PEARSON ALWAYS LEARNING

Jane B. Reece • Lisa A. Urry • Michael L. Cain
Steven A. Wasserman • Peter V. Minorsky • Robert B. Jackson

Campbell Biology

Custom Edition for Radford University
BIOL 132

Taken from:
Campbell Biology, Ninth Edition
by Jane B. Reece, Lisa A. Urry, Michael L. Cain,
Steven A. Wasserman, Peter V. Minorsky, and Robert B. Jackson

Cover Art: Courtesy of Photodisc/Getty Images

Taken from:

Campbell Biology, Ninth Edition
by Jane B. Reece, Lisa A. Urry, Michael L. Cain, Steven A. Wasserman, Peter V. Minorsky, and Robert B. Jackson

Published by Benjamin Cummings
San Francisco, California 94111

This special edition published in cooperation with Pearson Learning Solutions.

Pearson Learning Solutions, 501 Boylston Street, Suite 900, Boston, MA 02116
A Pearson Education Company
www.pearsoned.com

Printed in the United States of America

1 2 3 4 5 6 7 8 9 10 V092 16 15 14 13 12 11

000200010271275160

SS

PEARSON

ISBN 10: 1-256-49550-6
ISBN 13: 978-1-256-49550-5

Brief Contents

About the Authors

The Ninth Edition author team's contributions reflect their biological expertise as researchers and teaching sensibilities gained from years of experience as instructors at diverse institutions. The team's highly collaborative style continues to be evident in the cohesiveness and consistency of the Ninth Edition.

Jane B. Reece

The head of the Ninth Edition author team, Jane Reece was Neil Campbell's longtime collaborator. She has participated in every edition of *BIOLOGY*. Earlier, Jane taught biology at Middlesex County College and Queensborough Community College. She holds an A.B. in Biology from Harvard University, an M.S. in Microbiology from Rutgers University, and a Ph.D. in Bacteriology from the University of California, Berkeley. Jane's research as a doctoral student and postdoctoral fellow focused on genetic recombination in bacteria. Besides her work on *CAMPBELL BIOLOGY*, she has been a coauthor of *Biology: Concepts & Connections*, *Essential Biology*, and *The World of the Cell*.

Lisa A. Urry

Lisa Urry (Chapter 1 and Units 1–3) is a professor and developmental biologist and recent Chair of the Biology Department at Mills College. After graduating from Tufts University with a double major in Biology and French, Lisa completed her Ph.D. in molecular and developmental biology at Massachusetts Institute of Technology (MIT). She has published a number of research papers, most of them focused on gene expression during embryonic and larval development in sea urchins. Lisa is also deeply committed to promoting opportunities for women in science education and research.

Michael L. Cain

Michael Cain (Units 4 and 5) is an ecologist and evolutionary biologist who is now writing full-time. Michael earned a joint degree in Biology and Math at Bowdoin College, an M.Sc. from Brown University, and a Ph.D. in Ecology and Evolutionary Biology from Cornell University. As a faculty member at New Mexico State University and Rose-Hulman Institute of Technology, he taught a wide range of courses, including introductory biology, ecology, evolution, botany, and conservation biology. Michael is the author of dozens of scientific papers on topics that include foraging behavior in insects and plants, long-distance seed dispersal, and speciation in crickets. In addition to his work on *CAMPBELL BIOLOGY*, Michael is also the lead author of an ecology textbook.

Steven A. Wasserman

Steve Wasserman (Unit 7) is a professor at the University of California, San Diego (UCSD). He earned his A.B. in Biology from Harvard University and his Ph.D. in Biological Sciences from MIT. Through his research on regulatory pathway mechanisms in the fruit fly *Drosophila*, Steve has contributed to the fields of developmental biology, reproduction, and immunity. As a faculty member at the University of Texas Southwestern Medical Center and UCSD, he has taught genetics, development, and physiology to undergraduate, graduate, and medical students. He has also served as the research mentor for more than a dozen doctoral students and more than 50 aspiring scientists at the undergraduate and high school levels. Steve has been the recipient of distinguished scholar awards from both the Markey Charitable Trust and the David and Lucille Packard Foundation. In 2007, he received UCSD's Distinguished Teaching Award for undergraduate teaching.

Peter V. Minorsky

Peter Minorsky (Unit 6) is a professor at Mercy College in New York, where he teaches evolution, ecology, botany, and introductory biology. He received his B.A. in Biology from Vassar College and his Ph.D. in Plant Physiology from Cornell University. He is also the science writer for the journal *Plant Physiology*. After a postdoctoral fellowship at the University of Wisconsin at Madison, Peter taught at Kenyon College, Union College, Western Connecticut State University, and Vassar College. He is an electrophysiologist who studies responses of plants to stress. Peter received the 2008 Award for Teaching Excellence at Mercy College.

Robert B. Jackson

Rob Jackson (Unit 8) is a professor of biology and Nicholas Chair of Environmental Sciences at Duke University. Rob holds a B.S. in Chemical Engineering from Rice University, as well as M.S. degrees in Ecology and Statistics and a Ph.D. in Ecology from Utah State University. Rob directed Duke's Program in Ecology for many years and just finished a term as the Vice President of Science for the Ecological Society of America. Rob has received numerous awards, including a Presidential Early Career Award in Science and Engineering from the National Science Foundation. He also enjoys popular writing, having published a trade book about the environment, *The Earth Remains Forever*, and two books of poetry for children, *Animal Mischief* and *Weekend Mischief*.

Neil A. Campbell

Neil Campbell combined the investigative nature of a research scientist with the soul of an experienced and caring teacher. He earned his M.A. in Zoology from UCLA and his Ph.D. in Plant Biology from the University of California, Riverside, where he received the Distinguished Alumnus Award in 2001. Neil published numerous research articles on desert and coastal plants and how the sensitive plant (*Mimosa*) and other legumes move their leaves. His 30 years of teaching in diverse environments included general biology courses at Cornell University, Pomona College, and San Bernardino Valley College, where he received the college's first Outstanding Professor Award in 1986. Neil was a visiting scholar in the Department of Botany and Plant Sciences at the University of California, Riverside. In addition to his authorship of this book, he coauthored *Biology: Concepts & Connections* and *Essential Biology* with Jane Reece. For the Ninth Edition of this book, we honor Neil's contributions to biology education by adopting the title *CAMPBELL BIOLOGY*.

Preface

Biology is an enormous subject, one that can seem overwhelming to students and scientists alike. Moreover, discoveries are being made at an unprecedented pace—from new kinds of small RNA molecules to the Neanderthal genome, from new biofuels to communities of organisms thriving beneath vast glaciers, from emerging infectious diseases to cancer vaccines. As a result, a general biology course faces a daunting challenge: to keep students from suffocating under an avalanche of information. *CAMPBELL BIOLOGY* addresses this challenge by providing a strong foundation for understanding both current knowledge and new developments in the context of underlying biological concepts.

Key Concepts and Unifying Themes

In each chapter of this textbook, a framework of three to six carefully chosen **Key Concepts** provide context for supporting details, helping students distinguish the "forest" from the "trees." The numbered Key Concepts are presented at the beginning of the chapter and then serve as headings for each chapter section. *Concept Check Questions* at the end of each section provide a hierarchical framework for self-assessment that builds students' confidence and then challenges them to push the limits of their understanding with several types of critical thinking questions. The *Summary of Key Concepts* at the end of the chapter refocuses students on the main points. *CAMPBELL BIOLOGY* also helps students organize and make sense of what they learn on a grander scale by emphasizing **evolution and other unifying themes** that pervade biology. These themes are introduced in Chapter 1 and integrated throughout the book.

New to This Edition: An Emphasis on Making Connections

In addition to Key Concepts and themes, we've created new features for the Ninth Edition that help students see the big picture by *making connections*. These include the following:

New Make Connections Questions: Making connections across chapters

New **Make Connections Questions** help students see how different areas of biology tie together, helping them overcome the tendency to compartmentalize information. Each question challenges students to move beyond memorization and gain a deeper understanding of biological principles by asking them to relate chapter content to material they learned earlier in the course. For example, we ask students to connect

- DNA replication (Chapter 16, see p. 319) to the cell cycle (Chapter 12);
- Soil formation (Chapter 37, see p. 789) to the properties of water (Chapter 3); and
- Aquatic biomes (Chapter 52, see p. 1163) to osmoregulation (Chapter 44).

At least three Make Connections Questions appear in each chapter. In addition, online *Make Connections Tutorials* in MasteringBiology® (see p. xi) connect content from two different chapters using figures from the book.

Expanded Evolution Coverage: Making connections to evolution in every chapter

Evolution is the core theme of biology, and in this edition it is more evident than ever. At least one **Evolution section in every chapter** focuses on evolutionary aspects of the chapter material, highlighted by a new Evolution banner. See, for example, the new discussions of enzyme evolution (p. 157), coevolution of flowers and pollinators (p. 806), and evolution of hormone function in animals (pp. 988–989).

New Impact Figures: Making connections between scientific advances and the real world

Our new **Impact Figures** motivate students by highlighting the dramatic impact of recent discoveries in biology. These figures feature high-interest topics such as induced pluripotent stem cells and regenerative medicine (Chapter 20, p. 417), the discovery of *Tiktaalik* (Chapter 34, p. 710), and the use of forensic ecology to track elephant poaching (Chapter 56, p. 1243). The *Why It Matters* section of each figure explains the relevance of the research to students' lives, global problems, or the field of biology itself. Each Impact Figure ends with a suggestion for *Further Reading* and a *What If?* or *Make Connections Question* to develop critical thinking skills.

New Visual Organizers and 3-D Art: Making connections visually

The new **Visual Organizer** format highlights the main parts of a figure, helping students see the key categories at a glance. See, for instance, Figure 17.24 on types of small-scale mutations (p. 345) or Figure 27.3, Gram staining (p. 557). Throughout the book, selected figures have been rendered in a more **3-D art style** while keeping an appropriate balance between realism and teaching effectiveness. Figure 52.3, Exploring Global Climate Patterns (p. 1146), is one example.

Restructured Chapter Reviews: Making connections at a higher level

In the chapter summaries, each concept section now concludes with a new **Summary of Key Concepts Question** that is tied to a major learning goal. Also, this edition increases student awareness of different levels of thinking by organizing the end-of-chapter questions into three levels based on **Bloom's taxonomy**, which classifies types of thinking that are important in learning. Our levels are (1) Knowledge/Comprehension, (2) Application/Analysis, and (3) Synthesis/Evaluation. (These same levels are used in the Campbell Test Bank.) The range of question types helps students develop critical thinking skills and prepare for the kinds of questions they'll encounter on exams. New **Write About a Theme Questions** give students practice writing short, coherent essays that connect the chapter's content to one of the book's themes. (A suggested grading rubric can be found on p. xv, and sample answers are provided for instructors in the MasteringBiology Instructor Resources area.) A new **MasteringBiology preview section** at the end of each chapter lists *Assignments*—tutorials, activities, and questions that instructors can assign. This section also directs students to the *eText* and *Study Area* for online self study.

New Content: Making connections to advances in science

As in each new edition, the Ninth Edition incorporates **new scientific content** and **organizational improvements**. These are summarized on pp. viii–ix, following this Preface.

MasteringBiology®: Making connections outside of class

MasteringBiology, the most widely used online assessment and tutorial program for biology, provides an extensive library of homework assignments that are graded automatically. In addition to BioFlix® Tutorials, other Tutorials, Activities, Reading Quiz Questions in every chapter, and 4,500 Test Bank Questions, MasteringBiology for the Ninth Edition features an **improved user interface** and the following **new Tutorials and Questions**: *Make Connections Tutorials, Student Misconceptions Questions* for every chapter, *Data Analysis Tutorials, Experimental Inquiry Tutorials, Video Tutor Sessions,* and *MasteringBiology: Virtual Biology Labs*. For more information, see pp. xvi–xix and www.masteringbiology.com.

Our Hallmark Features

Besides our Key Concepts and unifying themes, several other features have contributed to the success of *Campbell Biology*. Because text and illustrations are equally important for learning biology, **integration of figures and text** has been a hallmark of this book since its inception. Our popular *Exploring Figures* on selected topics epitomize this approach: Each is a learning unit of core content that brings together related illustrations and text. Another example is our *Guided Tour Figures*, which use descriptions in blue type to walk students through complex figures like an instructor would, pointing out key structures, functions, and steps of processes.

To encourage **active learning**, recent editions have incorporated new types of questions: *What If? Questions, Figure Legend Questions*, and *Draw It Questions* that ask students to sketch a structure, annotate a figure, or graph data. In the Ninth Edition, these questions are augmented by the new *Make Connections Questions*. Online, the highly interactive *MasteringBiology tutorials* are sophisticated active-learning tools.

Finally, *Campbell Biology* features **scientific inquiry**, an essential component of any biology course. Complementing stories of scientific discovery in the text narrative and the unit-opening interviews, *Inquiry Figures* help students understand "how we know what we know" and provide a model of how to think like a scientist. Each one begins with a research question and then describes how researchers designed an experiment, interpreted their results, and drew conclusions. The source article is referenced, and a What If? Question asks students to consider an alternative scenario. Selected Inquiry Figures invite students to read and analyze the original research article in the supplement *Inquiry in Action: Interpreting Scientific Papers* (see p. xxi). At the end of each chapter, *Scientific Inquiry Questions* give students additional opportunities to practice critical thinking by developing hypotheses, designing experiments, and analyzing real research data. Beyond the book, activities involving scientific inquiry are featured in MasteringBiology and other supplements, both print and electronic (see pp. xviii–xxi).

Our Partnership with Instructors

A core value underlying all our work as authors is our belief in the importance of our partnership with instructors. Our primary way of serving instructors, of course, is providing a textbook, supplements, and media resources that serve their students well. In addition, Benjamin Cummings makes available a rich variety of instructor resources, in both print and electronic form (see p. xx). In our continuing efforts to improve the book and its supplements, we benefit tremendously from instructor feedback, not only in formal reviews from hundreds of scientists, but also via informal communication, often by e-mail.

The real test of any textbook is how well it helps instructors teach and students learn. We welcome comments from the students and instructors who use *Campbell Biology*. Please address your suggestions to any of us:

Jane Reece at janereece@cal.berkeley.edu
Lisa Urry (Chapter 1 and Units 1–3) at lurry@mills.edu
Michael Cain (Units 4 and 5) at mcain@bowdoin.edu
Peter Minorsky (Unit 6) at pminorsky@mercy.edu
Steve Wasserman (Unit 7) at stevenw@ucsd.edu
Rob Jackson (Unit 8) at jackson@duke.edu

New Content

This section provides just a few highlights of new content and organizational improvements in *CAMPBELL BIOLOGY*, Ninth Edition.

CHAPTER 1 Introduction: Themes in the Study of Life

We have added a separate new theme on energy flow while retaining a theme on environmental interactions. Concept 1.3, on the scientific method, has been reframed to more accurately reflect the scientific process, with a focus on observations and hypotheses. A new Concept 1.4 discusses the value of technology to society while emphasizing the cooperative nature of science and the value of diversity among scientists.

UNIT ONE The Chemistry of Life

For this edition, the basic chemistry is enlivened by new content connecting it to evolution, ecology, and other areas of biology. Examples of new material include omega-3 fatty acids, the isomeric forms of methamphetamine, arsenic contamination of groundwater, and the basis of mad cow disease. The burgeoning importance of nucleic acids throughout biology has prompted us to expand our coverage of DNA and RNA structures in this first unit. In fact, a general aim for the first two units is to infuse the chapters with more detail about nucleic acids, genes, and related topics. Another enhancement, in this and the next two units, is the inclusion of more computer models of important proteins in contexts where they support students' understanding of molecular function.

UNIT TWO The Cell

For Chapter 6, we developed an Exploring Figure on microscopy, which includes new types of microscopy, and we added micrographs of various cell types to the Exploring Figure on eukaryotic cells. We also expanded our description of chromosome composition, with the goal of preempting some common student misconceptions about chromosomes and DNA. New connections to evolution include an introduction to the endosymbiont theory in Chapter 6 and some interesting evolutionary adaptations of cell membranes in Chapter 7. We've added a new section to Chapter 8 on the evolution of enzymes with new functions, which not only strengthens enzyme coverage but also provides an early introduction to the concept that mutations contribute to molecular evolution. In Chapter 9, we simplified the glycolysis figure and emphasized pyruvate oxidation as a separate step to help students focus on the main ideas. In keeping with our increased focus on global issues in the Ninth Edition, Chapter 10 has an Impact Figure on biofuels and a discussion of the possible effect of climate change on the distribution of C_3 and C_4 plants. In Chapter 11, we have added an Impact Figure to highlight the importance and medical relevance of G protein-coupled receptors.

UNIT THREE Genetics

In Chapters 13–17, we have added material to stimulate student interest—for example, a new Impact Figure on genetic testing for disease-associated mutations. As done throughout the Ninth Edition, we ask students to make connections between chapters so that they avoid the trap of compartmentalizing the information in each chapter. For instance, Chapter 15 discusses the Philadelphia chromosome associated with chronic myelogenous leukemia and asks students to connect this information to what they learned about signaling in the cell cycle in Chapter 12. Also, we encourage students to connect what they learn about DNA replication and chromosome structure in Chapter 16 to the material on chromosome behavior during the cell cycle in Chapter 12. Chapter 16 has a new figure showing a current 3-D model of the DNA replication complex, with the lagging strand looping back through it.

Chapters 18–21 are extensively updated, with the changes dominated by new genomic sequence data and discoveries about the regulation of gene expression. (The introduction to genes, genomes, and gene expression in Units One and Two should help prepare students for these revisions.) Chapter 18 includes a new section on nuclear architecture, which describes the organization of chromatin in the nucleus in relation to gene expression. The roles of various types of RNA molecules in regulation also receive special attention. In the section on cancer, we describe how technical advances can contribute to personalized cancer treatments based on the molecular characteristics of an individual's tumor. Chapter 19 discusses the 2009 H1N1 flu pandemic. Chapter 20 includes advances in techniques for DNA sequencing and for obtaining induced pluripotent stem (iPS) cells. Finally, the heavily revised Chapter 21 describes what has been learned from the sequencing of many genomes, including those of a number of human individuals.

UNIT FOUR Mechanisms of Evolution

For this edition, we have continued to bolster our presentation of the vast evidence for evolution by adding new examples and figures that illustrate key conceptual points throughout the unit. For example, Chapter 22 now presents research data on adaptive evolution in soapberry bugs, fossil findings that shed light on the origins of cetaceans, and an Impact Figure on the rise of methicillin-resistant *Staphylococcus aureus*. Chapter 23 examines gene flow and adaptation in songbird populations. Chapter 24 incorporates several new examples of speciation research, including reproductive isolation in mosquitofish, speciation in shrimp, and hybridization of bear species. Other changes strengthen the storyline of the unit, ensuring that the chapters flow smoothly and build to a clear overall picture of what evolution is and how it works. For instance, new connections between Chapters 24 and 25 illustrate how differences in speciation and extinction rates shape the broad patterns in the history of life. We've also added earlier and more discussion of "tree thinking," the interpretation and application of phylogenetic trees, beginning in Chapter 22.

Unit Five The Evolutionary History of Biological Diversity

One of our goals for the diversity unit was to expand the coverage of the scientific evidence underlying the evolutionary story told in the chapters. So, for example, Chapter 27 now presents new findings on the evolutionary origin of bacterial flagella. In keeping with our increased emphasis on big-picture "tree thinking," we've added an "evogram" on tetrapod evolution in Chapter 34. (An evogram is a diagram illustrating the multiple lines of evidence that support the hypothesis shown in an evolutionary tree.) In addition, to help engage students, we've included new applications and woven more ecological information into our discussions of groups of organisms. Examples include new material on global growth of photosynthetic protists (Chapter 28), endangered molluscs (Chapter 33), and the impact of a pathogenic chytrid fungus on amphibian population declines (Chapters 31 and 34).

Unit Six Plant Form and Function

Plant biology is in a transitional phase; some professors prefer strong coverage of classical botany while others seek more in-depth coverage of the molecular biology of plants. In developing the Ninth Edition, we have continued to balance the old and the new to provide students with a basic understanding of plant anatomy and function while highlighting dynamic areas of plant research and the many important connections between plants and other organisms. One major revision goal was to provide more explicit discussion of the evolutionary aspects of plant biology, such as the coevolution of flowering plants and pollinators (Chapter 38). Updates include new findings in plant development in Concept 35.5 and new material on the dynamism of plant architecture as it relates to resource acquisition in Chapter 36.

Unit Seven Animal Form and Function

In revising this unit, we strove to introduce physiological systems through a comparative approach that underscores how adaptations are linked to shared physiological challenges. In particular, we have highlighted the interrelationship of the endocrine and nervous systems at multiple points in the unit, helping students appreciate how these two forms of communication link tissues, organs, and individuals. Other revisions aim to keep students focused on fundamental concepts amid the details of complex systems. For example, many figures have been reconceived to emphasize key information, including new figures comparing single and double circulation (Chapter 42) and examining the function of antigen receptors (Chapter 43), as well as new Exploring Figures on the vertebrate kidney (Chapter 44) and the structure and function of the human eye (Chapter 50). Chapter 43 has been significantly revised to support students' conceptual understanding of basic immunological responses and the key cellular players. Throughout the unit, new state-of-the-art images and material on current and compelling topics—such as circadian rhythms (Chapter 40), novel strains of influenza (Chapter 43), the effects of climate change on animal reproductive cycles (Chapter 46), and advances in understanding brain plasticity and function (Chapter 49)—will help engage students and encourage them to make connections beyond the text.

Unit Eight Ecology

Our revision was informed by the fact that biologists are increasingly asked to apply their knowledge to help solve global problems, such as climate change, that already are profoundly affecting life on Earth. As part of our increased emphasis on global ecology in this edition, we have made significant changes to Unit Eight's organization and content. The organizational changes begin with the introductory chapter of the unit (Chapter 52), which includes a new Key Concept 52.1: "Earth's climate varies by latitude and season and is changing rapidly." Introducing the global nature of climate and its effects on life at the beginning of the chapter provides a logical foundation for the rest of the material. New content in Chapters 53 and 54 highlights factors that limit population growth, the ecological importance of disease, positive interactions among organisms, and biodiversity. Chapter 55 now explores restoration ecology together with ecosystem ecology because successful restoration efforts depend on understanding ecosystem structure and function. Finally, the new title of the unit's capstone, Chapter 56, reflects its emphasis on the combined importance of conservation and our changing Earth: "Conservation Biology and Global Change." Several new Impact Figures in the unit show students how ecologists apply biological knowledge and ecological theory at all scales to understand and solve problems in the world around them.

To the Student: How to Use This Book

Focus on the Key Concepts.

Each chapter is organized around a framework of 3 to 6 **Key Concepts** that will help you stay focused on the big picture and give you a context for the supporting details.

52

An Introduction to Ecology and the Biosphere

▲ **Figure 52.1** **What threatens this amphibian's survival?**

Before you begin reading the chapter, get oriented by reading the **list of Key Concepts,** which introduces the big ideas covered in the chapter.

KEY CONCEPTS

52.1 Earth's climate varies by latitude and season and is changing rapidly

52.2 The structure and distribution of terrestrial biomes are controlled by climate and disturbance

52.3 Aquatic biomes are diverse and dynamic systems that cover most of Earth

52.4 Interactions between organisms and the environment limit the distribution of species

Each **Key Concept** serves as the heading for a major section of the chapter.

OVERVIEW

Discovering Ecology

When University of Delaware undergraduate Justin Yeager spent his summer abroad in Costa Rica, all he wanted was to see the tropical rain forest and to practice his Spanish. Instead, he rediscovered the variable harlequin toad (*Atelopus varius*), a species thought to be extinct in the mountain slopes of Costa Rica and Panama where it once lived **(Figure 52.1)**. During the 1980s and 1990s, roughly two-thirds of the 82 known species of harlequin toads vanished. Scientists think that a disease-causing chytrid fungus, *Batrachochytrium dendrobatidis* (see Figure 31.26), contributed to many of these extinctions. Why was the fungus suddenly thriving in the rain forest? Cloudier days and warmer nights associated with global warming appear to have created an environment ideal for its success. As of 2009, the species that Yeager found was surviving as a single known population of fewer than 100 individuals.

What environmental factors limit the geographic distribution of harlequin toads? How do variations in their food supply or interactions with other species, such as pathogens, affect the size of their population? Questions like these are the subject of **ecology** (from the Greek *oikos*, home, and *logos*, study), the scientific study of the interactions between organisms and the environment. Ecological interactions occur at a hierarchy of scales that ecologists study, from single organisms to the globe **(Figure 52.2)**.

Ecology's roots are in our basic human interest in observing other organisms. Naturalists, including Aristotle and Darwin, have long studied the living world and systematically recorded their observations. However, modern ecology involves more than observation. It is a rigorous experimental science that requires a breadth of biological knowledge. Ecologists generate hypotheses, manipulate environmental variables, and observe the outcome. In this unit, you will encounter many examples of ecological experiments, whose complex challenges have made ecologists innovators in experimental design and statistical inference.

In addition to providing a conceptual framework for understanding the field of ecology, Figure 52.2 provides the organizational framework for our final unit. In this chapter, we first describe Earth's climate and the importance of climate and other physical factors in determining the location of major life zones on land and in the oceans. We then examine how ecologists determine what controls the distribution and abundance of individual species. The next three chapters investigate population, community, and ecosystem ecology in detail, including approaches for restoring degraded ecosystems. The final chapter explores conservation biology and global ecology as we consider how ecologists apply biological knowledge to predict the global consequences of human activities and to conserve Earth's biodiversity.

CONCEPT 52.1

Earth's climate varies by latitude and season and is changing rapidly

The most significant influence on the distribution of organisms on land and in the oceans is **climate**, the long-term, prevailing weather conditions in a given area. Four physical

After reading a concept section, check your understanding using the **Concept Check Questions** at the end of the section. Work through these questions on your own or in a study group—they're good practice for the kinds of questions you might be asked on an exam.

What If? Questions ask you to apply what you've learned. New **Make Connections Questions** ask you to relate content in the chapter to a concept you learned earlier in the course.

If you can answer these questions (see Appendix A to check your work), you're ready to move on.

CONCEPT CHECK 52.1

1. Explain how the sun's unequal heating of Earth's surface leads to the development of deserts around 30° north and south of the equator.
2. What are some of the differences in microclimate between an unplanted agricultural field and a nearby stream corridor with trees?
3. WHAT IF? Changes in Earth's climate at the end of the last ice age happened gradually, taking centuries to thousands of years. If the current global warming happens very quickly, as predicted, how may this rapid climate change affect the ability of long-lived trees to evolve, compared with annual plants, which have much shorter generation times?
4. MAKE CONNECTIONS In Concept 10.4 (pp. 199–201), you learned about the important differences between C_3 and C_4 plants. Focusing just on the effects of temperature, would you expect the global distribution of C_4 plants to expand or contract as Earth becomes warmer? Why?

For suggested answers, see Appendix A.

Make connections across biology.

By relating the content of a chapter to material you learned earlier in the course, new **Make Connections Questions** help you develop a deeper understanding of biological principles.

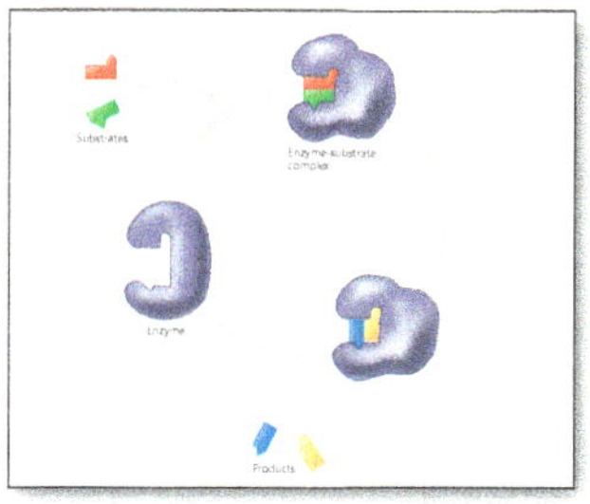

CONCEPT CHECK 41.1

2. **MAKE CONNECTIONS** Review the discussion of enzymes in metabolic reactions in Concept 8.4 (pp. 152–156). Then explain why vitamins are required in very small amounts in the diet.

Enzymes (Chapter 8) ⟷ **Animal nutrition** (Chapter 41)

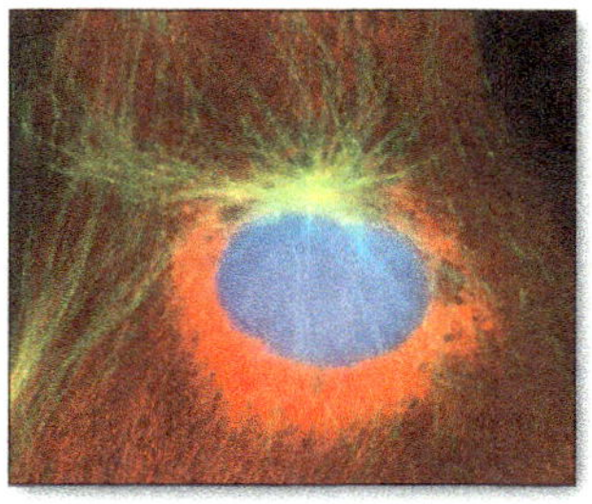

CONCEPT CHECK 16.2

3. **MAKE CONNECTIONS** What is the relationship between DNA replication and the S phase of the cell cycle? See Figure 12.6, page 231.

Cell cycle (Chapter 12) ⟷ **DNA replication** (Chapter 16)

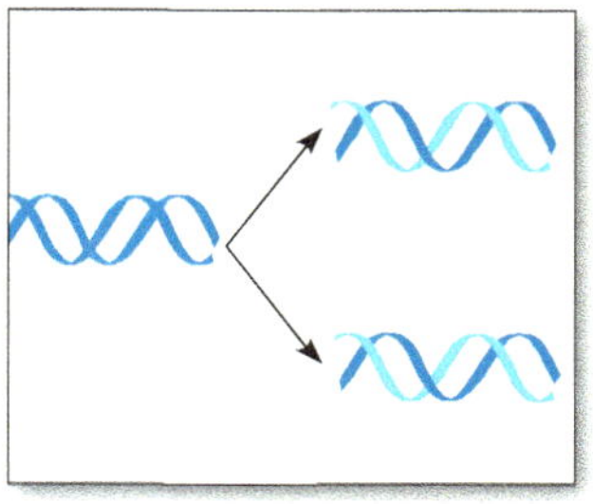

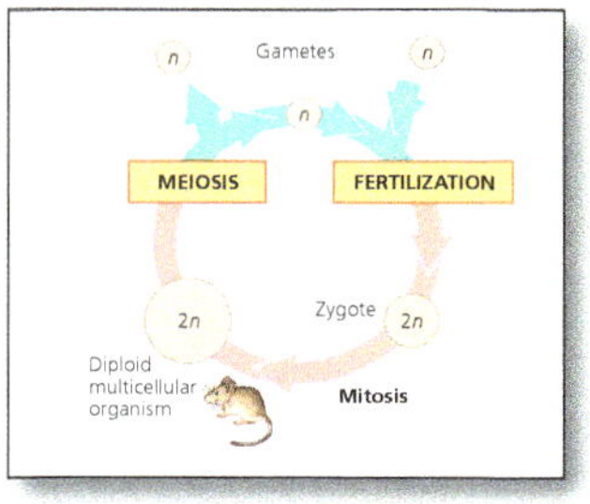

CONCEPT CHECK 31.2

1. **MAKE CONNECTIONS** Compare Figure 31.5 with Figure 13.6 (p. 252). In terms of haploidy versus diploidy, how do the life cycles of fungi and animals differ?

Meiosis (Chapter 13) ⟷ **Fungi** (Chapter 31)

MasteringBIOLOGY®
www.masteringbiology.com

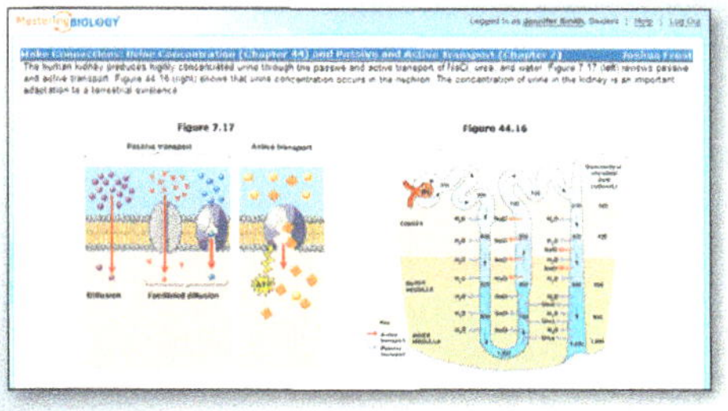

New **Make Connections Tutorials** help you connect biological concepts across chapters in an interactive way.

Make connections to **evolution,** the fundamental theme of biology.

Look for new **Evolution banners** highlighting sections in each chapter that focus on evolutionary aspects of the topic.

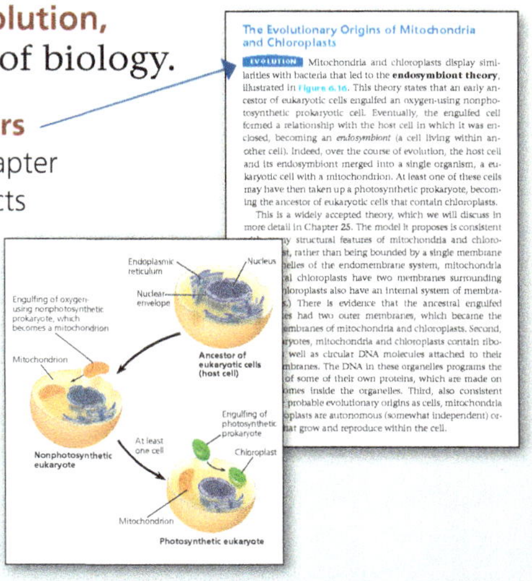

Practice thinking like a scientist.

New **Impact Figures** demonstrate the dramatic impact of recent discoveries in biology and show that biology is constantly changing as new discoveries add to our understanding.

▼ Figure 56.9

IMPACT

Forensic Ecology and Elephant Poaching

This array of severed tusks is part of an illegal shipment of 6,000 kg of ivory intercepted on its way from Africa to Singapore in 2002. Investigators wondered whether the elephants slaughtered for the ivory—perhaps as many as 6,500—were killed in the area where the shipment originated, primarily Zambia, or instead were killed across Africa, indicating a broader smuggling ring. Samuel Wasser, of the University of Washington, and colleagues amplified specific segments of DNA from the tusks using the polymerase chain reaction (PCR). These segments included stretches of DNA containing short tandem repeats (STRs; see Concept 20.4, pp. 420–421), the number of which varies among different elephant populations. The researchers then compared alleles at seven or more loci with a reference DNA database they had generated for elephants of known geographic origin. Their results showed conclusively that the elephants came from a narrow east-west band centered on Zambia rather than from across Africa.

WHY IT MATTERS The DNA analyses suggested that poaching rates were 30 times higher in Zambia than previously estimated. This news led to improved antipoaching efforts by the Zambian government. Techniques like those used in this study are being employed by conservation biologists to track the harvesting of many endangered species, including whales, sharks, and orchids.

FURTHER READING S. K. Wasser et al., Forensic tools battle ivory poachers, *Scientific American* 399:68–76 (2009); S. K. Wasser et al., Using DNA to track the origin of the largest ivory seizure since the 1989 trade ban, *Proceedings of the National Academy of Sciences USA* 104:4228–4233 (2007).

MAKE CONNECTIONS Figure 26.6 (p. 539) describes another example in which conservation biologists used DNA analyses to compare harvested samples with a reference DNA database. How are these examples similar, and how are they different? What limitations might there be to using such forensic methods in other suspected cases of poaching?

Why It Matters explains the relevance of the research.

Further Reading directs you to articles to explore.

A **Make Connections** or **What If? Question** encourages critical thinking.

Inquiry Figures reveal "how we know what we know" by highlighting how researchers designed an experiment, interpreted their results, and drew conclusions.

▼ Figure 37.14 **INQUIRY**

Does the invasive weed garlic mustard disrupt mutualistic associations between native tree seedlings and arbuscular mycorrhizal fungi?

EXPERIMENT Kristina Stinson, of Harvard University, and colleagues investigated the effect of invasive garlic mustard on the growth of native tree seedlings and associated mycorrhizal fungi. In one experiment, they grew seedlings of three North American trees—sugar maple, red maple, and white ash—in four different soils. Two of the soil samples were collected from a location where garlic mustard was growing, and one of these samples was sterilized. The other two soil samples were collected from a location devoid of garlic mustard, and one was then sterilized. After four months of growth, the researchers harvested the shoots and roots and determined the dried biomass. The roots were also analyzed for percent colonization by arbuscular mycorrhizal fungi.

RESULTS Native tree seedlings grew more slowly and were less able to form mycorrhizal associations when grown either in sterilized soil or in unsterilized soil collected from a location that had been invaded by garlic mustard.

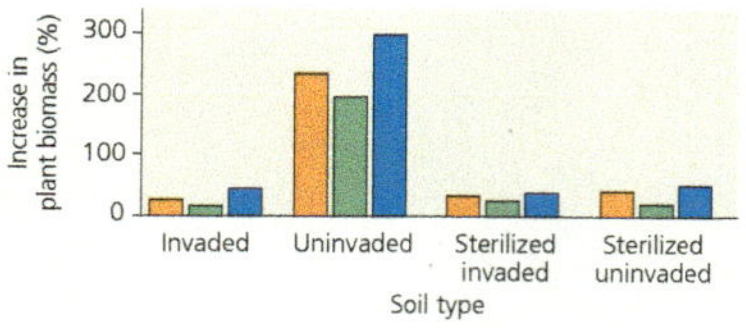

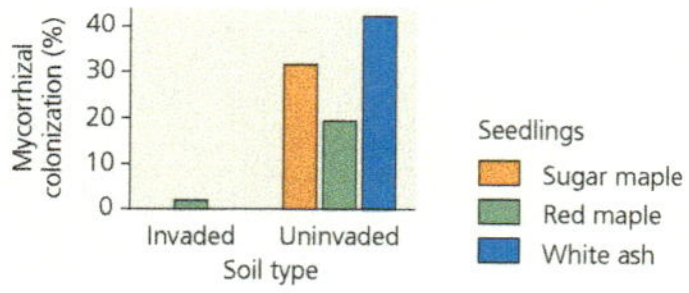

CONCLUSION The data support the hypothesis that garlic mustard suppresses growth of native trees by affecting the soil in a way that disrupts mutualistic associations between the trees and arbuscular mycorrhizal fungi.

SOURCE K. A. Stinson et al., Invasive plant suppresses the growth of native tree seedlings by disrupting belowground mutualisms, *PLoS Biol (Public Library of Science: Biology)* 4(5): e140 (2006).

INQUIRY IN ACTION Read and analyze the original paper in *Inquiry in Action: Interpreting Scientific Papers.*

WHAT IF? What effect would applying inorganic phosphate to soil invaded by garlic mustard have on the plant's ability to outcompete native species?

Some **Inquiry Figures** invite you to read and analyze the original research paper in its complete form. You can find the journal article, along with a worksheet guiding you through it, in the separate book ***Inquiry in Action: Interpreting Scientific Papers.***

After exploring the featured experiment, test your analytical skills by answering the **What If? Question**. Suggested answers are provided in Appendix A to help you gauge your understanding.

MasteringBIOLOGY®
www.masteringbiology.com

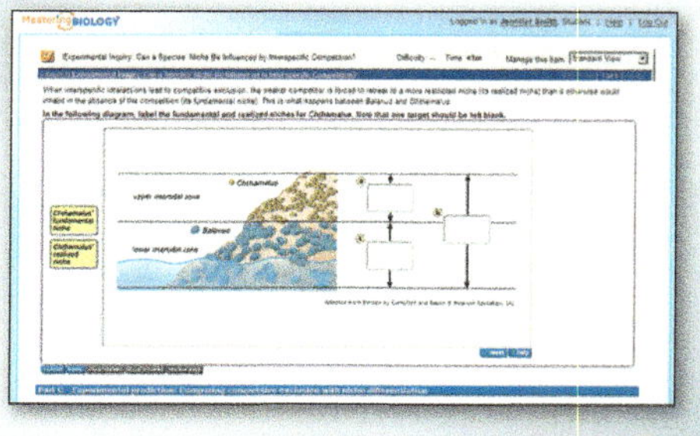

New **Experimental Inquiry Tutorials** give you practice analyzing experimental design and data and drawing conclusions.

Study the figures as you read the text.

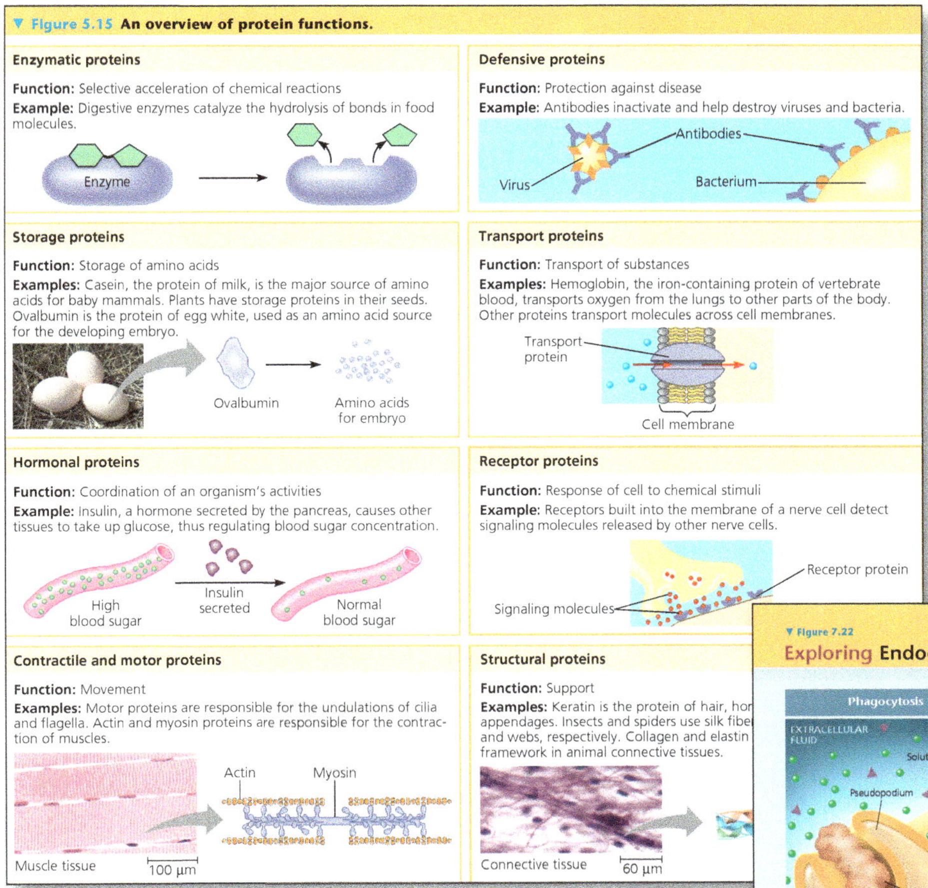
▼ **Figure 5.15 An overview of protein functions.**

Enzymatic proteins **Function:** Selective acceleration of chemical reactions **Example:** Digestive enzymes catalyze the hydrolysis of bonds in food molecules.	**Defensive proteins** **Function:** Protection against disease **Example:** Antibodies inactivate and help destroy viruses and bacteria.
Storage proteins **Function:** Storage of amino acids **Examples:** Casein, the protein of milk, is the major source of amino acids for baby mammals. Plants have storage proteins in their seeds. Ovalbumin is the protein of egg white, used as an amino acid source for the developing embryo.	**Transport proteins** **Function:** Transport of substances **Examples:** Hemoglobin, the iron-containing protein of vertebrate blood, transports oxygen from the lungs to other parts of the body. Other proteins transport molecules across cell membranes.
Hormonal proteins **Function:** Coordination of an organism's activities **Example:** Insulin, a hormone secreted by the pancreas, causes other tissues to take up glucose, thus regulating blood sugar concentration.	**Receptor proteins** **Function:** Response of cell to chemical stimuli **Example:** Receptors built into the membrane of a nerve cell detect signaling molecules released by other nerve cells.
Contractile and motor proteins **Function:** Movement **Examples:** Motor proteins are responsible for the undulations of cilia and flagella. Actin and myosin proteins are responsible for the contraction of muscles.	**Structural proteins** **Function:** Support **Examples:** Keratin is the protein of hair, hor... appendages. Insects and spiders use silk fibe... and webs, respectively. Collagen and elastin ... framework in animal connective tissues.

◀ New **Visual Organizers** help you to see important categories at a glance.

By integrating text and visuals, **Exploring Figures** help you learn more efficiently.
▼

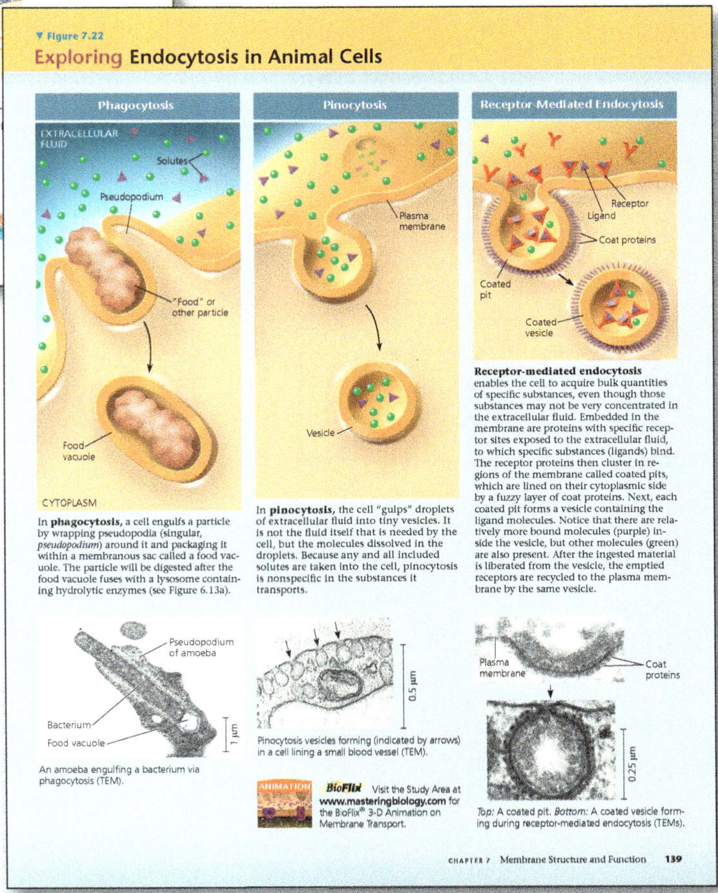
▼ Figure 7.22
Exploring Endocytosis in Animal Cells

In **phagocytosis**, a cell engulfs a particle by wrapping pseudopodia (singular, *pseudopodium*) around it and packaging it within a membranous sac called a food vacuole. The particle will be digested after the food vacuole fuses with a lysosome containing hydrolytic enzymes (see Figure 6.13a).

In **pinocytosis**, the cell "gulps" droplets of extracellular fluid into tiny vesicles. It is not the fluid itself that is needed by the cell, but the molecules dissolved in the droplets. Because any and all included solutes are taken into the cell, pinocytosis is nonspecific in the substances it transports.

Receptor-mediated endocytosis enables the cell to acquire bulk quantities of specific substances, even though those substances may not be very concentrated in the extracellular fluid. Embedded in the membrane are proteins with specific receptor sites exposed to the extracellular fluid, to which specific substances (ligands) bind. The receptor proteins then cluster in regions of the membrane called coated pits, which are lined on their cytoplasmic side by a fuzzy layer of coat proteins. Next, each coated pit forms a vesicle containing the ligand molecules. Notice that there are relatively more bound molecules (purple) inside the vesicle, but other molecules (green) are also present. After the ingested material is liberated from the vesicle, the emptied receptors are recycled to the plasma membrane by the same vesicle.

An amoeba engulfing a bacterium via phagocytosis (TEM).

Pinocytosis vesicles forming (indicated by arrows) in a cell lining a small blood vessel (TEM).

ANIMATION **BioFlix** Visit the Study Area at **www.masteringbiology.com** for the BioFlix® 3-D Animation on Membrane Transport.

Top: A coated pit. *Bottom:* A coated vesicle forming during receptor-mediated endocytosis (TEMs).

CHAPTER 7 Membrane Structure and Function 139

In selected illustrations, **a three-dimensional art style** helps you visualize biological structures. ▶

▲
BioFlix® icons direct you to high-impact 3-D animations in the Study Area at www.masteringbiology.com.

MasteringBIOLOGY®
www.masteringbiology.com

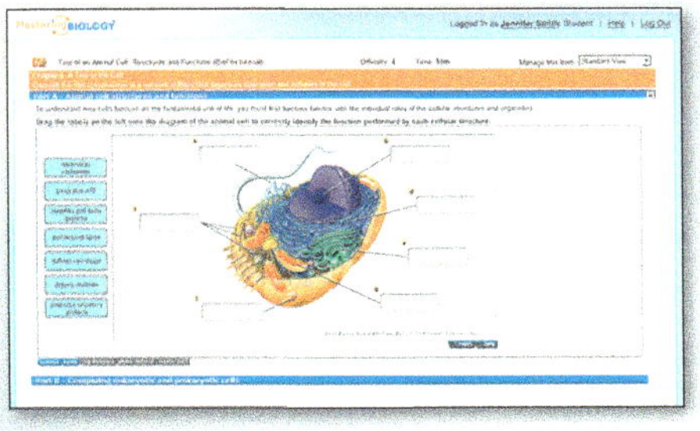

Many **Tutorials** and **Activities** integrate art from the textbook, providing a unified learning experience.

Review what you've learned.

Chapter Reviews help you efficiently master the chapter content by focusing on the main points of the chapter and offering opportunities to practice for exams.

Summary Figures present key information in a visual way.

23

The Evolution of Populations

▲ **Figure 23.1** **Is this finch evolving?**

EVOLUTION

KEY CONCEPTS

23.1 Genetic variation makes evolution possible
23.2 The Hardy-Weinberg equation can be used to test whether a population is evolving
23.3 Natural selection, genetic drift, and gene flow can alter allele frequencies in a population
23.4 Natural selection is the only mechanism that consistently causes adaptive evolution

OVERVIEW

The Smallest Unit of Evolution

One common misconception about evolution is that individual organisms evolve. It is true that natural selection acts on individuals: Each organism's traits affect its survival and reproductive success compared with other individuals. But the evolutionary impact of natural selection is only apparent in the changes in a *population* of organisms over time.

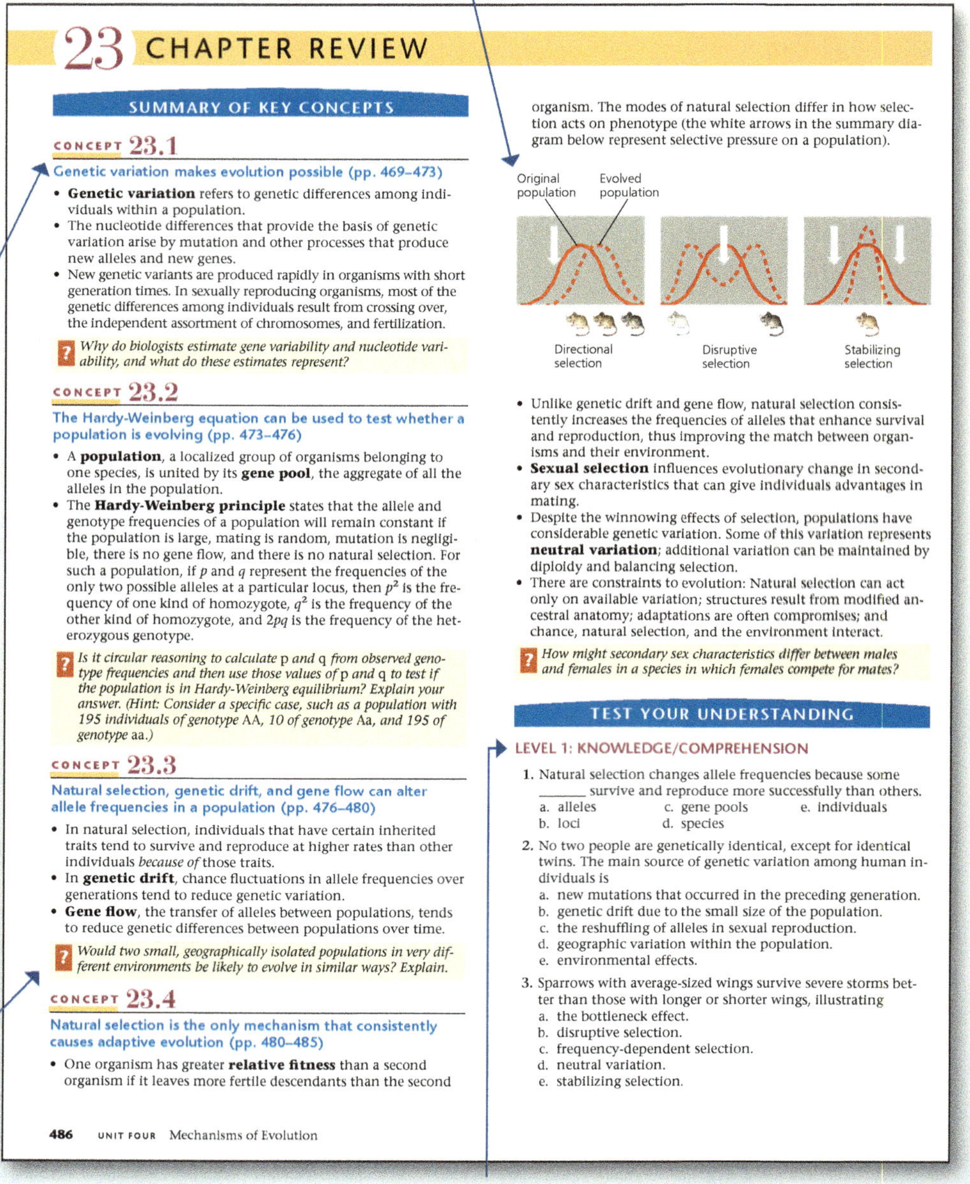

23 CHAPTER REVIEW

SUMMARY OF KEY CONCEPTS

CONCEPT 23.1

Genetic variation makes evolution possible (pp. 469–473)

- **Genetic variation** refers to genetic differences among individuals within a population.
- The nucleotide differences that provide the basis of genetic variation arise by mutation and other processes that produce new alleles and new genes.
- New genetic variants are produced rapidly in organisms with short generation times. In sexually reproducing organisms, most of the genetic differences among individuals result from crossing over, the independent assortment of chromosomes, and fertilization.

? *Why do biologists estimate gene variability and nucleotide variability, and what do these estimates represent?*

CONCEPT 23.2

The Hardy-Weinberg equation can be used to test whether a population is evolving (pp. 473–476)

- A **population**, a localized group of organisms belonging to one species, is united by its **gene pool**, the aggregate of all the alleles in the population.
- The **Hardy-Weinberg principle** states that the allele and genotype frequencies of a population will remain constant if the population is large, mating is random, mutation is negligible, there is no gene flow, and there is no natural selection. For such a population, if p and q represent the frequencies of the only two possible alleles at a particular locus, then p^2 is the frequency of one kind of homozygote, q^2 is the frequency of the other kind of homozygote, and $2pq$ is the frequency of the heterozygous genotype.

? *Is it circular reasoning to calculate* p *and* q *from observed genotype frequencies and then use those values of* p *and* q *to test if the population is in Hardy-Weinberg equilibrium? Explain your answer. (Hint: Consider a specific case, such as a population with 195 individuals of genotype AA, 10 of genotype Aa, and 195 of genotype aa.)*

CONCEPT 23.3

Natural selection, genetic drift, and gene flow can alter allele frequencies in a population (pp. 476–480)

- In natural selection, individuals that have certain inherited traits tend to survive and reproduce at higher rates than other individuals *because of* those traits.
- In **genetic drift**, chance fluctuations in allele frequencies over generations tend to reduce genetic variation.
- **Gene flow**, the transfer of alleles between populations, tends to reduce genetic differences between populations over time.

? *Would two small, geographically isolated populations in very different environments be likely to evolve in similar ways? Explain.*

CONCEPT 23.4

Natural selection is the only mechanism that consistently causes adaptive evolution (pp. 480–485)

- One organism has greater **relative fitness** than a second organism if it leaves more fertile descendants than the second organism. The modes of natural selection differ in how selection acts on phenotype (the white arrows in the summary diagram below represent selective pressure on a population).

Original population | Evolved population

Directional selection | Disruptive selection | Stabilizing selection

- Unlike genetic drift and gene flow, natural selection consistently increases the frequencies of alleles that enhance survival and reproduction, thus improving the match between organisms and their environment.
- **Sexual selection** influences evolutionary change in secondary sex characteristics that can give individuals advantages in mating.
- Despite the winnowing effects of selection, populations have considerable genetic variation. Some of this variation represents **neutral variation**; additional variation can be maintained by diploidy and balancing selection.
- There are constraints to evolution: Natural selection can act only on available variation; structures result from modified ancestral anatomy; adaptations are often compromises; and chance, natural selection, and the environment interact.

? *How might secondary sex characteristics differ between males and females in a species in which females compete for mates?*

TEST YOUR UNDERSTANDING

LEVEL 1: KNOWLEDGE/COMPREHENSION

1. Natural selection changes allele frequencies because some _______ survive and reproduce more successfully than others.
   a. alleles b. loci c. gene pools d. species e. individuals
2. No two people are genetically identical, except for identical twins. The main source of genetic variation among human individuals is
   a. new mutations that occurred in the preceding generation.
   b. genetic drift due to the small size of the population.
   c. the reshuffling of alleles in sexual reproduction.
   d. geographic variation within the population.
   e. environmental effects.
3. Sparrows with average-sized wings survive severe storms better than those with longer or shorter wings, illustrating
   a. the bottleneck effect.
   b. disruptive selection.
   c. frequency-dependent selection.
   d. neutral variation.
   e. stabilizing selection.

486 UNIT FOUR Mechanisms of Evolution

Key Concepts, which were introduced in the beginning of the chapter and developed in the text, are summarized in the Chapter Review.

New **Summary of Key Concepts Questions** appear at the end of each concept summary. Check your answers using Appendix A.

To help you prepare for the various kinds of questions that may appear on a test, the end-of-chapter questions are now organized into three levels based on Bloom's Taxonomy:

Level 1: Knowledge/Comprehension
Level 2: Application/Analysis
Level 3: Synthesis/Evaluation

Evolution Connection Questions in the Chapter Review ask you to think critically about how an aspect of the chapter relates to evolution. ▶

Scientific Inquiry ▶ **Questions** at the end of each chapter give you opportunities to practice scientific thinking by developing hypotheses, designing experiments, and analyzing real research data.

Draw It Exercises in each chapter ask you to put pencil to paper and draw a structure, annotate a figure, or graph experimental data.

LEVEL 2: APPLICATION/ANALYSIS

4. If the nucleotide variability of a locus equals 0%, what is the gene variability and number of alleles at that locus?
   a. gene variability = 0%; number of alleles = 0
   b. gene variability = 0%; number of alleles = 1
   c. gene variability = 0%; number of alleles = 2
   d. gene variability > 0%; number of alleles = 2
   e. Without more information, gene variability and number of alleles cannot be determined.

5. There are 40 individuals in population 1, all with genotype *A1A1*, and there are 25 individuals in population 2, all with genotype *A2A2*. Assume that these populations are located far from each other and that their environmental conditions are very similar. Based on the information given here, the observed genetic variation is most likely an example of
   a. genetic drift.
   b. gene flow.
   c. disruptive selection.
   d. discrete variation.
   e. directional selection.

6. A fruit fly population has a gene with two alleles, *A1* and *A2*. Tests show that 70% of the gametes produced in the population contain the *A1* allele. If the population is in Hardy-Weinberg equilibrium, what proportion of the flies carry both *A1* and *A2*?
   a. 0.7 b. 0.49 c. 0.21 d. 0.42 e. 0.09

LEVEL 3: SYNTHESIS/EVALUATION

7. EVOLUTION CONNECTION
   How is the process of evolution revealed by the imperfections of living organisms?

8. SCIENTIFIC INQUIRY
   DRAW IT Richard Koehn, of the State University of New York, Stony Brook, and Thomas Hilbish, of the University of South Carolina, studied genetic variation in the marine mussel *Mytilus edulis* around Long Island, New York. They measured the frequency of a particular allele (lap^{94}) for an enzyme involved in regulating the mussel's internal saltwater balance. The researchers presented their data as a series of pie charts linked to sampling sites within Long Island Sound, where the salinity is highly variable, and along the coast of the open ocean, where salinity is constant:

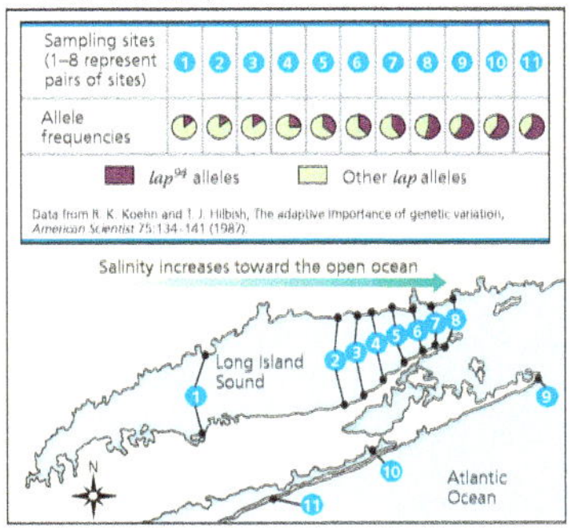

(*Question 8, continued*)
Create a data table for the 11 sampling sites by estimating the frequency of lap^{94} from the pie charts. (*Hint*: Think of each pie chart as a clock face to help you estimate the proportion of the shaded area.) Then graph the frequencies for sites 1–8 to show how the frequency of this allele changes with increasing salinity in Long Island Sound (from southwest to northeast). How do the data from sites 9–11 compare with the data from the sites within the Sound?

Construct a hypothesis that explains the patterns you observe in the data and that accounts for the following observations: (1) the lap^{94} allele helps mussels maintain osmotic balance in water with a high salt concentration but is costly to use in less salty water; and (2) mussels produce larvae that can disperse long distances before they settle on rocks and grow into adults.

9. WRITE ABOUT A THEME
   Emergent Properties Heterozygotes at the sickle-cell locus produce both normal and abnormal (sickle-cell) hemoglobin (see Concept 14.4). When hemoglobin molecules are packed into a heterozygote's red blood cells, some cells receive relatively large quantities of abnormal hemoglobin, making these cells prone to sickling. In a short essay (approximately 100–150 words), explain how these molecular and cellular events lead to emergent properties at the individual and population levels of biological organization.

For selected answers, see Appendix A.

MasteringBIOLOGY www.masteringbiology.com

1. MasteringBiology® Assignments

Make Connections Tutorial Hardy-Weinberg Principle (Chapter 23) and Inheritance of Alleles (Chapter 14)
Experimental Inquiry Tutorial Did Natural Selection of Ground Finches Occur When the Environment Changed?
BioFlix **Tutorial** Mechanisms of Evolution
Tutorial Hardy-Weinberg Principle
Activities Genetic Variation from Sexual Recombination • The Hardy-Weinberg Principle • Causes of Evolutionary Change • Three Modes of Natural Selection
Questions Student Misconceptions • Reading Quiz • Multiple Choice • End-of-Chapter

2. eText

Read your book online, search, take notes, highlight text, and more.

3. The Study Area

Practice Tests • Cumulative Test • *BioFlix* 3-D Animations • MP3 Tutor Sessions • Videos • Activities • Investigations • Lab Media • Audio Glossary • Word Study Tools • Art

CHAPTER 23 The Evolution of Populations 487

◀ New **Write About a Theme Questions** give you practice writing a short essay that connects the chapter's content to one of the bookwide themes introduced in Chapter 1.

◀ Each chapter now ends with a preview of the **MasteringBiology®** resources that can help you succeed in the course.

◀ This **Writing Rubric** explains criteria on which your writing may be graded. The rubric and tips for writing good short-answer essays can be found in the Study Area at www.masteringbiology.com.

Suggested Grading Rubric for "Write About a Theme" Short-Answer Essays

	Understanding of Theme and Relationship to Topic	Use of Supporting Examples or Details	Appropriate Use of Terminology	Quality of Writing
4	Evidence of full and complete understanding	Examples well chosen, details accurate and applied to theme	Accurate scientific terminology enhances the essay	Excellent organization, sentence structure, and grammar
3	Evidence of good understanding	Examples or details are generally well applied to theme	Terminology is correctly used	Good sentence flow, sentence structure, and grammar
2	Evidence of a basic understanding	Supporting examples and details are adequate	Terminology used is not totally accurate or appropriate	Some organizational and grammatical problems
1	Evidence of limited understanding	Examples and details are minimal	Appropriate terminology is not present	Poorly organized. Grammatical and spelling errors detract from essay
0	Essay shows no understanding of theme	Examples lacking or incorrect	Terminology lacking or incorrect	Essay is very poorly written

To the Student: How to Effectively

MasteringBIOLOGY®
www.masteringbiology.com

The Mastering system empowers you to take charge of your learning—at your convenience, 24/7.

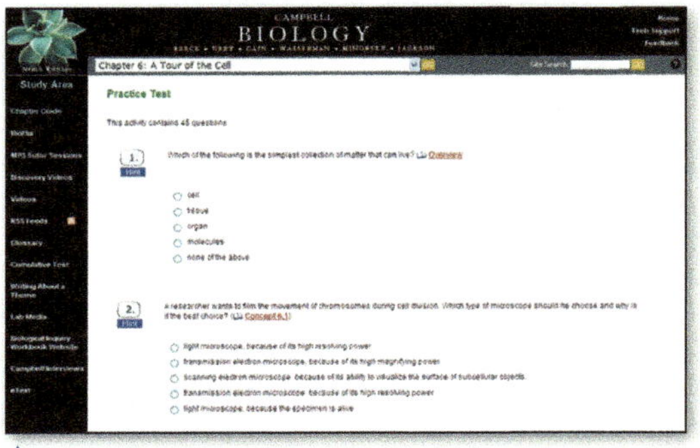

▲ **Practice Tests** help you assess your understanding of each chapter, providing feedback for right and wrong answers.

Use the Study Area on your own or in a study group.

*Bio***Flix**

▲ **BioFlix® 3-D animations** explore the most difficult biology topics, reinforced with tutorials, quizzes, and more.

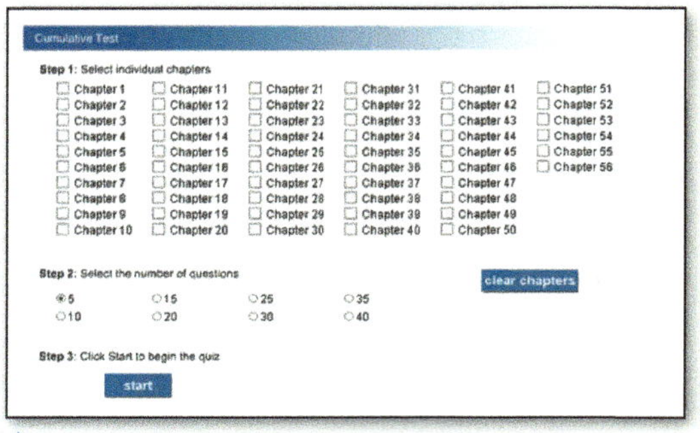

▲ The **Cumulative Test** allows you to build a practice test with questions from multiple chapters.

Access your book online.

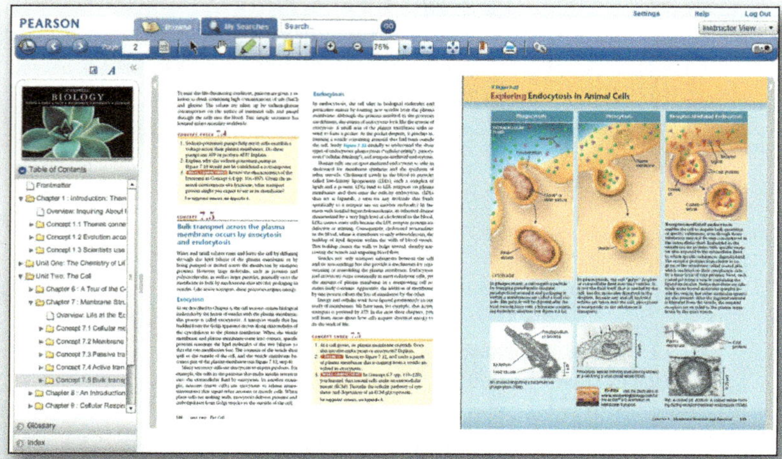

◀ The **Pearson eText** gives you access to the text whenever and wherever you can access the Internet. The eText includes powerful interactive and customization functions:

- write notes
- highlight text
- bookmark pages
- zoom
- click hyperlinked words to view definitions
- search
- link to media activities and quizzes

Your professor can also write notes for the class and highlight important material using a new tool that works like an electronic pen on a whiteboard.

Get personalized coaching & feedback.

Your instructor may assign self-paced **MasteringBiology® tutorials** that provide individualized coaching with specific hints and feedback on the toughest topics in the course.

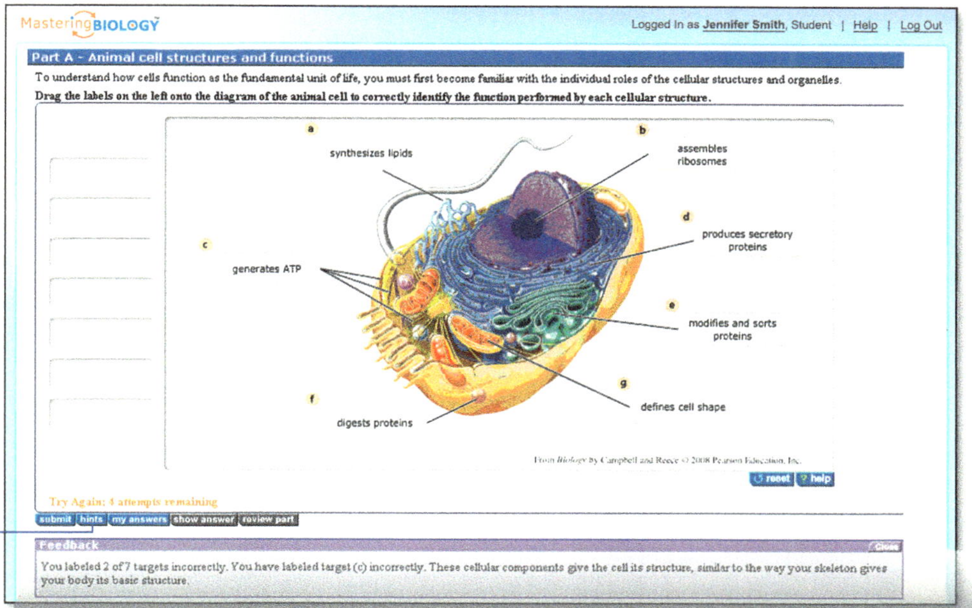

Feedback Close

You labeled 2 of 7 targets incorrectly. You have labeled target (c) incorrectly. These cellular components give the cell its structure, similar to the way your skeleton gives your body its basic structure.

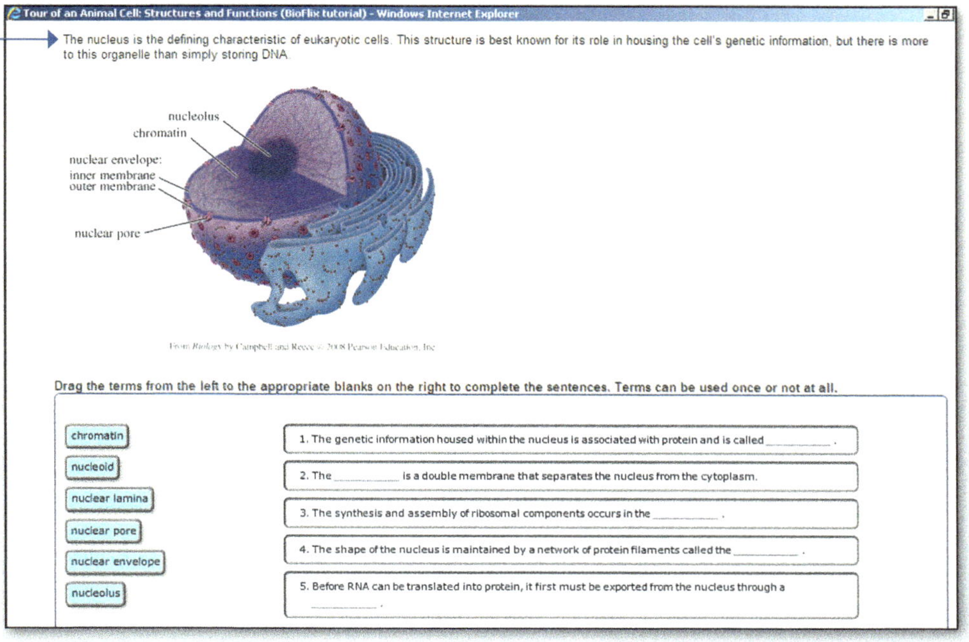

1 If you get stuck…

2 Specific wrong-answer **feedback** appears in the purple feedback box.

3 You are offered **hints** to coach you to the correct response.

To the Instructor: New Content in

MasteringBIOLOGY®
www.masteringbiology.com

MasteringBiology is the most effective and widely used online science tutorial, homework, and assessment system available.

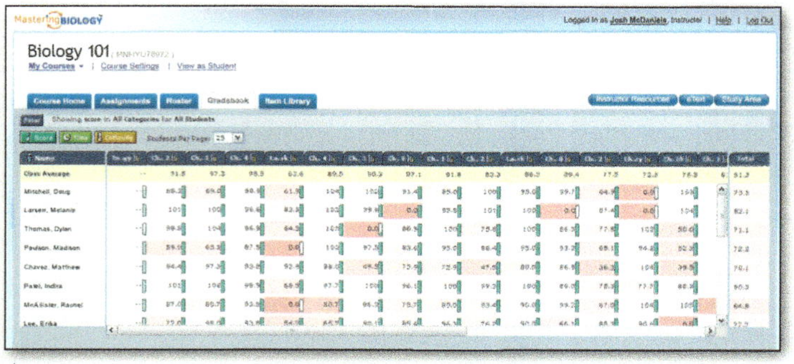

▲ The MasteringBiology **gradebook** provides you with quick results and easy-to-interpret insights into student performance. Every assignment is **automatically graded** and shades of red highlight vulnerable students and challenging assignments.

BioFlix

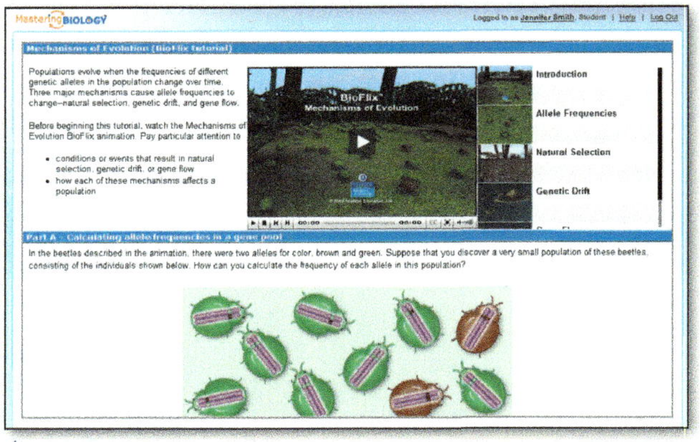

▲ **BioFlix® Tutorials** use 3-D, movie-quality animations and coaching exercises to help students master tough topics outside of class. Tutorials and animations include:

- A Tour of the Animal Cell
- A Tour of the Plant Cell
- Membrane Transport
- Cellular Respiration
- Photosynthesis
- Mitosis
- Meiosis
- DNA Replication
- Protein Synthesis
- Mechanisms of Evolution
- Water Transport in Plants
- Homeostasis: Regulating Blood Sugar
- Gas Exchange
- How Neurons Work
- How Synapses Work
- Muscle Contraction
- Population Ecology
- The Carbon Cycle

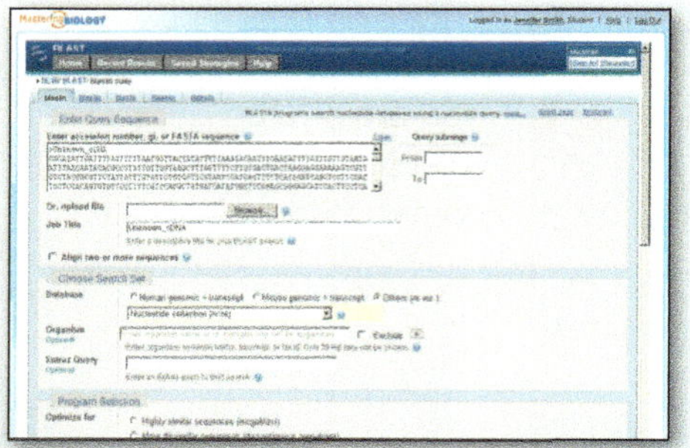

▲ New **Data Analysis Tutorials** allow students to analyze real data from online databases.

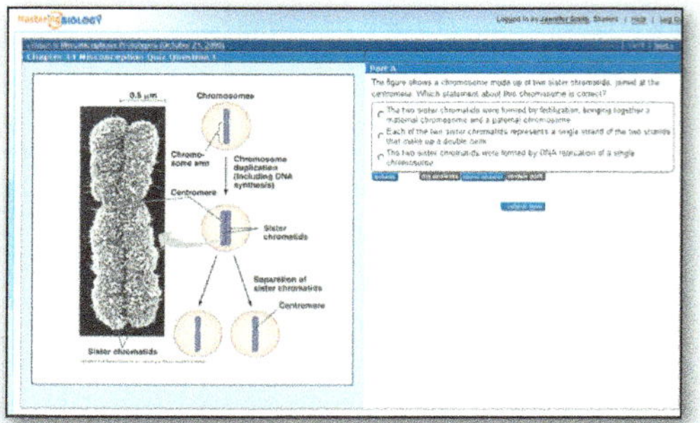

▲ New **Student Misconceptions Questions** provide assignable quizzes for each chapter based on common student misconceptions. Students are provided with feedback, and the instructor is provided with in-class strategies for overcoming these misconceptions.

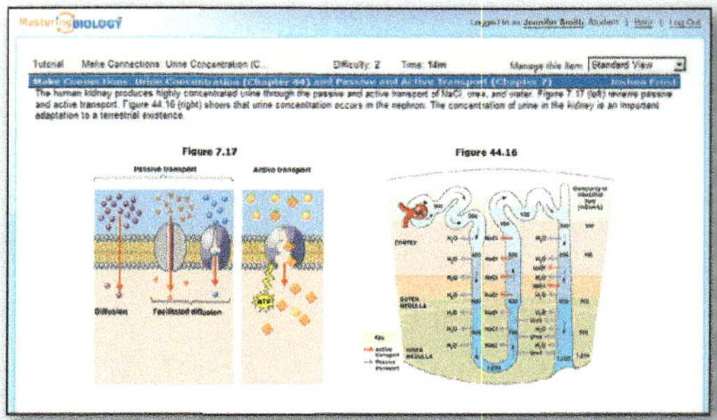

▲ New **Make Connections Tutorials** help students connect what they are learning in one chapter with material they learned in an earlier chapter.

MasteringBiology®

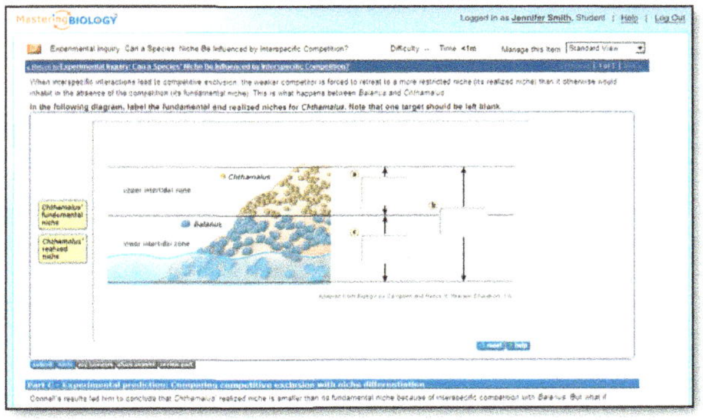

◀ New **Experimental Inquiry Tutorials,** based on some of biology's most influential experiments, give students practice analyzing experimental design and data, and help students understand how to reach conclusions based on collected data. Topics include:

- What Can You Learn About the Process of Science from Investigating a Cricket's Chirp?
- Which Wavelengths of Light Drive Photosynthesis?
- What Is the Inheritance Pattern of Sex-Linked Traits?
- Does DNA Replication Follow the Conservative, Semiconservative, or Dispersive Model?
- Did Natural Selection of Ground Finches Occur When the Environment Changed?
- What Effect Does Auxin Have on Coleoptile Growth?
- What Role Do Genes Play in Appetite Regulation?
- How Do Calcium Ions Help to Prevent Polyspermy During Egg Fertilization?
- Can a Species' Niche Be Influenced by Interspecific Competition?
- What Factors Influence the Loss of Nutrients from a Forest Ecosystem?

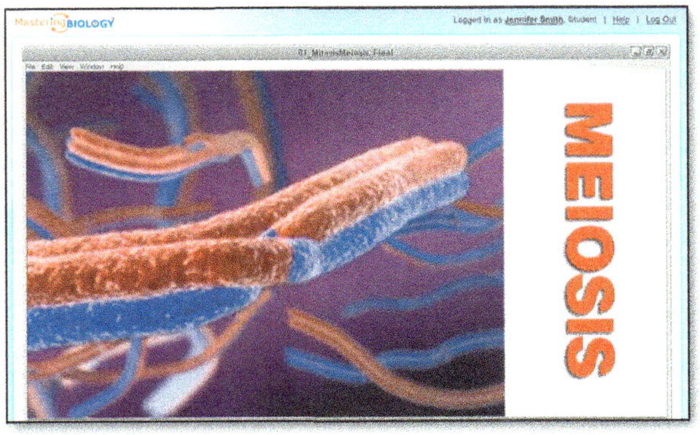

◀ The New **Video Tutor Sessions** walk students through tough topics with clearly explained visuals and demonstrations. Topics include:

- Mitosis and Meiosis
- Sex-Linked Pedigrees
- DNA Structure
- DNA Profiling Techniques
- Biodiversity
- Phylogenetic Trees

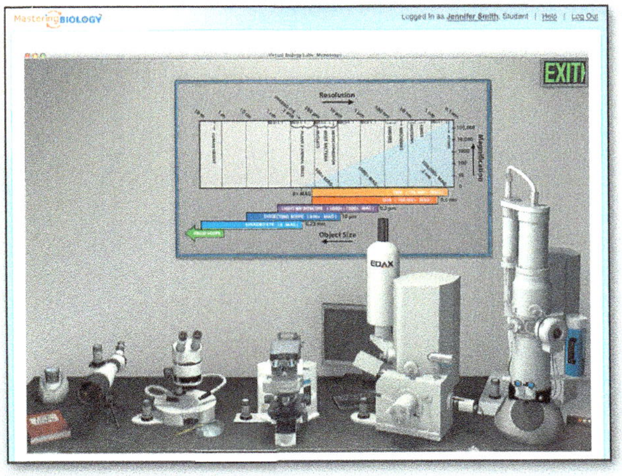

◀ The new **MasteringBiology: Virtual Biology Labs** online environment promotes critical thinking skills using virtual experiments and explorations that may be difficult to perform in a wet lab environment due to time, cost, or safety concerns.

- MasteringBiology: Virtual Biology Labs offer unique learning experiences in microscopy, molecular biology, genetics, ecology, and systematics.
- Choose from 20–30 automatically graded, "pre-set" lab activities that are ready to assign to students, or create your own from scratch.
- Each "pre-set" lab provides an assignable lab report with questions that are automatically graded and recorded in the MasteringBiology gradebook.
- Student subscriptions are available standalone or packaged with the *Campbell Biology* textbook.

Supplements

For Instructors

Instructor Resource DVD

978-0-321-67786-0 • 0-321-67786-2

Assets for each chapter include:

- All figures, photos, and tables in JPEG and PowerPoint®
- Prepared PowerPoint Presentations for each chapter, with lecture notes, editable figures from the text, and links to animations and videos
- Clicker Questions in PowerPoint
- 250+ Instructor Animations, including 3-D BioFlix®
- Discovery Channel™ Videos
- Test Bank questions in TestGen® software and Microsoft® Word

BioFlix® animations invigorate classroom lectures with 3-minute "movie quality" 3-D graphics (see list on p. xviii).

Customizable PowerPoints provide a jumpstart for each lecture.

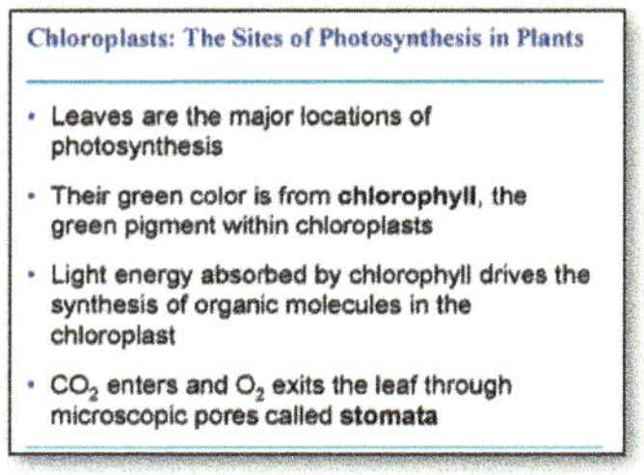

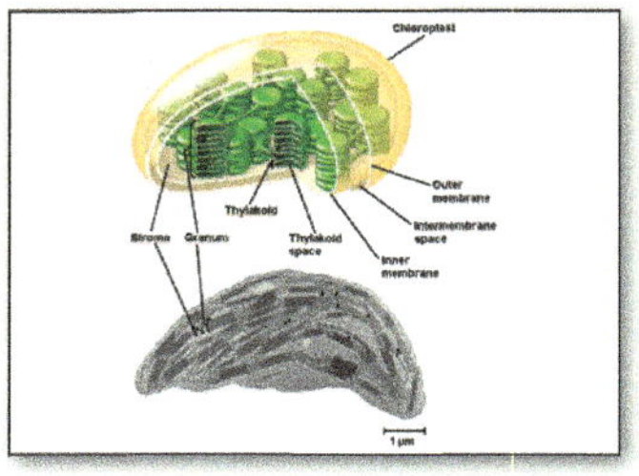

Clicker Questions can be used to stimulate effective classroom discussions (for use with or without clickers).

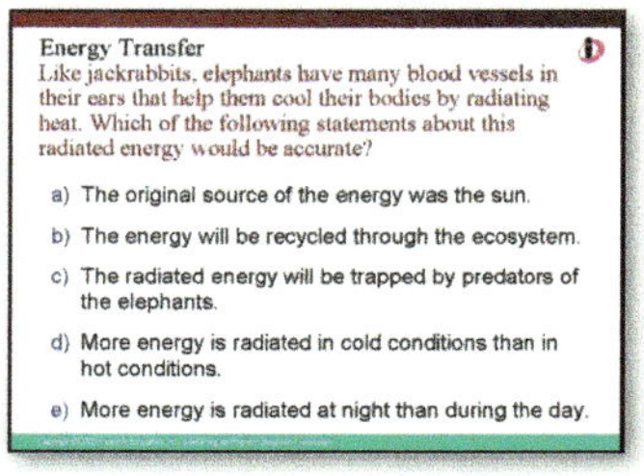

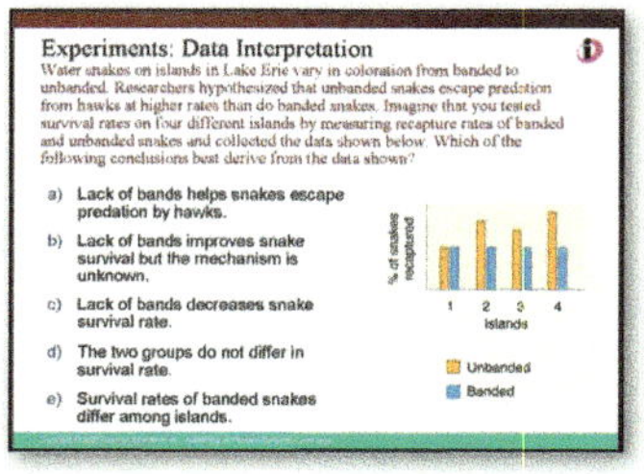

All of the art and photos from the book are provided with customizable labels. More than 1,600 photos from the text and other sources are included.

Printed Test Bank

by Louise Paquin, *McDaniel College*, Michael Dini, *Texas Tech University*, John Lepri, *University of North Carolina, Greensboro*, Jung Choi, *Georgia Institute of Technology*, John Zarnetske, *Hoosick Falls Central School*, and Ronald Balsamo, *Villanova University*

978-0-321-69729-5 • 0-321-69729-4

For the Ninth Edition, the Test Bank authors have increased the number of high-level application and synthesis questions.

Transparency Acetates

978-0-321-69708-0 • 0-321-69708-1

Course Management Systems

Test Bank questions, quizzes, and selected content from the Study Area of MasteringBiology are available in these popular course management systems:

CourseCompass™ (www.pearsonhighered.com)

Blackboard (www.pearsonhighered.com)

Instructor Resources Area in MasteringBiology

This area includes:

- PowerPoint® Lectures
- Clicker Questions
- JPEG Images
- Animations
- Videos
- Lecture Outlines
- Learning Objectives
- Strategies for Overcoming Common Student Misconceptions
- Instructor Guides for Supplements
- Write About a Theme: Tips for Grading Short-Answer Essays
- Suggested Answers for Essay Questions
- Test Bank Files
- Lab Media

For Students

***Study Guide*, Ninth Edition**
by Martha R. Taylor, *Cornell University*
978-0-321-62992-0 • 0-321-62992-2

This popular study guide helps students extract key ideas from the textbook and organize their knowledge of biology. Exercises include concept maps, chapter summaries, word roots, chapter tests, and a variety of interactive questions in various formats.

Inquiry in Action: Interpreting Scientific Papers*, Second Edition
by Ruth Buskirk, *University of Texas at Austin*, and Christopher M. Gillen, *Kenyon College*
978-0-321-68336-6 • 0-321-68336-6

Selected Inquiry Figures in the Ninth Edition direct students to read and analyze the complete original research paper. In this supplement, those articles are reprinted and accompanied by questions that help students analyze the article. The Inquiry Figures from the book are reprinted in the supplement.

Practicing Biology: A Student Workbook*, Fourth Edition
by Jean Heitz and Cynthia Giffen, *University of Wisconsin, Madison*
978-0-321-68328-1 • 0-321-68328-5

This workbook offers a variety of activities to suit different learning styles. Activities such as modeling and mapping allow students to visualize and understand biological processes. Other activities focus on basic skills, such as reading and drawing graphs.

Biological Inquiry: A Workbook of Investigative Cases*, Third Edition
by Margaret Waterman, *Southeast Missouri State University*, and Ethel Stanley, *BioQUEST Curriculum Consortium and Beloit College*
978-0-321-68320-5 • 0-321-68320-X

This workbook offers ten investigative cases. A student website is in the Study Area at **www.masteringbiology.com.**

***Study Card*, Ninth Edition**
978-0-321-68322-9 • 0-321-68322-6

This quick-reference card provides an overview of the entire field of biology and helps students see the connections between topics.

***Spanish Glossary*, Ninth Edition**
by Laura P. Zanello, *University of California, Riverside*
978-0-321-68321-2 • 0-321-68321-8

This resource provides definitions in Spanish for all the glossary terms.

Into the Jungle: Great Adventures in the Search for Evolution
by Sean B. Carroll, *University of Wisconsin, Madison*
978-0-321-55671-4 • 0-321-55671-2

These nine short tales vividly depict key discoveries in evolutionary biology and the excitement of the scientific process. Online resources available at www.aw-bc.com/carroll.

Get Ready for Biology
978-0-321-50057-1 • 0-321-50057-1

This engaging workbook helps students brush up on important math and study skills and get up to speed on biological terminology and the basics of chemistry and cell biology.

***A Short Guide to Writing About Biology*, Seventh Edition**
by Jan A. Pechenik, *Tufts University*
978-0-321-66838-7 • 0-321-66838-3

This best-selling writing guide teaches students to think as biologists and to express ideas clearly and concisely through their writing.

***An Introduction to Chemistry for Biology Students*, Ninth Edition**
by George I. Sackheim, *University of Illinois, Chicago*
978-0- 8053-9571-6 • 0-8053-9571-7

This text/workbook helps students review and master all the basic facts, concepts, and terminology of chemistry that they need for their life science course.

***The Chemistry of Life CD-ROM, Biology Version*, Second Edition**
by Robert M. Thornton, *University of California, Davis*
978-0-8053-3063-2 • 0-8053-3063-1

This CD-ROM uses animations, interactive simulations, and quizzes with feedback to help students learn or review the chemistry needed to succeed in introductory biology.

For Lab

***Investigating Biology Laboratory Manual*, Seventh Edition**
by Judith G. Morgan, *Emory University*, and M. Eloise Brown Carter, *Oxford College of Emory University*
978-0-321-66821-9 • 0-321-66821-9

The Seventh Edition emphasizes connections to recurring themes in biology, including structure and function, unity and diversity, and the overarching theme of evolution.

***Annotated Instructor Edition for Investigating Biology Laboratory Manual*, Seventh Edition**
by Judith G. Morgan, *Emory University*, and M. Eloise Brown Carter, *Oxford College of Emory University*
978-0-321-67668-9 • 0-321-67668-8

***Preparation Guide for Investigating Biology Laboratory Manual*, Seventh Edition**
by Judith G. Morgan, *Emory University*, and M. Eloise Brown Carter, *Oxford College of Emory University*
978-0-321-67669-6 • 0-321-67669-6

Symbiosis: The Benjamin Cummings Custom Laboratory Program for the Biological Sciences

www.pearsoncustom.com/database/symbiosis/bc.html

MasteringBiology®: Virtual Biology Labs

www.masteringbiology.com
This online environment promotes critical thinking skills using virtual experiments and explorations that may be difficult to perform in a wet lab environment due to time, cost, or safety concerns. Designed to supplement or substitute for existing wet labs, this product offers students unique learning experiences and critical thinking exercises in the areas of microscopy, molecular biology, genetics, ecology, and systematics.

*** An Instructor Guide is available for download in the Instructor Resources Area at www.masteringbiology.com.**

Featured Figures

Impact Figures

Exploring Figures

Inquiry Figures

* The Inquiry Figure, original research paper, and a worksheet to guide you through the paper are provided in *Inquiry in Action: Interpreting Scientific Papers*, Second Edition.

† See the related Experimental Inquiry Tutorial in MasteringBiology® (www.masteringbiology.com).

Research Method Figures

Interviews

UNIT 1

The Chemistry of Life 28

Susan Solomon

National Oceanic and Atmospheric Administration, Boulder, Colorado

UNIT 2

The Cell 92

Bonnie L. Bassler

Princeton University

UNIT 3

Genetics 246

Joan A. Steitz

Yale University

UNIT 4

Mechanisms of Evolution 450

Geerat J. Vermeij

University of California, Davis

UNIT 5

The Evolutionary History of Biological Diversity 534

W. Ford Doolittle

Dalhousie University, Canada

UNIT 6

Plant Form and Function 736

Luis Herrera-Estrella

National Polytechnic Institute, Mexico

UNIT 7

Animal Form and Function 850

Baldomero M. Olivera

University of Utah

UNIT 8

Ecology 1142

Camille Parmesan

University of Texas, Austin

Acknowledgments

The authors wish to express their gratitude to the global community of instructors, researchers, students, and publishing professionals who have contributed to this edition.

As authors of this text, we are mindful of the daunting challenge of keeping up to date in all areas of our rapidly expanding subject. We are grateful to the many scientists who helped shape this edition by discussing their research fields with us, answering specific questions in their areas of expertise, and sharing their ideas about biology education. We are especially grateful to the following, listed alphabetically: John Archibald, John Armour, Kristian Axelsen, Scott Bowling, Barbara Bowman, Andy Cameron, Scott Carroll, Amy Cheng-Vollmer, Michele Clamp, David DeRosier, Doug DeSimone, Binh An Diep, David Ehrhardt, Robert Fowler, Peter Fraser, Matt Friedman, Tom Gingeras, Anita Gondor, Ken Halanych, Jeff Hardin, Catherine Hurlbut, Adam Johnson, Dale Kaiser, Patrick Keeling, Emir Khatipov, Chris Killian, Andrew Knoll, Nikos Kyrpides, Teri Liegler, Zhe-Xi Luo, Kent MacDonald, Nick Matzke, Melissa Michael, Nadia Naffakh, Rolf Ohlsson, Aharon Oren, Tom Owens, Kevin Padian, Nathalie Pardigon, Bruce Pavlik, Kevin J. Peterson, Michael Pollock, Rebekah Rasooly, Andrew Roger, Ole Seehausen, Alastair Simpson, Betty Smocovitis, Frank Solomon, Pam Soltis, Hans Thewissen, Mark Uhen, Vance Vredenburg, Elisabeth Wade, Phil Zamore, and Christine Zardecki. In addition, a total of 168 biologists, listed on pages xxvi–xxvii, provided detailed reviews of chapters for this edition, helping us ensure the book's scientific accuracy and improve its pedagogical effectiveness. And finally, we thank Marty Taylor, author of all nine editions of the *Student Study Guide*, for her many contributions to the accuracy, clarity, and consistency of the book.

Thanks also to the other professors and students, from all over the world, who offered suggestions directly to the authors. We alone bear the responsibility for any errors that remain in the text, but the dedication of our consultants, reviewers, and other correspondents makes us confident in the accuracy and effectiveness of this edition.

Interviews with prominent scientists have been a hallmark of *Campbell Biology* since its inception, and conducting these interviews was again one of the great pleasures of revising the book. To open the eight units of this edition, we are proud to include interviews with Susan Solomon, Bonnie Bassler, Joan Steitz, Geerat Vermeij, Ford Doolittle, Luis Herrera-Estrella, Toto Olivera, and Camille Parmesan.

The value of *Campbell Biology* as a learning tool is greatly enhanced by the supplementary materials that have been created for instructors and students. We recognize that the dedicated authors of these materials are essentially writing mini (and not so mini) books. We much appreciate the hard work and creativity of all the authors listed, with their creations, on pages xx–xxi. In addition, we are grateful to Joan Sharp (*Lecture Outlines*, *Learning Objectives*, and *Student Misconceptions*) and Erin Barley and Kathleen Fitzpatrick (*PowerPoint® Lectures*).

The electronic media for this text are invaluable teaching and learning aids. We thank the hardworking and creative authors of the new material for this edition: Tom Owens, Joan Sharp, and Jennifer Yeh (*MasteringBiology®*); Richard Cowlishaw, Tod Duncan, and Stephanie Pandolfi (*Study Area Practice Tests*); and Eric Simon (*VideoTutors*). And we thank Brian White for his work. Also, we are grateful to the many other people—biology instructors, editors, artists, production experts, and narrators—who are listed in the credits for these and other elements of the electronic media that accompany the book. And we thank the reviewers and class testers of BioFlix® and MasteringBiology.

Campbell Biology, Ninth Edition, results from an unusually strong synergy between a team of scientists and a team of publishing professionals. Our editorial team at Benjamin Cummings again demonstrated unmatched talents, commitment, and pedagogical insights. Our new Acquisitions Editor, Josh Frost, brought publishing savvy, intelligence, and a much appreciated level head to leading the whole team. The clarity and effectiveness of every page owe much to our extraordinary Supervising Editors Pat Burner and Beth Winickoff, who headed the top-notch development team: Developmental Editors John Burner, Matt Lee, and Mary Catherine Hager; and Developmental Artists Hilair Chism, Carla Simmons, Andrew Recher, and Jay McElroy. Our unsurpassed Senior Editorial Manager Ginnie Simione Jutson, Executive Director of Development Deborah Gale, Assistant Editor Logan Triglia, and Editor-in-Chief Beth Wilbur were indispensable in moving the project in the right direction. We also want to thank Robin Heyden for organizing the annual Biology Leadership Conferences and keeping us in touch with the world of AP Biology.

You would not have this beautiful book in your hands today if not for the work of the book production team: Executive Managing Editor Erin Gregg; Managing Editor Michael Early; Senior Production Project Manager Shannon Tozier; Senior Photo Editor Donna Kalal; Photo Researcher Maureen Spuhler; Copy Editor Janet Greenblatt; Art Editor Laura Murray; Proofreaders Joanna Dinsmore and Pete Shanks; Permissions Editors Sue Ewing and Beth Keister; Senior Project Editor Emily Bush, S4Carlisle; Composition Manager Holly Paige, S4Carlisle; Art Production Manager Kristina Seymour, Precision Graphics; Design Director Mark Ong; Designer Gary Hespenheide; and Manufacturing Buyer Michael Penne. We also thank those who worked on the book's supplements: Susan Berge, Nina Lewallen Hufford, Brady Golden, Jane Brundage, James Bruce, and John Hammett. And for creating the wonderful package of electronic media that accompanies the book, we are grateful to Tania Mlawer, Director of Editorial Content for MasteringBiology, and Jonathan Ballard, Deb Greco, Sarah Jensen, Mary Catherine Hager, Alice Fugate, Juliana Tringali, Josh Gentry, Steve Wright, Kristen Sutton, Katie Foley, Karen Sheh, and David Kokorowski, as well as Director of Media Development Lauren Fogel and Director of Media Strategy Stacy Treco.

For their important roles in marketing the book, we thank Christy Lawrence, Lauren Harp, Scott Dustan, Lillian Carr, Jane Campbell, Jessica Perry, Nora Massuda, and Jessica Tree.

We are grateful to Linda Davis, President of Pearson Math and Science, who has shared our commitment to excellence and provided strong support for five editions now. Moreover, we thank Paul Corey, President of Pearson Benjamin Cummings, and Editorial Director Frank Ruggirello for their enthusiasm, encouragement, and support.

The Pearson sales team, which represents *Campbell Biology* on campus, is an essential link to the users of the text. They tell us what you like and don't like about the book, communicate the features of the book, and provide prompt service. We thank them for their hard work and professionalism. For representing our book to our international audience, we thank our sales and marketing partners throughout the world. They are all strong allies in biology education.

Finally, we wish to thank our families and friends for their encouragement and patience throughout this long project. Our special thanks to Paul, Dan, Maria, Armelle, and Sean (J.B.R.); Lily, Ross, Lily-too, and Alex (L.A.U.); Debra and Hannah (M.L.C.); Harry, Elga, Aaron, Sophie, Noah, and Gabriele (S.A.W.); Natalie (P.V.M.); and Sally, Robert, David, and Will (R.B.J.). And, as always, thanks to Rochelle, Allison, Jason, and McKay.

Jane Reece, Lisa Urry, Michael Cain,
Steve Wasserman, Peter Minorsky, and Rob Jackson

Reviewers

Ninth Edition Reviewers

Ann Aguanno, *Marymount Manhattan College*
Marc Albrecht, *University of Nebraska*
John Alcock, *Arizona State University*
Eric Alcorn, *Acadia University*
Terry Austin, *Temple College*
Brian Bagatto, *University of Akron*
Virginia Baker, *Chipola College*
Bonnie Baxter, *Westminster College*
Marilee Benore, *University of Michigan, Dearborn*
Catherine Black, *Idaho State University*
William Blaker, *Furman University*
Edward Blumenthal, *Marquette University*
David Bos, *Purdue University*
Scott Bowling, *Auburn University*
Beth Burch, *Huntington University*
Ragan Callaway, *The University of Montana*
Kenneth M. Cameron, *University of Wisconsin, Madison*
Patrick Canary, *Northland Pioneer College*
Cheryl Keller Capone, *Pennsylvania State University*
Karen I. Champ, *Central Florida Community College*
David Champlin, *University of Southern Maine*
Brad Chandler, *Palo Alto College*
Wei-Jen Chang, *Hamilton College*
Jung Choi, *Georgia Institute of Technology*
Steve Christensen, *Brigham Young University, Idaho*
James T. Colbert, *Iowa State University*
William Cushwa, *Clark College*
Shannon Datwyler, *California State University, Sacramento*
Eugene Delay, *University of Vermont*
Daniel DerVartanian, *University of Georgia*
Janet De Souza-Hart, *Massachusetts College of Pharmacy & Health Sciences*
Kathryn A. Durham, *Lorain Community College*
Curt Elderkin, *College of New Jersey*
Mary Ellard-Ivey, *Pacific Lutheran University*
George Ellmore, *Tufts University*
Robert C. Evans, *Rutgers University, Camden*
Sam Fan, *Bradley University*
Paul Farnsworth, *University of New Mexico*
Myriam Alhadeff Feldman, *Cascadia Community College*
Teresa Fischer, *Indian River Community College*
David Fitch, *New York University*
T. Fleming, *Bradley University*
Robert Fowler, *San Jose State University*
Robert Franklin, *College of Charleston*
Art Fredeen, *University of Northern British Columbia*
Matt Friedman, *University of Chicago*
Cynthia M. Galloway, *Texas A&M University, Kingsville*
Simon Gilroy, *University of Wisconsin, Madison*
Jim Goetze, *Laredo Community College*
Lynda Goff, *University of California, Santa Cruz*
Roy Golsteyn, *University of Lethbridge*
Barbara E. Goodman, *University of South Dakota*
David Grise, *Texas A&M University, Corpus Christi*
Devney Hamilton, *Stanford University (student)*
Matthew B. Hamilton, *Georgetown University*
Jeanne M. Harris, *University of Vermont*
Stephanie Harvey, *Georgia Southwestern State University*
Bernard Hauser, *University of Florida*
Andreas Hejnol, *Sars International Centre for Marine Molecular Biology*
Jason Hodin, *Stanford University*
Sara Huang, *Los Angeles Valley College*
Catherine Hurlbut, *Florida State College, Jacksonville*
Diane Husic, *Moravian College*
Thomas Jacobs, *University of Illinois*
Mark Jaffe, *Nova Southeastern University*
Douglas Jensen, *Converse College*
Lance Johnson, *Midland Lutheran College*
Cheryl Jorcyk, *Boise State University*
Caroline Kane, *University of California, Berkeley*
Jennifer Katcher, *Pima Community College*
Eric G. Keeling, *Cary Institute of Ecosystem Studies*
Chris Kennedy, *Simon Fraser University*
Hillar Klandorf, *West Virginia University*
Mark Knauss, *Georgia Highlands College*
Roger Koeppe, *University of Arkansas*
Peter Kourtev, *Central Michigan University*
Eliot Krause, *Seton Hall University*
Steven Kristoff, *Ivy Tech Community College*
William Kroll, *Loyola University*
Rukmani Kuppuswami, *Laredo Community College*
Lee Kurtz, *Georgia Gwinnett College*
Michael P. Labare, *United States Military Academy, West Point*
Ellen Lamb, *University of North Carolina, Greensboro*
William Lamberts, *College of St Benedict and St John's University*
Tali D. Lee, *University of Wisconsin, Eau Claire*
Hugh Lefcort, *Gonzaga University*
Alcinda Lewis, *University of Colorado, Boulder*
Graeme Lindbeck, *Valenica Community College*
Hannah Lui, *University of California, Irvine*
Cindy Malone, *California State University, Northridge*
Julia Marrs, *Barnard College (student)*
Kathleen Marrs, *Indiana University-Purdue University, Indianapolis*
Mike Mayfield, *Ball State University*
Kamau Mbuthia, *Bowling Green State University*
Tanya McGhee, *Craven Community College*
Darcy Medica, *Pennsylvania State University*
Susan Meiers, *Western Illinois University*
Alex Mills, *University of Windsor*
Eli Minkoff, *Bates College*
Subhash Minocha, *University of New Hampshire*
Ivona Mladenovic, *Simon Fraser University*
Courtney Murren, *College of Charleston*
Kimberlyn Nelson, *Pennsylvania State University*
Jacalyn Newman, *University of Pittsburgh*
Kathleen Nolta, *University of Michigan*
Aharon Oren, *The Hebrew University*
Henry R. Owen, *Eastern Illinois University*
Stephanie Pandolfi, *Michigan State University*
Nathalie Pardigon, *Institut Pasteur*
Cindy Paszkowski, *University of Alberta*
Andrew Pease, *Stevenson University*
Nancy Pelaez, *Purdue University*
Irene Perry, *University of Texas of the Permian Basin*
Roger Persell, *Hunter College*
Mark Pilgrim, *College of Coastal Georgia*
Vera M. Piper, *Shenadoah University*
Crima Pogge, *City College of San Francisco*
Michael Pollock, *Mount Royal University*
Roberta Pollock, *Occidental College*
Therese M. Poole, *Georgia State University*
Angela R. Porta, *Kean University*
Robert Powell, *Avila University*
Elena Pravosudova, *University of Nevada, Reno*
Terrell Pritts, *University of Arkansas, Little Rock*
Monica Ranes-Goldberg, *University of California, Berkeley*
Robert S. Rawding, *Gannon University*
Sarah Richart, *Azusa Pacific University*
Kenneth Robinson, *Purdue University*
Heather Roffey, *Marianopolis College*
Patricia Rugaber, *College of Coastal Georgia*
Scott Russell, *University of Oklahoma*
Louis Santiago, *University of California, Riverside*
Tom Sawicki, *Spartanburg Community College*
Thomas W. Schoener, *University of California, Davis*
Patricia Schulte, *University of British Colombia*
Brenda Schumpert, *Valencia Community College*
David Schwartz, *Houston Community College*
Brent Selinger, *University of Lethbridge*
Alison M. Shakarian, *Salve Regina University*
Robin L. Sherman, *Nova Southeastern University*
Sedonia Sipes, *Southern Illinois University, Carbondale*
Joel Stafstrom, *Northern Illinois University*
Alam Stam, *Capital University*
Judy Stone, *Colby College*
Cynthia Surmacz, *Bloomsburg University*
David Tam, *University of North Texas*
Yves Tan, *Cabrillo College*
Emily Taylor, *California Polytechnic State University*
Franklyn Tan Te, *Miami Dade College*
Kent Thomas, *Wichita State University*

Saba Valadkhan, *Center for RNA Molecular Biology*
Sarah VanVickle-Chavez, *Washington University, St. Louis*
William Velhagen, *New York University*
Janice Voltzow, *University of Scranton*
Margaret Voss, *Penn State Erie*
Charles Wade, *C.S. Mott Community College*
Claire Walczak, *Indiana University*
Jerry Waldvogel, *Clemson University*
Robert Lee Wallace, *Ripon College*
Fred Wasserman, *Boston University*
John Weishampel, *University of Central Florida*
Susan Whittemore, *Keene State College*
Janet Wolkenstein, *Hudson Valley Community College*
Grace Wyngaard, *James Madison University*
Paul Yancey, *Whitman College*
Anne D. Yoder, *Duke University*
Nina Zanetti, *Siena College*
Sam Zeveloff, *Weber State University*
Theresa Zucchero, *Methodist University*

Reviewers of Previous Editions

Kenneth Able *(State University of New York, Albany)*
Thomas Adams *(Michigan State University)*
Martin Adamson *(University of British Columbia)*
Dominique Adriaens *(Ghent University)*
Shylaja Akkaraju *(Bronx Community College of CUNY)*
John Alcock *(Arizona State University)*
George R. Aliaga *(Tarrant County College)*
Richard Almon *(State University of New York, Buffalo)*
Bonnie Amos *(Angelo State University)*
Katherine Anderson *(University of California, Berkeley)*
Richard J. Andren *(Montgomery County Community College)*
Estry Ang *(University of Pittsburgh, Greensburg)*
Jeff Appling *(Clemson University)*
J. David Archibald *(San Diego State University)*
David Armstrong *(University of Colorado, Boulder)*
Howard J. Arnott *(University of Texas, Arlington)*
Mary Ashley *(University of Illinois, Chicago)*
Angela S. Aspbury *(Texas State University)*
Robert Atherton *(University of Wyoming)*
Karl Aufderheide *(Texas A&M University)*
Leigh Auleb *(San Francisco State University)*
P. Stephen Baenziger *(University of Nebraska)*
Ellen Baker *(Santa Monica College)*
Katherine Baker *(Millersville University)*
William Barklow *(Framingham State College)*
Susan Barman *(Michigan State University)*
Steven Barnhart *(Santa Rosa Junior College)*
Andrew Barton *(University of Maine Farmington)*
Rebecca A. Bartow *(Western Kentucky University)*
Ron Basmajian *(Merced College)*
David Bass *(University of Central Oklahoma)*
Bonnie Baxter *(Hobart & William Smith Colleges)*
Tim Beagley *(Salt Lake Community College)*
Margaret E. Beard *(College of the Holy Cross)*
Tom Beatty *(University of British Columbia)*
Chris Beck *(Emory University)*
Wayne Becker *(University of Wisconsin, Madison)*
Patricia Bedinger *(Colorado State University)*
Jane Beiswenger *(University of Wyoming)*
Anne Bekoff *(University of Colorado, Boulder)*
Marc Bekoff *(University of Colorado, Boulder)*
Tania Beliz *(College of San Mateo)*
Adrianne Bendich *(Hoffman-La Roche, Inc.)*
Barbara Bentley *(State University of New York, Stony Brook)*
Darwin Berg *(University of California, San Diego)*
Werner Bergen *(Michigan State University)*
Gerald Bergstrom *(University of Wisconsin, Milwaukee)*
Anna W. Berkovitz *(Purdue University)*
Dorothy Berner *(Temple University)*
Annalisa Berta *(San Diego State University)*
Paulette Bierzychudek *(Pomona College)*
Charles Biggers *(Memphis State University)*
Kenneth Birnbaum *(New York University)*
Michael W. Black *(California Polytechnic State University, San Luis Obispo)*
Robert Blanchard *(University of New Hampshire)*
Andrew R. Blaustein *(Oregon State University)*
Judy Bluemer *(Morton College)*
Edward Blumenthal *(Marquette University)*
Robert Blystone *(Trinity University)*
Robert Boley *(University of Texas, Arlington)*
Jason E. Bond *(East Carolina University)*
Eric Bonde *(University of Colorado, Boulder)*
Cornelius Bondzi *(Hampton University)*
Richard Boohar *(University of Nebraska, Omaha)*
Carey L. Booth *(Reed College)*
Allan Bornstein *(Southeast Missouri State University)*
Oliver Bossdorf *(State University of New York, Stony Book)*
James L. Botsford *(New Mexico State University)*
Lisa Boucher *(University of Nebraska, Omaha)*
J. Michael Bowes *(Humboldt State University)*
Richard Bowker *(Alma College)*
Robert Bowker *(Glendale Community College, Arizona)*
Barbara Bowman *(Mills College)*
Barry Bowman *(University of California, Santa Cruz)*
Deric Bownds *(University of Wisconsin, Madison)*
Robert Boyd *(Auburn University)*
Sunny Boyd *(University of Notre Dame)*
Jerry Brand *(University of Texas, Austin)*
Edward Braun *(Iowa State University)*
Theodore A. Bremner *(Howard University)*
James Brenneman *(University of Evansville)*
Charles H. Brenner *(Berkeley, California)*
Lawrence Brewer *(University of Kentucky)*
Donald P. Briskin *(University of Illinois, Urbana)*
Paul Broady *(University of Canterbury)*
Chad Brommer *(Emory University)*
Judith L. Bronstein *(University of Arizona)*
Danny Brower *(University of Arizona)*
Carole Browne *(Wake Forest University)*
Mark Browning *(Purdue University)*
David Bruck *(San Jose State University)*
Robb T. Brumfield *(Louisiana State University)*
Herbert Bruneau *(Oklahoma State University)*
Gary Brusca *(Humboldt State University)*
Richard C. Brusca *(University of Arizona, Arizona-Sonora Desert Museum)*
Alan H. Brush *(University of Connecticut, Storrs)*
Howard Buhse *(University of Illinois, Chicago)*
Arthur Buikema *(Virginia Tech)*
Al Burchsted *(College of Staten Island)*
Meg Burke *(University of North Dakota)*
Edwin Burling *(De Anza College)*
William Busa *(Johns Hopkins University)*
Jorge Busciglio *(University of California, Irvine)*
John Bushnell *(University of Colorado)*
Linda Butler *(University of Texas, Austin)*
David Byres *(Florida Community College, Jacksonville)*
Guy A. Caldwell *(University of Alabama)*
Jane Caldwell *(West Virginia University)*
Kim A. Caldwell *(University of Alabama)*
R. Andrew Cameron *(California Institute of Technology)*
Alison Campbell *(University of Waikato)*
Iain Campbell *(University of Pittsburgh)*
W. Zacheus Cande *(University of California, Berkeley)*
Robert E. Cannon *(University of North Carolina, Greensboro)*
Deborah Canington *(University of California, Davis)*
Frank Cantelmo *(St. John's University)*
John Capeheart *(University of Houston, Downtown)*
Gregory Capelli *(College of William and Mary)*
Richard Cardullo *(University of California, Riverside)*
Nina Caris *(Texas A&M University)*
Jeffrey Carmichael *(University of North Dakota)*
Robert Carroll *(East Carolina University)*
Laura L. Carruth *(Georgia State University)*
J. Aaron Cassill *(University of Texas, San Antonio)*
David Champlin *(University of Southern Maine)*
Bruce Chase *(University of Nebraska, Omaha)*
P. Bryant Chase *(Florida State University)*
Doug Cheeseman *(De Anza College)*
Shepley Chen *(University of Illinois, Chicago)*
Giovina Chinchar *(Tougaloo College)*
Joseph P. Chinnici *(Virginia Commonwealth University)*
Jung H. Choi *(Georgia Institute of Technology)*
Geoffrey Church *(Fairfield University)*
Henry Claman *(University of Colorado Health Science Center)*
Anne Clark *(Binghamton University)*
Greg Clark *(University of Texas)*
Patricia J. Clark *(Indiana University-Purdue University, Indianapolis)*
Ross C. Clark *(Eastern Kentucky University)*
Lynwood Clemens *(Michigan State University)*
Janice J. Clymer *(San Diego Mesa College)*
William P. Coffman *(University of Pittsburgh)*
Austin Randy Cohen *(California State University, Northridge)*
J. John Cohen *(University of Colorado Health Science Center)*
Jim Colbert *(Iowa State University)*
Jan Colpaert *(Hasselt University)*

Robert Colvin *(Ohio University)*
Jay Comeaux *(McNeese State University)*
David Cone *(Saint Mary's University)*
Elizabeth Connor *(University of Massachusetts)*
Joanne Conover *(University of Connecticut)*
Gregory Copenhaver *(University of North Carolina, Chapel Hill)*
John Corliss *(University of Maryland)*
James T. Costa *(Western Carolina University)*
Stuart J. Coward *(University of Georgia)*
Charles Creutz *(University of Toledo)*
Bruce Criley *(Illinois Wesleyan University)*
Norma Criley *(Illinois Wesleyan University)*
Joe W. Crim *(University of Georgia)*
Greg Crowther *(University of Washington)*
Karen Curto *(University of Pittsburgh)*
Anne Cusic *(University of Alabama, Birmingham)*
Richard Cyr *(Pennsylvania State University)*
Marymegan Daly *(The Ohio State University)*
W. Marshall Darley *(University of Georgia)*
Cynthia Dassler *(The Ohio State University)*
Marianne Dauwalder *(University of Texas, Austin)*
Larry Davenport *(Samford University)*
Bonnie J. Davis *(San Francisco State University)*
Jerry Davis *(University of Wisconsin, La Crosse)*
Michael A. Davis *(Central Connecticut State University)*
Thomas Davis *(University of New Hampshire)*
John Dearn *(University of Canberra)*
Maria E. de Bellard *(California State University, Northridge)*
Teresa DeGolier *(Bethel College)*
James Dekloe *(University of California, Santa Cruz)*
Patricia A. DeLeon *(University of Delaware)*
Veronique Delesalle *(Gettysburg College)*
T. Delevoryas *(University of Texas, Austin)*
Roger Del Moral *(University of Washington)*
Charles F. Delwiche *(University of Maryland)*
Diane C. DeNagel *(Northwestern University)*
William L. Dentler *(University of Kansas)*
Daniel Dervartanian *(University of Georgia)*
Jean DeSaix *(University of North Carolina, Chapel Hill)*
Biao Ding *(Ohio State University)*
Michael Dini *(Texas Tech University)*
Andrew Dobson *(Princeton University)*
Stanley Dodson *(University of Wisconsin, Madison)*
Mark Drapeau *(University of California, Irvine)*
John Drees *(Temple University School of Medicine)*
Charles Drewes *(Iowa State University)*
Marvin Druger *(Syracuse University)*
Gary Dudley *(University of Georgia)*
Susan Dunford *(University of Cincinnati)*
Betsey Dyer *(Wheaton College)*
Robert Eaton *(University of Colorado)*
Robert S. Edgar *(University of California, Santa Cruz)*
Douglas J. Eernisse *(California State University, Fullerton)*
Betty J. Eidemiller *(Lamar University)*
Brad Elder *(Doane College)*
William D. Eldred *(Boston University)*
Michelle Elekonich *(University of Nevada, Las Vegas)*
Mary Ellard-Ivey *(Pacific Lutheran University)*
Norman Ellstrand *(University of California, Riverside)*
Johnny El-Rady *(University of South Florida)*
Dennis Emery *(Iowa State University)*
John Endler *(University of California, Santa Barbara)*
Margaret T. Erskine *(Lansing Community College)*
Gerald Esch *(Wake Forest University)*
Frederick B. Essig *(University of South Florida)*
Mary Eubanks *(Duke University)*
David Evans *(University of Florida)*
Robert C. Evans *(Rutgers University, Camden)*
Sharon Eversman *(Montana State University)*
Olukemi Fadayomi *(Ferris State University)*
Lincoln Fairchild *(Ohio State University)*
Peter Fajer *(Florida State University)*
Bruce Fall *(University of Minnesota)*
Lynn Fancher *(College of DuPage)*
Ellen H. Fanning *(Vanderbilt University)*
Paul Farnsworth *(University of Texas, San Antonio)*
Larry Farrell *(Idaho State University)*
Jerry F. Feldman *(University of California, Santa Cruz)*
Lewis Feldman *(University of California, Berkeley)*
Eugene Fenster *(Longview Community College)*
Russell Fernald *(University of Oregon)*
Rebecca Ferrell *(Metropolitan State College of Denver)*
Kim Finer *(Kent State University)*
Milton Fingerman *(Tulane University)*
Barbara Finney *(Regis College)*
Frank Fish *(West Chester University)*
David Fisher *(University of Hawaii, Manoa)*
Jonathan S. Fisher *(St. Louis University)*
Steven Fisher *(University of California, Santa Barbara)*
Kirk Fitzhugh *(Natural History Museum of Los Angeles County)*
Lloyd Fitzpatrick *(University of North Texas)*
William Fixsen *(Harvard University)*
Abraham Flexer *(Manuscript Consultant, Boulder, Colorado)*
Kerry Foresman *(University of Montana)*
Norma Fowler *(University of Texas, Austin)*
Robert G. Fowler *(San Jose State University)*
David Fox *(University of Tennessee, Knoxville)*
Carl Frankel *(Pennsylvania State University, Hazleton)*
James Franzen *(University of Pittsburgh)*
Bill Freedman *(Dalhousie University)*
Otto Friesen *(University of Virginia)*
Frank Frisch *(Chapman University)*
Virginia Fry *(Monterey Peninsula College)*
Bernard Frye *(University of Texas, Arlington)*
Jed Fuhrman *(University of Southern California)*
Alice Fulton *(University of Iowa)*
Chandler Fulton *(Brandeis University)*
Sara Fultz *(Stanford University)*
Berdell Funke *(North Dakota State University)*
Anne Funkhouser *(University of the Pacific)*
Zofia E. Gagnon *(Marist College)*
Michael Gaines *(University of Miami)*
Arthur W. Galston *(Yale University)*
Stephen Gammie *(University of Wisconsin, Madison)*
Carl Gans *(University of Michigan)*
John Gapter *(University of Northern Colorado)*
Andrea Gargas *(University of Wisconsin, Madison)*
Lauren Garner *(California Polytechnic State University, San Luis Obispo)*
Reginald Garrett *(University of Virginia)*
Patricia Gensel *(University of North Carolina)*
Chris George *(California Polytechnic State University, San Luis Obispo)*
Robert George *(University of Wyoming)*
J. Whitfield Gibbons *(University of Georgia)*
J. Phil Gibson *(Agnes Scott College)*
Frank Gilliam *(Marshall University)*
Simon Gilroy *(Pennsylvania State University)*
Alan D. Gishlick *(Gustavus Adolphus College)*
Todd Gleeson *(University of Colorado)*
Jessica Gleffe *(University of California, Irvine)*
John Glendinning *(Barnard College)*
David Glenn-Lewin *(Wichita State University)*
William Glider *(University of Nebraska)*
Tricia Glidewell *(Marist School)*
Elizabeth A. Godrick *(Boston University)*
Lynda Goff *(University of California, Santa Cruz)*
Elliott Goldstein *(Arizona State University)*
Paul Goldstein *(University of Texas, El Paso)*
Sandra Gollnick *(State University of New York, Buffalo)*
Anne Good *(University of California, Berkeley)*
Judith Goodenough *(University of Massachusetts, Amherst)*
Wayne Goodey *(University of British Columbia)*
Robert Goodman *(University of Wisconsin, Madison)*
Ester Goudsmit *(Oakland University)*
Linda Graham *(University of Wisconsin, Madison)*
Robert Grammer *(Belmont University)*
Joseph Graves *(Arizona State University)*
Phyllis Griffard *(University of Houston, Downtown)*
A. J. F. Griffiths *(University of British Columbia)*
William Grimes *(University of Arizona)*
Mark Gromko *(Bowling Green State University)*
Serine Gropper *(Auburn University)*
Katherine L. Gross *(Ohio State University)*
Gary Gussin *(University of Iowa)*
Mark Guyer *(National Human Genome Research Institute)*
Ruth Levy Guyer *(Bethesda, Maryland)*
R. Wayne Habermehl *(Montgomery County Community College)*
Mac Hadley *(University of Arizona)*
Joel Hagen *(Radford University)*
Jack P. Hailman *(University of Wisconsin)*
Leah Haimo *(University of California, Riverside)*
Ken Halanych *(Auburn University)*
Jody Hall *(Brown University)*
Douglas Hallett, *(Northern Arizona University)*
Rebecca Halyard *(Clayton State College)*
E. William Hamilton *(Washington and Lee University)*
Sam Hammer *(Boston University)*
Penny Hanchey-Bauer *(Colorado State University)*
William F. Hanna *(Massasoit Community College)*

Laszlo Hanzely *(Northern Illinois University)*
Jeff Hardin *(University of Wisconsin, Madison)*
Lisa Harper *(University of California, Berkeley)*
Richard Harrison *(Cornell University)*
Carla Hass *(Pennsylvania State University)*
Chris Haufler *(University of Kansas)*
Bernard A. Hauser *(University of Florida)*
Chris Haynes *(Shelton State Community College)*
Evan B. Hazard *(Bemidji State University, Emeritus)*
H. D. Heath *(California State University, East Bay)*
George Hechtel *(State University of New York, Stony Brook)*
S. Blair Hedges *(Pennsylvania State University)*
Brian Hedlund *(University of Nevada, Las Vegas)*
David Heins *(Tulane University)*
Jean Heitz *(University of Wisconsin, Madison)*
John D. Helmann *(Cornell University)*
Colin Henderson *(University of Montana)*
Susan Hengeveld *(Indiana University)*
Michelle Henricks *(University of California, Los Angeles)*
Caroll Henry *(Chicago State University)*
Frank Heppner *(University of Rhode Island)*
Albert Herrera *(University of Southern California)*
Scott Herrick *(Missouri Western State College)*
Ira Herskowitz *(University of California, San Francisco)*
Paul E. Hertz *(Barnard College)*
David Hibbett *(Clark University)*
R. James Hickey *(Miami University)*
William Hillenius *(College of Charleston)*
Kenneth Hillers *(California Polytechnic State University, San Luis Obispo)*
Ralph Hinegardner *(University of California, Santa Cruz)*
William Hines *(Foothill College)*
Robert Hinrichsen *(Indiana University of Pennsylvania)*
Helmut Hirsch *(State University of New York, Albany)*
Tuan-hua David Ho *(Washington University)*
Carl Hoagstrom *(Ohio Northern University)*
James Hoffman *(University of Vermont)*
A. Scott Holaday *(Texas Tech University)*
N. Michele Holbrook *(Harvard University)*
James Holland *(Indiana State University, Bloomington)*
Charles Holliday *(Lafayette College)*
Lubbock Karl Holte *(Idaho State University)*
Alan R. Holyoak *(Brigham Young University, Idaho)*
Laura Hoopes *(Occidental College)*
Nancy Hopkins *(Massachusetts Institute of Technology)*
Sandra Horikami *(Daytona Beach Community College)*
Kathy Hornberger *(Widener University)*
Pius F. Horner *(San Bernardino Valley College)*
Becky Houck *(University of Portland)*
Margaret Houk *(Ripon College)*
Daniel J. Howard *(New Mexico State University)*
Ronald R. Hoy *(Cornell University)*
Cristin Hulslander *(University of Oregon)*
Donald Humphrey *(Emory University School of Medicine)*
Robert J. Huskey *(University of Virginia)*
Steven Hutcheson *(University of Maryland, College Park)*
Sandra Hsu *(Skyline College)*
Linda L. Hyde *(Gordon College)*
Bradley Hyman *(University of California, Riverside)*
Mark Iked *(San Bernardino Valley College)*
Jeffrey Ihara *(Mira Costa College)*
Cheryl Ingram-Smith *(Clemson University)*
Alice Jacklet *(State University of New York, Albany)*
John Jackson *(North Hennepin Community College)*
John C. Jahoda *(Bridgewater State College)*
Dan Johnson *(East Tennessee State University)*
Lee Johnson *(The Ohio State University)*
Randall Johnson *(University of California, San Diego)*
Stephen Johnson *(William Penn University)*
Wayne Johnson *(Ohio State University)*
Kenneth C. Jones *(California State University, Northridge)*
Russell Jones *(University of California, Berkeley)*
Chad Jordan *(North Carolina State University)*
Alan Journet *(Southeast Missouri State University)*
Walter Judd *(University of Florida)*
Thomas W. Jurik *(Iowa State University)*
Caroline M. Kane *(University of California, Berkeley)*
Thomas C. Kane *(University of Cincinnati)*
Tamos Kapros *(University of Missouri)*
E. L. Karlstrom *(University of Puget Sound)*
Jennifer Katcher *(Pima Community College)*
Laura A. Katz *(Smith College)*
Maureen Kearney *(Field Museum of Natural History)*
Patrick Keeling *(University of British Columbia)*
Elizabeth A. Kellogg *(University of Missouri, St. Louis)*
Norm Kenkel *(University of Manitoba)*
Chris Kennedy *(Simon Fraser University)*
George Khoury *(National Cancer Institute)*
Rebecca T. Kimball *(University of Florida)*
Mark Kirk *(University of Missouri, Columbia)*
Robert Kitchin *(University of Wyoming)*
Attila O. Klein *(Brandeis University)*
Daniel Klionsky *(University of Michigan)*
Jennifer Knight *(University of Colorado)*
Ned Knight *(Linfield College)*
David Kohl *(University of California, Santa Barbara)*
Greg Kopf *(University of Pennsylvania School of Medicine)*
Thomas Koppenheffer *(Trinity University)*
Margareta Krabbe *(Uppsala University)*
Anselm Kratochwil *(Universität Osnabrück)*
Deborah M. Kristan *(California State University, San Marcos)*
William Kroll *(Loyola University, Chicago)*
Janis Kuby *(San Francisco State University)*
Justin P. Kumar *(Indiana University)*
David Kurijaka *(Ohio University)*
Marc-André Lachance *(University of Western Ontario)*
J. A. Lackey *(State University of New York, Oswego)*
Elaine Lai *(Brandeis University)*
Mohamed Lakrim *(Kingsborough Community College)*
Lynn Lamoreux *(Texas A&M University)*
William L'Amoreaux *(College of Staten Island)*
Carmine A. Lanciani *(University of Florida)*
Kenneth Lang *(Humboldt State University)*
Dominic Lannutti *(El Paso Community College)*
Allan Larson *(Washington University)*
John Latto *(University of California, Santa Barbara)*
Diane K. Lavett *(State University of New York, Cortland, and Emory University)*
Charles Leavell *(Fullerton College)*
C. S. Lee *(University of Texas)*
Daewoo Lee *(Ohio University)*
Robert Leonard *(University of California, Riverside)*
Michael R. Leonardo *(Coe College)*
John Lepri *(University of North Carolina, Greensboro)*
Donald Levin *(University of Texas, Austin)*
Mike Levine *(University of California, Berkeley)*
Joseph Levine *(Boston College)*
Bill Lewis *(Shoreline Community College)*
John Lewis *(Loma Linda University)*
Lorraine Lica *(California State University, East Bay)*
Harvey Liftin *(Broward Community College)*
Harvey Lillywhite *(University of Florida, Gainesville)*
Graeme Lindbeck *(Valencia Community College)*
Clark Lindgren *(Grinnell College)*
Diana Lipscomb *(George Washington University)*
Christopher Little *(The University of Texas, Pan American)*
Kevin D. Livingstone *(Trinity University)*
Andrea Lloyd *(Middlebury College)*
Sam Loker *(University of New Mexico)*
Christopher A. Loretz *(State University of New York, Buffalo)*
Jane Lubchenco *(Oregon State University)*
Douglas B. Luckie *(Michigan State University)*
Margaret A. Lynch *(Tufts University)*
Steven Lynch *(Louisiana State University, Shreveport)*
Richard Machemer Jr. *(St. John Fisher College)*
Elizabeth Machunis-Masuoka *(University of Virginia)*
James MacMahon *(Utah State University)*
Christine R. Maher *(University of Southern Maine)*
Linda Maier *(University of Alabama, Huntsville)*
Jose Maldonado *(El Paso Community College)*
Richard Malkin *(University of California, Berkeley)*
Charles Mallery *(University of Miami)*
Keith Malmos *(Valencia Community College, East Campus)*
Cindy Malone *(California State University, Northridge)*
Carol Mapes *(Kutztown University of Pennsylvania)*
William Margolin *(University of Texas Medical School)*
Lynn Margulis *(Boston University)*
Kathleen A. Marrs *(Indiana University-Purdue University, Indianapolis)*
Edith Marsh *(Angelo State University)*
Diane L. Marshall *(University of New Mexico)*
Linda Martin Morris *(University of Washington)*
Karl Mattox *(Miami University of Ohio)*
Joyce Maxwell *(California State University, Northridge)*
Jeffrey D. May *(Marshall University)*
Lee McClenaghan *(San Diego State University)*
Richard McCracken *(Purdue University)*
Andrew McCubbin *(Washington State University)*
Kerry McDonald *(University of Missouri, Columbia)*
Jacqueline McLaughlin *(Pennsylvania State University, Lehigh Valley)*
Neal McReynolds *(Texas A&M International)*

Lisa Marie Meffert *(Rice University)*
Michael Meighan *(University of California, Berkeley)*
Scott Meissner *(Cornell University)*
Paul Melchior *(North Hennepin Community College)*
Phillip Meneely *(Haverford College)*
John Merrill *(Michigan State University)*
Brian Metscher *(University of California, Irvine)*
Ralph Meyer *(University of Cincinnati)*
James Mickle *(North Carolina State University)*
Roger Milkman *(University of Iowa)*
Helen Miller *(Oklahoma State University)*
John Miller *(University of California, Berkeley)*
Kenneth R. Miller *(Brown University)*
John E. Minnich *(University of Wisconsin, Milwaukee)*
Michael J. Misamore *(Texas Christian University)*
Kenneth Mitchell *(Tulane University School of Medicine)*
Alan Molumby *(University of Illinois, Chicago)*
Nicholas Money *(Miami University)*
Russell Monson *(University of Colorado, Boulder)*
Joseph P. Montoya *(Georgia Institute of Technology)*
Frank Moore *(Oregon State University)*
Janice Moore *(Colorado State University)*
Randy Moore *(Wright State University)*
William Moore *(Wayne State University)*
Carl Moos *(Veterans Administration Hospital, Albany, New York)*
Michael Mote *(Temple University)*
Alex Motten *(Duke University)*
Jeanette Mowery *(Madison Area Technical College)*
Deborah Mowshowitz *(Columbia University)*
Rita Moyes *(Texas A&M College Station)*
Darrel L. Murray *(University of Illinois, Chicago)*
John Mutchmor *(Iowa State University)*
Elliot Myerowitz *(California Institute of Technology)*
Gavin Naylor *(Iowa State University)*
John Neess *(University of Wisconsin, Madison)*
Tom Neils *(Grand Rapids Community College)*
Raymond Neubauer *(University of Texas, Austin)*
Todd Newbury *(University of California, Santa Cruz)*
James Newcomb *(New England College)*
Harvey Nichols *(University of Colorado, Boulder)*
Deborah Nickerson *(University of South Florida)*
Bette Nicotri *(University of Washington)*
Caroline Niederman *(Tomball College)*
Maria Nieto *(California State University, East Bay)*
Anders Nilsson *(University of Umeå)*
Greg Nishiyama *(College of the Canyons)*
Charles R. Noback *(College of Physicians and Surgeons, Columbia University)*
Jane Noble-Harvey *(Delaware University)*
Mary C. Nolan *(Irvine Valley College)*
Peter Nonacs *(University of California, Los Angeles)*
Mohamed A. F. Noor *(Duke University)*
Shawn Nordell *(St. Louis University)*
Richard S. Norman *(University of Michigan, Dearborn, Emeritus)*
David O. Norris *(University of Colorado, Boulder)*
Steven Norris *(California State University, Channel Islands)*
Gretchen North *(Occidental College)*
Cynthia Norton *(University of Maine, Augusta)*
Steve Norton *(East Carolina University)*
Steve Nowicki *(Duke University)*
Bette H. Nybakken *(Hartnell College)*
Brian O'Conner *(University of Massachusetts, Amherst)*
Gerard O'Donovan *(University of North Texas)*
Eugene Odum *(University of Georgia)*
Mark P. Oemke *(Alma College)*
Linda Ogren *(University of California, Santa Cruz)*
Patricia O'Hern *(Emory University)*
Nathan O. Okia *(Auburn University, Montgomery)*
Jeanette Oliver *(St. Louis Community College, Florissant Valley)*
Gary P. Olivetti *(University of Vermont)*
John Olsen *(Rhodes College)*
Laura J. Olsen *(University of Michigan)*
Sharman O'Neill *(University of California, Davis)*
Wan Ooi *(Houston Community College)*
John Oross *(University of California, Riverside)*
Gay Ostarello *(Diablo Valley College)*
Catherine Ortega *(Fort Lewis College)*
Charissa Osborne *(Butler University)*
Thomas G. Owens *(Cornell University)*
Penny Padgett *(University of North Carolina, Chapel Hill)*
Kevin Padian *(University of California, Berkeley)*
Dianna Padilla *(State University of New York, Stony Brook)*
Anthony T. Paganini *(Michigan State University)*
Barry Palevitz *(University of Georgia)*
Michael A. Palladino *(Monmouth University)*
Daniel Papaj *(University of Arizona)*
Peter Pappas *(County College of Morris)*
Bulah Parker *(North Carolina State University)*
Stanton Parmeter *(Chemeketa Community College)*
Robert Patterson *(San Francisco State University)*
Ronald Patterson *(Michigan State University)*
Crellin Pauling *(San Francisco State University)*
Kay Pauling *(Foothill Community College)*
Daniel Pavuk *(Bowling Green State University)*
Debra Pearce *(Northern Kentucky University)*
Patricia Pearson *(Western Kentucky University)*
Shelley Penrod *(North Harris College)*
Imara Y. Perera *(North Carolina State University)*
Beverly Perry *(Houston Community College)*
David Pfennig *(University of North Carolina, Chapel Hill)*
David S. Pilliod *(California Polytechnic State University, San Luis Obispo)*
J. Chris Pires *(University of Missouri, Columbia)*
Bob Pittman *(Michigan State University)*
James Platt *(University of Denver)*
Martin Poenie *(University of Texas, Austin)*
Scott Poethig *(University of Pennsylvania)*
Jeffrey Pommerville *(Texas A&M University)*
Angela R. Porta *(Kean University)*
Warren Porter *(University of Wisconsin)*
Daniel Potter *(University of California, Davis)*
Donald Potts *(University of California, Santa Cruz)*
Andy Pratt *(University of Canterbury)*
David Pratt *(University of California, Davis)*
Halina Presley *(University of Illinois, Chicago)*
Mary V. Price *(University of California, Riverside)*
Mitch Price *(Pennsylvania State University)*
Rong Sun Pu *(Kean University)*
Rebecca Pyles *(East Tennessee State University)*
Scott Quackenbush *(Florida International University)*
Ralph Quatrano *(Oregon State University)*
Peter Quinby *(University of Pittsburgh)*
Val Raghavan *(Ohio State University)*
Deanna Raineri *(University of Illinois, Champaign-Urbana)*
Talitha Rajah *(Indiana University Southeast)*
Charles Ralph *(Colorado State University)*
Thomas Rand *(Saint Mary's University)*
Robert H. Reavis *(Glendale Community College)*
Kurt Redborg *(Coe College)*
Ahnya Redman *(Pennsylvania State University)*
Brian Reeder *(Morehead State University)*
Bruce Reid *(Kean University)*
David Reid *(Blackburn College)*
C. Gary Reiness *(Lewis & Clark College)*
Charles Remington *(Yale University)*
Erin Rempala *(San Diego Mesa College)*
David Reznick *(University of California, Riverside)*
Douglas Rhoads *(University of Arkansas)*
Fred Rhoades *(Western Washington State University)*
Eric Ribbens *(Western Illinois University)*
Christina Richards *(New York University)*
Christopher Riegle *(Irvine Valley College)*
Loren Rieseberg *(University of British Columbia)*
Bruce B. Riley *(Texas A&M University)*
Donna Ritch *(Pennsylvania State University)*
Carol Rivin *(Oregon State University East)*
Laurel Roberts *(University of Pittsburgh)*
Thomas Rodella *(Merced College)*
Rodney Rogers *(Drake University)*
William Roosenburg *(Ohio University)*
Mike Rosenzweig *(Virginia Polytechnic Institute and State University)*
Wayne Rosing *(Middle Tennessee State University)*
Thomas Rost *(University of California, Davis)*
Stephen I. Rothstein *(University of California, Santa Barbara)*
John Ruben *(Oregon State University)*
Albert Ruesink *(Indiana University)*
Neil Sabine *(Indiana University)*
Tyson Sacco *(Cornell University)*
Rowan F. Sage *(University of Toronto)*
Tammy Lynn Sage *(University of Toronto)*
Don Sakaguchi *(Iowa State University)*
Walter Sakai *(Santa Monica College)*
Mark F. Sanders *(University of California, Davis)*
Ted Sargent *(University of Massachusetts, Amherst)*
K. Sathasivan *(University of Texas, Austin)*
Gary Saunders *(University of New Brunswick)*
Thomas R. Sawicki *(Spartanburg Community College)*
Inder Saxena *(University of Texas, Austin)*
Carl Schaefer *(University of Connecticut)*
Maynard H. Schaus *(Virginia Wesleyan College)*

Renate Scheibe *(University of Osnabrück)*
David Schimpf *(University of Minnesota, Duluth)*
William H. Schlesinger *(Duke University)*
Mark Schlissel *(University of California, Berkeley)*
Christopher J. Schneider *(Boston University)*
Thomas W. Schoener *(University of California, Davis)*
Robert Schorr *(Colorado State University)*
Patricia M. Schulte *(University of British Columbia)*
Karen S. Schumaker *(University of Arizona)*
David J. Schwartz *(Houston Community College)*
Christa Schwintzer *(University of Maine)*
Erik P. Scully *(Towson State University)*
Robert W. Seagull *(Hofstra University)*
Edna Seaman *(Northeastern University)*
Duane Sears *(University of California, Santa Barbara)*
Orono Shukdeb Sen *(Bethune-Cookman College)*
Wendy Sera *(Seton Hill University)*
Timothy E. Shannon *(Francis Marion University)*
Joan Sharp *(Simon Fraser University)*
Victoria C. Sharpe *(Blinn College)*
Elaine Shea *(Loyola College, Maryland)*
Stephen Sheckler *(Virginia Polytechnic Institute and State University)*
Richard Sherwin *(University of Pittsburgh)*
Lisa Shimeld *(Crafton Hills College)*
James Shinkle *(Trinity University)*
Barbara Shipes *(Hampton University)*
Richard M. Showman *(University of South Carolina)*
Peter Shugarman *(University of Southern California)*
Alice Shuttey *(DeKalb Community College)*
James Sidie *(Ursinus College)*
Daniel Simberloff *(Florida State University)*
Rebecca Simmons *(University of North Dakota)*
Anne Simon *(University of Maryland, College Park)*
Robert Simons *(University of California, Los Angeles)*
Alastair Simpson *(Dalhousie University)*
Susan Singer *(Carleton College)*
Roger Sloboda *(Dartmouth University)*
John Smarrelli *(Le Moyne College)*
Andrew T. Smith *(Arizona State University)*
Kelly Smith *(University of North Florida)*
Nancy Smith-Huerta *(Miami Ohio University)*
John Smol *(Queen's University)*
Andrew J. Snope *(Essex Community College)*
Julio G. Soto *(San Jose State University)*
Mitchell Sogin *(Woods Hole Marine Biological Laboratory)*
Susan Sovonick-Dunford *(University of Cincinnati)*
Frederick W. Spiegel *(University of Arkansas)*
John Stachowicz *(University of California, Davis)*
Amanda Starnes *(Emory University)*
Karen Steudel *(University of Wisconsin)*
Barbara Stewart *(Swarthmore College)*
Gail A. Stewart *(Camden County College)*
Cecil Still *(Rutgers University, New Brunswick)*
Margery Stinson *(Southwestern College)*
James Stockand *(University of Texas Health Science Center, San Antonio)*
John Stolz *(California Institute of Technology)*
Richard D. Storey *(Colorado College)*
Stephen Strand *(University of California, Los Angeles)*
Eric Strauss *(University of Massachusetts, Boston)*
Antony Stretton *(University of Wisconsin, Madison)*
Russell Stullken *(Augusta College)*
Mark Sturtevant *(University of Michigan, Flint)*
John Sullivan *(Southern Oregon State University)*
Gerald Summers *(University of Missouri)*
Judith Sumner *(Assumption College)*
Marshall D. Sundberg *(Emporia State University)*
Lucinda Swatzell *(Southeast Missouri State University)*
Daryl Sweeney *(University of Illinois, Champaign-Urbana)*
Samuel S. Sweet *(University of California, Santa Barbara)*
Janice Swenson *(University of North Florida)*
Michael A. Sypes *(Pennsylvania State University)*
Lincoln Taiz *(University of California, Santa Cruz)*
Samuel Tarsitano *(Southwest Texas State University)*
David Tauck *(Santa Clara University)*
Emily Taylor *(California Polytechnic State University, San Luis Obispo)*
James Taylor *(University of New Hampshire)*
John W. Taylor *(University of California, Berkeley)*
Martha R. Taylor *(Cornell University)*
Thomas Terry *(University of Connecticut)*
Roger Thibault *(Bowling Green State University)*
William Thomas *(Colby-Sawyer College)*
Cyril Thong *(Simon Fraser University)*
John Thornton *(Oklahoma State University)*
Robert Thornton *(University of California, Davis)*
William Thwaites *(Tillamook Bay Community College)*
Stephen Timme *(Pittsburg State University)*
Eric Toolson *(University of New Mexico)*
Leslie Towill *(Arizona State University)*
James Traniello *(Boston University)*
Paul Q. Trombley *(Florida State University)*
Nancy J. Trun *(Duquesne University)*
Constantine Tsoukas *(San Diego State University)*
Marsha Turell *(Houston Community College)*
Robert Tuveson *(University of Illinois, Urbana)*
Maura G. Tyrrell *(Stonehill College)*
Catherine Uekert *(Northern Arizona University)*
Claudia Uhde-Stone *(California State University, East Bay)*
Gordon Uno *(University of Oklahoma)*
Lisa A. Urry *(Mills College)*
Saba Valadkhan *(Case Western Reserve University School of Medicine)*
James W. Valentine *(University of California, Santa Barbara)*
Joseph Vanable *(Purdue University)*
Theodore Van Bruggen *(University of South Dakota)*
Kathryn VandenBosch *(Texas A&M University)*
Gerald Van Dyke *(North Carolina State University)*
Brandi Van Roo *(Framingham State College)*
Moira Van Staaden *(Bowling Green State University)*
Steven D. Verhey *(Central Washington University)*
Kathleen Verville *(Washington College)*
Sara Via *(University of Maryland)*
Frank Visco *(Orange Coast College)*
Laurie Vitt *(University of California, Los Angeles)*
Neal Voelz *(St. Cloud State University)*
Thomas J. Volk *(University of Wisconsin, La Crosse)*
Leif Asbjørn Vøllestad *(University of Oslo)*
Susan D. Waaland *(University of Washington)*
William Wade *(Dartmouth Medical College)*
D. Alexander Wait *(Southwest Missouri State University)*
John Waggoner *(Loyola Marymount University)*
Jyoti Wagle *(Houston Community College)*
Edward Wagner *(University of California, Irvine)*
Dan Walker *(San Jose State University)*
Robert L. Wallace *(Ripon College)*
Jeffrey Walters *(North Carolina State University)*
Linda Walters *(University of Central Florida)*
Nickolas M. Waser *(University of California, Riverside)*
Margaret Waterman *(University of Pittsburgh)*
Charles Webber *(Loyola University of Chicago)*
Peter Webster *(University of Massachusetts, Amherst)*
Terry Webster *(University of Connecticut, Storrs)*
Beth Wee *(Tulane University)*
Andrea Weeks *(George Mason University)*
Peter Wejksnora *(University of Wisconsin, Milwaukee)*
Kentwood Wells *(University of Connecticut)*
David J. Westenberg, *(University of Missouri, Rolla)*
Richard Wetts *(University of California, Irvine)*
Matt White *(Ohio University)*
Ernest H. Williams *(Hamilton College)*
Kathy Williams *(San Diego State University)*
Stephen Williams *(Glendale Community College)*
Elizabeth Willott *(University of Arizona)*
Christopher Wills *(University of California, San Diego)*
Paul Wilson *(California State University, Northridge)*
Fred Wilt *(University of California, Berkeley)*
Peter Wimberger *(University of Puget Sound)*
Robert Winning *(Eastern Michigan University)*
E. William Wischusen *(Louisiana State University)*
Susan Whittemore *(Keene State College)*
Clarence Wolfe *(Northern Virginia Community College)*
Vickie L. Wolfe *(Marshall University)*
Robert T. Woodland *(University of Massachusetts Medical School)*
Joseph Woodring *(Louisiana State University)*
Denise Woodward *(Pennsylvania State University)*
Patrick Woolley *(East Central College)*
Sarah E. Wyatt *(Ohio University)*
Ramin Yadegari *(University of Arizona)*
Paul Yancey *(Whitman College)*
Philip Yant *(University of Michigan)*
Linda Yasui *(Northern Illinois University)*
Hideo Yonenaka *(San Francisco State University)*
Gina M. Zainelli *(Loyola University, Chicago)*
Edward Zalisko *(Blackburn College)*
Zai Ming Zhao *(University of Texas, Austin)*
John Zimmerman *(Kansas State University)*
Miriam Zolan *(Indiana University)*
Uko Zylstra *(Calvin College)*

Detailed Contents

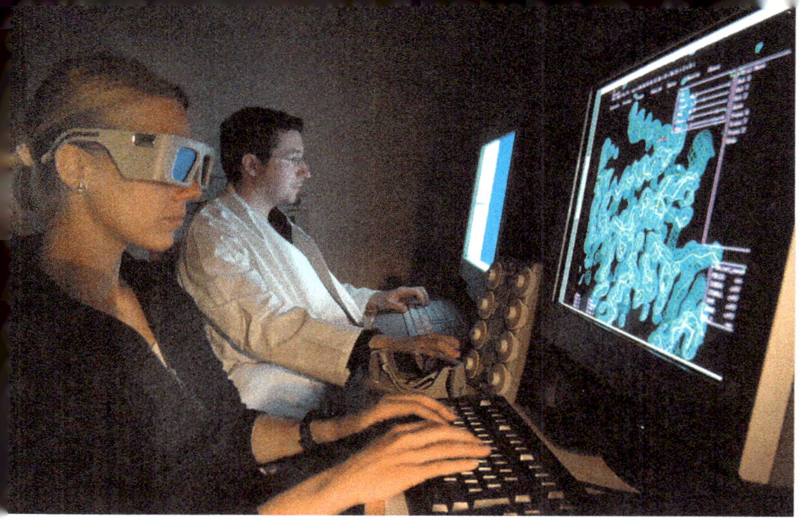

UNIT 2

The Cell 92

Interview: Bonnie L. Bassler

6 A Tour of the Cell 94

7 Membrane Structure and Function 125

8 An Introduction to Metabolism 142

9 Cellular Respiration and Fermentation 163

10 Photosynthesis 184

11 Cell Communication 206

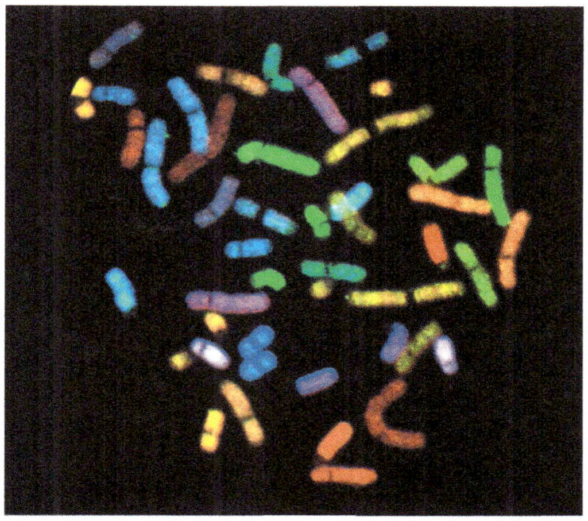

UNIT

7 Animal Form and Function 850

Interview: Baldomero M. Olivera

43 The Immune System 929

UNIT 1

The Chemistry of Life

An Interview with

Susan Solomon

Although Susan Solomon is not a biologist, her research as an atmospheric chemist has profound implications for life on Earth. Since earning degrees from the Illinois Institute of Technology and the University of California, Berkeley, Dr. Solomon has been a leader in determining the cause of the Antarctic ozone hole and in producing the 2007 report of the United Nations Intergovernmental Panel on Climate Change (IPCC), which concluded that warming of Earth's climate is unequivocal. These activities have given her a public role in communicating science to policymakers and society at large. In recognition of her scientific accomplishments, she has been awarded the U.S. National Medal of Science, the Blue Planet Prize, and, with Al Gore and the other IPCC members, the Nobel Peace Prize. A member of the U.S. National Academy of Sciences, the European Academy of Sciences, the Academy of Sciences of France, and the Royal Society of the United Kingdom, she works for the National Oceanic and Atmospheric Administration in Boulder, Colorado.

How is Earth's atmosphere important to life?

Life on Earth today could not have evolved without an atmosphere. We all know that we and many other organisms require oxygen (O_2) from the atmosphere, and plants use carbon dioxide (CO_2) to grow. The atmosphere also contains a form of oxygen called ozone that has three oxygen atoms (O_3) instead of two. Organisms would never have been able to leave the ocean and survive on land without the development of an ozone layer in the upper atmosphere. Ozone has the important property of absorbing ultraviolet (UV) light, which would otherwise cause DNA damage. Damage from UV can lead to skin cancer and cataracts; it can also harm many crops and even phytoplankton [small photosynthetic aquatic organisms].

Early in your career, you led an expedition to make measurements of the atmosphere in Antarctica. Tell us about that.

In 1985, the British Antarctic Survey reported a surprising discovery: that the springtime ozone at their station in Antarctica had fallen by 30–50% since the late 1970s, resulting in an "ozone hole"! People had begun to be worried about whether the ozone layer might be vulnerable to changes caused by human activity, but only very minor changes had been expected. In 1986, I had the chance to lead a new Antarctic expedition to help confirm the British data and to study the problem further. We didn't just measure ozone; we measured about a dozen other atmospheric molecules that allowed us to tell *why* the ozone was being destroyed.

What did you find out?

It turns out that the ozone chemistry in Antarctica is extremely different from what it is anywhere else. That's because Antarctica is very cold—it really is the coldest place on Earth. It's so cold that clouds form in its upper stratosphere, about 10–30 kilometers above sea level, and those clouds help convert chemicals from chlorofluorocarbons (CFCs) to ozone-damaging substances.

CFCs are synthetic compounds, made only by humans. They were used back then for a variety of purposes—for example, in refrigeration, as solvents, and as propellants for sprays. Many tons of CFCs were emitted every year. I came up with the idea that the reason an ozone hole developed in Antarctica had to do with chemical reactions that happen between a gas and a surface and that the surface in this case was the small particles that make up those stratospheric clouds. Our data supported this hypothesis. The reactions on those particles make the CFCs hundreds of times more damaging than they would be otherwise. The absence of such clouds in most other parts of the world is why we don't have ozone holes elsewhere, although stratospheric clouds form occasionally in the Arctic and there is significant ozone loss there. Scientists had been concerned since the mid-1970s that human use of CFCs might cause some ozone depletion, but they had expected a loss of only about 3–5% in 100 years.

How do CFCs destroy ozone?

When CFCs arrive at the stratosphere, which typically takes a few years, high-energy radiation up there can break them down, releasing chlorine atoms. The chlorine atoms destroy the ozone catalytically, which means that the atoms don't get used up in the process. So even if only a small amount of CFC is broken down, the tiny bit of chlorine produced can destroy an enormous amount of ozone.

In the U.N. Montreal Protocol of 1987, the nations of the world agreed to stop producing CFCs. However, the CFCs in the atmosphere disappear only very slowly; typically they hang around for 50 to 100 years. What that means is that the CFCs we've already put in the atmosphere will continue to produce an ozone hole for many decades to come, even though we're not using these substances any more. Global emissions are very near zero now, and we're beginning to see the ozone hole slowly diminish. But it will probably not go away completely until around 2060.

While the ozone hole remains, it continues to cause damage. For example, there is evidence that the phytoplankton in the Antarctic Ocean are being affected by increased UV, and the phytoplankton are the base of the main Antarctic food chain: They feed the krill, which feed the fish, which feed the penguins, seals, and whales.

Let's talk about an effect that other atmospheric changes are having—climate change.

There's no question that the planet is getting warmer. We know that, on average, our planet is about 1.4°F (0.8°C) warmer than it was 100 years ago, and this past decade has been the warmest decade in at least the last 100 years. We also know that glaciers worldwide are retreating and that sea level is rising. There's a breadth of scientific data, acquired by different techniques, that tells us that global warming really is unequivocal.

The warming has to do with the greenhouse effect, right?

We're lucky that this planet has a greenhouse effect, because if it didn't, we would be very cold indeed! Our planet is heated by the sun, and much of the infrared radiation (heat) that would otherwise be released back into space is trapped by "greenhouse gases" in the atmosphere. This makes the planet about 30°C hotter than it would be otherwise. But of course anything can be bad if you have too much of it, and what we're doing now is increasing the greenhouse effect of our atmosphere beyond its natural state. If we keep emitting the greenhouse gases that are causing the warming, then we will see some very significant warming in the coming century.

The main greenhouse gas we're adding to the atmosphere is CO_2, from burning fossil fuel and to a lesser extent deforestation. We have increased the atmospheric concentration of CO_2 by about 30% compared to any value that has been found for the last 800,000 years. This has been determined by digging up ice cores in Antarctica and measuring the gases in the air bubbles trapped in the ice. So we know that we have perturbed the atmosphere in a way that the planet hasn't seen in at least 800,000 years.

The CFCs we discussed earlier are actually the third most important greenhouse gas at present, after CO_2 and methane. Pound for pound, CFCs are much more potent as greenhouse gases than CO_2. The phase-out of CFCs since the signing of the Montreal Protocol has not only avoided a lot of ozone destruction that would otherwise have happened, but has also reduced our input of gases that cause climate change.

How is life on Earth being affected by climate change?

There are some things that we can already begin to see and talk about, but there's an enormous amount that we still don't know. We do know that the oceans are getting more acidic because CO_2 is taken up by the ocean and converted to carbonic acid, which can affect the ability of shellfish to make their shells. Other ocean life is also likely to be harmed by the increased acidity, such as the organisms of coral reefs. But there's also emerging evidence that some other marine organisms may do better—lobsters, perhaps.

As a westerner I'm extremely concerned about the greatly increased population of pine beetles in the western United States. These beetles are killing pine trees in unprecedented numbers. There's good evidence that a contributing factor to this explosion of pine beetles is global warming. I think we're going to see more of this kind of thing. Also, it is clear that bird migration is already being affected by global warming. Whether global warming will lead to extinction of some animals is an important question. The signature extinction issue is the polar bear; as the sea ice of the Arctic decreases, the polar bear could become extinct. We don't really know yet how much biological adaptation is possible in the time available. We'll probably find out that there are some winner species out there and some loser species. In agriculture, many crops are sensitive to increasing temperatures. One of the relevant findings about corn is that for every degree of warming, about 10% of crop production is lost—a big change.

"There's a breadth of scientific data, acquired by different techniques, that tells us that global warming really is unequivocal."

Does less precipitation always go along with higher temperatures?

In some places there will probably be less precipitation and in other places more. There's a band of subtropical and tropical regions where we are pretty confident that it will get drier—for example, Mexico, the Mediterranean region, parts of Australia. In the higher latitudes, places like Canada and Norway will likely get wetter. In between, it's harder to predict.

Tell us about the IPCC and your work on it.

The IPCC is fundamentally a mechanism for the communication of information about climate change from the science community to the policy community. It was set up in 1988 when people were beginning to recognize that climate change was a real possibility. Policymakers decided that they needed to get reliable scientific information so they could begin to talk about what to do, if anything. Every six or seven years, scientists are asked by their governments to get together and assess what we know and don't know on the basis of the published scientific literature.

I have been involved in the IPCC since 1992, and in 2001 I was elected by the panel, representing over 100 governments, to co-chair the scientific assessment team. In a process lasting several years, we generated a detailed report summarizing the state of climate science. Our report was then reviewed by dozens of governments and more than 600 scientists. The report itself and every one of their 30,000 comments are available on the Internet. We refined and refined the draft in consideration of those comments and finalized the document in 2007.

What were the main conclusions of your 2007 report?

The first conclusion, based on many independent lines of evidence, was that the Earth is warming. There's no doubt we are now living on a planet that is warmer than a century ago. The second main conclusion was that most—more than half—of the warming is very likely due to increases in greenhouse gases, primarily CO_2. We did a careful analysis of the uncertainties: When we say "very likely," we mean that there's a 90% chance or greater that most of the warming is due to emissions of greenhouse gases by human activity.

What have you learned about working at the interface of science and policy?

It's one of the most difficult things a scientist can do. Science normally takes us into a laboratory or out into the field or into scholarly discussions with colleagues. Getting involved with policy is quite different: It takes us out of the lab and makes us much more aware of the strong emotions around many issues. In that sense, it's a bit daunting. But it's uplifting to see how valuable science can be in helping society make more informed choices. Scientists can help make sure that whatever it is we choose to do as a society we're doing knowingly, not in ignorance. I appreciate all the reasons why people ask tough questions about the science. How much do we know? What really are the uncertainties? Yes, there's a lot at stake here, all the more reason why there has to be really good science going into it.

Susan Solomon (right) with Jane Reece

2

The Chemical Context of Life

▲ **Figure 2.1 Who tends this "garden"?**

KEY CONCEPTS

2.1 Matter consists of chemical elements in pure form and in combinations called compounds
2.2 An element's properties depend on the structure of its atoms
2.3 The formation and function of molecules depend on chemical bonding between atoms
2.4 Chemical reactions make and break chemical bonds

OVERVIEW

A Chemical Connection to Biology

The Amazon rain forest in South America is a showcase for the diversity of life on Earth. Colorful birds, insects, and other animals live in a densely-packed environment of trees, shrubs, vines, and wildflowers, and an excursion along a waterway or a forest path typically reveals a lush variety of plant life. Visitors traveling near the Amazon's headwaters in Peru are therefore surprised to come across tracts of forest like that seen in the foreground of the photo in **Figure 2.1**. This patch is almost completely dominated by a single plant species—a small flowering tree called *Duroia hirsuta*. Travelers may wonder if the plot of land is planted and maintained by local people, but the indigenous people are as mystified as the visitors. They call these stands of *Duroia* trees "devil's gardens," from a legend attributing them to an evil forest spirit.

Seeking a scientific explanation, a research team at Stanford University recently solved the "devil's garden" mystery. **Figure 2.2** describes their main experiment. The researchers showed that the "farmers" who create and maintain these gardens are actually ants that live in the hollow stems of the *Duroia* trees. The ants do not plant the *Duroia* trees, but they prevent other plant species from growing in the garden by injecting intruders with a poisonous chemical. In this way, the ants create space for the growth of the *Duroia* trees that serve as their home. With the ability to maintain and expand its habitat, a single colony of devil's garden ants can live for hundreds of years.

The chemical used by the ants to weed their garden turns out to be formic acid. This substance is produced by many species of ants and in fact got its name from the Latin word for ant, *formica*. For many ant species, the formic acid probably serves as a disinfectant that protects the ants against microbial parasites. The devil's garden ant is the first ant species found to use formic acid as an herbicide, an important addition to the list of functions mediated by chemicals in the insect world. Scientists have long known that chemicals play a major role in insect communication, attraction of mates, and defense against predators.

Research on devil's gardens is only one example of the relevance of chemistry to the study of life. Unlike a list of college courses, nature is not neatly packaged into the individual natural sciences—biology, chemistry, physics, and so forth. Biologists specialize in the study of life, but organisms and their environments are natural systems to which the concepts of chemistry and physics apply. Biology is a multidisciplinary science.

This unit of chapters introduces some basic concepts of chemistry that apply to the study of life. We will make many connections to the themes introduced in Chapter 1. One of these themes is the organization of life into a hierarchy of structural levels, with additional properties emerging at each successive level. In this unit, we will see how emergent properties are apparent at the lowest levels of biological organization—such as the ordering of atoms into molecules and the interactions of those molecules within cells. Somewhere in the transition from molecules to cells, we will cross the blurry boundary between nonlife and life. This chapter focuses on the chemical components that make up all matter.

▼ Figure 2.2 INQUIRY

What creates "devil's gardens" in the rain forest?

EXPERIMENT Working under Deborah Gordon and with Michael Greene, graduate student Megan Frederickson sought the cause of "devil's gardens," stands of a single species of tree, *Duroia hirsuta*. One hypothesis was that ants living in these trees, *Myrmelachista schumanni*, produce a poisonous chemical that kills trees of other species; another was that the *Duroia* trees themselves kill competing trees, perhaps by means of a chemical.

To test these hypotheses, Frederickson did field experiments in Peru. Two saplings of a local nonhost tree species, *Cedrela odorata*, were planted inside each of ten devil's gardens. At the base of one sapling, a sticky insect barrier was applied; the other was unprotected. Two more *Cedrela* saplings, with and without barriers, were planted about 50 meters outside each garden.

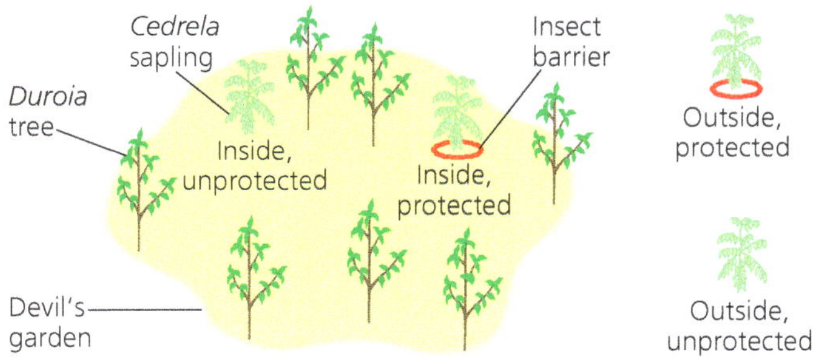

The researchers observed ant activity on the *Cedrela* leaves and measured areas of dead leaf tissue after one day. They also chemically analyzed contents of the ants' poison glands.

RESULTS The ants made injections from the tips of their abdomens into leaves of unprotected saplings in their gardens (see photo). Within one day, these leaves developed dead areas (see graph). The protected saplings were uninjured, as were the saplings planted outside the gardens. Formic acid was the only chemical detected in the poison glands of the ants.

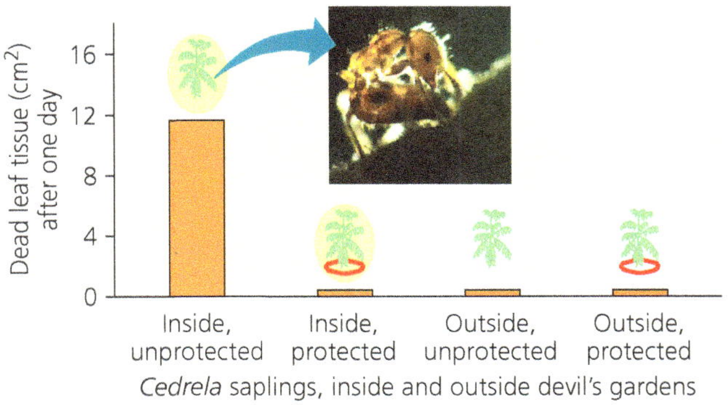

CONCLUSION Ants of the species *Myrmelachista schumanni* kill nonhost trees by injecting the leaves with formic acid, thus creating hospitable habitats (devil's gardens) for the ant colony.

SOURCE M. E. Frederickson, M. J. Greene, and D. M. Gordon, "Devil's gardens" bedevilled by ants, *Nature* 437:495–496 (2005).

INQUIRY IN ACTION Read and analyze the original paper in *Inquiry in Action: Interpreting Scientific Papers*.

WHAT IF? What would be the results if the unprotected saplings' inability to grow in the devil's gardens was caused by a chemical released by the *Duroia* trees rather than by the ants?

CONCEPT **2.1**

Matter consists of chemical elements in pure form and in combinations called compounds

Organisms are composed of **matter**, which is defined as anything that takes up space and has mass.* Matter exists in many diverse forms. Rocks, metals, oils, gases, and humans are just a few examples of what seems an endless assortment of matter.

Elements and Compounds

Matter is made up of elements. An **element** is a substance that cannot be broken down to other substances by chemical reactions. Today, chemists recognize 92 elements occurring in nature; gold, copper, carbon, and oxygen are examples. Each element has a symbol, usually the first letter or two of its name. Some symbols are derived from Latin or German; for instance, the symbol for sodium is Na, from the Latin word *natrium*.

A **compound** is a substance consisting of two or more different elements combined in a fixed ratio. Table salt, for example, is sodium chloride (NaCl), a compound composed of the elements sodium (Na) and chlorine (Cl) in a 1:1 ratio. Pure sodium is a metal, and pure chlorine is a poisonous gas. When chemically combined, however, sodium and chlorine form an edible compound. Water (H_2O), another compound, consists of the elements hydrogen (H) and oxygen (O) in a 2:1 ratio. These are simple examples of organized matter having emergent properties: A compound has characteristics different from those of its elements (**Figure 2.3**).

▲ **Figure 2.3 The emergent properties of a compound.** The metal sodium combines with the poisonous gas chlorine, forming the edible compound sodium chloride, or table salt.

*Sometimes we substitute the term weight for mass, although the two are not identical. Mass is the amount of matter in an object, whereas the weight of an object is how strongly that mass is pulled by gravity. The weight of an astronaut walking on the moon is approximately 1/6 the astronaut's weight on Earth, but his or her mass is the same. However, as long as we are earthbound, the weight of an object is a measure of its mass; in everyday language, therefore, we tend to use the terms interchangeably.

The Elements of Life

Of the 92 natural elements, about 20–25% are **essential elements** that an organism needs to live a healthy life and reproduce. The essential elements are similar among organisms, but there is some variation—for example, humans need 25 elements, but plants need only 17.

Just four elements—oxygen (O), carbon (C), hydrogen (H), and nitrogen (N)—make up 96% of living matter. Calcium (Ca), phosphorus (P), potassium (K), sulfur (S), and a few other elements account for most of the remaining 4% of an organism's mass. **Trace elements** are required by an organism in only minute quantities. Some trace elements, such as iron (Fe), are needed by all forms of life; others are required only by certain species. For example, in vertebrates (animals with backbones), the element iodine (I) is an essential ingredient of a hormone produced by the thyroid gland. A daily intake of only 0.15 milligram (mg) of iodine is adequate for normal activity of the human thyroid. An iodine deficiency in the diet causes the thyroid gland to grow to abnormal size, a condition called goiter. Where it is available, eating seafood or iodized salt reduces the incidence of goiter. All the elements needed by the human body are listed in **Table 2.1**.

Some naturally occurring elements are toxic to organisms. In humans, for instance, the element arsenic has been linked to numerous diseases and can be lethal. In some areas of the world, arsenic occurs naturally and can make its way into the groundwater. As a result of using water from drilled wells in southern Asia, millions of people have been inadvertently exposed to arsenic-laden water. Efforts are under way to reduce arsenic levels in their water supply.

Table 2.1 Elements in the Human Body

Element	Symbol	Percentage of Body Mass (including water)	
Oxygen	O	65.0%	96.3%
Carbon	C	18.5%	
Hydrogen	H	9.5%	
Nitrogen	N	3.3%	
Calcium	Ca	1.5%	3.7%
Phosphorus	P	1.0%	
Potassium	K	0.4%	
Sulfur	S	0.3%	
Sodium	Na	0.2%	
Chlorine	Cl	0.2%	
Magnesium	Mg	0.1%	

Trace elements (less than 0.01% of mass): Boron (B), chromium (Cr), cobalt (Co), copper (Cu), fluorine (F), iodine (I), iron (Fe), manganese (Mn), molybdenum (Mo), selenium (Se), silicon (Si), tin (Sn), vanadium (V), zinc (Zn)

▲ **Figure 2.4 Serpentine plant community.** The plants in the large photo are growing on serpentine soil, which contains elements that are usually toxic to plants. The insets show a close-up of serpentine rock and one of the plants, a Tiburon Mariposa lily.

Case Study: Evolution of Tolerance to Toxic Elements

EVOLUTION Some species have become adapted to environments containing elements that are usually toxic. A compelling example is found in serpentine plant communities. Serpentine is a jade-like mineral that contains toxic elements such as chromium, nickel, and cobalt. Although most plants cannot survive in soil that forms from serpentine rock, a small number of plant species have adaptations that allow them to do so **(Figure 2.4)**. Presumably, variants of ancestral, nonserpentine species arose that could survive in serpentine soils, and subsequent natural selection resulted in the distinctive array of species we see in these areas today.

CONCEPT CHECK 2.1

1. **MAKE CONNECTIONS** Review the discussion of emergent properties in Chapter 1 (p. 3). Explain how table salt has emergent properties.
2. Is a trace element an essential element? Explain.
3. In humans, iron is a trace element required for the proper functioning of hemoglobin, the molecule that carries oxygen in red blood cells. What might be the effects of an iron deficiency?
4. **MAKE CONNECTIONS** Review the discussion of natural selection in Chapter 1 (pp. 14–16) and explain how natural selection might have played a role in the evolution of species that are tolerant of serpentine soils.

For suggested answers, see Appendix A.

CONCEPT **2.2**

An element's properties depend on the structure of its atoms

Each element consists of a certain type of atom that is different from the atoms of any other element. An **atom** is the smallest unit of matter that still retains the properties of an element. Atoms are so small that it would take about a million of them to stretch across the period printed at the end of this sentence. We symbolize atoms with the same abbreviation used for the element that is made up of those atoms. For example, the symbol C stands for both the element carbon and a single carbon atom.

Subatomic Particles

Although the atom is the smallest unit having the properties of an element, these tiny bits of matter are composed of even smaller parts, called *subatomic particles*. Physicists have split the atom into more than a hundred types of particles, but only three kinds of particles are relevant here: **neutrons**, **protons**, and **electrons**. Protons and electrons are electrically charged. Each proton has one unit of positive charge, and each electron has one unit of negative charge. A neutron, as its name implies, is electrically neutral.

Protons and neutrons are packed together tightly in a dense core, or **atomic nucleus**, at the center of an atom; protons give the nucleus a positive charge. The electrons form a sort of cloud of negative charge around the nucleus, and it is the attraction between opposite charges that keeps the electrons in the vicinity of the nucleus. **Figure 2.5** shows two commonly used models of the structure of the helium atom as an example.

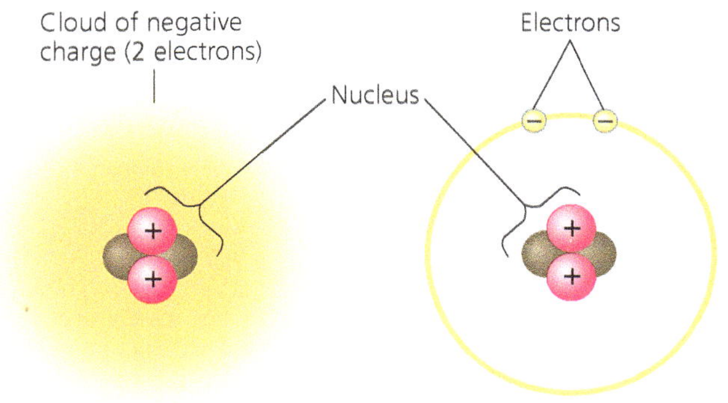

(a) This model represents the two electrons as a cloud of negative charge.

(b) In this more simplified model, the electrons are shown as two small yellow spheres on a circle around the nucleus.

▲ **Figure 2.5 Simplified models of a helium (He) atom.** The helium nucleus consists of 2 neutrons (brown) and 2 protons (pink). Two electrons (yellow) exist outside the nucleus. These models are not to scale; they greatly overestimate the size of the nucleus in relation to the electron cloud.

The neutron and proton are almost identical in mass, each about 1.7×10^{24} gram (g). Grams and other conventional units are not very useful for describing the mass of objects so minuscule. Thus, for atoms and subatomic particles (and for molecules, too), we use a unit of measurement called the **dalton**, in honor of John Dalton, the British scientist who helped develop atomic theory around 1800. (The dalton is the same as the *atomic mass unit*, or *amu*, a unit you may have encountered elsewhere.) Neutrons and protons have masses close to 1 dalton. Because the mass of an electron is only about 1/2,000 that of a neutron or proton, we can ignore electrons when computing the total mass of an atom.

Atomic Number and Atomic Mass

Atoms of the various elements differ in their number of subatomic particles. All atoms of a particular element have the same number of protons in their nuclei. This number of protons, which is unique to that element, is called the **atomic number** and is written as a subscript to the left of the symbol for the element. The abbreviation $_{2}He$, for example, tells us that an atom of the element helium has 2 protons in its nucleus. Unless otherwise indicated, an atom is neutral in electrical charge, which means that its protons must be balanced by an equal number of electrons. Therefore, the atomic number tells us the number of protons and also the number of electrons in an electrically neutral atom.

We can deduce the number of neutrons from a second quantity, the **mass number**, which is the sum of protons plus neutrons in the nucleus of an atom. The mass number is written as a superscript to the left of an element's symbol. For example, we can use this shorthand to write an atom of helium as $^{4}_{2}He$. Because the atomic number indicates how many protons there are, we can determine the number of neutrons by subtracting the atomic number from the mass number: The helium atom, $^{4}_{2}He$, has 2 neutrons. For sodium (Na):

Mass number = number of protons + neutrons
= 23 for sodium

$^{23}_{11}Na$

Atomic number = number of protons
= number of electrons in a neutral atom
= 11 for sodium

Number of neutrons = mass number − atomic number
= 23 − 11 = 12 for sodium

The simplest atom is hydrogen, $^{1}_{1}H$, which has no neutrons; it consists of a single proton with a single electron.

As mentioned earlier, the contribution of electrons to mass is negligible. Therefore, almost all of an atom's mass is concentrated in its nucleus. Because neutrons and protons each have a mass very close to 1 dalton, the mass number is an approximation of the total mass of an atom, called its **atomic mass**. So we might say that the atomic mass of sodium ($^{23}_{11}Na$) is 23 daltons, although more precisely it is 22.9898 daltons.

Isotopes

All atoms of a given element have the same number of protons, but some atoms have more neutrons than other atoms of the same element and therefore have greater mass. These different atomic forms of the same element are called **isotopes** of the element. In nature, an element occurs as a mixture of its isotopes. For example, consider the three isotopes of the element carbon, which has the atomic number 6. The most common isotope is carbon-12, ${}^{12}_{6}C$, which accounts for about 99% of the carbon in nature. The isotope ${}^{12}_{6}C$ has 6 neutrons. Most of the remaining 1% of carbon consists of atoms of the isotope ${}^{13}_{6}C$, with 7 neutrons. A third, even rarer isotope, ${}^{14}_{6}C$, has 8 neutrons. Notice that all three isotopes of carbon have 6 protons; otherwise, they would not be carbon. Although the isotopes of an element have slightly different masses, they behave identically in chemical reactions. (The number usually given as the atomic mass of an element, such as 22.9898 daltons for sodium, is actually an average of the atomic masses of all the element's naturally occurring isotopes.)

Both ${}^{12}C$ and ${}^{13}C$ are stable isotopes, meaning that their nuclei do not have a tendency to lose particles. The isotope ${}^{14}C$, however, is unstable, or radioactive. A **radioactive isotope** is one in which the nucleus decays spontaneously, giving off particles and energy. When the decay leads to a change in the number of protons, it transforms the atom to an atom of a different element. For example, when a radioactive carbon atom decays, it becomes an atom of nitrogen.

Radioactive isotopes have many useful applications in biology. In Chapter 25, you will learn how researchers use measurements of radioactivity in fossils to date these relics of past life. As shown in **Figure 2.6**, radioactive isotopes are also useful as tracers to follow atoms through metabolism, the chemical processes of an organism. Cells use the radioactive atoms as they would use nonradioactive isotopes of the same element, but the radioactive tracers can be readily detected.

Radioactive tracers are important diagnostic tools in medicine. For example, certain kidney disorders can be diagnosed by injecting small doses of substances containing radioactive isotopes into the blood and then measuring the amount of tracer excreted in the urine. Radioactive tracers are also used in combination with sophisticated imaging instruments. PET scanners, for instance, can monitor chemical processes, such as those involved in cancerous growth, as they actually occur in the body **(Figure 2.7)**.

Although radioactive isotopes are very useful in biological research and medicine, radiation from decaying isotopes also poses a hazard to life by damaging cellular molecules. The severity of this damage depends on the type and amount of radiation an organism absorbs. One of the most serious environmental threats is radioactive fallout from nuclear accidents. The doses of most isotopes used in medical diagnosis, however, are relatively safe.

▼ **Figure 2.6** RESEARCH METHOD

Radioactive Tracers

APPLICATION Scientists use radioactive isotopes to label certain chemical compounds, creating tracers that allow them to follow a metabolic process or locate the compound within an organism. In this example, radioactive tracers are utilized to determine the effect of temperature on the rate at which cells make copies of their DNA.

TECHNIQUE

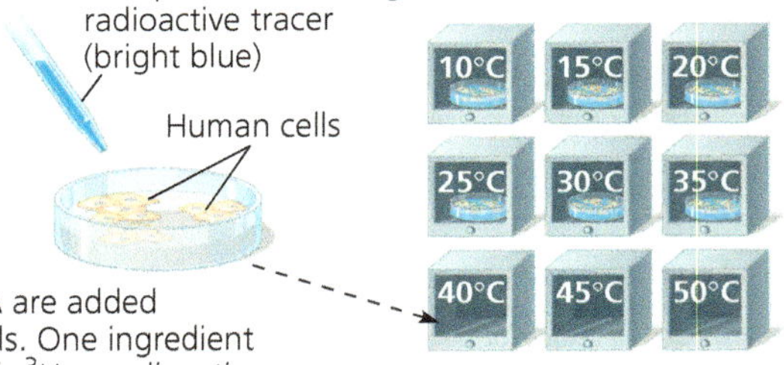

1 Compounds used by cells to make DNA are added to human cells. One ingredient is labeled with ${}^{3}H$, a radioactive isotope of hydrogen. Nine dishes of cells are incubated at different temperatures. The cells make new DNA, incorporating the radioactive tracer.

2 Cells from each incubator are placed in tubes; their DNA is isolated; and unused labeled compounds are removed.

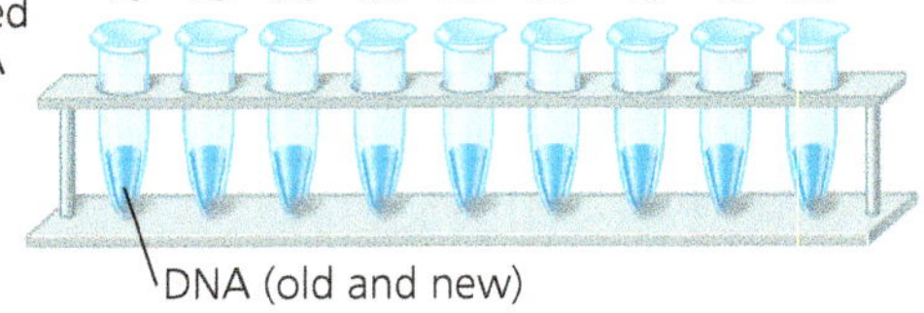

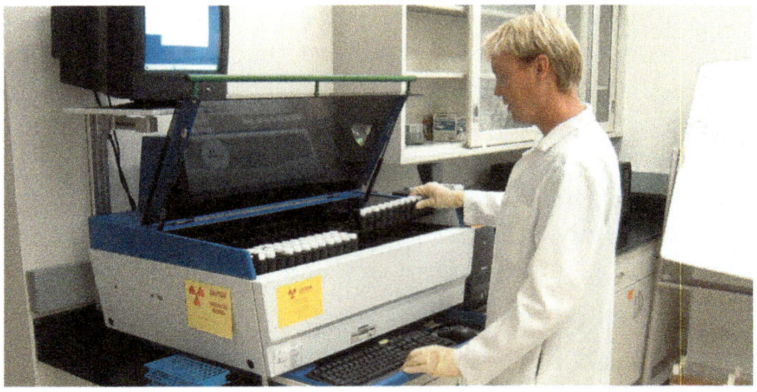

3 A solution called scintillation fluid is added to the samples, which are then placed in a scintillation counter. As the ${}^{3}H$ in the newly made DNA decays, it emits radiation that excites chemicals in the scintillation fluid, causing them to give off light. Flashes of light are recorded by the scintillation counter.

RESULTS The frequency of flashes, which is recorded as counts per minute, is proportional to the amount of the radioactive tracer present, indicating the amount of new DNA. In this experiment, when the counts per minute are plotted against temperature, it is clear that temperature affects the rate of DNA synthesis; the most DNA was made at 35°C.

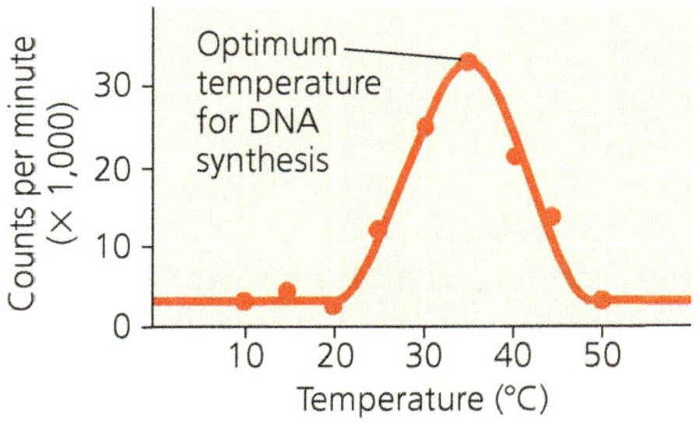

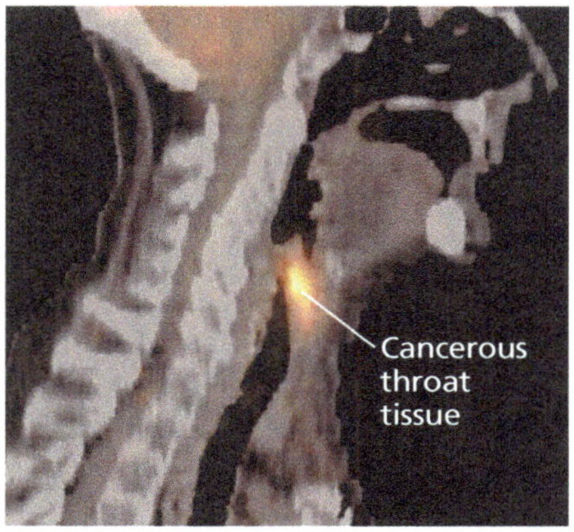

◀ **Figure 2.7 A PET scan, a medical use for radioactive isotopes.** PET, an acronym for positron-emission tomography, detects locations of intense chemical activity in the body. The bright yellow spot marks an area with an elevated level of radioactively labeled glucose, which in turn indicates high metabolic activity, a hallmark of cancerous tissue.

The Energy Levels of Electrons

The simplified models of the atom in Figure 2.5 greatly exaggerate the size of the nucleus relative to the volume of the whole atom. If an atom of helium were the size of a typical football stadium, the nucleus would be the size of a pencil eraser in the center of the field. Moreover, the electrons would be like two tiny gnats buzzing around the stadium. Atoms are mostly empty space.

When two atoms approach each other during a chemical reaction, their nuclei do not come close enough to interact. Of the three kinds of subatomic particles we have discussed, only electrons are directly involved in the chemical reactions between atoms.

An atom's electrons vary in the amount of energy they possess. **Energy** is defined as the capacity to cause change—for instance, by doing work. **Potential energy** is the energy that matter possesses because of its location or structure. For example, water in a reservoir on a hill has potential energy because of its altitude. When the gates of the reservoir's dam are opened and the water runs downhill, the energy can be used to do work, such as turning generators. Because energy has been expended, the water has less energy at the bottom of the hill than it did in the reservoir. Matter has a natural tendency to move to the lowest possible state of potential energy; in this example, the water runs downhill. To restore the potential energy of a reservoir, work must be done to elevate the water against gravity.

The electrons of an atom have potential energy because of how they are arranged in relation to the nucleus. The negatively charged electrons are attracted to the positively charged nucleus. It takes work to move a given electron farther away from the nucleus, so the more distant an electron is from the nucleus, the greater its potential energy. Unlike the continuous flow of water downhill, changes in the potential energy of electrons can occur only in steps of fixed amounts. An electron having a certain amount of energy is something like a ball on a staircase (**Figure 2.8a**). The ball can have different amounts of potential energy, depending on which step it is on, but it cannot spend much time between the steps. Similarly, an electron's potential energy is determined by its energy level. An electron cannot exist between energy levels.

(a) A ball bouncing down a flight of stairs provides an analogy for energy levels of electrons, because the ball can come to rest only on each step, not between steps.

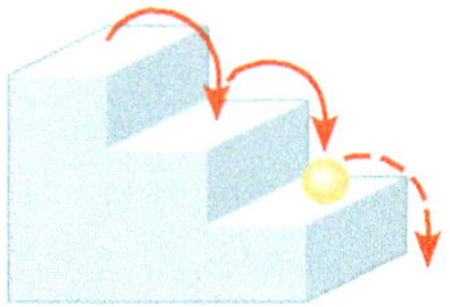

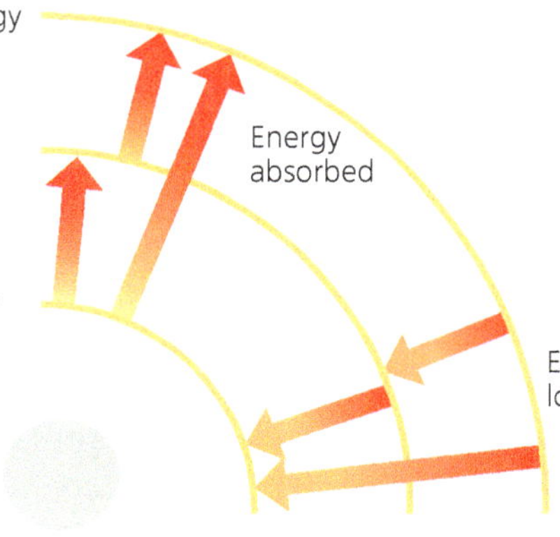

(b) An electron can move from one shell to another only if the energy it gains or loses is exactly equal to the difference in energy between the energy levels of the two shells. Arrows in this model indicate some of the stepwise changes in potential energy that are possible.

▲ **Figure 2.8 Energy levels of an atom's electrons.** Electrons exist only at fixed levels of potential energy called electron shells.

An electron's energy level is correlated with its average distance from the nucleus. Electrons are found in different **electron shells**, each with a characteristic average distance and energy level. In diagrams, shells can be represented by concentric circles (**Figure 2.8b**). The first shell is closest to the nucleus, and electrons in this shell have the lowest potential energy. Electrons in the second shell have more energy, and electrons in the third shell even more energy. An electron can change the shell it occupies, but only by absorbing or losing an amount of energy equal to the difference in potential energy between its position in the old shell and that in the new shell. When an electron absorbs energy, it moves to a shell farther out from the nucleus. For example, light energy can excite an electron to a higher energy level. (Indeed, this is the first step taken when plants harness the energy of sunlight for photosynthesis, the process that produces food from carbon dioxide and water.) When an electron loses energy, it "falls back" to a shell closer to the nucleus, and the lost energy is usually released to the environment as heat. For example, sunlight excites electrons in the surface of a car to higher energy levels. When the electrons fall back to their original levels, the car's surface heats up. This thermal energy can be transferred to the air or to your hand if you touch the car.

Electron Distribution and Chemical Properties

The chemical behavior of an atom is determined by the distribution of electrons in the atom's electron shells. Beginning with hydrogen, the simplest atom, we can imagine building the atoms of the other elements by adding 1 proton and 1 electron at a time (along with an appropriate number of neutrons). **Figure 2.9**, an abbreviated version of what is called the *periodic table of the elements*, shows this distribution of electrons for the first 18 elements, from hydrogen ($_1$H) to argon ($_{18}$Ar). The elements are arranged in three rows, or periods, corresponding to the number of electron shells in their atoms. The left-to-right sequence of elements in each row corresponds to the sequential addition of electrons and protons. (See Appendix B for the complete periodic table.)

Hydrogen's 1 electron and helium's 2 electrons are located in the first shell. Electrons, like all matter, tend to exist in the lowest available state of potential energy. In an atom, this state is in the first shell. However, the first shell can hold no more than 2 electrons; thus, hydrogen and helium are the only elements in the first row of the table. An atom with more than 2 electrons must use higher shells because the first shell is full. The next element, lithium, has 3 electrons. Two of these electrons fill the first shell, while the third electron occupies the second shell. The second shell holds a maximum of 8 electrons. Neon, at the end of the second row, has 8 electrons in the second shell, giving it a total of 10 electrons.

The chemical behavior of an atom depends mostly on the number of electrons in its *outermost* shell. We call those outer electrons **valence electrons** and the outermost electron shell the **valence shell**. In the case of lithium, there is only 1 valence electron, and the second shell is the valence shell. Atoms with the same number of electrons in their valence shells exhibit similar chemical behavior. For example, fluorine (F) and chlorine (Cl) both have 7 valence electrons, and both form compounds when combined with the element sodium (see Figure 2.3). An atom with a completed valence shell is unreactive; that is, it will not interact readily with other atoms. At the far right of the periodic table are helium, neon, and argon, the only three elements shown in Figure 2.9 that have full valence shells. These elements are said to be *inert*, meaning chemically unreactive. All the other atoms in Figure 2.9 are chemically reactive because they have incomplete valence shells.

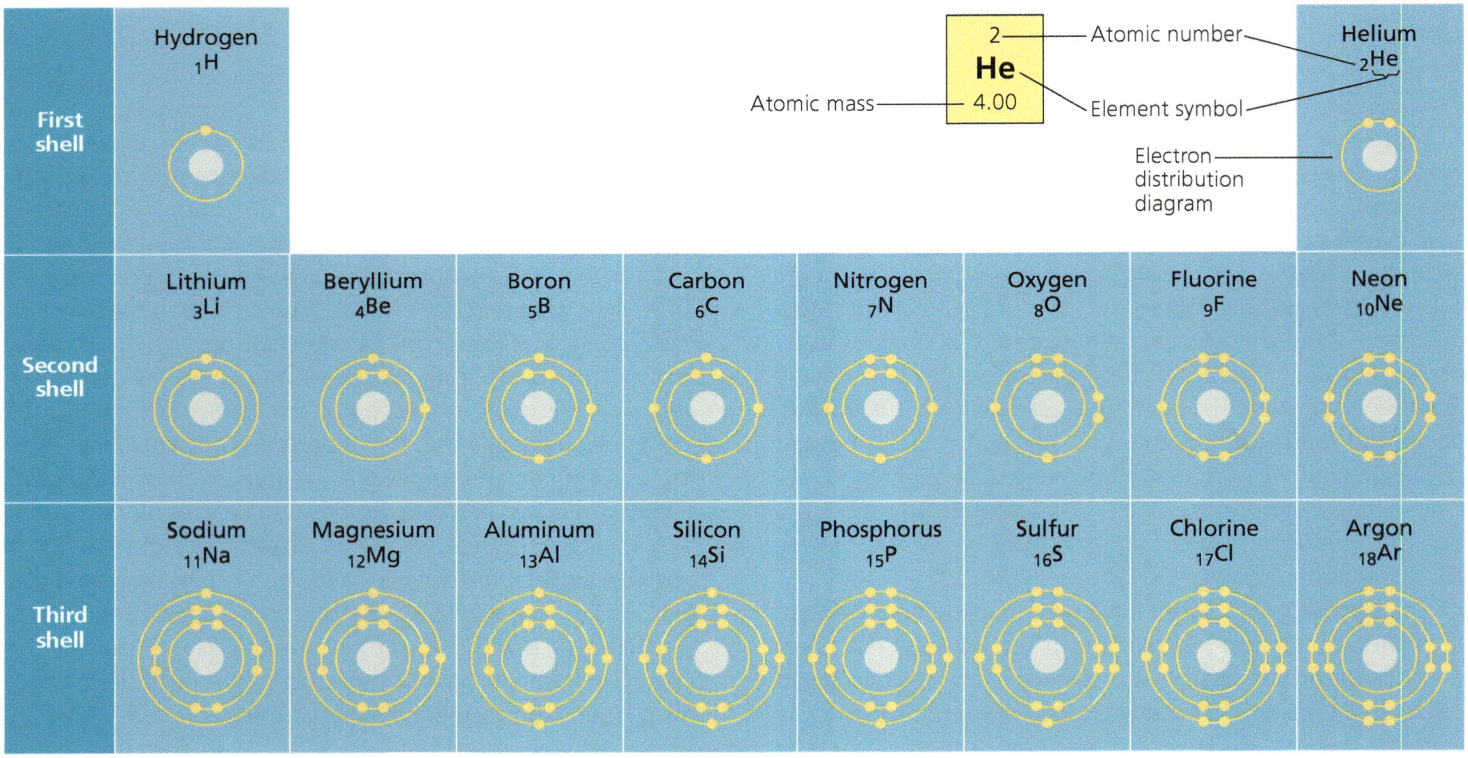

▲ **Figure 2.9 Electron distribution diagrams for the first 18 elements in the periodic table.** In a standard periodic table (see Appendix B), information for each element is presented as shown for helium in the inset. In the diagrams in this table, electrons are represented as yellow dots and electron shells as concentric circles. These diagrams are a convenient way to picture the distribution of an atom's electrons among its electron shells, but these simplified models do not accurately represent the shape of the atom or the location of its electrons. The elements are arranged in rows, each representing the filling of an electron shell. As electrons are added, they occupy the lowest available shell.

? *What is the atomic number of magnesium? How many protons and electrons does it have? How many electron shells? How many valence electrons?*

Electron Orbitals

In the early 1900s, the electron shells of an atom were visualized as concentric paths of electrons orbiting the nucleus, somewhat like planets orbiting the sun. It is still convenient to use two-dimensional concentric-circle diagrams, as in Figure 2.9, to symbolize three-dimensional electron shells. However, you need to remember that each concentric circle represents only the *average* distance between an electron in that shell and the nucleus. Accordingly, the concentric-circle diagrams do not give a real picture of an atom. In reality, we can never know the exact location of an electron. What we can do instead is describe the space in which an electron spends most of its time. The three-dimensional space where an electron is found 90% of the time is called an **orbital**.

Each electron shell contains electrons at a particular energy level, distributed among a specific number of orbitals of distinctive shapes and orientations. **Figure 2.10** shows the orbitals of neon as an example, with its electron distribution diagram for reference. You can think of an orbital as a component of an electron shell. The first electron shell has only one spherical *s* orbital (called 1*s*), but the second shell has four orbitals: one large spherical *s* orbital (called 2*s*) and three dumbbell-shaped *p* orbitals (called 2*p* orbitals). (The third shell and other higher electron shells also have *s* and *p* orbitals, as well as orbitals of more complex shapes.)

No more than 2 electrons can occupy a single orbital. The first electron shell can therefore accommodate up to 2 electrons in its *s* orbital. The lone electron of a hydrogen atom occupies the 1*s* orbital, as do the 2 electrons of a helium atom. The four orbitals of the second electron shell can hold up to 8 electrons, 2 in each orbital. Electrons in each of the four orbitals have nearly the same energy, but they move in different volumes of space.

The reactivity of atoms arises from the presence of unpaired electrons in one or more orbitals of their valence shells. As you will see in the next section, atoms interact in a way that completes their valence shells. When they do so, it is the *unpaired* electrons that are involved.

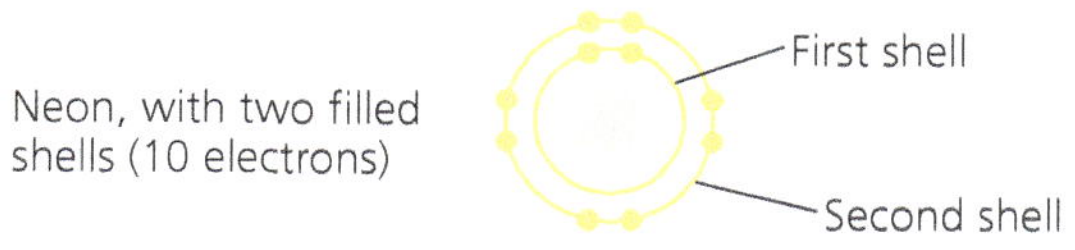

(a) Electron distribution diagram. An electron distribution diagram is shown here for a neon atom, which has a total of 10 electrons. Each concentric circle represents an electron shell, which can be subdivided into electron orbitals.

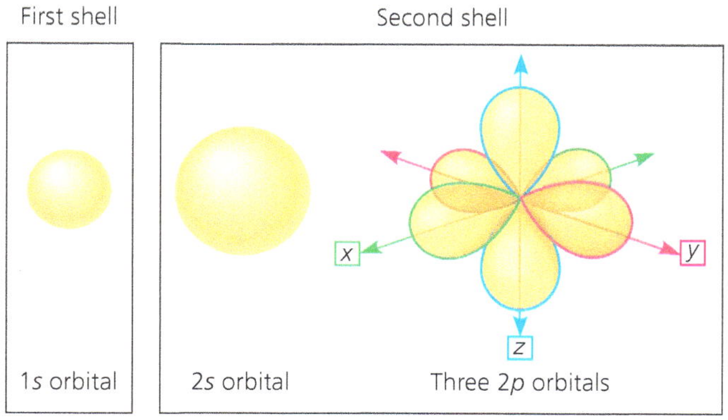

(b) Separate electron orbitals. The three-dimensional shapes represent electron orbitals—the volumes of space where the electrons of an atom are most likely to be found. Each orbital holds a maximum of 2 electrons. The first electron shell, on the left, has one spherical (*s*) orbital, designated 1*s*. The second shell, on the right, has one larger *s* orbital (designated 2*s* for the second shell) plus three dumbbell-shaped orbitals called *p* orbitals (2*p* for the second shell). The three 2*p* orbitals lie at right angles to one another along imaginary *x*-, *y*-, and *z*-axes of the atom. Each 2*p* orbital is outlined here in a different color.

(c) Superimposed electron orbitals. To reveal the complete picture of the electron orbitals of neon, we superimpose the 1*s* orbital of the first shell and the 2*s* and three 2*p* orbitals of the second shell.

▲ **Figure 2.10 Electron orbitals.**

CONCEPT CHECK 2.2

1. A lithium atom has 3 protons and 4 neutrons. What is its atomic mass in daltons?
2. A nitrogen atom has 7 protons, and the most common isotope of nitrogen has 7 neutrons. A radioactive isotope of nitrogen has 8 neutrons. Write the atomic number and mass number of this radioactive nitrogen as a chemical symbol with a subscript and superscript.
3. How many electrons does fluorine have? How many electron shells? Name the orbitals that are occupied. How many electrons are needed to fill the valence shell?
4. **WHAT IF?** In Figure 2.9, if two or more elements are in the same row, what do they have in common? If two or more elements are in the same column, what do they have in common?

For suggested answers, see Appendix A.

CONCEPT 2.3

The formation and function of molecules depend on chemical bonding between atoms

Now that we have looked at the structure of atoms, we can move up the hierarchy of organization and see how atoms combine to form molecules and ionic compounds. Atoms with incomplete valence shells can interact with certain other atoms in such a way that each partner completes its valence shell: The atoms either share or transfer valence electrons. These interactions usually result in atoms staying close together, held by attractions called **chemical bonds**. The strongest kinds of chemical bonds are covalent bonds and ionic bonds.

Covalent Bonds

A **covalent bond** is the sharing of a pair of valence electrons by two atoms. For example, let's consider what happens when two hydrogen atoms approach each other. Recall that hydrogen has 1 valence electron in the first shell, but the shell's capacity is 2 electrons. When the two hydrogen atoms come close enough for their 1*s* orbitals to overlap, they can share their electrons (**Figure 2.11**). Each hydrogen atom is now associated with 2 electrons in what amounts to a completed valence shell. Two or more atoms held together by covalent bonds constitute a **molecule**, in this case a hydrogen molecule.

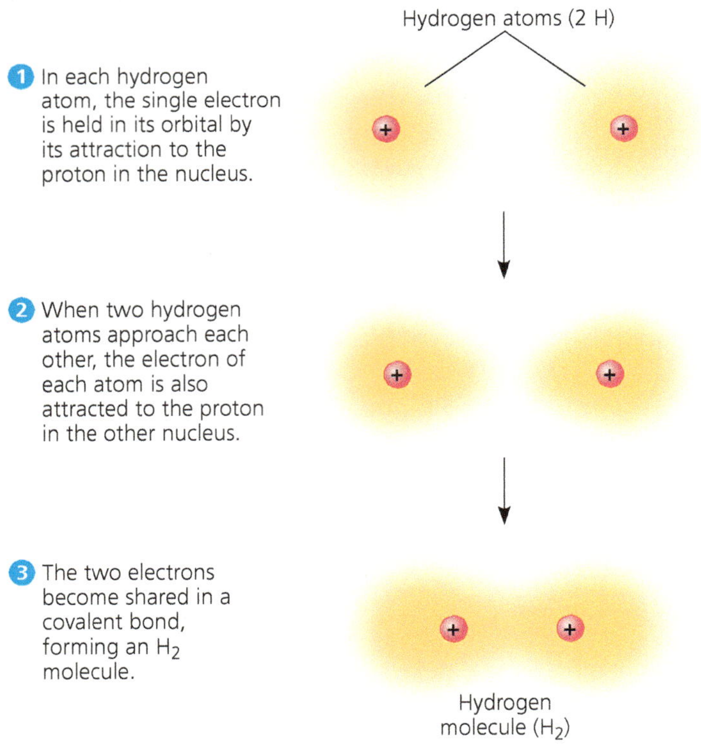

▲ **Figure 2.11 Formation of a covalent bond.**

Figure 2.12a shows several ways of representing a hydrogen molecule. Its *molecular formula*, H_2, simply indicates that the molecule consists of two atoms of hydrogen. Electron sharing can be depicted by an electron distribution diagram or by a *Lewis dot structure*, in which element symbols are surrounded by dots that represent the valence electrons (H:H). We can also use a *structural formula*, H—H, where the line represents a **single bond**, a pair of shared electrons. A space-filling model comes closest to representing the actual shape of the molecule.

Oxygen has 6 electrons in its second electron shell and therefore needs 2 more electrons to complete its valence shell. Two oxygen atoms form a molecule by sharing *two* pairs of valence electrons (**Figure 2.12b**). The atoms are thus joined by a **double bond** (O=O).

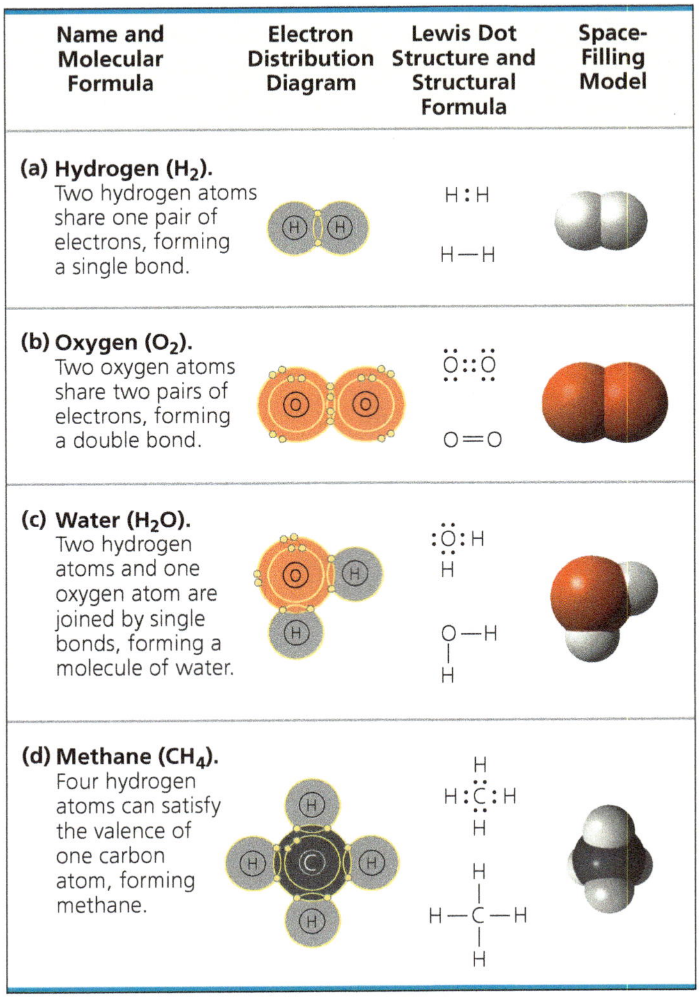

▲ **Figure 2.12 Covalent bonding in four molecules.** The number of electrons required to complete an atom's valence shell generally determines how many covalent bonds that atom will form. This figure shows several ways of indicating covalent bonds.

Each atom that can share valence electrons has a bonding capacity corresponding to the number of covalent bonds the atom can form. When the bonds form, they give the atom a full complement of electrons in the valence shell. The bonding capacity of oxygen, for example, is 2. This bonding capacity is called the atom's **valence** and usually equals the number of unpaired electrons required to complete the atom's outermost (valence) shell. See if you can determine the valences of hydrogen, oxygen, nitrogen, and carbon by studying the electron distribution diagrams in Figure 2.9. You can see that the valence of hydrogen is 1; oxygen, 2; nitrogen, 3; and carbon, 4. However, the situation is more complicated for elements in the third row of the periodic table. Phosphorus, for example, can have a valence of 3, as we would predict from the presence of 3 unpaired electrons in its valence shell. In some molecules that are biologically important, however, phosphorus can form three single bonds and one double bond. Therefore, it can also have a valence of 5.

The molecules H_2 and O_2 are pure elements rather than compounds because a compound is a combination of two or more *different* elements. Water, with the molecular formula H_2O, is a compound. Two atoms of hydrogen are needed to satisfy the valence of one oxygen atom. **Figure 2.12c** shows the structure of a water molecule. Water is so important to life that Chapter 3 is devoted entirely to its structure and behavior.

Methane, the main component of natural gas, is a compound with the molecular formula CH_4. It takes four hydrogen atoms, each with a valence of 1, to complement one atom of carbon, with its valence of 4 **(Figure 2.12d)**. We will look at many other compounds of carbon in Chapter 4.

Atoms in a molecule attract shared electrons to varying degrees, depending on the element. The attraction of a particular atom for the electrons of a covalent bond is called its **electronegativity**. The more electronegative an atom is, the more strongly it pulls shared electrons toward itself. In a covalent bond between two atoms of the same element, the electrons are shared equally because the two atoms have the same electronegativity—the tug-of-war is at a standoff. Such a bond is called a **nonpolar covalent bond**. For example, the single bond of H_2 is nonpolar, as is the double bond of O_2. However, when one atom is bonded to a more electronegative atom, the electrons of the bond are not shared equally. This type of bond is called a **polar covalent bond**. Such bonds vary in their polarity, depending on the relative electronegativity of the two atoms. For example, the bonds between the oxygen and hydrogen atoms of a water molecule are quite polar **(Figure 2.13)**.

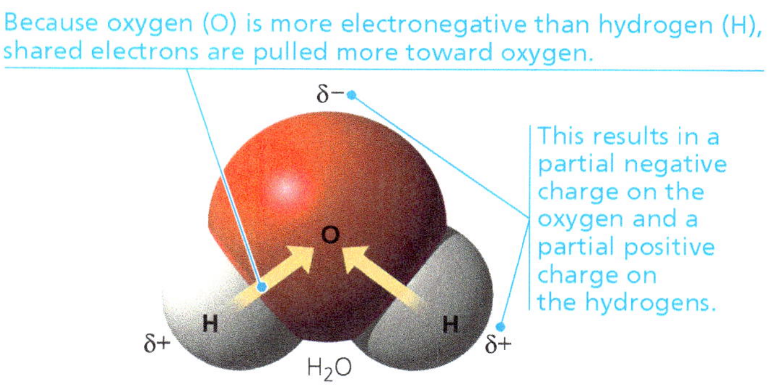

▲ **Figure 2.13 Polar covalent bonds in a water molecule.**

Oxygen is one of the most electronegative of all the elements, attracting shared electrons much more strongly than hydrogen does. In a covalent bond between oxygen and hydrogen, the electrons spend more time near the oxygen nucleus than they do near the hydrogen nucleus. Because electrons have a negative charge and are pulled toward oxygen in a water molecule, the oxygen atom has a partial negative charge (indicated by the Greek letter δ with a minus sign, δ−, or "delta minus"), and each hydrogen atom has a partial positive charge (δ+, or "delta plus"). In contrast, the individual bonds of methane (CH_4) are much less polar because the electronegativities of carbon and hydrogen are similar.

Ionic Bonds

In some cases, two atoms are so unequal in their attraction for valence electrons that the more electronegative atom strips an electron completely away from its partner. This is what happens when an atom of sodium ($_{11}Na$) encounters an atom of chlorine ($_{17}Cl$) **(Figure 2.14)**. A sodium atom has a total of 11 electrons, with its single valence electron in the third electron shell. A chlorine atom has a total of 17 electrons,

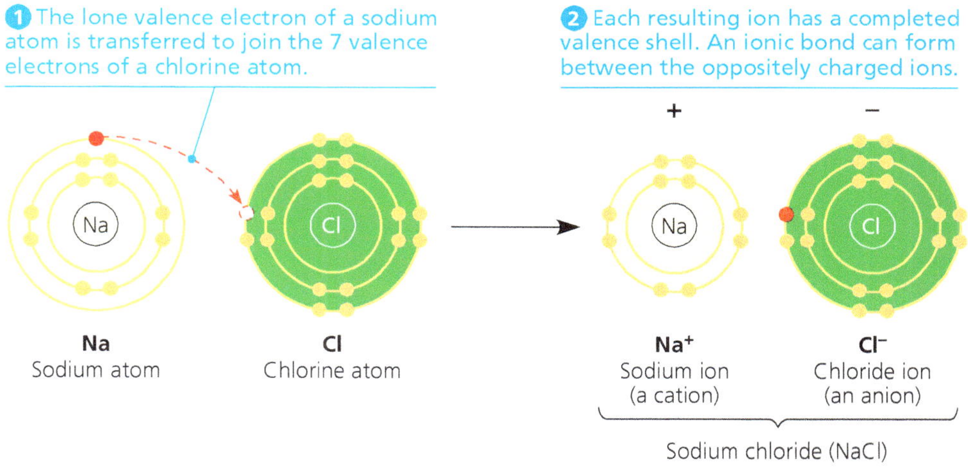

▲ **Figure 2.14 Electron transfer and ionic bonding.** The attraction between oppositely charged atoms, or ions, is an ionic bond. An ionic bond can form between any two oppositely charged ions, even if they have not been formed by transfer of an electron from one to the other.

with 7 electrons in its valence shell. When these two atoms meet, the lone valence electron of sodium is transferred to the chlorine atom, and both atoms end up with their valence shells complete. (Because sodium no longer has an electron in the third shell, the second shell is now the valence shell.)

The electron transfer between the two atoms moves one unit of negative charge from sodium to chlorine. Sodium, now with 11 protons but only 10 electrons, has a net electrical charge of 1+. A charged atom (or molecule) is called an **ion**. When the charge is positive, the ion is specifically called a **cation**; the sodium atom has become a cation. Conversely, the chlorine atom, having gained an extra electron, now has 17 protons and 18 electrons, giving it a net electrical charge of 1–. It has become a chloride ion—an **anion**, or negatively charged ion. Because of their opposite charges, cations and anions attract each other; this attraction is called an **ionic bond**. The transfer of an electron is not the formation of a bond; rather, it allows a bond to form because it results in two ions of opposite charge. Any two ions of opposite charge can form an ionic bond. The ions do not need to have acquired their charge by an electron transfer with each other.

Compounds formed by ionic bonds are called **ionic compounds**, or **salts**. We know the ionic compound sodium chloride (NaCl) as table salt **(Figure 2.15)**. Salts are often found in nature as crystals of various sizes and shapes. Each salt crystal is an aggregate of vast numbers of cations and anions bonded by their electrical attraction and arranged in a three-dimensional lattice. Unlike a covalent compound, which consists of molecules having a definite size and number of atoms, an ionic compound does not consist of molecules. The formula for an ionic compound, such as NaCl, indicates only the ratio of elements in a crystal of the salt. "NaCl" by itself is not a molecule.

Not all salts have equal numbers of cations and anions. For example, the ionic compound magnesium chloride ($MgCl_2$) has two chloride ions for each magnesium ion. Magnesium ($_{12}Mg$) must lose 2 outer electrons if the atom is to have a complete valence shell, so it tends to become a cation with a net charge of 2+ (Mg^{2+}). One magnesium cation can therefore form ionic bonds with two chloride anions.

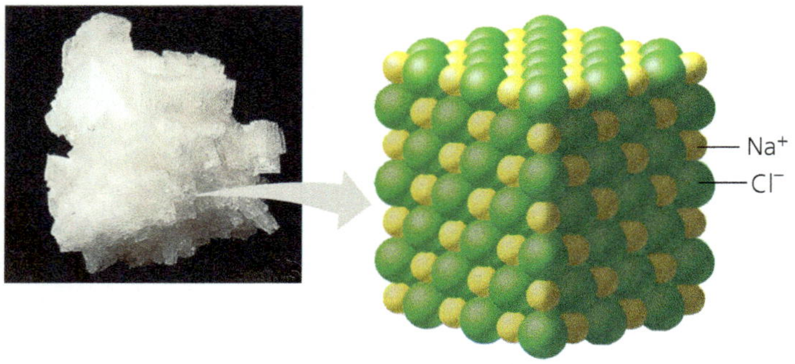

▲ **Figure 2.15 A sodium chloride (NaCl) crystal.** The sodium ions (Na^+) and chloride ions (Cl^-) are held together by ionic bonds. The formula NaCl tells us that the ratio of Na^+ to Cl^- is 1:1.

The term *ion* also applies to entire molecules that are electrically charged. In the salt ammonium chloride (NH_4Cl), for instance, the anion is a single chloride ion (Cl^-), but the cation is ammonium (NH_4^+), a nitrogen atom with four covalently bonded hydrogen atoms. The whole ammonium ion has an electrical charge of 1+ because it is 1 electron short.

Environment affects the strength of ionic bonds. In a dry salt crystal, the bonds are so strong that it takes a hammer and chisel to break enough of them to crack the crystal in two. If the same salt crystal is dissolved in water, however, the ionic bonds are much weaker because each ion is partially shielded by its interactions with water molecules. Most drugs are manufactured as salts because they are quite stable when dry but can dissociate (come apart) easily in water. In the next chapter, you will learn how water dissolves salts.

Weak Chemical Bonds

In organisms, most of the strongest chemical bonds are covalent bonds, which link atoms to form a cell's molecules. But weaker bonding within and between molecules is also indispensable in the cell, contributing greatly to the emergent properties of life. Many large biological molecules are held in their functional form by weak bonds. In addition, when two molecules in the cell make contact, they may adhere temporarily by weak bonds. The reversibility of weak bonding can be an advantage: Two molecules can come together, respond to one another in some way, and then separate.

Several types of weak chemical bonds are important in organisms. One is the ionic bond as it exists between ions dissociated in water, which we just discussed. Hydrogen bonds and van der Waals interactions are also crucial to life.

Hydrogen Bonds

Among the various kinds of weak chemical bonds, hydrogen bonds are so important in the chemistry of life that they deserve special attention. The partial positive charge on a hydrogen atom that is covalently bonded to an electronegative atom allows the hydrogen to be attracted to a different electronegative atom nearby. This noncovalent attraction between a hydrogen and an electronegative atom is called a **hydrogen bond**. In living cells, the electronegative partners are usually oxygen or nitrogen atoms. Refer to **Figure 2.16** to examine the simple case of hydrogen bonding between water (H_2O) and ammonia (NH_3).

Van der Waals Interactions

Even a molecule with nonpolar covalent bonds may have positively and negatively charged regions. Electrons are not always symmetrically distributed in such a molecule; at any

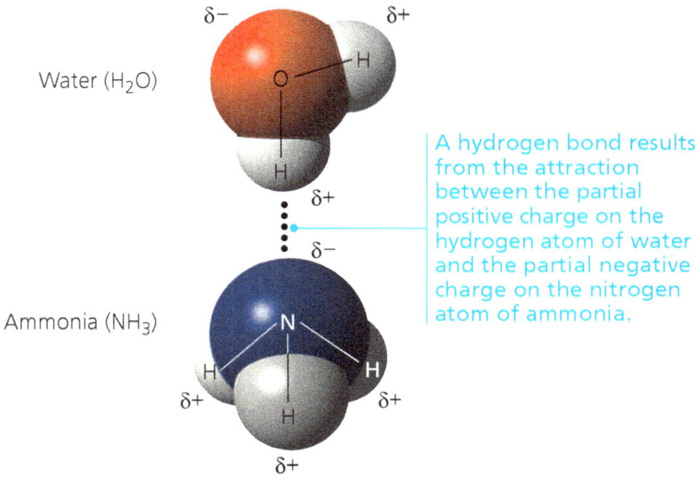

▲ **Figure 2.16 A hydrogen bond.**

DRAW IT *Draw five water molecules using structural formulas and indicating partial charges, and show how they can make hydrogen bonds with each other.*

instant, they may accumulate by chance in one part of the molecule or another. The results are ever-changing regions of positive and negative charge that enable all atoms and molecules to stick to one another. These **van der Waals interactions** are individually weak and occur only when atoms and molecules are very close together. When many such interactions occur simultaneously, however, they can be powerful: Van der Waals interactions are the reason a gecko lizard (right) can walk straight up a wall! Each gecko toe has hundreds of thousands of tiny hairs, with multiple projections at each hair's tip that increase surface area. Apparently, the van der Waals interactions between the hair tip molecules and the molecules of the wall's surface are so numerous that despite their individual weakness, together they can support the gecko's body weight.

Van der Waals interactions, hydrogen bonds, ionic bonds in water, and other weak bonds may form not only between molecules but also between parts of a large molecule, such as a protein. The cumulative effect of weak bonds is to reinforce the three-dimensional shape of the molecule. You will learn more about the very important biological roles of weak bonds in Chapter 5.

Molecular Shape and Function

A molecule has a characteristic size and shape. The precise shape of a molecule is usually very important to its function in the living cell.

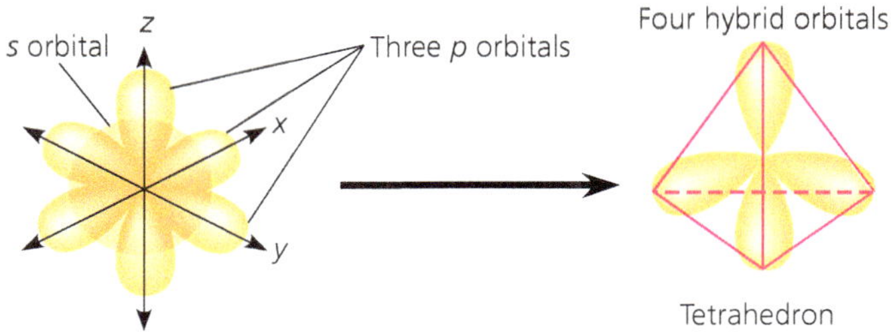

(a) Hybridization of orbitals. The single *s* and three *p* orbitals of a valence shell involved in covalent bonding combine to form four teardrop-shaped hybrid orbitals. These orbitals extend to the four corners of an imaginary tetrahedron (outlined in pink).

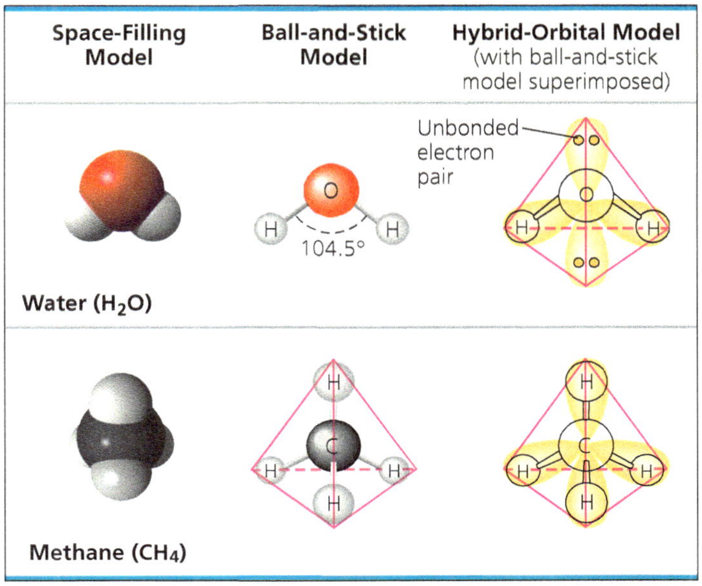

(b) Molecular-shape models. Three models representing molecular shape are shown for water and methane. The positions of the hybrid orbitals determine the shapes of the molecules.

▲ **Figure 2.17 Molecular shapes due to hybrid orbitals.**

A molecule consisting of two atoms, such as H_2 or O_2, is always linear, but most molecules with more than two atoms have more complicated shapes. These shapes are determined by the positions of the atoms' orbitals. When an atom forms covalent bonds, the orbitals in its valence shell undergo rearrangement. For atoms with valence electrons in both *s* and *p* orbitals (review Figure 2.10), the single *s* and three *p* orbitals form four new hybrid orbitals shaped like identical teardrops extending from the region of the atomic nucleus **(Figure 2.17a)**. If we connect the larger ends of the teardrops with lines, we have the outline of a geometric shape called a tetrahedron, a pyramid with a triangular base.

For the water molecule (H_2O), two of the hybrid orbitals in the oxygen atom's valence shell are shared with hydrogen atoms **(Figure 2.17b)**. The result is a molecule shaped roughly like a V, with its two covalent bonds spread apart at an angle of 104.5°.

The methane molecule (CH_4) has the shape of a completed tetrahedron because all four hybrid orbitals of the carbon atom are shared with hydrogen atoms (see Figure 2.17b). The carbon nucleus is at the center, with its four covalent bonds radiating to hydrogen nuclei at the corners of the tetrahedron. Larger molecules containing multiple carbon atoms, including many of the molecules that make up living matter, have more complex overall shapes. However, the tetrahedral shape of a carbon atom bonded to four other atoms is often a repeating motif within such molecules.

Molecular shape is crucial in biology because it determines how biological molecules recognize and respond to one another with specificity. Biological molecules often bind temporarily to each other by forming weak bonds, but this can happen only if their shapes are complementary. We can see this specificity in the effects of opiates, drugs derived from opium. Opiates, such as morphine and heroin, relieve pain and alter mood by weakly binding to specific receptor molecules on the surfaces of brain cells. Why would brain cells carry receptors for opiates, compounds that are not made by our bodies? The discovery of endorphins in 1975 answered this question. Endorphins are signaling molecules made by the pituitary gland that bind to the receptors, relieving pain and producing euphoria during times of stress, such as intense exercise. It turns out that opiates have shapes similar to endorphins and mimic them by binding to endorphin receptors in the brain. That is why opiates (such as morphine) and endorphins have similar effects (**Figure 2.18**). The role of molecular shape in brain chemistry illustrates the relationship between structure and function, one of biology's unifying themes.

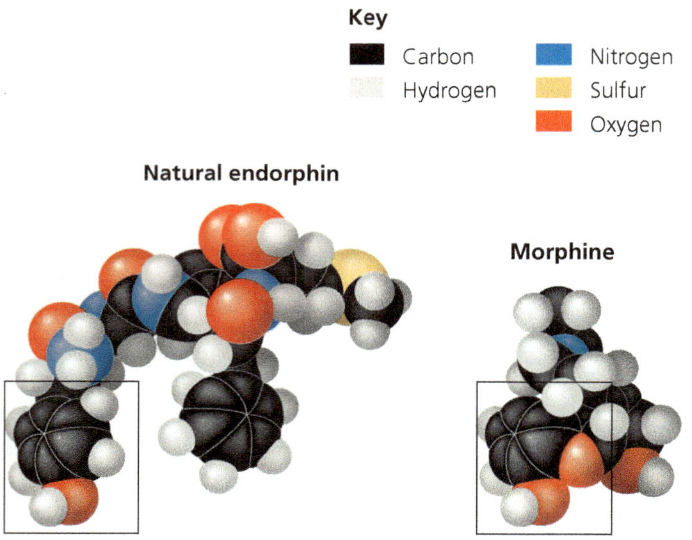

(a) Structures of endorphin and morphine. The boxed portion of the endorphin molecule (left) binds to receptor molecules on target cells in the brain. The boxed portion of the morphine molecule (right) is a close match.

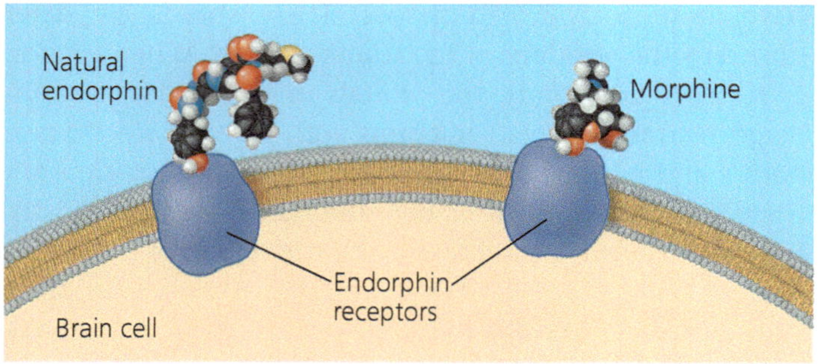

(b) Binding to endorphin receptors. Both endorphin and morphine can bind to endorphin receptors on the surface of a brain cell.

▲ **Figure 2.18 A molecular mimic.** Morphine affects pain perception and emotional state by mimicking the brain's natural endorphins.

CONCEPT CHECK 2.3

1. Why does the structure H—C═C—H fail to make sense chemically?
2. What holds the atoms together in a crystal of magnesium chloride ($MgCl_2$)?
3. **WHAT IF?** If you were a pharmaceutical researcher, why would you want to learn the three-dimensional shapes of naturally occurring signaling molecules?

For suggested answers, see Appendix A.

CONCEPT 2.4

Chemical reactions make and break chemical bonds

The making and breaking of chemical bonds, leading to changes in the composition of matter, are called **chemical reactions**. An example is the reaction between hydrogen and oxygen molecules that forms water:

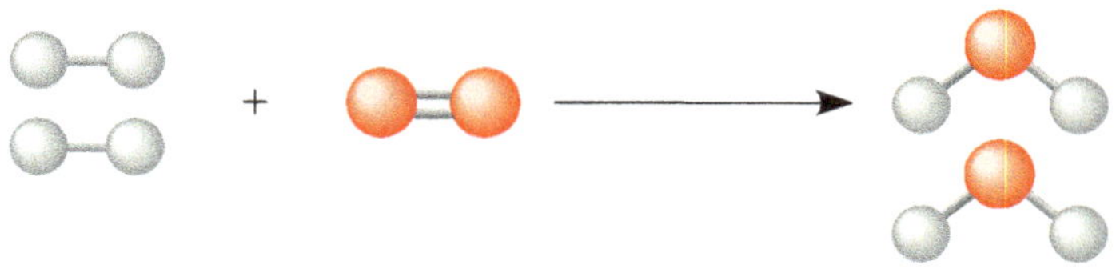

$$2\,H_2 + O_2 \longrightarrow 2\,H_2O$$

Reactants **Reaction** **Products**

This reaction breaks the covalent bonds of H_2 and O_2 and forms the new bonds of H_2O. When we write a chemical reaction, we use an arrow to indicate the conversion of the starting materials, called the **reactants**, to the **products**. The coefficients indicate the number of molecules involved; for example, the coefficient 2 in front of the H_2 means that

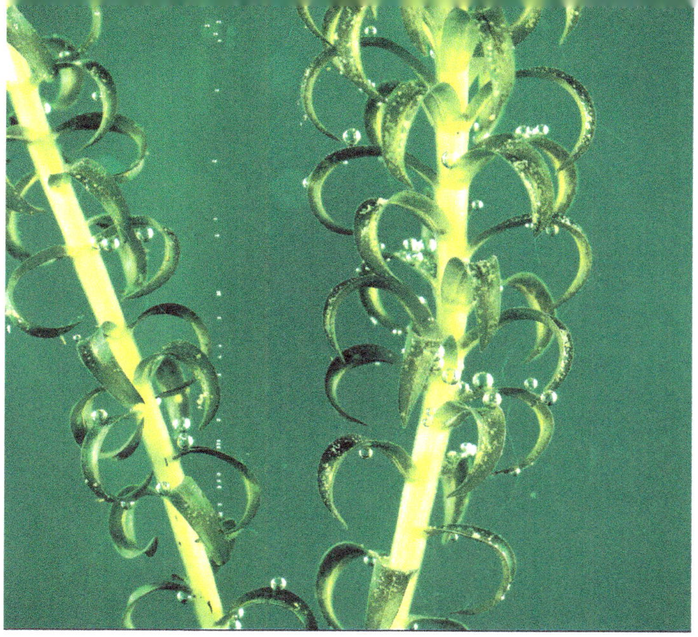

▲ **Figure 2.19 Photosynthesis: a solar-powered rearrangement of matter.** *Elodea*, a freshwater plant, produces sugar by rearranging the atoms of carbon dioxide and water in the chemical process known as photosynthesis, which is powered by sunlight. Much of the sugar is then converted to other food molecules. Oxygen gas (O_2) is a by-product of photosynthesis; notice the bubbles of oxygen escaping from the leaves in the photo.

? *Explain how this photo relates to the reactants and products in the equation for photosynthesis given in the text. (You will learn more about photosynthesis in Chapter 10.)*

the reaction starts with two molecules of hydrogen. Notice that all atoms of the reactants must be accounted for in the products. Matter is conserved in a chemical reaction: Reactions cannot create or destroy matter but can only rearrange it.

Photosynthesis, which takes place within the cells of green plant tissues, is a particularly important example of how chemical reactions rearrange matter. Humans and other animals ultimately depend on photosynthesis for food and oxygen, and this process is at the foundation of almost all ecosystems. The following chemical shorthand summarizes the process of photosynthesis:

$$6\ CO_2 + 6\ H_2O \rightarrow C_6H_{12}O_6 + 6\ O_2$$

The raw materials of photosynthesis are carbon dioxide (CO_2), which is taken from the air, and water (H_2O), which is absorbed from the soil. Within the plant cells, sunlight powers the conversion of these ingredients to a sugar called glucose ($C_6H_{12}O_6$) and oxygen molecules (O_2), a by-product that the plant releases into the surroundings (**Figure 2.19**). Although photosynthesis is actually a sequence of many chemical reactions, we still end up with the same number and types of atoms that we had when we started. Matter has simply been rearranged, with an input of energy provided by sunlight.

All chemical reactions are reversible, with the products of the forward reaction becoming the reactants for the reverse reaction. For example, hydrogen and nitrogen molecules can combine to form ammonia, but ammonia can also decompose to regenerate hydrogen and nitrogen:

$$3\ H_2 + N_2 \rightleftharpoons 2\ NH_3$$

The two opposite-headed arrows indicate that the reaction is reversible.

One of the factors affecting the rate of a reaction is the concentration of reactants. The greater the concentration of reactant molecules, the more frequently they collide with one another and have an opportunity to react and form products. The same holds true for products. As products accumulate, collisions resulting in the reverse reaction become more frequent. Eventually, the forward and reverse reactions occur at the same rate, and the relative concentrations of products and reactants stop changing. The point at which the reactions offset one another exactly is called **chemical equilibrium**. This is a dynamic equilibrium; reactions are still going on, but with no net effect on the concentrations of reactants and products. Equilibrium does *not* mean that the reactants and products are equal in concentration, but only that their concentrations have stabilized at a particular ratio. The reaction involving ammonia reaches equilibrium when ammonia decomposes as rapidly as it forms. In some chemical reactions, the equilibrium point may lie so far to the right that these reactions go essentially to completion; that is, virtually all the reactants are converted to products.

We will return to the subject of chemical reactions after more detailed study of the various types of molecules that are important to life. In the next chapter, we focus on water, the substance in which all the chemical processes of organisms occur.

CONCEPT CHECK 2.4

1. **MAKE CONNECTIONS** Consider the reaction between hydrogen and oxygen that forms water, shown with ball-and-stick models on page 42. Study Figure 2.12 and draw the Lewis dot structures representing this reaction.
2. Which type of chemical reaction occurs faster at equilibrium, the formation of products from reactants or reactants from products?
3. **WHAT IF?** Write an equation that uses the products of photosynthesis as reactants and the reactants of photosynthesis as products. Add energy as another product. This new equation describes a process that occurs in your cells. Describe this equation in words. How does this equation relate to breathing?

For suggested answers, see Appendix A.

2 CHAPTER REVIEW

SUMMARY OF KEY CONCEPTS

CONCEPT 2.1

Matter consists of chemical elements in pure form and in combinations called compounds (pp. 31–32)

- **Elements** cannot be broken down chemically to other substances. A **compound** contains two or more different elements in a fixed ratio. Oxygen, carbon, hydrogen, and nitrogen make up approximately 96% of living matter.

? *In what way does the need for iodine or iron in your diet differ from your need for calcium or phosphorus?*

CONCEPT 2.2

An element's properties depend on the structure of its atoms (pp. 33–37)

- An **atom**, the smallest unit of an element, has the following components:

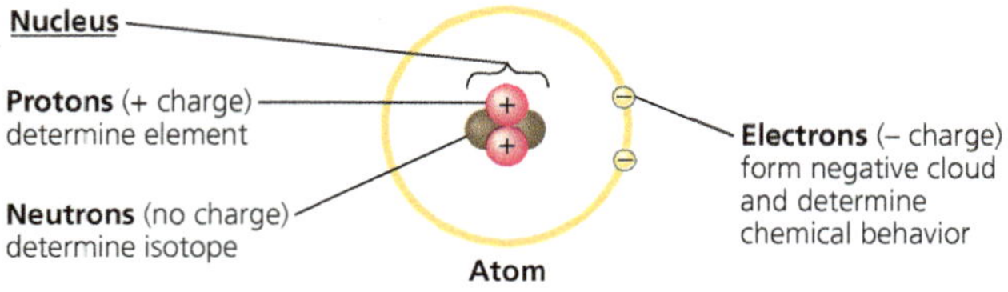

- An electrically neutral atom has equal numbers of electrons and protons; the number of protons determines the **atomic number**. The **atomic mass** is measured in **daltons** and is roughly equal to the sum of protons plus neutrons. **Isotopes** of an element differ from each other in neutron number and therefore mass. Unstable isotopes give off particles and energy as radioactivity.
- In an atom, electrons occupy specific **electron shells**; the electrons in a shell have a characteristic energy level. Electron distribution in shells determines the chemical behavior of an atom. An atom that has an incomplete outer shell, the **valence shell**, is reactive.
- Electrons exist in **orbitals**, three-dimensional spaces with specific shapes that are components of electron shells.

DRAW IT *Draw the electron distribution diagrams for neon ($_{10}$Ne) and argon ($_{18}$Ar). Use these diagrams to explain why these elements are chemically unreactive.*

CONCEPT 2.3

The formation and function of molecules depend on chemical bonding between atoms (pp. 38–42)

- **Chemical bonds** form when atoms interact and complete their valence shells. **Covalent bonds** form when pairs of electrons are shared.

- **Molecules** consist of two or more covalently bonded atoms. The attraction of an atom for the electrons of a covalent bond is its **electronegativity**. If both atoms are the same, they have the same electronegativity and share a **nonpolar covalent bond**. Electrons of a **polar covalent bond** are pulled closer to the more electronegative atom.
- An **ion** forms when an atom or molecule gains or loses an electron and becomes charged. An **ionic bond** is the attraction between two oppositely charged ions.

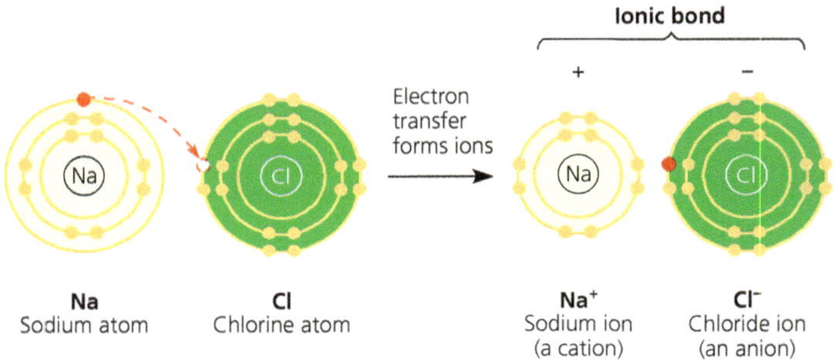

- Weak bonds reinforce the shapes of large molecules and help molecules adhere to each other. A **hydrogen bond** is an attraction between a hydrogen atom carrying a partial positive charge ($\delta+$) and an electronegative atom ($\delta-$). **Van der Waals interactions** occur between transiently positive and negative regions of molecules.
- A molecule's shape is determined by the positions of its atoms' valence orbitals. Covalent bonds result in hybrid orbitals, which are responsible for the shapes of H_2O, CH_4, and many more complex biological molecules. Shape is usually the basis for the recognition of one biological molecule by another.

? *In terms of electron sharing between atoms, compare nonpolar covalent bonds, polar covalent bonds, and the formation of ions.*

CONCEPT 2.4

Chemical reactions make and break chemical bonds (pp. 42–43)

- **Chemical reactions** change **reactants** into **products** while conserving matter. All chemical reactions are theoretically reversible. **Chemical equilibrium** is reached when the forward and reverse reaction rates are equal.

? *What would happen to the concentration of products if more reactants were added to a reaction that was in chemical equilibrium? How would this addition affect the equilibrium?*

TEST YOUR UNDERSTANDING

LEVEL 1: KNOWLEDGE/COMPREHENSION

1. In the term *trace element*, the adjective *trace* means that
   a. the element is required in very small amounts.
   b. the element can be used as a label to trace atoms through an organism's metabolism.
   c. the element is very rare on Earth.
   d. the element enhances health but is not essential for the organism's long-term survival.
   e. the element passes rapidly through the organism.

2. Compared with ^{31}P, the radioactive isotope ^{32}P has
   a. a different atomic number.
   b. a different charge.
   c. one more proton.
   d. one more electron.
   e. one more neutron.

3. The reactivity of an atom arises from
   a. the average distance of the outermost electron shell from the nucleus.
   b. the existence of unpaired electrons in the valence shell.
   c. the sum of the potential energies of all the electron shells.
   d. the potential energy of the valence shell.
   e. the energy difference between the *s* and *p* orbitals.

4. Which statement is true of all atoms that are anions?
   a. The atom has more electrons than protons.
   b. The atom has more protons than electrons.
   c. The atom has fewer protons than does a neutral atom of the same element.
   d. The atom has more neutrons than protons.
   e. The net charge is 1−.

5. Which of the following statements correctly describes any chemical reaction that has reached equilibrium?
   a. The concentrations of products and reactants are equal.
   b. The reaction is now irreversible.
   c. Both forward and reverse reactions have halted.
   d. The rates of the forward and reverse reactions are equal.
   e. No reactants remain.

LEVEL 2: APPLICATION/ANALYSIS

6. We can represent atoms by listing the number of protons, neutrons, and electrons—for example, $2p^+$, $2n^0$, $2e^-$ for helium. Which of the following represents the ^{18}O isotope of oxygen?
   a. $6p^+$, $8n^0$, $6e^-$
   b. $8p^+$, $10n^0$, $8e^-$
   c. $9p^+$, $9n^0$, $9e^-$
   d. $7p^+$, $2n^0$, $9e^-$
   e. $10p^+$, $8n^0$, $9e^-$

7. The atomic number of sulfur is 16. Sulfur combines with hydrogen by covalent bonding to form a compound, hydrogen sulfide. Based on the number of valence electrons in a sulfur atom, predict the molecular formula of the compound.
   a. HS b. HS_2 c. H_2S d. H_3S_2 e. H_4S

8. What coefficients must be placed in the following blanks so that all atoms are accounted for in the products?

   $C_6H_{12}O_6 \rightarrow$ _____ C_2H_6O + _____ CO_2

   a. 1; 2 b. 3; 1 c. 1; 3 d. 1; 1 e. 2; 2

9. **DRAW IT** Draw Lewis dot structures for each hypothetical molecule shown below, using the correct number of valence electrons for each atom. Determine which molecule makes sense because each atom has a complete valence shell and each bond has the correct number of electrons. Explain what makes the other molecules nonsensical, considering the number of bonds each type of atom can make.

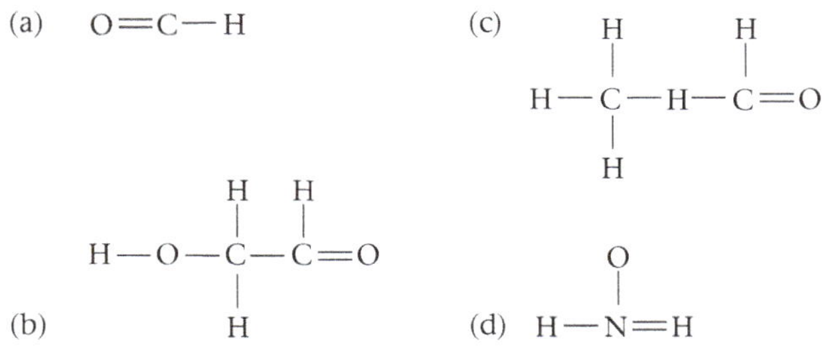

LEVEL 3: SYNTHESIS/EVALUATION

10. **EVOLUTION CONNECTION**

    The percentages of naturally occurring elements making up the human body (see Table 2.1) are similar to the percentages of these elements found in other organisms. How could you account for this similarity among organisms?

11. **SCIENTIFIC INQUIRY**

    Female silkworm moths (*Bombyx mori*) attract males by emitting chemical signals that spread through the air. A male hundreds of meters away can detect these molecules and fly toward their source. The sensory organs responsible for this behavior are the comblike antennae visible in the photograph shown here. Each filament of an antenna is equipped with thousands of receptor cells that detect the sex attractant. Based on what you learned in this chapter, propose a hypothesis to account for the ability of the male moth to detect a specific molecule in the presence of many other molecules in the air. What predictions does your hypothesis make? Design an experiment to test one of these predictions.

12. **WRITE ABOUT A THEME**

    Emergent Properties While waiting at an airport, Neil Campbell once overheard this claim: "It's paranoid and ignorant to worry about industry or agriculture contaminating the environment with their chemical wastes. After all, this stuff is just made of the same atoms that were already present in our environment." Drawing on your knowledge of electron distribution, bonding, and the theme of emergent properties (pp. 3–5), write a short essay (100–150 words) countering this argument.

For selected answers, see Appendix A.

MasteringBIOLOGY www.masteringbiology.com

1. MasteringBiology® Assignments

Tutorials The Anatomy of Atoms • Atomic Number and Mass Number
Activities Structure of the Atomic Nucleus • Electron Arrangement • Covalent Bonds • Nonpolar and Polar Molecules • Ionic Bonds • Hydrogen Bonds
Questions Student Misconceptions • Reading Quiz • Multiple Choice • End-of-Chapter

2. eText

Read your book online, search, take notes, highlight text, and more.

3. The Study Area

Practice Tests • Cumulative Test • *BioFlix* 3-D Animations • MP3 Tutor Sessions • Videos • Activities • Investigations • Lab Media • Audio Glossary • Word Study Tools • Art

3

Water and Life

▲ **Figure 3.1 How does the habitat of a polar bear depend on the chemistry of water?**

KEY CONCEPTS

3.1 Polar covalent bonds in water molecules result in hydrogen bonding

3.2 Four emergent properties of water contribute to Earth's suitability for life

3.3 Acidic and basic conditions affect living organisms

OVERVIEW

The Molecule That Supports All of Life

As astronomers study newly discovered planets orbiting distant stars, they hope to find evidence of water on these far-off celestial bodies, for water is the substance that makes possible life as we know it here on Earth. All organisms familiar to us are made mostly of water and live in an environment dominated by water. Water is the biological medium here on Earth, and possibly on other planets as well.

Three-quarters of Earth's surface is covered by water. Although most of this water is in liquid form, water is also present on Earth as a solid (ice) and a gas (water vapor). Water is the only common substance to exist in the natural environment in all three physical states of matter. Furthermore, the solid state of water floats on the liquid, a rare property emerging from the chemistry of the water molecule. Ice can thus provide a hunting platform for the polar bear in **Figure 3.1**.

The abundance of water is a major reason Earth is habitable. In a classic book called *The Fitness of the Environment*, ecologist Lawrence Henderson highlighted the importance of water to life. While acknowledging that life adapts to its environment through natural selection, Henderson emphasized that for life to exist at all, the environment must first be suitable.

Life on Earth began in water and evolved there for 3 billion years before spreading onto land. Modern life, even terrestrial (land-dwelling) life, remains tied to water. All living organisms require water more than any other substance. Human beings, for example, can survive for quite a few weeks without food, but only a week or so without water. Molecules of water participate in many chemical reactions necessary to sustain life. Most cells are surrounded by water, and cells themselves are about 70–95% water.

What properties of the simple water molecule make it so indispensable to life on Earth? In this chapter, you will learn how the structure of a water molecule allows it to interact with other molecules, including other water molecules. This ability leads to water's unique emergent properties that help make Earth suitable for life.

CONCEPT 3.1

Polar covalent bonds in water molecules result in hydrogen bonding

Water is so common that it is easy to overlook the fact that it is an exceptional substance with many extraordinary qualities. Following the theme of emergent properties, we can trace water's unique behavior to the structure and interactions of its molecules.

Studied on its own, the water molecule is deceptively simple. It is shaped like a wide V, with its two hydrogen atoms joined to the oxygen atom by single covalent bonds. Oxygen is more electronegative than hydrogen, so the electrons of the covalent bonds spend more time closer to oxygen than to hydrogen; these are **polar covalent bonds** (see Figure 2.13). This unequal sharing of electrons and water's V-like shape make it a **polar molecule**, meaning that its overall charge is unevenly distributed: The oxygen region of the molecule has a partial negative charge ($\delta-$), and each hydrogen has a partial positive charge ($\delta+$).

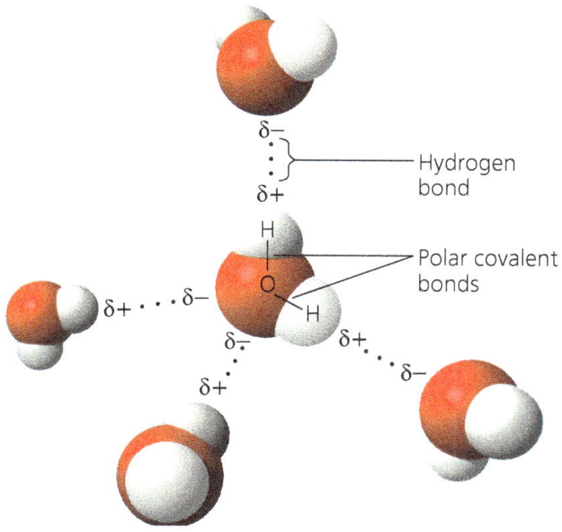

▲ **Figure 3.2 Hydrogen bonds between water molecules.** The charged regions in a water molecule are due to its polar covalent bonds. Oppositely charged regions of neighboring water molecules are attracted to each other, forming hydrogen bonds. Each molecule can hydrogen-bond to multiple partners, and these associations are constantly changing.

DRAW IT *Draw partial charges on all the atoms of the water molecule on the far left above, and draw two more water molecules hydrogen-bonded to it.*

The properties of water arise from attractions between oppositely charged atoms of different water molecules: The slightly positive hydrogen of one molecule is attracted to the slightly negative oxygen of a nearby molecule. The two molecules are thus held together by a hydrogen bond (**Figure 3.2**). When water is in its liquid form, its hydrogen bonds are very fragile, each about 1/20 as strong as a covalent bond. The hydrogen bonds form, break, and re-form with great frequency. Each lasts only a few trillionths of a second, but the molecules are constantly forming new hydrogen bonds with a succession of partners. Therefore, at any instant, a substantial percentage of all the water molecules are hydrogen-bonded to their neighbors. The extraordinary qualities of water are emergent properties resulting in large part from the hydrogen bonding that organizes water molecules into a higher level of structural order.

CONCEPT CHECK 3.1

1. **MAKE CONNECTIONS** What is electronegativity, and how does it affect interactions between water molecules? Review p. 39 and Figure 2.13.
2. Why is it unlikely that two neighboring water molecules would be arranged like this?

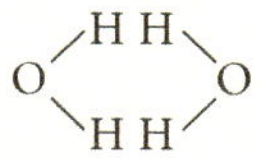

3. **WHAT IF?** What would be the effect on the properties of the water molecule if oxygen and hydrogen had equal electronegativity?

For suggested answers, see Appendix A.

CONCEPT 3.2

Four emergent properties of water contribute to Earth's suitability for life

We will examine four emergent properties of water that contribute to Earth's suitability as an environment for life: cohesive behavior, ability to moderate temperature, expansion upon freezing, and versatility as a solvent.

Cohesion of Water Molecules

Water molecules stay close to each other as a result of hydrogen bonding. Although the arrangement of molecules in a sample of liquid water is constantly changing, at any given moment many of the molecules are linked by multiple hydrogen bonds. These linkages make water more structured than most other liquids. Collectively, the hydrogen bonds hold the substance together, a phenomenon called **cohesion**.

Cohesion due to hydrogen bonding contributes to the transport of water and dissolved nutrients against gravity in plants (**Figure 3.3**). Water from the roots reaches the leaves through a network of water-conducting cells. As water evaporates from a

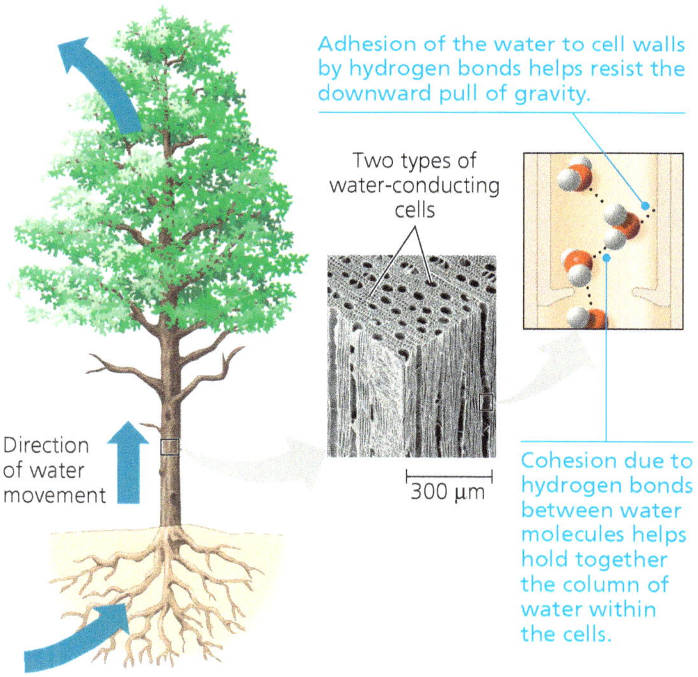

▲ **Figure 3.3 Water transport in plants.** Evaporation from leaves pulls water upward from the roots through water-conducting cells. Because of the properties of cohesion and adhesion, the tallest trees can transport water more than 100 m upward—approximately one-quarter the height of the Empire State Building in New York City.

BioFlix Visit the Study Area at **www.masteringbiology.com** for the BioFlix® 3-D Animation on Water Transport in Plants.

▲ **Figure 3.4 Walking on water.** The high surface tension of water, resulting from the collective strength of its hydrogen bonds, allows this raft spider to walk on the surface of a pond.

leaf, hydrogen bonds cause water molecules leaving the veins to tug on molecules farther down, and the upward pull is transmitted through the water-conducting cells all the way to the roots. **Adhesion**, the clinging of one substance to another, also plays a role. Adhesion of water to cell walls by hydrogen bonds helps counter the downward pull of gravity (see Figure 3.3).

Related to cohesion is **surface tension**, a measure of how difficult it is to stretch or break the surface of a liquid. Water has a greater surface tension than most other liquids. At the interface between water and air is an ordered arrangement of water molecules, hydrogen-bonded to one another and to the water below. This makes the water behave as though coated with an invisible film. You can observe the surface tension of water by slightly overfilling a drinking glass; the water will stand above the rim. In a more biological example, some animals can stand, walk, or run on water without breaking the surface **(Figure 3.4)**.

Moderation of Temperature by Water

Water moderates air temperature by absorbing heat from air that is warmer and releasing the stored heat to air that is cooler. Water is effective as a heat bank because it can absorb or release a relatively large amount of heat with only a slight change in its own temperature. To understand this capability of water, we must first look briefly at heat and temperature.

Heat and Temperature

Anything that moves has **kinetic energy**, the energy of motion. Atoms and molecules have kinetic energy because they are always moving, although not necessarily in any particular direction. The faster a molecule moves, the greater its kinetic energy. **Heat** is a form of energy. For a given body of matter, the amount of heat is a measure of the matter's *total* kinetic energy due to motion of its molecules; thus, heat depends in part on the matter's volume. Although heat is related to temperature, they are not the same thing. **Temperature** is a measure of heat intensity that represents the *average* kinetic energy of the molecules, regardless of volume. When water is heated in a coffeemaker, the average speed of the molecules increases, and the thermometer records this as a rise in temperature of the liquid. The amount of heat also increases in this case. Note, however, that although the pot of coffee has a much higher temperature than, say, the water in a swimming pool, the swimming pool contains more heat because of its much greater volume.

Whenever two objects of different temperature are brought together, heat passes from the warmer to the cooler object until the two are the same temperature. Molecules in the cooler object speed up at the expense of the kinetic energy of the warmer object. An ice cube cools a drink not by adding coldness to the liquid, but by absorbing heat from the liquid as the ice itself melts.

In general, we will use the **Celsius scale** to indicate temperature. (Celsius degrees are abbreviated °C; Appendix C shows how to convert between Celsius and Fahrenheit.) At sea level, water freezes at 0°C and boils at 100°C. The temperature of the human body averages 37°C, and comfortable room temperature is about 20–25°C.

One convenient unit of heat used in this book is the **calorie (cal)**. A calorie is the amount of heat it takes to raise the temperature of 1 g of water by 1°C. Conversely, a calorie is also the amount of heat that 1 g of water releases when it cools by 1°C. A **kilocalorie (kcal)**, 1,000 cal, is the quantity of heat required to raise the temperature of 1 kilogram (kg) of water by 1°C. (The "calories" on food packages are actually kilocalories.) Another energy unit used in this book is the **joule (J)**. One joule equals 0.239 cal; one calorie equals 4.184 J.

Water's High Specific Heat

The ability of water to stabilize temperature stems from its relatively high specific heat. The **specific heat** of a substance is defined as the amount of heat that must be absorbed or lost for 1 g of that substance to change its temperature by 1°C. We already know water's specific heat because we have defined a calorie as the amount of heat that causes 1 g of water to change its temperature by 1°C. Therefore, the specific heat of water is 1 calorie per gram and per degree Celsius, abbreviated as 1 cal/g·°C. Compared with most other substances, water has an unusually high specific heat. For example, ethyl alcohol, the type of alcohol in alcoholic beverages, has a specific heat of 0.6 cal/g·°C; that is, only 0.6 cal is required to raise the temperature of 1 g of ethyl alcohol by 1°C.

Because of the high specific heat of water relative to other materials, water will change its temperature less when it absorbs or loses a given amount of heat. The reason you can burn your fingers by touching the side of an iron pot on the stove when the water in the pot is still lukewarm is that the specific heat of water is ten times greater than that of iron.

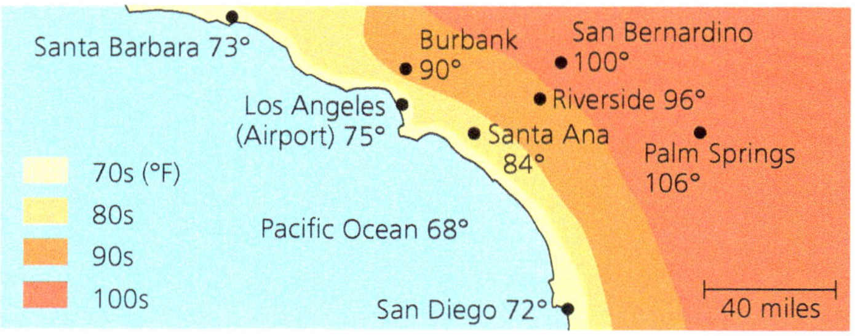

▲ **Figure 3.5 Effect of a large body of water on climate.** By absorbing or releasing heat, oceans moderate coastal climates. In this example from an August day in Southern California, the relatively cool ocean reduces coastal air temperatures by absorbing heat.

In other words, the same amount of heat will raise the temperature of 1 g of the iron much faster than it will raise the temperature of 1 g of the water. Specific heat can be thought of as a measure of how well a substance resists changing its temperature when it absorbs or releases heat. Water resists changing its temperature; when it does change its temperature, it absorbs or loses a relatively large quantity of heat for each degree of change.

We can trace water's high specific heat, like many of its other properties, to hydrogen bonding. Heat must be absorbed in order to break hydrogen bonds; by the same token, heat is released when hydrogen bonds form. A calorie of heat causes a relatively small change in the temperature of water because much of the heat is used to disrupt hydrogen bonds before the water molecules can begin moving faster. And when the temperature of water drops slightly, many additional hydrogen bonds form, releasing a considerable amount of energy in the form of heat.

What is the relevance of water's high specific heat to life on Earth? A large body of water can absorb and store a huge amount of heat from the sun in the daytime and during summer while warming up only a few degrees. At night and during winter, the gradually cooling water can warm the air. This is the reason coastal areas generally have milder climates than inland regions **(Figure 3.5)**. The high specific heat of water also tends to stabilize ocean temperatures, creating a favorable environment for marine life. Thus, because of its high specific heat, the water that covers most of Earth keeps temperature fluctuations on land and in water within limits that permit life. Also, because organisms are made primarily of water, they are better able to resist changes in their own temperature than if they were made of a liquid with a lower specific heat.

Evaporative Cooling

Molecules of any liquid stay close together because they are attracted to one another. Molecules moving fast enough to overcome these attractions can depart the liquid and enter the air as a gas. This transformation from a liquid to a gas is called vaporization, or *evaporation*. Recall that the speed of molecular movement varies and that temperature is the *average* kinetic energy of molecules. Even at low temperatures, the speediest molecules can escape into the air. Some evaporation occurs at any temperature; a glass of water at room temperature, for example, will eventually evaporate completely. If a liquid is heated, the average kinetic energy of molecules increases and the liquid evaporates more rapidly.

Heat of vaporization is the quantity of heat a liquid must absorb for 1 g of it to be converted from the liquid to the gaseous state. For the same reason that water has a high specific heat, it also has a high heat of vaporization relative to most other liquids. To evaporate 1 g of water at 25°C, about 580 cal of heat is needed—nearly double the amount needed to vaporize a gram of alcohol or ammonia. Water's high heat of vaporization is another emergent property resulting from the strength of its hydrogen bonds, which must be broken before the molecules can make their exodus from the liquid.

The high amount of energy required to vaporize water has a wide range of effects. On a global scale, for example, it helps moderate Earth's climate. A considerable amount of solar heat absorbed by tropical seas is consumed during the evaporation of surface water. Then, as moist tropical air circulates poleward, it releases heat as it condenses and forms rain. On an organismal level, water's high heat of vaporization accounts for the severity of steam burns. These burns are caused by the heat energy released when steam condenses into liquid on the skin.

As a liquid evaporates, the surface of the liquid that remains behind cools down. This **evaporative cooling** occurs because the "hottest" molecules, those with the greatest kinetic energy, are the most likely to leave as gas. It is as if the hundred fastest runners at a college transferred to another school; the average speed of the remaining students would decline.

Evaporative cooling of water contributes to the stability of temperature in lakes and ponds and also provides a mechanism that prevents terrestrial organisms from overheating. For example, evaporation of water from the leaves of a plant helps keep the tissues in the leaves from becoming too warm in the sunlight. Evaporation of sweat from human skin dissipates body heat and helps prevent overheating on a hot day or when excess heat is generated by strenuous activity. High humidity on a hot day increases discomfort because the high concentration of water vapor in the air inhibits the evaporation of sweat from the body.

Floating of Ice on Liquid Water

Water is one of the few substances that are less dense as a solid than as a liquid. In other words, ice floats on liquid water. While other materials contract and become denser when they solidify, water expands. The cause of this exotic behavior is, once again, hydrogen bonding. At temperatures above

▶ **Figure 3.6 Ice: crystalline structure and floating barrier.** In ice, each molecule is hydrogen-bonded to four neighbors in a three-dimensional crystal. Because the crystal is spacious, ice has fewer molecules than an equal volume of liquid water. In other words, ice is less dense than liquid water. Floating ice becomes a barrier that protects the liquid water below from the colder air. The marine organism shown here is a type of shrimp called krill; it was photographed beneath floating ice in the Southern Ocean near Antarctica.

WHAT IF? *If water did not form hydrogen bonds, what would happen to the shrimp's environment?*

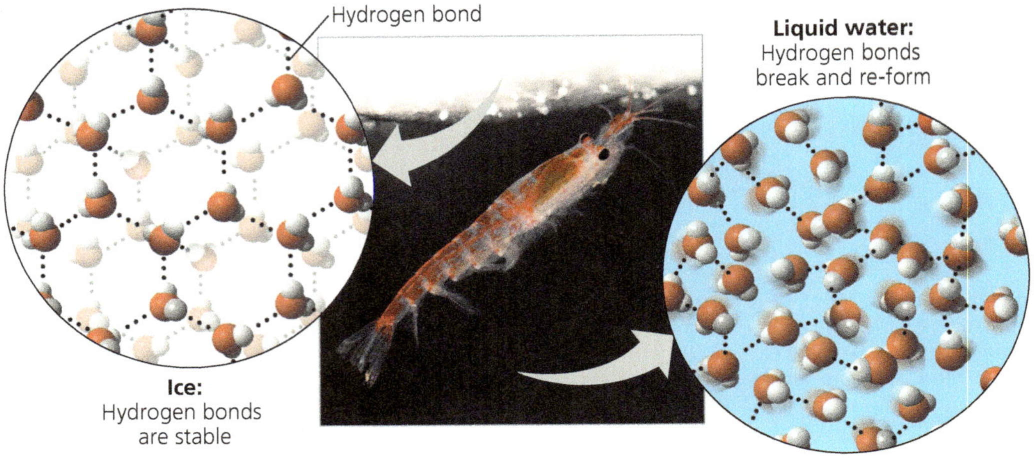

4°C, water behaves like other liquids, expanding as it warms and contracting as it cools. As the temperature falls from 4°C to 0°C, water begins to freeze because more and more of its molecules are moving too slowly to break hydrogen bonds. At 0°C, the molecules become locked into a crystalline lattice, each water molecule hydrogen-bonded to four partners (Figure 3.6). The hydrogen bonds keep the molecules at "arm's length," far enough apart to make ice about 10% less dense (10% fewer molecules for the same volume) than liquid water at 4°C. When ice absorbs enough heat for its temperature to rise above 0°C, hydrogen bonds between molecules are disrupted. As the crystal collapses, the ice melts, and molecules are free to slip closer together. Water reaches its greatest density at 4°C and then begins to expand as the molecules move faster. Even in liquid water, many of the molecules are connected by hydrogen bonds, though only transiently: The hydrogen bonds are constantly breaking and re-forming.

The ability of ice to float due to its lower density is an important factor in the suitability of the environment for life. If ice sank, then eventually all ponds, lakes, and even oceans would freeze solid, making life as we know it impossible on Earth. During summer, only the upper few inches of the ocean would thaw. Instead, when a deep body of water cools, the floating ice insulates the liquid water below, preventing it from freezing and allowing life to exist under the frozen surface, as shown in the photo in Figure 3.6. Besides insulating the water below, ice also provides solid habitat for some animals, such as polar bears and seals (see Figure 3.1).

Along with many other scientists, Susan Solomon, the interviewee for this unit (see pp. 28-29), is worried that these bodies of ice are at risk of disappearing. Global warming, which is caused by carbon dioxide and other "greenhouse" gases in the atmosphere, is having a profound effect on icy environments around the globe. In the Arctic, the average air temperature has risen 1.4°C just since 1961. This temperature increase has affected the seasonal balance between Arctic sea ice and liquid water, causing ice to form later in the year, to melt earlier, and to cover a smaller area. The alarming rate at which glaciers and Arctic sea ice are disappearing is posing an extreme challenge to animals that depend on ice for their survival.

Water: The Solvent of Life

A sugar cube placed in a glass of water will dissolve. The glass will then contain a uniform mixture of sugar and water; the concentration of dissolved sugar will be the same everywhere in the mixture. A liquid that is a completely homogeneous mixture of two or more substances is called a **solution**. The dissolving agent of a solution is the **solvent**, and the substance that is dissolved is the **solute**. In this case, water is the solvent and sugar is the solute. An **aqueous solution** is one in which water is the solvent.

The medieval alchemists tried to find a universal solvent, one that would dissolve anything. They learned that nothing works better than water. Yet, water is not a universal solvent; if it were, it would dissolve any container in which it was stored, including our cells. Water is a very versatile solvent, however, a quality we can trace to the polarity of the water molecule.

Suppose, for example, that a spoonful of table salt, the ionic compound sodium chloride (NaCl), is placed in water (Figure 3.7). At the surface of each grain, or crystal, of salt, the sodium and chloride ions are exposed to the solvent. These ions and the water molecules have a mutual affinity owing to the attraction between opposite charges. The oxygen regions of the water molecules are negatively charged and are attracted to sodium cations. The hydrogen regions are positively charged and are attracted to chloride anions. As a result, water molecules surround the individual sodium and chloride ions, separating and shielding them from one another. The sphere of water molecules around each dissolved ion is called a **hydration shell**. Working inward from the surface of each salt crystal, water eventually dissolves all the ions. The result is a solution of two solutes, sodium cations and chloride anions, homogeneously mixed with water, the solvent. Other ionic compounds also dissolve in water. Seawater, for instance, contains a great variety of dissolved ions, as do living cells.

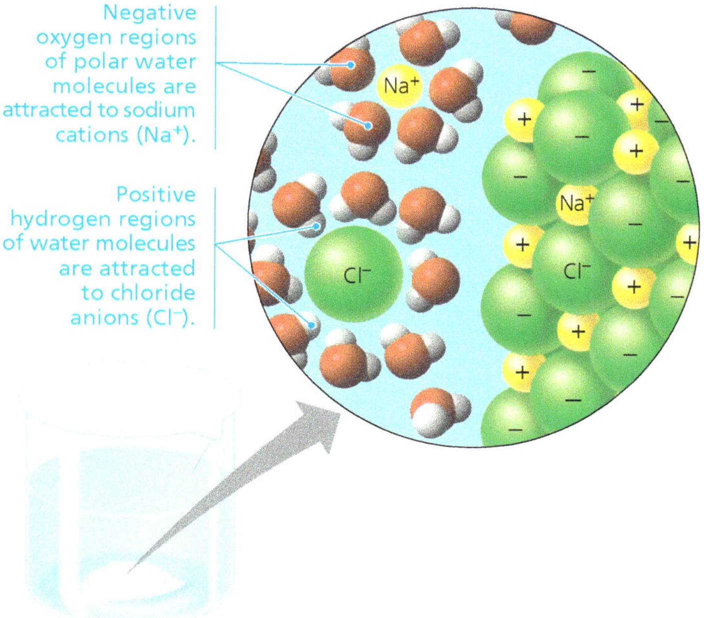

▲ **Figure 3.7 Table salt dissolving in water.** A sphere of water molecules, called a hydration shell, surrounds each solute ion.

WHAT IF? *What would happen if you heated this solution for a long time?*

A compound does not need to be ionic to dissolve in water; many compounds made up of nonionic polar molecules, such as sugars, are also water-soluble. Such compounds dissolve when water molecules surround each of the solute molecules, forming hydrogen bonds with them. Even molecules as large as proteins can dissolve in water if they have ionic and polar regions on their surface (**Figure 3.8**). Many different kinds of polar compounds are dissolved (along with ions) in the water of such biological fluids as blood, the sap of plants, and the liquid within all cells. Water is the solvent of life.

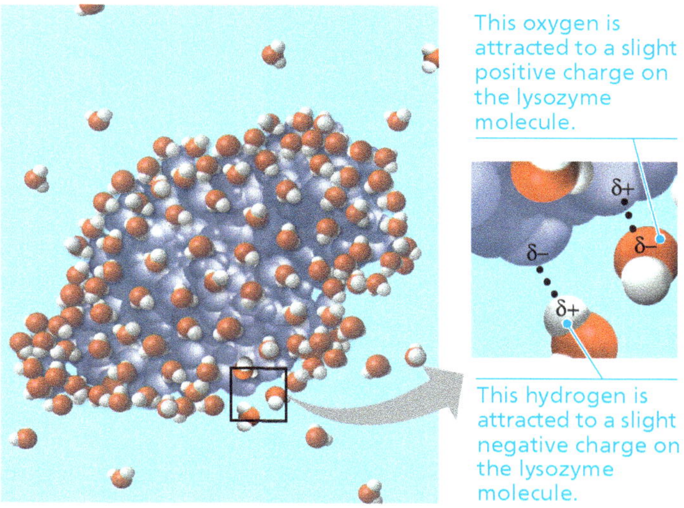

▲ **Figure 3.8 A water-soluble protein.** Human lysozyme is a protein found in tears and saliva that has antibacterial action. This model shows the lysozyme molecule (purple) in an aqueous environment. Ionic and polar regions on the protein's surface attract water molecules.

Hydrophilic and Hydrophobic Substances

Any substance that has an affinity for water is said to be **hydrophilic** (from the Greek *hydro*, water, and *philios*, loving). In some cases, substances can be hydrophilic without actually dissolving. For example, some molecules in cells are so large that they do not dissolve. Instead, they remain suspended in the aqueous liquid of the cell. Such a mixture is an example of a **colloid**, a stable suspension of fine particles in a liquid. Another example of a hydrophilic substance that does not dissolve is cotton, a plant product. Cotton consists of giant molecules of cellulose, a compound with numerous regions of partial positive and partial negative charges that can form hydrogen bonds with water. Water adheres to the cellulose fibers. Thus, a cotton towel does a great job of drying the body, yet it does not dissolve in the washing machine. Cellulose is also present in the walls of water-conducting cells in a plant; you read earlier how the adhesion of water to these hydrophilic walls allows water transport to occur.

There are, of course, substances that do not have an affinity for water. Substances that are nonionic and nonpolar (or otherwise cannot form hydrogen bonds) actually seem to repel water; these substances are said to be **hydrophobic** (from the Greek *phobos*, fearing). An example from the kitchen is vegetable oil, which, as you know, does not mix stably with water-based substances such as vinegar. The hydrophobic behavior of the oil molecules results from a prevalence of relatively nonpolar covalent bonds, in this case bonds between carbon and hydrogen, which share electrons almost equally. Hydrophobic molecules related to oils are major ingredients of cell membranes. (Imagine what would happen to a cell if its membrane dissolved!)

Solute Concentration in Aqueous Solutions

Biological chemistry is "wet" chemistry. Most of the chemical reactions in organisms involve solutes dissolved in water. To understand such reactions, we must know how many atoms and molecules are involved and be able to calculate the concentration of solutes in an aqueous solution (the number of solute molecules in a volume of solution).

When carrying out experiments, we use mass to calculate the number of molecules. We know the mass of each atom in a given molecule, so we can calculate the **molecular mass**, which is simply the sum of the masses of all the atoms in a molecule. As an example, let's calculate the molecular mass of table sugar (sucrose), which has the molecular formula $C_{12}H_{22}O_{11}$. In round numbers of daltons, the mass of a carbon atom is 12, the mass of a hydrogen atom is 1, and the mass of an oxygen atom is 16. Thus, sucrose has a molecular mass of $(12 \times 12) + (22 \times 1) + (11 \times 16) = 342$ daltons. Of course, weighing out small numbers of molecules is not practical. For this reason, we usually measure substances in units called moles. Just as a dozen always means 12 objects, a **mole (mol)** represents an exact number of objects: 6.02×10^{23}, which is called Avogadro's number. Because of the way in which Avogadro's number and the unit *dalton* were originally defined, there are 6.02×10^{23} daltons in 1 g. This is significant because once we determine the molecular mass of a molecule such as sucrose, we can use the same number (342), but with the unit *gram*, to represent the mass of 6.02×10^{23} molecules of sucrose, or 1 mol of sucrose (this is sometimes called the *molar mass*). To obtain 1 mol of sucrose in the lab, therefore, we weigh out 342 g.

The practical advantage of measuring a quantity of chemicals in moles is that a mole of one substance has exactly the same number of molecules as a mole of any other substance. If the molecular mass of substance A is 342 daltons and that of substance B is 10 daltons, then 342 g of A will have the same number of molecules as 10 g of B. A mole of ethyl alcohol (C_2H_6O) also contains 6.02×10^{23} molecules, but its mass is only 46 g because the mass of a molecule of ethyl alcohol is less than that of a molecule of sucrose. Measuring in moles makes it convenient for scientists working in the laboratory to combine substances in fixed ratios of molecules.

How would we make a liter (L) of solution consisting of 1 mol of sucrose dissolved in water? We would measure out 342 g of sucrose and then gradually add water, while stirring, until the sugar was completely dissolved. We would then add enough water to bring the total volume of the solution up to 1 L. At that point, we would have a 1-molar (1 *M*) solution of sucrose. **Molarity**—the number of moles of solute per liter of solution—is the unit of concentration most often used by biologists for aqueous solutions.

Water's capacity as a versatile solvent complements the other properties discussed in this chapter. Since these remarkable properties allow water to support life on Earth so well, scientists who seek life elsewhere in the universe look for water as a sign that a planet might sustain life.

Possible Evolution of Life on Other Planets with Water

EVOLUTION Humans have probably always gazed skyward, wondering whether other living beings exist beyond Earth. And if life has arisen on other planets, into what form or forms has it evolved? Biologists who look for life elsewhere in the universe (known as *astrobiologists*) have concentrated their search on planets that might have water. To date, more than 200 planets have been found outside our solar system, and there is evidence for the presence of water vapor on one or two of them. In our own solar system, Mars has been most compelling to astrobiologists as a focus of study.

Like Earth, Mars has an ice cap at both poles. And in the decades since the age of space exploration began, scientists have found intriguing signs that water may exist elsewhere on Mars. Finally, in 2008, the robotic spacecraft *Phoenix* landed on Mars and began to sample its surface. Years of debate were resolved by the images sent back from *Phoenix*: Ice is definitely present just under Mars's surface, and enough water vapor is in the Martian atmosphere for frost to form (**Figure 3.9**). This exciting finding has reinvigorated the search for signs of life, past or present, on Mars and other planets. If any life-forms or fossils are found, their study will shed light on the process of evolution from an entirely new perspective.

◄ **Figure 3.9 Subsurface ice and morning frost on Mars.** This photograph was taken by the Mars lander *Phoenix* in 2008. The trench was scraped by a robotic arm, uncovering ice (white in rectangle near bottom) below the surface material. Frost also appears as a white coating in several places in the upper half of the image. This photograph was colorized by NASA to highlight the ice.

CONCEPT CHECK 3.2

1. Describe how properties of water contribute to the upward movement of water in a tree.
2. Explain the saying "It's not the heat; it's the humidity."
3. How can the freezing of water crack boulders?
4. The concentration of the appetite-regulating hormone ghrelin is about 1.3×10^{-10} *M* in a fasting person. How many molecules of ghrelin are in 1 L of blood?
5. **WHAT IF?** A water strider (which can walk on water) has legs that are coated with a hydrophobic substance. What might be the benefit? What would happen if the substance were hydrophilic?

For suggested answers, see Appendix A.

CONCEPT 3.3

Acidic and basic conditions affect living organisms

Occasionally, a hydrogen atom participating in a hydrogen bond between two water molecules shifts from one molecule to the other. When this happens, the hydrogen atom leaves its electron behind, and what is actually transferred is a **hydrogen ion** (H^+), a single proton with a charge of 1+. The water molecule that lost a proton is now a **hydroxide ion** (OH^-), which has a charge of 1−. The proton binds to the other water molecule, making that molecule a **hydronium ion** (H_3O^+). We can picture the chemical reaction as shown at the top of the next page.

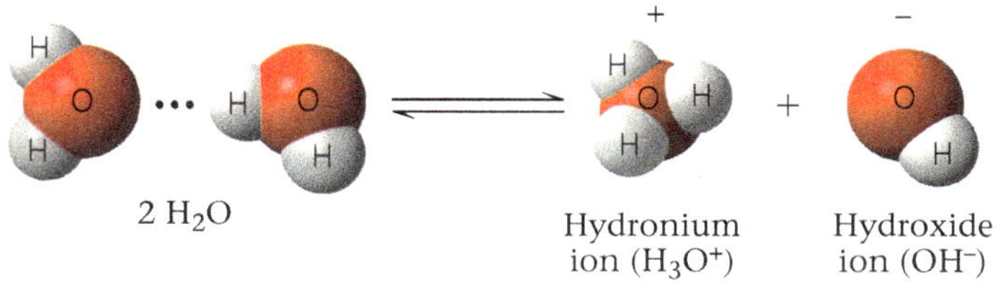

By convention, H^+ (the hydrogen ion) is used to represent H_3O^+ (the hydronium ion), and we follow that practice here. Keep in mind, though, that H^+ does not exist on its own in an aqueous solution. It is always associated with another water molecule in the form of H_3O^+.

As indicated by the double arrows, this is a reversible reaction that reaches a state of dynamic equilibrium when water molecules dissociate at the same rate that they are being re-formed from H^+ and OH^-. At this equilibrium point, the concentration of water molecules greatly exceeds the concentrations of H^+ and OH^-. In pure water, only one water molecule in every 554 million is dissociated; the concentration of each ion in pure water is 10^{-7} *M* (at 25°C). This means there is only one ten-millionth of a mole of hydrogen ions per liter of pure water and an equal number of hydroxide ions.

Although the dissociation of water is reversible and statistically rare, it is exceedingly important in the chemistry of life. H^+ and OH^- are very reactive. Changes in their concentrations can drastically affect a cell's proteins and other complex molecules. As we have seen, the concentrations of H^+ and OH^- are equal in pure water, but adding certain kinds of solutes, called acids and bases, disrupts this balance. Biologists use something called the pH scale to describe how acidic or basic (the opposite of acidic) a solution is. In the remainder of this chapter, you will learn about acids, bases, and pH and why changes in pH can adversely affect organisms.

Acids and Bases

What would cause an aqueous solution to have an imbalance in H^+ and OH^- concentrations? When acids dissolve in water, they donate additional H^+ to the solution. An **acid** is a substance that increases the hydrogen ion concentration of a solution. For example, when hydrochloric acid (HCl) is added to water, hydrogen ions dissociate from chloride ions:

$$HCl \rightarrow H^+ + Cl^-$$

This source of H^+ (dissociation of water is the other source) results in an acidic solution—one having more H^+ than OH^-.

A substance that reduces the hydrogen ion concentration of a solution is called a **base**. Some bases reduce the H^+ concentration directly by accepting hydrogen ions. Ammonia (NH_3), for instance, acts as a base when the unshared electron pair in nitrogen's valence shell attracts a hydrogen ion from the solution, resulting in an ammonium ion (NH_4^+):

$$NH_3 + H^+ \rightleftharpoons NH_4^+$$

Other bases reduce the H^+ concentration indirectly by dissociating to form hydroxide ions, which combine with hydrogen ions and form water. One such base is sodium hydroxide (NaOH), which in water dissociates into its ions:

$$NaOH \rightarrow Na^+ + OH^-$$

In either case, the base reduces the H^+ concentration. Solutions with a higher concentration of OH^- than H^+ are known as basic solutions. A solution in which the H^+ and OH^- concentrations are equal is said to be neutral.

Notice that single arrows were used in the reactions for HCl and NaOH. These compounds dissociate completely when mixed with water, so hydrochloric acid is called a strong acid and sodium hydroxide a strong base. In contrast, ammonia is a relatively weak base. The double arrows in the reaction for ammonia indicate that the binding and release of hydrogen ions are reversible reactions, although at equilibrium there will be a fixed ratio of NH_4^+ to NH_3.

There are also weak acids, which reversibly release and accept back hydrogen ions. An example is carbonic acid:

$$\underset{\text{Carbonic acid}}{H_2CO_3} \rightleftharpoons \underset{\text{Bicarbonate ion}}{HCO_3^-} + \underset{\text{Hydrogen ion}}{H^+}$$

Here the equilibrium so favors the reaction in the left direction that when carbonic acid is added to pure water, only 1% of the molecules are dissociated at any particular time. Still, that is enough to shift the balance of H^+ and OH^- from neutrality.

The pH Scale

In any aqueous solution at 25°C, the *product* of the H^+ and OH^- concentrations is constant at 10^{-14}. This can be written

$$[H^+][OH^-] = 10^{-14}$$

In such an equation, brackets indicate molar concentration. In a neutral solution at room temperature (25°C), $[H^+] = 10^{-7}$ and $[OH^-] = 10^{-7}$, so in this case, 10^{-14} is the product of $10^{-7} \times 10^{-7}$. If enough acid is added to a solution to increase $[H^+]$ to 10^{-5} *M*, then $[OH^-]$ will decline by an equivalent amount to 10^{-9} *M* (note that $10^{-5} \times 10^{-9} = 10^{-14}$). This constant relationship expresses the behavior of acids and bases in an aqueous solution. An acid not only adds hydrogen ions to a solution, but also removes hydroxide ions because of the tendency for H^+ to combine with OH^-, forming water. A base has the opposite effect, increasing OH^- concentration but also reducing H^+ concentration by the formation of water. If enough of a base is added to raise the OH^- concentration to 10^{-4} *M*, it will cause the H^+ concentration to drop to 10^{-10} *M*. Whenever we know the concentration of either H^+ or OH^- in an aqueous solution, we can deduce the concentration of the other ion.

Because the H^+ and OH^- concentrations of solutions can vary by a factor of 100 trillion or more, scientists have

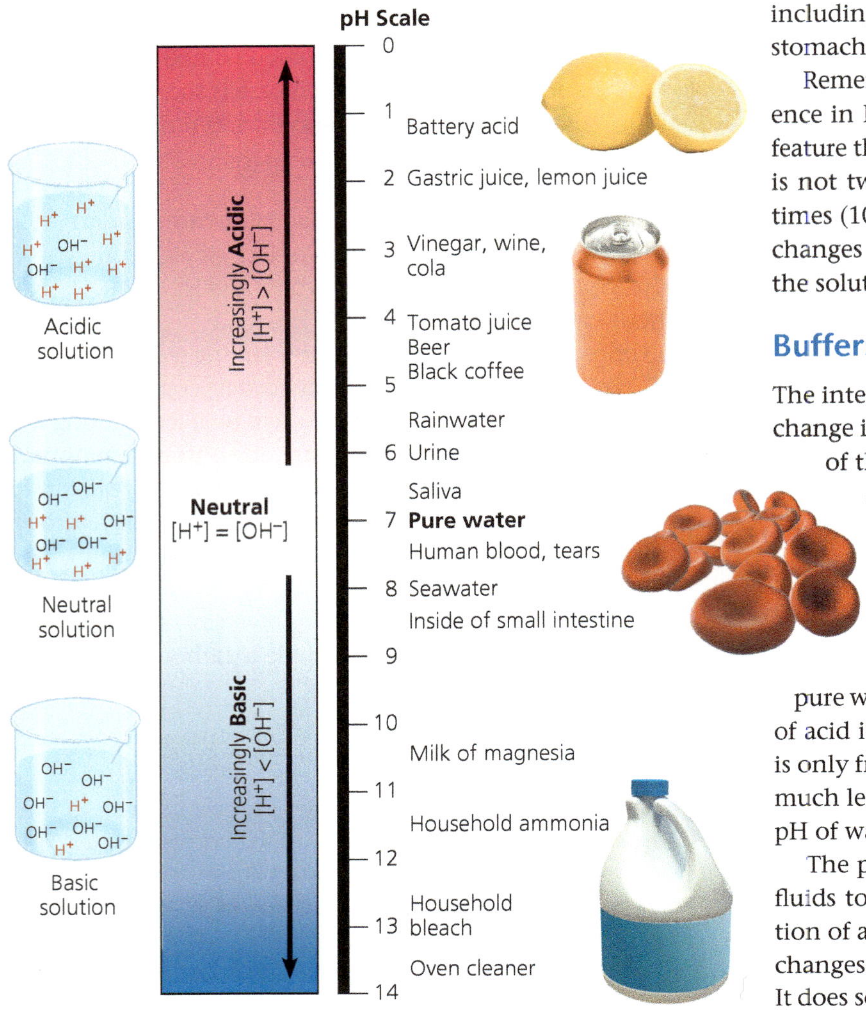

▲ **Figure 3.10 The pH scale and pH values of some aqueous solutions.**

developed a way to express this variation more conveniently than in moles per liter. The pH scale **(Figure 3.10)** compresses the range of H^+ and OH^- concentrations by employing logarithms. The **pH** of a solution is defined as the negative logarithm (base 10) of the hydrogen ion concentration:

$$pH = -\log [H^+]$$

For a neutral aqueous solution, $[H^+]$ is 10^{-7} *M*, giving us

$$-\log 10^{-7} = -(-7) = 7$$

Notice that pH *declines* as H^+ concentration *increases*. Notice, too, that although the pH scale is based on H^+ concentration, it also implies OH^- concentration. A solution of pH 10 has a hydrogen ion concentration of 10^{-10} *M* and a hydroxide ion concentration of 10^{-4} *M*.

The pH of a neutral aqueous solution at 25°C is 7, the midpoint of the pH scale. A pH value less than 7 denotes an acidic solution; the lower the number, the more acidic the solution. The pH for basic solutions is above 7. Most biological fluids are within the range pH 6–8. There are a few exceptions, however, including the strongly acidic digestive juice of the human stomach, which has a pH of about 2.

Remember that each pH unit represents a tenfold difference in H^+ and OH^- concentrations. It is this mathematical feature that makes the pH scale so compact. A solution of pH 3 is not twice as acidic as a solution of pH 6, but a thousand times ($10 \times 10 \times 10$) more acidic. When the pH of a solution changes slightly, the actual concentrations of H^+ and OH^- in the solution change substantially.

Buffers

The internal pH of most living cells is close to 7. Even a slight change in pH can be harmful, because the chemical processes of the cell are very sensitive to the concentrations of hydrogen and hydroxide ions. The pH of human blood is very close to 7.4, or slightly basic. A person cannot survive for more than a few minutes if the blood pH drops to 7 or rises to 7.8, and a chemical system exists in the blood that maintains a stable pH. If you add 0.01 mol of a strong acid to a liter of pure water, the pH drops from 7.0 to 2.0. If the same amount of acid is added to a liter of blood, however, the pH decrease is only from 7.4 to 7.3. Why does the addition of acid have so much less of an effect on the pH of blood than it does on the pH of water?

The presence of substances called buffers allows biological fluids to maintain a relatively constant pH despite the addition of acids or bases. A **buffer** is a substance that minimizes changes in the concentrations of H^+ and OH^- in a solution. It does so by accepting hydrogen ions from the solution when they are in excess and donating hydrogen ions to the solution when they have been depleted. Most buffer solutions contain a weak acid and its corresponding base, which combine reversibly with hydrogen ions.

There are several buffers that contribute to pH stability in human blood and many other biological solutions. One of these is carbonic acid (H_2CO_3), formed when CO_2 reacts with water in blood plasma. As mentioned earlier, carbonic acid dissociates to yield a bicarbonate ion (HCO_3^-) and a hydrogen ion (H^+):

$$\underset{\substack{H^+ \text{ donor} \\ \text{(acid)}}}{H_2CO_3} \underset{\substack{\text{Response to} \\ \text{a drop in pH}}}{\overset{\substack{\text{Response to} \\ \text{a rise in pH}}}{\rightleftharpoons}} \underset{\substack{H^+ \text{ acceptor} \\ \text{(base)}}}{HCO_3^-} + \underset{\substack{\text{Hydrogen} \\ \text{ion}}}{H^+}$$

The chemical equilibrium between carbonic acid and bicarbonate acts as a pH regulator, the reaction shifting left or right as other processes in the solution add or remove hydrogen ions. If the H^+ concentration in blood begins to fall (that is, if pH rises), the reaction proceeds to the right and more carbonic acid dissociates, replenishing hydrogen ions. But when H^+ concentration in blood begins to rise (when pH drops), the reaction proceeds to the left, with HCO_3^- (the base) removing the hydrogen ions from the solution and forming H_2CO_3. Thus, the carbonic acid–bicarbonate buffering system consists of an acid and a base in equilibrium with each other. Most other buffers are also acid-base pairs.

Acidification: A Threat to Water Quality

Among the many threats to water quality posed by human activities is the burning of fossil fuels, which releases gaseous compounds into the atmosphere. When certain of these compounds react with water, the water becomes more acidic, altering the delicate balance of conditions for life on Earth.

Carbon dioxide is the main product of fossil fuel combustion. About 25% of human-generated CO_2 is absorbed by the oceans. In spite of the huge volume of water in the oceans, scientists worry that the absorption of so much CO_2 will harm marine ecosystems.

Recent data have shown that such fears are well founded. When CO_2 dissolves in seawater, it reacts with water to form carbonic acid, which lowers ocean pH, a process known as **ocean acidification**. Based on measurements of CO_2 levels in air bubbles trapped in ice over thousands of years, scientists calculate that the pH of the oceans is 0.1 pH unit lower now than at any time in the past 420,000 years. Recent studies predict that it will drop another 0.3–0.5 pH unit by the end of this century.

As seawater acidifies, the extra hydrogen ions combine with carbonate ions (CO_3^{2-}) to form bicarbonate ions (HCO_3^-), thereby reducing the carbonate concentration **(Figure 3.11)**.

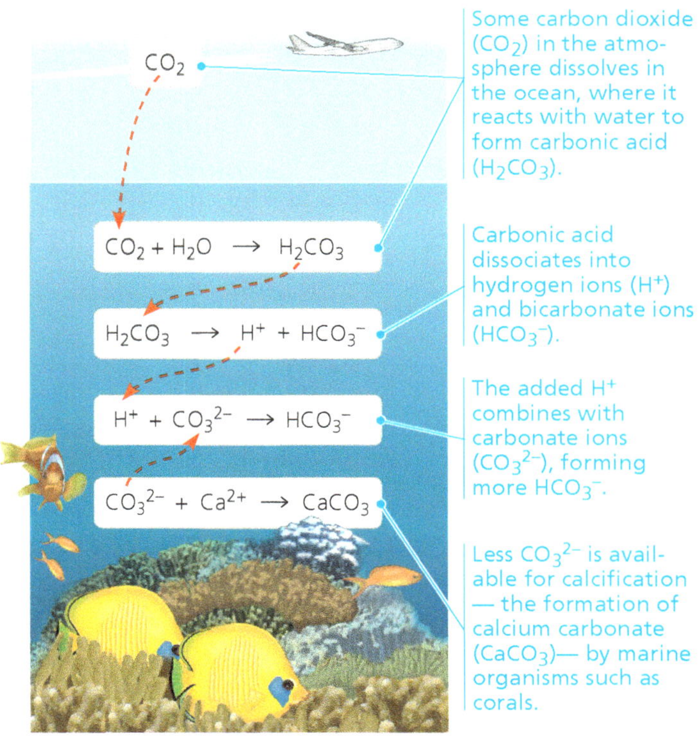

▲ **Figure 3.11 Atmospheric CO_2 from human activities and its fate in the ocean.**

▼ **Figure 3.12**

IMPACT

The Threat of Ocean Acidification to Coral Reef Ecosystems

Recently, scientists have sounded the alarm about the effects of ocean acidification, the process in which oceans become more acidic due to increased atmospheric carbon dioxide levels (see Figure 3.11). They predict that the resulting decrease in the concentration of carbonate ion (CO_3^{2-}) will take a serious toll on coral reef calcification. Taking many studies into account, and including the effects of ocean warming as well, one group of scientists defined three scenarios for coral reefs during this century, depending on whether the concentration of atmospheric CO_2 (a) stays at today's level, (b) increases at the current rate, or (c) increases more rapidly. The photographs below show coral reefs resembling those predicted under each scenario.

(a) **(b)** **(c)**

The healthy coral reef in (a) supports a highly diverse group of species and bears little resemblance to the damaged coral reef in (c).

WHY IT MATTERS The disappearance of coral reef ecosystems would be a tragic loss of biological diversity. In addition, coral reefs provide shoreline protection, a feeding ground for many commercial fishery species, and a popular tourist draw, so coastal human communities would suffer from greater wave damage, collapsed fisheries, and reduced tourism.

FURTHER READING O. Hoegh-Guldberg et al., Coral reefs under rapid climate change and ocean acidification, *Science* 318:1737–1742 (2007). S. C. Doney, The dangers of ocean acidification, *Scientific American*, March 2006, 58–65.

WHAT IF? Would lowering the ocean's carbonate concentration have any effect, even indirectly, on organisms that don't form $CaCO_3$? Explain.

Scientists predict that ocean acidification will cause the carbonate concentration to decrease by 40% by the year 2100. This is of great concern because carbonate is required for calcification, the production of calcium carbonate ($CaCO_3$) by many marine organisms, including reef-building corals and animals that build shells. Coral reefs are sensitive ecosystems that act as havens for a great diversity of marine life **(Figure 3.12)**.

The burning of fossil fuels is also a major source of sulfur oxides and nitrogen oxides. These compounds react with water in the air to form strong acids, which fall to Earth with rain or snow. **Acid precipitation** refers to rain, snow, or fog with a pH lower (more acidic) than 5.2. (Uncontaminated rain has a pH of about 5.6, which is slightly acidic due to the formation of carbonic acid from CO_2 and water.) Acid precipitation can damage life in lakes and streams, and it adversely affects plants on land by changing soil chemistry. To address this problem, the U.S. Congress amended the Clean Air Act in 1990, and the mandated improvements in industrial technologies have been largely responsible for improving the health of most North American lakes and forests.

If there is any reason for optimism about the future quality of water resources on our planet, it is that we have made progress in learning about the delicate chemical balances in oceans, lakes, and rivers. Continued progress can come only from the actions of informed individuals, like yourselves, who are concerned about environmental quality. This requires understanding the crucial role that water plays in the suitability of the environment for continued life on Earth.

CONCEPT CHECK 3.3

1. Compared with a basic solution at pH 9, the same volume of an acidic solution at pH 4 has ____ times as many hydrogen ions (H^+).
2. HCl is a strong acid that dissociates in water: $HCl \rightarrow H^+ + Cl^-$. What is the pH of 0.01 *M* HCl?
3. Acetic acid (CH_3COOH) can be a buffer, similar to carbonic acid. Write the dissociation reaction, identifying the acid, base, H^+ acceptor, and H^+ donor.
4. **WHAT IF?** Given a liter of pure water and a liter solution of acetic acid, what would happen to the pH if you added 0.01 mol of a strong acid to each? Use the reaction equation from question 3 to explain the result.

For suggested answers, see Appendix A.

3 CHAPTER REVIEW

SUMMARY OF KEY CONCEPTS

CONCEPT 3.1

Polar covalent bonds in water molecules result in hydrogen bonding (pp. 46–47)

- A hydrogen bond forms when the slightly negatively charged oxygen of one water molecule is attracted to the slightly positively charged hydrogen of a nearby water molecule. Hydrogen bonding between water molecules is the basis for water's properties.

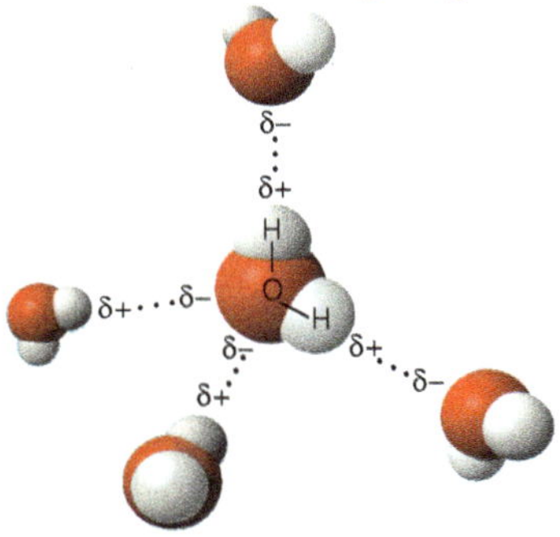

DRAW IT *Label a hydrogen bond and a polar covalent bond in this figure. How many hydrogen bonds can each water molecule make?*

CONCEPT 3.2

Four emergent properties of water contribute to Earth's suitability for life (pp. 47–52)

- Hydrogen bonding keeps water molecules close to each other, and this **cohesion** helps pull water upward in the microscopic water-conducting cells of plants. Hydrogen bonding is also responsible for water's **surface tension**.
- Water has a high **specific heat**: Heat is absorbed when hydrogen bonds break and is released when hydrogen bonds form. This helps keep temperatures relatively steady, within limits that permit life. **Evaporative cooling** is based on water's high **heat of vaporization**. The evaporative loss of the most energetic water molecules cools a surface.
- Ice floats because it is less dense than liquid water. This allows life to exist under the frozen surfaces of lakes and polar seas.

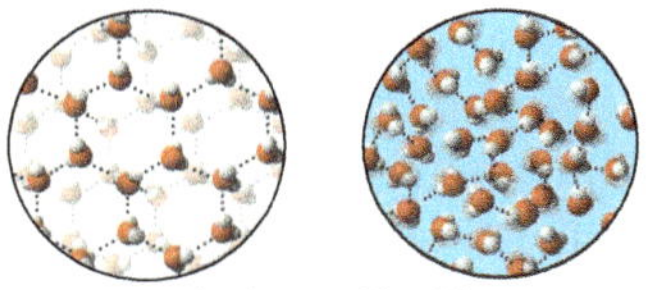

Ice: stable hydrogen bonds

Liquid water: transient hydrogen bonds

- Water is an unusually versatile **solvent** because its polar molecules are attracted to charged and polar substances capable of forming hydrogen bonds. **Hydrophilic** substances have an affinity for water; **hydrophobic** substances do not. **Molarity**, the number of moles of **solute** per liter of **solution**, is used as a measure of solute concentration in solutions. A **mole** is a certain number of molecules of a substance. The mass of a mole of a substance in grams is the same as the **molecular mass** in daltons.
- The emergent properties of water support life on Earth and may contribute to the potential for life to have evolved on other planets.

? *Describe how different types of solutes dissolve in water. Explain the difference between a solution and a colloid.*

CONCEPT 3.3

Acidic and basic conditions affect living organisms (pp. 52–56)

- A water molecule can transfer an H^+ to another water molecule to form H_3O^+ (represented simply by H^+) and OH^-.
- The concentration of H^+ is expressed as **pH**; $pH = -\log [H^+]$. **Buffers** in biological fluids resist changes in pH. A buffer consists of an acid-base pair that combines reversibly with hydrogen ions.
- The burning of fossil fuels increases the amount of CO_2 in the atmosphere. Some CO_2 dissolves in the oceans, causing **ocean acidification**, which has potentially grave consequences for coral reefs. The burning of fossil fuels also releases oxides of sulfur and nitrogen, leading to **acid precipitation**.

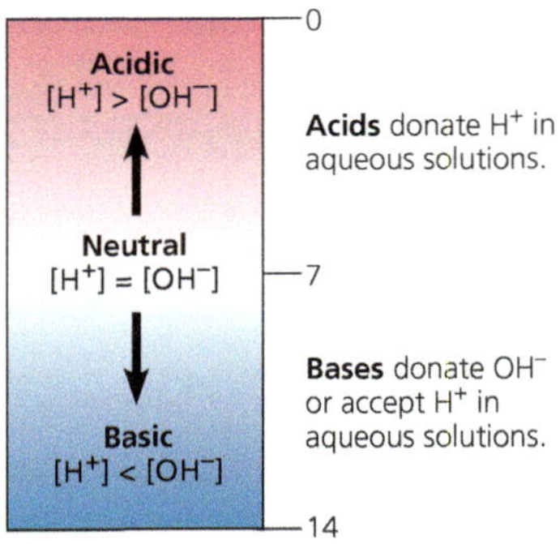

? *Explain how increasing amounts of CO_2 dissolving in the ocean leads to ocean acidification. How does this change in pH affect carbonate ion concentration and the rate of calcification?*

TEST YOUR UNDERSTANDING

LEVEL 1: KNOWLEDGE/COMPREHENSION

1. Many mammals control their body temperature by sweating. Which property of water is most directly responsible for the ability of sweat to lower body temperature?
   a. water's change in density when it condenses
   b. water's ability to dissolve molecules in the air
   c. the release of heat by the formation of hydrogen bonds
   d. the absorption of heat by the breaking of hydrogen bonds
   e. water's high surface tension

2. The bonds that are broken when water vaporizes are
   a. ionic bonds.
   b. hydrogen bonds between water molecules.
   c. covalent bonds between atoms within water molecules.
   d. polar covalent bonds.
   e. nonpolar covalent bonds.

3. Which of the following is a hydrophobic material?
   a. paper
   b. table salt
   c. wax
   d. sugar
   e. pasta

4. We can be sure that a mole of table sugar and a mole of vitamin C are equal in their
   a. mass in daltons.
   b. mass in grams.
   c. volume.
   d. number of atoms.
   e. number of molecules.

5. Measurements show that the pH of a particular lake is 4.0. What is the hydrogen ion concentration of the lake?
   a. 4.0 *M* b. 10^{-10} *M* c. 10^{-4} *M* d. 10^{4} *M* e. 4%

6. What is the *hydroxide* ion concentration of the lake described in question 5?
   a. 10^{-10} *M* b. 10^{-4} *M* c. 10^{-7} *M* d. 10^{-14}*M* e. 10 *M*

LEVEL 2: APPLICATION/ANALYSIS

7. A slice of pizza has 500 kcal. If we could burn the pizza and use all the heat to warm a 50-L container of cold water, what would be the approximate increase in the temperature of the water? (*Note*: A liter of cold water weighs about 1 kg.)
   a. 50°C b. 5°C c. 1°C d. 100°C e. 10°C

8. How many grams of acetic acid ($C_2H_4O_2$) would you use to make 10 L of a 0.1 *M* aqueous solution of acetic acid? (*Note*: The atomic masses, in daltons, are approximately 12 for carbon, 1 for hydrogen, and 16 for oxygen.)
   a. 10 g b. 0.1 g c. 6.0 g d. 60 g e. 0.6 g

9. **DRAW IT** Draw the hydration shells that form around a potassium ion and a chloride ion when potassium chloride (KCl) dissolves in water. Label the positive, negative, and partial charges on the atoms.

10. **MAKE CONNECTIONS** What do global warming (see Chapter 1, p. 6) and ocean acidification have in common?

LEVEL 3: SYNTHESIS/EVALUATION

11. In agricultural areas, farmers pay close attention to the weather forecast. Right before a predicted overnight freeze, farmers spray water on crops to protect the plants. Use the properties of water to explain how this method works. Be sure to mention why hydrogen bonds are responsible for this phenomenon.

12. **EVOLUTION CONNECTION**
This chapter explains how the emergent properties of water contribute to the suitability of the environment for life. Until fairly recently, scientists assumed that other physical requirements for life included a moderate range of temperature, pH, atmospheric pressure, and salinity, as well as low levels of toxic chemicals. That view has changed with the discovery of organisms known as extremophiles, which have been found flourishing in hot, acidic sulfur springs, around hydrothermal vents deep in the ocean, and in soils with high levels of toxic metals. Why would astrobiologists be interested in studying extremophiles? What does the existence of life in such extreme environments say about the possibility of life on other planets?

13. **SCIENTIFIC INQUIRY**
Design a controlled experiment to test the hypothesis that acid precipitation inhibits the growth of *Elodea*, a common freshwater plant (see Figure 2.19, p. 43).

14. **SCIENTIFIC INQUIRY**
In a study reported in 2000, C. Langdon and colleagues used an artificial coral reef system to test the effect of carbonate concentration on the rate of calcification by reef organisms. The graph on the right presents one set of their results. Describe what these data show. How do these results relate to the ocean acidification that is associated with increasing atmospheric CO_2 levels?

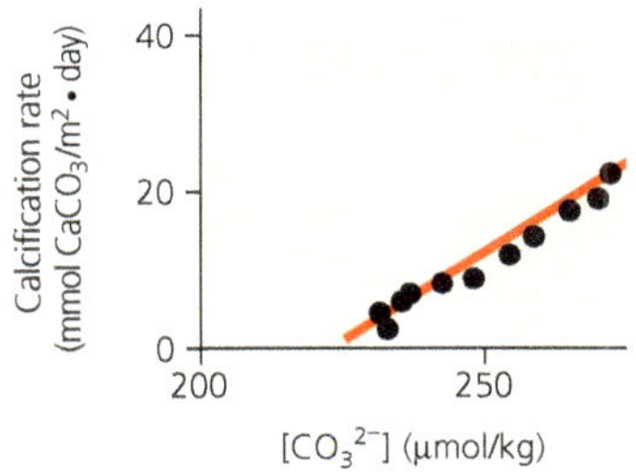

15. **SCIENCE, TECHNOLOGY, AND SOCIETY**
Agriculture, industry, and the growing populations of cities all compete, through political influence, for water. If you were in charge of water resources in an arid region, what would your priorities be for allocating the limited water supply for various uses? How would you try to build consensus among the different special-interest groups?

16. **WRITE ABOUT A THEME**
Emergent Properties Several emergent properties of water contribute to the suitability of the environment for life. In a short essay (100–150 words), describe how the ability of water to function as a versatile solvent arises from the structure of water molecules.

For selected answers, see Appendix A.

MasteringBIOLOGY **www.masteringbiology.com**

1. MasteringBiology® Assignments

Tutorials Hydrogen Bonding and Water • The pH Scale
Activities The Polarity of Water • Cohesion of Water • Dissociation of Water Molecules • Acids, Bases, and pH
Questions Student Misconceptions • Reading Quiz • Multiple Choice • End-of-Chapter

2. eText

Read your book online, search, take notes, highlight text, and more.

3. The Study Area

Practice Tests • Cumulative Test • ***BioFlix*** 3-D Animations • MP3 Tutor Sessions • Videos • Activities • Investigations • Lab Media • Audio Glossary • Word Study Tools • Art

4

Carbon and the Molecular Diversity of Life

▲ **Figure 4.1** **What properties make carbon the basis of all life?**

KEY CONCEPTS

4.1 Organic chemistry is the study of carbon compounds
4.2 Carbon atoms can form diverse molecules by bonding to four other atoms
4.3 A few chemical groups are key to the functioning of biological molecules

OVERVIEW

Carbon: The Backbone of Life

Water is the universal medium for life on Earth, but living organisms, such as the plants and Roosevelt elk in Figure 4.1, are made up of chemicals based mostly on the element carbon. Carbon enters the biosphere through the action of plants. Plants use solar energy to transform atmospheric CO_2 into the molecules of life, which are then taken in by plant-eating animals.

Of all chemical elements, carbon is unparalleled in its ability to form molecules that are large, complex, and varied, making possible the diversity of organisms that have evolved on Earth. Proteins, DNA, carbohydrates, and other molecules that distinguish living matter from inanimate material are all composed of carbon atoms bonded to one another and to atoms of other elements. Hydrogen (H), oxygen (O), nitrogen (N), sulfur (S), and phosphorus (P) are other common ingredients of these compounds, but it is the element carbon (C) that accounts for the enormous variety of biological molecules.

Large biological molecules, such as proteins, are the main focus of Chapter 5. In this chapter, we investigate the properties of smaller molecules. We will use these small molecules to illustrate concepts of molecular architecture that will help explain why carbon is so important to life, at the same time highlighting the theme that emergent properties arise from the organization of matter in living organisms.

CONCEPT **4.1**

Organic chemistry is the study of carbon compounds

For historical reasons, compounds containing carbon are said to be organic, and the branch of chemistry that specializes in the study of carbon compounds is called **organic chemistry**. Organic compounds range from simple molecules, such as methane (CH_4), to colossal ones, such as proteins, with thousands of atoms. Most organic compounds contain hydrogen atoms in addition to carbon atoms.

The overall percentages of the major elements of life—C, H, O, N, S, and P—are quite uniform from one organism to another. Because of carbon's versatility, however, this limited assortment of atomic building blocks can be used to build an inexhaustible variety of organic molecules. Different species of organisms, and different individuals within a species, are distinguished by variations in their organic molecules.

Since the dawn of human history, people have used other organisms as sources of valued substances—from foods and medicines to fabrics. The science of organic chemistry originated in attempts to purify and improve the yield of such products. By the early 1800s, chemists had learned to make many simple compounds in the laboratory by combining elements under the right conditions. Artificial synthesis of the complex molecules extracted from living matter seemed impossible, however. At that time, the Swedish chemist Jöns Jakob Berzelius made the distinction between organic compounds, those thought to arise only in living organisms, and inorganic compounds, those found only in the nonliving world. *Vitalism*, the belief in a life force outside the jurisdiction of physical and chemical laws, provided the foundation for the new discipline of organic chemistry.

Chemists began to chip away at the support for vitalism when they finally learned to synthesize organic compounds in the laboratory. In 1828, Friedrich Wöhler, a German chemist who had studied with Berzelius, tried to make an "inorganic" salt, ammonium cyanate, by mixing solutions of ammonium ions (NH_4^+) and cyanate ions (CNO^-). Wöhler was astonished to find that instead he had made urea, an organic compound present in the urine of animals. Wöhler challenged the vitalists when he wrote, "I must tell you that I can prepare urea without requiring a kidney or an animal, either man or dog." However, one of the ingredients used in the synthesis, the cyanate, had been extracted from animal blood, and the vitalists were not swayed by Wöhler's discovery. A few years later, however, Hermann Kolbe, a student of Wöhler's, made the organic compound acetic acid from inorganic substances that could be prepared directly from pure elements. Vitalism crumbled completely after several decades of laboratory synthesis of increasingly complex organic compounds.

Organic Molecules and the Origin of Life on Earth

EVOLUTION In 1953, Stanley Miller, a graduate student of Harold Urey's at the University of Chicago, helped bring the abiotic (nonliving) synthesis of organic compounds into the context of evolution. Study **Figure 4.2** to learn about his classic experiment. From his results, Miller concluded that complex organic molecules could arise spontaneously under conditions thought to have existed on the early Earth. Miller also performed experiments designed to mimic volcanic conditions, with roughly similar results. In 2008, a former graduate student of Miller's discovered some samples from these experiments. Reanalyzing them using modern equipment, he identified additional organic compounds that had not been found by Miller. Although the jury is still out, these experiments support the idea that abiotic synthesis of organic compounds, perhaps near volcanoes, could have been an early stage in the origin of life (see Chapter 25).

The pioneers of organic chemistry helped shift the mainstream of biological thought from vitalism to *mechanism*, the view that physical and chemical laws govern all natural phenomena, including the processes of life. Organic chemistry was redefined as the study of carbon compounds, regardless of origin. Organisms produce most of the naturally occurring organic compounds, and these molecules represent a diversity and range of complexity unrivaled by inorganic compounds. However, the rules of chemistry apply to all molecules. The foundation of organic chemistry is not some intangible life force, but the unique chemical versatility of the element carbon.

CONCEPT CHECK 4.1

1. Why was Wöhler astonished to find he had made urea?
2. **WHAT IF?** When Miller tried his experiment without the electrical discharge, no organic compounds were found. What might explain this result?

For suggested answers, see Appendix A.

▼ Figure 4.2 INQUIRY

Can organic molecules form under conditions estimated to simulate those on the early Earth?

EXPERIMENT In 1953, Stanley Miller set up a closed system to mimic conditions thought to have existed on the early Earth. A flask of water simulated the primeval sea. The water was heated so that some vaporized and moved into a second, higher flask containing the "atmosphere"—a mixture of gases. Sparks were discharged in the synthetic atmosphere to mimic lightning.

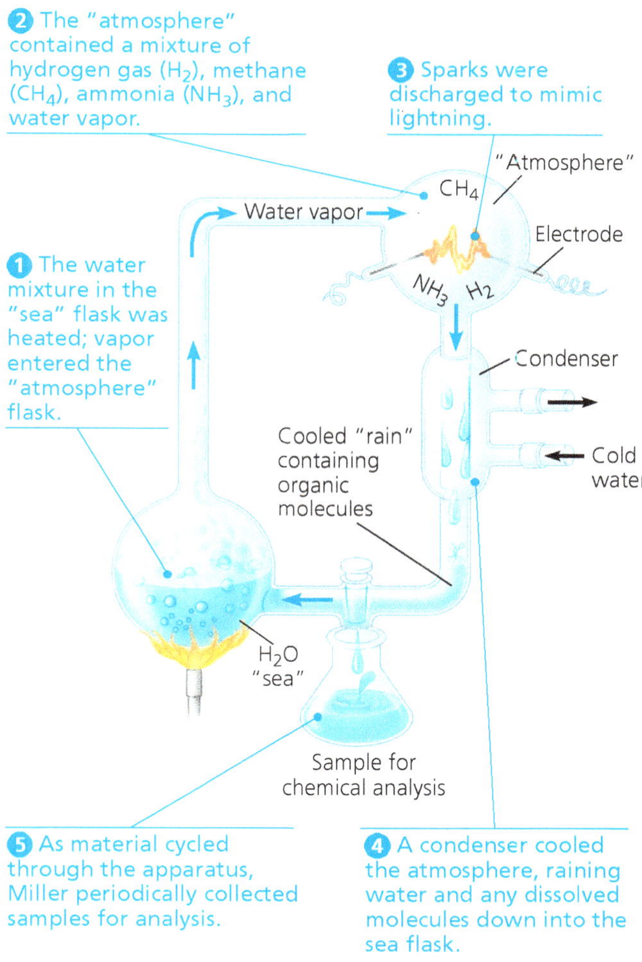

RESULTS Miller identified a variety of organic molecules that are common in organisms. These included simple compounds, such as formaldehyde (CH_2O) and hydrogen cyanide (HCN), and more complex molecules, such as amino acids and long chains of carbon and hydrogen known as hydrocarbons.

CONCLUSION Organic molecules, a first step in the origin of life, may have been synthesized abiotically on the early Earth. (We will explore this hypothesis in more detail in Chapter 25.)

SOURCE S. L. Miller, A production of amino acids under possible primitive Earth conditions, *Science* 117:528–529 (1953).

WHAT IF? If Miller had increased the concentration of NH_3 in his experiment, how might the relative amounts of the products HCN and CH_2O have differed?

CONCEPT 4.2

Carbon atoms can form diverse molecules by bonding to four other atoms

The key to an atom's chemical characteristics is its electron configuration. This configuration determines the kinds and number of bonds an atom will form with other atoms.

The Formation of Bonds with Carbon

Carbon has 6 electrons, with 2 in the first electron shell and 4 in the second shell; thus, it has 4 valence electrons in a shell that holds 8 electrons. A carbon atom usually completes its valence shell by sharing its 4 electrons with other atoms so that 8 electrons are present. Each pair of shared electrons constitutes a covalent bond (see Figure 2.12d). In organic molecules, carbon usually forms single or double covalent bonds. Each carbon atom acts as an intersection point from which a molecule can branch off in as many as four directions. This ability is one facet of carbon's versatility that makes large, complex molecules possible.

When a carbon atom forms four single covalent bonds, the arrangement of its four hybrid orbitals causes the bonds to angle toward the corners of an imaginary tetrahedron (see Figure 2.17b). The bond angles in methane (CH_4) are 109.5° **(Figure 4.3a)**, and they are roughly the same in any group of atoms where carbon has four single bonds. For example, ethane (C_2H_6) is shaped like two overlapping tetrahedrons **(Figure 4.3b)**. In molecules with more carbons, every grouping of a carbon bonded to four other atoms has a tetrahedral shape. But when two carbon atoms are joined by a double bond, as in ethene (C_2H_4), the atoms joined to those carbons are in the same plane as the carbons **(Figure 4.3c)**. We find it convenient to write molecules as structural formulas, as if the molecules being represented are two-dimensional, but keep in mind that molecules are three-dimensional and that the shape of a molecule often determines its function.

The electron configuration of carbon gives it covalent compatibility with many different elements. **Figure 4.4** shows the valences of carbon and its most frequent partners—hydrogen, oxygen, and nitrogen. These are the four major atomic components of organic molecules. These valences are the basis for the rules of covalent bonding in organic chemistry—the building code for the architecture of organic molecules.

Let's consider how the rules of covalent bonding apply to carbon atoms with partners other than hydrogen. We'll look at two examples, the simple molecules carbon dioxide and urea.

In the carbon dioxide molecule (CO_2), a single carbon atom is joined to two atoms of oxygen by double covalent bonds. The structural formula for CO_2 is shown here:

$$O{=}C{=}O$$

Each line in a structural formula represents a pair of shared electrons. Thus, the two double bonds in CO_2 have the same number of shared electrons as four single bonds. The arrangement completes the valence shells of all atoms in the molecule.

Name and Comment	Molecular Formula	Structural Formula	Ball-and-Stick Model (molecular shape in pink)	Space-Filling Model
(a) Methane. When a carbon atom has four single bonds to other atoms, the molecule is tetrahedral.	CH_4	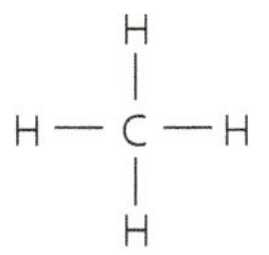	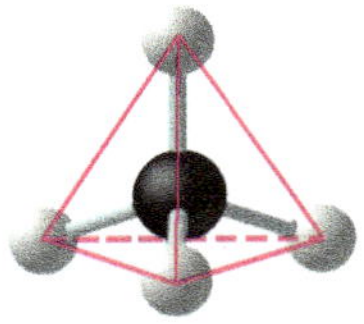	
(b) Ethane. A molecule may have more than one tetrahedral group of single-bonded atoms. (Ethane consists of two such groups.)	C_2H_6	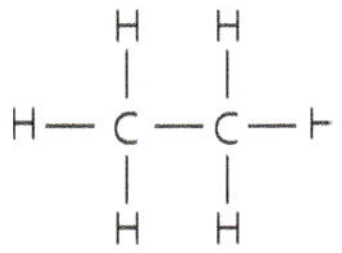	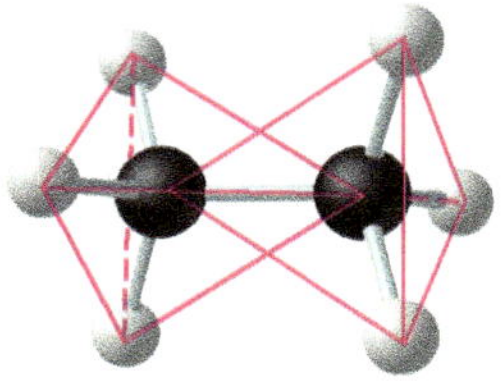	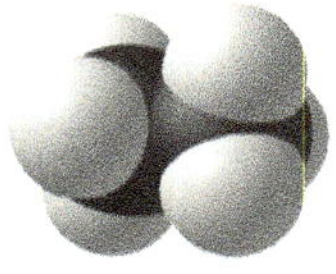
(c) Ethene (ethylene). When two carbon atoms are joined by a double bond, all atoms attached to those carbons are in the same plane; the molecule is flat.	C_2H_4	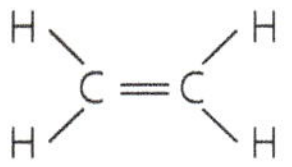	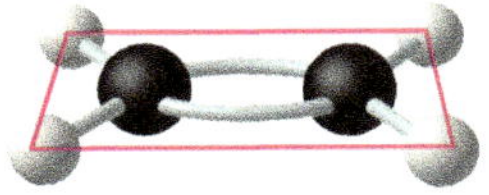	

▲ **Figure 4.3 The shapes of three simple organic molecules.**

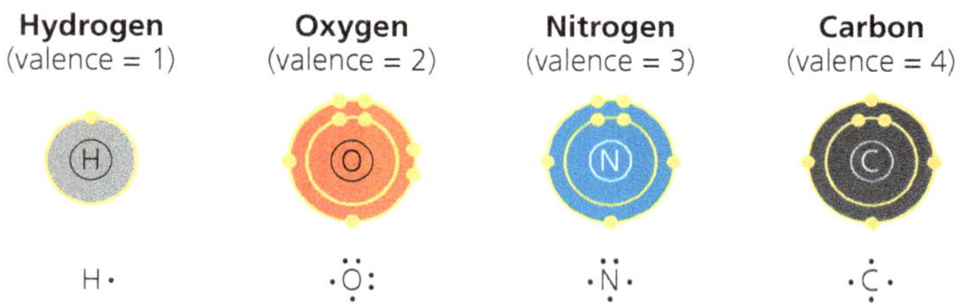

▲ **Figure 4.4 Valences of the major elements of organic molecules.** Valence is the number of covalent bonds an atom can form. It is generally equal to the number of electrons required to complete the valence (outermost) shell (see Figure 2.9). All the electrons are shown for each atom in the electron distribution diagrams (top). Only the valence shell electrons are shown in the Lewis dot structures (bottom). Note that carbon can form four bonds.

MAKE CONNECTIONS *Refer to Figure 2.9 (p. 36) and draw the Lewis dot structures for sodium, phosphorus, sulfur, and chlorine.*

Because CO_2 is a very simple molecule and lacks hydrogen, it is often considered inorganic, even though it contains carbon. Whether we call CO_2 organic or inorganic, however, it is clearly important to the living world as the source of carbon for all organic molecules in organisms.

Urea, $CO(NH_2)_2$, is the organic compound found in urine that Wöhler synthesized in the early 1800s. Again, each atom has the required number of covalent bonds. In this case, one carbon atom participates in both single and double bonds.

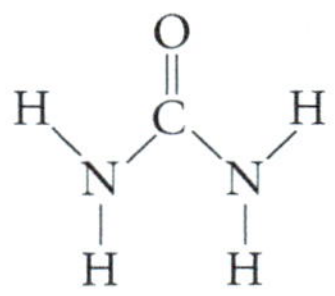

Urea

Urea and carbon dioxide are molecules with only one carbon atom. But as Figure 4.3 shows, a carbon atom can also use one or more valence electrons to form covalent bonds to other carbon atoms, linking the atoms into chains of seemingly infinite variety.

Molecular Diversity Arising from Carbon Skeleton Variation

Carbon chains form the skeletons of most organic molecules. The skeletons vary in length and may be straight, branched, or arranged in closed rings (**Figure 4.5**). Some carbon skeletons have double bonds, which vary in number and location. Such variation in carbon skeletons is one important source of the molecular complexity and diversity that characterize living matter. In addition, atoms of other elements can be bonded to the skeletons at available sites.

Hydrocarbons

All of the molecules shown in Figures 4.3 and 4.5 are **hydrocarbons**, organic molecules consisting of only carbon and hydrogen. Atoms of hydrogen are attached to the carbon skeleton wherever electrons are available for covalent bonding. Hydrocarbons are the major components of petroleum, which is called a fossil fuel because it consists of the partially decomposed remains of organisms that lived millions of years ago.

▼ **Figure 4.5 Four ways that carbon skeletons can vary.**

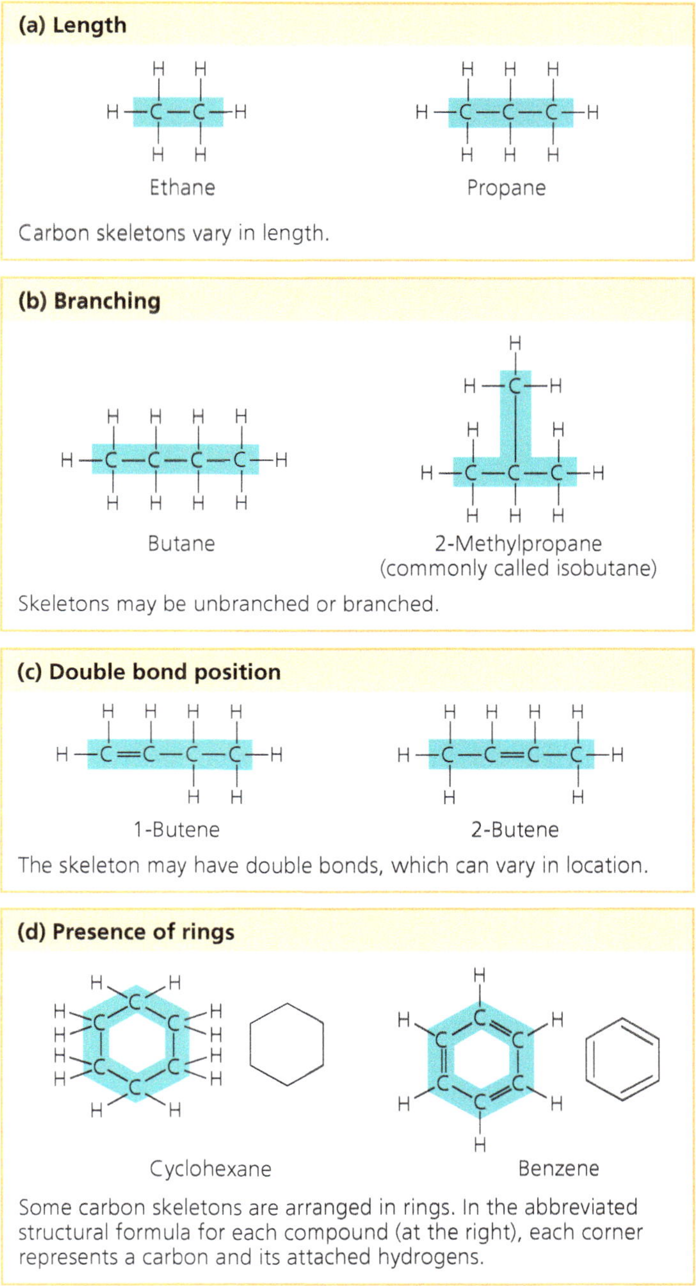

Although hydrocarbons are not prevalent in most living organisms, many of a cell's organic molecules have regions consisting of only carbon and hydrogen. For example, the molecules known as fats have long hydrocarbon tails attached to a nonhydrocarbon component (**Figure 4.6**, on the next page). Neither petroleum nor fat dissolves in water; both are hydrophobic compounds because the great majority of their bonds are relatively nonpolar carbon-to-hydrogen linkages. Another characteristic of hydrocarbons is that they can undergo reactions that release a relatively large amount of energy. The gasoline that fuels a car consists of hydrocarbons, and the hydrocarbon tails of fats serve as stored fuel for animals.

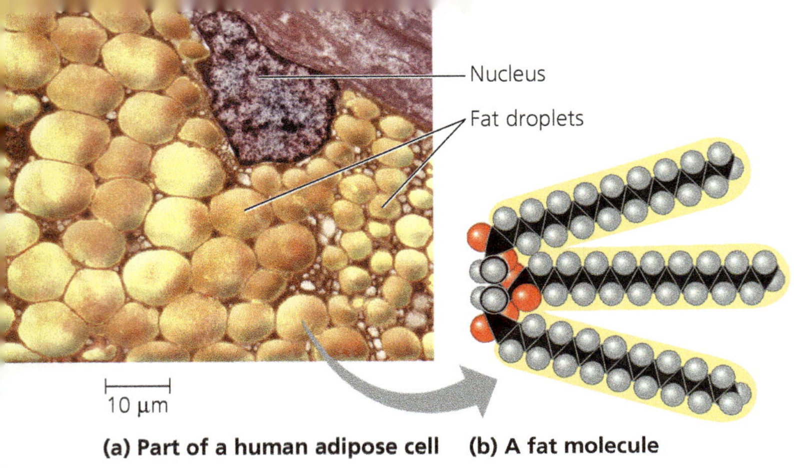

▲ **Figure 4.6 The role of hydrocarbons in fats.** **(a)** Mammalian adipose cells stockpile fat molecules as a fuel reserve. This colorized micrograph shows part of a human adipose cell with many fat droplets, each containing a large number of fat molecules. **(b)** A fat molecule consists of a small, nonhydrocarbon component joined to three hydrocarbon tails that account for the hydrophobic behavior of fats. The tails can be broken down to provide energy. (Black = carbon; gray = hydrogen; red = oxygen.)

MAKE CONNECTIONS *How do the tails account for the hydrophobic nature of fats? (See Concept 3.2, p. 51.)*

Isomers

Variation in the architecture of organic molecules can be seen in **isomers**, compounds that have the same numbers of atoms of the same elements but different structures and hence different properties. We will examine three types of isomers: structural isomers, *cis-trans* isomers, and enantiomers.

Structural isomers differ in the covalent arrangements of their atoms. Compare, for example, the two five-carbon compounds in **Figure 4.7a**. Both have the molecular formula C_5H_{12}, but they differ in the covalent arrangement of their carbon skeletons. The skeleton is straight in one compound but branched in the other. The number of possible isomers increases tremendously as carbon skeletons increase in size. There are only three forms of C_5H_{12} (two of which are shown in Figure 4.7a), but there are 18 variations of C_8H_{18} and 366,319 possible structural isomers of $C_{20}H_{42}$. Structural isomers may also differ in the location of double bonds.

In ***cis-trans* isomers** (formerly called *geometric isomers*), carbons have covalent bonds to the same atoms, but these atoms differ in their spatial arrangements due to the inflexibility of double bonds. Single bonds allow the atoms they join to rotate freely about the bond axis without changing the compound. In contrast, double bonds do not permit such rotation. If a double bond joins two carbon atoms, and each C also has two different atoms (or groups of atoms) attached to it, then two distinct *cis-trans* isomers are possible. Consider a simple molecule with two double-bonded carbons, each of which has an H and an X attached to it (**Figure 4.7b**). The arrangement with both Xs on the same side of the double bond is called a *cis isomer*, and that with the Xs on opposite sides is called a *trans* isomer. The subtle difference in shape between such isomers can dramatically affect the biological activities of organic molecules. For example, the biochemistry of vision involves a light-induced change of retinal, a chemical compound in the eye, from the *cis* isomer to the *trans* isomer (see Figure 50.17). Another example involves *trans* fats, which are discussed in Chapter 5.

▼ **Figure 4.7 Three types of isomers, compounds with the same molecular formula but different structures.**

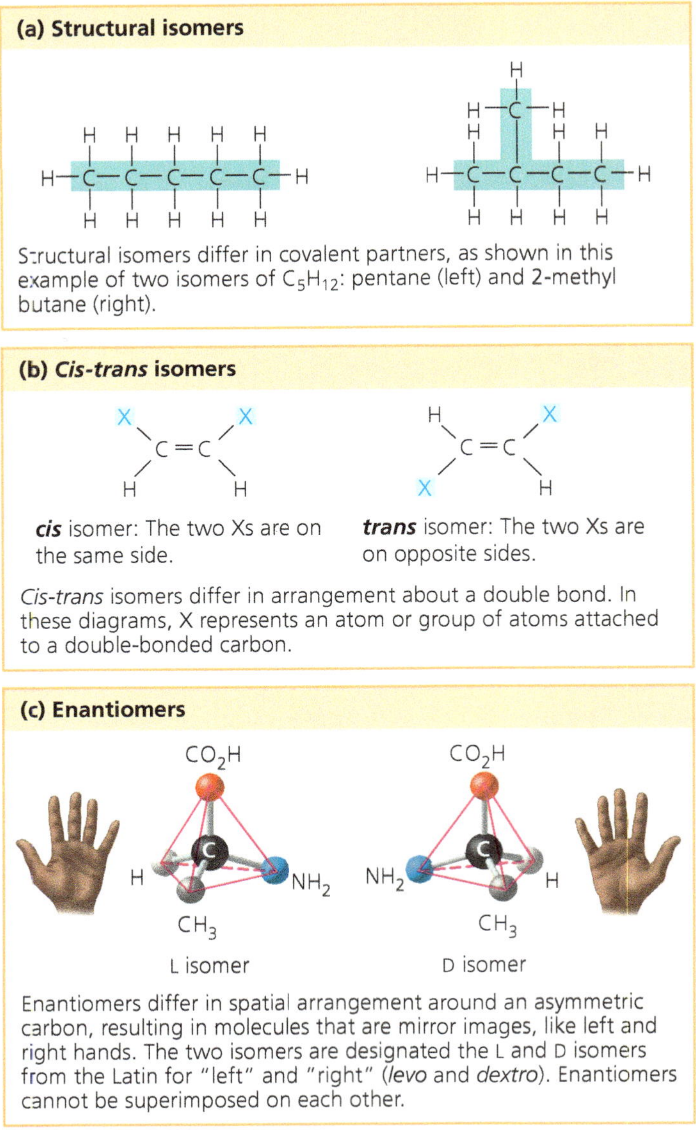

(a) Structural isomers differ in covalent partners, as shown in this example of two isomers of C_5H_{12}: pentane (left) and 2-methyl butane (right).

(b) *Cis-trans* isomers differ in arrangement about a double bond. In these diagrams, X represents an atom or group of atoms attached to a double-bonded carbon.

(c) Enantiomers differ in spatial arrangement around an asymmetric carbon, resulting in molecules that are mirror images, like left and right hands. The two isomers are designated the L and D isomers from the Latin for "left" and "right" (*levo* and *dextro*). Enantiomers cannot be superimposed on each other.

DRAW IT *There are three structural isomers of C_5H_{12}; draw the one not shown in (a).*

Enantiomers are isomers that are mirror images of each other and that differ in shape due to the presence of an *asymmetric carbon*, one that is attached to four different atoms or groups of atoms. (See the middle carbon in the ball-and-stick models shown in **Figure 4.7c**.) The four groups can be arranged in space around the asymmetric carbon in two different ways that are mirror images. Enantiomers are, in a way, left-handed and right-handed versions of the molecule. Just as your right hand won't fit into a left-handed glove, a "right-handed" molecule won't fit into the same space as the "left-handed" version.

Drug	Condition	Effective Enantiomer	Ineffective Enantiomer
Ibuprofen	Pain; inflammation	 *S*-Ibuprofen	 *R*-Ibuprofen
Albuterol	Asthma	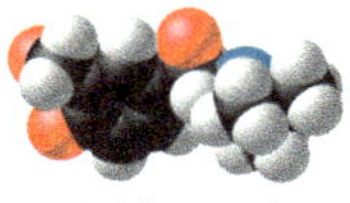 *R*-Albuterol	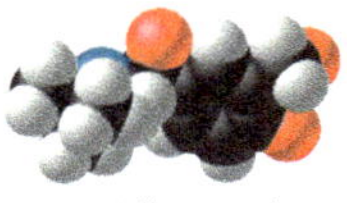 *S*-Albuterol

▲ **Figure 4.8 The pharmacological importance of enantiomers.** Ibuprofen and albuterol are examples of drugs whose enantiomers have different effects. (*S* and *R* are letters used in one system to distinguish between enantiomers.) Ibuprofen reduces inflammation and pain. It is commonly sold as a mixture of the two enantiomers. The *S* enantiomer is 100 times more effective than the other. Albuterol is used to relax bronchial muscles, improving airflow in asthma patients. Only *R*-albuterol is synthesized and sold as a drug; the *S* form counteracts the active *R* form.

Usually, only one isomer is biologically active because only that form can bind to specific molecules in an organism.

The concept of enantiomers is important in the pharmaceutical industry because the two enantiomers of a drug may not be equally effective, as is the case for both ibuprofen and the asthma medication albuterol (**Figure 4.8**). Methamphetamine also occurs in two enantiomers that have very different effects. One enantiomer is the highly addictive stimulant drug known as "crank," sold illegally in the street drug trade. The other has a much weaker effect and is even found as an ingredient in an over-the-counter vapor inhaler for treatment of nasal congestion! The differing effects of enantiomers in the body demonstrate that organisms are sensitive to even the most subtle variations in molecular architecture. Once again, we see that molecules have emergent properties that depend on the specific arrangement of their atoms.

CONCEPT CHECK 4.2

1. **DRAW IT** Draw a structural formula for C_2H_4.
2. Which molecules in Figure 4.5 are isomers? For each pair, identify the type of isomer.
3. How are gasoline and fat chemically similar?
4. **WHAT IF?** Can propane (C_3H_8) form isomers?

For suggested answers, see Appendix A.

CONCEPT 4.3

A few chemical groups are key to the functioning of biological molecules

The distinctive properties of an organic molecule depend not only on the arrangement of its carbon skeleton but also on the chemical groups attached to that skeleton. We can think of hydrocarbons, the simplest organic molecules, as the underlying framework for more complex organic molecules. A number of chemical groups can replace one or more of the hydrogens bonded to the carbon skeleton of the hydrocarbon. (Some groups include atoms of the carbon skeleton, as we will see.) These groups may participate in chemical reactions or may contribute to function indirectly by their effects on molecular shape. The number and arrangement of the groups help give each molecule its unique properties.

The Chemical Groups Most Important in the Processes of Life

Consider the differences between estradiol (a type of estrogen) and testosterone. These compounds are female and male sex hormones, respectively, in humans and other vertebrates. Both are steroids, organic molecules with a common carbon skeleton in the form of four fused rings. These sex hormones differ only in the chemical groups attached to the rings (shown here in abbreviated form); the distinctions in molecular architecture are shaded in blue:

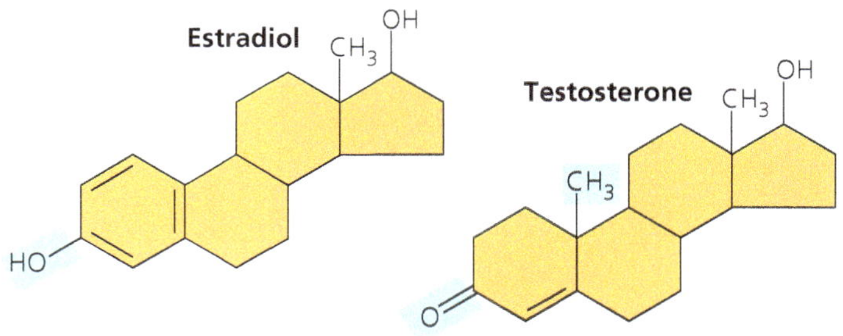

The different actions of these two molecules on many targets throughout the body help produce the contrasting anatomical and physiological features of male and female vertebrates. Thus, even our sexuality has its biological basis in variations of molecular architecture.

In the example of sex hormones, different chemical groups contribute to function by affecting the molecule's shape. In other cases, the chemical groups affect molecular function by being directly involved in chemical reactions; these important chemical groups are known as **functional groups**. Each functional group participates in chemical reactions in a characteristic way from one organic molecule to another.

The seven chemical groups most important in biological processes are the hydroxyl, carbonyl, carboxyl, amino, sulfhydryl, phosphate, and methyl groups. The first six groups can act as functional groups; they are also hydrophilic and thus increase the solubility of organic compounds in water. The methyl group is not reactive, but instead often serves as a recognizable tag on biological molecules. Before reading further, study **Figure 4.9** on the next two pages to familiarize yourself with these biologically important chemical groups.

▼ **Figure 4.9**

Exploring Some Biologically Important Chemical Groups

CHEMICAL GROUP	Hydroxyl	Carbonyl	Carboxyl
STRUCTURE	 (may be written HO—) In a **hydroxyl group** (—OH), a hydrogen atom is bonded to an oxygen atom, which in turn is bonded to the carbon skeleton of the organic molecule. (Do not confuse this functional group with the hydroxide ion, OH^-.)	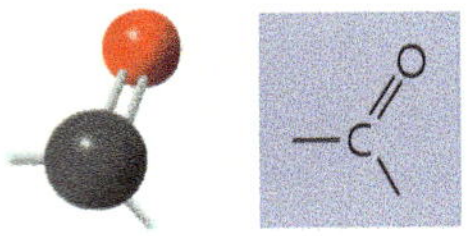 The **carbonyl group** (>CO) consists of a carbon atom joined to an oxygen atom by a double bond.	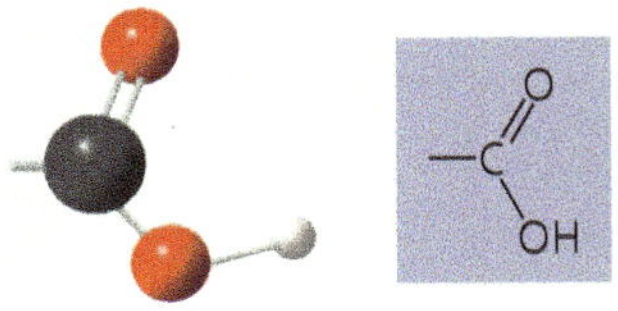 When an oxygen atom is double-bonded to a carbon atom that is also bonded to an —OH group, the entire assembly of atoms is called a **carboxyl group** (—COOH).
NAME OF COMPOUND	**Alcohols** (Their specific names usually end in *-ol*.)	**Ketones** if the carbonyl group is within a carbon skeleton **Aldehydes** if the carbonyl group is at the end of the carbon skeleton	**Carboxylic acids**, or organic acids
EXAMPLE	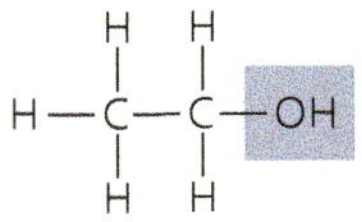 **Ethanol**, the alcohol present in alcoholic beverages	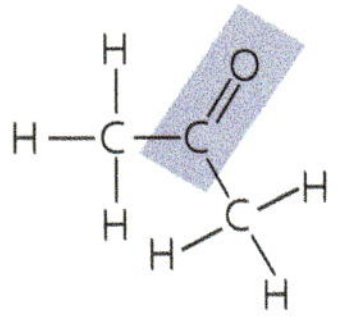 **Acetone**, the simplest ketone 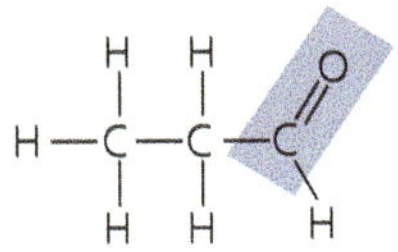 **Propanal**, an aldehyde	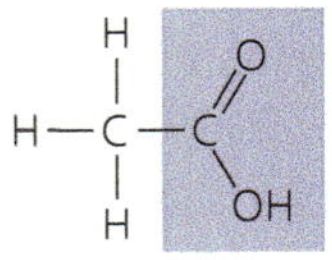 **Acetic acid**, which gives vinegar its sour taste
FUNCTIONAL PROPERTIES	• Is polar as a result of the electrons spending more time near the electronegative oxygen atom. • Can form hydrogen bonds with water molecules, helping dissolve organic compounds such as sugars. (Sugars are shown in Figure 5.3.)	• A ketone and an aldehyde may be structural isomers with different properties, as is the case for acetone and propanal. • Ketone and aldehyde groups are also found in sugars, giving rise to two major groups of sugars: ketoses (containing ketone groups) and aldoses (containing aldehyde groups).	• Acts as an acid; can donate an H^+ because the covalent bond between oxygen and hydrogen is so polar: 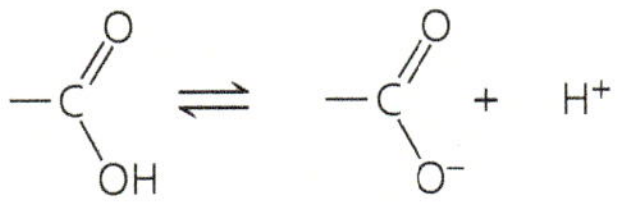 Nonionized Ionized • Found in cells in the ionized form with a charge of 1– and called a carboxylate ion.

Amino

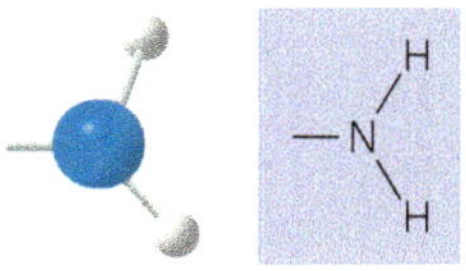

The **amino group** (—NH_2) consists of a nitrogen atom bonded to two hydrogen atoms and to the carbon skeleton.

Amines

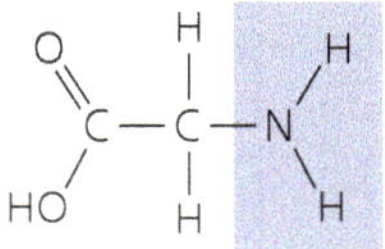

Glycine, a compound that is both an amine and a carboxylic acid because it has both an amino group and a carboxyl group; compounds with both groups are called **amino acids**

- Acts as a base; can pick up an H^+ from the surrounding solution (water, in living organisms):

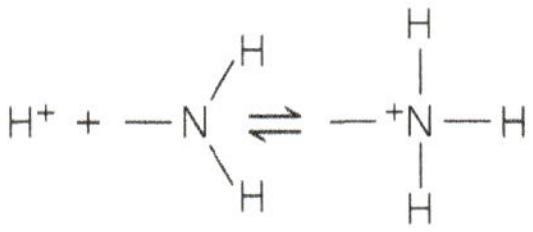

Nonionized   Ionized

- Found in cells in the ionized form with a charge of 1+.

Sulfhydryl

(may be written HS—)

The **sulfhydryl group** (—SH) consists of a sulfur atom bonded to an atom of hydrogen; it resembles a hydroxyl group in shape.

Thiols

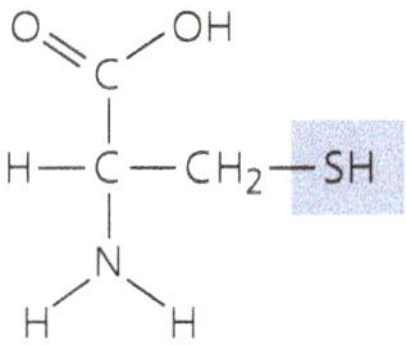

Cysteine, an important sulfur-containing amino acid

- Two sulfhydryl groups can react, forming a covalent bond. This "cross-linking" helps stabilize protein structure (see Figure 5.20, Tertiary Structure).
- Cross-linking of cysteines in hair proteins maintains the curliness or straightness of hair. Straight hair can be "permanently" curled by shaping it around curlers and then breaking and re-forming the cross-linking bonds.

Phosphate

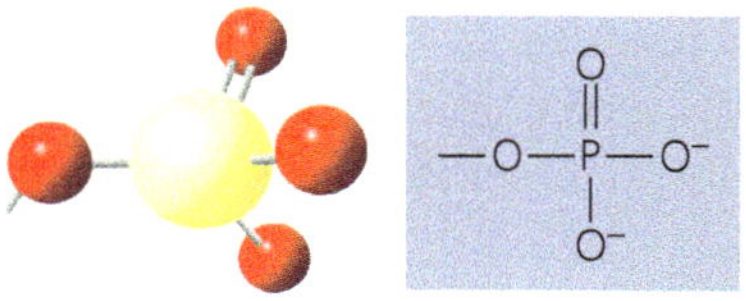

In the **phosphate group** shown here, a phosphorus atom is bonded to four oxygen atoms; one oxygen is bonded to the carbon skeleton; two oxygens carry negative charges (—OPO_3^{2-}). In this text, Ⓟ represents an attached phosphate group.

Organic phosphates

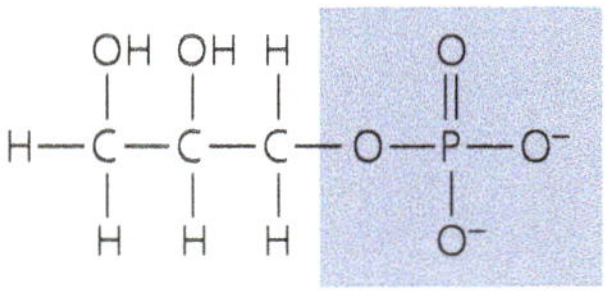

Glycerol phosphate, which takes part in many important chemical reactions in cells; glycerol phosphate also provides the backbone for phospholipids, the most prevalent molecules in cell membranes

- Contributes negative charge to the molecule of which it is a part (2− when at the end of a molecule, as above; 1− when located internally in a chain of phosphates).
- Molecules containing phosphate groups have the potential to react with water, releasing energy.

Methyl

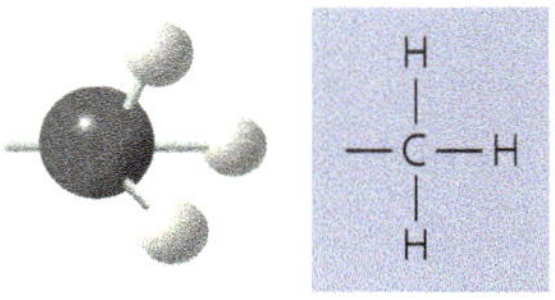

A **methyl group** (—CH_3) consists of a carbon bonded to three hydrogen atoms. The carbon of a methyl group may be attached to a carbon or to a different atom.

Methylated compounds

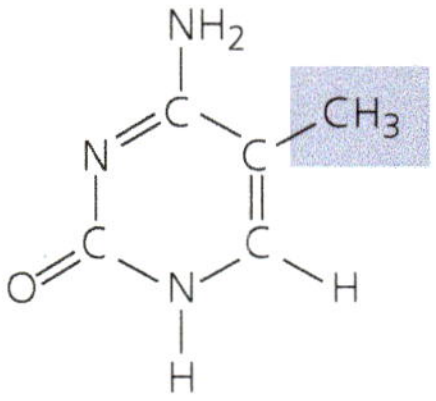

5-Methyl cytidine, a component of DNA that has been modified by addition of a methyl group

- Addition of a methyl group to DNA, or to molecules bound to DNA, affects the expression of genes.
- Arrangement of methyl groups in male and female sex hormones affects their shape and function (see p. 63).

MAKE CONNECTIONS *Given the information in this figure and what you know about the electronegativity of oxygen (see Concept 2.3, p. 39), predict which of the following molecules would be the stronger acid (see Concept 3.3, p. 53). Explain your answer.*

ATP: An Important Source of Energy for Cellular Processes

The "Phosphate" column in Figure 4.9 shows a simple example of an organic phosphate molecule. A more complicated organic phosphate, **adenosine triphosphate**, or **ATP**, is worth mentioning here because its function in the cell is so important. ATP consists of an organic molecule called adenosine attached to a string of three phosphate groups:

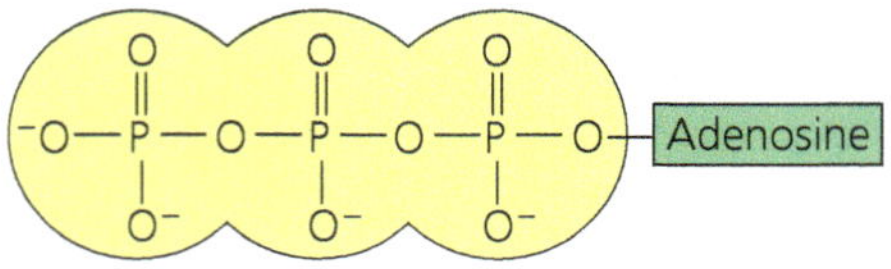

Where three phosphates are present in series, as in ATP, one phosphate may be split off as a result of a reaction with water. This inorganic phosphate ion, $HOPO_3^{2-}$, is often abbreviated Ⓟ$_i$ in this book. Having lost one phosphate, ATP becomes adenosine *di*phosphate, or ADP. Although ATP is sometimes said to store energy, it is more accurate to think of it as storing the potential to react with water. This reaction releases energy that can be used by the cell. You will learn about this in more detail in Chapter 8.

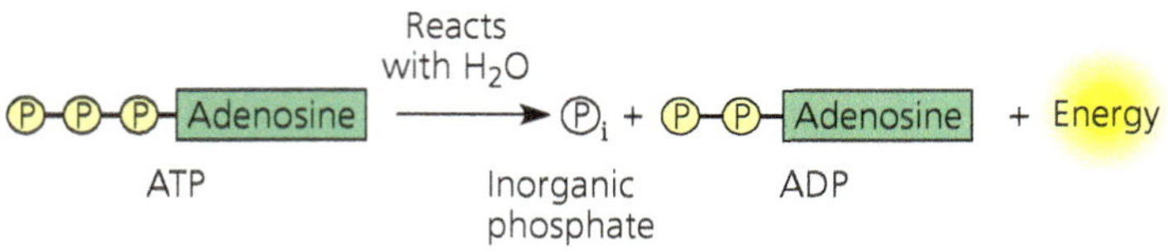

CONCEPT CHECK 4.3

1. What does the term *amino acid* signify about the structure of such a molecule?
2. What chemical change occurs to ATP when it reacts with water and releases energy?
3. **WHAT IF?** Suppose you had an organic molecule such as cysteine (see Figure 4.9, sulfhydryl group example), and you chemically removed the $—NH_2$ group and replaced it with —COOH. Draw the structural formula for this molecule and speculate about its chemical properties. Is the central carbon asymmetric before the change? After?

For suggested answers, see Appendix A.

The Chemical Elements of Life: *A Review*

Living matter, as you have learned, consists mainly of carbon, oxygen, hydrogen, and nitrogen, with smaller amounts of sulfur and phosphorus. These elements all form strong covalent bonds, an essential characteristic in the architecture of complex organic molecules. Of all these elements, carbon is the virtuoso of the covalent bond. The versatility of carbon makes possible the great diversity of organic molecules, each with particular properties that emerge from the unique arrangement of its carbon skeleton and the chemical groups appended to that skeleton. At the foundation of all biological diversity lies this variation at the molecular level.

4 CHAPTER REVIEW

SUMMARY OF KEY CONCEPTS

CONCEPT 4.1

Organic chemistry is the study of carbon compounds (pp. 58–59)

- Living matter is made mostly of carbon, oxygen, hydrogen, and nitrogen, with some sulfur and phosphorus. Biological diversity has its molecular basis in carbon's ability to form a huge number of molecules with particular shapes and chemical properties.
- Organic compounds were once thought to arise only within living organisms, but this idea (vitalism) was disproved when chemists were able to synthesize organic compounds in the laboratory.

? *How did Stanley Miller's experiments extend the idea of mechanism to the origin of life?*

CONCEPT 4.2

Carbon atoms can form diverse molecules by bonding to four other atoms (pp. 60–63)

- Carbon, with a valence of 4, can bond to various other atoms, including O, H, and N. Carbon can also bond to other carbon atoms, forming the carbon skeletons of organic compounds. These skeletons vary in length and shape and have bonding sites for atoms of other elements. **Hydrocarbons** consist only of carbon and hydrogen.
- **Isomers** are compounds with the same molecular formula but different structures and properties. Three types of isomers are **structural isomers**, ***cis-trans*** **isomers**, and **enantiomers**.

? *Refer back to Figure 4.9. What type of isomers are acetone and propanal? How many asymmetric carbons are present in acetic acid, glycine, and glycerol phosphate? Can these three molecules exist as forms that are enantiomers?*

CONCEPT 4.3

A few chemical groups are key to the functioning of biological molecules (pp. 63–66)

- Chemical groups attached to the carbon skeletons of organic molecules participate in chemical reactions (**functional groups**) or contribute to function by affecting molecular shape (see Figure 4.9).
- **ATP (adenosine triphosphate)** consists of adenosine attached to three phosphate groups. ATP can react with water, forming inorganic phosphate and ADP (adenosine diphosphate). This reaction releases energy that can be used by the cell (see the equation at the top of the next page).

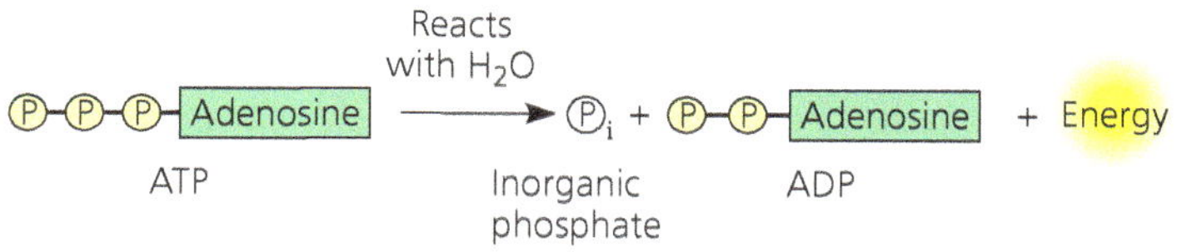

? *In what ways does a methyl group differ chemically from the other six important chemical groups shown in Figure 4.9?*

TEST YOUR UNDERSTANDING

LEVEL 1: KNOWLEDGE/COMPREHENSION

1. Organic chemistry is currently defined as
   a. the study of compounds made only by living cells.
   b. the study of carbon compounds.
   c. the study of vital forces.
   d. the study of natural (as opposed to synthetic) compounds.
   e. the study of hydrocarbons.
2. Which functional group is *not* present in this molecule?
   a. carboxyl
   b. sulfhydryl
   c. hydroxyl
   d. amino

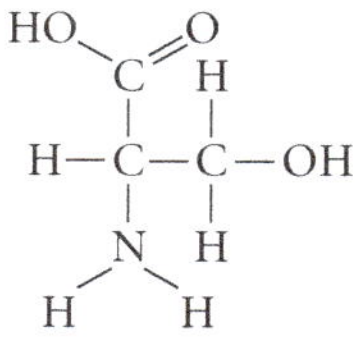

3. **MAKE CONNECTIONS** Which chemical group is most likely to be responsible for an organic molecule behaving as a base (see Concept 3.3, p. 53)?
   a. hydroxyl
   b. carbonyl
   c. carboxyl
   d. amino
   e. phosphate

LEVEL 2: APPLICATION/ANALYSIS

4. Which of the following hydrocarbons has a double bond in its carbon skeleton?
   a. C_3H_8 b. C_2H_6 c. CH_4 d. C_2H_4 e. C_2H_2
5. Choose the term that correctly describes the relationship between these two sugar molecules:
   a. structural isomers
   b. *cis-trans* isomers
   c. enantiomers
   d. isotopes

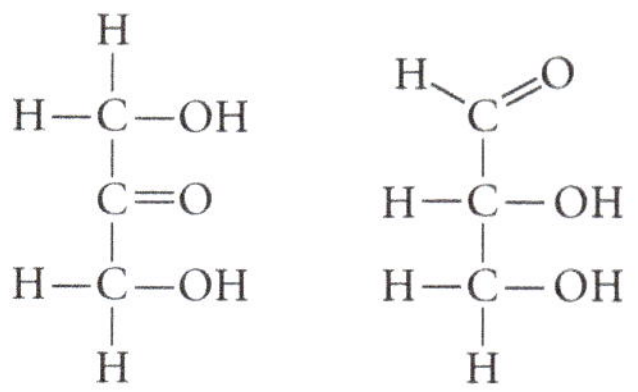

6. Identify the asymmetric carbon in this molecule:

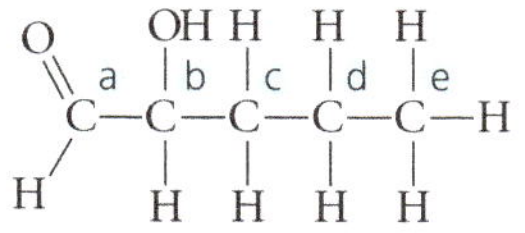

7. Which action could produce a carbonyl group?
   a. the replacement of the —OH of a carboxyl group with hydrogen
   b. the addition of a thiol to a hydroxyl
   c. the addition of a hydroxyl to a phosphate
   d. the replacement of the nitrogen of an amine with oxygen
   e. the addition of a sulfhydryl to a carboxyl
8. Which of the molecules shown in question 5 has an asymmetric carbon? Which carbon is asymmetric?

LEVEL 3: SYNTHESIS/EVALUATION

9. **EVOLUTION CONNECTION**

DRAW IT Some scientists think that life elsewhere in the universe might be based on the element silicon, rather than on carbon, as on Earth. Look at the electron distribution diagram for silicon in Figure 2.9 and draw the Lewis dot structure for silicon. What properties does silicon share with carbon that would make silicon-based life more likely than, say, neon-based life or aluminum-based life?

10. **SCIENTIFIC INQUIRY**

Thalidomide achieved notoriety 50 years ago because of a wave of birth defects among children born to women who took this drug during pregnancy as a treatment for morning sickness. Thalidomide is a mixture of two enantiomers; one reduces morning sickness, but the other causes severe birth defects. (Although the beneficial enantiomer can be synthesized and given to patients, it is converted in the body to the harmful enantiomer.) The U.S. Food and Drug Administration (FDA) withheld approval of thalidomide in 1960. Since then, however, the FDA has approved this drug for the treatment of conditions associated with Hansen's disease (leprosy) and newly diagnosed multiple myeloma, a blood and bone marrow cancer. In clinical trials, thalidomide also shows promise as a treatment for AIDS, tuberculosis, inflammatory diseases, and some other types of cancer. Assuming that molecules related to thalidomide could be synthesized in the laboratory, describe in a broad way the type of experiments you would do to improve the benefits of this drug and minimize its harmful effects.

11. **WRITE ABOUT A THEME**

Structure and Function In 1918, an epidemic of sleeping sickness caused an unusual rigid paralysis in some survivors, similar to symptoms of advanced Parkinson's disease. Years later, L-dopa (below, left), a chemical used to treat Parkinson's disease, was given to some of these patients, as dramatized in the movie *Awakenings*, starring Robin Williams. L-dopa was remarkably effective at eliminating the paralysis, at least temporarily. However, its enantiomer, D-dopa (right), was subsequently shown to have no effect at all, as is the case for Parkinson's disease. In a short essay (100–150 words), discuss how the effectiveness of one enantiomer and not the other illustrates the theme of structure and function.

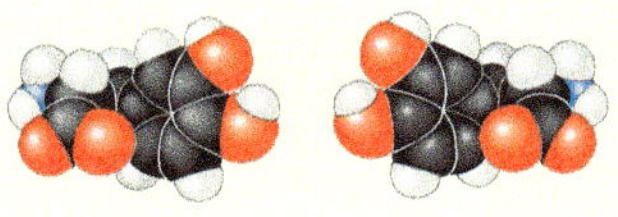

L-dopa **D-dopa**

For selected answers, see Appendix A.

MasteringBIOLOGY www.masteringbiology.com

1. MasteringBiology® Assignments

Tutorial Carbon Bonding and Functional Groups
Activities Diversity of Carbon-Based Molecules • Isomers • Functional Groups
Questions Student Misconceptions • Reading Quiz • Multiple Choice • End-of-Chapter

2. eText

Read your book online, search, take notes, highlight text, and more.

3. The Study Area

Practice Tests • Cumulative Test • *BioFlix* 3-D Animations • MP3 Tutor Sessions • Videos • Activities • Investigations • Lab Media • Audio Glossary • Word Study Tools • Art

5

The Structure and Function of Large Biological Molecules

▲ **Figure 5.1 Why do scientists study the structures of macromolecules?**

KEY CONCEPTS

5.1 Macromolecules are polymers, built from monomers
5.2 Carbohydrates serve as fuel and building material
5.3 Lipids are a diverse group of hydrophobic molecules
5.4 Proteins include a diversity of structures, resulting in a wide range of functions
5.5 Nucleic acids store, transmit, and help express hereditary information

OVERVIEW

The Molecules of Life

Given the rich complexity of life on Earth, we might expect organisms to have an enormous diversity of molecules. Remarkably, however, the critically important large molecules of all living things—from bacteria to elephants—fall into just four main classes: carbohydrates, lipids, proteins, and nucleic acids. On the molecular scale, members of three of these classes—carbohydrates, proteins, and nucleic acids—are huge and are therefore called **macromolecules**. For example, a protein may consist of thousands of atoms that form a molecular colossus with a mass well over 100,000 daltons. Considering the size and complexity of macromolecules, it is noteworthy that biochemists have determined the detailed structure of so many of them. The scientist in the foreground of **Figure 5.1** is using 3-D glasses to help her visualize the structure of the protein displayed on her screen.

The architecture of a large biological molecule helps explain how that molecule works. Like water and simple organic molecules, large biological molecules exhibit unique emergent properties arising from the orderly arrangement of their atoms. In this chapter, we'll first consider how macromolecules are built. Then we'll examine the structure and function of all four classes of large biological molecules: carbohydrates, lipids, proteins, and nucleic acids.

CONCEPT 5.1

Macromolecules are polymers, built from monomers

The macromolecules in three of the four classes of life's organic compounds—carbohydrates, proteins, and nucleic acids—are chain-like molecules called polymers (from the Greek *polys*, many, and *meros*, part). A **polymer** is a long molecule consisting of many similar or identical building blocks linked by covalent bonds, much as a train consists of a chain of cars. The repeating units that serve as the building blocks of a polymer are smaller molecules called **monomers** (from the Greek *monos*, single). Some of the molecules that serve as monomers also have other functions of their own.

The Synthesis and Breakdown of Polymers

Although each class of polymer is made up of a different type of monomer, the chemical mechanisms by which cells make and break down polymers are basically the same in all cases. In cells, these processes are facilitated by **enzymes**, specialized macromolecules that speed up chemical reactions. Monomers are connected by a reaction in which two molecules are covalently bonded to each other, with the loss of a water molecule; this is known as a **dehydration reaction** (**Figure 5.2a**). When a bond forms between two monomers, each monomer contributes part of the water molecule that is released during the reaction: One monomer provides a hydroxyl group (—OH), while the other provides a hydrogen (—H). This reaction is repeated as monomers are added to the chain one by one, making a polymer.

Polymers are disassembled to monomers by **hydrolysis**, a process that is essentially the reverse of the dehydration reac-

Figure 5.2 The synthesis and breakdown of polymers.

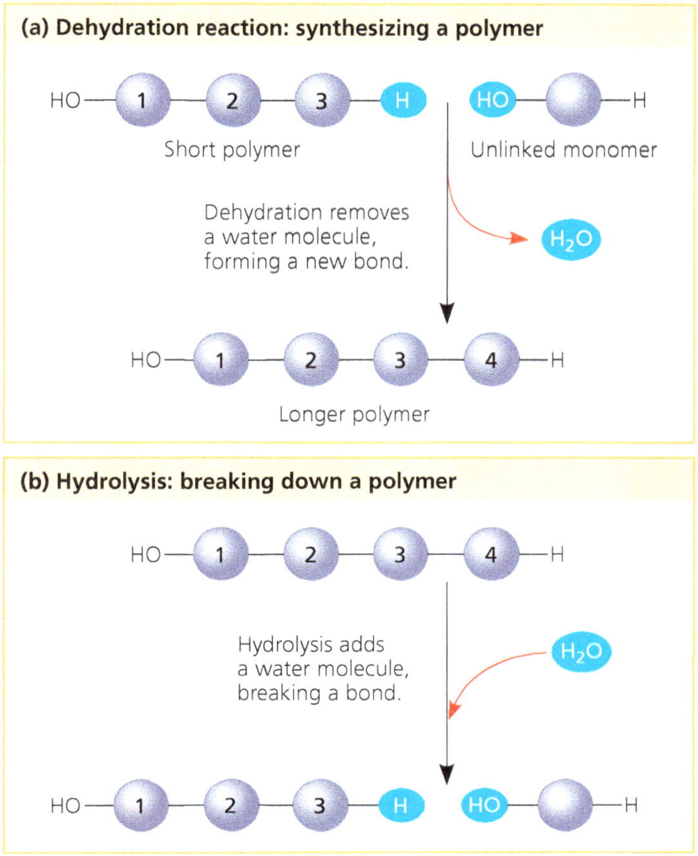

tion **(Figure 5.2b)**. Hydrolysis means to break using water (from the Greek *hydro*, water, and *lysis*, break). The bond between the monomers is broken by the addition of a water molecule, with the hydrogen from the water attaching to one monomer and the hydroxyl group attaching to the adjacent monomer. An example of hydrolysis working within our bodies is the process of digestion. The bulk of the organic material in our food is in the form of polymers that are much too large to enter our cells. Within the digestive tract, various enzymes attack the polymers, speeding up hydrolysis. The released monomers are then absorbed into the bloodstream for distribution to all body cells. Those cells can then use dehydration reactions to assemble the monomers into new, different polymers that can perform specific functions required by the cell.

The Diversity of Polymers

Each cell has thousands of different macromolecules; the collection varies from one type of cell to another even in the same organism. The inherent differences between human siblings reflect small variations in polymers, particularly DNA and proteins. Molecular differences between unrelated individuals are more extensive and those between species greater still. The diversity of macromolecules in the living world is vast, and the possible variety is effectively limitless.

What is the basis for such diversity in life's polymers? These molecules are constructed from only 40 to 50 common monomers and some others that occur rarely. Building a huge variety of polymers from such a limited number of monomers is analogous to constructing hundreds of thousands of words from only 26 letters of the alphabet. The key is arrangement—the particular linear sequence that the units follow. However, this analogy falls far short of describing the great diversity of macromolecules because most biological polymers have many more monomers than the number of letters in the longest word. Proteins, for example, are built from 20 kinds of amino acids arranged in chains that are typically hundreds of amino acids long. The molecular logic of life is simple but elegant: Small molecules common to all organisms are ordered into unique macromolecules.

Despite this immense diversity, molecular structure and function can still be grouped roughly by class. Let's examine each of the four major classes of large biological molecules. For each class, the large molecules have emergent properties not found in their individual building blocks.

CONCEPT CHECK 5.1

1. What are the four main classes of large biological molecules? Which class does not consist of polymers?
2. How many molecules of water are needed to completely hydrolyze a polymer that is ten monomers long?
3. **WHAT IF?** Suppose you eat a serving of fish. What reactions must occur for the amino acid monomers in the protein of the fish to be converted to new proteins in your body?

For suggested answers, see Appendix A.

CONCEPT 5.2

Carbohydrates serve as fuel and building material

Carbohydrates include both sugars and polymers of sugars. The simplest carbohydrates are the monosaccharides, or simple sugars; these are the monomers from which more complex carbohydrates are constructed. Disaccharides are double sugars, consisting of two monosaccharides joined by a covalent bond. Carbohydrates also include macromolecules called polysaccharides, polymers composed of many sugar building blocks.

Sugars

Monosaccharides (from the Greek *monos*, single, and *sacchar*, sugar) generally have molecular formulas that are some multiple of the unit CH_2O. Glucose ($C_6H_{12}O_6$), the most common monosaccharide, is of central importance in the chemistry

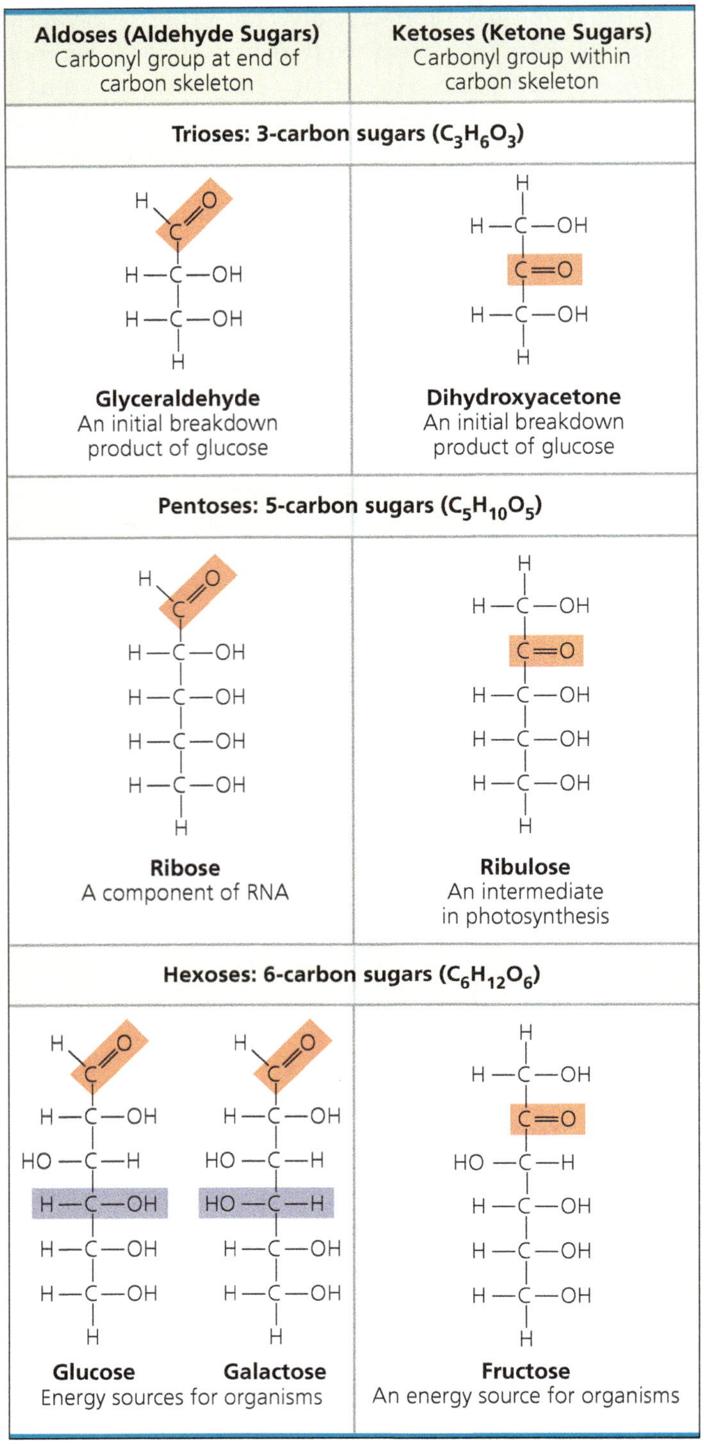

▲ **Figure 5.3 The structure and classification of some monosaccharides.** Sugars vary in the location of their carbonyl groups (orange), the length of their carbon skeletons, and the spatial arrangement around asymmetric carbons (compare, for example, the purple portions of glucose and galactose).

MAKE CONNECTIONS *In the 1970s, a process was developed that converts the glucose in corn syrup to its sweeter isomer, fructose. High-fructose corn syrup, a common ingredient in soft drinks and processed food, is a mixture of glucose and fructose. What type of isomers are glucose and fructose? See Figure 4.7, p. 62.*

of life. In the structure of glucose, we can see the trademarks of a sugar: The molecule has a carbonyl group (C=O) and multiple hydroxyl groups (—OH) **(Figure 5.3)**. Depending on the location of the carbonyl group, a sugar is either an aldose (aldehyde sugar) or a ketose (ketone sugar). Glucose, for example, is an aldose; fructose, an isomer of glucose, is a ketose. (Most names for sugars end in *-ose.*) Another criterion for classifying sugars is the size of the carbon skeleton, which ranges from three to seven carbons long. Glucose, fructose, and other sugars that have six carbons are called hexoses. Trioses (three-carbon sugars) and pentoses (five-carbon sugars) are also common.

Still another source of diversity for simple sugars is in the spatial arrangement of their parts around asymmetric carbons. (Recall that an asymmetric carbon is a carbon attached to four different atoms or groups of atoms.) Glucose and galactose, for example, differ only in the placement of parts around one asymmetric carbon (see the purple boxes in Figure 5.3). What seems like a small difference is significant enough to give the two sugars distinctive shapes and behaviors.

Although it is convenient to draw glucose with a linear carbon skeleton, this representation is not completely accurate. In aqueous solutions, glucose molecules, as well as most other five- and six-carbon sugars, form rings **(Figure 5.4)**.

Monosaccharides, particularly glucose, are major nutrients for cells. In the process known as cellular respiration, cells extract energy in a series of reactions starting with glucose molecules. Simple-sugar molecules are not only a major fuel for cellular work, but their carbon skeletons also serve as raw material for the synthesis of other types of small organic molecules, such as amino acids and fatty acids. Sugar molecules that are not immediately used in these ways are generally incorporated as monomers into disaccharides or polysaccharides.

A **disaccharide** consists of two monosaccharides joined by a **glycosidic linkage**, a covalent bond formed between two monosaccharides by a dehydration reaction. For example, maltose is a disaccharide formed by the linking of two molecules of glucose **(Figure 5.5a)**. Also known as malt sugar, maltose is an ingredient used in brewing beer. The most prevalent disaccharide is sucrose, which is table sugar. Its two monomers are glucose and fructose **(Figure 5.5b)**. Plants generally transport carbohydrates from leaves to roots and other nonphotosynthetic organs in the form of sucrose. Lactose, the sugar present in milk, is another disaccharide, in this case a glucose molecule joined to a galactose molecule.

Polysaccharides

Polysaccharides are macromolecules, polymers with a few hundred to a few thousand monosaccharides joined by glycosidic linkages. Some polysaccharides serve as storage material, hydrolyzed as needed to provide sugar for cells. Other polysaccharides serve as building material for structures that

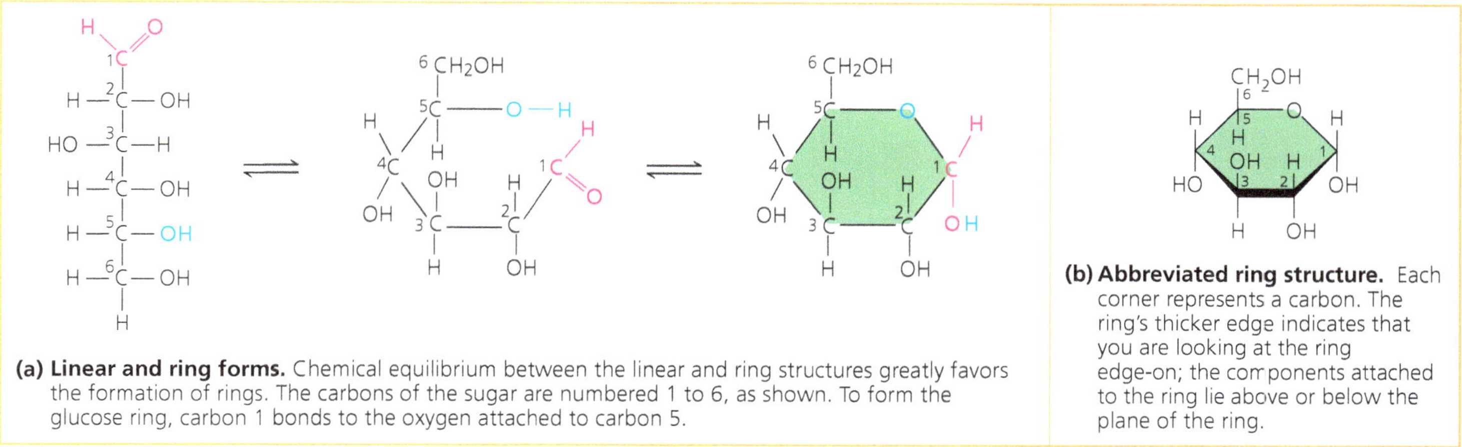

(a) **Linear and ring forms.** Chemical equilibrium between the linear and ring structures greatly favors the formation of rings. The carbons of the sugar are numbered 1 to 6, as shown. To form the glucose ring, carbon 1 bonds to the oxygen attached to carbon 5.

(b) **Abbreviated ring structure.** Each corner represents a carbon. The ring's thicker edge indicates that you are looking at the ring edge-on; the components attached to the ring lie above or below the plane of the ring.

▲ **Figure 5.4 Linear and ring forms of glucose.**

DRAW IT *Start with the linear form of fructose (see Figure 5.3) and draw the formation of the fructose ring in two steps. First, number the carbons starting at the top of the linear structure. Then attach carbon 5 via its oxygen to carbon 2. Compare the number of carbons in the fructose and glucose rings.*

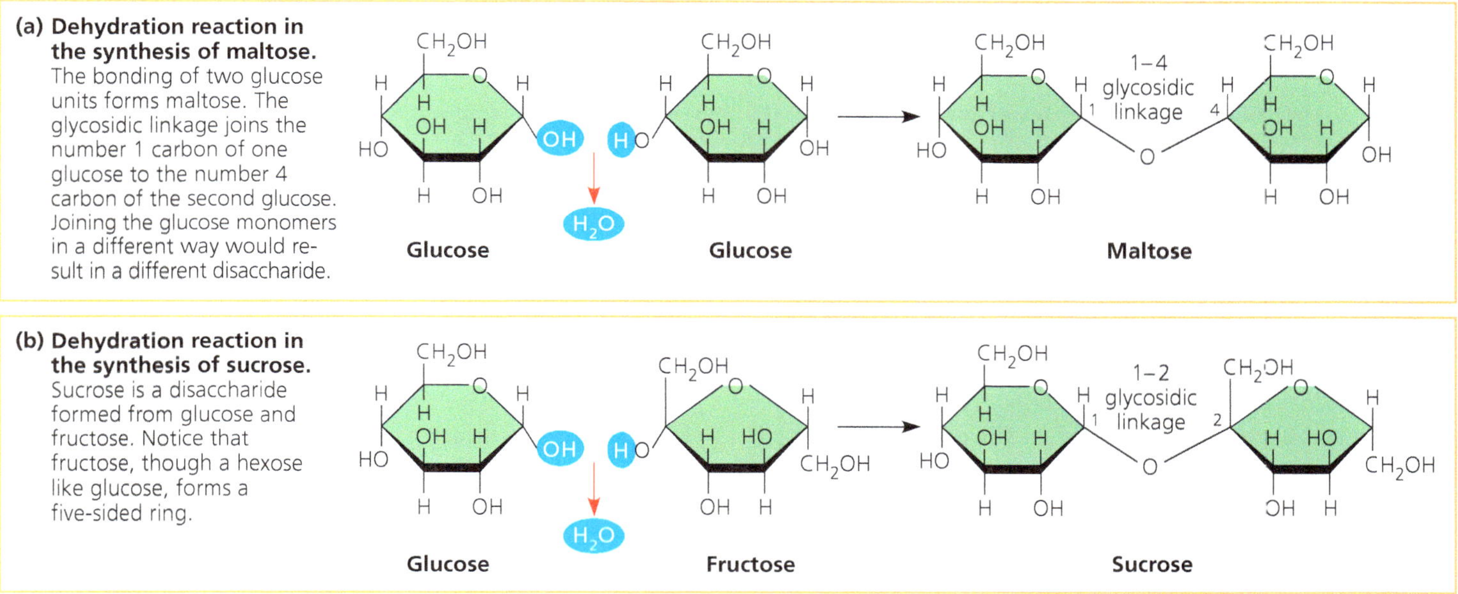

(a) **Dehydration reaction in the synthesis of maltose.** The bonding of two glucose units forms maltose. The glycosidic linkage joins the number 1 carbon of one glucose to the number 4 carbon of the second glucose. Joining the glucose monomers in a different way would result in a different disaccharide.

(b) **Dehydration reaction in the synthesis of sucrose.** Sucrose is a disaccharide formed from glucose and fructose. Notice that fructose, though a hexose like glucose, forms a five-sided ring.

▲ **Figure 5.5 Examples of disaccharide synthesis.**

DRAW IT *Referring to Figure 5.4, number the carbons in each sugar in this figure. Show how the numbering is consistent with the name of the glycosidic linkage in each disaccharide.*

protect the cell or the whole organism. The architecture and function of a polysaccharide are determined by its sugar monomers and by the positions of its glycosidic linkages.

Storage Polysaccharides

Both plants and animals store sugars for later use in the form of storage polysaccharides. Plants store **starch**, a polymer of glucose monomers, as granules within cellular structures known as plastids, which include chloroplasts. Synthesizing starch enables the plant to stockpile surplus glucose. Because glucose is a major cellular fuel, starch represents stored energy. The sugar can later be withdrawn from this carbohydrate "bank" by hydrolysis, which breaks the bonds between the glucose monomers. Most animals, including humans, also have enzymes that can hydrolyze plant starch, making glucose available as a nutrient for cells. Potato tubers and grains—the fruits of wheat, maize (corn), rice, and other grasses—are the major sources of starch in the human diet.

Most of the glucose monomers in starch are joined by 1–4 linkages (number 1 carbon to number 4 carbon), like the glucose units in maltose (see Figure 5.5a). The simplest form of starch, amylose, is unbranched. Amylopectin, a more complex

(a) Starch: a plant polysaccharide. This micrograph shows part of a plant cell with a chloroplast, the cellular organelle where glucose is made and then stored as starch granules. Amylose (unbranched) and amylopectin (branched) are two forms of starch.

(b) Glycogen: an animal polysaccharide. Animal cells stockpile glycogen as dense clusters of granules within liver and muscle cells, as shown in this micrograph of part of a liver cell. Mitochondria are cellular organelles that help break down glucose released from glycogen. Note that glycogen is more branched than amylopectin starch.

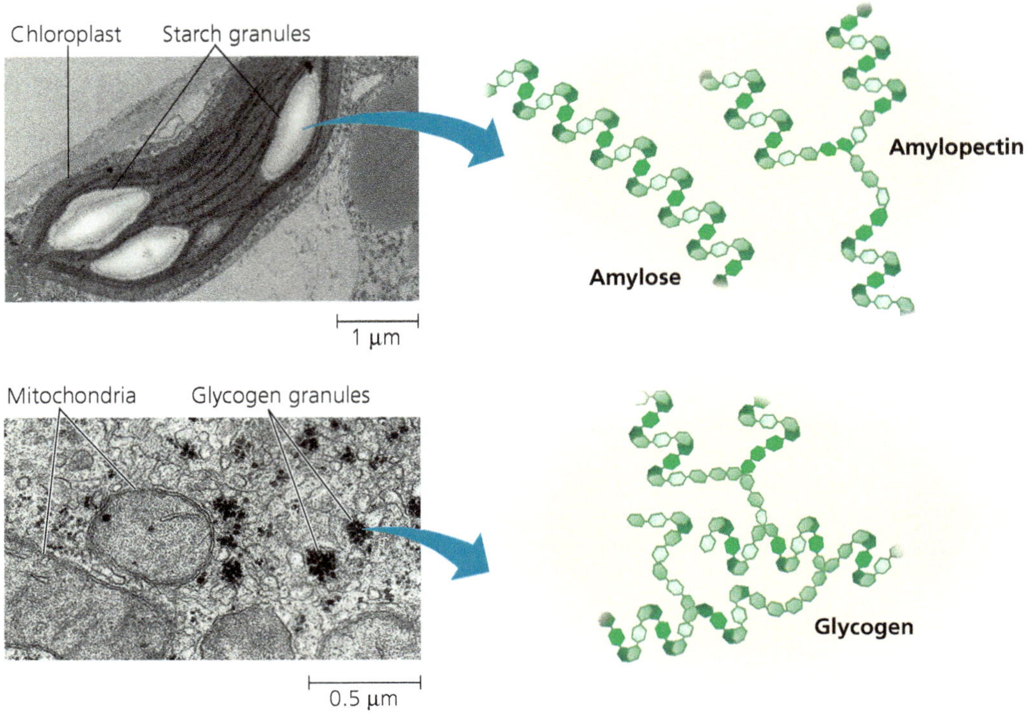

▲ **Figure 5.6 Storage polysaccharides of plants and animals.** These examples, starch and glycogen, are composed entirely of glucose monomers, represented here by hexagons. Because of the angle of the 1–4 linkages, the polymer chains tend to form helices in unbranched regions.

starch, is a branched polymer with 1–6 linkages at the branch points. Both of these starches are shown in **Figure 5.6a**.

Animals store a polysaccharide called **glycogen**, a polymer of glucose that is like amylopectin but more extensively branched **(Figure 5.6b)**. Humans and other vertebrates store glycogen mainly in liver and muscle cells. Hydrolysis of glycogen in these cells releases glucose when the demand for sugar increases. This stored fuel cannot sustain an animal for long, however. In humans, for example, glycogen stores are depleted in about a day unless they are replenished by consumption of food. This is an issue of concern in low-carbohydrate diets.

Structural Polysaccharides

Organisms build strong materials from structural polysaccharides. For example, the polysaccharide called **cellulose** is a major component of the tough walls that enclose plant cells. On a global scale, plants produce almost 10^{14} kg (100 billion tons) of cellulose per year; it is the most abundant organic compound on Earth. Like starch, cellulose is a polymer of glucose, but the glycosidic linkages in these two polymers differ. The difference is based on the fact that there are actually two slightly different ring structures for glucose **(Figure 5.7a)**. When glucose forms a ring, the hydroxyl group attached to the number 1 carbon is positioned either below or above the plane of the ring. These two ring forms for glucose are called alpha (α) and beta (β), respectively. In starch, all the glucose monomers are in the α configuration **(Figure 5.7b)**, the arrangement we saw in Figures 5.4 and 5.5. In contrast, the glucose monomers of cellulose are all in the β configuration, making every glucose monomer "upside down" with respect to its neighbors **(Figure 5.7c)**.

The differing glycosidic linkages in starch and cellulose give the two molecules distinct three-dimensional shapes. Whereas certain starch molecules are largely helical, a cellulose molecule is straight. Cellulose is never branched, and some hydroxyl groups on its glucose monomers are free to hydrogen-bond with the hydroxyls of other cellulose molecules lying parallel to it. In plant cell walls, parallel cellulose molecules held together in this way are grouped into units called microfibrils **(Figure 5.8)**. These cable-like microfibrils are a strong building material for plants and an important substance for humans because cellulose is the major constituent of paper and the only component of cotton.

Enzymes that digest starch by hydrolyzing its α linkages are unable to hydrolyze the β linkages of cellulose because of the distinctly different shapes of these two molecules. In fact, few organisms possess enzymes that can digest cellulose. Animals, including humans, do not; the cellulose in our food passes through the digestive tract and is eliminated with the feces. Along the way, the cellulose abrades the wall of the digestive tract and stimulates the lining to secrete mucus, which aids in the smooth passage of food through the tract. Thus, although cellulose is not a nutrient for humans, it is an important part of a healthful diet. Most fresh fruits, vegetables, and whole grains are rich in cellulose. On food packages, "insoluble fiber" refers mainly to cellulose.

(a) α and β glucose ring structures. These two interconvertible forms of glucose differ in the placement of the hydroxyl group (highlighted in blue) attached to the number 1 carbon.

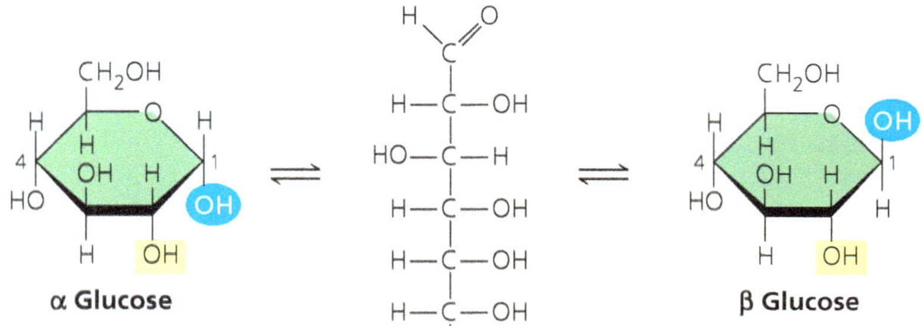

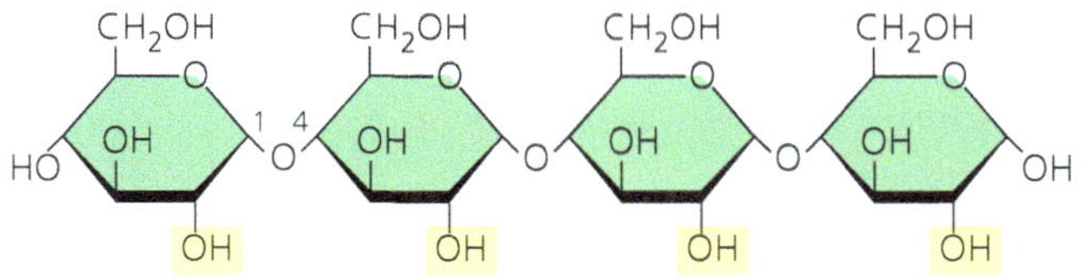

(b) Starch: 1–4 linkage of α glucose monomers. All monomers are in the same orientation. Compare the positions of the —OH groups highlighted in yellow with those in cellulose (c).

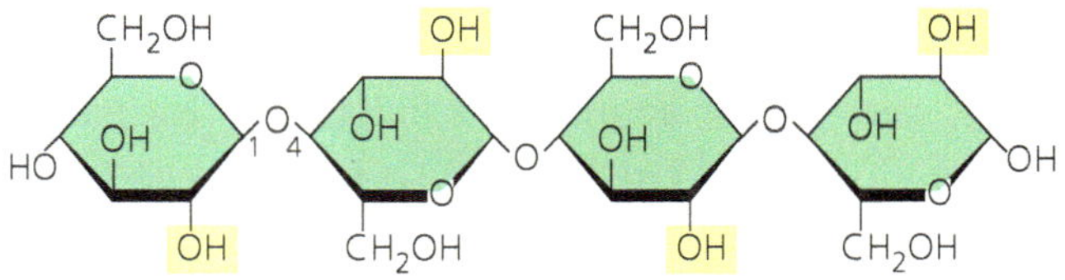

(c) Cellulose: 1–4 linkage of β glucose monomers. In cellulose, every β glucose monomer is upside down with respect to its neighbors.

▲ **Figure 5.7 Starch and cellulose structures.**

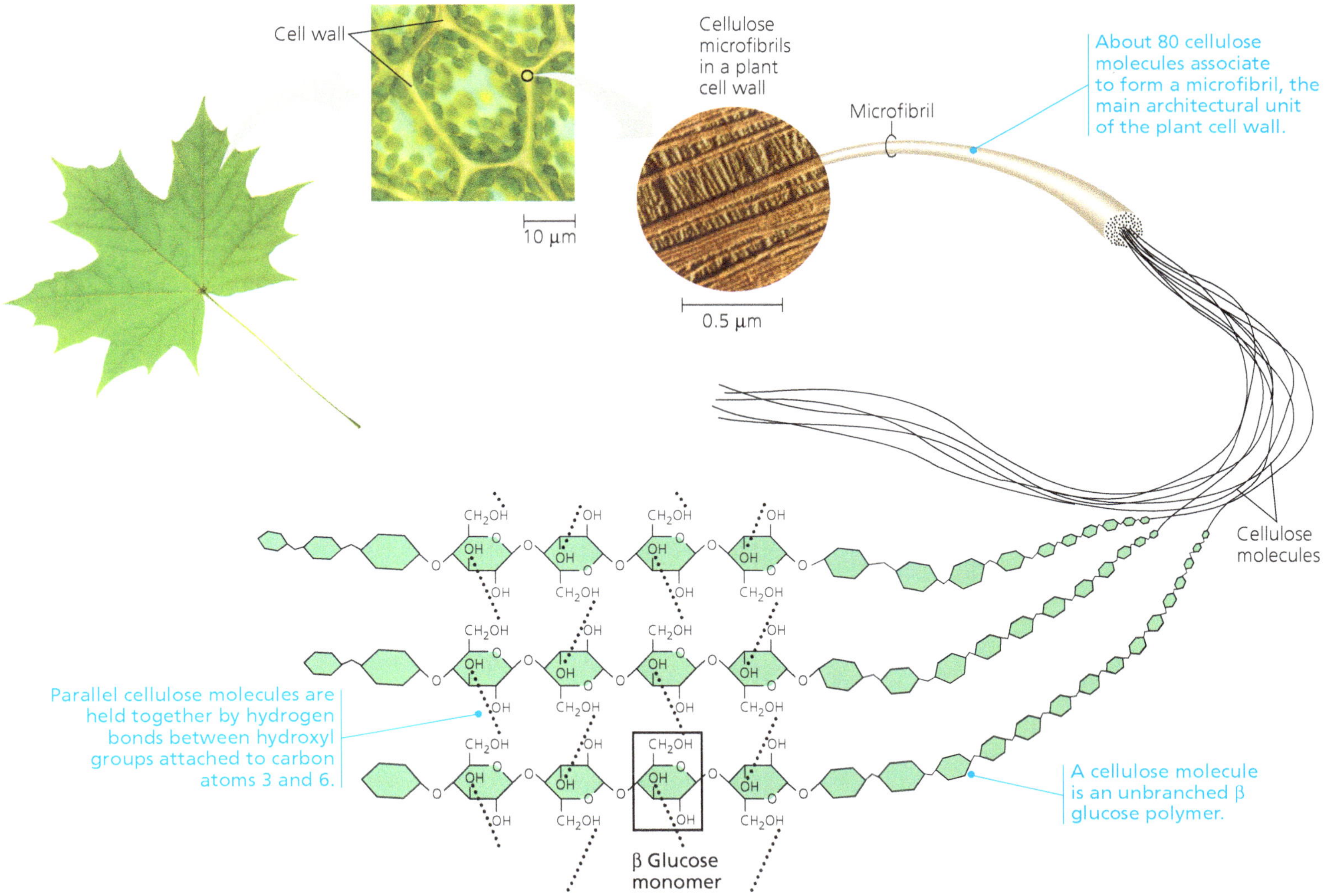

▲ **Figure 5.8 The arrangement of cellulose in plant cell walls.**

Some microorganisms can digest cellulose, breaking it down into glucose monomers. A cow harbors cellulose-digesting prokaryotes and protists in its stomach. These microbes hydrolyze the cellulose of hay and grass and convert the glucose to other compounds that nourish the cow. Similarly, a termite, which is unable to digest cellulose by itself, has prokaryotes or protists living in its gut that can make a meal of wood. Some fungi can also digest cellulose, thereby helping recycle chemical elements within Earth's ecosystems.

Another important structural polysaccharide is **chitin**, the carbohydrate used by arthropods (insects, spiders, crustaceans, and related animals) to build their exoskeletons (**Figure 5.9**). An exoskeleton is a hard case that surrounds the soft parts of an animal. Pure chitin is leathery and flexible, but it becomes hardened when encrusted with calcium carbonate, a salt. Chitin is also found in many fungi, which use this polysaccharide rather than cellulose as the building material for their cell walls. Chitin is similar to cellulose, with β linkages, except that the glucose monomer of chitin has a nitrogen-containing appendage (see Figure 5.9, top right).

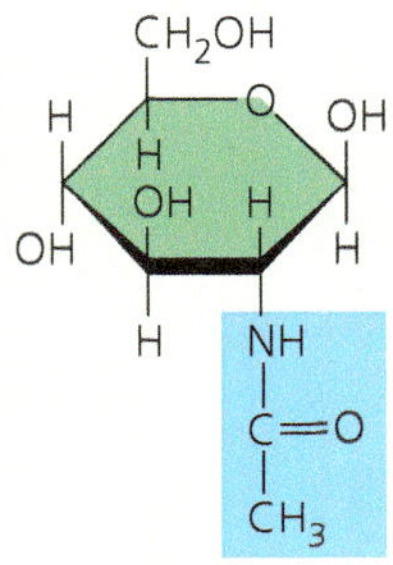

◄ The structure of the chitin monomer

◄ Chitin forms the exoskeleton of arthropods. This cicada is molting, shedding its old exoskeleton and emerging in adult form.

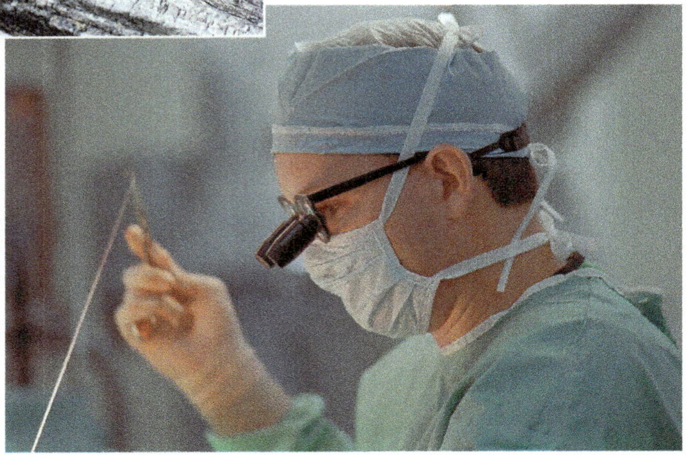

▲ Chitin is used to make a strong and flexible surgical thread that decomposes after the wound or incision heals.

▲ **Figure 5.9 Chitin, a structural polysaccharide.**

CONCEPT CHECK 5.2

1. Write the formula for a monosaccharide that has three carbons.
2. A dehydration reaction joins two glucose molecules to form maltose. The formula for glucose is $C_6H_{12}O_6$. What is the formula for maltose?
3. WHAT IF? After a cow is given antibiotics to treat an infection, a vet gives the animal a drink of "gut culture" containing various prokaryotes. Why is this necessary?

For suggested answers, see Appendix A.

CONCEPT 5.3

Lipids are a diverse group of hydrophobic molecules

Lipids are the one class of large biological molecules that does not include true polymers, and they are generally not big enough to be considered macromolecules. The compounds called **lipids** are grouped together because they share one important trait: They mix poorly, if at all, with water. The hydrophobic behavior of lipids is based on their molecular structure. Although they may have some polar bonds associated with oxygen, lipids consist mostly of hydrocarbon regions. Lipids are varied in form and function. They include waxes and certain pigments, but we will focus on the most biologically important types of lipids: fats, phospholipids, and steroids.

Fats

Although fats are not polymers, they are large molecules assembled from smaller molecules by dehydration reactions. A **fat** is constructed from two kinds of smaller molecules: glycerol and fatty acids (**Figure 5.10a**). Glycerol is an alcohol; each of its three carbons bears a hydroxyl group. A **fatty acid** has a long carbon skeleton, usually 16 or 18 carbon atoms in length. The carbon at one end of the skeleton is part of a carboxyl group, the functional group that gives these molecules the name fatty *acid*. The rest of the skeleton consists of a hydrocarbon chain. The relatively nonpolar C—H bonds in the hydrocarbon chains of fatty acids are the reason fats are hydrophobic. Fats separate from water because the water molecules hydrogen-bond to one another and exclude the fats. This is the reason that vegetable oil (a liquid fat) separates from the aqueous vinegar solution in a bottle of salad dressing.

In making a fat, three fatty acid molecules are each joined to glycerol by an ester linkage, a bond between a hydroxyl group and a carboxyl group. The resulting fat, also called a **triacylglycerol**, thus consists of three fatty acids linked to one glycerol molecule. (Still another name for a fat is

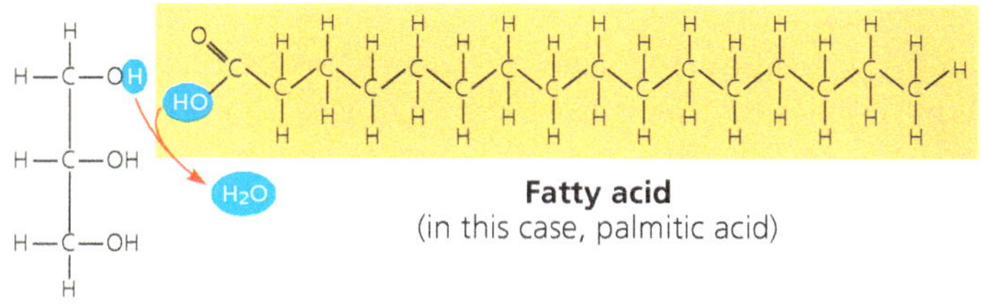

(a) One of three dehydration reactions in the synthesis of a fat

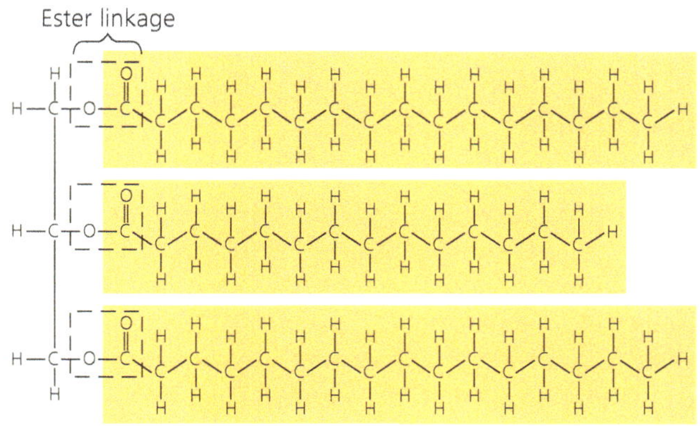

(b) Fat molecule (triacylglycerol)

▲ **Figure 5.10 The synthesis and structure of a fat, or triacylglycerol.** The molecular building blocks of a fat are one molecule of glycerol and three molecules of fatty acids. **(a)** One water molecule is removed for each fatty acid joined to the glycerol. **(b)** A fat molecule with three fatty acid units, two of them identical. The carbons of the fatty acids are arranged zigzag to suggest the actual orientations of the four single bonds extending from each carbon (see Figure 4.3a).

triglyceride, a word often found in the list of ingredients on packaged foods.) The fatty acids in a fat can be the same, or they can be of two or three different kinds, as in **Figure 5.10b**.

The terms *saturated fats* and *unsaturated fats* are commonly used in the context of nutrition **(Figure 5.11)**. These terms refer to the structure of the hydrocarbon chains of the fatty acids. If there are no double bonds between carbon atoms composing a chain, then as many hydrogen atoms as possible are bonded to the carbon skeleton. Such a structure is said to be *saturated* with hydrogen, and the resulting fatty acid therefore called a **saturated fatty acid** **(Figure 5.11a)**. An **unsaturated fatty acid** has one or more double bonds, with one fewer hydrogen atom on each double-bonded carbon. Nearly all double bonds in naturally occurring fatty acids are *cis* double bonds, which cause a kink in the hydrocarbon chain wherever they occur **(Figure 5.11b)**. (See Figure 4.7 to remind yourself about *cis* and *trans* double bonds.)

A fat made from saturated fatty acids is called a saturated fat. Most animal fats are saturated: The hydrocarbon chains of their fatty acids—the "tails" of the fat molecules—lack double bonds, and their flexibility allows the fat molecules to pack together tightly. Saturated animal fats—such as lard and butter—are solid at room temperature. In contrast, the fats of plants

▼ **Figure 5.11 Saturated and unsaturated fats and fatty acids.**

(a) Saturated fat

At room temperature, the molecules of a saturated fat, such as the fat in butter, are packed closely together, forming a solid.

Structural formula of a saturated fat molecule (Each hydrocarbon chain is represented as a zigzag line, where each bend represents a carbon atom and hydrogens are not shown.)

Space-filling model of stearic acid, a saturated fatty acid (red = oxygen, black = carbon, gray = hydrogen)

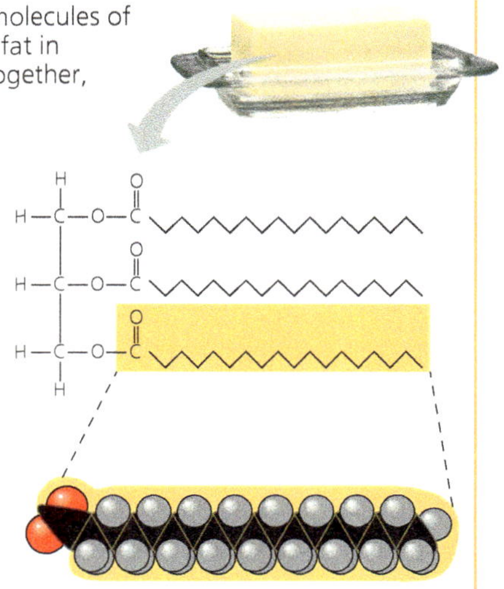

(b) Unsaturated fat

At room temperature, the molecules of an unsaturated fat such as olive oil cannot pack together closely enough to solidify because of the kinks in some of their fatty acid hydrocarbon chains.

Structural formula of an unsaturated fat molecule

Space-filling model of oleic acid, an unsaturated fatty acid

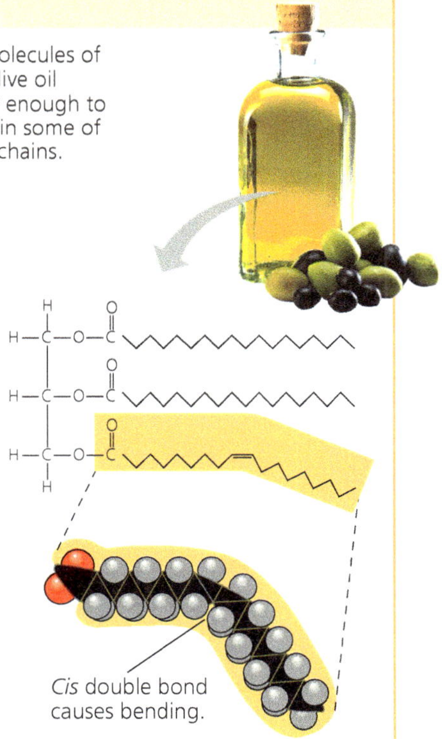

and fishes are generally unsaturated, meaning that they are built of one or more types of unsaturated fatty acids. Usually liquid at room temperature, plant and fish fats are referred to as oils—olive oil and cod liver oil are examples. The kinks where the *cis* double bonds are located prevent the molecules from packing together closely enough to solidify at room temperature. The phrase "hydrogenated vegetable oils" on food labels means that unsaturated fats have been synthetically converted to saturated fats by adding hydrogen. Peanut butter, margarine, and many other products are hydrogenated to prevent lipids from separating out in liquid (oil) form.

A diet rich in saturated fats is one of several factors that may contribute to the cardiovascular disease known as atherosclerosis. In this condition, deposits called plaques develop within the walls of blood vessels, causing inward bulges that impede blood flow and reduce the resilience of the vessels. Recent studies have shown that the process of hydrogenating vegetable oils produces not only saturated fats but also unsaturated fats with *trans* double bonds. These **trans fats** may contribute more than saturated fats to atherosclerosis (see Chapter 42) and other problems. Because trans fats are especially common in baked goods and processed foods, the U.S. Department of Agriculture requires nutritional labels to include information on trans fat content. Some U.S. cities and at least one country—Denmark—have even banned the use of trans fats in restaurants.

Certain unsaturated fatty acids must be supplied in the human diet because they cannot be synthesized in the body. These essential fatty acids include the omega-3 fatty acids, which are required for normal growth in children and appear to protect against cardiovascular disease in adults. Fatty fish and certain nuts and vegetable oils are rich in omega-3 fatty acids (so named because they have a double bond at the third carbon-carbon bond from the end of the hydrocarbon chain).

The major function of fats is energy storage. The hydrocarbon chains of fats are similar to gasoline molecules and just as rich in energy. A gram of fat stores more than twice as much energy as a gram of a polysaccharide, such as starch. Because plants are relatively immobile, they can function with bulky energy storage in the form of starch. (Vegetable oils are generally obtained from seeds, where more compact storage is an asset to the plant.) Animals, however, must carry their energy stores with them, so there is an advantage to having a more compact reservoir of fuel—fat. Humans and other mammals stock their long-term food reserves in adipose cells (see Figure 4.6a), which swell and shrink as fat is deposited and withdrawn from storage. In addition to storing energy, adipose tissue also cushions such vital organs as the kidneys, and a layer of fat beneath the skin insulates the body. This subcutaneous layer is especially thick in whales, seals, and most other marine mammals, protecting them from cold ocean water.

Phospholipids

Cells could not exist without another type of lipid—**phospholipids** (**Figure 5.12**). Phospholipids are essential for cells because they make up cell membranes. Their structure provides a classic example of how form fits function at the molecular level. As shown in Figure 5.12, a phospholipid is similar to a fat molecule but has only two fatty acids attached to glycerol rather than three. The third hydroxyl group of glycerol is joined to a phosphate group, which has a negative electrical charge in the cell. Additional small molecules, which are usually charged or polar, can be linked to the phosphate group to form a variety of phospholipids.

The two ends of phospholipids show different behavior toward water. The hydrocarbon tails are hydrophobic and are excluded from water. However, the phosphate group and its attachments form a hydrophilic head that has an affinity for water. When phospholipids are added to water, they self-assemble into double-layered structures called "bilayers," shielding their hydrophobic portions from water (**Figure 5.13**).

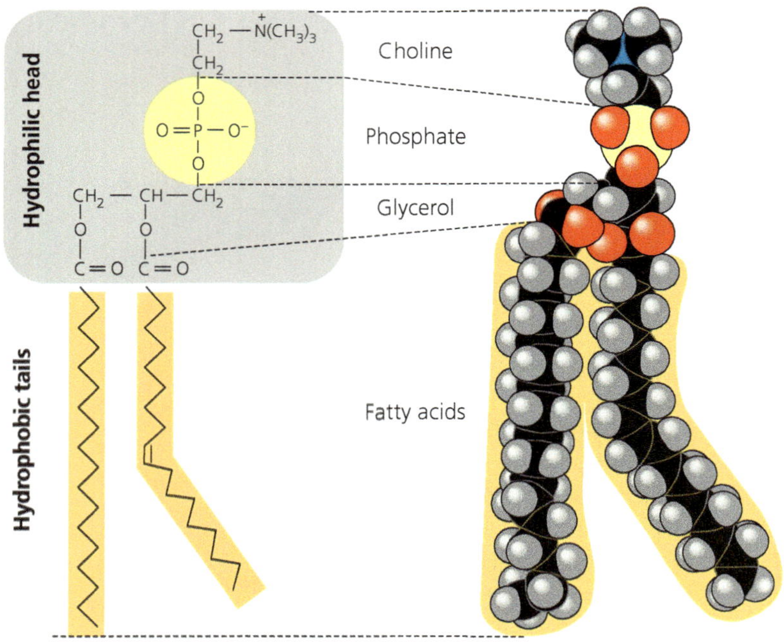

(a) Structural formula **(b) Space-filling model** **(c) Phospholipid symbol**

◄ **Figure 5.12 The structure of a phospholipid.** A phospholipid has a hydrophilic (polar) head and two hydrophobic (nonpolar) tails. Phospholipid diversity is based on differences in the two fatty acids and in the groups attached to the phosphate group of the head. This particular phospholipid, called a phosphatidylcholine, has an attached choline group. The kink in one of its tails is due to a *cis* double bond. Shown here are **(a)** the structural formula, **(b)** the space-filling model (yellow = phosphorus, blue = nitrogen), and **(c)** the symbol for a phospholipid that will appear throughout this book. (In most figures, this symbol will be used to represent a phospholipid with either saturated or unsaturated tails.)

DRAW IT *Draw an oval around the hydrophilic head of the space-filling model.*

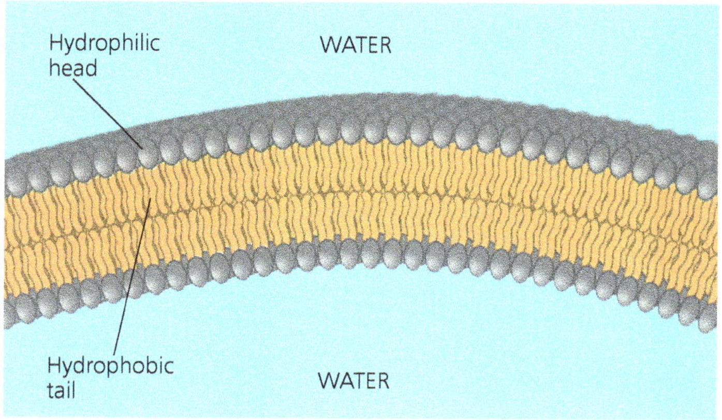

▲ **Figure 5.13 Bilayer structure formed by self-assembly of phospholipids in an aqueous environment.** The phospholipid bilayer shown here is the main fabric of biological membranes. Note that the hydrophilic heads of the phospholipids are in contact with water in this structure, whereas the hydrophobic tails are in contact with each other and remote from water.

At the surface of a cell, phospholipids are arranged in a similar bilayer. The hydrophilic heads of the molecules are on the outside of the bilayer, in contact with the aqueous solutions inside and outside of the cell. The hydrophobic tails point toward the interior of the bilayer, away from the water. The phospholipid bilayer forms a boundary between the cell and its external environment; in fact, cells could not exist without phospholipids.

Steroids

Steroids are lipids characterized by a carbon skeleton consisting of four fused rings. Different steroids, such as cholesterol and the vertebrate sex hormones, are distinguished by the particular chemical groups attached to this ensemble of rings (**Figure 5.14**). **Cholesterol** is a crucial molecule in animals. It is a common component of animal cell membranes and is also the precursor from which other steroids are synthesized. In vertebrates, cholesterol is synthesized in the liver and obtained from the diet. A high level of cholesterol in the blood may contribute to atherosclerosis. In fact, both saturated fats and trans fats exert their negative impact on health by affecting cholesterol levels.

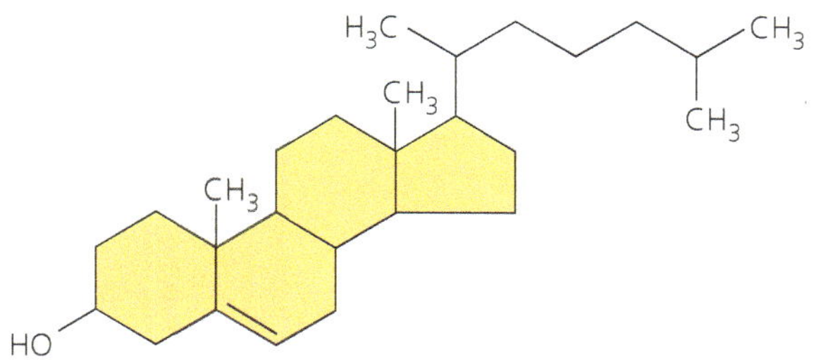

▲ **Figure 5.14 Cholesterol, a steroid.** Cholesterol is the molecule from which other steroids, including the sex hormones, are synthesized. Steroids vary in the chemical groups attached to their four interconnected rings (shown in gold).

MAKE CONNECTIONS *Compare cholesterol with the sex hormones shown in Concept 4.3 on p. 63. Circle the chemical groups that cholesterol has in common with estradiol; put a square around the chemical groups that cholesterol has in common with testosterone.*

CONCEPT CHECK 5.3

1. Compare the structure of a fat (triglyceride) with that of a phospholipid.
2. Why are human sex hormones considered lipids?
3. **WHAT IF?** Suppose a membrane surrounded an oil droplet, as it does in the cells of plant seeds. Describe and explain the form it might take.

For suggested answers, see Appendix A.

CONCEPT 5.4

Proteins include a diversity of structures, resulting in a wide range of functions

Nearly every dynamic function of a living being depends on proteins. In fact, the importance of proteins is underscored by their name, which comes from the Greek word *proteios*, meaning "first," or "primary." Proteins account for more than 50% of the dry mass of most cells, and they are instrumental in almost everything organisms do. Some proteins speed up chemical reactions, while others play a role in defense, storage, transport, cellular communication, movement, or structural support. **Figure 5.15**, on the next page, shows examples of proteins with these functions, which you'll learn more about in later chapters.

Life would not be possible without enzymes, most of which are proteins. Enzymatic proteins regulate metabolism by acting as **catalysts**, chemical agents that selectively speed up chemical reactions without being consumed by the reaction. Because an enzyme can perform its function over and over again, these molecules can be thought of as workhorses that keep cells running by carrying out the processes of life.

A human has tens of thousands of different proteins, each with a specific structure and function; proteins, in fact, are the most structurally sophisticated molecules known. Consistent with their diverse functions, they vary extensively in structure, each type of protein having a unique three-dimensional shape.

Polypeptides

Diverse as proteins are, they are all unbranched polymers constructed from the same set of 20 amino acids. Polymers of amino acids are called **polypeptides**. A **protein** is a biologically functional molecule that consists of one or more polypeptides, each folded and coiled into a specific three-dimensional structure.

▼ Figure 5.15 An overview of protein functions.

Enzymatic proteins

Function: Selective acceleration of chemical reactions

Example: Digestive enzymes catalyze the hydrolysis of bonds in food molecules.

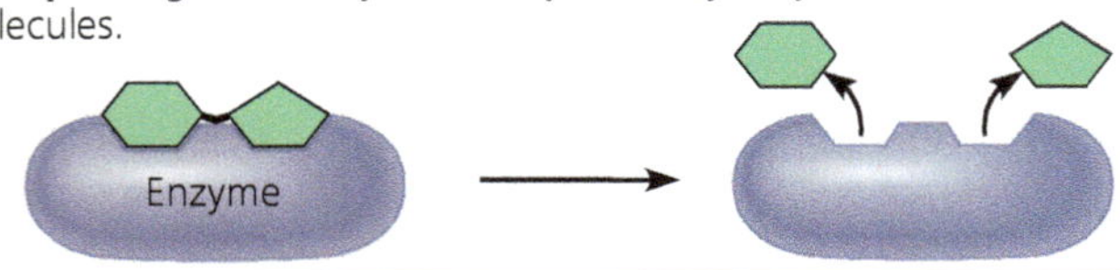

Defensive proteins

Function: Protection against disease

Example: Antibodies inactivate and help destroy viruses and bacteria.

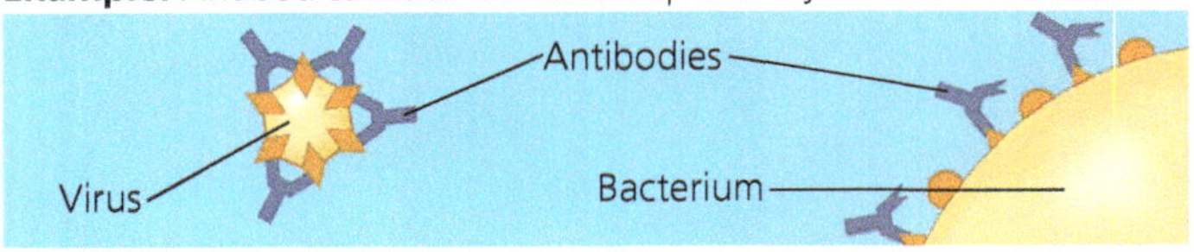

Storage proteins

Function: Storage of amino acids

Examples: Casein, the protein of milk, is the major source of amino acids for baby mammals. Plants have storage proteins in their seeds. Ovalbumin is the protein of egg white, used as an amino acid source for the developing embryo.

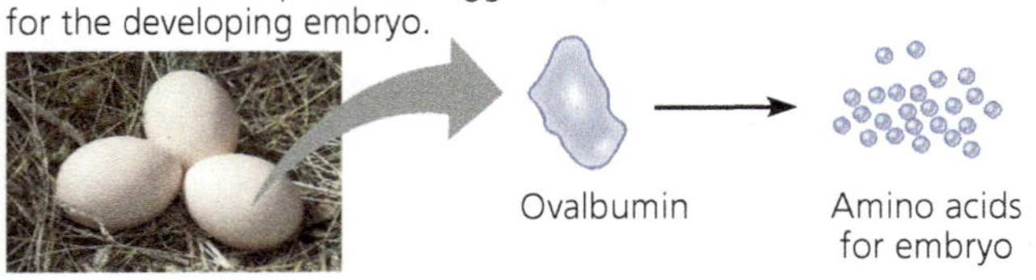

Transport proteins

Function: Transport of substances

Examples: Hemoglobin, the iron-containing protein of vertebrate blood, transports oxygen from the lungs to other parts of the body. Other proteins transport molecules across cell membranes.

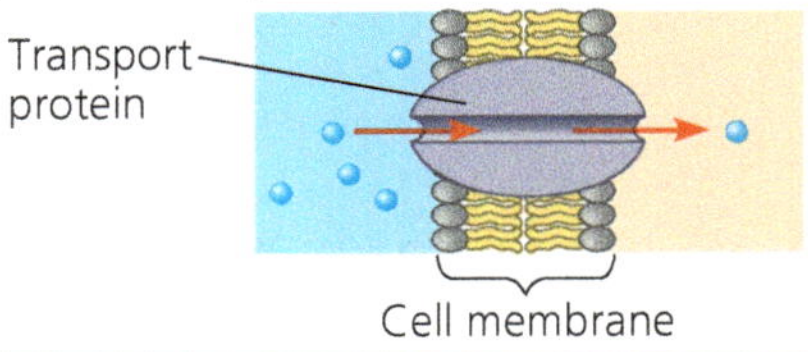

Hormonal proteins

Function: Coordination of an organism's activities

Example: Insulin, a hormone secreted by the pancreas, causes other tissues to take up glucose, thus regulating blood sugar concentration.

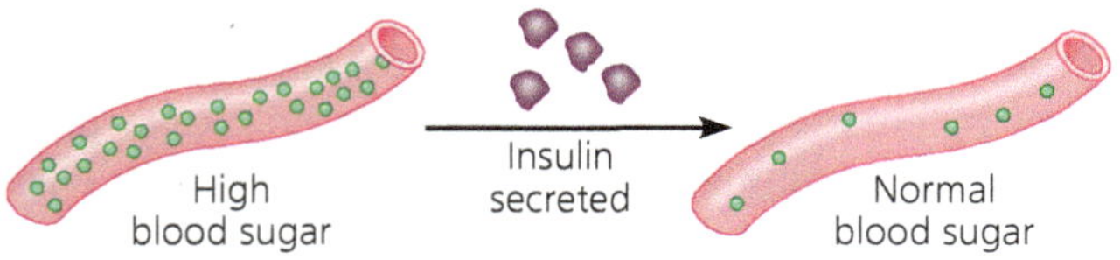

Receptor proteins

Function: Response of cell to chemical stimuli

Example: Receptors built into the membrane of a nerve cell detect signaling molecules released by other nerve cells.

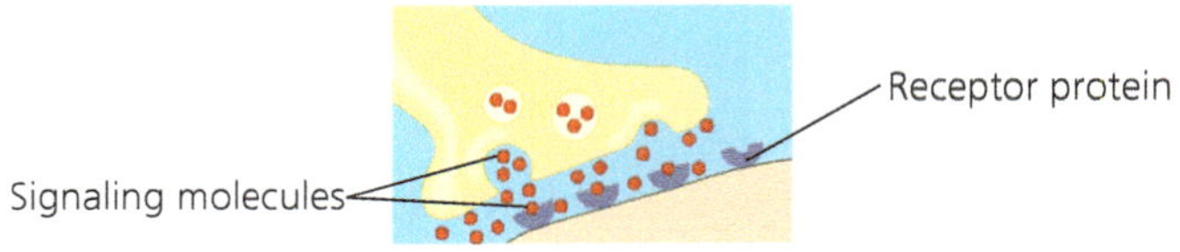

Contractile and motor proteins

Function: Movement

Examples: Motor proteins are responsible for the undulations of cilia and flagella. Actin and myosin proteins are responsible for the contraction of muscles.

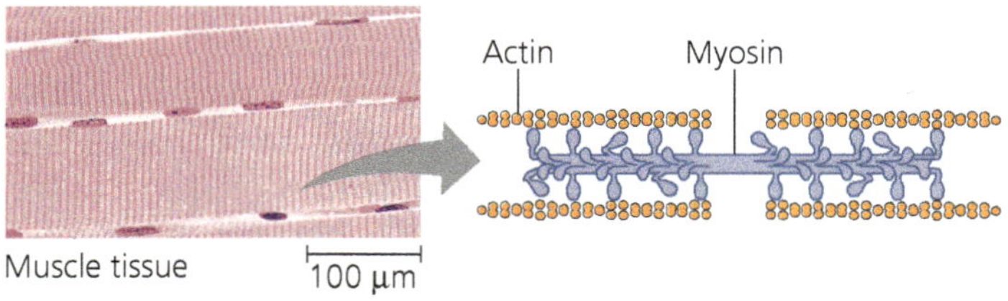

Structural proteins

Function: Support

Examples: Keratin is the protein of hair, horns, feathers, and other skin appendages. Insects and spiders use silk fibers to make their cocoons and webs, respectively. Collagen and elastin proteins provide a fibrous framework in animal connective tissues.

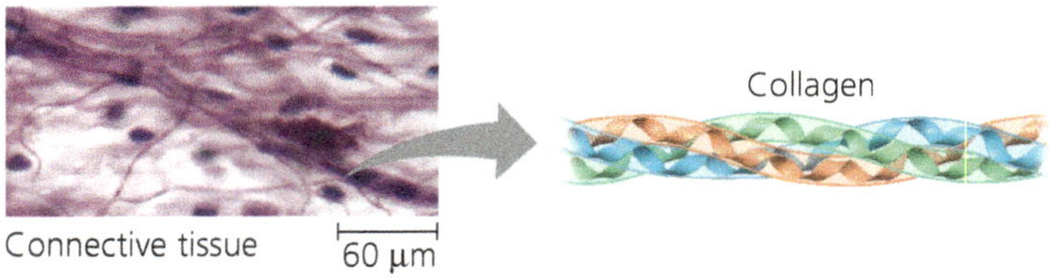

Amino Acid Monomers

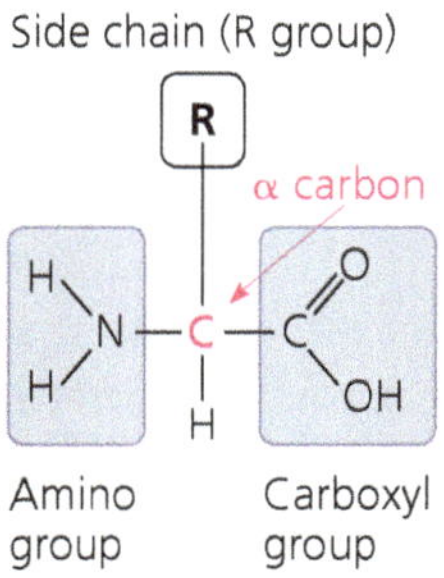

All amino acids share a common structure. An **amino acid** is an organic molecule possessing both an amino group and a carboxyl group (see Figure 4.9). The illustration at the right shows the general formula for an amino acid. At the center of the amino acid is an asymmetric carbon atom called the *alpha (α) carbon*. Its four different partners are an amino group, a carboxyl group, a hydrogen atom, and a variable group symbolized by R. The R group, also called the side chain, differs with each amino acid.

Figure 5.16 shows the 20 amino acids that cells use to build their thousands of proteins. Here the amino groups and carboxyl groups are all depicted in ionized form, the way they usually exist at the pH found in a cell. The side chain (R group) may be as simple as a hydrogen atom, as in the amino acid glycine, or it may be a carbon skeleton with various functional groups attached, as in glutamine.

▼ **Figure 5.16 The 20 amino acids of proteins.** The amino acids are grouped here according to the properties of their side chains (R groups) and shown in their prevailing ionic forms at pH 7.2, the pH within a cell. The three-letter and one-letter abbreviations for the amino acids are in parentheses. All amino acids used in proteins are L enantiomers, the form shown here (see Figure 4.7).

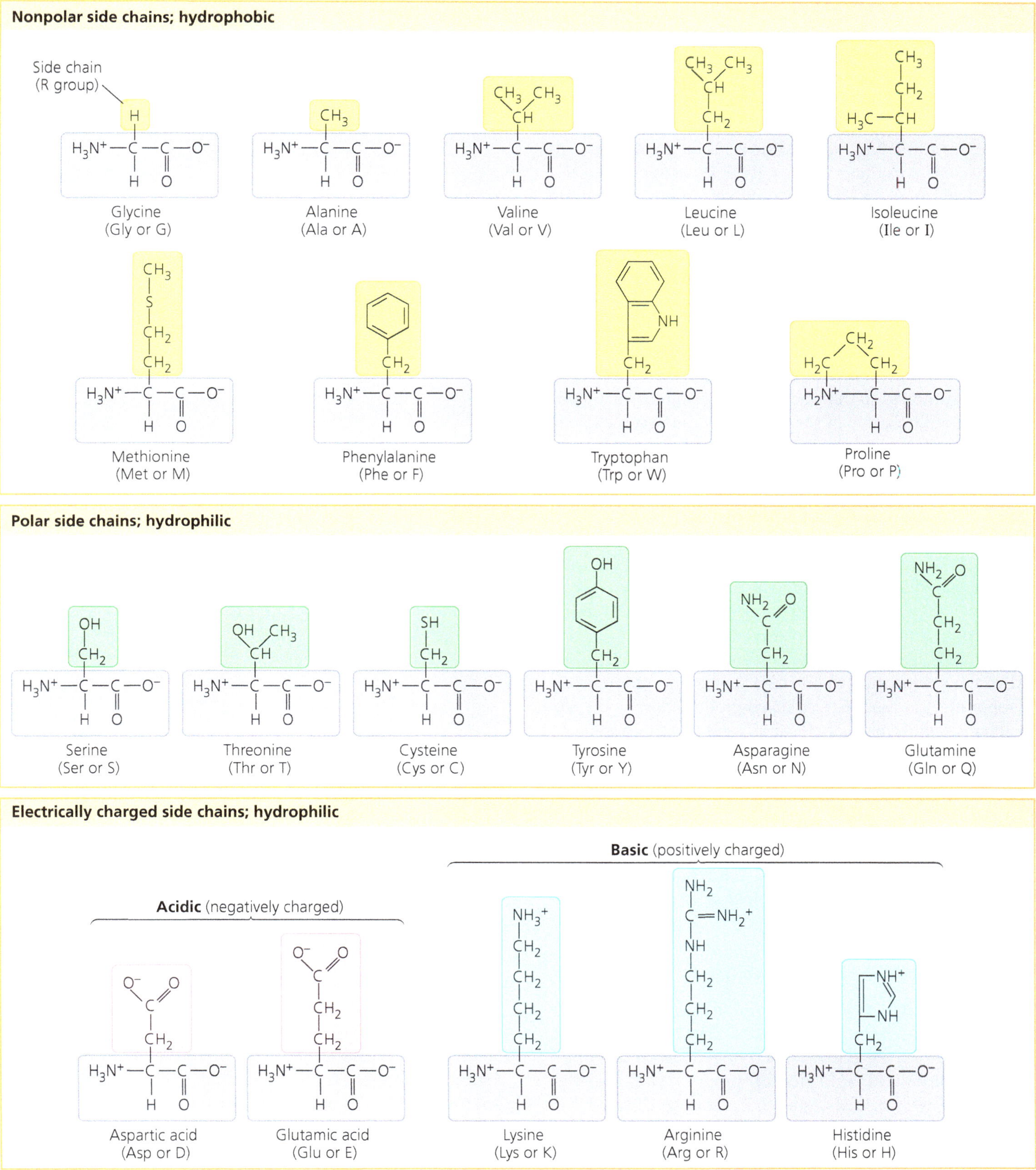

The physical and chemical properties of the side chain determine the unique characteristics of a particular amino acid, thus affecting its functional role in a polypeptide. In Figure 5.16, the amino acids are grouped according to the properties of their side chains. One group consists of amino acids with nonpolar side chains, which are hydrophobic. Another group consists of amino acids with polar side chains, which are hydrophilic. Acidic amino acids are those with side chains that are generally negative in charge owing to the presence of a carboxyl group, which is usually dissociated (ionized) at cellular pH. Basic amino acids have amino groups in their side chains that are generally positive in charge. (Notice that *all* amino acids have carboxyl groups and amino groups; the terms *acidic* and *basic* in this context refer only to groups on the side chains.) Because they are charged, acidic and basic side chains are also hydrophilic.

Amino Acid Polymers

Now that we have examined amino acids, let's see how they are linked to form polymers **(Figure 5.17)**. When two amino acids are positioned so that the carboxyl group of one is adjacent to the amino group of the other, they can become joined by a dehydration reaction, with the removal of a water molecule. The resulting covalent bond is called a **peptide bond**. Repeated over and over, this process yields a polypeptide, a polymer of many amino acids linked by peptide bonds.

The repeating sequence of atoms highlighted in purple in Figure 5.17 is called the polypeptide backbone. Extending from this backbone are the different side chains (R groups) of the amino acids. Polypeptides range in length from a few amino acids to a thousand or more. Each specific polypeptide has a unique linear sequence of amino acids. Note that one end of the polypeptide chain has a free amino group, while the opposite end has a free carboxyl group. Thus, a polypeptide of any length has a single amino end (N-terminus) and a single carboxyl end (C-terminus). In a polypeptide of any significant size, the side chains far outnumber the terminal groups, so the chemical nature of the molecule as a whole is determined by the kind and sequence of the side chains. The immense variety of polypeptides in nature illustrates an important concept introduced earlier—that cells can make many different polymers by linking a limited set of monomers into diverse sequences.

Protein Structure and Function

The specific activities of proteins result from their intricate three-dimensional architecture, the simplest level of which is the sequence of their amino acids. The pioneer in determining the amino acid sequence of proteins was Frederick Sanger, who, with his colleagues at Cambridge University in England, worked on the hormone insulin in the late 1940s and early 1950s. He used agents that break polypeptides at specific places, followed by chemical methods to determine the amino acid sequence in these small fragments. Sanger and his co-workers were able, after years of effort, to reconstruct the complete amino acid sequence of insulin. Since then, most of the steps involved in sequencing a polypeptide have been automated.

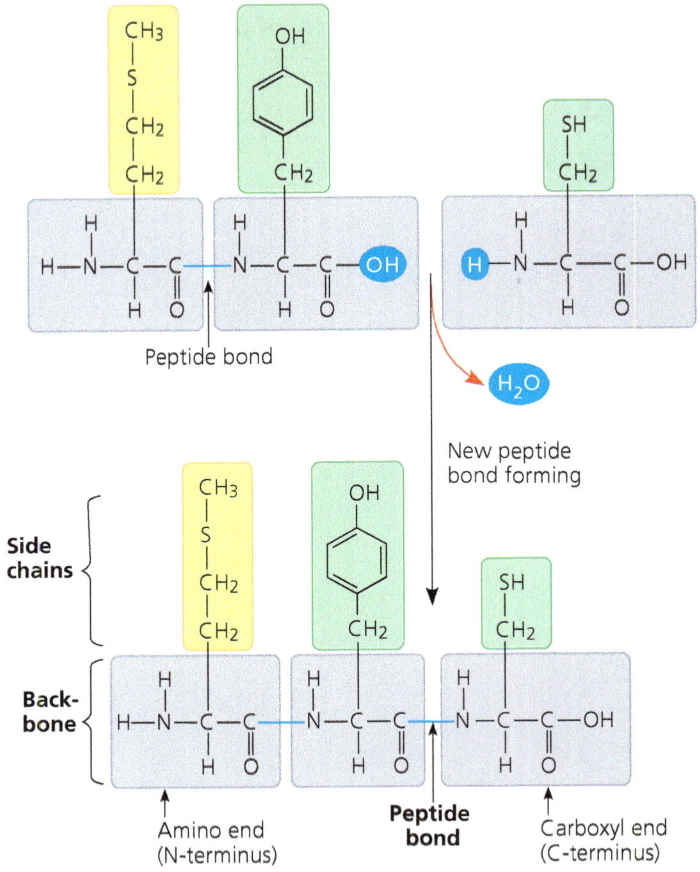

▲ **Figure 5.17 Making a polypeptide chain.** Peptide bonds are formed by dehydration reactions, which link the carboxyl group of one amino acid to the amino group of the next. The peptide bonds are formed one at a time, starting with the amino acid at the amino end (N-terminus). The polypeptide has a repetitive backbone (purple) to which the amino acid side chains (yellow and green) are attached.

DRAW IT *Circle and label the carboxyl and amino groups that will form the new peptide bond.*

Once we have learned the amino acid sequence of a polypeptide, what can it tell us about the three-dimensional structure (commonly referred to simply as the "structure") of the protein and its function? The term *polypeptide* is not synonymous with the term *protein*. Even for a protein consisting of a single polypeptide, the relationship is somewhat analogous to that between a long strand of yarn and a sweater of particular size and shape that can be knit from the yarn. A functional protein is not *just* a polypeptide chain, but one or more polypeptides precisely twisted, folded, and coiled into a molecule of unique shape **(Figure 5.18)**. And it is the amino acid sequence of each polypeptide that determines what

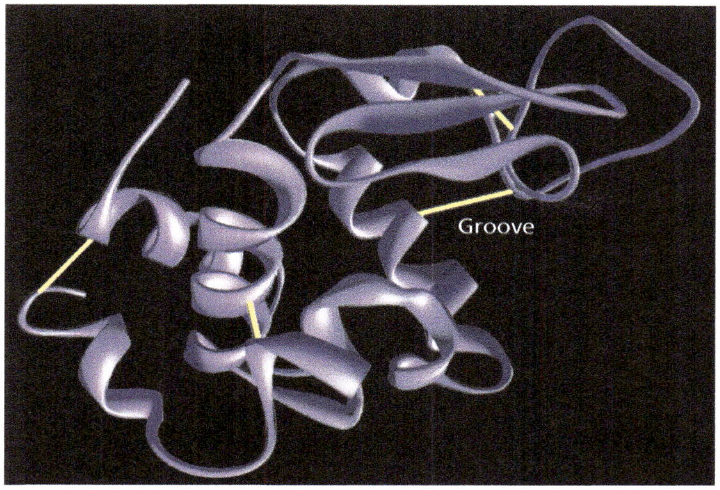

(a) A **ribbon model** shows how the single polypeptide chain folds and coils to form the functional protein. (The yellow lines represent disulfide bridges that stabilize the protein's shape; see Figure 5.20.)

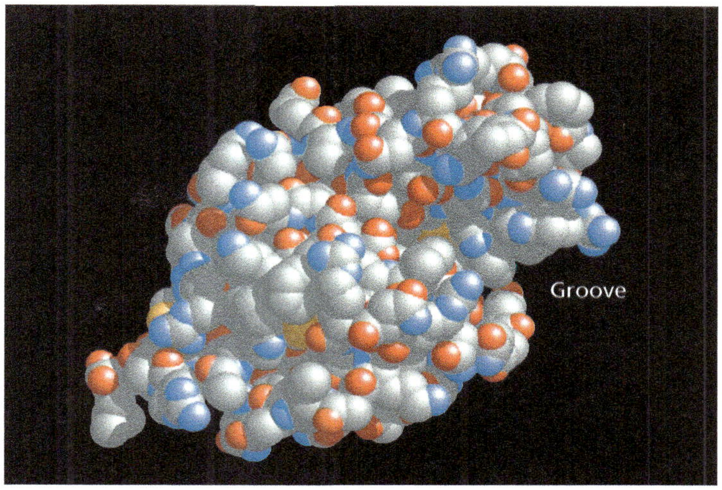

(b) A **space-filling model** shows more clearly the globular shape seen in many proteins, as well as the specific three-dimensional structure unique to lysozyme.

▲ **Figure 5.18 Structure of a protein, the enzyme lysozyme.** Present in our sweat, tears, and saliva, lysozyme is an enzyme that helps prevent infection by binding to and destroying specific molecules on the surface of many kinds of bacteria. The groove is the part of the protein that recognizes and binds to the target molecules on bacterial walls.

three-dimensional structure the protein will have under normal cellular conditions.

When a cell synthesizes a polypeptide, the chain generally folds spontaneously, assuming the functional structure for that protein. This folding is driven and reinforced by the formation of a variety of bonds between parts of the chain, which in turn depends on the sequence of amino acids. Many proteins are roughly spherical (*globular proteins*), while others are shaped like long fibers (*fibrous proteins*). Even within these broad categories, countless variations exist.

A protein's specific structure determines how it works. In almost every case, the function of a protein depends on its ability to recognize and bind to some other molecule. In an especially striking example of the marriage of form and function, **Figure 5.19** shows the exact match of shape between an antibody (a protein in the body) and the particular foreign substance on a flu virus that the antibody binds to and marks for destruction. In Chapter 43, you'll learn more about how the immune system generates antibodies that match the shapes of specific foreign molecules so well. Also, you may recall from Chapter 2 that natural signaling molecules called endorphins bind to specific receptor proteins on the surface of brain cells in humans, producing euphoria and relieving pain. Morphine, heroin, and other opiate drugs are able to mimic endorphins because they all share a similar shape with endorphins and can thus fit into and bind to endorphin receptors in the brain. This fit is very specific, something like a lock and key (see Figure 2.18). Thus, the function of a protein—for instance, the ability of a receptor protein to bind to a particular pain-relieving signaling molecule—is an emergent property resulting from exquisite molecular order.

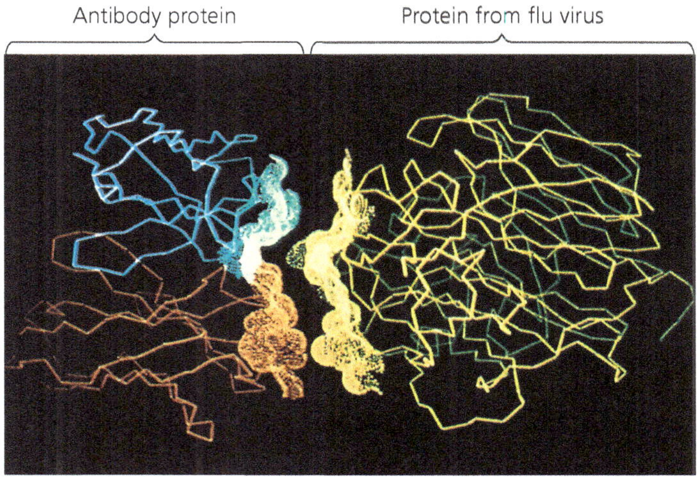

▲ **Figure 5.19 An antibody binding to a protein from a flu virus.** A technique called X-ray crystallography was used to generate a computer model of an antibody protein (blue and orange, left) bound to a flu virus protein (green and yellow, right). Computer software was then used to back the images away from each other, revealing the exact complementarity of shape between the two protein surfaces.

Four Levels of Protein Structure

With the goal of understanding the function of a protein, learning about its structure is often productive. In spite of their great diversity, all proteins share three superimposed levels of structure, known as primary, secondary, and tertiary structure. A fourth level, quaternary structure, arises when a protein consists of two or more polypeptide chains. **Figure 5.20**, on the following two pages, describes these four levels of protein structure. Be sure to study this figure thoroughly before going on to the next section.

▼ **Figure 5.20**

Exploring Levels of Protein Structure

Primary Structure

Linear chain of amino acids

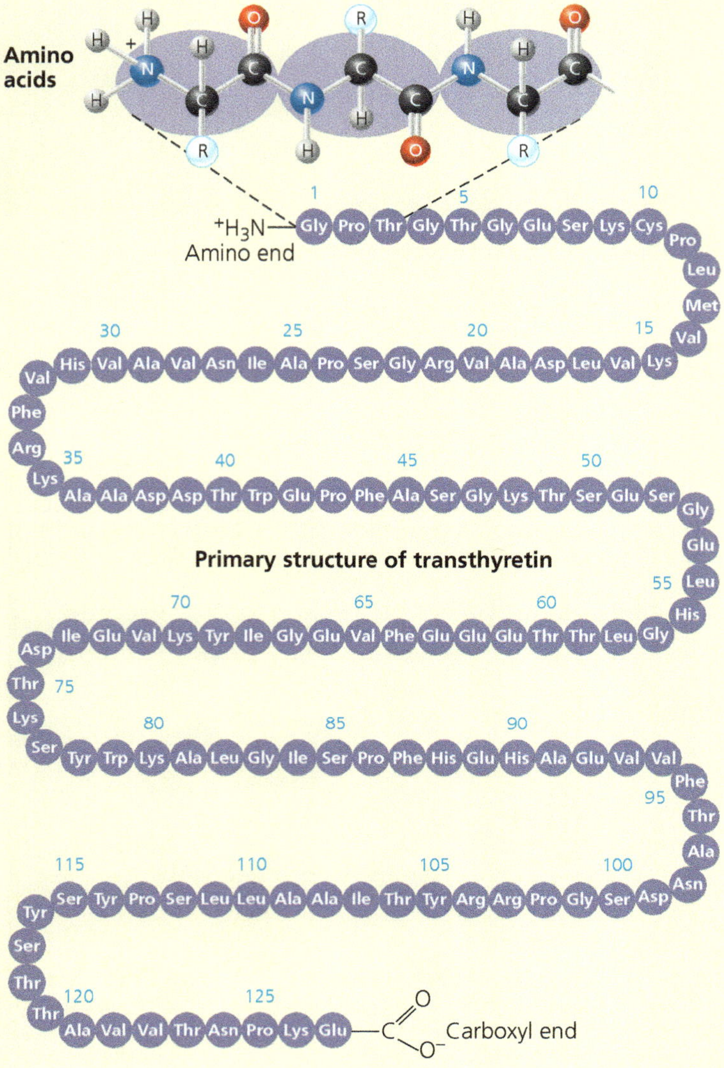

The **primary structure** of a protein is a linked series of amino acids with a unique sequence. As an example, let's consider transthyretin, a globular blood protein that transports vitamin A and one of the thyroid hormones throughout the body. Transthyretin is made up of four identical polypeptide chains, each composed of 127 amino acids. Shown here is one of these chains unraveled for a closer look at its primary structure. Each of the 127 positions along the chain is occupied by one of the 20 amino acids, indicated here by its three-letter abbreviation.

The primary structure is like the order of letters in a very long word. If left to chance, there would be 20^{127} different ways of making a polypeptide chain 127 amino acids long. However, the precise primary structure of a protein is determined not by the random linking of amino acids, but by inherited genetic information. The primary structure in turn dictates secondary and tertiary structure, due to the chemical nature of the backbone and the side chains (R groups) of the amino acids positioned along the chain.

Secondary Structure

Regions stabilized by hydrogen bonds between atoms of the polypeptide backbone

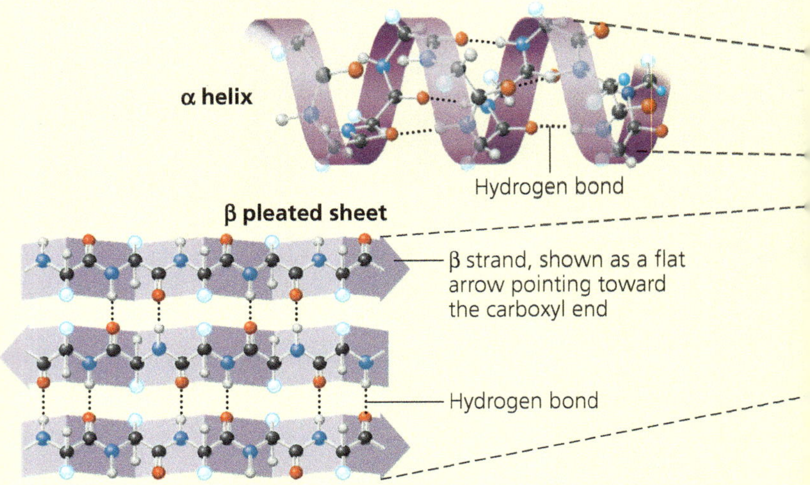

Most proteins have segments of their polypeptide chains repeatedly coiled or folded in patterns that contribute to the protein's overall shape. These coils and folds, collectively referred to as **secondary structure**, are the result of hydrogen bonds between the repeating constituents of the polypeptide backbone (not the amino acid side chains). Within the backbone, the oxygen atoms have a partial negative charge, and the hydrogen atoms attached to the nitrogens have a partial positive charge (see Figure 2.16); therefore, hydrogen bonds can form between these atoms. Individually, these hydrogen bonds are weak, but because they are repeated many times over a relatively long region of the polypeptide chain, they can support a particular shape for that part of the protein.

One such secondary structure is the **α helix**, a delicate coil held together by hydrogen bonding between every fourth amino acid, shown above. Although each transthyretin polypeptide has only one α helix region (see tertiary structure on the next page), other globular proteins have multiple stretches of α helix separated by nonhelical regions (see hemoglobin on the next page). Some fibrous proteins, such as α-keratin, the structural protein of hair, have the α helix formation over most of their length.

The other main type of secondary structure is the **β pleated sheet**. As shown above, in this structure two or more strands of the polypeptide chain lying side by side (called β strands) are connected by hydrogen bonds between parts of the two parallel polypeptide backbones. β pleated sheets make up the core of many globular proteins, as is the case for transthyretin (see tertiary structure on the next page), and dominate some fibrous proteins, including the silk protein of a spider's web. The teamwork of so many hydrogen bonds makes each spider silk fiber stronger than a steel strand of the same weight.

▼ Spiders secrete silk fibers made of a structural protein containing β pleated sheets, which allow the spider web to stretch and recoil.

Tertiary Structure

Three-dimensional shape stabilized by interactions between side chains

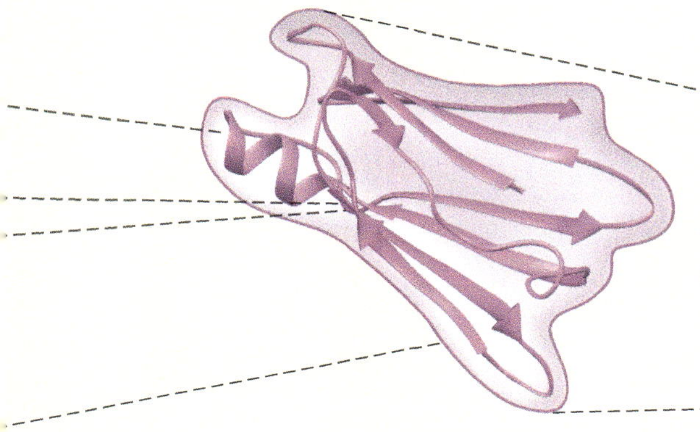

Superimposed on the patterns of secondary structure is a protein's tertiary structure, shown above in a ribbon model of the transthyretin polypeptide. While secondary structure involves interactions between backbone constituents, **tertiary structure** is the overall shape of a polypeptide resulting from interactions between the side chains (R groups) of the various amino acids. One type of interaction that contributes to tertiary structure is—somewhat misleadingly—called a **hydrophobic interaction**. As a polypeptide folds into its functional shape, amino acids with hydrophobic (nonpolar) side chains usually end up in clusters at the core of the protein, out of contact with water. Thus, a "hydrophobic interaction" is actually caused by the exclusion of nonpolar substances by water molecules. Once nonpolar amino acid side chains are close together, van der Waals interactions help hold them together. Meanwhile, hydrogen bonds between polar side chains and ionic bonds between positively and negatively charged side chains also help stabilize tertiary structure. These are all weak interactions in the aqueous cellular environment, but their cumulative effect helps give the protein a unique shape.

Covalent bonds called **disulfide bridges** may further reinforce the shape of a protein. Disulfide bridges form where two cysteine monomers, which have sulfhydryl groups (—SH) on their side chains (see Figure 4.9), are brought close together by the folding of the protein. The sulfur of one cysteine bonds to the sulfur of the second, and the disulfide bridge (—S—S—) rivets parts of the protein together (see yellow lines in Figure 5.18a). All of these different kinds of interactions can contribute to the tertiary structure of a protein, as shown here in a small part of a hypothetical protein:

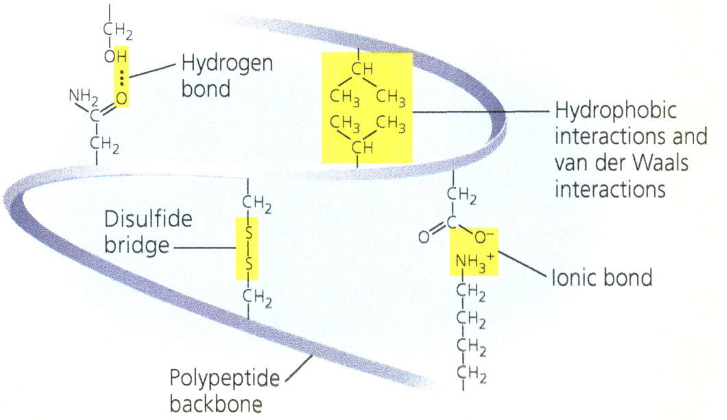

Quaternary Structure

Association of multiple polypeptides, forming a functional protein

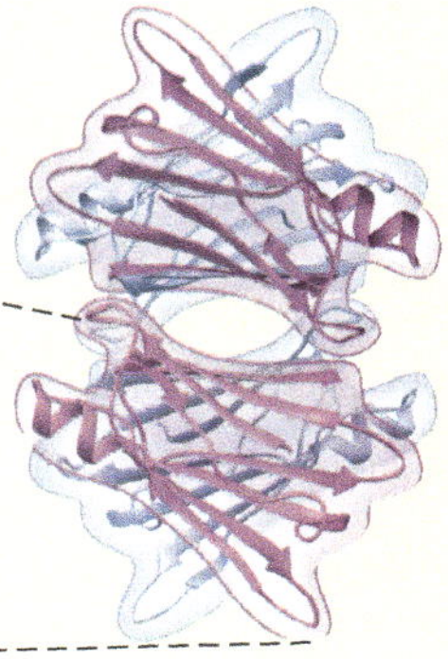

Some proteins consist of two or more polypeptide chains aggregated into one functional macromolecule. **Quaternary structure** is the overall protein structure that results from the aggregation of these polypeptide subunits. For example, shown above is the complete globular transthyretin protein, made up of its four polypeptides.

Another example is collagen, shown below, which is a fibrous protein that has three identical helical polypeptides intertwined into a larger triple helix, giving the long fibers great strength. This suits collagen fibers to their function as the girders of connective tissue in skin, bone, tendons, ligaments, and other body parts. (Collagen accounts for 40% of the protein in a human body.)

Collagen

Hemoglobin, the oxygen-binding protein of red blood cells shown below, is another example of a globular protein with quaternary structure. It consists of four polypeptide subunits, two of one kind (α) and two of another kind (β). Both α and β subunits consist primarily of α-helical secondary structure. Each subunit has a nonpolypeptide component, called heme, with an iron atom that binds oxygen.

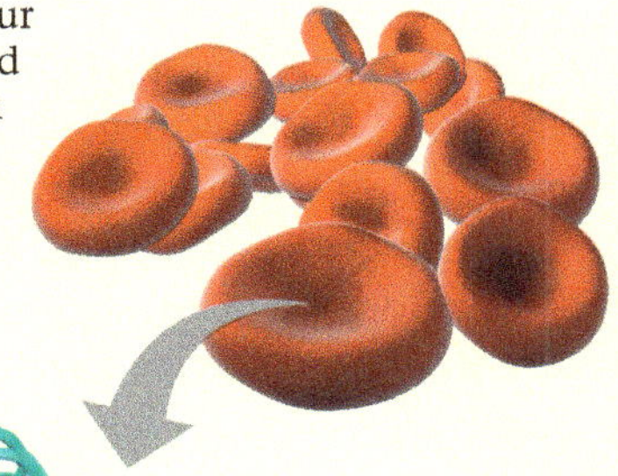

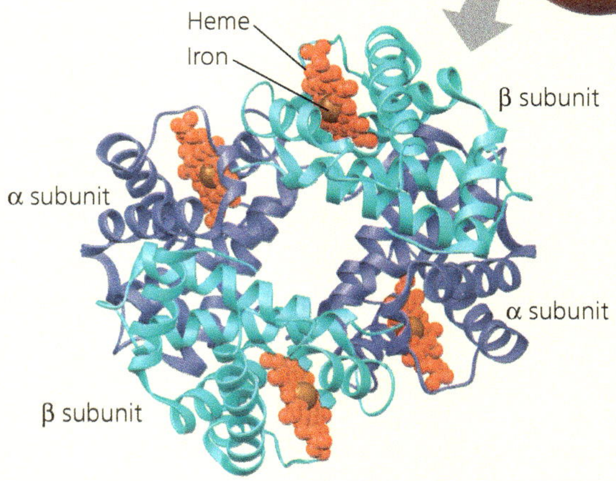

Hemoglobin

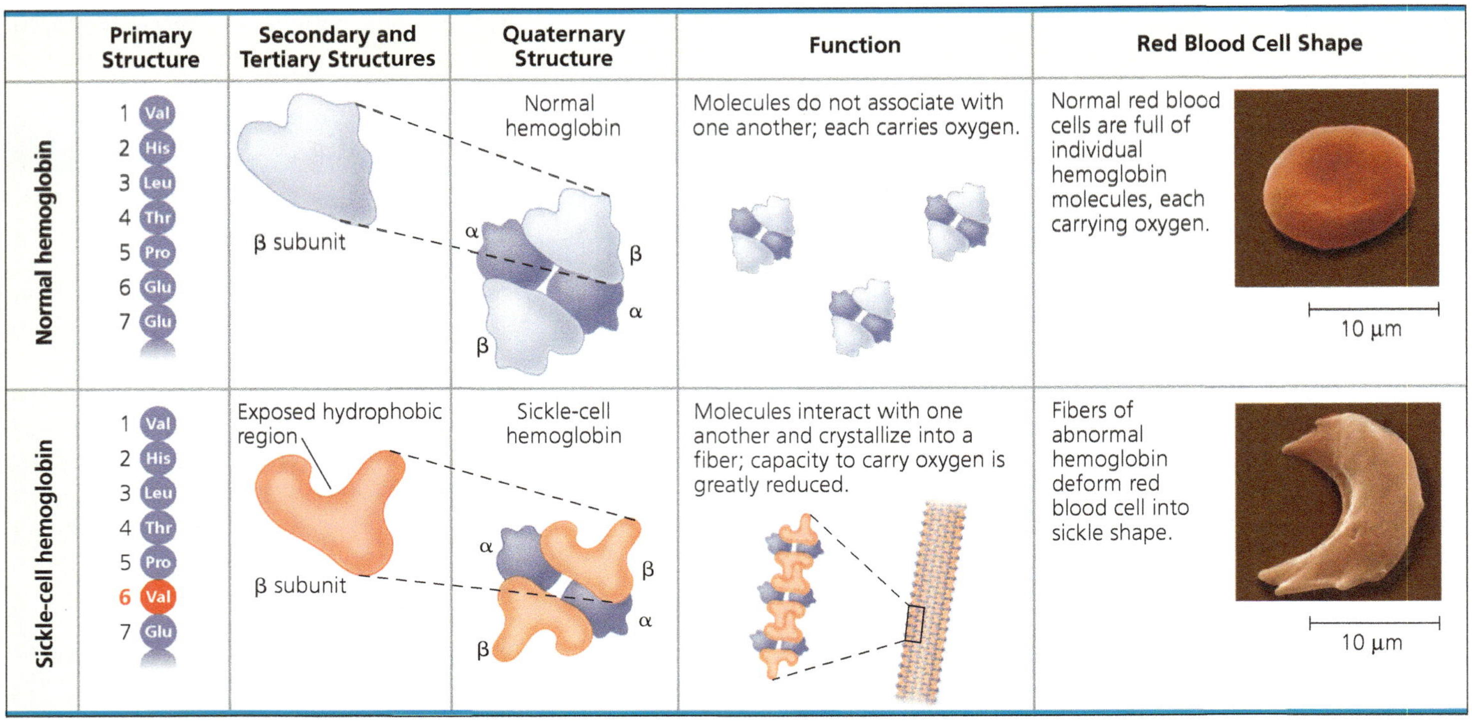

▲ **Figure 5.21 A single amino acid substitution in a protein causes sickle-cell disease.**

MAKE CONNECTIONS *Considering the chemical characteristics of the amino acids valine and glutamic acid (see Figure 5.16), propose a possible explanation for the dramatic effect on protein function that occurs when valine is substituted for glutamic acid.*

Sickle-Cell Disease: A Change in Primary Structure

Even a slight change in primary structure can affect a protein's shape and ability to function. For instance, **sickle-cell disease**, an inherited blood disorder, is caused by the substitution of one amino acid (valine) for the normal one (glutamic acid) at a particular position in the primary structure of hemoglobin, the protein that carries oxygen in red blood cells. Normal red blood cells are disk-shaped, but in sickle-cell disease, the abnormal hemoglobin molecules tend to crystallize, deforming some of the cells into a sickle shape **(Figure 5.21)**. A person with the disease has periodic "sickle-cell crises" when the angular cells clog tiny blood vessels, impeding blood flow. The toll taken on such patients is a dramatic example of how a simple change in protein structure can have devastating effects on protein function.

What Determines Protein Structure?

You've learned that a unique shape endows each protein with a specific function. But what are the key factors determining protein structure? You already know most of the answer: A polypeptide chain of a given amino acid sequence can spontaneously arrange itself into a three-dimensional shape determined and maintained by the interactions responsible for secondary and tertiary structure. This folding normally occurs as the protein is being synthesized in the crowded environment within a cell, aided by other proteins. However, protein structure also depends on the physical and chemical conditions of the protein's environment. If the pH, salt concentration, temperature, or other aspects of its environment are altered, the weak chemical bonds and interactions within a protein may be destroyed, causing the protein to unravel and lose its native shape, a change called **denaturation** **(Figure 5.22)**. Because it is misshapen, the denatured protein is biologically inactive.

Most proteins become denatured if they are transferred from an aqueous environment to a nonpolar solvent, such as

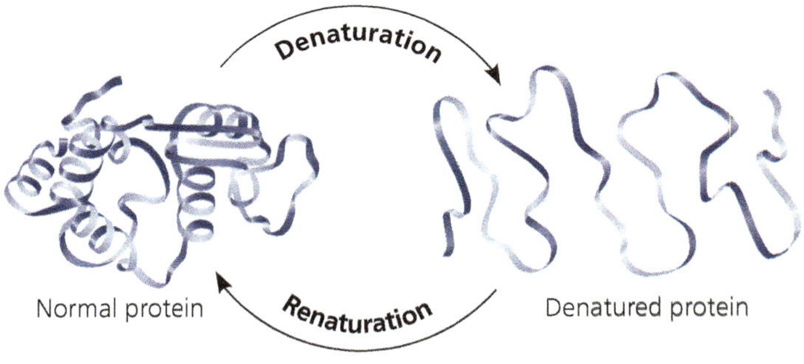

▲ **Figure 5.22 Denaturation and renaturation of a protein.** High temperatures or various chemical treatments will denature a protein, causing it to lose its shape and hence its ability to function. If the denatured protein remains dissolved, it can often renature when the chemical and physical aspects of its environment are restored to normal.

ether or chloroform; the polypeptide chain refolds so that its hydrophobic regions face outward toward the solvent. Other denaturation agents include chemicals that disrupt the hydrogen bonds, ionic bonds, and disulfide bridges that maintain a protein's shape. Denaturation can also result from excessive heat, which agitates the polypeptide chain enough to overpower the weak interactions that stabilize the structure. The white of an egg becomes opaque during cooking because the denatured proteins are insoluble and solidify. This also explains why excessively high fevers can be fatal: Proteins in the blood can denature at very high body temperatures.

When a protein in a test-tube solution has been denatured by heat or chemicals, it can sometimes return to its functional shape when the denaturing agent is removed. We can conclude that the information for building specific shape is intrinsic to the protein's primary structure. The sequence of amino acids determines the protein's shape—where an α helix can form, where β pleated sheets can exist, where disulfide bridges are located, where ionic bonds can form, and so on. But how does protein folding occur in the cell?

Protein Folding in the Cell

Biochemists now know the amino acid sequence for more than 10 million proteins and the three-dimensional shape for more than 20,000. Researchers have tried to correlate the primary structure of many proteins with their three-dimensional structure to discover the rules of protein folding. Unfortunately, however, the protein-folding process is not that simple. Most proteins probably go through several intermediate structures on their way to a stable shape, and looking at the mature structure does not reveal the stages of folding required to achieve that form. However, biochemists have developed methods for tracking a protein through such stages.

Crucial to the folding process are **chaperonins** (also called chaperone proteins), protein molecules that assist in the proper folding of other proteins (**Figure 5.23**). Chaperonins do not specify the final structure of a polypeptide. Instead, they keep the new polypeptide segregated from "bad influences" in the cytoplasmic environment while it folds spontaneously. The chaperonin shown in Figure 5.23, from the bacterium *E. coli*, is a giant multiprotein complex shaped like a hollow cylinder. The cavity provides a shelter for folding polypeptides. In the past decade, researchers have discovered molecular systems that interact with chaperonins and check whether proper folding has occurred. Such systems either refold the misfolded proteins correctly or mark them for destruction.

Misfolding of polypeptides is a serious problem in cells. Many diseases, such as Alzheimer's, Parkinson's, and mad cow disease, are associated with an accumulation of misfolded proteins. In fact, misfolded versions of the transthyretin protein featured in Figure 5.20 have been implicated in several diseases, including one form of senile dementia.

Even when scientists have a correctly folded protein in hand, determining its exact three-dimensional structure is not simple, for a single protein molecule has thousands of atoms. The first 3-D structures were worked out in 1959 for hemoglobin and a related protein. The method that made these feats possible was **X-ray crystallography**, which has since been used to determine the 3-D structure of many other proteins. In a recent example, Roger Kornberg and his colleagues at Stanford University used this method to elucidate the structure of RNA polymerase, an enzyme that plays a crucial role in the expression of genes (**Figure 5.24**, on the next page). Another method for analyzing protein structure is nuclear magnetic resonance (NMR) spectroscopy, which does not require protein crystallization. A still newer approach employs bioinformatics (see Chapter 1) to predict the 3-D structure of polypeptides from their amino acid sequence. X-ray crystallography, NMR spectroscopy, and bioinformatics are complementary approaches to understanding protein structure and function.

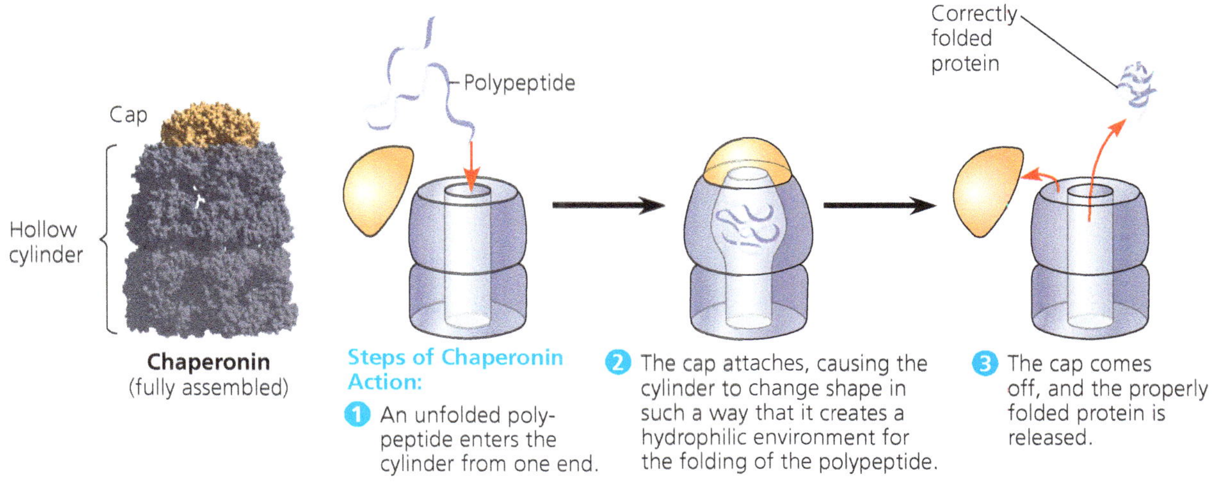

▶ **Figure 5.23 A chaperonin in action.** The computer graphic (left) shows a large chaperonin protein complex from the bacterium *E. coli*. It has an interior space that provides a shelter for the proper folding of newly made polypeptides. The complex consists of two proteins: One protein is a hollow cylinder; the other is a cap that can fit on either end.

▼ **Figure 5.24** INQUIRY

What can the 3-D shape of the enzyme RNA polymerase II tell us about its function?

EXPERIMENT In 2006, Roger Kornberg was awarded the Nobel Prize in Chemistry for using X-ray crystallography to determine the 3-D shape of RNA polymerase II, which binds to the DNA double helix and synthesizes RNA. After crystallizing a complex of all three components, Kornberg and his colleagues aimed an X-ray beam through the crystal. The atoms of the crystal diffracted (bent) the X-rays into an orderly array that a digital detector recorded as a pattern of spots called an X-ray diffraction pattern.

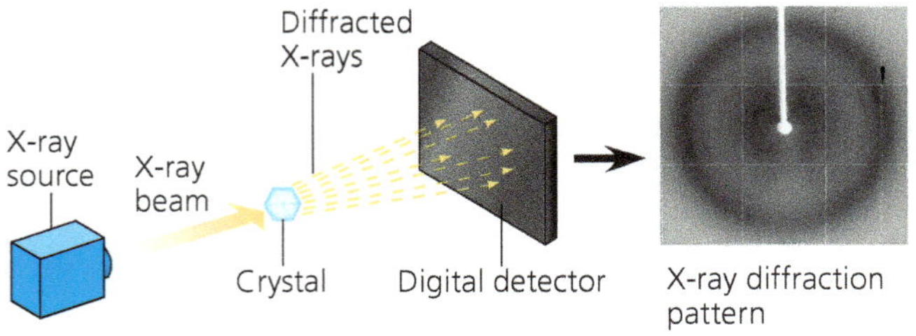

RESULTS Using data from X-ray diffraction patterns, as well as the amino acid sequence determined by chemical methods, Kornberg and colleagues built a 3-D model of the complex with the help of computer software.

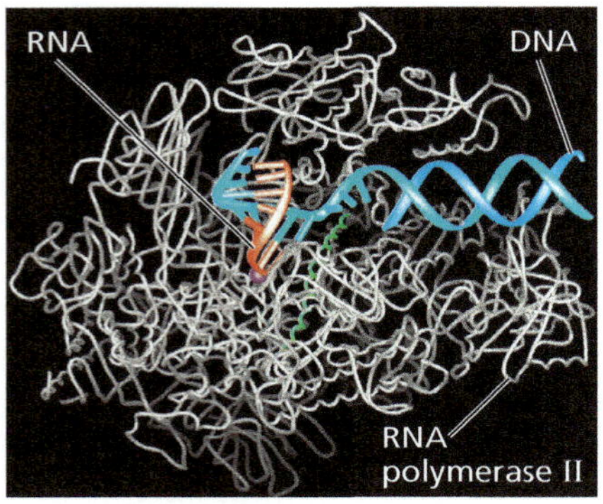

CONCLUSION By analyzing their model, the researchers developed a hypothesis about the functions of different regions of RNA polymerase II. For example, the region above the DNA may act as a clamp that holds the nucleic acids in place. (You'll learn more about this enzyme in Chapter 17.)

SOURCE A. L. Gnatt et al., Structural basis of transcription: an RNA polymerase II elongation complex at 3.3Å, *Science* 292:1876–1882 (2001).

WHAT IF? If you were an author of the paper and were describing the model, what type of protein structure would you call the small polypeptide spirals in RNA polymerase II?

CONCEPT CHECK 5.4

1. Why does a denatured protein no longer function normally?
2. What parts of a polypeptide participate in the bonds that hold together secondary structure? Tertiary structure?
3. **WHAT IF?** Where would you expect a polypeptide region that is rich in the amino acids valine, leucine, and isoleucine to be located in the folded polypeptide? Explain.

For suggested answers, see Appendix A.

CONCEPT 5.5

Nucleic acids store, transmit, and help express hereditary information

If the primary structure of polypeptides determines a protein's shape, what determines primary structure? The amino acid sequence of a polypeptide is programmed by a discrete unit of inheritance known as a **gene**. Genes consist of DNA, which belongs to the class of compounds called nucleic acids. **Nucleic acids** are polymers made of monomers called *nucleotides*.

The Roles of Nucleic Acids

The two types of nucleic acids, **deoxyribonucleic acid (DNA)** and **ribonucleic acid (RNA)**, enable living organisms to reproduce their complex components from one generation to the next. Unique among molecules, DNA provides directions for its own replication. DNA also directs RNA synthesis and, through RNA, controls protein synthesis (**Figure 5.25**).

DNA is the genetic material that organisms inherit from their parents. Each chromosome contains one long DNA molecule, usually carrying several hundred or more genes. When a cell reproduces itself by dividing, its DNA molecules are copied and passed along from one generation of cells to the next. Encoded in the structure of DNA is the information that

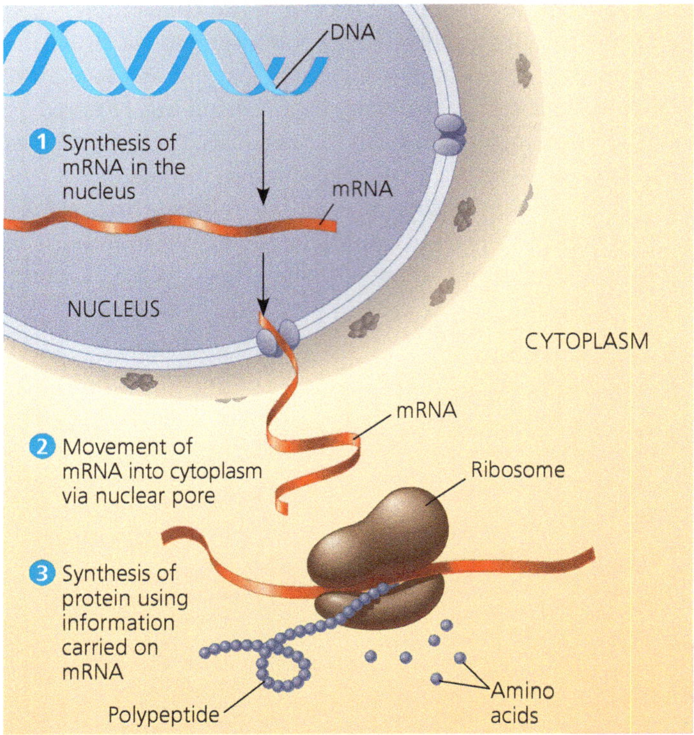

▲ **Figure 5.25 DNA → RNA → protein.** In a eukaryotic cell, DNA in the nucleus programs protein production in the cytoplasm by dictating synthesis of messenger RNA (mRNA). (The cell nucleus is actually much larger relative to the other elements of this figure.)

programs all the cell's activities. The DNA, however, is not directly involved in running the operations of the cell, any more than computer software by itself can print a bank statement or read the bar code on a box of cereal. Just as a printer is needed to print out a statement and a scanner is needed to read a bar code, proteins are required to implement genetic programs. The molecular hardware of the cell—the tools for biological functions—consists mostly of proteins. For example, the oxygen carrier in red blood cells is the protein hemoglobin, not the DNA that specifies its structure.

How does RNA, the other type of nucleic acid, fit into gene expression, the flow of genetic information from DNA to proteins? Each gene along a DNA molecule directs synthesis of a type of RNA called *messenger RNA* (*mRNA*). The mRNA molecule interacts with the cell's protein-synthesizing machinery to direct production of a polypeptide, which folds into all or part of a protein. We can summarize the flow of genetic information as DNA → RNA → protein (see Figure 5.25). The sites of protein synthesis are tiny structures called ribosomes. In a eukaryotic cell, ribosomes are in the cytoplasm, but DNA resides in the nucleus. Messenger RNA conveys genetic instructions for building proteins from the nucleus to the cytoplasm. Prokaryotic cells lack nuclei but still use mRNA to convey a message from the DNA to ribosomes and other cellular equipment that translate the coded information into amino acid sequences. In recent years, the spotlight has been turned on other, previously unknown types of RNA that play many other roles in the cell. As is so often true in biology, the story is still being written! You'll hear more about the newly discovered functions of RNA molecules in Chapter 18.

The Components of Nucleic Acids

Nucleic acids are macromolecules that exist as polymers called **polynucleotides** (**Figure 5.26a**). As indicated by the name, each polynucleotide consists of monomers called **nucleotides**. A nucleotide, in general, is composed of three parts: a nitrogen-containing (nitrogenous) base, a five-carbon sugar (a pentose), and one or more phosphate groups (**Figure 5.26b**). In a polynucleotide, each monomer has only one phosphate group. The portion of a nucleotide without any phosphate groups is called a *nucleoside*.

To build a nucleotide, let's first consider the nitrogenous bases (**Figure 5.26c**). Each nitrogenous base has one or two rings that include nitrogen atoms. (They are called nitrogenous *bases* because the nitrogen atoms tend to take up H^+

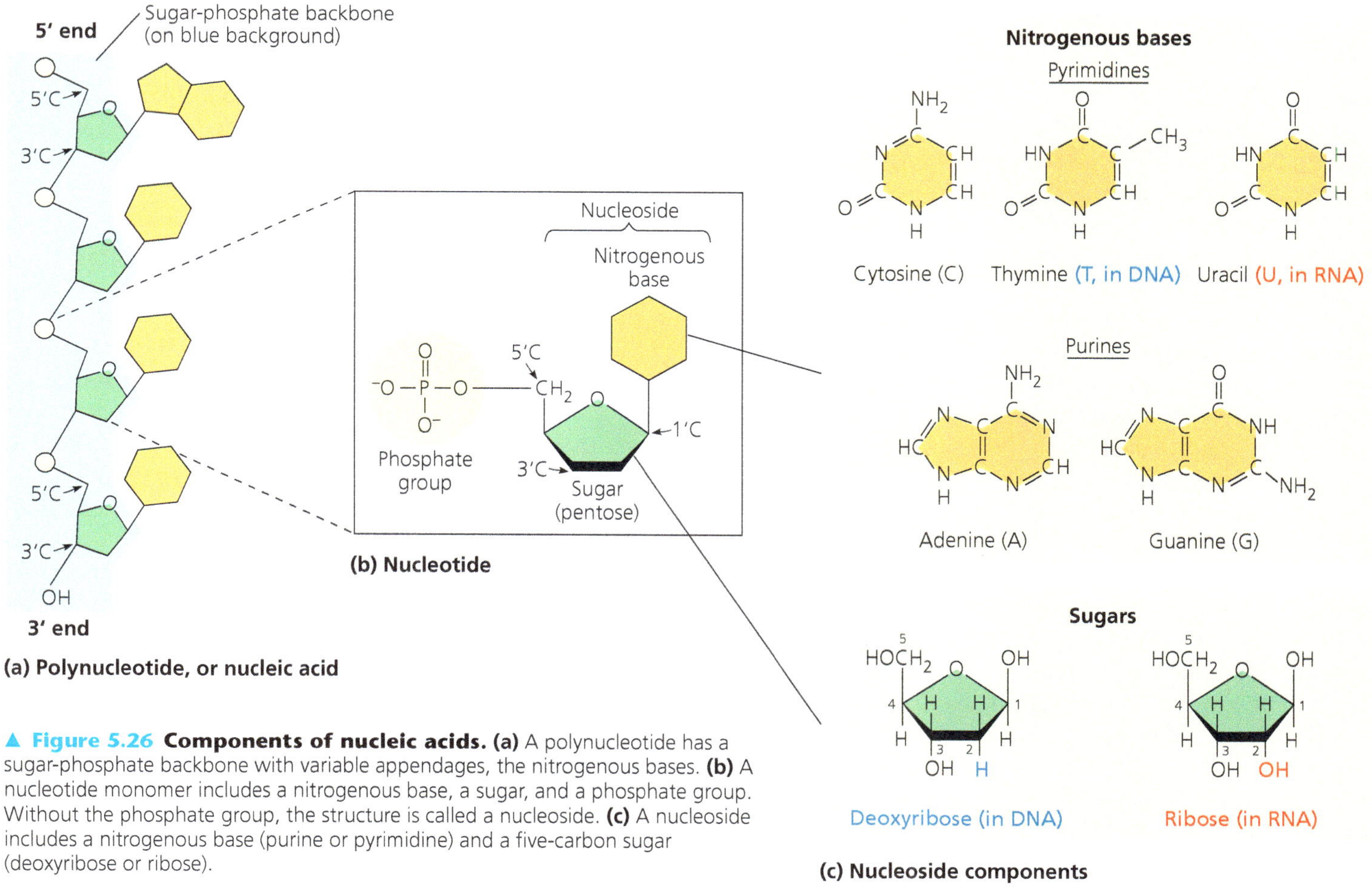

▲ **Figure 5.26 Components of nucleic acids. (a)** A polynucleotide has a sugar-phosphate backbone with variable appendages, the nitrogenous bases. **(b)** A nucleotide monomer includes a nitrogenous base, a sugar, and a phosphate group. Without the phosphate group, the structure is called a nucleoside. **(c)** A nucleoside includes a nitrogenous base (purine or pyrimidine) and a five-carbon sugar (deoxyribose or ribose).

from solution, thus acting as bases.) There are two families of nitrogenous bases: pyrimidines and purines. A **pyrimidine** has one six-membered ring of carbon and nitrogen atoms. The members of the pyrimidine family are cytosine (C), thymine (T), and uracil (U). **Purines** are larger, with a six-membered ring fused to a five-membered ring. The purines are adenine (A) and guanine (G). The specific pyrimidines and purines differ in the chemical groups attached to the rings. Adenine, guanine, and cytosine are found in both DNA and RNA; thymine is found only in DNA and uracil only in RNA.

Now let's add a sugar to the nitrogenous base. In DNA the sugar is **deoxyribose**; in RNA it is **ribose** (see Figure 5.26c). The only difference between these two sugars is that deoxyribose lacks an oxygen atom on the second carbon in the ring; hence the name *deoxy*ribose. To distinguish the numbers of the sugar carbons from those used for the ring atoms of the attached nitrogenous base, the sugar carbon numbers of a nucleoside or nucleotide have a prime (′) after them. Thus, the second carbon in the sugar ring is the 2′ ("2 prime") carbon, and the carbon that sticks up from the ring is called the 5′ carbon.

So far, we have built a nucleoside (nitrogenous base plus sugar). To complete the construction of a nucleotide, we attach a phosphate group to the 5′ carbon of the sugar (see Figure 5.26b). The molecule is now a nucleoside monophosphate, better known as a nucleotide.

Nucleotide Polymers

Now we can see how these nucleotides are linked together to build a polynucleotide. Adjacent nucleotides are joined by a phosphodiester linkage, which consists of a phosphate group that links the sugars of two nucleotides. This bonding results in a backbone with a repeating pattern of sugar-phosphate units (see Figure 5.26a). (Note that the nitrogenous bases are not part of the backbone.) The two free ends of the polymer are distinctly different from each other. One end has a phosphate attached to a 5′ carbon, and the other end has a hydroxyl group on a 3′ carbon; we refer to these as the 5′ end and the 3′ end, respectively. We can say that a polynucleotide has a built-in directionality along its sugar-phosphate backbone, from 5′ to 3′, somewhat like a one-way street. All along this sugar-phosphate backbone are appendages consisting of the nitrogenous bases.

The sequence of bases along a DNA (or mRNA) polymer is unique for each gene and provides very specific information to the cell. Because genes are hundreds to thousands of nucleotides long, the number of possible base sequences is effectively limitless. A gene's meaning to the cell is encoded in its specific sequence of the four DNA bases. For example, the sequence 5′-AGGTAACTT-3′ means one thing, whereas the sequence 5′-CGCTTTAAC-3′ has a different meaning. (Entire genes, of course, are much longer.) The linear order of bases in a gene specifies the amino acid sequence—the primary structure—of a protein, which in turn specifies that protein's three-dimensional structure and its function in the cell.

The Structures of DNA and RNA Molecules

RNA molecules usually exist as single polynucleotide chains like the one shown in Figure 5.26a. In contrast, DNA molecules have two polynucleotides, or "strands," that spiral around an imaginary axis, forming a **double helix** **(Figure 5.27a)**. The two sugar-phosphate backbones run in opposite 5′ → 3′ directions from each other; this arrangement is referred to as **antiparallel**, somewhat like a divided highway. The sugar-phosphate backbones are on the outside of the helix, and the nitrogenous bases are paired in the interior of the helix. The two strands are held together by hydrogen bonds between the paired bases (see Figure 5.27a). Most DNA molecules are very long, with thousands or even millions of base pairs. One long DNA double helix includes many genes, each one a particular segment of the molecule.

Only certain bases in the double helix are compatible with each other. Adenine (A) always pairs with thymine (T), and guanine (G) always pairs with cytosine (C). If we were to read the sequence of bases along one strand of the double helix, we would know the sequence of bases along the other strand. If a stretch of one strand has the base sequence 5′-AGGTCCG-3′, then the base-pairing rules tell us that the same stretch of the other strand must have the sequence 3′-TCCAGGC-5′. The two strands of the double helix are *complementary*, each the predictable counterpart of the other. It is this feature of DNA that makes it possible to generate two identical copies of each DNA molecule in a cell that is preparing to divide. When the cell divides, the copies are distributed to the daughter cells, making them genetically identical to the parent cell. Thus, the structure of DNA accounts for its function of transmitting genetic information whenever a cell reproduces.

Complementary base pairing can also occur between parts of two RNA molecules or even between two stretches of nucleotides in the *same* RNA molecule. In fact, base pairing within an RNA molecule allows it to take on the particular three-dimensional shape necessary for its function. Consider, for example, the type of RNA called *transfer RNA* (*tRNA*), which brings amino acids to the ribosome during the synthesis of a polypeptide. A tRNA molecule is about 80 nucleotides in length. Its functional shape results from base pairing between nucleotides where complementary stretches of the molecule run antiparallel to each other **(Figure 5.27b)**.

Note that in RNA, adenine (A) pairs with uracil (U); thymine (T) is not present in RNA. Another difference between RNA and DNA is that DNA almost always exists as a double helix, whereas RNA molecules are more variable in shape. This variability arises because the extent and location of complementary base pairing within an RNA molecule differs in different types of RNA, as you will see in Chapter 17.

▶ **Figure 5.27 The structures of DNA and tRNA molecules. (a)** The DNA molecule is usually a double helix, with the sugar-phosphate backbones of the antiparallel polynucleotide strands (symbolized here by blue ribbons) on the outside of the helix. Holding the two strands together are pairs of nitrogenous bases attached to each other by hydrogen bonds. As illustrated here with symbolic shapes for the bases, adenine (A) can pair only with thymine (T), and guanine (G) can pair only with cytosine (C). Each DNA strand in this figure is the structural equivalent of the polynucleotide diagrammed in Figure 5.26a. **(b)** A tRNA molecule has a roughly L-shaped structure, with complementary base pairing of antiparallel stretches of RNA. In RNA, A pairs with U.

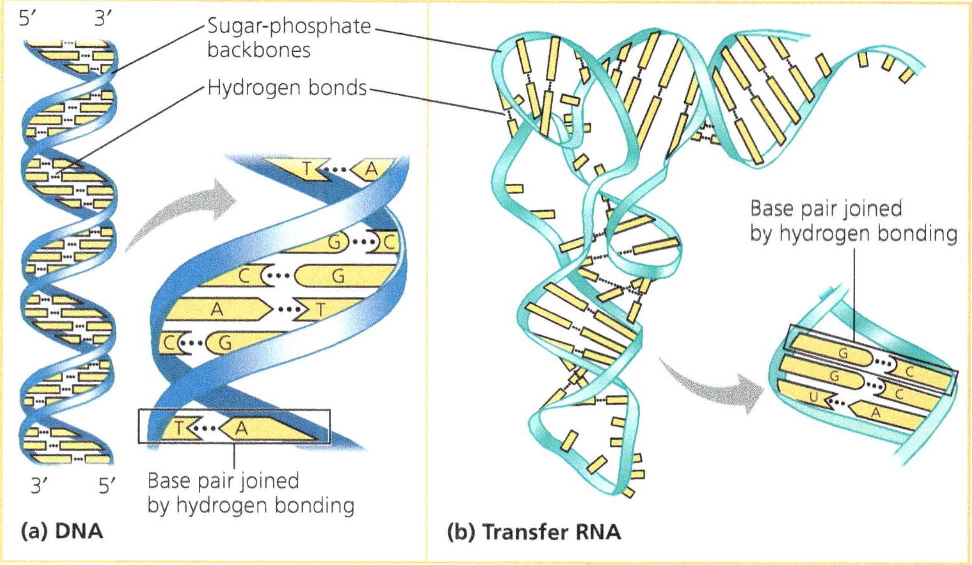

DNA and Proteins as Tape Measures of Evolution

EVOLUTION We are accustomed to thinking of shared traits, such as hair and milk production in mammals, as evidence of shared ancestors. Because we now understand that DNA carries heritable information in the form of genes, we can see that genes and their products (proteins) document the hereditary background of an organism. The linear sequences of nucleotides in DNA molecules are passed from parents to offspring; these sequences determine the amino acid sequences of proteins. Siblings have greater similarity in their DNA and proteins than do unrelated individuals of the same species. If the evolutionary view of life is valid, we should be able to extend this concept of "molecular genealogy" to relationships between species: We should expect two species that appear to be closely related based on fossil and anatomical evidence to also share a greater proportion of their DNA and protein sequences than do more distantly related species. In fact, that is the case. An example is the comparison of the β polypeptide chain of human hemoglobin with the corresponding hemoglobin polypeptide in other vertebrates. In this chain of 146 amino acids, humans and gorillas differ in just 1 amino acid, while humans and frogs differ in 67 amino acids. Molecular biology has added a new tape measure to the toolkit biologists use to assess evolutionary kinship.

The Theme of Emergent Properties in the Chemistry of Life: *A Review*

Recall that life is organized along a hierarchy of structural levels (see Figure 1.4). With each increasing level of order, new properties emerge. In Chapters 2–5, we have dissected the chemistry of life. But we have also begun to develop a more integrated view of life, exploring how properties emerge with increasing order.

We have seen that water's behavior results from the interactions of its molecules, each an ordered arrangement of hydrogen and oxygen atoms. We reduced the complexity and diversity of organic compounds to carbon skeletons and appended chemical groups. We saw that macromolecules are assembled from small organic molecules, taking on new properties. By completing our overview with an introduction to macromolecules and lipids, we have built a bridge to Unit Two, where we will study cell structure and function. We will keep a balance between the need to reduce life to simpler processes and the ultimate satisfaction of viewing those processes in their integrated context.

CONCEPT CHECK 5.5

1. **DRAW IT** Go to Figure 5.26a and, for the top three nucleotides, number all the carbons in the sugars, circle the nitrogenous bases, and star the phosphates.
2. **DRAW IT** In a DNA double helix, a region along one DNA strand has this sequence of nitrogenous bases: 5′-TAGGCCT-3′. Copy this sequence, and write down its complementary strand, clearly indicating the 5′ and 3′ ends of the complementary strand.
3. **WHAT IF?** (a) Suppose a substitution occurred in one DNA strand of the double helix in question 2, resulting in

   5′-TAAGCCT-3′
   3′-ATCCGGA-5′

   Copy these two strands, and circle and label the mismatched bases. (b) If the modified top strand is used by the cell to construct a complementary strand, what would that matching strand be?

For suggested answers, see Appendix A.

5 CHAPTER REVIEW

SUMMARY OF KEY CONCEPTS

CONCEPT 5.1

Macromolecules are polymers, built from monomers (pp. 68–69)

- Carbohydrates, proteins, and nucleic acids are **polymers**, chains of **monomers**. The components of lipids vary.

Monomers form larger molecules by **dehydration reactions**, in which water molecules are released. Polymers can disassemble by the reverse process, **hydrolysis**. An immense variety of polymers can be built from a small set of monomers.

? *What is the fundamental basis for the differences between carbohydrates, proteins, and nucleic acids?*

Large Biological Molecules	Components	Examples	Functions
CONCEPT 5.2 **Carbohydrates serve as fuel and building material (pp. 69–74)** ? *Compare the composition, structure, and function of starch and cellulose. What role do starch and cellulose play in the human body?*	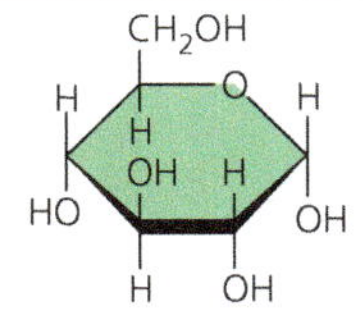 Monosaccharide monomer	**Monosaccharides:** glucose, fructose	Fuel; carbon sources that can be converted to other molecules or combined into polymers
		Disaccharides: lactose, sucrose	
		Polysaccharides: • Cellulose (plants) • Starch (plants) • Glycogen (animals) • Chitin (animals and fungi)	• Strengthens plant cell walls • Stores glucose for energy • Stores glucose for energy • Strengthens exoskeletons and fungal cell walls
CONCEPT 5.3 **Lipids are a diverse group of hydrophobic molecules (pp. 74–77)** ? *Why are lipids not considered to be macromolecules or polymers?*	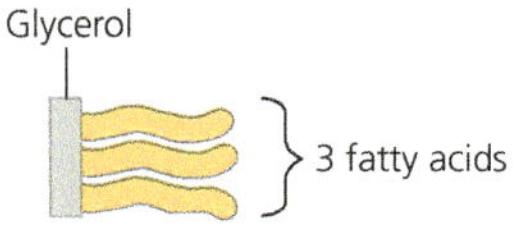	**Triacylglycerols** (fats or oils): glycerol + 3 fatty acids	Important energy source 
	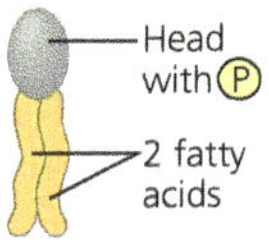	**Phospholipids:** phosphate group + 2 fatty acids	Lipid bilayers of membranes 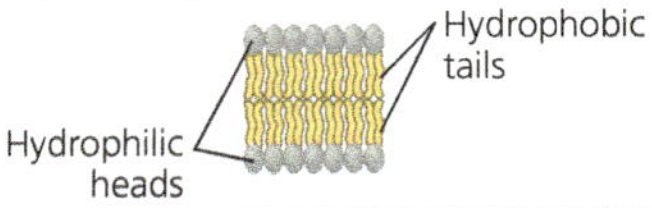
	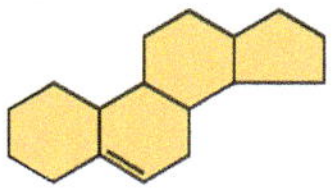 Steroid backbone	**Steroids:** four fused rings with attached chemical groups	• Component of cell membranes (cholesterol) • Signaling molecules that travel through the body (hormones)
CONCEPT 5.4 **Proteins include a diversity of structures, resulting in a wide range of functions (pp. 77–86)** ? *Proteins are the most structurally and functionally diverse class of biological molecules. Explain the basis for this diversity.*	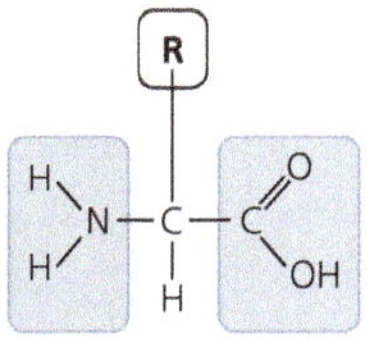 Amino acid monomer (20 types)	• Enzymes • Structural proteins • Storage proteins • Transport proteins • Hormones • Receptor proteins • Motor proteins • Defensive proteins	• Catalyze chemical reactions • Provide structural support • Store amino acids • Transport substances • Coordinate organismal responses • Receive signals from outside cell • Function in cell movement • Protect against disease
CONCEPT 5.5 **Nucleic acids store, transmit, and help express hereditary information (pp. 86–89)** ? *What role does complementary base pairing play in the functions of nucleic acids?*	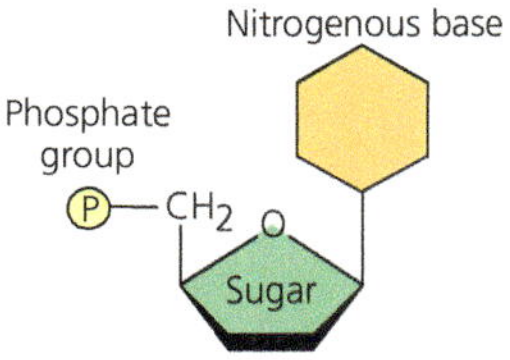	**DNA:**  • Sugar = deoxyribose • Nitrogenous bases = C, G, A, T • Usually double-stranded	Stores hereditary information
		RNA:  • Sugar = ribose • Nitrogenous bases = C, G, A, U • Usually single-stranded	Various functions during gene expression, including carrying instructions from DNA to ribosomes

TEST YOUR UNDERSTANDING

LEVEL 1: KNOWLEDGE/COMPREHENSION

1. Which of the following categories includes all others in the list?
   a. monosaccharide
   b. disaccharide
   c. starch
   d. carbohydrate
   e. polysaccharide

2. The enzyme amylase can break glycosidic linkages between glucose monomers only if the monomers are in the α form. Which of the following could amylase break down?
   a. glycogen, starch, and amylopectin
   b. glycogen and cellulose
   c. cellulose and chitin
   d. starch and chitin
   e. starch, amylopectin, and cellulose

3. Which of the following statements concerning *unsaturated* fats is true?
   a. They are more common in animals than in plants.
   b. They have double bonds in the carbon chains of their fatty acids.
   c. They generally solidify at room temperature.
   d. They contain more hydrogen than do saturated fats having the same number of carbon atoms.
   e. They have fewer fatty acid molecules per fat molecule.

4. The structural level of a protein *least* affected by a disruption in hydrogen bonding is the
   a. primary level.
   b. secondary level.
   c. tertiary level
   d. quaternary level.
   e. All structural levels are equally affected.

5. Enzymes that break down DNA catalyze the hydrolysis of the covalent bonds that join nucleotides together. What would happen to DNA molecules treated with these enzymes?
   a. The two strands of the double helix would separate.
   b. The phosphodiester linkages of the polynucleotide backbone would be broken.
   c. The purines would be separated from the deoxyribose sugars.
   d. The pyrimidines would be separated from the deoxyribose sugars.
   e. All bases would be separated from the deoxyribose sugars.

LEVEL 2: APPLICATION/ANALYSIS

6. The molecular formula for glucose is $C_6H_{12}O_6$. What would be the molecular formula for a polymer made by linking ten glucose molecules together by dehydration reactions?
   a. $C_{60}H_{120}O_{60}$
   b. $C_6H_{12}O_6$
   c. $C_{60}H_{102}O_{51}$
   d. $C_{60}H_{100}O_{50}$
   e. $C_{60}H_{111}O_{51}$

7. Which of the following pairs of base sequences could form a short stretch of a normal double helix of DNA?
   a. 5′-purine-pyrimidine-purine-pyrimidine-3′ with 3′-purine-pyrimidine-purine-pyrimidine-5′
   b. 5′-AGCT-3′ with 5′-TCGA-3′
   c. 5′-GCGC-3′ with 5′-TATA-3′
   d. 5′-ATGC-3′ with 5′-GCAT-3′
   e. All of these pairs are correct.

8. Construct a table that organizes the following terms, and label the columns and rows.

phosphodiester linkages	polypeptides	monosaccharides
peptide bonds	triacylglycerols	nucleotides
glycosidic linkages	polynucleotides	amino acids
ester linkages	polysaccharides	fatty acids

9. **DRAW IT** Copy the polynucleotide strand in Figure 5.26a and label the bases G, T, C, and T, starting from the 5′ end. Assuming this is a DNA polynucleotide, now draw the complementary strand, using the same symbols for phosphates (circles), sugars (pentagons), and bases. Label the bases. Draw arrows showing the 5′ → 3′ direction of each strand. Use the arrows to make sure the second strand is antiparallel to the first. *Hint*: After you draw the first strand vertically, turn the paper upside down; it is easier to draw the second strand from the 5′ toward the 3′ direction as you go from top to bottom.

LEVEL 3: SYNTHESIS/EVALUATION

10. **EVOLUTION CONNECTION**
    Comparisons of amino acid sequences can shed light on the evolutionary divergence of related species. If you were comparing two living species, would you expect all proteins to show the same degree of divergence? Why or why not?

11. **SCIENTIFIC INQUIRY**
    Suppose you are a research assistant in a lab studying DNA-binding proteins. You have been given the amino acid sequences of all the proteins encoded by the genome of a certain species and have been asked to find candidate proteins that could bind DNA. What type of amino acids would you expect to see in such proteins? Why?

12. **SCIENCE, TECHNOLOGY, AND SOCIETY**
    Some amateur and professional athletes take anabolic steroids to help them "bulk up" or build strength. The health risks of this practice are extensively documented. Apart from health considerations, how do you feel about the use of chemicals to enhance athletic performance? Is an athlete who takes anabolic steroids cheating, or is such use part of the preparation required to succeed in competition? Explain.

13. **WRITE ABOUT A THEME**
    Structure and Function Proteins, which have diverse functions in a cell, are all polymers of the same subunits—amino acids. Write a short essay (100–150 words) that discusses how the structure of amino acids allows this one type of polymer to perform so many functions.

For selected answers, see Appendix A.

UNIT 2

The Cell

An Interview with

Bonnie L. Bassler

Bonnie Bassler loves her life as a biologist scrutinizing the secret lives of bacteria. For the past 20 years or so, Bonnie and her lab (her "gang," as she calls them) have made momentous discoveries about how bacterial cells use chemicals to communicate with each other in a process called quorum sensing. Dr. Bassler has a B.S. in Biochemistry from the University of California at Davis and a Ph.D. in Biochemistry from The Johns Hopkins University. Among her many awards and honors, she has received a MacArthur Foundation Fellowship and is a member of the National Academy of Sciences. She is the 2010–2011 President of the American Society for Microbiology, the largest specialized life science organization in the world. At Princeton University since 1994, she is currently the Squibb Professor in Molecular Biology and an Investigator of the Howard Hughes Medical Institute.

How did you get started in science?

I've always been interested in nature and animals, and in puzzles and mystery books—I really like figuring things out. As an undergraduate at UC Davis, I worked in a lab on a bacterial project while taking courses in both biochemistry and genetics. Then, as a graduate student at Johns Hopkins, I learned a lot of biochemistry while studying marine bacteria. The bacteria belong to the genus *Vibrio*, and I was working on chemotaxis, movement by cells toward food or away from noxious chemicals in the environment.

What are the advantages of using bacteria for research in cell biology?

Bacteria have been the foundation of molecular biology for the last 100 years because they're accessible. They grow fast, they form clones of identical cells, and they're amenable to biochemical and genetic analyses. Most of what we initially learned about molecular biology—about genes and proteins and other biomolecules—came from work done on bacteria. Because of evolutionary history, the most basic and ancient life processes that happen in bacteria also happen in humans and other higher organisms. Humans have more bells and whistles, more proteins, more sophistication, more complexity. But if you want to understand the basic components of a process, very often you can use bacteria to do that. Also, working with bacteria fits my personality. I prefer having 10 billion bacterial offspring the day after an experiment to having to wait weeks or months for a small number of baby mice. Every morning there's a surprise waiting for me in the incubator!

What is quorum sensing, and how did you first hear about it?

When I was finishing my graduate work, I heard a talk by Mike Silverman, a scientist with the Agouron Institute in San Diego, about how bacteria "talk" to each other, "count" their own numbers, and coordinate their behavior. Mike had been working on a light-producing (bioluminescent) marine bacterium called *Vibrio fischeri* that lives symbiotically inside a variety of marine animals. The animal provides nutrients for the bacteria, which live in an enclosed space within the animal's body. In return, the bacteria provide light that benefits the animal—by scaring away predators or attracting prey or a mate. But if only a small number of bacteria are present, they do not make light—producing light would waste energy because the light wouldn't be visible. The word "quorum" means "the number needed to do something," and bacteria can sense whether there is a quorum or not and act accordingly.

The way quorum sensing works is that bacteria release certain signaling chemicals into the environment. As the bacterial cells increase in number, the molecules reach a concentration at which many of them bind to receptor proteins on the surface of or inside the bacteria. The signaling molecule fits together with the receptor like a key in a lock. In the case of the surface receptors, each receptor molecule has a part on the outside of the cell and a part on the inside. The signaling chemical binds to the outer part of the receptor, "tickling" the protein so that it makes something happen inside the cell. For instance, in *Vibrio fischeri*, binding of signaling molecules ultimately turns on genes that code for enzymes that make light. Mike had worked out this mechanism of how cells of *Vibrio fischeri* turn on light in synchrony.

It's important to understand that back then, we just didn't think about bacteria like that—we thought bacteria ignored each other and did their own thing as solitary cells. I was totally fascinated. I thought, "He's either crazy or he's brilliant—but I just have to work on that." I went up to the podium after his talk and begged him to let me be his postdoc. Finally he said yes, even though he was a geneticist and I was a biochemist! He took a chance on me.

How does a genetic approach differ from a biochemical one?

Geneticists make lots of mutant organisms, then think up clever strategies to find the ones with mutations in the genes they're interested in. In the case of quorum sensing in bioluminescent bacteria, you look for cells that remain dark. If you have mutated genes involved in quorum sensing, you would expect the bacteria not to make light because light emission depends on the cells communicating with each other. Eventually, you would hope to identify the components that function in normal light-emitting bacteria but not in the mutants. Biochemists, on the other hand, start by isolating molecules and studying their properties directly. Genetics and biochemistry are complementary approaches. I'm glad I know both because the combination is more effective than either approach by itself.

What did you learn about quorum sensing as a postdoc?

In Mike's lab, I worked on another species of bioluminescent *Vibrio* called *Vibrio harveyi*. Because these bacteria are free-living in the ocean, we thought their quorum-sensing molecular circuitry might be more complicated than that of *Vibrio fischeri*. What I found was that *Vibrio harveyi* has two parallel systems for quorum sensing, one that senses cells of the same species and one that counts bacteria of other species. Fast-forwarding a decade or so, this second system seems to be present in *many* bacteria, and the second signaling molecule appears to be universal. So, apparently bacteria can measure the ratios of these two signals, and they're saying, "How many of us and how many of you are there?" Then they do different things, depending on who is in the majority. And this isn't just restricted to bioluminescence. Other bacterial behaviors are also controlled by quorum sensing, such as forming an organized thin layer (called a biofilm) on your teeth or coordinating a virulent infection.

Tell us more about biofilms.

We used to think that most bacteria lived as individual cells suspended in liquid environments. But we now understand that in the wild, they live attached to surfaces in biofilms, and they secrete carbohydrates and other molecules that form a protective slime on the biofilm surface. Most of us have noticed the biofilm coating our teeth every morning. Believe it or not, there are about 600 bacterial species in that biofilm just trying to make a living, getting nutrients from us, but the side effect is that we get cavities. And when someone has a lung infection or an implant or heart valve that harbors an infection, the bacteria are growing as a biofilm in the lungs or on the introduced device. So we now understand why these infections are so hard to treat: It's because the slime on the biofilm is providing a protective shield that antibiotics can't penetrate.

What questions are you and your lab asking now?

My group is interested in how information outside an organism gets inside so that the organism does the right thing at the right time. We work on bacteria because they're simple, but we hope that we will have insights for people working on higher organisms. And we're curious about how collective behaviors first evolved on Earth. How did multicellularity come about? We know that the first organisms were bacteria, but how did they begin to do things together? How did groups of cells in your body come to act like a liver or a heart? We're very interested in how the flow of information through networks facilitates multicellularity.

Are there applications for the basic research you do?

When you're asking fundamental questions, you hope that surprises, things you never thought of, will come out of it. Now that we know that bacteria talk to each other and perform group activities, the question is whether we could interfere with these conversations for therapeutic purposes. Could we make molecules that keep bacteria from "talking" or "hearing"? Maybe these would be new antibiotics. Biofilms are a terrible problem in medicine and dental health, and now that we are starting to know the molecular basis for their formation, maybe we can learn how to prevent them from forming.

Bacteria get a lot of bad press for the negative things they do. On the other hand, bacteria also do many miraculous things that keep us alive; they are working for us every instant of our lives. You are covered with a bacterial biofilm that acts as invisible body armor—these good bacteria occupy all the spaces on your skin, preventing invading bacteria from attaching. Throughout your gut you have a huge mass of bacteria, and they're making vitamins for you and helping you digest your food. So all biofilms aren't bad—and for the good biofilms, what if we could find a molecule to make quorum sensing better? Rather than an antibiotic, this would be a probiotic.

Where do you think this field is going?

I think we'll be turning our attention to the possibility of communication between organisms from different kingdoms and different domains. Bacteria have been around for over 4 billion years and have probably been living with multicellular eukaryotic hosts for hundreds of millions of years. So why wouldn't these hosts have evolved strategies to listen in, say, to the conversation being carried out by a group of pathogenic bacteria? Does our immune system "hear" bacterial signaling molecules? Do hosts actively prevent quorum sensing among pathogenic bacteria? Do they tune in and help the good bacteria? I think this is going to be a dialogue, not a monologue.

What do you enjoy most about your life as a scientist?

I love what I work on. I figured out as a postdoc how much fun this life in science is—that it is not about me against other scientists or who is going to discover something first. Instead, it's me against this bacterium, and we are in it head-to-head for the rest of our lives, in a contest of wills between bacteria trying to keep their secrets and me trying to discover them. Also, the basic question of how groups work together fascinates me. I work with a fantastic group of students and we share everything—everybody gives everybody everything, and then we all get more. That's quorum sensing! Both my molecular and nonmolecular lives involve getting the group to do more than the individual. I love that parallelism! My gang of students show me their data, and it's my job to help them figure out the science and get on with their careers. I'm so lucky—having 24 hands and 12 brains is so much better than two hands and one brain. The science is always changing, and trying to keep up with these young and tireless people is hugely challenging and rewarding.

What is your advice to an undergraduate who is considering a career in biology?

For undergraduates who are considering a life in science, my advice is to work on something that you are passionate about. Don't be limited by thinking that bench science is the only thing a scientist can do. There are so many potential careers for a biologist. You could work on Capitol Hill as a scientific advisor or policymaker. You could teach. You could be a lawyer. You could be a writer who helps the public understand science. You could work on science education at the kindergarten level. Figure out your particular combination of personality traits and what you really love doing as a scientist; then make that niche for yourself and bring science to that career. The sky's the limit for biologists because biology is the science of the 21st century and it touches every part of our lives.

"The sky's the limit for biologists because biology is the science of the 21st century and it touches every part of our lives."

Bonnie Bassler (left) with Lisa Urry (center) and Jane Reece

6

A Tour of the Cell

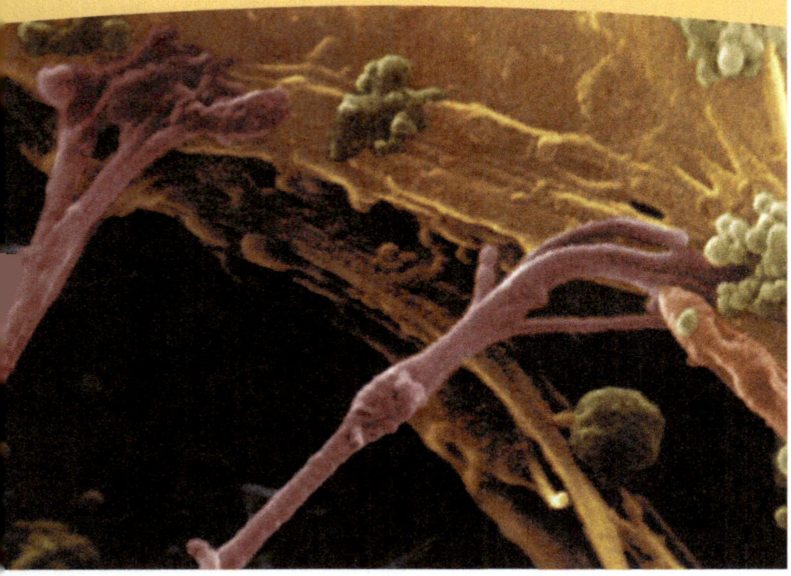

▲ **Figure 6.1 How do your brain cells help you learn about biology?**

KEY CONCEPTS

6.1 Biologists use microscopes and the tools of biochemistry to study cells

6.2 Eukaryotic cells have internal membranes that compartmentalize their functions

6.3 The eukaryotic cell's genetic instructions are housed in the nucleus and carried out by the ribosomes

6.4 The endomembrane system regulates protein traffic and performs metabolic functions in the cell

6.5 Mitochondria and chloroplasts change energy from one form to another

6.6 The cytoskeleton is a network of fibers that organizes structures and activities in the cell

6.7 Extracellular components and connections between cells help coordinate cellular activities

OVERVIEW

The Fundamental Units of Life

Given the scope of biology, you may wonder sometimes how you will ever learn all the material in this course! The answer involves cells, which are as fundamental to the living systems of biology as the atom is to chemistry. The contraction of muscle cells moves your eyes as you read this sentence. The words on the page are translated into signals that nerve cells carry to your brain. **Figure 6.1** shows extensions from one nerve cell (purple) making contact with another nerve cell (orange) in the brain. As you study, your goal is to make connections like these that solidify memories and permit learning to occur.

All organisms are made of cells. In the hierarchy of biological organization, the cell is the simplest collection of matter that can be alive. Indeed, many forms of life exist as single-celled organisms. More complex organisms, including plants and animals, are multicellular; their bodies are cooperatives of many kinds of specialized cells that could not survive for long on their own. Even when cells are arranged into higher levels of organization, such as tissues and organs, the cell remains the organism's basic unit of structure and function.

All cells are related by their descent from earlier cells. However, they have been modified in many different ways during the long evolutionary history of life on Earth. But although cells can differ substantially from one another, they share common features. In this chapter, we'll first examine the tools and techniques that allow us to understand cells, then tour the cell and become acquainted with its components.

CONCEPT 6.1

Biologists use microscopes and the tools of biochemistry to study cells

How can cell biologists investigate the inner workings of a cell, usually too small to be seen by the unaided eye? Before we tour the cell, it will be helpful to learn how cells are studied.

Microscopy

The development of instruments that extend the human senses has gone hand in hand with the advance of science. The discovery and early study of cells progressed with the invention of microscopes in 1590 and their refinement during the 1600s. Cell walls were first seen by Robert Hooke in 1665 as he looked through a microscope at dead cells from the bark of an oak tree. But it took the wonderfully crafted lenses of Antoni van Leeuwenhoek to visualize living cells. Imagine Hooke's awe when he visited van Leeuwenhoek in 1674 and the world of microorganisms—what his host called "very little animalcules"—was revealed to him.

The microscopes first used by Renaissance scientists, as well as the microscopes you are likely to use in the laboratory, are all light microscopes. In a **light microscope (LM)**, visible light is passed through the specimen and then through glass lenses. The lenses refract (bend) the light in such a way that the image of the specimen is magnified as it is projected into the eye or into a camera (see Appendix D).

Three important parameters in microscopy are magnification, resolution, and contrast. *Magnification* is the ratio of an object's image size to its real size. Light microscopes can magnify effectively to about 1,000 times the actual size of the specimen; at greater magnifications, additional details cannot be seen clearly. *Resolution* is a measure of the clarity of the image; it is the minimum distance two points can be separated and still be distinguished as two points. For example, what appears to the unaided eye as one star in the sky may be resolved as twin stars with a telescope, which has a higher resolving ability than the eye. Similarly, using standard techniques, the light microscope cannot resolve detail finer than about 0.2 micrometer (µm), or 200 nanometers (nm), regardless of the magnification (**Figure 6.2**). The third parameter, *contrast*, accentuates differences in parts of the sample. Improvements in light microscopy have included new methods for enhancing contrast, such as staining or labeling cell components to stand out visually. **Figure 6.3**, on the next page, shows different types of microscopy; study this figure as you read the rest of this section.

Until recently, the resolution barrier prevented cell biologists from using standard light microscopy to study **organelles**, the membrane-enclosed structures within eukaryotic cells. To see these structures in any detail required the development of a new instrument. In the 1950s, the electron microscope was introduced to biology. Rather than light, the **electron microscope (EM)** focuses a beam of electrons through the specimen or onto its surface (see Appendix D). Resolution is inversely related to the wavelength of the radiation a microscope uses for imaging, and electron beams have much shorter wavelengths than visible light. Modern electron microscopes can theoretically achieve a resolution of about 0.002 nm, though in practice they usually cannot resolve structures smaller than about 2 nm across. Still, this is a hundredfold improvement over the standard light microscope.

The **scanning electron microscope (SEM)** is especially useful for detailed study of the topography of a specimen (see Figure 6.3). The electron beam scans the surface of the sample, usually coated with a thin film of gold. The beam excites electrons on the surface, and these secondary electrons are detected by a device that translates the pattern of electrons into an electronic signal to a video screen. The result is an image of the specimen's surface that appears three-dimensional.

The **transmission electron microscope (TEM)** is used to study the internal structure of cells (see Figure 6.3). The TEM aims an electron beam through a very thin section of the specimen, similar to the way a light microscope transmits light through a slide. The specimen has been stained with atoms of heavy metals, which attach to certain cellular structures, thus enhancing the electron density of some parts of the cell more than others. The electrons passing through the specimen are scattered more in the denser regions, so fewer are transmitted. The image displays the pattern of transmitted electrons. Instead of using glass lenses, the TEM uses electromagnets as lenses to bend the paths of the electrons, ultimately focusing the image onto a monitor for viewing.

Electron microscopes have revealed many organelles and other subcellular structures that were impossible to resolve with the light microscope. But the light microscope offers advantages, especially in studying living cells. A disadvantage of

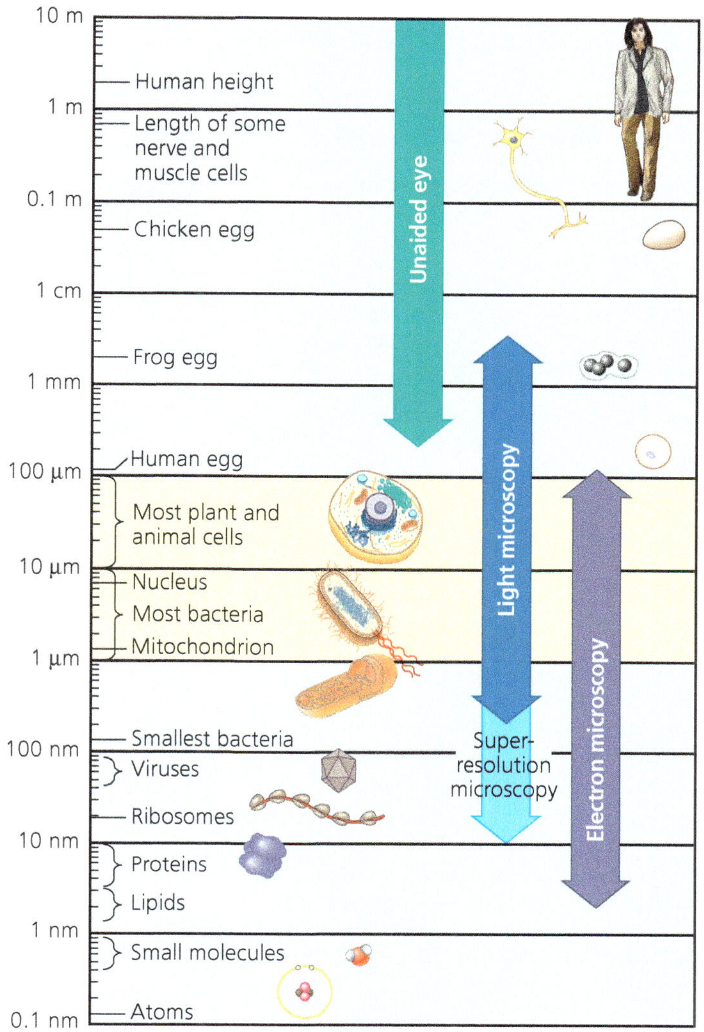

1 centimeter (cm) = 10^{-2} meter (m) = 0.4 inch
1 millimeter (mm) = 10^{-3} m
1 micrometer (µm) = 10^{-3} mm = 10^{-6} m
1 nanometer (nm) = 10^{-3} µm = 10^{-9} m

▲ **Figure 6.2 The size range of cells.** Most cells are between 1 and 100 µm in diameter (yellow region of chart) and are therefore visible only under a microscope. Notice that the scale along the left side is logarithmic to accommodate the range of sizes shown. Starting at the top of the scale with 10 m and going down, each reference measurement marks a tenfold decrease in diameter or length. For a complete table of the metric system, see Appendix C.

▼ **Figure 6.3**

Exploring **Microscopy**

Light Microscopy (LM)

Brightfield (unstained specimen). Light passes directly through the specimen. Unless the cell is naturally pigmented or artificially stained, the image has little contrast. (The first four light micrographs show human cheek epithelial cells; the scale bar pertains to all four micrographs.)

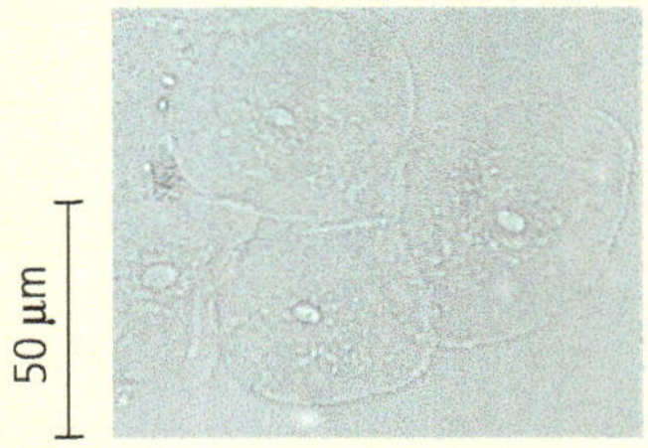

Brightfield (stained specimen). Staining with various dyes enhances contrast. Most staining procedures require that cells be fixed (preserved).

Phase-contrast. Variations in density within the specimen are amplified to enhance contrast in unstained cells, which is especially useful for examining living, unpigmented cells.

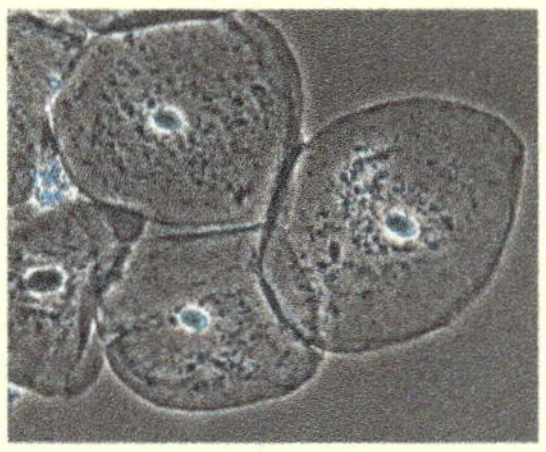

Differential-interference-contrast (Nomarski). As in phase-contrast microscopy, optical modifications are used to exaggerate differences in density, making the image appear almost 3-D.

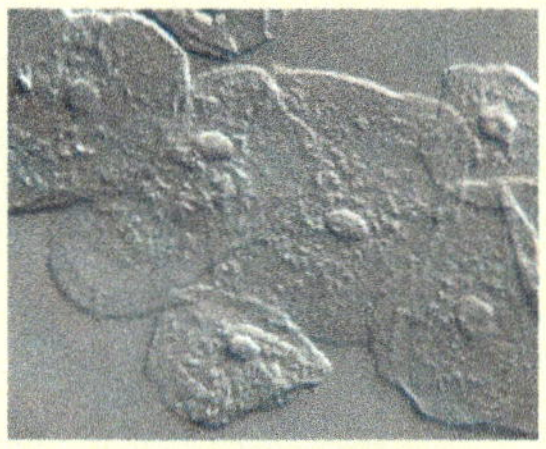

Fluorescence. The locations of specific molecules in the cell can be revealed by labeling the molecules with fluorescent dyes or antibodies; some cells have molecules that fluoresce on their own. Fluorescent substances absorb ultraviolet radiation and emit visible light. In this fluorescently labeled uterine cell, nuclear material is blue, organelles called mitochondria are orange, and the cell's "skeleton" is green.

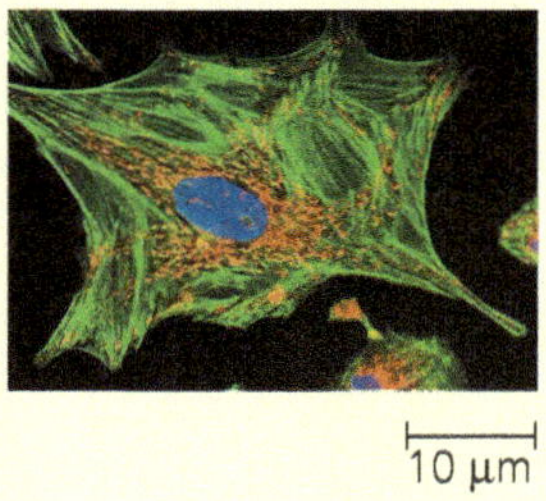

Confocal. The top image is a standard fluorescence micrograph of fluorescently labeled nervous tissue (nerve cells are green, support cells are orange, and regions of overlap are yellow); below it is a confocal image of the same tissue. Using a laser, this "optical sectioning" technique eliminates out-of-focus light from a thick sample, creating a single plane of fluorescence in the image. By capturing sharp images at many different planes, a 3-D reconstruction can be created. The standard image is blurry because out-of-focus light is not excluded.

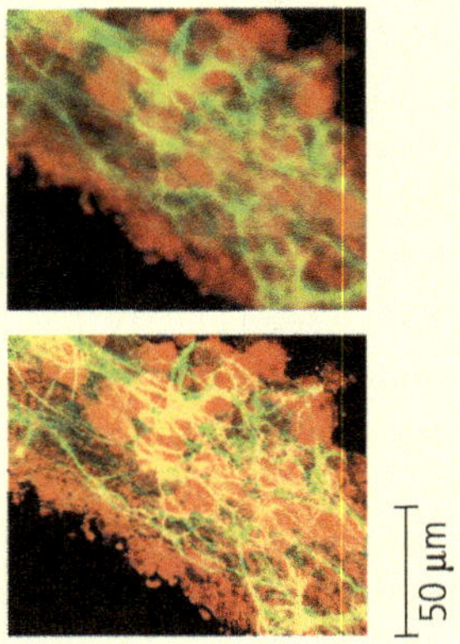

Deconvolution. The top of this split image is a compilation of standard fluorescence micrographs through the depth of a white blood cell. Below is an image of the same cell reconstructed from many blurry images at different planes, each of which was processed using deconvolution software. This process digitally removes out-of-focus light and reassigns it to its source, creating a much sharper 3-D image.

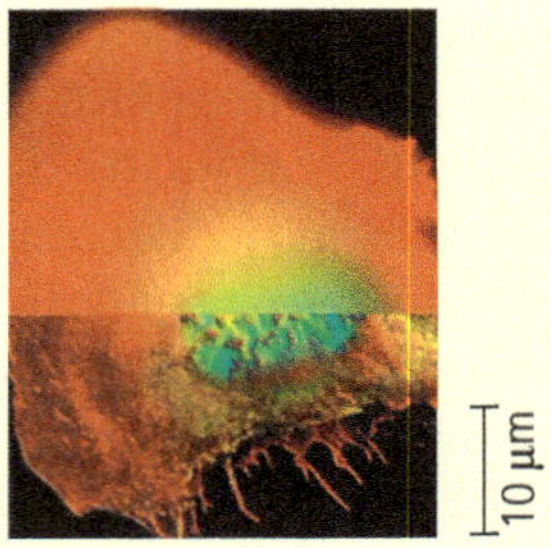

Super-resolution. On the top is a confocal image of part of a nerve cell, using a fluorescent label that binds to a molecule clustered in small sacs in the cell (vesicles) that are 40 nm in diameter. The greenish-yellow spots are blurry because 40 nm is below the 200-nm limit of resolution for standard light microscopy. Below is an image of the same part of the cell, seen using a new "super-resolution" technique. Sophisticated equipment is used to light up individual fluorescent molecules and record their position. Combining information from many molecules in different places "breaks" the limit of resolution, resulting in the sharp greenish-yellow dots seen here. (Each dot is a 40-nm vesicle.)

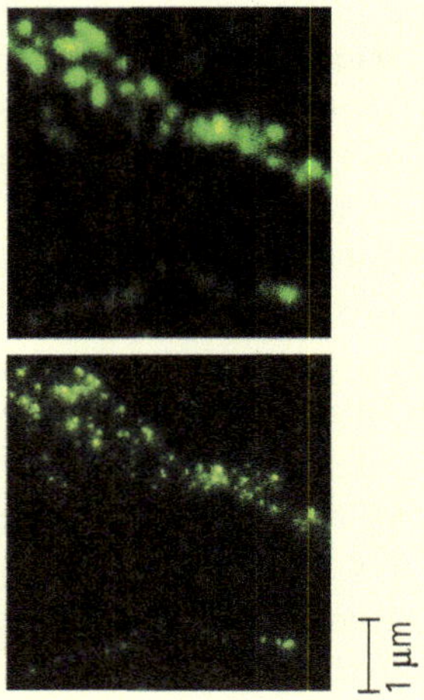

Electron Microscopy (EM)

Scanning electron microscopy (SEM). Micrographs taken with a scanning electron microscope show a 3-D image of the surface of a specimen. This SEM shows the surface of a cell from a trachea (windpipe) covered with cilia. Beating of the cilia helps move inhaled debris upward toward the throat. The SEM and TEM shown here have been artificially colorized. (Electron micrographs are black and white, but are often artificially colorized to highlight particular structures.)

Abbreviations used in this book:
LM = Light Micrograph
SEM = Scanning Electron Micrograph
TEM = Transmission Electron Micrograph

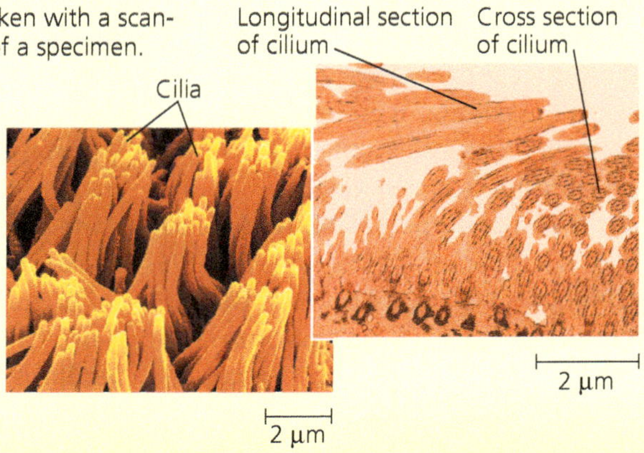

Transmission electron microscopy (TEM). A transmission electron microscope profiles a thin section of a specimen. Here we see a section through a tracheal cell, revealing its internal structure. In preparing the TEM, some cilia were cut along their lengths, creating longitudinal sections, while other cilia were cut straight across, creating cross sections.

electron microscopy is that the methods used to prepare the specimen kill the cells. For all microscopy techniques, in fact, specimen preparation can introduce artifacts, structural features seen in micrographs that do not exist in the living cell.

In the past several decades, light microscopy has been revitalized by major technical advances (see Figure 6.3). Labeling individual cellular molecules or structures with fluorescent markers has made it possible to see such structures with increasing detail. In addition, both confocal and deconvolution microscopy have sharpened images of three-dimensional tissues and cells. Finally, over the past ten years, a group of new techniques and labeling molecules have allowed researchers to "break" the resolution barrier and distinguish subcellular structures as small as 10–20 nm across. As this "super-resolution microscopy" becomes more widespread, the images we'll see of living cells may well be as awe-inspiring to us as van Leeuwenhoek's were to Robert Hooke 350 years ago.

Microscopes are the most important tools of *cytology*, the study of cell structure. To understand the function of each structure, however, required the integration of cytology and *biochemistry*, the study of the chemical processes (metabolism) of cells.

Cell Fractionation

A useful technique for studying cell structure and function is **cell fractionation**, which takes cells apart and separates major organelles and other subcellular structures from one another (Figure 6.4). The instrument used is the centrifuge, which spins test tubes holding mixtures of disrupted cells at a series of increasing speeds. At each speed, the resulting force causes a fraction of the cell components to settle to the bottom of the tube, forming a pellet. At lower speeds, the pellet consists of larger components, and higher speeds yield a pellet with smaller components.

Cell fractionation enables researchers to prepare specific cell components in bulk and identify their functions, a task not usually possible with intact cells. For example, on one of the cell fractions, biochemical tests showed the presence of enzymes involved in cellular respiration, while electron microscopy revealed large numbers of the organelles called mitochondria. Together, these data helped biologists determine that mitochondria are the sites of cellular respiration. Biochemistry and cytology thus complement each other in correlating cell function with structure.

CONCEPT CHECK 6.1

1. How do stains used for light microscopy compare with those used for electron microscopy?
2. WHAT IF? Which type of microscope would you use to study (a) the changes in shape of a living white blood cell and (b) the details of surface texture of a hair?

For suggested answers, see Appendix A.

▼ **Figure 6.4** RESEARCH METHOD

Cell Fractionation

APPLICATION Cell fractionation is used to isolate (fractionate) cell components based on size and density.

TECHNIQUE Cells are homogenized in a blender to break them up. The resulting mixture (homogenate) is centrifuged. The supernatant (liquid) is poured into another tube and centrifuged at a higher speed for a longer time. This process is repeated several times. This "differential centrifugation" results in a series of pellets, each containing different cell components.

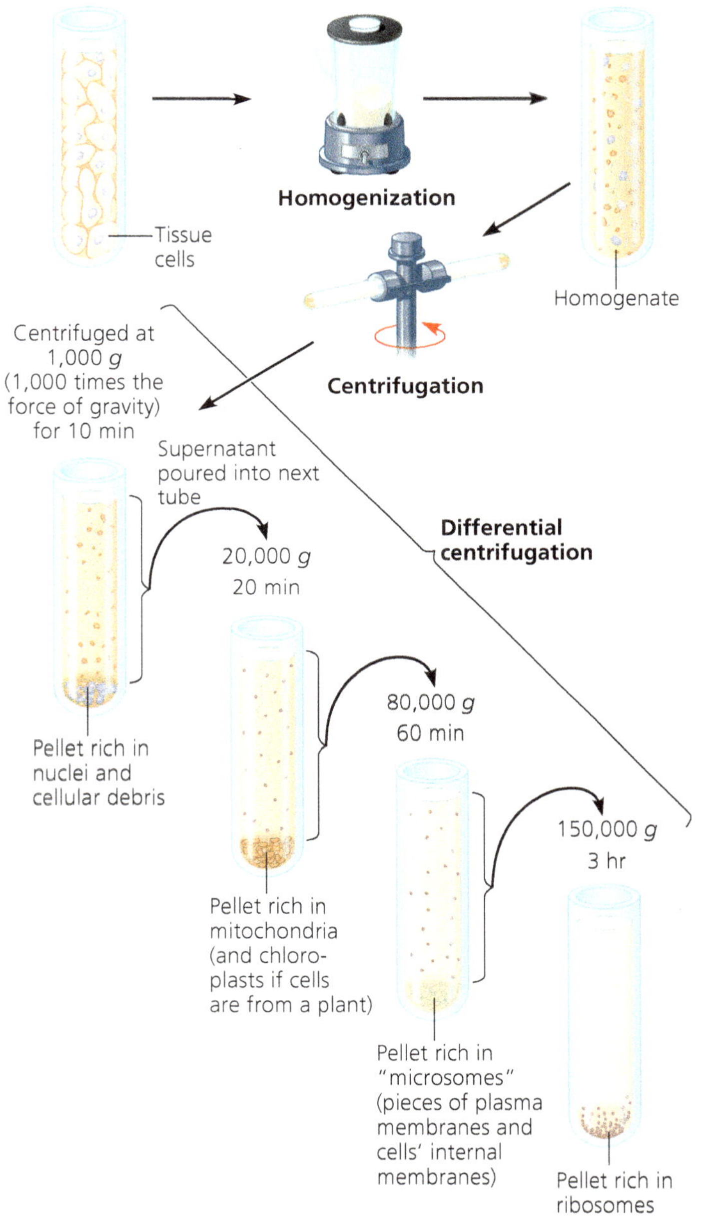

RESULTS In early experiments, researchers used microscopy to identify the organelles in each pellet and biochemical methods to determine their metabolic functions. These identifications established a baseline for this method, enabling today's researchers to know which cell fraction they should collect in order to isolate and study particular organelles.

CONCEPT **6.2**

Eukaryotic cells have internal membranes that compartmentalize their functions

Cells—the basic structural and functional units of every organism—are of two distinct types: prokaryotic and eukaryotic. Organisms of the domains Bacteria and Archaea consist of prokaryotic cells. Protists, fungi, animals, and plants all consist of eukaryotic cells.

Comparing Prokaryotic and Eukaryotic Cells

All cells share certain basic features: They are all bounded by a selective barrier, called the *plasma membrane*. Inside all cells is a semifluid, jellylike substance called **cytosol**, in which subcellular components are suspended. All cells contain *chromosomes*, which carry genes in the form of DNA. And all cells have *ribosomes*, tiny complexes that make proteins according to instructions from the genes.

A major difference between prokaryotic and eukaryotic cells is the location of their DNA. In a **eukaryotic cell**, most of the DNA is in an organelle called the *nucleus*, which is bounded by a double membrane (see Figure 6.8, on pp. 100–101). In a **prokaryotic cell**, the DNA is concentrated in a region that is not membrane-enclosed, called the **nucleoid** (**Figure 6.5**). The word *eukaryotic* means "true nucleus" (from the Greek *eu*, true, and *karyon*, kernel, here referring to the nucleus), and the word *prokaryotic* means "before nucleus" (from the Greek *pro*, before), reflecting the fact that prokaryotic cells evolved before eukaryotic cells.

The interior of either type of cell is called the **cytoplasm**; in eukaryotic cells, this term refers only to the region between the nucleus and the plasma membrane. Within the cytoplasm of a eukaryotic cell, suspended in cytosol, are a variety of organelles of specialized form and function. These membrane-bounded structures are absent in prokaryotic cells. Thus, the presence or absence of a true nucleus is just one aspect of the disparity in structural complexity between the two types of cells.

Eukaryotic cells are generally much larger than prokaryotic cells (see Figure 6.2). Size is a general feature of cell structure that relates to function. The logistics of carrying out cellular metabolism sets limits on cell size. At the lower limit, the smallest cells known are bacteria called mycoplasmas, which have diameters between 0.1 and 1.0 µm. These are perhaps the smallest packages with enough DNA to program metabolism and enough enzymes and other cellular equipment to carry out the activities necessary for a cell to sustain itself and reproduce. Typical bacteria are 1–5 µm in diameter, about ten times the size of mycoplasmas. Eukaryotic cells are typically 10–100 µm in diameter.

Metabolic requirements also impose theoretical upper limits on the size that is practical for a single cell. At the boundary of every cell, the **plasma membrane** functions as a selective barrier that allows passage of enough oxygen, nutrients, and wastes to service the entire cell (**Figure 6.6**). For each square micrometer of membrane, only a limited amount of a particular

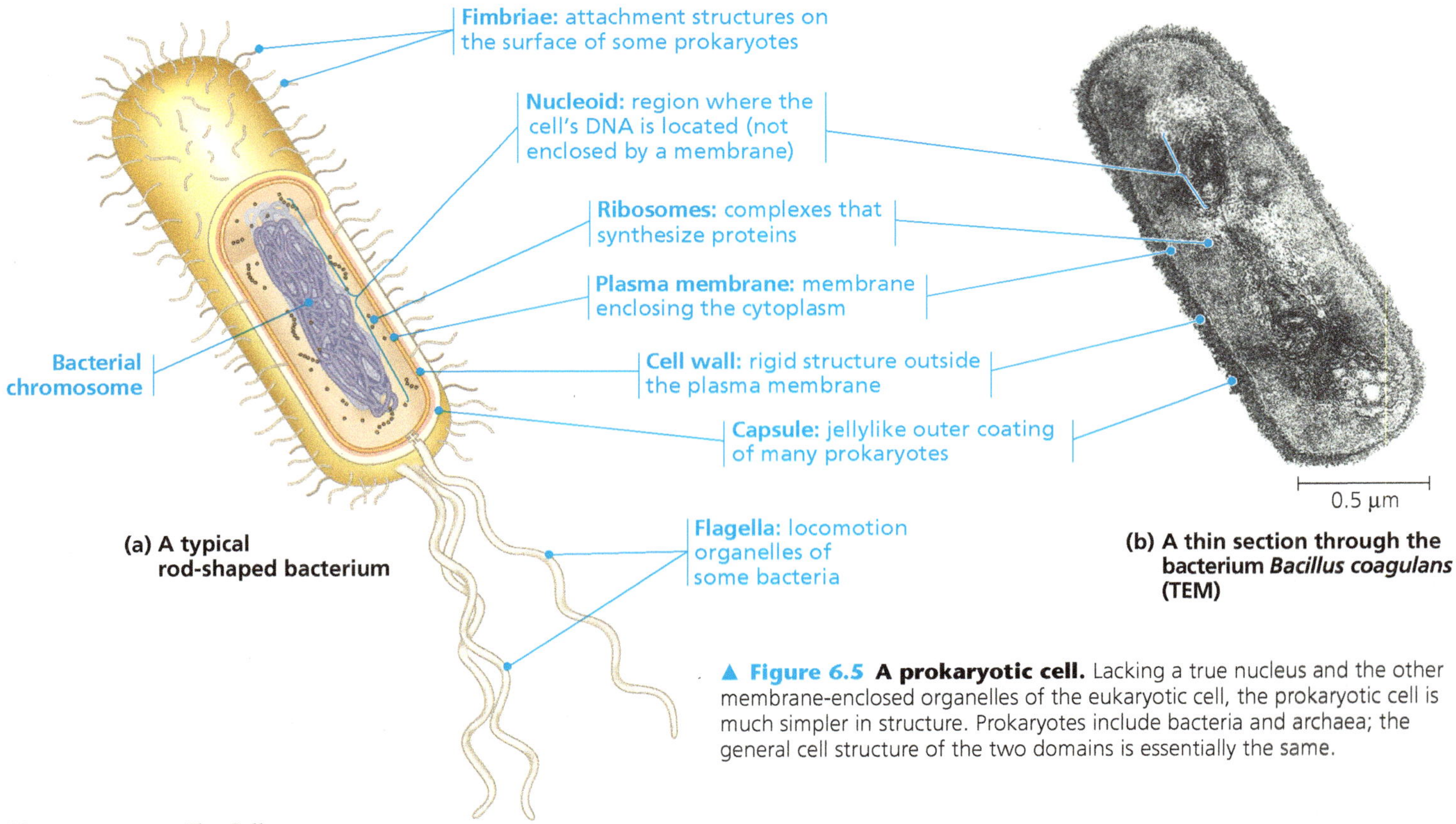

▲ **Figure 6.5 A prokaryotic cell.** Lacking a true nucleus and the other membrane-enclosed organelles of the eukaryotic cell, the prokaryotic cell is much simpler in structure. Prokaryotes include bacteria and archaea; the general cell structure of the two domains is essentially the same.

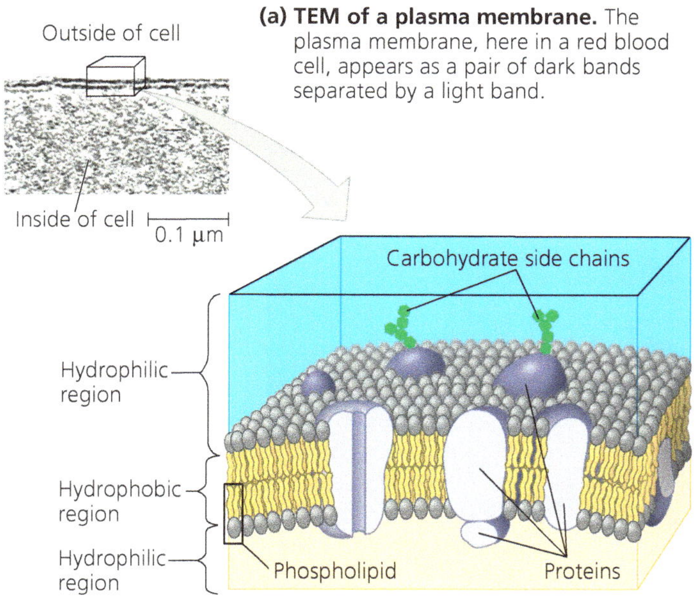

(a) TEM of a plasma membrane. The plasma membrane, here in a red blood cell, appears as a pair of dark bands separated by a light band.

(b) Structure of the plasma membrane

▲ **Figure 6.6 The plasma membrane.** The plasma membrane and the membranes of organelles consist of a double layer (bilayer) of phospholipids with various proteins attached to or embedded in it. The hydrophobic parts, including phospholipid tails and interior portions of membrane proteins, are found in the interior of the membrane. The hydrophilic parts, including phospholipid heads, exterior portions of proteins, and channels of proteins, are in contact with the aqueous solution. Carbohydrate side chains may be attached to proteins or lipids on the outer surface of the plasma membrane.

MAKE CONNECTIONS *Review Figure 5.12 (p. 76) and describe the characteristics of a phospholipid that allow it to function as the major component in the plasma membrane.*

substance can cross per second, so the ratio of surface area to volume is critical. As a cell (or any other object) increases in size, its volume grows proportionately more than its surface area. (Area is proportional to a linear dimension squared, whereas volume is proportional to the linear dimension cubed.) Thus, a smaller object has a greater ratio of surface area to volume **(Figure 6.7)**.

The need for a surface area sufficiently large to accommodate the volume helps explain the microscopic size of most cells and the narrow, elongated shapes of others, such as nerve cells. Larger organisms do not generally have *larger* cells than smaller organisms—they simply have *more* cells (see Figure 6.7). A sufficiently high ratio of surface area to volume is especially important in cells that exchange a lot of material with their surroundings, such as intestinal cells. Such cells may have many long, thin projections from their surface called *microvilli*, which increase surface area without an appreciable increase in volume.

The evolutionary relationships between prokaryotic and eukaryotic cells will be discussed later in this chapter, and prokaryotic cells will be described in detail in Chapter 27. Most of the discussion of cell structure that follows in this chapter applies to eukaryotic cells.

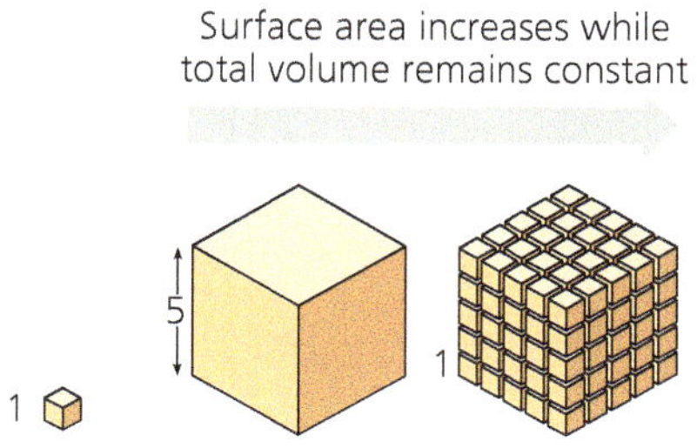

Total surface area [sum of the surface areas (height × width) of all box sides × number of boxes]	6	150	750
Total volume [height × width × length × number of boxes]	1	125	125
Surface-to-volume (S-to-V) ratio [surface area ÷ volume]	6	1.2	6

▲ **Figure 6.7 Geometric relationships between surface area and volume.** In this diagram, cells are represented as boxes. Using arbitrary units of length, we can calculate the cell's surface area (in square units, or units^2), volume (in cubic units, or units^3), and ratio of surface area to volume. A high surface-to-volume ratio facilitates the exchange of materials between a cell and its environment.

A Panoramic View of the Eukaryotic Cell

In addition to the plasma membrane at its outer surface, a eukaryotic cell has extensive and elaborately arranged internal membranes that divide the cell into compartments—the organelles mentioned earlier. The cell's compartments provide different local environments that facilitate specific metabolic functions, so incompatible processes can go on simultaneously inside a single cell. The plasma membrane and organelle membranes also participate directly in the cell's metabolism, because many enzymes are built right into the membranes.

Because membranes are so fundamental to the organization of the cell, Chapter 7 will discuss them in detail. The basic fabric of most biological membranes is a double layer of phospholipids and other lipids. Embedded in this lipid bilayer or attached to its surfaces are diverse proteins (see Figure 6.6). However, each type of membrane has a unique composition of lipids and proteins suited to that membrane's specific functions. For example, enzymes embedded in the membranes of the organelles called mitochondria function in cellular respiration.

Before continuing with this chapter, examine the eukaryotic cells in **Figure 6.8**, on the next two pages. The generalized diagrams of an animal cell and a plant cell introduce the various organelles and highlight the key differences between animal and plant cells. The micrographs at the bottom of the figure give you a glimpse of cells from different types of eukaryotic organisms.

▼ Figure 6.8

Exploring Eukaryotic Cells

Animal Cell (cutaway view of generalized cell)

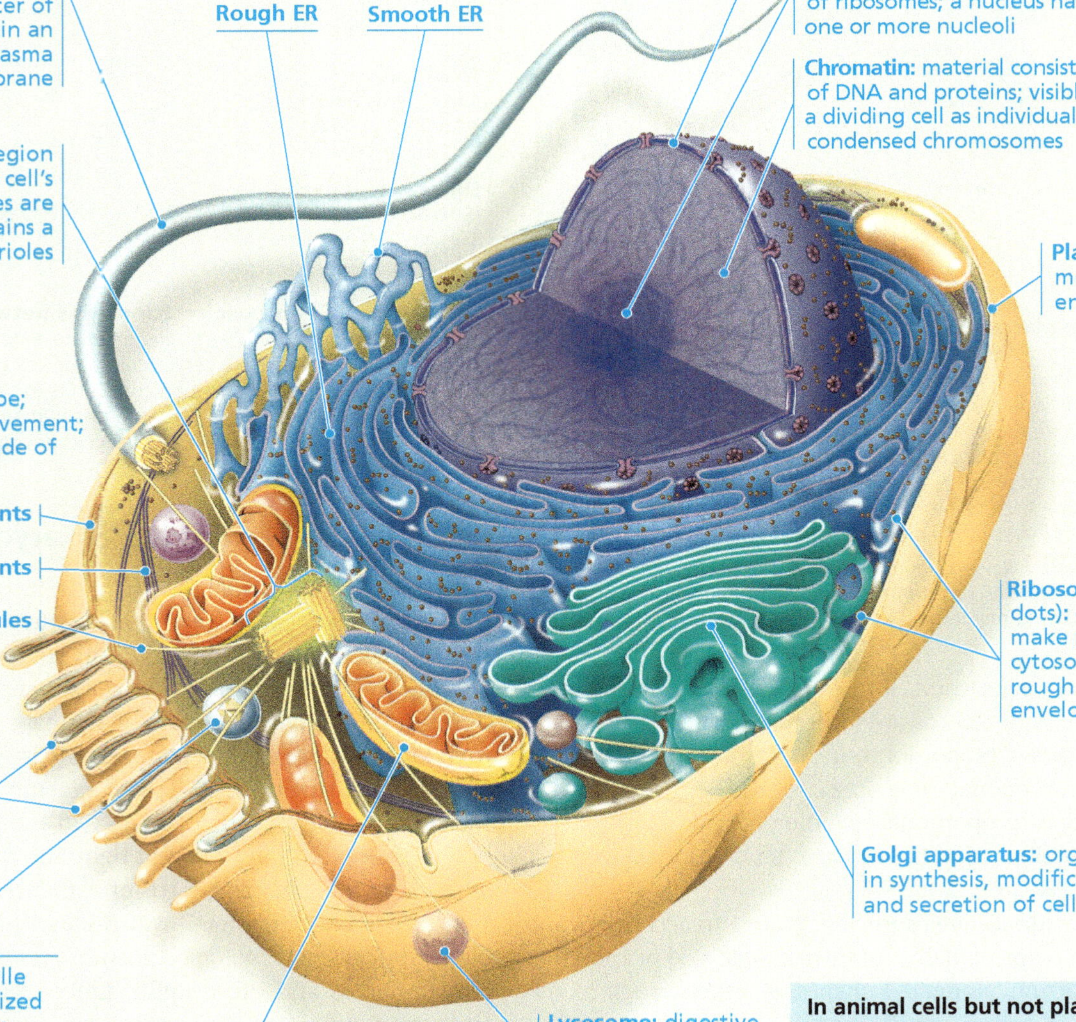

In animal cells but not plant cells:
Lysosomes
Centrosomes, with centrioles
Flagella (but present in some plant sperm)

Animal Cells

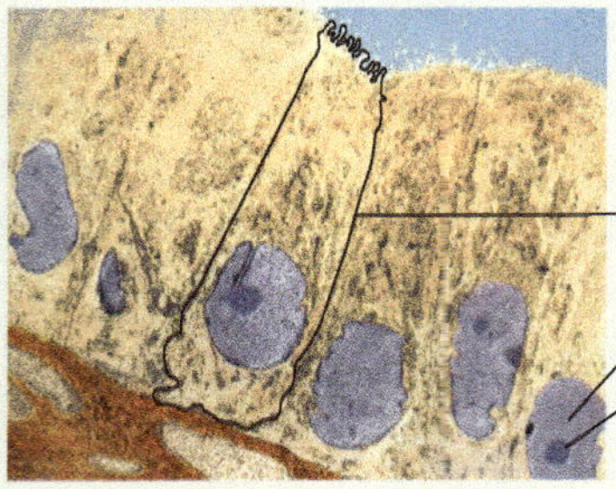

Human cells from lining of uterus (colorized TEM)

Fungal Cells

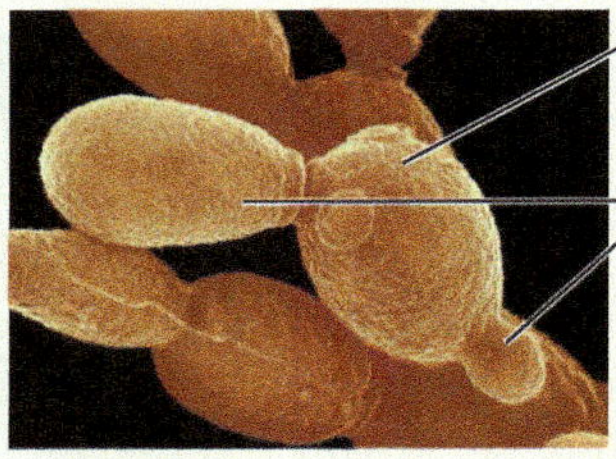

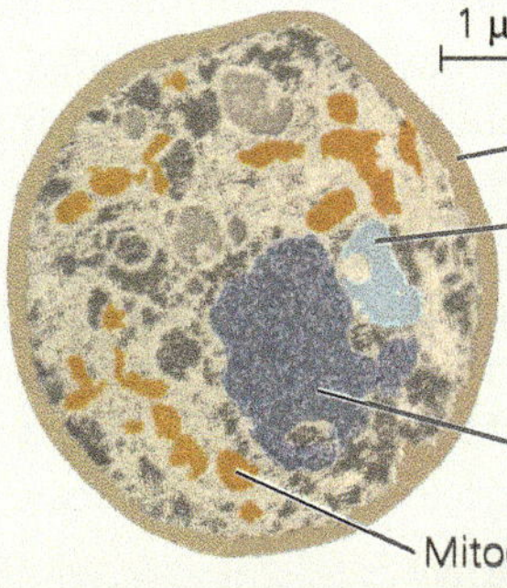

Yeast cells: reproducing by budding (above, colorized SEM) and a single cell (right, colorized TEM)

Plant Cell (cutaway view of generalized cell)

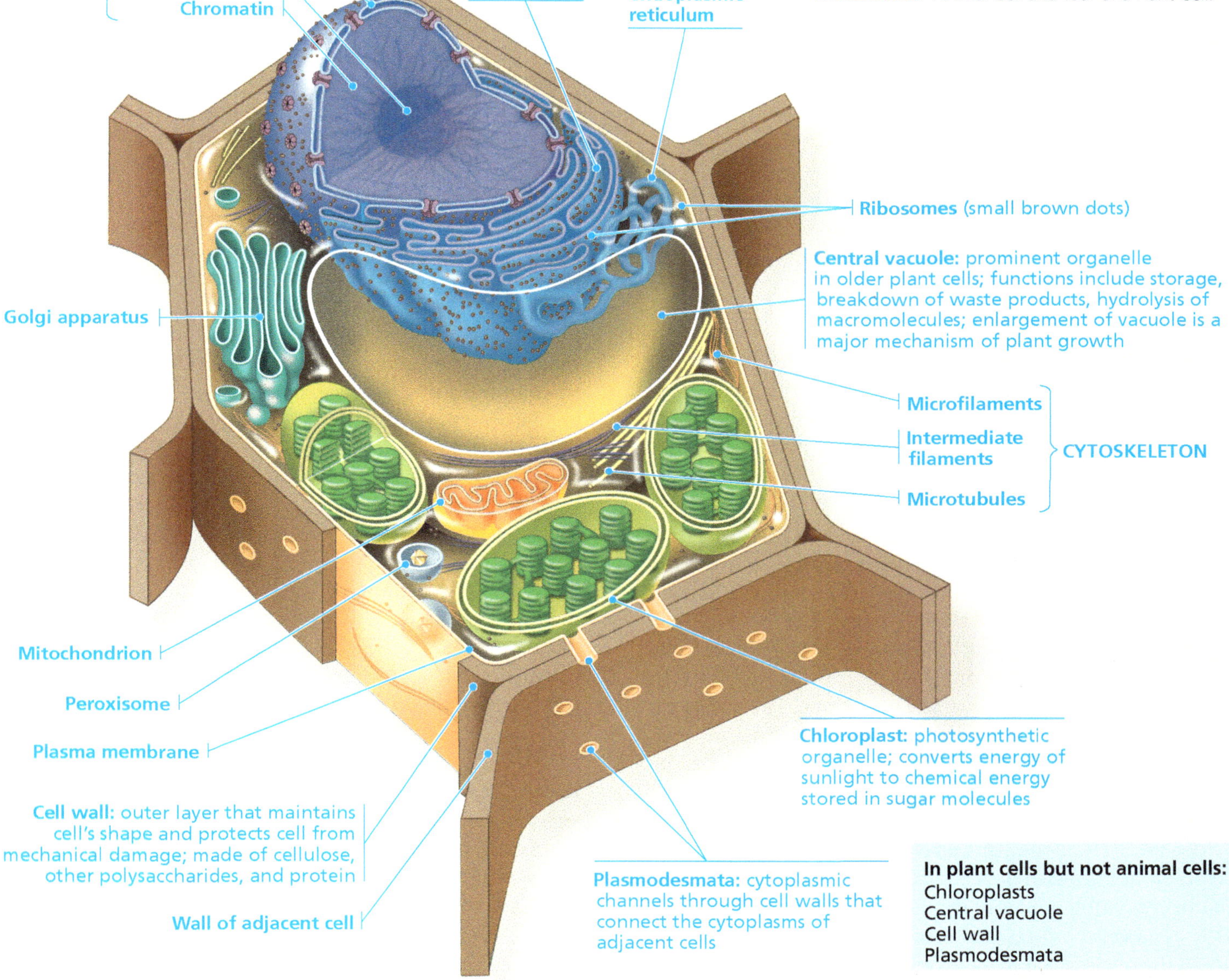

BioFlix Visit the Study Area at **www.masteringbiology.com** for the BioFlix® 3-D Animations Tour of an Animal Cell and Tour of a Plant Cell.

In plant cells but not animal cells:
Chloroplasts
Central vacuole
Cell wall
Plasmodesmata

Plant Cells

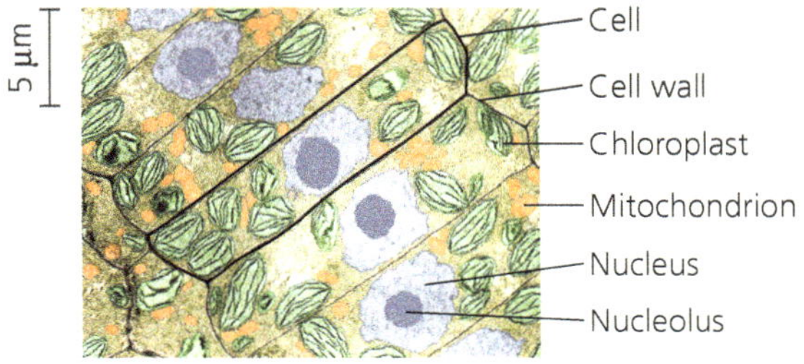

Cells from duckweed (*Spirodela oligorrhiza*), a floating plant (colorized TEM)

Protistan Cells

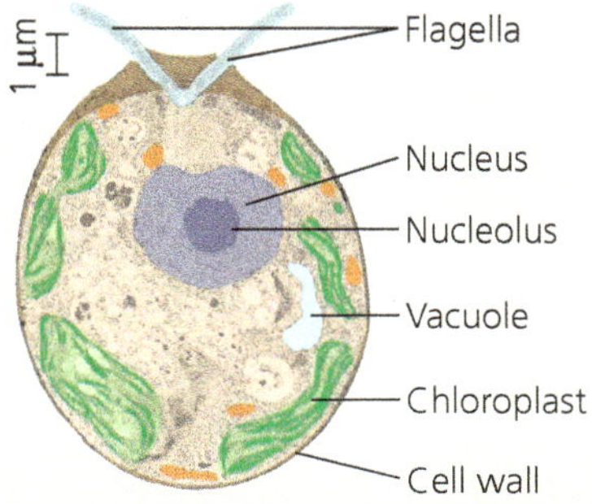

Unicellular green alga *Chlamydomonas* (above, colorized SEM; right, colorized TEM)

CONCEPT CHECK 6.2

1. After carefully reviewing Figure 6.8, briefly describe the structure and function of the nucleus, the mitochondrion, the chloroplast, and the endoplasmic reticulum.
2. **WHAT IF?** Imagine an elongated cell (such as a nerve cell) that measures $125 \times 1 \times 1$ arbitrary units. Predict how its surface-to-volume ratio would compare with those in Figure 6.7. Then calculate the ratio and check your prediction.

For suggested answers, see Appendix A.

CONCEPT 6.3

The eukaryotic cell's genetic instructions are housed in the nucleus and carried out by the ribosomes

On the first stop of our detailed tour of the cell, let's look at two cellular components involved in the genetic control of the cell: the nucleus, which houses most of the cell's DNA, and the ribosomes, which use information from the DNA to make proteins.

The Nucleus: Information Central

The **nucleus** contains most of the genes in the eukaryotic cell. (Some genes are located in mitochondria and chloroplasts.) It is generally the most conspicuous organelle in a eukaryotic cell, averaging about 5 μm in diameter. The **nuclear envelope** encloses the nucleus (**Figure 6.9**), separating its contents from the cytoplasm.

The nuclear envelope is a *double* membrane. The two membranes, each a lipid bilayer with associated proteins, are separated by a space of 20–40 nm. The envelope is perforated by pore structures that are about 100 nm in diameter. At the lip of each pore, the inner and outer membranes of the nuclear envelope are continuous. An intricate protein structure called a *pore complex* lines each pore and plays an important role in the cell by regulating the entry and exit of proteins and RNAs, as well as large complexes of macromolecules. Except at the pores, the nuclear side of the envelope is lined by the **nuclear lamina**, a netlike array of protein filaments that maintains the shape of the nucleus by mechanically supporting the nuclear envelope. There is also much evidence for a *nuclear matrix*, a framework of protein fibers extending throughout the nuclear interior. The nuclear lamina and matrix may help organize the genetic material so it functions efficiently.

Within the nucleus, the DNA is organized into discrete units called **chromosomes**, structures that carry the genetic information. Each chromosome contains one long DNA molecule associated with many proteins. Some of the proteins help coil the DNA molecule of each chromosome, reducing its length and allowing it to fit into the nucleus. The complex of DNA and proteins making up chromosomes is called **chromatin**. When a cell is not dividing, stained chromatin appears as a diffuse mass in micrographs, and the chromosomes cannot be distinguished from one another, even though discrete chromosomes are present. As a cell prepares to divide, however, the chromosomes coil (condense) further, becoming thick enough to be distinguished as separate structures. Each eukaryotic species has a characteristic number of chromosomes. For example, a typical human cell has 46 chromosomes in its nucleus; the exceptions are the sex cells (eggs and sperm), which have only 23 chromosomes in humans. A fruit fly cell has 8 chromosomes in most cells and 4 in the sex cells.

A prominent structure within the nondividing nucleus is the **nucleolus** (plural, *nucleoli*), which appears through the electron microscope as a mass of densely stained granules and fibers adjoining part of the chromatin. Here a type of RNA called *ribosomal RNA* (rRNA) is synthesized from instructions in the DNA. Also in the nucleolus, proteins imported from the cytoplasm are assembled with rRNA into large and small subunits of ribosomes. These subunits then exit the nucleus through the nuclear pores to the cytoplasm, where a large and a small subunit can assemble into a ribosome. Sometimes there are two or more nucleoli; the number depends on the species and the stage in the cell's reproductive cycle.

As we saw in Figure 5.25, the nucleus directs protein synthesis by synthesizing messenger RNA (mRNA) according to instructions provided by the DNA. The mRNA is then transported to the cytoplasm via the nuclear pores. Once an mRNA molecule reaches the cytoplasm, ribosomes translate the mRNA's genetic message into the primary structure of a specific polypeptide. This process of transcribing and translating genetic information is described in detail in Chapter 17.

Ribosomes: Protein Factories

Ribosomes, which are complexes made of ribosomal RNA and protein, are the cellular components that carry out protein synthesis (**Figure 6.10**). Cells that have high rates of protein synthesis have particularly large numbers of ribosomes. For example, a human pancreas cell has a few million ribosomes. Not surprisingly, cells active in protein synthesis also have prominent nucleoli.

Ribosomes build proteins in two cytoplasmic locales. At any given time, *free ribosomes* are suspended in the cytosol, while *bound ribosomes* are attached to the outside of the endoplasmic reticulum or nuclear envelope (see Figure 6.10). Bound and free ribosomes are structurally identical, and ribosomes can alternate between the two roles. Most of the proteins made on free ribosomes function within the cytosol; examples are enzymes that catalyze the first steps of sugar breakdown. Bound ribosomes generally make proteins that are destined for insertion into membranes, for packaging

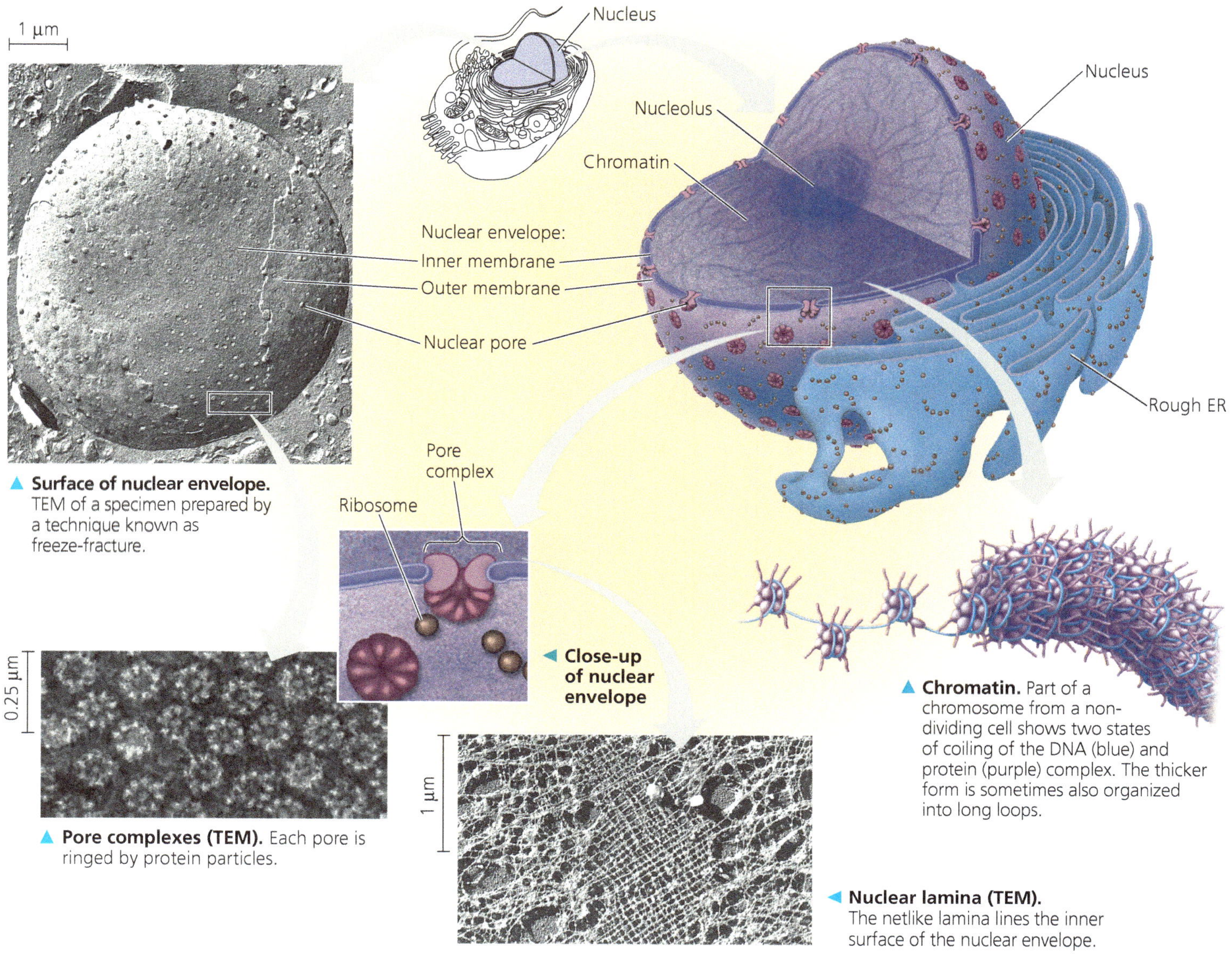

▲ **Surface of nuclear envelope.** TEM of a specimen prepared by a technique known as freeze-fracture.

◄ **Close-up of nuclear envelope**

▲ **Chromatin.** Part of a chromosome from a non-dividing cell shows two states of coiling of the DNA (blue) and protein (purple) complex. The thicker form is sometimes also organized into long loops.

▲ **Pore complexes (TEM).** Each pore is ringed by protein particles.

◄ **Nuclear lamina (TEM).** The netlike lamina lines the inner surface of the nuclear envelope.

▲ **Figure 6.9 The nucleus and its envelope.** Within the nucleus are the chromosomes, which appear as a mass of chromatin (DNA and associated proteins), and one or more nucleoli (singular, *nucleolus*), which function in ribosome synthesis. The nuclear envelope, which consists of two membranes separated by a narrow space, is perforated with pores and lined by the nuclear lamina.

MAKE CONNECTIONS *Since the chromosomes contain the genetic material and reside in the nucleus, how does the rest of the cell get access to the information they carry? See Figure 5.25, page 86.*

► **Figure 6.10 Ribosomes.** This electron micrograph of part of a pancreas cell shows many ribosomes, both free (in the cytosol) and bound (to the endoplasmic reticulum). The simplified diagram of a ribosome shows its two subunits.

DRAW IT *After you have read the section on ribosomes, circle a ribosome in the micrograph that might be making a protein that will be secreted.*

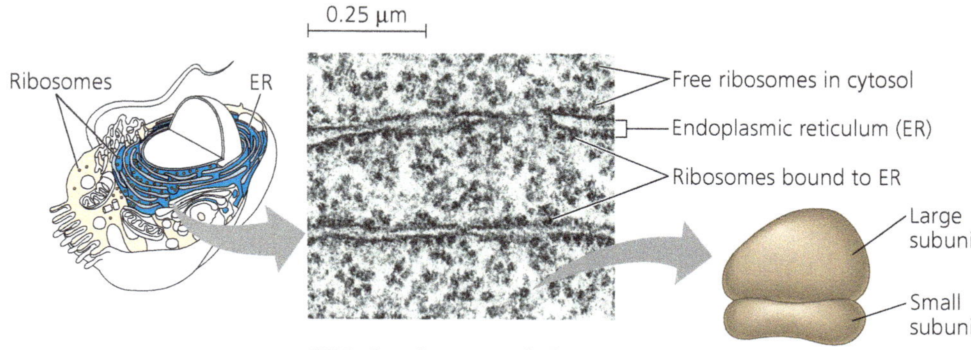

TEM showing ER and ribosomes **Diagram of a ribosome**

within certain organelles such as lysosomes (see Figure 6.8), or for export from the cell (secretion). Cells that specialize in protein secretion—for instance, the cells of the pancreas that secrete digestive enzymes—frequently have a high proportion of bound ribosomes. You will learn more about ribosome structure and function in Chapter 17.

CONCEPT CHECK 6.3

1. What role do ribosomes play in carrying out genetic instructions?
2. Describe the molecular composition of nucleoli and explain their function.
3. **WHAT IF?** As a cell begins the process of dividing, its chromatin becomes more and more condensed. Does the number of chromosomes change during this process? Explain.

For suggested answers, see Appendix A.

CONCEPT 6.4

The endomembrane system regulates protein traffic and performs metabolic functions in the cell

Many of the different membranes of the eukaryotic cell are part of the **endomembrane system**, which includes the nuclear envelope, the endoplasmic reticulum, the Golgi apparatus, lysosomes, various kinds of vesicles and vacuoles, and the plasma membrane. This system carries out a variety of tasks in the cell, including synthesis of proteins, transport of proteins into membranes and organelles or out of the cell, metabolism and movement of lipids, and detoxification of poisons. The membranes of this system are related either through direct physical continuity or by the transfer of membrane segments as tiny **vesicles** (sacs made of membrane). Despite these relationships, the various membranes are not identical in structure and function. Moreover, the thickness, molecular composition, and types of chemical reactions carried out in a given membrane are not fixed, but may be modified several times during the membrane's life. Having already discussed the nuclear envelope, we will now focus on the endoplasmic reticulum and the other endomembranes to which the endoplasmic reticulum gives rise.

The Endoplasmic Reticulum: Biosynthetic Factory

The **endoplasmic reticulum (ER)** is such an extensive network of membranes that it accounts for more than half the total membrane in many eukaryotic cells. (The word *endoplasmic* means "within the cytoplasm," and *reticulum* is Latin for "little net.") The ER consists of a network of membranous tubules and sacs called cisternae (from the Latin *cisterna*, a reservoir for a liquid). The ER membrane separates the internal compartment of the ER, called the ER lumen (cavity) or cisternal space, from the cytosol. And because the ER membrane is continuous with the nuclear envelope, the space between the two membranes of the envelope is continuous with the lumen of the ER **(Figure 6.11)**.

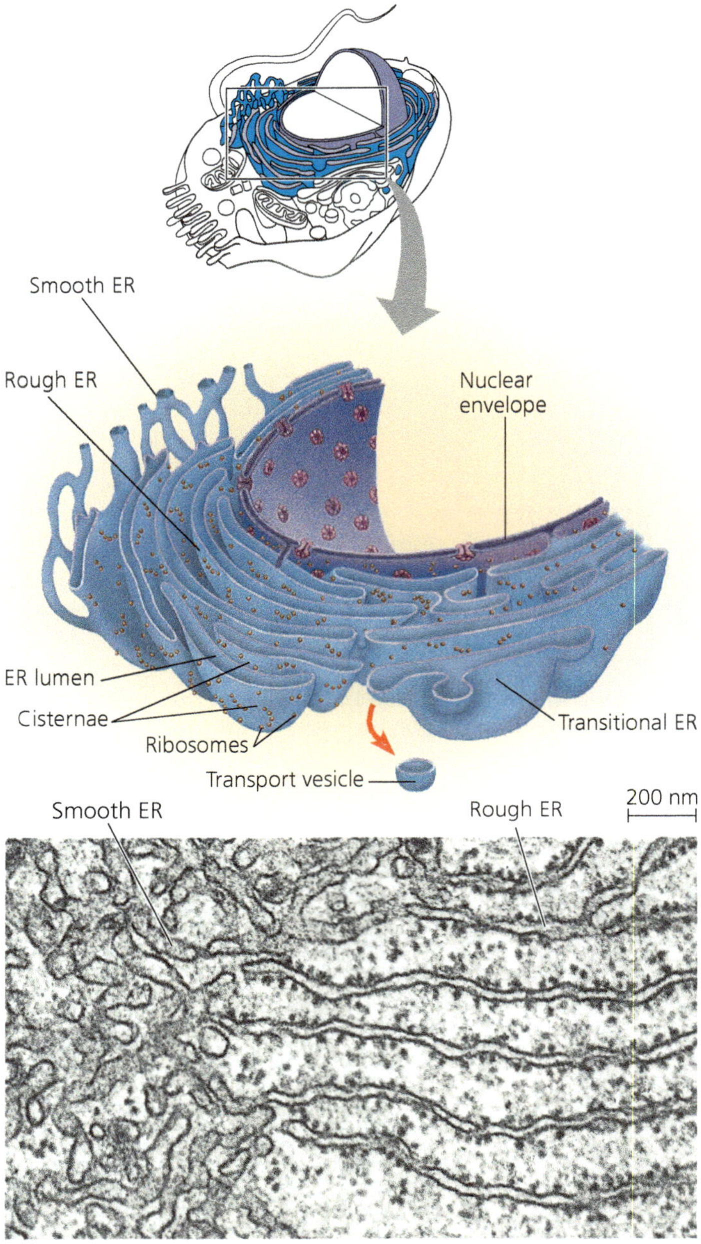

▲ **Figure 6.11 Endoplasmic reticulum (ER).** A membranous system of interconnected tubules and flattened sacs called cisternae, the ER is also continuous with the nuclear envelope. (The drawing is a cutaway view.) The membrane of the ER encloses a continuous compartment called the ER lumen (or cisternal space). Rough ER, which is studded on its outer surface with ribosomes, can be distinguished from smooth ER in the electron micrograph (TEM). Transport vesicles bud off from a region of the rough ER called transitional ER and travel to the Golgi apparatus and other destinations.

There are two distinct, though connected, regions of the ER that differ in structure and function: smooth ER and rough ER. **Smooth ER** is so named because its outer surface lacks ribosomes. **Rough ER** is studded with ribosomes on the outer surface of the membrane and thus appears rough through the electron microscope. As already mentioned, ribosomes are also attached to the cytoplasmic side of the nuclear envelope's outer membrane, which is continuous with rough ER.

Functions of Smooth ER

The smooth ER functions in diverse metabolic processes, which vary with cell type. These processes include synthesis of lipids, metabolism of carbohydrates, detoxification of drugs and poisons, and storage of calcium ions.

Enzymes of the smooth ER are important in the synthesis of lipids, including oils, phospholipids, and steroids. Among the steroids produced by the smooth ER in animal cells are the sex hormones of vertebrates and the various steroid hormones secreted by the adrenal glands. The cells that synthesize and secrete these hormones—in the testes and ovaries, for example—are rich in smooth ER, a structural feature that fits the function of these cells.

Other enzymes of the smooth ER help detoxify drugs and poisons, especially in liver cells. Detoxification usually involves adding hydroxyl groups to drug molecules, making them more soluble and easier to flush from the body. The sedative phenobarbital and other barbiturates are examples of drugs metabolized in this manner by smooth ER in liver cells. In fact, barbiturates, alcohol, and many other drugs induce the proliferation of smooth ER and its associated detoxification enzymes, thus increasing the rate of detoxification. This, in turn, increases tolerance to the drugs, meaning that higher doses are required to achieve a particular effect, such as sedation. Also, because some of the detoxification enzymes have relatively broad action, the proliferation of smooth ER in response to one drug can increase tolerance to other drugs as well. Barbiturate abuse, for example, can decrease the effectiveness of certain antibiotics and other useful drugs.

The smooth ER also stores calcium ions. In muscle cells, for example, the smooth ER membrane pumps calcium ions from the cytosol into the ER lumen. When a muscle cell is stimulated by a nerve impulse, calcium ions rush back across the ER membrane into the cytosol and trigger contraction of the muscle cell. In other cell types, calcium ion release from the smooth ER triggers different responses, such as secretion of vesicles carrying newly synthesized proteins.

Functions of Rough ER

Many types of cells secrete proteins produced by ribosomes attached to rough ER. For example, certain pancreatic cells synthesize the protein insulin in the ER and secrete this hormone into the bloodstream. As a polypeptide chain grows from a bound ribosome, the chain is threaded into the ER lumen through a pore formed by a protein complex in the ER membrane. As the new polypeptide enters the ER lumen, it folds into its native shape. Most secretory proteins are **glycoproteins**, proteins that have carbohydrates covalently bonded to them. The carbohydrates are attached to the proteins in the ER by enzymes built into the ER membrane.

After secretory proteins are formed, the ER membrane keeps them separate from proteins that are produced by free ribosomes and that will remain in the cytosol. Secretory proteins depart from the ER wrapped in the membranes of vesicles that bud like bubbles from a specialized region called transitional ER (see Figure 6.11). Vesicles in transit from one part of the cell to another are called **transport vesicles**; we will discuss their fate shortly.

In addition to making secretory proteins, rough ER is a membrane factory for the cell; it grows in place by adding membrane proteins and phospholipids to its own membrane. As polypeptides destined to be membrane proteins grow from the ribosomes, they are inserted into the ER membrane itself and anchored there by their hydrophobic portions. Like the smooth ER, the rough ER also makes membrane phospholipids; enzymes built into the ER membrane assemble phospholipids from precursors in the cytosol. The ER membrane expands and portions of it are transferred in the form of transport vesicles to other components of the endomembrane system.

The Golgi Apparatus: Shipping and Receiving Center

After leaving the ER, many transport vesicles travel to the **Golgi apparatus**. We can think of the Golgi as a warehouse for receiving, sorting, shipping, and even some manufacturing. Here, products of the ER, such as proteins, are modified and stored and then sent to other destinations. Not surprisingly, the Golgi apparatus is especially extensive in cells specialized for secretion.

The Golgi apparatus consists of flattened membranous sacs—cisternae—looking like a stack of pita bread (**Figure 6.12**, on the next page). A cell may have many, even hundreds, of these stacks. The membrane of each cisterna in a stack separates its internal space from the cytosol. Vesicles concentrated in the vicinity of the Golgi apparatus are engaged in the transfer of material between parts of the Golgi and other structures.

A Golgi stack has a distinct structural directionality, with the membranes of cisternae on opposite sides of the stack differing in thickness and molecular composition. The two sides of a Golgi stack are referred to as the *cis* face and the *trans* face; these act, respectively, as the receiving and shipping departments of the Golgi apparatus. The *cis* face is usually located near the ER. Transport vesicles move material from the ER to the Golgi apparatus. A vesicle that buds from the ER can add its membrane and the contents of its lumen to the *cis* face by fusing with a Golgi membrane. The *trans* face gives rise to vesicles that pinch off and travel to other sites.

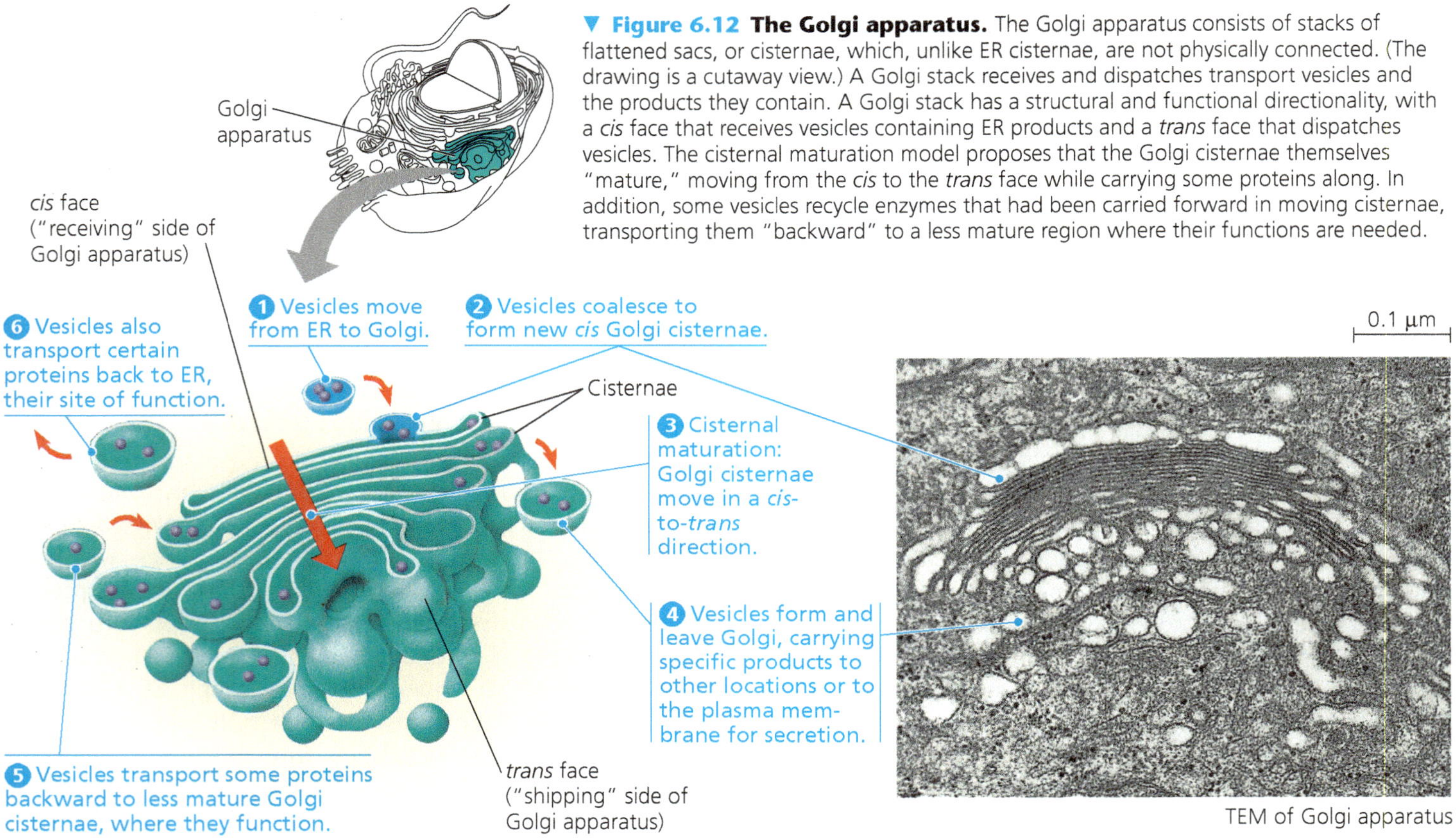

▼ **Figure 6.12 The Golgi apparatus.** The Golgi apparatus consists of stacks of flattened sacs, or cisternae, which, unlike ER cisternae, are not physically connected. (The drawing is a cutaway view.) A Golgi stack receives and dispatches transport vesicles and the products they contain. A Golgi stack has a structural and functional directionality, with a *cis* face that receives vesicles containing ER products and a *trans* face that dispatches vesicles. The cisternal maturation model proposes that the Golgi cisternae themselves "mature," moving from the *cis* to the *trans* face while carrying some proteins along. In addition, some vesicles recycle enzymes that had been carried forward in moving cisternae, transporting them "backward" to a less mature region where their functions are needed.

Products of the endoplasmic reticulum are usually modified during their transit from the *cis* region to the *trans* region of the Golgi apparatus. For example, glycoproteins formed in the ER have their carbohydrates modified, first in the ER itself, then as they pass through the Golgi. The Golgi removes some sugar monomers and substitutes others, producing a large variety of carbohydrates. Membrane phospholipids may also be altered in the Golgi.

In addition to its finishing work, the Golgi apparatus also manufactures some macromolecules. Many polysaccharides secreted by cells are Golgi products. For example, pectins and certain other noncellulose polysaccharides are made in the Golgi of plant cells and then incorporated along with cellulose into their cell walls. Like secretory proteins, nonprotein Golgi products that will be secreted depart from the *trans* face of the Golgi inside transport vesicles that eventually fuse with the plasma membrane.

The Golgi manufactures and refines its products in stages, with different cisternae containing unique teams of enzymes. Until recently, biologists viewed the Golgi as a static structure, with products in various stages of processing transferred from one cisterna to the next by vesicles. While this may occur, recent research has given rise to a new model of the Golgi as a more dynamic structure. According to the *cisternal maturation model*, the cisternae of the Golgi actually progress forward from the *cis* to the *trans* face, carrying and modifying their cargo as they move. Figure 6.12 shows the details of this model.

Before a Golgi stack dispatches its products by budding vesicles from the *trans* face, it sorts these products and targets them for various parts of the cell. Molecular identification tags, such as phosphate groups added to the Golgi products, aid in sorting by acting like ZIP codes on mailing labels. Finally, transport vesicles budded from the Golgi may have external molecules on their membranes that recognize "docking sites" on the surface of specific organelles or on the plasma membrane, thus targeting the vesicles appropriately.

Lysosomes: Digestive Compartments

A **lysosome** is a membranous sac of hydrolytic enzymes that an animal cell uses to digest (hydrolyze) macromolecules. Lysosomal enzymes work best in the acidic environment found in lysosomes. If a lysosome breaks open or leaks its contents, the released enzymes are not very active because the cytosol has a neutral pH. However, excessive leakage from a large number of lysosomes can destroy a cell by self-digestion.

Hydrolytic enzymes and lysosomal membrane are made by rough ER and then transferred to the Golgi apparatus for further processing. At least some lysosomes probably arise by budding from the *trans* face of the Golgi apparatus (see Figure 6.12). How are the proteins of the inner surface of the lysosomal membrane and the digestive enzymes themselves spared from destruction? Apparently, the three-dimensional shapes of these proteins protect vulnerable bonds from enzymatic attack.

Lysosomes carry out intracellular digestion in a variety of circumstances. Amoebas and many other protists eat by engulfing smaller organisms or food particles, a process called **phagocytosis** (from the Greek *phagein*, to eat, and *kytos*, vessel, referring here to the cell). The *food vacuole* formed in this way then fuses with a lysosome, whose enzymes digest the food (**Figure 6.13a**, bottom). Digestion products, including simple sugars, amino acids, and other monomers, pass into the cytosol and become nutrients for the cell. Some human cells also carry out phagocytosis. Among them are macrophages, a type of white blood cell that helps defend the body by engulfing and destroying bacteria and other invaders (see Figure 6.13a, top, and Figure 6.33).

Lysosomes also use their hydrolytic enzymes to recycle the cell's own organic material, a process called *autophagy*. During autophagy, a damaged organelle or small amount of cytosol becomes surrounded by a double membrane (of unknown origin), and a lysosome fuses with the outer membrane of this vesicle **(Figure 6.13b)**. The lysosomal enzymes dismantle the enclosed material, and the organic monomers are returned to the cytosol for reuse. With the help of lysosomes, the cell continually renews itself. A human liver cell, for example, recycles half of its macromolecules each week.

The cells of people with inherited lysosomal storage diseases lack a functioning hydrolytic enzyme normally present in lysosomes. The lysosomes become engorged with indigestible substrates, which begin to interfere with other cellular activities. In Tay-Sachs disease, for example, a lipid-digesting enzyme is missing or inactive, and the brain becomes impaired by an accumulation of lipids in the cells. Fortunately, lysosomal storage diseases are rare in the general population.

Vacuoles: Diverse Maintenance Compartments

Vacuoles are large vesicles derived from the endoplasmic reticulum and Golgi apparatus. Thus, vacuoles are an integral part of a cell's endomembrane system. Like all cellular membranes, the vacuolar membrane is selective in transporting solutes; as a result, the solution inside a vacuole differs in composition from the cytosol.

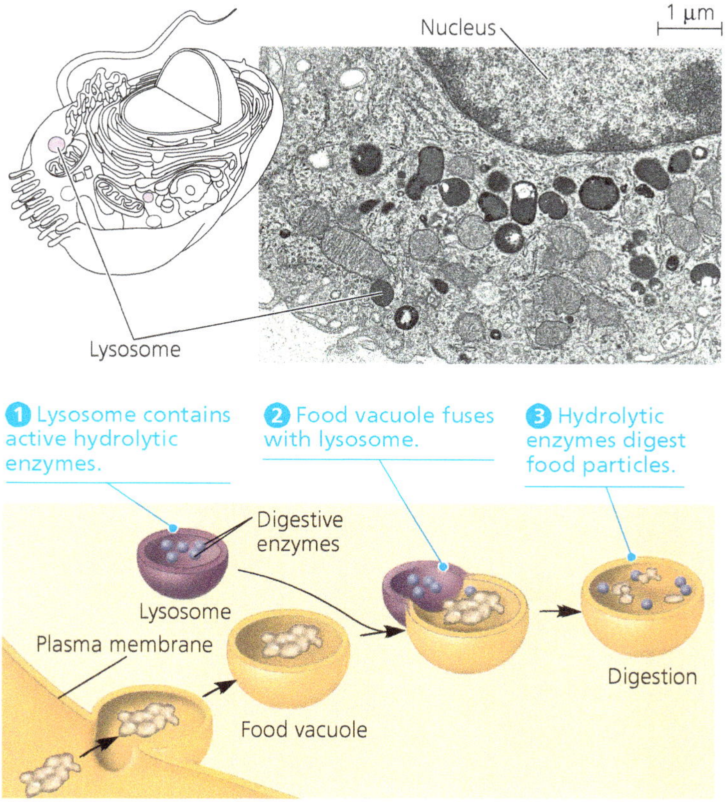

(a) Phagocytosis: lysosome digesting food

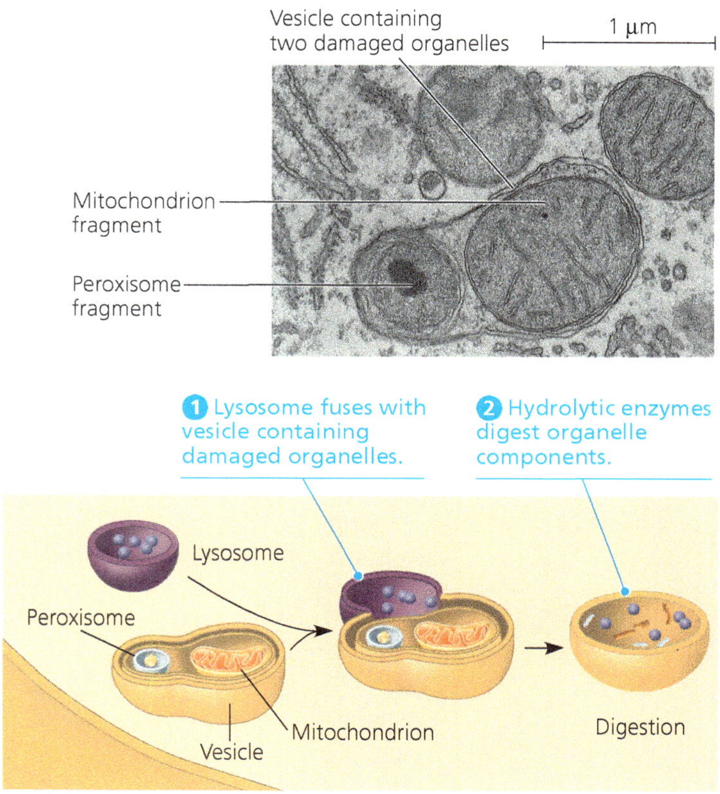

(b) Autophagy: lysosome breaking down damaged organelles

▲ **Figure 6.13 Lysosomes.** Lysosomes digest (hydrolyze) materials taken into the cell and recycle intracellular materials. **(a)** *Top*: In this macrophage (a type of white blood cell) from a rat, the lysosomes are very dark because of a stain that reacts with one of the products of digestion within the lysosome (TEM). Macrophages ingest bacteria and viruses and destroy them using lysosomes. *Bottom*: This diagram shows one lysosome fusing with a food vacuole during the process of phagocytosis by a protist. **(b)** *Top*: In the cytoplasm of this rat liver cell is a vesicle containing two disabled organelles; the vesicle will fuse with a lysosome in the process of autophagy (TEM). *Bottom*: This diagram shows fusion of such a vesicle with a lysosome. This type of vesicle has a double membrane of unknown origin. The outer membrane fuses with the lysosome, and the inner membrane is degraded along with the damaged organelles.

Vacuoles perform a variety of functions in different kinds of cells. **Food vacuoles**, formed by phagocytosis, have already been mentioned (see Figure 6.13a). Many freshwater protists have **contractile vacuoles** that pump excess water out of the cell, thereby maintaining a suitable concentration of ions and molecules inside the cell (see Figure 7.16). In plants and fungi, certain vacuoles carry out enzymatic hydrolysis, a function shared by lysosomes in animal cells. (In fact, some biologists consider these hydrolytic vacuoles to be a type of lysosome.) In plants, smaller vacuoles can hold reserves of important organic compounds, such as the proteins stockpiled in the storage cells in seeds. Vacuoles may also help protect the plant against herbivores by storing compounds that are poisonous or unpalatable to animals. Some plant vacuoles contain pigments, such as the red and blue pigments of petals that help attract pollinating insects to flowers.

Mature plant cells generally contain a large **central vacuole** (**Figure 6.14**), which develops by the coalescence of smaller vacuoles. The solution inside the central vacuole, called cell sap, is the plant cell's main repository of inorganic ions, including potassium and chloride. The central vacuole plays a major role in the growth of plant cells, which enlarge as the vacuole absorbs water, enabling the cell to become larger with a minimal investment in new cytoplasm. The cytosol often occupies only a thin layer between the central vacuole and the plasma membrane, so the ratio of plasma membrane surface to cytosolic volume is sufficient, even for a large plant cell.

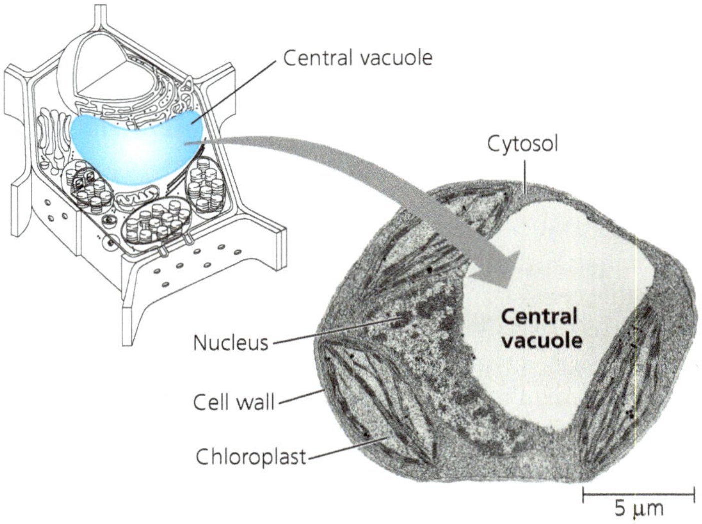

▲ **Figure 6.14 The plant cell vacuole.** The central vacuole is usually the largest compartment in a plant cell; the rest of the cytoplasm is often confined to a narrow zone between the vacuolar membrane and the plasma membrane (TEM).

The Endomembrane System: *A Review*

Figure 6.15 reviews the endomembrane system, showing the flow of membrane lipids and proteins through the various organelles. As the membrane moves from the ER to the Golgi and then elsewhere, its molecular composition and metabolic functions are modified, along with those of its contents. The

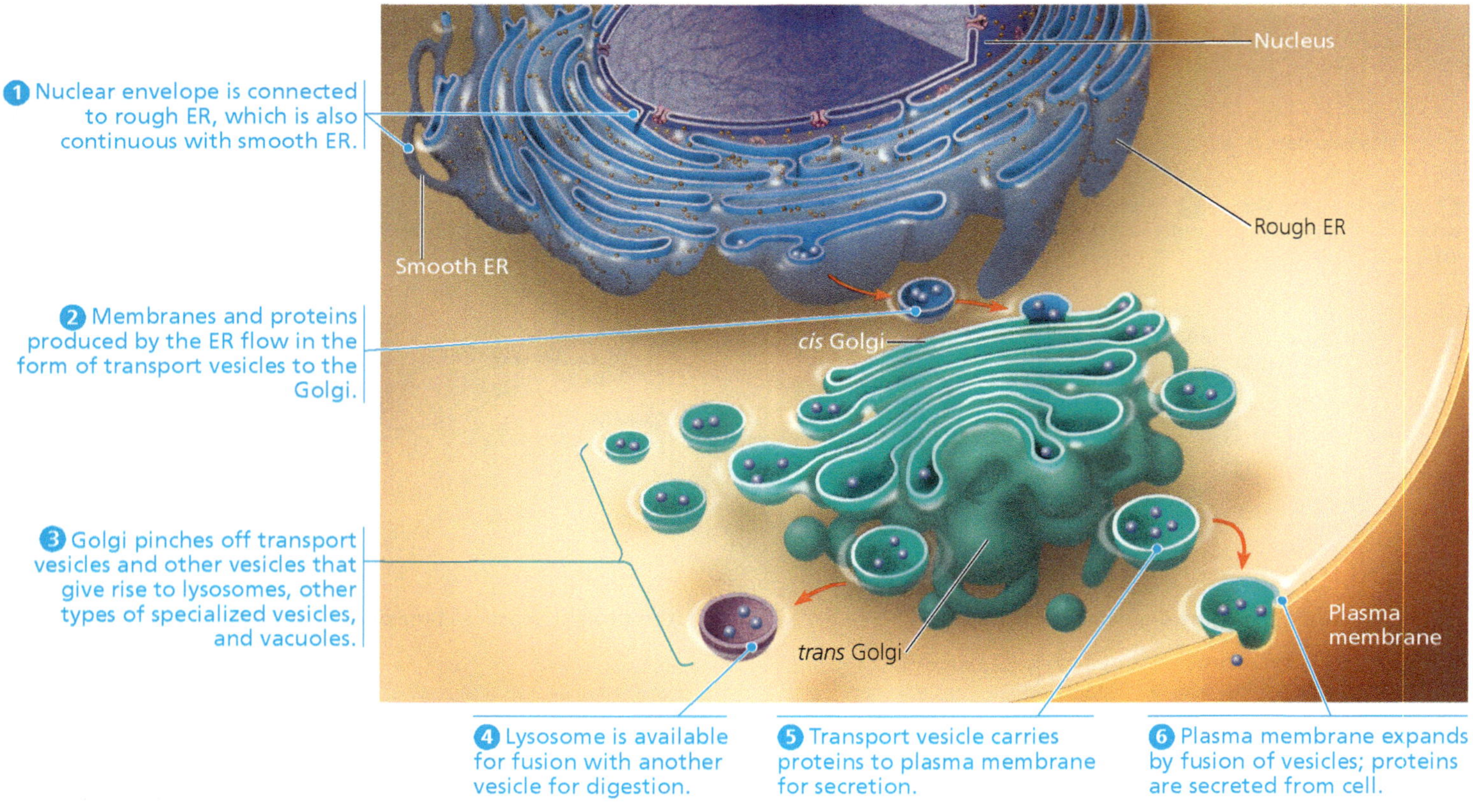

▲ **Figure 6.15 Review: relationships among organelles of the endomembrane system.** The red arrows show some of the migration pathways for membranes and the materials they enclose.

endomembrane system is a complex and dynamic player in the cell's compartmental organization.

We'll continue our tour of the cell with some organelles that are not closely related to the endomembrane system but play crucial roles in the energy transformations carried out by cells.

CONCEPT CHECK 6.4

1. Describe the structural and functional distinctions between rough and smooth ER.
2. Describe how transport vesicles integrate the endomembrane system.
3. **WHAT IF?** Imagine a protein that functions in the ER but requires modification in the Golgi apparatus before it can achieve that function. Describe the protein's path through the cell, starting with the mRNA molecule that specifies the protein.

For suggested answers, see Appendix A.

CONCEPT 6.5

Mitochondria and chloroplasts change energy from one form to another

Organisms transform the energy they acquire from their surroundings. In eukaryotic cells, mitochondria and chloroplasts are the organelles that convert energy to forms that cells can use for work. **Mitochondria** (singular, *mitochondrion*) are the sites of cellular respiration, the metabolic process that uses oxygen to generate ATP by extracting energy from sugars, fats, and other fuels. **Chloroplasts**, found in plants and algae, are the sites of photosynthesis. These organelles convert solar energy to chemical energy by absorbing sunlight and using it to drive the synthesis of organic compounds such as sugars from carbon dioxide and water.

In addition to having related functions, mitochondria and chloroplasts share a similar evolutionary origin, something we'll discuss briefly before describing their structure. In this section, we will also consider the peroxisome, an oxidative organelle. The evolutionary origin of the peroxisome, as well as its relation to other organelles, is still under debate.

The Evolutionary Origins of Mitochondria and Chloroplasts

EVOLUTION Mitochondria and chloroplasts display similarities with bacteria that led to the **endosymbiont theory**, illustrated in Figure 6.16. This theory states that an early ancestor of eukaryotic cells engulfed an oxygen-using nonphotosynthetic prokaryotic cell. Eventually, the engulfed cell formed a relationship with the host cell in which it was enclosed, becoming an *endosymbiont* (a cell living within another cell). Indeed, over the course of evolution, the host cell and its endosymbiont merged into a single organism, a eukaryotic cell with a mitochondrion. At least one of these cells may have then taken up a photosynthetic prokaryote, becoming the ancestor of eukaryotic cells that contain chloroplasts.

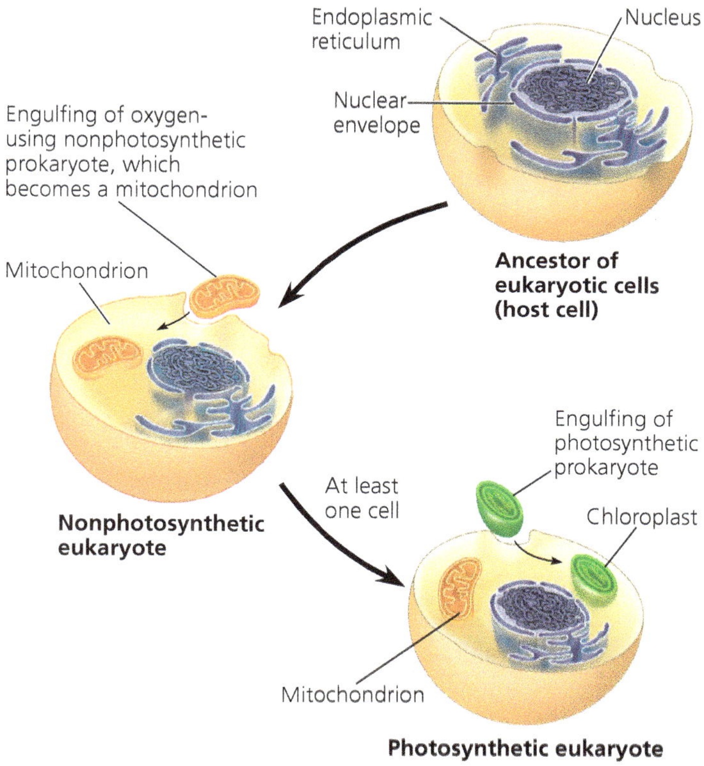

▲ **Figure 6.16 The endosymbiont theory of the origin of mitochondria and chloroplasts in eukaryotic cells.** According to this theory, the proposed ancestors of mitochondria were oxygen-using nonphotosynthetic prokaryotes, while the proposed ancestors of chloroplasts were photosynthetic prokaryotes. The large arrows represent change over evolutionary time; the small arrows inside the cells show the process of the endosymbiont becoming an organelle.

This is a widely accepted theory, which we will discuss in more detail in Chapter 25. The model it proposes is consistent with many structural features of mitochondria and chloroplasts. First, rather than being bounded by a single membrane like organelles of the endomembrane system, mitochondria and typical chloroplasts have two membranes surrounding them. (Chloroplasts also have an internal system of membranous sacs.) There is evidence that the ancestral engulfed prokaryotes had two outer membranes, which became the double membranes of mitochondria and chloroplasts. Second, like prokaryotes, mitochondria and chloroplasts contain ribosomes, as well as circular DNA molecules attached to their inner membranes. The DNA in these organelles programs the synthesis of some of their own proteins, which are made on the ribosomes inside the organelles. Third, also consistent with their probable evolutionary origins as cells, mitochondria and chloroplasts are autonomous (somewhat independent) organelles that grow and reproduce within the cell.

In Chapters 9 and 10, we will focus on how mitochondria and chloroplasts function as energy transformers. Here we are concerned mainly with their structures and their roles.

Mitochondria: Chemical Energy Conversion

Mitochondria are found in nearly all eukaryotic cells, including those of plants, animals, fungi, and most protists. Some cells have a single large mitochondrion, but more often a cell has hundreds or even thousands of mitochondria; the number correlates with the cell's level of metabolic activity. For example, cells that move or contract have proportionally more mitochondria per volume than less active cells.

The mitochondrion is enclosed by two membranes, each a phospholipid bilayer with a unique collection of embedded proteins (**Figure 6.17**). The outer membrane is smooth, but the inner membrane is convoluted, with infoldings called **cristae**. The inner membrane divides the mitochondrion into two internal compartments. The first is the intermembrane space, the narrow region between the inner and outer membranes. The second compartment, the **mitochondrial matrix**, is enclosed by the inner membrane. The matrix contains many different enzymes as well as the mitochondrial DNA and ribosomes. Enzymes in the matrix catalyze some of the steps of cellular respiration. Other proteins that function in respiration, including the enzyme that makes ATP, are built into the inner membrane. As highly folded surfaces, the cristae give the inner mitochondrial membrane a large surface area, thus enhancing the productivity of cellular respiration. This is another example of structure fitting function.

Mitochondria are generally in the range of 1–10 μm long. Time-lapse films of living cells reveal mitochondria moving around, changing their shapes, and fusing or dividing in two, unlike the static structures seen in electron micrographs of dead cells. These observations helped cell biologists understand that mitochondria in a living cell form a branched tubular network, seen in a whole cell in Figure 6.17.

Chloroplasts: Capture of Light Energy

Chloroplasts contain the green pigment chlorophyll, along with enzymes and other molecules that function in the photosynthetic production of sugar. These lens-shaped organelles, about 3–6 μm in length, are found in leaves and other green organs of plants and in algae (**Figure 6.18** and Figure 6.27c).

The contents of a chloroplast are partitioned from the cytosol by an envelope consisting of two membranes separated by a very narrow intermembrane space. Inside the chloroplast is another membranous system in the form of flattened, interconnected sacs called **thylakoids**. In some regions, thylakoids are stacked like poker chips; each stack is called a **granum** (plural, *grana*). The fluid outside the thylakoids is the **stroma**, which contains the chloroplast DNA and ribosomes as well as many enzymes. The membranes of the chloroplast divide the chloroplast space into three compartments: the intermembrane space, the stroma, and the thylakoid space. In Chapter 10, you will learn how this compartmental organization enables the chloroplast to convert light energy to chemical energy during photosynthesis.

As with mitochondria, the static and rigid appearance of chloroplasts in micrographs or schematic diagrams is not true

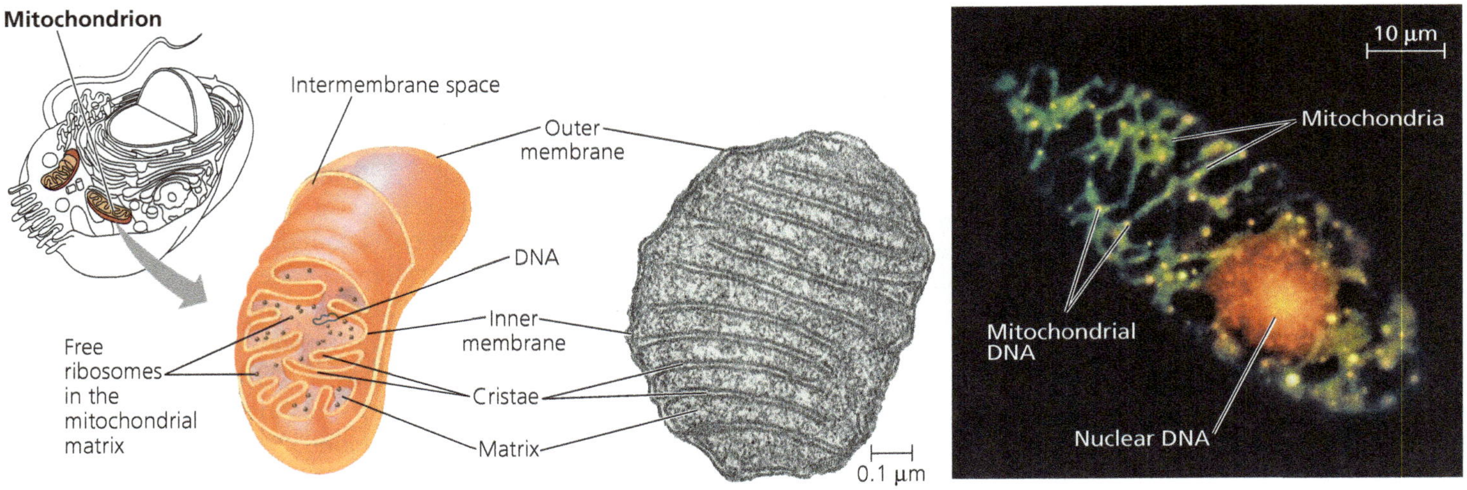

(a) Diagram and TEM of mitochondrion

(b) Network of mitochondria in a protist cell (LM)

▲ **Figure 6.17 The mitochondrion, site of cellular respiration. (a)** The inner and outer membranes of the mitochondrion are evident in the drawing and electron micrograph (TEM). The cristae are infoldings of the inner membrane, which increase its surface area. The cutaway drawing shows the two compartments bounded by the membranes: the intermembrane space and the mitochondrial matrix. Many respiratory enzymes are found in the inner membrane and the matrix. Free ribosomes are also present in the matrix. The DNA molecules are usually circular and are attached to the inner mitochondrial membrane. **(b)** The light micrograph shows an entire unicellular protist (*Euglena gracilis*) at a much lower magnification than the TEM. The mitochondrial matrix has been stained green. The mitochondria form a branched tubular network. The nuclear DNA is stained red, and the molecules of mitochondrial DNA appear as bright yellow spots.

▼ **Figure 6.18 The chloroplast, site of photosynthesis. (a)** Many plants have disk-shaped choloroplasts, as shown here. A typical chloroplast has three compartments: the intermembrane space, the stroma, and the thylakoid space. Free ribosomes are present in the stroma, as are copies of chloroplast DNA molecules. **(b)** This fluorescence micrograph shows a cell of the green alga *Spirogyra crassa*, which is named for its spiral chloroplasts. Under natural light the chloroplasts appear green, but under ultraviolet light they naturally fluoresce red, as shown here.

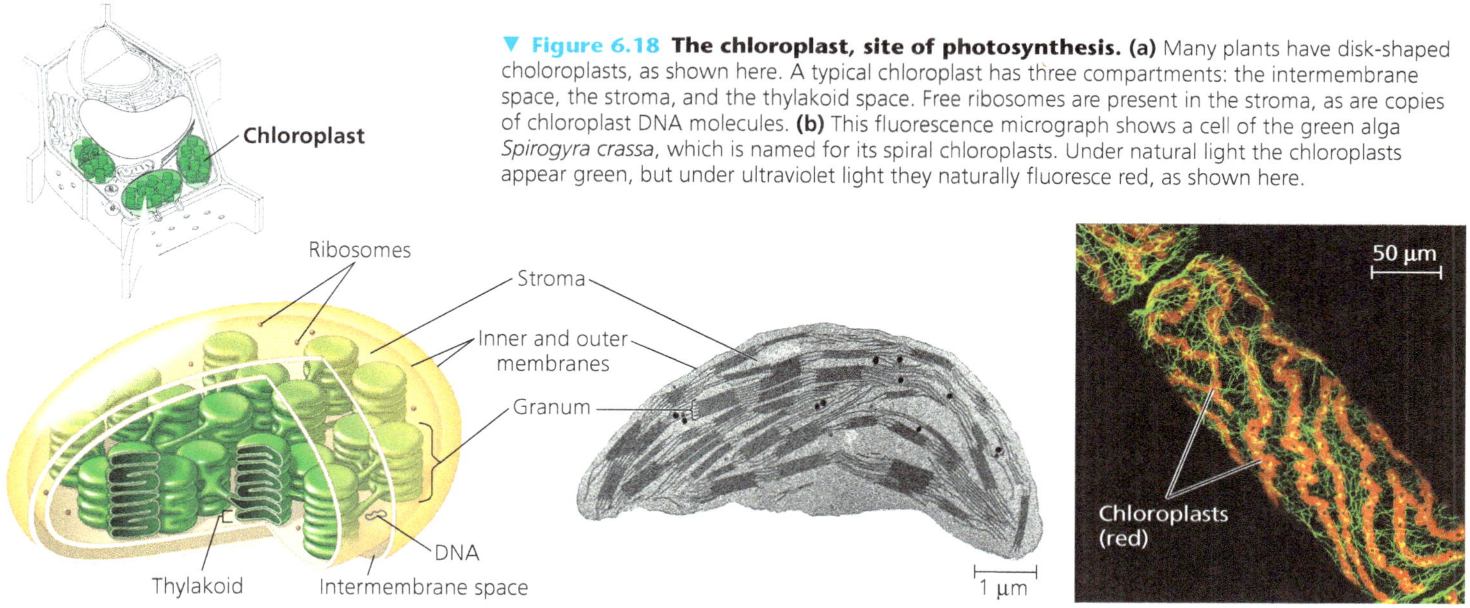

(a) Diagram and TEM of chloroplast

(b) Chloroplasts in an algal cell

to their dynamic behavior in the living cell. Their shape is changeable, and they grow and occasionally pinch in two, reproducing themselves. They are mobile and, with mitochondria and other organelles, move around the cell along tracks of the cytoskeleton, a structural network we will consider later in this chapter.

The chloroplast is a specialized member of a family of closely related plant organelles called **plastids**. One type of plastid, the *amyloplast*, is a colorless organelle that stores starch (amylose), particularly in roots and tubers. Another is the *chromoplast*, which has pigments that give fruits and flowers their orange and yellow hues.

Peroxisomes: Oxidation

The **peroxisome** is a specialized metabolic compartment bounded by a single membrane (**Figure 6.19**). Peroxisomes contain enzymes that remove hydrogen atoms from various substrates and transfer them to oxygen (O_2), thus producing hydrogen peroxide (H_2O_2) as a by-product (from which the organelle derives its name). These reactions have many different functions. Some peroxisomes use oxygen to break fatty acids down into smaller molecules that are transported to mitochondria and used as fuel for cellular respiration. Peroxisomes in the liver detoxify alcohol and other harmful compounds by transferring hydrogen from the poisons to oxygen. The H_2O_2 formed by peroxisomes is itself toxic, but the organelle also contains an enzyme that converts H_2O_2 to water. This is an excellent example of how the cell's compartmental structure is crucial to its functions: The enzymes that produce hydrogen peroxide and those that dispose of this toxic compound are sequestered away from other cellular components that could be damaged.

Specialized peroxisomes called *glyoxysomes* are found in the fat-storing tissues of plant seeds. These organelles contain enzymes that initiate the conversion of fatty acids to sugar, which the emerging seedling uses as a source of energy and carbon until it can produce its own sugar by photosynthesis.

How peroxisomes are related to other organelles is still an open question. They grow larger by incorporating proteins made in the cytosol and ER, as well as lipids made in the ER and within the peroxisome itself. Peroxisomes may increase in number by splitting in two when they reach a certain size, sparking the suggestion of an endosymbiotic evolutionary origin, but others argue against this scenario. The debate continues.

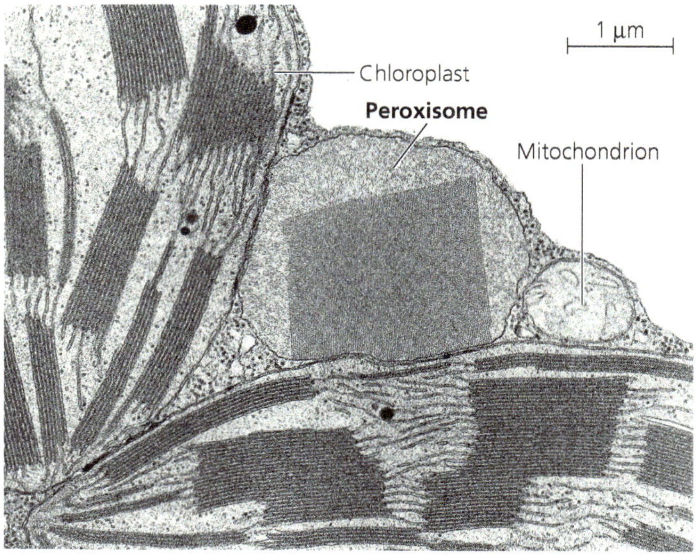

▲ **Figure 6.19 A peroxisome.** Peroxisomes are roughly spherical and often have a granular or crystalline core that is thought to be a dense collection of enzyme molecules. This peroxisome is in a leaf cell (TEM). Notice its proximity to two chloroplasts and a mitochondrion. These organelles cooperate with peroxisomes in certain metabolic functions.

CONCEPT CHECK 6.5

1. Describe two common characteristics of chloroplasts and mitochondria. Consider both function and membrane structure.
2. Do plant cells have mitochondria? Explain.
3. **WHAT IF?** A classmate proposes that mitochondria and chloroplasts should be classified in the endomembrane system. Argue against the proposal.

For suggested answers, see Appendix A.

CONCEPT 6.6

The cytoskeleton is a network of fibers that organizes structures and activities in the cell

In the early days of electron microscopy, biologists thought that the organelles of a eukaryotic cell floated freely in the cytosol. But improvements in both light microscopy and electron microscopy have revealed the **cytoskeleton**, a network of fibers extending throughout the cytoplasm **(Figure 6.20)**. The cytoskeleton, which plays a major role in organizing the structures and activities of the cell, is composed of three types of molecular structures: microtubules, microfilaments, and intermediate filaments.

Roles of the Cytoskeleton: Support and Motility

The most obvious function of the cytoskeleton is to give mechanical support to the cell and maintain its shape. This is especially important for animal cells, which lack walls. The remarkable strength and resilience of the cytoskeleton as a whole is based on its architecture. Like a dome tent, the cytoskeleton is stabilized by a balance between opposing forces exerted by its elements. And just as the skeleton of an animal helps fix the positions of other body parts, the cytoskeleton provides anchorage for many organelles and even cytosolic enzyme molecules. The cytoskeleton is more dynamic than an animal skeleton, however. It can be quickly dismantled in one part of the cell and reassembled in a new location, changing the shape of the cell.

Several types of cell motility (movement) also involve the cytoskeleton. The term *cell motility* encompasses both changes in cell location and more limited movements of parts of the cell. Cell motility generally requires the interaction of the cytoskeleton with **motor proteins**. Examples of such cell motility abound. Cytoskeletal elements and motor proteins work together with plasma membrane molecules to allow whole cells to move along fibers outside the cell. Motor proteins bring about the bending of cilia and flagella by gripping microtubules within those organelles and sliding them against each other. A similar mechanism involving microfilaments causes muscle cells to contract. Inside the cell, vesicles and other organelles often use motor protein "feet" to "walk" to their destinations along a track provided by the cytoskeleton. For example, this is how vesicles containing neurotransmitter molecules migrate to the tips of axons, the long extensions of nerve cells that release these molecules as chemical signals to adjacent nerve cells **(Figure 6.21)**. The vesicles that bud off

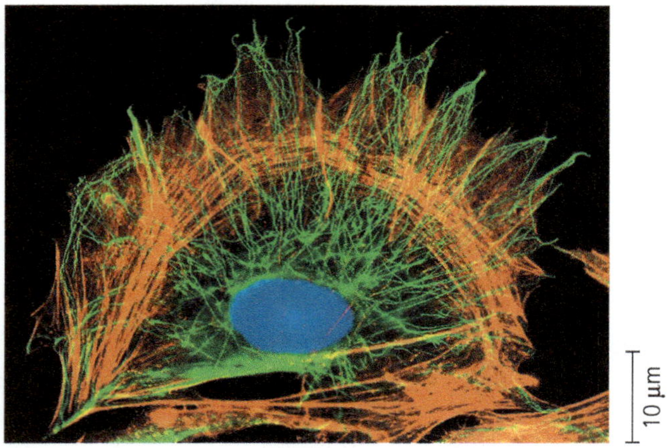

▲ **Figure 6.20 The cytoskeleton.** As shown in this fluorescence micrograph, the cytoskeleton extends throughout the cell. The cytoskeletal elements have been tagged with different fluorescent molecules: green for microtubules and red for microfilaments. A third component of the cytoskeleton, intermediate filaments, is not evident here. (The DNA in the nucleus is blue.)

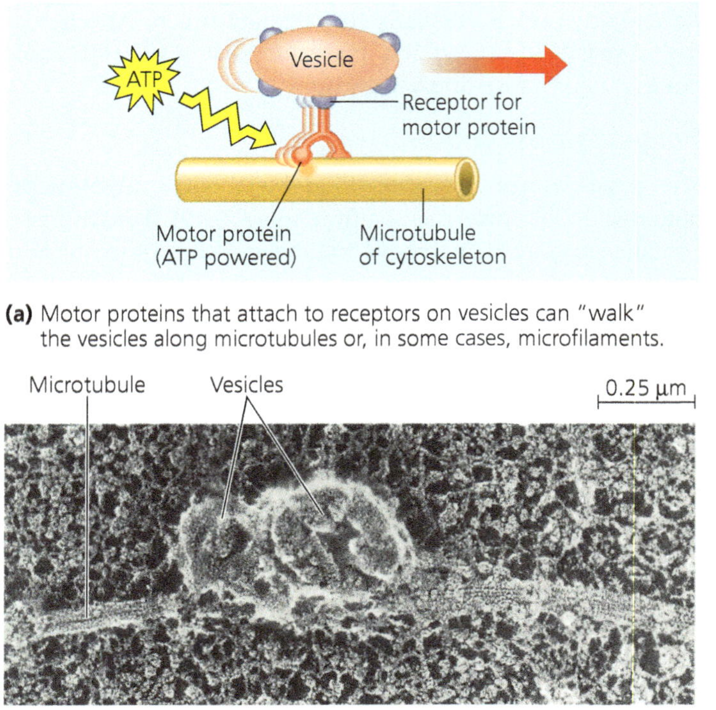

(a) Motor proteins that attach to receptors on vesicles can "walk" the vesicles along microtubules or, in some cases, microfilaments.

(b) In this SEM of a squid giant axon (a nerve cell extension), two vesicles containing neurotransmitters migrate toward the tip of the axon via the mechanism shown in (a).

▲ **Figure 6.21 Motor proteins and the cytoskeleton.**

Table 6.1 The Structure and Function of the Cytoskeleton

Property	Microtubules (Tubulin Polymers)	Microfilaments (Actin Filaments)	Intermediate Filaments
Structure	Hollow tubes; wall consists of 13 columns of tubulin molecules	Two intertwined strands of actin, each a polymer of actin subunits	Fibrous proteins supercoiled into thicker cables
Diameter	25 nm with 15-nm lumen	7 nm	8–12 nm
Protein subunits	Tubulin, a dimer consisting of α-tubulin and β-tubulin	Actin	One of several different proteins (such as keratins), depending on cell type
Main functions	Maintenance of cell shape (compression-resisting "girders") Cell motility (as in cilia or flagella) Chromosome movements in cell division Organelle movements	Maintenance of cell shape (tension-bearing elements) Changes in cell shape Muscle contraction Cytoplasmic streaming Cell motility (as in pseudopodia) Cell division (cleavage furrow formation)	Maintenance of cell shape (tension-bearing elements) Anchorage of nucleus and certain other organelles Formation of nuclear lamina

Fluorescence micrographs of fibroblasts, a favorite cell type for cell biology studies. In each, the structure of interest has been tagged with fluorescent molecules. In the first and third micrographs, the DNA in the nucleus has also been tagged (blue or orange).

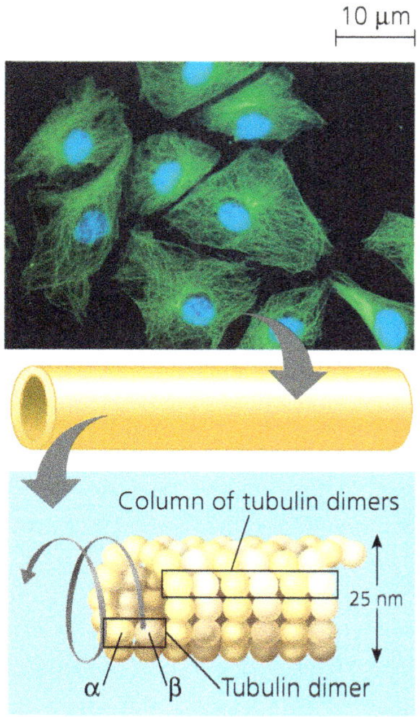

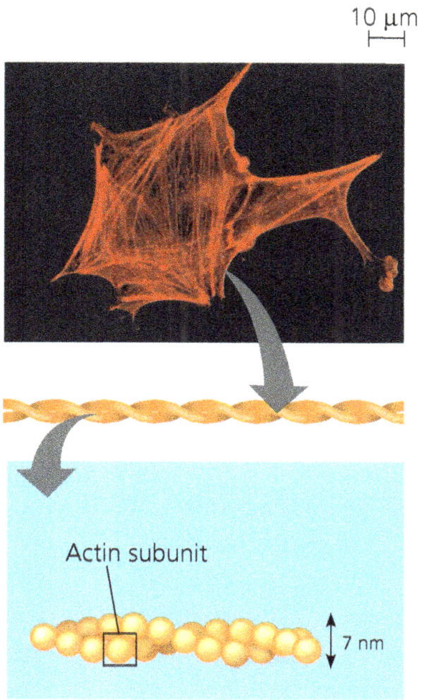

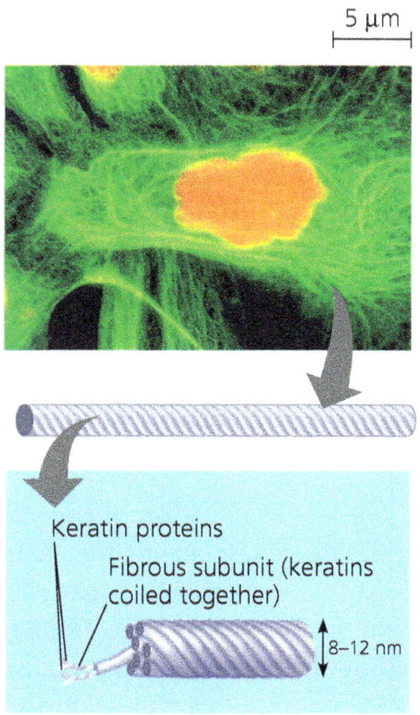

from the ER travel to the Golgi along cytoskeletal tracks. The cytoskeleton also manipulates the plasma membrane, making it bend inward to form food vacuoles or other phagocytic vesicles. And the streaming of cytoplasm that circulates materials within many large plant cells is yet another kind of cellular movement brought about by the cytoskeleton.

Components of the Cytoskeleton

Now let's look more closely at the three main types of fibers that make up the cytoskeleton: *Microtubules* are the thickest of the three types; *microfilaments* (also called actin filaments) are the thinnest; and *intermediate filaments* are fibers with diameters in a middle range (**Table 6.1**).

Microtubules

All eukaryotic cells have **microtubules**, hollow rods measuring about 25 nm in diameter and from 200 nm to 25 µm in length. The wall of the hollow tube is constructed from a globular protein called tubulin. Each tubulin protein is a *dimer*, a molecule made up of two subunits. A tubulin dimer consists of two slightly different polypeptides, α-tubulin and β-tubulin. Microtubules grow in length by adding tubulin dimers; they can also be disassembled and their tubulin used to build microtubules elsewhere in the cell. Because of the orientation of tubulin dimers, the two ends of a microtubule are slightly different. One end can accumulate or release tubulin dimers at a much higher rate than the other, thus growing and shrinking significantly during cellular activities. (This is called the "plus end," not because it can only add tubulin proteins but because it's the end where both "on" and "off" rates are much higher.)

Microtubules shape and support the cell and also serve as tracks along which organelles equipped with motor proteins can move. In addition to the example in Figure 6.21, microtubules guide secretory vesicles from the Golgi apparatus to the plasma membrane. Microtubules are also involved in the separation of chromosomes during cell division, which will be discussed in Chapter 12.

Centrosomes and Centrioles In animal cells, microtubules grow out from a **centrosome**, a region that is often located near the nucleus and is considered a "microtubule-organizing center." These microtubules function as compression-resisting girders of the cytoskeleton. Within the centrosome is a pair of **centrioles**, each composed of nine sets of triplet microtubules arranged in a ring **(Figure 6.22)**. Before an animal cell divides, the centrioles replicate. Although centrosomes with centrioles may help organize microtubule assembly in animal cells, they are not essential for this function in all eukaryotes; fungi and almost all plant cells lack centrosomes with centrioles but have well-organized microtubules. Apparently, other microtubule-organizing centers play the role of centrosomes in these cells.

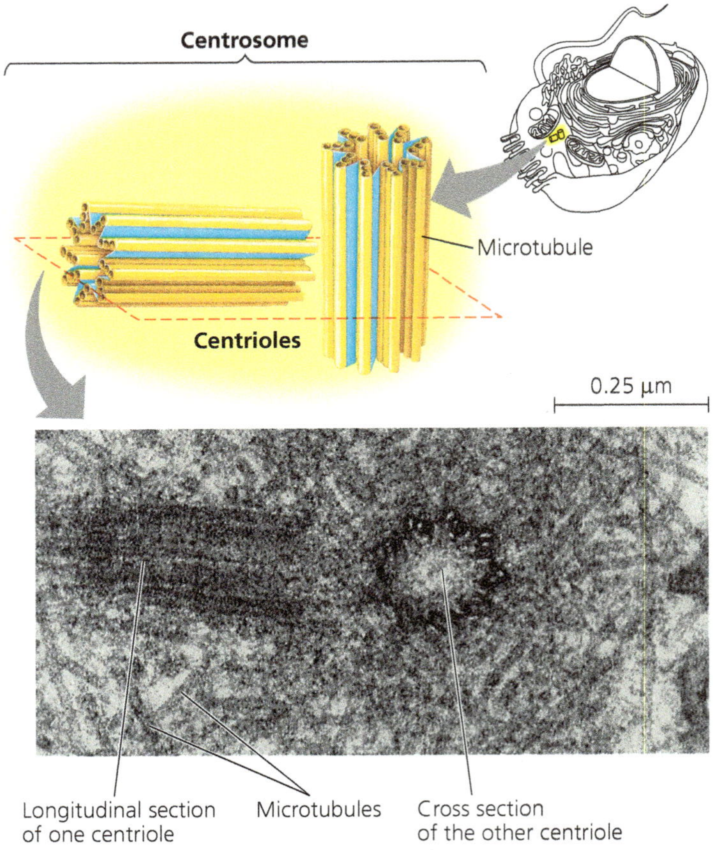

▲ **Figure 6.22 Centrosome containing a pair of centrioles.** Most animal cells have a centrosome, a region near the nucleus where the cell's microtubules are initiated. Within the centrosome is a pair of centrioles, each about 250 nm (0.25 µm) in diameter. The two centrioles are at right angles to each other, and each is made up of nine sets of three microtubules. The blue portions of the drawing represent nontubulin proteins that connect the microtubule triplets (TEM).

? *How many microtubules are in a centrosome? In the drawing, circle and label one microtubule and describe its structure. Circle and label a triplet.*

Cilia and Flagella In eukaryotes, a specialized arrangement of microtubules is responsible for the beating of **flagella** (singular, *flagellum*) and **cilia** (singular, *cilium*), microtubule-containing extensions that project from some cells. (The bacterial flagellum, shown in Figure 6.5, has a completely different structure.) Many unicellular eukaryotes are propelled through water by cilia or flagella that act as locomotor appendages, and the sperm of animals, algae, and some plants have flagella. When cilia or flagella extend from cells that are held in place as part of a tissue layer, they can move fluid over the surface of the tissue. For example, the ciliated lining of the trachea (windpipe) sweeps mucus containing trapped debris out of the lungs (see the EMs in Figure 6.3). In a woman's reproductive tract, the cilia lining the oviducts help move an egg toward the uterus.

Motile cilia usually occur in large numbers on the cell surface. They are about 0.25 µm in diameter and about 2–20 µm long. Flagella are the same diameter but longer, 10–200 µm. Also, flagella are usually limited to just one or a few per cell.

Flagella and cilia differ in their beating patterns **(Figure 6.23)**. A flagellum has an undulating motion that generates force in the same direction as the flagellum's axis, like the tail of a fish. In contrast, cilia work more like oars, with alternating power and recovery strokes generating force in a direction perpendicular to the cilium's axis, much as the oars of a racing crew boat extend outward at a right angle to the boat's forward movement.

A cilium may also act as a signal-receiving "antenna" for the cell. Cilia that have this function are generally nonmotile, and there is only one per cell. (In fact, in vertebrate animals, it appears that almost all cells have such a cilium, which is called a *primary cilium*.) Membrane proteins on this kind of cilium transmit molecular signals from the cell's environment to its interior, triggering signaling pathways that may lead to changes in the cell's activities. Cilium-based signaling appears to be crucial to brain function and to embryonic development.

Though different in length, number per cell, and beating pattern, motile cilia and flagella share a common structure. Each motile cilium and flagellum has a group of microtubules sheathed in an extension of the plasma membrane **(Figure 6.24)**. Nine doublets of microtubules are arranged in a ring; in the center of the ring are two single microtubules.

▶ Figure 6.23 A comparison of the beating of flagella and motile cilia.

(a) Motion of flagella. A flagellum usually undulates, its snakelike motion driving a cell in the same direction as the axis of the flagellum. Propulsion of a human sperm cell is an example of flagellate locomotion (LM).

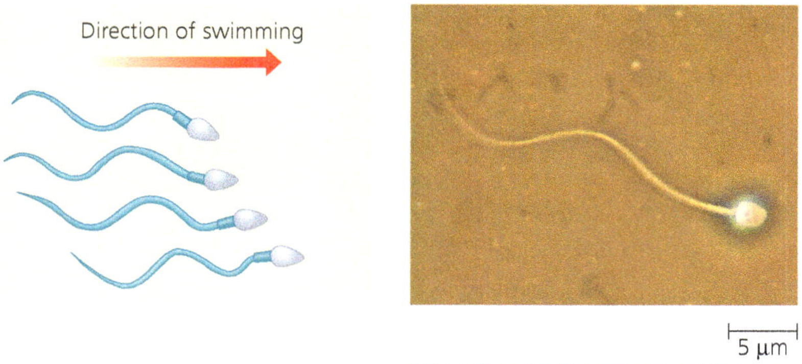

(b) Motion of cilia. Cilia have a back-and-forth motion. The rapid power stroke moves the cell in a direction perpendicular to the axis of the cilium. Then, during the slower recovery stroke, the cilium bends and sweeps sideways, closer to the cell surface. A dense nap of cilia, beating at a rate of about 40 to 60 strokes a second, covers this *Colpidium*, a freshwater protist (colorized SEM).

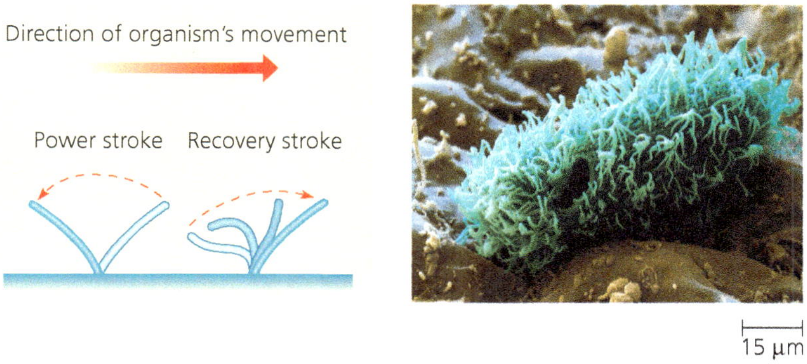

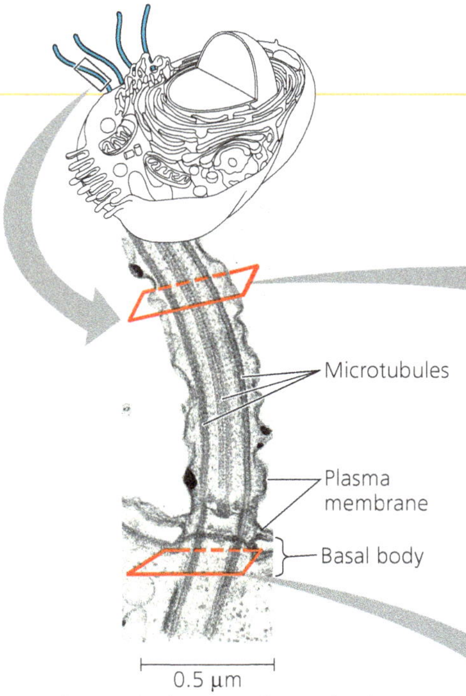

(a) A longitudinal section of a motile cilium shows microtubules running the length of the structure (TEM).

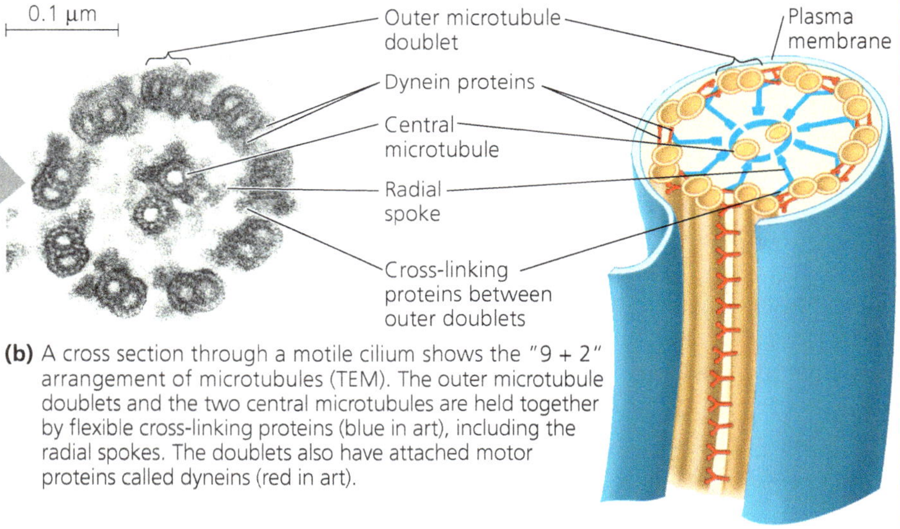

(b) A cross section through a motile cilium shows the "9 + 2" arrangement of microtubules (TEM). The outer microtubule doublets and the two central microtubules are held together by flexible cross-linking proteins (blue in art), including the radial spokes. The doublets also have attached motor proteins called dyneins (red in art).

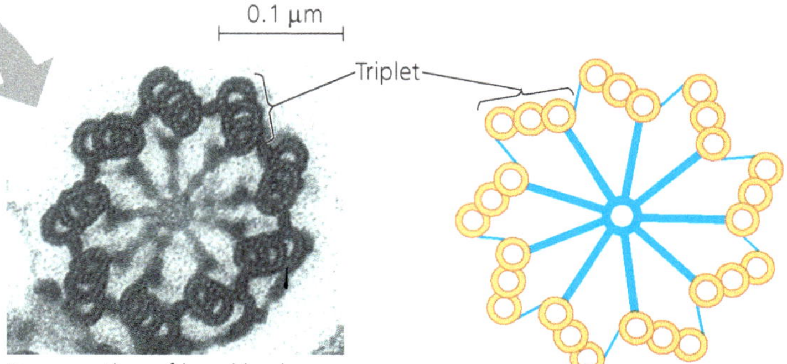

Cross section of basal body

(c) Basal body: The nine outer doublets of a cilium or flagellum extend into the basal body, where each doublet joins another microtubule to form a ring of nine triplets. Each triplet is connected to the next by nontubulin proteins (thinner blue lines in diagram). This is a "9 + 0" arrangement: the two central microtubules are not present because they terminate above the basal body (TEM).

▲ Figure 6.24 Structure of a flagellum or motile cilium.

DRAW IT *In (a), circle the central pair of microtubules. Show where they terminate, and explain why they aren't seen in the cross section of the basal body in (c).*

This arrangement, referred to as the "9 + 2" pattern, is found in nearly all eukaryotic flagella and motile cilia. (Nonmotile primary cilia have a "9 + 0" pattern, lacking the central pair of microtubules.) The microtubule assembly of a cilium or flagellum is anchored in the cell by a **basal body**, which is structurally very similar to a centriole, with microtubule triplets in a "9 + 0" pattern. In fact, in many animals (including humans), the basal body of the fertilizing sperm's flagellum enters the egg and becomes a centriole.

In flagella and motile cilia, flexible cross-linking proteins, evenly spaced along the length of the cilium or flagellum, connect the outer doublets to each other and to the two central microtubules. Each outer doublet also has pairs of protruding proteins spaced along its length and reaching toward the neighboring doublet; these are large motor proteins called **dyneins**, each composed of several polypeptides. Dyneins are responsible for the bending movements of the organelle. A dynein molecule performs a complex cycle of movements caused by changes in the shape of the protein, with ATP providing the energy for these changes **(Figure 6.25)**.

The mechanics of dynein-based bending involve a process that resembles walking. A typical dynein protein has two "feet" that "walk" along the microtubule of the adjacent doublet, one foot maintaining contact while the other releases and reattaches one step farther along the microtubule. Without any restraints on the movement of the microtubule doublets, one doublet would continue to "walk" along and slide past the surface of the other, elongating the cilium or flagellum rather than bending it (see Figure 6.25a). For lateral movement of a cilium or flagellum, the dynein "walking" must have something to pull against, as when the muscles in your leg pull against your bones to move your knee. In cilia and flagella, the microtubule doublets seem to be held in place by the cross-linking proteins just inside the outer doublets and by the radial spokes and other structural elements. Thus, neighboring doublets cannot slide past each other very far. Instead, the forces exerted by dynein "walking" cause the doublets to curve, bending the cilium or flagellum (see Figure 6.25b and c).

Microfilaments (Actin Filaments)

Microfilaments are solid rods about 7 nm in diameter. They are also called actin filaments because they are built from molecules of **actin**, a globular protein. A microfilament is a twisted double chain of actin subunits (see Table 6.1). Besides occurring as linear filaments, microfilaments can form structural networks when certain proteins bind along the side of an actin filament and allow a new filament to extend as a branch. Like microtubules, microfilaments seem to be present in all eukaryotic cells.

In contrast to the compression-resisting role of microtubules, the structural role of microfilaments in the cytoskeleton is to bear tension (pulling forces). A three-dimensional network formed by microfilaments just inside the plasma

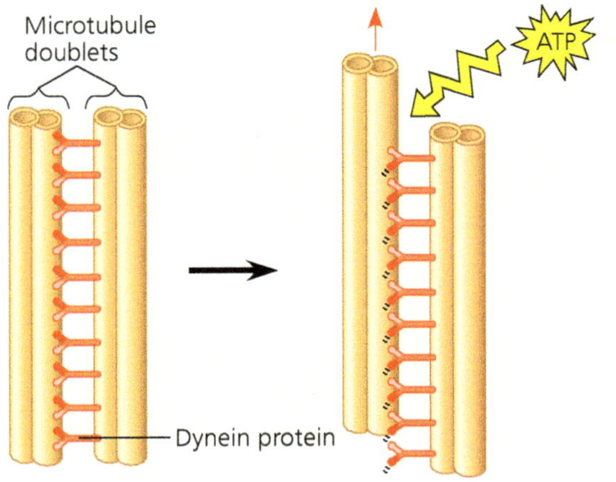

(a) Effect of unrestrained dynein movement. If a cilium or flagellum had no cross-linking proteins, the two feet of each dynein along one doublet (powered by ATP) would alternately grip and release the adjacent doublet. This "walking" motion would push the adjacent doublet up. Instead of bending, the doublets would slide past each other.

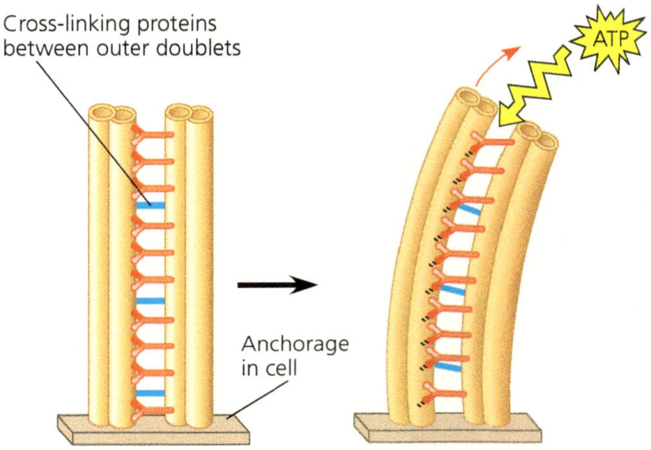

(b) Effect of cross-linking proteins. In a cilium or flagellum, two adjacent doublets cannot slide far because they are physically restrained by proteins, so they bend. (Only two of the nine outer doublets in Figure 6.24b are shown here.)

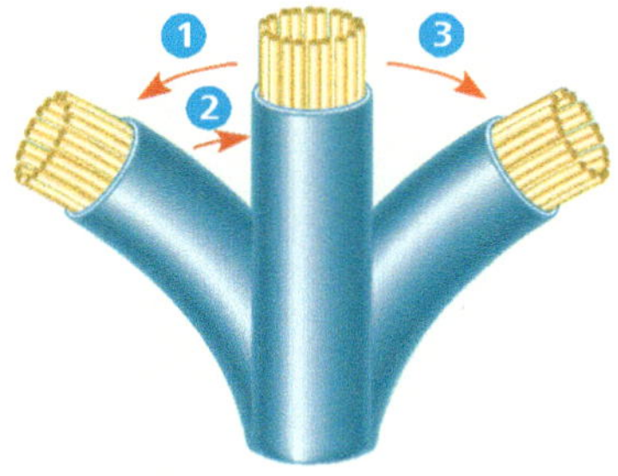

(c) Wavelike motion. Synchronized cycles of movement of many dyneins probably cause a bend to begin at the base of the cilium or flagellum and move outward toward the tip. Many successive bends, such as the ones shown here to the left and right, result in a wavelike motion. In this diagram, the two central microtubules and the cross-linking proteins are not shown.

▲ **Figure 6.25 How dynein "walking" moves flagella and cilia.**

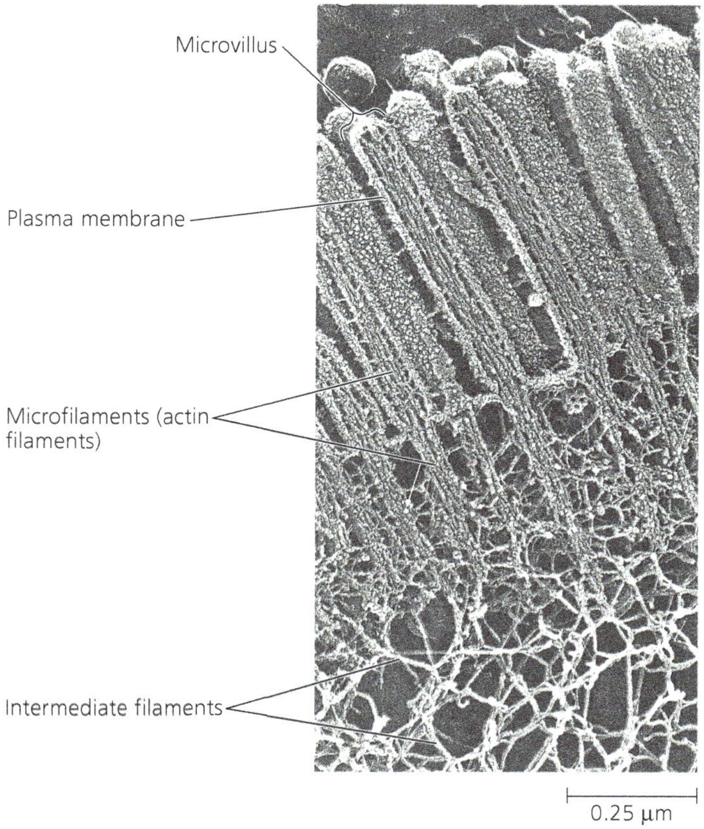

▲ **Figure 6.26 A structural role of microfilaments.** The surface area of this nutrient-absorbing intestinal cell is increased by its many microvilli (singular, *microvillus*), cellular extensions reinforced by bundles of microfilaments. These actin filaments are anchored to a network of intermediate filaments (TEM).

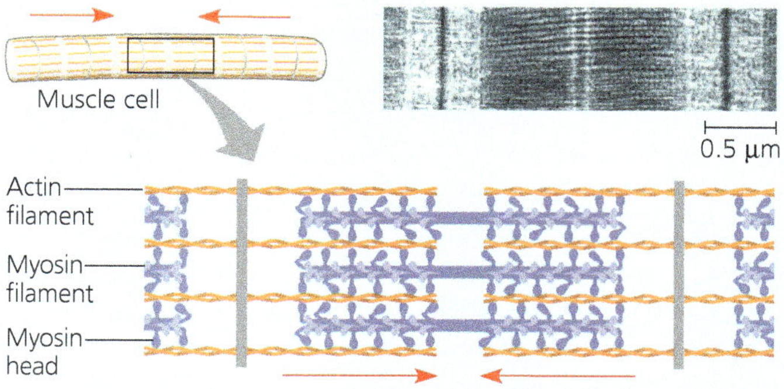

(a) Myosin motors in muscle cell contraction. The "walking" of myosin projections (the so-called heads) drives the parallel myosin and actin filaments past each other so that the actin filaments approach each other in the middle (red arrows). This shortens the muscle cell. Muscle contraction involves the shortening of many muscle cells at the same time (TEM).

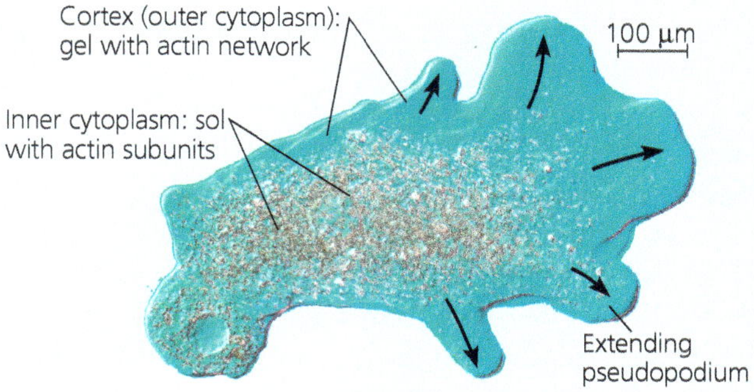

(b) Amoeboid movement. Interaction of actin filaments with myosin causes contraction of the cell, pulling the cell's trailing end (at left) forward (to the right) (LM).

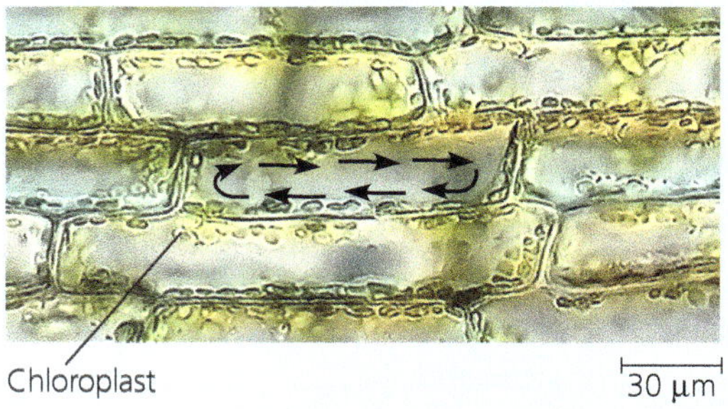

(c) Cytoplasmic streaming in plant cells. A layer of cytoplasm cycles around the cell, moving over a carpet of parallel actin filaments. Myosin motors attached to organelles in the fluid cytosol may drive the streaming by interacting with the actin (LM).

▲ **Figure 6.27 Microfilaments and motility.** In these three examples, interactions between actin filaments and motor proteins bring about cell movement.

membrane (*cortical microfilaments*) helps support the cell's shape (see Figure 6.8). This network gives the outer cytoplasmic layer of a cell, called the **cortex**, the semisolid consistency of a gel, in contrast with the more fluid (*sol*) state of the interior cytoplasm. In animal cells specialized for transporting materials across the plasma membrane, such as intestinal cells, bundles of microfilaments make up the core of microvilli, delicate projections that increase the cell's surface area (Figure 6.26).

Microfilaments are well known for their role in cell motility, particularly as part of the contractile apparatus of muscle cells. Thousands of actin filaments are arranged parallel to one another along the length of a muscle cell, interdigitated with thicker filaments made of a protein called **myosin** (Figure 6.27a). Like dynein when it interacts with microtubules, myosin acts as a motor protein by means of projections that "walk" along the actin filaments. Contraction of the muscle cell results from the actin and myosin filaments sliding past one another in this way, shortening the cell. In other kinds of cells, actin filaments are associated with myosin in miniature and less elaborate versions of the arrangement in muscle cells. These actin-myosin aggregates are responsible for localized contractions of cells. For example, a contracting belt of microfilaments forms a cleavage furrow that pinches a dividing animal cell into two daughter cells.

Localized contraction brought about by actin and myosin also plays a role in amoeboid movement (Figure 6.27b). A cell such as an amoeba crawls along a surface by extending cellular extensions called **pseudopodia** (from the Greek *pseudes*, false, and *pod*, foot), and moving toward them. Pseudopodia extend by assembly of actin subunits into microfilament networks that convert cytoplasm from a sol to a gel inside these cell projections. Cell surface proteins on the pseudopodium make strong attachments to the "road." Next, the interaction of microfilaments with myosin near the cell's trailing end causes contraction of that region, loosening its cell-surface attachments and pulling it forward toward the pseudopodia. Amoebae lacking myosin can still form pseudopodia, but forward movement is greatly slowed. Amoebas are not the only cells that move by crawling; so do many cells in the animal body, including some white blood cells.

In plant cells, both actin-myosin interactions and sol-gel transformations brought about by actin may be involved in **cytoplasmic streaming**, a circular flow of cytoplasm within cells **(Figure 6.27c)**. This movement, which is especially common in large plant cells, speeds the distribution of materials within the cell.

Intermediate Filaments

Intermediate filaments are named for their diameter, which, at 8–12 nm, is larger than the diameter of microfilaments but smaller than that of microtubules (see Table 6.1, p. 113). Specialized for bearing tension (like microfilaments), intermediate filaments are a diverse class of cytoskeletal elements. Each type is constructed from a particular molecular subunit belonging to a family of proteins whose members include the keratins. Microtubules and microfilaments, in contrast, are consistent in diameter and composition in all eukaryotic cells.

Intermediate filaments are more permanent fixtures of cells than are microfilaments and microtubules, which are often disassembled and reassembled in various parts of a cell. Even after cells die, intermediate filament networks often persist; for example, the outer layer of our skin consists of dead skin cells full of keratin proteins. Chemical treatments that remove microfilaments and microtubules from the cytoplasm of living cells leave a web of intermediate filaments that retains its original shape. Such experiments suggest that intermediate filaments are especially sturdy and that they play an important role in reinforcing the shape of a cell and fixing the position of certain organelles. For instance, the nucleus typically sits within a cage made of intermediate filaments, fixed in location by branches of the filaments that extend into the cytoplasm. Other intermediate filaments make up the nuclear lamina, which lines the interior of the nuclear envelope (see Figure 6.9). By supporting a cell's shape, intermediate filaments help the cell carry out its specific function. For example, the long extensions (axons) of nerve cells that transmit impulses are strengthened by intermediate filaments. Thus, the various kinds of intermediate filaments may function together as the permanent framework of the entire cell.

CONCEPT CHECK 6.6

1. Describe shared features of microtubule-based motion of flagella and microfilament-based muscle contraction.
2. How do cilia and flagella bend?
3. **WHAT IF?** Males afflicted with Kartagener's syndrome are sterile because of immotile sperm, and they tend to suffer from lung infections. This disorder has a genetic basis. Suggest what the underlying defect might be.

For suggested answers, see Appendix A.

CONCEPT 6.7

Extracellular components and connections between cells help coordinate cellular activities

Having crisscrossed the cell to explore its interior components, we complete our tour of the cell by returning to the surface of this microscopic world, where there are additional structures with important functions. The plasma membrane is usually regarded as the boundary of the living cell, but most cells synthesize and secrete materials that are extracellular, or external to the plasma membrane. Although these materials and the structures they form are outside the cell, their study is important to cell biology because they are involved in a great many cellular functions.

Cell Walls of Plants

The **cell wall** is an extracellular structure of plant cells that distinguishes them from animal cells (see Figure 6.8). The wall protects the plant cell, maintains its shape, and prevents excessive uptake of water. On the level of the whole plant, the strong walls of specialized cells hold the plant up against the force of gravity. Prokaryotes, fungi, and some protists also have cell walls, as you saw in Figures 6.5 and 6.8, but we will postpone discussion of them until Unit Five.

Plant cell walls are much thicker than the plasma membrane, ranging from 0.1 µm to several micrometers. The exact chemical composition of the wall varies from species to species and even from one cell type to another in the same plant, but the basic design of the wall is consistent. Microfibrils made of the polysaccharide cellulose (see Figure 5.8) are synthesized by an enzyme called cellulose synthase and secreted to the extracellular space, where they become embedded in a matrix of other polysaccharides and proteins. This combination of materials, strong fibers in a "ground substance" (matrix), is the same basic architectural design found in steel-reinforced concrete and in fiberglass.

A young plant cell first secretes a relatively thin and flexible wall called the **primary cell wall (Figure 6.28)**. In actively

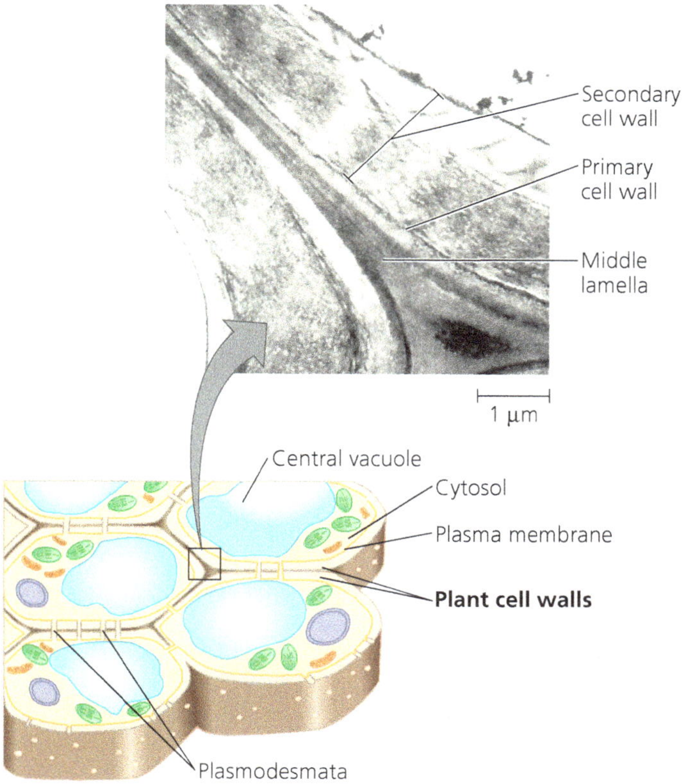

▲ **Figure 6.28 Plant cell walls.** The drawing shows several cells, each with a large vacuole, a nucleus, and several chloroplasts and mitochondria. The transmission electron micrograph shows the cell walls where two cells come together. The multilayered partition between plant cells consists of adjoining walls individually secreted by the cells.

growing cells, the cellulose fibrils are oriented at right angles to the direction of cell expansion. Researchers investigated the role of microtubules in orienting these cellulose fibrils (**Figure 6.29**). Their observations strongly support the idea that microtubules in the cell cortex guide cellulose synthase as it synthesizes and deposits cellulose fibrils. By orienting cellulose deposition, microtubules thus affect the growth pattern of the cells.

Between primary walls of adjacent cells is the **middle lamella**, a thin layer rich in sticky polysaccharides called pectins. The middle lamella glues adjacent cells together (see Figure 6.28). (Pectin is used as a thickening agent in jams and jellies.) When the cell matures and stops growing, it strengthens its wall. Some plant cells do this simply by secreting hardening substances into the primary wall. Other cells add a **secondary cell wall** between the plasma membrane and the primary wall. The secondary wall, often deposited in several laminated layers, has a strong and durable matrix that affords the cell protection and support. Wood, for example, consists mainly of secondary walls. Plant cell walls are usually perforated by channels between adjacent cells called plasmodesmata (see Figure 6.28), which will be discussed shortly.

▼ **Figure 6.29** INQUIRY

What role do microtubules play in orienting deposition of cellulose in cell walls?

EXPERIMENT Previous experiments on preserved plant tissues had shown alignment of microtubules in the cell cortex with cellulose fibrils in the cell wall. Also, drugs that disrupted microtubules were observed to cause disoriented cellulose fibrils. To further investigate the possible role of cortical microtubules in guiding cellulose fibril deposition, David Ehrhardt and colleagues at Stanford University used a type of confocal microscopy to study cell wall deposition in living cells. In these cells, they labeled both cellulose synthase and microtubules with fluorescent markers and observed them over time.

RESULTS Each fluorescence image below represents a combination of 30 images taken over a 5-minute period to detect the movement of cellulose synthase and microtubules. These two coincided highly over time. The labeling molecules caused cellulose synthase to fluoresce green and the microtubules to fluoresce red. The arrowheads indicate prominent areas where the two are seen to align.

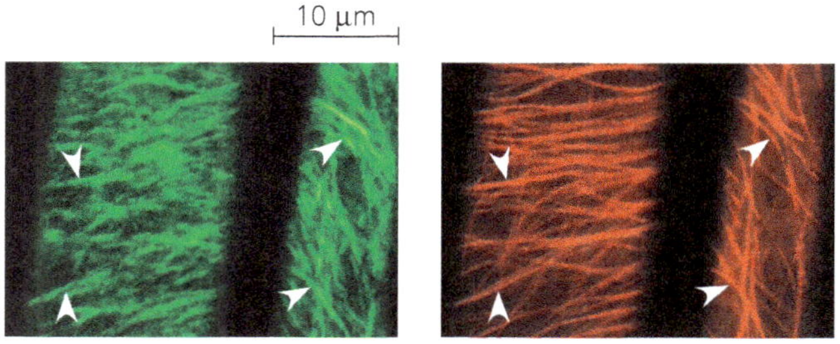

Distribution of cellulose synthase over time

Distribution of microtubules over time

CONCLUSION The organization of microtubules appears to directly guide the path of cellulose synthase as it lays down cellulose, thus determining the orientation of cellulose fibrils.

SOURCE A. R. Paradez et al., Visualization of cellulose synthase demonstrates functional association with microtubules, *Science* 312:1491–1495 (2006).

WHAT IF? In a second experiment, the researchers exposed the plant cells to blue light, previously shown to cause reorientation of microtubules. What events would you predict would follow blue light exposure?

The Extracellular Matrix (ECM) of Animal Cells

Although animal cells lack walls akin to those of plant cells, they do have an elaborate **extracellular matrix (ECM)**. The main ingredients of the ECM are glycoproteins and other carbohydrate-containing molecules secreted by the cells. (Recall that glycoproteins are proteins with covalently bonded carbohydrate, usually short chains of sugars.) The most abundant glycoprotein in the ECM of most animal cells is **collagen**, which forms strong fibers outside the cells (see Figure 5.20). In fact, collagen accounts for about 40% of the total protein in the human body. The collagen fibers are embedded in a network woven out of **proteoglycans** secreted

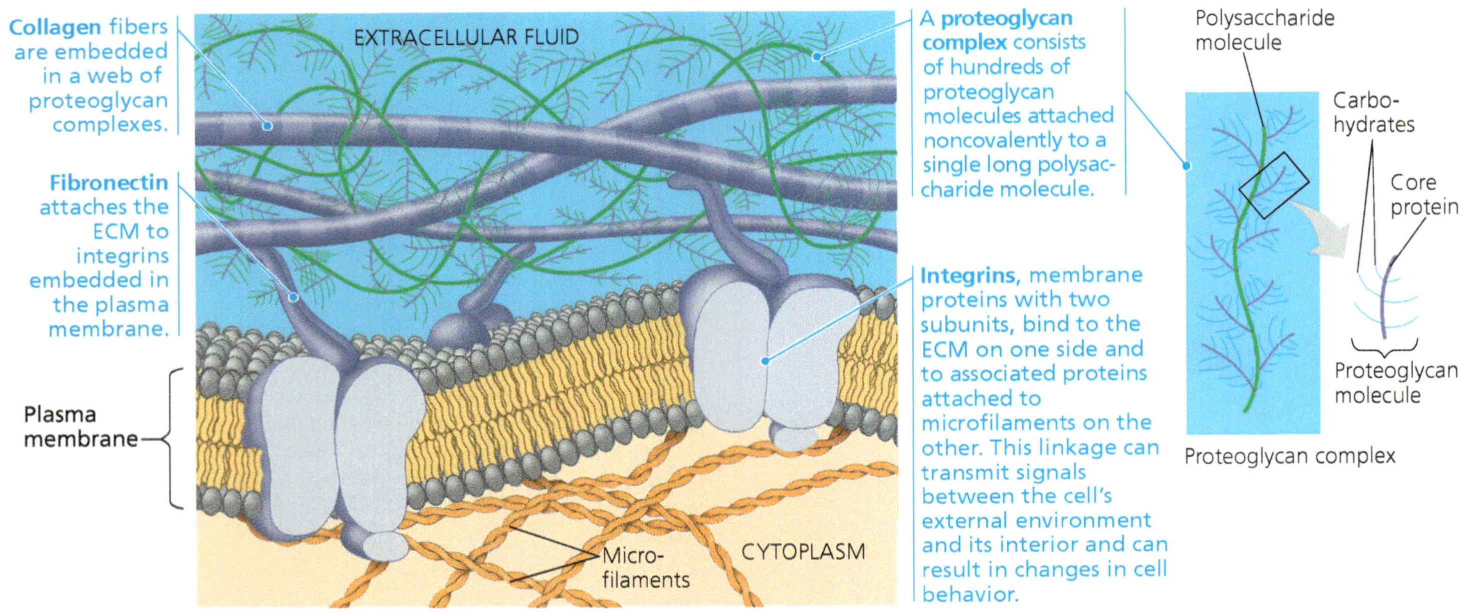

▲ **Figure 6.30 Extracellular matrix (ECM) of an animal cell.** The molecular composition and structure of the ECM vary from one cell type to another. In this example, three different types of ECM molecules are present: proteoglycans, collagen, and fibronectin.

by cells **(Figure 6.30)**. A proteoglycan molecule consists of a small core protein with many carbohydrate chains covalently attached, so that it may be up to 95% carbohydrate. Large proteoglycan complexes can form when hundreds of proteoglycan molecules become noncovalently attached to a single long polysaccharide molecule, as shown in Figure 6.30. Some cells are attached to the ECM by ECM glycoproteins such as **fibronectin**. Fibronectin and other ECM proteins bind to cell-surface receptor proteins called **integrins** that are built into the plasma membrane. Integrins span the membrane and bind on their cytoplasmic side to associated proteins attached to microfilaments of the cytoskeleton. The name *integrin* is based on the word *integrate*: Integrins are in a position to transmit signals between the ECM and the cytoskeleton and thus to integrate changes occurring outside and inside the cell.

Current research on fibronectin, other ECM molecules, and integrins is revealing the influential role of the extracellular matrix in the lives of cells. By communicating with a cell through integrins, the ECM can regulate a cell's behavior. For example, some cells in a developing embryo migrate along specific pathways by matching the orientation of their microfilaments to the "grain" of fibers in the extracellular matrix. Researchers have also learned that the extracellular matrix around a cell can influence the activity of genes in the nucleus. Information about the ECM probably reaches the nucleus by a combination of mechanical and chemical signaling pathways. Mechanical signaling involves fibronectin, integrins, and microfilaments of the cytoskeleton. Changes in the cytoskeleton may in turn trigger chemical signaling pathways inside the cell, leading to changes in the set of proteins being made by the cell and therefore changes in the cell's function. In this way, the extracellular matrix of a particular tissue may help coordinate the behavior of all the cells of that tissue. Direct connections between cells also function in this coordination, as we discuss next.

Cell Junctions

Cells in an animal or plant are organized into tissues, organs, and organ systems. Neighboring cells often adhere, interact, and communicate via sites of direct physical contact.

Plasmodesmata in Plant Cells

It might seem that the nonliving cell walls of plants would isolate plant cells from one another. But in fact, as shown in **Figure 6.31**, cell walls are perforated with **plasmodesmata** (singular, *plasmodesma*; from the Greek *desmos*, to bind),

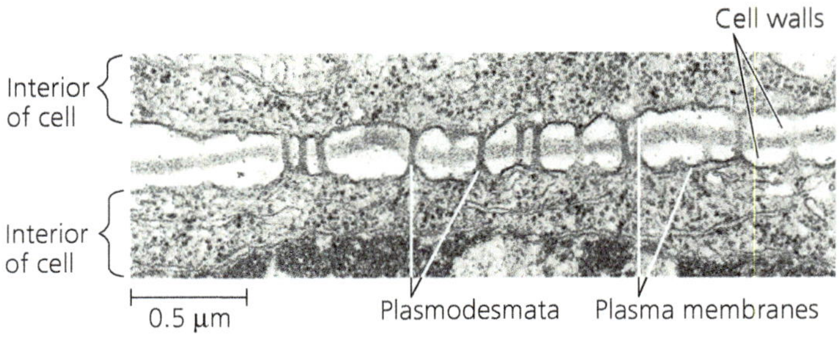

▲ **Figure 6.31 Plasmodesmata between plant cells.** The cytoplasm of one plant cell is continuous with the cytoplasm of its neighbors via plasmodesmata, cytoplasmic channels through the cell walls (TEM).

membrane-lined channels filled with cytoplasm. Cytosol passes through the plasmodesmata and joins the internal chemical environments of adjacent cells. These connections unify most of the plant into one living continuum. The plasma membranes of adjacent cells line the channel of each plasmodesma and thus are continuous. Water and small solutes can pass freely from cell to cell, and recent experiments have shown that in some circumstances, certain proteins and RNA molecules can also do this (see Concept 36.6). The macromolecules transported to neighboring cells appear to reach the plasmodesmata by moving along fibers of the cytoskeleton.

Tight Junctions, Desmosomes, and Gap Junctions in Animal Cells

In animals, there are three main types of cell junctions: *tight junctions*, *desmosomes*, and *gap junctions*. (Gap junctions are most like the plasmodesmata of plants, although gap junction pores are not lined with membrane.) All three types of cell junctions are especially common in epithelial tissue, which lines the external and internal surfaces of the body. **Figure 6.32** uses epithelial cells of the intestinal lining to illustrate these junctions.

▼ **Figure 6.32**

Exploring Cell Junctions in Animal Tissues

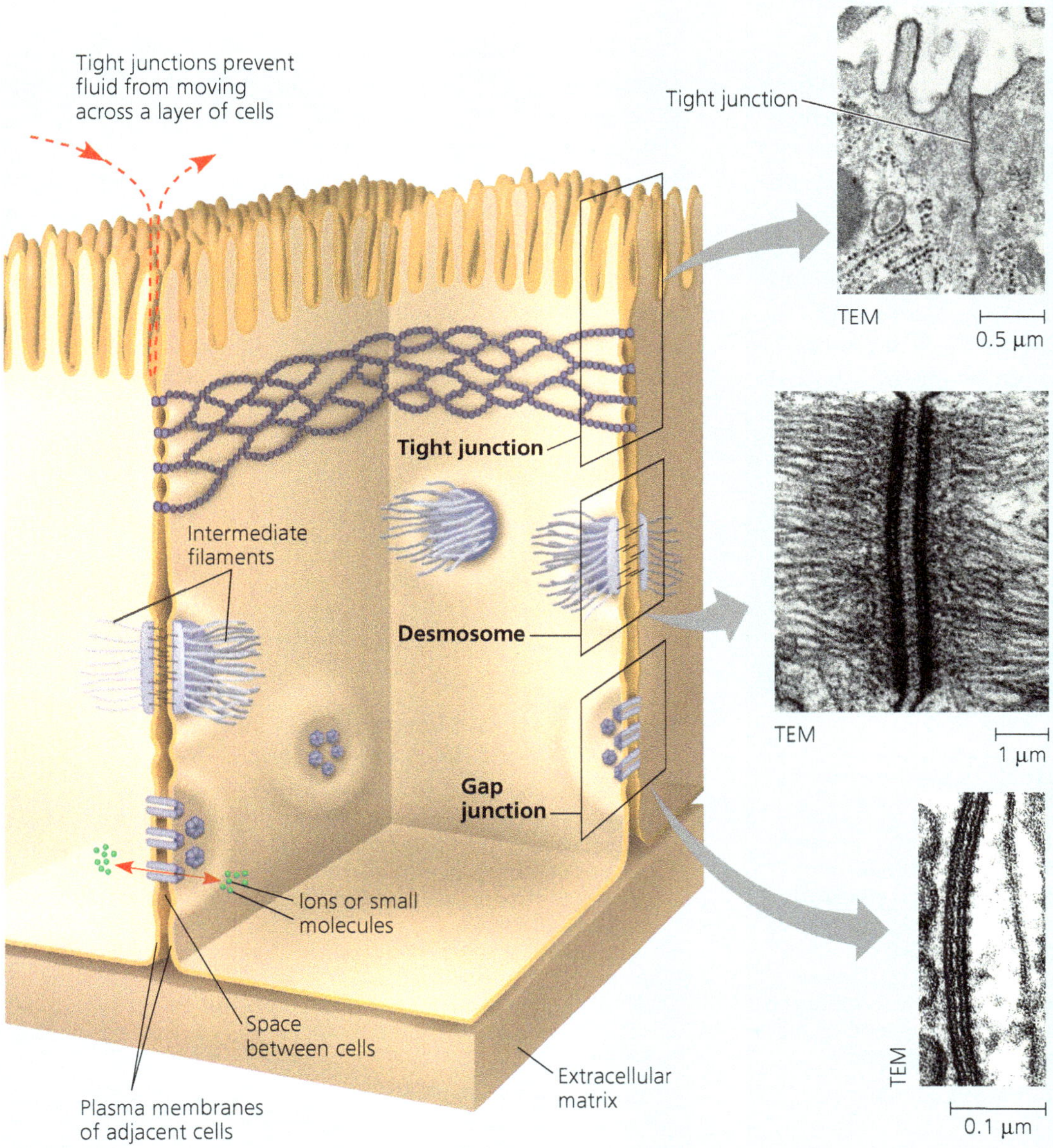

Tight Junctions

At **tight junctions**, the plasma membranes of neighboring cells are very tightly pressed against each other, bound together by specific proteins (purple). Forming continuous seals around the cells, tight junctions prevent leakage of extracellular fluid across a layer of epithelial cells. For example, tight junctions between skin cells make us watertight by preventing leakage between cells in our sweat glands.

Desmosomes

Desmosomes (also called *anchoring junctions*) function like rivets, fastening cells together into strong sheets. Intermediate filaments made of sturdy keratin proteins anchor desmosomes in the cytoplasm. Desmosomes attach muscle cells to each other in a muscle. Some "muscle tears" involve the rupture of desmosomes.

Gap Junctions

Gap junctions (also called *communicating junctions*) provide cytoplasmic channels from one cell to an adjacent cell and in this way are similar in their function to the plasmodesmata in plants. Gap junctions consist of membrane proteins that surround a pore through which ions, sugars, amino acids, and other small molecules may pass. Gap junctions are necessary for communication between cells in many types of tissues, such as heart muscle, and in animal embryos.

CONCEPT CHECK 6.7

1. In what way are the cells of plants and animals structurally different from single-celled eukaryotes?
2. **WHAT IF?** If the plant cell wall or the animal extracellular matrix were impermeable, what effect would this have on cell function?
3. **MAKE CONNECTIONS** The polypeptide chain that makes up a tight junction weaves back and forth through the membrane four times, with two extracellular loops, and one loop plus short C-terminal and N-terminal tails in the cytoplasm. Looking at Figure 5.16 (p. 79), what would you predict about the amino acid sequence of the tight-junction protein?

For suggested answers, see Appendix A.

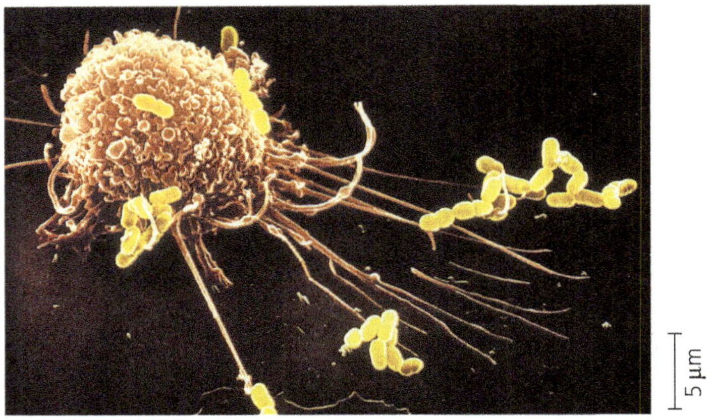

▲ **Figure 6.33 The emergence of cellular functions.** The ability of this macrophage (brown) to recognize, apprehend, and destroy bacteria (yellow) is a coordinated activity of the whole cell. Its cytoskeleton, lysosomes, and plasma membrane are among the components that function in phagocytosis (colorized SEM).

The Cell: A Living Unit Greater Than the Sum of Its Parts

From our panoramic view of the cell's compartmental organization to our close-up inspection of each organelle's architecture, this tour of the cell has provided many opportunities to correlate structure with function. (This would be a good time to review cell structure by returning to Figure 6.8, on pp. 100 and 101.) But even as we dissect the cell, remember that none of its components works alone. As an example of cellular integration, consider the microscopic scene in **Figure 6.33**. The large cell is a macrophage (see Figure 6.13a). It helps defend the mammalian body against infections by ingesting bacteria (the smaller cells) into phagocytic vesicles. The macrophage crawls along a surface and reaches out to the bacteria with thin pseudopodia (called filopodia). Actin filaments interact with other elements of the cytoskeleton in these movements. After the macrophage engulfs the bacteria, they are destroyed by lysosomes. The elaborate endomembrane system produces the lysosomes. The digestive enzymes of the lysosomes and the proteins of the cytoskeleton are all made on ribosomes. And the synthesis of these proteins is programmed by genetic messages dispatched from the DNA in the nucleus. All these processes require energy, which mitochondria supply in the form of ATP. Cellular functions arise from cellular order: The cell is a living unit greater than the sum of its parts.

6 CHAPTER REVIEW

SUMMARY OF KEY CONCEPTS

CONCEPT 6.1

Biologists use microscopes and the tools of biochemistry to study cells (pp. 94–97)

- Improvements in microscopy that affect the parameters of magnification, resolution, and contrast have catalyzed progress in the study of cell structure. **Light microscopy** (LM) and **electron microscopy** (EM), as well as other types, remain important tools.
- Cell biologists can obtain pellets enriched in particular cellular components by centrifuging disrupted cells at sequential speeds, a process known as **cell fractionation**. Larger cellular components are in the pellet after lower-speed centrifugation, and smaller components are in the pellet after higher-speed centrifugation.

? *How do microscopy and biochemistry complement each other to reveal cell structure and function?*

CONCEPT 6.2

Eukaryotic cells have internal membranes that compartmentalize their functions (pp. 98–102)

- All cells are bounded by a **plasma membrane**.
- **Prokaryotic cells** lack nuclei and other membrane-enclosed **organelles**, while **eukaryotic cells** have internal membranes that compartmentalize cellular functions.
- The surface-to-volume ratio is an important parameter affecting cell size and shape.
- Plant and animal cells have most of the same organelles: a nucleus, endoplasmic reticulum, Golgi apparatus, and mitochondria. Some organelles are found only in plant or in animal cells. Chloroplasts are present only in cells of photosynthetic eukaryotes.

? *Explain how the compartmental organization of a eukaryotic cell contributes to its biochemical functioning.*

	Cell Component	Structure	Function
CONCEPT 6.3 **The eukaryotic cell's genetic instructions are housed in the nucleus and carried out by the ribosomes (pp. 102–104)** ? *Describe the relationship between the nucleus and ribosomes.*	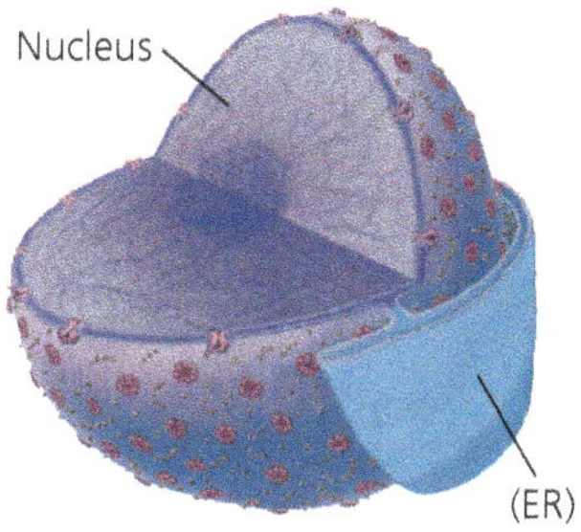	Surrounded by nuclear envelope (double membrane) perforated by nuclear pores; nuclear envelope continuous with endoplasmic reticulum (ER)	Houses chromosomes, which are made of chromatin (DNA and proteins); contains nucleoli, where ribosomal subunits are made; pores regulate entry and exit of materials
	Ribosome 	Two subunits made of ribosomal RNA and proteins; can be free in cytosol or bound to ER	Protein synthesis
CONCEPT 6.4 **The endomembrane system regulates protein traffic and performs metabolic functions in the cell (pp. 104–109)** ? *Describe the key role played by transport vesicles in the endomembrane system.*	Endoplasmic reticulum 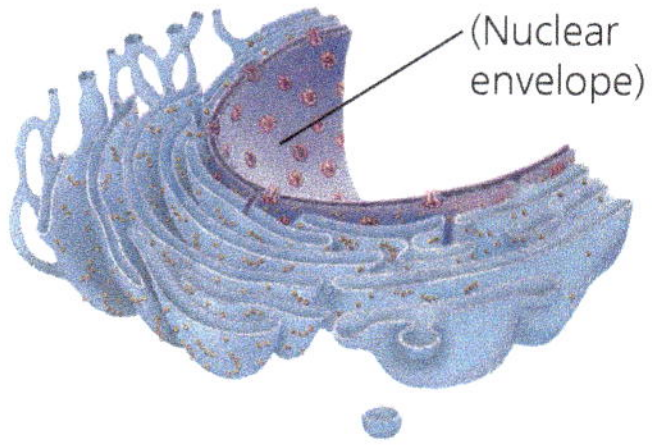	Extensive network of membrane-bounded tubules and sacs; membrane separates lumen from cytosol; continuous with nuclear envelope	Smooth ER: synthesis of lipids, metabolism of carbohydrates, Ca^{2+} storage, detoxification of drugs and poisons Rough ER: aids in synthesis of secretory and other proteins from bound ribosomes; adds carbohydrates to proteins to make glycoproteins; produces new membrane
	Golgi apparatus 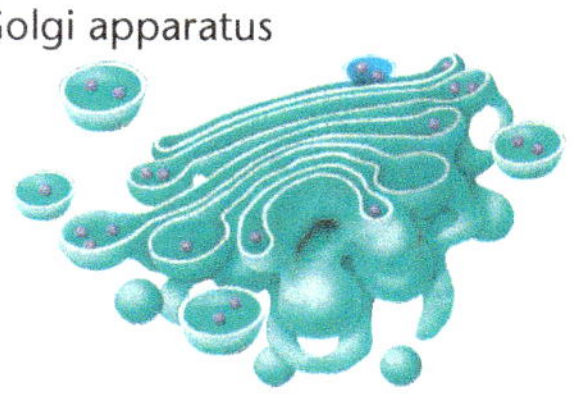	Stacks of flattened membranous sacs; has polarity (*cis* and *trans* faces)	Modification of proteins, carbohydrates on proteins, and phospholipids; synthesis of many polysaccharides; sorting of Golgi products, which are then released in vesicles
	Lysosome 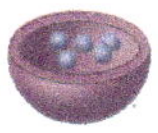	Membranous sac of hydrolytic enzymes (in animal cells)	Breakdown of ingested substances, cell macromolecules, and damaged organelles for recycling
	Vacuole 	Large membrane-bounded vesicle	Digestion, storage, waste disposal, water balance, cell growth, and protection
CONCEPT 6.5 **Mitochondria and chloroplasts change energy from one form to another (pp. 109–112)** ? *What is the endosymbiont theory?*	Mitochondrion 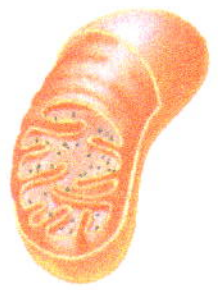	Bounded by double membrane; inner membrane has infoldings (cristae)	Cellular respiration
	Chloroplast 	Typically two membranes around fluid stroma, which contains thylakoids stacked into grana (in cells of photosynthetic eukaryotes, including plants)	Photosynthesis
	Peroxisome 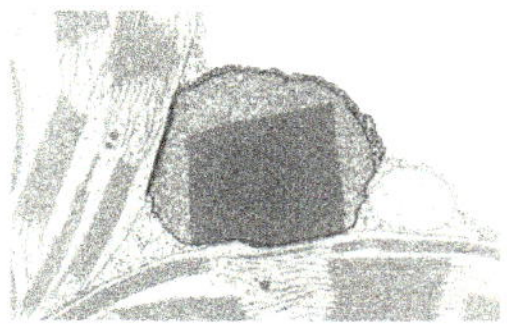	Specialized metabolic compartment bounded by a single membrane	Contains enzymes that transfer hydrogen atoms from substrates to oxygen, producing hydrogen peroxide (H_2O_2) as a by-product; H_2O_2 is converted to water by another enzyme

CONCEPT 6.6

The cytoskeleton is a network of fibers that organizes structures and activities in the cell (pp. 112–118)

- The **cytoskeleton** functions in structural support for the cell and in motility and signal transmission.
- **Microtubules** shape the cell, guide organelle movement, and separate chromosomes in dividing cells. **Cilia** and **flagella** are motile appendages containing microtubules. Primary cilia also play sensory and signaling roles. **Microfilaments** are thin rods functioning in muscle contraction, amoeboid movement, **cytoplasmic streaming**, and microvillus support. **Intermediate filaments** support cell shape and fix organelles in place.

? *Describe the role of motor proteins inside the eukaryotic cell and in whole-cell movement.*

CONCEPT 6.7

Extracellular components and connections between cells help coordinate cellular activities (pp. 118–122)

- Plant **cell walls** are made of cellulose fibers embedded in other polysaccharides and proteins. Cellulose deposition is oriented along microtubules.
- Animal cells secrete glycoproteins and proteoglycans that form the **extracellular matrix (ECM)**, which functions in support, adhesion, movement, and regulation.
- Cell junctions connect neighboring cells in plants and animals. Plants have **plasmodesmata** that pass through adjoining cell walls. Animal cells have **tight junctions**, **desmosomes**, and **gap junctions**.

? *Compare the composition and functions of a plant cell wall and the extracellular matrix of an animal cell.*

TEST YOUR UNDERSTANDING

LEVEL 1: KNOWLEDGE/COMPREHENSION

1. Which structure is *not* part of the endomembrane system?
   a. nuclear envelope
   b. chloroplast
   c. Golgi apparatus
   d. plasma membrane
   e. ER

2. Which structure is common to plant *and* animal cells?
   a. chloroplast
   b. wall made of cellulose
   c. central vacuole
   d. mitochondrion
   e. centriole

3. Which of the following is present in a prokaryotic cell?
   a. mitochondrion
   b. ribosome
   c. nuclear envelope
   d. chloroplast
   e. ER

4. Which structure-function pair is *mismatched*?
   a. nucleolus; production of ribosomal subunits
   b. lysosome; intracellular digestion
   c. ribosome; protein synthesis
   d. Golgi; protein trafficking
   e. microtubule; muscle contraction

LEVEL 2: APPLICATION/ANALYSIS

5. Cyanide binds to at least one molecule involved in producing ATP. If a cell is exposed to cyanide, most of the cyanide will be found within the
   a. mitochondria.
   b. ribosomes.
   c. peroxisomes.
   d. lysosomes.
   e. endoplasmic reticulum.

6. What is the most likely pathway taken by a newly synthesized protein that will be secreted by a cell?
   a. ER → Golgi → nucleus
   b. Golgi → ER → lysosome
   c. nucleus → ER → Golgi
   d. ER → Golgi → vesicles that fuse with plasma membrane
   e. ER → lysosomes → vesicles that fuse with plasma membrane

7. Which cell would be best for studying lysosomes?
   a. muscle cell
   b. nerve cell
   c. phagocytic white blood cell
   d. leaf cell of a plant
   e. bacterial cell

8. **DRAW IT** From memory, draw two eukaryotic cells, labeling the structures listed here and showing any physical connections between the internal structures of each cell: nucleus, rough ER, smooth ER, mitochondrion, centrosome, chloroplast, vacuole, lysosome, microtubule, cell wall, ECM, microfilament, Golgi apparatus, intermediate filament, plasma membrane, peroxisome, ribosome, nucleolus, nuclear pore, vesicle, flagellum, microvilli, plasmodesma.

LEVEL 3: SYNTHESIS/EVALUATION

9. **EVOLUTION CONNECTION**
   Which aspects of cell structure best reveal evolutionary unity? What are some examples of specialized modifications?

10. **SCIENTIFIC INQUIRY**
    Imagine protein X, destined to span the plasma membrane. Assume that the mRNA carrying the genetic message for protein X has already been translated by ribosomes in a cell culture. If you fractionate the cells (see Figure 6.4), in which fraction would you find protein X? Explain by describing its transit through the cell.

11. **WRITE ABOUT A THEME**
    Emergent Properties Considering some of the characteristics that define life and drawing on your new knowledge of cellular structures and functions, write a short essay (100–150 words) that discusses this statement: Life is an emergent property that appears at the level of the cell. (Review pp. 3–5 in Chapter 1.)

For selected answers, see Appendix A.

MasteringBIOLOGY www.masteringbiology.com

1. MasteringBiology® Assignments

BioFlix Tutorials Tour of an Animal Cell: The Endomembrane System • Structures and Functions; Tour of a Plant Cell: Structures and Functions
Tutorial Connections Between Cells
Activities Metric System Review • Prokaryotic Cell Structure and Function • Discovery Channel Video: Cells • Role of the Nucleus and Ribosomes in Protein Synthesis • Transport into the Nucleus • A Pulse Chase Experiment • The Endomembrane System • Cilia and Flagella • Cell Junctions • Review: Animal Cell Structure and Function
Questions Student Misconceptions • Reading Quiz • Multiple Choice • End-of-Chapter

2. eText

Read your book online, search, take notes, highlight text, and more.

3. The Study Area

Practice Tests • Cumulative Test • **BioFlix** 3-D Animations • MP3 Tutor Sessions • Videos • Activities • Investigations • Lab Media • Audio Glossary • Word Study Tools • Art

7

Membrane Structure and Function

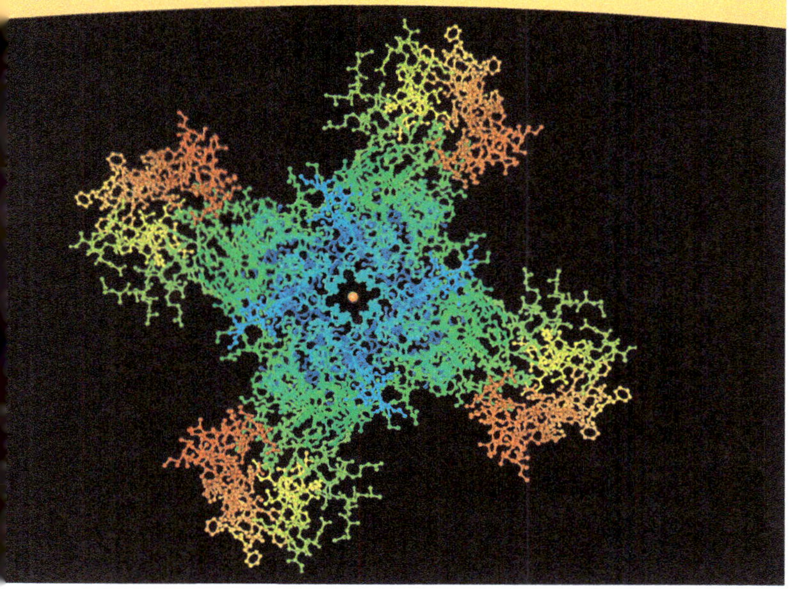

▲ **Figure 7.1 How do cell membrane proteins help regulate chemical traffic?**

KEY CONCEPTS

7.1 Cellular membranes are fluid mosaics of lipids and proteins
7.2 Membrane structure results in selective permeability
7.3 Passive transport is diffusion of a substance across a membrane with no energy investment
7.4 Active transport uses energy to move solutes against their gradients
7.5 Bulk transport across the plasma membrane occurs by exocytosis and endocytosis

OVERVIEW

Life at the Edge

The plasma membrane is the edge of life, the boundary that separates the living cell from its surroundings. A remarkable film only about 8 nm thick—it would take over 8,000 plasma membranes to equal the thickness of this page—the plasma membrane controls traffic into and out of the cell it surrounds. Like all biological membranes, the plasma membrane exhibits **selective permeability**; that is, it allows some substances to cross it more easily than others. One of the earliest episodes in the evolution of life may have been the formation of a membrane that enclosed a solution different from the surrounding solution while still permitting the uptake of nutrients and elimination of waste products. The ability of the cell to discriminate in its chemical exchanges with its environment is fundamental to life, and it is the plasma membrane and its component molecules that make this selectivity possible.

In this chapter, you will learn how cellular membranes control the passage of substances. The image in Figure 7.1 shows the elegant structure of a eukaryotic plasma membrane protein that plays a crucial role in nerve cell signaling. This protein provides a channel for a stream of potassium ions (K^+) to exit a nerve cell at a precise moment after nerve stimulation, restoring the cell's ability to fire again. (The orange ball in the center represents one potassium ion moving through the channel.) In this way, the plasma membrane and its proteins not only act as an outer boundary but also enable the cell to carry out its functions. The same applies to the many varieties of internal membranes that partition the eukaryotic cell: The molecular makeup of each membrane allows compartmentalized specialization in cells. To understand how membranes work, we'll begin by examining their architecture.

CONCEPT 7.1

Cellular membranes are fluid mosaics of lipids and proteins

Lipids and proteins are the staple ingredients of membranes, although carbohydrates are also important. The most abundant lipids in most membranes are phospholipids. The ability of phospholipids to form membranes is inherent in their molecular structure. A phospholipid is an **amphipathic** molecule, meaning it has both a hydrophilic region and a hydrophobic region (see Figure 5.12). Other types of membrane lipids are also amphipathic. Furthermore, most of the proteins within membranes have both hydrophobic and hydrophilic regions.

How are phospholipids and proteins arranged in the membranes of cells? In the **fluid mosaic model**, the membrane is a fluid structure with a "mosaic" of various proteins embedded in or attached to a double layer (bilayer) of phospholipids. Scientists propose models as hypotheses, ways of organizing and explaining existing information. Let's explore how the fluid mosaic model was developed.

Membrane Models: *Scientific Inquiry*

Scientists began building molecular models of the membrane decades before membranes were first seen with the electron microscope (in the 1950s). In 1915, membranes isolated from red blood cells were chemically analyzed and found to be composed of lipids and proteins. Ten years later, two Dutch scientists reasoned that cell membranes must be phospholipid bilayers. Such a double layer of molecules could exist as a stable boundary between two aqueous compartments because the molecular arrangement shelters the hydrophobic tails of the phospholipids from water while exposing the hydrophilic heads to water **(Figure 7.2)**.

If a phospholipid bilayer was the main fabric of a membrane, where were the proteins located? Although the heads of phospholipids are hydrophilic, the surface of a pure phospholipid bilayer adheres less strongly to water than does the surface of a biological membrane. Given this difference, Hugh Davson and James Danielli suggested in 1935 that the membrane might be coated on both sides with hydrophilic proteins. They proposed a sandwich model: a phospholipid bilayer between two layers of proteins.

When researchers first used electron microscopes to study cells in the 1950s, the pictures seemed to support the Davson-Danielli model. By the late 1960s, however, many cell biologists recognized two problems with the model. First, inspection of a variety of membranes revealed that membranes with different functions differ in structure and chemical composition. A second, more serious problem became apparent once membrane proteins were better characterized. Unlike proteins dissolved in the cytosol, membrane proteins are not very soluble in water because they are amphipathic. If such proteins were layered on the surface of the membrane, their hydrophobic parts would be in aqueous surroundings.

Taking these observations into account, S. J. Singer and G. Nicolson proposed in 1972 that membrane proteins reside in the phospholipid bilayer with their hydrophilic regions protruding **(Figure 7.3)**. This molecular arrangement would maximize contact of hydrophilic regions of proteins and phospholipids with water in the cytosol and extracellular fluid, while providing their hydrophobic parts with a nonaqueous environment. In this fluid mosaic model, the membrane is a mosaic of protein molecules bobbing in a fluid bilayer of phospholipids.

A method of preparing cells for electron microscopy called freeze-fracture has demonstrated visually that proteins are indeed embedded in the phospholipid bilayer of the membrane **(Figure 7.4)**. Freeze-fracture splits a membrane along the middle of the bilayer, somewhat like pulling apart a chunky peanut butter sandwich. When the membrane layers are viewed in the electron microscope, the interior of the

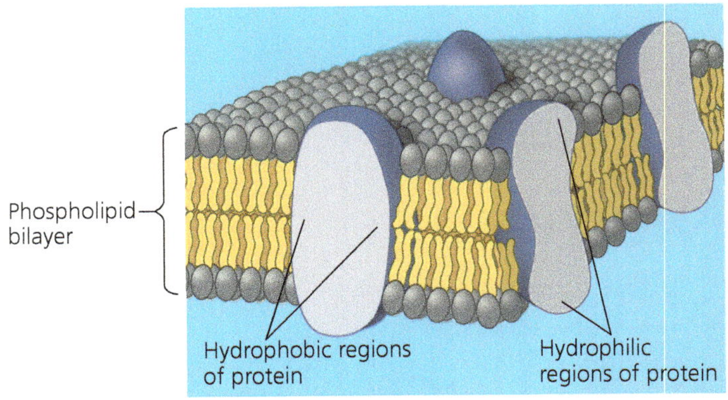

▲ **Figure 7.3 The original fluid mosaic model for membranes.**

▼ **Figure 7.2 Phospholipid bilayer (cross section).**

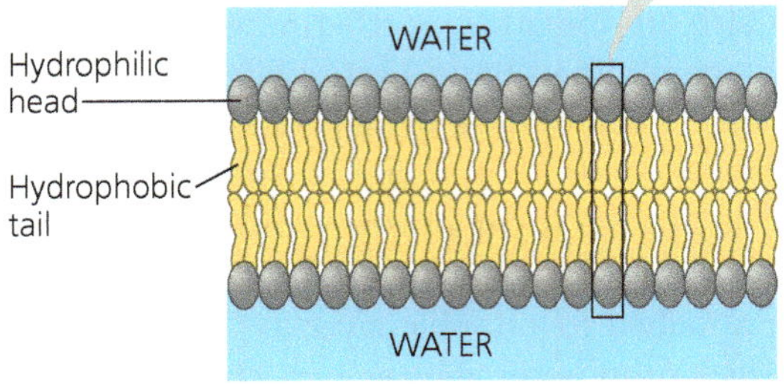

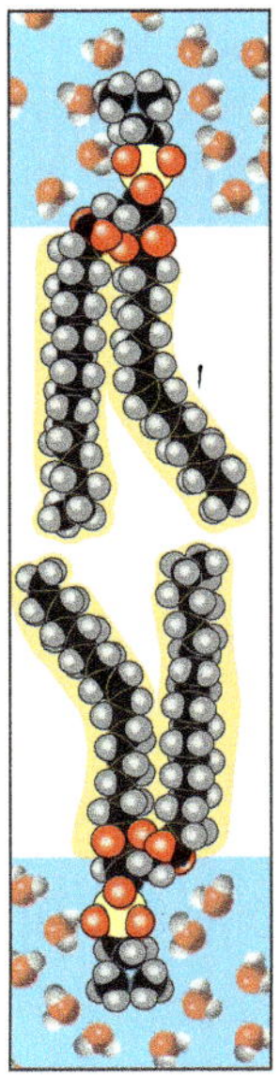

MAKE CONNECTIONS *Consulting Figure 5.12 (p. 76), circle the hydrophilic and hydrophobic portions of the enlarged phospholipids on the right. Explain what each portion contacts when the phospholipids are in the plasma membrane.*

▼ **Figure 7.4** **RESEARCH METHOD**

Freeze-fracture

APPLICATION A cell membrane can be split into its two layers, revealing the structure of the membrane's interior.

TECHNIQUE A cell is frozen and fractured with a knife. The fracture plane often follows the hydrophobic interior of a membrane, splitting the phospholipid bilayer into two separated layers. Each membrane protein goes wholly with one of the layers.

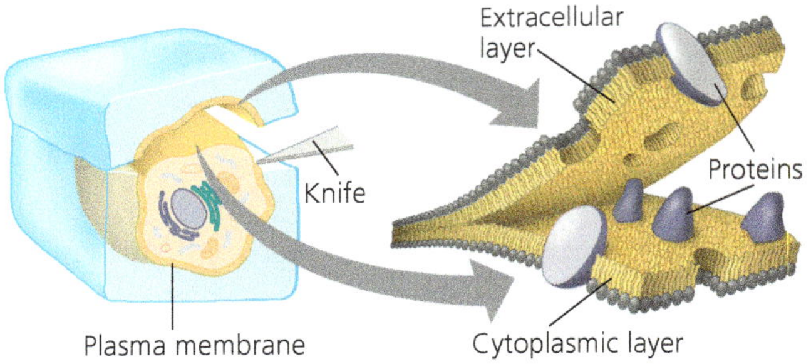

RESULTS These SEMs show membrane proteins (the "bumps") in the two layers, demonstrating that proteins are embedded in the phospholipid bilayer.

Inside of extracellular layer

Inside of cytoplasmic layer

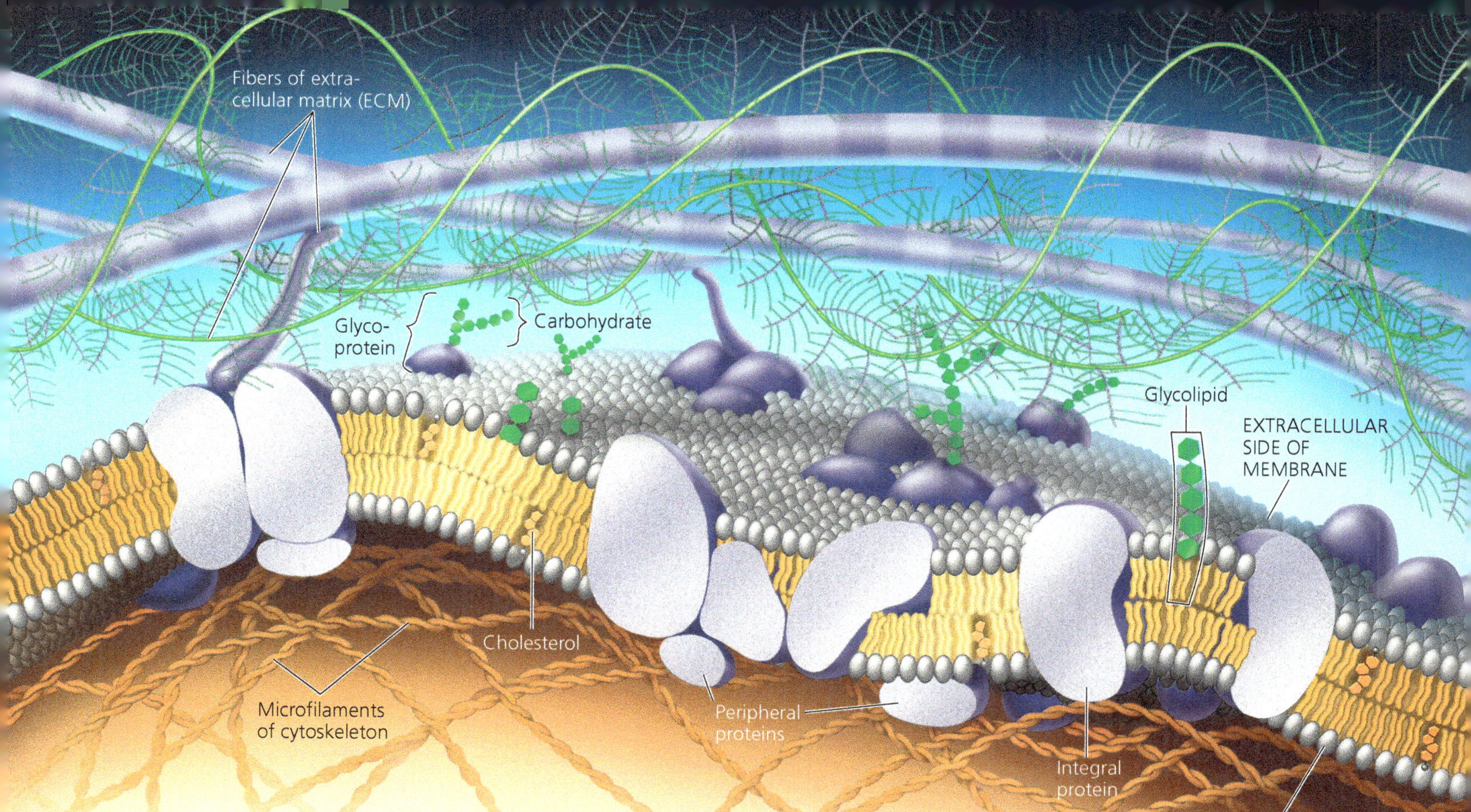

▲ **Figure 7.5 Updated model of an animal cell's plasma membrane (cutaway view).**

bilayer appears cobblestoned, with protein particles interspersed in a smooth matrix, in agreement with the fluid mosaic model. Some proteins remain attached to one layer or the other, like the peanut chunks in the sandwich.

Because models are hypotheses, replacing one model of membrane structure with another does not imply that the original model was worthless. The acceptance or rejection of a model depends on how well it fits observations and explains experimental results. New findings may make a model obsolete; even then, it may not be totally scrapped, but revised to incorporate the new observations. The fluid mosaic model is continually being refined. For example, groups of proteins are often found associated in long-lasting, specialized patches, where they carry out common functions. The lipids themselves appear to form defined regions as well. Also, the membrane may be much more packed with proteins than imagined in the classic fluid mosaic model—compare the updated model in **Figure 7.5** with the original model in Figure 7.3. Let's now take a closer look at membrane structure.

The Fluidity of Membranes

Membranes are not static sheets of molecules locked rigidly in place. A membrane is held together primarily by hydrophobic interactions, which are much weaker than covalent bonds (see Figure 5.20). Most of the lipids and some of the proteins can shift about laterally—that is, in the plane of the membrane, like partygoers elbowing their way through a crowded room **(Figure 7.6)**. It is quite rare, however, for a molecule to flip-flop transversely across the membrane, switching from one phospholipid layer to the other; to do so, the hydrophilic part of the molecule must cross the hydrophobic interior of the membrane.

The lateral movement of phospholipids within the membrane is rapid. Adjacent phospholipids switch positions about 10^7 times per second, which means that a phospholipid can travel about 2 μm—the length of many bacterial cells—in 1 second. Proteins are much larger than lipids and move more slowly, but some membrane proteins do drift, as shown in a classic experiment described in **Figure 7.7**, on the next page.

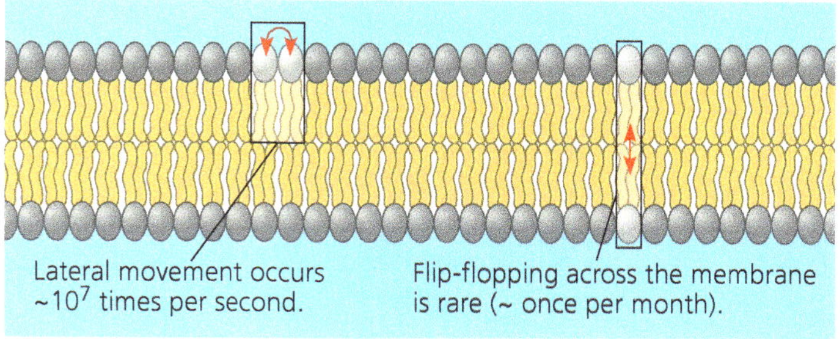

▲ **Figure 7.6 The movement of phospholipids.**

▼ **Figure 7.7** INQUIRY

Do membrane proteins move?

EXPERIMENT Larry Frye and Michael Edidin, at Johns Hopkins University, labeled the plasma membrane proteins of a mouse cell and a human cell with two different markers and fused the cells. Using a microscope, they observed the markers on the hybrid cell.

RESULTS

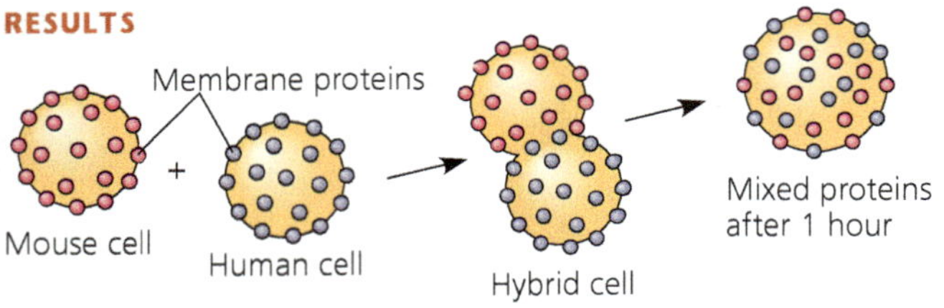

CONCLUSION The mixing of the mouse and human membrane proteins indicates that at least some membrane proteins move sideways within the plane of the plasma membrane.

SOURCE L. D. Frye and M. Edidin, The rapid intermixing of cell surface antigens after formation of mouse-human heterokaryons, *Journal of Cell Science* 7:319 (1970).

WHAT IF? Suppose the proteins did not mix in the hybrid cell, even many hours after fusion. Would you be able to conclude that proteins don't move within the membrane? What other explanation could there be?

And some membrane proteins seem to move in a highly directed manner, perhaps driven along cytoskeletal fibers by motor proteins connected to the membrane proteins' cytoplasmic regions. However, many other membrane proteins seem to be held immobile by their attachment to the cytoskeleton or to the extracellular matrix (see Figure 7.5).

A membrane remains fluid as temperature decreases until finally the phospholipids settle into a closely packed arrangement and the membrane solidifies, much as bacon grease forms lard when it cools. The temperature at which a membrane solidifies depends on the types of lipids it is made of. The membrane remains fluid to a lower temperature if it is rich in phospholipids with unsaturated hydrocarbon tails (see Figures 5.11 and 5.12). Because of kinks in the tails where double bonds are located, unsaturated hydrocarbon tails cannot pack together as closely as saturated hydrocarbon tails, and this makes the membrane more fluid **(Figure 7.8a)**.

The steroid cholesterol, which is wedged between phospholipid molecules in the plasma membranes of animal cells, has different effects on membrane fluidity at different temperatures **(Figure 7.8b)**. At relatively high temperatures—at 37°C, the body temperature of humans, for example—cholesterol makes the membrane less fluid by restraining phospholipid movement. However, because cholesterol also hinders the close packing of phospholipids, it lowers the temperature required for the membrane to solidify. Thus, cholesterol can be thought of as a "fluidity buffer" for the membrane, resisting changes in membrane fluidity that can be caused by changes in temperature.

Membranes must be fluid to work properly; they are usually about as fluid as salad oil. When a membrane solidifies, its permeability changes, and enzymatic proteins in the membrane may become inactive if their activity requires them to be able to move within the membrane. However, membranes that are too fluid cannot support protein function either. Therefore, extreme environments pose a challenge for life, resulting in evolutionary adaptations that include differences in membrane lipid composition.

Evolution of Differences in Membrane Lipid Composition

EVOLUTION Variations in the cell membrane lipid compositions of many species appear to be evolutionary adaptations that maintain the appropriate membrane fluidity under specific environmental conditions. For instance, fishes that live in extreme cold have membranes with a high proportion of unsaturated hydrocarbon tails, enabling their membranes to remain fluid (see Figure 7.8a). At the other extreme, some bacteria and archaea thrive at temperatures greater than 90°C (194°F) in thermal hot springs and geysers. Their membranes include unusual lipids that may prevent excessive fluidity at such high temperatures.

The ability to change the lipid composition of cell membranes in response to changing temperatures has evolved in organisms that live where temperatures vary. In many plants that tolerate extreme cold, such as winter wheat, the percentage of unsaturated phospholipids increases in autumn, an adjustment that keeps the membranes from solidifying during winter. Certain bacteria and archaea can also change the proportion of unsaturated phospholipids in their cell membranes, depending on the temperature at which they are growing. Overall, natural selection has apparently favored organisms whose mix of membrane lipids ensures an appropriate level of membrane fluidity for their environment.

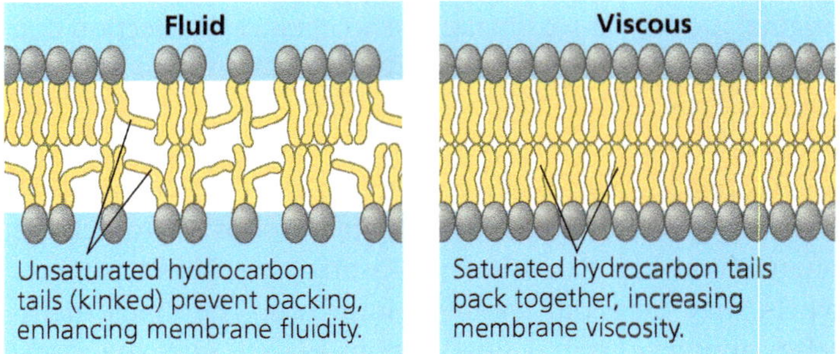

(a) Unsaturated versus saturated hydrocarbon tails.

(b) Cholesterol within the animal cell membrane. Cholesterol reduces membrane fluidity at moderate temperatures by reducing phospholipid movement, but at low temperatures it hinders solidification by disrupting the regular packing of phospholipids.

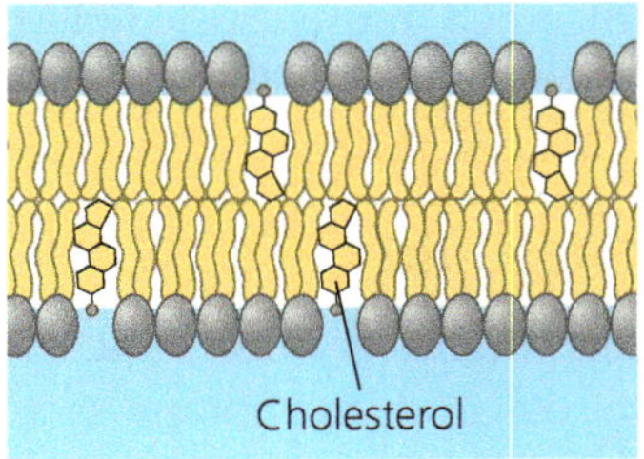

▲ **Figure 7.8 Factors that affect membrane fluidity.**

Membrane Proteins and Their Functions

Now we come to the *mosaic* aspect of the fluid mosaic model. Somewhat like a tile mosaic, a membrane is a collage of different proteins, often clustered together in groups, embedded in the fluid matrix of the lipid bilayer (see Figure 7.5). More than 50 kinds of proteins have been found so far in the plasma membrane of red blood cells, for example. Phospholipids form the main fabric of the membrane, but proteins determine most of the membrane's functions. Different types of cells contain different sets of membrane proteins, and the various membranes within a cell each have a unique collection of proteins.

Notice in Figure 7.5 that there are two major populations of membrane proteins: integral proteins and peripheral proteins. **Integral proteins** penetrate the hydrophobic interior of the lipid bilayer. The majority are *transmembrane proteins*, which span the membrane; other integral proteins extend only partway into the hydrophobic interior. The hydrophobic regions of an integral protein consist of one or more stretches of nonpolar amino acids (see Figure 5.16), usually coiled into α helices (**Figure 7.9**). The hydrophilic parts of the molecule are exposed to the aqueous solutions on either side of the membrane. Some proteins also have a hydrophilic channel through their center that allows passage of hydrophilic substances (see Figure 7.1). **Peripheral proteins** are not embedded in the lipid bilayer at all; they are appendages loosely bound to the surface of the membrane, often to exposed parts of integral proteins (see Figure 7.5).

On the cytoplasmic side of the plasma membrane, some membrane proteins are held in place by attachment to the cytoskeleton. And on the extracellular side, certain membrane proteins are attached to fibers of the extracellular matrix (see Figure 6.30; *integrins* are one type of integral protein). These attachments combine to give animal cells a stronger framework than the plasma membrane alone could provide.

Figure 7.10 gives an overview of six major functions performed by proteins of the plasma membrane. A single cell

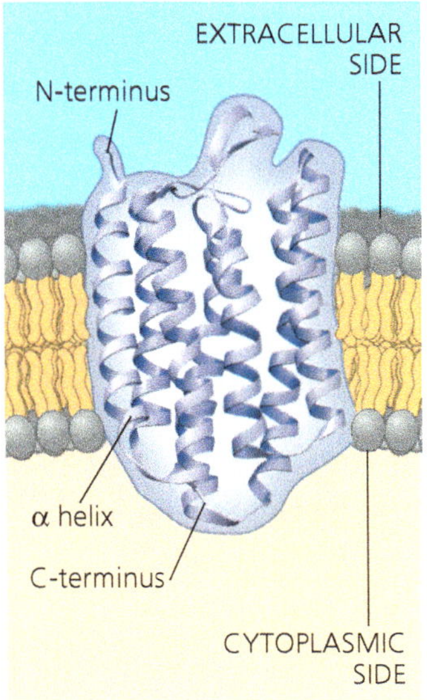

◀ **Figure 7.9 The structure of a transmembrane protein.** Bacteriorhodopsin (a bacterial transport protein) has a distinct orientation in the membrane, with its N-terminus outside the cell and its C-terminus inside. This ribbon model highlights the α-helical secondary structure of the hydrophobic parts, which lie mostly within the hydrophobic interior of the membrane. The protein includes seven transmembrane helices. The nonhelical hydrophilic segments are in contact with the aqueous solutions on the extracellular and cytoplasmic sides of the membrane.

(a) Transport. *Left:* A protein that spans the membrane may provide a hydrophilic channel across the membrane that is selective for a particular solute. *Right:* Other transport proteins shuttle a substance from one side to the other by changing shape (see Figure 7.17). Some of these proteins hydrolyze ATP as an energy source to actively pump substances across the membrane.

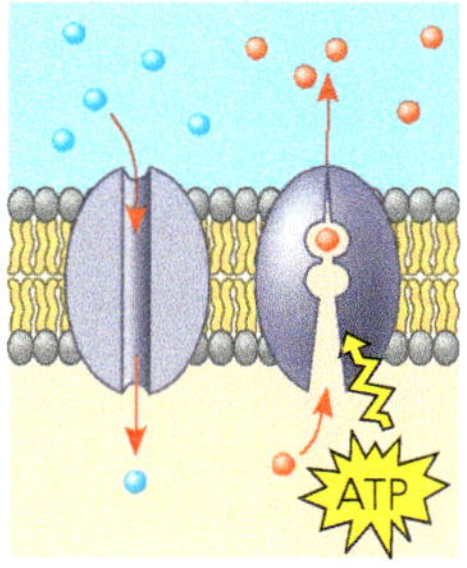

(b) Enzymatic activity. A protein built into the membrane may be an enzyme with its active site exposed to substances in the adjacent solution. In some cases, several enzymes in a membrane are organized as a team that carries out sequential steps of a metabolic pathway.

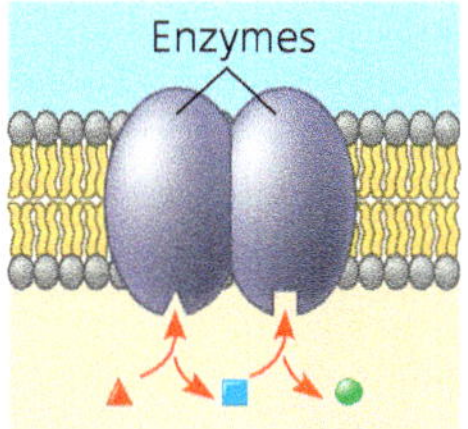

(c) Signal transduction. A membrane protein (receptor) may have a binding site with a specific shape that fits the shape of a chemical messenger, such as a hormone. The external messenger (signaling molecule) may cause the protein to change shape, allowing it to relay the message to the inside of the cell, usually by binding to a cytoplasmic protein (see Figure 11.6).

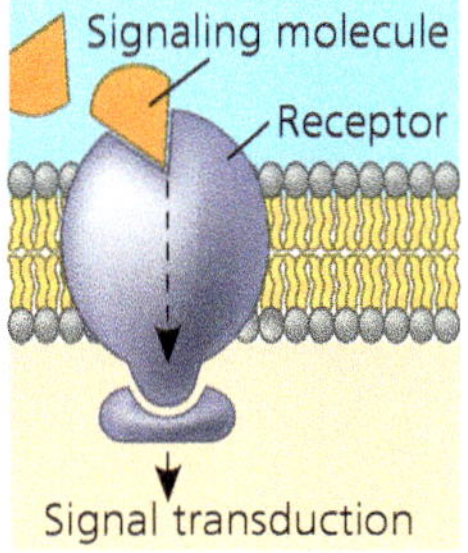

(d) Cell-cell recognition. Some glycoproteins serve as identification tags that are specifically recognized by membrane proteins of other cells. This type of cell-cell binding is usually short-lived compared to that shown in (e).

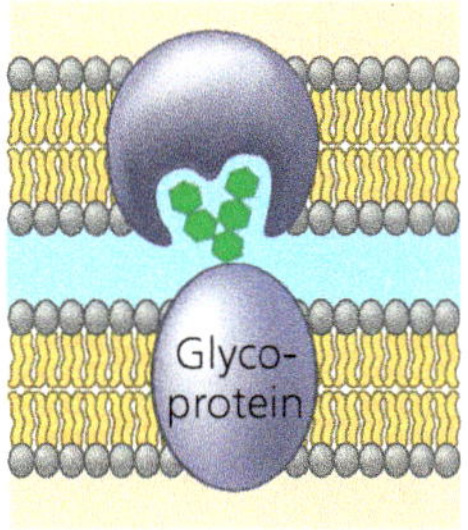

(e) Intercellular joining. Membrane proteins of adjacent cells may hook together in various kinds of junctions, such as gap junctions or tight junctions (see Figure 6.32). This type of binding is more long-lasting than that shown in (d).

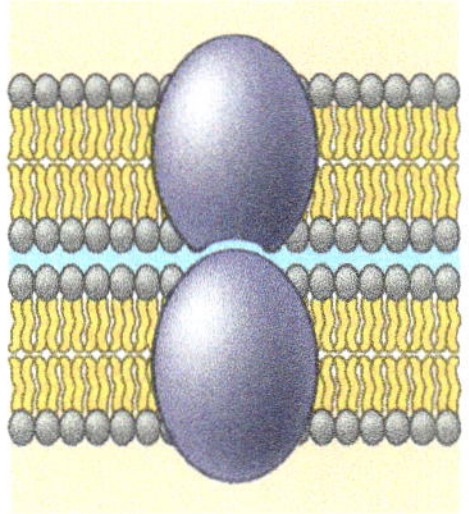

(f) Attachment to the cytoskeleton and extracellular matrix (ECM). Microfilaments or other elements of the cytoskeleton may be noncovalently bound to membrane proteins, a function that helps maintain cell shape and stabilizes the location of certain membrane proteins. Proteins that can bind to ECM molecules can coordinate extracellular and intracellular changes (see Figure 6.30).

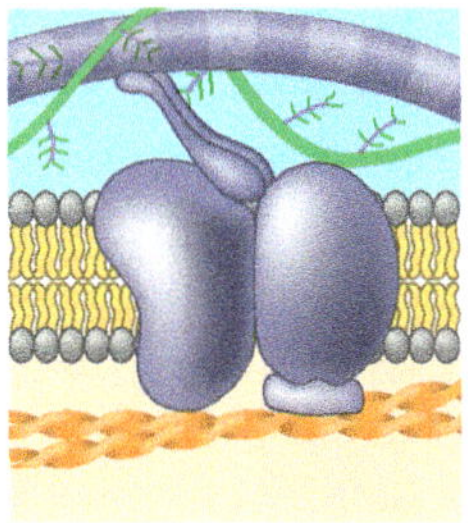

▲ **Figure 7.10 Some functions of membrane proteins.** In many cases, a single protein performs multiple tasks.

? *Some transmembrane proteins can bind to a particular ECM molecule and, when bound, transmit a signal into the cell. Use the proteins shown here to explain how this might occur.*

▼ Figure 7.11

IMPACT

Blocking HIV Entry into Cells as a Treatment for HIV Infections

Despite multiple exposures to HIV, a small number of people do not develop AIDS and show no evidence of HIV-infected cells. Comparing their genes with the genes of infected individuals, researchers discovered that resistant individuals have an unusual form of a gene that codes for an immune cell-surface protein called CCR5. Further work showed that HIV binds to a main protein receptor (CD4) on an immune cell, but on most cell types, HIV also needs to bind to CCR5 as a "co-receptor" to infect the cell (below, left). An absence of CCR5 on the cells of resistant individuals, due to the gene alteration, prevents the virus from entering the cells (below, right).

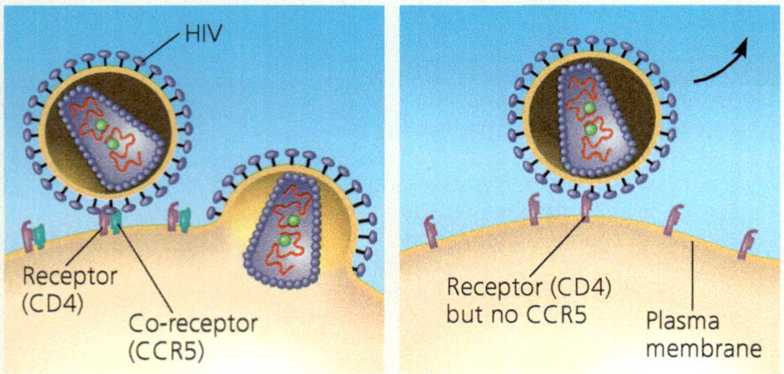

HIV can infect a cell that has CCR5 on its surface, as in most people.

HIV cannot infect a cell lacking CCR5 on its surface, as in resistant individuals.

WHY IT MATTERS Researchers have been searching for drugs to block cell-surface receptors involved in HIV infection. The main receptor protein, CD4, performs many important functions for cells, so interfering with it could cause dangerous side effects. Discovery of the CCR5 co-receptor provided a safer target for development of drugs that mask CCR5 and block HIV entry. One such drug, maraviroc (brand name Selzentry), was approved for treatment of HIV infection in 2007.

FURTHER READING T. Kenakin, New bull's-eyes for drugs, *Scientific American* 293(4):50–57 (2005).

MAKE CONNECTIONS Study Figures 2.18 (p. 42) and 5.19 (p. 81), both of which show pairs of molecules binding to each other. What would you predict about CCR5 that would allow HIV to bind to it? How could a drug molecule interfere with this binding?

may have membrane proteins carrying out several of these functions, and a single membrane protein may have multiple functions. In this way, the membrane is a functional mosaic as well as a structural one.

Proteins on the surface of a cell are important in the medical field because some proteins can help outside agents invade the cell. For example, cell-surface proteins help the human immunodeficiency virus (HIV) infect immune system cells, leading to acquired immune deficiency syndrome (AIDS). (You'll read more about HIV in Chapter 19.) Learning about the proteins that HIV binds to on immune cells has been central to developing a treatment for HIV infection **(Figure 7.11)**.

The Role of Membrane Carbohydrates in Cell-Cell Recognition

Cell-cell recognition, a cell's ability to distinguish one type of neighboring cell from another, is crucial to the functioning of an organism. It is important, for example, in the sorting of cells into tissues and organs in an animal embryo. It is also the basis for the rejection of foreign cells by the immune system, an important line of defense in vertebrate animals (see Chapter 43). Cells recognize other cells by binding to molecules, often containing carbohydrates, on the extracellular surface of the plasma membrane (see Figure 7.10d).

Membrane carbohydrates are usually short, branched chains of fewer than 15 sugar units. Some are covalently bonded to lipids, forming molecules called **glycolipids**. (Recall that *glyco* refers to the presence of carbohydrate.) However, most are covalently bonded to proteins, which are thereby **glycoproteins** (see Figure 7.5).

The carbohydrates on the extracellular side of the plasma membrane vary from species to species, among individuals of the same species, and even from one cell type to another in a single individual. The diversity of the molecules and their location on the cell's surface enable membrane carbohydrates to function as markers that distinguish one cell from another. For example, the four human blood types designated A, B, AB, and O reflect variation in the carbohydrate part of glycoproteins on the surface of red blood cells.

Synthesis and Sidedness of Membranes

Membranes have distinct inside and outside faces. The two lipid layers may differ in specific lipid composition, and each protein has directional orientation in the membrane (see Figure 7.9). **Figure 7.12** shows how membrane sidedness arises: The asymmetrical arrangement of proteins, lipids, and their associated carbohydrates in the plasma membrane is determined as the membrane is being built by the endoplasmic reticulum (ER) and Golgi apparatus.

CONCEPT CHECK 7.1

1. The carbohydrates attached to some proteins and lipids of the plasma membrane are added as the membrane is made and refined in the ER and Golgi apparatus. The new membrane then forms transport vesicles that travel to the cell surface. On which side of the vesicle membrane are the carbohydrates?
2. **WHAT IF?** The soil immediately around hot springs is much warmer than that in neighboring regions. Two closely related species of native grasses are found, one in the warmer region and one in the cooler region. If you analyzed their membrane lipid compositions, what would you expect to find? Explain.

For suggested answers, see Appendix A.

▼ **Figure 7.12 Synthesis of membrane components and their orientation in the membrane.** The cytoplasmic (orange) face of the plasma membrane differs from the extracellular (aqua) face. The latter arises from the inside face of ER, Golgi, and vesicle membranes.

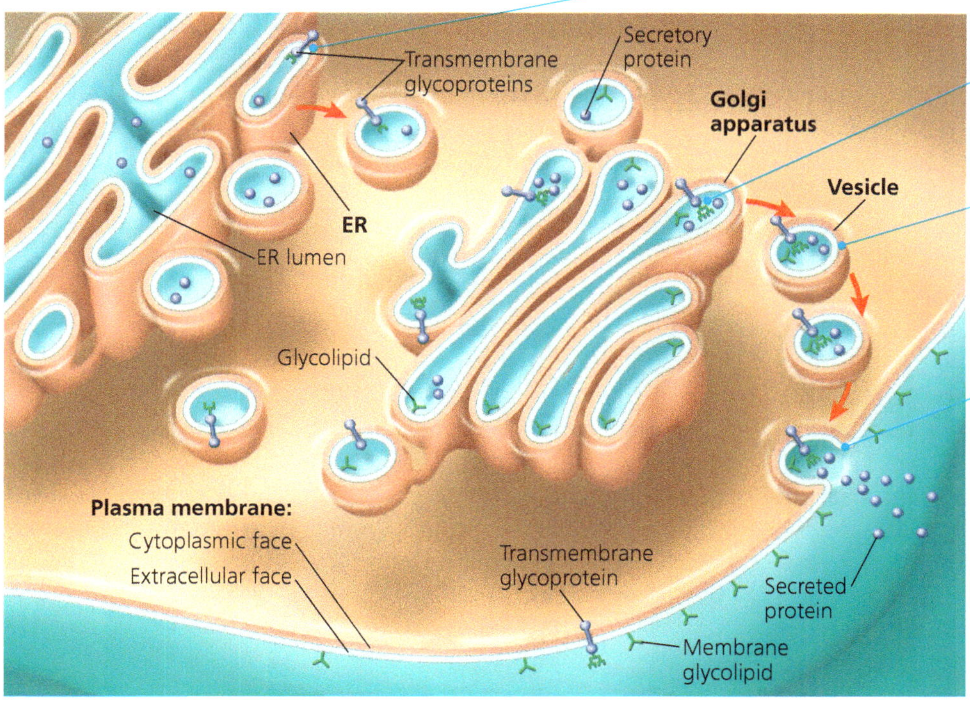

1 Membrane proteins and lipids are synthesized in the endoplasmic reticulum (ER). Carbohydrates (green) are added to the transmembrane proteins (purple dumbbells), making them glycoproteins. The carbohydrate portions may then be modified.

2 Inside the Golgi apparatus, the glycoproteins undergo further carbohydrate modification, and lipids acquire carbohydrates, becoming glycolipids.

3 The glycoproteins, glycolipids, and secretory proteins (purple spheres) are transported in vesicles to the plasma membrane.

4 As vesicles fuse with the plasma membrane, the outside face of the vesicle becomes continuous with the inside (cytoplasmic) face of the plasma membrane. This releases the secretory proteins from the cell, a process called *exocytosis*, and positions the carbohydrates of membrane glycoproteins and glycolipids on the outside (extracellular) face of the plasma membrane.

DRAW IT *Draw an integral membrane protein extending from partway through the ER membrane into the ER lumen. Next, draw the protein where it would be located in a series of numbered steps ending at the plasma membrane. Would the protein contact the cytoplasm or the extracellular fluid?*

CONCEPT 7.2

Membrane structure results in selective permeability

The biological membrane is an exquisite example of a supramolecular structure—many molecules ordered into a higher level of organization—with emergent properties beyond those of the individual molecules. The remainder of this chapter focuses on one of the most important of those properties: the ability to regulate transport across cellular boundaries, a function essential to the cell's existence. We will see once again that form fits function: The fluid mosaic model helps explain how membranes regulate the cell's molecular traffic.

A steady traffic of small molecules and ions moves across the plasma membrane in both directions. Consider the chemical exchanges between a muscle cell and the extracellular fluid that bathes it. Sugars, amino acids, and other nutrients enter the cell, and metabolic waste products leave it. The cell takes in O_2 for use in cellular respiration and expels CO_2. Also, the cell regulates its concentrations of inorganic ions, such as Na^+, K^+, Ca^{2+}, and Cl^-, by shuttling them one way or the other across the plasma membrane. In spite of heavy traffic through them, cell membranes are selectively permeable, and substances do not cross the barrier indiscriminately. The cell is able to take up some small molecules and ions and exclude others. Also, substances that move through the membrane do so at different rates.

The Permeability of the Lipid Bilayer

Nonpolar molecules, such as hydrocarbons, carbon dioxide, and oxygen, are hydrophobic and can therefore dissolve in the lipid bilayer of the membrane and cross it easily, without the aid of membrane proteins. However, the hydrophobic interior of the membrane impedes the direct passage of ions and polar molecules, which are hydrophilic, through the membrane. Polar molecules such as glucose and other sugars pass only slowly through a lipid bilayer, and even water, an extremely small polar molecule, does not cross very rapidly. A charged atom or molecule and its surrounding shell of water (see Figure 3.7) find the hydrophobic interior of the membrane even more difficult to penetrate. Furthermore, the lipid bilayer is only one aspect of the gatekeeper system responsible for the selective permeability of a cell. Proteins built into the membrane play key roles in regulating transport.

Transport Proteins

Cell membranes *are* permeable to specific ions and a variety of polar molecules. These hydrophilic substances can avoid contact with the lipid bilayer by passing through **transport proteins** that span the membrane.

Some transport proteins, called *channel proteins*, function by having a hydrophilic channel that certain molecules or atomic ions use as a tunnel through the membrane (see Figure 7.10a, left). For example, the passage of water molecules through the membrane in certain cells is greatly facilitated by channel proteins known as **aquaporins**. Each aquaporin allows entry of up to *3 billion* (3×10^9) water molecules per second, passing single file through its central channel, which fits ten at a time. Without aquaporins, only a tiny fraction of these water molecules would pass through the same area of the cell membrane in a second, so the channel protein brings about a tremendous increase in rate. Other transport proteins, called *carrier proteins*, hold onto their passengers and change shape in a way that shuttles them across the membrane (see Figure 7.10a, right). A transport protein is specific for the substance it translocates (moves), allowing only a certain substance (or a small group of related substances) to cross the membrane. For example, a specific carrier protein in the plasma membrane of red blood cells transports glucose across the membrane 50,000 times faster than glucose can pass through on its own. This "glucose transporter" is so selective that it even rejects fructose, a structural isomer of glucose.

Thus, the selective permeability of a membrane depends on both the discriminating barrier of the lipid bilayer and the specific transport proteins built into the membrane. But what establishes the *direction* of traffic across a membrane? At a given time, what determines whether a particular substance will enter the cell or leave the cell? And what mechanisms actually drive molecules across membranes? We will address these questions next as we explore two modes of membrane traffic: passive transport and active transport.

CONCEPT CHECK 7.2

1. Two molecules that can cross a lipid bilayer without help from membrane proteins are O_2 and CO_2. What property allows this to occur?
2. Why is a transport protein needed to move water molecules rapidly and in large quantities across a membrane?
3. **MAKE CONNECTIONS** Aquaporins exclude passage of hydronium ions (H_3O^+; see pp. 52–53). Recent research on fat metabolism has shown that some aquaporins allow passage of glycerol, a three-carbon alcohol (see Figure 5.10, p. 75), as well as H_2O. Since H_3O^+ is much closer in size to water than is glycerol, what do you suppose is the basis of this selectivity?

For suggested answers, see Appendix A.

CONCEPT 7.3

Passive transport is diffusion of a substance across a membrane with no energy investment

Molecules have a type of energy called thermal energy (heat), due to their constant motion. One result of this motion is **diffusion**, the movement of molecules of any substance so that they spread out evenly into the available space. Each molecule moves randomly, yet diffusion of a *population* of molecules may be directional. To understand this process, let's imagine a synthetic membrane separating pure water from a solution of a dye in water. Study **Figure 7.13a** carefully to appreciate how diffusion would result in both solutions having equal concentrations of the dye molecules. Once that point is reached, there will be a dynamic equilibrium, with as many dye molecules crossing the membrane each second in one direction as in the other.

We can now state a simple rule of diffusion: In the absence of other forces, a substance will diffuse from where it is more concentrated to where it is less concentrated. Put another way, any substance will diffuse down its **concentration gradient**, the region along which the density of a chemical substance increases or decreases (in this case, decreases). No work must be done to make this happen; diffusion is a spontaneous process, needing no input of energy. Note that each substance diffuses down its *own* concentration gradient, unaffected by the concentration gradients of other substances **(Figure 7.13b)**.

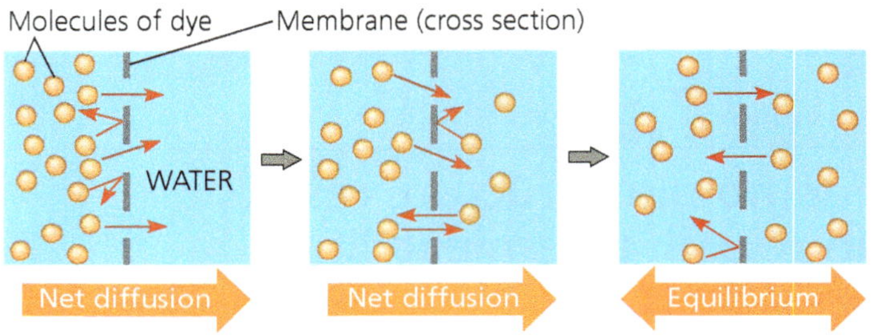

(a) Diffusion of one solute. The membrane has pores large enough for molecules of dye to pass through. Random movement of dye molecules will cause some to pass through the pores; this will happen more often on the side with more dye molecules. The dye diffuses from where it is more concentrated to where it is less concentrated (called diffusing down a concentration gradient). This leads to a dynamic equilibrium: The solute molecules continue to cross the membrane, but at equal rates in both directions.

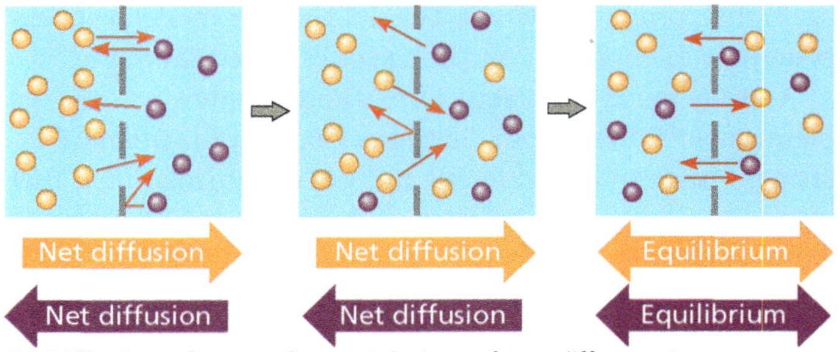

(b) Diffusion of two solutes. Solutions of two different dyes are separated by a membrane that is permeable to both. Each dye diffuses down its own concentration gradient. There will be a net diffusion of the purple dye toward the left, even though the *total* solute concentration was initially greater on the left side.

▲ **Figure 7.13 The diffusion of solutes across a synthetic membrane.** Each of the large arrows under the diagrams shows the net diffusion of the dye molecules of that color.

Much of the traffic across cell membranes occurs by diffusion. When a substance is more concentrated on one side of a membrane than on the other, there is a tendency for the substance to diffuse across the membrane down its concentration gradient (assuming that the membrane is permeable to that substance). One important example is the uptake of oxygen by a cell performing cellular respiration. Dissolved oxygen diffuses into the cell across the plasma membrane. As long as cellular respiration consumes the O_2 as it enters, diffusion into the cell will continue because the concentration gradient favors movement in that direction.

The diffusion of a substance across a biological membrane is called **passive transport** because the cell does not have to expend energy to make it happen. The concentration gradient itself represents potential energy (see Chapter 2, p. 35) and drives diffusion. Remember, however, that membranes are selectively permeable and therefore have different effects on the rates of diffusion of various molecules. In the case of water, aquaporins allow water to diffuse very rapidly across the membranes of certain cells. As we'll see next, the movement of water across the plasma membrane has important consequences for cells.

Effects of Osmosis on Water Balance

To see how two solutions with different solute concentrations interact, picture a U-shaped glass tube with a selectively permeable artificial membrane separating two sugar solutions (**Figure 7.14**). Pores in this synthetic membrane are too small for sugar molecules to pass through but large enough for water molecules. How does this affect the *water* concentration? It seems logical that the solution with the higher concentration of solute would have the lower concentration of water and that water would diffuse into it from the other side for that reason. However, for a dilute solution like most biological fluids, solutes do not affect the water concentration significantly. Instead, tight clustering of water molecules around the hydrophilic solute molecules makes some of the water unavailable to cross the membrane. It is the difference in *free* water concentration that is important. In the end, the effect is the same: Water diffuses across the membrane from the region of lower solute concentration (higher free water concentration) to that of higher solute concentration (lower free water concentration) until the solute concentrations on both sides of the membrane are equal. The diffusion of free water across a selectively permeable membrane, whether artificial or cellular, is called **osmosis**. The movement of water across cell membranes and the balance of water between the cell and its environment are crucial to organisms. Let's now apply to living cells what we have learned about osmosis in artificial systems.

Water Balance of Cells Without Walls

To explain the behavior of a cell in a solution, we must consider both solute concentration and membrane permeability.

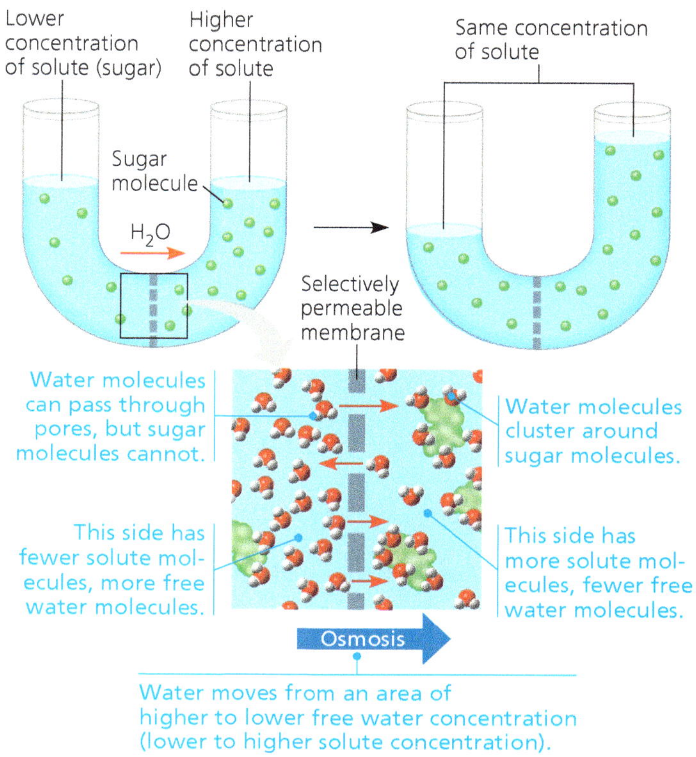

▲ **Figure 7.14 Osmosis.** Two sugar solutions of different concentrations are separated by a membrane that the solvent (water) can pass through but the solute (sugar) cannot. Water molecules move randomly and may cross in either direction, but overall, water diffuses from the solution with less concentrated solute to that with more concentrated solute. This diffusion of water, or osmosis, equalizes the sugar concentrations on both sides.

WHAT IF? *If an orange dye capable of passing through the membrane was added to the left side of the tube above, how would it be distributed at the end of the experiment? (See Figure 7.13.) Would the final solution levels in the tube be affected?*

Both factors are taken into account in the concept of **tonicity**, the ability of a surrounding solution to cause a cell to gain or lose water. The tonicity of a solution depends in part on its concentration of solutes that cannot cross the membrane (nonpenetrating solutes) relative to that inside the cell. If there is a higher concentration of nonpenetrating solutes in the surrounding solution, water will tend to leave the cell, and vice versa.

If a cell without a wall, such as an animal cell, is immersed in an environment that is **isotonic** to the cell (*iso* means "same"), there will be no *net* movement of water across the plasma membrane. Water diffuses across the membrane, but at the same rate in both directions. In an isotonic environment, the volume of an animal cell is stable (**Figure 7.15a**, on the next page).

Now let's transfer the cell to a solution that is **hypertonic** to the cell (*hyper* means "more," in this case referring to nonpenetrating solutes). The cell will lose water, shrivel, and probably die. This is one way an increase in the salinity (saltiness) of a lake can kill animals there; if the lake water becomes hypertonic to the animals' cells, the cells might shrivel and

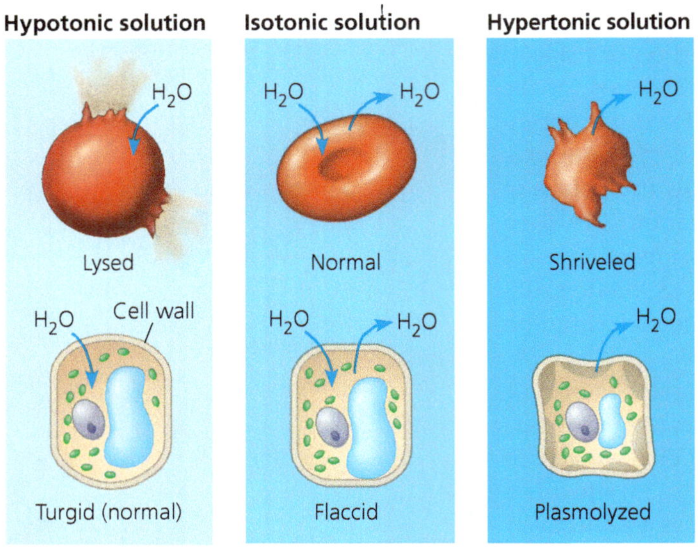

(a) Animal cell. An animal cell fares best in an isotonic environment unless it has special adaptations that offset the osmotic uptake or loss of water.

(b) Plant cell. Plant cells are turgid (firm) and generally healthiest in a hypotonic environment, where the uptake of water is eventually balanced by the wall pushing back on the cell.

▲ **Figure 7.15 The water balance of living cells.** How living cells react to changes in the solute concentration of their environment depends on whether or not they have cell walls. **(a)** Animal cells, such as this red blood cell, do not have cell walls. **(b)** Plant cells do. (Arrows indicate net water movement after the cells were first placed in these solutions.)

die. However, taking up too much water can be just as hazardous to an animal cell as losing water. If we place the cell in a solution that is **hypotonic** to the cell (*hypo* means "less"), water will enter the cell faster than it leaves, and the cell will swell and lyse (burst) like an overfilled water balloon.

A cell without rigid walls can tolerate neither excessive uptake nor excessive loss of water. This problem of water balance is automatically solved if such a cell lives in isotonic surroundings. Seawater is isotonic to many marine invertebrates. The cells of most terrestrial (land-dwelling) animals are bathed in an extracellular fluid that is isotonic to the cells. In hypertonic or hypotonic environments, however, organisms that lack rigid cell walls must have other adaptations for **osmoregulation**, the control of solute concentrations and water balance. For example, the unicellular protist *Paramecium caudatum* lives in pond water, which is hypotonic to the cell. *P. caudatum* has a plasma membrane that is much less permeable to water than the membranes of most other cells, but this only slows the uptake of water, which continually enters the cell. The *P. caudatum* cell doesn't burst because it is also equipped with a contractile vacuole, an organelle that functions as a bilge pump to force water out of the cell as fast as it enters by osmosis **(Figure 7.16)**. We will examine other evolutionary adaptations for osmoregulation in Chapter 44.

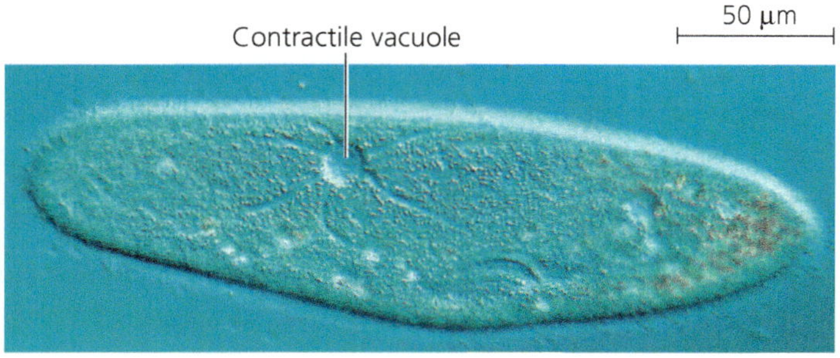

▲ **Figure 7.16 The contractile vacuole of *Paramecium caudatum*.** The vacuole collects fluid from a system of canals in the cytoplasm. When full, the vacuole and canals contract, expelling fluid from the cell (LM).

Water Balance of Cells with Walls

The cells of plants, prokaryotes, fungi, and some protists are surrounded by walls (see Figure 6.28). When such a cell is immersed in a hypotonic solution—bathed in rainwater, for example—the wall helps maintain the cell's water balance. Consider a plant cell. Like an animal cell, the plant cell swells as water enters by osmosis **(Figure 7.15b)**. However, the relatively inelastic wall will expand only so much before it exerts a back pressure on the cell, called *turgor pressure*, that opposes further water uptake. At this point, the cell is **turgid** (very firm), which is the healthy state for most plant cells. Plants that are not woody, such as most houseplants, depend for mechanical support on cells kept turgid by a surrounding hypotonic solution. If a plant's cells and their surroundings are isotonic, there is no net tendency for water to enter, and the cells become **flaccid** (limp).

However, a wall is of no advantage if the cell is immersed in a hypertonic environment. In this case, a plant cell, like an animal cell, will lose water to its surroundings and shrink. As the plant cell shrivels, its plasma membrane pulls away from the wall. This phenomenon, called **plasmolysis**, causes the plant to wilt and can lead to plant death. The walled cells of bacteria and fungi also plasmolyze in hypertonic environments.

Facilitated Diffusion: Passive Transport Aided by Proteins

Let's look more closely at how water and certain hydrophilic solutes cross a membrane. As mentioned earlier, many polar molecules and ions impeded by the lipid bilayer of the membrane diffuse passively with the help of transport proteins that span the membrane. This phenomenon is called **facilitated diffusion**. Cell biologists are still trying to learn exactly how various transport proteins facilitate diffusion. Most transport proteins are very specific: They transport some substances but not others.

As described earlier, the two types of transport proteins are channel proteins and carrier proteins. Channel proteins simply provide corridors that allow specific molecules or ions to cross the membrane **(Figure 7.17a)**. The hydrophilic passageways

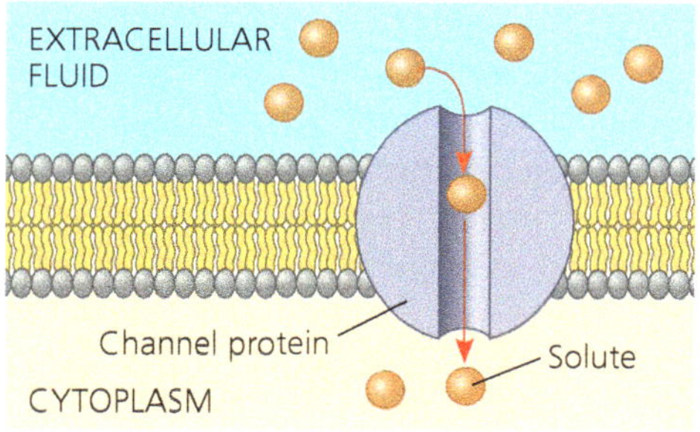

(a) A channel protein (purple) has a channel through which water molecules or a specific solute can pass.

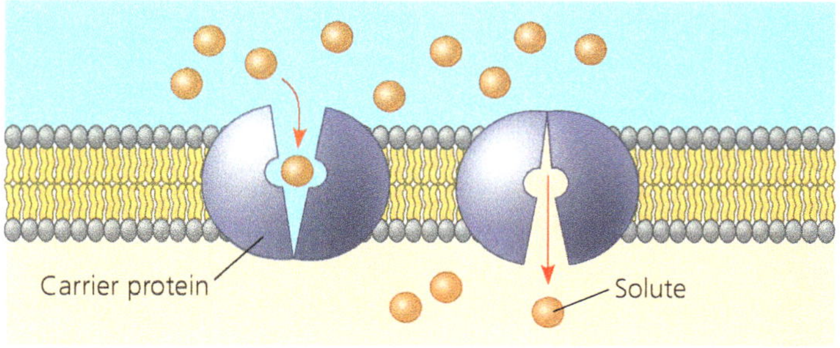

(b) A carrier protein alternates between two shapes, moving a solute across the membrane during the shape change.

▲ **Figure 7.17 Two types of transport proteins that carry out facilitated diffusion.** In both cases, the protein can transport the solute in either direction, but the net movement is down the concentration gradient of the solute.

provided by these proteins can allow water molecules or small ions to diffuse very quickly from one side of the membrane to the other. Aquaporins, the water channel proteins, facilitate the massive amounts of diffusion that occur in plant cells and in animal cells such as red blood cells (see Figure 7.15). Certain kidney cells also have a high number of aquaporins, allowing them to reclaim water from urine before it is excreted. If the kidneys did not perform this function, you would excrete about 180 L of urine per day—and have to drink an equal volume of water!

Channel proteins that transport ions are called **ion channels**. Many ion channels function as **gated channels**, which open or close in response to a stimulus. For some gated channels, the stimulus is electrical. The ion channel shown in Figure 7.1, for example, opens in response to an electrical stimulus, allowing potassium ions to leave the cell. Other gated channels open or close when a specific substance other than the one to be transported binds to the channel. Both types of gated channels are important in the functioning of the nervous system, as you'll learn in Chapter 48.

Carrier proteins, such as the glucose transporter mentioned earlier, seem to undergo a subtle change in shape that somehow translocates the solute-binding site across the membrane **(Figure 7.17b)**. Such a change in shape may be triggered by the binding and release of the transported molecule. Like ion channels, carrier proteins involved in facilitated diffusion result in the net movement of a substance down its concentration gradient. No energy input is thus required: This is passive transport.

In certain inherited diseases, specific transport systems are either defective or missing altogether. An example is cystinuria, a human disease characterized by the absence of a carrier protein that transports cysteine and some other amino acids across the membranes of kidney cells. Kidney cells normally reabsorb these amino acids from the urine and return them to the blood, but an individual afflicted with cystinuria develops painful stones from amino acids that accumulate and crystallize in the kidneys.

CONCEPT CHECK 7.3

1. How do you think a cell performing cellular respiration rids itself of the resulting CO_2?
2. In the supermarket, produce is often sprayed with water. Explain why this makes vegetables look crisp.
3. **WHAT IF?** If a *Paramecium caudatum* swims from a hypotonic to an isotonic environment, will its contractile vacuole become more active or less? Why?

For suggested answers, see Appendix A.

CONCEPT 7.4

Active transport uses energy to move solutes against their gradients

Despite the help of transport proteins, facilitated diffusion is considered passive transport because the solute is moving down its concentration gradient, a process that requires no energy. Facilitated diffusion speeds transport of a solute by providing efficient passage through the membrane, but it does not alter the direction of transport. Some transport proteins, however, can move solutes against their concentration gradients, across the plasma membrane from the side where they are less concentrated (whether inside or outside) to the side where they are more concentrated.

The Need for Energy in Active Transport

To pump a solute across a membrane against its gradient requires work; the cell must expend energy. Therefore, this type of membrane traffic is called **active transport**. The transport proteins that move solutes against their concentration gradients are all carrier proteins rather than channel proteins. This makes sense because when channel proteins are open, they merely allow solutes to diffuse down their concentration gradients rather than picking them up and transporting them against their gradients.

Active transport enables a cell to maintain internal concentrations of small solutes that differ from concentrations in its environment. For example, compared with its surroundings,

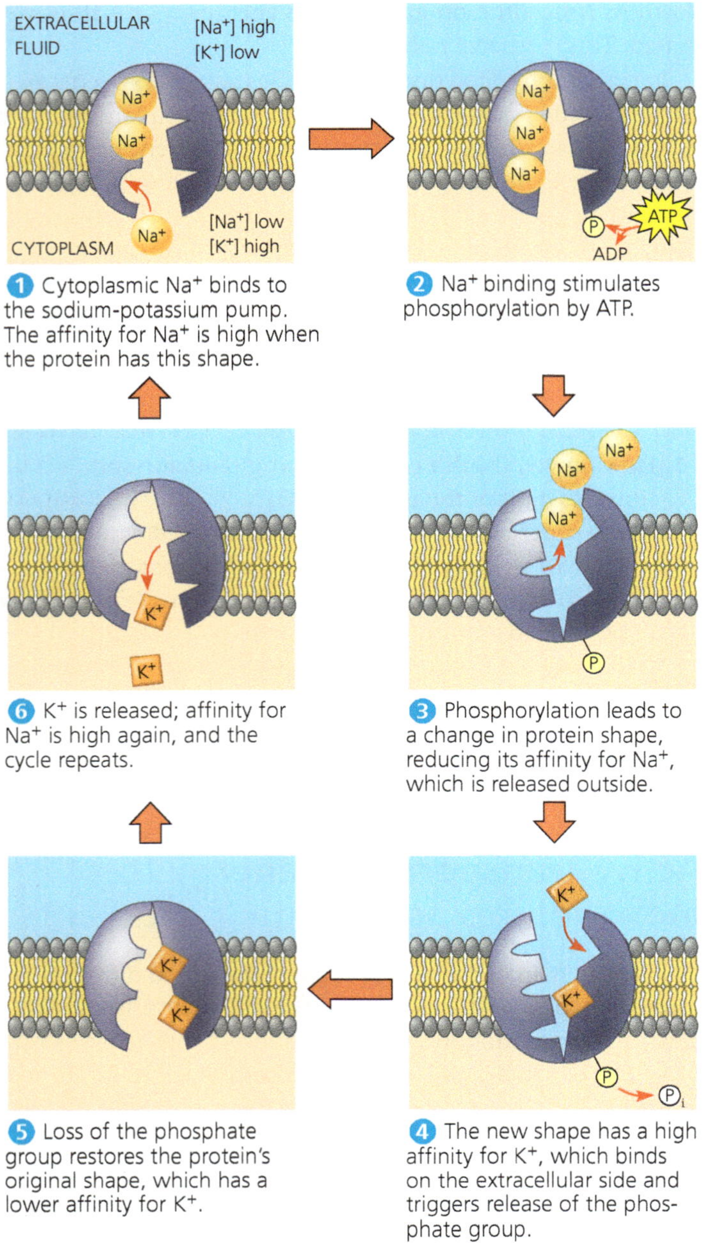

▲ **Figure 7.18 The sodium-potassium pump: a specific case of active transport.** This transport system pumps ions against steep concentration gradients: Sodium ion concentration ($[Na^+]$) is high outside the cell and low inside, while potassium ion concentration ($[K^+]$) is low outside the cell and high inside. The pump oscillates between two shapes in a cycle that moves 3 Na^+ out of the cell for every 2 K^+ pumped into the cell. The two shapes have different affinities for Na^+ and K^+. ATP powers the shape change by transferring a phosphate group to the transport protein (phosphorylating the protein).

an animal cell has a much higher concentration of potassium ions (K^+) and a much lower concentration of sodium ions (Na^+). The plasma membrane helps maintain these steep gradients by pumping Na^+ out of the cell and K^+ into the cell.

As in other types of cellular work, ATP supplies the energy for most active transport. One way ATP can power active transport is by transferring its terminal phosphate group directly to the transport protein. This can induce the protein to change its shape in a manner that translocates a solute bound to the protein across the membrane. One transport system that works this way is the **sodium-potassium pump**, which exchanges Na^+ for K^+ across the plasma membrane of animal cells (**Figure 7.18**). The distinction between passive transport and active transport is reviewed in **Figure 7.19**.

How Ion Pumps Maintain Membrane Potential

All cells have voltages across their plasma membranes. Voltage is electrical potential energy—a separation of opposite charges. The cytoplasmic side of the membrane is negative in charge relative to the extracellular side because of an unequal distribution of anions and cations on the two sides. The voltage across a membrane, called a **membrane potential**, ranges from about −50 to −200 millivolts (mV). (The minus sign indicates that the inside of the cell is negative relative to the outside.)

The membrane potential acts like a battery, an energy source that affects the traffic of all charged substances across the membrane. Because the inside of the cell is negative compared with the outside, the membrane potential favors the passive transport of cations into the cell and anions out of the cell. Thus, *two* forces drive the diffusion of ions across a membrane: a chemical force (the ion's concentration gradient) and an electrical force (the effect of the membrane potential on

▼ **Figure 7.19 Review: passive and active transport.**

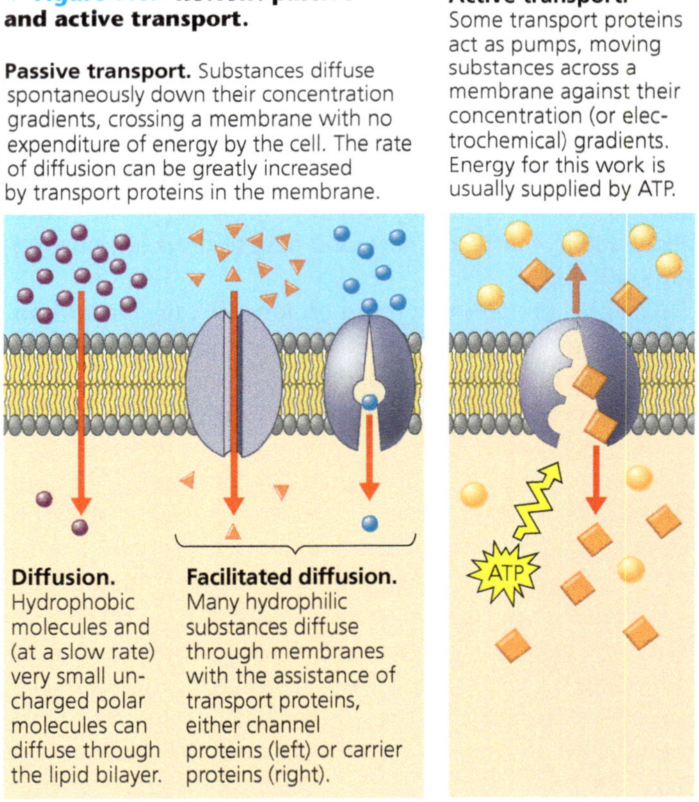

? *For each solute in the right panel, describe its direction of movement, and state whether it is going with or against its concentration gradient.*

the ion's movement). This combination of forces acting on an ion is called the **electrochemical gradient.**

In the case of ions, then, we must refine our concept of passive transport: An ion diffuses not simply down its *concentration* gradient but, more exactly, down its *electrochemical* gradient. For example, the concentration of Na^+ inside a resting nerve cell is much lower than outside it. When the cell is stimulated, gated channels open that facilitate Na^+ diffusion. Sodium ions then "fall" down their electrochemical gradient, driven by the concentration gradient of Na^+ and by the attraction of these cations to the negative side (inside) of the membrane. In this example, both electrical and chemical contributions to the electrochemical gradient act in the same direction across the membrane, but this is not always so. In cases where electrical forces due to the membrane potential oppose the simple diffusion of an ion down its concentration gradient, active transport may be necessary. In Chapter 48, you'll learn about the importance of electrochemical gradients and membrane potentials in the transmission of nerve impulses.

Some membrane proteins that actively transport ions contribute to the membrane potential. An example is the sodium-potassium pump. Notice in Figure 7.18 that the pump does not translocate Na^+ and K^+ one for one, but pumps three sodium ions out of the cell for every two potassium ions it pumps into the cell. With each "crank" of the pump, there is a net transfer of one positive charge from the cytoplasm to the extracellular fluid, a process that stores energy as voltage. A transport protein that generates voltage across a membrane is called an **electrogenic pump**. The sodium-potassium pump appears to be the major electrogenic pump of animal cells. The main electrogenic pump of plants, fungi, and bacteria is a **proton pump**, which actively transports protons (hydrogen ions, H^+) out of the cell. The pumping of H^+ transfers positive charge from the cytoplasm to the extracellular solution **(Figure 7.20)**. By generating voltage across membranes, electrogenic pumps help store energy that can be tapped for cellular work. One important use of proton gradients in the cell is for ATP synthesis during cellular respiration, as you will see in Chapter 9. Another is a type of membrane traffic called cotransport.

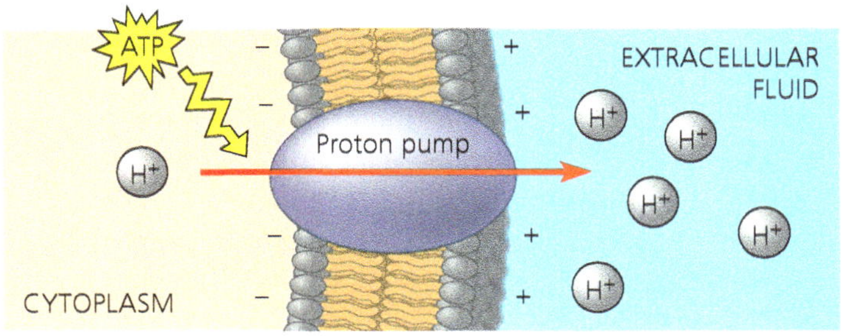

▲ **Figure 7.20 A proton pump.** Proton pumps are electrogenic pumps that store energy by generating voltage (charge separation) across membranes. A proton pump translocates positive charge in the form of hydrogen ions. The voltage and H^+ concentration gradient represent a dual energy source that can drive other processes, such as the uptake of nutrients. Most proton pumps are powered by ATP.

Cotransport: Coupled Transport by a Membrane Protein

A single ATP-powered pump that transports a specific solute can indirectly drive the active transport of several other solutes in a mechanism called **cotransport**. A substance that has been pumped across a membrane can do work as it moves back across the membrane by diffusion, analogous to water that has been pumped uphill and performs work as it flows back down. Another transport protein, a cotransporter separate from the pump, can couple the "downhill" diffusion of this substance to the "uphill" transport of a second substance against its own concentration (or electrochemical) gradient. For example, a plant cell uses the gradient of H^+ generated by its proton pumps to drive the active transport of amino acids, sugars, and several other nutrients into the cell. One transport protein couples the return of H^+ to the transport of sucrose into the cell **(Figure 7.21)**. This protein can translocate sucrose into the cell against a concentration gradient, but only if the sucrose molecule travels in the company of a hydrogen ion. The hydrogen ion uses the transport protein as an avenue to diffuse down the electrochemical gradient maintained by the proton pump. Plants use sucrose-H^+ cotransport to load sucrose produced by photosynthesis into cells in the veins of leaves. The vascular tissue of the plant can then distribute the sugar to nonphotosynthetic organs, such as roots.

What we know about cotransport proteins in animal cells has helped us find more effective treatments for diarrhea, a serious problem in developing countries. Normally, sodium in waste is reabsorbed in the colon, maintaining constant levels in the body, but diarrhea expels waste so rapidly that reabsorption is not possible, and sodium levels fall precipitously.

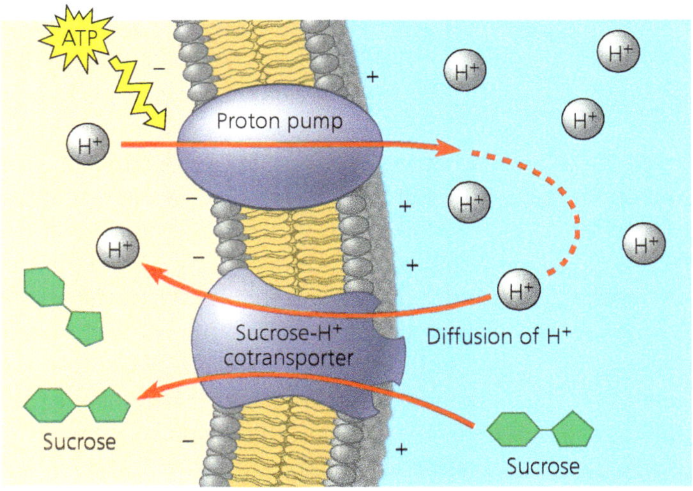

▲ **Figure 7.21 Cotransport: active transport driven by a concentration gradient.** A carrier protein, such as this sucrose-H^+ cotransporter in a plant cell, is able to use the diffusion of H^+ down its electrochemical gradient into the cell to drive the uptake of sucrose. The H^+ gradient is maintained by an ATP-driven proton pump that concentrates H^+ outside the cell, thus storing potential energy that can be used for active transport, in this case of sucrose. Thus, ATP indirectly provides the energy necessary for cotransport. (The cell wall is not shown.)

To treat this life-threatening condition, patients are given a solution to drink containing high concentrations of salt (NaCl) and glucose. The solutes are taken up by sodium-glucose cotransporters on the surface of intestinal cells and passed through the cells into the blood. This simple treatment has lowered infant mortality worldwide.

CONCEPT CHECK 7.4

1. Sodium-potassium pumps help nerve cells establish a voltage across their plasma membranes. Do these pumps use ATP or produce ATP? Explain.
2. Explain why the sodium-potassium pump in Figure 7.18 would not be considered a cotransporter.
3. **MAKE CONNECTIONS** Review the characteristics of the lysosome in Concept 6.4 (pp. 106–107). Given the internal environment of a lysosome, what transport protein might you expect to see in its membrane?

For suggested answers, see Appendix A.

CONCEPT 7.5

Bulk transport across the plasma membrane occurs by exocytosis and endocytosis

Water and small solutes enter and leave the cell by diffusing through the lipid bilayer of the plasma membrane or by being pumped or moved across the membrane by transport proteins. However, large molecules, such as proteins and polysaccharides, as well as larger particles, generally cross the membrane in bulk by mechanisms that involve packaging in vesicles. Like active transport, these processes require energy.

Exocytosis

As we described in Chapter 6, the cell secretes certain biological molecules by the fusion of vesicles with the plasma membrane; this process is called **exocytosis**. A transport vesicle that has budded from the Golgi apparatus moves along microtubules of the cytoskeleton to the plasma membrane. When the vesicle membrane and plasma membrane come into contact, specific proteins rearrange the lipid molecules of the two bilayers so that the two membranes fuse. The contents of the vesicle then spill to the outside of the cell, and the vesicle membrane becomes part of the plasma membrane (see Figure 7.12, step 4).

Many secretory cells use exocytosis to export products. For example, the cells in the pancreas that make insulin secrete it into the extracellular fluid by exocytosis. In another example, neurons (nerve cells) use exocytosis to release neurotransmitters that signal other neurons or muscle cells. When plant cells are making walls, exocytosis delivers proteins and carbohydrates from Golgi vesicles to the outside of the cell.

Endocytosis

In **endocytosis**, the cell takes in biological molecules and particulate matter by forming new vesicles from the plasma membrane. Although the proteins involved in the processes are different, the events of endocytosis look like the reverse of exocytosis. A small area of the plasma membrane sinks inward to form a pocket. As the pocket deepens, it pinches in, forming a vesicle containing material that had been outside the cell. Study **Figure 7.22** carefully to understand the three types of endocytosis: phagocytosis ("cellular eating"), pinocytosis ("cellular drinking"), and receptor-mediated endocytosis.

Human cells use receptor-mediated endocytosis to take in cholesterol for membrane synthesis and the synthesis of other steroids. Cholesterol travels in the blood in particles called low-density lipoproteins (LDLs), each a complex of lipids and a protein. LDLs bind to LDL receptors on plasma membranes and then enter the cells by endocytosis. (LDLs thus act as **ligands**, a term for any molecule that binds specifically to a receptor site on another molecule.) In humans with familial hypercholesterolemia, an inherited disease characterized by a very high level of cholesterol in the blood, LDLs cannot enter cells because the LDL receptor proteins are defective or missing. Consequently, cholesterol accumulates in the blood, where it contributes to early atherosclerosis, the buildup of lipid deposits within the walls of blood vessels. This buildup causes the walls to bulge inward, thereby narrowing the vessels and impeding blood flow.

Vesicles not only transport substances between the cell and its surroundings but also provide a mechanism for rejuvenating or remodeling the plasma membrane. Endocytosis and exocytosis occur continually in most eukaryotic cells, yet the amount of plasma membrane in a nongrowing cell remains fairly constant. Apparently, the addition of membrane by one process offsets the loss of membrane by the other.

Energy and cellular work have figured prominently in our study of membranes. We have seen, for example, that active transport is powered by ATP. In the next three chapters, you will learn more about how cells acquire chemical energy to do the work of life.

CONCEPT CHECK 7.5

1. As a cell grows, its plasma membrane expands. Does this involve endocytosis or exocytosis? Explain.
2. **DRAW IT** Return to Figure 7.12, and circle a patch of plasma membrane that is coming from a vesicle involved in exocytosis.
3. **MAKE CONNECTIONS** In Concept 6.7 (pp. 119–120), you learned that animal cells make an extracellular matrix (ECM). Describe the cellular pathway of synthesis and deposition of an ECM glycoprotein.

For suggested answers, see Appendix A.

▼ Figure 7.22

Exploring Endocytosis in Animal Cells

Phagocytosis

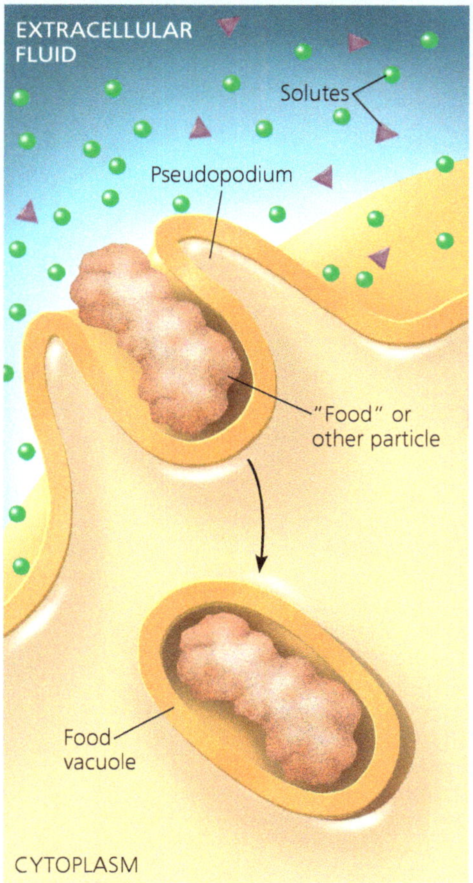

In **phagocytosis,** a cell engulfs a particle by wrapping pseudopodia (singular, *pseudopodium*) around it and packaging it within a membranous sac called a food vacuole. The particle will be digested after the food vacuole fuses with a lysosome containing hydrolytic enzymes (see Figure 6.13a).

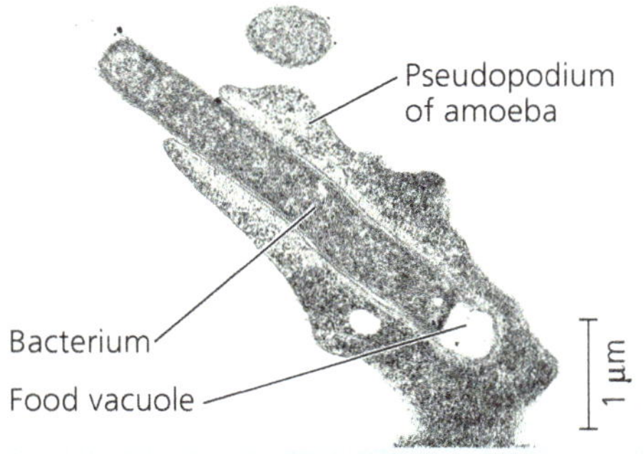

An amoeba engulfing a bacterium via phagocytosis (TEM).

Pinocytosis

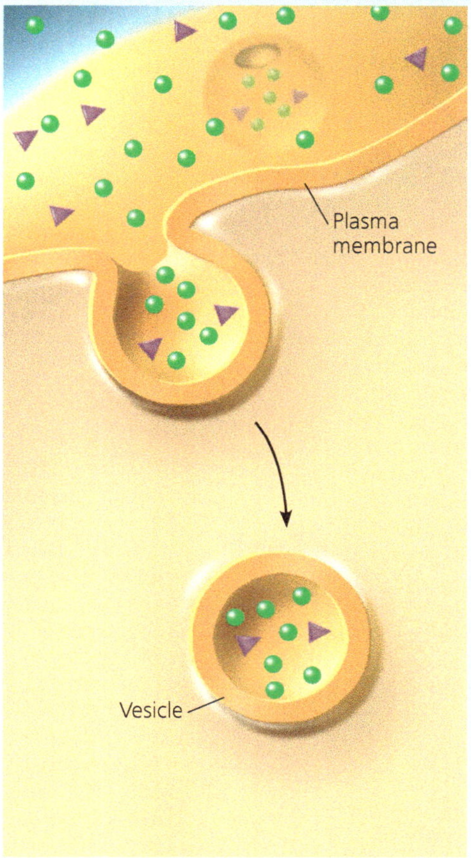

In **pinocytosis,** the cell "gulps" droplets of extracellular fluid into tiny vesicles. It is not the fluid itself that is needed by the cell, but the molecules dissolved in the droplets. Because any and all included solutes are taken into the cell, pinocytosis is nonspecific in the substances it transports.

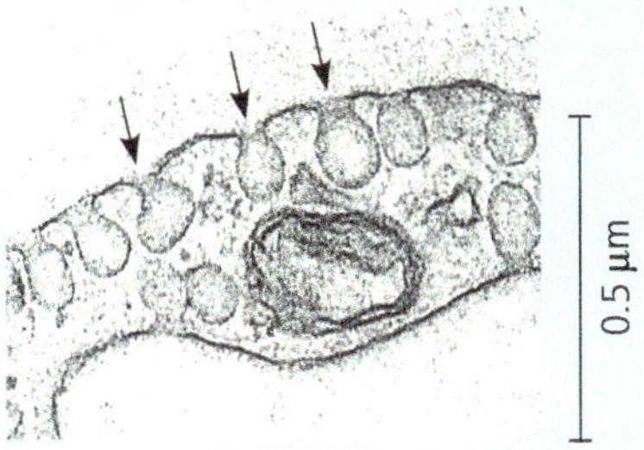

Pinocytosis vesicles forming (indicated by arrows) in a cell lining a small blood vessel (TEM).

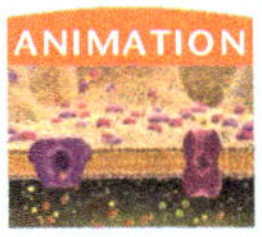

BioFlix Visit the Study Area at **www.masteringbiology.com** for the BioFlix® 3-D Animation on Membrane Transport.

Receptor-Mediated Endocytosis

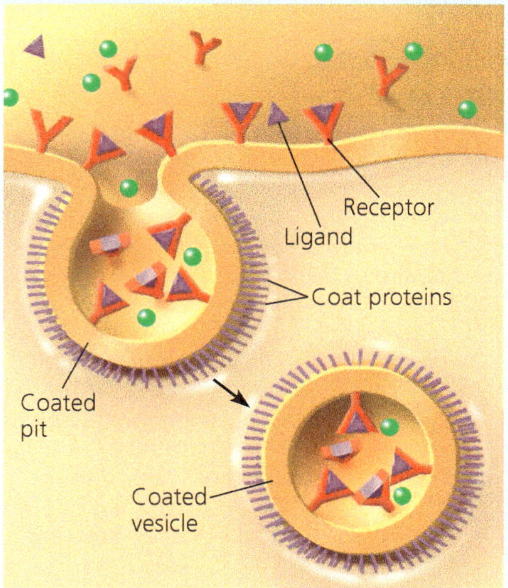

Receptor-mediated endocytosis enables the cell to acquire bulk quantities of specific substances, even though those substances may not be very concentrated in the extracellular fluid. Embedded in the membrane are proteins with specific receptor sites exposed to the extracellular fluid, to which specific substances (ligands) bind. The receptor proteins then cluster in regions of the membrane called coated pits, which are lined on their cytoplasmic side by a fuzzy layer of coat proteins. Next, each coated pit forms a vesicle containing the ligand molecules. Notice that there are relatively more bound molecules (purple) inside the vesicle, but other molecules (green) are also present. After the ingested material is liberated from the vesicle, the emptied receptors are recycled to the plasma membrane by the same vesicle.

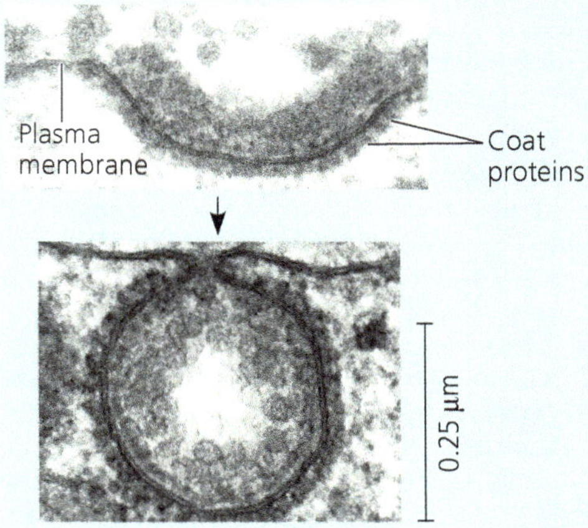

Top: A coated pit. *Bottom:* A coated vesicle forming during receptor-mediated endocytosis (TEMs).

7 CHAPTER REVIEW

SUMMARY OF KEY CONCEPTS

CONCEPT 7.1

Cellular membranes are fluid mosaics of lipids and proteins (pp. 125–131)

- The Davson-Danielli sandwich model of the membrane has been replaced by the **fluid mosaic model**, in which **amphipathic** proteins are embedded in the phospholipid bilayer. Proteins with related functions often cluster in patches.
- Phospholipids and some proteins move laterally within the membrane. The unsaturated hydrocarbon tails of some phospholipids keep membranes fluid at lower temperatures, while cholesterol helps membranes resist changes in fluidity caused by temperature changes. Differences in membrane lipid composition, as well as the ability to change lipid composition, are evolutionary adaptations that ensure membrane fluidity.
- **Integral proteins** are embedded in the lipid bilayer; **peripheral proteins** are attached to the membrane surface. The functions of membrane proteins include transport, enzymatic activity, signal transduction, cell-cell recognition, intercellular joining, and attachment to the cytoskeleton and extracellular matrix. Short chains of sugars linked to proteins (in **glycoproteins**) and lipids (in **glycolipids**) on the exterior side of the plasma membrane interact with surface molecules of other cells.
- Membrane proteins and lipids are synthesized in the ER and modified in the ER and Golgi apparatus. The inside and outside faces of membranes differ in molecular composition.

? *In what ways are membranes crucial to life?*

CONCEPT 7.2

Membrane structure results in selective permeability (pp. 131–132)

- A cell must exchange molecules and ions with its surroundings, a process controlled by the **selective permeability** of the plasma membrane. Hydrophobic substances are soluble in lipid and pass through membranes rapidly, whereas polar molecules and ions generally require specific **transport proteins** to cross the membrane.

? *How do **aquaporins** affect the permeability of a membrane?*

CONCEPT 7.3

Passive transport is diffusion of a substance across a membrane with no energy investment (pp. 132–135)

- **Diffusion** is the spontaneous movement of a substance down its **concentration gradient**. Water diffuses out through the permeable membrane of a cell (**osmosis**) if the solution outside has a higher solute concentration (**hypertonic**) than the cytosol; water enters the cell if the solution has a lower solute concentration (**hypotonic**). If the concentrations are equal (**isotonic**), no net osmosis occurs. Cell survival depends on balancing water uptake and loss. Cells lacking walls (as in animals and some protists) are isotonic with their environments or have adaptations for **osmoregulation**. Plants, prokaryotes, fungi, and some protists have relatively inelastic cell walls, so the cells don't burst in a hypotonic environment.
- In a type of **passive transport** called **facilitated diffusion**, a transport protein speeds the movement of water or a solute across a membrane down its concentration gradient. **Ion channels**, some of which are **gated channels**, facilitate the diffusion of ions across a membrane. Carrier proteins can undergo changes in shape that translocate bound solutes across the membrane.

Passive transport: Facilitated diffusion

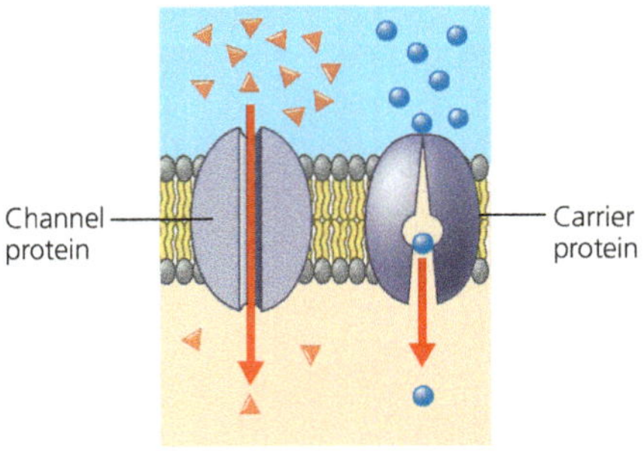

? *What happens to a cell placed in a hypertonic solution? Describe the free water concentration inside and out.*

CONCEPT 7.4

Active transport uses energy to move solutes against their gradients (pp. 135–138)

- Specific membrane proteins use energy, usually in the form of ATP, to do the work of **active transport**. The **sodium-potassium pump** is an example.
- Ions can have both a concentration (chemical) gradient and an electrical gradient (voltage). These gradients combine in the **electrochemical gradient**, which determines the net direction of ionic diffusion. **Electrogenic pumps**, such as the sodium-potassium pump and **proton pumps**, are transport proteins that contribute to electrochemical gradients.
- **Cotransport** of two solutes occurs when a membrane protein enables the "downhill" diffusion of one solute to drive the "uphill" transport of the other.

Active transport

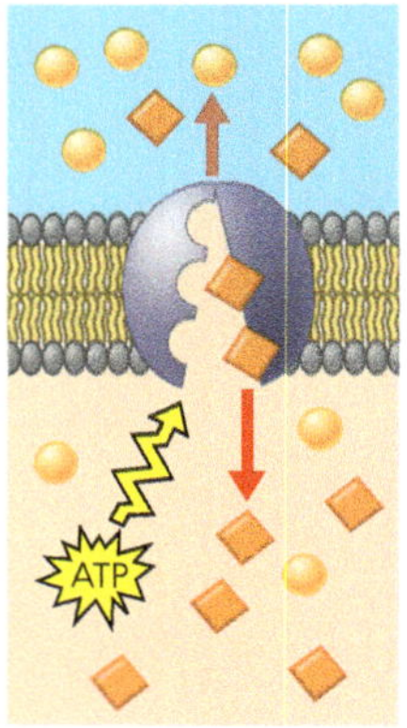

? *ATP is not directly involved in the functioning of a cotransporter. Why, then, is cotransport considered active transport?*

CONCEPT 7.5

Bulk transport across the plasma membrane occurs by exocytosis and endocytosis (p. 138)

- In **exocytosis**, transport vesicles migrate to the plasma membrane, fuse with it, and release their contents. In **endocytosis**, molecules enter cells within vesicles that pinch inward from the plasma membrane. The three types of endocytosis are **phagocytosis**, **pinocytosis**, and **receptor-mediated endocytosis**.

? *Which type of endocytosis involves ligands? What does this type of transport enable a cell to do?*

TEST YOUR UNDERSTANDING

LEVEL 1: KNOWLEDGE/COMPREHENSION

1. In what way do the membranes of a eukaryotic cell vary?
   a. Phospholipids are found only in certain membranes.
   b. Certain proteins are unique to each membrane.

c. Only certain membranes of the cell are selectively permeable.
d. Only certain membranes are constructed from amphipathic molecules.
e. Some membranes have hydrophobic surfaces exposed to the cytoplasm, while others have hydrophilic surfaces facing the cytoplasm.

2. According to the fluid mosaic model of membrane structure, proteins of the membrane are mostly
   a. spread in a continuous layer over the inner and outer surfaces of the membrane.
   b. confined to the hydrophobic interior of the membrane.
   c. embedded in a lipid bilayer.
   d. randomly oriented in the membrane, with no fixed inside-outside polarity.
   e. free to depart from the fluid membrane and dissolve in the surrounding solution.

3. Which of the following factors would tend to increase membrane fluidity?
   a. a greater proportion of unsaturated phospholipids
   b. a greater proportion of saturated phospholipids
   c. a lower temperature
   d. a relatively high protein content in the membrane
   e. a greater proportion of relatively large glycolipids compared with lipids having smaller molecular masses

LEVEL 2: APPLICATION/ANALYSIS

4. Which of the following processes includes all others?
   a. osmosis
   b. diffusion of a solute across a membrane
   c. facilitated diffusion
   d. passive transport
   e. transport of an ion down its electrochemical gradient

5. Based on Figure 7.21, which of these experimental treatments would increase the rate of sucrose transport into the cell?
   a. decreasing extracellular sucrose concentration
   b. decreasing extracellular pH
   c. decreasing cytoplasmic pH
   d. adding an inhibitor that blocks the regeneration of ATP
   e. adding a substance that makes the membrane more permeable to hydrogen ions

6. **DRAW IT** An artificial "cell" consisting of an aqueous solution enclosed in a selectively permeable membrane is immersed in a beaker containing a different solution, the "environment," as shown below. The membrane is permeable to water and to the simple sugars glucose and fructose but impermeable to the disaccharide sucrose.
   a. Draw solid arrows to indicate the net movement of solutes into and/or out of the cell.
   b. Is the solution outside the cell isotonic, hypotonic, or hypertonic?

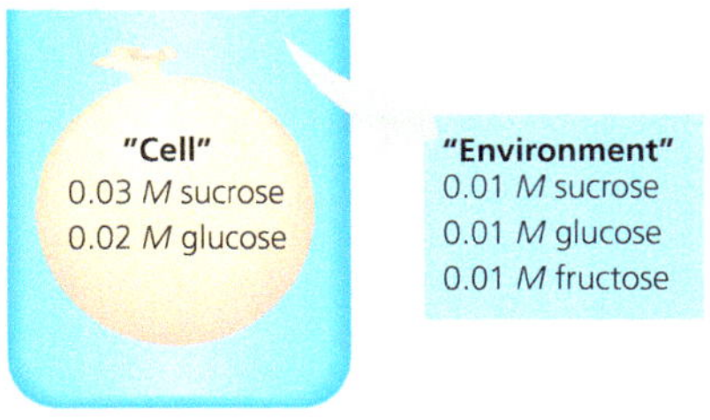

   c. Draw a dashed arrow to show the net osmosis, if any.
   d. Will the artificial cell become more flaccid, more turgid, or stay the same?
   e. Eventually, will the two solutions have the same or different solute concentrations?

LEVEL 3: SYNTHESIS/EVALUATION

7. **EVOLUTION CONNECTION**
   Paramecium and other protists that live in hypotonic environments have cell membranes that limit water uptake, while those living in isotonic environments have membranes that are more permeable to water. What water regulation adaptations might have evolved in protists in hypertonic habitats such as Great Salt Lake? In habitats with changing salt concentration?

8. **SCIENTIFIC INQUIRY**
   An experiment is designed to study the mechanism of sucrose uptake by plant cells. Cells are immersed in a sucrose solution, and the pH of the solution is monitored. Samples of the cells are taken at intervals, and their sucrose concentration is measured. After a decrease in the pH of the solution to a steady, slightly acidic level, sucrose uptake begins. Propose a hypothesis for these results. What do you think would happen if an inhibitor of ATP regeneration by the cell were added to the beaker once the pH is at a steady level? Explain.

9. **SCIENCE, TECHNOLOGY, AND SOCIETY**
   Extensive irrigation in arid regions causes salts to accumulate in the soil. (When water evaporates, salts that were dissolved in the water are left behind in the soil.) Based on what you learned about water balance in plant cells, explain why increased soil salinity (saltiness) might be harmful to crops. Suggest ways to minimize damage. What costs are attached to your solutions?

10. **WRITE ABOUT A THEME**
    Environmental Interactions A human pancreatic cell obtains O_2, fuel molecules such as glucose, and building materials such as amino acids and cholesterol from its environment, and it releases CO_2 as a waste product of cellular respiration. In response to hormonal signals, the cell secretes digestive enzymes. It also regulates its ion concentrations by exchange with its environment. Based on what you have just learned about the structure and function of cellular membranes, write a short essay (100–150 words) that describes how such a cell accomplishes these interactions with its environment.

For selected answers, see Appendix A.

MasteringBIOLOGY www.masteringbiology.com

1. MasteringBiology® Assignments

Make Connections Tutorial Plasma Membranes (Chapter 7) and Phospholipid Structure (Chapter 5)

BioFlix **Tutorials** Membrane Transport: Diffusion and Passive Transport • The Sodium-Potassium Pump • Cotransport • Bulk Transport
Tutorial Osmosis
Activities Membrane Structure • Selective Permeability of Membranes • Diffusion • Diffusion and Osmosis • Facilitated Diffusion • Membrane Transport Proteins • Osmosis and Water Balance in Cells • Active Transport • Exocytosis and Endocytosis
Questions Student Misconceptions • Reading Quiz • Multiple Choice • End-of-Chapter

2. eText

Read your book online, search, take notes, highlight text, and more.

3. The Study Area

Practice Tests • Cumulative Test • ***BioFlix*** 3-D Animations • MP3 Tutor Sessions • Videos • Activities • Investigations • Lab Media • Audio Glossary • Word Study Tools • Art

8

An Introduction to Metabolism

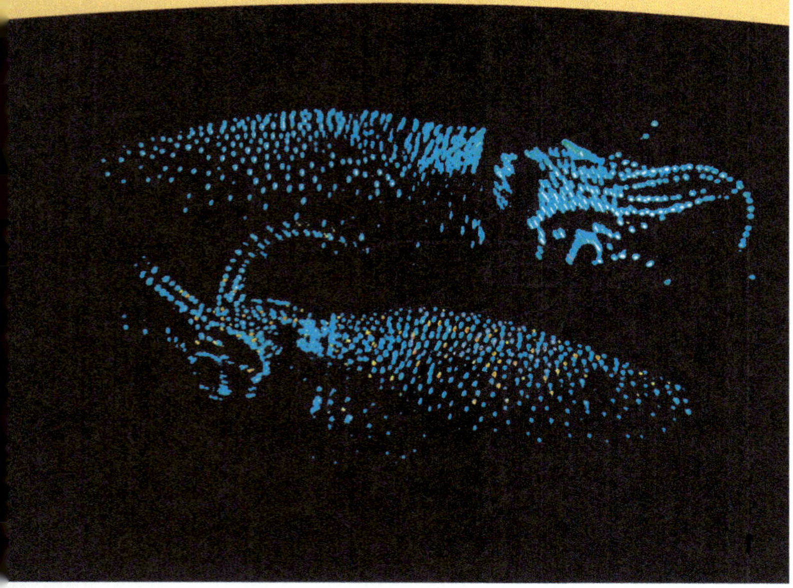

▲ **Figure 8.1 What causes these two squid to glow?**

KEY CONCEPTS

8.1 An organism's metabolism transforms matter and energy, subject to the laws of thermodynamics

8.2 The free-energy change of a reaction tells us whether or not the reaction occurs spontaneously

8.3 ATP powers cellular work by coupling exergonic reactions to endergonic reactions

8.4 Enzymes speed up metabolic reactions by lowering energy barriers

8.5 Regulation of enzyme activity helps control metabolism

OVERVIEW

The Energy of Life

The living cell is a chemical factory in miniature, where thousands of reactions occur within a microscopic space. Sugars can be converted to amino acids that are linked together into proteins when needed, and when food is digested, proteins are dismantled into amino acids that can be converted to sugars. Small molecules are assembled into polymers, which may be hydrolyzed later as the needs of the cell change. In multicellular organisms, many cells export chemical products that are used in other parts of the organism. The process called cellular respiration drives the cellular economy by extracting the energy stored in sugars and other fuels. Cells apply this energy to perform various types of work, such as the transport of solutes across the plasma membrane, which we discussed in Chapter 7. In a more exotic example, cells of the two firefly squid (*Watasenia scintillans*) shown mating in **Figure 8.1** convert the energy stored in certain organic molecules to light, a process called bioluminescence. (The light pattern aids in mate recognition and protection from predators lurking below.) Bioluminescence and other metabolic activities carried out by a cell are precisely coordinated and controlled. In its complexity, its efficiency, and its responsiveness to subtle changes, the cell is peerless as a chemical factory. The concepts of metabolism that you learn in this chapter will help you understand how matter and energy flow during life's processes and how that flow is regulated.

CONCEPT 8.1

An organism's metabolism transforms matter and energy, subject to the laws of thermodynamics

The totality of an organism's chemical reactions is called **metabolism** (from the Greek *metabole*, change). Metabolism is an emergent property of life that arises from orderly interactions between molecules.

Organization of the Chemistry of Life into Metabolic Pathways

We can picture a cell's metabolism as an elaborate road map of the thousands of chemical reactions that occur in a cell, arranged as intersecting metabolic pathways. A **metabolic pathway** begins with a specific molecule, which is then altered in a series of defined steps, resulting in a certain product. Each step of the pathway is catalyzed by a specific enzyme:

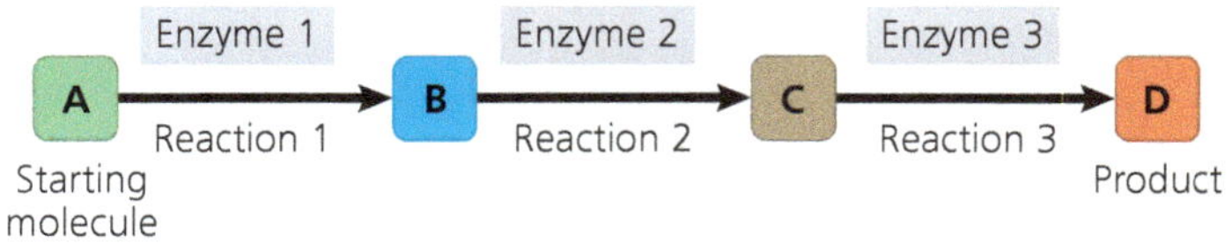

Analogous to the red, yellow, and green stoplights that control the flow of automobile traffic, mechanisms that regulate enzymes balance metabolic supply and demand.

Metabolism as a whole manages the material and energy resources of the cell. Some metabolic pathways release energy by breaking down complex molecules to simpler compounds. These degradative processes are called **catabolic pathways**, or breakdown pathways. A major pathway of catabolism is cellular respiration, in which the sugar glucose and other organic fuels are broken down in the presence of oxygen to carbon dioxide and water. (Pathways can have more than one starting molecule and/or product.) Energy that was stored in the organic molecules becomes available to do the work of the cell, such as ciliary beating or membrane transport. **Anabolic pathways**, in contrast, consume energy to build complicated molecules from simpler ones; they are sometimes called biosynthetic pathways. Examples of anabolism are the synthesis of an amino acid from simpler molecules and the synthesis of a protein from amino acids. Catabolic and anabolic pathways are the "downhill" and "uphill" avenues of the metabolic landscape. Energy released from the downhill reactions of catabolic pathways can be stored and then used to drive the uphill reactions of anabolic pathways.

In this chapter, we will focus on mechanisms common to metabolic pathways. Because energy is fundamental to all metabolic processes, a basic knowledge of energy is necessary to understand how the living cell works. Although we will use some nonliving examples to study energy, the concepts demonstrated by these examples also apply to **bioenergetics**, the study of how energy flows through living organisms.

Forms of Energy

Energy is the capacity to cause change. In everyday life, energy is important because some forms of energy can be used to do work—that is, to move matter against opposing forces, such as gravity and friction. Put another way, energy is the ability to rearrange a collection of matter. For example, you expend energy to turn the pages of this book, and your cells expend energy in transporting certain substances across membranes. Energy exists in various forms, and the work of life depends on the ability of cells to transform energy from one form to another.

Energy can be associated with the relative motion of objects; this energy is called **kinetic energy**. Moving objects can perform work by imparting motion to other matter: A pool player uses the motion of the cue stick to push the cue ball, which in turn moves the other balls; water gushing through a dam turns turbines; and the contraction of leg muscles pushes bicycle pedals. **Heat**, or **thermal energy**, is kinetic energy associated with the random movement of atoms or molecules. Light is also a type of energy that can be harnessed to perform work, such as powering photosynthesis in green plants.

An object not presently moving may still possess energy. Energy that is not kinetic is called **potential energy**; it is energy that matter possesses because of its location or structure. Water behind a dam, for instance, possesses energy because of its altitude above sea level. Molecules possess energy because of the arrangement of electrons in the bonds between their atoms. **Chemical energy** is a term used by biologists to refer to the potential energy available for release in a chemical reaction. Recall that catabolic pathways release energy by breaking down complex molecules. Biologists say that these complex molecules, such as glucose, are high in chemical energy. During a catabolic reaction, some bonds are broken and others formed, releasing energy and resulting in lower-energy breakdown products. This transformation also occurs, for example, in the engine of a car when the hydrocarbons of gasoline react explosively with oxygen, releasing the energy that pushes the pistons and producing exhaust. Although less explosive, a similar reaction of food molecules with oxygen provides chemical energy in biological systems, producing carbon dioxide and water as waste products. Biochemical pathways, carried out in the context of cellular structures, enable cells to release chemical energy from food molecules and use the energy to power life processes.

How is energy converted from one form to another? Consider the divers in **Figure 8.2**. The young woman climbing the ladder to the diving platform is releasing chemical energy from the food she ate for lunch and using some of that energy to perform the work of climbing. The kinetic energy of muscle movement is thus being transformed into potential energy due to her increasing height above the water. The young man diving is converting his potential energy to kinetic energy, which is then transferred to the water as he enters it. A small amount of energy is lost as heat due to friction.

▲ **Figure 8.2 Transformations between potential and kinetic energy.**

Now let's go back one step and consider the original source of the organic food molecules that provided the necessary chemical energy for the diver to climb the steps. This chemical energy was itself derived from light energy by plants during photosynthesis. Organisms are energy transformers.

The Laws of Energy Transformation

The study of the energy transformations that occur in a collection of matter is called **thermodynamics**. Scientists use the word *system* to denote the matter under study; they refer to the rest of the universe—everything outside the system—as the *surroundings*. An *isolated system*, such as that approximated by liquid in a thermos bottle, is unable to exchange either energy or matter with its surroundings. In an *open system*, energy and matter can be transferred between the system and its surroundings. Organisms are open systems. They absorb energy—for instance, light energy or chemical energy in the form of organic molecules—and release heat and metabolic waste products, such as carbon dioxide, to the surroundings. Two laws of thermodynamics govern energy transformations in organisms and all other collections of matter.

The First Law of Thermodynamics

According to the **first law of thermodynamics**, the energy of the universe is constant: *Energy can be transferred and transformed, but it cannot be created or destroyed.* The first law is also known as the *principle of conservation of energy*. The electric company does not make energy, but merely converts it to a form that is convenient for us to use. By converting sunlight to chemical energy, a plant acts as an energy transformer, not an energy producer.

The brown bear in **Figure 8.3a** will convert the chemical energy of the organic molecules in its food to kinetic and other forms of energy as it carries out biological processes. What happens to this energy after it has performed work? The second law of thermodynamics helps to answer this question.

The Second Law of Thermodynamics

If energy cannot be destroyed, why can't organisms simply recycle their energy over and over again? It turns out that during every energy transfer or transformation, some energy becomes unavailable to do work. In most energy transformations, more usable forms of energy are at least partly converted to heat, which is the energy associated with the random motion of atoms or molecules. Only a small fraction of the chemical energy from the food in Figure 8.3a is transformed into the motion of the brown bear shown in **Figure 8.3b**; most is lost as heat, which dissipates rapidly through the surroundings.

In the process of carrying out chemical reactions that perform various kinds of work, living cells unavoidably convert other forms of energy to heat. A system can put heat to work only when there is a temperature difference that results in the heat flowing from a warmer location to a cooler one. If temperature is uniform, as it is in a living cell, then the only use for heat energy generated during a chemical reaction is to warm a body of matter, such as the organism. (This can make a room crowded with people uncomfortably warm, as each person is carrying out a multitude of chemical reactions!)

A logical consequence of the loss of usable energy during energy transfer or transformation is that each such event makes the universe more disordered. Scientists use a quantity called **entropy** as a measure of disorder, or randomness.

(a) First law of thermodynamics: Energy can be transferred or transformed but neither created nor destroyed. For example, chemical reactions in this brown bear (*Ursus arctos*) will convert the chemical (potential) energy in the fish into the kinetic energy of running, shown in (b).

(b) Second law of thermodynamics: Every energy transfer or transformation increases the disorder (entropy) of the universe. For example, as it runs, disorder is increased around the bear by the release of heat and small molecules that are the by-products of metabolism. A brown bear can run at speeds up to 35 miles per hour (56 km/hr)—as fast as a racehorse.

▲ **Figure 8.3 The two laws of thermodynamics.**

The more randomly arranged a collection of matter is, the greater its entropy. We can now state the **second law of thermodynamics**: *Every energy transfer or transformation increases the entropy of the universe.* Although order can increase locally, there is an unstoppable trend toward randomization of the universe as a whole.

In many cases, increased entropy is evident in the physical disintegration of a system's organized structure. For example, you can observe increasing entropy in the gradual decay of an unmaintained building. Much of the increasing entropy of the universe is less apparent, however, because it appears as increasing amounts of heat and less ordered forms of matter. As the bear in Figure 8.3b converts chemical energy to kinetic energy, it is also increasing the disorder of its surroundings by producing heat and small molecules, such as the CO_2 it exhales, that are the breakdown products of food.

The concept of entropy helps us understand why certain processes occur without any input of energy. It turns out that for a process to occur on its own, without outside help, it must increase the entropy of the universe. A process that can occur without an input of energy is called a **spontaneous process**. Note that as we're using it here, the word *spontaneous* does not imply that such a process would occur quickly; rather, the word signifies that the process is energetically favorable. (In fact, it may be helpful for you to think of the phrase "energetically favorable" when you read the formal term "spontaneous.") Some spontaneous processes, such as an explosion, may be virtually instantaneous, while others, such as the rusting of an old car over time, are much slower. A process that cannot occur on its own is said to be nonspontaneous; it will happen only if energy is added to the system. We know from experience that certain events occur spontaneously and others do not. For instance, we know that water flows downhill spontaneously but moves uphill only with an input of energy, such as when a machine pumps the water against gravity. This understanding gives us another way to state the second law: *For a process to occur spontaneously, it must increase the entropy of the universe.*

Biological Order and Disorder

Living systems increase the entropy of their surroundings, as predicted by thermodynamic law. It is true that cells create ordered structures from less organized starting materials. For example, simpler molecules are ordered into the more complex structure of an amino acid, and amino acids are ordered into polypeptide chains. At the organismal level as well, complex and beautifully ordered structures result from biological processes that use simpler starting materials **(Figure 8.4)**. However, an organism also takes in organized forms of matter and energy from the surroundings and replaces them with less ordered forms. For example, an animal obtains starch, proteins, and other complex molecules from the food it eats. As catabolic pathways break these molecules down, the animal releases carbon dioxide and water—small molecules that possess less chemical energy than the food did. The depletion of chemical energy is accounted for by heat generated during metabolism. On a larger scale, energy flows into most ecosystems in the form of light and exits in the form of heat (see Figure 1.6).

▲ **Figure 8.4 Order as a characteristic of life.** Order is evident in the detailed structures of the sea urchin skeleton and the succulent plant shown here. As open systems, organisms can increase their order as long as the order of their surroundings decreases.

During the early history of life, complex organisms evolved from simpler ancestors. For example, we can trace the ancestry of the plant kingdom from much simpler organisms called green algae to more complex flowering plants. However, this increase in organization over time in no way violates the second law. The entropy of a particular system, such as an organism, may actually decrease as long as the total entropy of the *universe*—the system plus its surroundings—increases. Thus, organisms are islands of low entropy in an increasingly random universe. The evolution of biological order is perfectly consistent with the laws of thermodynamics.

CONCEPT CHECK 8.1

1. **MAKE CONNECTIONS** How does the second law of thermodynamics help explain the diffusion of a substance across a membrane? See Figure 7.13 on page 132.
2. Describe the forms of energy found in an apple as it grows on a tree, then falls, then is digested by someone who eats it.
3. **WHAT IF?** If you place a teaspoon of sugar in the bottom of a glass of water, it will dissolve completely over time. Left longer, eventually the water will disappear and the sugar crystals will reappear. Explain these observations in terms of entropy.

For suggested answers, see Appendix A.

CONCEPT 8.2

The free-energy change of a reaction tells us whether or not the reaction occurs spontaneously

The laws of thermodynamics that we've just discussed apply to the universe as a whole. As biologists, we want to understand the chemical reactions of life—for example, which reactions occur spontaneously and which ones require some input of energy from outside. But how can we know this without assessing the energy and entropy changes in the entire universe for each separate reaction?

Free-Energy Change, ΔG

Recall that the universe is really equivalent to "the system" plus "the surroundings." In 1878, J. Willard Gibbs, a professor at Yale, defined a very useful function called the Gibbs free energy of a system (without considering its surroundings), symbolized by the letter G. We'll refer to the Gibbs free energy simply as free energy. **Free energy** is the portion of a system's energy that can perform work when temperature and pressure are uniform throughout the system, as in a living cell. Let's consider how we determine the free-energy change that occurs when a system changes—for example, during a chemical reaction.

The change in free energy, ΔG, can be calculated for a chemical reaction by applying the following equation:

$$\Delta G = \Delta H - T\Delta S$$

This equation uses only properties of the system (the reaction) itself: ΔH symbolizes the change in the system's *enthalpy* (in biological systems, equivalent to total energy); ΔS is the change in the system's entropy; and T is the absolute temperature in Kelvin (K) units (K = °C + 273; see Appendix C).

Once we know the value of ΔG for a process, we can use it to predict whether the process will be spontaneous (that is, whether it is energetically favorable and will occur without an input of energy). More than a century of experiments has shown that only processes with a negative ΔG are spontaneous. For ΔG to be negative, either ΔH must be negative (the system gives up enthalpy and H decreases) or $T\Delta S$ must be positive (the system gives up order and S increases), or both: When ΔH and $T\Delta S$ are tallied, ΔG has a negative value ($\Delta G < 0$) for all spontaneous processes. In other words, every spontaneous process decreases the system's free energy, and processes that have a positive or zero ΔG are never spontaneous.

This information is immensely interesting to biologists, for it gives us the power to predict which kinds of change can happen without help. Such spontaneous changes can be harnessed to perform work. This principle is very important in the study of metabolism, where a major goal is to determine which reactions can supply energy for cellular work.

Free Energy, Stability, and Equilibrium

As we saw in the previous section, when a process occurs spontaneously in a system, we can be sure that ΔG is negative. Another way to think of ΔG is to realize that it represents the difference between the free energy of the final state and the free energy of the initial state:

$$\Delta G = G_{\text{final state}} - G_{\text{initial state}}$$

Thus, ΔG can be negative only when the process involves a loss of free energy during the change from initial state to final state. Because it has less free energy, the system in its final state is less likely to change and is therefore more stable than it was previously.

We can think of free energy as a measure of a system's instability—its tendency to change to a more stable state. Unstable systems (higher G) tend to change in such a way that they become more stable (lower G). For example, a diver on top of a platform is less stable (more likely to fall) than when floating in the water; a drop of concentrated dye is less stable (more likely to disperse) than when the dye is spread randomly through the liquid; and a glucose molecule is less stable (more likely to break down) than the simpler molecules into which it can be split **(Figure 8.5)**. Unless something prevents it, each of these systems will move toward greater stability: The diver falls, the solution becomes uniformly colored, and the glucose molecule is broken down.

Another term that describes a state of maximum stability is *equilibrium*, which you learned about in Chapter 2 in connection with chemical reactions. There is an important relationship between free energy and equilibrium, including chemical equilibrium. Recall that most chemical reactions are reversible and proceed to a point at which the forward and backward reactions occur at the same rate. The reaction is then said to be at chemical equilibrium, and there is no further net change in the relative concentration of products and reactants.

As a reaction proceeds toward equilibrium, the free energy of the mixture of reactants and products decreases. Free energy increases when a reaction is somehow pushed away from equilibrium, perhaps by removing some of the products (and thus changing their concentration relative to that of the reactants). For a system at equilibrium, G is at its lowest possible value in that system. We can think of the equilibrium state as a free-energy valley. Any change from the equilibrium position will have a positive ΔG and will not be spontaneous. For this reason, systems never spontaneously move away from equilibrium. Because a system at equilibrium cannot spontaneously change, it can do no work. *A process is spontaneous and can perform work only when it is moving toward equilibrium.*

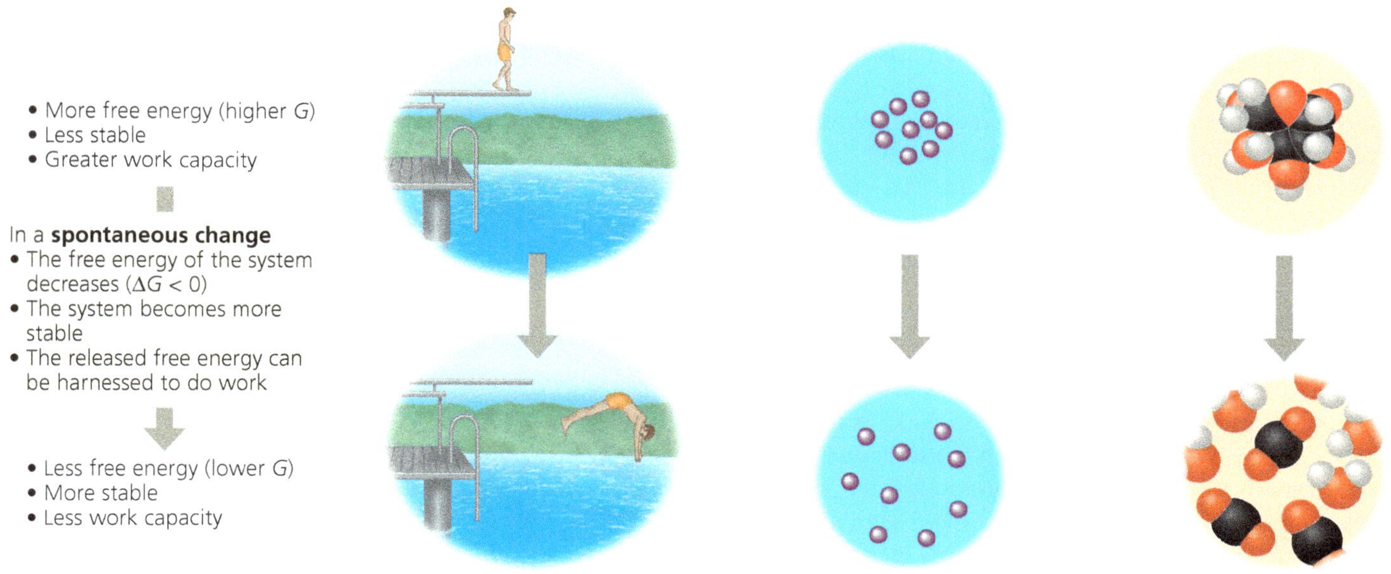

(a) Gravitational motion. Objects move spontaneously from a higher altitude to a lower one.

(b) Diffusion. Molecules in a drop of dye diffuse until they are randomly dispersed.

(c) Chemical reaction. In a cell, a glucose molecule is broken down into simpler molecules.

▲ **Figure 8.5 The relationship of free energy to stability, work capacity, and spontaneous change.** Unstable systems (top) are rich in free energy, *G*. They have a tendency to change spontaneously to a more stable state (bottom), and it is possible to harness this "downhill" change to perform work.

Free Energy and Metabolism

We can now apply the free-energy concept more specifically to the chemistry of life's processes.

Exergonic and Endergonic Reactions in Metabolism

Based on their free-energy changes, chemical reactions can be classified as either exergonic ("energy outward") or endergonic ("energy inward"). An **exergonic reaction** proceeds with a net release of free energy **(Figure 8.6a)**. Because the chemical mixture loses free energy (G decreases), ΔG is negative for an exergonic reaction. Using ΔG as a standard for spontaneity, exergonic reactions are those that occur spontaneously. (Remember, the word *spontaneous* implies that it is energetically favorable, not that it will occur rapidly.) The magnitude of ΔG for an exergonic reaction represents the maximum amount of work the reaction can perform.* The greater the decrease in free energy, the greater the amount of work that can be done.

We can use the overall reaction for cellular respiration as an example:

$$C_6H_{12}O_6 + 6\ O_2 \rightarrow 6\ CO_2 + 6\ H_2O$$
$$\Delta G = -686\ \text{kcal/mol}\ (-2{,}870\ \text{kJ/mol})$$

*The word *maximum* qualifies this statement, because some of the free energy is released as heat and cannot do work. Therefore, ΔG represents a theoretical upper limit of available energy.

▼ **Figure 8.6 Free energy changes (ΔG) in exergonic and endergonic reactions.**

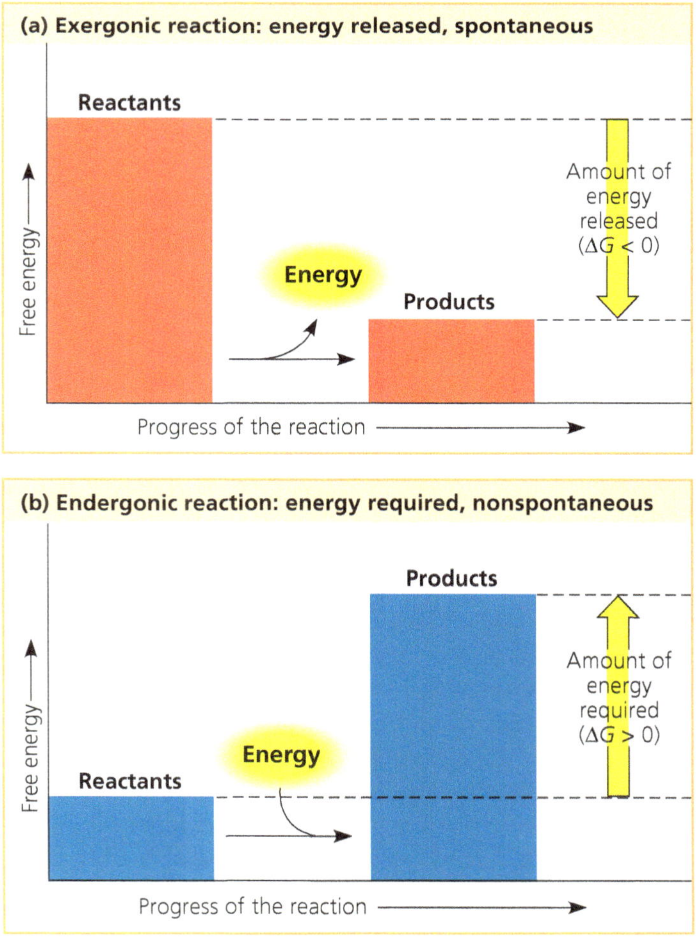

For each mole (180 g) of glucose broken down by respiration under what are called "standard conditions" (1 *M* of each reactant and product, 25°C, pH 7), 686 kcal (2,870 kJ) of energy are made available for work. Because energy must be conserved, the chemical products of respiration store 686 kcal less free energy per mole than the reactants. The products are, in a sense, the spent exhaust of a process that tapped the free energy stored in the bonds of the sugar molecules.

It is important to realize that the breaking of bonds does not release energy; on the contrary, as you will soon see, it requires energy. The phrase "energy stored in bonds" is shorthand for the potential energy that can be released when new bonds are formed after the original bonds break, as long as the products are of lower free energy than the reactants.

An **endergonic reaction** is one that absorbs free energy from its surroundings **(Figure 8.6b)**. Because this kind of reaction essentially *stores* free energy in molecules (G increases), ΔG is positive. Such reactions are nonspontaneous, and the magnitude of ΔG is the quantity of energy required to drive the reaction. If a chemical process is exergonic (downhill), releasing energy in one direction, then the reverse process must be endergonic (uphill), using energy. A reversible process cannot be downhill in both directions. If $\Delta G = -686$ kcal/mol for respiration, which converts glucose and oxygen to carbon dioxide and water, then the reverse process—the conversion of carbon dioxide and water to glucose and oxygen—must be strongly endergonic, with $\Delta G = +686$ kcal/mol. Such a reaction would never happen by itself.

How, then, do plants make the sugar that organisms use for energy? Plants get the required energy—686 kcal to make a mole of glucose—from the environment by capturing light and converting its energy to chemical energy. Next, in a long series of exergonic steps, they gradually spend that chemical energy to assemble glucose molecules.

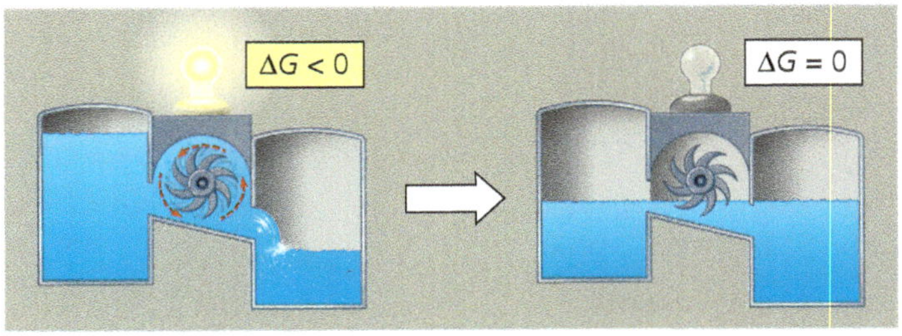

(a) An isolated hydroelectric system. Water flowing downhill turns a turbine that drives a generator providing electricity to a lightbulb, but only until the system reaches equilibrium.

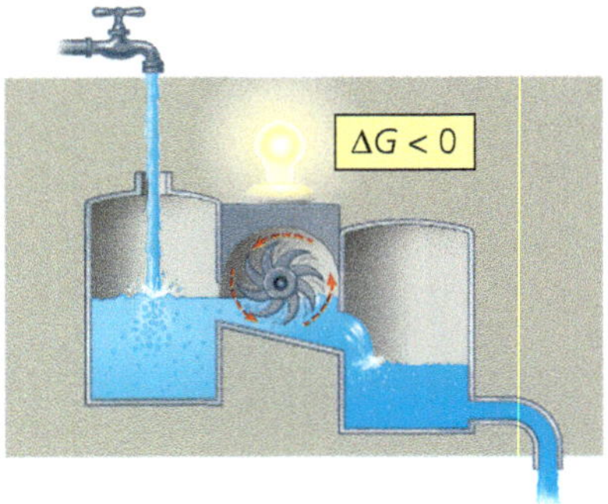

(b) An open hydroelectric system. Flowing water keeps driving the generator because intake and outflow of water keep the system from reaching equilibrium.

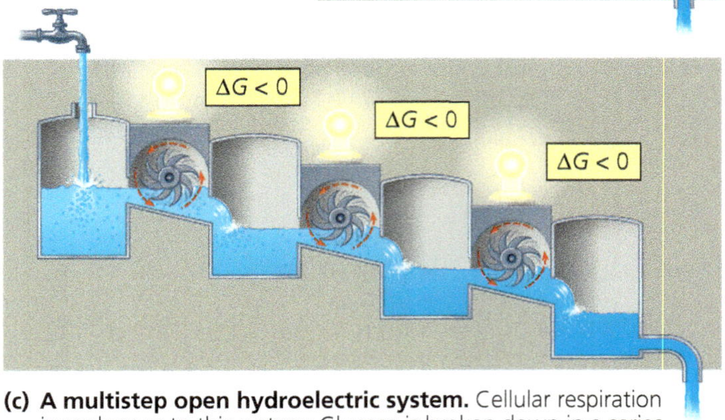

(c) A multistep open hydroelectric system. Cellular respiration is analogous to this system: Glucose is broken down in a series of exergonic reactions that power the work of the cell. The product of each reaction becomes the reactant for the next, so no reaction reaches equilibrium.

▲ **Figure 8.7 Equilibrium and work in isolated and open systems.**

Equilibrium and Metabolism

Reactions in an isolated system eventually reach equilibrium and can then do no work, as illustrated by the isolated hydroelectric system in **Figure 8.7a**. The chemical reactions of metabolism are reversible, and they, too, would reach equilibrium if they occurred in the isolation of a test tube. Because systems at equilibrium are at a minimum of G and can do no work, a cell that has reached metabolic equilibrium is dead! The fact that metabolism as a whole is never at equilibrium is one of the defining features of life.

Like most systems, a living cell is not in equilibrium. The constant flow of materials in and out of the cell keeps the metabolic pathways from ever reaching equilibrium, and the cell continues to do work throughout its life. This principle is illustrated by the open (and more realistic) hydroelectric system in **Figure 8.7b**. However, unlike this simple single-step system, a catabolic pathway in a cell releases free energy in a series of reactions. An example is cellular respiration, illustrated by analogy in **Figure 8.7c**. Some of the reversible reactions of respiration are constantly "pulled" in one direction—that is, they are kept out of equilibrium. The key to maintaining this lack of equilibrium is that the product of a reaction does not accumulate but instead becomes a reactant in the next step; finally, waste products are expelled from the cell. The overall sequence of reactions is kept going by the huge free-energy difference between glucose and oxygen at the top of the energy "hill" and carbon dioxide and water at the "downhill" end. As long as our cells have a steady supply of glucose or other fuels and oxygen and are able to expel waste products to the surroundings, their metabolic pathways never reach equilibrium and can continue to do the work of life.

We see once again how important it is to think of organisms as open systems. Sunlight provides a daily source of free energy for an ecosystem's plants and other photosynthetic organisms. Animals and other nonphotosynthetic organisms in an ecosystem must have a source of free energy in the form of the organic products of photosynthesis. Now that we have applied the free-energy concept to metabolism, we are ready to see how a cell actually performs the work of life.

CONCEPT CHECK 8.2

1. Cellular respiration uses glucose and oxygen, which have high levels of free energy, and releases CO_2 and water, which have low levels of free energy. Is cellular respiration spontaneous or not? Is it exergonic or endergonic? What happens to the energy released from glucose?
2. **MAKE CONNECTIONS** As you saw in Figure 7.20 on page 137, a key process in metabolism is the transport of hydrogen ions (H^+) across a membrane to create a concentration gradient. Other processes can result in an equal concentration of H^+ on each side. Which situation allows the H^+ to perform work in this system? How is the answer consistent with what is shown in regard to energy in Figure 7.20?
3. **WHAT IF?** Some night-time partygoers wear glow-in-the-dark necklaces. The necklaces start glowing once they are "activated," which usually involves snapping the necklace in a way that allows two chemicals to react and emit light in the form of chemiluminescence. Is the chemical reaction exergonic or endergonic? Explain your answer.

For suggested answers, see Appendix A.

CONCEPT 8.3

ATP powers cellular work by coupling exergonic reactions to endergonic reactions

A cell does three main kinds of work:

- *Chemical work*, the pushing of endergonic reactions that would not occur spontaneously, such as the synthesis of polymers from monomers (chemical work will be discussed further here and in Chapters 9 and 10)
- *Transport work*, the pumping of substances across membranes against the direction of spontaneous movement (see Chapter 7)
- *Mechanical work*, such as the beating of cilia (see Chapter 6), the contraction of muscle cells, and the movement of chromosomes during cellular reproduction

A key feature in the way cells manage their energy resources to do this work is **energy coupling**, the use of an exergonic process to drive an endergonic one. ATP is responsible for mediating most energy coupling in cells, and in most cases it acts as the immediate source of energy that powers cellular work.

The Structure and Hydrolysis of ATP

ATP (adenosine triphosphate) was introduced in Chapter 4 when we discussed the phosphate group as a functional group. ATP contains the sugar ribose, with the nitrogenous base adenine and a chain of three phosphate groups bonded to it (**Figure 8.8a**). In addition to its role in energy coupling, ATP is also one of the nucleoside triphosphates used to make RNA (see Figure 5.26).

The bonds between the phosphate groups of ATP can be broken by hydrolysis. When the terminal phosphate bond is broken by addition of a water molecule, a molecule of inorganic phosphate ($HOPO_3^{2-}$, abbreviated Ⓟ$_i$ throughout this book) leaves the ATP, which becomes adenosine diphosphate,

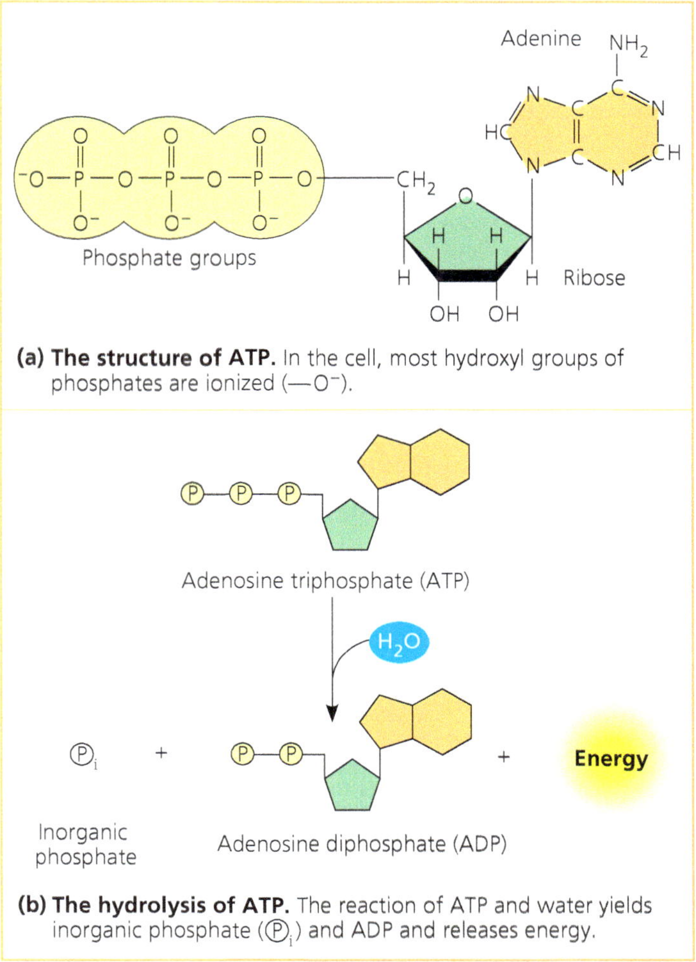

▲ **Figure 8.8 The structure and hydrolysis of adenosine triphosphate (ATP).**

or ADP **(Figure 8.8b)**. The reaction is exergonic and releases 7.3 kcal of energy per mole of ATP hydrolyzed:

$$ATP + H_2O \rightarrow ADP + Ⓟ_i$$
$$\Delta G = -7.3 \text{ kcal/mol } (-30.5 \text{ kJ/mol})$$

This is the free-energy change measured under standard conditions. In the cell, conditions do not conform to standard conditions, primarily because reactant and product concentrations differ from 1 *M*. For example, when ATP hydrolysis occurs under cellular conditions, the actual ΔG is about -13 kcal/mol, 78% greater than the energy released by ATP hydrolysis under standard conditions.

Because their hydrolysis releases energy, the phosphate bonds of ATP are sometimes referred to as high-energy phosphate bonds, but the term is misleading. The phosphate bonds of ATP are not unusually strong bonds, as "high-energy" may imply; rather, the reactants (ATP and water) themselves have high energy relative to the energy of the products (ADP and $Ⓟ_i$). The release of energy during the hydrolysis of ATP comes from the chemical change to a state of lower free energy, not from the phosphate bonds themselves.

ATP is useful to the cell because the energy it releases on losing a phosphate group is somewhat greater than the energy most other molecules could deliver. But why does this hydrolysis release so much energy? If we reexamine the ATP molecule in Figure 8.8a, we can see that all three phosphate groups are negatively charged. These like charges are crowded together, and their mutual repulsion contributes to the instability of this region of the ATP molecule. The triphosphate tail of ATP is the chemical equivalent of a compressed spring.

How the Hydrolysis of ATP Performs Work

When ATP is hydrolyzed in a test tube, the release of free energy merely heats the surrounding water. In an organism, this same generation of heat can sometimes be beneficial. For instance, the process of shivering uses ATP hydrolysis during muscle contraction to generate heat and warm the body. In most cases in the cell, however, the generation of heat alone would be an inefficient (and potentially dangerous) use of a valuable energy resource. Instead, the cell's proteins harness the energy released during ATP hydrolysis in several ways to perform the three types of cellular work—chemical, transport, and mechanical.

For example, with the help of specific enzymes, the cell is able to use the energy released by ATP hydrolysis directly to drive chemical reactions that, by themselves, are endergonic. If the ΔG of an endergonic reaction is less than the amount of energy released by ATP hydrolysis, then the two reactions can be coupled so that, overall, the coupled reactions are exergonic **(Figure 8.9)**. This usually involves the transfer of a phosphate

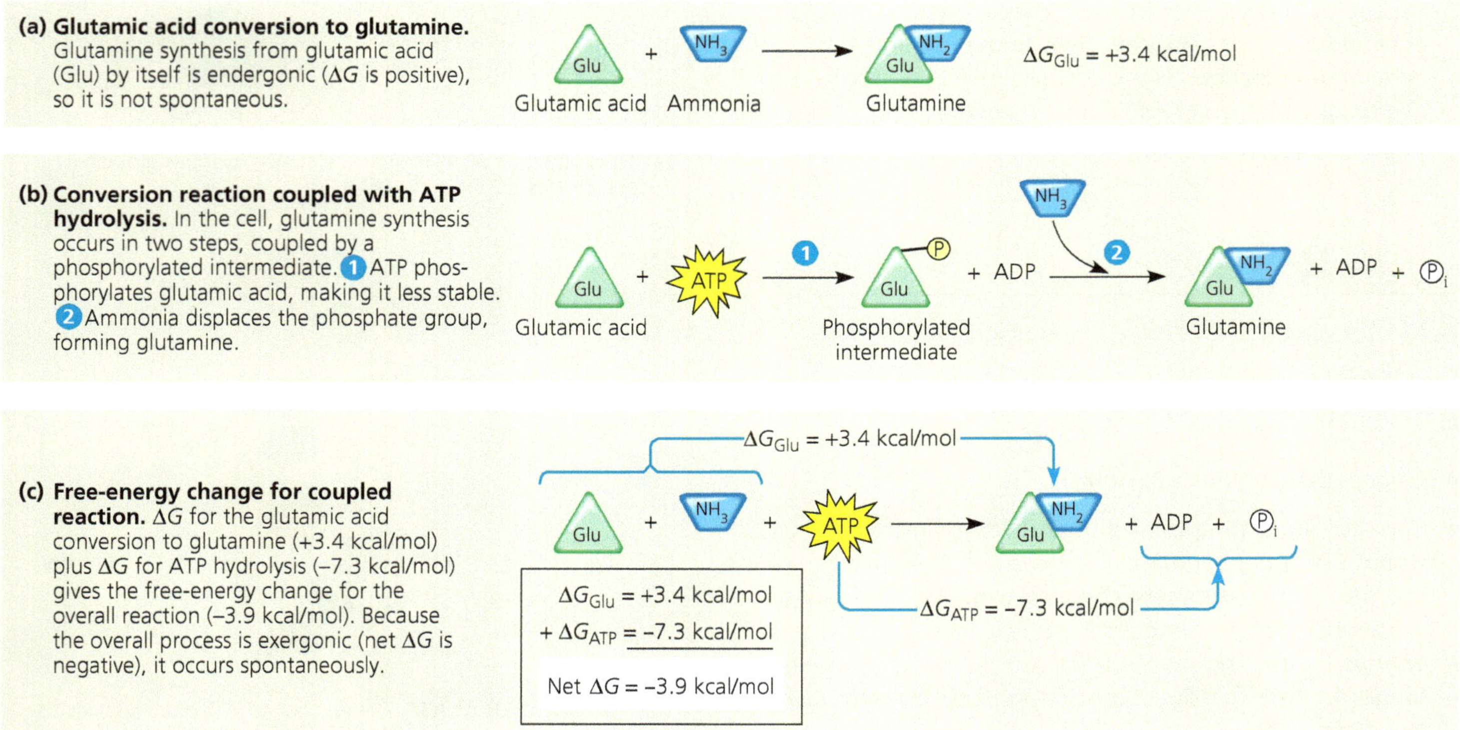

▲ **Figure 8.9 How ATP drives chemical work: Energy coupling using ATP hydrolysis.** In this example, the exergonic process of ATP hydrolysis is used to drive an endergonic process—the cellular synthesis of the amino acid glutamine from glutamic acid and ammonia.

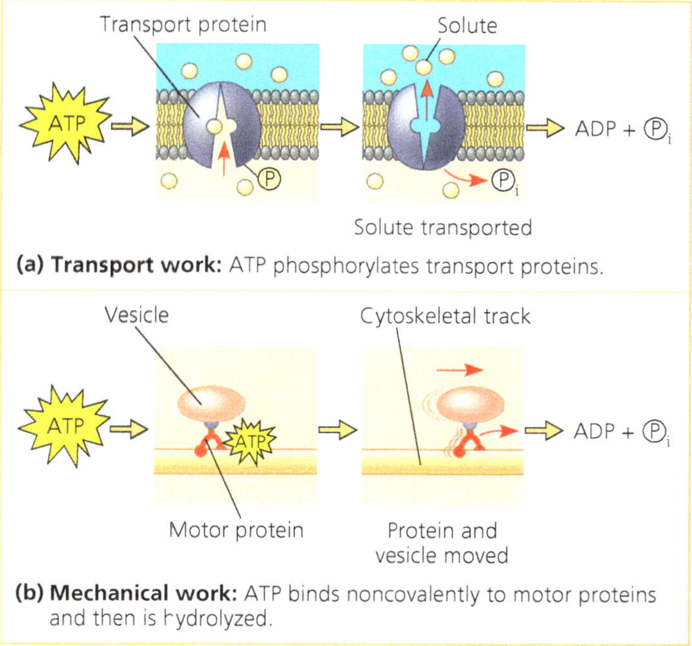

(a) Transport work: ATP phosphorylates transport proteins.

(b) Mechanical work: ATP binds noncovalently to motor proteins and then is hydrolyzed.

▲ **Figure 8.10 How ATP drives transport and mechanical work.** ATP hydrolysis causes changes in the shapes and binding affinities of proteins. This can occur either **(a)** directly, by phosphorylation, as shown for a membrane protein carrying out active transport of a solute (see also Figure 7.18), or **(b)** indirectly, via noncovalent binding of ATP and its hydrolytic products, as is the case for motor proteins that move vesicles (and other organelles) along cytoskeletal "tracks" in the cell (see also Figure 6.21).

group from ATP to some other molecule, such as the reactant. The recipient with the phosphate group covalently bonded to it is then called a **phosphorylated intermediate**. The key to coupling exergonic and endergonic reactions is the formation of this phosphorylated intermediate, which is more reactive (less stable) than the original unphosphorylated molecule.

Transport and mechanical work in the cell are also nearly always powered by the hydrolysis of ATP. In these cases, ATP hydrolysis leads to a change in a protein's shape and often its ability to bind another molecule. Sometimes this occurs via a phosphorylated intermediate, as seen for the transport protein in **Figure 8.10a**. In most instances of mechanical work involving motor proteins "walking" along cytoskeletal elements **(Figure 8.10b)**, a cycle occurs in which ATP is first bound noncovalently to the motor protein. Next, ATP is hydrolyzed, releasing ADP and Ⓟ$_i$. Another ATP molecule can then bind. At each stage, the motor protein changes its shape and ability to bind the cytoskeleton, resulting in movement of the protein along the cytoskeletal track.

The Regeneration of ATP

An organism at work uses ATP continuously, but ATP is a renewable resource that can be regenerated by the addition of phosphate to ADP **(Figure 8.11)**. The free energy required to

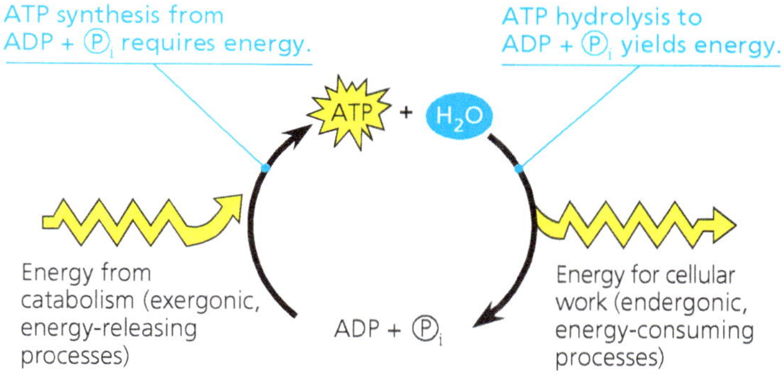

▲ **Figure 8.11 The ATP cycle.** Energy released by breakdown reactions (catabolism) in the cell is used to phosphorylate ADP, regenerating ATP. Chemical potential energy stored in ATP drives most cellular work.

phosphorylate ADP comes from exergonic breakdown reactions (catabolism) in the cell. This shuttling of inorganic phosphate and energy is called the ATP cycle, and it couples the cell's energy-yielding (exergonic) processes to the energy-consuming (endergonic) ones. The ATP cycle proceeds at an astonishing pace. For example, a working muscle cell recycles its entire pool of ATP in less than a minute. That turnover represents 10 million molecules of ATP consumed and regenerated per second per cell. If ATP could not be regenerated by the phosphorylation of ADP, humans would use up nearly their body weight in ATP each day.

Because both directions of a reversible process cannot be downhill, the regeneration of ATP from ADP and Ⓟ$_i$ is necessarily endergonic:

$$\text{ADP} + Ⓟ_i \rightarrow \text{ATP} + H_2O$$
$$\Delta G = +7.3 \text{ kcal/mol } (+30.5 \text{ kJ/mol}) \text{ (standard conditions)}$$

Since ATP formation from ADP and Ⓟ$_i$ is not spontaneous, free energy must be spent to make it occur. Catabolic (exergonic) pathways, especially cellular respiration, provide the energy for the endergonic process of making ATP. Plants also use light energy to produce ATP. Thus, the ATP cycle is a revolving door through which energy passes during its transfer from catabolic to anabolic pathways.

CONCEPT CHECK 8.3

1. How does ATP typically transfer energy from exergonic to endergonic reactions in the cell?
2. Which of the following combinations has more free energy: glutamic acid + ammonia + ATP, or glutamine + ADP + Ⓟ$_i$? Explain your answer.
3. **MAKE CONNECTIONS** Considering what you learned in Concepts 7.3 and 7.4 (pp. 134–136), does Figure 8.10a show passive or active transport? Explain.

For suggested answers, see Appendix A.

CONCEPT 8.4

Enzymes speed up metabolic reactions by lowering energy barriers

The laws of thermodynamics tell us what will and will not happen under given conditions but say nothing about the rate of these processes. A spontaneous chemical reaction occurs without any requirement for outside energy, but it may occur so slowly that it is imperceptible. For example, even though the hydrolysis of sucrose (table sugar) to glucose and fructose is exergonic, occurring spontaneously with a release of free energy ($\Delta G = -7$ kcal/mol), a solution of sucrose dissolved in sterile water will sit for years at room temperature with no appreciable hydrolysis. However, if we add a small amount of the enzyme sucrase to the solution, then all the sucrose may be hydrolyzed within seconds, as shown below:

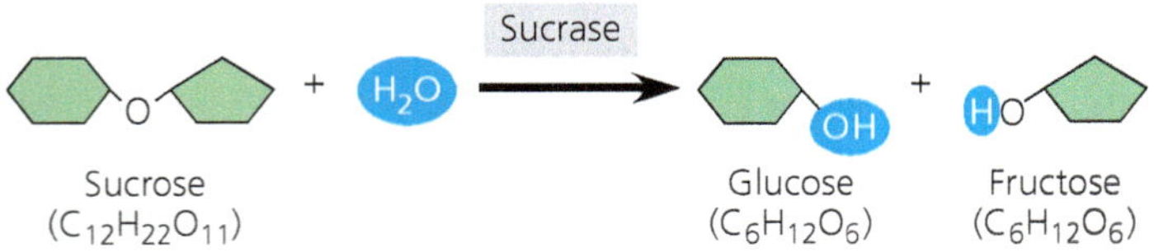

How does the enzyme do this?

An **enzyme** is a macromolecule that acts as a **catalyst**, a chemical agent that speeds up a reaction without being consumed by the reaction. (In this chapter, we are focusing on enzymes that are proteins. RNA enzymes, also called ribozymes, are discussed in Chapters 17 and 25.) Without regulation by enzymes, chemical traffic through the pathways of metabolism would become terribly congested because many chemical reactions would take such a long time. In the next two sections, we will see what prevents a spontaneous reaction from occurring faster and how an enzyme changes the situation.

The Activation Energy Barrier

Every chemical reaction between molecules involves both bond breaking and bond forming. For example, the hydrolysis of sucrose involves breaking the bond between glucose and fructose and one of the bonds of a water molecule and then forming two new bonds, as shown above. Changing one molecule into another generally involves contorting the starting molecule into a highly unstable state before the reaction can proceed. This contortion can be compared to the bending of a metal key ring when you pry it open to add a new key. The key ring is highly unstable in its opened form but returns to a stable state once the key is threaded all the way onto the ring. To reach the contorted state where bonds can change, reactant molecules must absorb energy from their surroundings. When the new bonds of the product molecules form, energy is released as heat, and the molecules return to stable shapes with lower energy than the contorted state.

The initial investment of energy for starting a reaction—the energy required to contort the reactant molecules so the bonds can break—is known as the *free energy of activation*, or **activation energy**, abbreviated E_A in this book. We can think of activation energy as the amount of energy needed to push the reactants to the top of an energy barrier, or uphill, so that the "downhill" part of the reaction can begin. Activation energy is often supplied in the form of thermal energy (heat) that the reactant molecules absorb from the surroundings. The absorption of thermal energy accelerates the reactant molecules, so they collide more often and more forcefully. It also agitates the atoms within the molecules, making the breakage of bonds more likely. When the molecules have absorbed enough energy for the bonds to break, the reactants are in an unstable condition known as the *transition state*.

Figure 8.12 graphs the energy changes for a hypothetical exergonic reaction that swaps portions of two reactant molecules:

$$\underset{\text{Reactants}}{AB + CD} \longrightarrow \underset{\text{Products}}{AC + BD}$$

The activation of the reactants is represented by the uphill portion of the graph, in which the free-energy content of the

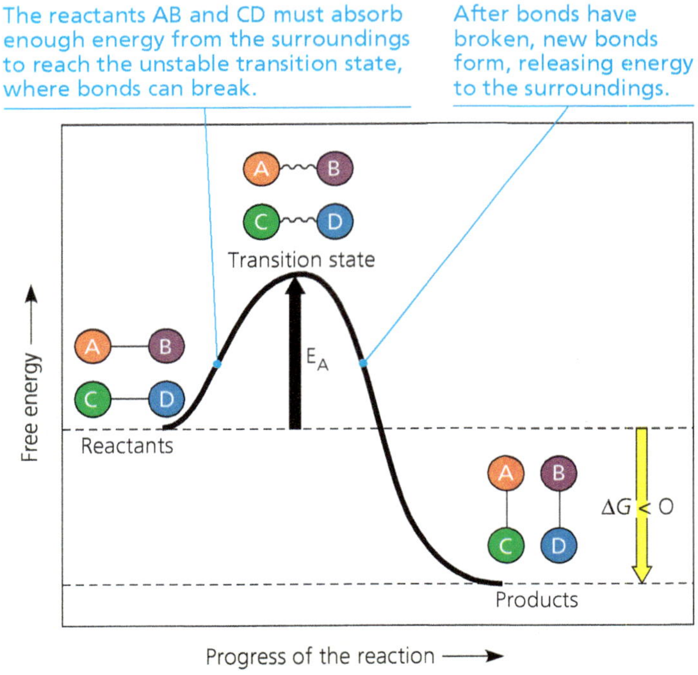

▲ **Figure 8.12 Energy profile of an exergonic reaction.** The "molecules" are hypothetical, with A, B, C, and D representing portions of the molecules. Thermodynamically, this is an exergonic reaction, with a negative ΔG, and the reaction occurs spontaneously. However, the activation energy (E_A) provides a barrier that determines the rate of the reaction.

DRAW IT *Graph the progress of an endergonic reaction in which EF and GH form products EG and FH, assuming that the reactants must pass through a transition state.*

reactant molecules is increasing. At the summit, when energy equivalent to E_A has been absorbed, the reactants are in the transition state: They are activated, and their bonds can be broken. As the atoms then settle into their new, more stable bonding arrangements, energy is released to the surroundings. This corresponds to the downhill part of the curve, which shows the loss of free energy by the molecules. The overall decrease in free energy means that E_A is repaid with interest, as the formation of new bonds releases more energy than was invested in the breaking of old bonds..

The reaction shown in Figure 8.12 is exergonic and occurs spontaneously ($\Delta G < 0$). However, the activation energy provides a barrier that determines the rate of the reaction. The reactants must absorb enough energy to reach the top of the activation energy barrier before the reaction can occur. For some reactions, E_A is modest enough that even at room temperature there is sufficient thermal energy for many of the reactant molecules to reach the transition state in a short time. In most cases, however, E_A is so high and the transition state is reached so rarely that the reaction will hardly proceed at all. In these cases, the reaction will occur at a noticeable rate only if the reactants are heated. For example, the reaction of gasoline and oxygen is exergonic and will occur spontaneously, but energy is required for the molecules to reach the transition state and react. Only when the spark plugs fire in an automobile engine can there be the explosive release of energy that pushes the pistons. Without a spark, a mixture of gasoline hydrocarbons and oxygen will not react because the E_A barrier is too high.

How Enzymes Lower the E_A Barrier

Proteins, DNA, and other complex molecules of the cell are rich in free energy and have the potential to decompose spontaneously; that is, the laws of thermodynamics favor their breakdown. These molecules persist only because at temperatures typical for cells, few molecules can make it over the hump of activation energy. However, the barriers for selected reactions must occasionally be surmounted for cells to carry out the processes needed for life. Heat speeds a reaction by allowing reactants to attain the transition state more often, but this solution would be inappropriate for biological systems. First, high temperature denatures proteins and kills cells. Second, heat would speed up *all* reactions, not just those that are needed. Instead of heat, organisms use catalysis to speed up reactions.

An enzyme catalyzes a reaction by lowering the E_A barrier (**Figure 8.13**), enabling the reactant molecules to absorb enough energy to reach the transition state even at moderate temperatures. An enzyme cannot change the ΔG for a reaction; it cannot make an endergonic reaction exergonic. Enzymes can only hasten reactions that would eventually occur anyway, but this function makes it possible for the cell to have a dynamic

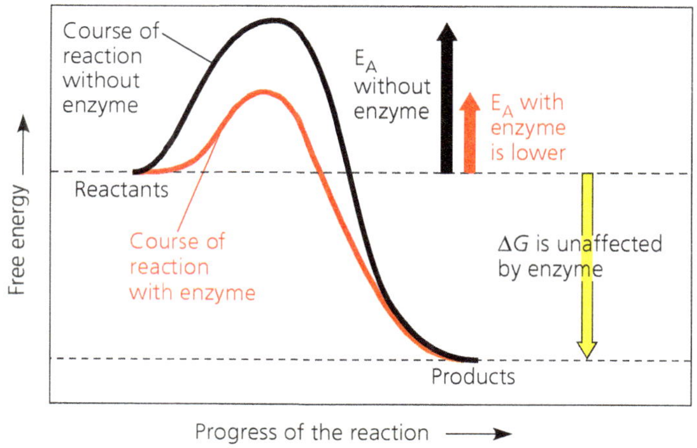

▲ **Figure 8.13 The effect of an enzyme on activation energy.** Without affecting the free-energy change (ΔG) for a reaction, an enzyme speeds the reaction by reducing its activation energy (E_A).

metabolism, routing chemicals smoothly through the cell's metabolic pathways. And because enzymes are very specific for the reactions they catalyze, they determine which chemical processes will be going on in the cell at any particular time.

Substrate Specificity of Enzymes

The reactant an enzyme acts on is referred to as the enzyme's **substrate**. The enzyme binds to its substrate (or substrates, when there are two or more reactants), forming an **enzyme-substrate complex**. While enzyme and substrate are joined, the catalytic action of the enzyme converts the substrate to the product (or products) of the reaction. The overall process can be summarized as follows:

$$\text{Enzyme + Substrate(s)} \rightleftharpoons \text{Enzyme-substrate complex} \rightleftharpoons \text{Enzyme + Product(s)}$$

For example, the enzyme sucrase (most enzyme names end in *-ase*) catalyzes the hydrolysis of the disaccharide sucrose into its two monosaccharides, glucose and fructose (see p. 152):

$$\text{Sucrase + Sucrose + } H_2O \rightleftharpoons \text{Sucrase-sucrose-}H_2O\text{ complex} \rightleftharpoons \text{Sucrase + Glucose + Fructose}$$

The reaction catalyzed by each enzyme is very specific; an enzyme can recognize its specific substrate even among closely related compounds. For instance, sucrase will act only on sucrose and will not bind to other disaccharides, such as maltose. What accounts for this molecular recognition? Recall that most enzymes are proteins, and proteins are macromolecules with unique three-dimensional configurations. The specificity of an enzyme results from its shape, which is a consequence of its amino acid sequence.

Only a restricted region of the enzyme molecule actually binds to the substrate. This region, called the **active site**, is typically a pocket or groove on the surface of the enzyme where catalysis occurs **(Figure 8.14a)**. Usually, the active site is formed by only a few of the enzyme's amino acids, with the rest of the protein molecule providing a framework that determines the configuration of the active site. The specificity of an enzyme is attributed to a compatible fit between the shape of its active site and the shape of the substrate.

An enzyme is not a stiff structure locked into a given shape. In fact, recent work by biochemists has shown clearly that enzymes (and other proteins as well) seem to "dance" between subtly different shapes in a dynamic equilibrium, with slight differences in free energy for each "pose." The shape that best fits the substrate isn't necessarily the one with the lowest energy, but during the very short time the enzyme takes on this shape, its active site can bind to the substrate. It has been known for more than 50 years that the active site itself is also not a rigid receptacle for the substrate. As the substrate enters the active site, the enzyme changes shape slightly due to interactions between the substrate's chemical groups and chemical groups on the side chains of the amino acids that form the active site. This shape change makes the active site fit even more snugly around the substrate **(Figure 8.14b)**. This **induced fit** is like a clasping handshake. Induced fit brings chemical groups of the active site into positions that enhance their ability to catalyze the chemical reaction.

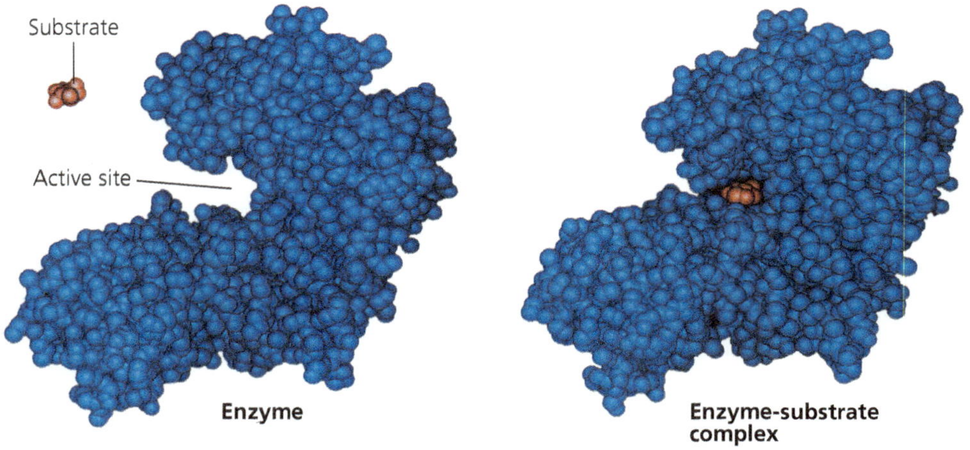

(a) In this computer graphic model, the active site of this enzyme (hexokinase, shown in blue) forms a groove on its surface. Its substrate is glucose (red).

(b) When the substrate enters the active site, it forms weak bonds with the enzyme, inducing a change in the shape of the protein. This change allows additional weak bonds to form, causing the active site to enfold the substrate and hold it in place.

▲ **Figure 8.14 Induced fit between an enzyme and its substrate.**

Catalysis in the Enzyme's Active Site

In most enzymatic reactions, the substrate is held in the active site by so-called weak interactions, such as hydrogen bonds and ionic bonds. R groups of a few of the amino acids that make up the active site catalyze the conversion of substrate to product, and the product departs from the active site. The enzyme is then free to take another substrate molecule into its active site. The entire cycle happens so fast that a single enzyme molecule typically acts on about a thousand substrate molecules per second. Some enzymes are much faster. Enzymes, like other catalysts, emerge from the reaction in their original form. Therefore, very small amounts of enzyme can have a huge metabolic impact by functioning over and over again in catalytic cycles. **Figure 8.15** shows a catalytic cycle involving two substrates and two products.

Most metabolic reactions are reversible, and an enzyme can catalyze either the forward or the reverse reaction, depending on which direction has a negative ΔG. This in turn depends mainly on the relative concentrations of reactants and products. The net effect is always in the direction of equilibrium.

Enzymes use a variety of mechanisms that lower activation energy and speed up a reaction (see Figure 8.15, step 3). First, in reactions involving two or more reactants, the active site provides a template on which the substrates can come together in the proper orientation for a reaction to occur between them. Second, as the active site of an enzyme clutches the bound substrates, the enzyme may stretch the substrate molecules toward their transition-state form, stressing and bending critical chemical bonds that must be broken during the reaction. Because E_A is proportional to the difficulty of breaking the bonds, distorting the substrate helps it approach the transition state and thus reduces the amount of free energy that must be absorbed to achieve that state.

Third, the active site may also provide a microenvironment that is more conducive to a particular type of reaction than the solution itself would be without the enzyme. For example, if the active site has amino acids with acidic R groups, the active site may be a pocket of low pH in an otherwise neutral cell. In such cases, an acidic amino acid may facilitate H^+ transfer to the substrate as a key step in catalyzing the reaction.

A fourth mechanism of catalysis is the direct participation of the active site in the chemical reaction. Sometimes this process even involves brief covalent bonding between the substrate and the side chain of an amino acid of the enzyme. Subsequent steps of the reaction restore the side chains to their original states, so that the active site is the same after the reaction as it was before.

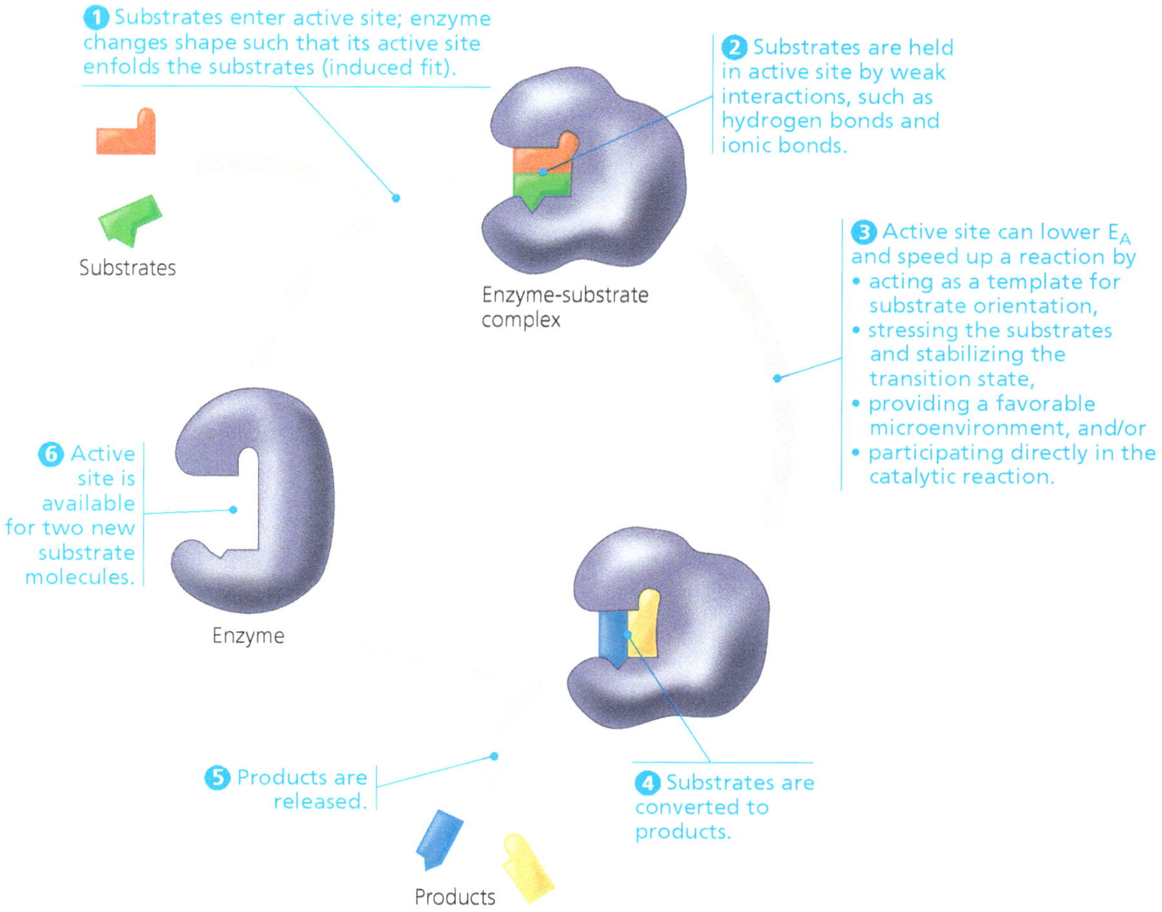

◀ **Figure 8.15 The active site and catalytic cycle of an enzyme.** An enzyme can convert one or more reactant molecules to one or more product molecules. The enzyme shown here converts two substrate molecules to two product molecules.

The rate at which a particular amount of enzyme converts substrate to product is partly a function of the initial concentration of the substrate: The more substrate molecules that are available, the more frequently they access the active sites of the enzyme molecules. However, there is a limit to how fast the reaction can be pushed by adding more substrate to a fixed concentration of enzyme. At some point, the concentration of substrate will be high enough that all enzyme molecules have their active sites engaged. As soon as the product exits an active site, another substrate molecule enters. At this substrate concentration, the enzyme is said to be *saturated*, and the rate of the reaction is determined by the speed at which the active site converts substrate to product. When an enzyme population is saturated, the only way to increase the rate of product formation is to add more enzyme. Cells often increase the rate of a reaction by producing more enzyme molecules.

Effects of Local Conditions on Enzyme Activity

The activity of an enzyme—how efficiently the enzyme functions—is affected by general environmental factors, such as temperature and pH. It can also be affected by chemicals that specifically influence that enzyme. In fact, researchers have learned much about enzyme function by employing such chemicals.

Effects of Temperature and pH

Recall from Chapter 5 that the three-dimensional structures of proteins are sensitive to their environment. As a consequence, each enzyme works better under some conditions than under other conditions, because these *optimal conditions* favor the most active shape for the enzyme molecule.

Temperature and pH are environmental factors important in the activity of an enzyme. Up to a point, the rate of an enzymatic reaction increases with increasing temperature, partly because substrates collide with active sites more frequently when the molecules move rapidly. Above that temperature, however, the speed of the enzymatic reaction drops sharply. The thermal agitation of the enzyme molecule disrupts the hydrogen bonds, ionic bonds, and other weak interactions that stabilize the active shape of the enzyme, and the protein molecule eventually denatures. Each enzyme has an optimal temperature at which its reaction rate is greatest. Without denaturing the enzyme, this temperature allows the greatest number of molecular collisions and the fastest conversion of the reactants to product molecules. Most human enzymes have optimal temperatures of about 35–40°C (close to human body temperature). The thermophilic bacteria that live in hot springs contain enzymes with optimal temperatures of 70°C or higher (**Figure 8.16a** on the next page).

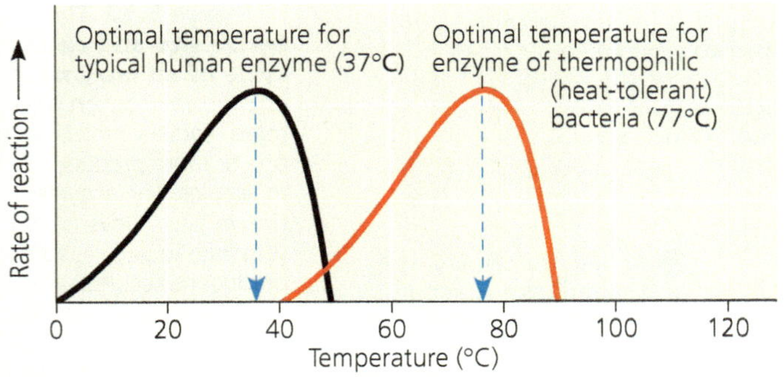

(a) Optimal temperature for two enzymes

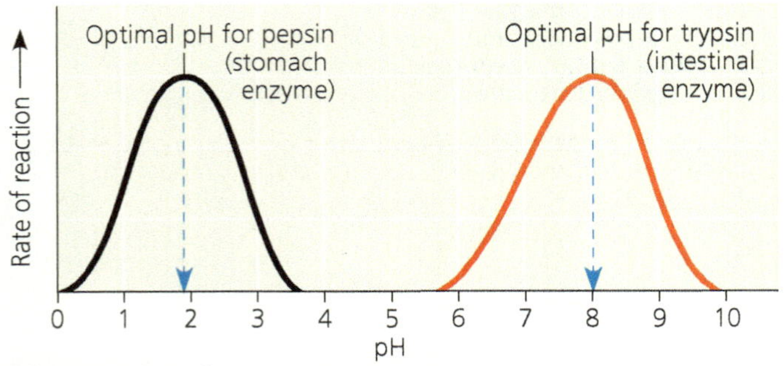

(b) Optimal pH for two enzymes

▲ **Figure 8.16 Environmental factors affecting enzyme activity.** Each enzyme has an optimal **(a)** temperature and **(b)** pH that favor the most active shape of the protein molecule.

DRAW IT *Given that a mature lysosome has an internal pH of around 4.5, draw a curve in (b) showing what you would predict for a lysosomal enzyme, labeling its optimal pH.*

Just as each enzyme has an optimal temperature, it also has a pH at which it is most active. The optimal pH values for most enzymes fall in the range of pH 6–8, but there are exceptions. For example, pepsin, a digestive enzyme in the human stomach, works best at pH 2. Such an acidic environment denatures most enzymes, but pepsin is adapted to maintain its functional three-dimensional structure in the acidic environment of the stomach. In contrast, trypsin, a digestive enzyme residing in the alkaline environment of the human intestine, has an optimal pH of 8 and would be denatured in the stomach **(Figure 8.16b)**.

Cofactors

Many enzymes require nonprotein helpers for catalytic activity. These adjuncts, called **cofactors**, may be bound tightly to the enzyme as permanent residents, or they may bind loosely and reversibly along with the substrate. The cofactors of some enzymes are inorganic, such as the metal atoms zinc, iron, and copper in ionic form. If the cofactor is an organic molecule, it is more specifically called a **coenzyme**. Most vitamins are important in nutrition because they act as coenzymes or raw materials from which coenzymes are made. Cofactors function in various ways, but in all cases where they are used, they perform a crucial chemical function in catalysis. You'll encounter examples of cofactors later in the book.

Enzyme Inhibitors

Certain chemicals selectively inhibit the action of specific enzymes, and we have learned a lot about enzyme function by studying the effects of these molecules. If the inhibitor attaches to the enzyme by covalent bonds, inhibition is usually irreversible.

Many enzyme inhibitors, however, bind to the enzyme by weak interactions, in which case inhibition is reversible. Some reversible inhibitors resemble the normal substrate molecule and compete for admission into the active site **(Figure 8.17a** and **b)**. These mimics, called **competitive inhibitors**, reduce

▼ **Figure 8.17 Inhibition of enzyme activity.**

(a) Normal binding

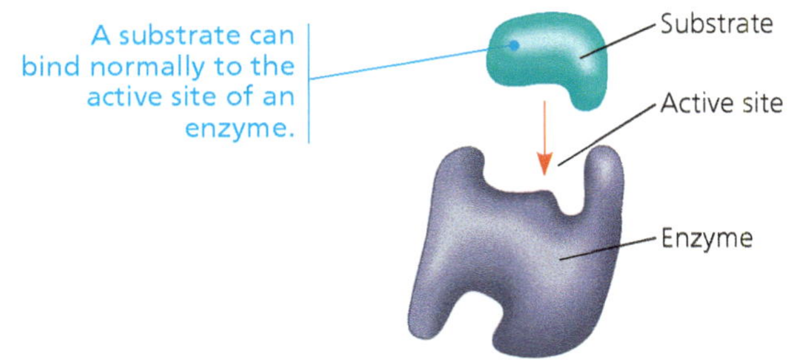

(b) Competitive inhibition

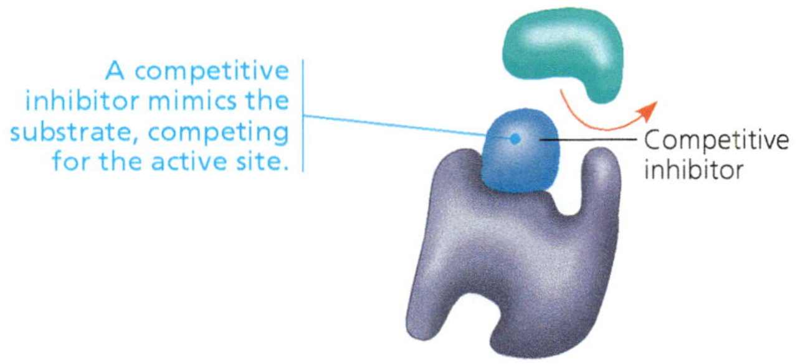

(c) Noncompetitive inhibition

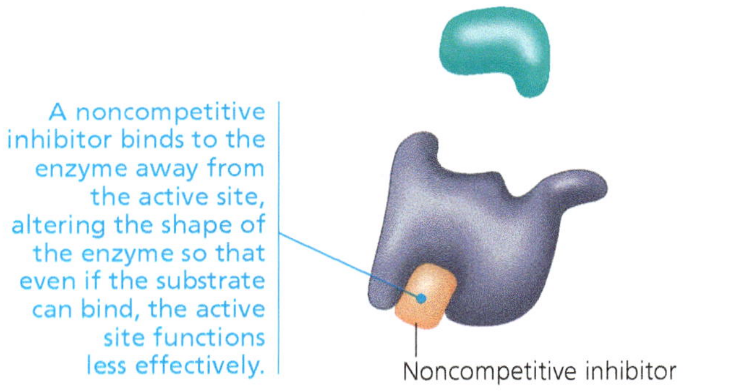

the productivity of enzymes by blocking substrates from entering active sites. This kind of inhibition can be overcome by increasing the concentration of substrate so that as active sites become available, more substrate molecules than inhibitor molecules are around to gain entry to the sites.

In contrast, **noncompetitive inhibitors** do not directly compete with the substrate to bind to the enzyme at the active site (**Figure 8.17c**). Instead, they impede enzymatic reactions by binding to another part of the enzyme. This interaction causes the enzyme molecule to change its shape in such a way that the active site becomes less effective at catalyzing the conversion of substrate to product.

Toxins and poisons are often irreversible enzyme inhibitors. An example is sarin, a nerve gas that caused the death of several people and injury to many others when it was released by terrorists in the Tokyo subway in 1995. This small molecule binds covalently to the R group on the amino acid serine, which is found in the active site of acetylcholinesterase, an enzyme important in the nervous system. Other examples include the pesticides DDT and parathion, inhibitors of key enzymes in the nervous system. Finally, many antibiotics are inhibitors of specific enzymes in bacteria. For instance, penicillin blocks the active site of an enzyme that many bacteria use to make their cell walls.

Citing enzyme inhibitors that are metabolic poisons may give the impression that enzyme inhibition is generally abnormal and harmful. In fact, molecules naturally present in the cell often regulate enzyme activity by acting as inhibitors. Such regulation—selective inhibition—is essential to the control of cellular metabolism, as we will discuss in Concept 8.5.

The Evolution of Enzymes

EVOLUTION Thus far, biochemists have discovered and named more than 4,000 different enzymes in various species, and this list probably represents the tip of the proverbial iceberg. How did this grand profusion of enzymes arise? Recall that most enzymes are proteins, and proteins are encoded by genes. A permanent change in a gene, known as a *mutation*, can result in a protein with one or more changed amino acids. In the case of an enzyme, if the changed amino acids are in the active site or some other crucial region, the altered enzyme might have a novel activity or might bind to a different substrate. Under environmental conditions where the new function benefits the organism, natural selection would tend to favor the mutated form of the gene, causing it to persist in the population. This simplified model is generally accepted as the main way in which the multitude of different enzymes arose over the past few billion years of life's history.

Data supporting this model have been collected by researchers using a lab procedure that mimics evolution in natural populations. One group tested whether the function of an enzyme called β-galactosidase could change over time in populations of the bacterium *Escherichia coli* (*E. coli*). β-galactosidase breaks down the disaccharide lactose into the simple sugars glucose and galactose. Using molecular techniques, the researchers introduced random mutations into *E. coli* genes and then tested the bacteria for their ability to break down a slightly different disaccharide (one that has the sugar fucose in place of galactose). They selected the mutant bacteria that could do this best and exposed them to another round of mutation and selection. After seven rounds, the "evolved" enzyme bound the new substrate several hundred times more strongly, and broke it down 10 to 20 times more quickly, than did the original enzyme.

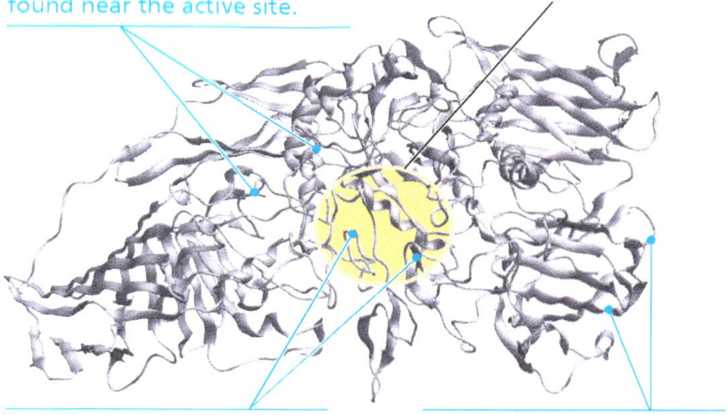

▲ **Figure 8.18 Mimicking evolution of an enzyme with a new function.** After seven rounds of mutation and selection in a lab, the enzyme β-galactosidase evolved into an enzyme specialized for breaking down a sugar different from lactose. This ribbon model shows one subunit of the altered enzyme; six amino acids were different.

The researchers found that six amino acids had changed in the enzyme altered in this experiment. Two of these changed amino acids were in the active site, two were nearby, and two were on the surface of the protein (**Figure 8.18**). This experiment and others like it strengthen the notion that a few changes can indeed alter enzyme function.

CONCEPT CHECK 8.4

1. Many spontaneous reactions occur very slowly. Why don't all spontaneous reactions occur instantly?
2. Why do enzymes act only on very specific substrates?
3. **WHAT IF?** Malonate is an inhibitor of the enzyme succinate dehydrogenase. How would you determine whether malonate is a competitive or noncompetitive inhibitor?
4. **MAKE CONNECTIONS** In nature, what conditions could lead to natural selection favoring bacteria with enzymes that could break down the fucose-containing disaccharide discussed above? See the discussion of natural selection in Concept 1.2, pages 14–16.

For suggested answers, see Appendix A.

CONCEPT 8.5

Regulation of enzyme activity helps control metabolism

Chemical chaos would result if all of a cell's metabolic pathways were operating simultaneously. Intrinsic to life's processes is a cell's ability to tightly regulate its metabolic pathways by controlling when and where its various enzymes are active. It does this either by switching on and off the genes that encode specific enzymes (as we will discuss in Unit Three) or, as we discuss here, by regulating the activity of enzymes once they are made.

Allosteric Regulation of Enzymes

In many cases, the molecules that naturally regulate enzyme activity in a cell behave something like reversible noncompetitive inhibitors (see Figure 8.17c): These regulatory molecules change an enzyme's shape and the functioning of its active site by binding to a site elsewhere on the molecule, via noncovalent interactions. **Allosteric regulation** is the term used to describe any case in which a protein's function at one site is affected by the binding of a regulatory molecule to a separate site. It may result in either inhibition or stimulation of an enzyme's activity.

Allosteric Activation and Inhibition

Most enzymes known to be allosterically regulated are constructed from two or more subunits, each composed of a polypeptide chain with its own active site. The entire complex oscillates between two different shapes, one catalytically active and the other inactive **(Figure 8.19a)**. In the simplest kind of allosteric regulation, an activating or inhibiting regulatory molecule binds to a regulatory site (sometimes called an allosteric site), often located where subunits join. The binding of an *activator* to a regulatory site stabilizes the shape that has functional active sites, whereas the binding of an *inhibitor* stabilizes the inactive form of the enzyme. The subunits of an allosteric enzyme fit together in such a way that a shape change in one subunit is transmitted to all others. Through this interaction of subunits, a single activator or inhibitor molecule that binds to one regulatory site will affect the active sites of all subunits.

Fluctuating concentrations of regulators can cause a sophisticated pattern of response in the activity of cellular enzymes. The products of ATP hydrolysis (ADP and Ⓟ$_i$), for example, play a complex role in balancing the flow of traffic between anabolic and catabolic pathways by their effects on key enzymes. ATP binds to several catabolic enzymes allosterically, lowering their affinity for substrate and thus inhibiting their activity. ADP, however, functions as an activator of the same enzymes. This is logical because catabolism functions

▼ **Figure 8.19 Allosteric regulation of enzyme activity.**

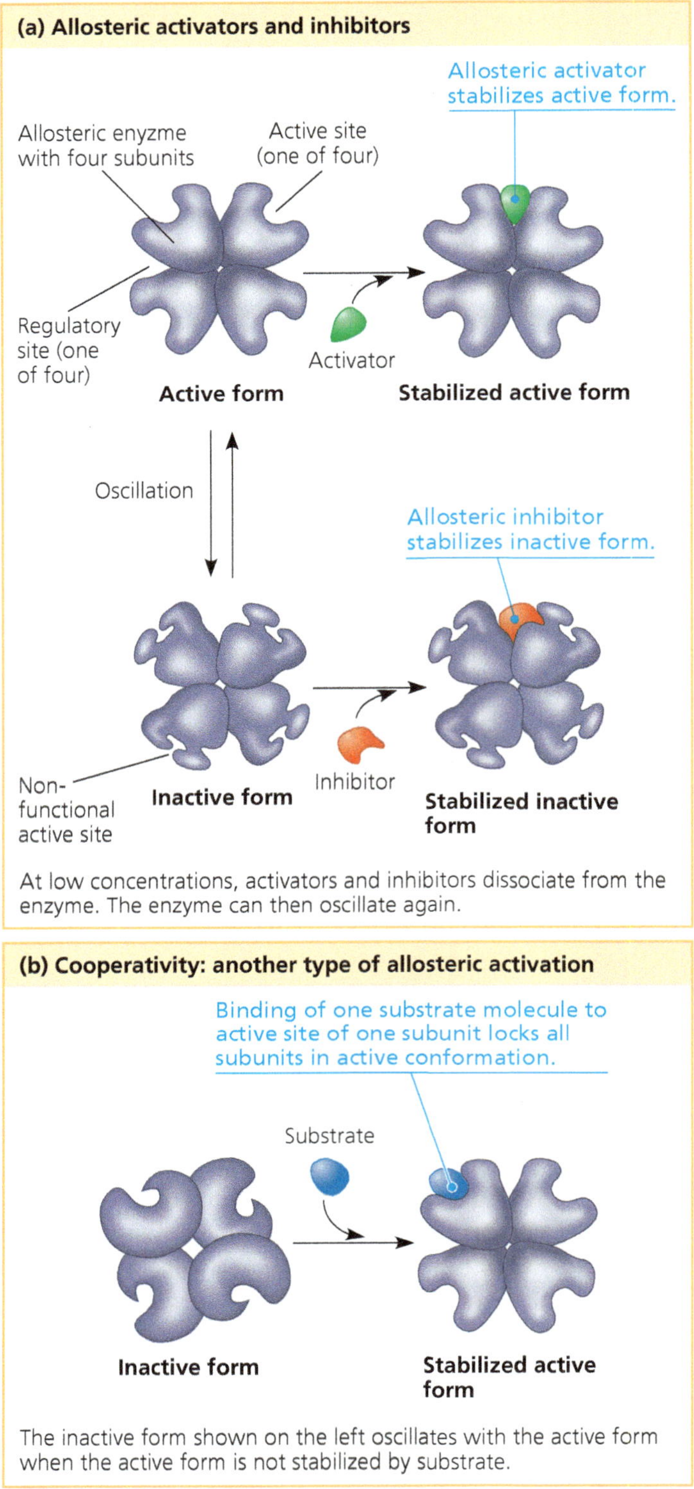

in regenerating ATP. If ATP production lags behind its use, ADP accumulates and activates the enzymes that speed up catabolism, producing more ATP. If the supply of ATP exceeds demand, then catabolism slows down as ATP molecules accumulate and bind to the same enzymes, inhibiting them. (You'll see specific examples of this type of regulation when you learn about cellular respiration in the next chapter.) ATP, ADP, and other related molecules also affect key enzymes in anabolic pathways. In this way, allosteric enzymes control the rates of important reactions in both sorts of metabolic pathways.

In another kind of allosteric activation, a *substrate* molecule binding to one active site in a multisubunit enzyme triggers a shape change in all the subunits, thereby increasing catalytic activity at the other active sites (**Figure 8.19b**). Called **cooperativity**, this mechanism amplifies the response of enzymes to substrates: One substrate molecule primes an enzyme to act on additional substrate molecules more readily. Cooperativity is considered "allosteric" regulation because binding of the substrate to one active site affects catalysis in another active site.

Although the vertebrate oxygen transport protein hemoglobin is not an enzyme, classic studies of cooperative binding in this protein have elucidated the principle of cooperativity. Hemoglobin is made up of four subunits, each of which has an oxygen-binding site (see Figure 5.20). The binding of an oxygen molecule to one binding site increases the affinity for oxygen of the remaining binding sites. Thus, where oxygen is at high levels, such as in the lungs or gills, hemoglobin's affinity for oxygen increases as more binding sites are filled. In oxygen-deprived tissues, however, the release of each oxygen molecule decreases the oxygen affinity of the other binding sites, resulting in the release of oxygen where it is most needed. Cooperativity works similarly in multisubunit enzymes that have been studied.

Identification of Allosteric Regulators

Although allosteric regulation is probably quite widespread, relatively few of the many known metabolic enzymes have been shown to be regulated in this way. Allosteric regulatory molecules are hard to characterize, in part because they tend to bind the enzyme at low affinity and are therefore hard to isolate. Recently, however, pharmaceutical companies have turned their attention to allosteric regulators. These molecules are attractive drug candidates for enzyme regulation because they exhibit higher specificity for particular enzymes than do inhibitors that bind to the active site. (An active site may be similar to the active site in another, related enzyme, whereas allosteric regulatory sites appear to be quite distinct between enzymes.)

Figure 8.20 describes a search for allosteric regulators, carried out as a collaboration between researchers at the University of California at San Francisco and a company called Sunesis Pharmaceuticals. The study was designed to find allosteric inhibitors of *caspases*, protein-digesting enzymes that play an active role in inflammation and cell death. (You'll learn more about caspases and cell death in Chapter 11.) By specifically regulating these enzymes, we may be able to better manage inappropriate inflammatory responses, such as those commonly seen in vascular and neurodegenerative diseases.

▼ Figure 8.20 **INQUIRY**

Are there allosteric inhibitors of caspase enzymes?

EXPERIMENT In an effort to identify allosteric inhibitors of caspases, Justin Scheer and co-workers screened close to 8,000 compounds for their ability to bind to a possible allosteric binding site in caspase 1 and inhibit the enzyme's activity. Each compound was designed to form a disulfide bond with a cysteine near the site in order to stabilize the low-affinity interaction that is expected of an allosteric inhibitor. As the caspases are known to exist in both active and inactive forms, the researchers hypothesized that this linkage might lock the enzyme in the inactive form.

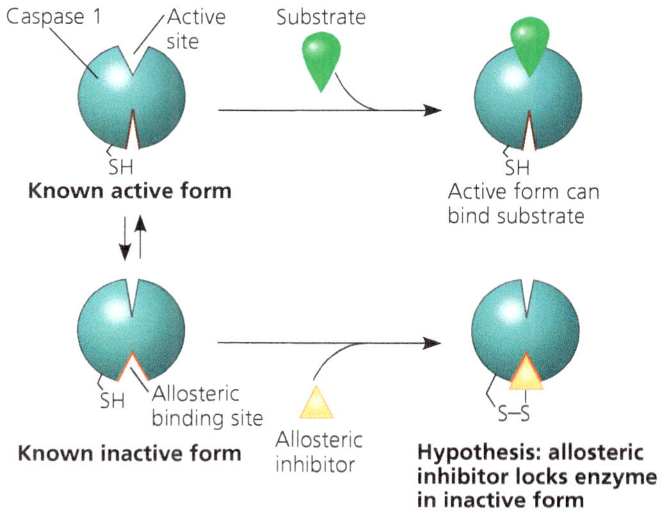

To test this model, X-ray diffraction analysis was used to determine the structure of caspase 1 when bound to one of the inhibitors and to compare it with the active and inactive structures.

RESULTS Fourteen compounds were identified that could bind to the proposed allosteric site (red) of caspase 1 and block enzymatic activity. The enzyme's shape when one such inhibitor was bound resembled the inactive caspase 1 more than the active form.

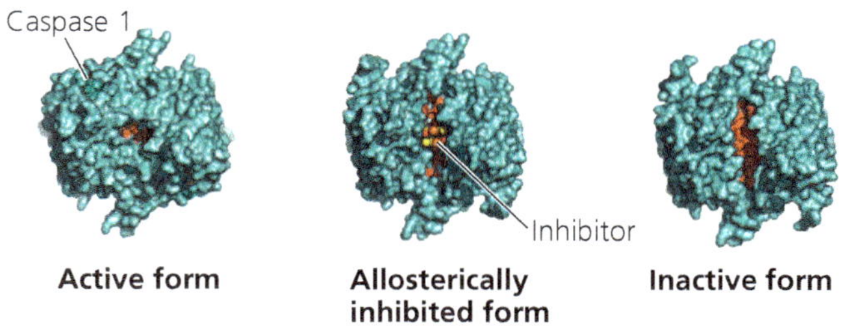

CONCLUSION That particular inhibitory compound apparently locks the enzyme in its inactive form, as expected for a true allosteric regulator. The data therefore support the existence of an allosteric inhibitory site on caspase 1 that can be used to control enzymatic activity.

SOURCE J. M. Scheer et al., A common allosteric site and mechanism in caspases, *Proceedings of the National Academy of Sciences* 103: 7595–7600 (2006).

WHAT IF? As a control, the researchers broke the disulfide linkage between one of the inhibitors and the caspase. Assuming that the experimental solution contains no other inhibitors, how would you expect the caspase 1 activity to be affected?

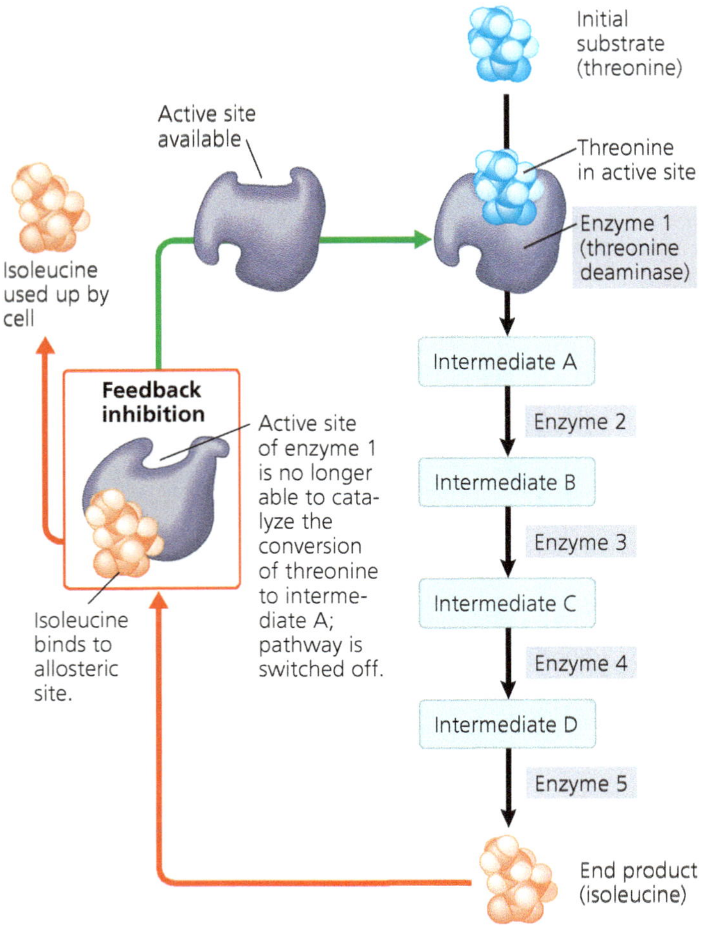

▲ **Figure 8.21 Feedback inhibition in isoleucine synthesis.**

Feedback Inhibition

When ATP allosterically inhibits an enzyme in an ATP-generating pathway, as we discussed earlier, the result is feedback inhibition, a common mode of metabolic control. In **feedback inhibition**, a metabolic pathway is switched off by the inhibitory binding of its end product to an enzyme that acts early in the pathway. **Figure 8.21** shows an example of this control mechanism operating on an anabolic pathway. Certain cells use this five-step pathway to synthesize the amino acid isoleucine from threonine, another amino acid. As isoleucine accumulates, it slows down its own synthesis by allosterically inhibiting the enzyme for the first step of the pathway. Feedback inhibition thereby prevents the cell from wasting chemical resources by making more isoleucine than is necessary.

Specific Localization of Enzymes Within the Cell

The cell is not just a bag of chemicals with thousands of different kinds of enzymes and substrates in a random mix. The cell is compartmentalized, and cellular structures help bring order to metabolic pathways. In some cases, a team of enzymes for several steps of a metabolic pathway are assembled into a multienzyme complex. The arrangement facilitates the sequence of reactions, with the product from the first enzyme becoming the substrate for an adjacent enzyme in the complex, and so on, until the end product is released. Some enzymes and enzyme complexes have fixed locations within the cell and act as structural components of particular membranes. Others are in solution within particular membrane-enclosed eukaryotic organelles, each with its own internal chemical environment. For example, in eukaryotic cells, the enzymes for cellular respiration reside in specific locations within mitochondria **(Figure 8.22)**.

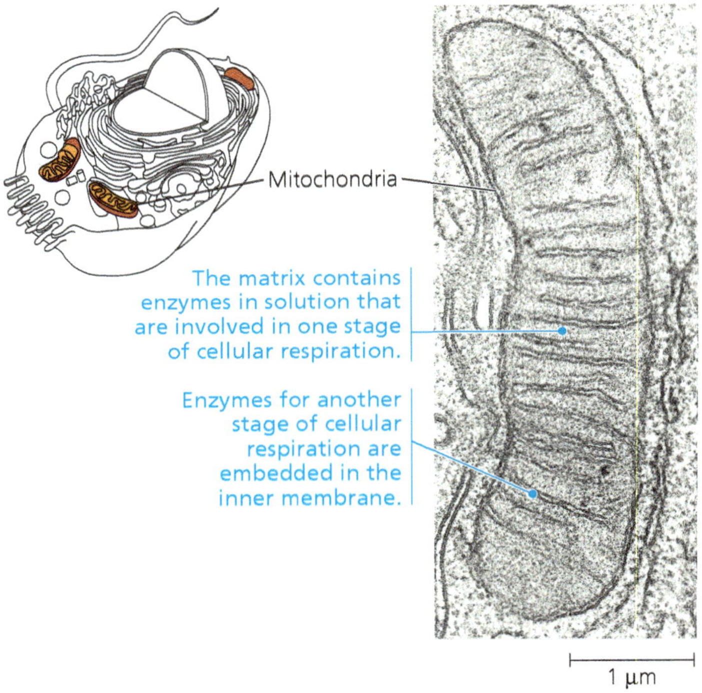

▲ **Figure 8.22 Organelles and structural order in metabolism.** Organelles such as the mitochondrion (TEM) contain enzymes that carry out specific functions, in this case cellular respiration.

In this chapter, you have learned that metabolism, the intersecting set of chemical pathways characteristic of life, is a choreographed interplay of thousands of different kinds of cellular molecules. In the next chapter, we explore cellular respiration, the major catabolic pathway that breaks down organic molecules, releasing energy for the crucial processes of life.

CONCEPT CHECK 8.5

1. How do an activator and an inhibitor have different effects on an allosterically regulated enzyme?
2. **WHAT IF?** Imagine you are a pharmacological researcher who wants to design a drug that inhibits a particular enzyme. Upon reading the scientific literature, you find that the enzyme's active site is similar to that of several other enzymes. What might be a good approach to developing your inhibitor drug?

For suggested answers, see Appendix A.

8 CHAPTER REVIEW

SUMMARY OF KEY CONCEPTS

CONCEPT 8.1

An organism's metabolism transforms matter and energy, subject to the laws of thermodynamics (pp. 142–145)

- **Metabolism** is the collection of chemical reactions that occur in an organism. **Enzymes** catalyze reactions in intersecting **metabolic pathways**, which may be **catabolic** (breaking down molecules, releasing energy) or **anabolic** (building molecules, consuming energy).
- **Energy** is the capacity to cause change; some forms of energy do work by moving matter. **Kinetic energy** is associated with motion and includes **thermal energy (heat)** associated with random motion of atoms or molecules. **Potential energy** is related to the location or structure of matter and includes **chemical energy** possessed by a molecule due to its structure.
- **The first law of thermodynamics**, conservation of energy, states that energy cannot be created or destroyed, only transferred or transformed. The **second law of thermodynamics** states that **spontaneous processes**, those requiring no outside input of energy, increase the **entropy** (disorder) of the universe.

? *Explain how the highly ordered structure of a cell does not conflict with the second law of thermodynamics.*

CONCEPT 8.2

The free-energy change of a reaction tells us whether or not the reaction occurs spontaneously (pp. 146–149)

- A living system's **free energy** is energy that can do work under cellular conditions. The change in free energy (ΔG) during a biological process is related directly to enthalpy change (ΔH) and to the change in entropy (ΔS): $\Delta G = \Delta H - T\Delta S$. Organisms live at the expense of free energy. During a spontaneous change, free energy decreases and the stability of a system increases. At maximum stability, the system is at equilibrium and can do no work.
- In an **exergonic** (spontaneous) chemical reaction, the products have less free energy than the reactants ($-\Delta G$). **Endergonic** (nonspontaneous) reactions require an input of energy ($+\Delta G$). The addition of starting materials and the removal of end products prevent metabolism from reaching equilibrium.

? *Explain the meaning of each component in the equation for the change in free energy of a spontaneous chemical reaction. Why are spontaneous reactions important in the metabolism of a cell?*

CONCEPT 8.3

ATP powers cellular work by coupling exergonic reactions to endergonic reactions (pp. 149–151)

- **ATP** is the cell's energy shuttle. Hydrolysis of its terminal phosphate yields ADP and Ⓟ$_i$ and releases free energy.
- Through **energy coupling**, the exergonic process of ATP hydrolysis drives endergonic reactions by transfer of a phosphate group to specific reactants, forming a **phosphorylated intermediate** that is more reactive. ATP hydrolysis (sometimes with protein phosphorylation) also causes changes in the shape and binding affinities of transport and motor proteins.
- Catabolic pathways drive regeneration of ATP from ADP + Ⓟ$_i$.

? *Describe the ATP cycle: How is ATP used and regenerated in a cell?*

CONCEPT 8.4

Enzymes speed up metabolic reactions by lowering energy barriers (pp. 152–157)

- In a chemical reaction, the energy necessary to break the bonds of the reactants is the **activation energy**, E_A.
- **Enzymes** lower the E_A barrier:

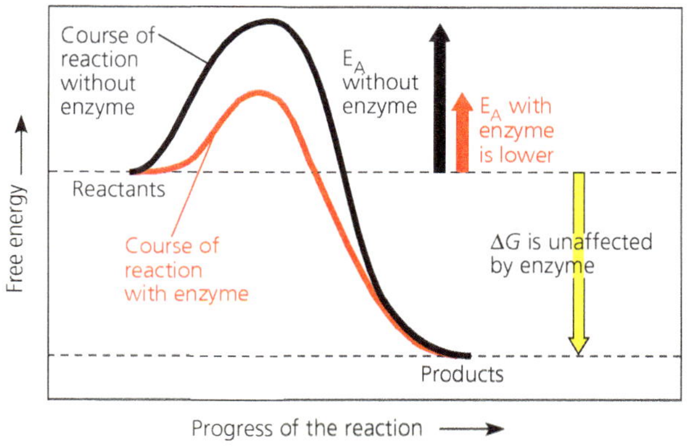

- Each type of enzyme has a unique **active site** that combines specifically with its **substrate(s)**, the reactant molecule(s) on which it acts. The enzyme changes shape slightly when it binds the substrate(s) (**induced fit**).
- The active site can lower an E_A barrier by orienting substrates correctly, straining their bonds, providing a favorable microenvironment, or even covalently bonding with the substrate.
- Each enzyme has an optimal temperature and pH. Inhibitors reduce enzyme function. A **competitive inhibitor** binds to the active site, whereas a **noncompetitive inhibitor** binds to a different site on the enzyme.
- Natural selection, acting on organisms with mutant genes encoding altered enzymes, is a major evolutionary force responsible for the diverse array of enzymes found in organisms.

? *How do both activation energy barriers and enzymes help maintain the structural and metabolic order of life?*

CONCEPT 8.5

Regulation of enzyme activity helps control metabolism (pp. 158–160)

- Many enzymes are subject to **allosteric regulation**: Regulatory molecules, either activators or inhibitors, bind to specific regulatory sites, affecting the shape and function of the enzyme. In **cooperativity**, binding of one substrate molecule can stimulate binding or activity at other active sites. In **feedback inhibition**, the end product of a metabolic pathway allosterically inhibits the enzyme for a previous step in the pathway.
- Some enzymes are grouped into complexes, some are incorporated into membranes, and some are contained inside organelles, increasing the efficiency of metabolic processes.

? *What roles do allosteric regulation and feedback inhibition play in the metabolism of a cell?*

TEST YOUR UNDERSTANDING

LEVEL 1: KNOWLEDGE/COMPREHENSION

1. Choose the pair of terms that correctly completes this sentence: Catabolism is to anabolism as ______ is to ______.
   a. exergonic; spontaneous
   b. exergonic; endergonic
   c. free energy; entropy
   d. work; energy
   e. entropy; enthalpy

2. Most cells cannot harness heat to perform work because
   a. heat is not a form of energy.
   b. cells do not have much heat; they are relatively cool.
   c. temperature is usually uniform throughout a cell.
   d. heat can never be used to do work.
   e. heat must remain constant during work.

3. Which of the following metabolic processes can occur without a net influx of energy from some other process?
   a. ADP + Ⓟ$_i \rightarrow$ ATP + H_2O
   b. $C_6H_{12}O_6 + 6\ O_2 \rightarrow 6\ CO_2 + 6\ H_2O$
   c. $6\ CO_2 + 6\ H_2O \rightarrow C_6H_{12}O_6 + 6\ O_2$
   d. amino acids $\rightarrow$ protein
   e. glucose + fructose $\rightarrow$ sucrose

4. If an enzyme in solution is saturated with substrate, the most effective way to obtain a faster yield of products is to
   a. add more of the enzyme.
   b. heat the solution to 90°C.
   c. add more substrate.
   d. add an allosteric inhibitor.
   e. add a noncompetitive inhibitor.

5. Some bacteria are metabolically active in hot springs because
   a. they are able to maintain a lower internal temperature.
   b. high temperatures make catalysis unnecessary.
   c. their enzymes have high optimal temperatures.
   d. their enzymes are completely insensitive to temperature.
   e. they use molecules other than proteins or RNAs as their main catalysts.

LEVEL 2: APPLICATION/ANALYSIS

6. If an enzyme is added to a solution where its substrate and product are in equilibrium, what will occur?
   a. Additional product will be formed.
   b. Additional substrate will be formed.
   c. The reaction will change from endergonic to exergonic.
   d. The free energy of the system will change.
   e. Nothing; the reaction will stay at equilibrium.

LEVEL 3: SYNTHESIS/EVALUATION

7. **DRAW IT** Using a series of arrows, draw the branched metabolic reaction pathway described by the following statements, and then answer the question at the end. Use red arrows and minus signs to indicate inhibition.

   L can form either M or N.
   M can form O.
   O can form either P or R.
   P can form Q.
   R can form S.
   O inhibits the reaction of L to form M.
   Q inhibits the reaction of O to form P.
   S inhibits the reaction of O to form R.

   Which reaction would prevail if both Q and S were present in the cell in high concentrations?
   a. L → M
   b. M → O
   c. L → N
   d. O → P
   e. R → S

8. **EVOLUTION CONNECTION**
   A recent revival of the antievolutionary "intelligent design" argument holds that biochemical pathways are too complex to have evolved, because all intermediate steps in a given pathway must be present to produce the final product. Critique this argument. How could you use the diversity of metabolic pathways that produce the same or similar products to support your case?

9. **SCIENTIFIC INQUIRY**
   DRAW IT A researcher has developed an assay to measure the activity of an important enzyme present in liver cells growing in culture. She adds the enzyme's substrate to a dish of cells and then measures the appearance of reaction products. The results are graphed as the amount of product on the *y*-axis versus time on the *x*-axis. The researcher notes four sections of the graph. For a short period of time, no products appear (section A). Then (section B) the reaction rate is quite high (the slope of the line is steep). Next, the reaction gradually slows down (section C). Finally, the graph line becomes flat (section D). Draw and label the graph, and propose a model to explain the molecular events occurring at each stage of this reaction profile.

10. **SCIENCE, TECHNOLOGY, AND SOCIETY**
    Organophosphates (organic compounds containing phosphate groups) are commonly used as insecticides to improve crop yield. Organophosphates typically interfere with nerve signal transmission by inhibiting the enzymes that degrade transmitter molecules. They affect humans and other vertebrates as well as insects. Thus, the use of organophosphate pesticides poses some health risks. On the other hand, these molecules break down rapidly upon exposure to air and sunlight. As a consumer, what level of risk are you willing to accept in exchange for an abundant and affordable food supply?

11. **WRITE ABOUT A THEME**
    Energy Transfer Life requires energy. In a short essay (100–150 words), describe the basic principles of bioenergetics in an animal cell. How is the flow and transformation of energy different in a photosynthesizing cell? Include the role of ATP and enzymes in your discussion.

For selected answers, see Appendix A.

9

Cellular Respiration and Fermentation

▲ **Figure 9.1** **How do these leaves power the work of life for this chimpanzee?**

KEY CONCEPTS

9.1 Catabolic pathways yield energy by oxidizing organic fuels

9.2 Glycolysis harvests chemical energy by oxidizing glucose to pyruvate

9.3 After pyruvate is oxidized, the citric acid cycle completes the energy-yielding oxidation of organic molecules

9.4 During oxidative phosphorylation, chemiosmosis couples electron transport to ATP synthesis

9.5 Fermentation and anaerobic respiration enable cells to produce ATP without the use of oxygen

9.6 Glycolysis and the citric acid cycle connect to many other metabolic pathways

OVERVIEW

Life Is Work

Living cells require transfusions of energy from outside sources to perform their many tasks—for example, assembling polymers, pumping substances across membranes, moving, and reproducing. The chimpanzee in **Figure 9.1** obtains energy for its cells by eating plants; some animals feed on other organisms that eat plants. The energy stored in the organic molecules of food ultimately comes from the sun. Energy flows into an ecosystem as sunlight and exits as heat; in contrast, the chemical elements essential to life are recycled **(Figure 9.2)**. Photosynthesis generates oxygen and organic molecules used by the mitochondria of eukaryotes (including plants and algae) as fuel for cellular respiration. Respiration breaks this fuel down, generating ATP. The waste products of this type of respiration, carbon dioxide and water, are the raw materials for photosynthesis. In this chapter, we consider how cells harvest the chemical energy stored in organic molecules and use it to generate ATP, the molecule that drives most cellular work. After presenting some basics about respiration, we will focus on three key pathways of respiration: glycolysis, the citric acid cycle, and oxidative phosphorylation. We'll also consider fermentation, a somewhat simpler pathway coupled to glycolysis that has deep evolutionary roots.

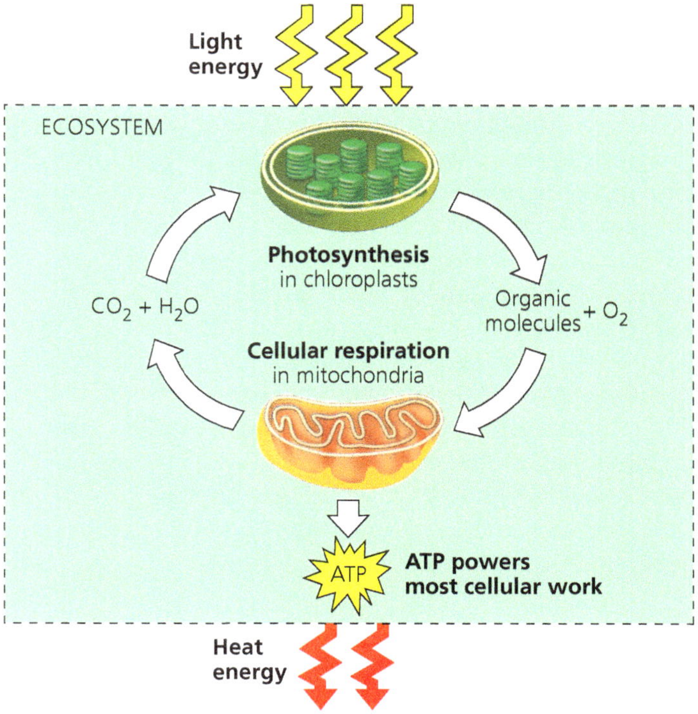

▲ **Figure 9.2** **Energy flow and chemical recycling in ecosystems.** Energy flows into an ecosystem as sunlight and ultimately leaves as heat, while the chemical elements essential to life are recycled.

BioFlix Visit the Study Area at **www.masteringbiology.com** for the BioFlix® 3-D Animation on The Carbon Cycle.

CONCEPT 9.1

Catabolic pathways yield energy by oxidizing organic fuels

As you learned in Chapter 8, metabolic pathways that release stored energy by breaking down complex molecules are called catabolic pathways. Electron transfer plays a major role in these pathways. In this section, we consider these processes, which are central to cellular respiration.

Catabolic Pathways and Production of ATP

Organic compounds possess potential energy as a result of the arrangement of electrons in the bonds between their atoms. Compounds that can participate in exergonic reactions can act as fuels. With the help of enzymes, a cell systematically degrades complex organic molecules that are rich in potential energy to simpler waste products that have less energy. Some of the energy taken out of chemical storage can be used to do work; the rest is dissipated as heat.

One catabolic process, **fermentation**, is a partial degradation of sugars or other organic fuel that occurs without the use of oxygen. However, the most prevalent and efficient catabolic pathway is **aerobic respiration**, in which oxygen is consumed as a reactant along with the organic fuel (*aerobic* is from the Greek *aer*, air, and *bios*, life). The cells of most eukaryotic and many prokaryotic organisms can carry out aerobic respiration. Some prokaryotes use substances other than oxygen as reactants in a similar process that harvests chemical energy without oxygen; this process is called *anaerobic respiration* (the prefix *an-* means "without"). Technically, the term **cellular respiration** includes both aerobic and anaerobic processes. However, it originated as a synonym for aerobic respiration because of the relationship of that process to organismal respiration, in which an animal breathes in oxygen. Thus, *cellular respiration* is often used to refer to the aerobic process, a practice we follow in most of this chapter.

Although very different in mechanism, aerobic respiration is in principle similar to the combustion of gasoline in an automobile engine after oxygen is mixed with the fuel (hydrocarbons). Food provides the fuel for respiration, and the exhaust is carbon dioxide and water. The overall process can be summarized as follows:

$$\text{Organic compounds} + \text{Oxygen} \rightarrow \text{Carbon dioxide} + \text{Water} + \text{Energy}$$

Although carbohydrates, fats, and proteins can all be processed and consumed as fuel, it is helpful to learn the steps of cellular respiration by tracking the degradation of the sugar glucose ($C_6H_{12}O_6$):

$$C_6H_{12}O_6 + 6\ O_2 \rightarrow 6\ CO_2 + 6\ H_2O + \text{Energy (ATP + heat)}$$

Glucose is the fuel that cells most often use; we will discuss other organic molecules contained in foods later in the chapter.

This breakdown of glucose is exergonic, having a free-energy change of −686 kcal (2,870 kJ) per mole of glucose decomposed ($\Delta G = -686$ kcal/mol). Recall that a negative ΔG indicates that the products of the chemical process store less energy than the reactants and that the reaction can happen spontaneously—in other words, without an input of energy.

Catabolic pathways do not directly move flagella, pump solutes across membranes, polymerize monomers, or perform other cellular work. Catabolism is linked to work by a chemical drive shaft—ATP, which you learned about in Chapter 8. To keep working, the cell must regenerate its supply of ATP from ADP and Ⓟi (see Figure 8.11). To understand how cellular respiration accomplishes this, let's examine the fundamental chemical processes known as oxidation and reduction.

Redox Reactions: Oxidation and Reduction

How do the catabolic pathways that decompose glucose and other organic fuels yield energy? The answer is based on the transfer of electrons during the chemical reactions. The relocation of electrons releases energy stored in organic molecules, and this energy ultimately is used to synthesize ATP.

The Principle of Redox

In many chemical reactions, there is a transfer of one or more electrons (e^-) from one reactant to another. These electron transfers are called oxidation-reduction reactions, or **redox reactions** for short. In a redox reaction, the loss of electrons from one substance is called **oxidation**, and the addition of electrons to another substance is known as **reduction**. (Note that *adding* electrons is called *reduction*; negatively charged electrons added to an atom *reduce* the amount of positive charge of that atom.) To take a simple, nonbiological example, consider the reaction between the elements sodium (Na) and chlorine (Cl) that forms table salt:

becomes oxidized (loses electron)

$$Na + Cl \longrightarrow Na^+ + Cl^-$$

becomes reduced (gains electron)

We could generalize a redox reaction this way:

becomes oxidized

$$Xe^- + Y \longrightarrow X + Ye^-$$

becomes reduced

In the generalized reaction, substance Xe^-, the electron donor, is called the **reducing agent**; it reduces Y, which accepts the donated electron. Substance Y, the electron acceptor, is the **oxidizing agent**; it oxidizes Xe^- by removing its electron. Because an electron transfer requires both a donor and an acceptor, oxidation and reduction always go together.

Not all redox reactions involve the complete transfer of electrons from one substance to another; some change the degree of electron sharing in covalent bonds. The reaction

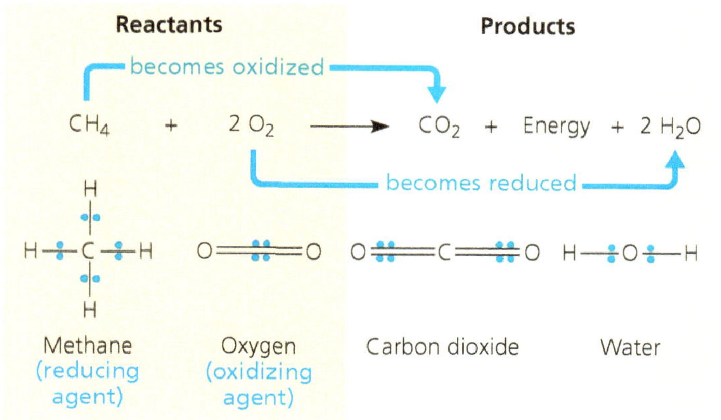

▲ **Figure 9.3 Methane combustion as an energy-yielding redox reaction.** The reaction releases energy to the surroundings because the electrons lose potential energy when they end up being shared unequally, spending more time near electronegative atoms such as oxygen.

between methane and oxygen, shown in **Figure 9.3**, is an example. As explained in Chapter 2, the covalent electrons in methane are shared nearly equally between the bonded atoms because carbon and hydrogen have about the same affinity for valence electrons; they are about equally electronegative. But when methane reacts with oxygen, forming carbon dioxide, electrons end up shared less equally between the carbon atom and its new covalent partners, the oxygen atoms, which are very electronegative. In effect, the carbon atom has partially "lost" its shared electrons; thus, methane has been oxidized.

Now let's examine the fate of the reactant O_2. The two atoms of the oxygen molecule (O_2) share their electrons equally. But when oxygen reacts with the hydrogen from methane, forming water, the electrons of the covalent bonds spend more time near the oxygen (see Figure 9.3). In effect, each oxygen atom has partially "gained" electrons, so the oxygen molecule has been reduced. Because oxygen is so electronegative, it is one of the most potent of all oxidizing agents.

Energy must be added to pull an electron away from an atom, just as energy is required to push a ball uphill. The more electronegative the atom (the stronger its pull on electrons), the more energy is required to take an electron away from it. An electron loses potential energy when it shifts from a less electronegative atom toward a more electronegative one, just as a ball loses potential energy when it rolls downhill. A redox reaction that moves electrons closer to oxygen, such as the burning (oxidation) of methane, therefore releases chemical energy that can be put to work.

Oxidation of Organic Fuel Molecules During Cellular Respiration

The oxidation of methane by oxygen is the main combustion reaction that occurs at the burner of a gas stove. The combustion of gasoline in an automobile engine is also a redox reaction; the energy released pushes the pistons. But the energy-yielding redox process of greatest interest to biologists is respiration: the oxidation of glucose and other molecules in food. Examine again the summary equation for cellular respiration, but this time think of it as a redox process:

becomes oxidized ($C_6H_{12}O_6 \rightarrow 6\,CO_2$)

$$C_6H_{12}O_6 + 6\,O_2 \longrightarrow 6\,CO_2 + 6\,H_2O + \text{Energy}$$

becomes reduced ($6\,O_2 \rightarrow 6\,H_2O$)

As in the combustion of methane or gasoline, the fuel (glucose) is oxidized and oxygen is reduced. The electrons lose potential energy along the way, and energy is released.

In general, organic molecules that have an abundance of hydrogen are excellent fuels because their bonds are a source of "hilltop" electrons, whose energy may be released as these electrons "fall" down an energy gradient when they are transferred to oxygen. The summary equation for respiration indicates that hydrogen is transferred from glucose to oxygen. But the important point, not visible in the summary equation, is that the energy state of the electron changes as hydrogen (with its electron) is transferred to oxygen. In respiration, the oxidation of glucose transfers electrons to a lower energy state, liberating energy that becomes available for ATP synthesis.

The main energy-yielding foods, carbohydrates and fats, are reservoirs of electrons associated with hydrogen. Only the barrier of activation energy holds back the flood of electrons to a lower energy state (see Figure 8.12). Without this barrier, a food substance like glucose would combine almost instantaneously with O_2. If we supply the activation energy by igniting glucose, it burns in air, releasing 686 kcal (2,870 kJ) of heat per mole of glucose (about 180 g). Body temperature is not high enough to initiate burning, of course. Instead, if you swallow some glucose, enzymes in your cells will lower the barrier of activation energy, allowing the sugar to be oxidized in a series of steps.

Stepwise Energy Harvest via NAD^+ and the Electron Transport Chain

If energy is released from a fuel all at once, it cannot be harnessed efficiently for constructive work. For example, if a gasoline tank explodes, it cannot drive a car very far. Cellular respiration does not oxidize glucose in a single explosive step either. Rather, glucose and other organic fuels are broken down in a series of steps, each one catalyzed by an enzyme. At key steps, electrons are stripped from the glucose. As is often the case in oxidation reactions, each electron travels with a proton—thus, as a hydrogen atom. The hydrogen atoms are not transferred directly to oxygen, but instead are usually passed first to an electron carrier, a coenzyme called **NAD^+** (nicotinamide adenine dinucleotide, a derivative of the vitamin niacin). NAD^+ is well suited as an electron carrier because it can cycle easily between oxidized (NAD^+) and reduced (NADH) states. As an electron acceptor, NAD^+ functions as an oxidizing agent during respiration.

How does NAD^+ trap electrons from glucose and other organic molecules? Enzymes called dehydrogenases remove a

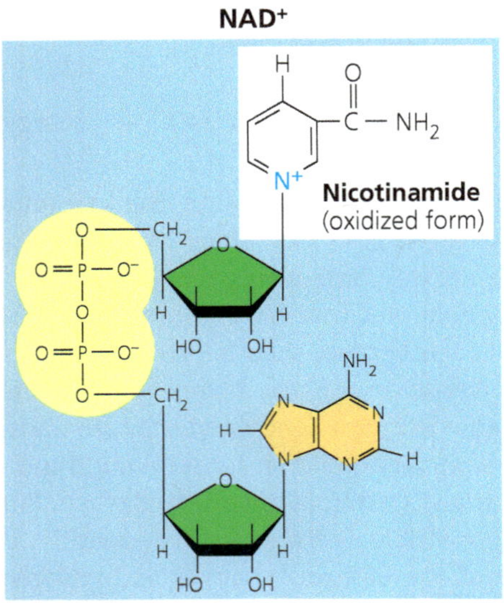

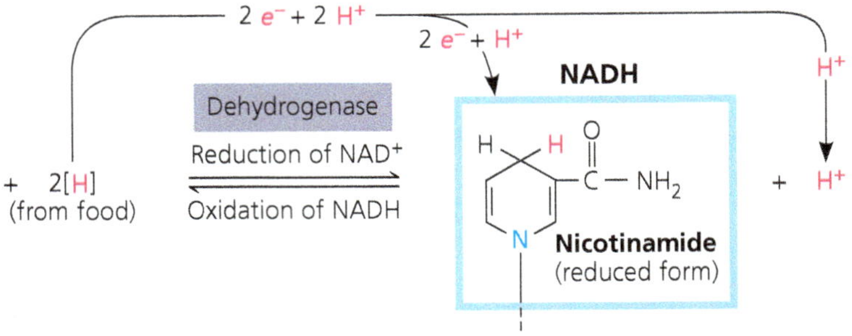

◀ **Figure 9.4 NAD$^+$ as an electron shuttle.** The full name for NAD$^+$, nicotinamide adenine dinucleotide, describes its structure: The molecule consists of two nucleotides joined together at their phosphate groups (shown in yellow). (Nicotinamide is a nitrogenous base, although not one that is present in DNA or RNA; see Figure 5.26.) The enzymatic transfer of 2 electrons and 1 proton (H^+) from an organic molecule in food to NAD$^+$ reduces the NAD$^+$ to NADH; the second proton (H^+) is released. Most of the electrons removed from food are transferred initially to NAD$^+$.

pair of hydrogen atoms (2 electrons and 2 protons) from the substrate (glucose, in this example), thereby oxidizing it. The enzyme delivers the 2 electrons along with 1 proton to its coenzyme, NAD$^+$ **(Figure 9.4)**. The other proton is released as a hydrogen ion (H^+) into the surrounding solution:

$$\text{H}-\overset{|}{\underset{|}{\text{C}}}-\text{OH} + \text{NAD}^+ \xrightarrow{\text{Dehydrogenase}} \overset{|}{\underset{|}{\text{C}}}=\text{O} + \text{NADH} + \text{H}^+$$

By receiving 2 negatively charged electrons but only 1 positively charged proton, NAD$^+$ has its charge neutralized when it is reduced to NADH. The name NADH shows the hydrogen that has been received in the reaction. NAD$^+$ is the most versatile electron acceptor in cellular respiration and functions in several of the redox steps during the breakdown of glucose.

Electrons lose very little of their potential energy when they are transferred from glucose to NAD$^+$. Each NADH molecule formed during respiration represents stored energy that can be tapped to make ATP when the electrons complete their "fall" down an energy gradient from NADH to oxygen.

How do electrons that are extracted from glucose and stored as potential energy in NADH finally reach oxygen? It will help to compare the redox chemistry of cellular respiration to a much simpler reaction: the reaction between hydrogen and oxygen to form water **(Figure 9.5a)**. Mix H_2 and O_2, provide a spark for activation energy, and the gases combine explosively. In fact, combustion of liquid H_2 and O_2 is harnessed to power the main engines of the space shuttle after it is launched, boosting it into orbit. The explosion represents a release of energy as the electrons of hydrogen "fall" closer to the electronegative oxygen atoms. Cellular respiration also brings hydrogen and oxygen together to form water, but there are two important differences. First, in cellular respiration, the hydrogen that reacts with oxygen is derived from organic molecules rather than H_2. Second, instead of occurring in one explosive reaction, respiration uses an

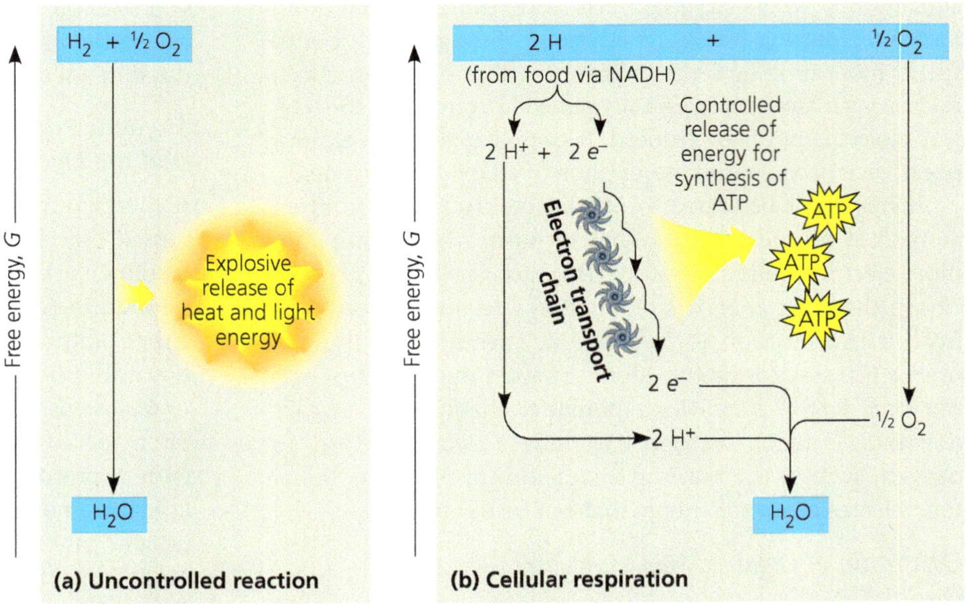

▲ **Figure 9.5 An introduction to electron transport chains. (a)** The one-step exergonic reaction of hydrogen with oxygen to form water releases a large amount of energy in the form of heat and light: an explosion. **(b)** In cellular respiration, the same reaction occurs in stages: An electron transport chain breaks the "fall" of electrons in this reaction into a series of smaller steps and stores some of the released energy in a form that can be used to make ATP. (The rest of the energy is released as heat.)

electron transport chain to break the fall of electrons to oxygen into several energy-releasing steps (**Figure 9.5b**). An **electron transport chain** consists of a number of molecules, mostly proteins, built into the inner membrane of the mitochondria of eukaryotic cells and the plasma membrane of aerobically respiring prokaryotes. Electrons removed from glucose are shuttled by NADH to the "top," higher-energy end of the chain. At the "bottom," lower-energy end, O_2 captures these electrons along with hydrogen nuclei (H^+), forming water.

Electron transfer from NADH to oxygen is an exergonic reaction with a free-energy change of −53 kcal/mol (−222 kJ/mol). Instead of this energy being released and wasted in a single explosive step, electrons cascade down the chain from one carrier molecule to the next in a series of redox reactions, losing a small amount of energy with each step until they finally reach oxygen, the terminal electron acceptor, which has a very great affinity for electrons. Each "downhill" carrier is more electronegative than, and thus capable of oxidizing, its "uphill" neighbor, with oxygen at the bottom of the chain. Therefore, the electrons removed from glucose by NAD^+ fall down an energy gradient in the electron transport chain to a far more stable location in the electronegative oxygen atom. Put another way, oxygen pulls electrons down the chain in an energy-yielding tumble analogous to gravity pulling objects downhill.

In summary, during cellular respiration, most electrons travel the following "downhill" route: glucose → NADH → electron transport chain → oxygen. Later in this chapter, you will learn more about how the cell uses the energy released from this exergonic electron fall to regenerate its supply of ATP. For now, having covered the basic redox mechanisms of cellular respiration, let's look at the entire process by which energy is harvested from organic fuels.

The Stages of Cellular Respiration: *A Preview*

The harvesting of energy from glucose by cellular respiration is a cumulative function of three metabolic stages:

1. Glycolysis (color-coded teal throughout the chapter)
2. Pyruvate oxidation and the citric acid cycle (color-coded salmon)
3. Oxidative phosphorylation: electron transport and chemiosmosis (color-coded violet)

Biochemists usually reserve the term *cellular respiration* for stages 2 and 3. We include glycolysis, however, because most respiring cells deriving energy from glucose use glycolysis to produce the starting material for the citric acid cycle.

As diagrammed in **Figure 9.6**, glycolysis and pyruvate oxidation followed by the citric acid cycle are the catabolic pathways that break down glucose and other organic fuels. **Glycolysis**, which occurs in the cytosol, begins the degradation process by breaking glucose into two molecules of a compound called pyruvate. In eukaryotes, pyruvate enters the mitochondrion and is oxidized to a compound called acetyl CoA, which enters the **citric acid cycle**. There, the breakdown of glucose to carbon dioxide is completed. (In prokaryotes, these processes take place in the cytosol.) Thus, the carbon dioxide produced by respiration represents fragments of oxidized organic molecules.

Some of the steps of glycolysis and the citric acid cycle are redox reactions in which dehydrogenases transfer electrons from substrates to NAD^+, forming NADH. In the third stage of respiration, the electron transport chain accepts electrons from the breakdown products of the first two stages (most often via NADH) and passes these electrons from one molecule to another. At the end of the chain, the electrons are combined with molecular oxygen and hydrogen ions (H^+), forming water (see

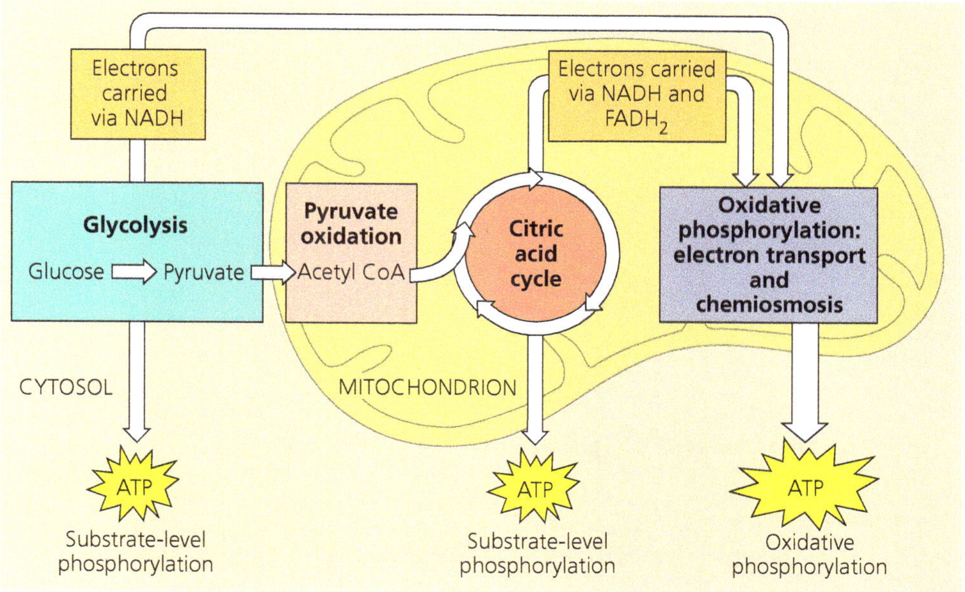

▶ **Figure 9.6 An overview of cellular respiration.** During glycolysis, each glucose molecule is broken down into two molecules of the compound pyruvate. In eukaryotic cells, as shown here, the pyruvate enters the mitochondrion. There it is oxidized to acetyl CoA, which is further oxidized to CO_2 in the citric acid cycle. NADH and a similar electron carrier, a coenzyme called $FADH_2$, transfer electrons derived from glucose to electron transport chains, which are built into the inner mitochondrial membrane. (In prokaryotes, the electron transport chains are located in the plasma membrane.) During oxidative phosphorylation, electron transport chains convert the chemical energy to a form used for ATP synthesis in the process called chemiosmosis.

BioFlix Visit the Study Area at **www.masteringbiology.com** for the BioFlix® 3-D Animation on Cellular Respiration.

Figure 9.5b). The energy released at each step of the chain is stored in a form the mitochondrion (or prokaryotic cell) can use to make ATP from ADP. This mode of ATP synthesis is called **oxidative phosphorylation** because it is powered by the redox reactions of the electron transport chain.

In eukaryotic cells, the inner membrane of the mitochondrion is the site of electron transport and chemiosmosis, the processes that together constitute oxidative phosphorylation. (In prokaryotes, these processes take place in the plasma membrane.) Oxidative phosphorylation accounts for almost 90% of the ATP generated by respiration. A smaller amount of ATP is formed directly in a few reactions of glycolysis and the citric acid cycle by a mechanism called **substrate-level phosphorylation** (**Figure 9.7**). This mode of ATP synthesis occurs when an enzyme transfers a phosphate group from a substrate molecule to ADP, rather than adding an inorganic phosphate to ADP as in oxidative phosphorylation. "Substrate molecule" here refers to an organic molecule generated as an intermediate during the catabolism of glucose.

For each molecule of glucose degraded to carbon dioxide and water by respiration, the cell makes up to about 32 molecules of ATP, each with 7.3 kcal/mol of free energy. Respiration cashes in the large denomination of energy banked in a single molecule of glucose (686 kcal/mol) for the small change of many molecules of ATP, which is more practical for the cell to spend on its work.

This preview has introduced you to how glycolysis, the citric acid cycle, and oxidative phosphorylation fit into the process of cellular respiration. We are now ready to take a closer look at each of these three stages of respiration.

CONCEPT CHECK 9.1

1. Compare and contrast aerobic and anaerobic respiration.
2. **WHAT IF?** If the following redox reaction occurred, which compound would be oxidized? Which reduced?

$$C_4H_6O_5 + NAD^+ \longrightarrow C_4H_4O_5 + NADH + H^+$$

For suggested answers, see Appendix A.

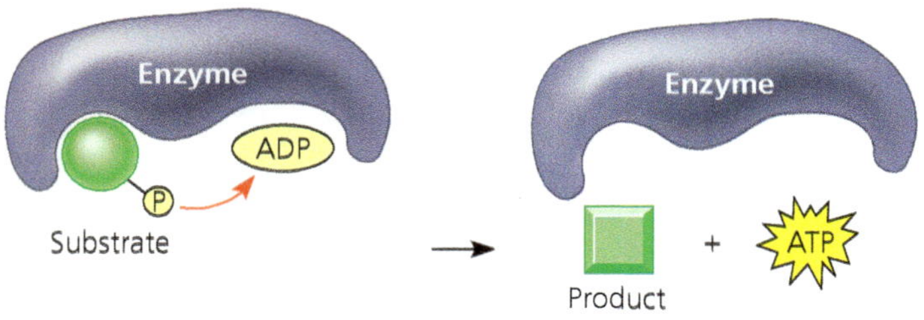

▲ **Figure 9.7 Substrate-level phosphorylation.** Some ATP is made by direct transfer of a phosphate group from an organic substrate to ADP by an enzyme. (For examples in glycolysis, see Figure 9.9, steps 7 and 10.)

MAKE CONNECTIONS *Review Figure 8.8 on page 149. Do you think the potential energy is higher for the reactants or the products in the reaction shown above? Explain.*

CONCEPT 9.2

Glycolysis harvests chemical energy by oxidizing glucose to pyruvate

The word *glycolysis* means "sugar splitting," and that is exactly what happens during this pathway. Glucose, a six-carbon sugar, is split into two three-carbon sugars. These smaller sugars are then oxidized and their remaining atoms rearranged to form two molecules of pyruvate. (Pyruvate is the ionized form of pyruvic acid.)

As summarized in **Figure 9.8**, glycolysis can be divided into two phases: energy investment and energy payoff. During the energy investment phase, the cell actually spends ATP. This investment is repaid with interest during the energy payoff phase, when ATP is produced by substrate-level phosphorylation and NAD^+ is reduced to NADH by electrons released from the oxidation of glucose. The net energy yield from glycolysis, per glucose molecule, is 2 ATP plus 2 NADH. The ten steps of the glycolytic pathway are shown in **Figure 9.9**.

All of the carbon originally present in glucose is accounted for in the two molecules of pyruvate; no carbon is released as CO_2 during glycolysis. Glycolysis occurs whether or not O_2 is present. However, if O_2 *is* present, the chemical energy stored in pyruvate and NADH can be extracted by pyruvate oxidation, the citric acid cycle, and oxidative phosphorylation.

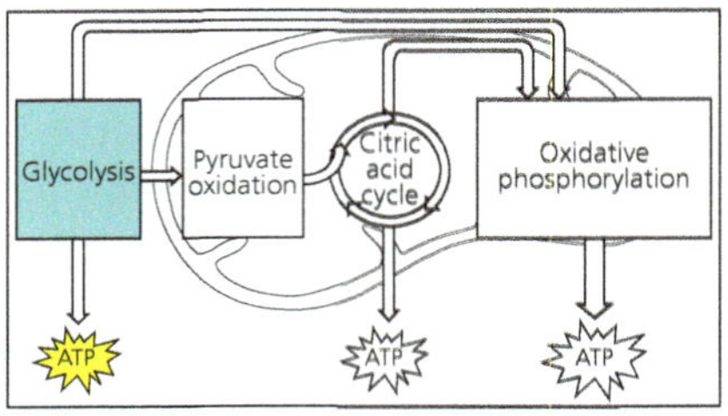

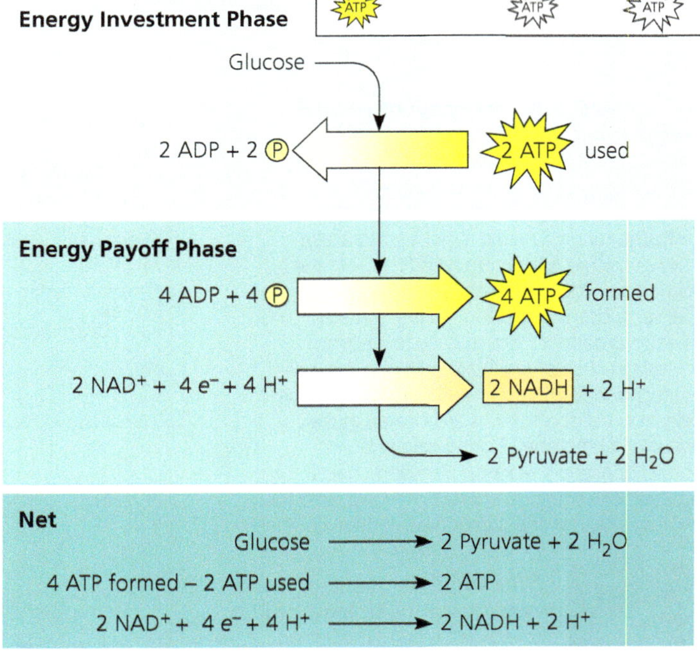

▲ **Figure 9.8 The energy input and output of glycolysis.**

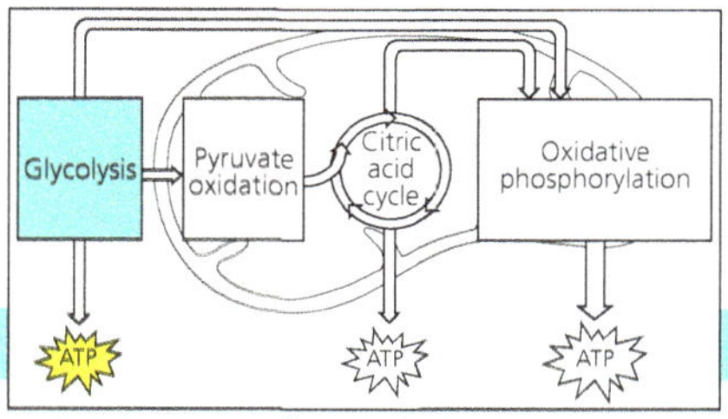

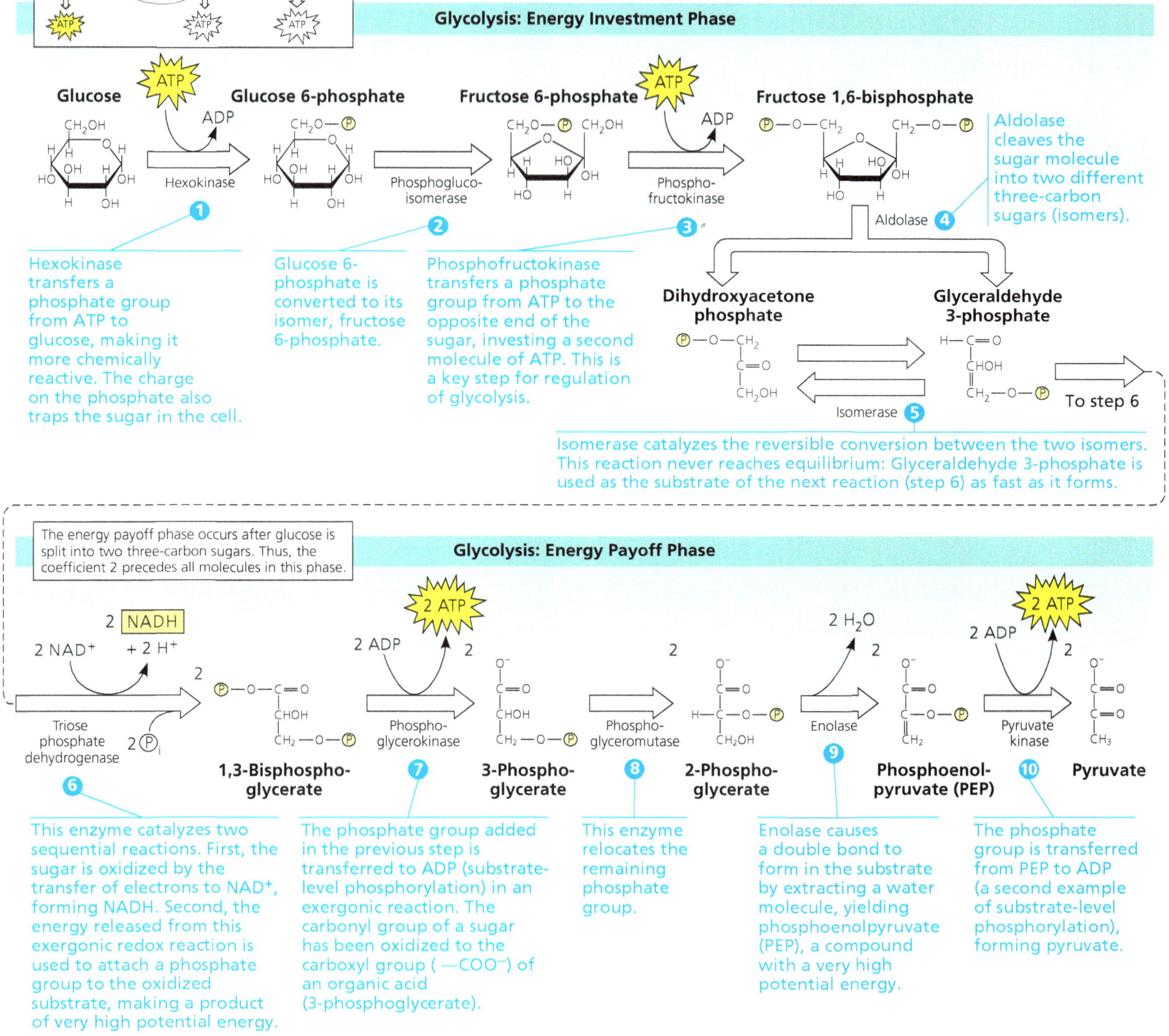

▼ **Figure 9.9 A closer look at glycolysis.** The orientation diagram on the left relates glycolysis to the entire process of respiration. Note that glycolysis is a source of ATP and NADH.

WHAT IF? *What would happen if you removed the dihydroxyacetone phosphate generated in step 4 as fast as it was produced?*

CONCEPT CHECK 9.2

1. During the redox reaction in glycolysis (step 6 in Figure 9.9), which molecule acts as the oxidizing agent? The reducing agent?
2. **MAKE CONNECTIONS** Step 3 in Figure 9.9 is a major point of regulation of glycolysis. The enzyme phosphofructokinase is allosterically regulated by ATP and related molecules (see Concept 8.5, p. 158). Considering the overall result of glycolysis, would you expect ATP to inhibit or stimulate activity of this enzyme? (*Hint*: Make sure you consider the role of ATP as an allosteric regulator, not as a substrate of the enzyme.)

For suggested answers, see Appendix A.

CONCEPT 9.3

After pyruvate is oxidized, the citric acid cycle completes the energy-yielding oxidation of organic molecules

Glycolysis releases less than a quarter of the chemical energy in glucose that can be released by cells; most of the energy remains stockpiled in the two molecules of pyruvate. If molecular oxygen is present, the pyruvate enters a mitochondrion (in eukaryotic cells), where the oxidation of glucose is completed. (In prokaryotic cells, this process occurs in the cytosol.)

Oxidation of Pyruvate to Acetyl CoA

Upon entering the mitochondrion via active transport, pyruvate is first converted to a compound called acetyl coenzyme A, or **acetyl CoA** (Figure 9.10). This step, linking glycolysis and the citric acid cycle, is carried out by a multienzyme complex that catalyzes three reactions: 1 Pyruvate's carboxyl group (—COO^-), which is already fully oxidized and thus has little chemical energy, is removed and given off as a molecule of CO_2. (This is the first step in which CO_2 is released during respiration.) 2 The remaining two-carbon fragment is oxidized, forming acetate (CH_3COO^-, the ionized form of acetic acid). The extracted electrons are transferred to NAD^+, storing energy in the form of NADH. 3 Finally, coenzyme A (CoA), a sulfur-containing compound derived from a B vitamin, is attached via its sulfur atom to the acetate, forming acetyl CoA, which has a high potential energy; in other words, the reaction of acetyl CoA to yield lower-energy products is highly exergonic. This molecule will now feed its acetyl group into the citric acid cycle for further oxidation.

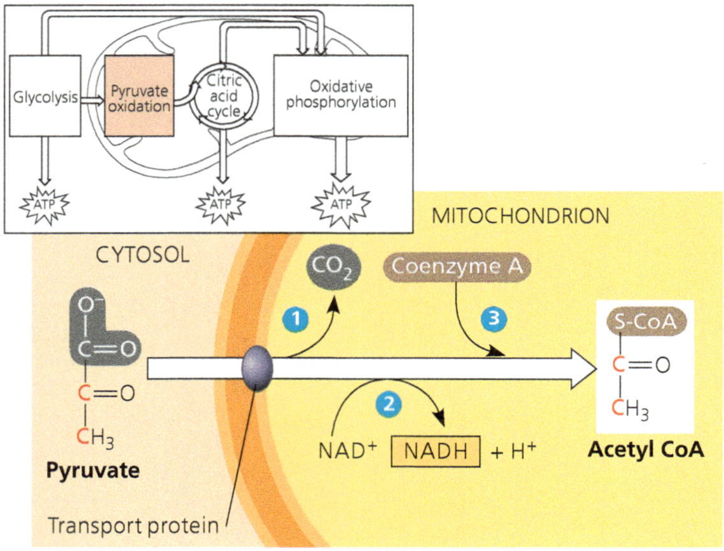

▲ **Figure 9.10 Oxidation of pyruvate to acetyl CoA, the step before the citric acid cycle.** Pyruvate is a charged molecule, so in eukaryotic cells it must enter the mitochondrion via active transport, with the help of a transport protein. Next, a complex of several enzymes (the pyruvate dehydrogenase complex) catalyzes the three numbered steps, which are described in the text. The acetyl group of acetyl CoA will enter the citric acid cycle. The CO_2 molecule will diffuse out of the cell. By convention, coenzyme A is abbreviated S-CoA when it is attached to a molecule, emphasizing the sulfur atom (S).

The Citric Acid Cycle

The citric acid cycle is also called the tricarboxylic acid cycle or the Krebs cycle, the latter honoring Hans Krebs, the German-British scientist who was largely responsible for working out the pathway in the 1930s. The cycle functions as a metabolic furnace that oxidizes organic fuel derived from pyruvate. Figure 9.11 summarizes the inputs and outputs as pyruvate is broken down to three CO_2 molecules, including the molecule of CO_2 released during the conversion of pyruvate to acetyl CoA. The cycle generates 1 ATP per turn by

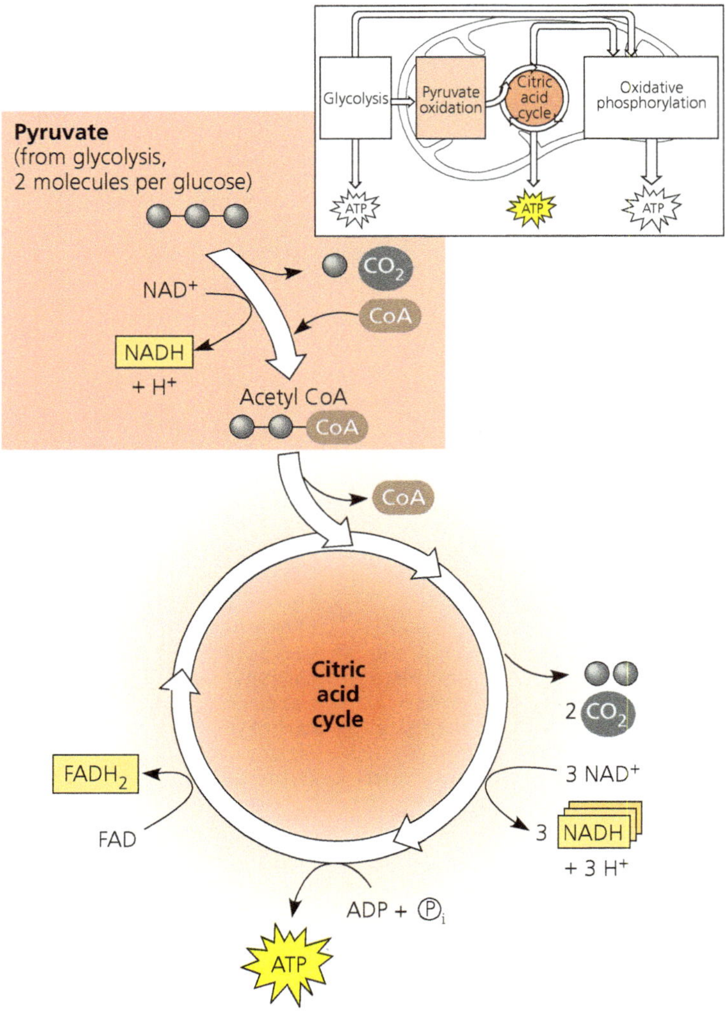

▲ **Figure 9.11 An overview of pyruvate oxidation and the citric acid cycle.** The inputs and outputs per pyruvate molecule are shown. To calculate on a per-glucose basis, multiply by 2, because each glucose molecule is split during glycolysis into two pyruvate molecules.

substrate-level phosphorylation, but most of the chemical energy is transferred to NAD^+ and a related electron carrier, the coenzyme FAD (flavin adenine dinucleotide, derived from riboflavin, a B vitamin), during the redox reactions. The reduced coenzymes, NADH and $FADH_2$, shuttle their cargo of high-energy electrons into the electron transport chain.

Now let's look at the citric acid cycle in more detail. The cycle has eight steps, each catalyzed by a specific enzyme. You can see in **Figure 9.12** that for each turn of the citric acid cycle, two carbons (red) enter in the relatively reduced form of an acetyl group (step 1), and two different carbons (blue) leave in the completely oxidized form of CO_2 molecules

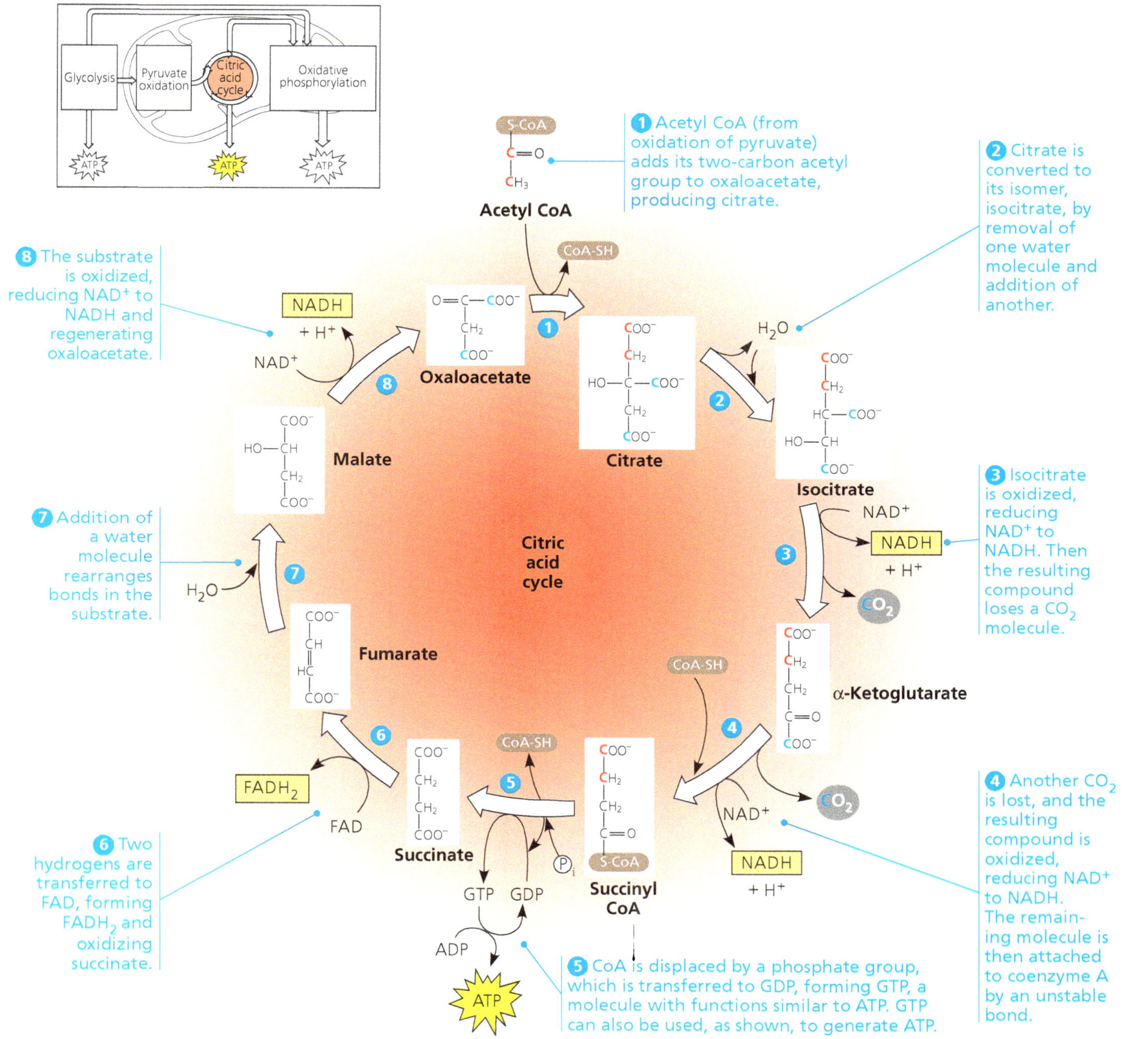

▲ **Figure 9.12 A closer look at the citric acid cycle.** In the chemical structures, red type traces the fate of the two carbon atoms that enter the cycle via acetyl CoA (step 1), and blue type indicates the two carbons that exit the cycle as CO_2 in steps 3 and 4. (The red labeling goes only through step 5 because the succinate molecule is symmetrical; the two ends cannot be distinguished from each other.) Notice that the carbon atoms that enter the cycle from acetyl CoA do not leave the cycle in the same turn. They remain in the cycle, occupying a different location in the molecules on their next turn, after another acetyl group is added. As a consequence, the oxaloacetate that is regenerated at step 8 is composed of different carbon atoms each time around. In eukaryotic cells, all the citric acid cycle enzymes are located in the mitochondrial matrix except for the enzyme that catalyzes step 6, which resides in the inner mitochondrial membrane. Carboxylic acids are represented in their ionized forms, as —COO^-, because the ionized forms prevail at the pH within the mitochondrion. For example, citrate is the ionized form of citric acid.

(steps 3 and 4). The acetyl group of acetyl CoA joins the cycle by combining with the compound oxaloacetate, forming citrate (step 1). (Citrate is the ionized form of citric acid, for which the cycle is named.) The next seven steps decompose the citrate back to oxaloacetate. It is this regeneration of oxaloacetate that makes this process a *cycle*.

Now let's tally the energy-rich molecules produced by the citric acid cycle. For each acetyl group entering the cycle, 3 NAD^+ are reduced to NADH (steps 3, 4, and 8). In step 6, electrons are transferred not to NAD^+, but to FAD, which accepts 2 electrons and 2 protons to become $FADH_2$. In many animal tissue cells, step 5 produces a guanosine triphosphate (GTP) molecule by substrate-level phosphorylation, as shown in Figure 9.12. GTP is a molecule similar to ATP in its structure and cellular function. This GTP may be used to make an ATP molecule (as shown) or directly power work in the cell. In the cells of plants, bacteria, and some animal tissues, step 5 forms an ATP molecule directly by substrate-level phosphorylation. The output from step 5 represents the only ATP generated during the citric acid cycle.

Most of the ATP produced by respiration results from oxidative phosphorylation, when the NADH and $FADH_2$ produced by the citric acid cycle relay the electrons extracted from food to the electron transport chain. In the process, they supply the necessary energy for the phosphorylation of ADP to ATP. We will explore this process in the next section.

CONCEPT CHECK 9.3

1. Name the molecules that conserve most of the energy from the citric acid cycle's redox reactions. How is this energy converted to a form that can be used to make ATP?
2. What processes in your cells produce the CO_2 that you exhale?
3. **WHAT IF?** The conversions shown in Figure 9.10 and step 4 of Figure 9.12 are each catalyzed by a large multienzyme complex. What similarities are there in the reactions that occur in these two cases?

For suggested answers, see Appendix A.

CONCEPT 9.4

During oxidative phosphorylation, chemiosmosis couples electron transport to ATP synthesis

Our main objective in this chapter is to learn how cells harvest the energy of glucose and other nutrients in food to make ATP. But the metabolic components of respiration we have dissected so far, glycolysis and the citric acid cycle, produce only 4 ATP molecules per glucose molecule, all by substrate-level phosphorylation: 2 net ATP from glycolysis and 2 ATP from the citric acid cycle. At this point, molecules of NADH (and $FADH_2$) account for most of the energy extracted from the glucose. These electron escorts link glycolysis and the citric acid cycle to the machinery of oxidative phosphorylation, which uses energy released by the electron transport chain to power ATP synthesis. In this section, you will learn first how the electron transport chain works and then how electron flow down the chain is coupled to ATP synthesis.

The Pathway of Electron Transport

The electron transport chain is a collection of molecules embedded in the inner membrane of the mitochondrion in eukaryotic cells. (In prokaryotes, these molecules reside in the plasma membrane.) The folding of the inner membrane to form cristae increases its surface area, providing space for thousands of copies of the chain in each mitochondrion. (Once again, we see that structure fits function.) Most components of the chain are proteins, which exist in multiprotein complexes numbered I through IV. Tightly bound to these proteins are *prosthetic groups*, nonprotein components essential for the catalytic functions of certain enzymes.

Figure 9.13 shows the sequence of electron carriers in the electron transport chain and the drop in free energy as electrons travel down the chain. During electron transport along the chain, electron carriers alternate between reduced and oxidized states as they accept and donate electrons. Each component of the chain becomes reduced when it accepts electrons from its "uphill" neighbor, which has a lower affinity for electrons (is less electronegative). It then returns to its oxidized form as it passes electrons to its "downhill," more electronegative neighbor.

Now let's take a closer look at the electron transport chain in Figure 9.13. We'll first describe the passage of electrons through complex I in some detail, as an illustration of the general principles involved in electron transport. Electrons removed from glucose by NAD^+, during glycolysis and the citric acid cycle, are transferred from NADH to the first molecule of the electron transport chain in complex I. This molecule is a flavoprotein, so named because it has a prosthetic group called flavin mononucleotide (FMN). In the next redox reaction, the flavoprotein returns to its oxidized form as it passes electrons to an iron-sulfur protein (Fe·S in complex I), one of a family of proteins with both iron and sulfur tightly bound. The iron-sulfur protein then passes the electrons to a compound called ubiquinone (Q in Figure 9.13). This electron carrier is a small hydrophobic molecule, the only member of the electron transport chain that is not a protein. Ubiquinone is individually mobile within the membrane rather than residing in a particular complex. (Another name for ubiquinone is coenzyme Q, or CoQ; you may have seen it sold as a nutritional supplement.)

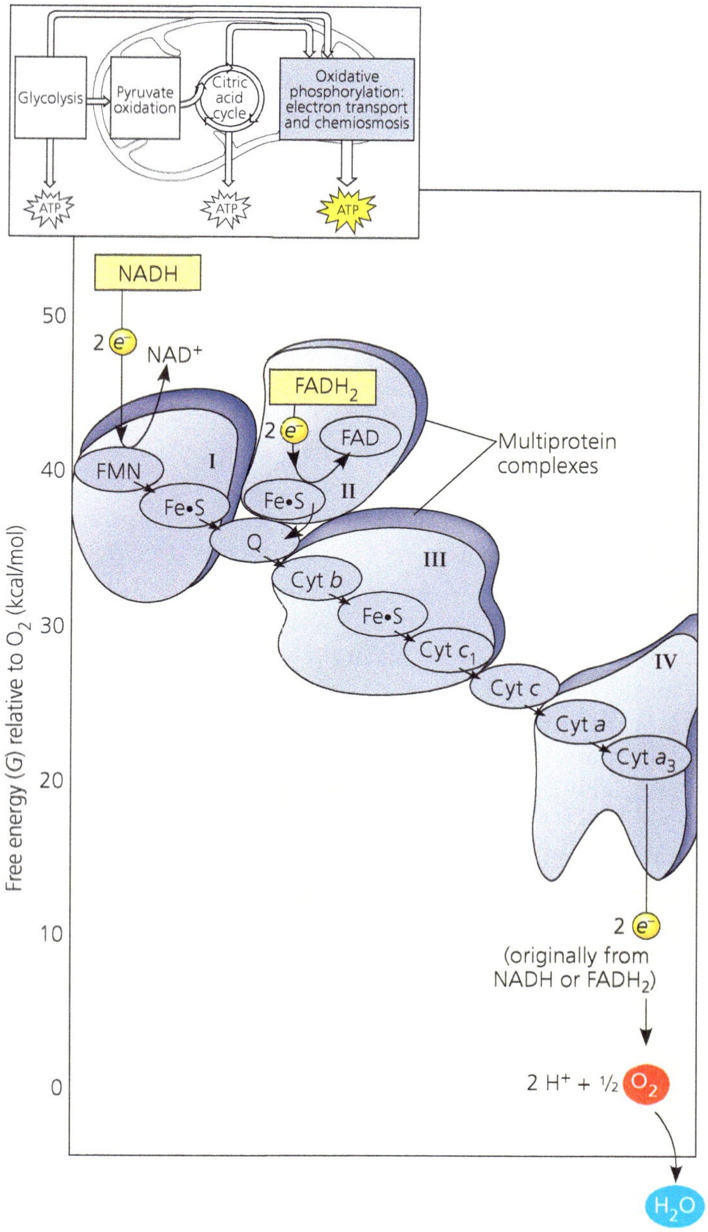

▲ **Figure 9.13 Free-energy change during electron transport.** The overall energy drop (ΔG) for electrons traveling from NADH to oxygen is 53 kcal/mol, but this "fall" is broken up into a series of smaller steps by the electron transport chain. (An oxygen atom is represented here as ½ O_2 to emphasize that the electron transport chain reduces molecular oxygen, O_2, not individual oxygen atoms.)

Most of the remaining electron carriers between ubiquinone and oxygen are proteins called **cytochromes**. Their prosthetic group, called a heme group, has an iron atom that accepts and donates electrons. (It is similar to the heme group in hemoglobin, the protein of red blood cells, except that the iron in hemoglobin carries oxygen, not electrons.) The electron transport chain has several types of cytochromes, each a different protein with a slightly different electron-carrying heme group. The last cytochrome of the chain, cyt a_3, passes its electrons to oxygen, which is *very* electronegative. Each oxygen atom also picks up a pair of hydrogen ions from the aqueous solution, forming water.

Another source of electrons for the transport chain is $FADH_2$, the other reduced product of the citric acid cycle. Notice in Figure 9.13 that $FADH_2$ adds its electrons to the electron transport chain from within complex II, at a lower energy level than NADH does. Consequently, although NADH and $FADH_2$ each donate an equivalent number of electrons (2) for oxygen reduction, the electron transport chain provides about one-third less energy for ATP synthesis when the electron donor is $FADH_2$ rather than NADH. We'll see why in the next section.

The electron transport chain makes no ATP directly. Instead, it eases the fall of electrons from food to oxygen, breaking a large free-energy drop into a series of smaller steps that release energy in manageable amounts. How does the mitochondrion (or the prokaryotic plasma membrane) couple this electron transport and energy release to ATP synthesis? The answer is a mechanism called chemiosmosis.

Chemiosmosis: The Energy-Coupling Mechanism

Populating the inner membrane of the mitochondrion or the prokaryotic plasma membrane are many copies of a protein complex called **ATP synthase**, the enzyme that actually makes ATP from ADP and inorganic phosphate. ATP synthase works like an ion pump running in reverse. Recall from Chapter 7 that ion pumps usually use ATP as an energy source to transport ions against their gradients. In fact, the proton pump shown in Figure 7.20 is an ATP synthase. As we mentioned in Chapter 8, enzymes can catalyze a reaction in either direction, depending on the ΔG for the reaction, which is affected by the local concentrations of reactants and products. Rather than hydrolyzing ATP to pump protons against their concentration gradient, under the conditions of cellular respiration ATP synthase uses the energy of an existing ion gradient to power ATP synthesis. The power source for the ATP synthase is a difference in the concentration of H^+ on opposite sides of the inner mitochondrial membrane. (We can also think of this gradient as a difference in pH, since pH is a measure of H^+ concentration.) This process, in which energy stored in the form of a hydrogen ion gradient across a membrane is used to drive cellular work such as the synthesis of ATP, is called **chemiosmosis** (from the Greek *osmos*, push). We have previously used the word *osmosis* in discussing water transport, but here it refers to the flow of H^+ across a membrane.

From studying the structure of ATP synthase, scientists have learned how the flow of H^+ through this large enzyme powers ATP generation. ATP synthase is a multisubunit complex with four main parts, each made up of multiple polypeptides. Protons move one by one into binding sites on one of the parts (the rotor), causing it to spin in a way that catalyzes ATP production from ADP and inorganic phosphate (**Figure 9.14**). The flow of protons thus behaves somewhat like a rushing stream that turns a waterwheel. ATP synthase is the smallest molecular rotary motor known in nature.

How does the inner mitochondrial membrane or the prokaryotic plasma membrane generate and maintain the H^+ gradient that drives ATP synthesis by the ATP synthase protein complex? Establishing the H^+ gradient is a major function of the electron transport chain, which is shown in its mitochondrial location in **Figure 9.15**. The chain is an energy converter that uses the exergonic flow of electrons from NADH and $FADH_2$ to pump H^+ across the membrane, from the mitochondrial matrix into the intermembrane space. The H^+ has a tendency to move back across the membrane, diffusing down its gradient. And the ATP synthases are the only sites that provide a route through the membrane for H^+. As we described previously, the passage of H^+ through ATP synthase uses the exergonic flow of H^+ to drive the phosphorylation of ADP. Thus, the energy stored in an H^+ gradient across a membrane couples the redox reactions of the electron transport chain to ATP synthesis, an example of chemiosmosis.

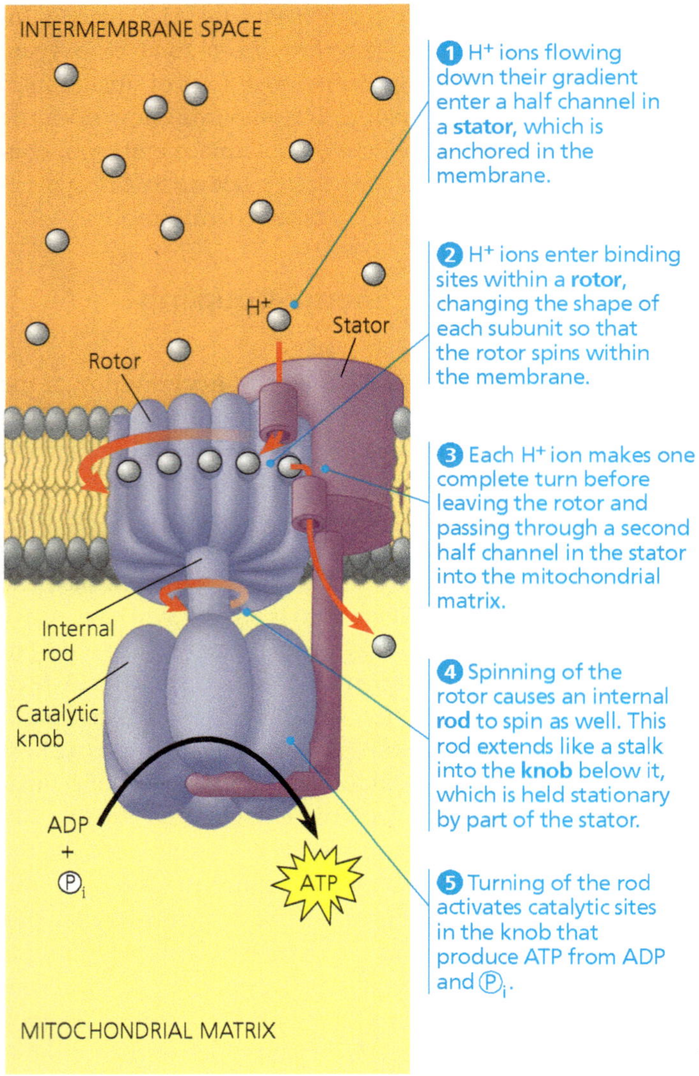

▲ **Figure 9.14 ATP synthase, a molecular mill.** The ATP synthase protein complex functions as a mill, powered by the flow of hydrogen ions. Multiple copies of this complex reside in mitochondrial and chloroplast membranes of eukaryotes and in the plasma membranes of prokaryotes. Each of the four parts of ATP synthase consists of a number of polypeptide subunits.

At this point, you may be wondering how the electron transport chain pumps hydrogen ions. Researchers have found that certain members of the electron transport chain accept and release protons (H^+) along with electrons. (The aqueous solutions inside and surrounding the cell are a ready source of H^+.) At certain steps along the chain, electron transfers cause H^+ to be taken up and released into the surrounding solution. In eukaryotic cells, the electron carriers are spatially arranged in the inner mitochondrial membrane in such a way that H^+ is accepted from the mitochondrial matrix and deposited in the intermembrane space (see Figure 9.15). The H^+ gradient that results is referred to as a **proton-motive force**, emphasizing the capacity of the gradient to perform work. The force drives H^+ back across the membrane through the H^+ channels provided by ATP synthases.

In general terms, *chemiosmosis is an energy-coupling mechanism that uses energy stored in the form of an H^+ gradient across a membrane to drive cellular work.* In mitochondria, the energy for gradient formation comes from exergonic redox reactions, and ATP synthesis is the work performed. But chemiosmosis also occurs elsewhere and in other variations. Chloroplasts use chemiosmosis to generate ATP during photosynthesis; in these organelles, light (rather than chemical energy) drives both electron flow down an electron transport chain and the resulting H^+ gradient formation. Prokaryotes, as already mentioned, generate H^+ gradients across their plasma membranes. They then tap the proton-motive force not only to make ATP inside the cell but also to rotate their flagella and to pump nutrients and waste products across the membrane. Because of its central importance to energy conversions in prokaryotes and eukaryotes, chemiosmosis has helped unify the study of bioenergetics. Peter Mitchell was awarded the Nobel Prize in 1978 for originally proposing the chemiosmotic model.

An Accounting of ATP Production by Cellular Respiration

In the last few sections, we have looked rather closely at the key processes of cellular respiration. Now let's take a step

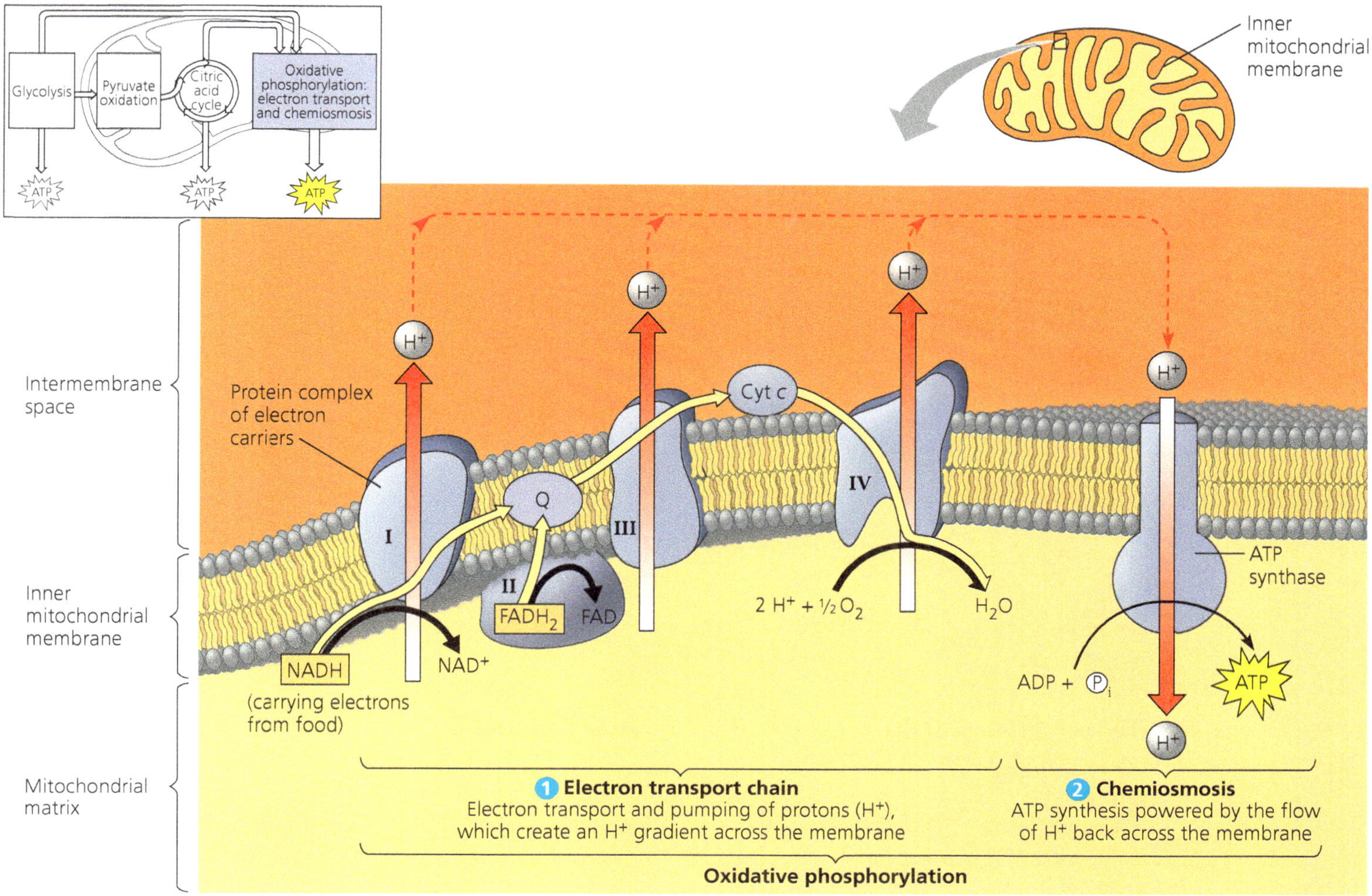

▲ **Figure 9.15 Chemiosmosis couples the electron transport chain to ATP synthesis.** ① NADH and $FADH_2$ shuttle high-energy electrons extracted from food during glycolysis and the citric acid cycle into an electron transport chain built into the inner mitochondrial membrane. The gold arrows trace the transport of electrons, which finally pass to oxygen at the "downhill" end of the chain, forming water. As Figure 9.13 showed, most of the electron carriers of the chain are grouped into four complexes. Two mobile carriers, ubiquinone (Q) and cytochrome *c* (Cyt *c*), move rapidly, ferrying electrons between the large complexes. As complexes I, III, and IV accept and then donate electrons, they pump protons from the mitochondrial matrix into the intermembrane space. (In prokaryotes, protons are pumped outside the plasma membrane.) Note that $FADH_2$ deposits its electrons via complex II and so results in fewer protons being pumped into the intermembrane space than occurs with NADH. Chemical energy originally harvested from food is transformed into a proton-motive force, a gradient of H^+ across the membrane. ② During chemiosmosis, the protons flow back down their gradient via ATP synthase, which is built into the membrane nearby. The ATP synthase harnesses the proton-motive force to phosphorylate ADP, forming ATP. Together, electron transport and chemiosmosis make up oxidative phosphorylation.

WHAT IF? *If complex IV were nonfunctional, could chemiosmosis produce any ATP, and if so, how would the rate of synthesis differ?*

back and remind ourselves of its overall function: harvesting the energy of glucose for ATP synthesis.

During respiration, most energy flows in this sequence: glucose → NADH → electron transport chain → proton-motive force → ATP. We can do some bookkeeping to calculate the ATP profit when cellular respiration oxidizes a molecule of glucose to six molecules of carbon dioxide. The three main departments of this metabolic enterprise are glycolysis, the citric acid cycle, and the electron transport chain, which drives oxidative phosphorylation. **Figure 9.16**, on the next page, gives a detailed accounting of the ATP yield per glucose molecule oxidized. The tally adds the 4 ATP produced directly by substrate-level phosphorylation during glycolysis and the citric acid cycle to the many more molecules of ATP generated by oxidative phosphorylation. Each NADH that transfers a pair of electrons from glucose to the electron transport chain contributes enough to the proton-motive force to generate a maximum of about 3 ATP.

Why are the numbers in Figure 9.16 inexact? There are three reasons we cannot state an exact number of ATP molecules generated by the breakdown of one molecule of glucose. First, phosphorylation and the redox reactions are not directly coupled to each other, so the ratio of the number of

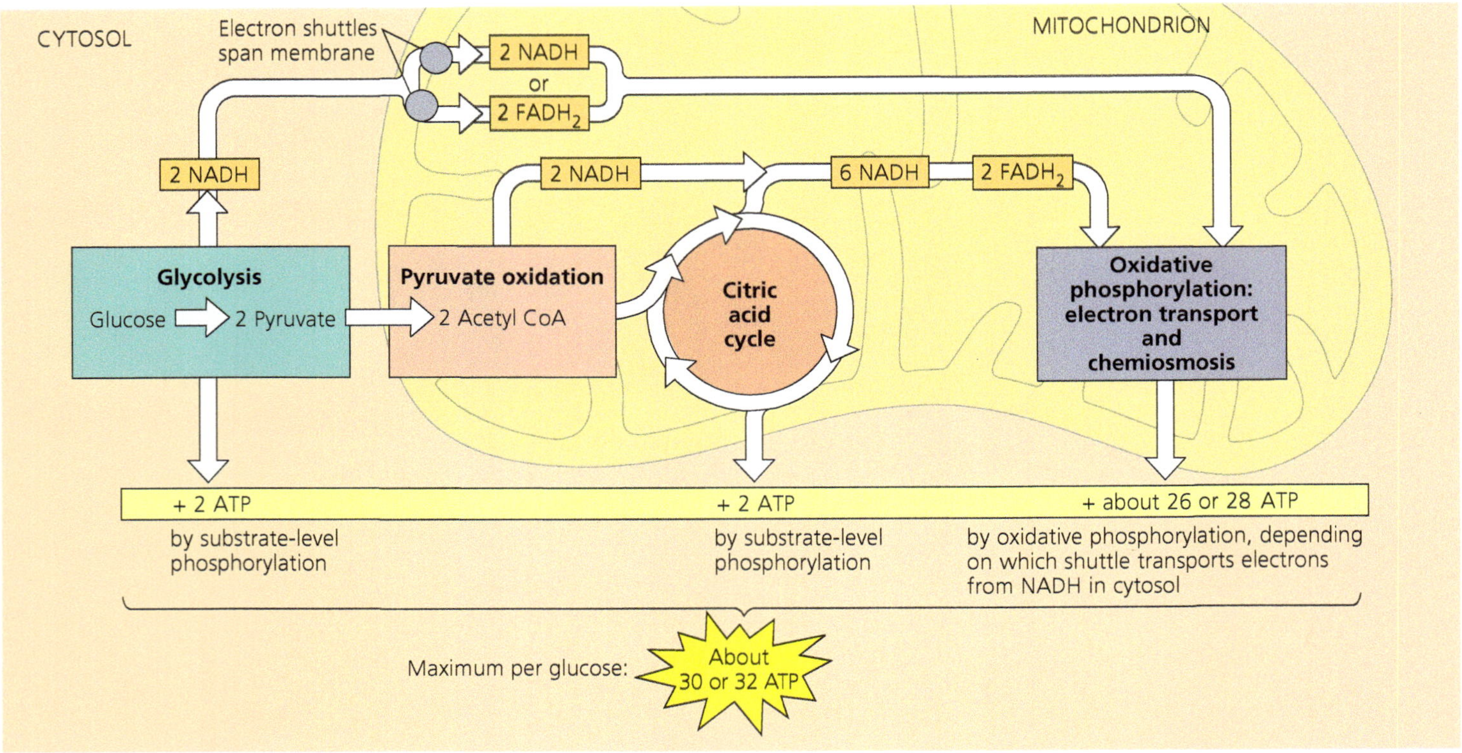

▲ **Figure 9.16 ATP yield per molecule of glucose at each stage of cellular respiration.**

? *Explain exactly how the numbers "26 or 28" were calculated.*

NADH molecules to the number of ATP molecules is not a whole number. We know that 1 NADH results in 10 H^+ being transported out across the inner mitochondrial membrane, but the exact number of H^+ that must reenter the mitochondrial matrix via ATP synthase to generate 1 ATP has long been debated. Based on experimental data, however, most biochemists now agree that the most accurate number is 4 H^+. Therefore, a single molecule of NADH generates enough proton-motive force for the synthesis of 2.5 ATP. The citric acid cycle also supplies electrons to the electron transport chain via $FADH_2$, but since its electrons enter later in the chain, each molecule of this electron carrier is responsible for transport of only enough H^+ for the synthesis of 1.5 ATP. These numbers also take into account the slight energetic cost of moving the ATP formed in the mitochondrion out into the cytosol, where it will be used.

Second, the ATP yield varies slightly depending on the type of shuttle used to transport electrons from the cytosol into the mitochondrion. The mitochondrial inner membrane is impermeable to NADH, so NADH in the cytosol is segregated from the machinery of oxidative phosphorylation. The 2 electrons of NADH captured in glycolysis must be conveyed into the mitochondrion by one of several electron shuttle systems. Depending on the kind of shuttle in a particular cell type, the electrons are passed either to NAD^+ or to FAD in the mitochondrial matrix (see Figure 9.16). If the electrons are passed to FAD, as in brain cells, only about 1.5 ATP can result from each NADH that was originally generated in the cytosol. If the electrons are passed to mitochondrial NAD^+, as in liver cells and heart cells, the yield is about 2.5 ATP per NADH.

A third variable that reduces the yield of ATP is the use of the proton-motive force generated by the redox reactions of respiration to drive other kinds of work. For example, the proton-motive force powers the mitochondrion's uptake of pyruvate from the cytosol. However, if *all* the proton-motive force generated by the electron transport chain were used to drive ATP synthesis, one glucose molecule could generate a maximum of 28 ATP produced by oxidative phosphorylation plus 4 ATP (net) from substrate-level phosphorylation to give a total yield of about 32 ATP (or only about 30 ATP if the less efficient shuttle were functioning).

We can now roughly estimate the efficiency of respiration—that is, the percentage of chemical energy in glucose that has been transferred to ATP. Recall that the complete oxidation of a mole of glucose releases 686 kcal of energy under standard conditions ($\Delta G = -686$ kcal/mol). Phosphorylation of ADP to form ATP stores at least 7.3 kcal per mole of ATP. Therefore, the efficiency of respiration is 7.3 kcal per mole of ATP times 32 moles of ATP per mole of glucose divided by 686 kcal per mole of glucose, which equals 0.34. Thus, about 34% of the potential chemical energy in glucose has been transferred to ATP; the actual percentage is bound to vary as ΔG varies under different cellular conditions. Cellular respiration is remarkably efficient in its energy conversion. By comparison, the most efficient automobile converts only about 25% of the energy stored in gasoline to energy that moves the car.

The rest of the energy stored in glucose is lost as heat. We humans use some of this heat to maintain our relatively high body temperature (37°C), and we dissipate the rest through sweating and other cooling mechanisms.

Under certain conditions, it may be beneficial to reduce the efficiency of cellular respiration. A remarkable adaptation is shown by hibernating mammals, which overwinter in a state of inactivity and lowered metabolism. Although their internal body temperature is lower than normal, it still must be kept significantly higher than the external air temperature. One type of tissue, called brown fat, is made up of cells packed full of mitochondria. The inner mitochondrial membrane contains a channel protein called the uncoupling protein, which allows protons to flow back down their concentration gradient without generating ATP. Activation of these proteins in hibernating mammals results in ongoing oxidation of stored fuel stores (fats), generating heat without any ATP production. In the absence of such an adaptation, the ATP level would build up to a point that cellular respiration would be shut down due to regulatory mechanisms to be discussed later.

CONCEPT CHECK 9.4

1. What effect would an absence of O_2 have on the process shown in Figure 9.15?
2. **WHAT IF?** In the absence of O_2, as in question 1, what do you think would happen if you decreased the pH of the intermembrane space of the mitochondrion? Explain your answer.
3. **MAKE CONNECTIONS** In Concept 7.1 (pp. 127–128), you learned that membranes must be fluid to function properly. How does the operation of the electron transport chain support that assertion?

For suggested answers, see Appendix A.

CONCEPT 9.5

Fermentation and anaerobic respiration enable cells to produce ATP without the use of oxygen

Because most of the ATP generated by cellular respiration is due to the work of oxidative phosphorylation, our estimate of ATP yield from aerobic respiration is contingent on an adequate supply of oxygen to the cell. Without the electronegative oxygen to pull electrons down the transport chain, oxidative phosphorylation eventually ceases. However, there are two general mechanisms by which certain cells can oxidize organic fuel and generate ATP *without* the use of oxygen: anaerobic respiration and fermentation. The distinction between these two is that an electron transport chain is used in anaerobic respiration but not in fermentation. (The electron transport chain is also called the respiratory chain because of its role in both types of cellular respiration.)

We have already mentioned anaerobic respiration, which takes place in certain prokaryotic organisms that live in environments without oxygen. These organisms have an electron transport chain but do not use oxygen as a final electron acceptor at the end of the chain. Oxygen performs this function very well because it is extremely electronegative, but other, less electronegative substances can also serve as final electron acceptors. Some "sulfate-reducing" marine bacteria, for instance, use the sulfate ion (SO_4^{2-}) at the end of their respiratory chain. Operation of the chain builds up a proton-motive force used to produce ATP, but H_2S (hydrogen sulfide) is produced as a by-product rather than water. The rotten-egg odor you may have smelled while walking through a salt marsh or a mudflat signals the presence of sulfate-reducing bacteria.

Fermentation is a way of harvesting chemical energy without using either oxygen or any electron transport chain—in other words, without cellular respiration. How can food be oxidized without cellular respiration? Remember, oxidation simply refers to the loss of electrons to an electron acceptor, so it does not need to involve oxygen. Glycolysis oxidizes glucose to two molecules of pyruvate. The oxidizing agent of glycolysis is NAD^+, and neither oxygen nor any electron transfer chain is involved. Overall, glycolysis is exergonic, and some of the energy made available is used to produce 2 ATP (net) by substrate-level phosphorylation. If oxygen *is* present, then additional ATP is made by oxidative phosphorylation when NADH passes electrons removed from glucose to the electron transport chain. But glycolysis generates 2 ATP whether oxygen is present or not—that is, whether conditions are aerobic or anaerobic.

As an alternative to respiratory oxidation of organic nutrients, fermentation is an extension of glycolysis that allows continuous generation of ATP by the substrate-level phosphorylation of glycolysis. For this to occur, there must be a sufficient supply of NAD^+ to accept electrons during the oxidation step of glycolysis. Without some mechanism to recycle NAD^+ from NADH, glycolysis would soon deplete the cell's pool of NAD^+ by reducing it all to NADH and would shut itself down for lack of an oxidizing agent. Under aerobic conditions, NAD^+ is recycled from NADH by the transfer of electrons to the electron transport chain. An anaerobic alternative is to transfer electrons from NADH to pyruvate, the end product of glycolysis.

Types of Fermentation

Fermentation consists of glycolysis plus reactions that regenerate NAD^+ by transferring electrons from NADH to pyruvate or derivatives of pyruvate. The NAD^+ can then be reused to oxidize sugar by glycolysis, which nets two molecules of ATP by substrate-level phosphorylation. There are many types of fermentation, differing in the end products formed from pyruvate. Two common types are alcohol fermentation and lactic acid fermentation.

In **alcohol fermentation** (Figure 9.17a), pyruvate is converted to ethanol (ethyl alcohol) in two steps. The first step releases carbon dioxide from the pyruvate, which is converted to the two-carbon compound acetaldehyde. In the second step, acetaldehyde is reduced by NADH to ethanol. This regenerates the supply of NAD^+ needed for the continuation of glycolysis. Many bacteria carry out alcohol fermentation under anaerobic conditions. Yeast (a fungus) also carries out alcohol fermentation. For thousands of years, humans have used yeast in brewing, winemaking, and baking. The CO_2 bubbles generated by baker's yeast during alcohol fermentation allow bread to rise.

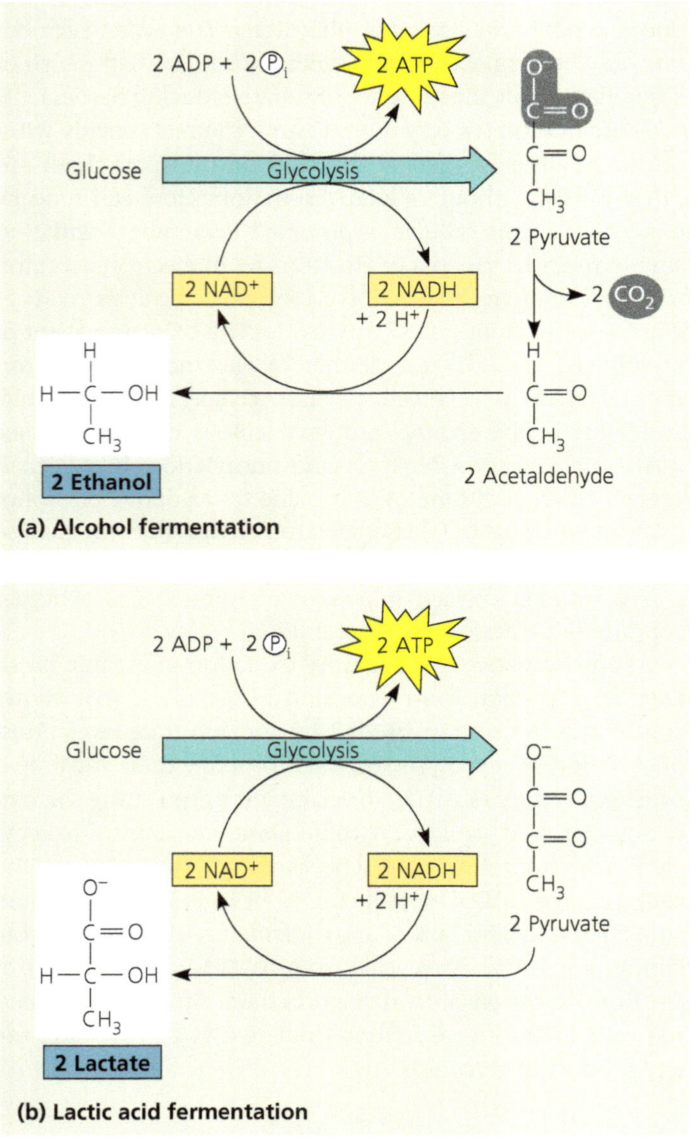

▲ **Figure 9.17 Fermentation.** In the absence of oxygen, many cells use fermentation to produce ATP by substrate-level phosphorylation. Pyruvate, the end product of glycolysis, serves as an electron acceptor for oxidizing NADH back to NAD^+, which can then be reused in glycolysis. Two of the common end products formed from fermentation are **(a)** ethanol and **(b)** lactate, the ionized form of lactic acid.

During **lactic acid fermentation** (Figure 9.17b), pyruvate is reduced directly by NADH to form lactate as an end product, with no release of CO_2. (Lactate is the ionized form of lactic acid.) Lactic acid fermentation by certain fungi and bacteria is used in the dairy industry to make cheese and yogurt.

Human muscle cells make ATP by lactic acid fermentation when oxygen is scarce. This occurs during strenuous exercise, when sugar catabolism for ATP production outpaces the muscle's supply of oxygen from the blood. Under these conditions, the cells switch from aerobic respiration to fermentation. The lactate that accumulates was previously thought to cause muscle fatigue and pain, but recent research suggests instead that increased levels of potassium ions (K^+) may be to blame, while lactate appears to enhance muscle performance. In any case, the excess lactate is gradually carried away by the blood to the liver, where it is converted back to pyruvate by liver cells. Because oxygen is available, this pyruvate can then enter the mitochondria in liver cells and complete cellular respiration.

Comparing Fermentation with Anaerobic and Aerobic Respiration

Fermentation, anaerobic respiration, and aerobic respiration are three alternative cellular pathways for producing ATP by harvesting the chemical energy of food. All three use glycolysis to oxidize glucose and other organic fuels to pyruvate, with a net production of 2 ATP by substrate-level phosphorylation. And in all three pathways, NAD^+ is the oxidizing agent that accepts electrons from food during glycolysis.

A key difference among the three pathways is the contrasting mechanisms for oxidizing NADH back to NAD^+, which is required to sustain glycolysis. In fermentation, the final electron acceptor is an organic molecule such as pyruvate (lactic acid fermentation) or acetaldehyde (alcohol fermentation). In cellular respiration, by contrast, electrons carried by NADH are transferred to an electron transport chain, where they move stepwise down a series of redox reactions to a final electron acceptor. In aerobic respiration, the final electron acceptor is oxygen; in anaerobic respiration, the final acceptor is another molecule that is electronegative (although invariably less so than oxygen). Passage of electrons from NADH to the electron transport chain not only regenerates the NAD^+ required for glycolysis but pays an ATP bonus when the stepwise electron transport from this NADH to oxygen drives oxidative phosphorylation. An even bigger ATP payoff comes from the oxidation of pyruvate in the mitochondrion, which is unique to respiration. Without an electron transport chain, the energy still stored in pyruvate is unavailable to most cells. Thus, cellular respiration harvests much more energy from each sugar molecule than fermentation can. In fact, aerobic respiration yields up to 16 times as much ATP per glucose molecule as does fermentation—up to 32 molecules of ATP for respiration, compared with 2 molecules of ATP produced by substrate-level phosphorylation in fermentation.

Some organisms, called **obligate anaerobes**, carry out only fermentation or anaerobic respiration. In fact, these organisms cannot survive in the presence of oxygen, some forms of which can actually be toxic if protective systems are not present in the cell. A few cell types, such as cells of the vertebrate brain, can carry out only aerobic oxidation of pyruvate, not fermentation. Other organisms, including yeasts and many bacteria, can make enough ATP to survive using either fermentation or respiration. Such species are called **facultative anaerobes**. On the cellular level, our muscle cells behave as facultative anaerobes. In such cells, pyruvate is a fork in the metabolic road that leads to two alternative catabolic routes (**Figure 9.18**). Under aerobic conditions, pyruvate can be converted to acetyl CoA, and oxidation continues in the citric acid cycle via aerobic respiration. Under anaerobic conditions, lactic acid fermentation occurs: Pyruvate is diverted from the citric acid cycle, serving instead as an electron acceptor to recycle NAD^+. To make the same amount of ATP, a facultative anaerobe has to consume sugar at a much faster rate when fermenting than when respiring.

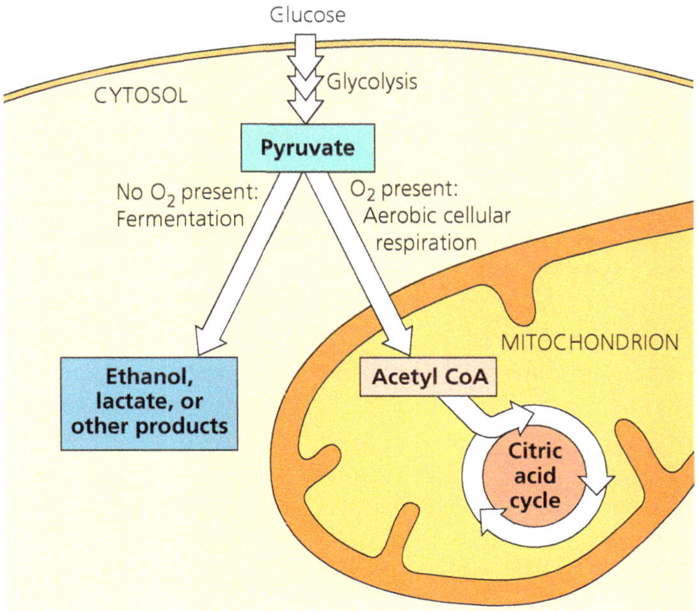

▲ **Figure 9.18 Pyruvate as a key juncture in catabolism.** Glycolysis is common to fermentation and cellular respiration. The end product of glycolysis, pyruvate, represents a fork in the catabolic pathways of glucose oxidation. In a facultative anaerobe or a muscle cell, which are capable of both aerobic cellular respiration and fermentation, pyruvate is committed to one of those two pathways, usually depending on whether or not oxygen is present.

The Evolutionary Significance of Glycolysis

EVOLUTION The role of glycolysis in both fermentation and respiration has an evolutionary basis. Ancient prokaryotes are thought to have used glycolysis to make ATP long before oxygen was present in Earth's atmosphere. The oldest known fossils of bacteria date back 3.5 billion years, but appreciable quantities of oxygen probably did not begin to accumulate in the atmosphere until about 2.7 billion years ago. Cyanobacteria produced this O_2 as a by-product of photosynthesis. Therefore, early prokaryotes may have generated ATP exclusively from glycolysis. The fact that glycolysis is today the most widespread metabolic pathway among Earth's organisms suggests that it evolved very early in the history of life. The cytosolic location of glycolysis also implies great antiquity; the pathway does not require any of the membrane-bounded organelles of the eukaryotic cell, which evolved approximately 1 billion years after the prokaryotic cell. Glycolysis is a metabolic heirloom from early cells that continues to function in fermentation and as the first stage in the breakdown of organic molecules by respiration.

CONCEPT CHECK 9.5

1. Consider the NADH formed during glycolysis. What is the final acceptor for its electrons during fermentation? What is the final acceptor for its electrons during aerobic respiration?
2. **WHAT IF?** A glucose-fed yeast cell is moved from an aerobic environment to an anaerobic one. How would its rate of glucose consumption change if ATP were to be generated at the same rate?

For suggested answers, see Appendix A.

CONCEPT 9.6

Glycolysis and the citric acid cycle connect to many other metabolic pathways

So far, we have treated the oxidative breakdown of glucose in isolation from the cell's overall metabolic economy. In this section, you will learn that glycolysis and the citric acid cycle are major intersections of the cell's catabolic and anabolic (biosynthetic) pathways.

The Versatility of Catabolism

Throughout this chapter, we have used glucose as the fuel for cellular respiration. But free glucose molecules are not common in the diets of humans and other animals. We obtain most of our calories in the form of fats, proteins, sucrose and other disaccharides, and starch, a polysaccharide. All these

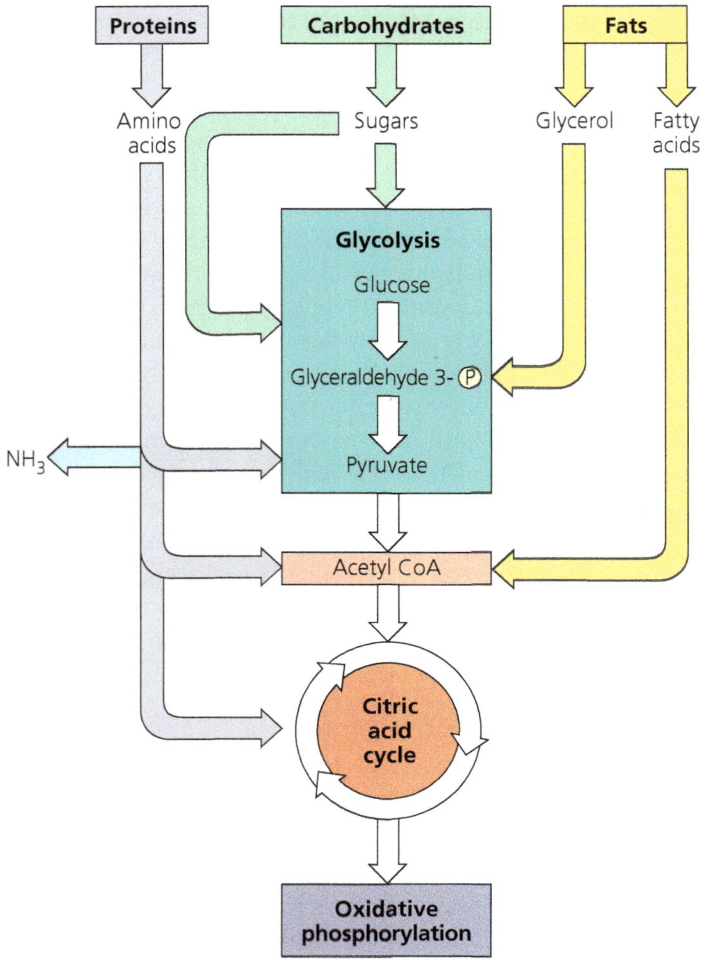

▲ **Figure 9.19 The catabolism of various molecules from food.** Carbohydrates, fats, and proteins can all be used as fuel for cellular respiration. Monomers of these molecules enter glycolysis or the citric acid cycle at various points. Glycolysis and the citric acid cycle are catabolic funnels through which electrons from all kinds of organic molecules flow on their exergonic fall to oxygen.

organic molecules in food can be used by cellular respiration to make ATP (**Figure 9.19**).

Glycolysis can accept a wide range of carbohydrates for catabolism. In the digestive tract, starch is hydrolyzed to glucose, which can then be broken down in the cells by glycolysis and the citric acid cycle. Similarly, glycogen, the polysaccharide that humans and many other animals store in their liver and muscle cells, can be hydrolyzed to glucose between meals as fuel for respiration. The digestion of disaccharides, including sucrose, provides glucose and other monosaccharides as fuel for respiration.

Proteins can also be used for fuel, but first they must be digested to their constituent amino acids. Many of the amino acids are used by the organism to build new proteins. Amino acids present in excess are converted by enzymes to intermediates of glycolysis and the citric acid cycle. Before amino acids can feed into glycolysis or the citric acid cycle, their amino groups must be removed, a process called *deamination*. The nitrogenous refuse is excreted from the animal in the form of ammonia (NH_3), urea, or other waste products.

Catabolism can also harvest energy stored in fats obtained either from food or from storage cells in the body. After fats are digested to glycerol and fatty acids, the glycerol is converted to glyceraldehyde 3-phosphate, an intermediate of glycolysis. Most of the energy of a fat is stored in the fatty acids. A metabolic sequence called **beta oxidation** breaks the fatty acids down to two-carbon fragments, which enter the citric acid cycle as acetyl CoA. NADH and $FADH_2$ are also generated during beta oxidation; they can enter the electron transport chain, leading to further ATP production. Fats make excellent fuel, in large part due to their chemical structure and the high energy level of their electrons (equally shared between carbon and hydrogen) compared to those of carbohydrates. A gram of fat oxidized by respiration produces more than twice as much ATP as a gram of carbohydrate. Unfortunately, this also means that a person trying to lose weight must work hard to use up fat stored in the body because so many calories are stockpiled in each gram of fat.

Biosynthesis (Anabolic Pathways)

Cells need substance as well as energy. Not all the organic molecules of food are destined to be oxidized as fuel to make ATP. In addition to calories, food must also provide the carbon skeletons that cells require to make their own molecules. Some organic monomers obtained from digestion can be used directly. For example, as previously mentioned, amino acids from the hydrolysis of proteins in food can be incorporated into the organism's own proteins. Often, however, the body needs specific molecules that are not present as such in food. Compounds formed as intermediates of glycolysis and the citric acid cycle can be diverted into anabolic pathways as precursors from which the cell can synthesize the molecules it requires. For example, humans can make about half of the 20 amino acids in proteins by modifying compounds siphoned away from the citric acid cycle; the rest are "essential amino acids" that must be obtained in the diet. Also, glucose can be made from pyruvate, and fatty acids can be synthesized from acetyl CoA. Of course, these anabolic, or biosynthetic, pathways do not generate ATP, but instead consume it.

In addition, glycolysis and the citric acid cycle function as metabolic interchanges that enable our cells to convert some kinds of molecules to others as we need them. For example, an intermediate compound generated during glycolysis, dihydroxyacetone phosphate (see Figure 9.9, step 5), can be converted to one of the major precursors of fats. If we eat more food than we need, we store fat even if our diet is fat-free. Metabolism is remarkably versatile and adaptable.

Regulation of Cellular Respiration via Feedback Mechanisms

Basic principles of supply and demand regulate the metabolic economy. The cell does not waste energy making more of a particular substance than it needs. If there is a glut of a certain amino acid, for example, the anabolic pathway that synthesizes that amino acid from an intermediate of the citric acid cycle is switched off. The most common mechanism for this control is feedback inhibition: The end product of the anabolic pathway inhibits the enzyme that catalyzes an early step of the pathway (see Figure 8.21). This prevents the needless diversion of key metabolic intermediates from uses that are more urgent.

The cell also controls its catabolism. If the cell is working hard and its ATP concentration begins to drop, respiration speeds up. When there is plenty of ATP to meet demand, respiration slows down, sparing valuable organic molecules for other functions. Again, control is based mainly on regulating the activity of enzymes at strategic points in the catabolic pathway. As shown in **Figure 9.20**, one important switch is phosphofructokinase, the enzyme that catalyzes step 3 of glycolysis (see Figure 9.9). That is the first step that commits the substrate irreversibly to the glycolytic pathway. By controlling the rate of this step, the cell can speed up or slow down the entire catabolic process. Phosphofructokinase can thus be considered the pacemaker of respiration.

Phosphofructokinase is an allosteric enzyme with receptor sites for specific inhibitors and activators. It is inhibited by ATP and stimulated by AMP (adenosine monophosphate), which the cell derives from ADP. As ATP accumulates, inhibition of the enzyme slows down glycolysis. The enzyme becomes active again as cellular work converts ATP to ADP (and AMP) faster than ATP is being regenerated. Phosphofructokinase is also sensitive to citrate, the first product of the citric acid cycle. If citrate accumulates in mitochondria, some of it passes into the cytosol and inhibits phosphofructokinase. This mechanism helps synchronize the rates of glycolysis and the citric acid cycle. As citrate accumulates, glycolysis slows down, and the supply of acetyl groups to the citric acid cycle decreases. If citrate consumption increases, either because of a demand for more ATP or because anabolic pathways are draining off intermediates of the citric acid cycle, glycolysis accelerates and meets the demand. Metabolic balance is augmented by the control of enzymes that catalyze other key steps of glycolysis and the citric acid cycle. Cells are thrifty, expedient, and responsive in their metabolism.

Cellular respiration and metabolic pathways play a role of central importance in organisms. Examine Figure 9.2 again to put cellular respiration into the broader context of energy flow and chemical cycling in ecosystems. The energy that keeps us alive is *released*, not *produced*, by cellular respiration. We are tapping energy that was stored in food by photosynthesis. In the next chapter, you will learn how photosynthesis captures light and converts it to chemical energy.

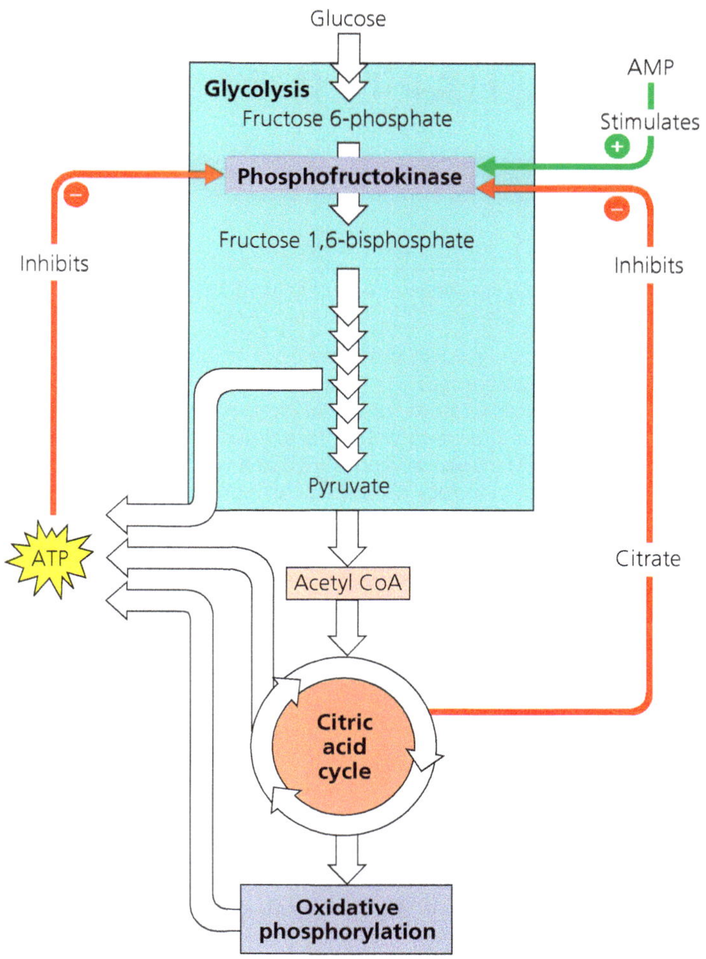

▲ **Figure 9.20 The control of cellular respiration.** Allosteric enzymes at certain points in the respiratory pathway respond to inhibitors and activators that help set the pace of glycolysis and the citric acid cycle. Phosphofructokinase, which catalyzes an early step in glycolysis (see Figure 9.9), is one such enzyme. It is stimulated by AMP (derived from ADP) but is inhibited by ATP and by citrate. This feedback regulation adjusts the rate of respiration as the cell's catabolic and anabolic demands change.

CONCEPT CHECK 9.6

1. **MAKE CONNECTIONS** Compare the structure of a fat (see Figure 5.10, p. 75) with that of a carbohydrate (see Figure 5.3, p. 70). What features of their structures make fat a much better fuel?
2. Under what circumstances might your body synthesize fat molecules?
3. **MAKE CONNECTIONS** Return to Figure 5.6b on page 72 and look at the arrangement of glycogen and mitochondria in the micrograph. What is the connection between glycogen and mitochondria?
4. **WHAT IF?** What will happen in a muscle cell that has used up its supply of oxygen and ATP? (Review Figures 9.18 and 9.20.)
5. **WHAT IF?** During intense exercise, can a muscle cell use fat as a concentrated source of chemical energy? Explain. (Review Figures 9.18 and 9.19.)

For suggested answers, see Appendix A.

9 CHAPTER REVIEW

SUMMARY OF KEY CONCEPTS

CONCEPT 9.1

Catabolic pathways yield energy by oxidizing organic fuels (pp. 164–168)

- Cells break down glucose and other organic fuels to yield chemical energy in the form of ATP. **Fermentation** is a partial degradation of glucose without the use of oxygen. **Cellular respiration** is a more complete breakdown of glucose; in **aerobic respiration**, oxygen is used as a reactant. The cell taps the energy stored in food molecules through **redox reactions**, in which one substance partially or totally shifts electrons to another. **Oxidation** is the loss of electrons from one substance, while **reduction** is the addition of electrons to the other.
- During aerobic respiration, glucose ($C_6H_{12}O_6$) is oxidized to CO_2, and O_2 is reduced to H_2O. Electrons lose potential energy during their transfer from glucose or other organic compounds to oxygen. Electrons are usually passed first to **NAD^+**, reducing it to NADH, and then from NADH to an **electron transport chain**, which conducts them to O_2 in energy-releasing steps. The energy is used to make ATP.
- Aerobic respiration occurs in three stages: (1) **glycolysis**, (2) pyruvate oxidation and the **citric acid cycle**, and (3) **oxidative phosphorylation** (electron transport and chemiosmosis).

? *Describe the difference between the two processes in cellular respiration that produce ATP: oxidative phosphorylation and substrate-level phosphorylation.*

CONCEPT 9.2

Glycolysis harvests chemical energy by oxidizing glucose to pyruvate (pp. 168–169)

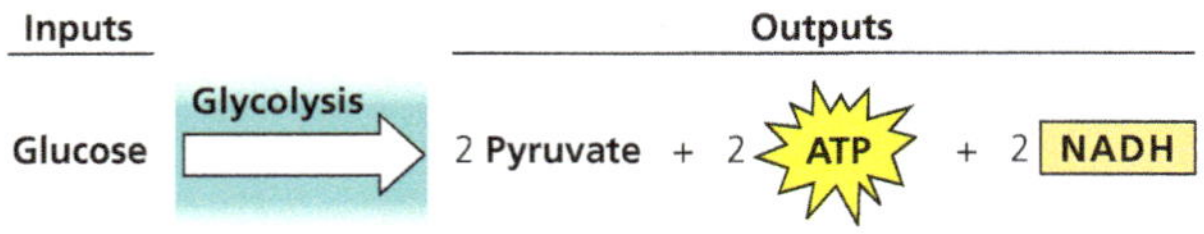

? *What is the source of energy for the formation of ATP and NADH in glycolysis?*

CONCEPT 9.3

After pyruvate is oxidized, the citric acid cycle completes the energy-yielding oxidation of organic molecules (pp. 170–172)

- In eukaryotic cells, pyruvate enters the mitochondrion and is oxidized to **acetyl CoA**, which is further oxidized in the citric acid cycle.

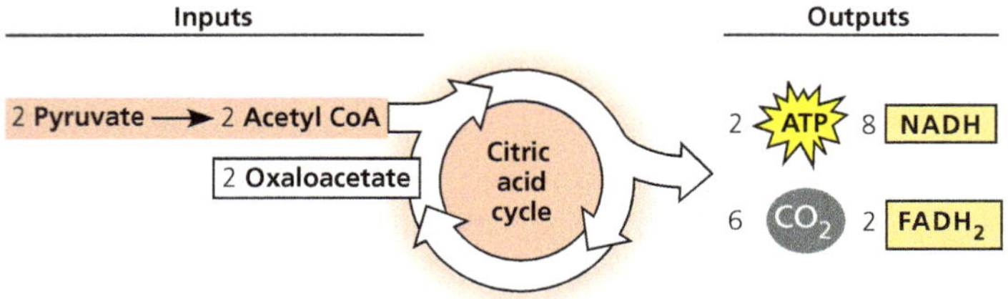

? *What molecular products indicate the complete oxidation of glucose during cellular respiration?*

CONCEPT 9.4

During oxidative phosphorylation, chemiosmosis couples electron transport to ATP synthesis (pp. 172–177)

- NADH and $FADH_2$ transfer electrons to the electron transport chain. Electrons move down the chain, losing energy in several energy-releasing steps. Finally, electrons are passed to O_2, reducing it to H_2O.

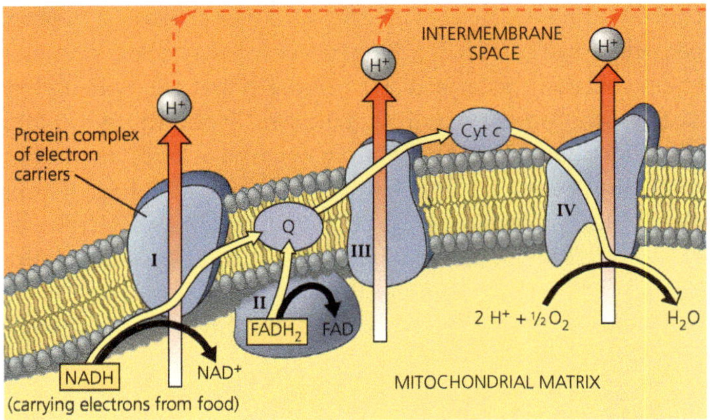

- At certain steps along the electron transport chain, electron transfer causes protein complexes to move H^+ from the mitochondrial matrix (in eukaryotes) to the intermembrane space, storing energy as a **proton-motive force** (H^+ gradient). As H^+ diffuses back into the matrix through **ATP synthase**, its passage drives the phosphorylation of ADP, a process called **chemiosmosis**.

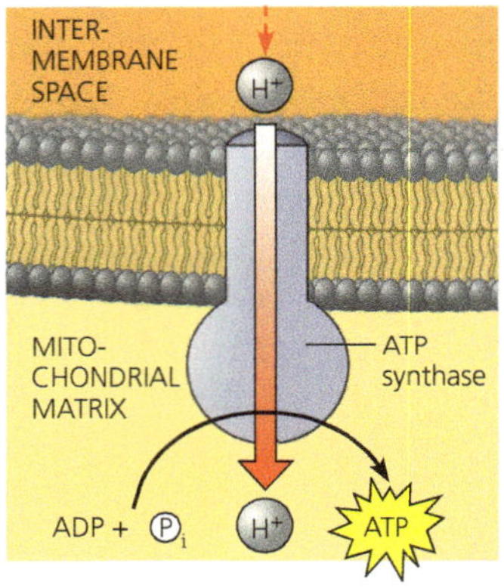

- About 34% of the energy stored in a glucose molecule is transferred to ATP during cellular respiration, producing a maximum of about 32 ATP.

? *Briefly explain the mechanism by which ATP synthase produces ATP. List three locations in which ATP synthases are found.*

CONCEPT 9.5

Fermentation and anaerobic respiration enable cells to produce ATP without the use of oxygen (pp. 177–179)

- Glycolysis nets 2 ATP by substrate-level phosphorylation, whether oxygen is present or not. Under anaerobic conditions, either anaerobic respiration or fermentation can take place. In anaerobic respiration, an electron transport chain is present with a final electron acceptor other than oxygen. In fermentation, the electrons from NADH are passed to pyruvate or a derivative of pyruvate, regenerating the NAD^+ required to oxidize more glucose. Two common types of fermentation are **alcohol fermentation** and **lactic acid fermentation**.
- Fermentation and anaerobic or aerobic respiration all use glycolysis to oxidize glucose, but they differ in their final electron acceptor and whether an electron transport chain is used (respiration) or not (fermentation). Respiration yields more ATP;

aerobic respiration, with O_2 as the final electron acceptor, yields about 16 times as much ATP as does fermentation.
- Glycolysis occurs in nearly all organisms and is thought to have evolved in ancient prokaryotes before there was O_2 in the atmosphere.

? *Which process yields more ATP, fermentation or anaerobic respiration? Explain.*

CONCEPT 9.6

Glycolysis and the citric acid cycle connect to many other metabolic pathways (pp. 179–181)

- Catabolic pathways funnel electrons from many kinds of organic molecules into cellular respiration. Many carbohydrates can enter glycolysis, most often after conversion to glucose. Amino acids of proteins must be deaminated before being oxidized. The fatty acids of fats undergo **beta oxidation** to two-carbon fragments and then enter the citric acid cycle as acetyl CoA. Anabolic pathways can use small molecules from food directly or build other substances using intermediates of glycolysis or the citric acid cycle.
- Cellular respiration is controlled by allosteric enzymes at key points in glycolysis and the citric acid cycle.

? *Describe how the catabolic pathways of glycolysis and the citric acid cycle intersect with anabolic pathways in the metabolism of a cell.*

TEST YOUR UNDERSTANDING

LEVEL 1: KNOWLEDGE/COMPREHENSION

1. The *immediate* energy source that drives ATP synthesis by ATP synthase during oxidative phosphorylation is the
   a. oxidation of glucose and other organic compounds.
   b. flow of electrons down the electron transport chain.
   c. affinity of oxygen for electrons.
   d. H^+ concentration across the membrane holding ATP synthase.
   e. transfer of phosphate to ADP.

2. Which metabolic pathway is common to both fermentation and cellular respiration of a glucose molecule?
   a. the citric acid cycle
   b. the electron transport chain
   c. glycolysis
   d. synthesis of acetyl CoA from pyruvate
   e. reduction of pyruvate to lactate

3. In mitochondria, exergonic redox reactions
   a. are the source of energy driving prokaryotic ATP synthesis.
   b. are directly coupled to substrate-level phosphorylation.
   c. provide the energy that establishes the proton gradient.
   d. reduce carbon atoms to carbon dioxide.
   e. are coupled via phosphorylated intermediates to endergonic processes.

4. The final electron acceptor of the electron transport chain that functions in aerobic oxidative phosphorylation is
   a. oxygen. b. water. c. NAD^+. d. pyruvate. e. ADP.

LEVEL 2: APPLICATION/ANALYSIS

5. What is the oxidizing agent in the following reaction?
   Pyruvate + NADH + H^+ → Lactate + NAD^+
   a. oxygen b. NADH c. NAD^+ d. lactate e. pyruvate

6. When electrons flow along the electron transport chains of mitochondria, which of the following changes occurs?
   a. The pH of the matrix increases.
   b. ATP synthase pumps protons by active transport.
   c. The electrons gain free energy.
   d. The cytochromes phosphorylate ADP to form ATP.
   e. NAD^+ is oxidized.

7. Most CO_2 from catabolism is released during
   a. glycolysis.
   b. the citric acid cycle.
   c. lactate fermentation.
   d. electron transport.
   e. oxidative phosphorylation.

LEVEL 3: SYNTHESIS/EVALUATION

8. **DRAW IT** The graph here shows the pH difference across the inner mitochondrial membrane over time in an actively respiring cell. At the time indicated by the vertical arrow, a metabolic poison is added that specifically and completely inhibits all function of mitochondrial ATP synthase. Draw what you would expect to see for the rest of the graphed line.

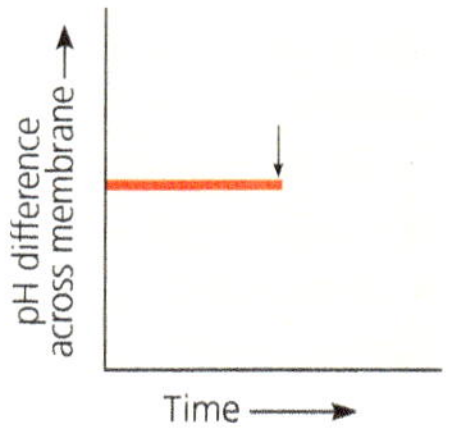

9. **EVOLUTION CONNECTION**
   ATP synthases are found in the prokaryotic plasma membrane and in mitochondria and chloroplasts. What does this suggest about the evolutionary relationship of these eukaryotic organelles to prokaryotes? How might the amino acid sequences of the ATP synthases from the different sources support or refute your hypothesis?

10. **SCIENTIFIC INQUIRY**
    In the 1930s, some physicians prescribed low doses of a compound called dinitrophenol (DNP) to help patients lose weight. This unsafe method was abandoned after some patients died. DNP uncouples the chemiosmotic machinery by making the lipid bilayer of the inner mitochondrial membrane leaky to H^+. Explain how this could cause weight loss and death.

11. **WRITE ABOUT A THEME**
    Emergent Properties In a short essay (100–150 words), explain how oxidative phosphorylation—the production of ATP using energy derived from the redox reactions of a spatially organized electron transport chain followed by chemiosmosis—is an example of how new properties emerge at each level of the biological hierarchy.

For selected answers, see Appendix A.

MasteringBIOLOGY www.masteringbiology.com

1. MasteringBiology® Assignments

BioFlix **Tutorials** Cellular Respiration: Inputs and Outputs • Glycolysis • Acetyl CoA Formation and the Citric Acid Cycle • Oxidative Phosphorylation • Summary
Tutorial Pathways for Pyruvate
Activities Build a Chemical Cycling System • Overview of Cellular Respiration • Redox Reactions • Glycolysis • The Citric Acid Cycle • Electron Transport • Fermentation • Glucose Metabolism • Discovery Channel Video: Space Plants
Questions Student Misconceptions • Reading Quiz • Multiple Choice • End-of-Chapter

2. eText

Read your book online, search, take notes, highlight text, and more.

3. The Study Area

Practice Tests • Cumulative Test • ***BioFlix*** 3-D Animations • MP3 Tutor Sessions • Videos • Activities • Investigations • Lab Media • Audio Glossary • Word Study Tools • Art

10

Photosynthesis

▲ **Figure 10.1 How can sunlight, seen here as a spectrum of colors in a rainbow, power the synthesis of organic substances?**

KEY CONCEPTS

10.1 Photosynthesis converts light energy to the chemical energy of food

10.2 The light reactions convert solar energy to the chemical energy of ATP and NADPH

10.3 The Calvin cycle uses the chemical energy of ATP and NADPH to reduce CO_2 to sugar

10.4 Alternative mechanisms of carbon fixation have evolved in hot, arid climates

OVERVIEW

The Process That Feeds the Biosphere

Life on Earth is solar powered. The chloroplasts of plants capture light energy that has traveled 150 million kilometers from the sun and convert it to chemical energy that is stored in sugar and other organic molecules. This conversion process is called **photosynthesis**. Let's begin by placing photosynthesis in its ecological context.

Photosynthesis nourishes almost the entire living world directly or indirectly. An organism acquires the organic compounds it uses for energy and carbon skeletons by one of two major modes: autotrophic nutrition or heterotrophic nutrition. **Autotrophs** are "self-feeders" (*auto-* means "self," and *trophos* means "feeder"); they sustain themselves without eating anything derived from other living beings. Autotrophs produce their organic molecules from CO_2 and other inorganic raw materials obtained from the environment. They are the ultimate sources of organic compounds for all nonautotrophic organisms, and for this reason, biologists refer to autotrophs as the *producers* of the biosphere.

Almost all plants are autotrophs; the only nutrients they require are water and minerals from the soil and carbon dioxide from the air. Specifically, plants are *photo*autotrophs, organisms that use light as a source of energy to synthesize organic substances (**Figure 10.1**). Photosynthesis also occurs in algae, certain other protists, and some prokaryotes (**Figure 10.2**). In this chapter, we will touch on these other groups in passing, but our emphasis will be on plants. Variations in autotrophic nutrition that occur in prokaryotes and algae will be described in Chapters 27 and 28.

Heterotrophs obtain their organic material by the second major mode of nutrition. Unable to make their own food, they live on compounds produced by other organisms (*hetero-* means "other"). Heterotrophs are the biosphere's *consumers*. The most obvious form of this "other-feeding" occurs when an animal eats plants or other animals. But heterotrophic nutrition may be more subtle. Some heterotrophs consume the remains of dead organisms by decomposing and feeding on organic litter such as carcasses, feces, and fallen leaves; they are known as decomposers. Most fungi and many types of prokaryotes get their nourishment this way. Almost all heterotrophs, including humans, are completely dependent, either directly or indirectly, on photoautotrophs for food—and also for oxygen, a by-product of photosynthesis.

The Earth's supply of fossil fuels was formed from remains of organisms that died hundreds of millions of years ago. In a sense, then, fossil fuels represent stores of the sun's energy from the distant past. Because these resources are being used at a much higher rate than they are replenished, researchers are exploring methods of capitalizing on the photosynthetic process to provide alternative fuels **(Figure 10.3)**.

In this chapter, you will learn how photosynthesis works. After a discussion of the general principles of photosynthesis, we will consider the two stages of photosynthesis: the light reactions, in which solar energy is captured and transformed into chemical energy; and the Calvin cycle, in which the chemical energy is used to make organic molecules of food. Finally, we will consider a few aspects of photosynthesis from an evolutionary perspective.

▲ **Figure 10.2 Photoautotrophs.** These organisms use light energy to drive the synthesis of organic molecules from carbon dioxide and (in most cases) water. They feed themselves and the entire living world. **(a)** On land, plants are the predominant producers of food. In aquatic environments, photoautotrophs include unicellular and **(b)** multicellular algae, such as this kelp; **(c)** some non-algal unicellular protists, such as *Euglena*; **(d)** the prokaryotes called cyanobacteria; and **(e)** other photosynthetic prokaryotes, such as these purple sulfur bacteria, which produce sulfur (the yellow globules within the cells) (c–e, LMs).

▼ **Figure 10.3**

IMPACT

Alternative Fuels from Plants and Algae

Biofuels from crops such as corn, soybeans, and cassava have been proposed as a supplement or even replacement for fossil fuels. To produce "bioethanol," the starch made naturally by the plants is simply converted to glucose and then fermented to ethanol by microorganisms. Alternatively, a simple chemical process can yield "biodiesel" from plant oils. Either product can be mixed with gasoline or used alone to power vehicles. Some species of unicellular algae are especially prolific oil producers, and they can be easily cultured in containers such as the tubular plastic bags shown below.

WHY IT MATTERS The rate of fossil fuel use by humans far outpaces its formation in the earth: Fossil fuels are a nonrenewable source of energy. Tapping the power of sunlight by using products of photosynthesis to generate energy is a sustainable alternative if cost-effective techniques can be developed. It is generally agreed that using algae is preferable to growing crops for this purpose because this use of cropland diminishes the food supply and drives up food prices.

FURTHER READING A. L. Haag, Algae bloom again, *Nature* 447:520–521 (2007).

WHAT IF? The main product of fossil fuel combustion is CO_2, and this combustion is the source of the increase in atmospheric CO_2 concentration. Scientists have proposed strategically situating containers of these algae near industrial plants, as shown above, or near highly congested city streets. Why does this arrangement make sense?

CONCEPT 10.1

Photosynthesis converts light energy to the chemical energy of food

The remarkable ability of an organism to harness light energy and use it to drive the synthesis of organic compounds emerges from structural organization in the cell: Photosynthetic enzymes and other molecules are grouped together in a biological membrane, enabling the necessary series of chemical reactions to be carried out efficiently. The process of photosynthesis most likely originated in a group of bacteria that had infolded regions of the plasma membrane containing clusters of such molecules. In existing photosynthetic bacteria, infolded photosynthetic membranes function similarly to the internal membranes of the chloroplast, a eukaryotic organelle. According to the endosymbiont theory, the original chloroplast was a photosynthetic prokaryote that lived inside an ancestor of eukaryotic cells. (You learned about this theory in Chapter 6 and it will be described more fully in Chapter 25.) Chloroplasts are present in a variety of photosynthesizing organisms (see Figure 10.2), but here we will focus on plants.

Chloroplasts: The Sites of Photosynthesis in Plants

All green parts of a plant, including green stems and unripened fruit, have chloroplasts, but the leaves are the major sites of photosynthesis in most plants **(Figure 10.4)**. There are about half a million chloroplasts in a chunk of leaf with a top surface area of 1 mm^2. Chloroplasts are found mainly in the cells of the **mesophyll**, the tissue in the interior of the leaf. Carbon dioxide enters the leaf, and oxygen exits, by way of microscopic pores called **stomata** (singular, *stoma*; from the Greek, meaning "mouth"). Water absorbed by the roots is delivered to the leaves in veins. Leaves also use veins to export sugar to roots and other nonphotosynthetic parts of the plant.

A typical mesophyll cell has about 30–40 chloroplasts, each organelle measuring about 2–4 μm by 4–7 μm. A chloroplast has an envelope of two membranes surrounding a dense fluid called the **stroma**. Suspended within the stroma is a third membrane system, made up of sacs called **thylakoids**, which segregates the stroma from the *thylakoid space* inside these sacs. In some places, thylakoid sacs are stacked in columns called *grana* (singular, *granum*). **Chlorophyll**, the green pigment that gives leaves their color, resides in the thylakoid membranes of the chloroplast. (The internal photosynthetic membranes of some prokaryotes are also called thylakoid membranes; see Figure 27.7b.) It is the light energy absorbed

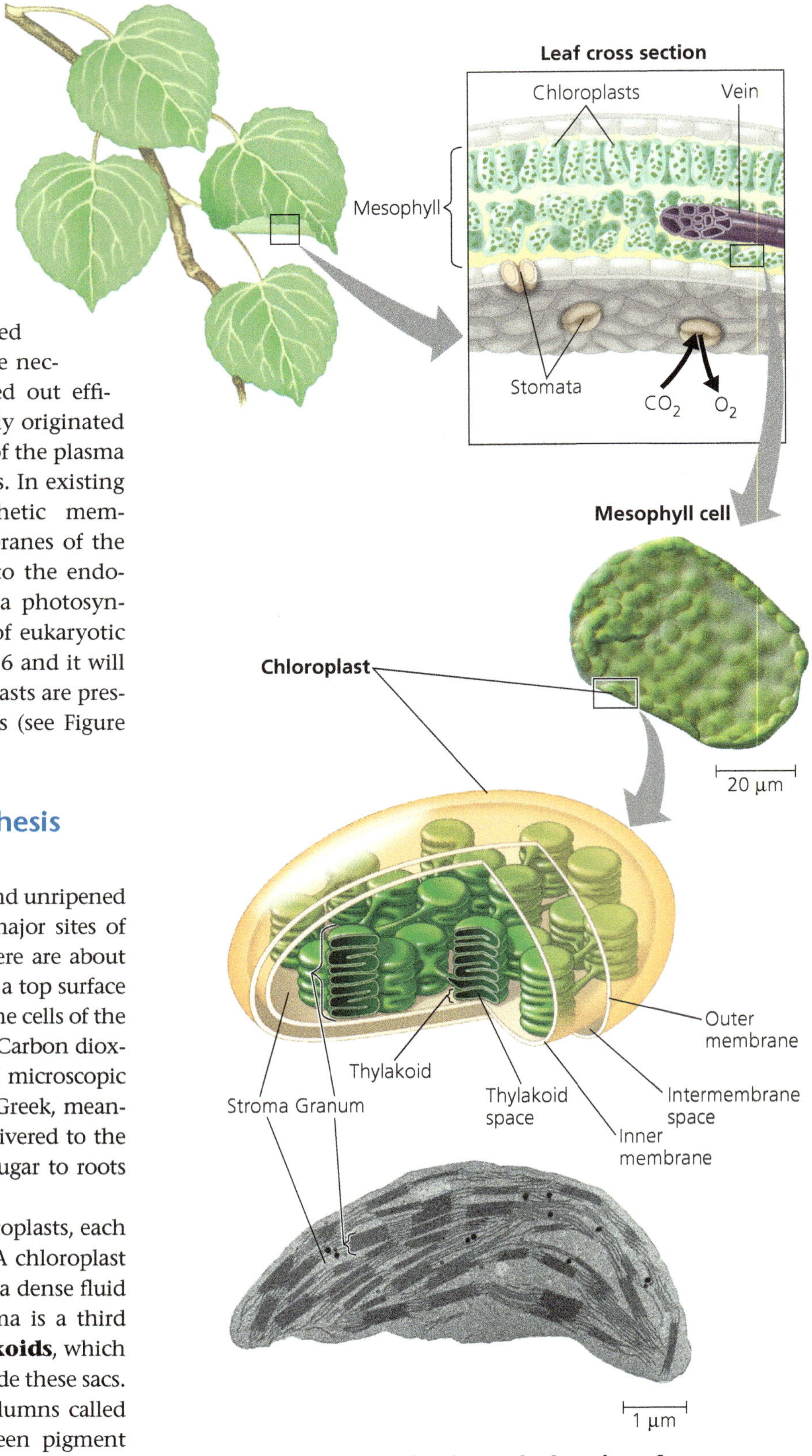

▲ **Figure 10.4 Zooming in on the location of photosynthesis in a plant.** Leaves are the major organs of photosynthesis in plants. These pictures take you into a leaf, then into a cell, and finally into a chloroplast, the organelle where photosynthesis occurs (middle, LM; bottom, TEM).

by chlorophyll that drives the synthesis of organic molecules in the chloroplast. Now that we have looked at the sites of photosynthesis in plants, we are ready to look more closely at the process of photosynthesis.

Tracking Atoms Through Photosynthesis: *Scientific Inquiry*

Scientists have tried for centuries to piece together the process by which plants make food. Although some of the steps are still not completely understood, the overall photosynthetic equation has been known since the 1800s: In the presence of light, the green parts of plants produce organic compounds and oxygen from carbon dioxide and water. Using molecular formulas, we can summarize the complex series of chemical reactions in photosynthesis with this chemical equation:

$$6\ CO_2 + 12\ H_2O + \text{Light energy} \rightarrow C_6H_{12}O_6 + 6\ O_2 + 6\ H_2O$$

We use glucose ($C_6H_{12}O_6$) here to simplify the relationship between photosynthesis and respiration, but the direct product of photosynthesis is actually a three-carbon sugar that can be used to make glucose. Water appears on both sides of the equation because 12 molecules are consumed and 6 molecules are newly formed during photosynthesis. We can simplify the equation by indicating only the net consumption of water:

$$6\ CO_2 + 6\ H_2O + \text{Light energy} \rightarrow C_6H_{12}O_6 + 6\ O_2$$

Writing the equation in this form, we can see that the overall chemical change during photosynthesis is the reverse of the one that occurs during cellular respiration. Both of these metabolic processes occur in plant cells. However, as you will soon learn, chloroplasts do not synthesize sugars by simply reversing the steps of respiration.

Now let's divide the photosynthetic equation by 6 to put it in its simplest possible form:

$$CO_2 + H_2O \rightarrow [CH_2O] + O_2$$

Here, the brackets indicate that CH_2O is not an actual sugar but represents the general formula for a carbohydrate. In other words, we are imagining the synthesis of a sugar molecule one carbon at a time. Six repetitions would theoretically produce a glucose molecule. Let's now use this simplified formula to see how researchers tracked the elements C, H, and O from the reactants of photosynthesis to the products.

The Splitting of Water

One of the first clues to the mechanism of photosynthesis came from the discovery that the O_2 given off by plants is derived from H_2O and not from CO_2. The chloroplast splits water into hydrogen and oxygen. Before this discovery, the prevailing hypothesis was that photosynthesis split carbon dioxide ($CO_2 \rightarrow C + O_2$) and then added water to the carbon ($C + H_2O \rightarrow [CH_2O]$). This hypothesis predicted that the O_2 released during photosynthesis came from CO_2. This idea was challenged in the 1930s by C. B. van Niel, of Stanford University. Van Niel was investigating photosynthesis in bacteria that make their carbohydrate from CO_2 but do not release O_2. He concluded that, at least in these bacteria, CO_2 is not split into carbon and oxygen. One group of bacteria used hydrogen sulfide (H_2S) rather than water for photosynthesis, forming yellow globules of sulfur as a waste product (these globules are visible in Figure 10.2e). Here is the chemical equation for photosynthesis in these sulfur bacteria:

$$CO_2 + 2\ H_2S \rightarrow [CH_2O] + H_2O + 2\ S$$

Van Niel reasoned that the bacteria split H_2S and used the hydrogen atoms to make sugar. He then generalized that idea, proposing that all photosynthetic organisms require a hydrogen source but that the source varies:

$$\begin{aligned} \text{Sulfur bacteria:}\quad & CO_2 + 2\ H_2S \rightarrow [CH_2O] + H_2O + 2\ S \\ \text{Plants:}\quad & CO_2 + 2\ H_2O \rightarrow [CH_2O] + H_2O + O_2 \\ \text{General:}\quad & CO_2 + 2\ H_2X \rightarrow [CH_2O] + H_2O + 2\ X \end{aligned}$$

Thus, van Niel hypothesized that plants split H_2O as a source of electrons from hydrogen atoms, releasing O_2 as a by-product.

Nearly 20 years later, scientists confirmed van Niel's hypothesis by using oxygen-18 (^{18}O), a heavy isotope, as a tracer to follow the fate of oxygen atoms during photosynthesis. The experiments showed that the O_2 from plants was labeled with ^{18}O *only* if water was the source of the tracer (experiment 1). If the ^{18}O was introduced to the plant in the form of CO_2, the label did not turn up in the released O_2 (experiment 2). In the following summary, red denotes labeled atoms of oxygen (^{18}O):

$$\begin{aligned} \text{Experiment 1:}\quad & CO_2 + 2\ H_2\color{red}{O} \rightarrow [CH_2O] + H_2O + \color{red}{O_2} \\ \text{Experiment 2:}\quad & C\color{red}{O_2} + 2\ H_2O \rightarrow [CH_2\color{red}{O}] + H_2\color{red}{O} + O_2 \end{aligned}$$

A significant result of the shuffling of atoms during photosynthesis is the extraction of hydrogen from water and its incorporation into sugar. The waste product of photosynthesis, O_2, is released to the atmosphere. **Figure 10.5** shows the fates of all atoms in photosynthesis.

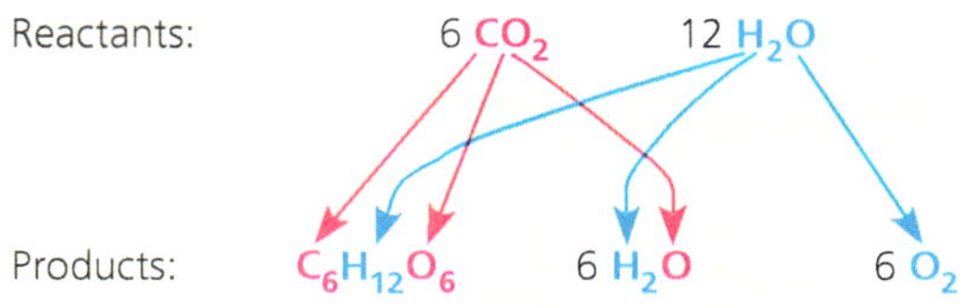

▲ **Figure 10.5 Tracking atoms through photosynthesis.** The atoms from CO_2 are shown in magenta, and the atoms from H_2O are shown in blue.

Photosynthesis as a Redox Process

Let's briefly compare photosynthesis with cellular respiration. Both processes involve redox reactions. During cellular respiration, energy is released from sugar when electrons associated with hydrogen are transported by carriers to oxygen, forming water as a by-product (see p. 164). The electrons lose potential energy as they "fall" down the electron transport chain toward electronegative oxygen, and the mitochondrion harnesses that energy to synthesize ATP (see Figure 9.15). Photosynthesis reverses the direction of electron flow. Water is split, and electrons are transferred along with hydrogen ions from the water to carbon dioxide, reducing it to sugar.

$$\text{Energy} + 6\,CO_2 + 6\,H_2O \longrightarrow C_6H_{12}O_6 + 6\,O_2$$

(becomes reduced: CO_2 → $C_6H_{12}O_6$; becomes oxidized: H_2O → O_2)

Because the electrons increase in potential energy as they move from water to sugar, this process requires energy—in other words is endergonic. This energy boost is provided by light.

The Two Stages of Photosynthesis: *A Preview*

The equation for photosynthesis is a deceptively simple summary of a very complex process. Actually, photosynthesis is not a single process, but two processes, each with multiple steps. These two stages of photosynthesis are known as the **light reactions** (the *photo* part of photosynthesis) and the **Calvin cycle** (the *synthesis* part) **(Figure 10.6)**.

The light reactions are the steps of photosynthesis that convert solar energy to chemical energy. Water is split, providing a source of electrons and protons (hydrogen ions, H^+) and giving off O_2 as a by-product. Light absorbed by chlorophyll drives a transfer of the electrons and hydrogen ions from water to an acceptor called **$NADP^+$** (nicotinamide adenine dinucleotide phosphate), where they are temporarily stored. The electron acceptor $NADP^+$ is first cousin to NAD^+, which functions as an electron carrier in cellular respiration; the two molecules differ only by the presence of an extra phosphate group in the $NADP^+$ molecule. The light reactions use solar power to reduce $NADP^+$ to NADPH by adding a pair of electrons along with an H^+. The light reactions also generate ATP, using chemiosmosis to power the addition of a phosphate group to ADP, a process called **photophosphorylation**. Thus, light energy is initially converted to chemical energy in the form of two compounds: NADPH, a source of electrons as "reducing power" that can be passed along to an electron acceptor, reducing it, and ATP, the versatile energy currency of cells. Notice that the light reactions produce no sugar; that happens in the second stage of photosynthesis, the Calvin cycle.

The Calvin cycle is named for Melvin Calvin, who, along with his colleagues, began to elucidate its steps in the late 1940s. The cycle begins by incorporating CO_2 from the air

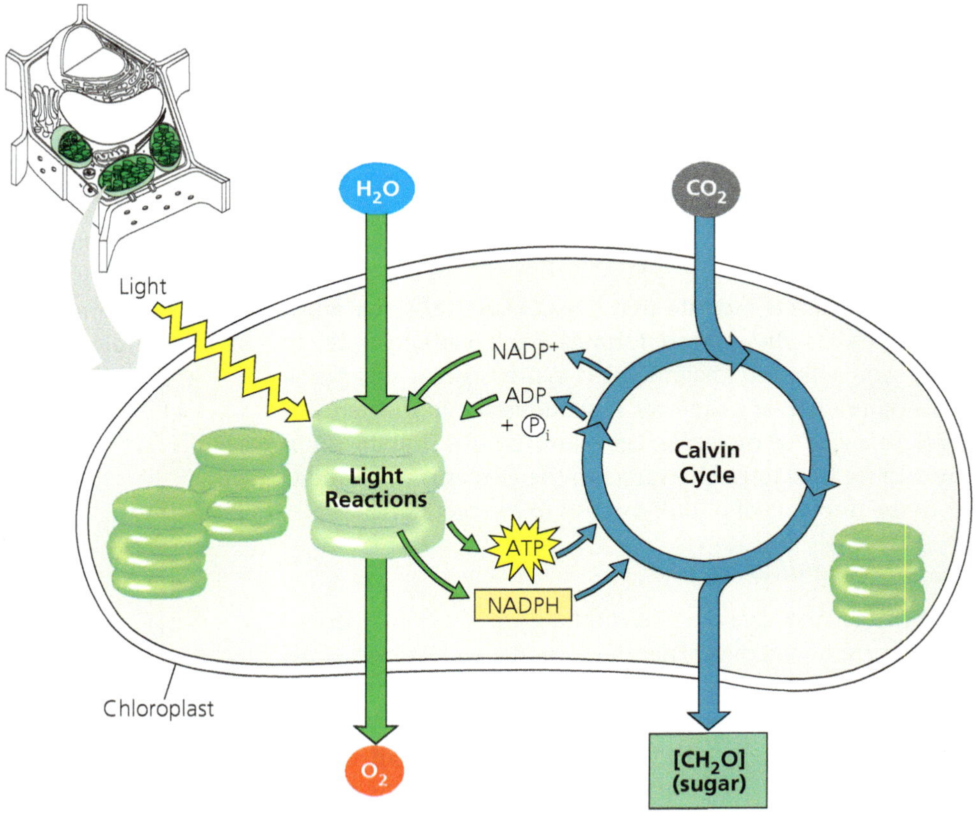

▶ **Figure 10.6 An overview of photosynthesis: cooperation of the light reactions and the Calvin cycle.** In the chloroplast, the thylakoid membranes are the sites of the light reactions, whereas the Calvin cycle occurs in the stroma. The light reactions use solar energy to make ATP and NADPH, which supply chemical energy and reducing power, respectively, to the Calvin cycle. The Calvin cycle incorporates CO_2 into organic molecules, which are converted to sugar. (Recall that most simple sugars have formulas that are some multiple of CH_2O.)

ANIMATION **BioFlix** Visit the Study Area at **www.masteringbiology.com** for the BioFlix® 3-D Animation on Photosynthesis.

into organic molecules already present in the chloroplast. This initial incorporation of carbon into organic compounds is known as **carbon fixation**. The Calvin cycle then reduces the fixed carbon to carbohydrate by the addition of electrons. The reducing power is provided by NADPH, which acquired its cargo of electrons in the light reactions. To convert CO_2 to carbohydrate, the Calvin cycle also requires chemical energy in the form of ATP, which is also generated by the light reactions. Thus, it is the Calvin cycle that makes sugar, but it can do so only with the help of the NADPH and ATP produced by the light reactions. The metabolic steps of the Calvin cycle are sometimes referred to as the dark reactions, or light-independent reactions, because none of the steps requires light *directly*. Nevertheless, the Calvin cycle in most plants occurs during daylight, for only then can the light reactions provide the NADPH and ATP that the Calvin cycle requires. In essence, the chloroplast uses light energy to make sugar by coordinating the two stages of photosynthesis.

As Figure 10.6 indicates, the thylakoids of the chloroplast are the sites of the light reactions, while the Calvin cycle occurs in the stroma. On the outside of the thylakoids, molecules of $NADP^+$ and ADP pick up electrons and phosphate, respectively, and NADPH and ATP are then released to the stroma, where they play crucial roles in the Calvin cycle. The two stages of photosynthesis are treated in this figure as metabolic modules that take in ingredients and crank out products. In the next two sections, we'll look more closely at how the two stages work, beginning with the light reactions.

CONCEPT CHECK 10.1

1. How do the reactant molecules of photosynthesis reach the chloroplasts in leaves?
2. How did the use of an oxygen isotope help elucidate the chemistry of photosynthesis?
3. **WHAT IF?** The Calvin cycle requires ATP and NADPH, products of the light reactions. If a classmate asserted that the light reactions don't depend on the Calvin cycle and, with continual light, could just keep on producing ATP and NADPH, how would you respond?

For suggested answers, see Appendix A.

CONCEPT 10.2

The light reactions convert solar energy to the chemical energy of ATP and NADPH

Chloroplasts are chemical factories powered by the sun. Their thylakoids transform light energy into the chemical energy of ATP and NADPH. To understand this conversion better, we need to know about some important properties of light.

The Nature of Sunlight

Light is a form of energy known as electromagnetic energy, also called electromagnetic radiation. Electromagnetic energy travels in rhythmic waves analogous to those created by dropping a pebble into a pond. Electromagnetic waves, however, are disturbances of electric and magnetic fields rather than disturbances of a material medium such as water.

The distance between the crests of electromagnetic waves is called the **wavelength**. Wavelengths range from less than a nanometer (for gamma rays) to more than a kilometer (for radio waves). This entire range of radiation is known as the **electromagnetic spectrum** (Figure 10.7). The segment most important to life is the narrow band from about 380 nm to 750 nm in wavelength. This radiation is known as **visible light** because it can be detected as various colors by the human eye.

The model of light as waves explains many of light's properties, but in certain respects light behaves as though it consists of discrete particles, called **photons**. Photons are not tangible objects, but they act like objects in that each of them has a fixed quantity of energy. The amount of energy is inversely related to the wavelength of the light: the shorter the wavelength, the greater the energy of each photon of that light. Thus, a photon of violet light packs nearly twice as much energy as a photon of red light.

Although the sun radiates the full spectrum of electromagnetic energy, the atmosphere acts like a selective window, allowing visible light to pass through while screening out a substantial fraction of other radiation. The part of the spectrum we can see—visible light—is also the radiation that drives photosynthesis.

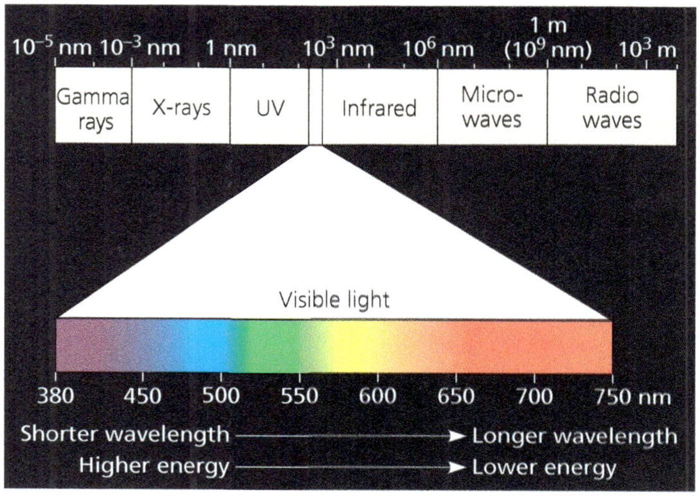

▲ **Figure 10.7 The electromagnetic spectrum.** White light is a mixture of all wavelengths of visible light. A prism can sort white light into its component colors by bending light of different wavelengths at different angles. (Droplets of water in the atmosphere can act as prisms, forming a rainbow; see Figure 10.1.) Visible light drives photosynthesis.

Photosynthetic Pigments: The Light Receptors

When light meets matter, it may be reflected, transmitted, or absorbed. Substances that absorb visible light are known as *pigments*. Different pigments absorb light of different wavelengths, and the wavelengths that are absorbed disappear. If a pigment is illuminated with white light, the color we see is the color most reflected or transmitted by the pigment. (If a pigment absorbs all wavelengths, it appears black.) We see green when we look at a leaf because chlorophyll absorbs violet-blue and red light while transmitting and reflecting green light **(Figure 10.8)**. The ability of a pigment to absorb various wavelengths of light can be measured with an instrument called a **spectrophotometer**. This machine directs beams of light of different wavelengths through a solution of the pigment and measures the fraction of the light transmitted at each wavelength. A graph plotting a pigment's light absorption versus wavelength is called an **absorption spectrum** **(Figure 10.9)**.

The absorption spectra of chloroplast pigments provide clues to the relative effectiveness of different wavelengths for driving photosynthesis, since light can perform work in chloroplasts only if it is absorbed. **Figure 10.10a** shows the absorption spectra of three types of pigments in chloroplasts: **chlorophyll *a***, which participates directly in the light reactions; the accessory pigment *chlorophyll b*; and a group of accessory pigments called carotenoids. The spectrum of chlorophyll *a* suggests that violet-blue and red light work best for photosynthesis, since they are absorbed, while green is the least effective color. This is confirmed by an **action spectrum** for photosynthesis **(Figure 10.10b)**, which profiles the relative effectiveness of different wavelengths of radiation in driving the process. An action spectrum is prepared by illuminating chloroplasts with light of different colors and then plotting wavelength against some measure of photosynthetic rate, such as CO_2 consumption or O_2 release. The action spectrum for photosynthesis was first demonstrated by Theodor W. Engelmann, a German botanist, in 1883. Before equipment for measuring O_2 levels had even been invented, Engelmann performed a

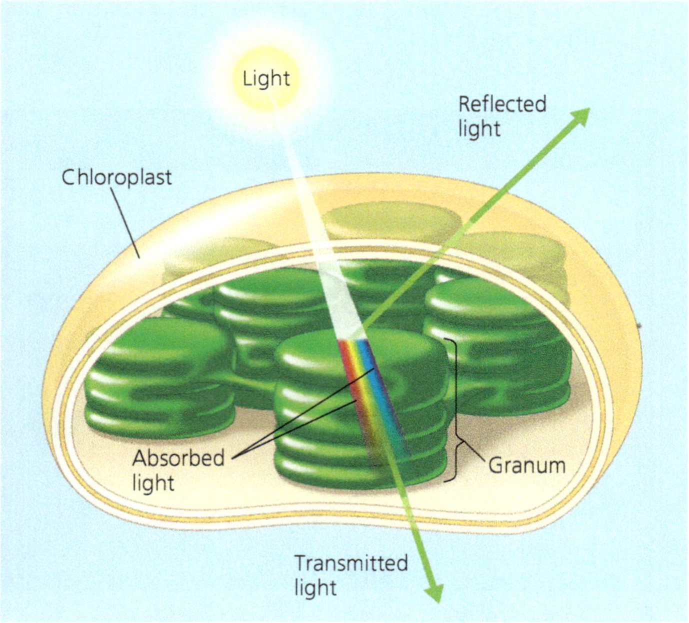

▲ **Figure 10.8 Why leaves are green: interaction of light with chloroplasts.** The chlorophyll molecules of chloroplasts absorb violet-blue and red light (the colors most effective in driving photosynthesis) and reflect or transmit green light. This is why leaves appear green.

▼ **Figure 10.9** **RESEARCH METHOD**

Determining an Absorption Spectrum

APPLICATION An absorption spectrum is a visual representation of how well a particular pigment absorbs different wavelengths of visible light. Absorption spectra of various chloroplast pigments help scientists decipher each pigment's role in a plant.

TECHNIQUE A spectrophotometer measures the relative amounts of light of different wavelengths absorbed and transmitted by a pigment solution.

1. White light is separated into colors (wavelengths) by a prism.
2. One by one, the different colors of light are passed through the sample (chlorophyll in this example). Green light and blue light are shown here.
3. The transmitted light strikes a photoelectric tube, which converts the light energy to electricity.
4. The electric current is measured by a galvanometer. The meter indicates the fraction of light transmitted through the sample, from which we can determine the amount of light absorbed.

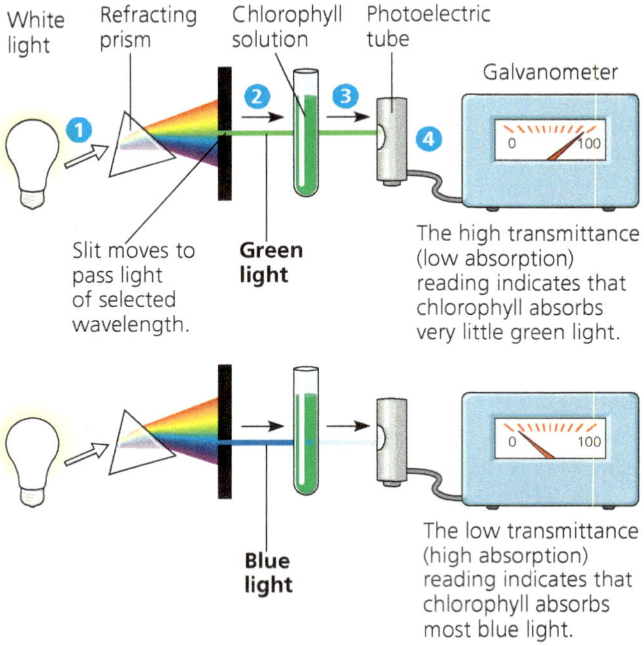

RESULTS See Figure 10.10a for absorption spectra of three types of chloroplast pigments.

▼ **Figure 10.10** INQUIRY

Which wavelengths of light are most effective in driving photosynthesis?

EXPERIMENT Absorption and action spectra, along with a classic experiment by Theodor W. Engelmann, reveal which wavelengths of light are photosynthetically important.

RESULTS

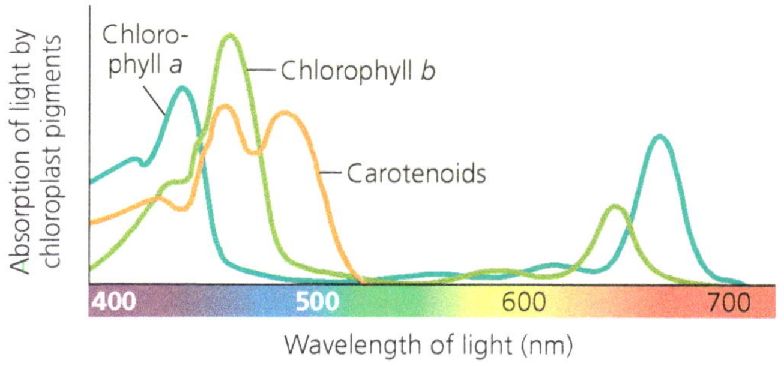

(a) Absorption spectra. The three curves show the wavelengths of light best absorbed by three types of chloroplast pigments.

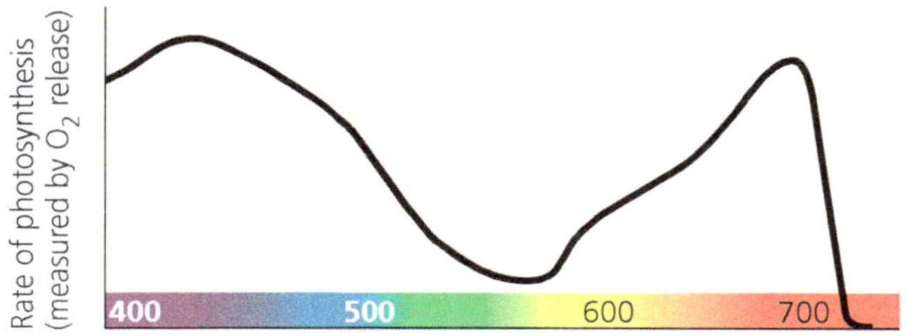

(b) Action spectrum. This graph plots the rate of photosynthesis versus wavelength. The resulting action spectrum resembles the absorption spectrum for chlorophyll *a* but does not match exactly (see part a). This is partly due to the absorption of light by accessory pigments such as chlorophyll *b* and carotenoids.

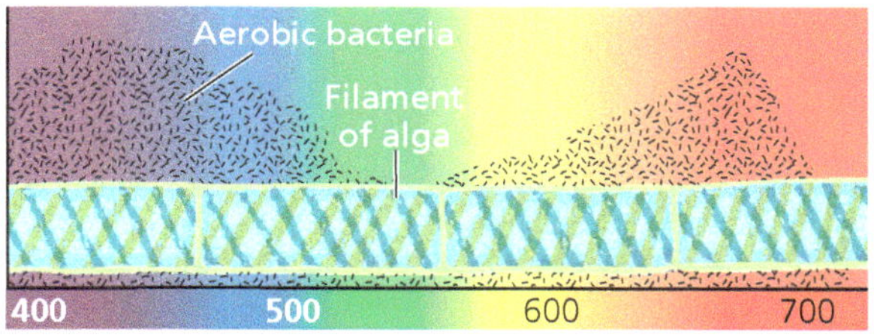

(c) Engelmann's experiment. In 1883, Theodor W. Engelmann illuminated a filamentous alga with light that had been passed through a prism, exposing different segments of the alga to different wavelengths. He used aerobic bacteria, which concentrate near an oxygen source, to determine which segments of the alga were releasing the most O_2 and thus photosynthesizing most. Bacteria congregated in greatest numbers around the parts of the alga illuminated with violet-blue or red light.

CONCLUSION Light in the violet-blue and red portions of the spectrum is most effective in driving photosynthesis.

SOURCE T. W. Engelmann, *Bacterium photometricum*. Ein Betrag zur vergleichenden Physiologie des Licht- und farbensinnes, *Archiv. für Physiologie* 30:95–124 (1883).

MB See the related Experimental Inquiry Tutorial in MasteringBiology.

WHAT IF? If Engelmann had used a filter that allowed only red light to pass through, how would the results have differed?

clever experiment in which he used bacteria to measure rates of photosynthesis in filamentous algae (**Figure 10.10c**). His results are a striking match to the modern action spectrum shown in Figure 10.10b.

Notice by comparing Figures 10.10a and 10.10b that the action spectrum for photosynthesis does not exactly match the absorption spectrum of chlorophyll *a*. The absorption spectrum of chlorophyll *a* alone underestimates the effectiveness of certain wavelengths in driving photosynthesis. This is partly because accessory pigments with different absorption spectra are also photosynthetically important in chloroplasts and broaden the spectrum of colors that can be used for photosynthesis. **Figure 10.11** shows the structure of chlorophyll *a* compared with that of **chlorophyll *b***. A slight structural difference between them is enough to cause the two pigments to absorb at slightly different wavelengths in the red and blue parts of the spectrum (see Figure 10.10a). As a result, chlorophyll *a* is blue green and chlorophyll *b* is olive green.

Other accessory pigments include **carotenoids**, hydrocarbons that are various shades of yellow and orange because they absorb violet and blue-green light (see Figure 10.10a). Carotenoids may broaden the spectrum of colors that can drive photosynthesis. However, a more important function of at least some carotenoids seems to be *photoprotection*: These

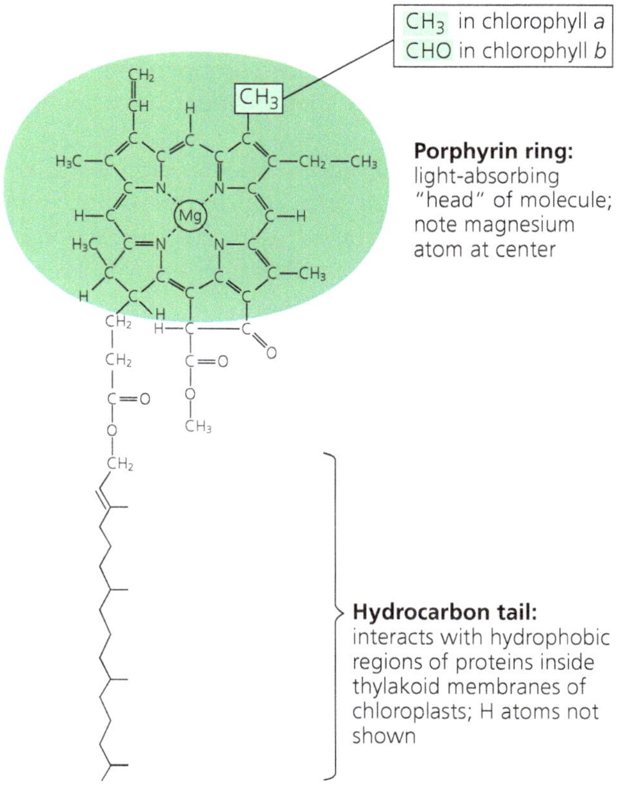

▲ **Figure 10.11 Structure of chlorophyll molecules in chloroplasts of plants.** Chlorophyll *a* and chlorophyll *b* differ only in one of the functional groups bonded to the porphyrin ring. (Also see the space-filling model of chlorophyll in Figure 1.4, p. 5.)

compounds absorb and dissipate excessive light energy that would otherwise damage chlorophyll or interact with oxygen, forming reactive oxidative molecules that are dangerous to the cell. Interestingly, carotenoids similar to the photoprotective ones in chloroplasts have a photoprotective role in the human eye. These and related molecules, often found in health food products, are valued as "phytochemicals" (from the Greek *phyton*, plant), compounds with antioxidant properties. Plants can synthesize all the antioxidants they require, but humans and other animals must obtain some of them from their diets.

Excitation of Chlorophyll by Light

What exactly happens when chlorophyll and other pigments absorb light? The colors corresponding to the absorbed wavelengths disappear from the spectrum of the transmitted and reflected light, but energy cannot disappear. When a molecule absorbs a photon of light, one of the molecule's electrons is elevated to an orbital where it has more potential energy. When the electron is in its normal orbital, the pigment molecule is said to be in its ground state. Absorption of a photon boosts an electron to an orbital of higher energy, and the pigment molecule is then said to be in an excited state. The only photons absorbed are those whose energy is exactly equal to the energy difference between the ground state and an excited state, and this energy difference varies from one kind of molecule to another. Thus, a particular compound absorbs only photons corresponding to specific wavelengths, which is why each pigment has a unique absorption spectrum.

Once absorption of a photon raises an electron from the ground state to an excited state, the electron cannot remain there long. The excited state, like all high-energy states, is unstable. Generally, when isolated pigment molecules absorb light, their excited electrons drop back down to the ground-state orbital in a billionth of a second, releasing their excess energy as heat. This conversion of light energy to heat is what makes the top of an automobile so hot on a sunny day. (White cars are coolest because their paint reflects all wavelengths of visible light, although it may absorb ultraviolet and other invisible radiation.) In isolation, some pigments, including chlorophyll, emit light as well as heat after absorbing photons. As excited electrons fall back to the ground state, photons are given off. This afterglow is called fluorescence. If a solution of chlorophyll isolated from chloroplasts is illuminated, it will fluoresce in the red-orange part of the spectrum and also give off heat **(Figure 10.12)**.

A Photosystem: A Reaction-Center Complex Associated with Light-Harvesting Complexes

Chlorophyll molecules excited by the absorption of light energy produce very different results in an intact chloroplast than they do in isolation (see Figure 10.12). In their native environment of the thylakoid membrane, chlorophyll molecules are organized along with other small organic molecules and proteins into complexes called photosystems.

A **photosystem** is composed of a **reaction-center complex** surrounded by several light-harvesting complexes **(Figure 10.13)**. The reaction-center complex is an organized association of proteins holding a special pair of chlorophyll *a* molecules. Each **light-harvesting complex** consists of various pigment molecules (which may include chlorophyll *a*, chlorophyll *b*, and carotenoids) bound to proteins. The number and variety of pigment molecules enable a photosystem to harvest light over a larger surface area and a larger portion of the spectrum than could any single pigment molecule alone. Together, these light-harvesting complexes act as an antenna for the reaction-center complex. When a pigment molecule

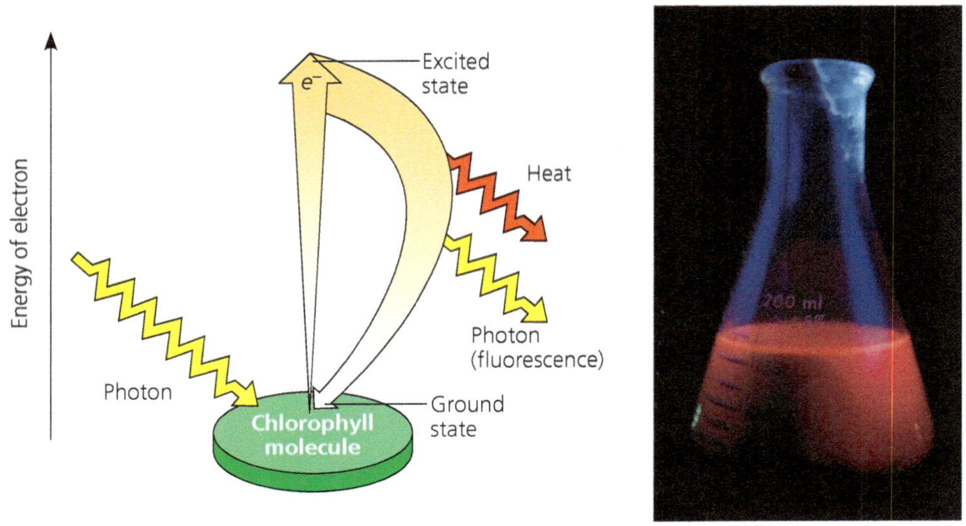

(a) Excitation of isolated chlorophyll molecule

(b) Fluorescence

▶ **Figure 10.12 Excitation of isolated chlorophyll by light. (a)** Absorption of a photon causes a transition of the chlorophyll molecule from its ground state to its excited state. The photon boosts an electron to an orbital where it has more potential energy. If the illuminated molecule exists in isolation, its excited electron immediately drops back down to the ground-state orbital, and its excess energy is given off as heat and fluorescence (light). **(b)** A chlorophyll solution excited with ultraviolet light fluoresces with a red-orange glow.

WHAT IF? *If a leaf containing a similar concentration of chlorophyll as the solution was exposed to the same ultraviolet light, no fluorescence would be seen. Explain the difference in fluorescence emission between the solution and the leaf.*

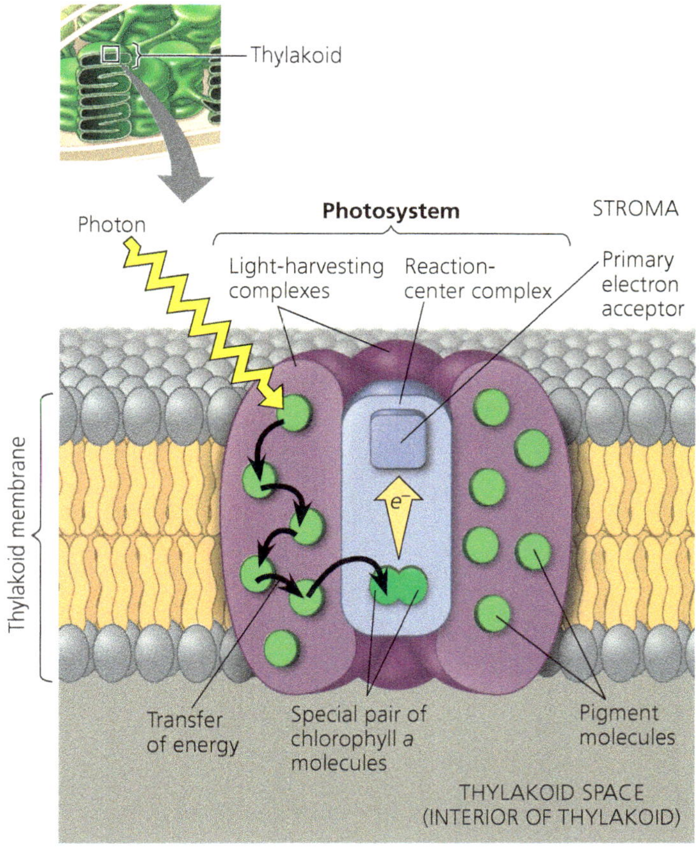

(a) How a photosystem harvests light. When a photon strikes a pigment molecule in a light-harvesting complex, the energy is passed from molecule to molecule until it reaches the reaction-center complex. Here, an excited electron from the special pair of chlorophyll *a* molecules is transferred to the primary electron acceptor.

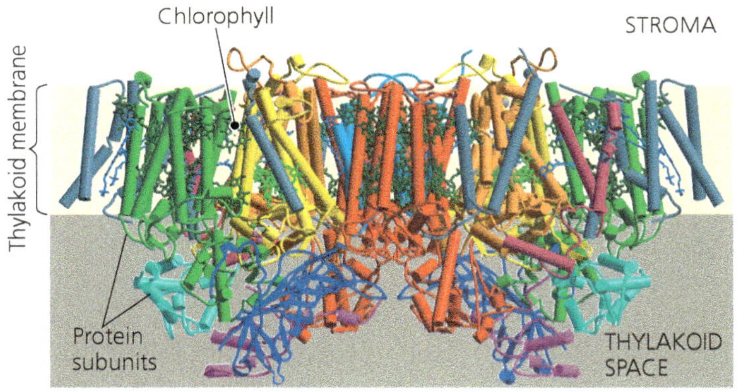

(b) Structure of photosystem II. This computer model of photosystem II, based on X-ray crystallography, shows two photosystem complexes side by side. Chlorophyll molecules (small green ball-and-stick models) are interspersed with protein subunits (cylinders and ribbons). For simplicity, photosystem II will be shown as a single complex in the rest of the chapter.

▲ **Figure 10.13 The structure and function of a photosystem.**

absorbs a photon, the energy is transferred from pigment molecule to pigment molecule within a light-harvesting complex, somewhat like a human "wave" at a sports arena, until it is passed into the reaction-center complex. The reaction-center complex also contains a molecule capable of accepting electrons and becoming reduced; this is called the **primary electron acceptor**. The pair of chlorophyll *a* molecules in the reaction-center complex are special because their molecular environment—their location and the other molecules with which they are associated—enables them to use the energy from light not only to boost one of their electrons to a higher energy level, but also to transfer it to a different molecule—the primary electron acceptor.

The solar-powered transfer of an electron from the reaction-center chlorophyll *a* pair to the primary electron acceptor is the first step of the light reactions. As soon as the chlorophyll electron is excited to a higher energy level, the primary electron acceptor captures it; this is a redox reaction. In the flask shown in Figure 10.12, isolated chlorophyll fluoresces because there is no electron acceptor, so electrons of photoexcited chlorophyll drop right back to the ground state. In the structured environment of a chloroplast, however, an electron acceptor is readily available, and the potential energy represented by the excited electron is not dissipated as light and heat. Thus, each photosystem—a reaction-center complex surrounded by light-harvesting complexes—functions in the chloroplast as a unit. It converts light energy to chemical energy, which will ultimately be used for the synthesis of sugar.

The thylakoid membrane is populated by two types of photosystems that cooperate in the light reactions of photosynthesis. They are called **photosystem II (PS II)** and **photosystem I (PS I)**. (They were named in order of their discovery, but photosystem II functions first in the light reactions.) Each has a characteristic reaction-center complex—a particular kind of primary electron acceptor next to a special pair of chlorophyll *a* molecules associated with specific proteins. The reaction-center chlorophyll *a* of photosystem II is known as P680 because this pigment is best at absorbing light having a wavelength of 680 nm (in the red part of the spectrum). The chlorophyll *a* at the reaction-center complex of photosystem I is called P700 because it most effectively absorbs light of wavelength 700 nm (in the far-red part of the spectrum). These two pigments, P680 and P700, are nearly identical chlorophyll *a* molecules. However, their association with different proteins in the thylakoid membrane affects the electron distribution in the two pigments and accounts for the slight differences in their light-absorbing properties. Now let's see how the two photosystems work together in using light energy to generate ATP and NADPH, the two main products of the light reactions.

Linear Electron Flow

Light drives the synthesis of ATP and NADPH by energizing the two photosystems embedded in the thylakoid membranes of chloroplasts. The key to this energy transformation is a flow of electrons through the photosystems and other molecular components built into the thylakoid membrane. This is called

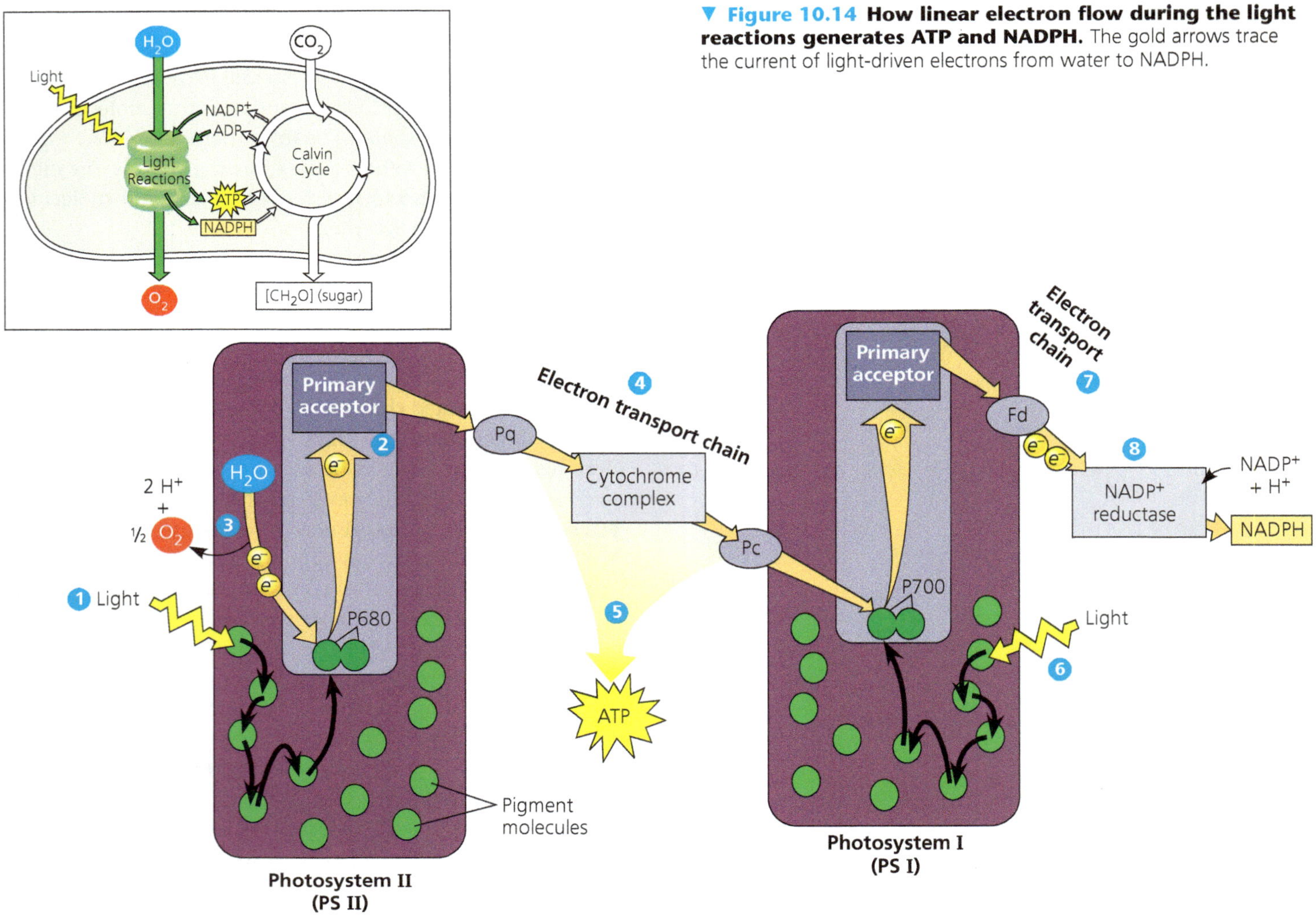

▼ **Figure 10.14 How linear electron flow during the light reactions generates ATP and NADPH.** The gold arrows trace the current of light-driven electrons from water to NADPH.

linear electron flow, and it occurs during the light reactions of photosynthesis, as shown in **Figure 10.14**. The following steps correspond to the numbered steps in the figure.

1. A photon of light strikes a pigment molecule in a light-harvesting complex of PS II, boosting one of its electrons to a higher energy level. As this electron falls back to its ground state, an electron in a nearby pigment molecule is simultaneously raised to an excited state. The process continues, with the energy being relayed to other pigment molecules until it reaches the P680 pair of chlorophyll *a* molecules in the PS II reaction-center complex. It excites an electron in this pair of chlorophylls to a higher energy state.
2. This electron is transferred from the excited P680 to the primary electron acceptor. We can refer to the resulting form of P680, missing an electron, as $P680^+$.
3. An enzyme catalyzes the splitting of a water molecule into two electrons, two hydrogen ions (H^+), and an oxygen atom. The electrons are supplied one by one to the $P680^+$ pair, each electron replacing one transferred to the primary electron acceptor. ($P680^+$ is the strongest biological oxidizing agent known; its electron "hole" must be filled. This greatly facilitates the transfer of electrons from the split water molecule.) The H^+ are released into the thylakoid lumen. The oxygen atom immediately combines with an oxygen atom generated by the splitting of another water molecule, forming O_2.
4. Each photoexcited electron passes from the primary electron acceptor of PS II to PS I via an electron transport chain, the components of which are similar to those of the electron transport chain that functions in cellular respiration. The electron transport chain between PS II and PS I is made up of the electron carrier plastoquinone (Pq), a cytochrome complex, and a protein called plastocyanin (Pc).
5. The exergonic "fall" of electrons to a lower energy level provides energy for the synthesis of ATP. As electrons pass through the cytochrome complex, H^+ are pumped into the thylakoid lumen, contributing to the proton gradient that is subsequently used in chemiosmosis.
6. Meanwhile, light energy has been transferred via light-harvesting complex pigments to the PS I reaction-center complex, exciting an electron of the P700 pair of chlorophyll *a* molecules located there. The photoexcited electron was then transferred to PS I's primary electron acceptor, creating an electron "hole" in the P700—which

we now can call P700$^+$. In other words, P700$^+$ can now act as an electron acceptor, accepting an electron that reaches the bottom of the electron transport chain from PS II.

7. Photoexcited electrons are passed in a series of redox reactions from the primary electron acceptor of PS I down a second electron transport chain through the protein ferredoxin (Fd). (This chain does not create a proton gradient and thus does not produce ATP.)

8. The enzyme NADP$^+$ reductase catalyzes the transfer of electrons from Fd to NADP$^+$. Two electrons are required for its reduction to NADPH. This molecule is at a higher energy level than water, and its electrons are more readily available for the reactions of the Calvin cycle than were those of water. This process also removes an H$^+$ from the stroma.

As complicated as the scheme shown in Figure 10.14 is, do not lose track of its functions. The light reactions use solar power to generate ATP and NADPH, which provide chemical energy and reducing power, respectively, to the carbohydrate-synthesizing reactions of the Calvin cycle. The energy changes of electrons during their linear flow through the light reactions are shown in a mechanical analogy in **Figure 10.15**.

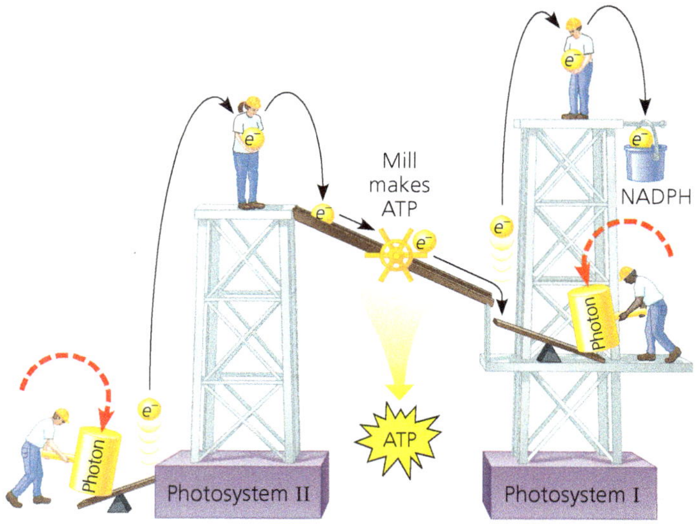

▲ **Figure 10.15 A mechanical analogy for linear electron flow during the light reactions.**

Cyclic Electron Flow

In certain cases, photoexcited electrons can take an alternative path called **cyclic electron flow**, which uses photosystem I but not photosystem II. You can see in **Figure 10.16** that cyclic flow is a short circuit: The electrons cycle back from ferredoxin (Fd) to the cytochrome complex and from there continue on to a P700 chlorophyll in the PS I reaction-center complex. There is no production of NADPH and no release of oxygen. Cyclic flow does, however, generate ATP.

Several of the currently existing groups of photosynthetic bacteria are known to have photosystem I but not photosystem II; for these species, which include the purple sulfur bacteria (see Figure 10.2e), cyclic electron flow is the sole means of generating ATP in photosynthesis. Evolutionary biologists hypothesize that these bacterial groups are descendants of the bacteria in which photosynthesis first evolved, in a form similar to cyclic electron flow.

Cyclic electron flow can also occur in photosynthetic species that possess both photosystems; this includes some prokaryotes, such as the cyanobacteria shown in Figure 10.2d, as well as the eukaryotic photosynthetic species that have been tested so far. Although the process is probably in part an "evolutionary leftover," it clearly plays at least one beneficial role for these organisms. Mutant plants that are not able to carry out cyclic electron flow are capable of growing well in low light, but do not grow well where light is intense. This is evidence for the idea that cyclic electron flow may be photoprotective. Later you'll learn more about cyclic electron flow as it relates to a particular adaptation of photosynthesis (C_4 plants; see Concept 10.4).

Whether ATP synthesis is driven by linear or cyclic electron flow, the actual mechanism is the same. Before we move on to consider the Calvin cycle, let's review chemiosmosis, the process that uses membranes to couple redox reactions to ATP production.

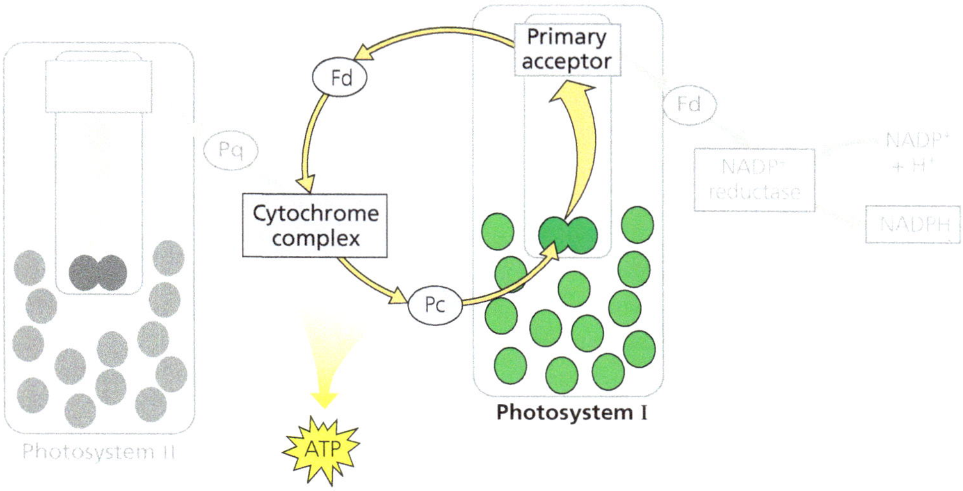

◄ **Figure 10.16 Cyclic electron flow.** Photoexcited electrons from PS I are occasionally shunted back from ferredoxin (Fd) to chlorophyll via the cytochrome complex and plastocyanin (Pc). This electron shunt supplements the supply of ATP (via chemiosmosis) but produces no NADPH. The "shadow" of linear electron flow is included in the diagram for comparison with the cyclic route. The two ferredoxin molecules shown in this diagram are actually one and the same—the final electron carrier in the electron transport chain of PS I.

? *Look at Figure 10.15, and explain how you would alter it to show a mechanical analogy for cyclic electron flow.*

A Comparison of Chemiosmosis in Chloroplasts and Mitochondria

Chloroplasts and mitochondria generate ATP by the same basic mechanism: chemiosmosis. An electron transport chain assembled in a membrane pumps protons across the membrane as electrons are passed through a series of carriers that are progressively more electronegative. In this way, electron transport chains transform redox energy to a proton-motive force, potential energy stored in the form of an H^+ gradient across a membrane. Built into the same membrane is an ATP synthase complex that couples the diffusion of hydrogen ions down their gradient to the phosphorylation of ADP. Some of the electron carriers, including the iron-containing proteins called cytochromes, are very similar in chloroplasts and mitochondria. The ATP synthase complexes of the two organelles are also very much alike. But there are noteworthy differences between oxidative phosphorylation in mitochondria and photophosphorylation in chloroplasts. In mitochondria, the high-energy electrons dropped down the transport chain are extracted from organic molecules (which are thus oxidized), while in chloroplasts, the source of electrons is water. Chloroplasts do not need molecules from food to make ATP; their photosystems capture light energy and use it to drive the electrons from water to the top of the transport chain. In other words, mitochondria use chemiosmosis to transfer chemical energy from food molecules to ATP, whereas chloroplasts transform light energy into chemical energy in ATP.

Although the spatial organization of chemiosmosis differs slightly between chloroplasts and mitochondria, it is easy to see similarities in the two **(Figure 10.17)**. The inner membrane of the mitochondrion pumps protons from the mitochondrial matrix out to the intermembrane space, which then serves as a reservoir of hydrogen ions. The thylakoid membrane of the chloroplast pumps protons from the stroma into the thylakoid space (interior of the thylakoid), which functions as the H^+ reservoir. If you imagine the cristae of mitochondria pinching off from the inner membrane, this may help you see how the thylakoid space and the intermembrane space are comparable spaces in the two organelles, while the mitochondrial matrix is analogous to the stroma of the chloroplast. In the mitochondrion, protons diffuse down their concentration gradient from the intermembrane space through ATP synthase to the matrix, driving ATP synthesis. In the chloroplast, ATP is synthesized as the hydrogen ions diffuse from the thylakoid space back to the stroma through ATP synthase complexes, whose catalytic knobs are on the stroma side of the membrane. Thus, ATP forms in the stroma, where it is used to help drive sugar synthesis during the Calvin cycle **(Figure 10.18)**.

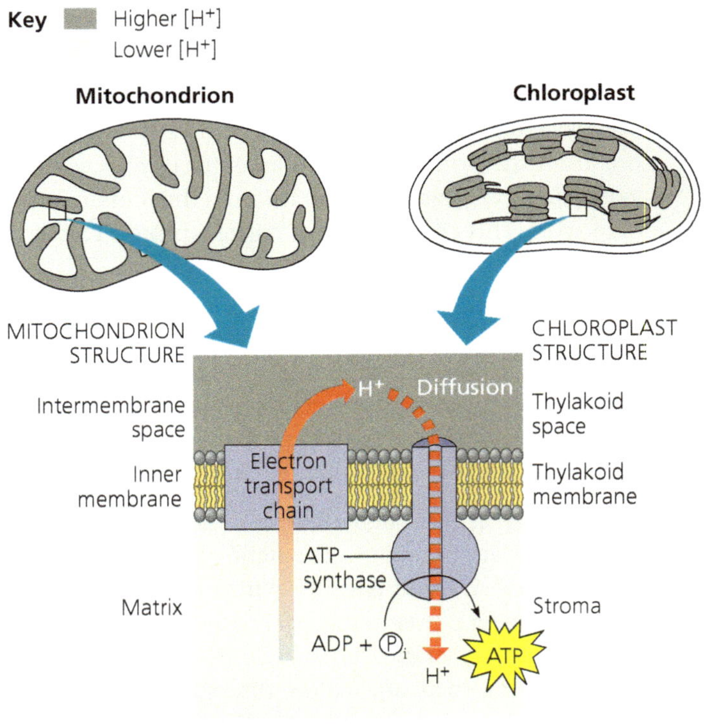

▲ **Figure 10.17 Comparison of chemiosmosis in mitochondria and chloroplasts.** In both kinds of organelles, electron transport chains pump protons (H^+) across a membrane from a region of low H^+ concentration (light gray in this diagram) to one of high H^+ concentration (dark gray). The protons then diffuse back across the membrane through ATP synthase, driving the synthesis of ATP.

The proton (H^+) gradient, or pH gradient, across the thylakoid membrane is substantial. When chloroplasts in an experimental setting are illuminated, the pH in the thylakoid space drops to about 5 (the H^+ concentration increases), and the pH in the stroma increases to about 8 (the H^+ concentration decreases). This gradient of three pH units corresponds to a thousandfold difference in H^+ concentration. If in the laboratory the lights are turned off, the pH gradient is abolished, but it can quickly be restored by turning the lights back on. Experiments such as this provided strong evidence in support of the chemiosmotic model.

Based on studies in several laboratories, Figure 10.18 shows a current model for the organization of the light-reaction "machinery" within the thylakoid membrane. Each of the molecules and molecular complexes in the figure is present in numerous copies in each thylakoid. Notice that NADPH, like ATP, is produced on the side of the membrane facing the stroma, where the Calvin cycle reactions take place.

Let's summarize the light reactions. Electron flow pushes electrons from water, where they are at a low state of potential energy, ultimately to NADPH, where they are stored at a high state of potential energy. The light-driven electron current also generates ATP. Thus, the equipment of the thylakoid membrane converts light energy to chemical energy stored in ATP and NADPH. (Oxygen is a by-product.) Let's now see how the Calvin cycle uses the products of the light reactions to synthesize sugar from CO_2.

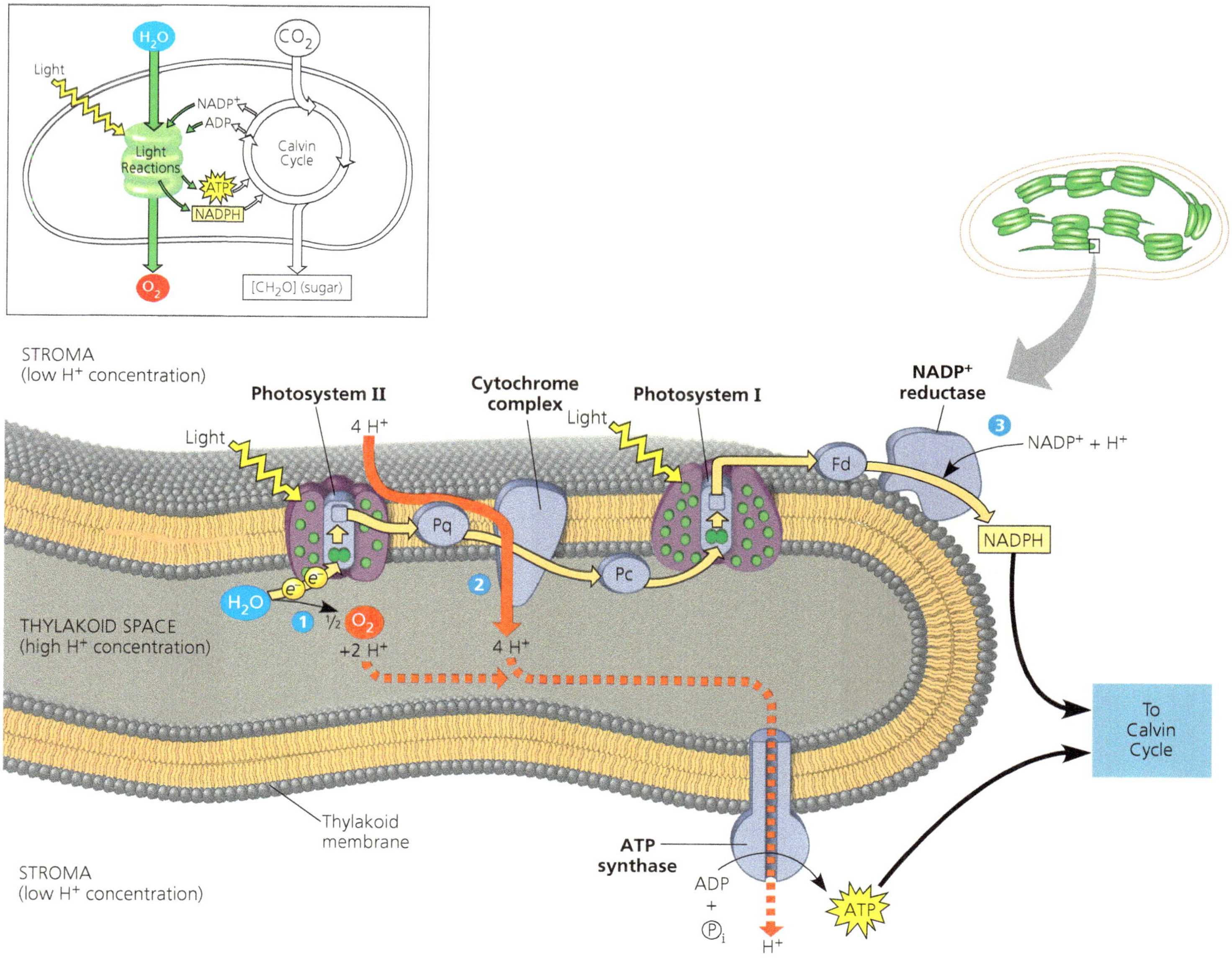

▲ **Figure 10.18 The light reactions and chemiosmosis: the organization of the thylakoid membrane.** This diagram shows a current model for the organization of the thylakoid membrane. The gold arrows track the linear electron flow outlined in Figure 10.14. As electrons pass from carrier to carrier in redox reactions, hydrogen ions removed from the stroma are deposited in the thylakoid space, storing energy as a proton-motive force (H^+ gradient). At least three steps in the light reactions contribute to the proton gradient: ① Water is split by photosystem II on the side of the membrane facing the thylakoid space; ② as plastoquinone (Pq), a mobile carrier, transfers electrons to the cytochrome complex, four protons are translocated across the membrane into the thylakoid space; and ③ a hydrogen ion is removed from the stroma when it is taken up by $NADP^+$. Notice that in step 2, hydrogen ions are being pumped from the stroma into the thylakoid space, as in Figure 10.17. The diffusion of H^+ from the thylakoid space back to the stroma (along the H^+ concentration gradient) powers the ATP synthase. These light-driven reactions store chemical energy in NADPH and ATP, which shuttle the energy to the carbohydrate-producing Calvin cycle.

CONCEPT CHECK 10.2

1. What color of light is *least* effective in driving photosynthesis? Explain.
2. Compared to a solution of isolated chlorophyll, why do intact chloroplasts release less heat and fluorescence when illuminated?
3. In the light reactions, what is the initial electron donor? Where do the electrons finally end up?
4. **WHAT IF?** In an experiment, isolated chloroplasts placed in an illuminated solution with the appropriate chemicals can carry out ATP synthesis. Predict what would happen to the rate of synthesis if a compound is added to the solution that makes membranes freely permeable to hydrogen ions.

For suggested answers, see Appendix A.

CONCEPT 10.3

The Calvin cycle uses the chemical energy of ATP and NADPH to reduce CO_2 to sugar

The Calvin cycle is similar to the citric acid cycle in that a starting material is regenerated after molecules enter and leave the cycle. However, while the citric acid cycle is catabolic, oxidizing acetyl CoA and using the energy to synthesize ATP, the Calvin cycle is anabolic, building carbohydrates from smaller molecules and consuming energy. Carbon enters the Calvin cycle in the form of CO_2 and leaves in the form of sugar. The cycle spends ATP as an energy source and consumes NADPH as reducing power for adding high-energy electrons to make the sugar.

As we mentioned previously, the carbohydrate produced directly from the Calvin cycle is actually not glucose, but a three-carbon sugar; the name of this sugar is **glyceraldehyde 3-phosphate (G3P)**. For the net synthesis of one molecule of G3P, the cycle must take place three times, fixing three molecules of CO_2. (Recall that carbon fixation refers to the initial incorporation of CO_2 into organic material.) As we trace the steps of the cycle, keep in mind that we are following three molecules of CO_2 through the reactions. **Figure 10.19** divides

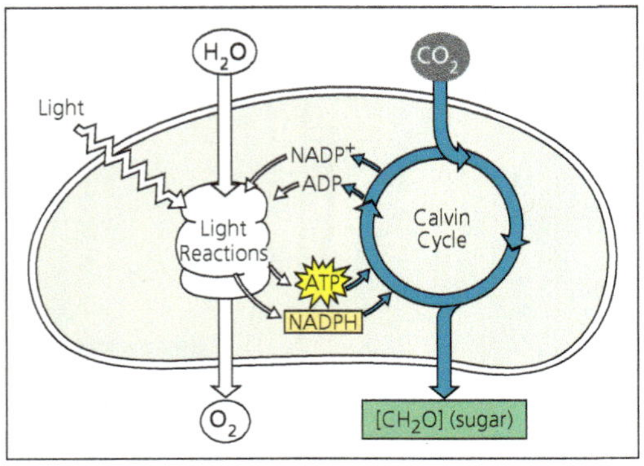

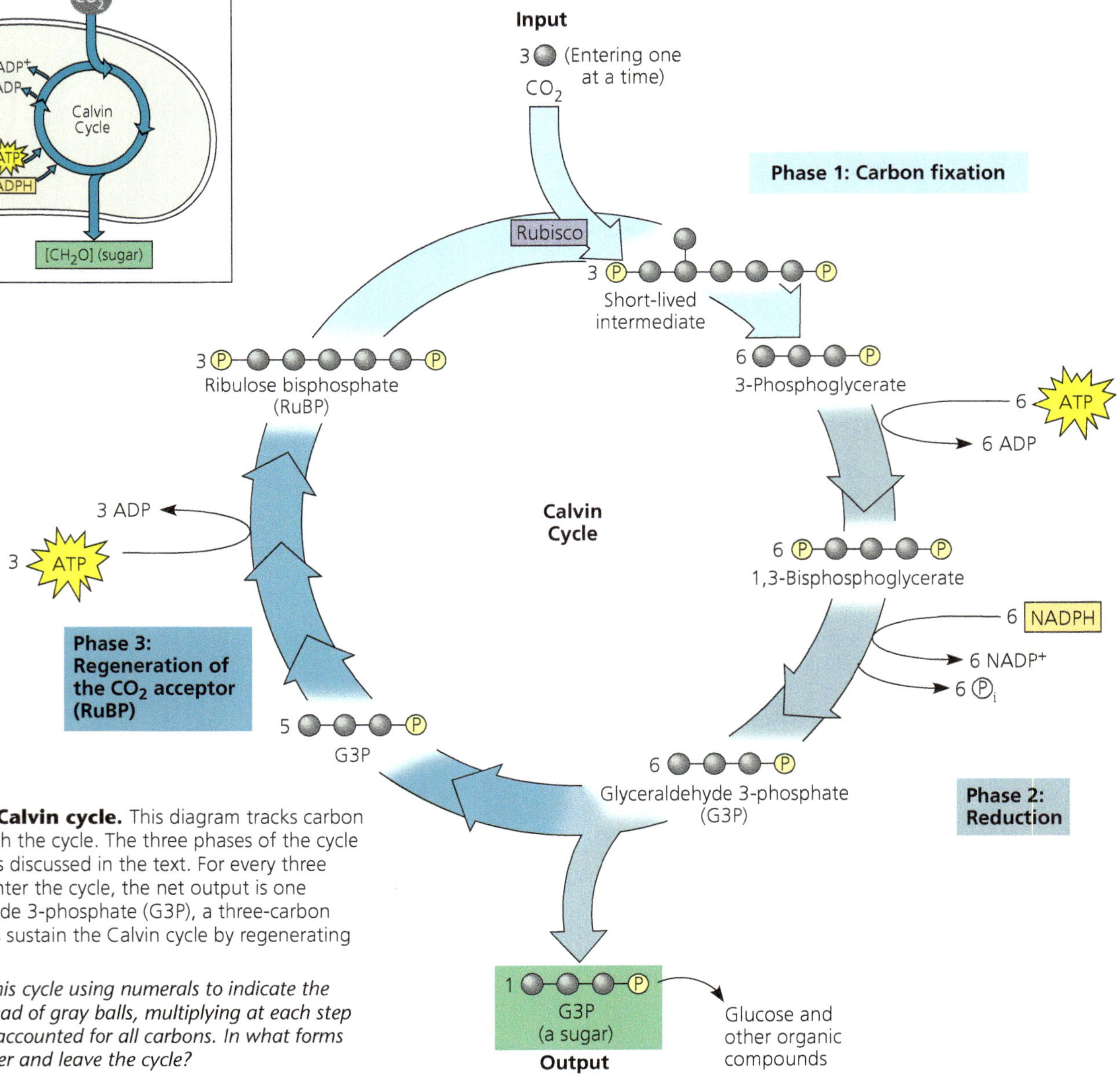

▲ **Figure 10.19 The Calvin cycle.** This diagram tracks carbon atoms (gray balls) through the cycle. The three phases of the cycle correspond to the phases discussed in the text. For every three molecules of CO_2 that enter the cycle, the net output is one molecule of glyceraldehyde 3-phosphate (G3P), a three-carbon sugar. The light reactions sustain the Calvin cycle by regenerating ATP and NADPH.

DRAW IT *Redraw this cycle using numerals to indicate the numbers of carbons instead of gray balls, multiplying at each step to ensure that you have accounted for all carbons. In what forms do the carbon atoms enter and leave the cycle?*

the Calvin cycle into three phases: carbon fixation, reduction, and regeneration of the CO_2 acceptor.

Phase 1: Carbon fixation. The Calvin cycle incorporates each CO_2 molecule, one at a time, by attaching it to a five-carbon sugar named ribulose bisphosphate (abbreviated RuBP). The enzyme that catalyzes this first step is RuBP carboxylase, or **rubisco**. (This is the most abundant protein in chloroplasts and is also thought to be the most abundant protein on Earth.) The product of the reaction is a six-carbon intermediate so unstable that it immediately splits in half, forming two molecules of 3-phosphoglycerate (for each CO_2 fixed).

Phase 2: Reduction. Each molecule of 3-phosphoglycerate receives an additional phosphate group from ATP, becoming 1,3-bisphosphoglycerate. Next, a pair of electrons donated from NADPH reduces 1,3-bisphosphoglycerate, which also loses a phosphate group, becoming G3P. Specifically, the electrons from NADPH reduce a carboyxl group on 1,3-bisphosphoglycerate to the aldehyde group of G3P, which stores more potential energy. G3P is a sugar—the same three-carbon sugar formed in glycolysis by the splitting of glucose (see Figure 9.9). Notice in Figure 10.19 that for every *three* molecules of CO_2 that enter the cycle, there are *six* molecules of G3P formed. But only one molecule of this three-carbon sugar can be counted as a net gain of carbohydrate. The cycle began with 15 carbons' worth of carbohydrate in the form of three molecules of the five-carbon sugar RuBP. Now there are 18 carbons' worth of carbohydrate in the form of six molecules of G3P. One molecule exits the cycle to be used by the plant cell, but the other five molecules must be recycled to regenerate the three molecules of RuBP.

Phase 3: Regeneration of the CO_2 acceptor (RuBP). In a complex series of reactions, the carbon skeletons of five molecules of G3P are rearranged by the last steps of the Calvin cycle into three molecules of RuBP. To accomplish this, the cycle spends three more molecules of ATP. The RuBP is now prepared to receive CO_2 again, and the cycle continues.

For the net synthesis of one G3P molecule, the Calvin cycle consumes a total of nine molecules of ATP and six molecules of NADPH. The light reactions regenerate the ATP and NADPH. The G3P spun off from the Calvin cycle becomes the starting material for metabolic pathways that synthesize other organic compounds, including glucose and other carbohydrates. Neither the light reactions nor the Calvin cycle alone can make sugar from CO_2. Photosynthesis is an emergent property of the intact chloroplast, which integrates the two stages of photosynthesis.

CONCEPT CHECK 10.3

1. To synthesize one glucose molecule, the Calvin cycle uses ________ molecules of CO_2, ________ molecules of ATP, and ________ molecules of NADPH.
2. Explain why the large numbers of ATP and NADPH molecules used during the Calvin cycle are consistent with the high value of glucose as an energy source.
3. WHAT IF? Explain why a poison that inhibits an enzyme of the Calvin cycle will also inhibit the light reactions.
4. MAKE CONNECTIONS Review Figures 9.9 (p. 169) and 10.19. Discuss the roles of intermediate and product played by glyceraldehyde 3-phosphate (G3P) in the two processes shown in these figures.

For suggested answers, see Appendix A.

CONCEPT 10.4

Alternative mechanisms of carbon fixation have evolved in hot, arid climates

EVOLUTION Ever since plants first moved onto land about 475 million years ago, they have been adapting to the problems of terrestrial life, particularly the problem of dehydration. In Chapters 29 and 36, we will consider anatomical adaptations that help plants conserve water, while in this chapter we are concerned with metabolic adaptations. The solutions often involve trade-offs. An important example is the compromise between photosynthesis and the prevention of excessive water loss from the plant. The CO_2 required for photosynthesis enters a leaf via stomata, the pores on the leaf surface (see Figure 10.4). However, stomata are also the main avenues of transpiration, the evaporative loss of water from leaves. On a hot, dry day, most plants close their stomata, a response that conserves water. This response also reduces photosynthetic yield by limiting access to CO_2. With stomata even partially closed, CO_2 concentrations begin to decrease in the air spaces within the leaf, and the concentration of O_2 released from the light reactions begins to increase. These conditions within the leaf favor an apparently wasteful process called photorespiration.

Photorespiration: An Evolutionary Relic?

In most plants, initial fixation of carbon occurs via rubisco, the Calvin cycle enzyme that adds CO_2 to ribulose bisphosphate. Such plants are called **C_3 plants** because the first organic product of carbon fixation is a three-carbon compound,

3-phosphoglycerate (see Figure 10.19). Rice, wheat, and soybeans are C_3 plants that are important in agriculture. When their stomata partially close on hot, dry days, C_3 plants produce less sugar because the declining level of CO_2 in the leaf starves the Calvin cycle. In addition, rubisco can bind O_2 in place of CO_2. As CO_2 becomes scarce within the air spaces of the leaf, rubisco adds O_2 to the Calvin cycle instead of CO_2. The product splits, and a two-carbon compound leaves the chloroplast. Peroxisomes and mitochondria rearrange and split this compound, releasing CO_2. The process is called **photorespiration** because it occurs in the light (*photo*) and consumes O_2 while producing CO_2 (*respiration*). However, unlike normal cellular respiration, photorespiration generates no ATP; in fact, photorespiration consumes ATP. And unlike photosynthesis, photorespiration produces no sugar. In fact, photorespiration *decreases* photosynthetic output by siphoning organic material from the Calvin cycle and releasing CO_2 that would otherwise be fixed.

How can we explain the existence of a metabolic process that seems to be counterproductive for the plant? According to one hypothesis, photorespiration is evolutionary baggage—a metabolic relic from a much earlier time when the atmosphere had less O_2 and more CO_2 than it does today. In the ancient atmosphere that prevailed when rubisco first evolved, the inability of the enzyme's active site to exclude O_2 would have made little difference. The hypothesis suggests that modern rubisco retains some of its chance affinity for O_2, which is now so concentrated in the atmosphere that a certain amount of photorespiration is inevitable.

We now know that, at least in some cases, photorespiration plays a protective role in plants. Plants that are impaired in their ability to carry out photorespiration (due to defective genes) are more susceptible to damage induced by excess light. Researchers consider this clear evidence that photorespiration acts to neutralize the otherwise damaging products of the light reactions, which build up when a low CO_2 concentration limits the progress of the Calvin cycle. Whether there are other benefits of photorespiration is still unknown. In many types of plants—including a significant number of crop plants—photorespiration drains away as much as 50% of the carbon fixed by the Calvin cycle. As heterotrophs that depend on carbon fixation in chloroplasts for our food, we naturally view photorespiration as wasteful. Indeed, if photorespiration could be reduced in certain plant species without otherwise affecting photosynthetic productivity, crop yields and food supplies might increase.

In some plant species, alternate modes of carbon fixation have evolved that minimize photorespiration and optimize the Calvin cycle—even in hot, arid climates. The two most important of these photosynthetic adaptations are C_4 photosynthesis and crassulacean acid metabolism (CAM).

C_4 Plants

The **C_4 plants** are so named because they preface the Calvin cycle with an alternate mode of carbon fixation that forms a four-carbon compound as its first product. Several thousand species in at least 19 plant families use the C_4 pathway. Among the C_4 plants important to agriculture are sugarcane and corn, members of the grass family.

A unique leaf anatomy is correlated with the mechanism of C_4 photosynthesis (**Figure 10.20**; compare with Figure 10.4). In C_4 plants, there are two distinct types of photosynthetic cells: bundle-sheath cells and mesophyll cells. **Bundle-sheath cells** are arranged into tightly packed sheaths around the veins of the leaf. Between the bundle sheath and the leaf surface are the more loosely arranged mesophyll cells. The Calvin cycle is confined to the chloroplasts of the bundle-sheath cells. However, the Calvin cycle is preceded by incorporation of CO_2 into organic compounds in the mesophyll cells. See the numbered steps in Figure 10.20, which are also described here:

1. The first step is carried out by an enzyme present only in mesophyll cells called **PEP carboxylase**. This enzyme adds CO_2 to phosphoenolpyruvate (PEP), forming the four-carbon product oxaloacetate. PEP carboxylase has a much higher affinity for CO_2 than does rubisco and no affinity for O_2. Therefore, PEP carboxylase can fix carbon efficiently when rubisco cannot—that is, when it is hot and dry and stomata are partially closed, causing CO_2 concentration in the leaf to fall and O_2 concentration to rise.
2. After the C_4 plant fixes carbon from CO_2, the mesophyll cells export their four-carbon products (malate in the example shown in Figure 10.20) to bundle-sheath cells through plasmodesmata (see Figure 6.31).
3. Within the bundle-sheath cells, the four-carbon compounds release CO_2, which is reassimilated into organic material by rubisco and the Calvin cycle. The same reaction regenerates pyruvate, which is transported to mesophyll cells. There, ATP is used to convert pyruvate to PEP, allowing the reaction cycle to continue; this ATP can be thought of as the "price" of concentrating CO_2 in the bundle-sheath cells. To generate this extra ATP, bundle-sheath cells carry out cyclic electron flow, the process described earlier in this chapter (see Figure 10.16). In fact, these cells contain PS I but no PS II, so cyclic electron flow is their only photosynthetic mode of generating ATP.

In effect, the mesophyll cells of a C_4 plant pump CO_2 into the bundle sheath, keeping the CO_2 concentration in the bundle-sheath cells high enough for rubisco to bind carbon

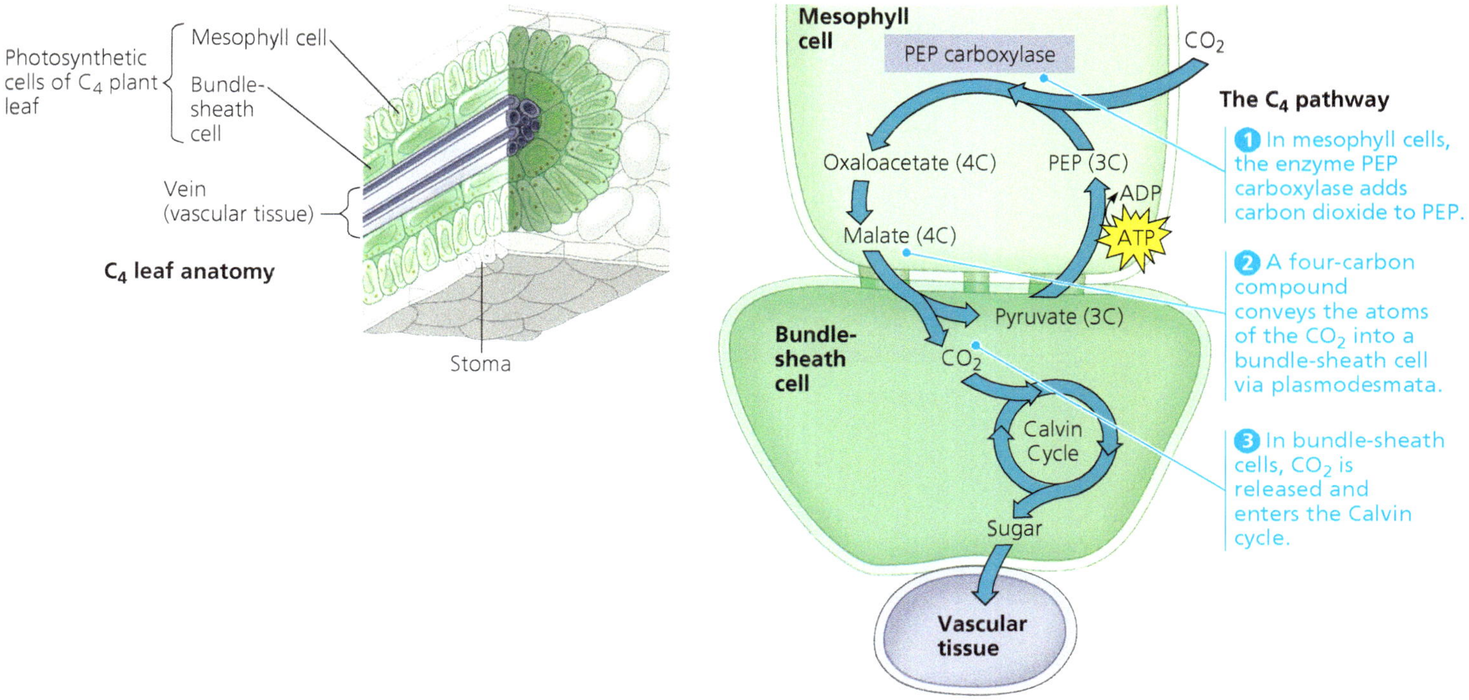

▲ **Figure 10.20** **C_4 leaf anatomy and the C_4 pathway.** The structure and biochemical functions of the leaves of C_4 plants are an evolutionary adaptation to hot, dry climates. This adaptation maintains a CO_2 concentration in the bundle sheath that favors photosynthesis over photorespiration.

dioxide rather than oxygen. The cyclic series of reactions involving PEP carboxylase and the regeneration of PEP can be thought of as a CO_2-concentrating pump that is powered by ATP. In this way, C_4 photosynthesis minimizes photorespiration and enhances sugar production. This adaptation is especially advantageous in hot regions with intense sunlight, where stomata partially close during the day, and it is in such environments that C_4 plants evolved and thrive today.

Since the Industrial Revolution began in the 1800s, human activities such as the burning of fossil fuels have drastically increased the concentration of CO_2 in the atmosphere. The resulting global climate change, including an increase in average temperatures around the planet, may have far-reaching effects on plant species. Scientists are concerned that increasing CO_2 concentration and temperature may affect C_3 and C_4 plants differently, thus changing the relative abundance of these species in a given plant community.

Which type of plant would stand to gain more from increasing CO_2 levels? Recall that in C_3 plants, the binding of O_2 rather than CO_2 by rubisco leads to photorespiration, lowering the efficiency of photosynthesis. C_4 plants overcome this problem by concentrating CO_2 in the bundle-sheath cells at the cost of ATP. Rising CO_2 levels should benefit C_3 plants by lowering the amount of photorespiration that occurs. At the same time, rising temperatures have the opposite effect, increasing photorespiration. (Other factors such as water availability may also come into play.) In contrast, many C_4 plants could be largely unaffected by increasing CO_2 levels or temperature. In different regions, the particular combination of these two factors is likely to alter the balance of C_3 and C_4 plants in varying ways. The effects of such a widespread and variable change in community structure are unpredictable and thus a cause of legitimate concern.

CAM Plants

A second photosynthetic adaptation to arid conditions has evolved in many succulent (water-storing) plants, numerous cacti, pineapples, and representatives of several other plant families. These plants open their stomata during the night and close them during the day, just the reverse of how other plants behave. Closing stomata during the day helps desert plants conserve water, but it also prevents CO_2 from entering the leaves. During the night, when their stomata are open, these plants take up CO_2 and incorporate it into a variety

▶ **Figure 10.21 C_4 and CAM photosynthesis compared.** Both adaptations are characterized by (1) preliminary incorporation of CO_2 into organic acids, followed by (2) transfer of CO_2 to the Calvin cycle. The C_4 and CAM pathways are two evolutionary solutions to the problem of maintaining photosynthesis with stomata partially or completely closed on hot, dry days.

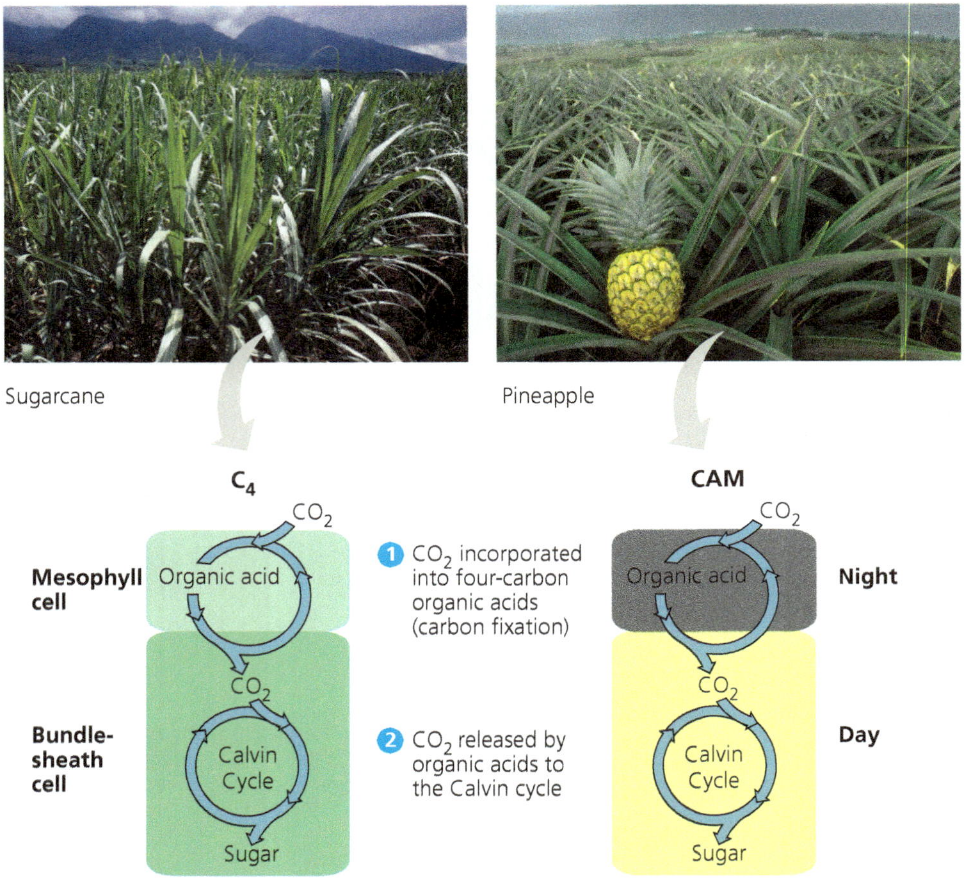

(a) Spatial separation of steps. In C_4 plants, carbon fixation and the Calvin cycle occur in different types of cells.

(b) Temporal separation of steps. In CAM plants, carbon fixation and the Calvin cycle occur in the same cells at different times.

of organic acids. This mode of carbon fixation is called **crassulacean acid metabolism**, or **CAM**, after the plant family Crassulaceae, the succulents in which the process was first discovered. The mesophyll cells of **CAM plants** store the organic acids they make during the night in their vacuoles until morning, when the stomata close. During the day, when the light reactions can supply ATP and NADPH for the Calvin cycle, CO_2 is released from the organic acids made the night before to become incorporated into sugar in the chloroplasts.

Notice in **Figure 10.21** that the CAM pathway is similar to the C_4 pathway in that carbon dioxide is first incorporated into organic intermediates before it enters the Calvin cycle. The difference is that in C_4 plants, the initial steps of carbon fixation are separated structurally from the Calvin cycle, whereas in CAM plants, the two steps occur at separate times but within the same cell. (Keep in mind that CAM, C_4, and C_3 plants all eventually use the Calvin cycle to make sugar from carbon dioxide.)

CONCEPT CHECK 10.4

1. Explain why photorespiration lowers photosynthetic output for plants.
2. The presence of only PS I, not PS II, in the bundle-sheath cells of C_4 plants has an effect on O_2 concentration. What is that effect, and how might that benefit the plant?
3. **MAKE CONNECTIONS** Refer to the discussion of ocean acidification in Concept 3.3 (p. 55). Ocean acidification and changes in the distribution of C_3 and C_4 plants may seem to be two very different problems, but what do they have in common? Explain.
4. **WHAT IF?** How would you expect the relative abundance of C_3 versus C_4 and CAM species to change in a geographic region whose climate becomes much hotter and drier, with no change in CO_2 concentration?

For suggested answers, see Appendix A.

The Importance of Photosynthesis: *A Review*

In this chapter, we have followed photosynthesis from photons to food. The light reactions capture solar energy and use it to make ATP and transfer electrons from water to $NADP^+$, forming NADPH. The Calvin cycle uses the ATP and NADPH to produce sugar from carbon dioxide. The energy that enters the chloroplasts as sunlight becomes stored as chemical energy in organic compounds. See **Figure 10.22** for a review of the entire process.

What are the fates of photosynthetic products? The sugar made in the chloroplasts supplies the entire plant with chemical energy and carbon skeletons for the synthesis of all the major organic molecules of plant cells. About 50% of the organic material made by photosynthesis is consumed as fuel for cellular respiration in the mitochondria of the plant cells. Sometimes there is a loss of photosynthetic products to photorespiration.

Technically, green cells are the only autotrophic parts of the plant. The rest of the plant depends on organic molecules exported from leaves via veins. In most plants, carbohydrate is transported out of the leaves in the form of sucrose, a disaccharide. After arriving at nonphotosynthetic cells, the sucrose provides raw material for cellular respiration and a multitude of anabolic pathways that synthesize proteins, lipids, and other products. A considerable amount of sugar in the form of glucose is linked together to make the polysaccharide cellulose, especially in plant cells that are still growing and maturing. Cellulose, the main ingredient of cell walls, is the most abundant organic molecule in the plant—and probably on the surface of the planet.

Most plants manage to make more organic material each day than they need to use as respiratory fuel and precursors for biosynthesis. They stockpile the extra sugar by synthesizing starch, storing some in the chloroplasts themselves and some in storage cells of roots, tubers, seeds, and fruits. In accounting for the consumption of the food molecules produced by photosynthesis, let's not forget that most plants lose leaves, roots, stems, fruits, and sometimes their entire bodies to heterotrophs, including humans.

On a global scale, photosynthesis is the process responsible for the presence of oxygen in our atmosphere. Furthermore, in terms of food production, the collective productivity of the minuscule chloroplasts is prodigious: Photosynthesis makes an estimated 160 billion metric tons of carbohydrate per year (a metric ton is 1,000 kg, about 1.1 tons). That's organic matter equivalent in mass to a stack of about 60 trillion copies of this textbook—17 stacks of books reaching from Earth to the sun! No other chemical process on the planet can match the output of photosynthesis. And as we mentioned earlier, researchers are seeking ways to capitalize on photosynthetic production to produce alternative fuels. No process is more important than photosynthesis to the welfare of life on Earth.

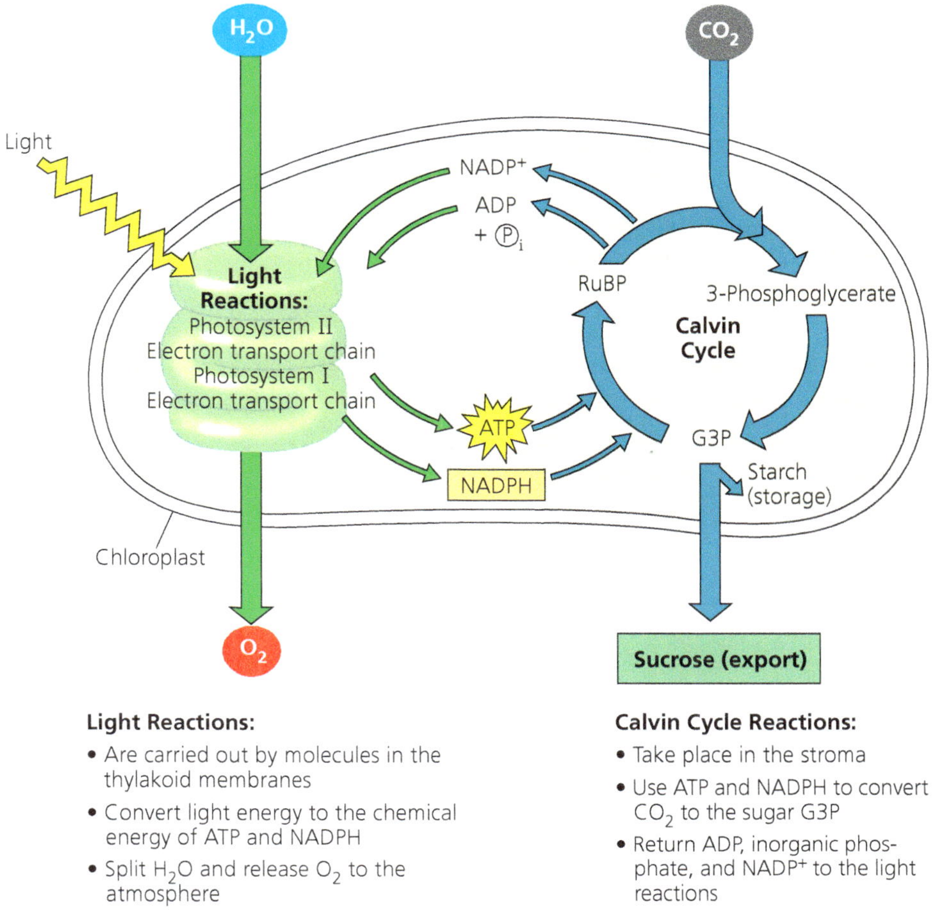

▶ **Figure 10.22 A review of photosynthesis.** This diagram outlines the main reactants and products of the light reactions and the Calvin cycle as they occur in the chloroplasts of plant cells. The entire ordered operation depends on the structural integrity of the chloroplast and its membranes. Enzymes in the chloroplast and cytosol convert glyceraldehyde 3-phosphate (G3P), the direct product of the Calvin cycle, to many other organic compounds.

MAKE CONNECTIONS *Return to the micrograph in Figure 5.6a, on page 72. Label and describe where the light reactions and the Calvin cycle take place. Also explain where the starch granules in the micrograph came from.*

10 CHAPTER REVIEW

SUMMARY OF KEY CONCEPTS

CONCEPT 10.1

Photosynthesis converts light energy to the chemical energy of food (pp. 186–189)

- In **autotrophic** eukaryotes, photosynthesis occurs in **chloroplasts**, organelles containing **thylakoids**. Stacks of thylakoids form grana. **Photosynthesis** is summarized as

  $6\ CO_2 + 12\ H_2O + \text{Light energy} \rightarrow C_6H_{12}O_6 + 6\ O_2 + 6\ H_2O.$

  Chloroplasts split water into hydrogen and oxygen, incorporating the electrons of hydrogen into sugar molecules. Photosynthesis is a redox process: H_2O is oxidized, and CO_2 is reduced. The **light reactions** in the thylakoid membranes split water, releasing O_2, producing ATP, and forming **NADPH**. The **Calvin cycle** in the **stroma** forms sugar from CO_2, using ATP for energy and NADPH for reducing power.

? *Compare and describe the roles of CO_2 and H_2O in respiration and photosynthesis.*

CONCEPT 10.2

The light reactions convert solar energy to the chemical energy of ATP and NADPH (pp. 189–197)

- Light is a form of electromagnetic energy. The colors we see as **visible light** include those **wavelengths** that drive photosynthesis. A pigment absorbs light of specific wavelengths; **chlorophyll *a*** is the main photosynthetic pigment in plants. Other accessory pigments absorb different wavelengths of light and pass the energy on to chlorophyll *a*.
- A pigment goes from a ground state to an excited state when a **photon** of light boosts one of the pigment's electrons to a higher-energy orbital. This excited state is unstable. Electrons from isolated pigments tend to fall back to the ground state, giving off heat and/or light.
- A **photosystem** is composed of a **reaction-center complex** surrounded by **light-harvesting complexes** that funnel the energy of photons to the reaction-center complex. When a special pair of reaction-center chlorophyll *a* molecules absorbs energy, one of its electrons is boosted to a higher energy level and transferred to the **primary electron acceptor**. **Photosystem II** contains P680 chlorophyll *a* molecules in the reaction-center complex; **photosystem I** contains P700 molecules.
- **Linear electron flow** during the light reactions uses both photosystems and produces NADPH, ATP, and oxygen:

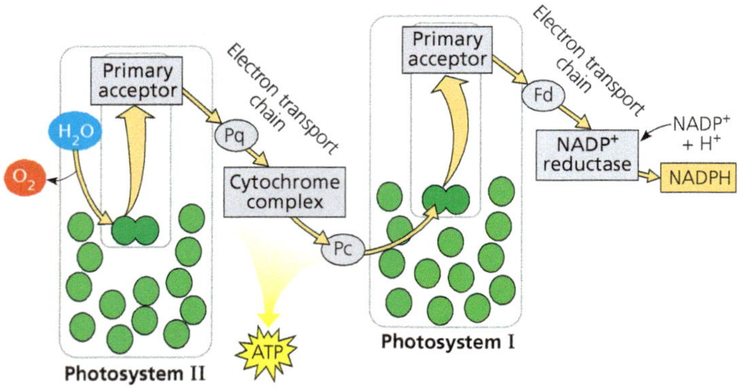

- **Cyclic electron flow** employs only photosystem I, producing ATP but no NADPH or O_2.
- During chemiosmosis in both mitochondria and chloroplasts, electron transport chains generate an H^+ gradient across a membrane. ATP synthase uses this proton-motive force to make ATP.

? *The absorption spectrum of chlorophyll a differs from the action spectrum of photosynthesis. Explain this observation.*

CONCEPT 10.3

The Calvin cycle uses the chemical energy of ATP and NADPH to reduce CO_2 to sugar (pp. 198–199)

- The Calvin cycle occurs in the stroma, using electrons from NADPH and energy from ATP. One molecule of **G3P** exits the cycle per three CO_2 molecules fixed and is converted to glucose and other organic molecules.

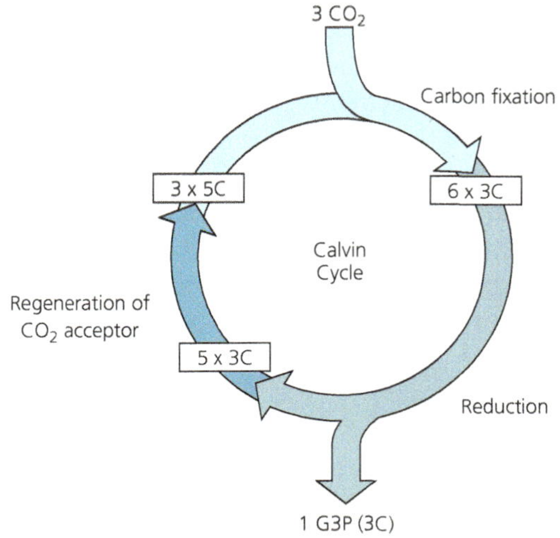

DRAW IT *On the diagram above, draw where ATP and NADPH are used and where rubisco functions. Describe these steps.*

CONCEPT 10.4

Alternative mechanisms of carbon fixation have evolved in hot, arid climates (pp. 199–202)

- On dry, hot days, **C_3 plants** close their stomata, conserving water. Oxygen from the light reactions builds up. In **photorespiration**, O_2 substitutes for CO_2 in the active site of rubisco. This process consumes organic fuel and releases CO_2 without producing ATP or carbohydrate. Photorespiration may be an evolutionary relic, and it may play a photoprotective role.
- **C_4 plants** minimize the cost of photorespiration by incorporating CO_2 into four-carbon compounds in mesophyll cells. These compounds are exported to **bundle-sheath cells**, where they release carbon dioxide for use in the Calvin cycle.
- **CAM plants** open their stomata at night, incorporating CO_2 into organic acids, which are stored in mesophyll cells. During the day, the stomata close, and the CO_2 is released from the organic acids for use in the Calvin cycle.
- Organic compounds produced by photosynthesis provide the energy and building material for ecosystems.

? *Why are C_4 and CAM photosynthesis more energetically expensive than C_3 photosynthesis? What climate conditions would favor C_4 and CAM plants?*

TEST YOUR UNDERSTANDING

LEVEL 1: KNOWLEDGE/COMPREHENSION

1. The light reactions of photosynthesis supply the Calvin cycle with
   a. light energy.
   b. CO_2 and ATP.
   c. H_2O and NADPH.
   d. ATP and NADPH.
   e. sugar and O_2.

2. Which of the following sequences correctly represents the flow of electrons during photosynthesis?
   a. NADPH → O_2 → CO_2
   b. H_2O → NADPH → Calvin cycle
   c. NADPH → chlorophyll → Calvin cycle
   d. H_2O → photosystem I → photosystem II
   e. NADPH → electron transport chain → O_2

3. How is photosynthesis similar in C_4 plants and CAM plants?
   a. In both cases, only photosystem I is used.
   b. Both types of plants make sugar without the Calvin cycle.
   c. In both cases, rubisco is not used to fix carbon initially.
   d. Both types of plants make most of their sugar in the dark.
   e. In both cases, thylakoids are not involved in photosynthesis.

4. Which of the following statements is a correct distinction between autotrophs and heterotrophs?
   a. Only heterotrophs require chemical compounds from the environment.
   b. Cellular respiration is unique to heterotrophs.
   c. Only heterotrophs have mitochondria.
   d. Autotrophs, but not heterotrophs, can nourish themselves beginning with CO_2 and other nutrients that are inorganic.
   e. Only heterotrophs require oxygen.

5. Which of the following does *not* occur during the Calvin cycle?
   a. carbon fixation
   b. oxidation of NADPH
   c. release of oxygen
   d. regeneration of the CO_2 acceptor
   e. consumption of ATP

LEVEL 2: APPLICATION/ANALYSIS

6. In mechanism, photophosphorylation is most similar to
   a. substrate-level phosphorylation in glycolysis.
   b. oxidative phosphorylation in cellular respiration.
   c. the Calvin cycle.
   d. carbon fixation.
   e. reduction of $NADP^+$.

7. Which process is most directly driven by light energy?
   a. creation of a pH gradient by pumping protons across the thylakoid membrane
   b. carbon fixation in the stroma
   c. reduction of $NADP^+$ molecules
   d. removal of electrons from chlorophyll molecules
   e. ATP synthesis

LEVEL 3: SYNTHESIS/EVALUATION

8. **EVOLUTION CONNECTION**
   Photorespiration can decrease soybeans' photosynthetic output by about 50%. Would you expect this figure to be higher or lower in wild relatives of soybeans? Why?

9. **SCIENTIFIC INQUIRY**
   MAKE CONNECTIONS **DRAW IT** The following diagram represents an experiment with isolated thylakoids. The thylakoids were first made acidic by soaking them in a solution at pH 4. After the thylakoid space reached pH 4, the thylakoids were transferred to a basic solution at pH 8. The thylakoids then made ATP in the dark. (See Concept 3.3, pp. 53–54, to review pH).

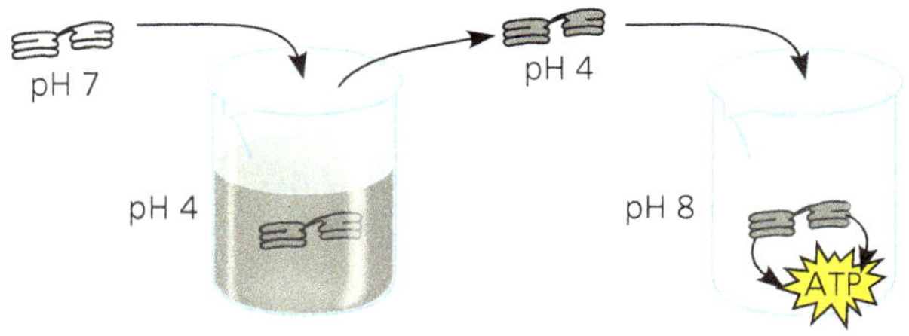

   Draw an enlargement of part of the thylakoid membrane in the beaker with the solution at pH 8. Draw ATP synthase. Label the areas of high H^+ concentration and low H^+ concentration. Show the direction protons flow through the enzyme, and show the reaction where ATP is synthesized. Would ATP end up in the thylakoid or outside of it? Explain why the thylakoids in the experiment were able to make ATP in the dark.

10. **SCIENCE, TECHNOLOGY, AND SOCIETY**
    Scientific evidence indicates that the CO_2 added to the air by the burning of wood and fossil fuels is contributing to global warming, a rise in global temperature. Tropical rain forests are estimated to be responsible for approximately 20% of global photosynthesis, yet the consumption of large amounts of CO_2 by living trees is thought to make little or no *net* contribution to reduction of global warming. Why might this be? (*Hint:* What processes in both living and dead trees produce CO_2?)

11. **WRITE ABOUT A THEME**
    Energy Transfer Life is solar powered. Almost all the producers of the biosphere depend on energy from the sun to produce the organic molecules that supply the energy and carbon skeletons needed for life. In a short essay (100–150 words), describe how the process of photosynthesis in the chloroplasts of plants transforms the energy of sunlight into the chemical energy of sugar molecules.

For selected answers, see Appendix A.

MasteringBIOLOGY® www.masteringbiology.com

1. MasteringBiology® Assignments

Experimental Inquiry Tutorial Which Wavelengths of Light Drive Photosynthesis?
BioFlix Tutorials Photosynthesis: Inputs, Outputs, and Chloroplast Structure • The Light Reactions • The Calvin Cycle
Tutorial Energy Flow in Plants—Concept Map
Activities Overview of Photosynthesis • The Sites of Photosynthesis • Chemiosmosis • Light Energy and Pigments • Photosynthesis • The Light Reactions • The Calvin Cycle • Photosynthesis in Dry Climates
Questions Student Misconceptions • Reading Quiz • Multiple Choice • End-of-Chapter

2. eText

Read your book online, search, take notes, highlight text, and more.

3. The Study Area

Practice Tests • Cumulative Test • BioFlix 3-D Animations • MP3 Tutor Sessions • Videos • Activities • Investigations • Lab Media • Audio Glossary • Word Study Tools • Art

11

Cell Communication

▲ **Figure 11.1 How does cell signaling trigger the desperate flight of this gazelle?**

KEY CONCEPTS

- **11.1** External signals are converted to responses within the cell
- **11.2** Reception: A signaling molecule binds to a receptor protein, causing it to change shape
- **11.3** Transduction: Cascades of molecular interactions relay signals from receptors to target molecules in the cell
- **11.4** Response: Cell signaling leads to regulation of transcription or cytoplasmic activities
- **11.5** Apoptosis integrates multiple cell-signaling pathways

OVERVIEW

Cellular Messaging

The Thomson's gazelle in **Figure 11.1** is fleeing for its life, seeking to escape the predatory cheetah nipping at its heels. The gazelle's heart is racing, its breathing accelerated and its muscles performing at their highest level. These physiological functions are all part of the "fight-or-flight" response, driven by hormones released from the adrenal glands at times of stress—in this case, when the gazelle first sensed the cheetah. Hormonal signaling and the subsequent response by cells and tissues throughout the gazelle's body illustrate how cell-to-cell communication allows the trillions of cells in a multicellular organism to "talk" to each other, coordinating their activities. Communication between cells is essential not only for multicellular organisms such as gazelles and oak trees but for many unicellular organisms as well.

In studying how cells signal to each other and how they interpret the signals they receive, biologists have discovered some universal mechanisms of cellular regulation, additional evidence for the evolutionary relatedness of all life. The same small set of cell-signaling mechanisms shows up again and again in diverse species, in biological processes ranging from hormone action to embryonic development to cancer. The signals received by cells, whether originating from other cells or from changes in the physical environment, take various forms, including light and touch. However, cells most often communicate with each other by chemical signals. For instance, the fight-or-flight response is triggered by a signaling molecule called epinephrine. In this chapter, we focus on the main mechanisms by which cells receive, process, and respond to chemical signals sent from other cells. We will also take a look at *apoptosis*, a type of programmed cell death that integrates input from multiple signaling pathways.

CONCEPT 11.1

External signals are converted to responses within the cell

What does a "talking" cell say to a "listening" cell, and how does the latter cell respond to the message? Let's approach these questions by first looking at communication among microorganisms, for microbes living today provide a glimpse into the role of cell signaling in the evolution of life on Earth.

Evolution of Cell Signaling

EVOLUTION One topic of cell "conversation" is sex—at least for the yeast *Saccharomyces cerevisiae*, which people have used for millennia to make bread, wine, and beer. Researchers have learned that cells of this yeast identify their mates by chemical signaling. There are two sexes, or mating types, called **a** and α (**Figure 11.2**). Cells of mating type **a** secrete a signaling

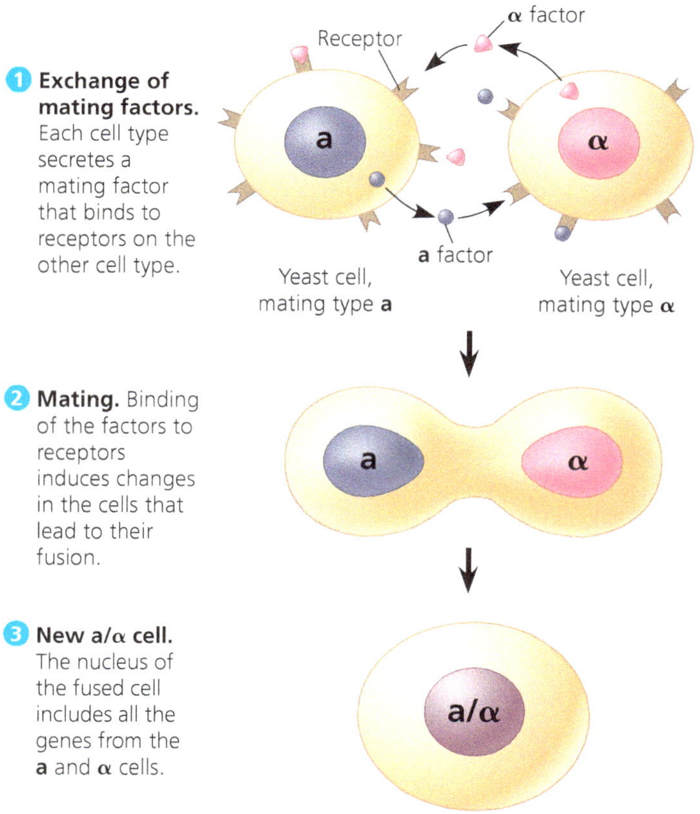

▲ **Figure 11.2 Communication between mating yeast cells.** *Saccharomyces cerevisiae* cells use chemical signaling to identify cells of opposite mating type and initiate the mating process. The two mating types and their corresponding chemical signaling molecules, or mating factors, are called **a** and **α**.

molecule called **a** factor, which can bind to specific receptor proteins on nearby **α** cells. At the same time, **α** cells secrete **α** factor, which binds to receptors on **a** cells. Without actually entering the cells, the two mating factors cause the cells to grow toward each other and also bring about other cellular changes. The result is the fusion, or mating, of two cells of opposite type. The new **a**/**α** cell contains all the genes of both original cells, a combination of genetic resources that provides advantages to the cell's descendants, which arise by subsequent cell divisions.

Once received at the yeast cell surface, how is the mating signal changed, or *transduced*, into a form that brings about the cellular response of mating? The received signal is converted to a specific cellular response in a series of steps called a **signal transduction pathway**. Many such pathways have been extensively studied in both yeast and animal cells. Amazingly, the molecular details of signal transduction in yeast and mammals are strikingly similar, even though the last common ancestor of these two groups of organisms lived over a billion years ago. These similarities—and others more recently uncovered between signaling systems in bacteria and plants—suggest that early versions of today's cell-signaling mechanisms evolved well before the first multicellular creatures appeared on Earth.

Scientists such as Bonnie Bassler, the interviewee for Unit 2 (see pp. 92–93), think that signaling mechanisms first evolved in ancient prokaryotes and single-celled eukaryotes, then were adopted for new uses by their multicellular descendants. Cell signaling is critical in the microbial world; a classic example in one bacterial species is shown in **Figure 11.3**. Bacterial cells secrete small molecules that can be detected by other bacterial cells. The concentration of such signaling molecules, sensed by the bacteria, allows them to monitor the local density of cells, a phenomenon called *quorum sensing*. Quorum sensing allows bacterial populations to coordinate their behaviors so they can carry out activities that are only productive when performed by a given number of cells in synchrony. One example is formation of a *biofilm*, an aggregation of bacterial cells adhered to a surface; the cells in the biofilm generally derive nutrition from the surface they are on. You have probably encountered biofilms many times, perhaps without realizing it. The slimy coating on a fallen log or on leaves lying on a forest path, or on your teeth each morning, are examples of bacterial biofilms. Biofilms are responsible for cavities—a good argument for daily tooth brushing and flossing to disrupt them!

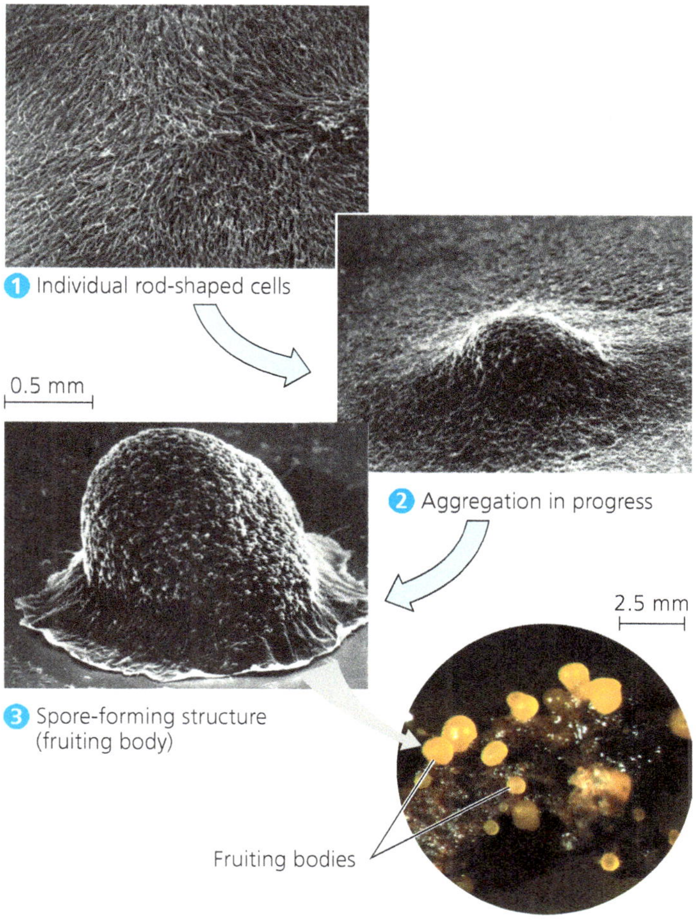

▲ **Figure 11.3 Communication among bacteria.** Soil-dwelling bacteria called myxobacteria ("slime bacteria") use chemical signals to share information about nutrient availability. When food is scarce, starving cells secrete a molecule that stimulates neighboring cells to aggregate. The cells form a structure, called a fruiting body, that produces thick-walled spores capable of surviving until the environment improves. The bacteria shown here are *Myxococcus xanthus* (steps 1–3, SEMs; lower photo, LM).

Local and Long-Distance Signaling

Like bacteria or yeast cells, cells in a multicellular organism usually communicate via chemical messengers targeted for cells that may or may not be immediately adjacent. As we saw in Chapters 6 and 7, eukaryotic cells may communicate by direct contact **(Figure 11.4)**, one type of local signaling. Both animals and plants have cell junctions that, where present, directly connect the cytoplasms of adjacent cells **(Figure 11.4a)**. In these cases, signaling substances dissolved in the cytosol can pass freely between adjacent cells. Moreover, animal cells may communicate via direct contact between membrane-bound cell-surface molecules in a process called cell-cell recognition **(Figure 11.4b)**. This sort of local signaling is important in embryonic development and the immune response.

In many other cases of local signaling, messenger molecules are secreted by the signaling cell. Some of these travel only short distances; such **local regulators** influence cells in the vicinity. One class of local regulators in animals, *growth factors*, consists of compounds that stimulate nearby target cells to grow and divide. Numerous cells can simultaneously receive and respond to the molecules of growth factor produced by a single cell in their vicinity. This type of local signaling in animals is called *paracrine signaling* **(Figure 11.5a)**.

Another, more specialized type of local signaling called *synaptic signaling* occurs in the animal nervous system **(Figure 11.5b)**. An electrical signal along a nerve cell triggers the secretion of neurotransmitter molecules carrying a chemical signal. These molecules diffuse across the synapse, the narrow space between the nerve cell and its target cell (often another nerve cell), triggering a response in the target cell.

Beyond communication through plasmodesmata (plant cell junctions), local signaling in plants is not as well understood. Because of their cell walls, plants use mechanisms somewhat different from those operating locally in animals.

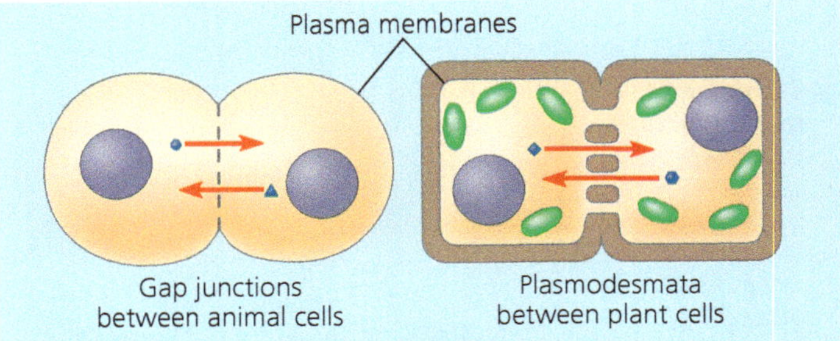

(a) Cell junctions. Both animals and plants have cell junctions that allow molecules to pass readily between adjacent cells without crossing plasma membranes.

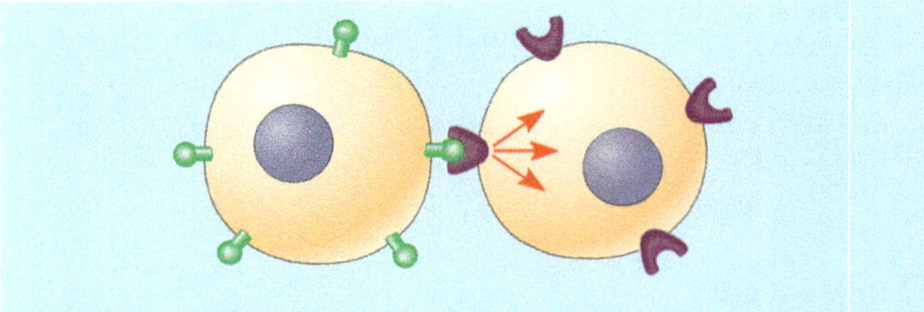
(b) Cell-cell recognition. Two cells in an animal may communicate by interaction between molecules protruding from their surfaces.

▲ **Figure 11.4 Communication by direct contact between cells.**

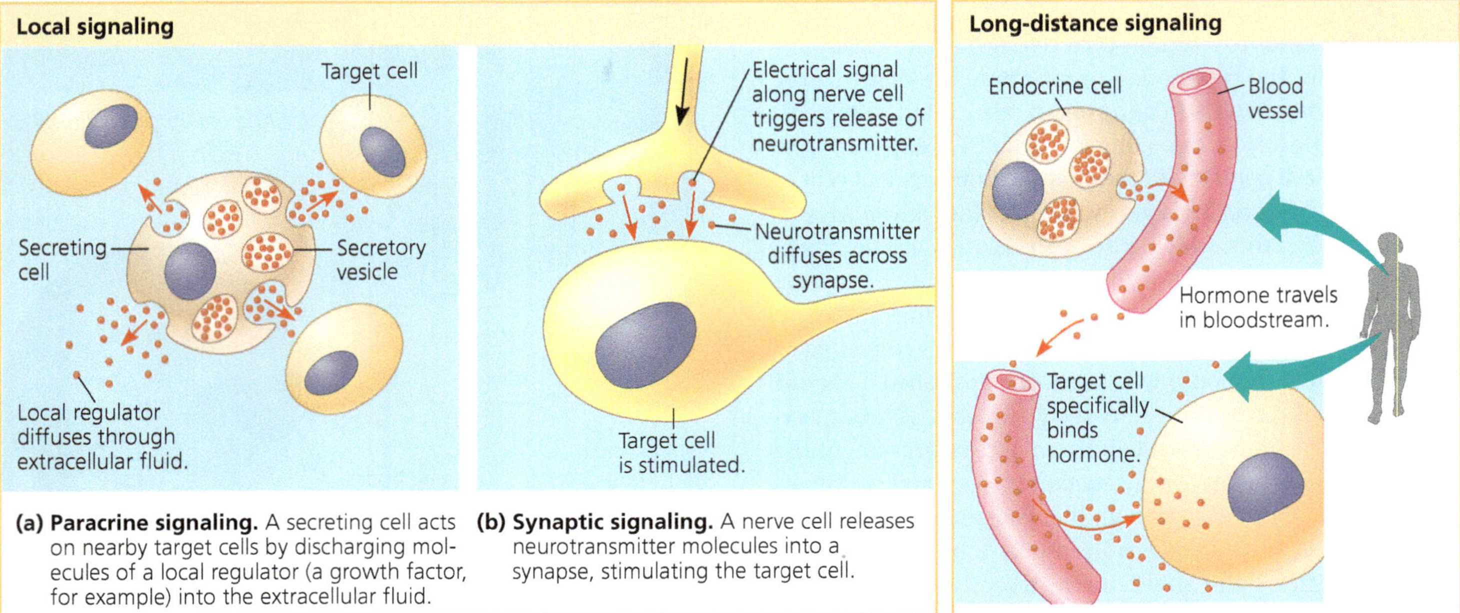

(a) Paracrine signaling. A secreting cell acts on nearby target cells by discharging molecules of a local regulator (a growth factor, for example) into the extracellular fluid.

(b) Synaptic signaling. A nerve cell releases neurotransmitter molecules into a synapse, stimulating the target cell.

(c) Endocrine (hormonal) signaling. Specialized endocrine cells secrete hormones into body fluids, often blood. Hormones reach virtually all body cells, but are bound only by some cells.

▲ **Figure 11.5 Local and long-distance cell signaling by secreted molecules in animals.** In both local and long-distance signaling, only specific target cells that can recognize a given signaling molecule will respond to it.

Both animals and plants use chemicals called **hormones** for long-distance signaling. In hormonal signaling in animals, also known as *endocrine signaling*, specialized cells release hormone molecules, which travel via the circulatory system to other parts of the body, where they reach target cells that can recognize and respond to the hormones **(Figure 11.5c)**. Plant hormones (often called *plant growth regulators*) sometimes travel in vessels but more often reach their targets by moving through cells or by diffusing through the air as a gas (see Chapter 39). Hormones vary widely in molecular size and type, as do local regulators. For instance, the plant hormone ethylene, a gas that promotes fruit ripening and helps regulate growth, is a hydrocarbon of only six atoms (C_2H_4), small enough to pass through cell walls. In contrast, the mammalian hormone insulin, which regulates sugar levels in the blood, is a protein with thousands of atoms.

The transmission of a signal through the nervous system can also be considered an example of long-distance signaling. An electrical signal travels the length of a nerve cell and is then converted back to a chemical signal when a signaling molecule is released and crosses the synapse to another nerve cell. Here it is converted back to an electrical signal. In this way, a nerve signal can travel along a series of nerve cells. Because some nerve cells are quite long, the nerve signal can quickly travel great distances—from your brain to your big toe, for example. This type of long-distance signaling will be covered in detail in Chapter 48.

What happens when a cell encounters a secreted signaling molecule? The ability of a cell to respond is determined by whether it has a specific receptor molecule that can bind to the signaling molecule. The information conveyed by this binding, the signal, must then be changed into another form—transduced—inside the cell before the cell can respond. The remainder of the chapter discusses this process, primarily as it occurs in animal cells.

The Three Stages of Cell Signaling: *A Preview*

Our current understanding of how chemical messengers act via signal transduction pathways had its origins in the pioneering work of Earl W. Sutherland, whose research led to a Nobel Prize in 1971. Sutherland and his colleagues at Vanderbilt University were investigating how the animal hormone epinephrine (also called adrenaline) stimulates the breakdown of the storage polysaccharide glycogen within liver cells and skeletal muscle cells. Glycogen breakdown releases the sugar glucose 1-phosphate, which the cell converts to glucose 6-phosphate. The cell (a liver cell, for example) can then use this compound, an early intermediate in glycolysis, for energy production. Alternatively, the compound can be stripped of phosphate and released from the liver cell into the blood as glucose, which can fuel cells throughout the body. Thus, one effect of epinephrine is the mobilization of fuel reserves, which can be used by the animal to either defend itself (fight) or escape whatever elicited a scare (flight). (The gazelle in Figure 11.1 is clearly engaged in the latter.)

Sutherland's research team discovered that epinephrine stimulates glycogen breakdown by somehow activating a cytosolic enzyme, glycogen phosphorylase. However, when epinephrine was added to a test-tube mixture containing the enzyme and its substrate, glycogen, no breakdown occurred. Epinephrine could activate glycogen phosphorylase only when the hormone was added to a solution containing *intact* cells. This result told Sutherland two things. First, epinephrine does not interact directly with the enzyme responsible for glycogen breakdown; an intermediate step or series of steps must be occurring inside the cell. Second, the plasma membrane is somehow involved in transmitting the signal.

Sutherland's early work suggested that the process going on at the receiving end of a cellular conversation can be dissected into three stages: reception, transduction, and response **(Figure 11.6)**:

1. **Reception.** Reception is the target cell's detection of a signaling molecule coming from outside the cell. A chemical signal is "detected" when the signaling molecule binds to a receptor protein located at the cell's surface or inside the cell.

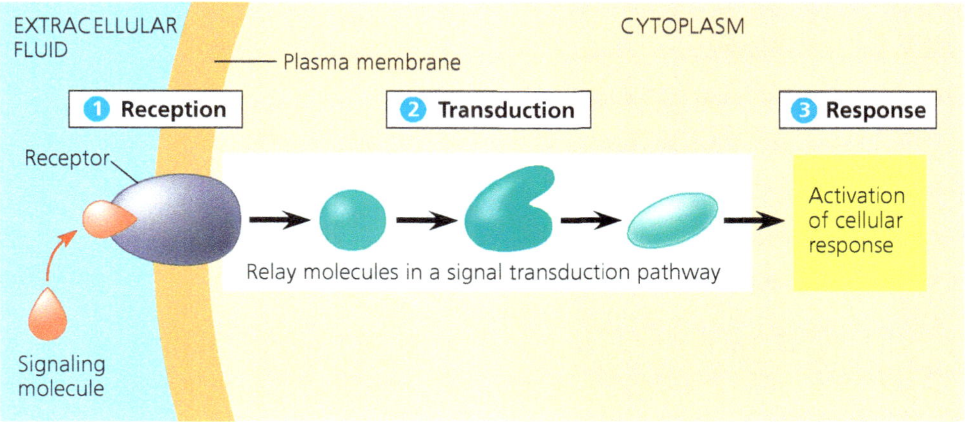

▶ **Figure 11.6 Overview of cell signaling.** From the perspective of the cell receiving the message, cell signaling can be divided into three stages: signal reception, signal transduction, and cellular response. When reception occurs at the plasma membrane, as shown here, the transduction stage is usually a pathway of several steps, with each relay molecule in the pathway bringing about a change in the next molecule. The final molecule in the pathway triggers the cell's response. The three stages are explained in more detail in the text.

? *How does the epinephrine in Sutherland's experiment fit into this diagram of cell signaling?*

2 **Transduction.** The binding of the signaling molecule changes the receptor protein in some way, initiating the process of transduction. The transduction stage converts the signal to a form that can bring about a specific cellular response. In Sutherland's system, the binding of epinephrine to a receptor protein in a liver cell's plasma membrane leads to activation of glycogen phosphorylase. Transduction sometimes occurs in a single step but more often requires a sequence of changes in a series of different molecules—a *signal transduction pathway*. The molecules in the pathway are often called relay molecules.

3 **Response.** In the third stage of cell signaling, the transduced signal finally triggers a specific cellular response. The response may be almost any imaginable cellular activity—such as catalysis by an enzyme (for example, glycogen phosphorylase), rearrangement of the cytoskeleton, or activation of specific genes in the nucleus. The cell-signaling process helps ensure that crucial activities like these occur in the right cells, at the right time, and in proper coordination with the activities of other cells of the organism. We'll now explore the mechanisms of cell signaling in more detail, including a discussion of fine-tuning and termination of the process.

CONCEPT CHECK 11.1

1. Explain how signaling is involved in ensuring that yeast cells fuse only with cells of the opposite mating type.
2. Explain how nerve cells provide examples of both local and long-distance signaling.
3. WHAT IF? When epinephrine is mixed with glycogen phosphorylase and glycogen in a test tube, is glucose 1-phosphate generated? Why or why not?
4. In liver cells, glycogen phosphorylase acts in which of the three stages of the signaling pathway associated with an epinephrine-initiated signal?

For suggested answers, see Appendix A.

CONCEPT 11.2

Reception: A signaling molecule binds to a receptor protein, causing it to change shape

A radio station broadcasts its signal indiscriminately, but it can only be picked up by radios tuned to the right wavelength: Reception of the signal depends on the receiver. Similarly, the signals emitted by an **a** yeast cell are "heard" only by its prospective mates, α cells. In the case of epinephrine, the hormone encounters many types of cells as it circulates in the blood, but only certain target cells detect and react to the hormone molecule. A receptor protein on or in the target cell allows the cell to "hear" the signal and respond to it. The signaling molecule is complementary in shape to a specific site on the receptor and attaches there, like a key in a lock or a substrate in the catalytic site of an enzyme. The signaling molecule behaves as a **ligand**, the term for a molecule that specifically binds to another molecule, often a larger one. Ligand binding generally causes a receptor protein to undergo a change in shape. For many receptors, this shape change directly activates the receptor, enabling it to interact with other cellular molecules. For other kinds of receptors, the immediate effect of ligand binding is to cause the aggregation of two or more receptor molecules, which leads to further molecular events inside the cell.

Most signal receptors are plasma membrane proteins. Their ligands are water-soluble and generally too large to pass freely through the plasma membrane. Other signal receptors, however, are located inside the cell. We discuss both of these types next.

Receptors in the Plasma Membrane

Most water-soluble signaling molecules bind to specific sites on receptor proteins that span the cell's plasma membrane. Such a transmembrane receptor transmits information from the extracellular environment to the inside of the cell by changing shape or aggregating when a specific ligand binds to it. We can see how cell-surface transmembrane receptors work by looking at three major types: G protein-coupled receptors, receptor tyrosine kinases, and ion channel receptors. These receptors are discussed and illustrated in **Figure 11.7**, on the next three pages; study this figure before going on.

Cell-surface receptor molecules play crucial roles in the biological systems of animals, and not surprisingly, their malfunctions are associated with many human diseases, including cancer, heart disease, and asthma. Working out the structure and function of these receptors will allow us to better understand and treat these conditions. Therefore, this endeavor has been a major focus of both university research teams and the pharmaceutical industry. In spite of this effort, and although cell-surface receptors make up 30% of all human proteins, they make up only 1% of the proteins whose structures have been determined by X-ray crystallography (see Figure 5.24): Their structures are very challenging to determine.

The largest family of human cell-surface receptors consists of the nearly 1,000 G protein-coupled receptors (GPCRs). After persistent efforts, researchers have made significant

▼ **Figure 11.7**

Exploring Cell-Surface Transmembrane Receptors

G Protein-Coupled Receptors

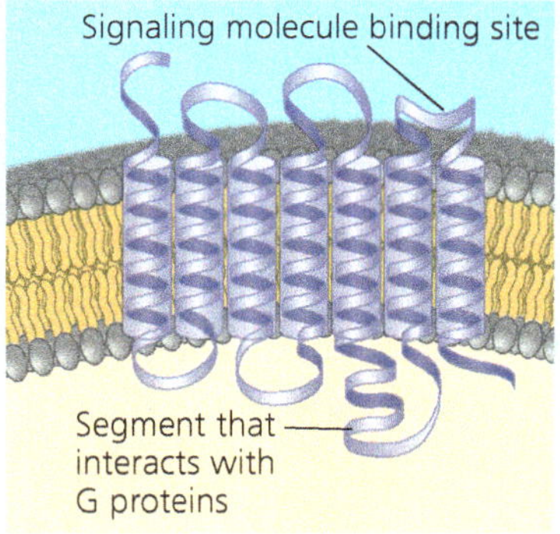

G protein-coupled receptor

A **G protein-coupled receptor** (GPCR) is a cell-surface transmembrane receptor that works with the help of a **G protein**, a protein that binds the energy-rich molecule GTP. Many different signaling molecules, including yeast mating factors, epinephrine and many other hormones, and neurotransmitters, use G protein-coupled receptors. These receptors vary in the binding sites for their signaling molecules (often referred to as their ligands) and also for different types of G proteins inside the cell. Nevertheless, G protein-coupled receptor proteins are all remarkably similar in structure.

In fact, they make up a large family of eukaryotic receptor proteins with a secondary structure in which the single polypeptide, represented here as a ribbon, has seven transmembrane α helices, outlined with cylinders and depicted in a row for clarity. Specific loops between the helices form binding sites for signaling and G protein molecules.

G protein-coupled receptor systems are extremely widespread and diverse in their functions, including roles in embryonic development and sensory reception. In humans, for example, vision, smell, and taste depend on such systems. Similarities in structure in G proteins and G protein-coupled receptors in diverse organisms suggest that G proteins and associated receptors evolved very early.

G protein systems are involved in many human diseases, including bacterial infections. The bacteria that cause cholera, pertussis (whooping cough), and botulism, among others, make their victims ill by producing toxins that interfere with G protein function. Pharmacologists now realize that up to 60% of all medicines used today exert their effects by influencing G protein pathways.

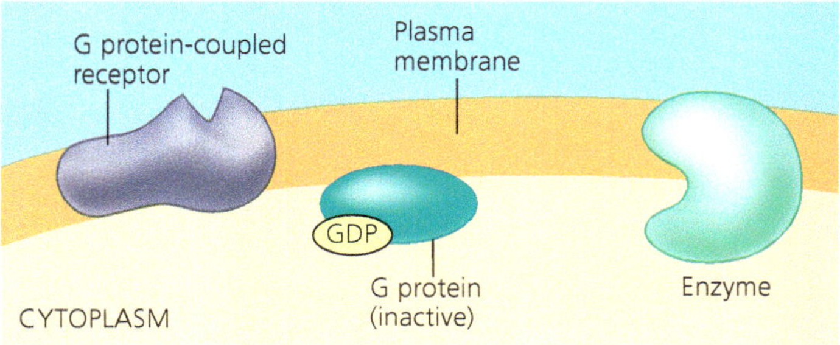

1 Loosely attached to the cytoplasmic side of the membrane, the G protein functions as a molecular switch that is either on or off, depending on which of two guanine nucleotides is attached, GDP or GTP—hence the term *G protein*. (GTP, or guanosine triphosphate, is similar to ATP.) When GDP is bound to the G protein, as shown above, the G protein is inactive. The receptor and G protein work together with another protein, usually an enzyme.

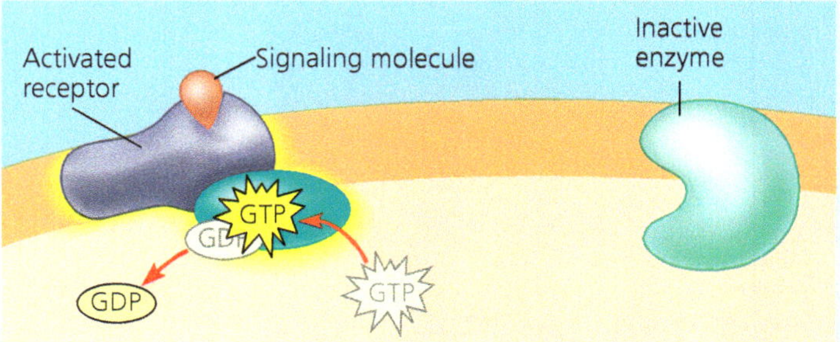

2 When the appropriate signaling molecule binds to the extracellular side of the receptor, the receptor is activated and changes shape. Its cytoplasmic side then binds an inactive G protein, causing a GTP to displace the GDP. This activates the G protein.

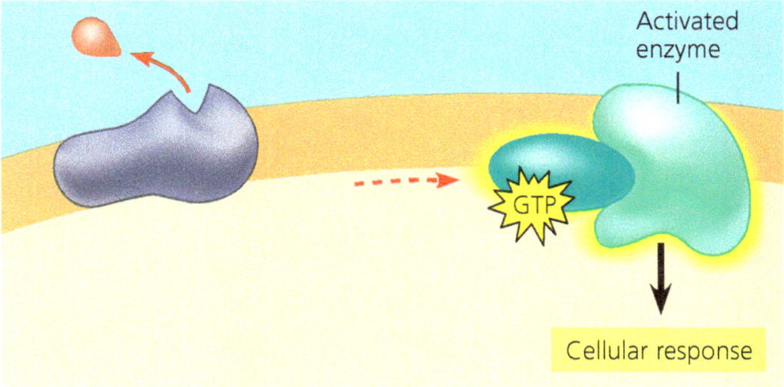

3 The activated G protein dissociates from the receptor, diffuses along the membrane, and then binds to an enzyme, altering the enzyme's shape and activity. Once activated, the enzyme can trigger the next step leading to a cellular response. (Binding of signaling molecules is reversible: Like other ligands, they bind and dissociate many times. The ligand concentration outside the cell determines how often a ligand is bound and causes signaling.)

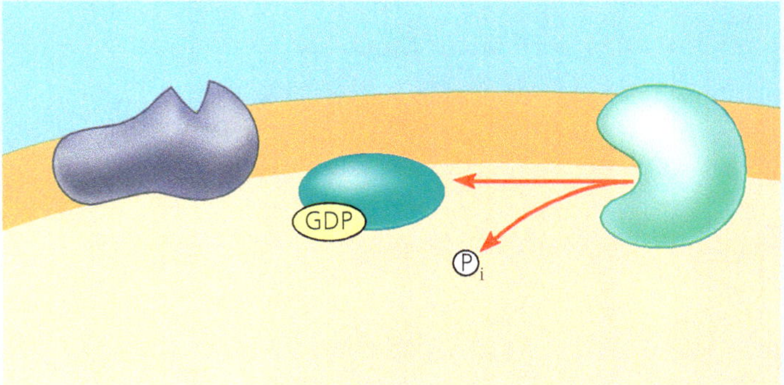

4 The changes in the enzyme and G protein are only temporary because the G protein also functions as a GTPase enzyme—in other words, it then hydrolyzes its bound GTP to GDP. Now inactive again, the G protein leaves the enzyme, which returns to its original state. The G protein is now available for reuse. The GTPase function of the G protein allows the pathway to shut down rapidly when the signaling molecule is no longer present.

Continued on next page

▼ **Figure 11.7 (continued)**

Exploring Cell-Surface Transmembrane Receptors

Receptor Tyrosine Kinases

Receptor tyrosine kinases (RTKs) belong to a major class of plasma membrane receptors characterized by having enzymatic activity. A *kinase* is an enzyme that catalyzes the transfer of phosphate groups. The part of the receptor protein extending into the cytoplasm functions as a tyrosine kinase, an enzyme that catalyzes the transfer of a phosphate group from ATP to the amino acid tyrosine on a substrate protein. Thus, receptor tyrosine kinases are membrane receptors that attach phosphates to tyrosines.

One receptor tyrosine kinase complex may activate ten or more different transduction pathways and cellular responses. Often, more than one signal transduction pathway can be triggered at once, helping the cell regulate and coordinate many aspects of cell growth and cell reproduction. The ability of a single ligand-binding event to trigger so many pathways is a key difference between receptor tyrosine kinases and G protein-coupled receptors. Abnormal receptor tyrosine kinases that function even in the absence of signaling molecules are associated with many kinds of cancer.

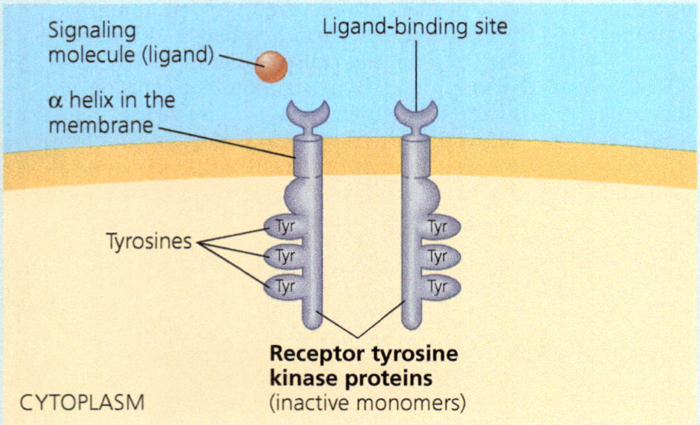

1 Many receptor tyrosine kinases have the structure depicted schematically here. Before the signaling molecule binds, the receptors exist as individual units referred to as monomers. Notice that each has an extracellular ligand-binding site, an α helix spanning the membrane, and an intracellular tail containing multiple tyrosines.

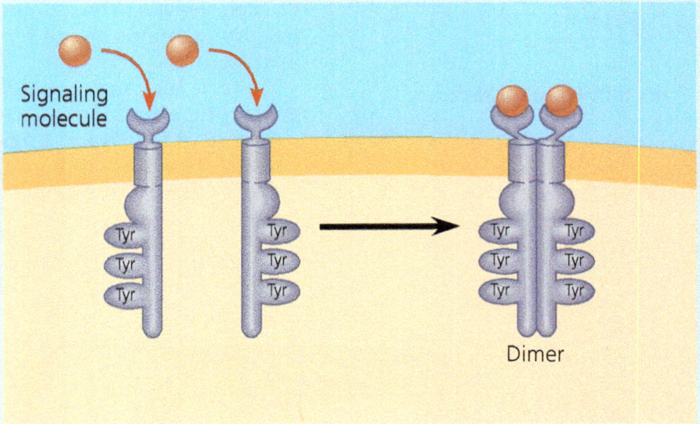

2 The binding of a signaling molecule (such as a growth factor) causes two receptor monomers to associate closely with each other, forming a complex known as a dimer (dimerization).

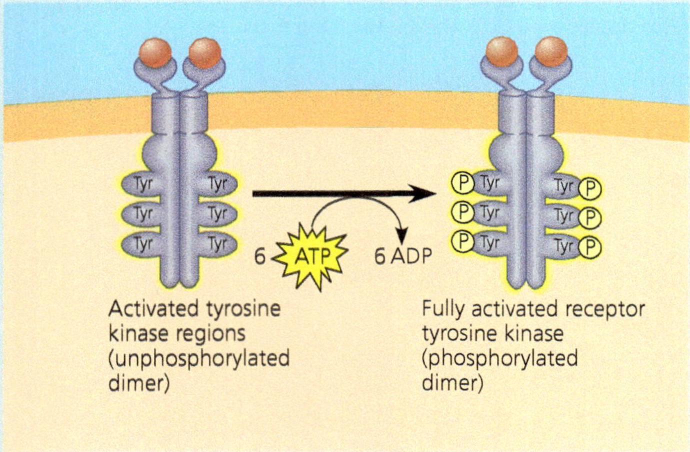

3 Dimerization activates the tyrosine kinase region of each monomer; each tyrosine kinase adds a phosphate from an ATP molecule to a tyrosine on the tail of the other monomer.

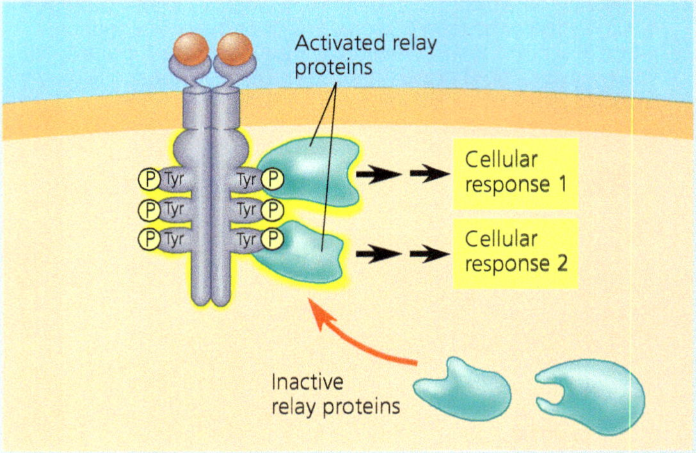

4 Now that the receptor is fully activated, it is recognized by specific relay proteins inside the cell. Each such protein binds to a specific phosphorylated tyrosine, undergoing a resulting structural change that activates the bound protein. Each activated protein triggers a transduction pathway, leading to a cellular response.

Ion Channel Receptors

A **ligand-gated ion channel** is a type of membrane receptor containing a region that can act as a "gate" when the receptor changes shape. When a signaling molecule binds as a ligand to the receptor protein, the gate opens or closes, allowing or blocking the flow of specific ions, such as Na^+ or Ca^{2+}, through a channel in the receptor. Like the other receptors we have discussed, these proteins bind the ligand at a specific site on their extracellular sides.

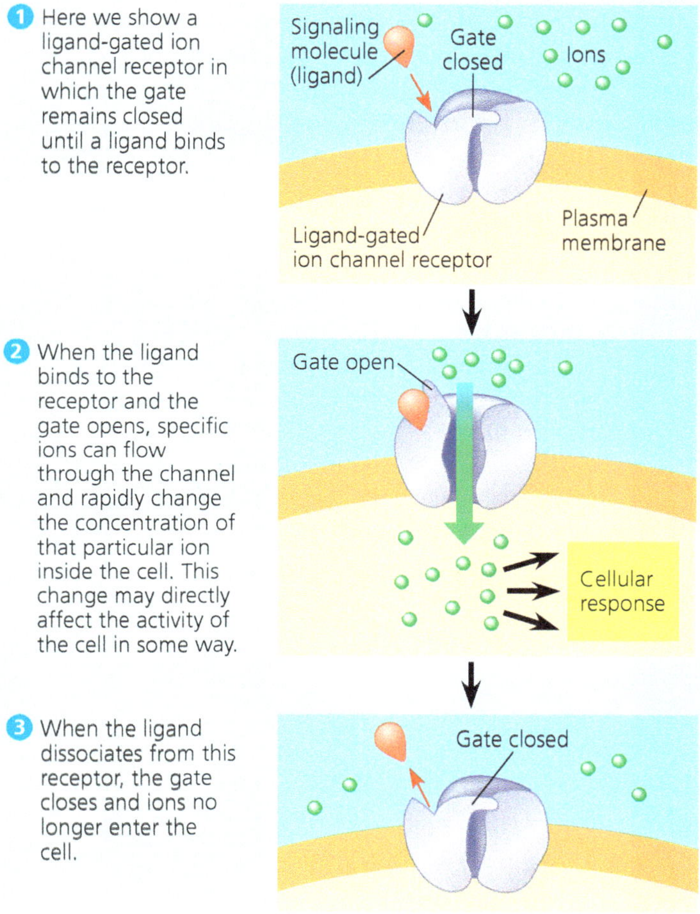

Ligand-gated ion channels are very important in the nervous system. For example, the neurotransmitter molecules released at a synapse between two nerve cells (see Figure 11.5b) bind as ligands to ion channels on the receiving cell, causing the channels to open. Ions flow in (or, in some cases, out), triggering an electrical signal that propagates down the length of the receiving cell. Some gated ion channels are controlled by electrical signals instead of ligands; these *voltage-gated ion channels* are also crucial to the functioning of the nervous system, as we will discuss in Chapter 48.

MAKE CONNECTIONS *Examine the ion channel protein in Figure 7.1 (p. 125) and the discussion of it on page 135. What type of stimulus opens that ion channel? According to the information given above, what type of ion channel is described?*

▼ **Figure 11.8**

IMPACT

Determining the Structure of a G Protein-Coupled Receptor (GPCR)

GPCRs are flexible and inherently unstable, so they have been difficult to crystallize, a required step in determining their structure by X-ray crystallography. Recently, however, researchers have crystallized the human β_2-adrenergic receptor in the presence of a ligand similar to the natural one (green in the model below) and cholesterol (orange), which stabilized the receptor enough for its structure to be determined. Two receptor molecules (blue) are shown here as ribbon models in a side view within the plasma membrane.

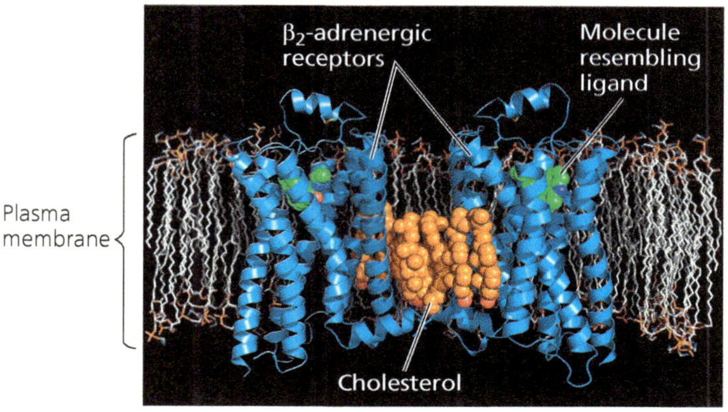

WHY IT MATTERS The β_2-adrenergic receptor is found on smooth muscle cells throughout the body, and abnormal forms of it are associated with diseases such as asthma, hypertension, and heart failure. Current drugs used for these conditions produce unwanted side effects, and further research may yield better drugs. Also, since GPCRs share structural similarities, this work on the β_2-adrenergic receptor will aid development of treatments for diseases associated with other GPCRs.

FURTHER READING R. Ranganathan, Signaling across the cell membrane, *Science* 318:1253–1254 (2007).

WHAT IF? The model shown above represents the receptor in an inactive state, not bound to a G protein. Can you suggest conditions for crystallizing the protein that would reveal the structure of the receptor while it is actively signaling to the inside of the cell?

breakthroughs in elucidating the structure of several G protein-coupled receptors over the past few years **(Figure 11.8)**.

Abnormal functioning of receptor tyrosine kinases (RTKs) is associated with many types of cancers. For example, patients with breast cancer cells that have excessive levels of a receptor tyrosine kinase called HER2 have a poor prognosis. Using molecular biological techniques, researchers have developed a protein called Herceptin that binds to HER2 on cells and inhibits their growth, thus thwarting further tumor development. In some clinical studies, treatment with Herceptin improved patient survival rates by more than one-third. One goal of ongoing research into these cell-surface receptors and other cell-signaling proteins is development of additional successful treatments.

Intracellular Receptors

Intracellular receptor proteins are found in either the cytoplasm or nucleus of target cells. To reach such a receptor, a chemical messenger passes through the target cell's plasma membrane. A number of important signaling molecules can do this because they are either hydrophobic enough or small enough to cross the hydrophobic interior of the membrane. Such hydrophobic chemical messengers include the steroid hormones and thyroid hormones of animals. Another chemical signaling molecule with an intracellular receptor is nitric oxide (NO), a gas; its very small molecules readily pass between the membrane phospholipids.

The behavior of testosterone is representative of steroid hormones. In males, the hormone is secreted by cells of the testes. It then travels through the blood and enters cells all over the body. However, only cells that contain receptor molecules for testosterone respond. In these cells, the hormone binds to the receptor protein, activating it **(Figure 11.9)**. With the hormone attached, the active form of the receptor protein then enters the nucleus and turns on specific genes that control male sex characteristics.

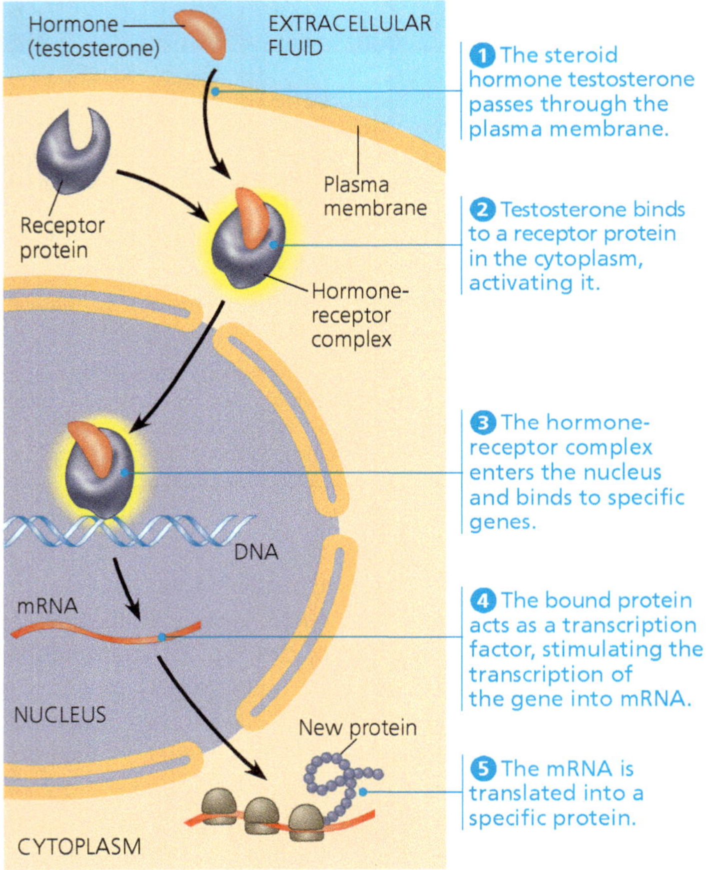

▲ **Figure 11.9 Steroid hormone interacting with an intracellular receptor.**

? *Why is a cell-surface receptor protein not required for this steroid hormone to enter the cell?*

How does the activated hormone-receptor complex turn on genes? Recall that the genes in a cell's DNA function by being transcribed and processed into messenger RNA (mRNA), which leaves the nucleus and is translated into a specific protein by ribosomes in the cytoplasm (see Figure 5.25). Special proteins called *transcription factors* control which genes are turned on—that is, which genes are transcribed into mRNA—in a particular cell at a particular time. The testosterone receptor, when activated, acts as a transcription factor that turns on specific genes.

By acting as a transcription factor, the testosterone receptor itself carries out the complete transduction of the signal. Most other intracellular receptors function in the same way, although many of them, such as the thyroid hormone receptor, are already in the nucleus before the signaling molecule reaches them. Interestingly, many of these intracellular receptor proteins are structurally similar, suggesting an evolutionary kinship.

CONCEPT CHECK 11.2

1. Nerve growth factor (NGF) is a water-soluble signaling molecule. Would you expect the receptor for NGF to be intracellular or in the plasma membrane? Why?
2. **WHAT IF?** What would the effect be if a cell made defective receptor tyrosine kinase proteins that were unable to dimerize?
3. **MAKE CONNECTIONS** How is ligand binding similar to the process of allosteric regulation of enzymes? See Figure 8.19 on page 158.

For suggested answers, see Appendix A.

CONCEPT 11.3

Transduction: Cascades of molecular interactions relay signals from receptors to target molecules in the cell

When receptors for signaling molecules are plasma membrane proteins, like most of those we have discussed, the transduction stage of cell signaling is usually a multistep pathway. Steps often include activation of proteins by addition or removal of phosphate groups or release of other small molecules or ions that act as messengers. One benefit of multiple steps is the possibility of greatly amplifying a signal. If some of the molecules in a pathway transmit the signal to numerous molecules at the next step in the series, the result can be a large number of activated molecules at the end of the pathway. Moreover, multistep pathways provide more opportunities for coordination and regulation than simpler systems do. This allows fine-tuning of the response, in both unicellular and multicellular organisms, as we'll discuss later in the chapter.

Signal Transduction Pathways

The binding of a specific signaling molecule to a receptor in the plasma membrane triggers the first step in the chain of molecular interactions—the signal transduction pathway—that leads to a particular response within the cell. Like falling dominoes, the signal-activated receptor activates another molecule, which activates yet another molecule, and so on, until the protein that produces the final cellular response is activated. The molecules that relay a signal from receptor to response, which we call relay molecules in this book, are often proteins. The interaction of proteins is a major theme of cell signaling. Indeed, protein interaction is a unifying theme of all regulation at the cellular level.

Keep in mind that the original signaling molecule is not physically passed along a signaling pathway; in most cases, it never even enters the cell. When we say that the signal is relayed along a pathway, we mean that certain information is passed on. At each step, the signal is transduced into a different form, commonly a shape change in a protein. Very often, the shape change is brought about by phosphorylation.

Protein Phosphorylation and Dephosphorylation

Previous chapters introduced the concept of activating a protein by adding one or more phosphate groups to it (see Figure 8.10a). In Figure 11.7, we have already seen how phosphorylation is involved in the activation of receptor tyrosine kinases. In fact, the phosphorylation and dephosphorylation of proteins is a widespread cellular mechanism for regulating protein activity. An enzyme that transfers phosphate groups from ATP to a protein is generally known as a **protein kinase**. Recall that a receptor tyrosine kinase phosphorylates tyrosines on the other receptor tyrosine kinase in a dimer. Most cytoplasmic protein kinases, however, act on proteins different from themselves. Another distinction is that most cytoplasmic protein kinases phosphorylate either of two other amino acids, serine or threonine, rather than tyrosine. Such serine/threonine kinases are widely involved in signaling pathways in animals, plants, and fungi.

Many of the relay molecules in signal transduction pathways are protein kinases, and they often act on other protein kinases in the pathway. **Figure 11.10** depicts a hypothetical

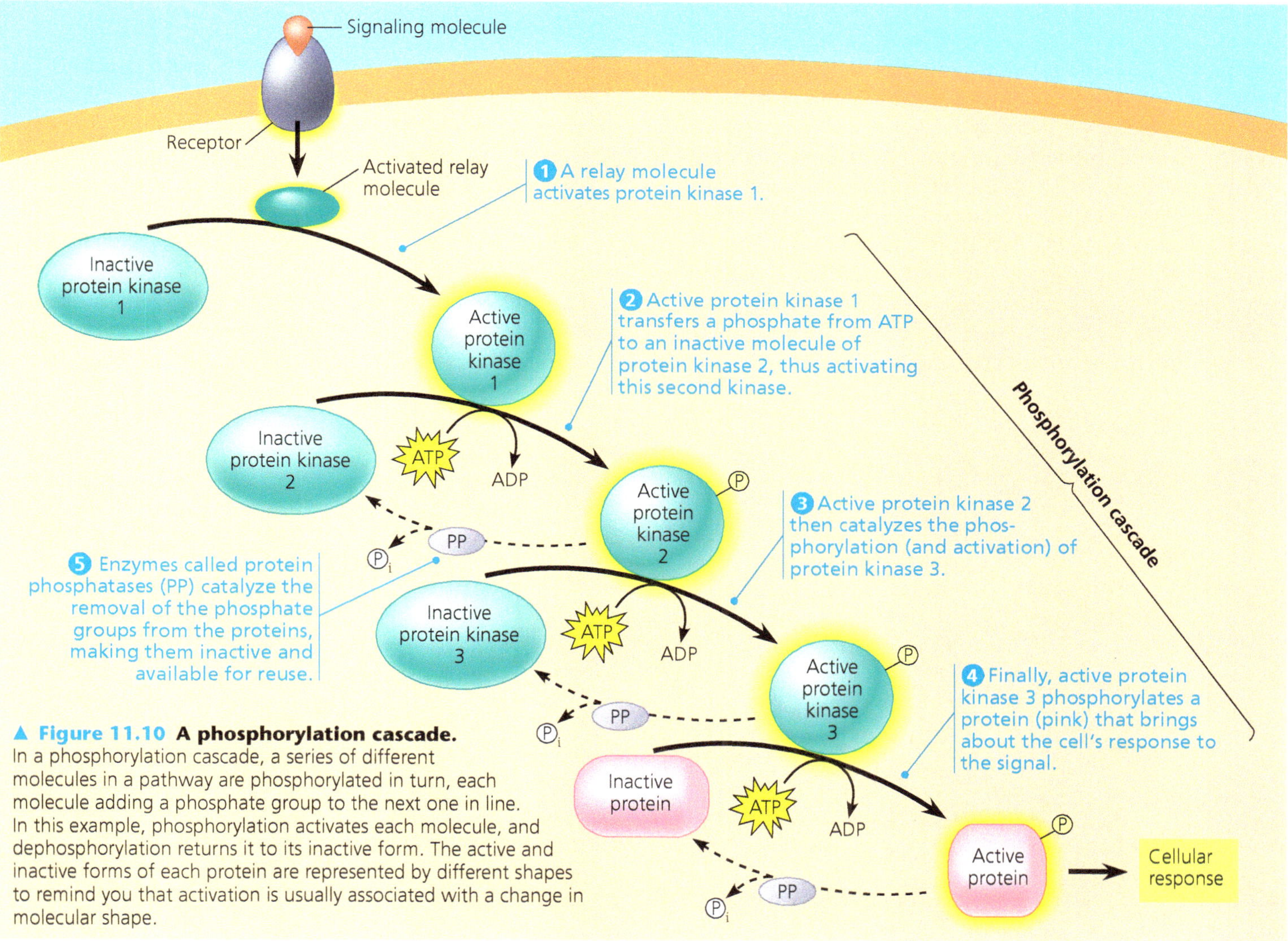

▲ **Figure 11.10 A phosphorylation cascade.** In a phosphorylation cascade, a series of different molecules in a pathway are phosphorylated in turn, each molecule adding a phosphate group to the next one in line. In this example, phosphorylation activates each molecule, and dephosphorylation returns it to its inactive form. The active and inactive forms of each protein are represented by different shapes to remind you that activation is usually associated with a change in molecular shape.

? *Which protein is responsible for activation of protein kinase 3?*

pathway containing three different protein kinases that create a "phosphorylation cascade." The sequence shown is similar to many known pathways, including those triggered in yeast by mating factors and in animal cells by many growth factors. The signal is transmitted by a cascade of protein phosphorylations, each bringing with it a shape change. Each such shape change results from the interaction of the newly added phosphate groups with charged or polar amino acids (see Figure 5.16). The addition of phosphate groups often changes a protein from an inactive form to an active form. In other cases, though, phosphorylation *decreases* the activity of the protein.

The importance of protein kinases can hardly be overstated. About 2% of our own genes are thought to code for protein kinases. A single cell may have hundreds of different kinds, each specific for a different substrate protein. Together, they probably regulate a large proportion of the thousands of proteins in a cell. Among these are most of the proteins that, in turn, regulate cell reproduction. Abnormal activity of such a kinase can cause abnormal cell growth and contribute to the development of cancer.

Equally important in the phosphorylation cascade are the **protein phosphatases**, enzymes that can rapidly remove phosphate groups from proteins, a process called dephosphorylation. By dephosphorylating and thus inactivating protein kinases, phosphatases provide the mechanism for turning off the signal transduction pathway when the initial signal is no longer present. Phosphatases also make the protein kinases available for reuse, enabling the cell to respond again to an extracellular signal. The phosphorylation-dephosphorylation system acts as a molecular switch in the cell, turning activities on or off, or up or down, as required. At any given moment, the activity of a protein regulated by phosphorylation depends on the balance in the cell between active kinase molecules and active phosphatase molecules.

Small Molecules and Ions as Second Messengers

Not all components of signal transduction pathways are proteins. Many signaling pathways also involve small, nonprotein, water-soluble molecules or ions called **second messengers**. (This term is used because the pathway's "first messenger" is considered to be the extracellular signaling molecule—the ligand—that binds to the membrane receptor.) Because second messengers are small and water-soluble, they can readily spread throughout the cell by diffusion. For example, as we'll see shortly, a second messenger called cyclic AMP carries the signal initiated by epinephrine from the plasma membrane of a liver or muscle cell into the cell's interior, where the signal eventually brings about glycogen breakdown. Second messengers participate in pathways that are initiated by both G protein-coupled receptors and receptor tyrosine kinases. The two most widely used second messengers are cyclic AMP and calcium ions, Ca^{2+}. A large variety of relay proteins are sensitive to the cytosolic concentration of one or the other of these second messengers.

Cyclic AMP

As discussed on page 209, Earl Sutherland established that epinephrine somehow causes glycogen breakdown without passing through the plasma membrane. This discovery prompted him to search for a second messenger that transmits the signal from the plasma membrane to the metabolic machinery in the cytoplasm.

Sutherland found that the binding of epinephrine to the plasma membrane of a liver cell elevates the cytosolic concentration of a compound called cyclic adenosine monophosphate, abbreviated as either **cyclic AMP** or **cAMP** (**Figure 11.11**). An enzyme embedded in the plasma

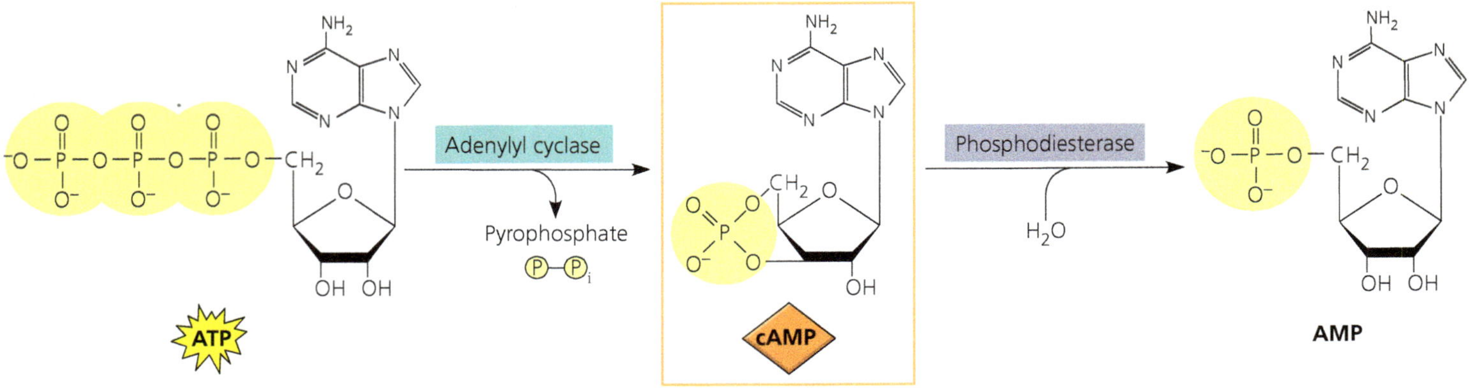

▲ **Figure 11.11 Cyclic AMP.** The second messenger cyclic AMP (cAMP) is made from ATP by adenylyl cyclase, an enzyme embedded in the plasma membrane. Cyclic AMP is inactivated by phosphodiesterase, an enzyme that converts it to AMP.

WHAT IF? *What would happen if a molecule that inactivated phosphodiesterase were introduced into the cell?*

membrane, **adenylyl cyclase**, converts ATP to cAMP in response to an extracellular signal—in this case, provided by epinephrine. But epinephrine doesn't stimulate adenylyl cyclase directly. When epinephrine outside the cell binds to a specific receptor protein, the protein activates adenylyl cyclase, which in turn can catalyze the synthesis of many molecules of cAMP. In this way, the normal cellular concentration of cAMP can be boosted 20-fold in a matter of seconds. The cAMP broadcasts the signal to the cytoplasm. It does not persist for long in the absence of the hormone because another enzyme, called phosphodiesterase, converts cAMP to AMP. Another surge of epinephrine is needed to boost the cytosolic concentration of cAMP again.

Subsequent research has revealed that epinephrine is only one of many hormones and other signaling molecules that trigger the formation of cAMP. It has also brought to light the other components of cAMP pathways, including G proteins, G protein-coupled receptors, and protein kinases **(Figure 11.12)**. The immediate effect of cAMP is usually the activation of a serine/threonine kinase called *protein kinase A*. The activated protein kinase A then phosphorylates various other proteins, depending on the cell type. (The complete pathway for epinephrine's stimulation of glycogen breakdown is shown later, in Figure 11.16.)

Further regulation of cell metabolism is provided by other G protein systems that *inhibit* adenylyl cyclase. In these systems, a different signaling molecule activates a different receptor, which in turn activates an *inhibitory* G protein.

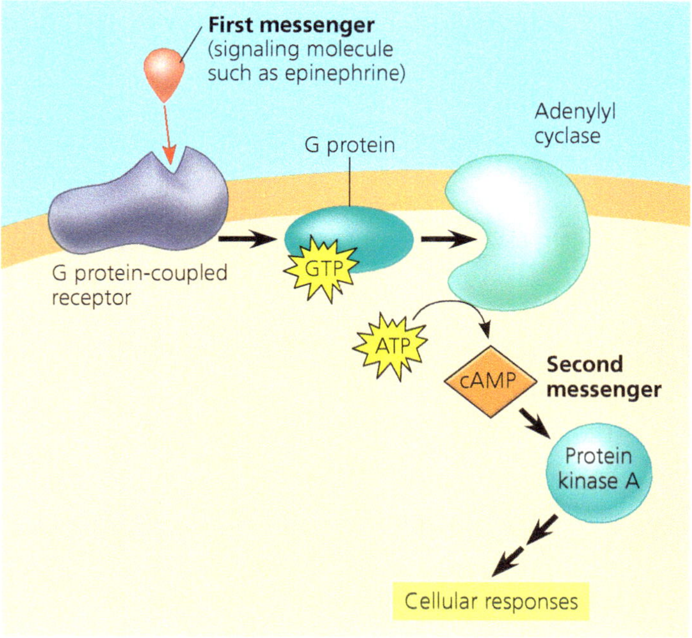

▲ **Figure 11.12 cAMP as a second messenger in a G protein signaling pathway.** The first messenger activates a G protein-coupled receptor, which activates a specific G protein. In turn, the G protein activates adenylyl cyclase, which catalyzes the conversion of ATP to cAMP. The cAMP then acts as a second messenger and activates another protein, usually protein kinase A, leading to cellular responses.

Now that we know about the role of cAMP in G protein signaling pathways, we can explain in molecular detail how certain microbes cause disease. Consider cholera, a disease that is frequently epidemic in places where the water supply is contaminated with human feces. People acquire the cholera bacterium, *Vibrio cholerae*, by drinking contaminated water. The bacteria form a biofilm on the lining of the small intestine and produce a toxin. The cholera toxin is an enzyme that chemically modifies a G protein involved in regulating salt and water secretion. Because the modified G protein is unable to hydrolyze GTP to GDP, it remains stuck in its active form, continuously stimulating adenylyl cyclase to make cAMP. The resulting high concentration of cAMP causes the intestinal cells to secrete large amounts of salts into the intestines, with water following by osmosis. An infected person quickly develops profuse diarrhea and if left untreated can soon die from the loss of water and salts.

Our understanding of signaling pathways involving cyclic AMP or related messengers has allowed us to develop treatments for certain conditions in humans. In one pathway, *cyclic GMP*, or *cGMP*, acts as a signaling molecule whose effects include relaxation of smooth muscle cells in artery walls. A compound that inhibits the hydrolysis of cGMP to GMP, thus prolonging the signal, was originally prescribed for chest pains because it increased blood flow to the heart muscle. Under the trade name Viagra, this compound is now widely used as a treatment for erectile dysfunction in human males. Because Viagra leads to dilation of blood vessels, it also allows increased blood flow to the penis, optimizing physiological conditions for penile erections.

Calcium Ions and Inositol Trisphosphate (IP_3)

Many signaling molecules in animals, including neurotransmitters, growth factors, and some hormones, induce responses in their target cells via signal transduction pathways that increase the cytosolic concentration of calcium ions (Ca^{2+}). Calcium is even more widely used than cAMP as a second messenger. Increasing the cytosolic concentration of Ca^{2+} causes many responses in animal cells, including muscle cell contraction, secretion of certain substances, and cell division. In plant cells, a wide range of hormonal and environmental stimuli can cause brief increases in cytosolic Ca^{2+} concentration, triggering various signaling pathways, such as the pathway for greening in response to light (see Figure 39.4). Cells use Ca^{2+} as a second messenger in both G protein and receptor tyrosine kinase pathways.

Although cells always contain some Ca^{2+}, this ion can function as a second messenger because its concentration in the cytosol is normally much lower than the concentration

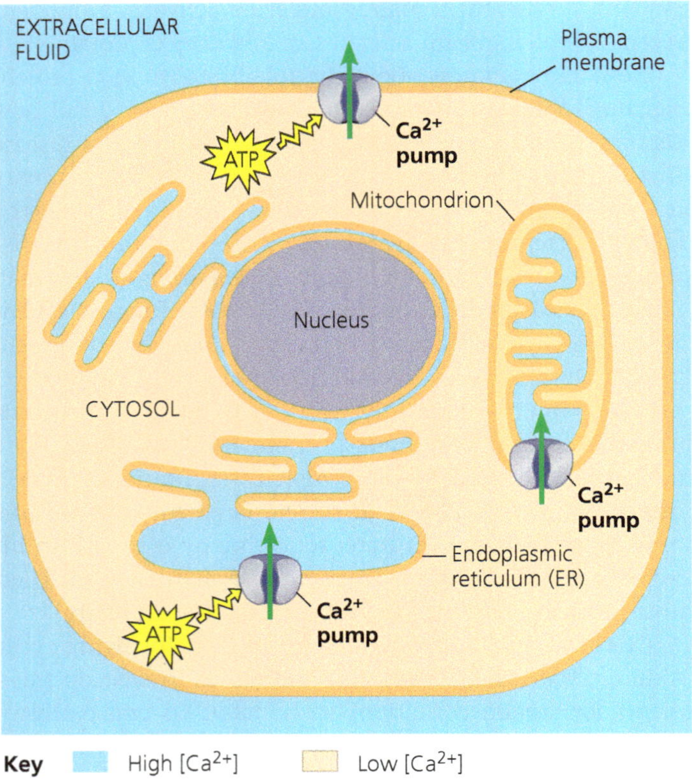

▲ **Figure 11.13 The maintenance of calcium ion concentrations in an animal cell.** The Ca^{2+} concentration in the cytosol is usually much lower (beige) than in the extracellular fluid and ER (blue). Protein pumps in the plasma membrane and the ER membrane, driven by ATP, move Ca^{2+} from the cytosol into the extracellular fluid and into the lumen of the ER. Mitochondrial pumps, driven by chemiosmosis (see Chapter 9), move Ca^{2+} into mitochondria when the calcium level in the cytosol rises significantly.

outside the cell **(Figure 11.13)**. In fact, the level of Ca^{2+} in the blood and extracellular fluid of an animal is often more than 10,000 times higher than that in the cytosol. Calcium ions are actively transported out of the cell and are actively imported from the cytosol into the endoplasmic reticulum (and, under some conditions, into mitochondria and chloroplasts) by various protein pumps. As a result, the calcium concentration in the ER is usually much higher than that in the cytosol. Because the cytosolic calcium level is low, a small change in absolute numbers of ions represents a relatively large percentage change in calcium concentration.

In response to a signal relayed by a signal transduction pathway, the cytosolic calcium level may rise, usually by a mechanism that releases Ca^{2+} from the cell's ER. The pathways leading to calcium release involve still other second messengers, **inositol trisphosphate (IP_3)** and **diacylglycerol (DAG)**. These two messengers are produced by cleavage of a certain kind of phospholipid in the plasma membrane. **Figure 11.14** shows how this occurs and how IP_3 stimulates the release of calcium from the ER. Because IP_3 acts before calcium in these pathways, calcium could be considered a "*third* messenger." However, scientists use the term *second messenger* for all small, nonprotein components of signal transduction pathways.

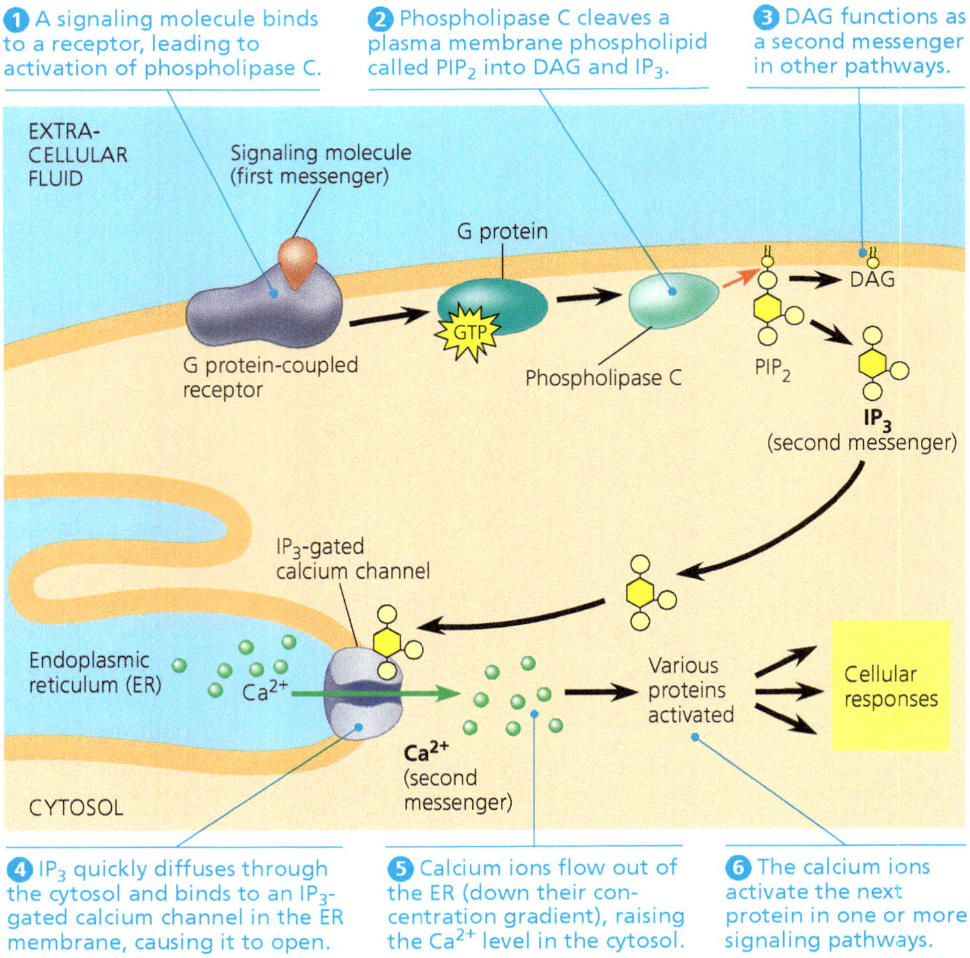

► **Figure 11.14 Calcium and IP_3 in signaling pathways.** Calcium ions (Ca^{2+}) and inositol trisphosphate (IP_3) function as second messengers in many signal transduction pathways. In this figure, the process is initiated by the binding of a signaling molecule to a G protein-coupled receptor. A receptor tyrosine kinase could also initiate this pathway by activating phospholipase C.

CONCEPT CHECK 11.3

1. What is a protein kinase, and what is its role in a signal transduction pathway?
2. When a signal transduction pathway involves a phosphorylation cascade, how does the cell's response get turned off?
3. What is the actual "signal" that is being transduced in any signal transduction pathway, such as those shown in Figures 11.6 and 11.10? In what way is this information being passed from the exterior to the interior of the cell?
4. **WHAT IF?** Upon activation of phospholipase C by the binding of a ligand to a receptor, what effect does the IP_3-gated calcium channel have on Ca^{2+} concentration in the cytosol?

For suggested answers, see Appendix A.

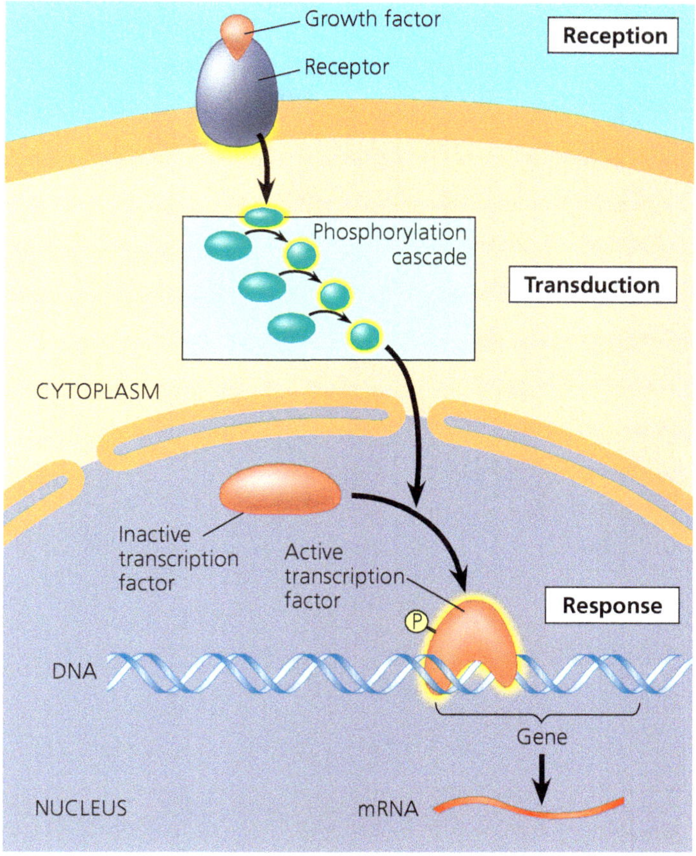

▲ **Figure 11.15 Nuclear responses to a signal: the activation of a specific gene by a growth factor.** This diagram is a simplified representation of a typical signaling pathway that leads to the regulation of gene activity in the cell nucleus. The initial signaling molecule, a local regulator called a growth factor, triggers a phosphorylation cascade, as in Figure 11.10. (The ATP molecules and phosphate groups are not shown.) Once phosphorylated, the last kinase in the sequence enters the nucleus and there activates a gene-regulating protein, a transcription factor. This protein stimulates transcription of a specific gene (or genes). The resulting mRNA then directs the synthesis of a particular protein in the cytoplasm.

CONCEPT 11.4

Response: Cell signaling leads to regulation of transcription or cytoplasmic activities

We now take a closer look at the cell's subsequent response to an extracellular signal—what some researchers call the "output response." What is the nature of the final step in a signaling pathway?

Nuclear and Cytoplasmic Responses

Ultimately, a signal transduction pathway leads to the regulation of one or more cellular activities. The response at the end of the pathway may occur in the nucleus of the cell or in the cytoplasm.

Many signaling pathways ultimately regulate protein synthesis, usually by turning specific genes on or off in the nucleus. Like an activated steroid receptor (see Figure 11.9), the final activated molecule in a signaling pathway may function as a transcription factor. **Figure 11.15** shows an example in which a signaling pathway activates a transcription factor that turns a gene on: The response to the growth factor signal is transcription, the synthesis of mRNA, which will be translated in the cytoplasm into a specific protein. In other cases, the transcription factor might regulate a gene by turning it off. Often a transcription factor regulates several different genes.

Sometimes a signaling pathway may regulate the *activity* of proteins rather than their *synthesis*, directly affecting proteins that function outside the nucleus. For example, a signal may cause the opening or closing of an ion channel in the plasma membrane or a change in cell metabolism. As we have seen, the response of liver cells to the hormone epinephrine helps regulate cellular energy metabolism by affecting the activity of an enzyme. The final step in the signaling pathway that begins with epinephrine binding activates the enzyme that catalyzes the breakdown of glycogen. **Figure 11.16**, on the next page, shows the complete pathway leading to the release of glucose 1-phosphate molecules from glycogen. Notice that as each molecule is activated, the response is amplified, a subject we'll return to shortly.

In addition to controlling enzymes, signaling events may regulate other cellular attributes, even activities of the cell as a whole. An example of the latter can be found in the processes leading to the mating of yeast cells (see Figure 11.2). Yeast cells are not motile; their mating process depends on the growth of localized projections of one cell toward a cell of the

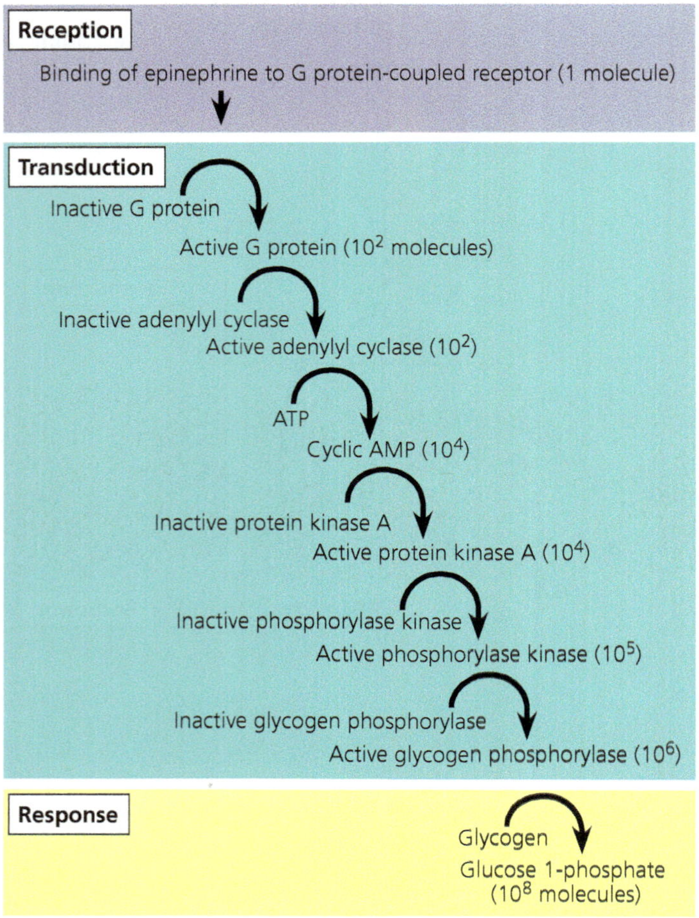

▲ **Figure 11.16 Cytoplasmic response to a signal: the stimulation of glycogen breakdown by epinephrine.** In this signaling system, the hormone epinephrine acts through a G protein-coupled receptor to activate a succession of relay molecules, including cAMP and two protein kinases (see also Figure 11.12). The final protein activated is the enzyme glycogen phosphorylase, which uses inorganic phosphate to release glucose monomers from glycogen in the form of glucose 1-phosphate molecules. This pathway amplifies the hormonal signal: One receptor protein can activate about 100 molecules of G protein, and each enzyme in the pathway, once activated, can act on many molecules of its substrate, the next molecule in the cascade. The number of activated molecules given for each step is approximate.

opposite mating type. As shown in **Figure 11.17**, binding of the mating factor causes this directional growth. When the mating factor binds, it activates signaling pathway kinases that affect the growth and orientation of cytoskeletal microfilaments. Because activation of signaling kinases is coupled in this way to cytoskeletal dynamics, cell projections emerge from regions of the plasma membrane exposed to the highest concentration of the mating factor. As a result, these projections are oriented toward the cell of the opposite mating type, which is the source of the signaling molecule.

The signal receptors, relay molecules, and second messengers introduced so far in this chapter participate in a variety of pathways, leading to both nuclear and cytoplasmic responses. Some of these pathways lead to cell division. The molecular messengers that initiate cell division pathways include growth factors and certain plant and animal hormones. Malfunctioning of growth factor pathways like the one in Figure 11.15 can contribute to the development of cancer, as we will see in Chapter 18.

Fine-Tuning of the Response

Regardless of whether the response occurs in the nucleus or in the cytoplasm, it is fine-tuned at multiple points rather than simply being turned "on" or "off." Here we'll consider four aspects of fine-tuning. First, as mentioned earlier, a signaling pathway with numerous steps between the initial signaling event at the cell surface and the cell's response results in amplification of the signal and thus the response. Second, such a multistep pathway has many different points at which the cell's response can be regulated, contributing to the specificity of the response and allowing coordination with other signaling pathways. Third, the overall efficiency of the response is enhanced by the presence of proteins known as scaffolding proteins. Finally, a crucial point in fine-tuning the response is the termination of the signal.

Signal Amplification

Elaborate enzyme cascades amplify the cell's response to a signal. At each catalytic step in the cascade, the number of activated products is much greater than in the preceding step. For example, in the epinephrine-triggered pathway in Figure 11.16, each adenylyl cyclase molecule catalyzes the formation of many cAMP molecules, each molecule of protein kinase A phosphorylates many molecules of the next kinase in the pathway, and so on. The amplification effect stems from the fact that these proteins persist in the active form long enough to process numerous molecules of substrate before they become inactive again. As a result of the signal's amplification, a small number of epinephrine molecules binding to receptors on the surface of a liver cell or muscle cell can lead to the release of hundreds of millions of glucose molecules from glycogen.

The Specificity of Cell Signaling and Coordination of the Response

Consider two different cells in your body—a liver cell and a heart muscle cell, for example. Both are in contact with your bloodstream and are therefore constantly exposed to many different hormone molecules, as well as to local regulators secreted by nearby cells. Yet the liver cell responds to some signals but ignores others, and the same is true for the heart cell. And some kinds of signals trigger responses in both cells—but different responses. For instance, epinephrine stimulates the liver cell to break down glycogen, but the main response of the heart cell to epinephrine is contraction, leading to a more rapid heartbeat. How do we account for this difference?

▼ Figure 11.17 INQUIRY

How do signals induce directional cell growth during mating in yeast?

EXPERIMENT When a yeast cell binds mating factor molecules from a cell of the opposite mating type, a signaling pathway causes it to grow a projection toward the potential mate. The cell with the projection is called a "shmoo" because it resembles a 1950s cartoon character by that name. Dina Matheos and colleagues in Mark Rose's lab at Princeton University sought to determine how mating factor signaling is linked to this asymmetrical growth. Previous work had shown that activation of Fus3, one of the kinases in the signaling cascade, caused it to move to the membrane near where the factor bound. Preliminary experiments by these researchers identified formin, a protein that directs the construction of microfilaments, as a phosphorylation target of Fus3 kinase. To examine the role of Fus3 and formin in shmoo formation, the researchers generated two mutant yeast strains: one that no longer had the kinase (this strain is called ΔFus3) and one that lacked the formin (Δformin). To observe the effects of these mutations on cell growth induced by the mating factor, the cell walls of each strain were first stained with a green fluorescent dye. These green-stained cells were then exposed to mating factor and stained with a red fluorescent dye that labeled new cell wall growth. Images taken of the cells after the staining procedure were then compared with a similarly treated strain that expressed Fus3 and formin (the wild type).

RESULTS The cells of the wild-type strain showed shmoo projections, whose walls were stained red, while the rest of their cell walls were green, indicating asymmetrical growth. Cells of both the ΔFus3 and Δformin strains showed no shmoo formation, and their cell walls were stained almost uniformly yellow. This color resulted from merged green and red stains, indicating symmetrical growth, characteristic of cells not exposed to mating factor.

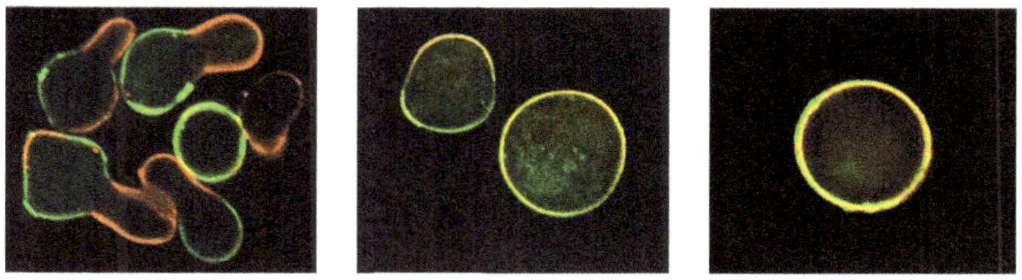

Wild type (with shmoos) ΔFus3 Δformin

CONCLUSION The similar defect (lack of ability to form shmoos) in strains lacking either Fus3 or formin suggests that both proteins are required for shmoo formation. These results led the investigators to propose the model shown here for the induction of asymmetrical growth in the receiving cell directed toward the cell of the opposite mating type.

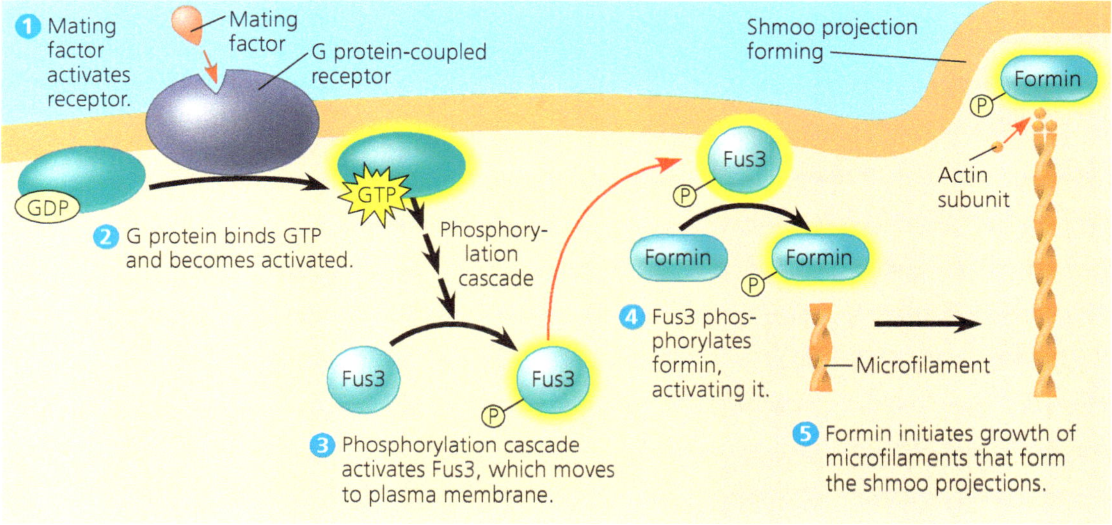

SOURCE D. Matheos et al., Pheromone-induced polarization is dependent on the Fus3p MAPK acting through the formin Bni1p, *Journal of Cell Biology* 165:99–109 (2004).

WHAT IF? Based on these results and the proposed model from this work, what would happen to a cell if its Fus3 kinase were not able to associate with the membrane upon activation?

The explanation for the specificity exhibited in cellular responses to signals is the same as the basic explanation for virtually all differences between cells: Because different kinds of cells turn on different sets of genes, *different kinds of cells have different collections of proteins* (**Figure 11.18**, on the next page). The response of a particular cell to a signal depends on its particular collection of signal receptor proteins, relay proteins, and proteins needed to carry out the response. A liver cell, for example, is poised to respond appropriately to epinephrine by having the proteins listed in Figure 11.16 as well as those needed to manufacture glycogen.

Thus, two cells that respond differently to the same signal differ in one or more of the proteins that handle and respond to the signal. Notice in Figure 11.18 that different pathways may have some molecules in common. For example, cells A, B, and C all use the same receptor protein for the red signaling molecule; differences in other proteins account for their differing responses. In cell D, a different receptor protein is used for the same signaling molecule, leading to yet another response. In cell B, a pathway that is triggered by a single kind of signal diverges to produce two responses; such branched pathways often involve receptor tyrosine kinases (which can activate multiple relay proteins) or second messengers (which can regulate numerous proteins). In cell C, two pathways triggered by separate signals converge to modulate a single response. Branching of pathways and "cross-talk" (interaction)

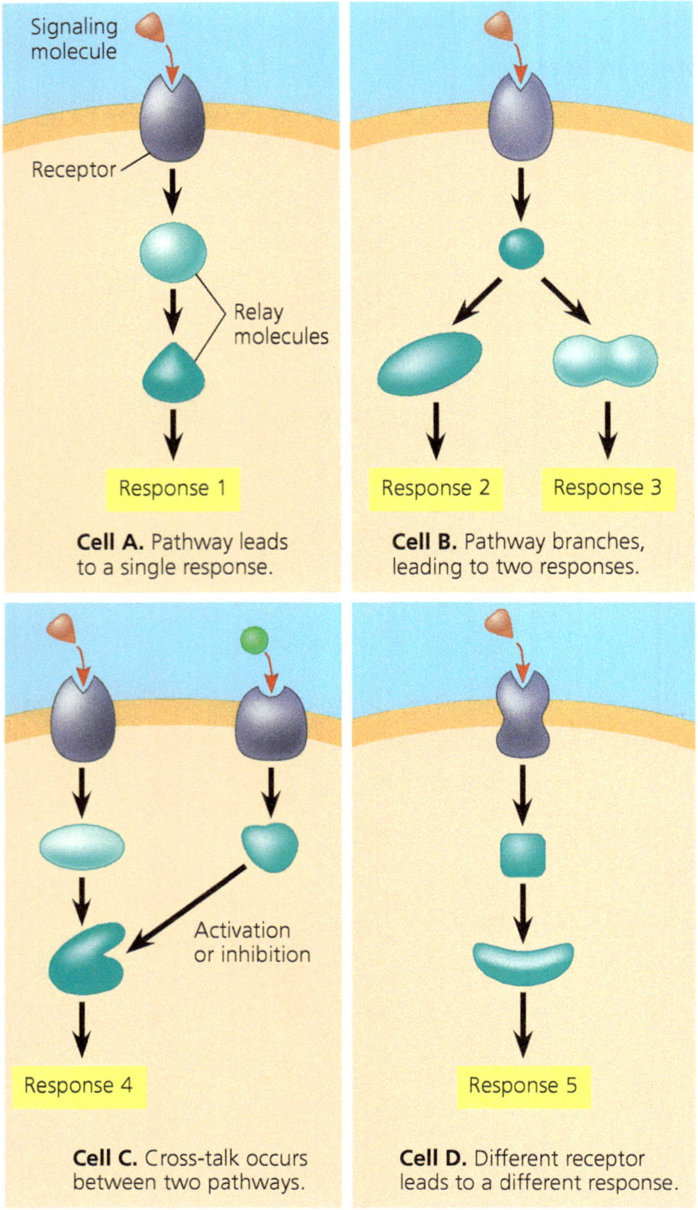

▲ **Figure 11.18 The specificity of cell signaling.** The particular proteins a cell possesses determine what signaling molecules it responds to and the nature of the response. The four cells in these diagrams respond to the same signaling molecule (red) in different ways because each has a different set of proteins (purple and teal). Note, however, that the same kinds of molecules can participate in more than one pathway.

MAKE CONNECTIONS *Study the signaling pathway shown in Figure 11.14 (p. 218), and explain how the situation pictured for cell B above could apply to that pathway.*

between pathways are important in regulating and coordinating a cell's responses to information coming in from different sources in the body. (You'll learn more about this coordination in Concept 11.5.) Moreover, the use of some of the same proteins in more than one pathway allows the cell to economize on the number of different proteins it must make.

Signaling Efficiency: Scaffolding Proteins and Signaling Complexes

The illustrations of signaling pathways in Figure 11.18 (as well as diagrams of other pathways in this chapter) are greatly simplified. The diagrams show only a few relay molecules and, for clarity's sake, display these molecules spread out in the cytosol. If this were true in the cell, signaling pathways would operate very inefficiently because most relay molecules are proteins, and proteins are too large to diffuse quickly through the viscous cytosol. How does a particular protein kinase, for instance, find its substrate?

In many cases, the efficiency of signal transduction is apparently increased by the presence of **scaffolding proteins**, large relay proteins to which several other relay proteins are simultaneously attached. For example, one scaffolding protein isolated from mouse brain cells holds three protein kinases and carries these kinases with it when it binds to an appropriately activated membrane receptor; it thus facilitates a specific phosphorylation cascade **(Figure 11.19)**. Researchers have found scaffolding proteins in brain cells that *permanently* hold together networks of signaling pathway proteins at synapses. This hardwiring enhances the speed and accuracy of signal transfer between cells, because the rate of protein-protein interaction is not limited by diffusion. Furthermore, in addition to this indirect role in activation of relay proteins, the scaffolding proteins themselves may more directly activate some of the other relay proteins.

When signaling pathways were first discovered, they were thought to be linear, independent pathways. Our understanding of cellular communication has benefited from the realization that signaling-pathway components interact with each other in various ways. As seen in Figure 11.18, some proteins may participate in more than one pathway, either in different cell types or in the same cell at different times or under different conditions. These observations underscore

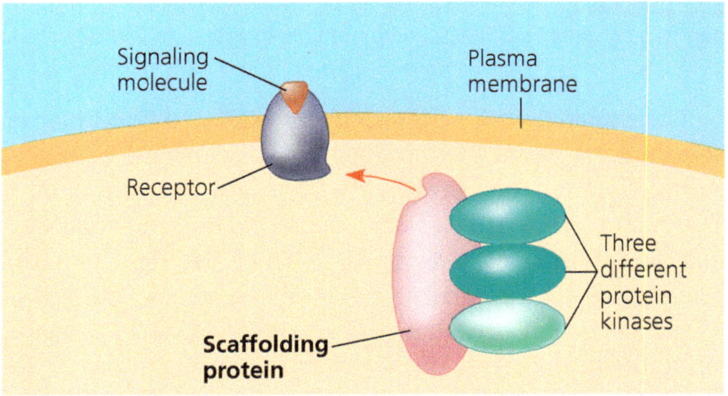

▲ **Figure 11.19 A scaffolding protein.** The scaffolding protein shown here (pink) simultaneously binds to a specific activated membrane receptor and three different protein kinases. This physical arrangement facilitates signal transduction by these molecules and may directly activate relay molecules in some cases.

the importance of transient—or, in some cases, permanent—protein complexes in the process of cell signaling.

The importance of the relay proteins that serve as points of branching or intersection in signaling pathways is highlighted by the problems arising when these proteins are defective or missing. For instance, in an inherited disorder called Wiskott-Aldrich syndrome (WAS), the absence of a single relay protein leads to such diverse effects as abnormal bleeding, eczema, and a predisposition to infections and leukemia. These symptoms are thought to arise primarily from the absence of the protein in cells of the immune system. By studying normal cells, scientists found that the WAS protein is located just beneath the cell surface. The protein interacts both with microfilaments of the cytoskeleton and with several different components of signaling pathways that relay information from the cell surface, including pathways regulating immune cell proliferation. This multifunctional relay protein is thus both a branch point and an important intersection point in a complex signal transduction network that controls immune cell behavior. When the WAS protein is absent, the cytoskeleton is not properly organized and signaling pathways are disrupted, leading to the WAS symptoms.

Termination of the Signal

To keep Figure 11.18 simple, we did not indicate the *inactivation* mechanisms that are an essential aspect of cell signaling. For a cell of a multicellular organism to remain capable of responding to incoming signals, each molecular change in its signaling pathways must last only a short time. As we saw in the cholera example, if a signaling pathway component becomes locked into one state, whether active or inactive, consequences for the organism can be dire.

The ability of a cell to receive new signals depends on reversibility of the changes produced by prior signals. The binding of signaling molecules to receptors is reversible. As the external concentration of signaling molecules falls, fewer receptors are bound at any given moment, and the unbound receptors revert to their inactive form. The cellular response occurs only when the concentration of receptors with bound signaling molecules is above a certain threshold. When the number of active receptors falls below that threshold, the cellular response ceases. Then, by a variety of means, the relay molecules return to their inactive forms: The GTPase activity intrinsic to a G protein hydrolyzes its bound GTP; the enzyme phosphodiesterase converts cAMP to AMP; protein phosphatases inactivate phosphorylated kinases and other proteins; and so forth. As a result, the cell is soon ready to respond to a fresh signal.

In this section, we explored the complexity of signaling initiation and termination in a single pathway, and we saw the potential for pathways to intersect with each other. In the next section, we'll consider one especially important network of interacting pathways in the cell.

CONCEPT CHECK 11.4

1. How can a target cell's response to a single hormone molecule result in a response that affects a million other molecules?
2. **WHAT IF?** If two cells have different scaffolding proteins, explain how they might behave differently in response to the same signaling molecule.
3. **MAKE CONNECTIONS** Review the discussion of protein phosphatases on page 216, and see Figure 11.10 on page 215. Some human diseases are associated with malfunctioning protein phosphatases. How would such proteins affect signaling pathways?

For suggested answers, see Appendix A.

CONCEPT 11.5

Apoptosis integrates multiple cell-signaling pathways

To be or not to be? One of the most elaborate networks of signaling pathways in the cell seems to ask and answer this question posed by Hamlet. Cells that are infected, damaged, or have reached the end of their functional life span often undergo "programmed cell death." The best-understood type of this controlled cell suicide is **apoptosis** (from the Greek, meaning "falling off," and used in a classic Greek poem to refer to leaves falling from a tree). During this process, cellular agents chop up the DNA and fragment the organelles and other cytoplasmic components. The cell shrinks and becomes lobed (a change called "blebbing"; **Figure 11.20**), and the cell's parts are packaged up in vesicles that are engulfed and digested by specialized scavenger cells, leaving no trace. Apoptosis protects neighboring cells from damage that they would otherwise suffer if a dying cell merely leaked out all its contents, including its many digestive enzymes.

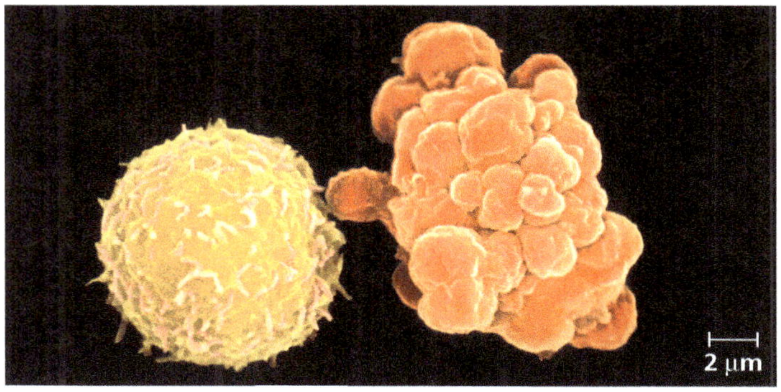

▲ **Figure 11.20 Apoptosis of a human white blood cell.** We can compare a normal white blood cell (left) with a white blood cell undergoing apoptosis (right). The apoptotic cell is shrinking and forming lobes ("blebs"), which eventually are shed as membrane-bounded cell fragments (colorized SEMs).

Apoptosis in the Soil Worm *Caenorhabditis elegans*

Embryonic development is a period during which apoptosis is widespread and plays a crucial role. The molecular mechanisms underlying apoptosis were worked out in detail by researchers studying embryonic development of a small soil worm, a nematode called *Caenorhabditis elegans*. Because the adult worm has only about a thousand cells, the researchers were able to work out the entire ancestry of each cell. The timely suicide of cells occurs exactly 131 times during normal development of *C. elegans*, at precisely the same points in the cell lineage of each worm. In worms and other species, apoptosis is triggered by signals that activate a cascade of "suicide" proteins in the cells destined to die.

Genetic research on *C. elegans* has revealed two key apoptosis genes, called *ced-3* and *ced-4* (*ced* stands for "cell death"), which encode proteins essential for apoptosis. The proteins are called Ced-3 and Ced-4, respectively. These and most other proteins involved in apoptosis are continually present in cells, but in inactive form; thus, regulation occurs at the level of protein activity rather than through gene activity and protein synthesis. In *C. elegans*, a protein in the outer mitochondrial membrane, called Ced-9 (the product of the *ced-9* gene), serves as a master regulator of apoptosis, acting as a brake in the absence of a signal promoting apoptosis **(Figure 11.21)**. When a death signal is received by the cell, it overrides the brake, and the apoptotic pathway activates proteases and nucleases, enzymes that cut up the proteins and DNA of the cell. The main proteases of apoptosis are called *caspases*; in the nematode, the chief caspase is Ced-3.

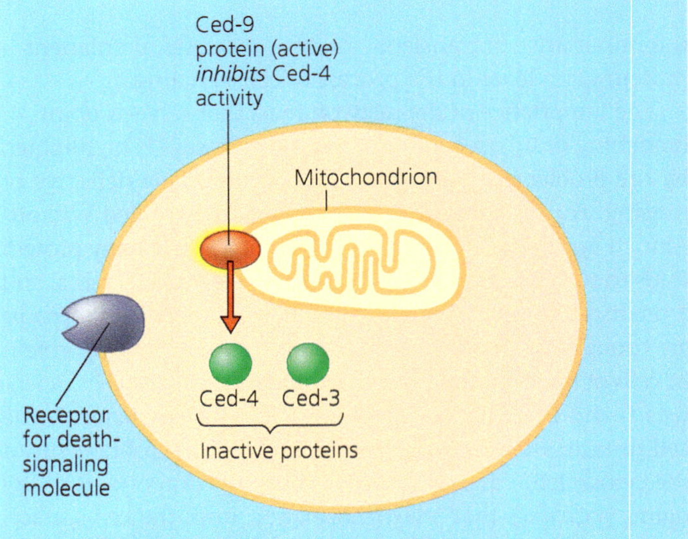

(a) No death signal. As long as Ced-9, located in the outer mitochondrial membrane, is active, apoptosis is inhibited, and the cell remains alive.

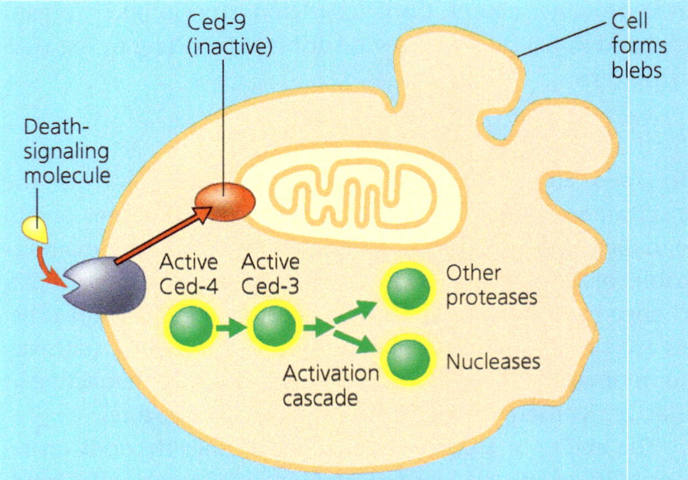

(b) Death signal. When a cell receives a death signal, Ced-9 is inactivated, relieving its inhibition of Ced-3 and Ced-4. Active Ced-3, a protease, triggers a cascade of reactions leading to activation of nucleases and other proteases. The action of these enzymes causes the changes seen in apoptotic cells and eventual cell death.

▲ **Figure 11.21 Molecular basis of apoptosis in *C. elegans*.** Three proteins, Ced-3, Ced-4, and Ced-9, are critical to apoptosis and its regulation in the nematode. Apoptosis is more complicated in mammals but involves proteins similar to those in the nematode.

Apoptotic Pathways and the Signals That Trigger Them

In humans and other mammals, several different pathways, involving about 15 different caspases, can carry out apoptosis. The pathway that is used depends on the type of cell and on the particular signal that initiates apoptosis. One major pathway involves certain mitochondrial proteins that are triggered to form molecular pores in the mitochondrial outer membrane, causing it to leak and release other proteins that promote apoptosis. Surprisingly, these latter include cytochrome *c*, which functions in mitochondrial electron transport in healthy cells (see Figure 9.15) but acts as a cell death factor when released from mitochondria. The process of mitochondrial apoptosis in mammals uses proteins similar to the nematode proteins Ced-3, Ced-4, and Ced-9. These can be thought of as relay proteins capable of transducing the apoptotic signal.

At key gateways into the apoptotic program, relay proteins integrate signals from several different sources and can send a cell down an apoptotic pathway. Often, the signal originates outside the cell, like the death-signaling molecule depicted in Figure 11.21b, which presumably was released by a neighboring cell. When a death-signaling ligand occupies a cell-surface receptor, this binding leads to activation of caspases and other enzymes that carry out apoptosis, without involving the mitochondrial pathway. This process of signal reception, transduction, and response is similar to what we discussed earlier in this chapter. In a twist on the classic scenario, two other types of alarm signals that can lead to apoptosis originate from *inside* the cell rather than from a cell-surface receptor.

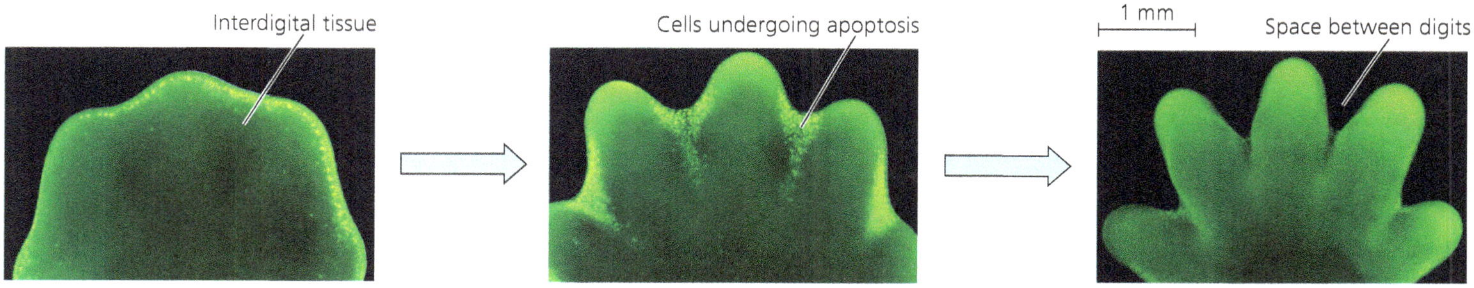

▲ **Figure 11.22 Effect of apoptosis during paw development in the mouse.** In mice, humans, other mammals, and land birds, the embryonic region that develops into feet or hands initially has a solid, platelike structure. Apoptosis eliminates the cells in the interdigital regions, thus forming the digits. The embryonic mouse paws shown in these fluorescence light micrographs are stained so that cells undergoing apoptosis appear a bright yellowish green. Apoptosis of cells begins at the margin of each interdigital region (left), peaks as the tissue in these regions is reduced (middle), and is no longer visible when the interdigital tissue has been eliminated (right).

One signal comes from the nucleus, generated when the DNA has suffered irreparable damage, and a second comes from the endoplasmic reticulum when excessive protein misfolding occurs. Mammalian cells make life-or-death "decisions" by somehow integrating the death signals and life signals they receive from these external and internal sources.

A built-in cell suicide mechanism is essential to development and maintenance in all animals. The similarities between apoptosis genes in nematodes and mammals, as well as the observation that apoptosis occurs in multicellular fungi and even in single-celled yeasts, indicate that the basic mechanism evolved early in the evolution of eukaryotes. In vertebrates, apoptosis is essential for normal development of the nervous system, for normal operation of the immune system, and for normal morphogenesis of hands and feet in humans and paws in other mammals (**Figure 11.22**). The level of apoptosis between the developing digits is lower in the webbed feet of ducks and other water birds than in the nonwebbed feet of land birds, such as chickens. In the case of humans, the failure of appropriate apoptosis can result in webbed fingers and toes.

Significant evidence points to the involvement of apoptosis in certain degenerative diseases of the nervous system, such as Parkinson's disease and Alzheimer's disease. Also, cancer can result from a failure of cell suicide; some cases of human melanoma, for example, have been linked to faulty forms of the human version of the *C. elegans* Ced-4 protein. It is not surprising, therefore, that the signaling pathways feeding into apoptosis are quite elaborate. After all, the life-or-death question is the most fundamental one imaginable for a cell.

This chapter has introduced you to many of the general mechanisms of cell communication, such as ligand binding, protein-protein interactions and shape changes, cascades of interactions, and protein phosphorylation. As you continue through the text, you will encounter numerous examples of cell signaling.

CONCEPT CHECK 11.5

1. Give an example of apoptosis during embryonic development, and explain its function in the developing embryo.
2. **WHAT IF?** What types of protein defects could result in apoptosis occurring when it should not? What types could result in apoptosis not occurring when it should?

For suggested answers, see Appendix A.

11 CHAPTER REVIEW

SUMMARY OF KEY CONCEPTS

CONCEPT 11.1

External signals are converted to responses within the cell (pp. 206–210)

- **Signal transduction pathways** are crucial for many processes, including the mating of yeast cells. In fact, signaling in microbes has much in common with processes in multicellular organisms, suggesting an early evolutionary origin of signaling mechanisms. Bacterial cells can sense the local density of bacterial cells (quorum sensing) by binding molecules secreted by other cells. In some cases, such signals lead to aggregation of these cells into biofilms.
- In local signaling, animal cells may communicate by direct contact or by secreting **local regulators**, such as growth factors or neurotransmitters. For long-distance signaling, both animals and plants use **hormones**; animals also pass signals electrically.

- Earl Sutherland discovered how the hormone epinephrine acts on cells. Like other hormones that bind to membrane receptors, it triggers a three-stage cell-signaling pathway:

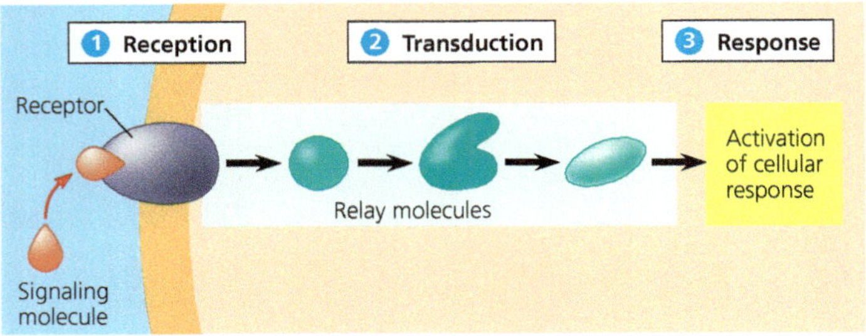

? *What determines* whether *a cell responds to a hormone such as epinephrine? What determines* how *a cell responds to such a hormone?*

CONCEPT 11.2

Reception: A signaling molecule binds to a receptor protein, causing it to change shape (pp. 210–214)

- The binding between signaling molecule (**ligand**) and receptor is highly specific. A specific shape change in a receptor is often the initial transduction of the signal.
- There are three major types of cell-surface transmembrane receptors: (1) **G protein-coupled receptors (GPCRs)** work with the help of cytoplasmic **G proteins**. Ligand binding activates the receptor, which then activates a specific G protein, which activates yet another protein, thus propagating the signal along a signal transduction pathway. (2) **Receptor tyrosine kinases (RTKs)** react to the binding of signaling molecules by forming dimers and then adding phosphate groups to tyrosines on the cytoplasmic part of the other monomer making up the dimer. Relay proteins in the cell can then be activated by binding to different phosphorylated tyrosines, allowing this receptor to trigger several pathways at once. (3) **Ligand-gated ion channels** open or close in response to binding by specific signaling molecules, regulating the flow of specific ions across the membrane.
- The activity of all three types of receptors is crucial to proper cell functioning, and abnormal GPCRs and RTKs are associated with many human diseases.
- Intracellular receptors are cytoplasmic or nuclear proteins. Signaling molecules that are hydrophobic or small enough to cross the plasma membrane bind to these receptors inside the cell.

? *How are the structures of a G protein-coupled receptor and a receptor tyrosine kinase similar? In what key way does the triggering of signal transduction pathways differ for these two types of receptors?*

CONCEPT 11.3

Transduction: Cascades of molecular interactions relay signals from receptors to target molecules in the cell (pp. 214–219)

- At each step in a signal transduction pathway, the signal is transduced into a different form, which commonly involves a shape change in a protein. Many signal transduction pathways include phosphorylation cascades, in which a series of **protein kinases** each add a phosphate group to the next one in line, activating it. Enzymes called **protein phosphatases** remove the phosphate groups. The balance between phosphorylation and dephosphorylation regulates the activity of proteins involved in the sequential steps of a signal transduction pathway.
- **Second messengers**, such as the small molecule **cyclic AMP (cAMP)** and the ion Ca^{2+}, diffuse readily through the cytosol and thus help broadcast signals quickly. Many G proteins activate **adenylyl cyclase**, which makes cAMP from ATP. Cells use Ca^{2+} as a second messenger in both G protein and tyrosine kinase pathways. The tyrosine kinase pathways can also involve two other second messengers, **diacylglycerol (DAG)** and **inositol trisphosphate (IP_3)**. IP_3 can trigger a subsequent increase in Ca^{2+} levels.

? *What is the difference between a protein kinase and a second messenger? Can both types of molecules operate in the same signal transduction pathway?*

CONCEPT 11.4

Response: Cell signaling leads to regulation of transcription or cytoplasmic activities (pp. 219–223)

- Some pathways lead to a nuclear response: Specific genes are turned on or off by activation of proteins called transcription factors. In other pathways, the response involves cytoplasmic regulation, including cytoskeletal rearrangement (which can lead to cell shape changes) or changes in enzyme activity.
- Cellular responses are not simply on or off; they are fine-tuned at many steps in the process. Each catalytic protein in a signaling pathway amplifies the signal by activating multiple copies of the next component of the pathway; for long pathways, the total amplification may be a millionfold or more. The particular combination of proteins in a cell gives the cell great specificity in both the signals it detects and the responses it carries out. **Scaffolding proteins** can increase signal transduction efficiency. Pathway branching and cross-talk further help the cell coordinate incoming signals and responses. Signal response is terminated quickly by the reversal of ligand binding.

? *What mechanisms in the cell terminate its response to a signal and maintain its ability to respond to new signals?*

CONCEPT 11.5

Apoptosis integrates multiple cell-signaling pathways (pp. 223–225)

- **Apoptosis** is a type of programmed cell death in which cell components are disposed of in an orderly fashion, without damage to neighboring cells. Studies of the soil worm *Caenorhabditis elegans* showed that apoptosis occurs at defined times during embryonic development and clarified molecular details of the signaling pathway involved in the process. A protein (Ced-9) in the mitochondrial membrane acts as a brake; when released by a death signal, it allows activation of caspases, the main proteases that carry out apoptosis, and nucleases.
- Several apoptotic signaling pathways exist in the cells of humans and other mammals, and these pathways may be triggered in several ways. A major pathway involves pore formation in the outer mitochondrial membrane, which leads to release of factors that activate caspases. Signals eliciting the apoptotic response can originate from outside or inside the cell.

? *What is an explanation for the similarities between genes in yeasts, nematodes, and mammals that control apoptosis?*

TEST YOUR UNDERSTANDING

LEVEL 1: KNOWLEDGE/COMPREHENSION

1. Phosphorylation cascades involving a series of protein kinases are useful for cellular signal transduction because
   a. they are species specific.
   b. they always lead to the same cellular response.
   c. they amplify the original signal manyfold.
   d. they counter the harmful effects of phosphatases.
   e. the number of molecules used is small and fixed.

2. Binding of a signaling molecule to which type of receptor leads directly to a change in the distribution of ions on opposite sides of the membrane?
   a. receptor tyrosine kinase
   b. G protein-coupled receptor
   c. phosphorylated receptor tyrosine kinase dimer
   d. ligand-gated ion channel
   e. intracellular receptor

3. The activation of receptor tyrosine kinases is characterized by
   a. dimerization and phosphorylation.
   b. dimerization and IP_3 binding.
   c. a phosphorylation cascade.
   d. GTP hydrolysis.
   e. channel protein shape change.

4. Lipid-soluble signaling molecules, such as testosterone, cross the membranes of all cells but affect only target cells because
   a. only target cells retain the appropriate DNA segments.
   b. intracellular receptors are present only in target cells.
   c. most cells lack the Y chromosome required.
   d. only target cells possess the cytosolic enzymes that transduce the testosterone.
   e. only in target cells is testosterone able to initiate the phosphorylation cascade leading to activated transcription factor.

5. Consider this pathway: epinephrine → G protein-coupled receptor → G protein → adenylyl cyclase → cAMP. Identify the second messenger.
   a. cAMP
   b. G protein
   c. GTP
   d. adenylyl cyclase
   e. G protein-coupled receptor

6. Apoptosis involves all but which of the following?
   a. fragmentation of the DNA
   b. cell-signaling pathways
   c. activation of cellular enzymes
   d. lysis of the cell
   e. digestion of cellular contents by scavenger cells

LEVEL 2: APPLICATION/ANALYSIS

7. Which observation suggested to Sutherland the involvement of a second messenger in epinephrine's effect on liver cells?
   a. Enzymatic activity was proportional to the amount of calcium added to a cell-free extract.
   b. Receptor studies indicated that epinephrine was a ligand.
   c. Glycogen breakdown was observed only when epinephrine was administered to intact cells.
   d. Glycogen breakdown was observed when epinephrine and glycogen phosphorylase were combined.
   e. Epinephrine was known to have different effects on different types of cells.

8. Protein phosphorylation is commonly involved with all of the following *except*
   a. regulation of transcription by extracellular signaling molecules.
   b. enzyme activation.
   c. activation of G protein-coupled receptors.
   d. activation of receptor tyrosine kinases.
   e. activation of protein kinase molecules.

LEVEL 3: SYNTHESIS/EVALUATION

9. **DRAW IT** Draw the following apoptotic pathway, which operates in human immune cells. A death signal is received when a molecule called Fas binds its cell-surface receptor. The binding of many Fas molecules to receptors causes receptor clustering. The intracellular regions of the receptors, when together, bind proteins called adaptor proteins. These in turn bind to inactive molecules of caspase-8, which become activated and then activate caspase-3. Once activated, caspase-3 initiates apoptosis.

10. **EVOLUTION CONNECTION**
    What evolutionary mechanisms might account for the origin and persistence of cell-to-cell signaling systems in unicellular prokaryotes?

11. **SCIENTIFIC INQUIRY**
    Epinephrine initiates a signal transduction pathway that involves production of cyclic AMP (cAMP) and leads to the breakdown of glycogen to glucose, a major energy source for cells. But glycogen breakdown is actually only part of the fight-or-flight response that epinephrine brings about; the overall effect on the body includes increased heart rate and alertness, as well as a burst of energy. Given that caffeine blocks the activity of cAMP phosphodiesterase, propose a mechanism by which caffeine ingestion leads to heightened alertness and sleeplessness.

12. **SCIENCE, TECHNOLOGY, AND SOCIETY**
    The aging process is thought to be initiated at the cellular level. Among the changes that can occur after a certain number of cell divisions is the loss of a cell's ability to respond to growth factors and other chemical signals. Much research into aging is aimed at understanding such losses, with the ultimate goal of significantly extending the human life span. Not everyone, however, agrees that this is a desirable goal. If life expectancy were greatly increased, what might be the social and ecological consequences?

13. **WRITE ABOUT A THEME**
    Emergent Properties The property of life emerges at the biological level of the cell. The highly regulated process of apoptosis is not simply the destruction of a cell; it is also an emergent property. Write a short essay (about 100–150 words) that briefly explains the role of apoptosis in the development and proper functioning of an animal and then describes how this form of programmed cell death is a process that emerges from the orderly integration of signaling pathways.

For selected answers, see Appendix A.

MasteringBIOLOGY® www.masteringbiology.com

1. MasteringBiology® Assignments

Tutorials Cell Signaling: Reception • Cell Signaling: Transduction and Response
Activities Overview of Cell Signaling • Reception • Signal Transduction Pathways • Cellular Responses • Build a Signaling Pathway
Questions Student Misconceptions • Reading Quiz • Multiple Choice • End-of-Chapter

2. eText

Read your book online, search, take notes, highlight text, and more.

3. The Study Area

Practice Tests • Cumulative Test • *BioFlix* 3-D Animations • MP3 Tutor Sessions • Videos • Activities • Investigations • Lab Media • Audio Glossary • Word Study Tools • Art

12

The Cell Cycle

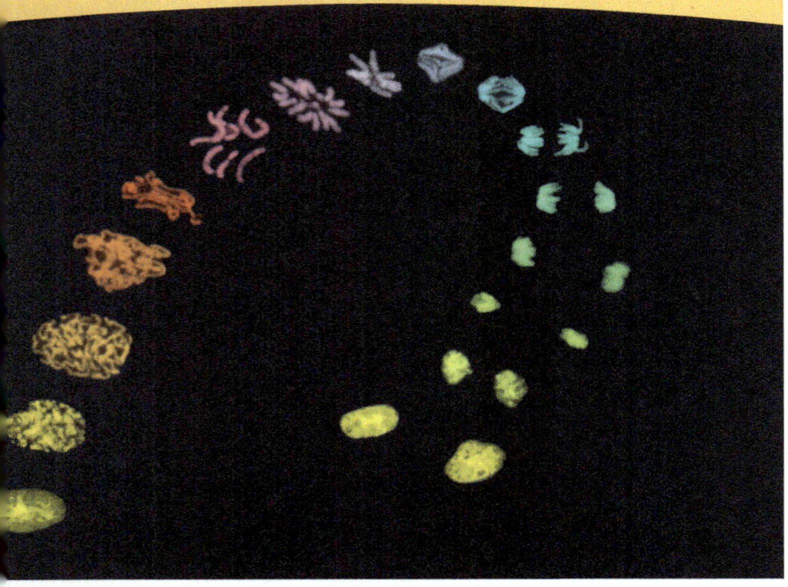

▲ **Figure 12.1 How do a cell's chromosomes change during cell division?**

KEY CONCEPTS

12.1 Most cell division results in genetically identical daughter cells

12.2 The mitotic phase alternates with interphase in the cell cycle

12.3 The eukaryotic cell cycle is regulated by a molecular control system

OVERVIEW

The Key Roles of Cell Division

The ability of organisms to produce more of their own kind is the one characteristic that best distinguishes living things from nonliving matter. This unique capacity to procreate, like all biological functions, has a cellular basis. Rudolf Virchow, a German physician, put it this way in 1855: "Where a cell exists, there must have been a preexisting cell, just as the animal arises only from an animal and the plant only from a plant." He summarized this concept with the Latin axiom "*Omnis cellula e cellula*," meaning "Every cell from a cell." The continuity of life is based on the reproduction of cells, or **cell division**. The series of fluorescence micrographs in **Figure 12.1** follows an animal cell's chromosomes, from lower left to lower right, as one cell divides into two.

Cell division plays several important roles in life. The division of one prokaryotic cell reproduces an entire organism. The same is true of a unicellular eukaryote **(Figure 12.2a)**. Cell division also enables multicellular eukaryotes to develop from a single cell, like the fertilized egg that gave rise to the two-celled embryo in **Figure 12.2b**. And after such an organism is fully grown, cell division continues to function in renewal and repair, replacing cells that die from normal wear and tear or accidents. For example, dividing cells in your bone marrow continuously make new blood cells **(Figure 12.2c)**.

The cell division process is an integral part of the **cell cycle**, the life of a cell from the time it is first formed from a dividing parent cell until its own division into two daughter cells. (Our use of the words *daughter* or *sister* in relation to cells is not meant to imply gender.) Passing identical genetic material to cellular offspring is a crucial function of cell division. In this chapter, you will learn how this process occurs. After studying the cellular mechanics of cell division in eukaryotes and bacteria, you will learn about the molecular control system that regulates progress through the eukaryotic cell cycle and what happens when the control system malfunctions. Because a breakdown in cell cycle control plays a major role in cancer development, this aspect of cell biology is an active area of research.

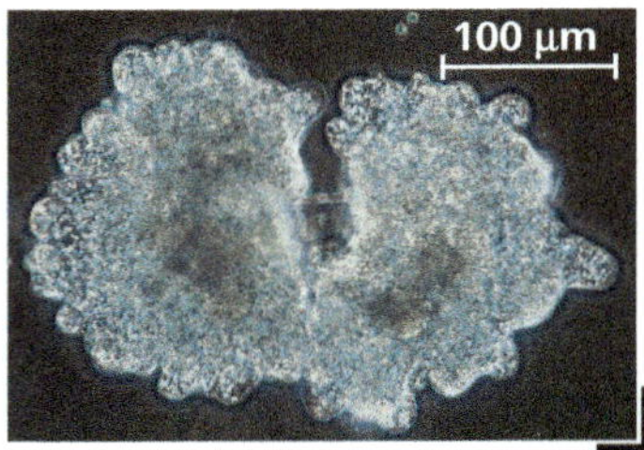

◀ **(a) Reproduction.** An amoeba, a single-celled eukaryote, is dividing into two cells. Each new cell will be an individual organism (LM).

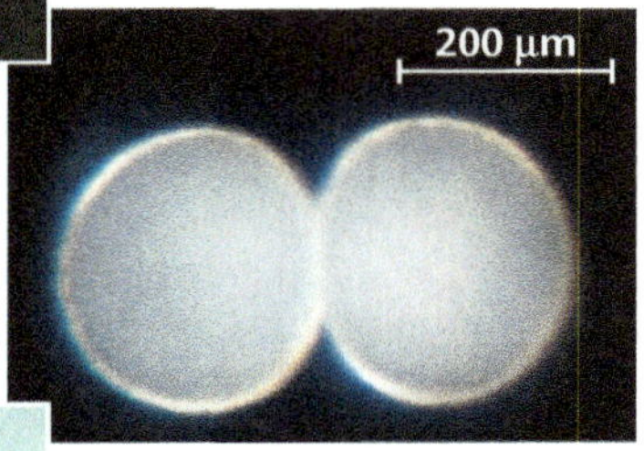

▶ **(b) Growth and development.** This micrograph shows a sand dollar embryo shortly after the fertilized egg divided, forming two cells (LM).

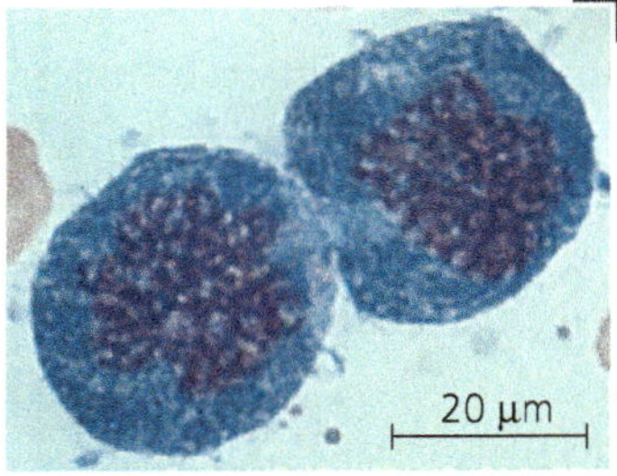

◀ **(c) Tissue renewal.** These dividing bone marrow cells will give rise to new blood cells (LM).

▲ **Figure 12.2 The functions of cell division.**

CONCEPT 12.1

Most cell division results in genetically identical daughter cells

The reproduction of an assembly as complex as a cell cannot occur by a mere pinching in half; a cell is not like a soap bubble that simply enlarges and splits in two. In both prokaryotes and eukaryotes, most cell division involves the distribution of identical genetic material—DNA—to two daughter cells. (The exception is meiosis, the special type of eukaryotic cell division that can produce sperm and eggs.) What is most remarkable about cell division is the fidelity with which the DNA is passed along from one generation of cells to the next. A dividing cell duplicates its DNA, allocates the two copies to opposite ends of the cell, and only then splits into daughter cells. After we describe the distribution of DNA during cell division in animal and plant cells, we'll consider the process in other eukaryotes as well as in bacteria.

Cellular Organization of the Genetic Material

A cell's endowment of DNA, its genetic information, is called its **genome**. Although a prokaryotic genome is often a single DNA molecule, eukaryotic genomes usually consist of a number of DNA molecules. The overall length of DNA in a eukaryotic cell is enormous. A typical human cell, for example, has about 2 m of DNA—a length about 250,000 times greater than the cell's diameter. Yet before the cell can divide to form genetically identical daughter cells, all of this DNA must be copied, or replicated, and then the two copies must be separated so that each daughter cell ends up with a complete genome.

The replication and distribution of so much DNA is manageable because the DNA molecules are packaged into structures called **chromosomes**, so named because they take up certain dyes used in microscopy (from the Greek *chroma*, color, and *soma*, body) (Figure 12.3). Each eukaryotic chromosome consists of one very long, linear DNA molecule associated with many proteins (see Figure 6.9). The DNA molecule carries several hundred to a few thousand genes, the units of information that specify an organism's inherited traits. The associated proteins maintain the structure of the chromosome and help control the activity of the genes. Together, the entire complex of DNA and proteins that is the building material of chromosomes is referred to as **chromatin**. As you will soon see, the chromatin of a chromosome varies in its degree of condensation during the process of cell division.

Every eukaryotic species has a characteristic number of chromosomes in each cell nucleus. For example, the nuclei of human **somatic cells** (all body cells except the reproductive cells) each contain 46 chromosomes, made up of two sets of 23, one set inherited from each parent. Reproductive cells, or **gametes**—sperm and eggs—have half as many chromosomes as somatic cells, or one set of 23 chromosomes in humans. The

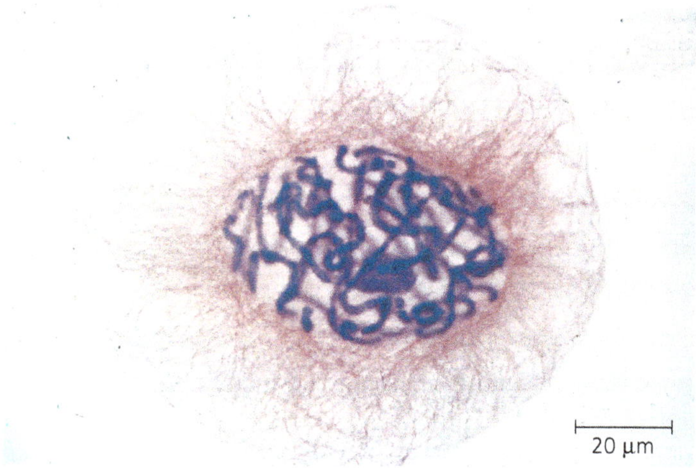

▲ **Figure 12.3 Eukaryotic chromosomes.** Chromosomes (stained purple) are visible within the nucleus of this cell from an African blood lily. The thinner red threads in the surrounding cytoplasm are the cytoskeleton. The cell is preparing to divide (LM).

number of chromosomes in somatic cells varies widely among species: 18 in cabbage plants, 48 in chimpanzees, 56 in elephants, 90 in hedgehogs, and 148 in one species of alga. We'll now consider how these chromosomes behave during cell division.

Distribution of Chromosomes During Eukaryotic Cell Division

When a cell is not dividing, and even as it replicates its DNA in preparation for cell division, each chromosome is in the form of a long, thin chromatin fiber. After DNA replication, however, the chromosomes condense as a part of cell division: Each chromatin fiber becomes densely coiled and folded, making the chromosomes much shorter and so thick that we can see them with a light microscope.

Each duplicated chromosome has two **sister chromatids**, which are joined copies of the original chromosome (Figure 12.4). The two chromatids, each containing an identical DNA molecule, are initially attached all along their lengths by protein complexes called *cohesins*; this attachment is known as *sister chromatid cohesion*. Each sister chromatid has a **centromere**, a region containing specific DNA sequences

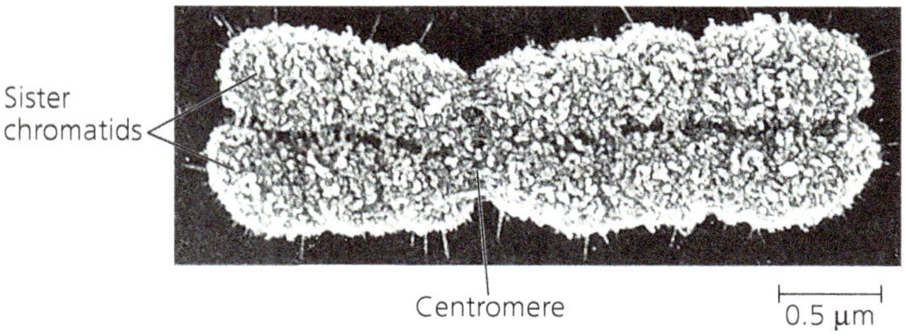

▲ **Figure 12.4 A highly condensed, duplicated human chromosome (SEM).**

DRAW IT *Circle one sister chromatid of the chromosome in this micrograph.*

▶ **Figure 12.5 Chromosome duplication and distribution during cell division.**

? *How many chromatid arms does the chromosome in ② have?*

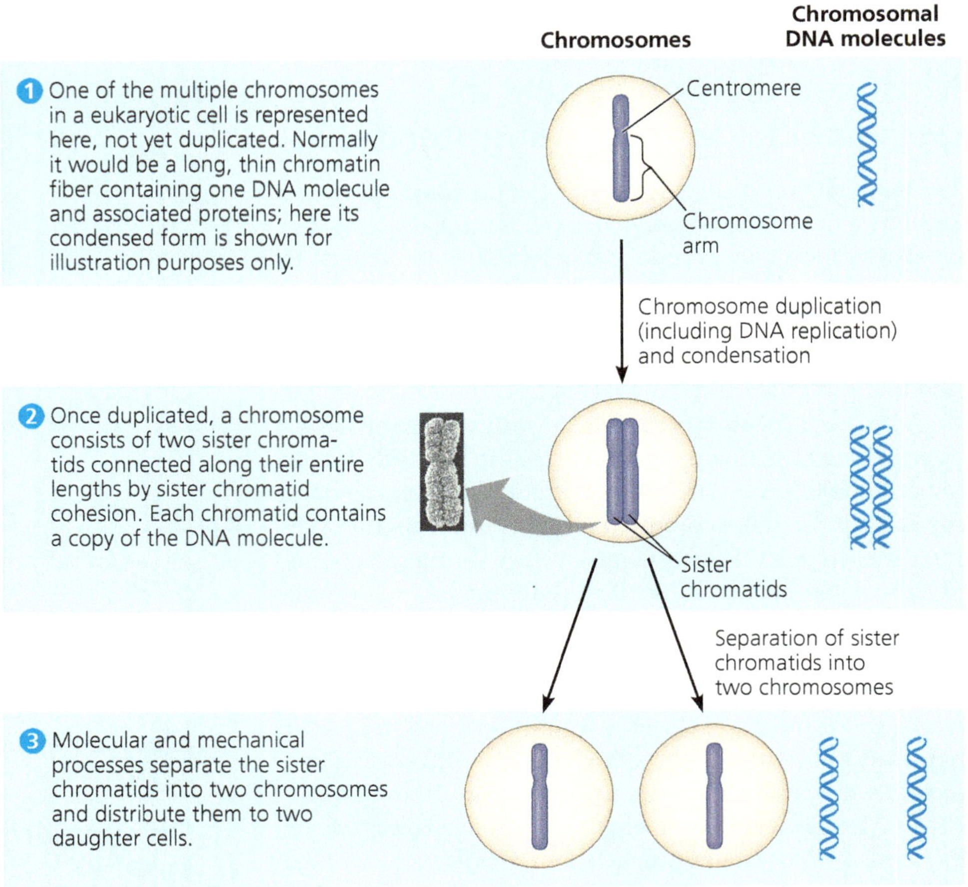

where the chromatid is attached most closely to its sister chromatid. This attachment is mediated by proteins bound to the centromeric DNA sequences and gives the condensed, duplicated chromosome a narrow "waist." The part of a chromatid on either side of the centromere is referred to as an *arm* of the chromatid. (An uncondensed, unduplicated chromosome has a single centromere and two arms.)

Later in the cell division process, the two sister chromatids of each duplicated chromosome separate and move into two new nuclei, one forming at each end of the cell. Once the sister chromatids separate, they are no longer called sister chromatids but are considered individual chromosomes. Thus, each new nucleus receives a collection of chromosomes identical to that of the parent cell (**Figure 12.5**). **Mitosis**, the division of the genetic material in the nucleus, is usually followed immediately by **cytokinesis**, the division of the cytoplasm. One cell has become two, each the genetic equivalent of the parent cell.

What happens to the chromosome number as we follow the human life cycle through the generations? You inherited 46 chromosomes, one set of 23 from each parent. They were combined in the nucleus of a single cell when a sperm from your father united with an egg from your mother, forming a fertilized egg, or zygote. Mitosis and cytokinesis produced the 200 trillion somatic cells that now make up your body, and the same processes continue to generate new cells to replace dead and damaged ones. In contrast, you produce gametes—eggs or sperm—by a variation of cell division called *meiosis*, which yields nonidentical daughter cells that have only one set of chromosomes, half as many chromosomes as the parent cell. Meiosis in humans occurs only in the gonads (ovaries or testes). In each generation, meiosis reduces the chromosome number from 46 (two sets of chromosomes) to 23 (one set). Fertilization fuses two gametes together and returns the chromosome number to 46, and mitosis conserves that number in every somatic cell nucleus of the new individual. In Chapter 13, we will examine the role of meiosis in reproduction and inheritance in more detail. In the remainder of this chapter, we focus on mitosis and the rest of the cell cycle in eukaryotes.

CONCEPT CHECK 12.1

1. How many chromatids are in a duplicated chromosome?
2. **WHAT IF?** A chicken has 78 chromosomes in its somatic cells. How many chromosomes did the chicken inherit from each parent? How many chromosomes are in each of the chicken's gametes? How many chromosomes will be in each somatic cell of the chicken's offspring?

For suggested answers, see Appendix A.

CONCEPT 12.2

The mitotic phase alternates with interphase in the cell cycle

In 1882, a German anatomist named Walther Flemming developed dyes that allowed him to observe, for the first time, the behavior of chromosomes during mitosis and cytokinesis. (In fact, Flemming coined the terms *mitosis* and *chromatin*.) During the period between one cell division and the next, it appeared to Flemming that the cell was simply growing larger. But we now know that many critical events occur during this stage in the life of a cell.

Phases of the Cell Cycle

Mitosis is just one part of the cell cycle (**Figure 12.6**). In fact, the **mitotic (M) phase**, which includes both mitosis and cytokinesis, is usually the shortest part of the cell cycle. Mitotic cell division alternates with a much longer stage called **interphase**, which often accounts for about 90% of the cycle. During interphase, a cell that is about to divide grows and copies its chromosomes in preparation for cell division. Interphase can be divided into subphases: the **G_1 phase** ("first gap"), the **S phase** ("synthesis"), and the **G_2 phase** ("second gap"). During all three subphases, a cell that will eventually divide grows by producing proteins and cytoplasmic organelles such as mitochondria and endoplasmic reticulum. However, chromosomes are duplicated only during the S phase. (We will discuss synthesis of DNA in Chapter 16.) Thus, a cell grows (G_1), continues to grow as it copies its chromosomes (S), grows more as it completes preparations for cell division (G_2), and divides (M). The daughter cells may then repeat the cycle.

A particular human cell might undergo one division in 24 hours. Of this time, the M phase would occupy less than 1 hour, while the S phase might occupy about 10–12 hours, or about half the cycle. The rest of the time would be apportioned between the G_1 and G_2 phases. The G_2 phase usually takes 4–6 hours; in our example, G_1 would occupy about 5–6 hours. G_1 is the most variable in length in different types of cells. Some cells in a multicellular organism divide very infrequently or not at all. These cells spend their time in G_1 (or a related phase called G_0) doing their job in the organism—a nerve cell carries impulses, for example.

Mitosis is conventionally broken down into five stages: **prophase**, **prometaphase**, **metaphase**, **anaphase**, and **telophase**. Overlapping with the latter stages of mitosis, cytokinesis completes the mitotic phase. **Figure 12.7**, on the next two pages, describes these stages in an animal cell. Study this figure thoroughly before progressing to the next two sections, which examine mitosis and cytokinesis more closely.

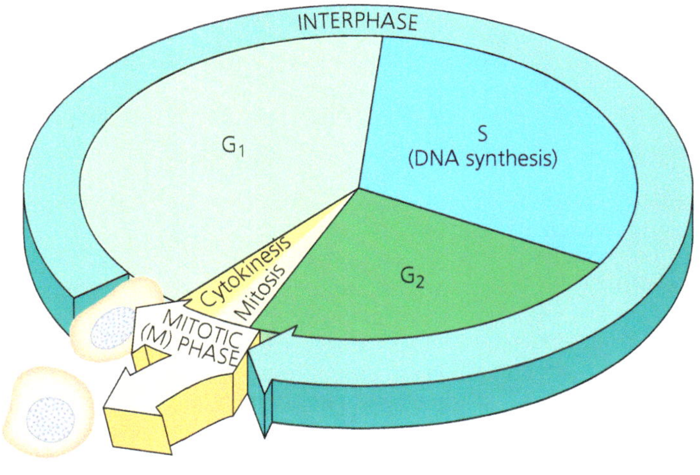

▲ **Figure 12.6 The cell cycle.** In a dividing cell, the mitotic (M) phase alternates with interphase, a growth period. The first part of interphase (G_1) is followed by the S phase, when the chromosomes duplicate; G_2 is the last part of interphase. In the M phase, mitosis distributes the daughter chromosomes to daughter nuclei, and cytokinesis divides the cytoplasm, producing two daughter cells. The relative durations of G_1, S, and G_2 may vary.

The Mitotic Spindle: *A Closer Look*

Many of the events of mitosis depend on the **mitotic spindle**, which begins to form in the cytoplasm during prophase. This structure consists of fibers made of microtubules and associated proteins. While the mitotic spindle assembles, the other microtubules of the cytoskeleton partially disassemble, providing the material used to construct the spindle. The spindle microtubules elongate (polymerize) by incorporating more subunits of the protein tubulin (see Table 6.1) and shorten (depolymerize) by losing subunits.

In animal cells, the assembly of spindle microtubules starts at the **centrosome**, a subcellular region containing material that functions throughout the cell cycle to organize the cell's microtubules. (It is also called the *microtubule-organizing center*.) A pair of centrioles is located at the center of the centrosome, but they are not essential for cell division: If the centrioles are destroyed with a laser microbeam, a spindle nevertheless forms during mitosis. In fact, centrioles are not even present in plant cells, which do form mitotic spindles.

During interphase in animal cells, the single centrosome duplicates, forming two centrosomes, which remain together near the nucleus. The two centrosomes move apart during prophase and prometaphase of mitosis as spindle microtubules grow out from them. By the end of prometaphase, the two centrosomes, one at each pole of the spindle, are at opposite ends of the cell. An **aster**, a radial array of short microtubules, extends from each centrosome. The spindle includes the centrosomes, the spindle microtubules, and the asters.

Each of the two sister chromatids of a duplicated chromosome has a **kinetochore**, a structure of proteins associated with specific sections of chromosomal DNA at each centromere. The chromosome's two kinetochores face in opposite directions. During prometaphase, some of the spindle microtubules attach to the kinetochores; these are called kinetochore microtubules. (The number of microtubules attached to a kinetochore varies among species, from one microtubule in yeast cells to 40 or so in some mammalian cells.) When one of a chromosome's kinetochores is "captured" by microtubules, the chromosome begins to move toward the pole from which those microtubules extend. However, this movement is checked as soon as microtubules from the opposite pole attach to the other kinetochore. What happens next is like a tug-of-war that ends in a draw. The chromosome moves first in one direction, then the other, back and forth, finally settling midway between the two ends of the cell. At metaphase, the centromeres of all the duplicated chromosomes are on a plane midway between the spindle's

▼ Figure 12.7

Exploring Mitosis in an Animal Cell

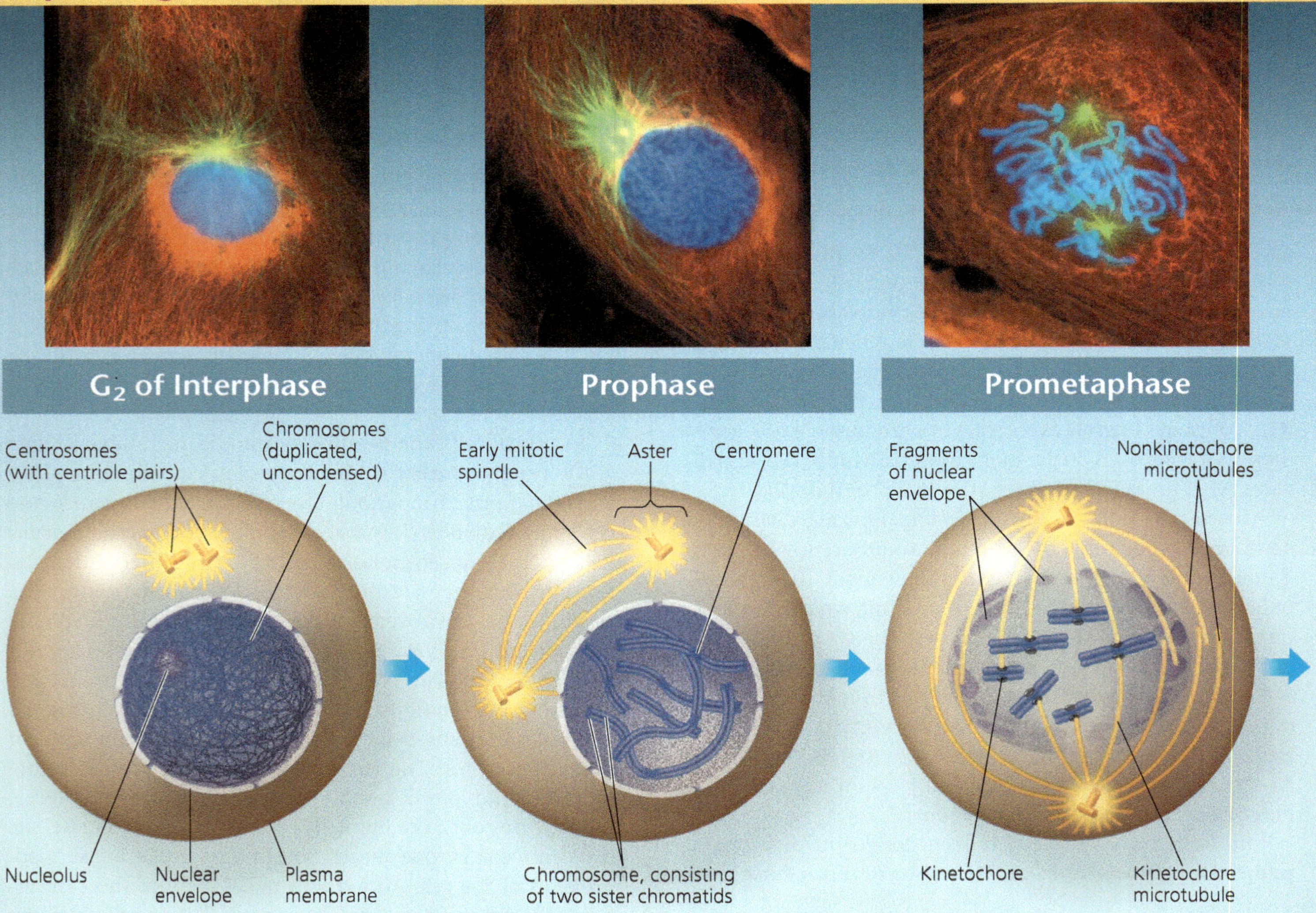

G_2 of Interphase

- A nuclear envelope encloses the nucleus.
- The nucleus contains one or more nucleoli (singular, *nucleolus*).
- Two centrosomes have formed by duplication of a single centrosome. Centrosomes are regions in animal cells that organize the microtubules of the spindle. Each centrosome contains two centrioles.
- Chromosomes, duplicated during S phase, cannot be seen individually because they have not yet condensed.

The light micrographs show dividing lung cells from a newt, which has 22 chromosomes in its somatic cells. Chromosomes appear blue, microtubules green, and intermediate filaments red. For simplicity, the drawings show only 6 chromosomes.

Prophase

- The chromatin fibers become more tightly coiled, condensing into discrete chromosomes observable with a light microscope.
- The nucleoli disappear.
- Each duplicated chromosome appears as two identical sister chromatids joined at their centromeres and, in some species, all along their arms by cohesins (sister chromatid cohesion).
- The mitotic spindle (named for its shape) begins to form. It is composed of the centrosomes and the microtubules that extend from them. The radial arrays of shorter microtubules that extend from the centrosomes are called asters ("stars").
- The centrosomes move away from each other, propelled partly by the lengthening microtubules between them.

Prometaphase

- The nuclear envelope fragments.
- The microtubules extending from each centrosome can now invade the nuclear area.
- The chromosomes have become even more condensed.
- Each of the two chromatids of each chromosome now has a kinetochore, a specialized protein structure at the centromere.
- Some of the microtubules attach to the kinetochores, becoming "kinetochore microtubules," which jerk the chromosomes back and forth.
- Nonkinetochore microtubules interact with those from the opposite pole of the spindle.

? *How many molecules of DNA are in the prometaphase drawing? How many molecules per chromosome? How many double helices are there per chromosome? Per chromatid?*

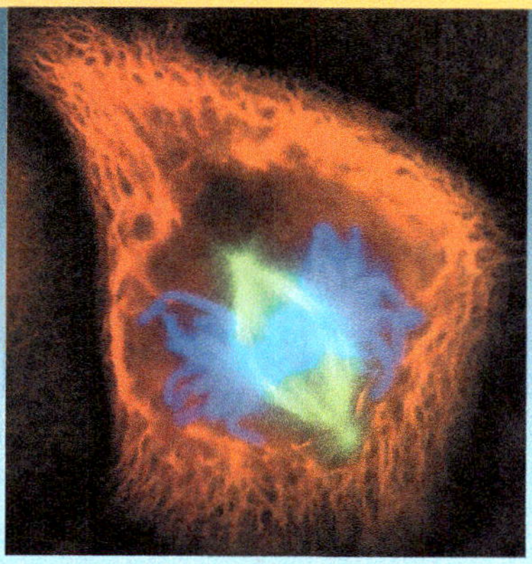

Metaphase

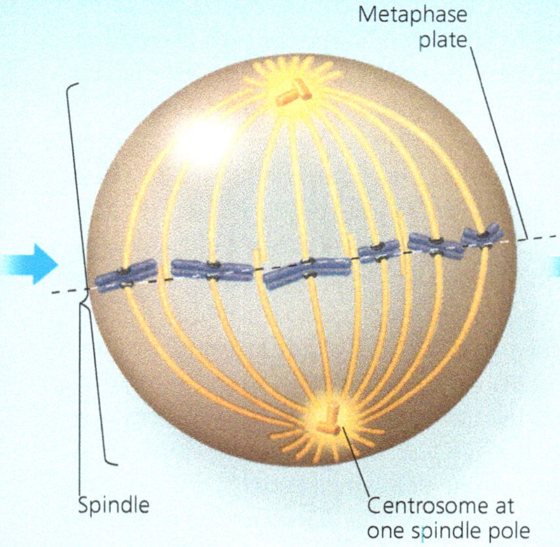

Metaphase

- The centrosomes are now at opposite poles of the cell.
- The chromosomes convene at the *metaphase plate*, a plane that is equidistant between the spindle's two poles. The chromosomes' centromeres lie at the metaphase plate.
- For each chromosome, the kinetochores of the sister chromatids are attached to kinetochore microtubules coming from opposite poles.

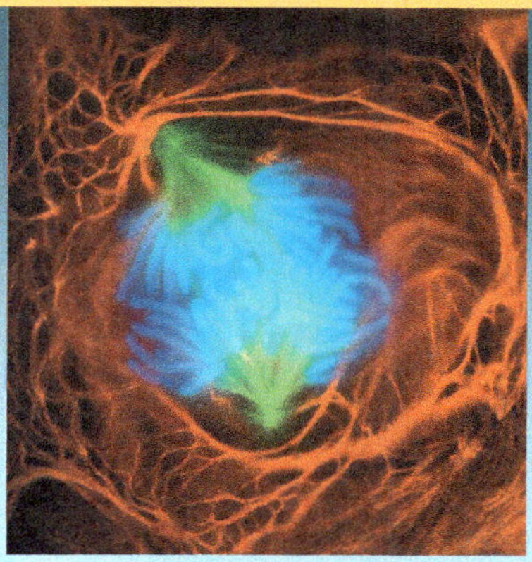

Anaphase

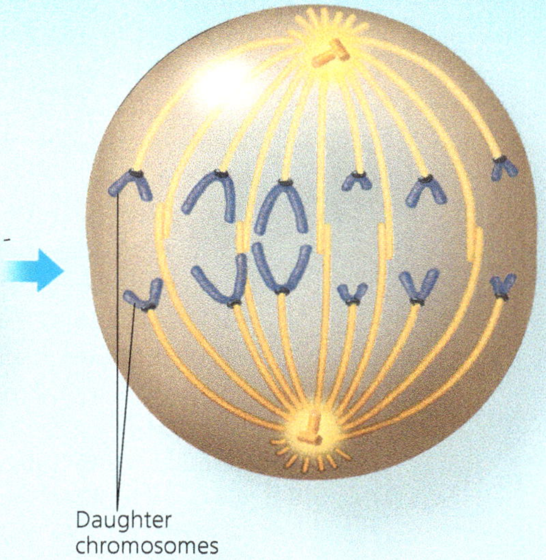

Anaphase

- Anaphase is the shortest stage of mitosis, often lasting only a few minutes.
- Anaphase begins when the cohesin proteins are cleaved. This allows the two sister chromatids of each pair to part suddenly. Each chromatid thus becomes a full-fledged chromosome.
- The two liberated daughter chromosomes begin moving toward opposite ends of the cell as their kinetochore microtubules shorten. Because these microtubules are attached at the centromere region, the chromosomes move centromere first (at about 1 μm/min).
- The cell elongates as the nonkinetochore microtubules lengthen.
- By the end of anaphase, the two ends of the cell have equivalent—and complete—collections of chromosomes.

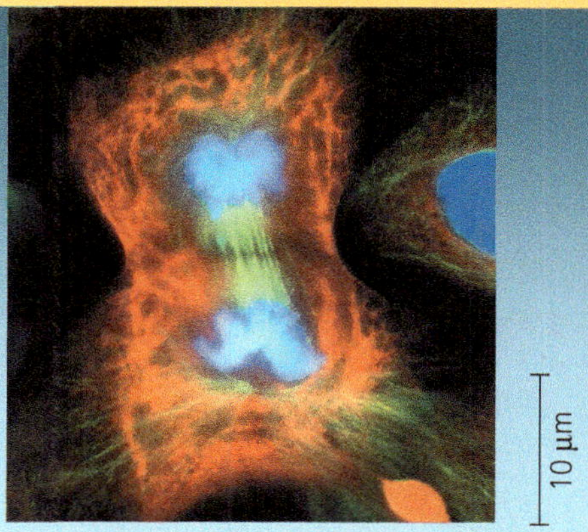

Telophase and Cytokinesis

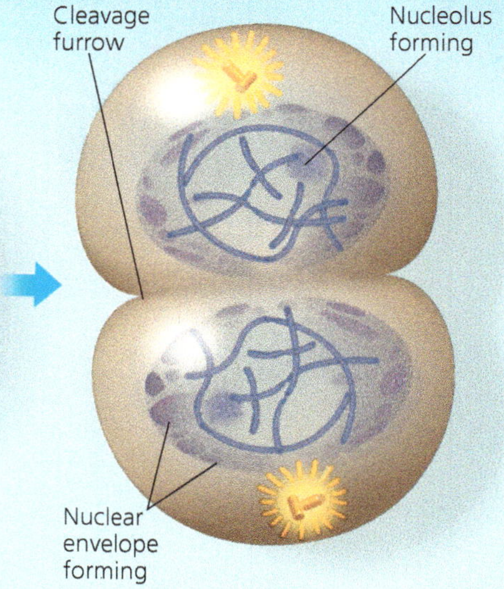

Telophase

- Two daughter nuclei form in the cell. Nuclear envelopes arise from the fragments of the parent cell's nuclear envelope and other portions of the endomembrane system.
- Nucleoli reappear.
- The chromosomes become less condensed.
- Any remaining spindle microtubules are depolymerized.
- Mitosis, the division of one nucleus into two genetically identical nuclei, is now complete.

Cytokinesis

- The division of the cytoplasm is usually well under way by late telophase, so the two daughter cells appear shortly after the end of mitosis.
- In animal cells, cytokinesis involves the formation of a cleavage furrow, which pinches the cell in two.

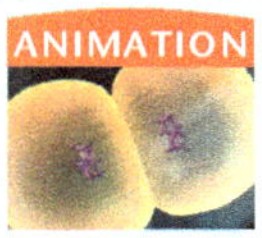

BioFlix Visit the Study Area at **www.masteringbiology.com** for the BioFlix® 3-D Animation on Mitosis.

two poles. This plane is called the **metaphase plate**, which is an imaginary rather than an actual cellular structure **(Figure 12.8)**. Meanwhile, microtubules that do not attach to kinetochores have been elongating, and by metaphase they overlap and interact with other nonkinetochore microtubules from the opposite pole of the spindle. (These are sometimes called "polar" microtubules.) By metaphase, the microtubules of the asters have also grown and are in contact with the plasma membrane. The spindle is now complete.

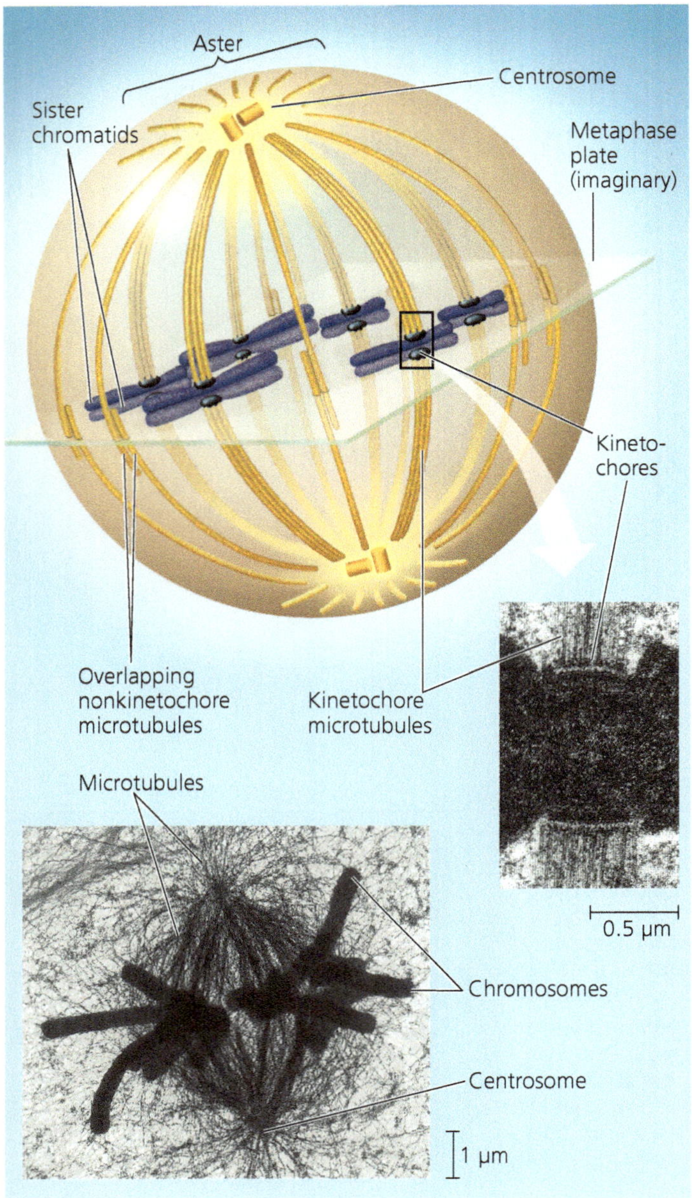

▲ **Figure 12.8 The mitotic spindle at metaphase.** The kinetochores of each chromosome's two sister chromatids face in opposite directions. Here, each kinetochore is attached to a *cluster* of kinetochore microtubules extending from the nearest centrosome. Nonkinetochore microtubules overlap at the metaphase plate (TEMs).

DRAW IT *On the lower micrograph, draw a line indicating the position of the metaphase plate. Circle an aster. Draw arrows indicating the directions of chromosome movement once anaphase begins.*

The structure of the completed spindle correlates well with its function during anaphase. Anaphase commences suddenly when the cohesins holding together the sister chromatids of each chromosome are cleaved by an enzyme called *separase*. Once the chromatids become separate, full-fledged chromosomes, they move toward opposite ends of the cell.

How do the kinetochore microtubules function in this poleward movement of chromosomes? Apparently, two mechanisms are in play, both involving motor proteins. (To review how motor proteins move an object along a microtubule, see Figure 6.21.) A clever experiment carried out in 1987 suggested that motor proteins on the kinetochores "walk" the chromosomes along the microtubules, which depolymerize at their kinetochore ends after the motor proteins have passed **(Figure 12.9)**. (This is referred to as the "Pacman" mechanism because of its resemblance to the arcade game character that moves by eating all the dots in its path.) However, other researchers, working with different cell types or cells from other species, have shown that chromosomes are "reeled in" by motor proteins at the spindle poles and that the microtubules depolymerize after they pass by these motor proteins. The general consensus now is that both mechanisms are used and that their relative contributions vary among cell types.

In a dividing animal cell, the nonkinetochore microtubules are responsible for elongating the whole cell during anaphase. Nonkinetochore microtubules from opposite poles overlap each other extensively during metaphase (see Figure 12.8). During anaphase, the region of overlap is reduced as motor proteins attached to the microtubules walk them away from one another, using energy from ATP. As the microtubules push apart from each other, their spindle poles are pushed apart, elongating the cell. At the same time, the microtubules lengthen somewhat by the addition of tubulin subunits to their overlapping ends. As a result, the microtubules continue to overlap.

At the end of anaphase, duplicate groups of chromosomes have arrived at opposite ends of the elongated parent cell. Nuclei re-form during telophase. Cytokinesis generally begins during anaphase or telophase, and the spindle eventually disassembles by depolymerization of microtubules.

Cytokinesis: *A Closer Look*

In animal cells, cytokinesis occurs by a process known as **cleavage**. The first sign of cleavage is the appearance of a **cleavage furrow**, a shallow groove in the cell surface near the old metaphase plate **(Figure 12.10a)**. On the cytoplasmic side of the furrow is a contractile ring of actin microfilaments associated with molecules of the protein myosin. The actin microfilaments interact with the myosin molecules, causing the ring to contract. The contraction of the dividing cell's ring of microfilaments is like the pulling of a drawstring. The cleavage furrow deepens until the parent cell is pinched in two, producing two completely separated cells, each with its own nucleus and share of cytosol, organelles, and other subcellular structures.

▼ Figure 12.9 **INQUIRY**

At which end do kinetochore microtubules shorten during anaphase?

EXPERIMENT Gary Borisy and colleagues at the University of Wisconsin wanted to determine whether kinetochore microtubules depolymerize at the kinetochore end or the pole end as chromosomes move toward the poles during mitosis. First they labeled the microtubules of a pig kidney cell in early anaphase with a yellow fluorescent dye.

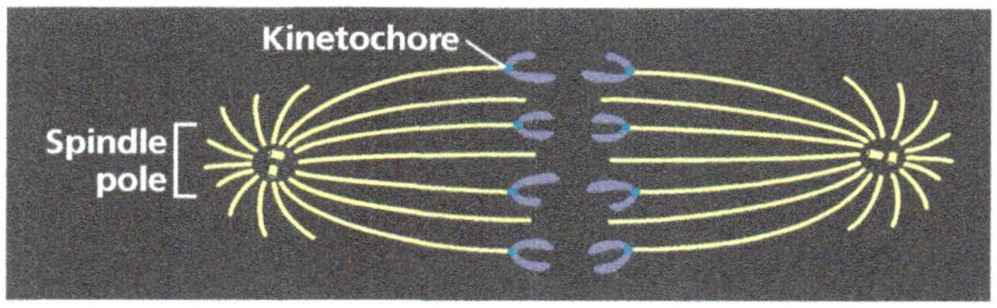

Then they marked a region of the kinetochore microtubules between one spindle pole and the chromosomes by using a laser to eliminate the fluorescence from that region, while leaving the microtubules intact (see below). As anaphase proceeded, they monitored the changes in microtubule length on either side of the mark.

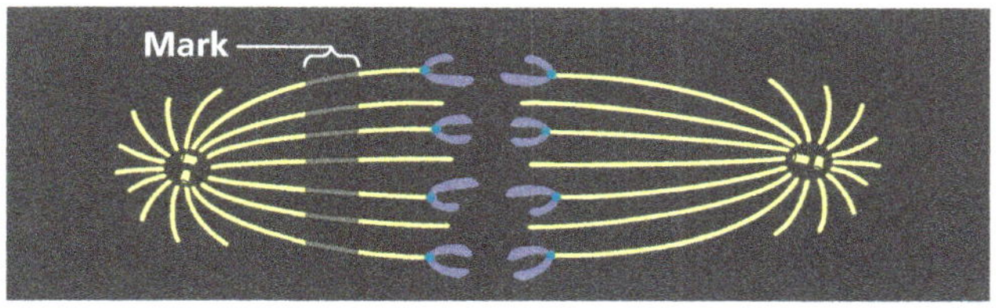

RESULTS As the chromosomes moved poleward, the microtubule segments on the kinetochore side of the mark shortened, while those on the spindle pole side stayed the same length.

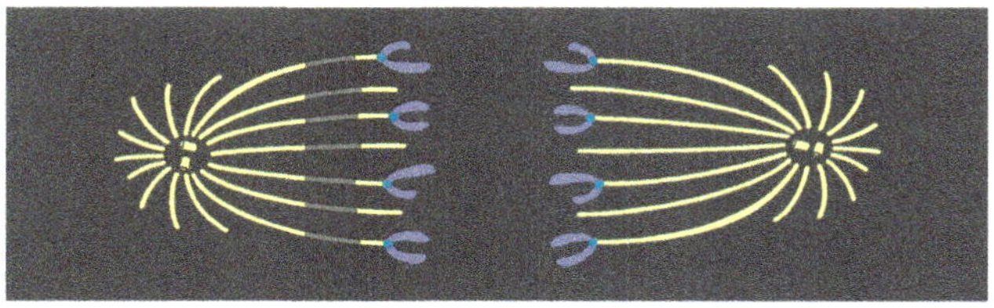

CONCLUSION During anaphase in this cell type, chromosome movement is correlated with kinetochore microtubules shortening at their kinetochore ends and not at their spindle pole ends. This experiment supports the hypothesis that during anaphase, a chromosome is walked along a microtubule as the microtubule depolymerizes at its kinetochore end, releasing tubulin subunits.

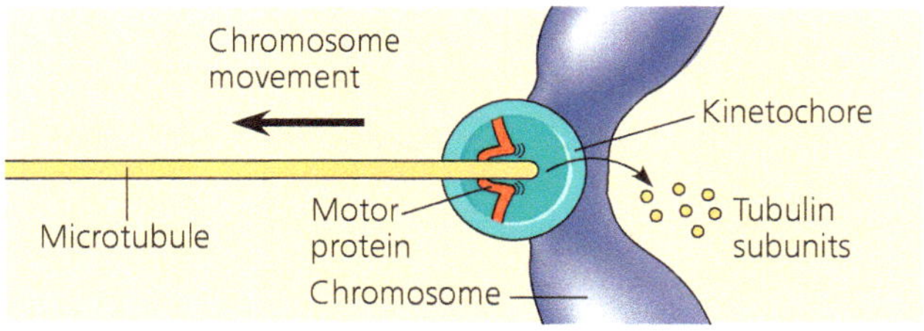

SOURCE G. J. Gorbsky, P. J. Sammak, and G. G. Borisy, Chromosomes move poleward in anaphase along stationary microtubules that coordinately disassemble from their kinetochore ends, *Journal of Cell Biology* 104:9–18 (1987).

WHAT IF? If this experiment had been done on a cell type in which "reeling in" at the poles was the main cause of chromosome movement, how would the mark have moved relative to the poles? How would the microtubule lengths have changed?

▼ Figure 12.10 **Cytokinesis in animal and plant cells.**

(a) Cleavage of an animal cell (SEM)

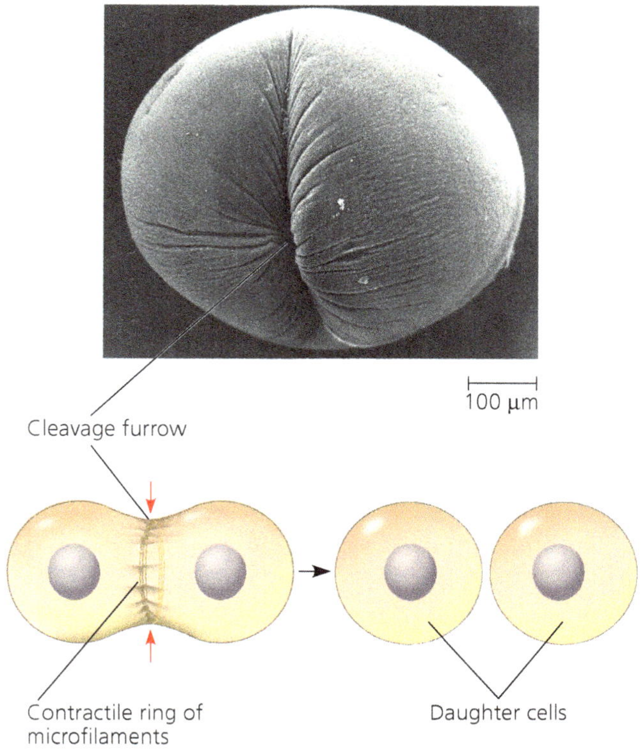

(b) Cell plate formation in a plant cell (TEM)

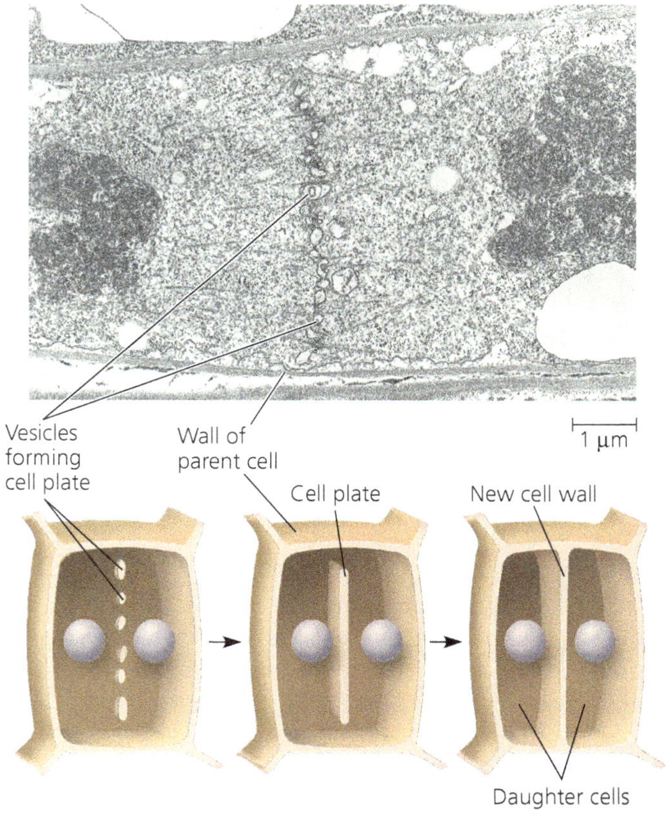

1 Prophase. The chromosomes are condensing and the nucleolus is beginning to disappear. Although not yet visible in the micrograph, the mitotic spindle is starting to form.

2 Prometaphase. Discrete chromosomes are now visible; each consists of two aligned, identical sister chromatids. Later in prometaphase, the nuclear envelope will fragment.

3 Metaphase. The spindle is complete, and the chromosomes, attached to microtubules at their kinetochores, are all at the metaphase plate.

4 Anaphase. The chromatids of each chromosome have separated, and the daughter chromosomes are moving to the ends of the cell as their kinetochore microtubules shorten.

5 Telophase. Daughter nuclei are forming. Meanwhile, cytokinesis has started: The cell plate, which will divide the cytoplasm in two, is growing toward the perimeter of the parent cell.

▲ **Figure 12.11 Mitosis in a plant cell.** These light micrographs show mitosis in cells of an onion root.

Cytokinesis in plant cells, which have cell walls, is markedly different. There is no cleavage furrow. Instead, during telophase, vesicles derived from the Golgi apparatus move along microtubules to the middle of the cell, where they coalesce, producing a **cell plate** (Figure 12.10b). Cell wall materials carried in the vesicles collect in the cell plate as it grows. The cell plate enlarges until its surrounding membrane fuses with the plasma membrane along the perimeter of the cell. Two daughter cells result, each with its own plasma membrane. Meanwhile, a new cell wall arising from the contents of the cell plate has formed between the daughter cells.

Figure 12.11 is a series of micrographs of a dividing plant cell. Examining this figure will help you review mitosis and cytokinesis.

Binary Fission in Bacteria

Prokaryotes (bacteria and archaea) can undergo a type of reproduction in which the cell grows to roughly double its size and then divides to form two cells. The term **binary fission**, meaning "division in half," refers to this process and to the asexual reproduction of single-celled eukaryotes, such as the amoeba in Figure 12.2a. However, the process in eukaryotes involves mitosis, while that in prokaryotes does not.

In bacteria, most genes are carried on a single *bacterial chromosome* that consists of a circular DNA molecule and associated proteins. Although bacteria are smaller and simpler than eukaryotic cells, the challenge of replicating their genomes in an orderly fashion and distributing the copies equally to two daughter cells is still formidable. The chromosome of the bacterium *Escherichia coli*, for example, when it is fully stretched out, is about 500 times as long as the cell. For such a long chromosome to fit within the cell requires that it be highly coiled and folded.

In *E. coli*, the process of cell division is initiated when the DNA of the bacterial chromosome begins to replicate at a specific place on the chromosome called the **origin of replication**, producing two origins. As the chromosome continues to replicate, one origin moves rapidly toward the opposite end of the cell (Figure 12.12). While the chromosome is replicating, the cell elongates. When replication is complete and the bacterium has reached about twice its initial size, its plasma membrane pinches inward, dividing the parent *E. coli* cell into two daughter cells. In this way, each cell inherits a complete genome.

Using the techniques of modern DNA technology to tag the origins of replication with molecules that glow green in fluorescence microscopy (see Figure 6.3), researchers have directly observed the movement of bacterial chromosomes. This movement is reminiscent of the poleward movements of the centromere regions of eukaryotic chromosomes during anaphase of mitosis, but bacteria don't have visible mitotic spindles or even microtubules. In most bacterial species studied, the two origins of replication end up at opposite ends of the cell or in some other very specific location, possibly anchored there by one or more proteins. How bacterial chromosomes move and how their specific location is established and maintained are still not fully understood. However, several proteins have been identified that play important roles: One resembling eukaryotic actin apparently functions in bacterial chromosome movement during cell division, and another that is related to tubulin seems to help pinch the plasma membrane inward, separating the two bacterial daughter cells.

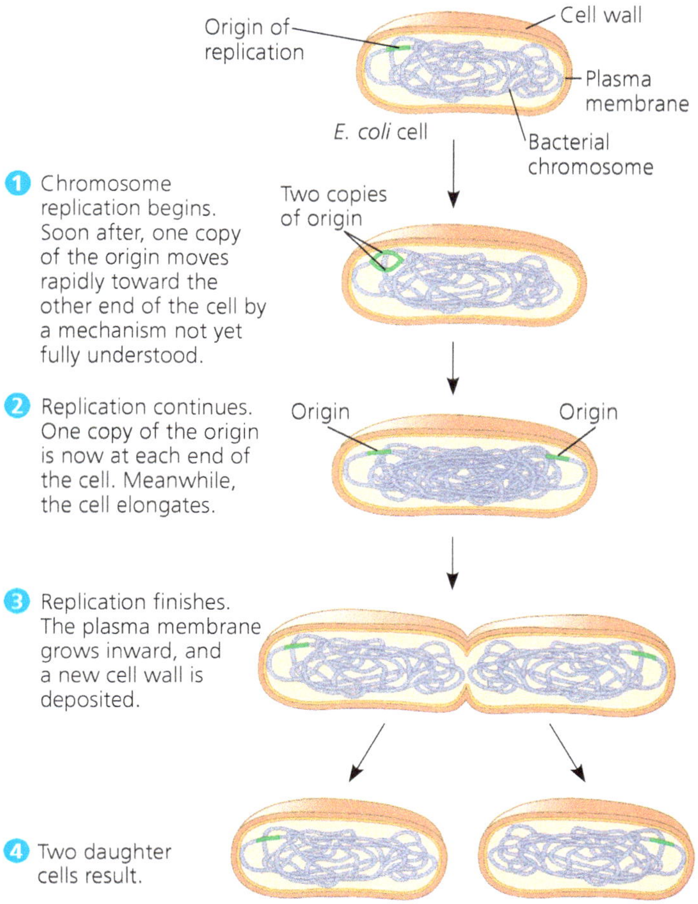

▲ **Figure 12.12 Bacterial cell division by binary fission.** The bacterium *E. coli*, shown here, has a single, circular chromosome.

The Evolution of Mitosis

EVOLUTION Given that prokaryotes preceded eukaryotes on Earth by more than a billion years, we might hypothesize that mitosis evolved from simpler prokaryotic mechanisms of cell reproduction. The fact that some of the proteins involved in bacterial binary fission are related to eukaryotic proteins that function in mitosis supports that hypothesis.

As eukaryotes evolved, along with their larger genomes and nuclear envelopes, the ancestral process of binary fission, seen today in bacteria, somehow gave rise to mitosis. **Figure 12.13** shows some variations on cell division in different groups of organisms. These processes may be similar to mechanisms used by ancestral species and thus may resemble steps in the evolution of mitosis from a binary fission-like process presumably carried out by very early bacteria. Possible intermediate stages are suggested by two unusual types of nuclear division found today in certain unicellular eukaryotes—dinoflagellates, diatoms, and some yeasts. These two modes of nuclear division are thought to be cases where ancestral mechanisms have remained relatively unchanged over evolutionary time. In both types, the nuclear envelope remains intact, in contrast to what happens in most eukaryotic cells.

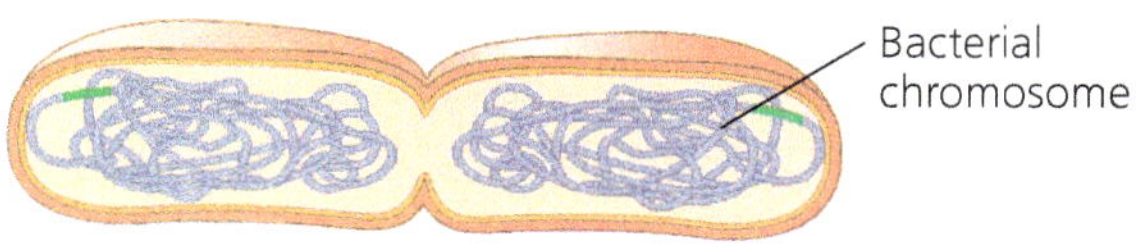

(a) Bacteria. During binary fission in bacteria, the origins of the daughter chromosomes move to opposite ends of the cell. The mechanism is not fully understood, but proteins may anchor the daughter chromosomes to specific sites on the plasma membrane.

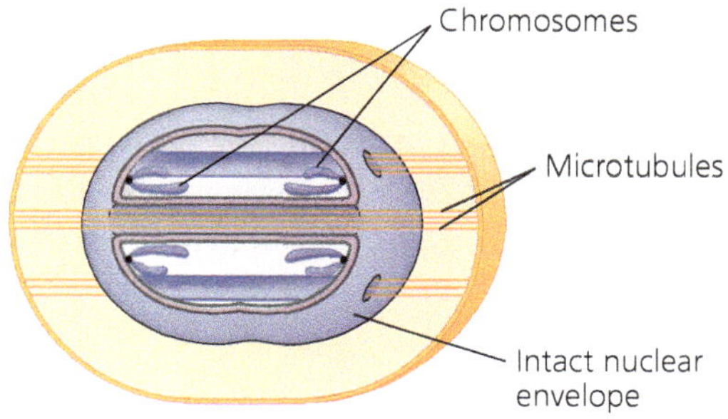

(b) Dinoflagellates. In unicellular protists called dinoflagellates, the chromosomes attach to the nuclear envelope, which remains intact during cell division. Microtubules pass through the nucleus inside cytoplasmic tunnels, reinforcing the spatial orientation of the nucleus, which then divides in a process reminiscent of bacterial binary fission.

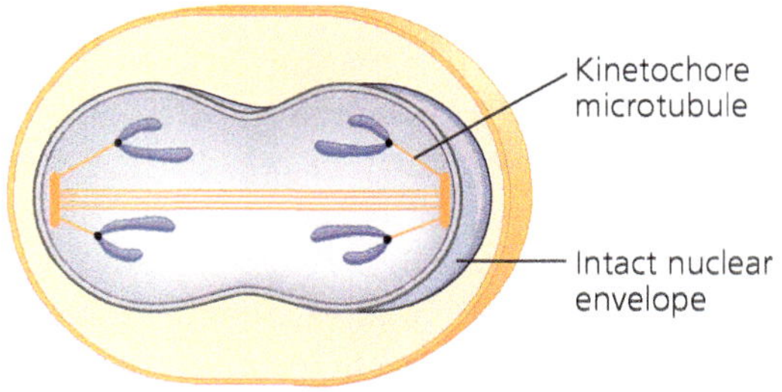

(c) Diatoms and some yeasts. In two other groups of unicellular protists, diatoms and some yeasts, the nuclear envelope also remains intact during cell division. In these organisms, the microtubules form a spindle *within* the nucleus. Microtubules separate the chromosomes, and the nucleus splits into two daughter nuclei.

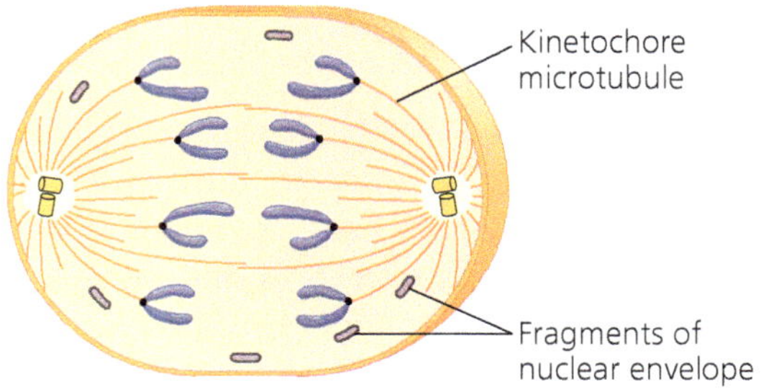

(d) Most eukaryotes. In most other eukaryotes, including plants and animals, the spindle forms outside the nucleus, and the nuclear envelope breaks down during mitosis. Microtubules separate the chromosomes, and two nuclear envelopes then form.

▲ **Figure 12.13 Mechanisms of cell division in several groups of organisms.** Some unicellular eukaryotes existing today have mechanisms of cell division that may resemble intermediate steps in the evolution of mitosis. Except for (a), these schematic diagrams do not show cell walls.

CONCEPT CHECK 12.2

1. How many chromosomes are shown in the diagram in Figure 12.8? Are they duplicated? How many chromatids are shown?
2. Compare cytokinesis in animal cells and plant cells.
3. What is the function of nonkinetochore microtubules?
4. Compare the roles of tubulin and actin during eukaryotic cell division with the roles of tubulin-like and actin-like proteins during bacterial binary fission.
5. **MAKE CONNECTIONS** What other functions do actin and tubulin carry out? Name the proteins they interact with to do so. (Review Figures 6.21a and 6.27a.)
6. **WHAT IF?** During which stages of the cell cycle does a chromosome consist of two identical chromatids?

For suggested answers, see Appendix A.

CONCEPT 12.3

The eukaryotic cell cycle is regulated by a molecular control system

The timing and rate of cell division in different parts of a plant or animal are crucial to normal growth, development, and maintenance. The frequency of cell division varies with the type of cell. For example, human skin cells divide frequently throughout life, whereas liver cells maintain the ability to divide but keep it in reserve until an appropriate need arises—say, to repair a wound. Some of the most specialized cells, such as fully formed nerve cells and muscle cells, do not divide at all in a mature human. These cell cycle differences result from regulation at the molecular level. The mechanisms of this regulation are of intense interest, not only for understanding the life cycles of normal cells but also for understanding how cancer cells manage to escape the usual controls.

Evidence for Cytoplasmic Signals

What controls the cell cycle? One reasonable hypothesis might be that each event in the cell cycle merely leads to the next, as in a simple metabolic pathway. According to this hypothesis, the replication of chromosomes in the S phase, for example, might cause cell growth during the G_2 phase, which might in turn lead inevitably to the onset of mitosis. However, this hypothesis, which proposes a pathway that is not subject to either internal or external regulation, turns out to be incorrect.

In the early 1970s, a variety of experiments led to an alternative hypothesis: that the cell cycle is driven by specific signaling molecules present in the cytoplasm. Some of the first strong evidence for this hypothesis came from experiments with mammalian cells grown in culture. In these experiments, two cells in different phases of the cell cycle were fused to form a single cell with two nuclei. If one of the original cells was in the S phase and the other was in G_1, the G_1 nucleus immediately entered the S phase, as though stimulated by signaling molecules present in the cytoplasm of the first cell. Similarly, if a cell undergoing mitosis (M phase) was fused with another cell in any stage of its cell cycle, even G_1, the second nucleus immediately entered mitosis, with condensation of the chromatin and formation of a mitotic spindle **(Figure 12.14)**.

▼ Figure 12.14 INQUIRY

Do molecular signals in the cytoplasm regulate the cell cycle?

EXPERIMENT Researchers at the University of Colorado wondered whether a cell's progression through the cell cycle is controlled by cytoplasmic molecules. To investigate this, they selected cultured mammalian cells that were at different phases of the cell cycle and induced them to fuse. Two such experiments are shown here.

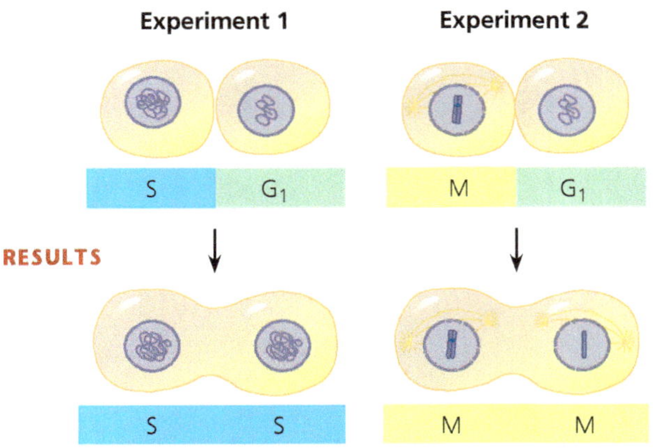

When a cell in the S phase was fused with a cell in G_1, the G_1 nucleus immediately entered the S phase—DNA was synthesized.

When a cell in the M phase was fused with a cell in G_1, the G_1 nucleus immediately began mitosis—a spindle formed and the chromosomes condensed, even though the chromosomes had not been duplicated.

CONCLUSION The results of fusing a G_1 cell with a cell in the S or M phase of the cell cycle suggest that molecules present in the cytoplasm during the S or M phase control the progression to those phases.

SOURCE R. T. Johnson and P. N. Rao, Mammalian cell fusion: Induction of premature chromosome condensation in interphase nuclei, *Nature* 226:717–722 (1970).

WHAT IF? If the progression of phases did not depend on cytoplasmic molecules and each phase began when the previous one was complete, how would the results have differed?

The Cell Cycle Control System

The experiment shown in Figure 12.14 and other experiments on animal cells and yeasts demonstrated that the sequential events of the cell cycle are directed by a distinct **cell cycle control system**, a cyclically operating set of molecules in the cell that both triggers and coordinates key events

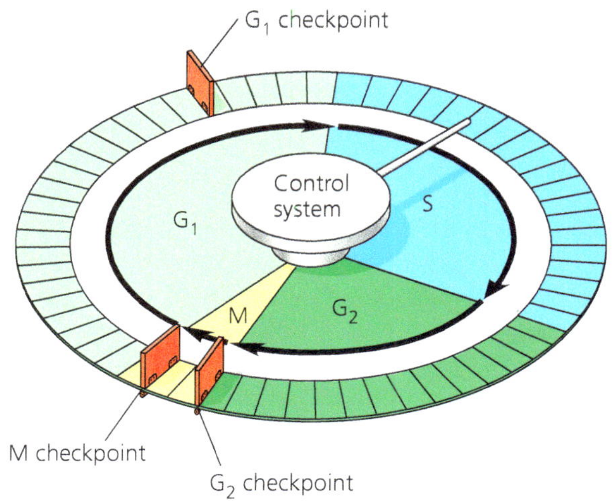

▲ **Figure 12.15 Mechanical analogy for the cell cycle control system.** In this diagram of the cell cycle, the flat "stepping stones" around the perimeter represent sequential events. Like the control device of an automatic washer, the cell cycle control system proceeds on its own, driven by a built-in clock. However, the system is subject to internal and external regulation at various checkpoints, of which three are shown (red).

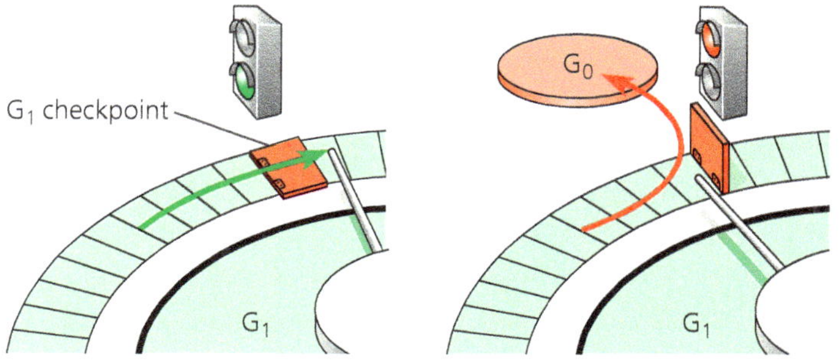

(a) If a cell receives a go-ahead signal at the G_1 checkpoint, the cell continues on in the cell cycle.

(b) If a cell does not receive a go-ahead signal at the G_1 checkpoint, the cell exits the cell cycle and goes into G_0, a nondividing state.

▲ **Figure 12.16 The G_1 checkpoint.**

WHAT IF? *What might be the result if the cell ignored the checkpoint and progressed through the cell cycle?*

in the cell cycle. The cell cycle control system has been compared to the control device of an automatic washing machine (Figure 12.15). Like the washer's timing device, the cell cycle control system proceeds on its own, according to a built-in clock. However, just as a washer's cycle is subject to both internal control (such as the sensor that detects when the tub is filled with water) and external adjustment (such as activation of the start mechanism), the cell cycle is regulated at certain checkpoints by both internal and external signals.

A **checkpoint** in the cell cycle is a control point where stop and go-ahead signals can regulate the cycle. (The signals are transmitted within the cell by the kinds of signal transduction pathways discussed in Chapter 11.) Animal cells generally have built-in stop signals that halt the cell cycle at checkpoints until overridden by go-ahead signals. Many signals registered at checkpoints come from cellular surveillance mechanisms inside the cell. These signals report whether crucial cellular processes that should have occurred by that point have in fact been completed correctly and thus whether or not the cell cycle should proceed. Checkpoints also register signals from outside the cell, as we will discuss later. Three major checkpoints are found in the G_1, G_2, and M phases (see Figure 12.15).

For many cells, the G_1 checkpoint—dubbed the "restriction point" in mammalian cells—seems to be the most important. If a cell receives a go-ahead signal at the G_1 checkpoint, it will usually complete the G_1, S, G_2, and M phases and divide. If it does not receive a go-ahead signal at that point, it will exit the cycle, switching into a nondividing state called the **G_0 phase** (Figure 12.16). Most cells of the human body are actually in the G_0 phase. As mentioned earlier, mature nerve cells and muscle cells never divide. Other cells, such as liver cells, can be "called back" from the G_0 phase to the cell cycle by external cues, such as growth factors released during injury.

To understand how cell cycle checkpoints work, we first need to see what kinds of molecules make up the cell cycle control system (the molecular basis for the cell cycle clock) and how a cell progresses through the cycle. Then we will consider the internal and external checkpoint signals that can make the clock pause or continue.

The Cell Cycle Clock: Cyclins and Cyclin-Dependent Kinases

Rhythmic fluctuations in the abundance and activity of cell cycle control molecules pace the sequential events of the cell cycle. These regulatory molecules are mainly proteins of two types: protein kinases and cyclins. Protein kinases are enzymes that activate or inactivate other proteins by phosphorylating them (see Chapter 11). Particular protein kinases give the go-ahead signals at the G_1 and G_2 checkpoints.

Many of the kinases that drive the cell cycle are actually present at a constant concentration in the growing cell, but much of the time they are in an inactive form. To be active, such a kinase must be attached to a **cyclin**, a protein that gets its name from its cyclically fluctuating concentration in the cell. Because of this requirement, these kinases are called **cyclin-dependent kinases**, or **Cdks**. The activity of a Cdk rises and falls with changes in the concentration of its cyclin partner. Figure 12.17a, on the next page, shows the fluctuating activity of **MPF**, the cyclin-Cdk complex that was discovered first (in frog eggs). Note that the peaks of MPF activity correspond to the peaks of cyclin concentration. The cyclin level rises during the S and G_2 phases and then falls abruptly during M phase.

The initials MPF stand for "maturation-promoting factor," but we can think of MPF as "M-phase-promoting factor" because it triggers the cell's passage past the G_2 checkpoint into

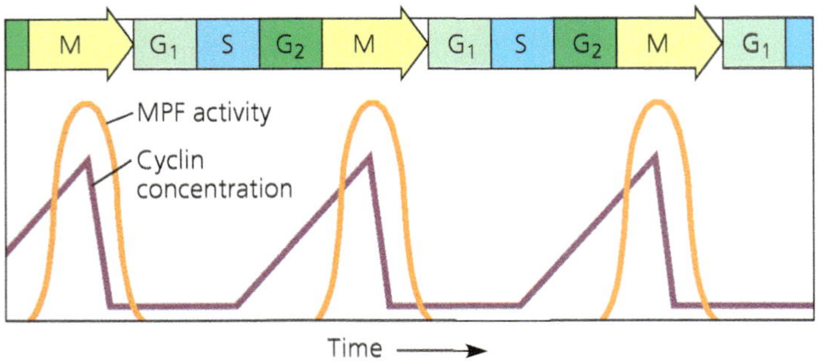

(a) Fluctuation of MPF activity and cyclin concentration during the cell cycle

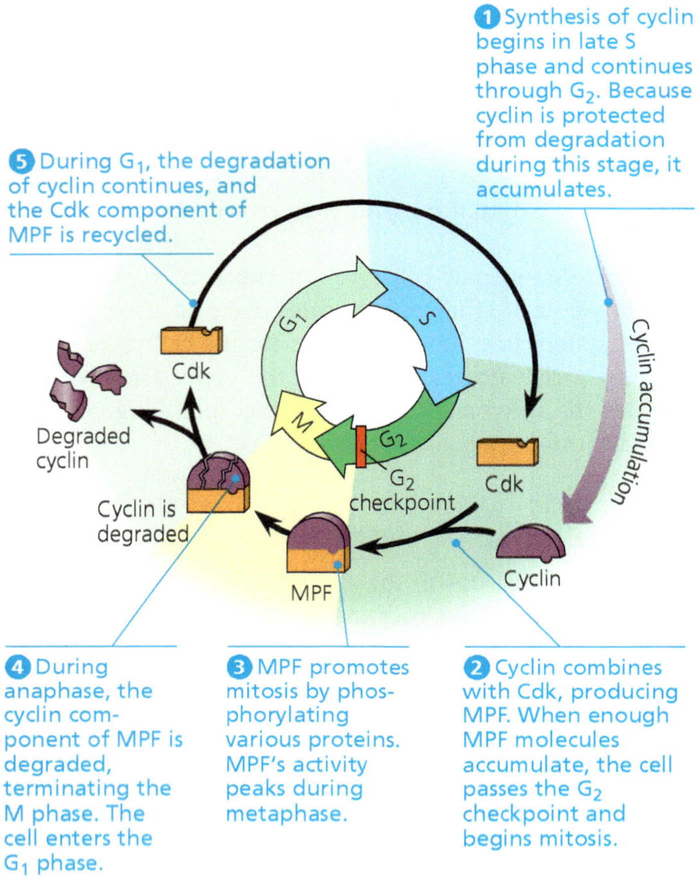

(b) Molecular mechanisms that help regulate the cell cycle

▲ **Figure 12.17 Molecular control of the cell cycle at the G_2 checkpoint.** The steps of the cell cycle are timed by rhythmic fluctuations in the activity of cyclin-dependent kinases (Cdks). Here we focus on a cyclin-Cdk complex in animal cells called MPF, which acts at the G_2 checkpoint as a go-ahead signal, triggering the events of mitosis.

? *Explain how the events in the diagram in (b) are related to the "Time" axis of the graph in (a).*

M phase (**Figure 12.17b**). When cyclins that accumulate during G_2 associate with Cdk molecules, the resulting MPF complex phosphorylates a variety of proteins, initiating mitosis. MPF acts both directly as a kinase and indirectly by activating other kinases. For example, MPF causes phosphorylation of various proteins of the nuclear lamina (see Figure 6.9), which promotes fragmentation of the nuclear envelope during prometaphase of mitosis. There is also evidence that MPF contributes to molecular events required for chromosome condensation and spindle formation during prophase.

During anaphase, MPF helps switch itself off by initiating a process that leads to the destruction of its own cyclin. The noncyclin part of MPF, the Cdk, persists in the cell, inactive until it becomes part of MPF again by associating with new cyclin molecules synthesized during the S and G_2 phases of the next round of the cycle.

Cell behavior at the G_1 checkpoint is also regulated by the activity of cyclin-Cdk protein complexes. Animal cells appear to have at least three Cdk proteins and several different cyclins that operate at this checkpoint. The fluctuating activities of different cyclin-Cdk complexes are of major importance in controlling all the stages of the cell cycle.

Stop and Go Signs: Internal and External Signals at the Checkpoints

Research scientists are currently working out the pathways that link signals originating inside and outside the cell with the responses by cyclin-dependent kinases and other proteins. An example of an internal signal occurs at the third important checkpoint, the M phase checkpoint. Anaphase, the separation of sister chromatids, does not begin until all the chromosomes are properly attached to the spindle at the metaphase plate. Researchers have learned that as long as some kinetochores are unattached to spindle microtubules, the sister chromatids remain together, delaying anaphase. Only when the kinetochores of all the chromosomes are properly attached to the spindle does the appropriate regulatory protein complex become activated. (In this case, the regulatory molecule is not a cyclin-Cdk complex but, instead, a different complex made up of several proteins.) Once activated, the complex sets off a chain of molecular events that activates the enzyme separase, which cleaves the cohesins, allowing the sister chromatids to separate. This mechanism ensures that daughter cells do not end up with missing or extra chromosomes.

Studies using animal cells in culture have led to the identification of many external factors, both chemical and physical, that can influence cell division. For example, cells fail to divide if an essential nutrient is lacking in the culture medium. (This is analogous to trying to run an automatic washing machine without the water supply hooked up; an internal sensor won't allow the machine to continue past the point where water is needed.) And even if all other conditions are favorable, most types of mammalian cells divide in culture only if the growth medium includes specific growth factors. As mentioned in Chapter 11, a **growth factor** is a protein released by certain cells that stimulates other cells to divide. Researchers have discovered more than 50 growth factors. Different cell types respond specifically to different growth factors or combinations of growth factors.

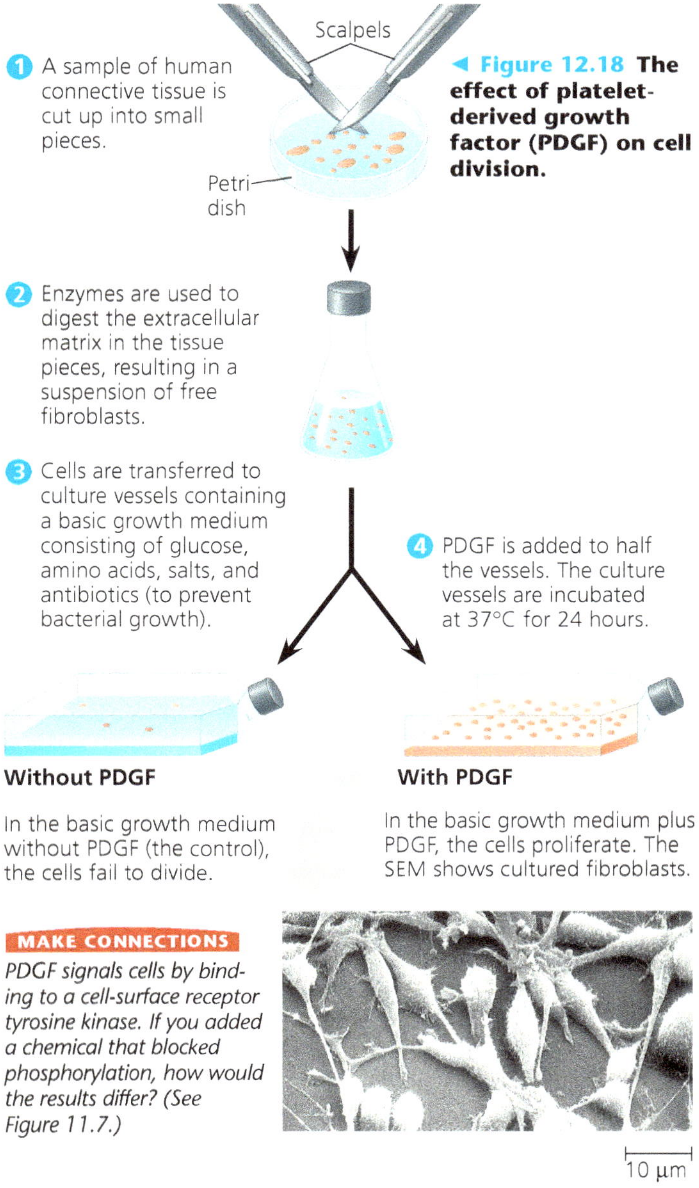

◀ **Figure 12.18 The effect of platelet-derived growth factor (PDGF) on cell division.**

MAKE CONNECTIONS

PDGF signals cells by binding to a cell-surface receptor tyrosine kinase. If you added a chemical that blocked phosphorylation, how would the results differ? (See Figure 11.7.)

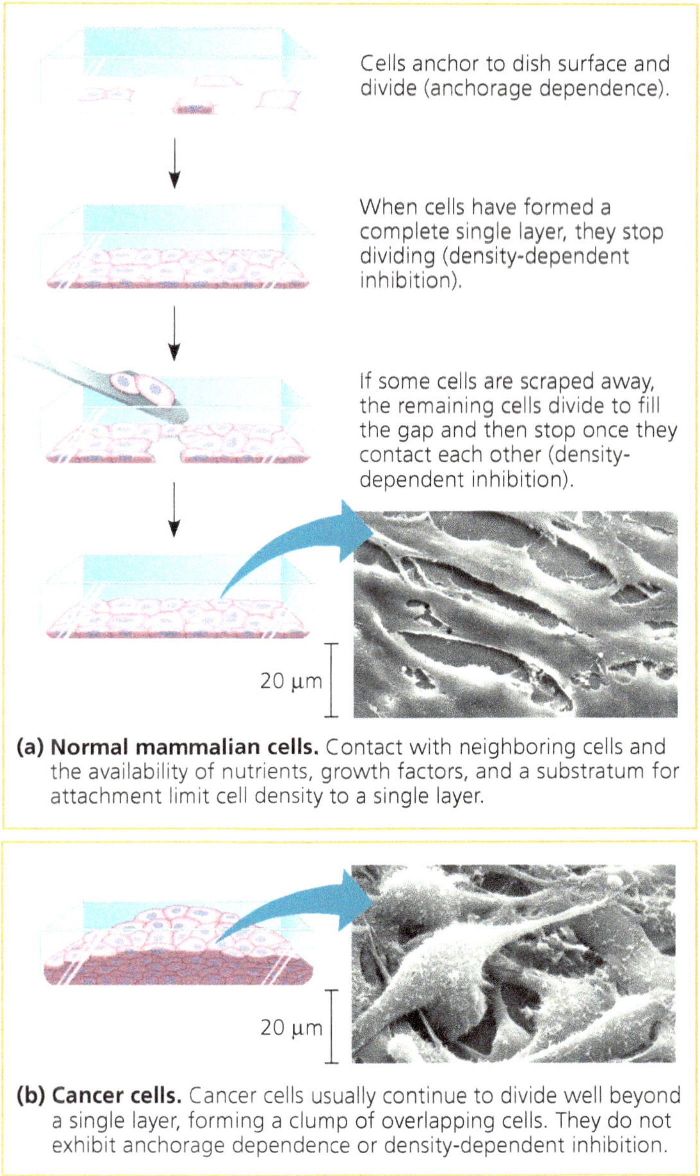

(a) Normal mammalian cells. Contact with neighboring cells and the availability of nutrients, growth factors, and a substratum for attachment limit cell density to a single layer.

(b) Cancer cells. Cancer cells usually continue to divide well beyond a single layer, forming a clump of overlapping cells. They do not exhibit anchorage dependence or density-dependent inhibition.

▲ **Figure 12.19 Density-dependent inhibition and anchorage dependence of cell division.** Individual cells are shown disproportionately large in the drawings.

Consider, for example, *platelet-derived growth factor (PDGF)*, which is made by blood cell fragments called platelets. The experiment illustrated in **Figure 12.18** demonstrates that PDGF is required for the division of cultured fibroblasts, a type of connective tissue cell. Fibroblasts have PDGF receptors on their plasma membranes. The binding of PDGF molecules to these receptors (which are receptor tyrosine kinases; see Chapter 11) triggers a signal transduction pathway that allows the cells to pass the G_1 checkpoint and divide. PDGF stimulates fibroblast division not only in the artificial conditions of cell culture, but also in an animal's body. When an injury occurs, platelets release PDGF in the vicinity. The resulting proliferation of fibroblasts helps heal the wound.

The effect of an external physical factor on cell division is clearly seen in **density-dependent inhibition**, a phenomenon in which crowded cells stop dividing **(Figure 12.19a)**. As first observed many years ago, cultured cells normally divide until they form a single layer of cells on the inner surface of the culture container, at which point the cells stop dividing. If some cells are removed, those bordering the open space begin dividing again and continue until the vacancy is filled. Follow-up studies revealed that the binding of a cell-surface protein to its counterpart on an adjoining cell sends a growth-inhibiting signal to both cells, preventing them from moving forward in the cell cycle, even in the presence of growth factors.

Most animal cells also exhibit **anchorage dependence** (see Figure 12.19a). To divide, they must be attached to a substratum, such as the inside of a culture jar or the extracellular matrix of a tissue. Experiments suggest that like cell density, anchorage is signaled to the cell cycle control system via pathways involving plasma membrane proteins and elements of the cytoskeleton linked to them.

Density-dependent inhibition and anchorage dependence appear to function in the body's tissues as well as in cell culture, checking the growth of cells at some optimal density and location. Cancer cells, which we discuss next, exhibit neither density-dependent inhibition nor anchorage dependence (**Figure 12.19b**).

Loss of Cell Cycle Controls in Cancer Cells

Cancer cells do not heed the normal signals that regulate the cell cycle. They divide excessively and invade other tissues. If unchecked, they can kill the organism.

Cancer cells in culture do not stop dividing when growth factors are depleted. A logical hypothesis is that cancer cells do not need growth factors in their culture medium to grow and divide. They may make a required growth factor themselves, or they may have an abnormality in the signaling pathway that conveys the growth factor's signal to the cell cycle control system even in the absence of that factor. Another possibility is an abnormal cell cycle control system. In all of these scenarios, the underlying basis of the abnormality is almost always a change in one or more genes that alters the function of their protein products, resulting in faulty cell cycle control. You will learn more in Chapter 18 about the genetic bases of these changes and how these conditions may lead to cancer.

There are other important differences between normal cells and cancer cells that reflect derangements of the cell cycle. If and when they stop dividing, cancer cells do so at random points in the cycle, rather than at the normal checkpoints. Moreover, cancer cells can go on dividing indefinitely in culture if they are given a continual supply of nutrients; in essence, they are "immortal." A striking example is a cell line that has been reproducing in culture since 1951. Cells of this line are called HeLa cells because their original source was a tumor removed from a woman named Henrietta Lacks. By contrast, nearly all normal mammalian cells growing in culture divide only about 20 to 50 times before they stop dividing, age, and die. (We'll see a possible reason for this phenomenon when we discuss DNA replication in Chapter 16.) Finally, cancer cells evade the normal controls that trigger a cell to undergo apoptosis when something is wrong—for example, when an irreparable mistake has occurred during DNA replication preceding mitosis.

The abnormal behavior of cancer cells can be catastrophic when it occurs in the body. The problem begins when a single cell in a tissue undergoes **transformation**, the process that converts a normal cell to a cancer cell. The body's immune system normally recognizes a transformed cell as an insurgent and destroys it. However, if the cell evades destruction, it may proliferate and form a tumor, a mass of abnormal cells within otherwise normal tissue. The abnormal cells may remain at the original site if they have too few genetic and cellular changes to survive at another site. In that case, the tumor is called a **benign tumor**. Most benign tumors do not cause serious problems and can be completely removed by surgery. In contrast, a **malignant tumor** includes cells whose genetic and cellular changes enable them to spread to new tissues and impair the functions of one or more organs. An individual with a malignant tumor is said to have cancer; **Figure 12.20** shows the development of breast cancer.

The changes that have occurred in cells of malignant tumors show up in many ways besides excessive proliferation. These cells may have unusual numbers of chromosomes,

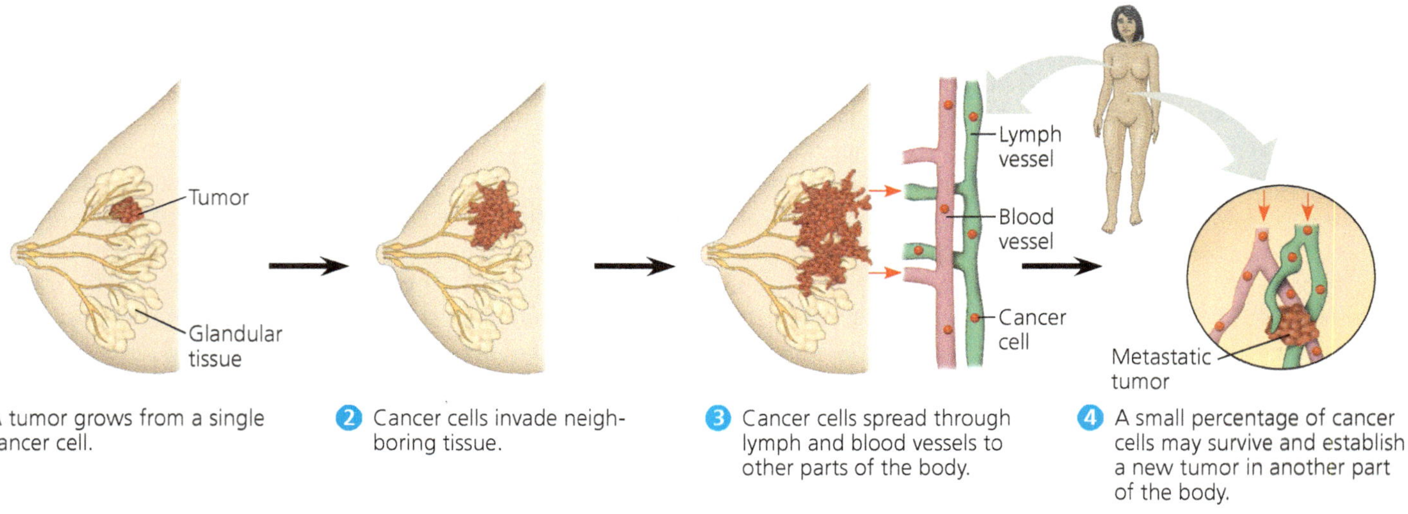

▲ **Figure 12.20 The growth and metastasis of a malignant breast tumor.** The cells of malignant (cancerous) tumors grow in an uncontrolled way and can spread to neighboring tissues and, via lymph and blood vessels, to other parts of the body. The spread of cancer cells beyond their original site is called metastasis.

though whether this is a cause or an effect of transformation is a current topic of debate. Their metabolism may be disabled, and they may cease to function in any constructive way. Abnormal changes on the cell surface cause cancer cells to lose attachments to neighboring cells and the extracellular matrix, allowing them to spread into nearby tissues. Cancer cells may also secrete signaling molecules that cause blood vessels to grow toward the tumor. A few tumor cells may separate from the original tumor, enter blood vessels and lymph vessels, and travel to other parts of the body. There, they may proliferate and form a new tumor. This spread of cancer cells to locations distant from their original site is called **metastasis** (see Figure 12.20).

A tumor that appears to be localized may be treated with high-energy radiation, which damages DNA in cancer cells much more than it does in normal cells, apparently because the majority of cancer cells have lost the ability to repair such damage. To treat known or suspected metastatic tumors, chemotherapy is used, in which drugs that are toxic to actively dividing cells are administered through the circulatory system. As you might expect, chemotherapeutic drugs interfere with specific steps in the cell cycle. For example, the drug Taxol freezes the mitotic spindle by preventing microtubule depolymerization, which stops actively dividing cells from proceeding past metaphase. The side effects of chemotherapy are due to the drugs' effects on normal cells that divide often. For example, nausea results from chemotherapy's effects on intestinal cells, hair loss from effects on hair follicle cells, and susceptibility to infection from effects on immune system cells.

Over the past several decades, researchers have produced a flood of valuable information about cell-signaling pathways and how their malfunction contributes to the development of cancer through effects on the cell cycle. Coupled with new molecular techniques, such as the ability to rapidly sequence the DNA of cells in a particular tumor, medical treatments for cancer are beginning to become more "personalized" to a particular patient's tumor. Breast cancer provides a good example. Basic research on the processes described in Chapters 11 and 12 has augmented our understanding of the molecular events underlying development of breast cancer. Proteins functioning in cell signaling pathways that affect the cell cycle are often found to be altered in breast cancer cells. Analyzing the level and sequences of such proteins has allowed physicians to better tailor the treatment to the cancers of some individuals, as shown in **Figure 12.21**.

One of the big lessons we've learned about the development of cancer, though, is how very complex the process is. There are many areas that remain to be explored. Perhaps the reason we have so many unanswered questions about cancer cells is that there is still so much to learn about how normal cells function. The cell, life's basic unit of structure and function, holds enough secrets to engage researchers well into the future.

▼ **Figure 12.21**

IMPACT

Advances in Treatment of Breast Cancer

Cancer cells, such as the breast cancer cell shown below, are analyzed by DNA sequencing and other molecular techniques to look for alterations in the level or sequence of specific proteins associated with cancer. For example, the cells of roughly 20–25% of breast cancer tumors show abnormally high amounts of a cell-surface receptor tyrosine kinase called HER2, and many show an increase in the number of estrogen receptor (ER) molecules, intracellular receptors that can trigger cell division. Based on lab findings, a physician can prescribe chemotherapy with a molecule that blocks the function of the specific protein (Herceptin for HER2 and tamoxifen for ERs). Treatment using these agents, when appropriate, has led to increased survival rates and fewer cancer recurrences.

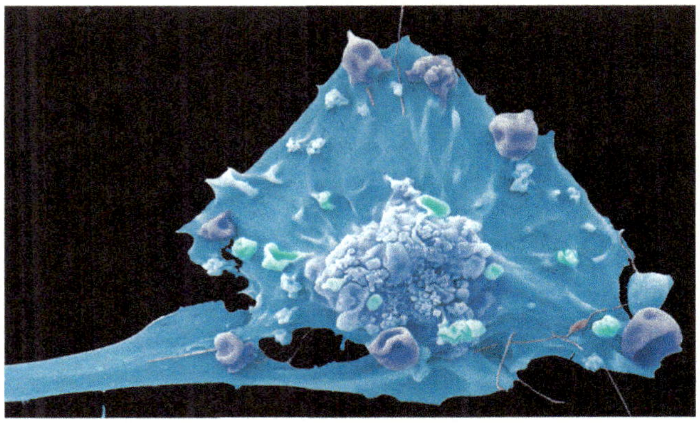

WHY IT MATTERS Approximately one out of every eight women will develop breast cancer, the most common cancer among women. Worldwide, the incidence of breast cancer has been increasing annually. However, the mortality rate from this disease is falling in the United States and elsewhere, probably a result of earlier detection and improved treatment. Furthermore, what we are learning from the study of breast cancer also enhances our understanding of the development and treatment of other types of cancer.

FURTHER READING F. J. Esteva and G. N. Hortobagyi, Gaining ground on breast cancer, *Scientific American* 298:58–65 (2008).

MAKE CONNECTIONS Review the material in Chapter 11 on receptor tyrosine kinases and intracellular receptors (Figures 11.7 and 11.9 on pp. 212–214). Explain in general how these receptors might function in triggering cell division.

CONCEPT CHECK 12.3

1. In Figure 12.14, why do the nuclei resulting from experiment 2 contain different amounts of DNA?
2. How does MPF allow a cell to pass the G_2 phase checkpoint and enter mitosis? (See Figure 12.17.)
3. What phase are most of your body cells in?
4. Compare and contrast a benign tumor and a malignant tumor.
5. **WHAT IF?** What would happen if you performed the experiment in Figure 12.18 with cancer cells?

For suggested answers, see Appendix A.

12 CHAPTER REVIEW

SUMMARY OF KEY CONCEPTS

- Unicellular organisms reproduce by **cell division**; multicellular organisms depend on cell division for their development from a fertilized egg and for growth and repair. Cell division is part of the **cell cycle**, an ordered sequence of events in the life of a cell from its origin until it divides into daughter cells.

CONCEPT 12.1

Most cell division results in genetically identical daughter cells (pp. 229–230)

- The genetic material (DNA) of a cell—its **genome**—is partitioned among **chromosomes**. Each eukaryotic chromosome consists of one DNA molecule associated with many proteins that maintain chromosome structure and help control the activity of genes. Together, the complex of DNA and associated proteins is called **chromatin**. The chromatin of a chromosome exists in different states of condensation at different times. In animals, **gametes** have one set of chromosomes and **somatic cells** have two sets.
- Cells replicate their genetic material before they divide, ensuring that each daughter cell can receive a copy of the DNA. In preparation for cell division, chromosomes are duplicated, each one then consisting of two identical **sister chromatids** joined along their lengths by sister chromatid cohesion and held most tightly together at a constricted region at the **centromeres** of the chromatids. When this cohesion is broken, the chromatids separate during cell division, becoming the chromosomes of the new daughter cells. Eukaryotic cell division consists of **mitosis** (division of the nucleus) and **cytokinesis** (division of the cytoplasm).

? *Differentiate between these terms: chromosome, chromatin, and chromatid.*

CONCEPT 12.2

The mitotic phase alternates with interphase in the cell cycle (pp. 230–238)

- Between divisions, a cell is in **interphase**: the $\mathbf{G_1}$, **S**, and $\mathbf{G_2}$ phases. The cell grows throughout interphase, but DNA is replicated only during the synthesis (S) phase. Mitosis and cytokinesis make up the **mitotic (M) phase** of the cell cycle.

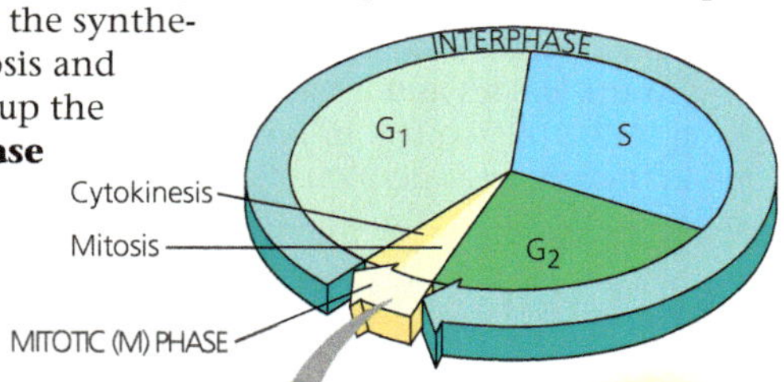

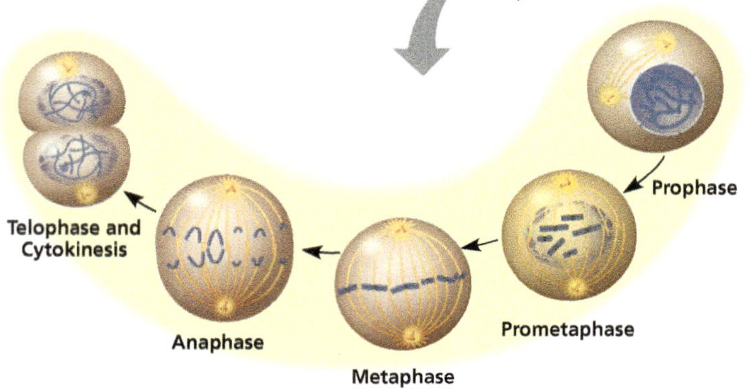

- The **mitotic spindle** is an apparatus of microtubules that controls chromosome movement during mitosis. In animal cells, the spindle arises from the **centrosomes** and includes spindle microtubules and **asters**. Some spindle microtubules attach to the **kinetochores** of chromosomes and move the chromosomes to the **metaphase plate**. In anaphase, sister chromatids separate, and motor proteins move them along the kinetochore microtubules toward opposite ends of the cell. Meanwhile, motor proteins push nonkinetochore microtubules from opposite poles away from each other, elongating the cell. In telophase, genetically identical daughter nuclei form at opposite ends of the cell.
- Mitosis is usually followed by cytokinesis. Animal cells carry out cytokinesis by **cleavage**, and plant cells form a **cell plate**.
- During **binary fission** in bacteria, the chromosome replicates and the two daughter chromosomes actively move apart. Some of the proteins involved in bacterial binary fission are related to eukaryotic actin and tubulin.
- Since prokaryotes preceded eukaryotes by more than a billion years, it is likely that mitosis evolved from prokaryotic cell division. Certain unicellular eukaryotes exhibit mechanisms of cell division that may be similar to those of ancestors of existing eukaryotes. Such mechanisms might have been intermediate steps in the evolution of mitosis from bacterial binary fission.

? *In which of the three subphases of interphase and the stages of mitosis do chromosomes exist as single DNA molecules?*

CONCEPT 12.3

The eukaryotic cell cycle is regulated by a molecular control system (pp. 238–243)

- Signaling molecules present in the cytoplasm regulate progress through the cell cycle.
- The **cell cycle control system** is molecularly based. Cyclic changes in regulatory proteins work as a cell cycle clock. The key molecules are **cyclins** and **cyclin-dependent kinases (Cdks)**. The clock has specific **checkpoints** where the cell cycle stops until a go-ahead signal is received. Cell culture has enabled researchers to study the molecular details of cell division. Both internal signals and external signals control the cell cycle checkpoints via signal transduction pathways. Most cells exhibit **density-dependent inhibition** of cell division as well as **anchorage dependence**.
- Cancer cells elude normal cell cycle regulation and divide out of control, forming tumors. **Malignant tumors** invade surrounding tissues and can undergo **metastasis**, exporting cancer cells to other parts of the body, where they may form secondary tumors. Recent advances in understanding the cell cycle and cell signaling, as well as techniques for sequencing DNA, have allowed improvements in cancer treatment.

? *Explain the significance of the G_1, G_2, and M checkpoints and the go-ahead signals involved in the cell cycle control system.*

TEST YOUR UNDERSTANDING

LEVEL 1: KNOWLEDGE/COMPREHENSION

1. Through a microscope, you can see a cell plate beginning to develop across the middle of a cell and nuclei forming on either side of the cell plate. This cell is most likely
   a. an animal cell in the process of cytokinesis.
   b. a plant cell in the process of cytokinesis.

c. an animal cell in the S phase of the cell cycle.
d. a bacterial cell dividing.
e. a plant cell in metaphase.

2. Vinblastine is a standard chemotherapeutic drug used to treat cancer. Because it interferes with the assembly of microtubules, its effectiveness must be related to
a. disruption of mitotic spindle formation.
b. inhibition of regulatory protein phosphorylation.
c. suppression of cyclin production.
d. myosin denaturation and inhibition of cleavage furrow formation.
e. inhibition of DNA synthesis.

3. One difference between cancer cells and normal cells is that cancer cells
a. are unable to synthesize DNA.
b. are arrested at the S phase of the cell cycle.
c. continue to divide even when they are tightly packed together.
d. cannot function properly because they are affected by density-dependent inhibition.
e. are always in the M phase of the cell cycle.

4. The decline of MPF activity at the end of mitosis is due to
a. the destruction of the protein kinase Cdk.
b. decreased synthesis of Cdk.
c. the degradation of cyclin.
d. the accumulation of cyclin.
e. synthesis of DNA.

5. In the cells of some organisms, mitosis occurs without cytokinesis. This will result in
a. cells with more than one nucleus.
b. cells that are unusually small.
c. cells lacking nuclei.
d. destruction of chromosomes.
e. cell cycles lacking an S phase.

6. Which of the following does *not* occur during mitosis?
a. condensation of the chromosomes
b. replication of the DNA
c. separation of sister chromatids
d. spindle formation
e. separation of the spindle poles

LEVEL 2: APPLICATION/ANALYSIS

7. In the light micrograph below of dividing cells near the tip of an onion root, identify a cell in each of the following stages: prophase, prometaphase, metaphase, anaphase, and telophase. Describe the major events occurring at each stage.

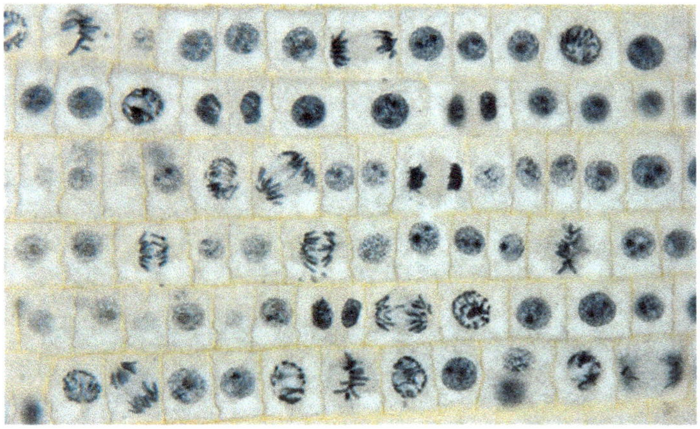

8. A particular cell has half as much DNA as some other cells in a mitotically active tissue. The cell in question is most likely in
a. G_1.
b. G_2.
c. prophase.
d. metaphase.
e. anaphase.

9. The drug cytochalasin B blocks the function of actin. Which of the following aspects of the animal cell cycle would be most disrupted by cytochalasin B?
a. spindle formation
b. spindle attachment to kinetochores
c. DNA synthesis
d. cell elongation during anaphase
e. cleavage furrow formation and cytokinesis

10. **DRAW IT** Draw one eukaryotic chromosome as it would appear during interphase, during each of the stages of mitosis, and during cytokinesis. Also draw and label the nuclear envelope and any microtubules attached to the chromosome(s).

LEVEL 3: SYNTHESIS/EVALUATION

11. **EVOLUTION CONNECTION**
The result of mitosis is that the daughter cells end up with the same number of chromosomes that the parent cell had. Another way to maintain the number of chromosomes would be to carry out cell division first and then duplicate the chromosomes in each daughter cell. Do you think this would be an equally good way of organizing the cell cycle? Why do you suppose that evolution has not led to this alternative?

12. **SCIENTIFIC INQUIRY**
Although both ends of a microtubule can gain or lose subunits, one end (called the plus end) polymerizes and depolymerizes at a higher rate than the other end (the minus end). For spindle microtubules, the plus ends are in the center of the spindle, and the minus ends are at the poles. Motor proteins that move along microtubules specialize in walking either toward the plus end or toward the minus end; the two types are called plus end–directed and minus end–directed motor proteins, respectively. Given what you know about chromosome movement and spindle changes during anaphase, predict which type of motor proteins would be present on (a) kinetochore microtubules and (b) nonkinetochore microtubules.

13. **WRITE ABOUT A THEME**
The Genetic Basis of Life The continuity of life is based on heritable information in the form of DNA. In a short essay (100–150 words), explain how the process of mitosis faithfully parcels out exact copies of this heritable information in the production of genetically identical daughter cells.

For selected answers, see Appendix A.

UNIT 5

The Evolutionary History of Biological Diversity

An Interview with

W. Ford Doolittle

Brimming with curiosity about nature, W. Ford Doolittle is a passionate advocate for both science and scientists. Dr. Doolittle holds degrees in molecular biology from Harvard and Stanford Universities. Since 1971 he has worked at Dalhousie University in Halifax, Canada, where he is now a Professor Emeritus. The author of more than 270 scientific articles, Dr. Doolittle is well known for his work on the molecular biology of cyanobacteria and archaea, early steps in the evolution of eukaryotes, and the importance of horizontal gene transfer in prokaryotes. He has also nurtured dozens of young scientists, making Dalhousie a world-renowned center for the study of evolution. The recipient of numerous awards and honorary degrees, Dr. Doolittle is an elected member of the Royal Society of Canada and the U.S. National Academy of Sciences. He is also an avid photographer, and in 2011 he will receive a B.F.A. in photography from the Nova Scotia College of Art and Design. Surrounded by stacks of journal articles, photography projects, and the bustle of a busy laboratory, Michael Cain and Jane Reece spoke with Dr. Doolittle in his office at Dalhousie.

How did you get started in science?

I hadn't planned on studying science until, as a high school student, I got a job washing glassware in Sol Spiegelman's lab at the University of Illinois. [Spiegelman was a molecular biologist whose work on RNA played a key role in developing understanding of biology's "central dogma" (see Chapter 17).] In college, I continued working in Sol's lab during the summer. It was a very exciting place to be, kind of like a hospital emergency ward where everything was important, everything was urgent. Some people worked all day, some people worked all night, some people worked all day and night! The lab was abuzz with simple and elegant experiments in molecular biology. I got the feeling that science was important, which I have never lost. And I got the feeling that science was important for its own sake, which I also have never lost. Doing science seems to fill our human need to know, to understand what the universe is like.

Did you become interested in evolution while working in Sol's lab?

Well, that came a little later. When I arrived at Dalhousie in 1971, I began studying blue-green algae (now called cyanobacteria) because they were a pretty color and few people were working on them. So the field was wide open. Lynn Margulis had just published her book *Origin of Eukaryotic Cells*, which put forward a startling idea—the endosymbiont hypothesis—positing that mitochondria and chloroplasts were formerly aerobic bacteria and cyanobacteria, respectively, that began living within larger cells, ultimately leading to the origin of eukaryotes. Then a fortuitous thing happened: I hired Linda Bonen, who brought a new technique to our lab, called RNA cataloging, from which you could get RNA sequence information. It's not a full sequence, but it was enough to start comparing ribosomal RNA molecules (rRNAs). I realized that we could use this technique to test the endosymbiont hypothesis for the origin of chloroplasts. We compared the sequences of rRNAs from chloroplasts with those from cyanobacteria and found they were similar, just as Margulis had predicted! From there, I became more and more interested in evolution, which continues to fascinate me today.

What are the main questions you are asking in your research?

I've had a long-standing interest in the evolution of genes and genomes. Most of my current work is on prokaryotes, but I've also been interested in questions about eukaryotes, such as, Where did introns come from? How did eukaryotic genomes originate? In recent years, my students and I have been occupied with two main research areas. One is *metagenomics*, where we collect environmental samples, such as mud from the bottom of harbors, and then sequence the prokaryotic DNA contained in the samples. We use these DNA sequences to shed light on how microbes interact with each other and their environment. The other main area of my current research is *horizontal gene transfer*, the transfer of genetic information—usually whole genes but not necessarily—from one species to another.

How can genes move from one prokaryotic species to another?

This can happen in a variety of ways, and such transfers turn out to be surprisingly common. Mechanisms have evolved in some bacteria enabling them to take up DNA molecules directly from the environment—and these DNA molecules may have come from individuals of other species that have died and broken apart. When this happens, sometimes the DNA is incorporated into the bacterial chromosomes and hence becomes part of their own genetic heritage. Many prokaryotes can also form temporary bridges that connect two or more individuals to one another, enabling genes to move from one species to another. And as we mentioned earlier, sometimes entire genomes are transferred in endosymbiosis. These and other processes that result in horizontal gene transfer have occurred throughout the nearly 4 billion years of life on Earth—and they continue to occur today.

What impact has horizontal gene transfer had on the evolutionary history of life?

Horizontal gene transfer, or HGT for short, has been a major force in the history of life. For instance, the endosymbiotic origins of mitochondria and chloroplasts were profoundly important events in the origin of eukaryotic cells. We humans wouldn't be here if it weren't for mitochondria, and of course without chloroplasts there wouldn't be any plants. Furthermore, HGT has been a dominant force in prokaryotic evolution—probably *the* dominant force. Genes that control key metabolic processes, such as photosynthesis and nitrogen fixation, can be transferred readily between species, and this has had an enormous impact on the ecological and evolutionary success of prokaryotes. And while one can construct evolutionary trees that identify major groups of prokaryotes, such trees can obscure the fact that there has been a tremendous amount of HGT, affecting the majority of the genes.

Tell us more about evolutionary trees.

An evolutionary tree (or *phylogenetic tree*) is a branching diagram that represents the evolutionary history of a group of organisms. For example, we might use morphological and genetic data to figure out a phylogenetic tree of animals. Such a tree can provide a huge amount of information. For any particular group of animals, whether they be weevils, whales, or worms, our tree could identify the ancestors and closest relatives of the group. If we traced the history of animals all the way back, we could use the tree to help us answer questions such as, What did the earliest animals look like? What features did they pass on to all their descendants, including us? Phylogenetic trees also have great practical value. The same techniques we use to reconstruct evolutionary history have been used in forensics, where phylogenetic trees have helped solve criminal cases, and epidemiology, where trees have been used to estimate when and where diseases such as AIDS originated.

What have you learned during your career about the pattern and process of evolution?

We can document the *pattern* of evolution, the "what happened" part, with data from many scientific disciplines. For example, if we measure genetic change in a population, we are documenting an evolutionary pattern. The *process* of evolution is the "what caused it" part, which focuses on the mechanisms that produce the observed pattern. Sixty years ago, many scientists thought that natural selection was *the* mechanism that explained most evolutionary patterns. But we don't think that anymore; we now know that genetic drift and catastrophic events such as an asteroid striking Earth can explain many observed evolutionary patterns. I call this "process pluralism" to indicate that evolutionary patterns can result from many processes. Results from my research suggest that we should be considering "pattern pluralism" as well. Consider the tree of life, a phylogenetic tree that represents the history of all life as a series of branching events, in which a common ancestor gives rise to two descendant groups of organisms. This approach works well for organisms such as animals, but for prokaryotes the importance of HGT leads me to question whether the tree of life pattern is real. For prokaryotes, the pattern of evolution may be more like a tangled network of organisms exchanging DNA, not a simple branching tree.

"For prokaryotes, the pattern of evolution may be more like a tangled network of organisms exchanging DNA, not a simple branching tree."

Does this mean that Charles Darwin was wrong when he introduced the idea of a tree of life in *The Origin of Species*?

Darwin set out in his book to document the fact of evolution—the pattern of evolutionary change over time—and to show that natural selection could account for observed patterns of change. He did a great job on both accounts. We now recognize that well-understood genetic and ecological processes operating over nearly 4 billion years of evolution are adequate to explain diversity and adaptation in the living world. As for his ideas about the tree of life, Darwin was focused on plants and animals, two groups for which the tree of life idea works well. Remember, when Darwin formulated his ideas, many scientists still thought that bacteria formed spontaneously from rotten meat. It wasn't even clear that bacteria were organisms. So I would say that Darwin's idea about the tree of life is still true, but that certain restrictions apply. It is similar to the situation in physics: Newton's laws work well under certain conditions, but not as well under other conditions. The tree of life still works well for plants and animals, but less well for prokaryotes due to HGT. And after all, most of life is prokaryotic.

Have recent discoveries altered our understanding of the evolutionary history of life?

Certainly. For one thing, our view of genomes and genome evolution has changed greatly in recent years. Who would have thought that we humans only have about the same number of genes as a 1-mm-long worm? It's astonishing. Not that long ago, little was known about endosymbiosis or HGT or the astoundingly rapid rates at which bacterial populations can evolve. Overall, the importance of prokaryotes and protists (single-celled eukaryotes) has really become clear. I mentioned metagenomics earlier, an approach that has already revealed previously unsuspected levels of microbial diversity. I think metagenomics will play an increasingly important role in future discoveries. It might even help us understand the consequences of increased atmospheric carbon dioxide (CO_2) and global warming, among the most important problems we face today. Microorganisms absorb vast amounts of CO_2 from the atmosphere. How will ongoing global warming affect microorganisms? Will they absorb even more CO_2, thus reducing effects of global warming? Or will the opposite occur? We don't know, but we can use metagenomics to figure out how climate change is currently affecting microbial communities. Once we understand what is going on, perhaps that can help us fix it.

You've devoted much time and effort to mentoring others. Can you comment on that?

Working in Sol Spiegelman's lab when I was young was a big turning point in my life. And as I gained experience, I tried to provide similar opportunities for others. One way I've done that is to establish programs that provide opportunities for young scientists to work together and to learn from each other and from more experienced scientists. I've always thought that it really helps to get experience early in your career, so I welcome people seeking such experience. One of the best ways to find what you really love doing is to try different things and see what most excites you. I did not plan to be a scientist, but I started doing science and I loved it. I'm a big believer in hands-on experience as a way of learning.

W. Ford Doolittle (center) with Michael Cain (left) and Jane Reece

27

Bacteria and Archaea

▲ **Figure 27.1 Why is this lake's water pink?**

EVOLUTION

KEY CONCEPTS

27.1 **Structural and functional adaptations contribute to prokaryotic success**
27.2 **Rapid reproduction, mutation, and genetic recombination promote genetic diversity in prokaryotes**
27.3 **Diverse nutritional and metabolic adaptations have evolved in prokaryotes**
27.4 **Molecular systematics is illuminating prokaryotic phylogeny**
27.5 **Prokaryotes play crucial roles in the biosphere**
27.6 **Prokaryotes have both beneficial and harmful impacts on humans**

OVERVIEW

Masters of Adaptation

In the heat of summer, parts of Utah's Great Salt Lake turn pink **(Figure 27.1)**, a sign of waters so salty that they would dehydrate your skin if you took a dip. The salt concentration can reach 32%, nearly ten times that of seawater. Yet despite these harsh conditions, the dramatic color of these waters is caused not by minerals or other nonliving sources, but by living things. What organisms can live in such an inhospitable environment, and how do they do it?

The pink color in the Great Salt Lake comes from trillions of prokaryotes in the domains Archaea and Bacteria, including archaea in the genus *Halobacterium*. These archaea have red membrane pigments, some of which capture the light energy that drives ATP synthesis. *Halobacterium* species are among the most salt-tolerant organisms on Earth; they thrive in salinities that dehydrate and kill other cells. *Halobacterium* compensates for water lost through osmosis by pumping potassium ions (K^+) into the cell until the ionic concentration inside the cell matches the concentration outside.

Like *Halobacterium*, many other prokaryotes can tolerate extreme conditions. Examples include *Deinococcus radiodurans*, which can survive 3 million rads of radiation (3,000 times the dose fatal to humans), and *Picrophilus oshimae*, which can grow at a pH of 0.03 (acidic enough to dissolve metal). Other prokaryotes live in environments that are too cold or too hot for most other organisms, and some have even been found living in rocks 3.2 km (2 miles) below Earth's surface.

Prokaryotic species are also very well adapted to more "normal" habitats—the lands and waters in which most other species are found. Their ability to adapt to a broad range of habitats helps explain why prokaryotes are the most abundant organisms on Earth: The number of prokaryotes in a handful of fertile soil is greater than the number of people who have ever lived. In this chapter, we'll examine the adaptations, diversity, and enormous ecological impact of these tiny organisms.

CONCEPT 27.1

Structural and functional adaptations contribute to prokaryotic success

As you read in Chapter 25, the first organisms to inhabit Earth were likely prokaryotes. Throughout their long evolutionary history, prokaryotic populations have been (and continue to be) subjected to natural selection in all kinds of environments, resulting in their enormous diversity today.

We'll begin by describing prokaryotes. Most prokaryotes are unicellular, although the cells of some species remain attached to each other after cell division. Prokaryotic cells typically have diameters of 0.5–5 µm, much smaller than the

10–100 μm diameter of many eukaryotic cells. (One notable exception, *Thiomargarita namibiensis*, can be 750 μm across—bigger than the dot on this i.) Prokaryotic cells have a variety of shapes (**Figure 27.2**). Finally, although they are unicellular and small, prokaryotes are well organized, achieving all of an organism's life functions within a single cell.

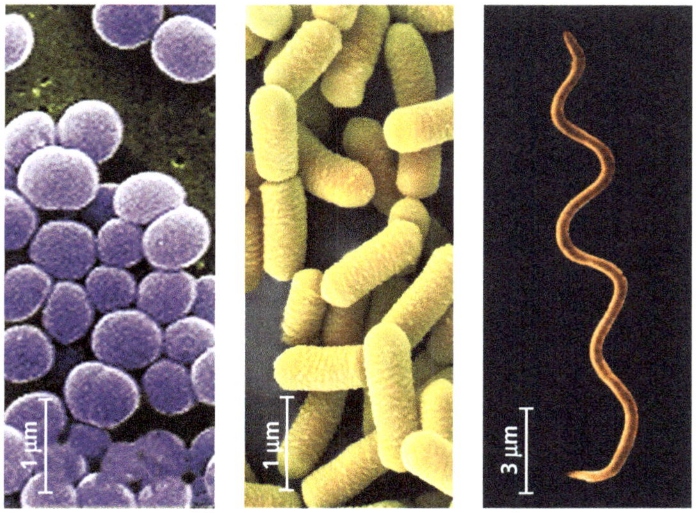

(a) Spherical **(b) Rod-shaped** **(c) Spiral**

▲ **Figure 27.2 The most common shapes of prokaryotes.** **(a)** Cocci (singular, *coccus*) are spherical prokaryotes. They occur singly, in pairs (diplococci), in chains of many cells (streptococci), and in clusters resembling bunches of grapes (staphylococci). **(b)** Bacilli (singular, *bacillus*) are rod-shaped prokaryotes. They are usually solitary, but in some forms the rods are arranged in chains (streptobacilli). **(c)** Spiral prokaryotes include spirilla, which range from comma-like shapes to loose coils, and spirochetes (shown here), which are corkscrew-shaped (colorized SEMs).

Cell-Surface Structures

A key feature of nearly all prokaryotic cells is the cell wall, which maintains cell shape, protects the cell, and prevents it from bursting in a hypotonic environment (see Chapter 7). In a hypertonic environment, most prokaryotes lose water and shrink away from their wall (plasmolyze), like other walled cells. Such water losses can inhibit cell reproduction. Thus, salt can be used to preserve foods because it causes prokaryotes to lose water, preventing them from rapidly multiplying.

The cell walls of prokaryotes differ in structure from those of eukaryotes. In eukaryotes that have cell walls, such as plants and fungi, the walls are usually made of cellulose or chitin (see Chapter 5). In contrast, most bacterial cell walls contain **peptidoglycan**, a polymer composed of modified sugars cross-linked by short polypeptides. This molecular fabric encloses the entire bacterium and anchors other molecules that extend from its surface. Archaeal cell walls contain a variety of polysaccharides and proteins but lack peptidoglycan.

Using a technique called the **Gram stain**, developed by the nineteenth-century Danish physician Hans Christian Gram, scientists can classify many bacterial species into two groups based on differences in cell wall composition. Samples are first stained with crystal violet dye and iodine, then rinsed in alcohol, and finally stained with a red dye such as safranin. The structure of a bacterium's cell wall determines the staining response (**Figure 27.3**). **Gram-positive** bacteria have simpler walls with a relatively large amount of peptidoglycan. **Gram-negative** bacteria have less peptidoglycan and are structurally more complex, with an outer membrane that contains lipopolysaccharides (carbohydrates bonded to lipids).

▼ **Figure 27.3 Gram staining.**

(a) Gram-positive bacteria

(b) Gram-negative bacteria

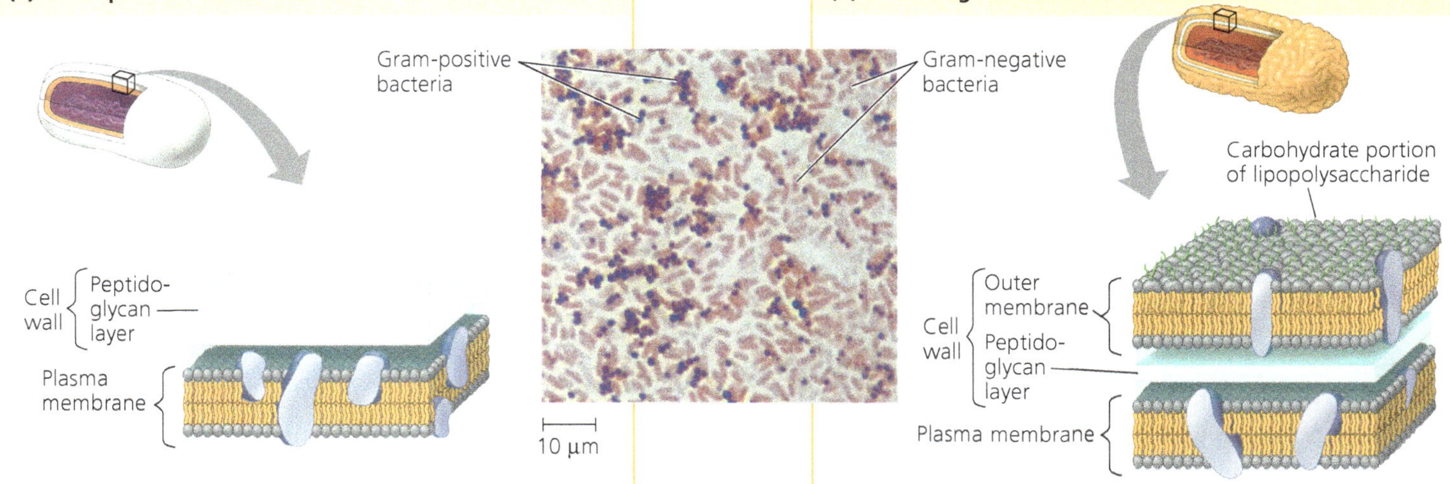

Gram-positive bacteria have a thick cell wall made of peptidoglycan, which traps the crystal violet in the cytoplasm. The alcohol rinse does not remove the crystal violet, which masks the red safranin dye.

Gram-negative bacteria have a thinner layer of peptidoglycan, and it is located in a layer between the plasma membrane and an outer membrane. The crystal violet is easily rinsed from the cytoplasm, and the cell appears pink or red from the dye.

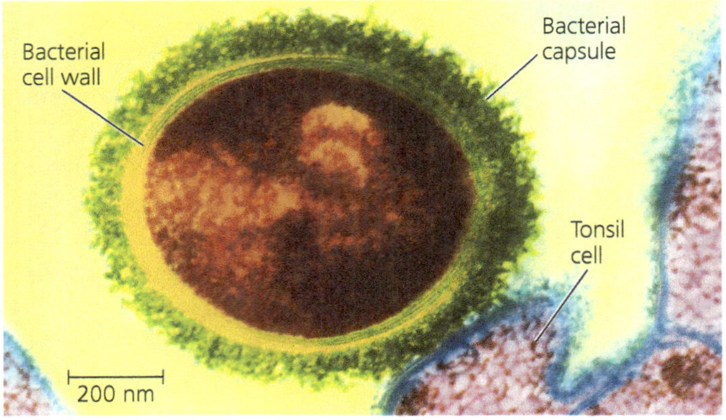

▲ **Figure 27.4 Capsule.** The polysaccharide capsule around this *Streptococcus* bacterium enables the prokaryote to attach to cells in the respiratory tract—in this colorized TEM, a tonsil cell.

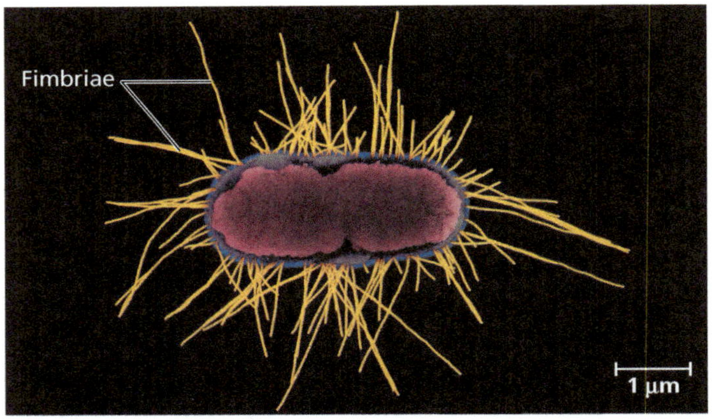

▲ **Figure 27.5 Fimbriae.** These numerous protein-containing appendages enable some prokaryotes to attach to surfaces or to other cells (colorized TEM).

Gram staining is a valuable tool in medicine for quickly determining if a patient's infection is due to gram-negative or to gram-positive bacteria. This information has treatment implications. The lipid portions of the lipopolysaccharides in the walls of many gram-negative bacteria are toxic, causing fever or shock. Furthermore, the outer membrane of a gram-negative bacterium helps protect it from the body's defenses. Gram-negative bacteria also tend to be more resistant than gram-positive species to antibiotics because the outer membrane impedes entry of the drugs. However, certain gram-positive species have virulent strains that are resistant to one or more antibiotics. (Figure 22.14 discusses one example: multidrug-resistant *Staphylococcus aureus*, which can cause lethal skin infections.)

The effectiveness of certain antibiotics, such as penicillin, derives from their inhibition of peptidoglycan cross-linking. The resulting cell wall may not be functional, particularly in gram-positive bacteria. Such drugs destroy many species of pathogenic bacteria without adversely affecting human cells, which do not have peptidoglycan.

The cell wall of many prokaryotes is surrounded by a sticky layer of polysaccharide or protein. This layer is called a **capsule** if it is dense and well-defined **(Figure 27.4)** or a *slime layer* if it is less well organized. Both kinds of sticky outer layers enable prokaryotes to adhere to their substrate or to other individuals in a colony. Some capsules and slime layers protect against dehydration, and some shield pathogenic prokaryotes from attacks by their host's immune system.

Some prokaryotes stick to their substrate or to one another by means of hairlike appendages called **fimbriae** (singular, *fimbria*) **(Figure 27.5)**. For example, the bacterium that causes gonorrhea, *Neisseria gonorrhoeae*, uses fimbriae to fasten itself to the mucous membranes of its host. Fimbriae are usually shorter and more numerous than **pili** (singular, *pilus*), appendages that pull two cells together prior to DNA transfer from one cell to the other (see Figure 27.12); pili are sometimes referred to as *sex pili*.

Motility

About half of all prokaryotes are capable of **taxis**, a directed movement toward or away from a stimulus (from the Greek *taxis*, to arrange). For example, prokaryotes that exhibit *chemotaxis* change their movement pattern in response to chemicals. They may move *toward* nutrients or oxygen (positive chemotaxis) or *away from* a toxic substance (negative chemotaxis). Some species can move at velocities exceeding 50 µm/sec—up to 50 times their body length per second. For perspective, consider that a person 1.7 m tall moving that fast would be running 306 km (190 miles) per hour!

Of the various structures that enable prokaryotes to move, the most common are flagella **(Figure 27.6)**. Flagella (singular, *flagellum*) may be scattered over the entire surface of the cell or concentrated at one or both ends. Prokaryotic flagella differ greatly from eukaryotic flagella: They are one-tenth the width and are not covered by an extension of the plasma membrane (see Figure 6.24). The flagella of prokaryotes are also very different in their molecular composition and their mechanism of propulsion. Among prokaryotes, bacterial and archaeal flagella are similar in size and rotation mechanism, but they are composed of different proteins. Overall, these structural and molecular comparisons suggest that the flagella of bacteria, archaea, and eukaryotes arose independently. Since the flagella of organisms in the three domains perform similar functions but probably are not related by common descent, it is likely that they are analogous, not homologous, structures.

Evolutionary Origins of Bacterial Flagella

The bacterial flagellum shown in Figure 27.6 has three main parts (the motor, hook, and filament) that are themselves composed of 42 different kinds of proteins. How could such a complex structure evolve? In fact, much evidence indicates that bacterial flagella originated as simpler structures that were modified in a stepwise fashion over time. As in the case of the

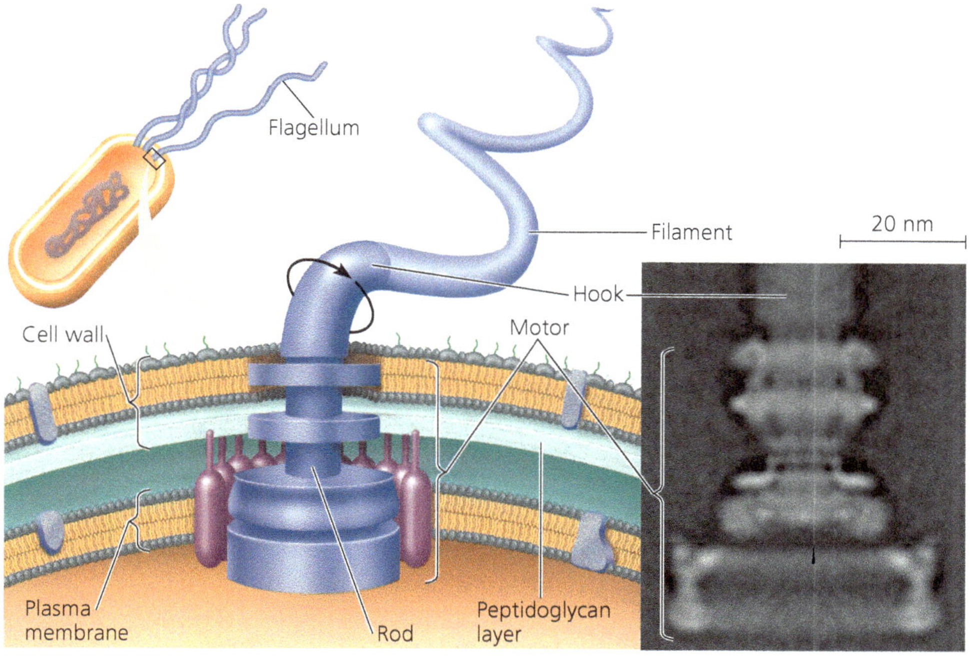

▲ **Figure 27.6 A prokaryotic flagellum.** The motor of a prokaryotic flagellum consists of a system of rings embedded in the cell wall and plasma membrane (TEM). ATP-driven pumps in the motor transport protons out of the cell. The diffusion of protons back into the cell provides the force that turns a curved hook and thereby causes the attached filament to rotate and propel the cell. (This diagram shows flagellar structures characteristic of gram-negative bacteria.)

human eye (see Concept 25.6), biologists asked whether a less complex version of the flagellum could still benefit its owner. Analyses of hundreds of bacterial genomes indicate that only half of the flagellum's protein components appear to be necessary for it to function; the others are inessential or not encoded in the genomes of some species. Of the 21 proteins required by all species studied to date, 19 are modified versions of proteins that perform other tasks in bacteria. For example, a set of 10 proteins in the motor are homologous to 10 similar proteins in a secretory system found in bacteria. (A secretory system is a protein complex that enables a cell to secrete certain macromolecules.) Two other proteins in the motor are homologous to proteins that function in ion transport. The proteins that comprise the rod, hook, and filament are all related to each other and are descended from an ancestral protein that formed a pilus-like tube. These findings suggest that the bacterial flagellum evolved as other proteins were added to an ancestral secretory system. This is an example of *exaptation*, the process in which existing structures take on new functions through descent with modification.

Internal Organization and DNA

The cells of prokaryotes are simpler than those of eukaryotes in both their internal structure and the physical arrangement of their DNA (see Figure 6.5). Prokaryotic cells lack the complex compartmentalization found in eukaryotic cells. However, some prokaryotic cells do have specialized membranes that perform metabolic functions **(Figure 27.7)**. These membranes are usually infoldings of the plasma membrane.

The genome of a prokaryote is structurally different from a eukaryotic genome and in most cases has considerably less DNA. In the majority of prokaryotes, the genome consists of a circular chromosome with many fewer proteins than found in the linear chromosomes of eukaryotes **(Figure 27.8)**. Also

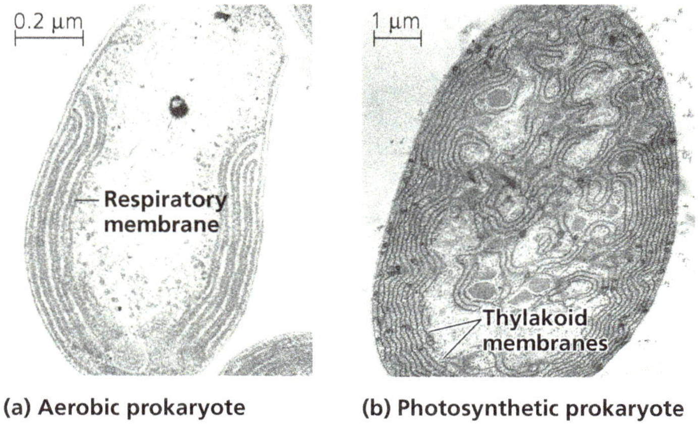

▲ **Figure 27.7 Specialized membranes of prokaryotes.**
(a) Infoldings of the plasma membrane, reminiscent of the cristae of mitochondria, function in cellular respiration in some aerobic prokaryotes (TEM). **(b)** Photosynthetic prokaryotes called cyanobacteria have thylakoid membranes, much like those in chloroplasts (TEM).

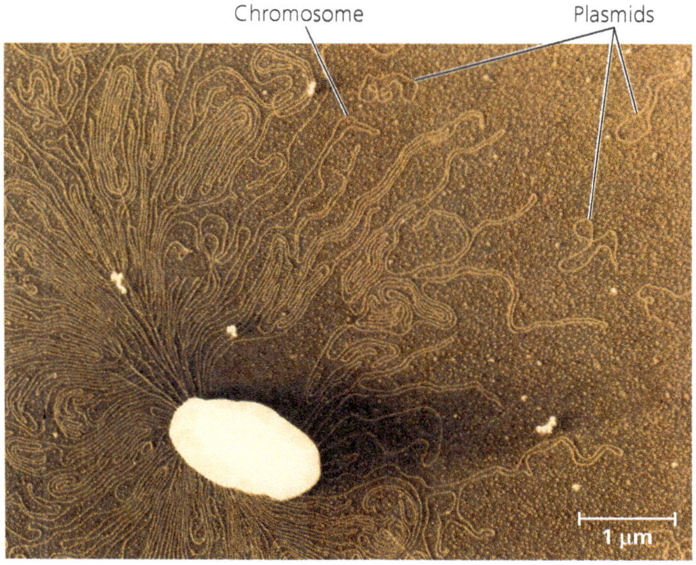

▲ **Figure 27.8 A prokaryotic chromosome and plasmids.** The thin, tangled loops surrounding this ruptured *E. coli* cell are parts of the cell's large, circular chromosome (colorized TEM). Three of the cell's plasmids, the much smaller rings of DNA, are also shown.

unlike eukaryotes, prokaryotes lack a membrane-bounded nucleus; their chromosome is located in the **nucleoid**, a region of cytoplasm that appears lighter than the surrounding cytoplasm in electron micrographs. In addition to its single chromosome, a typical prokaryotic cell may also have much smaller rings of independently replicating DNA molecules called **plasmids** (see Figure 27.8), most carrying only a few genes.

As explained in Chapters 16 and 17, DNA replication, transcription, and translation are fundamentally similar processes in prokaryotes and eukaryotes, although there are some differences. For example, prokaryotic ribosomes are slightly smaller than eukaryotic ribosomes and differ in their protein and RNA content. These differences allow certain antibiotics, such as erythromycin and tetracycline, to bind to ribosomes and block protein synthesis in prokaryotes but not in eukaryotes. As a result, people can use these antibiotics to kill or inhibit the growth of bacteria without harming themselves.

Reproduction and Adaptation

Prokaryotes are highly successful in part because of their potential to reproduce quickly in a favorable environment. By *binary fission* (see Figure 12.12), a single prokaryotic cell divides into 2 cells, which then divide into 4, 8, 16, and so on. Under optimal conditions, many prokaryotes can divide every 1–3 hours; some species can produce a new generation in only 20 minutes. If reproduction continued unchecked at this rate, a single prokaryotic cell could give rise to a colony outweighing Earth in only two days!

In reality, of course, prokaryotic reproduction is limited. The cells eventually exhaust their nutrient supply, poison themselves with metabolic wastes, face competition from other microorganisms, or are consumed by other organisms. For example, the well-studied bacterium *Escherichia coli* can divide every 20 minutes under ideal lab conditions, one reason it is used as a model organism in research. However, when growing in a human intestine, one of its natural environments, *E. coli* cells divide only once every 12–24 hours. But whether cell division occurs every 20 minutes or every few days, reproduction in prokaryotes draws attention to three key features of their biology: *They are small, they reproduce by binary fission, and they have short generation times.* As a result, prokaryotic populations can consist of many trillions of individuals—far more than populations of multicellular eukaryotes, such as plants and animals.

The ability of some prokaryotes to withstand harsh conditions also contributes to their success. Some, like *Halobacterium*, can survive in harsh environments because of particular biochemical adaptations; others, because of particular structural adaptations. Certain bacteria, for example, develop resistant cells called **endospores** when they lack an essential nutrient **(Figure 27.9)**. The original cell produces a copy of its chromosome and surrounds it with a tough multilayered structure,

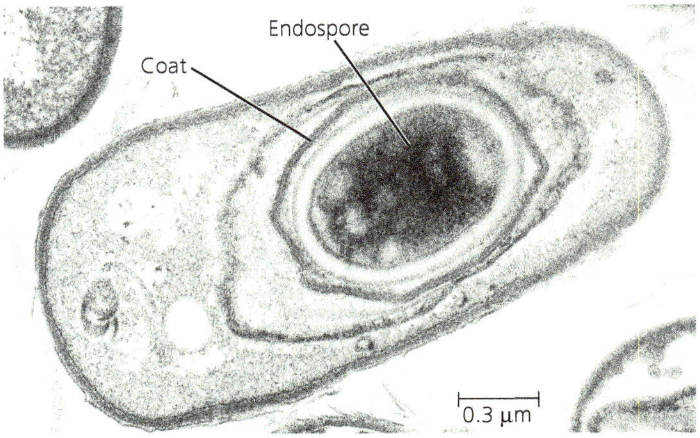

▲ **Figure 27.9 An endospore.** *Bacillus anthracis*, the bacterium that causes the disease anthrax, produces endospores (TEM). An endospore's protective, multilayered coat helps it survive in the soil for years.

forming the endospore. Water is removed from the endospore, and its metabolism halts. The original cell then lyses, releasing the endospore. Most endospores are so durable that they can survive in boiling water; killing them requires heating lab equipment to 121°C under high pressure. In less hostile environments, endospores can remain dormant but viable for centuries, able to rehydrate and resume metabolism when their environment improves.

Finally, in part because of their short generation times, prokaryotic populations can evolve substantially in short periods of time. For example, in a remarkable study that spanned 20,000 generations (roughly eight years) of evolution, researchers at Michigan State University documented adaptive evolution in bacterial populations **(Figure 27.10)**. The ability of prokaryotes to adapt rapidly to new conditions highlights the point that although the structure of their cells is simpler than that of eukaryotic cells, prokaryotes are not "primitive" or "inferior" in an evolutionary sense. They are, in fact, highly evolved: For over 3.5 billion years, prokaryotic populations have responded successfully to many different types of environmental challenges. As we will see, one reason for this is that their populations harbor high levels of genetic diversity on which selection can act.

CONCEPT CHECK 27.1

1. Identify and explain at least two adaptations that enable prokaryotes to survive in environments too harsh for other organisms.
2. Contrast the cellular and DNA structures of prokaryotes and eukaryotes.
3. **MAKE CONNECTIONS** Suggest a hypothesis to explain why the thylakoid membranes of chloroplasts resemble those of cyanobacteria. Refer to Figure 6.18 (p. 111) and Figure 26.21 (p. 552).

For suggested answers, see Appendix A.

▼ Figure 27.10 INQUIRY

Can prokaryotes evolve rapidly in response to environmental change?

EXPERIMENT Vaughn Cooper and Richard Lenski, of Michigan State University, tested the ability of *E. coli* populations to adapt to a new environment. They established 12 populations, each founded by a single cell from an *E. coli* strain, and followed these populations for 20,000 generations (3,000 days). To maintain a continual supply of resources, each day the researchers performed a *serial transfer*: They transferred 0.1 mL of each population to a new tube containing 9.9 mL of fresh growth medium. The growth medium used throughout the experiment provided a challenging environment that contained only low levels of glucose and other resources needed for growth.

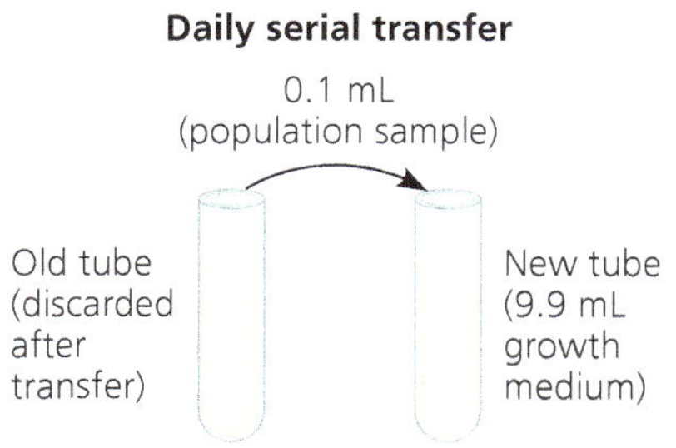

Samples were periodically removed from the 12 populations and grown in competition with the common ancestral strain in the experimental (low-glucose) environment.

RESULTS The fitness of the experimental populations, as measured by the rate at which each population grew, increased rapidly for the first 5,000 generations (two years) and more slowly for the next 15,000 generations. The graph below shows the averages for the 12 populations.

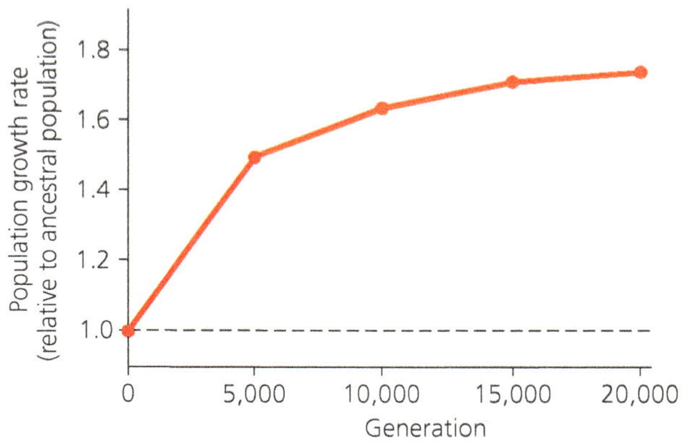

CONCLUSION Populations of *E. coli* continued to accumulate beneficial mutations for 20,000 generations, allowing rapid evolution of improved performance in their new environment.

SOURCE V. S. Cooper and R. E. Lenski, The population genetics of ecological specialization in evolving *Escherichia coli* populations, *Nature* 407:736–739 (2000).

WHAT IF? Suggest possible functions of the genes whose sequence or expression was altered as the experimental populations evolved in the low-glucose environment.

CONCEPT 27.2

Rapid reproduction, mutation, and genetic recombination promote genetic diversity in prokaryotes

As we discussed in Unit Four, genetic variation is a prerequisite for natural selection to occur in a population. The diverse adaptations exhibited by prokaryotes suggest that their populations must have considerable genetic variation—and they do. For example, a ribosomal RNA gene can differ more between two strains of *E. coli* than it does between a human and a platypus. In this section, we'll examine three factors that give rise to high levels of genetic diversity in prokaryotes: rapid reproduction, mutation, and genetic recombination.

Rapid Reproduction and Mutation

In sexually reproducing species, the generation of a novel allele by a new mutation is rare for any particular gene. Instead, most of the genetic variation in sexual populations results from the way existing alleles are arranged in new combinations during meiosis and fertilization (see Chapter 13). Prokaryotes do not reproduce sexually, so at first glance their extensive genetic variation may seem puzzling. In fact, this variation can result from prokaryotes' rapid reproduction and mutation.

Consider a prokaryote reproducing by binary fission. After repeated rounds of division, most of the offspring cells are genetically identical to the original parent cell. However, if errors occur during DNA replication—such as insertions, deletions, or base-pair substitutions—some of the offspring cells may differ genetically. The probability of a spontaneous mutation occurring in a given *E. coli* gene averages only about one in 10 million (1×10^{-7}) per cell division. But among the 2×10^{10} new *E. coli* cells that arise each day in a person's intestine, there will be approximately $(2 \times 10^{10}) \times (1 \times 10^{-7}) = 2{,}000$ bacteria that have a mutation in that gene. The total number of mutations when all 4,300 *E. coli* genes are considered is about $4{,}300 \times 2{,}000 = 9$ million per day per human host.

The key point is that new mutations, though rare, can increase genetic diversity quickly in species with short generation times and large populations. This diversity, in turn, can lead to rapid evolution: Individuals that are genetically better equipped for their environment tend to survive and reproduce more prolifically than less fit individuals (see Figure 27.10).

Genetic Recombination

Although new mutations are a major source of variation in prokaryotic populations, additional diversity arises from *genetic recombination*, the combining of DNA from two sources. In eukaryotes, the sexual processes of meiosis and fertilization combine DNA from two individuals in a single zygote. But meiosis and fertilization do not occur in prokaryotes. Instead, three other mechanisms—transformation, transduction, and conjugation—can bring together prokaryotic DNA from different individuals (that is, cells). When the individuals are members of different species, this movement of genes from one organism to another is called *horizontal gene transfer*. Although scientists have found evidence that each of these mechanisms can transfer DNA within and between species in both domain Bacteria and domain Archaea, to date most of our knowledge comes from research on bacteria.

Transformation and Transduction

In **transformation**, the genotype and possibly phenotype of a prokaryotic cell are altered by the uptake of foreign DNA from its surroundings. For example, a harmless strain of *Streptococcus pneumoniae* can be transformed into pneumonia-causing cells if the cells are placed in a medium containing DNA from a pathogenic strain (see p. 306). This transformation occurs when a nonpathogenic cell takes up a piece of DNA carrying the allele for pathogenicity and replaces its own allele with the foreign allele, an exchange of homologous DNA segments. The cell is now a recombinant: Its chromosome contains DNA derived from two different cells.

For many years after transformation was discovered in laboratory cultures, most biologists thought the process to be too rare and haphazard to play an important role in natural bacterial populations. But researchers have since learned that many bacteria have cell-surface proteins that recognize DNA from closely related species and transport it into the cell. Once inside the cell, the foreign DNA can be incorporated into the genome by homologous DNA exchange.

In **transduction**, phages (from "bacteriophages," the viruses that infect bacteria) carry prokaryotic genes from one host cell to another. In most cases, transduction results from accidents that occur during the phage replicative cycle (**Figure 27.11**). A virus that carries prokaryotic DNA may not be able to replicate because it lacks some or all of its own genetic material. However, the virus can attach to another prokaryotic cell (a recipient) and inject prokaryotic DNA acquired from the first cell (the donor). If some of this DNA is then incorporated into the recipient cell's chromosome by DNA recombination, a recombinant cell is formed.

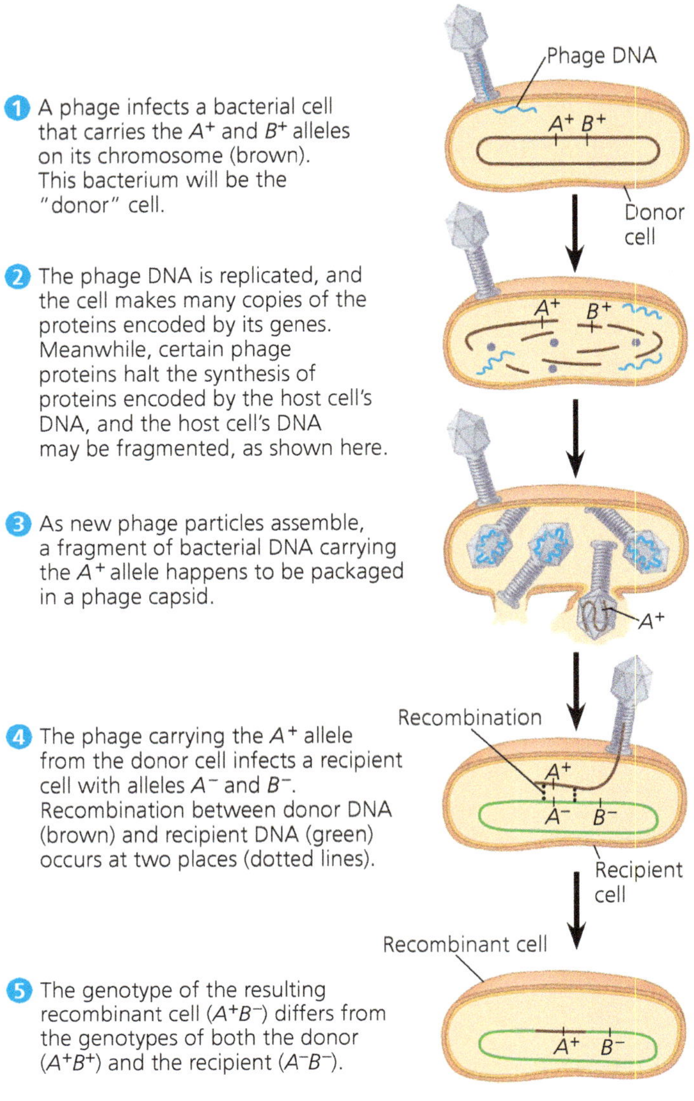

▲ **Figure 27.11 Transduction.** Phages may carry pieces of a bacterial chromosome from one cell (the donor) to another (the recipient). If recombination occurs after the transfer, genes from the donor may be incorporated into the recipient's genome.

? *Under what circumstances would transduction result in horizontal gene transfer?*

Conjugation and Plasmids

In a process called **conjugation**, DNA is transferred between two prokaryotic cells (usually of the same species) that are temporarily joined. In bacteria, the DNA transfer is always one-way: One cell donates the DNA, and the other receives it. The best-understood mechanism is that used by *E. coli*, and we will focus on this organism for the rest of this section. In *E. coli*, a pilus of the donor cell attaches to the recipient (**Figure 27.12**). The pilus then retracts, pulling the two cells together, much like a grappling hook. The next step is thought

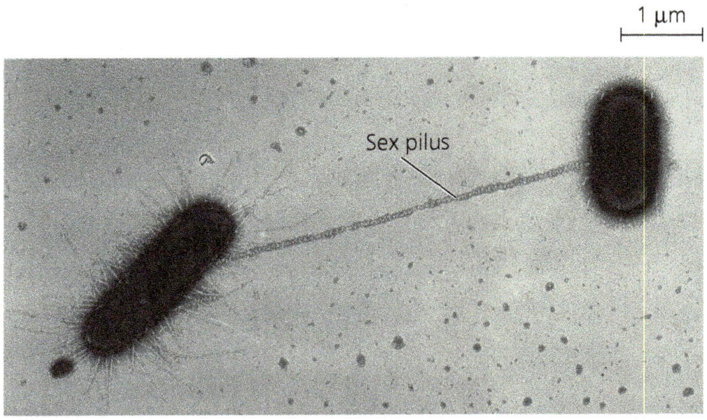

▲ **Figure 27.12 Bacterial conjugation.** The *E. coli* donor cell (left) extends a pilus that attaches to a recipient cell, a key first step in the transfer of DNA. The pilus is a flexible tube of protein subunits (TEM).

to be the formation of a temporary "mating bridge" between the two cells, through which the donor may transfer DNA to the recipient. This is an unsettled issue, however, and recent evidence indicates that DNA may pass directly through the pilus, which is hollow.

In either case, the ability to form pili and donate DNA during conjugation results from the presence of a particular piece of DNA called the **F factor** (F for *f*ertility). The F factor of *E. coli* consists of about 25 genes, most required for the production of pili. The F factor can exist either as a plasmid or as a segment of DNA within the bacterial chromosome.

The F Factor as a Plasmid The F factor in its plasmid form is called the **F plasmid**. Cells containing the F plasmid, designated F^+ cells, function as DNA donors during conjugation. Cells lacking the F factor, designated F^-, function as DNA recipients during conjugation. The F^+ condition is transferable in the sense that an F^+ cell converts an F^- cell to F^+ if a copy of the entire F^+ plasmid is transferred **(Figure 27.13a)**.

The F Factor in the Chromosome Chromosomal genes can be transferred during conjugation when the donor cell's F factor is integrated into the chromosome. A cell with the F factor built into its chromosome is called an *Hfr cell* (for *h*igh *f*requency of *r*ecombination). Like an F^+ cell, an Hfr cell functions as a donor during conjugation with an F^- cell **(Figure 27.13b)**. When chromosomal DNA from an Hfr cell enters an F^- cell, homologous regions of the Hfr and F^- chromosomes may align, allowing segments of their DNA to be exchanged. This results in the production of a recombinant bacterium that has genes derived from two different cells—a new genetic variant on which evolution can act.

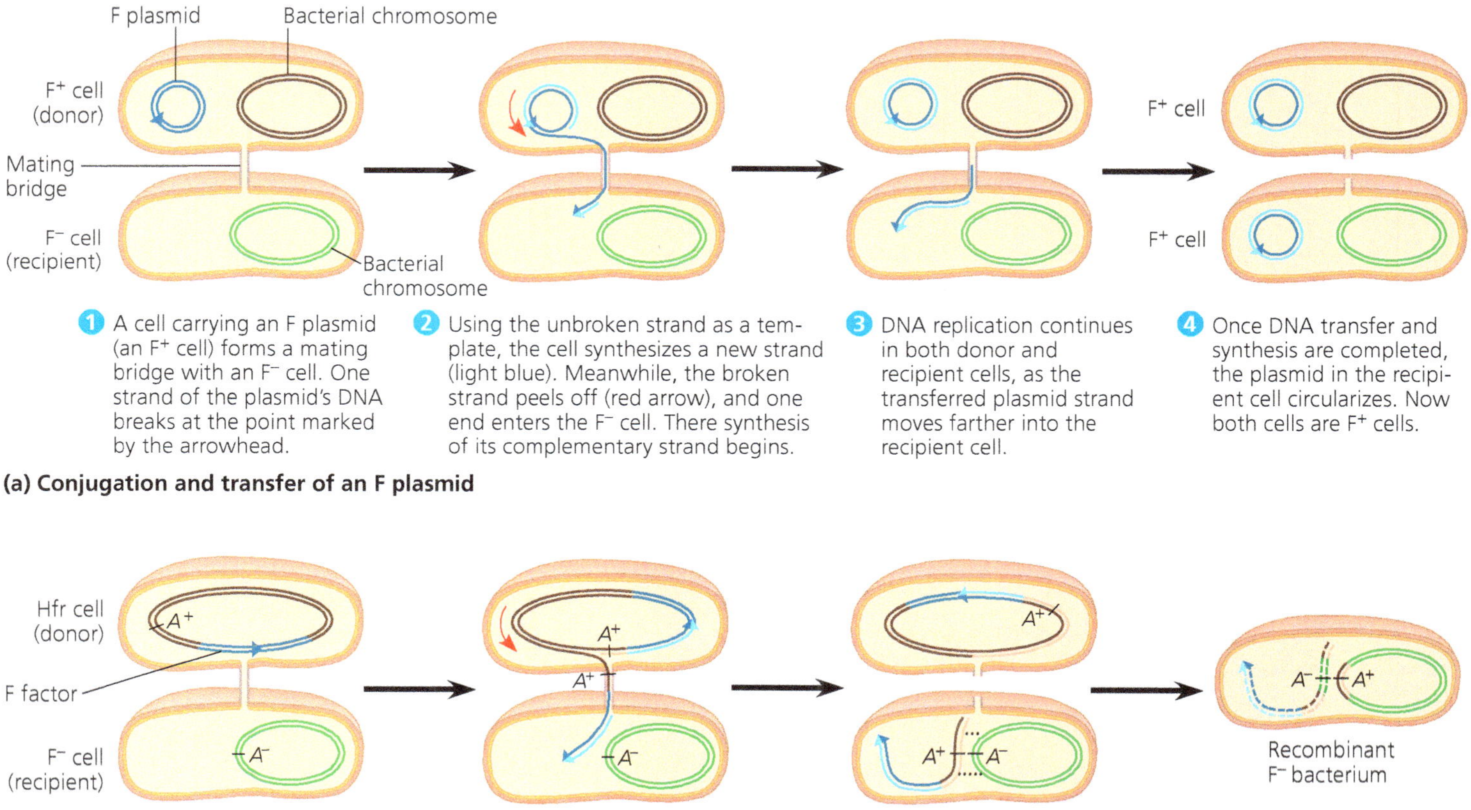

1. A cell carrying an F plasmid (an F^+ cell) forms a mating bridge with an F^- cell. One strand of the plasmid's DNA breaks at the point marked by the arrowhead.
2. Using the unbroken strand as a template, the cell synthesizes a new strand (light blue). Meanwhile, the broken strand peels off (red arrow), and one end enters the F^- cell. There synthesis of its complementary strand begins.
3. DNA replication continues in both donor and recipient cells, as the transferred plasmid strand moves farther into the recipient cell.
4. Once DNA transfer and synthesis are completed, the plasmid in the recipient cell circularizes. Now both cells are F^+ cells.

(a) Conjugation and transfer of an F plasmid

1. In an Hfr cell, the F factor (dark blue) is integrated into the bacterial chromosome. Since an Hfr cell has all of the F-factor genes, it can form a mating bridge with an F^- cell and transfer DNA.
2. A single strand of the F factor breaks and begins to move through the bridge. DNA replication occurs in both donor and recipient cells, resulting in double-stranded DNA (daughter strands shown in lighter color).
3. The mating bridge usually breaks before the entire chromosome is transferred. DNA recombination (indicated by dotted lines) can result in the exchange of homologous genes between the transferred DNA (brown) and the recipient's chromosome (green).
4. Cellular enzymes degrade any linear DNA not incorporated into the chromosome. The recipient cell, with a new combination of genes but no F factor, is now a recombinant F^- cell.

(b) Conjugation and transfer of part of an Hfr bacterial chromosome, resulting in recombination

▲ **Figure 27.13 Conjugation and recombination in *E. coli*.** The DNA replication that accompanies transfer of an F plasmid or part of an Hfr bacterial chromosome is called *rolling circle replication*. In effect, the intact circular parental DNA strand "rolls" as its other strand peels off and a new complementary strand is synthesized.

R Plasmids and Antibiotic Resistance During the 1950s in Japan, physicians started noticing that some hospital patients with bacterial dysentery, which produces severe diarrhea, did not respond to antibiotics that had generally been effective in the past. Apparently, resistance to these antibiotics had evolved in certain strains of *Shigella*, the bacterium that causes the disease.

Eventually, researchers began to identify the specific genes that confer antibiotic resistance in *Shigella* and other pathogenic bacteria. Sometimes, mutation in a chromosomal gene of the pathogen can confer resistance. For example, a mutation in one gene may make it less likely that the pathogen will transport a particular antibiotic into its cell. Mutation in a different gene may alter the intracellular target protein for an antibiotic molecule, reducing its inhibitory effect. In other cases, bacteria have "resistance genes," which code for enzymes that specifically destroy or otherwise hinder the effectiveness of certain antibiotics, such as tetracycline or ampicillin. Such resistance genes are carried by plasmids known as **R plasmids** (R for *r*esistance).

Exposing a bacterial population to a specific antibiotic, whether in a laboratory culture or within a host organism, will kill antibiotic-sensitive bacteria but not those that happen to have R plasmids with genes that counter the antibiotic. Under these circumstances, we would predict that natural selection would cause the fraction of the bacterial population carrying genes for antibiotic resistance to increase, and that is exactly what happens. The medical consequences are also predictable: As you've read, resistant strains of pathogens are becoming more common, making the treatment of certain bacterial infections more difficult. The problem is compounded by the fact that many R plasmids, like F plasmids, have genes that encode pili and enable DNA transfer from one bacterial cell to another by conjugation. Making the problem still worse, some R plasmids carry as many as ten genes for resistance to that many antibiotics.

CONCEPT CHECK 27.2

1. What features of prokaryotes make it likely that considerable genetic variation will be added to their populations in each generation?
2. Distinguish between the three mechanisms of transferring DNA from one bacterial cell to another.
3. In a rapidly changing environment, which bacterial population would likely be more successful, one that includes individuals capable of conjugation or one that does not? Explain.
4. WHAT IF? If a nonpathogenic bacterium were to acquire resistance to antibiotics, could this strain pose a health risk to people? Explain. In general, how does DNA transfer among bacteria affect the spread of resistance genes?

For suggested answers, see Appendix A.

CONCEPT 27.3

Diverse nutritional and metabolic adaptations have evolved in prokaryotes

The extensive genetic variation found in prokaryotic populations is reflected in the diverse nutritional adaptations of prokaryotes. Like all organisms, prokaryotes can be categorized by how they obtain energy and the carbon used in building the organic molecules that make up cells. Every type of nutrition observed in eukaryotes is represented among prokaryotes, along with some nutritional modes unique to prokaryotes. In fact, prokaryotes have an astounding range of metabolic adaptations, much broader than that found in eukaryotes.

Organisms that obtain energy from light are called *phototrophs*, and those that obtain energy from chemicals are called *chemotrophs*. Organisms that need only CO_2 in some form as a carbon source are called *autotrophs*. In contrast, *heterotrophs* require at least one organic nutrient, such as glucose, to make other organic compounds. Combining possible energy sources and carbon sources results in four major modes of nutrition, summarized in Table 27.1.

The Role of Oxygen in Metabolism

Prokaryotic metabolism also varies with respect to oxygen (O_2). **Obligate aerobes** must use O_2 for cellular respiration (see Chapter 9) and cannot grow without it. **Obligate anaerobes**, on the other hand, are poisoned by O_2. Some obligate anaerobes live exclusively by fermentation; others extract chemical energy by **anaerobic respiration**, in which substances other than O_2, such as nitrate ions (NO_3^-) or sulfate ions (SO_4^{2-}), accept electrons at the "downhill" end of electron transport chains. **Facultative anaerobes** use O_2 if it is present but can also carry out fermentation or anaerobic respiration in an anaerobic environment.

Nitrogen Metabolism

Nitrogen is essential for the production of amino acids and nucleic acids in all organisms. Whereas eukaryotes can obtain nitrogen from only a limited group of nitrogen compounds, prokaryotes can metabolize nitrogen in a wide variety of forms. For example, some cyanobacteria and some methanogens (a group of archaea) convert atmospheric nitrogen (N_2) to ammonia (NH_3), a process called **nitrogen fixation**. The cells can then incorporate this "fixed" nitrogen into amino acids and other organic molecules. In terms of their nutrition, nitrogen-fixing cyanobacteria are some of the most self-sufficient organisms, since they need only light, CO_2, N_2, water, and some minerals to grow.

Nitrogen fixation by prokaryotes has a large impact on other organisms. For example, nitrogen-fixing prokaryotes can increase the nitrogen available to plants, which cannot use

Table 27.1 Major Nutritional Modes

Mode	Energy Source	Carbon Source	Types of Organisms
AUTOTROPH			
Photoautotroph	Light	CO_2, HCO_3^-, or related compound	Photosynthetic prokaryotes (for example, cyanobacteria); plants; certain protists (for example, algae)
Chemoautotroph	Inorganic chemicals (such as H_2S, NH_3, or Fe^{2+})	CO_2, HCO_3^-, or related compound	Unique to certain prokaryotes (for example, *Sulfolobus*)
HETEROTROPH			
Photoheterotroph	Light	Organic compounds	Unique to certain aquatic and salt-loving prokaryotes (for example, *Rhodobacter*, *Chloroflexus*)
Chemoheterotroph	Organic compounds	Organic compounds	Many prokaryotes (for example, *Clostridium*) and protists; fungi; animals; some plants

atmospheric nitrogen but can use the nitrogen compounds that the prokaryotes produce from ammonia. Chapter 55 discusses this and other essential roles that prokaryotes play in the nitrogen cycles of ecosystems.

Metabolic Cooperation

Cooperation between prokaryotic cells allows them to use environmental resources they could not use as individual cells. In some cases, this cooperation takes place between specialized cells of a filament. For instance, the cyanobacterium *Anabaena* has genes that encode proteins for photosynthesis and for nitrogen fixation, but a single cell cannot carry out both processes at the same time. The reason is that photosynthesis produces O_2, which inactivates the enzymes involved in nitrogen fixation. Instead of living as isolated cells, *Anabaena* forms filamentous chains (**Figure 27.14**). Most cells in a filament carry out only photosynthesis, while a few specialized cells called **heterocysts** (sometimes called *heterocytes*) carry out only nitrogen fixation. Each heterocyst is surrounded by a thickened cell wall that restricts entry of O_2 produced by neighboring photosynthetic cells. Intercellular connections allow heterocysts to transport fixed nitrogen to neighboring cells and to receive carbohydrates.

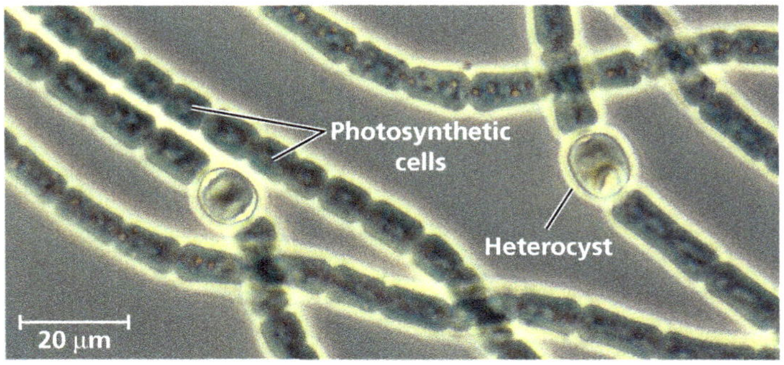

▲ **Figure 27.14 Metabolic cooperation in a prokaryote.** In the filamentous cyanobacterium *Anabaena*, cells called heterocysts fix nitrogen, while the other cells carry out photosynthesis (LM). *Anabaena* is found in many freshwater lakes.

Metabolic cooperation between different prokaryotic species often occurs in surface-coating colonies known as **biofilms**. Cells in a biofilm secrete signaling molecules that recruit nearby cells, causing the colonies to grow. The cells also produce polysaccharides and proteins that stick the cells to the substrate and to one another. Channels in the biofilm allow nutrients to reach cells in the interior and wastes to be expelled. Biofilms are common in nature, but they can cause problems by contaminating industrial products and medical equipment and contributing to tooth decay and more serious health problems. Altogether, damage caused by biofilms costs billions of dollars annually.

In another example of cooperation between prokaryotes, sulfate-consuming bacteria coexist with methane-consuming archaea in ball-shaped aggregates on the ocean floor. The bacteria appear to use the archaea's waste products, such as organic compounds and hydrogen. In turn, the bacteria produce sulfur compounds that the archaea use as oxidizing agents when they consume methane in the absence of oxygen. This partnership has global ramifications: Each year, these archaea consume an estimated 300 billion kilograms of methane, a major contributor to the greenhouse effect (see Chapter 55).

CONCEPT CHECK 27.3

1. Distinguish between the four major modes of nutrition, noting which are unique to prokaryotes.
2. A bacterium requires only the amino acid methionine as an organic nutrient and lives in lightless caves. What mode of nutrition does it employ? Explain.
3. **WHAT IF?** Describe what you might eat for a typical meal if humans, like cyanobacteria, could fix nitrogen.

For suggested answers, see Appendix A.

CONCEPT **27.4**

Molecular systematics is illuminating prokaryotic phylogeny

Until the late 20th century, systematists based prokaryotic taxonomy on phenotypic criteria such as shape, motility, nutritional mode, and response to Gram staining. These criteria are still valuable in certain contexts, such as the rapid identification of pathogenic bacteria cultured from a patient's blood. But when it comes to prokaryotic phylogeny, comparing these characteristics does not reveal a clear evolutionary history. Applying molecular systematics to the investigation of prokaryotic phylogeny, however, has led to some dramatic conclusions.

Lessons from Molecular Systematics

As discussed in Chapter 26, microbiologists began comparing the sequences of prokaryotic genes in the 1970s. Using small-subunit ribosomal RNA as a marker for evolutionary relationships, Carl Woese and his colleagues concluded that many prokaryotes once classified as bacteria are actually more closely related to eukaryotes and belong in a domain of their own: Archaea. Microbiologists have since analyzed larger amounts of genetic data—including hundreds of entire genomes—and have concluded that a few traditional taxonomic groups, such as cyanobacteria, do appear to be monophyletic. However, other groups, such as gram-negative bacteria, are scattered throughout several lineages. **Figure 27.15** shows one phylogenetic hypothesis for some of the major taxa of prokaryotes based on molecular systematics.

One lesson from studying prokaryotic phylogeny is that the genetic diversity of prokaryotes is immense. When researchers began to sequence the genes of prokaryotes, they could investigate only the small fraction of species that could be cultured in the laboratory. In the 1980s, researchers began using the polymerase chain reaction (PCR; see Chapter 20) to analyze the genes of prokaryotes collected from the environment (such as from soil or water samples). Such "genetic prospecting" is now widely used; in fact, today entire prokaryotic genomes can be obtained from environmental samples using *metagenomics* (see Chapter 21). Each year these techniques add new branches to the tree of life. While only about 7,800 prokaryotic species have been assigned scientific names, a single handful of soil could contain 10,000 prokaryotic species by some estimates. Taking full stock of this diversity will require many years of research.

Another important lesson from molecular systematics is the apparent significance of horizontal gene transfer in the evolution of prokaryotes. Over hundreds of millions of years, prokaryotes have acquired genes from even distantly related species, and they continue to do so today. As a result, significant portions of the genomes of many prokaryotes are actually mosaics of genes imported from other species. As we saw in Chapter 26, such gene transfers can make it difficult to determine the root of the tree of life. Still, it is clear that for billions of years, the prokaryotes have evolved in two separate lineages, the archaea and the bacteria (see Figure 27.15).

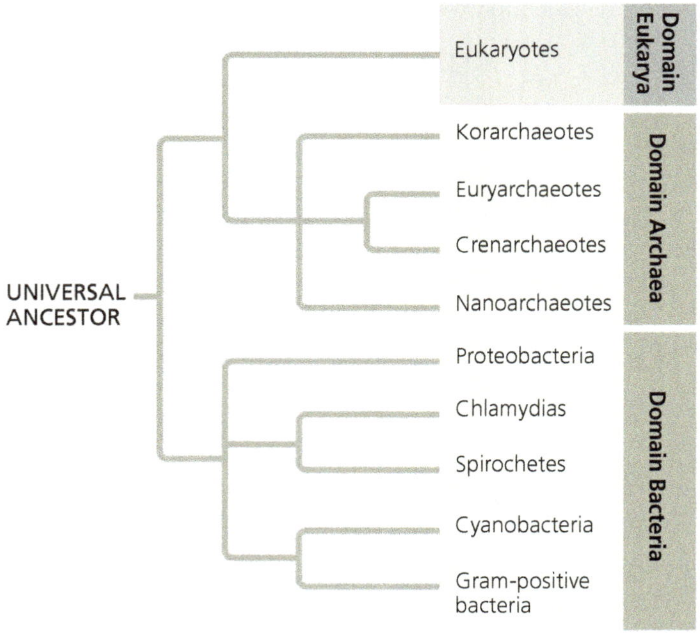

▲ **Figure 27.15 A simplified phylogeny of prokaryotes.** This phylogenetic tree based on molecular data shows one of several debated hypotheses of the relationships between the major prokaryotic groups discussed in this chapter. Within Archaea, the placement of the korarchaeotes and nanoarchaeotes remains unclear.

Archaea

Archaea share certain traits with bacteria and other traits with eukaryotes **(Table 27.2)**. However, archaea also have many unique characteristics, as we would expect in a taxon that has followed a separate evolutionary path for so long.

Table 27.2 A Comparison of the Three Domains of Life

CHARACTERISTIC	DOMAIN		
	Bacteria	Archaea	Eukarya
Nuclear envelope	Absent	Absent	Present
Membrane-enclosed organelles	Absent	Absent	Present
Peptidoglycan in cell wall	Present	Absent	Absent
Membrane lipids	Unbranched hydrocarbons	Some branched hydrocarbons	Unbranched hydrocarbons
RNA polymerase	One kind	Several kinds	Several kinds
Initiator amino acid for protein synthesis	Formyl-methionine	Methionine	Methionine
Introns in genes	Very rare	Present in some genes	Present in many genes
Response to the antibiotics streptomycin and chloramphenicol	Growth inhibited	Growth not inhibited	Growth not inhibited
Histones associated with DNA	Absent	Present in some species	Present
Circular chromosome	Present	Present	Absent
Growth at temperatures > 100°C	No	Some species	No

The first prokaryotes assigned to domain Archaea live in environments so extreme that few other organisms can survive there. Such organisms are called **extremophiles**, meaning "lovers" of extreme conditions (from the Greek *philos*, lover), and include extreme halophiles and extreme thermophiles.

Extreme halophiles (from the Greek *halo*, salt) live in highly saline environments, such as the Great Salt Lake and the Dead Sea (see Figure 27.1). Some species merely tolerate salinity, while others require an environment that is several times saltier than seawater (which has a salinity of 3.5%). For example, the proteins and cell wall of *Halobacterium* have unusual features that improve function in extremely salty environments but render these organisms incapable of survival if the salinity drops below 9%.

Extreme thermophiles (from the Greek *thermos*, hot) thrive in very hot environments **(Figure 27.16)**. For example, archaea in the genus *Sulfolobus* live in sulfur-rich volcanic springs as hot as 90°C. At temperatures this high, the cells of most organisms die because, for example, their DNA does not remain in a double helix and many of their proteins denature. *Sulfolobus* and other extreme thermophiles avoid this fate because their DNA and proteins have adaptations that make them stable at high temperatures. One extreme thermophile that lives near deep-sea hot springs called *hydrothermal vents* is informally known as "strain 121," since it can reproduce even at 121°C. Another extreme thermophile, *Pyrococcus furiosus*, is used in biotechnology as a source of DNA polymerase for the PCR technique (see Chapter 20).

Other archaea live in more moderate environments. Consider the **methanogens**, archaea that release methane as a by-product of their unique ways of obtaining energy. Many methanogens use CO_2 to oxidize H_2, a process that produces

▲ **Figure 27.16 Extreme thermophiles.** Orange and yellow colonies of thermophilic prokaryotes grow in the hot water of a Nevada geyser.

MAKE CONNECTIONS *Review the discussion of enzymes in Concept 8.4 (pp. 155–156). How might the enzymes of thermophiles differ from those of other organisms?*

both energy and methane waste. Among the strictest of anaerobes, methanogens are poisoned by O_2. Although some methanogens live in extreme environments, such as under kilometers of ice in Greenland, others live in swamps and marshes where other microorganisms have consumed all the O_2. The "marsh gas" found in such environments is the methane released by these archaea. Other species of methanogens inhabit the anaerobic environment within the guts of cattle, termites, and other herbivores, playing an essential role in the nutrition of these animals. Methanogens also have an important application as decomposers in sewage treatment facilities.

Many extreme halophiles and all known methanogens are archaea in the clade Euryarchaeota (from the Greek *eurys*, broad, a reference to the habitat range of these prokaryotes). The euryarchaeotes also include some extreme thermophiles, though most thermophilic species belong to a second clade, Crenarchaeota (*cren* means "spring," such as a hydrothermal spring). Recently, genetic prospecting has revealed many species of euryarchaeotes and crenarchaeotes that are not extremophiles. These archaea exist in habitats ranging from farm soils to lake sediments to the surface waters of the open ocean.

New findings continue to update the picture of archaeal phylogeny. In 1996, researchers sampling a hot spring in Yellowstone National Park discovered archaea that do not appear to belong to either Euryarchaeota or Crenarchaeota. They placed these archaea in a new clade, Korarchaeota (from the Greek *koron*, young man). In 2002, researchers exploring hydrothermal vents off the coast of Iceland discovered archaeal cells only 0.4 μm in diameter attached to a much larger crenarchaeote. The genome of the smaller archaean is one of the smallest known of any organism, containing only 500,000 base pairs. Genetic analysis indicates that this prokaryote belongs to a fourth archaeal clade, Nanoarchaeota (from the Greek *nanos*, dwarf). Within a year after this clade was named, three other DNA sequences from nanoarchaeote species were isolated: one from Yellowstone's hot springs, one from hot springs in Siberia, and one from a hydrothermal vent in the Pacific. As prospecting continues, it seems likely that the tree in Figure 27.15 will undergo further changes.

Bacteria

Bacteria include the vast majority of prokaryotic species of which most people are aware, from the pathogenic species that cause strep throat and tuberculosis to the beneficial species used to make Swiss cheese and yogurt. Every major mode of nutrition and metabolism is represented among bacteria, and even a small taxonomic group of bacteria may contain species exhibiting many different nutritional modes. As we'll see, the diverse nutritional and metabolic capabilities of bacteria—and archaea—are behind the great impact of these tiny organisms on Earth and its life. Examine **Figure 27.17**, on the following two pages, for a closer look at several major groups of bacteria.

▼ Figure 27.17

Exploring Major Groups of Bacteria

Proteobacteria

This large and diverse clade of gram-negative bacteria includes photoautotrophs, chemoautotrophs, and heterotrophs. Some proteobacteria are anaerobic, while others are aerobic. Molecular systematists currently recognize five subgroups of proteobacteria; the phylogenetic tree at right shows their relationships based on molecular data.

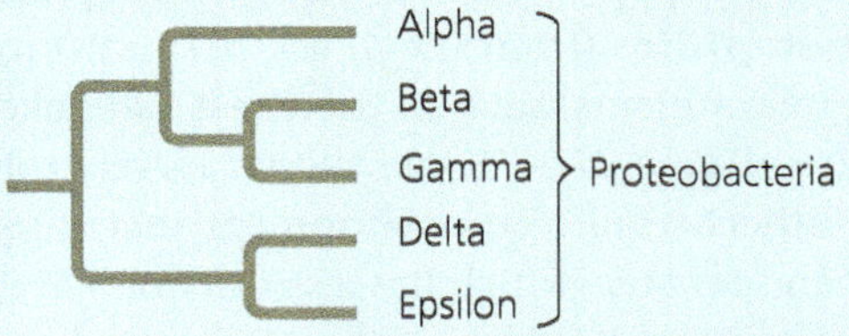

Subgroup: Alpha Proteobacteria

Many of the species in this subgroup are closely associated with eukaryotic hosts. For example, *Rhizobium* species live in nodules within the roots of legumes (plants of the pea/bean family), where the bacteria convert atmospheric N_2 to compounds the host plant can use to make proteins. Species in the genus *Agrobacterium* produce tumors in plants; genetic engineers use these bacteria to carry foreign DNA into the genomes of crop plants (see Figure 20.26). As explained in Chapter 25, scientists hypothesize that mitochondria evolved from aerobic alpha proteobacteria through endosymbiosis.

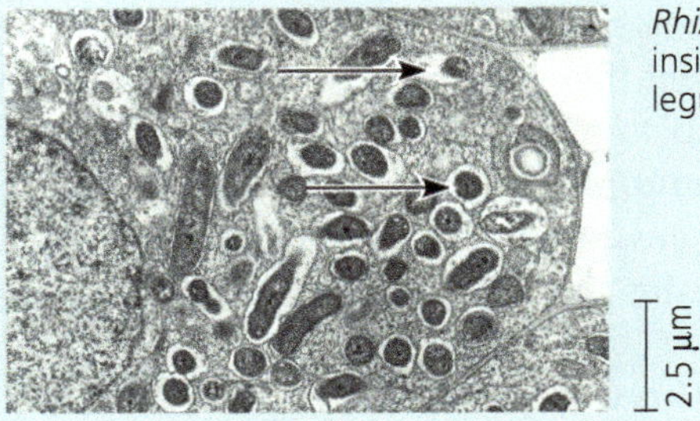

Rhizobium (arrows) inside a root cell of a legume (TEM)

Subgroup: Beta Proteobacteria

This nutritionally diverse subgroup includes *Nitrosomonas*, a genus of soil bacteria that play an important role in nitrogen recycling by oxidizing ammonium (NH_4^+), producing nitrite (NO_2^-) as a waste product.

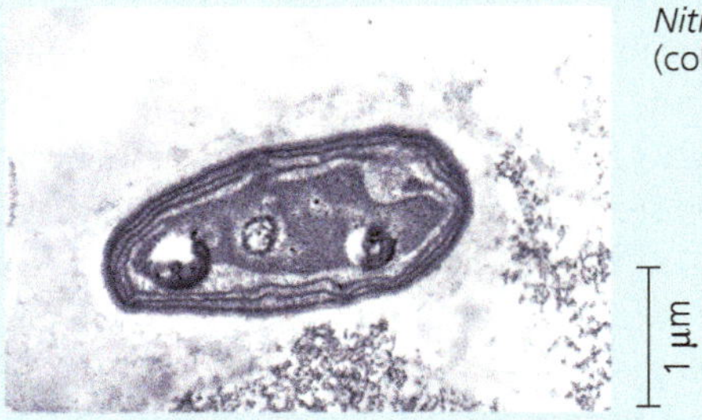

Nitrosomonas (colorized TEM)

Subgroup: Gamma Proteobacteria

This subgroup's autotrophic members include sulfur bacteria such as *Thiomargarita namibiensis* (see p. 557), which obtain energy by oxidizing H_2S, producing sulfur as a waste product (the small globules in the photograph at right). Some heterotrophic gamma proteobacteria are pathogens; for example, *Legionella* causes Legionnaires' disease, *Salmonella* is responsible for some cases of food poisoning, and *Vibrio cholerae* causes cholera. *Escherichia coli*, a common resident of the intestines of humans and other mammals, normally is not pathogenic.

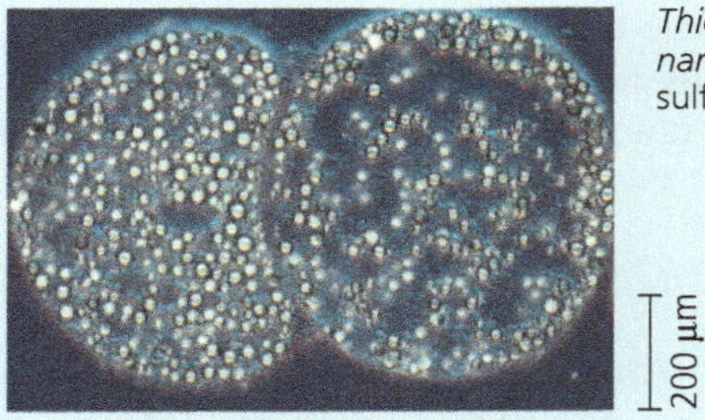

Thiomargarita namibiensis containing sulfur wastes (LM)

Subgroup: Delta Proteobacteria

This subgroup includes the slime-secreting myxobacteria. When the soil dries out or food is scarce, the cells congregate into a fruiting body that releases resistant "myxospores." These cells found new colonies in favorable environments. Another group of delta proteobacteria, the bdellovibrios, attack other bacteria, charging at up to 100 μm/sec (comparable to a human running 240 km/hr). The attack begins when a bdellovibrio attaches to specific molecules found on the outer covering of some bacterial species. The bdellovibrio then drills into its prey by using digestive enzymes and spinning at 100 revolutions per second.

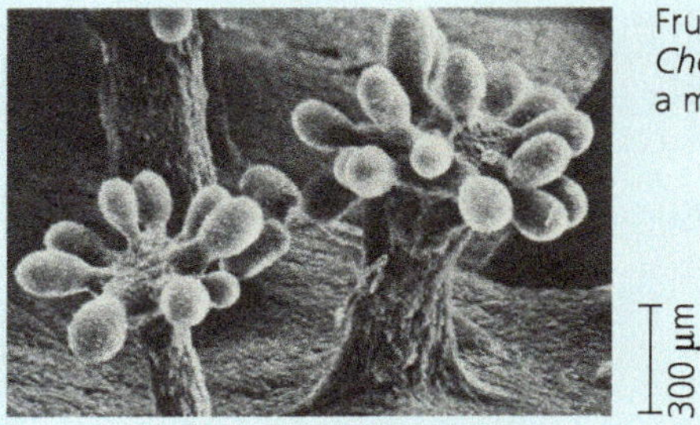

Fruiting bodies of *Chondromyces crocatus*, a myxobacterium (SEM)

Subgroup: Epsilon Proteobacteria

Most species in this subgroup are pathogenic to humans or other animals. Epsilon proteobacteria include *Campylobacter*, which causes blood poisoning and intestinal inflammation, and *Helicobacter pylori*, which causes stomach ulcers.

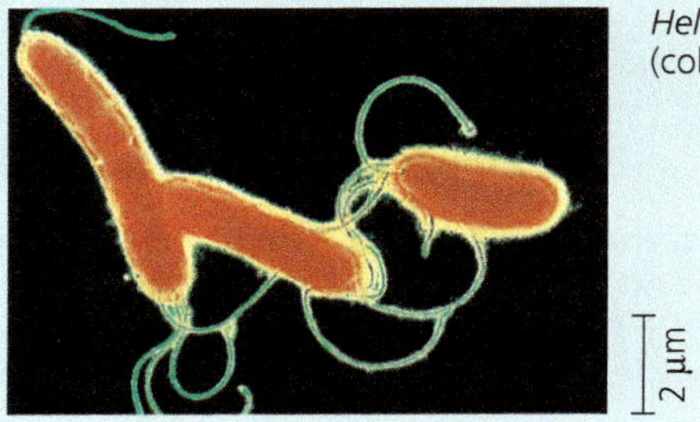

Helicobacter pylori (colorized TEM)

Chlamydias

These parasites can survive only within animal cells, depending on their hosts for resources as basic as ATP. The gram-negative walls of chlamydias are unusual in that they lack peptidoglycan. One species, *Chlamydia trachomatis*, is the most common cause of blindness in the world and also causes nongonococcal urethritis, the most common sexually transmitted disease in the United States.

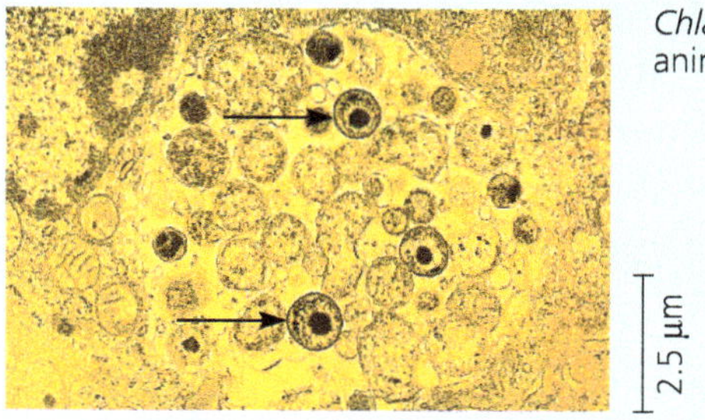

Chlamydia (arrows) inside an animal cell (colorized TEM)

Spirochetes

These helical heterotrophs spiral through their environment by means of rotating, internal, flagellum-like filaments. Many spirochetes are free-living, but others are notorious pathogenic parasites: *Treponema pallidum* causes syphilis, and *Borrelia burgdorferi* causes Lyme disease (see Figure 27.20).

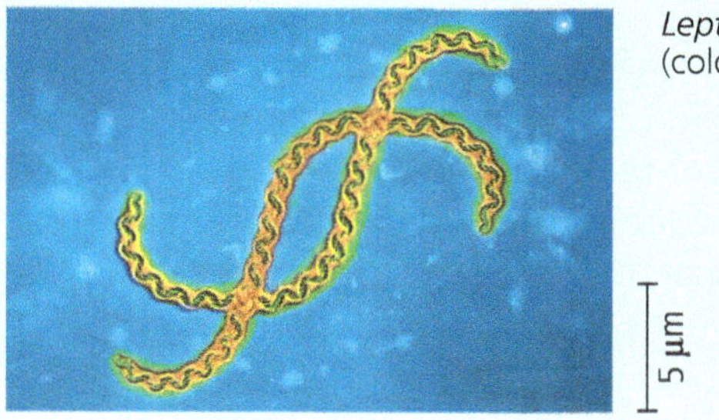

Leptospira, a spirochete (colorized TEM)

Cyanobacteria

These photoautotrophs are the only prokaryotes with plantlike, oxygen-generating photosynthesis. (In fact, as we'll discuss in Chapter 28, chloroplasts likely evolved from an endosymbiotic cyanobacterium.) Both solitary and filamentous cyanobacteria are abundant components of freshwater and marine *phytoplankton*, the collection of photosynthetic organisms that drift near the water's surface. Some filaments have cells specialized for nitrogen fixation, the process that incorporates atmospheric N_2 into inorganic compounds that can be used in the synthesis of amino acids and other organic molecules (see Figure 27.14).

Oscillatoria, a filamentous cyanobacterium

Gram-Positive Bacteria

Gram-positive bacteria rival the proteobacteria in diversity. Species in one subgroup, the actinomycetes (from the Greek *mykes*, fungus, for which these bacteria were once mistaken), form colonies containing branched chains of cells. Two species of actinomycetes cause tuberculosis and leprosy. However, most actinomycetes are free-living species that help decompose the organic matter in soil; their secretions are partly responsible for the "earthy" odor of rich soil. Soil-dwelling species in the genus *Streptomyces* (top) are cultured by pharmaceutical companies as a source of many antibiotics, including streptomycin.

Gram-positive bacteria include many solitary species, such as *Bacillus anthracis* (see Figure 27.9), which causes anthrax, and *Clostridium botulinum*, which causes botulism. The various species of *Staphylococcus* and *Streptococcus* are also gram-positive bacteria.

Mycoplasmas (bottom) are the only bacteria known to lack cell walls. They are also the tiniest known cells, with diameters as small as 0.1 µm, only about five times as large as a ribosome. Mycoplasmas have small genomes—*Mycoplasma genitalium* has only 517 genes, for example. Many mycoplasmas are free-living soil bacteria, but others are pathogens.

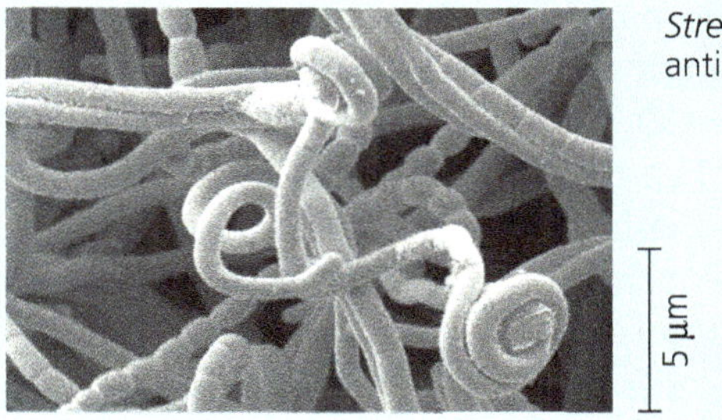

Streptomyces, the source of many antibiotics (SEM)

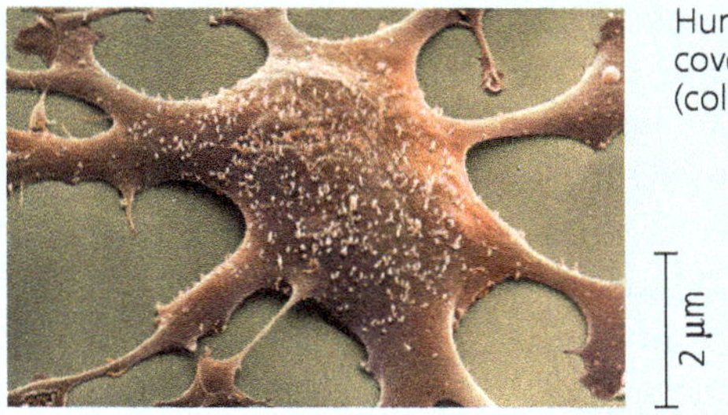

Hundreds of mycoplasmas covering a human fibroblast cell (colorized SEM)

CONCEPT CHECK 27.4

1. Explain how molecular systematics has contributed to our understanding of prokaryotic phylogeny.
2. How has genetic prospecting contributed to our understanding of prokaryotic diversity and phylogeny?
3. **WHAT IF?** What would the discovery of a bacterial species that is a methanogen imply about the evolution of the methane-producing pathway?

For suggested answers, see Appendix A.

CONCEPT 27.5

Prokaryotes play crucial roles in the biosphere

If humans were to disappear from the planet tomorrow, life on Earth would change for many species, but few would be driven to extinction. In contrast, prokaryotes are so important to the biosphere that if they were to disappear, the prospects of survival for many other species would be dim.

Chemical Recycling

The atoms that make up the organic molecules in all living things were at one time part of inorganic substances in the soil, air, and water. Sooner or later, those atoms will return there. Ecosystems depend on the continual recycling of chemical elements between the living and nonliving components of the environment, and prokaryotes play a major role in this process. For example, chemoheterotrophic prokaryotes function as **decomposers**, breaking down dead organisms as well as waste products and thereby unlocking supplies of carbon, nitrogen, and other elements. Without the actions of prokaryotes and other decomposers such as fungi, all life would cease. (See Chapter 55 for a detailed discussion of chemical cycles.)

Prokaryotes also convert some molecules to forms that can be taken up by other organisms. Cyanobacteria and other autotrophic prokaryotes use CO_2 to make organic compounds such as sugars, which are then passed up through food chains. Cyanobacteria also produce atmospheric O_2, and a variety of prokaryotes fix atmospheric nitrogen (N_2) into forms that other organisms can use to make the building blocks of proteins and nucleic acids. Under some conditions, prokaryotes can increase the availability of nutrients that plants require for growth, such as nitrogen, phosphorus, and potassium **(Figure 27.18)**. Prokaryotes can also *decrease* the availability of key plant nutrients; this occurs when prokaryotes "immobilize" nutrients by using them to synthesize molecules that remain within their cells. Thus, prokaryotes can have complex effects on soil nutrient concentrations. In marine environments, a 2005 study found that an archaean from the clade Crenarchaeota can perform nitrification, a key step in the nitrogen cycle (see Figure 55.14). Crenarchaeotes dominate the oceans by numbers, comprising an estimated 10^{28} cells. The sheer abundance of these organisms suggests that they may have a large impact on the global nitrogen cycle; scientists are investigating this possibility.

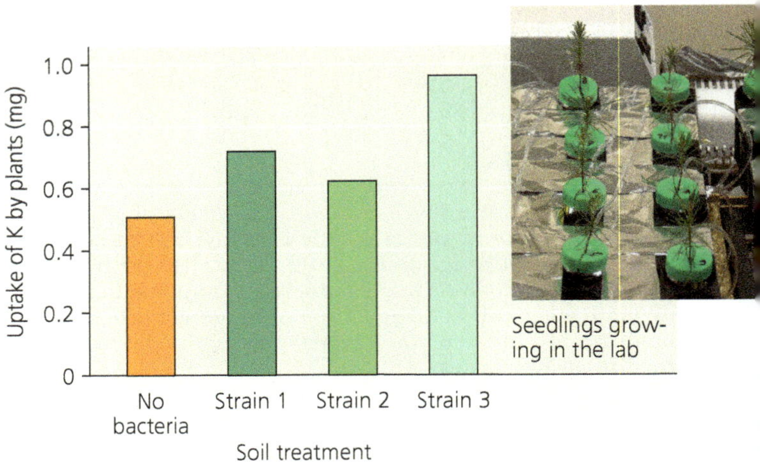

▲ **Figure 27.18 Impact of bacteria on soil nutrient availability.** Pine seedlings grown in sterile soils to which one of three strains of the bacterium *Burkholderia glathei* had been added absorbed more potassium (K) than did seedlings grown in soil without any bacteria. Other results (not shown) demonstrated that strain 3 increased the amount of K released from mineral crystals to the soil.

WHAT IF? *Estimate the average uptake of K for seedlings in soils with bacteria. What would you expect this average to be if bacteria had no effect on nutrient availability?*

Ecological Interactions

Prokaryotes play a central role in many ecological interactions. Consider **symbiosis** (from a Greek word meaning "living together"), an ecological relationship in which two species live in close contact with each other. Prokaryotes often form symbiotic associations with much larger organisms. In general, the larger organism in a symbiotic relationship is known as the **host**, and the smaller is known as the **symbiont**. There are many cases in which a prokaryote and its host participate in **mutualism**, an ecological interaction between two species in which both benefit **(Figure 27.19)**. Other interactions take the form of **commensalism**, an ecological relationship in which one species benefits while the other is not harmed or helped in any significant way. For example, more than 150 bacterial species live on the surface of your body, covering portions of your skin with up to 10 million cells per square centimeter. Some of these species are commensalists: You provide them with food, such as the oils that exude from your pores, and a place to live, while they do not harm or benefit you. Finally, some prokaryotes engage in **parasitism**, an ecological relationship in which a **parasite** eats the cell contents, tissues, or body fluids of its host; as a group, parasites harm but usually do not kill their host, at least not immediately (unlike a predator). Parasites that cause disease are known as **pathogens**, many of which are prokaryotic.

▲ **Figure 27.19 Mutualism: bacterial "headlights."** The glowing oval below the eye of the flashlight fish (*Photoblepharon palpebratus*) is an organ harboring bioluminescent bacteria. The fish uses the light to attract prey and to signal potential mates. The bacteria receive nutrients from the fish.

(We'll discuss mutualism, commensalism, and parasitism in greater detail in Chapter 54.)

The very existence of an ecosystem can depend on prokaryotes. For example, consider the diverse ecological communities found at hydrothermal vents. These communities are densely populated by many different kinds of animals, including worms, clams, crabs, and fishes. But since sunlight does not penetrate to the deep ocean floor, the community does not include photosynthetic organisms. Instead, the energy that supports the community is derived from the metabolic activities of chemoautotrophic bacteria. These bacteria harvest chemical energy from compounds such as hydrogen sulfide (H_2S) that are released from the vent. An active hydrothermal vent may support hundreds of eukaryotic species, but when the vent stops releasing chemicals, the chemoautotrophic bacteria cannot survive. As a result, the entire vent community collapses.

CONCEPT CHECK 27.5

1. Explain how prokaryotes, though small, can be considered giants in their collective impact on Earth and its life.
2. **MAKE CONNECTIONS** After reviewing photosynthesis in Figure 10.6 (p. 188), summarize the main steps by which cyanobacteria produce O_2 and use CO_2 to make organic compounds.

For suggested answers, see Appendix A.

CONCEPT 27.6

Prokaryotes have both beneficial and harmful impacts on humans

Though the best-known prokaryotes tend to be the bacteria that cause illness in humans, these pathogens represent only a small fraction of prokaryotic species. Many other prokaryotes have positive interactions with humans, and some play essential roles in agriculture and industry.

Mutualistic Bacteria

As is true for many other eukaryotes, human well-being can depend on mutualistic prokaryotes. For example, our intestines are home to an estimated 500–1,000 species of bacteria; their cells outnumber all human cells in the body by a factor of ten. Different species live in different portions of the intestines, and they vary in their ability to process different foods. Many of these species are mutualists, digesting food that our own intestines cannot break down. In 2003, scientists published the first complete genome of one of these gut mutualists, *Bacteroides thetaiotaomicron*. The genome includes a large array of genes involved in synthesizing carbohydrates, vitamins, and other nutrients needed by humans. Signals from the bacterium activate human genes that build the network of intestinal blood vessels necessary to absorb nutrient molecules. Other signals induce human cells to produce antimicrobial compounds to which *B. thetaiotaomicron* is not susceptible. This action may reduce the population sizes of other, competing species, thus potentially benefiting both *B. thetaiotaomicron* and its human host.

Pathogenic Bacteria

All the pathogenic prokaryotes known to date are bacteria, and they deserve their negative reputation. Bacteria cause about half of all human diseases. Roughly 2 million people die each year of the lung disease tuberculosis, caused by *Mycobacterium tuberculosis*. And another 2 million people die each year from diarrheal diseases caused by various bacteria.

Some bacterial diseases are transmitted by other species, such as fleas or ticks. In the United States, the most widespread pest-carried disease is Lyme disease, which infects 15,000 to 20,000 people each year (**Figure 27.20**). Caused by a bacterium carried by ticks that live on deer and field mice,

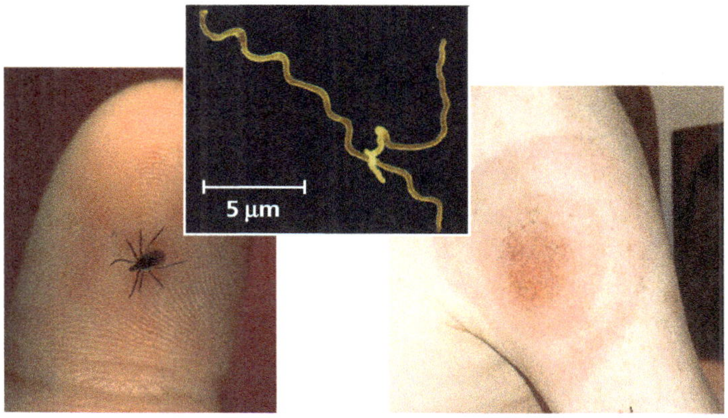

▲ **Figure 27.20 Lyme disease.** Ticks in the genus *Ixodes* spread the disease by transmitting the spirochete *Borrelia burgdorferi* (colorized SEM). A rash may develop at the site of the tick's bite; the rash may be large and ring-shaped (as shown) or much less distinctive.

Lyme disease can result in debilitating arthritis, heart disease, nervous disorders, and death if untreated.

Pathogenic prokaryotes usually cause illness by producing poisons, which are classified as exotoxins or endotoxins. **Exotoxins** are proteins secreted by certain bacteria and other organisms. Cholera, a dangerous diarrheal disease, is caused by an exotoxin secreted by the proteobacterium *Vibrio cholerae*. The exotoxin stimulates intestinal cells to release chloride ions into the gut, and water follows by osmosis. In another example, the potentially fatal disease botulism is caused by botulinum toxin, an exotoxin secreted by the gram-positive bacterium *Clostridium botulinum* as it ferments various foods, including improperly canned meat, seafood, and vegetables. Like other exotoxins, the botulinum toxin can produce disease even if the bacteria that manufacture it are not present. In one such case, eight people contracted botulism after eating salted fish that did not contain any *C. botulinum* bacteria, but did contain the botulinum toxin. Even though the bacterium was no longer present, at some point in the fish preparation process, the bacterium had been able to grow and secrete the toxin.

Endotoxins are lipopolysaccharide components of the outer membrane of gram-negative bacteria. In contrast to exotoxins, endotoxins are released only when the bacteria die and their cell walls break down. Endotoxin-producing bacteria include species in the genus *Salmonella*, such as *Salmonella typhi*, which causes typhoid fever. You might have heard of food poisoning caused by other *Salmonella* species that are frequently found in poultry.

Since the 19th century, improved sanitation systems in the industrialized world have greatly reduced the threat of pathogenic bacteria. Antibiotics have saved a great many lives and reduced the incidence of disease. However, resistance to antibiotics is currently evolving in many bacterial strains. As you read earlier, the rapid reproduction of bacteria enables cells carrying resistance genes to quickly give rise to large populations as a result of natural selection, and these genes can also spread to other species by horizontal gene transfer.

Horizontal gene transfer can also spread genes associated with virulence, turning normally harmless bacteria into potent pathogens. *E. coli*, for instance, is ordinarily a harmless symbiont in the human intestines, but pathogenic strains that cause bloody diarrhea have emerged. One of the most dangerous strains, called O157:H7, is a global threat; in the United States alone, there are 75,000 cases of O157:H7 infection per year, often from contaminated beef or produce. In 2001, scientists sequenced the genome of O157:H7 and compared it with the genome of a harmless strain of *E. coli* called K-12. They discovered that 1,387 out of the 5,416 genes in O157:H7 have no counterpart in K-12. Many of these 1,387 genes are found in chromosomal regions that include phage DNA. This result suggests that at least some of the 1,387 genes were incorporated into the genome of O157:H7 through phage-mediated horizontal gene transfer (transduction). Some of the genes found only in O157:H7 are associated with virulence, including genes that code for adhesive fimbriae that enable O157:H7 to attach itself to the intestinal wall and extract nutrients.

Pathogenic bacteria also pose a potential threat as weapons of bioterrorism. For example, endospores of *Bacillus anthracis* sent through the mail in 2001 caused 18 people to develop inhalation anthrax, which was fatal in 5 of the cases. Such scenarios have stimulated more research on pathogenic prokaryotic species in the hope of developing new vaccines and antibiotics.

Prokaryotes in Research and Technology

On a positive note, we reap many benefits from the metabolic capabilities of both bacteria and archaea. For example, humans have long used bacteria to convert milk to cheese and yogurt. In recent years, our greater understanding of prokaryotes has led to an explosion of new applications in biotechnology; two examples are the use of *E. coli* in gene cloning (see Figure 20.2) and the use of *Agrobacterium tumefaciens* in producing transgenic plants such as Golden Rice (see Figure 20.26 and p. 816).

Bacteria may soon figure prominently in a major industry: plastics. Globally, each year about 350 billion pounds of plastic are produced from petroleum and used to make toys, storage containers, soft drink bottles, and many other items. These products degrade slowly, creating environmental problems. Bacteria can now be used to make natural plastics **(Figure 27.21a)**. For example, some bacteria synthesize a type of polymer known as PHA (polyhydroxyalkanoate), which they use to store chemical energy. The PHA they produce can be extracted, formed into pellets, and used to make durable, biodegradable plastics.

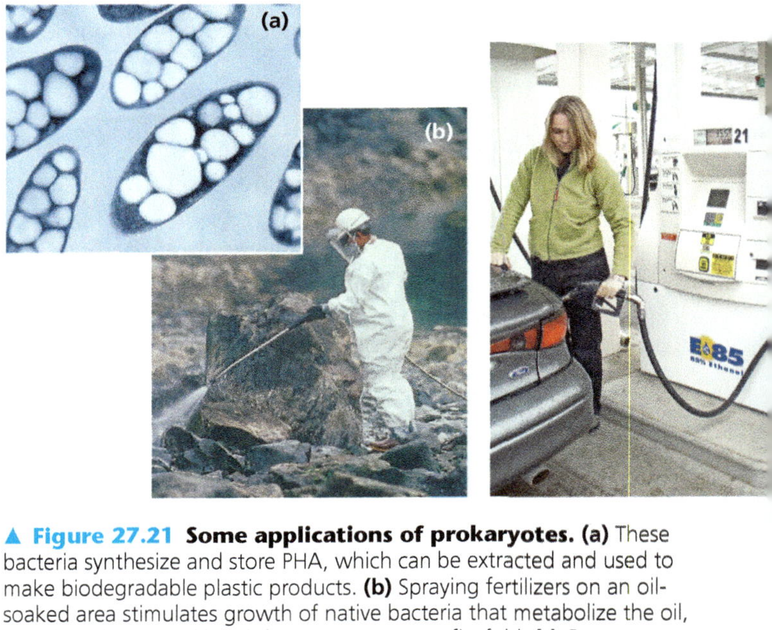

▲ **Figure 27.21 Some applications of prokaryotes. (a)** These bacteria synthesize and store PHA, which can be extracted and used to make biodegradable plastic products. **(b)** Spraying fertilizers on an oil-soaked area stimulates growth of native bacteria that metabolize the oil, speeding the natural breakdown process up to fivefold. **(c)** Current research seeks to develop bacteria that produce ethanol (E-85) fuel efficiently from renewable plant products.

Another way to harness prokaryotes is in **bioremediation**, the use of organisms to remove pollutants from soil, air, or water. For example, anaerobic bacteria and archaea decompose the organic matter in sewage, converting it to material that can be used as landfill or fertilizer after chemical sterilization. Other bioremediation applications include cleaning up oil spills (**Figure 27.21b**) and precipitating radioactive material (such as uranium) out of groundwater.

Through genetic engineering, humans can now modify bacteria to produce vitamins, antibiotics, hormones, and other products (see Chapter 20). Researchers are seeking to reduce fossil fuel use by engineering bacteria that can produce ethanol from various forms of biomass, including agricultural waste, switchgrass, municipal waste (such as paper products that are not recycled), and corn (**Figure 27.21c**).

The usefulness of prokaryotes largely derives from their diverse forms of nutrition and metabolism. All this metabolic versatility evolved prior to the appearance of the structural novelties that heralded the evolution of eukaryotic organisms, to which we devote the remainder of this unit.

CONCEPT CHECK 27.6

1. Identify at least two ways that prokaryotes have affected you positively today.
2. A pathogenic bacterium's toxin causes symptoms that increase the bacterium's chance of spreading from host to host. Does this information indicate whether the poison is an exotoxin or endotoxin? Explain.
3. **WHAT IF?** How might a sudden and dramatic change in your diet affect the diversity of prokaryotic species that live in your digestive tract?

For suggested answers, see Appendix A.

27 CHAPTER REVIEW

SUMMARY OF KEY CONCEPTS

CONCEPT 27.1

Structural and functional adaptations contribute to prokaryotic success (pp. 556–561)

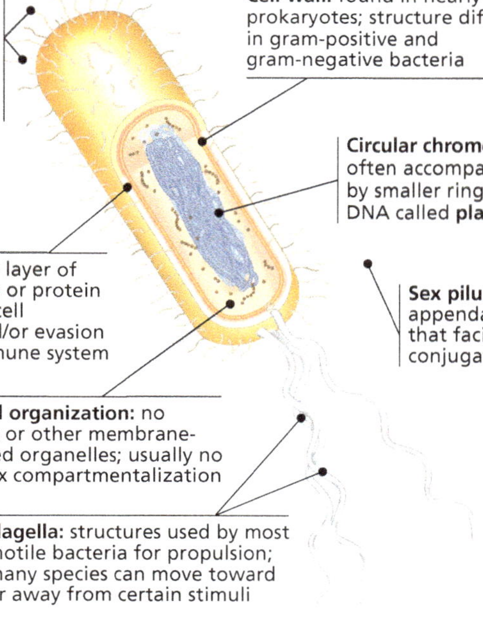

- Prokaryotes can reproduce quickly by binary fission. Some form endospores, which can remain viable in harsh conditions for centuries. Prokaryotic populations can evolve in short periods of time in response to changing environmental conditions.

? *Describe features of prokaryotes that enable them to thrive in many different environments.*

CONCEPT 27.2

Rapid reproduction, mutation, and genetic recombination promote genetic diversity in prokaryotes (pp. 561–564)

- Because prokaryotes can often proliferate rapidly, mutations can quickly increase a population's genetic variation, making adaptive evolution possible.
- Genetic diversity in prokaryotes also can arise by recombination of the DNA from two different cells (via transformation, transduction, or conjugation). By transferring advantageous alleles, such as ones for antibiotic resistance, genetic recombination can promote adaptive evolution in prokaryotic populations.

? *Mutations are rare and prokaryotes reproduce asexually; yet their populations can have high genetic diversity. Explain how this can occur.*

CONCEPT 27.3

Diverse nutritional and metabolic adaptations have evolved in prokaryotes (pp. 564–565)

- Nutritional diversity is much greater in prokaryotes than in eukaryotes. As a group, prokaryotes perform all four modes of nutrition: **photoautotrophy**, **chemoautotrophy**, **photoheterotrophy**, and **chemoheterotrophy**.
- Among prokaryotes, **obligate aerobes** require O_2, **obligate anaerobes** are poisoned by O_2, and **facultative anaerobes** can survive with or without O_2.
- Unlike eukaryotes, prokaryotes can metabolize nitrogen in many different forms. Some can convert atmospheric nitrogen to ammonia, a process called **nitrogen fixation**.
- Prokaryotic cells and even species may cooperate metabolically. In *Anabaena*, photosynthetic cells and nitrogen-fixing cells exchange metabolic products. Metabolic cooperation also occurs in surface-coating **biofilms** that include different species.

? *Describe the range of prokaryotic metabolic adaptations.*

CONCEPT 27.4

Molecular systematics is illuminating prokaryotic phylogeny (pp. 565–570)

- Molecular systematics is leading to a phylogenetic classification of prokaryotes, allowing systematists to identify major new clades.
- Some archaea, such as extreme thermophiles and extreme halophiles, live in extreme environments. Other archaea live in moderate environments such as soils and lakes.
- Diverse nutritional types are scattered among the major groups of bacteria. The two largest groups are the proteobacteria and the gram-positive bacteria.

? *What impact have molecular data had on constructing prokaryotic phylogeny?*

CONCEPT 27.5

Prokaryotes play crucial roles in the biosphere (pp. 570–571)

- Decomposition by heterotrophic prokaryotes and the synthetic activities of autotrophic and nitrogen-fixing prokaryotes contribute to the recycling of elements in ecosystems.
- Many prokaryotes have a symbiotic relationship with a host; the relationships between prokaryotes and their hosts range from mutualism to commensalism to parasitism.

? *In what ways are prokaryotes key to the survival of many species?*

CONCEPT 27.6

Prokaryotes have both beneficial and harmful impacts on humans (pp. 571–573)

- Humans depend on mutualistic prokaryotes, including hundreds of species that live in our intestines and help digest food.
- Pathogenic bacteria typically cause disease by releasing **exotoxins** or **endotoxins** and are potential weapons of bioterrorism. Horizontal gene transfer can spread genes associated with virulence to harmless species or strains.
- Experiments with bacteria such as *E. coli* have led to important advances in DNA technology. Prokaryotes can be used in bioremediation, production of biodegradable plastics, and the synthesis of vitamins, antibiotics, and other products.

? *Describe beneficial and harmful impacts of prokaryotes on humans.*

TEST YOUR UNDERSTANDING

LEVEL 1: KNOWLEDGE/COMPREHENSION

1. Genetic variation in bacterial populations cannot result from
   a. transduction.
   b. transformation.
   c. conjugation.
   d. mutation.
   e. meiosis.

2. Photoautotrophs use
   a. light as an energy source and CO_2 as a carbon source.
   b. light as an energy source and methane as a carbon source.
   c. N_2 as an energy source and CO_2 as a carbon source.
   d. CO_2 as both an energy source and a carbon source.
   e. H_2S as an energy source and CO_2 as a carbon source.

3. Which of the following statements is *not* true?
   a. Archaea and bacteria have different membrane lipids.
   b. Both archaea and bacteria generally lack membrane-enclosed organelles.
   c. The cell walls of archaea lack peptidoglycan.
   d. Only bacteria have histones associated with DNA.
   e. Only some archaea use CO_2 to oxidize H_2, releasing methane.

4. Which of the following involves metabolic cooperation among prokaryotic cells?
   a. binary fission
   b. endospore formation
   c. endotoxin release
   d. biofilms
   e. photoautotrophy

5. Bacteria perform the following ecological roles. Which role typically does *not* involve symbiosis?
   a. skin commensalist
   b. decomposer
   c. aggregate with methane-consuming archaea
   d. gut mutualist
   e. pathogen

6. Plantlike photosynthesis that releases O_2 occurs in
   a. cyanobacteria.
   b. chlamydias.
   c. archaea.
   d. actinomycetes.
   e. chemoautotrophic bacteria.

LEVEL 2: APPLICATION/ANALYSIS

7. **EVOLUTION CONNECTION**
   In patients infected with nonresistant strains of the tuberculosis bacterium, antibiotics can relieve symptoms in a few weeks. However, it takes much longer to halt the infection, and patients may discontinue treatment while bacteria are still present. How might this result in the evolution of drug-resistant pathogens?

LEVEL 3: SYNTHESIS/EVALUATION

8. **SCIENTIFIC INQUIRY**
   DRAW IT The nitrogen-fixing bacterium *Rhizobium* infects the roots of some plant species, forming a mutualism in which the bacterium provides nitrogen, and the plant provides carbohydrates. Scientists measured the 12-week growth of one such plant species (*Acacia irrorata*) when infected by six different *Rhizobium* strains. (a) Graph the data. (b) Interpret your graph.

Rhizobium strain	1	2	3	4	5	6
Plant mass (g)	0.91	0.06	1.56	1.72	0.14	1.03

Source: J. J. Burdon et al., Variation in the effectiveness of symbiotic associations between native rhizobia and temperate Australian *Acacia*: within species interactions, *Journal of Applied Ecology* 36:398–408 (1999).
Note: Without *Rhizobium*, after 12 weeks, *Acacia* plants have a mass of about 0.1 g.

9. **WRITE ABOUT A THEME**
   Energy Transfer In a short essay (about 100–150 words), discuss how prokaryotes and other members of hydrothermal vent communities transfer and transform energy.

For selected answers, see Appendix A.

MasteringBIOLOGY www.masteringbiology.com

1. MasteringBiology® Assignments

Make Connections Tutorial Bacterial Conjugation (Chapter 27) and Binary Fission (Chapter 12)
Tutorial Diversity in Bacteria
Activities Classification of Prokaryotes • The Tree of Life • Discovery Channel Videos: Bacteria; Antibiotics; Tasty Bacteria
Questions Student Misconceptions • Reading Quiz • Multiple Choice • End-of-Chapter

2. eText

Read your book online, search, take notes, highlight text, and more.

3. The Study Area

Practice Tests • Cumulative Test • *BioFlix* 3-D Animations • MP3 Tutor Sessions • Videos • Activities • Investigations • Lab Media • Audio Glossary • Word Study Tools • Art

28

Protists

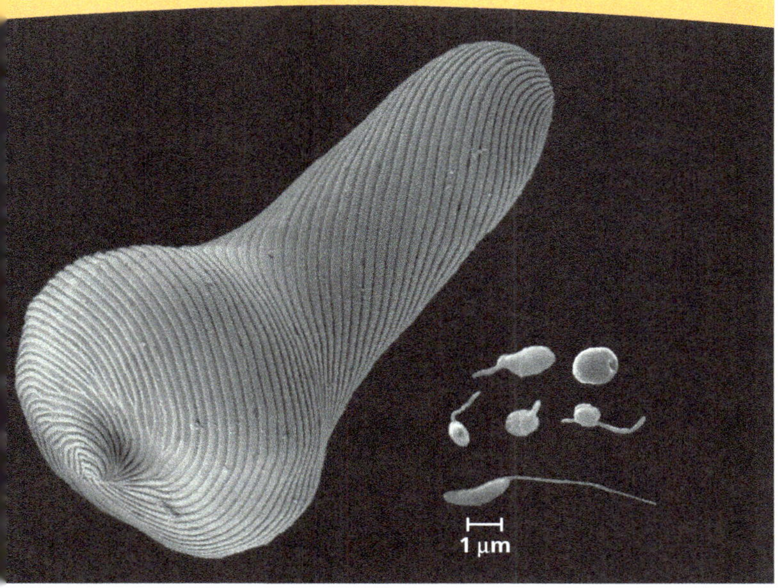

▲ **Figure 28.1 Which of these organisms are prokaryotes and which are eukaryotes?**

EVOLUTION

KEY CONCEPTS

28.1 Most eukaryotes are single-celled organisms
28.2 Excavates include protists with modified mitochondria and protists with unique flagella
28.3 Chromalveolates may have originated by secondary endosymbiosis
28.4 Rhizarians are a diverse group of protists defined by DNA similarities
28.5 Red algae and green algae are the closest relatives of land plants
28.6 Unikonts include protists that are closely related to fungi and animals
28.7 Protists play key roles in ecological communities

OVERVIEW

Living Small

Knowing that most prokaryotes are extremely small organisms, you might assume that **Figure 28.1** depicts six prokaryotes and one much larger eukaryote. But in fact, the only prokaryote is the organism immediately above the scale bar. The other six organisms are members of diverse, mostly unicellular groups of eukaryotes informally known as **protists**. Very small eukaryotes have intrigued scientists for more than 300 years, ever since the Dutch microscopist Antoni van Leeuwenhoek first laid eyes on them. As he discovered, viewing a drop of pond water under a light microscope can reveal a fascinating world of unicellular protists and prokaryotes. Some protists propel themselves with whipping flagella, while others creep along by means of blob-like appendages. Some are shaped like tiny trumpets; others resemble miniature jewelry. Recalling his observations, van Leeuwenhoek wrote, "No more pleasant sight has met my eye than this, of so many thousands of living creatures in one small drop of water."

Until recently, biologists thought that 300 years of observation had uncovered a representative sample of living protist species. But in the last decade, genetic prospecting has turned up a treasure trove of previously unknown protists within the world of microscopic life. Many of these newly discovered organisms are just 0.5–2 µm in diameter—as small as many prokaryotes.

The surprising discovery of many species of minuscule protists followed close on the heels of recent findings regarding protist phylogeny. All protists were once classified in a single kingdom, Protista, but advances in eukaryotic systematics have caused the kingdom to crumble. It has become clear that the kingdom Protista is in fact polyphyletic (see Figure 26.10): Some protists are more closely related to plants, fungi, or animals than they are to other protists. As a result, the kingdom Protista has been abandoned, and various lineages of protists are now recognized as kingdoms in their own right. Most biologists still use the term *protist*, but only as a convenient way to refer to eukaryotes that are not plants, animals, or fungi.

In this chapter, you will become acquainted with some of the most significant groups of protists. You will learn about their structural and biochemical adaptations as well as their enormous impact on ecosystems, agriculture, industry, and human health.

CONCEPT 28.1

Most eukaryotes are single-celled organisms

Protists, along with plants, animals, and fungi, are classified as eukaryotes; they are in domain Eukarya, one of the three domains of life. Unlike the cells of prokaryotes, eukaryotic cells have a nucleus and other membrane-bounded organelles, such as mitochondria and the Golgi apparatus. Such organelles provide specific locations in which particular cellular functions are accomplished, making the structure and organization of eukaryotic cells more complex than those of prokaryotic cells.

We'll survey the diversity of eukaryotes throughout the rest of this unit, beginning in this chapter with the protists. As you explore this material, bear in mind that

- the organisms in most eukaryotic lineages are protists, and
- most protists are unicellular.

Thus, life differs greatly from how most of us commonly think of it. The large, multicellular organisms that we know best (plants, animals, and fungi) are the tips of just a few branches on the great tree of life (see Figure 26.21).

Structural and Functional Diversity in Protists

Given the polyphyletic nature of the group once called Protista, it isn't surprising that few general characteristics of protists can be cited without exceptions. In fact, protists exhibit more structural and functional diversity than any other group of eukaryotes.

Most protists are unicellular, although there are some colonial and multicellular species. Single-celled protists are justifiably considered the simplest eukaryotes, but at the cellular level, many protists are very complex—the most elaborate of all cells. In multicellular organisms, essential biological functions are carried out by organs. Unicellular protists carry out the same essential functions, but they do so using subcellular organelles, not multicellular organs. The organelles that protists use are mostly those discussed in Chapter 6, including the nucleus, endoplasmic reticulum, Golgi apparatus, and lysosomes. Certain protists also rely on organelles not found in most other eukaryotic cells, such as contractile vacuoles that pump excess water from the protistan cell (see Figure 7.16).

Protists are more nutritionally diverse than other eukaryote groups. Some protists are photoautotrophs and contain chloroplasts. Some are heterotrophs, absorbing organic molecules or ingesting larger food particles. Still other protists, called **mixotrophs**, combine photosynthesis *and* heterotrophic nutrition. Photoautotrophy, heterotrophy, and mixotrophy have all arisen independently in many protist lineages.

Reproduction and life cycles also are highly varied among protists. Some protists are only known to reproduce asexually; others can also reproduce sexually or at least employ the sexual processes of meiosis and fertilization. All three basic types of sexual life cycles (see Figure 13.6) are represented among protists, along with some variations that do not quite fit any of these types. We will examine the life cycles of several protist groups later in this chapter.

Endosymbiosis in Eukaryotic Evolution

What gave rise to the enormous diversity of protists that exist today? There is abundant evidence that much of protist diversity has its origins in **endosymbiosis**, the process in which certain unicellular organisms engulf other cells, which become endosymbionts and ultimately organelles in the host cell. For example, as we discussed in Chapter 25, structural, biochemical, and DNA sequence data indicate that the first eukaryotes acquired mitochondria by engulfing an aerobic prokaryote (specifically, an alpha proteobacterium). The early origin of mitochondria is supported by the fact that all eukaryotes studied so far have either mitochondria or modified versions of them.

Much evidence also indicates that later in eukaryotic history, a lineage of heterotrophic eukaryotes acquired an additional endosymbiont—a photosynthetic cyanobacterium—that then evolved into plastids. As the hypothesis illustrated in **Figure 28.2** shows, this plastid-bearing lineage gave rise to two lineages of photosynthetic protists, or **algae:** red algae and green algae. This hypothesis is supported by the observation that the DNA of plastid genes in red algae and green algae closely resembles the DNA of cyanobacteria. In addition, plastids in red algae and green algae are surrounded by two membranes. Transport proteins in these membranes are homologous to proteins in the inner and outer membranes of cyanobacterial endosymbionts, providing further support for the hypothesis.

On several occasions during eukaryotic evolution, red algae and green algae underwent **secondary endosymbiosis**: They were ingested in the food vacuoles of heterotrophic eukaryotes and became endosymbionts themselves. For example, protists known as chlorarachniophytes likely evolved when a heterotrophic eukaryote engulfed a green alga. Evidence for this process can be found within the engulfed cell, which contains a tiny vestigial nucleus, called a *nucleomorph*. Genes from the nucleomorph are still transcribed, and their DNA sequences indicate the engulfed cell was a green alga. Also consistent with the hypothesis that chlorarachniophytes evolved from a eukaryote that engulfed another eukaryote, their plastids are surrounded by *four* membranes. The two inner membranes originated as the inner and outer membranes of the ancient cyanobacterium. The third membrane is derived from the engulfed alga's plasma membrane, and the outermost membrane is derived from the heterotrophic eukaryote's food vacuole. In some other protists, plastids acquired by secondary endosymbiosis are surrounded by three membranes, indicating that one of the original four membranes was lost during the course of evolution.

Five Supergroups of Eukaryotes

Our understanding of the evolutionary history of protists has been in a state of flux in recent years. Not only has kingdom Protista been abandoned, but a variety of other hypotheses have been discarded as well. For example, in the early 1990s,

▼ Figure 28.2 Diversity of plastids produced by endosymbiosis. Studies of plastid-bearing eukaryotes suggest that plastids evolved from a gram-negative cyanobacterium that was engulfed by an ancestral heterotrophic eukaryote (primary endosymbiosis). That ancestor then diversified into red algae and green algae, some of which were subsequently engulfed by other eukaryotes (secondary endosymbiosis).

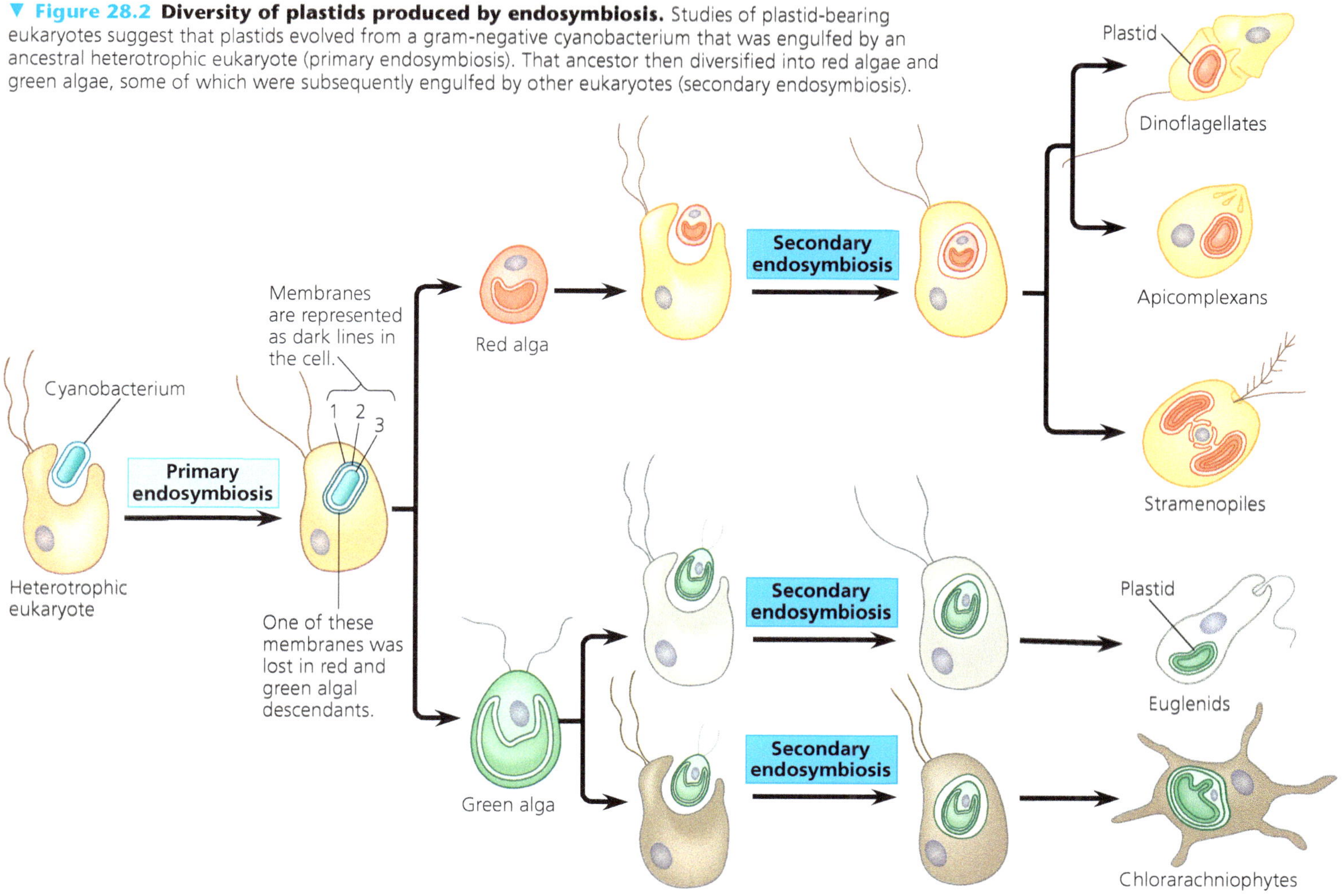

many biologists thought that the oldest lineage of living eukaryotes consisted of the *amitochondriate protists,* organisms without conventional mitochondria and with fewer membrane-bounded organelles than other protist groups. But recent structural and DNA data have undermined this hypothesis. Many of the so-called amitochondriate protists have been shown to have mitochondria—though reduced ones—and some of these organisms are now classified in entirely different groups. For example, microsporidians, once considered amitochondriate protists, are now classified as fungi.

The ongoing changes in our understanding of the phylogeny of protists pose challenges to students and instructors alike. Hypotheses about these relationships are a focus of scientific activity, changing rapidly as new data cause previous ideas to be modified or discarded. In this chapter, our discussion is organized around one current hypothesis: the five supergroups of eukaryotes shown in **Figure 28.3**, on the next two pages. Because the root of the eukaryotic tree is not known, all five supergroups are shown as diverging simultaneously from a common ancestor. We know that is not correct, but we do not know which organisms were the first to diverge from the others. In addition, while some of the groups in Figure 28.3 are well supported by morphological and DNA data, others are more controversial. As you read this chapter, it may be helpful to focus less on the specific names of groups of organisms and more on why the organisms are important and how ongoing research is elucidating their evolutionary relationships.

CONCEPT CHECK 28.1

1. Cite at least four examples of structural and functional diversity among protists.
2. Summarize the role of endosymbiosis in eukaryotic evolution.
3. **WHAT IF?** After studying Figure 28.3, on the next two pages, draw a simplified version of the phylogenetic tree that shows only the five supergroups of eukaryotes. Now sketch how the tree would look if the unikonts were the first group of eukaryotes to diverge from other eukaryotes.

For suggested answers, see Appendix A.

▼ Figure 28.3

Exploring Protistan Diversity

The tree below represents a phylogenetic hypothesis for the relationships among all the eukaryotes on Earth today. The eukaryotic groups at the branch tips are related in larger "supergroups," labeled vertically at the far right of the tree. The kingdoms Plantae (land plants), Fungi, and Animalia (animals) have survived from the five-kingdom system of classification. Groups that were formerly classified in the kingdom Protista are listed in beige boxes. Dotted lines indicate evolutionary relationships that are uncertain and proposed clades that are under active debate.

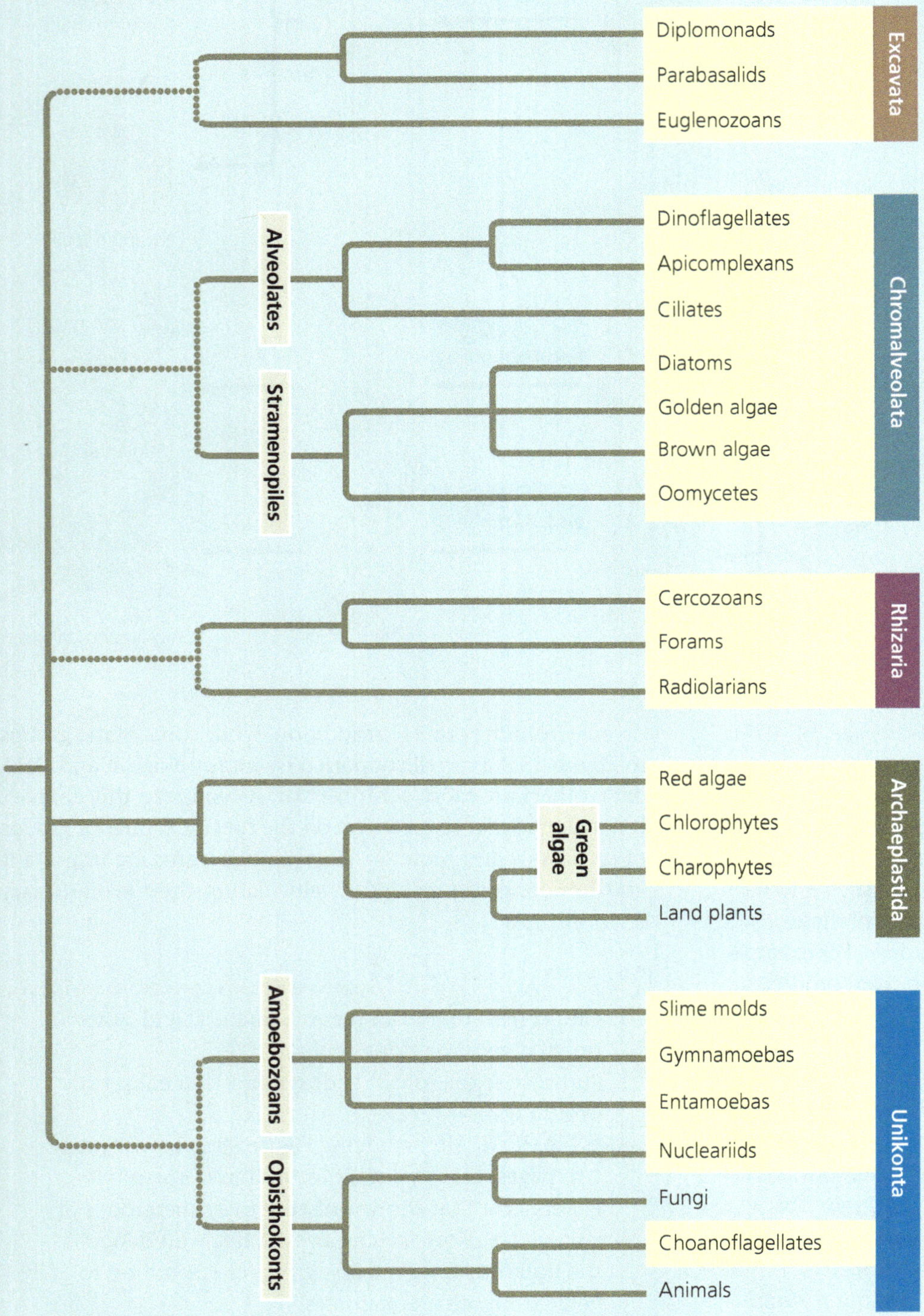

Excavata

Some members of this supergroup have an "excavated" groove on one side of the cell body. Two major clades (the parabasalids and diplomonads) have modified mitochondria; others (the euglenozoans) have flagella that differ in structure from those of other organisms. Excavates include parasites such as *Giardia*, as well as many predatory and photosynthetic species.

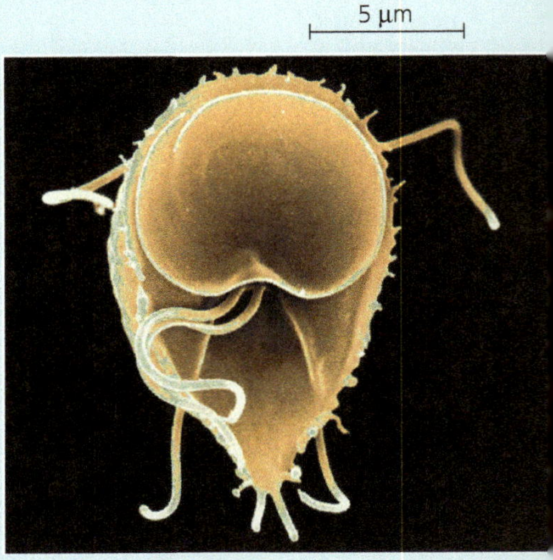

***Giardia intestinalis*, a diplomonad parasite.** This diplomonad (colorized SEM), which lacks the characteristic surface groove of the Excavata, can infect people when they drink water contaminated with feces containing *Giardia* cysts. Drinking such water—even from a seemingly pristine stream—can cause severe diarrhea. Boiling the water kills the parasite.

Chromalveolata

This group may have originated by an ancient secondary endosymbiosis event. Chromalveolates include some of the most important photosynthetic organisms on Earth, such as the diatoms shown here. The group also includes the brown algae that form underwater kelp "forests," as well as important pathogens, such as *Plasmodium*, which causes malaria, and *Phytophthora*, which caused the devastating potato famine in 19th-century Ireland.

Diatom diversity. These beautiful single-celled protists are important photosynthetic organisms in aquatic communities (LM).

Rhizaria

This group contains many species of amoebas, most of which have pseudopodia that are threadlike in shape. Pseudopodia are extensions that can bulge from any portion of the cell; they are used in movement and in the capture of prey. Several recent molecular phylogenetic studies have suggested that Rhizaria should be nested within Chromalveolata; this hypothesis is currently being tested by other research groups.

***Globigerina*, a foram in the supergroup Rhizaria.** Threadlike pseudopodia extend through pores in the shell, or *test* (LM). The inset SEM shows a foram test, which is hardened by calcium carbonate.

Archaeplastida

This group of eukaryotes includes red algae and green algae, along with land plants (kingdom Plantae, discussed in Chapters 29 and 30). Red algae and green algae include unicellular species, colonial species (such as the green alga *Volvox*), and multicellular species. Many of the large algae known informally as "seaweeds" are multicellular red or green algae. Protists in Archaeplastida include key photosynthetic species that form the base of the food web in some aquatic communities.

***Volvox*, a colonial freshwater green alga.** The colony is a hollow ball whose wall is composed of hundreds of biflagellated cells (see inset LM) embedded in a gelatinous matrix. The cells are usually connected by cytoplasmic strands; if isolated, these cells cannot reproduce. The large colonies seen here will eventually release the small "daughter" colonies within them (LM).

Unikonta

This group of eukaryotes includes amoebas that have lobe- or tube-shaped pseudopodia, as well as animals, fungi, and non-amoeba protists that are closely related to animals or fungi. According to one current hypothesis, the unikonts may have been the first group of eukaryotes to diverge from other eukaryotes (see Figure 28.23); however, this hypothesis has yet to be widely accepted.

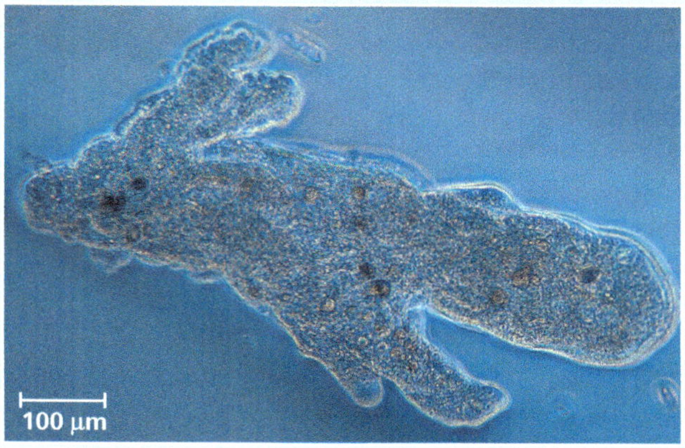

A unikont amoeba. This amoeba (*Amoeba proteus*) is using its pseudopodia to move.

CONCEPT 28.2

Excavates include protists with modified mitochondria and protists with unique flagella

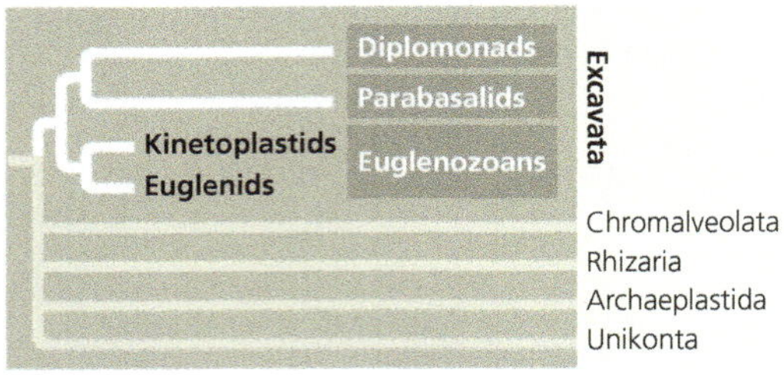

Now that we have examined some of the broad patterns in eukaryotic evolution, we will look more closely at the five main groups of protists shown in Figure 28.3.

We begin this tour with **Excavata** (the excavates), a clade recently proposed based on morphological studies of the cytoskeleton. Some members of this diverse group also have an "excavated" feeding groove on one side of the cell body.

The excavates include the diplomonads, the parabasalids, and the euglenozoans. Molecular data indicate that each of these three groups is monophyletic, but the data have neither confirmed nor strongly refuted the monophyly of the excavate supergroup. Although many excavates share certain unique cytoskeletal features, we cannot yet tell whether that is because the excavates are monophyletic or because the common ancestor of eukaryotes had those features. Overall, support for the excavate clade is relatively weak, making it one of the more controversial of the five supergroups.

Diplomonads and Parabasalids

The protists in these two groups lack plastids and have modified mitochondria (until recently, they were thought to lack mitochondria altogether). Most diplomonads and parabasalids are found in anaerobic environments.

Diplomonads have modified mitochondria called *mitosomes*. These organelles lack functional electron transport chains and hence cannot use oxygen to help extract energy from carbohydrates and other organic molecules. Instead, diplomonads get the energy they need from anaerobic biochemical pathways.

Structurally, diplomonads have two equal-sized nuclei and multiple flagella. Recall that eukaryotic flagella are extensions of the cytoplasm, consisting of bundles of microtubules covered by the cell's plasma membrane (see Figure 6.24). They are quite different from prokaryotic flagella, which are filaments composed of the globular protein flagellin attached to the cell surface (see Figure 27.6).

Many diplomonads are parasites. An infamous example is *Giardia intestinalis* (also known as *Giardia lamblia*; see Figure 28.3), which inhabits the intestines of mammals.

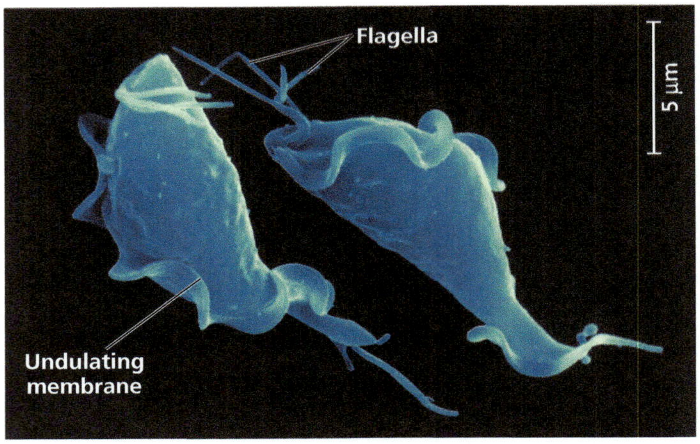

▲ **Figure 28.4 The parabasalid *Trichomonas vaginalis*** (colorized SEM).

Parabasalids also have reduced mitochondria; called *hydrogenosomes*, these organelles generate some energy anaerobically, releasing hydrogen gas as a by-product. The best-known parabasalid is *Trichomonas vaginalis*, a sexually transmitted parasite that infects some 5 million people each year. *T. vaginalis* travels along the mucus-coated lining of the human reproductive and urinary tracts by moving its flagella and by undulating part of its plasma membrane (**Figure 28.4**). In females, if the vagina's normal acidity is disturbed, *T. vaginalis* can outcompete beneficial microorganisms there and infect the vagina. (*Trichomonas* infections also can occur in the urethra of males, though often without symptoms.) *T. vaginalis* has a gene that allows it to feed on the vaginal lining, promoting infection. Studies suggest that the protist acquired this gene by horizontal gene transfer from bacterial parasites in the vagina.

Euglenozoans

Protists called **euglenozoans** belong to a diverse clade that includes predatory heterotrophs, photosynthetic autotrophs, and parasites. The main morphological feature that distinguishes protists in this clade is the presence of a rod with either a spiral or a crystalline structure inside each of their flagella (**Figure 28.5**). The two best-studied groups of euglenozoans are the kinetoplastids and the euglenids.

Kinetoplastids

Protists called **kinetoplastids** have a single, large mitochondrion that contains an organized mass of DNA called a *kinetoplast*. These protists include species that feed on prokaryotes in freshwater, marine, and moist terrestrial ecosystems, as well as species that parasitize animals, plants, and other protists. For example, kinetoplastids in the genus *Trypanosoma* infect humans and cause sleeping sickness, a neurological disease that is invariably fatal if not treated. The infection occurs via the bite of a vector (carrier) organism, the African tsetse fly (**Figure 28.6**). Trypanosomes also cause Chagas' disease, which

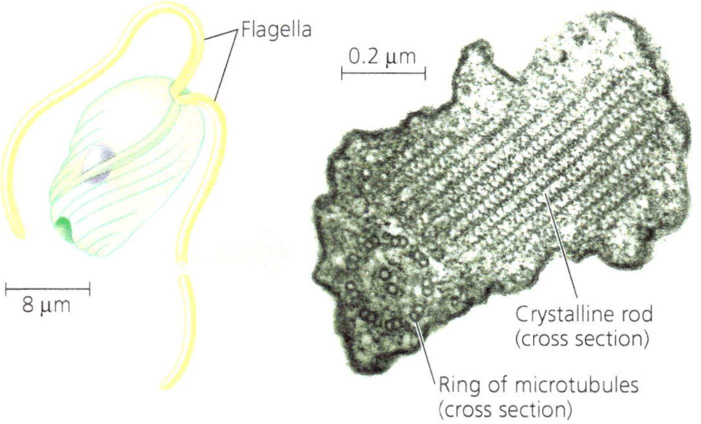

▲ **Figure 28.5 Euglenozoan flagellum.** Most euglenozoans have a crystalline rod inside one of their flagella (the TEM is a flagellum shown in cross section). The rod lies alongside the 9 + 2 ring of microtubules found in all eukaryotic flagella (compare with Figure 6.24).

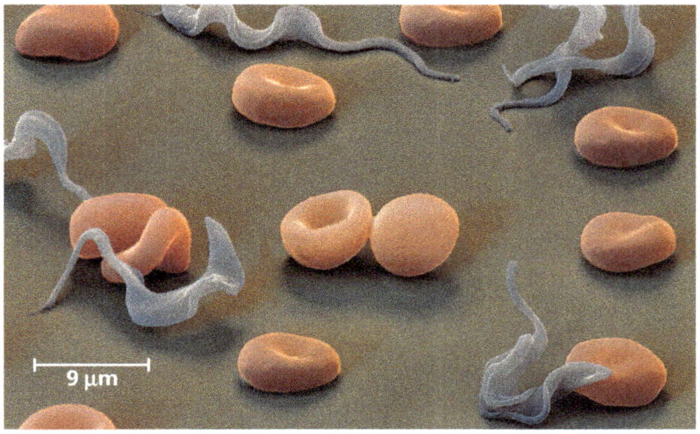

▲ **Figure 28.6 *Trypanosoma*, the kinetoplastid that causes sleeping sickness.** The purple, ribbon-shaped cells among these red blood cells are the trypanosomes (colorized SEM).

is transmitted by bloodsucking insects and can lead to congestive heart failure.

Trypanosomes evade immune responses with an effective "bait-and-switch" defense. The surface of a trypanosome is coated with millions of copies of a single protein. However, before the host's immune system can recognize the protein and mount an attack, new generations of the parasite switch to another surface protein with a different molecular structure. Frequent changes in the surface protein prevent the host from developing immunity (see Figure 43.24). About a third of *Trypanosoma*'s genome is dedicated to producing these surface proteins.

Euglenids

A **euglenid** has a pocket at one end of the cell from which one or two flagella emerge **(Figure 28.7)**. Many species of the euglenid *Euglena* are mixotrophs: In sunlight they are autotrophic, but when sunlight is unavailable, they can become heterotrophic, absorbing organic nutrients from their environment. Many other euglenids engulf prey by phagocytosis.

CONCEPT CHECK 28.2

1. Why do some biologists describe the mitochondria of diplomonads and parabasalids as "highly reduced"?
2. **WHAT IF?** DNA sequence data for a diplomonad, a euglenid, a plant, and an unidentified protist suggest that the unidentified species is most closely related to the diplomonad. Further studies reveal that the unknown species has fully functional mitochondria. Based on these data, at what point on the phylogenetic tree in Figure 28.3 did the mystery protist's lineage probably diverge from other eukaryote lineages? Explain.

For suggested answers, see Appendix A.

▼ **Figure 28.7 *Euglena*, a euglenid commonly found in pond water.**

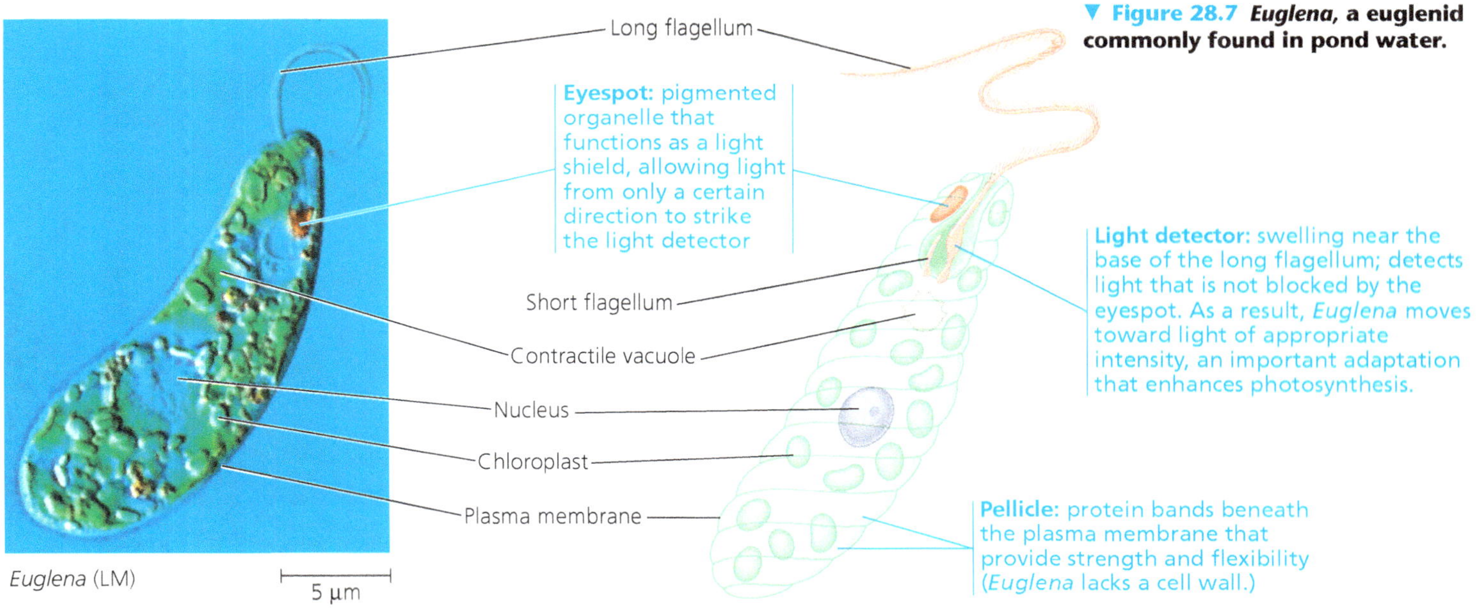

CONCEPT 28.3

Chromalveolates may have originated by secondary endosymbiosis

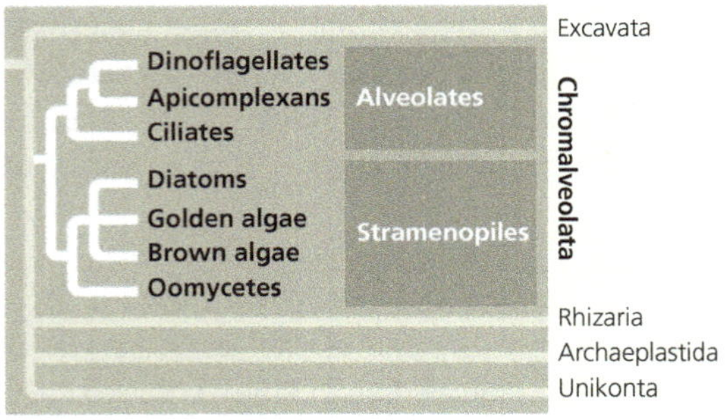

The supergroup **Chromalveolata** (the chromalveolates), a large, extremely diverse clade of protists, has recently been proposed based on two lines of evidence. First, some (though not all) DNA sequence data suggest that the chromalveolates form a monophyletic group. Second, some data support the hypothesis that the chromalveolates originated more than a billion years ago, when a common ancestor of the group engulfed a single-celled, photosynthetic red alga. Because red algae are thought to have originated by primary endosymbiosis (see Figure 28.2), such an origin for the chromalveolates is referred to as secondary endosymbiosis.

How strong is the evidence that the chromalveolates originated by secondary endosymbiosis? Many species in the clade have plastids whose structure and DNA indicate that they are of red algal origin. Others have reduced plastids that seem to be derived from a red algal endosymbiont. Still other species lack plastids altogether, yet some of these species have plastid genes in their nuclear DNA. Such data have led researchers to suggest that the common ancestor of the chromalveolates had plastids of red algal origin, but that later, some evolutionary lineages within the group lost the plastids. Others question this idea, based on the absence of plastid genes in the genomes of several chromalveolates that lack plastids. Overall, the endosymbiotic origin of the chromalveolates is an interesting idea, but like any scientific hypothesis, new data may show it to be incorrect.

The chromalveolates are perhaps the most controversial of the five supergroups we describe in this chapter. Even so, for many scientists, this supergroup represents the best current hypothesis for the phylogeny of the two large protist clades to which we now turn: the alveolates and the stramenopiles.

Alveolates

The **alveolates** are a group of protists whose monophyly is well supported by molecular systematics. Structurally, species in this group have membrane-bounded sacs (alveoli) just under the plasma membrane (**Figure 28.8**). The function of the alveoli is unknown; researchers hypothesize that they may help stabilize the cell surface or regulate the cell's water and ion content.

The alveolates include three subgroups: a group of flagellates (the dinoflagellates), a group of parasites (the apicomplexans), and a group of protists that move using cilia (the ciliates).

Dinoflagellates

The **dinoflagellates** are characterized by cells that are reinforced by cellulose plates. Two flagella located in grooves in this "armor" make dinoflagellates (from the Greek *dinos*, whirling) spin as they move through the water (**Figure 28.9**). Dinoflagellates are abundant components of both marine and freshwater plankton, communities of mostly microscopic organisms that drift in currents near the water's surface. These dinoflagellates include some of the most important species of *phytoplankton* (photosynthetic plankton, which include photosynthetic bacteria as well as algae). However, many photosynthetic dinoflagellates are mixotrophic, and roughly half of all dinoflagellates are purely heterotrophic.

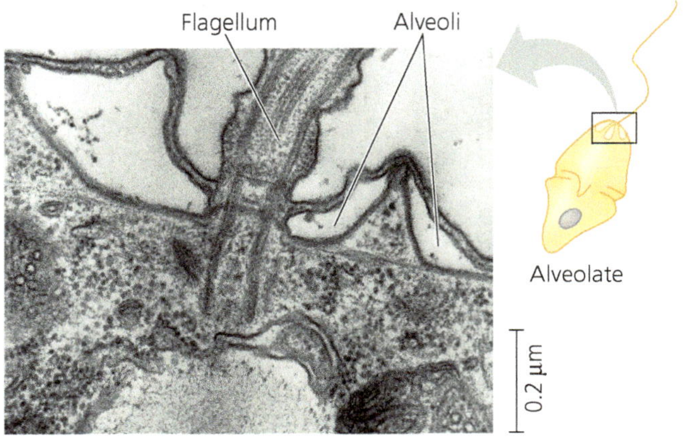

▲ **Figure 28.8 Alveoli.** These sacs under the plasma membrane are a characteristic that distinguishes alveolates from other eukaryotes (TEM).

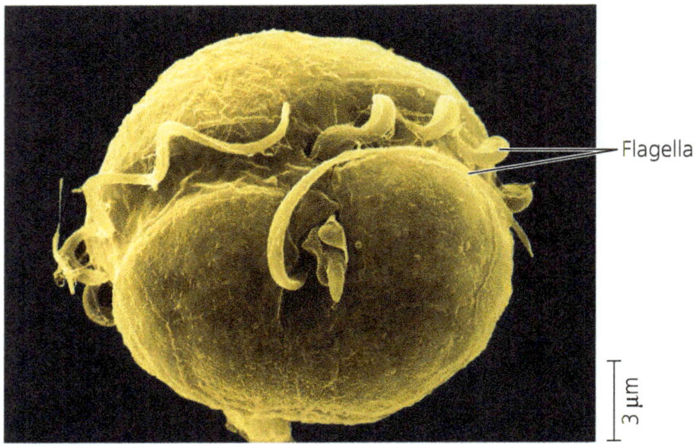

▲ **Figure 28.9 *Pfiesteria shumwayae,* a dinoflagellate.** Beating of the spiral flagellum, which lies in a groove that encircles the cell, makes this alveolate spin (colorized SEM).

Episodes of explosive population growth, or *blooms*, in dinoflagellates sometimes cause a phenomenon called "red tide." The blooms make coastal waters appear brownish red or pink because of the presence of carotenoids, the most common pigments in dinoflagellate plastids. Toxins produced by certain dinoflagellates (such as *Karenia brevis*, which inhabits the Gulf of Mexico) have caused massive kills of invertebrates and fishes. Humans who eat molluscs that have accumulated the toxins are affected as well, sometimes fatally.

Apicomplexans

Nearly all **apicomplexans** are parasites of animals, and some cause serious human diseases. The parasites spread through their host as tiny infectious cells called *sporozoites*. Apicomplexans are so named because one end (the *apex*) of the sporozoite cell contains a *complex* of organelles specialized for penetrating host cells and tissues. Although apicomplexans are not photosynthetic, recent data show that they retain a modified plastid (apicoplast), most likely of red algal origin.

Most apicomplexans have intricate life cycles with both sexual and asexual stages. Those life cycles often require two or more host species for completion. For example, *Plasmodium*, the parasite that causes malaria, lives in both mosquitoes and humans (**Figure 28.10**).

Historically, malaria has rivaled tuberculosis as the leading cause of human death by infectious disease. The incidence of malaria was greatly diminished in the 1960s by insecticides that reduced carrier populations of *Anopheles* mosquitoes and by drugs that killed *Plasmodium* in humans. But the emergence of resistant varieties of both *Anopheles* and *Plasmodium* has led

▼ **Figure 28.10 The two-host life cycle of *Plasmodium*, the apicomplexan that causes malaria.**

? *Are morphological differences between sporozoites, merozoites, and gametocytes caused by different genomes or by differences in gene expression? Explain.*

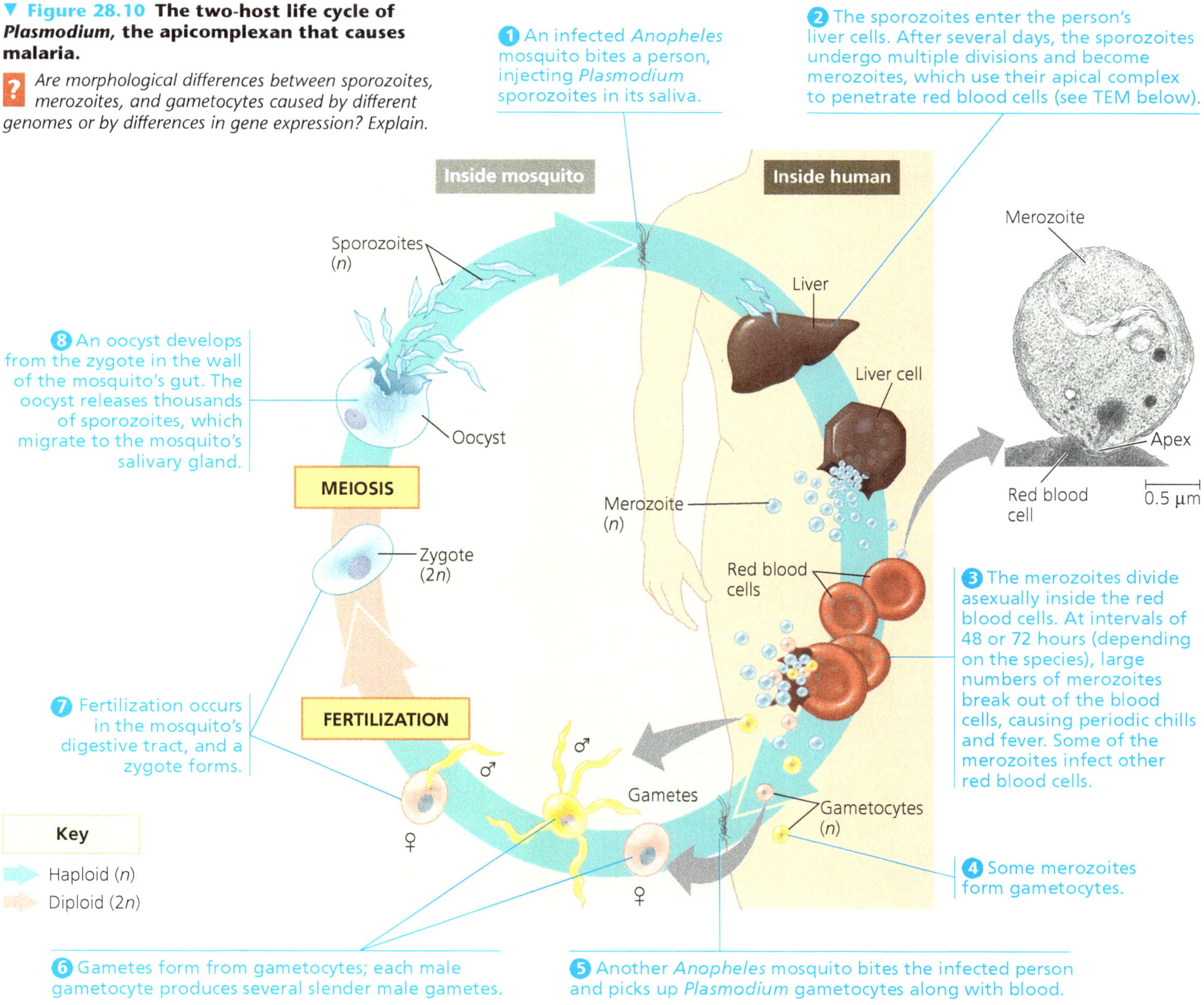

to a resurgence of malaria. About 250 million people in the tropics are currently infected, and 900,000 die each year.

The search for malarial vaccines has been hampered by the fact that *Plasmodium* lives mainly inside cells, hidden from the host's immune system. And, like trypanosomes, *Plasmodium* continually changes its surface proteins. The urgent need for treatments has inspired the sequencing of several *Plasmodium* genomes. Furthermore, researchers have now tracked the expression of most of the parasite's genes at numerous points in its life cycle. This research could help identify vaccine targets. Drugs that target the apicoplast are also in development. This approach may be effective because the apicoplast, derived by secondary endosymbiosis from a prokaryote, has metabolic pathways different from those in humans.

Ciliates

Ciliates are a large, varied group of protists named for their use of cilia to move and feed **(Figure 28.11a)**. The cilia may

▼ Figure 28.11 Structure and function in the ciliate *Paramecium caudatum*.

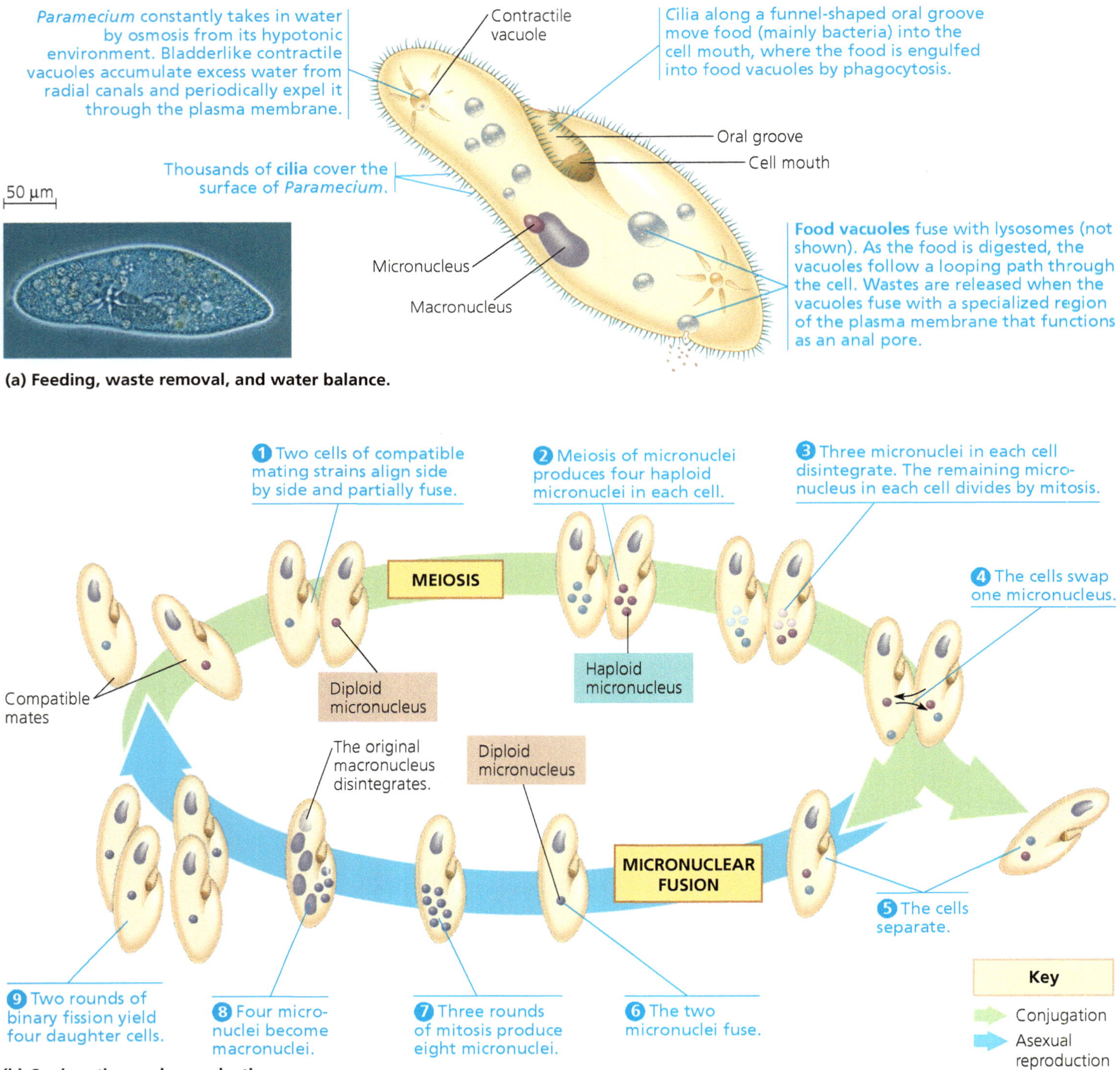

(b) Conjugation and reproduction.

completely cover the cell surface or may be clustered in a few rows or tufts. In certain species, rows of tightly packed cilia function collectively in locomotion. Other ciliates scurry about on leg-like structures constructed from many cilia bonded together.

A distinctive feature of ciliates is the presence of two types of nuclei: tiny micronuclei and large macronuclei. A cell has one or more nuclei of each type. Genetic variation results from **conjugation**, a sexual process in which two individuals exchange haploid micronuclei but do not reproduce (**Figure 28.11b**). Ciliates generally reproduce asexually by binary fission, during which the existing macronucleus disintegrates and a new one is formed from the cell's micronuclei. Each macronucleus typically contains multiple copies of the ciliate's genome. Genes in the macronucleus control the everyday functions of the cell, such as feeding, waste removal, and maintaining water balance (see Figure 28.11a).

Stramenopiles

Another major subgroup of the chromalveolates is the **stramenopiles**, protists that include some of the most important photosynthetic organisms on the planet, as well as several clades of heterotrophs. Their name (from the Latin *stramen*, straw, and *pilos*, hair) refers to their characteristic flagellum, which has numerous fine, hairlike projections. In most stramenopiles, this "hairy" flagellum is paired with a shorter "smooth" (nonhairy) flagellum (**Figure 28.12**). Here we'll focus on four groups of stramenopiles: diatoms, golden algae, brown algae, and oomycetes.

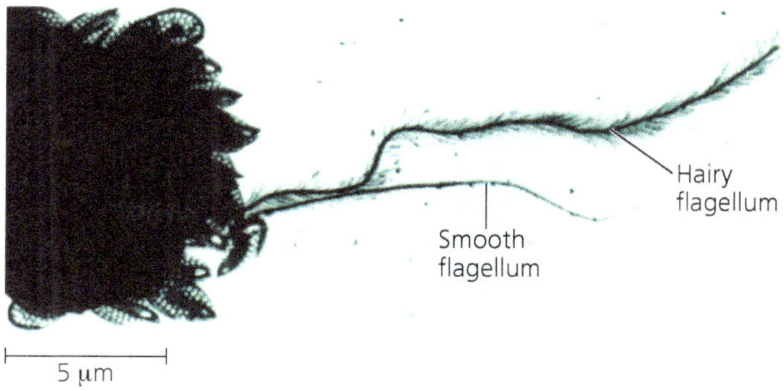

▲ **Figure 28.12 Stramenopile flagella.** Most stramenopiles, such as *Synura petersenii*, have two flagella: one covered with fine, stiff hairs and a shorter one that is smooth.

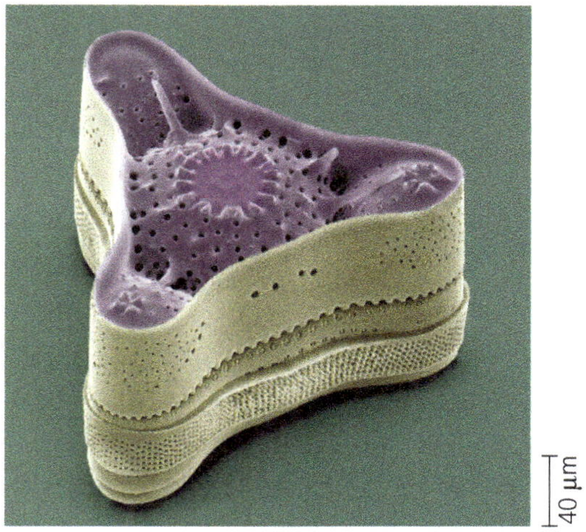

▲ **Figure 28.13 The diatom *Triceratium morlandii*** (colorized SEM).

Diatoms

Diatoms are unicellular algae that have a unique glass-like wall made of hydrated silica (silicon dioxide) embedded in an organic matrix. The wall consists of two parts that overlap like a shoe box and its lid. These walls provide effective protection from the crushing jaws of predators: Live diatoms can withstand pressures as great as 1.4 million kg/m^2, equal to the pressure under each leg of a table supporting an elephant! Much of the diatoms' strength comes from the delicate lacework of holes and grooves in their walls (**Figure 28.13**); if the walls were smooth, it would take 60% less force to crush them.

With an estimated 100,000 living species, diatoms are a highly diverse group of protists (see Figure 28.3). They are a major component of phytoplankton both in the ocean and in lakes: One bucket of water scooped from the surface of the sea may contain millions of these microscopic algae. The abundance of diatoms in the past is also evident in the fossil record, where massive accumulations of fossilized diatom walls are major constituents of sediments known as *diatomaceous earth*. These sediments are mined for their quality as a filtering medium and for many other uses.

You might expect that since diatoms are so widespread and abundant, their photosynthetic activity would affect global carbon dioxide levels, and this is indeed the case. Diatoms have this effect in part because of the chain of events that follows their rapid population growth (a bloom) when ample nutrients are available. Typically, diatoms are eaten by a variety of protists and invertebrates, but during a bloom, many escape this fate. When these uneaten diatoms die, their bodies sink to the ocean floor. Diatoms that sink to the ocean floor are not very likely to be broken down by bacteria and other decomposers. Hence, the carbon in their bodies remains there, rather than being released as carbon dioxide as the decomposers respire. The overall effect of these events is that carbon dioxide absorbed by diatoms during photosynthesis is transported, or "pumped," to the ocean floor. With an eye toward reducing global warming by lowering atmospheric carbon dioxide levels, some scientists advocate promoting diatom blooms by fertilizing the ocean with essential nutrients such as iron. Other scientists question this strategy, noting that small-scale tests of this idea have yielded mixed results and that it is difficult to predict the effects of large-scale manipulations of ecological communities.

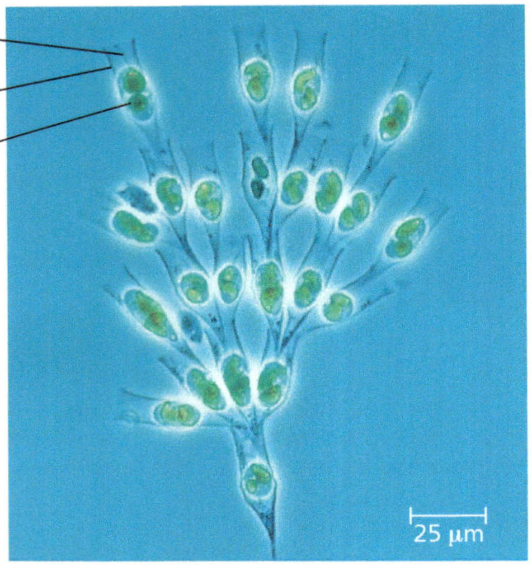

▲ **Figure 28.14** ***Dinobryon*, a colonial golden alga found in fresh water** (LM).

▲ **Figure 28.15 Seaweeds: adapted to life at the ocean's margins.** The sea palm (*Postelsia*) lives on rocks along the coast of the northwestern United States and western Canada. The thallus of this brown alga is well adapted to maintaining a firm foothold despite the crashing surf.

Golden Algae

The characteristic color of **golden algae** results from their yellow and brown carotenoids. The cells of golden algae are typically biflagellated, with both flagella attached near one end of the cell.

Many golden algae are components of freshwater and marine plankton. While all golden algae are photosynthetic, some species are mixotrophic. These mixotrophs can absorb dissolved organic compounds or ingest food particles, including living cells, by phagocytosis. Most species are unicellular, but some, such as those in the freshwater genus *Dinobryon*, are colonial **(Figure 28.14)**. If environmental conditions deteriorate, many species form protective cysts that can survive for decades.

Brown Algae

The largest and most complex algae are **brown algae**. All are multicellular, and most are marine. Brown algae are especially common along temperate coasts, where the water is cool. They owe their characteristic brown or olive color to the carotenoids in their plastids.

Many of the species commonly called "seaweeds" are brown algae. (Some large, multicellular species of red and green algae are also referred to as seaweeds. We will examine them later in this chapter.) Brown algae include species that have the most complex multicellular anatomy of all algae; some even have specialized tissues and organs that resemble those in plants. But morphological and DNA evidence indicates that the similarities evolved independently in the algal and plant lineages and are thus analogous, not homologous.

The term **thallus** (plural, *thalli*; from the Greek *thallos*, sprout) refers to an algal body that is plantlike. Unlike the body of a plant, however, a thallus lacks true roots, stems, and leaves. A typical thallus consists of a rootlike **holdfast**, which anchors the alga, and a stemlike **stipe**, which supports leaflike **blades** **(Figure 28.15)**. The blades provide most of the alga's photosynthetic surface. Some brown algae are equipped with gas-filled, bubble-shaped floats, which help keep the blades up near the water surface. Beyond the intertidal zone in deeper waters live giant seaweeds known as kelps. The stipes of these brown algae may be as long as 60 m, more than half the length of a football field.

Brown algae that inhabit the intertidal zone must cope with water churned by waves and wind, along with low tides that expose the algae to the drying atmosphere and intense rays of the sun. Unique adaptations enable these seaweeds to survive. For example, their cell walls are composed of cellulose and gel-forming polysaccharides that help cushion the thalli from waves and reduce drying when the algae are exposed.

Brown algae are important commodities for humans. Some species are eaten, such as *Laminaria* (Japanese "kombu"), which is used in soups. In addition, the gel-forming substance in the cell walls of brown algae, called algin, is used to thicken many processed foods, including pudding and salad dressing.

Alternation of Generations

A variety of life cycles have evolved among the multicellular algae. The most complex life cycles include an **alternation of generations**, the alternation of multicellular haploid and diploid forms. Although haploid and diploid conditions alternate in *all* sexual life cycles—human gametes, for example, are haploid—the term *alternation of generations* applies only to life cycles in which both haploid and diploid stages are multicellular. As you will read in Chapter 29, alternation of generations also evolved in plants.

The complex life cycle of the brown alga *Laminaria* provides an example of alternation of generations (**Figure 28.16**). The diploid individual is called the *sporophyte* because it produces spores. The spores are haploid and move by means of flagella; they are called zoospores. The zoospores develop into haploid, multicellular male and female *gametophytes*, which produce gametes. The union of two gametes (fertilization, or syngamy) results in a diploid zygote, which matures and gives rise to a new multicellular sporophyte.

In *Laminaria*, the two generations are **heteromorphic**, meaning that the sporophytes and gametophytes are structurally different. Other algal life cycles have an alternation of **isomorphic** generations, in which the sporophytes and gametophytes look similar to each other, although they differ in chromosome number.

Oomycetes (Water Molds and Their Relatives)

Oomycetes include the water molds, the white rusts, and the downy mildews. Based on their morphology, these organisms were previously classified as fungi (in fact, *oomycete* means

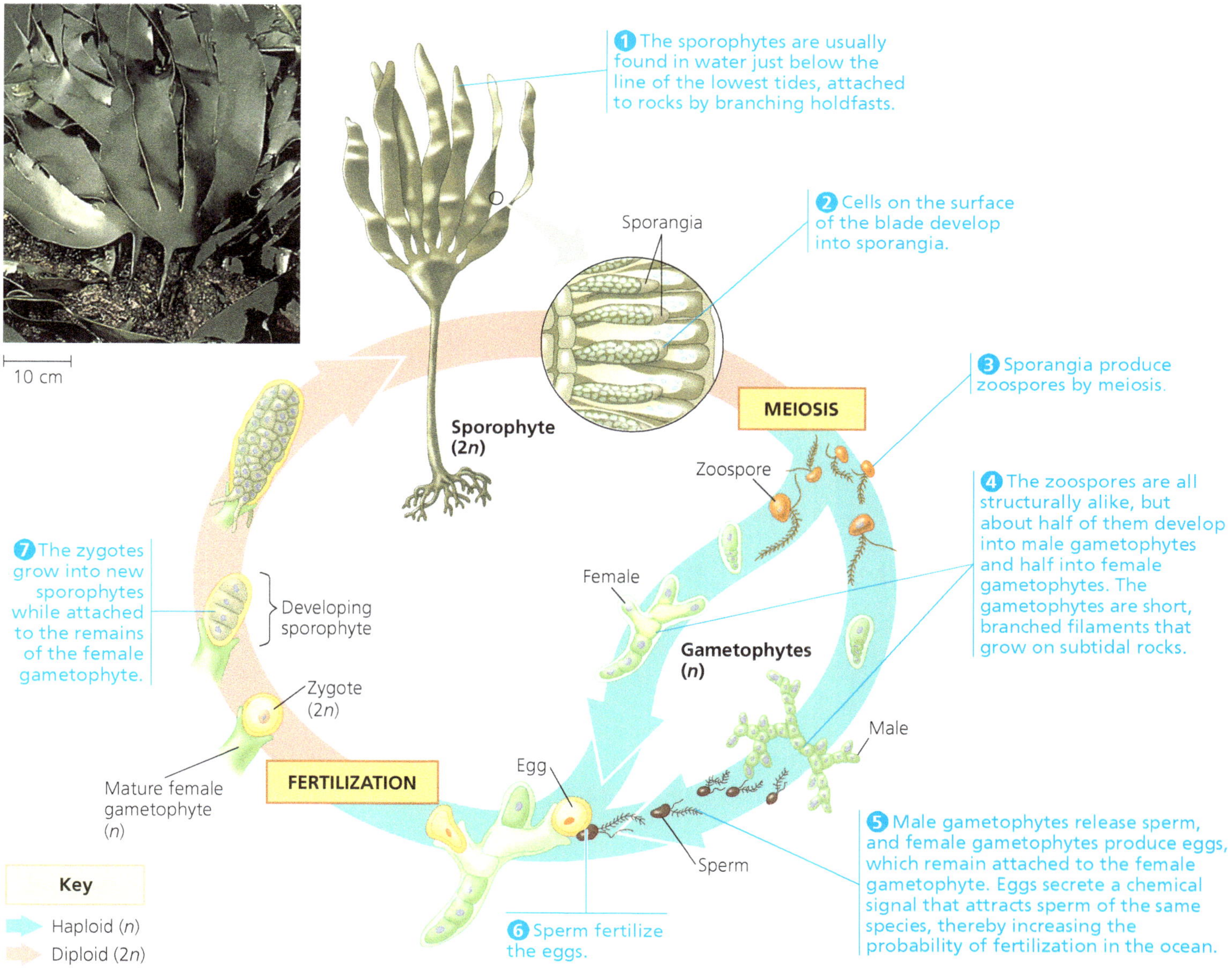

▲ **Figure 28.16 The life cycle of the brown alga *Laminaria*: an example of alternation of generations.**

? *Are the sperm shown in 5 genetically identical to one another? Are they genetically identical to the egg in 6? Explain.*

"egg fungus"). For example, many oomycetes have multinucleate filaments (hyphae) that resemble fungal hyphae **(Figure 28.17)**. However, there are key differences between oomycetes and fungi. Among the differences, oomycetes typically have cell walls made of cellulose, whereas the walls of fungi consist mainly of another polysaccharide, chitin. Data from molecular systematics have confirmed that oomycetes are not closely related to fungi. Their superficial similarity is a case of convergent evolution. In both oomycetes and fungi, the high surface-to-volume ratio of filamentous structures enhances the uptake of nutrients from the environment.

Although oomycetes descended from plastid-bearing ancestors, they no longer have plastids and do not perform photosynthesis. Instead, they typically acquire nutrients as decomposers or parasites. Most water molds are decomposers that grow as cottony masses on dead algae and animals, mainly in freshwater habitats (see Figure 28.17). White rusts and downy mildews generally live on land as plant parasites.

The ecological impact of oomycetes can be significant. For example, the oomycete *Phytophthora infestans* causes potato late blight, which turns the stalk and stem of potato plants to black slime. Late blight contributed to the devastating Irish famine of the 19th century, in which a million people died and at least that many were forced to leave Ireland. The disease remains a major problem today, causing crop losses of typically 15% in North America and as high as 70% in some parts of Russia where farmers cannot afford pesticides.

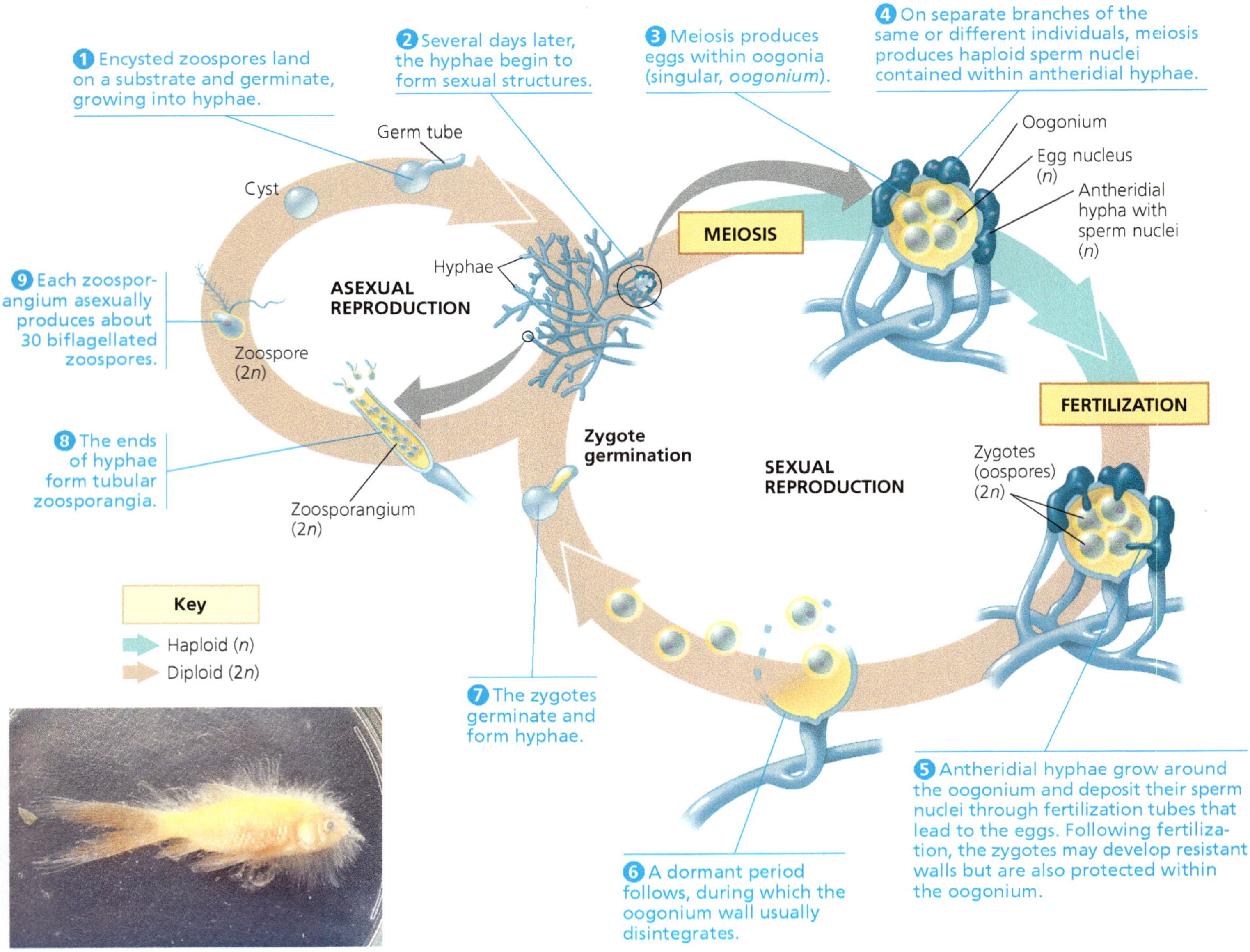

▲ **Figure 28.17 The life cycle of a water mold.** Water molds help decompose dead insects, fishes, and other animals in fresh water. (Note the hyphal mass growing on the goldfish in the photograph.)

To understand this pathogen better, molecular biologists have isolated DNA from a specimen of *P. infestans* preserved from the Irish potato blight of the 1840s. Genetic studies show that in recent decades, this oomycete has acquired genes that make it more aggressive and more resistant to pesticides. Scientists are looking within the genomes of both *Phytophthora* and potatoes to identify new weapons against the disease. Already, researchers have transferred genes from a blight-resistant strain of wild potato into domestic potatoes to produce a resistant crop strain.

CONCEPT CHECK 28.3

1. Summarize the evidence for and against the hypothesis that the species currently classified as chromalveolates are members of a single clade.
2. **WHAT IF?** Would you expect the plastid DNA of photosynthetic dinoflagellates, diatoms, and golden algae to be more similar to the nuclear DNA of plants (domain Eukarya) or to the chromosomal DNA of cyanobacteria (domain Bacteria)? Explain.
3. **MAKE CONNECTIONS** Which of the three life cycles in Figure 13.6 (p. 252) exhibits alternation of generations? How does it differ from the other two?

For suggested answers, see Appendix A.

CONCEPT 28.4

Rhizarians are a diverse group of protists defined by DNA similarities

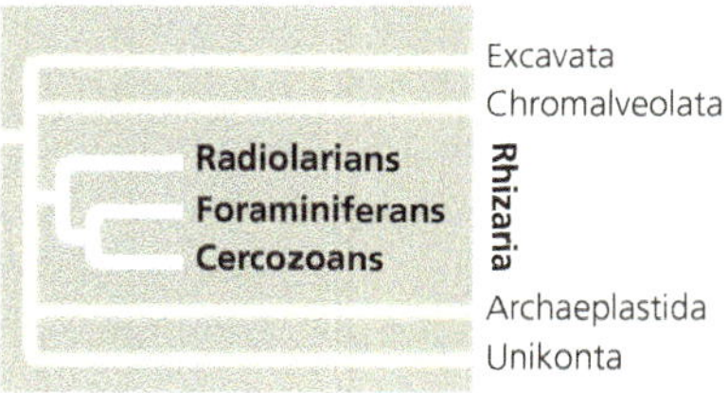

The clade **Rhizaria** has recently been proposed based on results from molecular systematics. Although its members vary greatly in morphology, DNA evidence suggests that rhizarians are a monophyletic group. According to some recent phylogenetic studies, Rhizaria should be nested within Chromalveolata, as discussed in Figure 28.3.

Many species in Rhizaria are among the organisms referred to as amoebas. **Amoebas** were formerly defined as protists that move and feed by means of **pseudopodia**, extensions that may bulge from almost anywhere on the cell surface. An amoeba moves by extending a pseudopodium and anchoring the tip; more cytoplasm then streams into the pseudopodium. However, based on molecular systematics, it is now clear that amoebas do not constitute a monophyletic group but are dispersed across many distantly related eukaryotic taxa. Most of those that belong to the clade Rhizaria are distinguished morphologically from other amoebas by having threadlike pseudopodia.

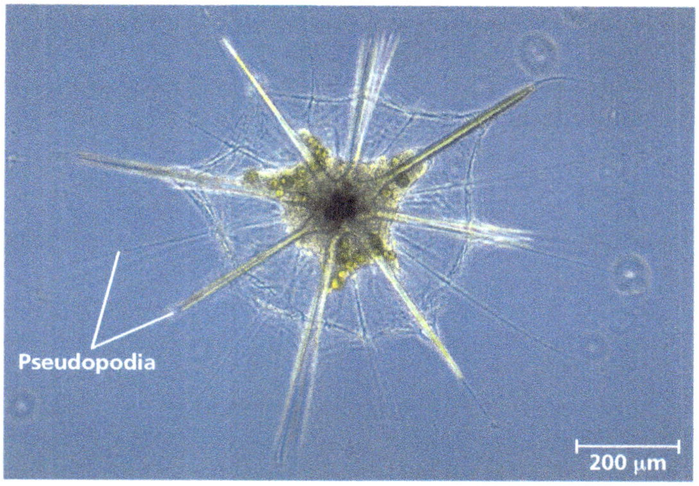

▲ **Figure 28.18 A radiolarian.** Numerous threadlike pseudopodia radiate from the central body of this radiolarian, which is found in the Red Sea (LM).

Rhizarians include three groups that we'll examine here: radiolarians, forams, and cercozoans.

Radiolarians

The protists called **radiolarians** have delicate, intricately symmetrical internal skeletons that are generally made of silica. The pseudopodia of these mostly marine protists radiate from the central body (Figure 28.18) and are reinforced by bundles of microtubules. The microtubules are covered by a thin layer of cytoplasm, which engulfs smaller microorganisms that become attached to the pseudopodia. Cytoplasmic streaming then carries the captured prey into the main part of the cell. After radiolarians die, their skeletons settle to the seafloor, where they have accumulated as an ooze that is hundreds of meters thick in some locations.

Forams

The protists called **foraminiferans** (from the Latin *foramen*, little hole, and *ferre*, to bear), or **forams**, are named for their porous shells, called **tests** (see Figure 28.3). Foram tests consist of a single piece of organic material hardened with calcium carbonate. The pseudopodia that extend through the pores function in swimming, test formation, and feeding. Many forams also derive nourishment from the photosynthesis of symbiotic algae that live within the tests.

Forams are found in both the ocean and fresh water. Most species live in sand or attach themselves to rocks or algae, but some are abundant in plankton. The largest forams, though single-celled, grow to a diameter of several centimeters.

Ninety percent of all identified species of forams are known from fossils. Along with the calcium-containing remains of other protists, the fossilized tests of forams are part of marine sediments, including sedimentary rocks that are now land formations. Foram fossils are excellent markers for correlating the ages of sedimentary rocks in different parts of the world.

Cercozoans

First identified in molecular phylogenies, the **cercozoans** form a large group that contains most of the amoeboid and flagellated protists that feed with threadlike pseudopodia. Cercozoan protists are common in marine, freshwater, and soil ecosystems.

Most cercozoans are heterotrophs. Many are parasites of plants, animals, or other protists; many others are predators. The predators include the most important consumers of bacteria in aquatic and soil ecosystems, along with species that eat other protists, fungi, and even small animals. One small group of cercozoans, the chlorarachniophytes (mentioned earlier in the discussion of secondary endosymbiosis), are mixotrophic: These organisms ingest smaller protists and bacteria as well as perform photosynthesis. At least one other cercozoan, *Paulinella chromatophora*, is an autotroph, deriving its energy from light and its carbon from carbon dioxide. This species has a distinctive sausage-shaped internal structure where photosynthesis is performed **(Figure 28.19)**. Genetic and morphological analyses indicate that these structures were derived from a cyanobacterium, although not the same cyanobacterium from which all other plastids were derived. As such, *Paulinella* appears to represent an intriguing additional evolutionary example of a eukaryotic lineage that obtained its photosynthetic apparatus directly from a cyanobacterium.

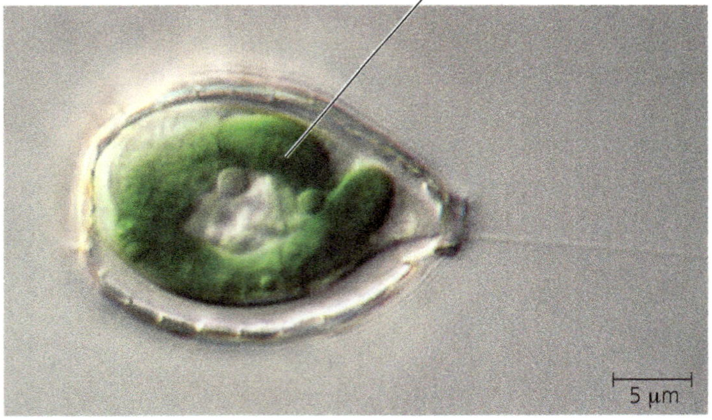

▲ **Figure 28.19 A second case of primary endosymbiosis?** The cercozoan *Paulinella* conducts photosynthesis in a unique structure called a chromatophore (LM). Chromatophores are surrounded by a membrane with a peptidoglycan layer, suggesting that they are derived from a bacterium. DNA evidence indicates that chromatophores are derived from a different cyanobacterium than that from which other plastids are derived.

CONCEPT CHECK 28.4

1. Explain why forams have such a well-preserved fossil record.
2. **WHAT IF?** Suppose DNA evidence suggests that a newly discovered amoeba is in the clade Rhizaria, yet its morphology is more similar to amoebas in other eukaryotic groups. Suggest an explanation that could account for these conflicting observations.
3. **MAKE CONNECTIONS** Review Figures 9.2 (p. 163) and 10.6 (p. 188), and then summarize how CO_2 and O_2 are both used and produced by chlorarachniophytes and other aerobic algae.

For suggested answers, see Appendix A.

CONCEPT 28.5

Red algae and green algae are the closest relatives of land plants

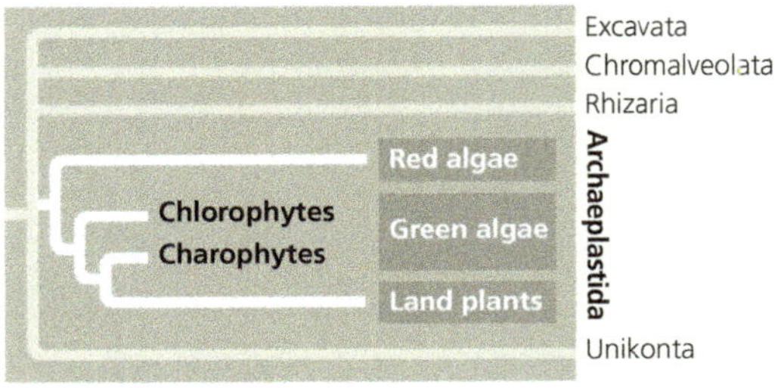

As we described in Chapter 25, molecular systematics and studies of cell structure support the following scenario: More than a billion years ago, a heterotrophic protist acquired a cyanobacterial endosymbiont, and the photosynthetic descendants of this ancient protist evolved into red algae and green algae. At least 475 million years ago, the lineage that produced green algae gave rise to land plants. Together, red algae, green algae, and land plants make up the fourth eukaryotic supergroup, which is called **Archaeplastida**. Archaeplastida is a monophyletic group that descended from the ancient protist that engulfed a cyanobacterium. We will examine land plants in Chapters 29 and 30; here we will look at the diversity of their closest algal relatives, red algae and green algae.

Red Algae

Many of the 6,000 known species of **red algae** (rhodophytes, from the Greek *rhodos*, red) are reddish, owing to a photosynthetic accessory pigment called phycoerythrin, which masks the green of chlorophyll **(Figure 28.20)**. However, species adapted to more shallow water have less phycoerythrin. As a result, red algal species may be greenish red in very shallow

► ***Bonnemaisonia hamifera.*** This red alga has a filamentous form.

◄ **Dulse (*Palmaria palmata*).** This edible species has a "leafy" form.

▼ **Nori.** The red alga *Porphyra* is the source of a traditional Japanese food.

▲ **Figure 28.20 Red algae.**

water, bright red at moderate depths, and almost black in deep water. Some species lack pigmentation altogether and function heterotrophically as parasites on other red algae.

Red algae are the most abundant large algae in the warm coastal waters of tropical oceans. Their accessory pigments, including phycoerythrin, allow them to absorb blue and green light, which penetrate relatively far into the water. A species of red alga has been discovered near the Bahamas at a depth of more than 260 m. There are also a small number of freshwater and terrestrial species.

Most red algae are multicellular. Although none are as big as the giant brown kelps, the largest multicellular red algae are included in the informal designation "seaweeds." You may have eaten one of these multicellular red algae, *Porphyra* (Japanese "nori"), as crispy sheets or as a wrap for sushi (see Figure 28.20). Red algae have especially diverse life cycles, and alternation of generations is common. But unlike other algae, they have no flagellated stages in their life cycle and depend on water currents to bring gametes together for fertilization.

Green Algae

The grass-green chloroplasts of **green algae** have a structure and pigment composition much like the chloroplasts of land plants. Molecular systematics and cellular morphology leave little doubt that green algae and land plants are closely related. In fact, some systematists now advocate including green algae in an expanded "plant" kingdom, Viridiplantae (from the Latin *viridis*, green). Phylogenetically, this change makes sense, since otherwise the green algae are a paraphyletic group.

Green algae are divided into two main groups, the charophytes and the chlorophytes. The charophytes are the algae most closely related to land plants, and so we will discuss them along with plants in Chapter 29.

The second group, the chlorophytes (from the Greek *chloros*, green), includes more than 7,000 species. Most live in fresh water, but there are also many marine and some terrestrial species. The simplest chlorophytes are unicellular organisms such as *Chlamydomonas*, which resemble gametes or zoospores of more complex chlorophytes. Various species of unicellular chlorophytes live in aquatic habitats as phytoplankton or inhabit damp soil. Some live symbiotically within other eukaryotes, contributing part of their photosynthetic output to the food supply of their hosts. Some chlorophytes have even adapted to one of the last habitats you might expect to find them: snow. These chlorophytes carry out photosynthesis despite subfreezing temperatures and intense visible and ultraviolet radiation. They are protected by the snow itself, which acts as a shield, and by radiation-blocking compounds in their cytoplasm. Other chlorophytes contain similar protective compounds in their cell wall or in a durable coat that surrounds the zygote.

Larger size and greater complexity evolved in chlorophytes by three different mechanisms:

1. The formation of colonies of individual cells, as seen in *Volvox* (see Figure 28.3) and in filamentous forms that contribute to the stringy masses known as pond scum

2. The formation of true multicellular bodies by cell division and differentiation, as in *Ulva* **(Figure 28.21a)**
3. The repeated division of nuclei with no cytoplasmic division, as in *Caulerpa* **(Figure 28.21b)**

Most chlorophytes have complex life cycles, with both sexual and asexual reproductive stages. Nearly all species of chlorophytes reproduce sexually by means of biflagellated gametes that have cup-shaped chloroplasts **(Figure 28.22)**. Alternation of generations has evolved in some chlorophytes, including *Ulva*.

CONCEPT CHECK 28.5

1. Contrast red algae and brown algae.
2. Why is it accurate to say that *Ulva* is truly multicellular but *Caulerpa* is not?
3. **WHAT IF?** Suggest a possible reason why species in the green algal lineage may have been more likely to colonize land than species in the red algal lineage.

For suggested answers, see Appendix A.

(a) ***Ulva*, or sea lettuce.** This edible chlorophyte has a multicellular thallus differentiated into leaflike blades. Its rootlike holdfast anchors the alga.

(b) ***Caulerpa*, an intertidal chlorophyte.** The branched filaments lack cross-walls and thus are multinucleate. In effect, the thallus is one huge "supercell."

▲ **Figure 28.21 Multicellular chlorophytes.**

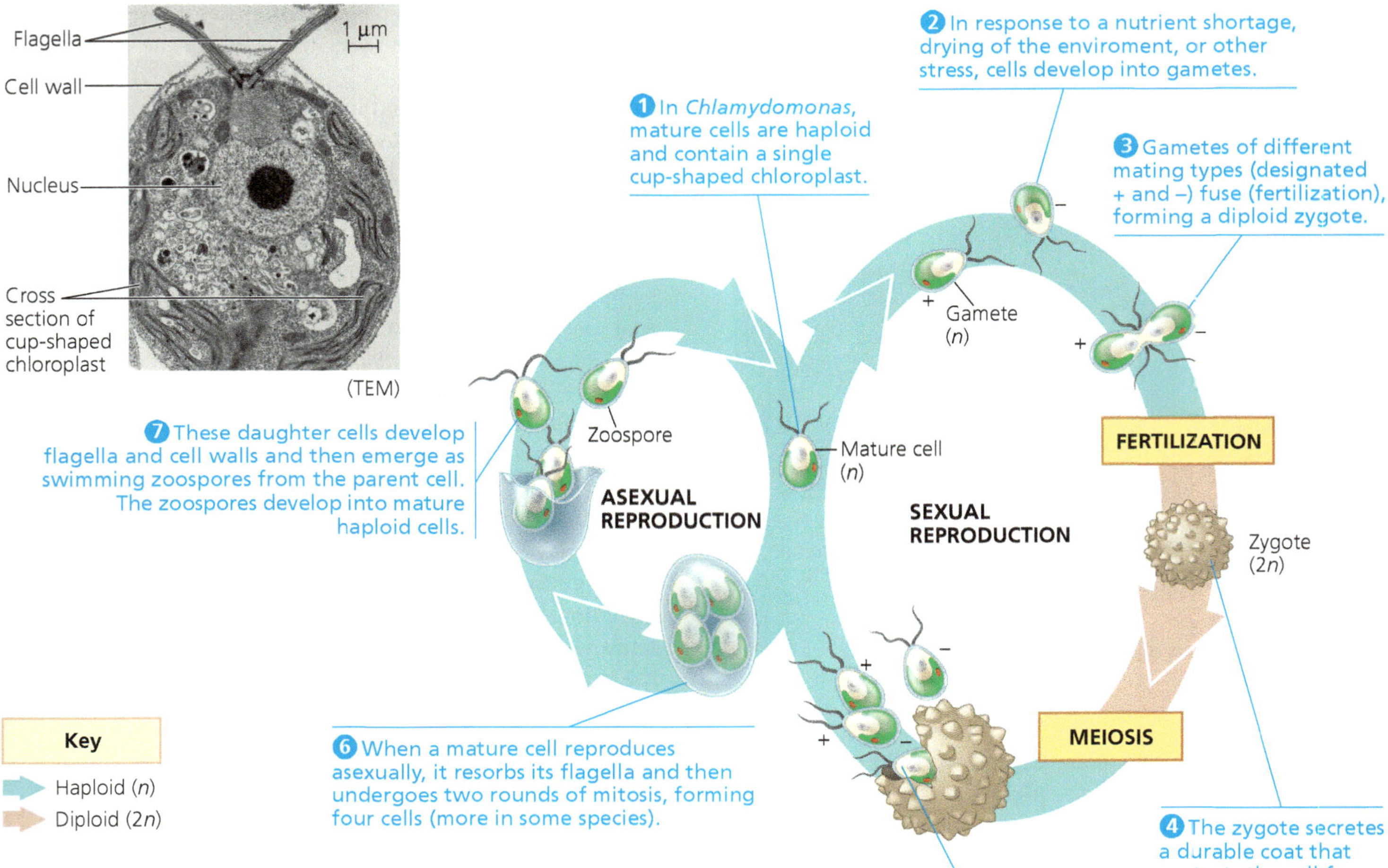

▲ **Figure 28.22 The life cycle of *Chlamydomonas*, a unicellular chlorophyte.**

DRAW IT *Circle the stage(s) in the diagram in which clones are formed, producing additional new daughter cells that are genetically identical to the parent cell(s).*

CONCEPT **28.6**

Unikonts include protists that are closely related to fungi and animals

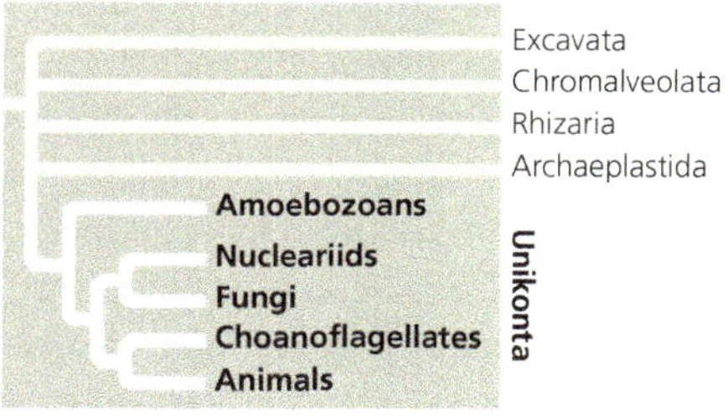

Unikonta is a recently proposed, extremely diverse supergroup of eukaryotes that includes animals, fungi, and some protists. There are two major clades of unikonts, the amoebozoans and the opisthokonts (animals, fungi, and closely related protist groups). Each of these two major clades is strongly supported by molecular systematics. The close relationship between amoebozoans and opisthokonts is more controversial. Support for this close relationship is provided by comparisons of myosin proteins and by several studies based on hundreds of genes, but not by other studies based on single genes.

Another controversy involving the unikonts concerns the root of the eukaryotic tree. Recall that the root of a phylogenetic tree anchors the tree in time: Branch points close to the root are the oldest. At present, the root of the eukaryotic tree is uncertain; thus, we do not know which group of eukaryotes was the first to diverge from other eukaryotes. Some hypotheses, such as the amitochondriate hypothesis described earlier, have been abandoned, but researchers have yet to agree on an alternative. If the root of the eukaryotic tree were known, scientists could infer characteristics of the common ancestor of all eukaryotes. Such information could help resolve some of the current debates about the five supergroups of eukaryotes.

In trying to determine the root of the eukaryotic tree, researchers have based their phylogenies on different sets of genes, producing conflicting results. Researchers at Oxford University have described a different approach, results from which support a radical new hypothesis about the root of the eukaryotic tree (**Figure 28.23**). According to their hypothesis, the unikonts were the first eukaryotes to diverge from other eukaryotes. This hypothesis proposes that animals and fungi belong to an early-diverging group of eukaryotes, while protists that lack typical mitochondria (such as the diplomonads and parabasalids) diverged much later in the history of life. This idea remains controversial and will require more supporting evidence to be widely accepted.

Amoebozoans

As already mentioned, **amoebozoans** form a clade that is well supported by molecular data. This clade includes many species

▼ **Figure 28.23** INQUIRY

What is the root of the eukaryotic tree?

EXPERIMENT Responding to the difficulty in determining the root of the eukaryotic phylogenetic tree, Alexandra Stechmann and Thomas Cavalier-Smith, of Oxford University, proposed a new approach. They studied two genes, one for the enzyme dihydrofolate reductase (DHFR), the other for the enzyme thymidylate synthase (TS). Their approach took advantage of a rare evolutionary event: In some organisms, the DHFR and TS genes have fused, leading to the production of a single protein with both enzyme activities. Stechmann and Cavalier-Smith amplified (using PCR; see Figure 20.8) and sequenced DHFR and TS genes in nine species (one choanoflagellate; two amoebozoans; one euglenozoan; two chromalveolates; and three rhizarians). They combined their data with previously published data for species of bacteria, animals, plants, and fungi.

RESULTS The bacteria studied all have separate DHFR and TS genes, suggesting that this is the ancestral condition (red dot on the tree below). Other taxa with separate genes are denoted by red type. Fused genes are a derived character, found in certain members (blue type) of the supergroups Excavata, Chromalveolata, Rhizaria, and Archaeplastida:

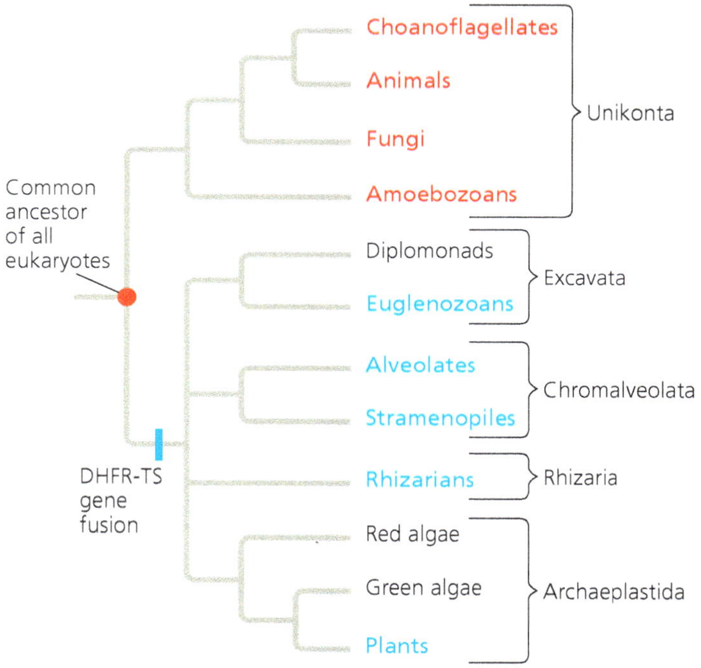

CONCLUSION These results support the hypothesis that the root of the tree is located between the unikonts and all other eukaryotes, suggesting that the unikonts were the first group of eukaryotes to diverge. Because support for this hypothesis is based on only one trait—the fusion of the DHFR and TS genes—more data are needed to evaluate its validity.

SOURCE A. Stechmann and T. Cavalier-Smith, Rooting the eukaryote tree by using a derived gene fusion, *Science* 297:89–91 (2002).

WHAT IF? Stechmann and Cavalier-Smith wrote that their conclusions are "valid only if the genes fused just once and were never secondarily split." Why is this assumption critical to their approach?

of amoebas that have lobe- or tube-shaped, rather than thread-like, pseudopodia. Amoebozoans include slime molds, gymnamoebas, and entamoebas.

Slime Molds

Slime molds, or mycetozoans (from the Latin, meaning "fungus animals"), were once thought to be fungi because, like fungi, they produce fruiting bodies that aid in spore dispersal. However, the resemblance between slime molds and fungi appears to be another example of evolutionary convergence. Molecular systematics places slime molds in Amoebozoa and suggests that they descended from unicellular ancestors. Slime molds have diverged into two main branches, plasmodial slime molds and cellular slime molds, distinguished in part by their unique life cycles.

Plasmodial Slime Molds Many **plasmodial slime molds** are brightly colored, often yellow or orange **(Figure 28.24)**. At one stage in their life cycle, they form a mass called a **plasmodium**, which may grow to a diameter of many centimeters. (Don't confuse a slime mold's plasmodium with the genus *Plasmodium*, which includes the parasitic apicomplexan that causes malaria.) Despite its size, the plasmodium is not multicellular; it is a single mass of cytoplasm that is undivided by plasma membranes and that contains many nuclei. This "supercell" is the product of mitotic nuclear divisions that are not followed by cytokinesis.

Within the plasmodium, cytoplasm streams first one way, then the other, in pulsing flows that are beautiful to watch through a microscope. This cytoplasmic streaming apparently helps distribute nutrients and oxygen. The plasmodium extends pseudopodia through moist soil, leaf mulch, or rotting

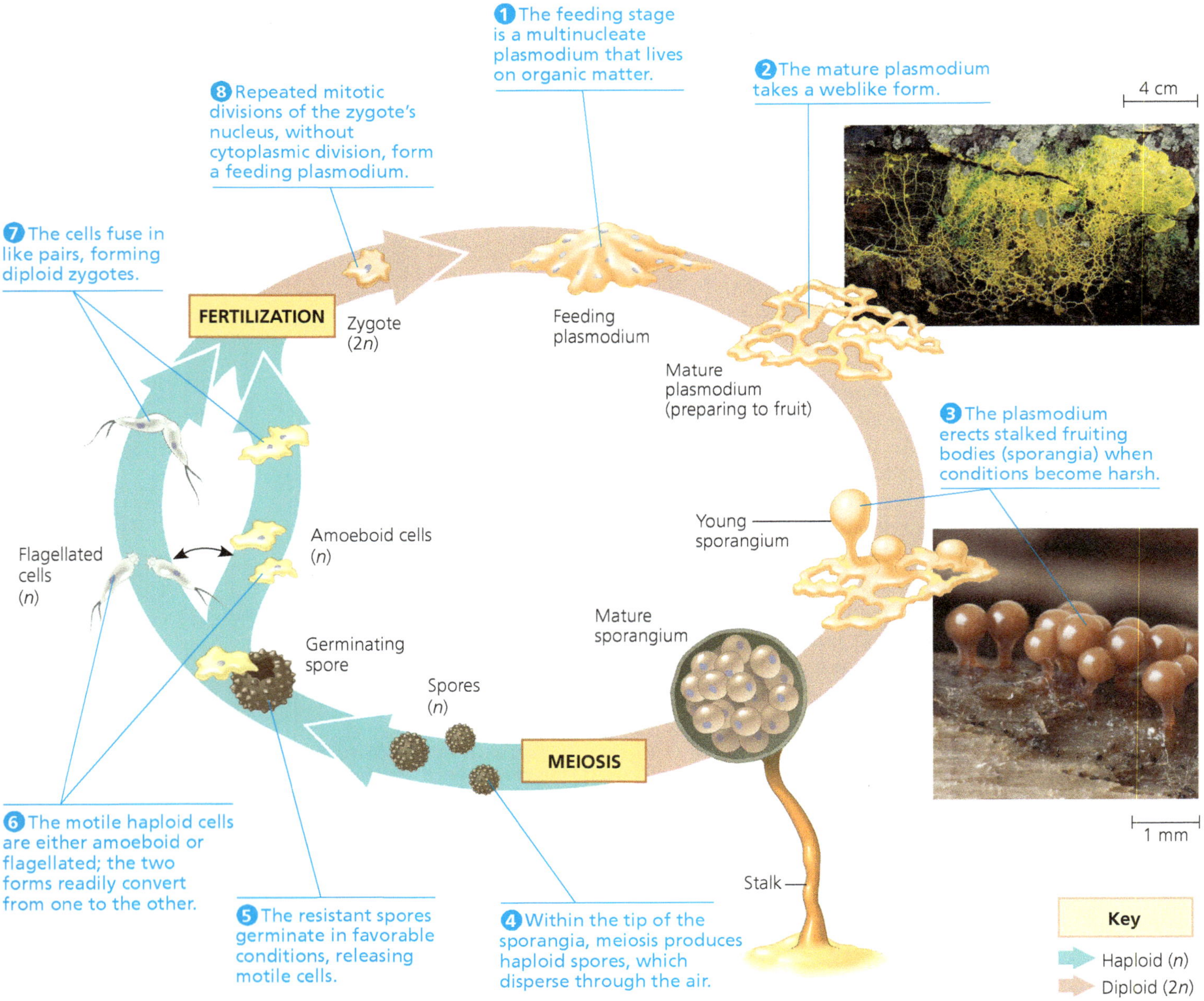

▲ **Figure 28.24 The life cycle of a plasmodial slime mold.**

logs, engulfing food particles by phagocytosis as it grows. If the habitat begins to dry up or there is no food left, the plasmodium stops growing and differentiates into fruiting bodies, which function in sexual reproduction.

Cellular Slime Molds The life cycle of the protists called **cellular slime molds** can prompt us to question what it means to be an individual organism. The feeding stage of these organisms consists of solitary cells that function individually, but when food is depleted, the cells form an aggregate that functions as a unit (**Figure 28.25**). Although this mass of cells superficially resembles a plasmodial slime mold, the cells remain separated by their individual plasma membranes. Cellular slime molds also differ from plasmodial slime molds in being haploid organisms (only the zygote is diploid) and in having fruiting bodies that function in asexual rather than sexual reproduction.

Dictyostelium discoideum, a cellular slime mold commonly found on forest floors, has become a model organism for studying the evolution of multicellularity. One line of research has focused on the slime mold's fruiting body stage. During this stage, the cells that form the stalk die as they dry out, while the spore cells at the top survive and have the potential to reproduce. Scientists have found that mutations in a single gene can turn individual *Dictyostelium* cells into "cheaters" that never become part of the stalk. Because these mutants gain a strong reproductive advantage over noncheaters, why don't all *Dictyostelium* cells cheat?

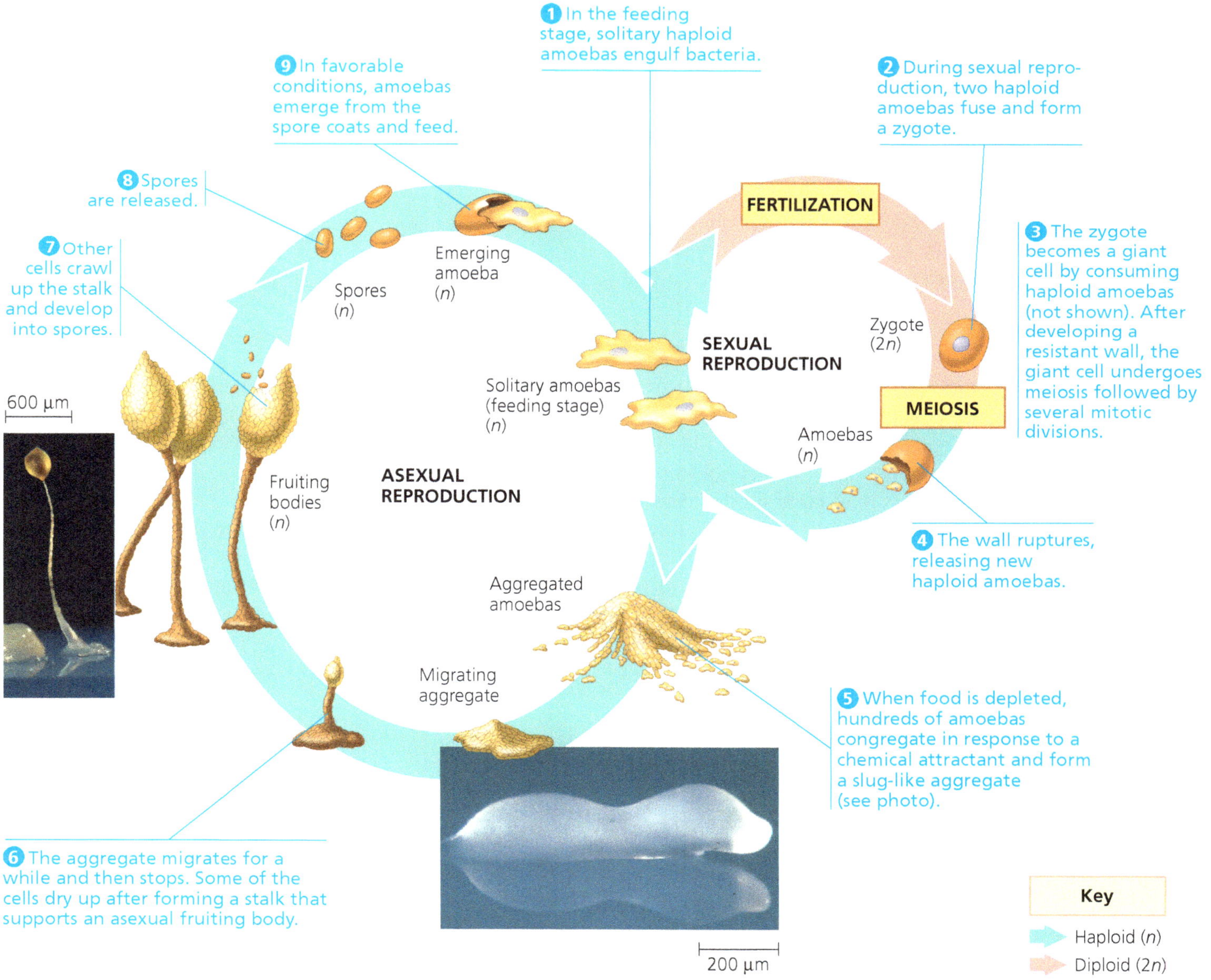

▲ **Figure 28.25 The life cycle of *Dictyostelium*, a cellular slime mold.**

Recent discoveries suggest an answer to this question. Cheating mutants lack a protein on their cell surface, and noncheating cells can recognize this difference. Noncheaters preferentially aggregate with other noncheaters, thus depriving cheaters of the opportunity to exploit them. Such a recognition system may have been important in the evolution of multicellular eukaryotes such as animals and plants.

Gymnamoebas

Gymnamoebas constitute a large and varied group of amoebozoans. These unicellular protists are ubiquitous in soil as well as freshwater and marine environments. Most are heterotrophs that actively seek and consume bacteria and other protists (see Figure 28.3). Some gymnamoebas also feed on detritus (nonliving organic matter).

Entamoebas

Whereas most amoebozoans are free-living, those that belong to the genus *Entamoeba* are parasites. They infect all classes of vertebrate animals as well as some invertebrates. Humans are host to at least six species of *Entamoeba*, but only one, *E. histolytica*, is known to be pathogenic. *E. histolytica* causes amebic dysentery and is spread via contaminated drinking water, food, or eating utensils. Responsible for up to 100,000 deaths worldwide every year, the disease is the third-leading cause of death due to eukaryotic parasites, after malaria (see Figure 28.10) and schistosomiasis (see Figure 33.11).

Opisthokonts

Opisthokonts are an extremely diverse group of eukaryotes that includes animals, fungi, and several groups of protists. We will discuss the evolutionary history of fungi and animals in Chapters 31–34. Of the opisthokont protists, we will discuss the nucleariids in Chapter 31 because they are more closely related to fungi than they are to other protists. Similarly, we will discuss choanoflagellates in Chapter 32, since they are more closely related to animals than they are to other protists. The nucleariids and choanoflagellates illustrate why scientists have abandoned the former kingdom Protista: A monophyletic group that included these single-celled eukaryotes would also have to include the multicellular animals and fungi that are closely related to them.

CONCEPT CHECK 28.6

1. Contrast the pseudopodia of amoebozoans and forams.
2. In what sense is "fungus animal" a fitting description of a slime mold? In what sense is it not fitting?
3. **WHAT IF?** If further evidence indicates that the root of the eukaryotic tree is as shown in Figure 28.23, would this evidence support, contradict, or have no bearing on the hypothesis that Excavata is monophyletic?

For suggested answers, see Appendix A.

CONCEPT 28.7

Protists play key roles in ecological communities

Most protists are aquatic, and they are found almost anywhere there is water, including moist terrestrial habitats such as damp soil and leaf litter. In oceans, ponds, and lakes, many protists are bottom-dwellers that attach to rocks and other substrates or creep through the sand and silt. Other protists are important constituents of plankton. We'll focus here on two key roles that protists play in the varied habitats in which they live: that of symbiont and that of producer.

Symbiotic Protists

Many protists form symbiotic associations with other species. For example, photosynthetic dinoflagellates are food-providing symbiotic partners of the coral polyps that build coral reefs. Coral reefs are highly diverse ecological communities. That diversity ultimately depends on corals—and on the mutualistic protist symbionts that nourish them. Corals support reef diversity by providing food to some species and habitat to many others.

Another example is the wood-digesting protists that inhabit the gut of many termite species (**Figure 28.26**). Unaided, termites cannot digest wood, and they rely on protistan or prokaryotic symbionts to do so. Termites cause over $3.5 billion in damage annually to wooden homes in the United States.

Symbiotic protists also include parasites that have compromised the economies of entire countries. Consider the malaria-causing protist *Plasmodium*: Income levels in countries hard hit by malaria are 33% lower than in similar countries free of the disease. Protists can have devastating effects on other species too. Massive fish kills have been attributed to *Pfiesteria shumwayae* (see Figure 28.9), a dinoflagellate parasite that attaches to its victims and eats their skin. Among species that parasitize plants, the oomycete protist *Phytophthora ramorum* has emerged as a major new forest pathogen. This species causes sudden oak death (SOD), a disease that has killed millions of oaks and other trees in California and Oregon (see Chapter 54).

▶ **Figure 28.26 A symbiotic protist.** This organism is a hypermastigote, a member of a group of parabasalids that live in the gut of termites and certain cockroaches and enable the hosts to digest wood (SEM).

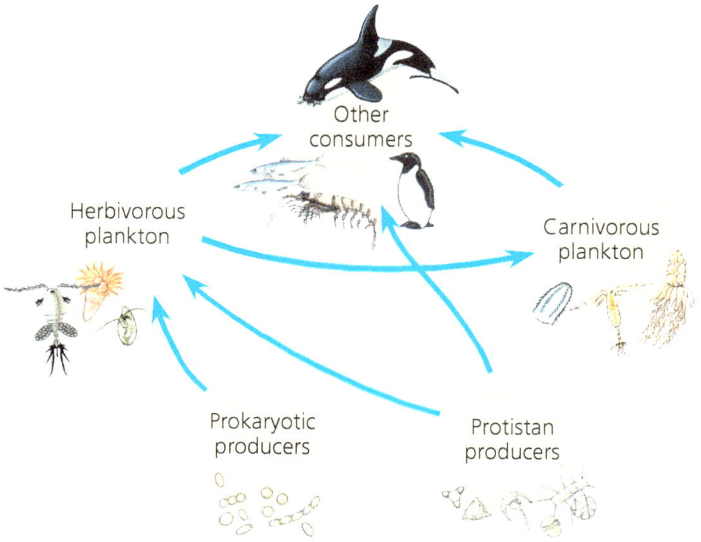

▲ **Figure 28.27 Protists: key producers in aquatic communities.** Arrows in this simplified food web lead from food sources to the organisms that eat them.

Photosynthetic Protists

Many protists are important **producers**, organisms that use energy from light (or inorganic chemicals) to convert carbon dioxide to organic compounds. Producers form the base of ecological food webs. In aquatic communities, the main producers are photosynthetic protists and prokaryotes. All other organisms in the community depend on them for food, either directly (by eating them) or indirectly (by eating an organism that ate a producer; **Figure 28.27**). Scientists estimate that roughly 30% of the world's photosynthesis is performed by diatoms, dinoflagellates, multicellular algae, and other aquatic protists. Photosynthetic prokaryotes contribute another 20%, and land plants are responsible for the remaining 50%.

Because producers form the foundation of food webs, factors that affect producers can dramatically affect their entire community. In aquatic environments, photosynthetic protists are often held in check by low concentrations of nitrogen, phosphorus, or iron. Various human actions can increase the concentrations of these elements in aquatic communities. For example, when fertilizer is applied to a field, some of the fertilizer may be washed by rainfall into a river that drains into a lake or ocean. When people add nutrients to aquatic communities in this or other ways, the abundance of photosynthetic protists can increase spectacularly. Such increases can alter the abundance of other species in the community, as we'll see in Chapter 55.

A pressing question is how global warming will affect protists and other producers. Satellite data indicate that the growth and biomass of photosynthetic protists and prokaryotes have declined in many regions as sea surface temperatures have increased **(Figure 28.28)**. If sustained, these changes would likely have far-reaching effects on marine ecosystems, fishery yields, and the global carbon cycle (see Chapter 55). Global warming can also affect producers on land, but there the base of food webs is occupied not by protists but by land plants, which we will discuss in Chapters 29 and 30.

▼ **Figure 28.28**

IMPACT

Marine Protists in a Warmer World

Photosynthetic protists are important components of marine food webs, each day converting millions of tons of carbon in CO_2 to organic molecules on which other organisms depend. How has global warming affected these key marine producers? Satellite data indicate that the growth and biomass of marine producers are negatively correlated to sea surface temperature (SST) across much of the tropical and midlatitude oceans (between the heavy black lines on the map below). In regions where SSTs have risen and growth has declined, the available nutrient supply may have been reduced by the formation of a light, warm layer of water that acts as a barrier preventing the rise, or upwelling, of cold, nutrient-rich waters from below.

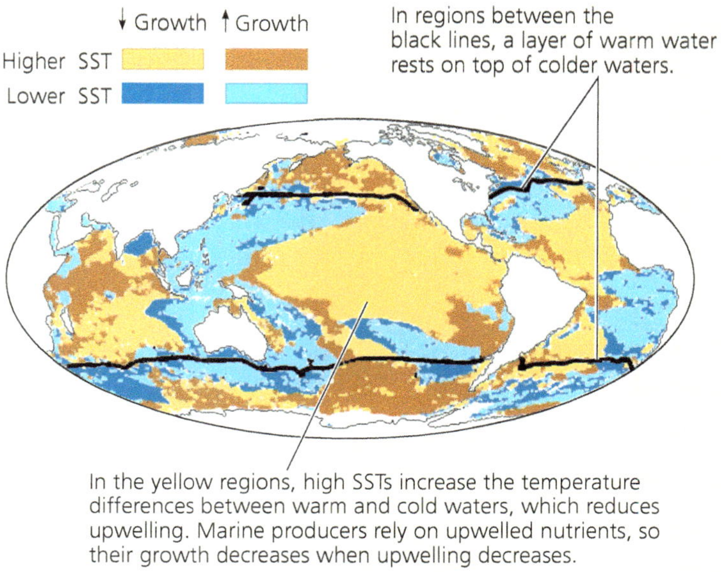

WHY IT MATTERS Major changes to marine ecosystems are expected if the growth and biomass of producers decrease as SSTs increase due to global warming. A decrease in diatom biomass, for example, would likely reduce both the amount of carbon pumped to the ocean floor and the catch of economically important fishes such as salmon and anchovies that feed on phytoplankton.

FURTHER READING M. J. Behrenfeld et al., Climate-driven trends in contemporary ocean productivity, *Nature* 444:752–755 (2006).

WHAT IF? If diatom populations continue to drop as the oceans warm, how might climate be affected in the future?

CONCEPT CHECK 28.7

1. Justify the claim that photosynthetic protists are among the biosphere's most important organisms.
2. Describe three symbioses that include protists.
3. **WHAT IF?** High water temperatures and pollution can cause corals to expel their dinoflagellate symbionts. Predict how such "coral bleaching" would affect corals and other species in the community.

For suggested answers, see Appendix A.

28 CHAPTER REVIEW

SUMMARY OF KEY CONCEPTS

CONCEPT 28.1

Most eukaryotes are single-celled organisms (pp. 575–579)

- Protists are more diverse than all other eukaryotes and are no longer classified in a single kingdom. Most are unicellular. Protists include photoautotrophs, heterotrophs, and mixotrophs. Protists are characterized by a wide diversity of life cycles.
- Mitochondria and plastids are thought to be descendants of bacteria that were engulfed by other cells and became endosymbionts. The plastid-bearing lineage eventually evolved into red algae and green algae. Other protist groups evolved from secondary endosymbiosis events in which red algae or green algae were themselves engulfed.
- In one hypothesis, eukaryotes are grouped into five supergroups, each a monophyletic clade: Excavata, Chromalveolata, Rhizaria, Archaeplastida, and Unikonta.

? *Describe similarities and differences between protists and other eukaryotes.*

Key Concept/Eukaryote Supergroup	Major Groups	Key Morphological Characteristics	Specific Examples
CONCEPT 28.2 **Excavates include protists with modified mitochondria and protists with unique flagella (pp. 580–581)** **?** *What evidence indicates that the excavates form a clade?*	**Diplomonads and parabasalids**	Modified mitochondria	*Giardia, Trichomonas* 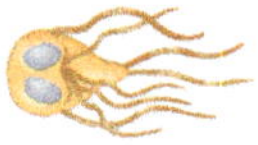
	Euglenozoans Kinetoplastids Euglenids	Spiral or crystalline rod inside flagella	*Trypanosoma, Euglena* 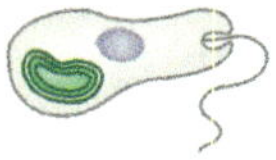
CONCEPT 28.3 **Chromalveolates may have originated by secondary endosymbiosis (pp. 582–589)** **?** *If Chromalveolata originated by secondary endosymbiosis, what can be inferred about the plastids of its members? Explain.*	**Alveolates** Dinoflagellates Apicomplexans Ciliates	Membrane-bounded sacs (alveoli) beneath plasma membrane	*Pfiesteria, Plasmodium, Paramecium* 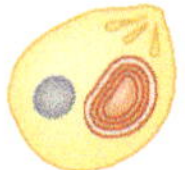
	Stramenopiles Diatoms Golden algae Brown algae Oomycetes	Hairy and smooth flagella	*Phytophthora, Laminaria* 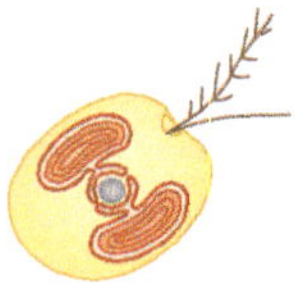
CONCEPT 28.4 **Rhizarians are a diverse group of protists defined by DNA similarities (pp. 589–590)** **?** *What are the main subgroups of rhizarians, and what unites these subgroups as a clade?*	**Radiolarians**	Amoebas with threadlike pseudopodia radiating from central body	*Hexacontium* 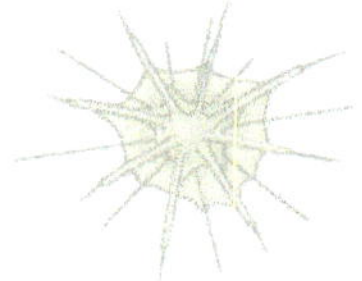
	Forams	Amoebas with threadlike pseudopodia and a porous shell	*Globigerina* 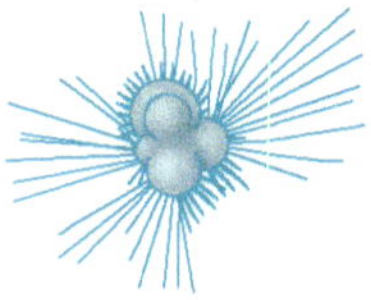
	Cercozoans	Amoebas and flagellated protists with threadlike pseudopodia	*Paulinella*
CONCEPT 28.5 **Red algae and green algae are the closest relatives of land plants (pp. 590–592) (Archaeplastida)** **?** *On what basis do some systematists place land plants in the same supergroup (Archaeplastida) as red and green algae?*	**Red algae**	Phycoerythrin (accessory pigment)	*Porphyra* 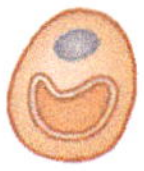
	Green algae	Plant-type chloroplasts	*Chlamydomonas, Ulva* 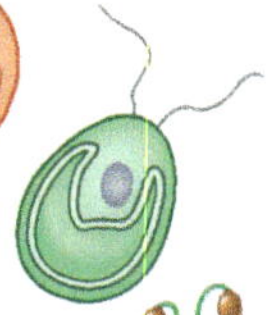
	Land plants	(See Chapters 29 and 30.)	Mosses, ferns, conifers, flowering plants 
CONCEPT 28.6 **Unikonts include protists that are closely related to fungi and animals (pp. 593–596)** **?** *Describe a key feature for each of the main protist subgroups of Unikonta.*	**Amoebozoans** Slime molds Gymnamoebas Entamoebas	Amoebas with lobe-shaped pseudopodia	*Amoeba, Entamoeba, Dictyostelium* 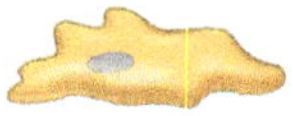
	Opisthokonts	(Highly variable; see Chapters 31–34.)	Nucleariids, choanoflagellates, animals, fungi 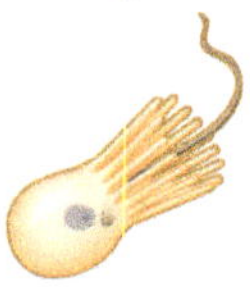

CONCEPT 28.7

Protists play key roles in ecological communities (pp. 596–597)

- Protists form a wide range of mutualistic and parasitic relationships that affect their symbiotic partners and many other members of the community.
- Photosynthetic protists are among the most important producers in aquatic communities. Because they are at the base of the food web, factors that affect photosynthetic protists affect many other species in the community.

? *Describe several protists that are ecologically important.*

TEST YOUR UNDERSTANDING

LEVEL 1: KNOWLEDGE/COMPREHENSION

1. Plastids that are surrounded by more than two membranes are evidence of
   a. evolution from mitochondria.
   b. fusion of plastids.
   c. origin of the plastids from archaea.
   d. secondary endosymbiosis.
   e. budding of the plastids from the nuclear envelope.

2. Biologists suspect that endosymbiosis gave rise to mitochondria before plastids partly because
   a. the products of photosynthesis could not be metabolized without mitochondrial enzymes.
   b. all eukaryotes have mitochondria (or their remnants), whereas many eukaryotes do not have plastids.
   c. mitochondrial DNA is less similar to prokaryotic DNA than is plastid DNA.
   d. without mitochondrial CO_2 production, photosynthesis could not occur.
   e. mitochondrial proteins are synthesized on cytosolic ribosomes, whereas plastids utilize their own ribosomes.

3. Which group is *incorrectly* paired with its description?
   a. rhizarians—morphologically diverse group defined by DNA similarities
   b. diatoms—important producers in aquatic communities
   c. red algae—acquired plastids by secondary endosymbiosis
   d. apicomplexans—parasites with intricate life cycles
   e. diplomonads—protists with modified mitochondria

4. Which protists are in the same eukaryotic supergroup as land plants?
   a. green algae
   b. dinoflagellates
   c. red algae
   d. brown algae
   e. both a and c

5. In life cycles with an alternation of generations, multicellular haploid forms alternate with
   a. unicellular haploid forms.
   b. unicellular diploid forms.
   c. multicellular haploid forms.
   d. multicellular diploid forms.
   e. multicellular polyploid forms.

LEVEL 2: APPLICATION/ANALYSIS

6. Based on the phylogenetic tree in Figure 28.3, which of the following statements is correct?
   a. The most recent common ancestor of Excavata is older than that of Chromalveolata.
   b. The most recent common ancestor of Chromalveolata is older than that of Rhizaria.
   c. The most recent common ancestor of red algae and land plants is older than that of nucleariids and fungi.
   d. The most basal (first to diverge) eukaryotic supergroup cannot be determined.
   e. Excavata is the most basal eukaryotic supergroup.

7. **EVOLUTION CONNECTION**

   DRAW IT Medical researchers seek to develop drugs that can kill or restrict the growth of human pathogens yet have few harmful effects on patients. These drugs often work by disrupting the metabolism of the pathogen or by targeting its structural features.

   Draw and label a phylogenetic tree that includes an ancestral prokaryote and the following groups of organisms: Excavata, Chromalveolata, Rhizaria, Archaeplastida, Unikonta, and, within Unikonta, amoebozoans, animals, choanoflagellates, fungi, and nucleariids. Based on this tree, hypothesize whether it would be most difficult to develop drugs to combat human pathogens that are prokaryotes, protists, animals, or fungi. (You do not need to consider the evolution of drug resistance by the pathogen.)

LEVEL 3: SYNTHESIS/EVALUATION

8. **SCIENTIFIC INQUIRY**

   Applying the "If . . . then" logic of science (see Chapter 1), what are a few of the predictions that arise from the hypothesis that plants evolved from green algae? Put another way, how could you test this hypothesis?

9. **WRITE ABOUT A THEME**

   Ecological Interactions Organisms interact with each other and the physical environment. In a short essay (100–150 words), explain how the response of diatom populations to a drop in nutrient availability can affect both other organisms and aspects of the physical environment (such as carbon dioxide concentrations).

For selected answers, see Appendix A.

UNIT 7

Animal Form and Function

An Interview with

Baldomero M. Olivera

Growing up in the Philippines, Baldomero (Toto) Olivera collected the shells of venomous cone snails as a hobby. Today, he is a leading expert on the use of cone snail toxins for exploring and modifying the activity of the mammalian nervous system. After graduating summa cum laude from the University of the Philippines, he earned a Ph.D. in Chemistry from the California Institute of Technology (Caltech). In 1970, Dr. Olivera joined the faculty at the University of Utah, where he is now a Distinguished Professor of Biology. In 2009, he became the first Filipino to be elected to the U.S. National Academy of Sciences.

How did you get started in science?

In high school I had a great teacher in both biology and chemistry. After I finished college, she encouraged me to go on to graduate school. She put an article in front of me about Caltech and told me to apply there. Fortunately, they accepted me.

When I got to Caltech, I assumed that since I'd taken five courses at a time as an undergraduate, I'd be expected to take even more in graduate school, so in my first semester I was taking seven courses. One was the only statistical mechanics course at Caltech, so not only chemists like me, but also physicists took it. The first day, the professor, Norman Davidson, reviewed all of classical mechanics, which I had never had. The second day he reviewed all of quantum mechanics. After the fourth lecture I went to him and said, "You know, Professor, I haven't understood a word you've said." He said, "Don't worry about it. Just hang in there." I did, and I ended up doing my Ph.D. research in his lab, studying DNA.

After graduate school I went to Stanford and continued to work on DNA as a postdoctoral fellow. I then took a faculty job in the Philippines, but soon found out that I probably would have no large lab equipment for at least a year. I decided to look for a project that we could start immediately with no equipment and, if at all possible, that would have some local advantages. Since I'd collected shells as a hobby, I knew that some of the local snails, called cone snails, had a venomous sting that could kill people. So we set out to purify the substances in the venom that were lethal to mammals.

Why do cone snails produce venom?

Not much was known about this until 1956, when a Yale graduate student, Alan Kohn, discovered that a cone snail he had in an aquarium hunted fish. (Kohn went on to become a leading expert on the ecology and taxonomy of cone snails.) We now know that about 100 of the 500–700 species of cone snails hunt fish. A similar number hunt other molluscs, and the rest hunt marine worms.

When a cone snail hunts, it first sniffs the fish by means of chemoreceptors, specialized cells that function in sensing smell or taste. Chemoreception by a cone snail in the marine environment is amazing. For example, fish-hunting cone snails in a large aquarium react the moment you drop in a fish. They move their siphon, which is sensory, back and forth so they can locate the fish. When the siphon touches fish skin, a disposable, harpoon-like tooth shoots out of the snail's proboscis and pierces the scales of the fish. The venom flows through the tooth into the fish. So the tooth, which is barbed, is both a harpoon and a hypodermic needle. The venom very rapidly immobilizes the harpooned fish, which the cone snail can then consume.

Some other animals hunt cone snails, which explains their cone-like shape. Among the major predators are crabs. If you're a crab, you grab the snail's shell and try to break it. But if the snail is cone-shaped, the crab claw slips, making it hard to break the shell. So the cone shape is a defensive adaptation. However, a cone-shaped snail shell that narrows to a point has an opening too small to allow a large prey animal to be pulled into the shell. So fish-hunting cone snails tend to be more cylindrical.

What did you hope to learn by studying cone snail venom?

We knew that venoms could be useful because the toxins of the pufferfish—*fugu* in Japanese—and the krait, a cobra-like snake, had been used to study ion channels in the nervous system. Ion channels are found on the surface of neurons, and they let ions in and out of the cell. This movement of ions mediates the electrical signaling within neurons and controls the transfer of a signal from one neuron to another. Venoms that interfere with that ion movement wipe out the prey animal's nervous system—that's why they're deadly.

The toxins in venoms are quite specific. The toxin of the pufferfish targets a sodium ion channel. The toxin of the krait targets a different type of ion channel, called a nicotinic receptor. By using these toxins in the lab to interfere with the function of particular kinds of ion channels, you can learn a great deal about how the nervous system normally functions.

How did you study the cone snail toxins?

In the Philippines, people eat marine snails, and the shells are by-products. Since I'd collected shells, I knew the people who sold them commercially and asked them to get me some live snails.

When you work with cone snails, you of course don't want to be stung. We put the snails on ice and they became very lethargic. We then dissected out the venom duct and pressed out the venom. The next step was to measure the activity of the venom. The medical literature said people stung by cone snails died because their diaphragms became paralyzed. So we assayed (tested) for paralysis. We would inject mice with various fractions of the venom and then immediately put them on a wire screen, upside down. Mice can actually stay attached this way indefinitely, but when paralysis sets in, they fall off. All we did was measure the time until they fell. It turns out that the falling time changes with the amount of toxin in a very reproducible way.

Using this falling-time assay, we purified an active peptide from two different fractions of the venom of the species *Conus geographus*. One of these peptides has only 13 amino acids; the other has 22. We found out that the peptide with 13 amino acids bound to exactly the same site on an ion channel as the krait toxin, which has about 80 amino acids. It was remarkable that a toxin had evolved in snails that was so much smaller but had the same activity.

How did the big breakthrough in your cone snail studies come about?

When I started as a full-time faculty member here at Utah, all our funding was still for experiments with DNA, and my graduate students all studied DNA. So we started recruiting undergraduates to work on the cone snail project—and that led to a very lucky break. One of the undergraduates, Craig Clark, said, "You guys are injecting into the body cavity of the mice. I think we should inject directly into the central nervous system." That was purely his idea. I thought it would probably just kill the mice.

At the time, we were beginning to use new techniques to separate the venom into many fractions, each containing different peptides. Using our falling-time assay, most fractions were inactive. But when Craig started injecting fractions into the central nervous system, the results were truly astonishing. Almost every fraction changed the mice's behavior in some way, and many of the effects were really bizarre. Some fractions made mice run around in circles. Others made mice scratch themselves. Still others made mice get up on their hind legs and move their front legs as if they were boxing. There was even a fraction that put juvenile, but not adult, mice into a sleeplike state. Craig purified the peptide with that activity, which he called the "sleeper peptide." I think the reason why university research is typically the most creative is that students sometimes do what they want, not what their professors tell them to do!

From then on, we let undergraduates pick which active fraction they wanted to study. Their job was to purify the peptide, which we would then synthesize and work on further.

Why and how did so much toxin diversity evolve in cone snails?

My suspicion is that if your survival depends on influencing the behavior of another animal and your strategy is purely chemical, it takes a very complex strategy for that to be successful in evolution. The adaptations that evolved in cone snails are in essence what we're just discovering in modern medical practice. To treat cancer or AIDS effectively, you can't use just one chemical; you use combination drug therapy. I think the evolution of venoms in cone snails followed the same principle.

For the cone snail toxins there are a few gene superfamilies, and they are subject to very accelerated evolution. There are parallels to the vertebrate adaptive immune system in that there are conserved and highly variable regions in the genes. The conserved regions are mostly all cut out. So, if you like, a cone snail peptide is equivalent to the highly variable part that makes an antibody very specific.

> "The adaptations that evolved in cone snails are in essence what we're just discovering in modern medical practice."

How did your studies with snails lead to a discovery that was medically important?

After Craig discovered that you could study the effects of toxins by injecting them into the central nervous system of mice, he recruited a high school student, Michael McIntosh, to the lab. Mike decided to work with a different snail, called *Conus magus*, or the magician's cone. One of the most striking discoveries he made was a peptide that gave mice a characteristic tremor, or shake.

When Mike purified the "shaker peptide," we learned that it blocks a certain type of calcium channel. In fish these calcium channels control the release of the neurotransmitter that carries signals from nerve to muscle. When shaker peptide blocks these channels, the neurotransmitter is not released, and the fish becomes paralyzed. In mammals, this type of calcium channel is found only in the pain circuitry. When you experience severe pain, pain fibers send signals to a nerve cell in your spinal cord that takes the pain signal up to your brain. The pain fibers in mammals release signals to the nerves using the same type of calcium channel that in fish carries signals to muscles to direct movement. This difference in receptor location and function is why the shaker peptide paralyzes and kills fish, but not mammals.

The shaker peptide is now an approved drug. If a patient has very severe pain and the shaker peptide is injected where the pain fiber connects to the nerve, the perception of pain stops because the pain fiber no longer releases the chemical signal to the nerve. Doctors are using the shaker peptide in this way for people who are suffering from severe pain and no longer get pain relief from morphine due to tolerance built up over time. A surgically implanted pump introduces the shaker peptide to the spinal cord at the site where the pain fiber makes its connection.

What other discoveries do you hope to see come out of studying cone snails?

There are 500–700 species of cone snails, but the total biodiversity of molluscs with venom is probably 12,000 species. The biggest group is the turrids. You don't find many of them in shallow water, but at 400 meters, where there are no cone snails, you find hundreds of species of turrids. Many of them are just a few millimeters in length, and nobody knows anything about their biology. Fortunately, hobbyists are extremely interested in their tiny shells, so fishermen in the Philippines gather them. They tie all their broken fine-mesh fishing nets into bundles, which they sink in water for 3 to 6 months. When they lift these bundles, each will contain several thousand marine animals, almost all of them alive. For gastropods alone, a net will hold something like 250 different species, of which 40 are venomous. Each venom has 200 compounds. So this is a vast new world.

What advice would you give students interested in biology or medicine?

One piece of advice is to get as good a basic science background as you possibly can. I started out with very basic chemistry and very basic biology, and then of course the science grew as my research career grew. Having strong basic science training allows you to feel confident that even if you don't understand a problem at first (like quantum mechanics), you can still tackle it. A second thing is that you ought to know what it is in science that really turns you on. Third, keep in mind that it is always fun to learn something new.

Baldomero Olivera (center) with Jane Reece and Steve Wasserman (right)

43

The Immune System

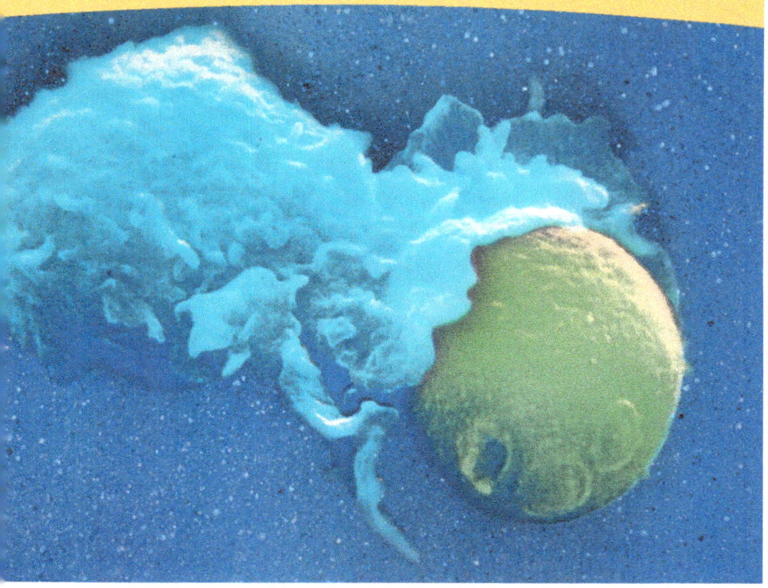

▲ **Figure 43.1 How do an animal's immune cells recognize foreign cells?**

KEY CONCEPTS

43.1 In innate immunity, recognition and response rely on traits common to groups of pathogens

43.2 In adaptive immunity, receptors provide pathogen-specific recognition

43.3 Adaptive immunity defends against infection of body fluids and body cells

43.4 Disruptions in immune system function can elicit or exacerbate disease

OVERVIEW

Recognition and Response

Pathogens, agents that cause disease, infect a wide range of animals. For a virus, bacterium, fungus, or other pathogen, the internal environment of an animal is a nearly ideal habitat. The animal body offers a ready source of nutrients, a protected setting for growth and reproduction, and a means of transport to new environments. From the perspective of a cold or flu virus, we are wonderful hosts. From our vantage point, things are not so ideal. Fortunately, adaptations have arisen over the course of evolution that protect animals against many invaders.

Dedicated immune cells in the body fluids and tissues of most animals specifically interact with and destroy pathogens. As shown in Figure 43.1 (a colorized scanning electron micrograph), an immune cell called a macrophage (blue) can engulf a yeast cell (green). Additional responses to infection take many forms, including proteins that punch holes in bacterial membranes or block viruses from entering body cells. These and other defenses make up the **immune system**, which enables an animal to avoid or limit many infections. A foreign molecule or cell doesn't have to be pathogenic to elicit an immune response, but we'll focus here on the immune system's role in defending against pathogens.

All animals have **innate immunity**, a defense that is active immediately upon infection and is the same whether or not the pathogen has been encountered previously. Innate immunity includes an outer covering, such as a skin or shell, that provides a significant barrier to entry by microbes. Sealing off the entire body surface is impossible, however, because gas exchange, nutrition, and reproduction require openings to the environment. Chemical secretions that trap or kill microbes guard the body's entrances and exits, while the linings of the digestive tract, airway, and other exchange surfaces provide additional barriers to infection.

If a pathogen breaches barrier defenses and enters the body, the problem of how to fend off attack changes substantially. Housed within the body fluids and tissues, the invader is no longer an outsider. To fight infections, an animal's immune system must detect foreign particles and cells within the body. In other words, a properly functioning immune system distinguishes nonself from self. Detection of nonself is accomplished by *molecular recognition*, in which receptor molecules bind specifically to molecules from foreign cells or viruses.

In innate immunity, a small preset group of receptor proteins bind to molecules or structures that are absent from animal bodies but common to a group of viruses, bacteria, or other microbes. Binding of an innate immune receptor to a foreign molecule activates internal defenses, enabling responses to a very broad range of pathogens.

A different type of molecular recognition provides the basis for **adaptive immunity**, a defense found only in vertebrates. Animals with adaptive immunity produce a vast arsenal of receptors, each of which recognizes a feature typically found only on a particular part of a particular molecule in a particular pathogen. As a result, recognition and response in adaptive immunity occur with tremendous specificity.

The adaptive immune response, also known as the acquired immune response, is activated after the innate immune response and develops more slowly. The names *adaptive* and

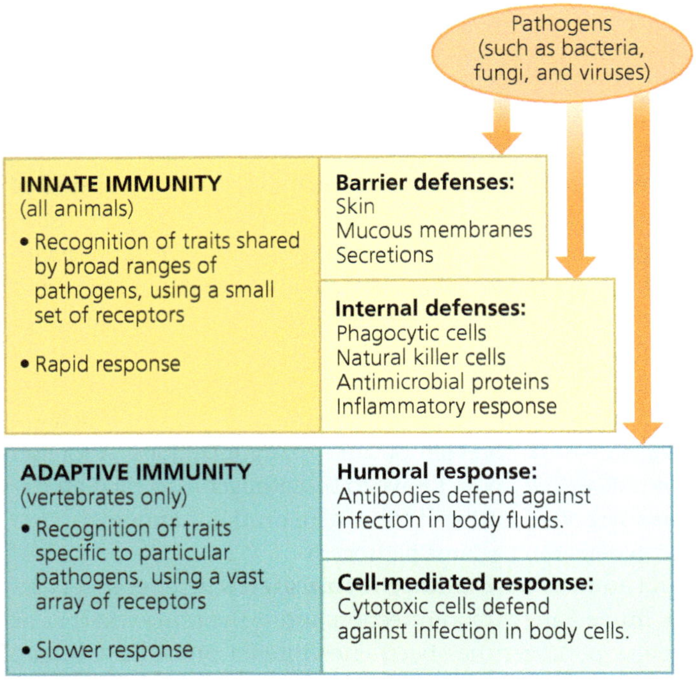

▲ **Figure 43.2 Overview of animal immunity.** Immune responses in animals can be divided into innate and adaptive immunity. Some components of innate immunity contribute to activation of adaptive immune defenses.

acquired reflect the fact that this immune response is enhanced by previous exposure to the infecting pathogen. Examples of adaptive responses include the synthesis of proteins that inactivate a bacterial toxin and the targeted killing of a virus-infected body cell.

Figure 43.2 provides an overview of the basic components of innate and adaptive immunity. In this chapter, you will learn how each type of immunity protects animals from disease. You will also examine how pathogens can avoid or overwhelm the immune system and how defects in the immune system can imperil an animal's health.

CONCEPT **43.1**

In innate immunity, recognition and response rely on traits common to groups of pathogens

Innate immunity is found in all animals (as well as in plants). In exploring innate immunity, we'll begin with invertebrates, which repel and fight infection with only this type of immunity. We'll then turn to vertebrates, in which innate immunity serves both as an immediate defense against infection and as the foundation for adaptive immune defenses.

Innate Immunity of Invertebrates

The great success of insects in terrestrial and freshwater habitats teeming with diverse microbes highlights the effectiveness of invertebrate innate immunity. In each of these environments, insects rely on their exoskeleton as a first line of defense against infection. Composed largely of the polysaccharide chitin, the exoskeleton provides an effective barrier defense against most pathogens. A chitin-based barrier is also present in the insect intestine, where it blocks infection by many pathogens ingested with food. **Lysozyme**, an enzyme that breaks down bacterial cell walls, further protects the insect digestive system.

Any pathogen that breaches an insect's barrier defenses encounters a number of internal immune defenses. Immune cells called *hemocytes* travel throughout the body in the hemolymph, the insect circulatory fluid. Some hemocytes carry out a defense called **phagocytosis**, the cellular ingestion and digestion of bacteria and other foreign substances **(Figure 43.3)**. Other hemocytes trigger the production of chemicals that kill pathogens and help entrap large parasites, such as *Plasmodium*, the parasite of mosquitoes that causes malaria. In addition, encounters with pathogens in the hemolymph cause hemocytes and certain other cells to secrete *antimicrobial peptides*, which are short chains of amino acids. The antimicrobial peptides circulate throughout the body of the insect **(Figure 43.4)** and inactivate or kill fungi and bacteria by disrupting their plasma membranes.

Immune cells of insects bind to molecules found only in the outer layers of fungi or bacteria. Fungal cell walls contain certain unique polysaccharides, whereas bacterial cell walls have polymers containing combinations of sugars and amino

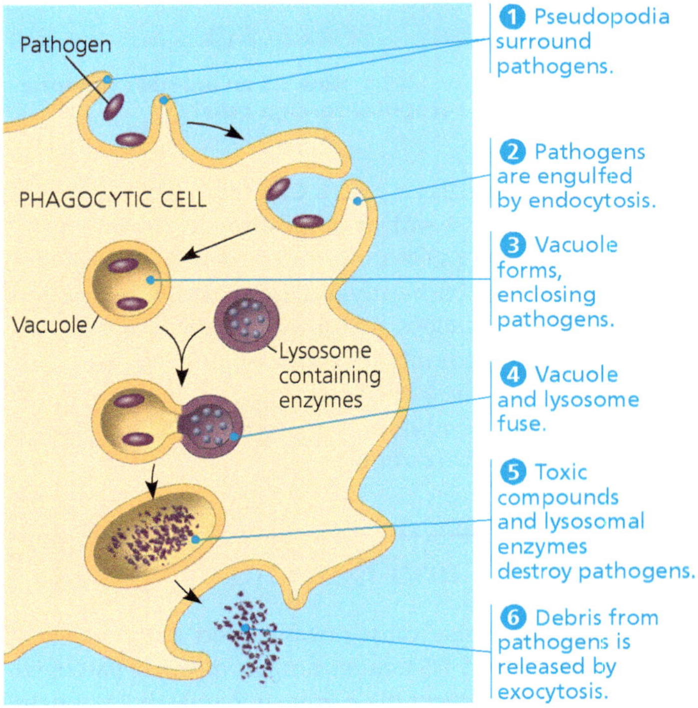

▲ **Figure 43.3 Phagocytosis.** This schematic depicts events in the ingestion and destruction of a microbe by a typical phagocytic cell.

▲ **Figure 43.4 An inducible innate immune response.** These fruit flies were engineered to express the green fluorescent protein (GFP) gene upon activation of the innate immune response. The fly on the top was injected with bacteria; the fly on the bottom was not. Only the infected fly activates antimicrobial peptide genes, produces GFP, and glows a bright green under fluorescent light.

acids not found in animal cells. Such macromolecules serve as identity tags in the process of pathogen recognition. Insect immune cells secrete specialized recognition proteins, each of which binds to a macromolecule characteristic of fungi or a broad class of bacteria.

Innate immune responses are distinct for different classes of pathogens. For example, when the fungus *Neurospora crassa* infects a fruit fly, pieces of the fungal cell wall bind a recognition protein. Together, the complex activates the protein Toll, a receptor on the surface of hemocytes. Signal transduction from the Toll receptor to the cell nucleus leads to synthesis of a set of antimicrobial peptides active against fungi. If the fly is instead infected by the bacterium *Micrococcus luteus*, a different recognition protein is activated, and the fly produces a different set of antimicrobial peptides effective against *M. luteus* and many related bacteria.

Because fruit flies secrete many distinct antimicrobial peptides in response to a single infection, it is difficult to study the activity of any one peptide. To get around this problem, Bruno Lemaitre and fellow researchers used modern genetic techniques to reprogram the fly immune system (Figure 43.5). They found that the synthesis of a single type of antimicrobial peptide in the fly's body could provide an effective immune defense. They also showed that particular antimicrobial peptides act against different kinds of pathogens.

▼ **Figure 43.5** INQUIRY

Can a single antimicrobial peptide protect fruit flies against infection?

EXPERIMENT In 2002, Bruno Lemaitre and colleagues in France devised a novel strategy to test the function of a single antimicrobial peptide. They began with a mutant fruit fly strain in which pathogens are recognized but the signaling that would normally trigger innate immune responses is blocked. As a result, the mutant flies do not make any antimicrobial peptides. The researchers then genetically engineered some of the mutant fruit flies to express significant amounts of a single antimicrobial peptide, either drosomycin or defensin. The scientists infected the various flies with the fungus *Neurospora crassa* and monitored survival over a five-day period. They repeated the procedure for infection by the bacterium *Micrococcus luteus*.

RESULTS

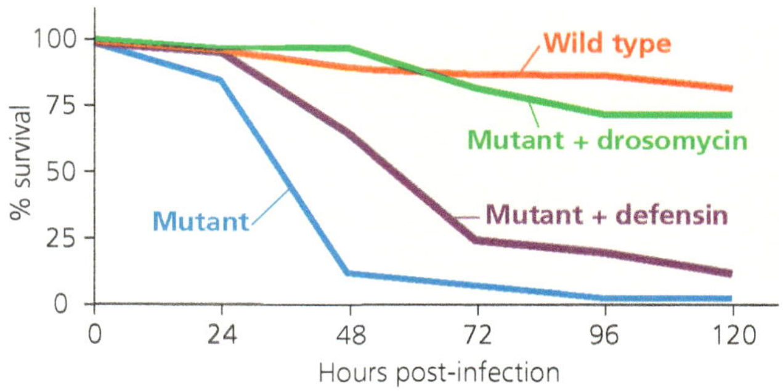

Fruit fly survival after infection by *N. crassa* fungi

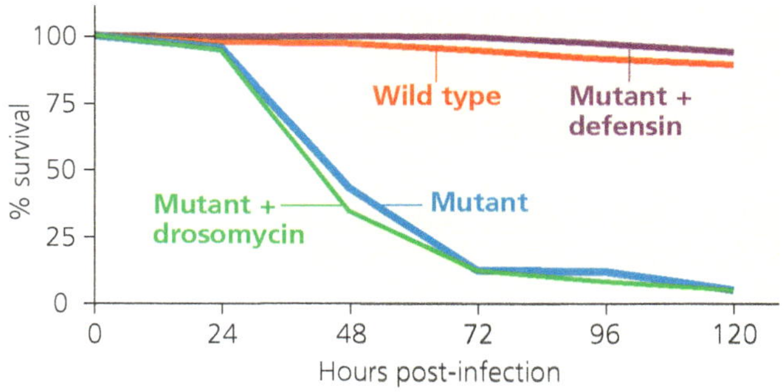

Fruit fly survival after infection by *M. luteus* bacteria

CONCLUSION Each of the two antimicrobial peptides provided a protective immune response. Furthermore, the different peptides defended against different pathogens. Drosomycin was effective against *N. crassa*, and defensin was effective against *M. luteus*.

SOURCE P. Tzou, J. Reichhart, and B. Lemaitre, Constitutive expression of a single antimicrobial peptide can restore wild-type resistance to infection in immunodeficient *Drosophila* mutants, *Proceedings of the National Academy of Sciences USA* 99:2152–2157 (2002).

WHAT IF? Even if a particular antimicrobial peptide showed no beneficial effect in such an experiment, why might it still be beneficial to flies?

Innate Immunity of Vertebrates

Among vertebrates, innate immune defenses coexist with the more recently evolved system of adaptive immunity. Because most of the recent discoveries regarding vertebrate innate immunity have come from studies of mice and humans, we'll focus here on mammals. We'll consider the innate defenses that are similar to those found among invertebrates: barrier defenses, phagocytosis, and antimicrobial peptides. We'll also examine some unique aspects of vertebrate innate immunity, such as natural killer cells, interferons, and the inflammatory response.

Barrier Defenses

In mammals, epithelial tissues block the entry of many pathogens. These barrier defenses include not only the skin but also the mucous membranes lining the digestive, respiratory, urinary, and reproductive tracts. Certain cells of the mucous membranes produce *mucus*, a viscous fluid that enhances defenses by trapping microbes and other particles. In the trachea, ciliated epithelial cells sweep mucus and any entrapped microbes upward, helping prevent infection of the lungs. Saliva, tears, and mucous secretions that bathe various exposed epithelia provide a washing action that also inhibits colonization by fungi and bacteria.

Beyond their physical role in inhibiting microbial entry, body secretions create an environment that is hostile to many microbes. Lysozyme in tears, saliva, and mucous secretions destroys the cell walls of susceptible bacteria as they enter the openings around the eyes or the upper respiratory tract. Microbes in food or water and those in swallowed mucus must also contend with the acidic environment of the stomach, which kills most of them before they can enter the intestines. Similarly, secretions from oil and sweat glands give human skin a pH ranging from 3 to 5, acidic enough to prevent the growth of many bacteria.

Cellular Innate Defenses

Pathogens entering the mammalian body are subject to phagocytosis. Phagocytic cells detect fungal or bacterial components using several types of receptors, some of which are very similar to the Toll receptor of insects. Each mammalian **Toll-like receptor (TLR)** binds to fragments of molecules characteristic of a set of pathogens **(Figure 43.6)**. For example, TLR3, on the inner surface of vesicles formed by endocytosis, is the sensor for double-stranded RNA, a form of nucleic acid characteristic of certain viruses. Similarly, TLR4, located on immune cell plasma membranes, recognizes lipopolysaccharide, a type of molecule found on the surface of many bacteria; and TLR5 recognizes flagellin, the main protein of bacterial flagella. In each case, the recognized macromolecule is normally absent from the vertebrate body and is an essential component of certain groups of pathogens.

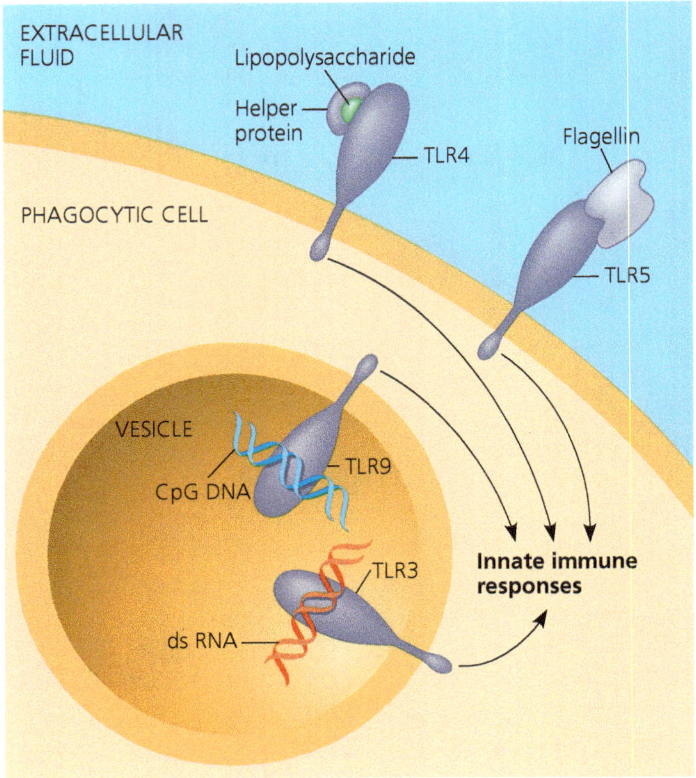

▲ **Figure 43.6 TLR signaling.** Each mammalian Toll-like receptor (TLR) recognizes a molecular pattern characteristic of a group of pathogens. Lipopolysaccharide, flagellin, CpG DNA (DNA containing unmethylated CG sequences), and double-stranded (ds) RNA are all found in bacteria, fungi, or viruses, but not in animal cells. Together with other recognition and response factors, TLR proteins trigger internal innate immune defenses.

? *Some TLR proteins are on the cell surface, whereas others are inside vesicles. Suggest a possible benefit of this distribution.*

After detecting invading pathogens, a phagocytic cell engulfs them, trapping them in a vacuole. The vacuole then fuses with a lysosome (see Figure 43.3), leading to destruction of the invaders in two ways. First, gases produced in the lysosome poison the engulfed pathogens. Second, lysozyme and other enzymes in the lysosome degrade the components of the pathogens.

The two main types of phagocytic cells in the mammalian body are neutrophils and macrophages. **Neutrophils**, which circulate in the blood, are attracted by signals from infected tissues and then engulf and destroy the infecting pathogens. **Macrophages** ("big eaters"), like the one shown in Figure 43.1, are larger phagocytic cells. Some migrate throughout the body, whereas others reside permanently in organs and tissues where they are likely to encounter pathogens. For example, some macrophages are located in the spleen, where pathogens in the blood become trapped.

Two other types of phagocytic cells—dendritic cells and eosinophils—provide additional functions in innate defense. **Dendritic cells** mainly populate tissues, such as skin, that contact the environment. They stimulate adaptive immunity against pathogens they encounter and engulf, as we'll explore shortly. *Eosinophils*, often found beneath mucosal surfaces, have low phagocytic activity but are important in defending against multicellular invaders, such as parasitic worms. Upon encountering such parasites, eosinophils discharge destructive enzymes.

Cellular innate defenses in vertebrates also involve **natural killer cells**. These cells circulate through the body and detect the abnormal array of surface proteins characteristic of some virus-infected and cancerous cells. Natural killer cells do not engulf stricken cells. Instead, they release chemicals that lead to cell death, inhibiting further spread of the virus or cancer.

Many cellular innate defenses of vertebrates involve the lymphatic system, a network that distributes the fluid called lymph throughout the body **(Figure 43.7)**. Some macrophages reside in the structures called lymph nodes, where they engulf pathogens that have flowed from the interstitial fluid into the lymph. Dendritic cells reside outside the lymphatic system but migrate to lymph nodes after interaction with pathogens. Within the lymph nodes, dendritic cells interact with other immune cells, stimulating adaptive immunity.

Antimicrobial Peptides and Proteins

In mammals, pathogen recognition triggers the production and release of a variety of peptides and proteins that attack pathogens or impede their reproduction. Some of these defense molecules function like the antimicrobial peptides of insects, damaging broad groups of pathogens by disrupting membrane integrity. Others, including the interferons and complement proteins, are unique to vertebrate immune systems.

Interferons are proteins that provide innate defense by interfering with viral infections. Virus-infected body cells secrete interferons, which induce nearby uninfected cells to produce substances that inhibit viral reproduction. In this way, interferons limit the cell-to-cell spread of viruses in the body, helping control viral infections such as colds and influenza. Some white blood cells secrete a different type of interferon that helps activate macrophages, enhancing their phagocytic ability. Pharmaceutical companies now use recombinant DNA technology to mass-produce interferons to help treat certain viral infections, such as hepatitis C.

The infection-fighting **complement system** consists of roughly 30 proteins in blood plasma. These proteins circulate in an inactive state and are activated by substances on the surface of many microbes. Activation results in a cascade of biochemical reactions that can lead to lysis (bursting) of invading cells. The complement system also functions in the inflammatory response, our next topic, as well as in the adaptive defenses discussed later in the chapter.

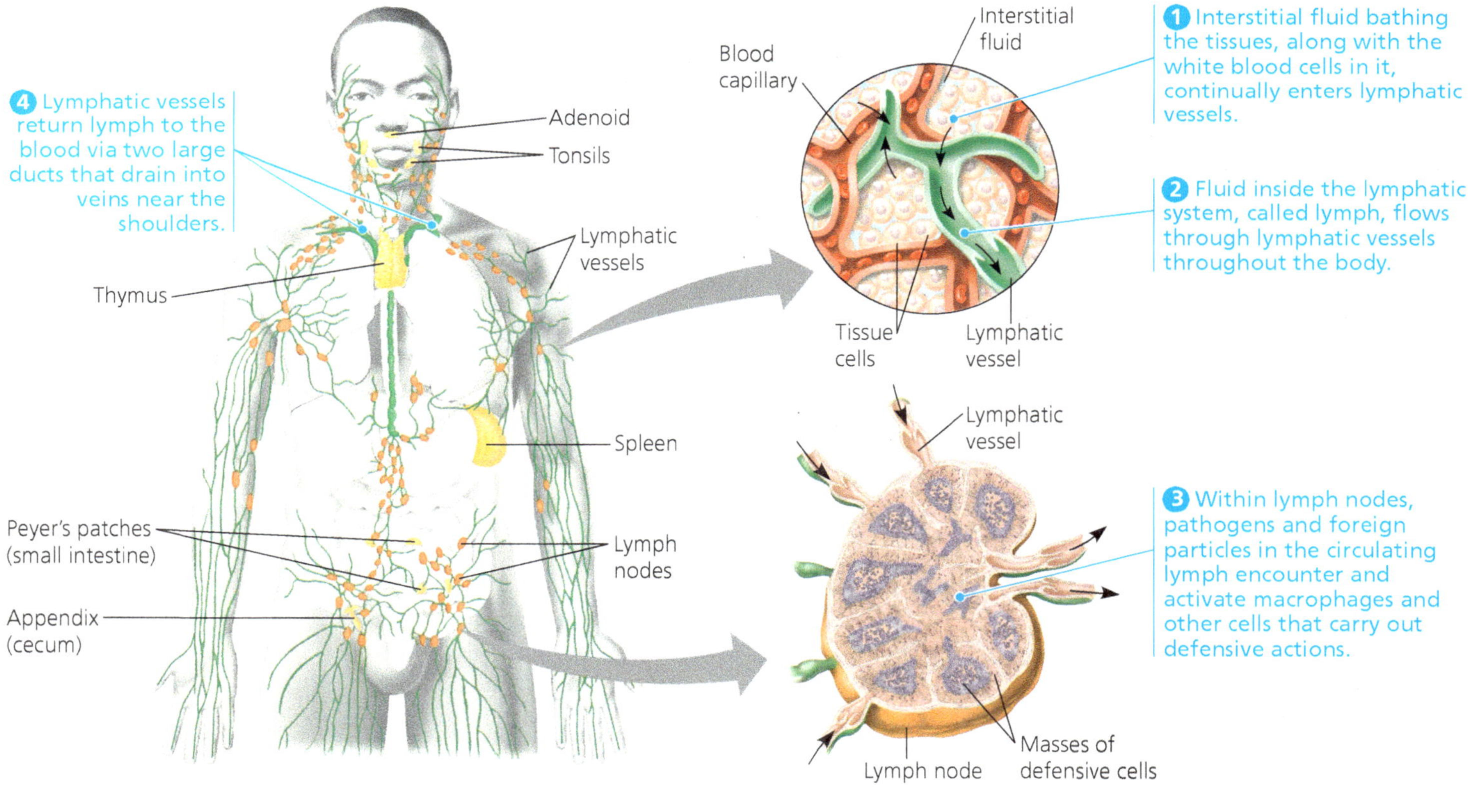

▲ **Figure 43.7 The human lymphatic system.** The lymphatic system consists of lymphatic vessels (shown in green), through which lymph travels, and structures that trap foreign substances. These structures include lymph nodes (orange) and lymphoid organs (yellow): the adenoids, tonsils, spleen, Peyer's patches, and appendix. Steps 1–4 trace the flow of lymph and illustrate the critical role of lymph nodes in activating adaptive immunity. (See also p. 909 for a description of the relationship between the lymphatic and circulatory systems.)

Inflammatory Response

The pain and swelling that alert you to a splinter under your skin are the result of a local **inflammatory response**, the changes brought about by signaling molecules released upon injury or infection (**Figure 43.8**). One important inflammatory signaling molecule is **histamine**, which is stored in the granules (vesicles) of **mast cells**, found in connective tissue. Histamine released at sites of damage triggers nearby blood vessels to dilate and become more permeable. Activated macrophages and neutrophils discharge **cytokines**, signaling molecules that enhance an immune response. These cytokines promote blood flow to the site of injury or infection. The increase in local blood supply causes the redness and increased skin temperature typical of the inflammatory response (from the Latin *inflammare*, to set on fire). Blood-engorged capillaries leak fluid into neighboring tissues, causing swelling.

During inflammation, cycles of signaling and response transform the site. Activated complement proteins promote further release of histamine, attracting more phagocytic cells that enter injured tissues (see Figure 43.8) and carry out additional phagocytosis. At the same time, enhanced blood flow to the site helps deliver antimicrobial peptides. The result is an accumulation of *pus*, a fluid rich in white blood cells, dead pathogens, and cell debris from damaged tissue.

A minor injury or infection causes a local inflammatory response, but severe tissue damage or infection may lead to a response that is systemic (throughout the body). Cells in injured or infected tissue often secrete molecules that stimulate the release of additional neutrophils from the bone marrow. In a severe infection, such as meningitis or appendicitis, the number of white blood cells in the blood may increase severalfold within a few hours.

Another systemic inflammatory response is fever. In response to certain pathogens, substances released by activated macrophages cause the body's thermostat to reset to a higher temperature (see Chapter 40). The benefits of the resulting fever are still a subject of debate. One of several competing hypotheses is that an elevated body temperature may enhance phagocytosis and, by speeding up chemical reactions, accelerate tissue repair.

Certain bacterial infections can induce an overwhelming systemic inflammatory response, leading to a life-threatening condition called *septic shock*. Characterized by very high fever, low blood pressure, and poor blood flow through capillaries, septic shock occurs most often in the very old and the very young. It is fatal in more than one-third of cases and kills more than 90,000 people each year in the United States alone.

Chronic (ongoing) inflammation can also threaten human health. For example, millions of individuals worldwide suffer from Crohn's disease and ulcerative colitis, often debilitating disorders in which an unregulated inflammatory response disrupts intestinal function.

Evasion of Innate Immunity by Pathogens

Adaptations have evolved in some pathogens that enable them to avoid destruction by phagocytic cells. For example, the outer capsule that surrounds certain bacteria interferes with molecular recognition and phagocytosis. One such bacterium, *Streptococcus pneumoniae*, played a critical role in the discovery that DNA can convey genetic information (see

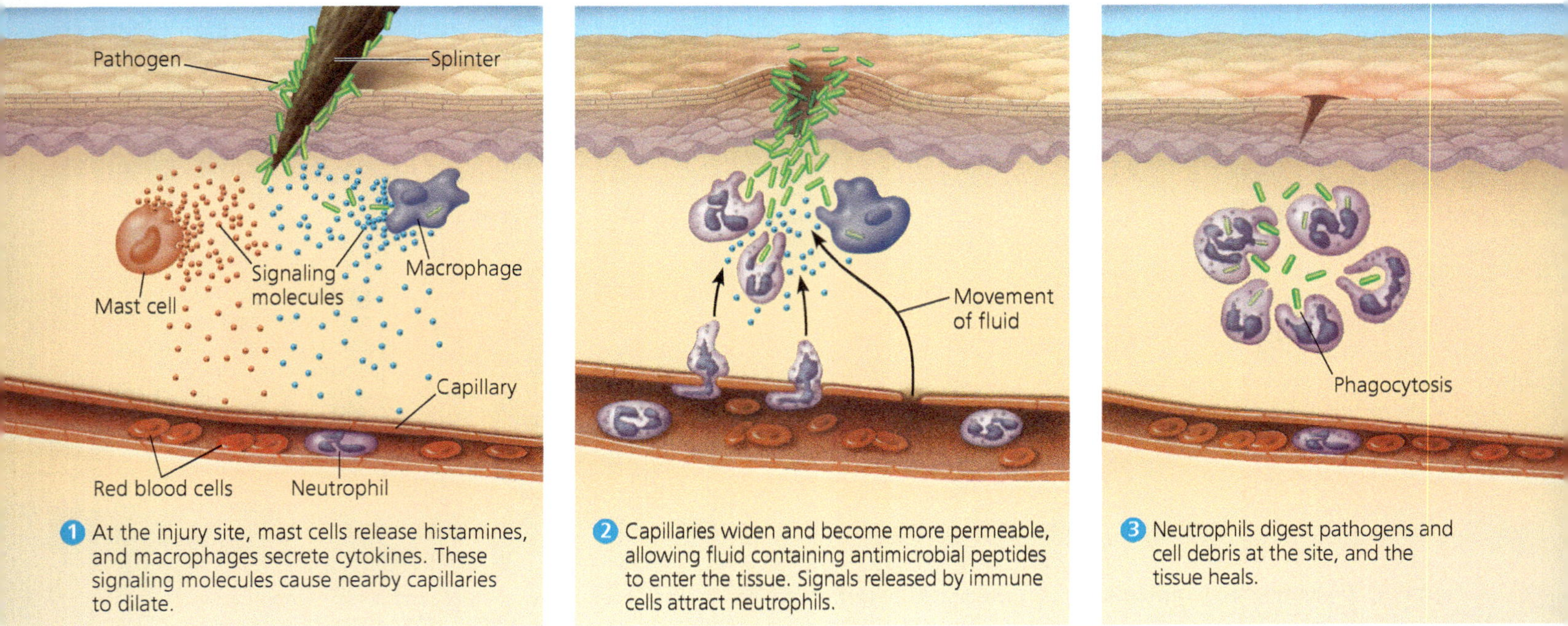

▲ **Figure 43.8 Major events in a local inflammatory response.**

Figure 16.2). Some bacteria, after being engulfed by a host cell, resist breakdown within lysosomes. An example is the bacterium that causes tuberculosis (TB). Rather than being destroyed within host cells, this bacterium grows and reproduces, effectively hidden from the body's innate immune defenses. These and other mechanisms that prevent destruction by the innate immune system make certain fungi and bacteria substantial pathogenic threats. Indeed, TB kills more than a million people a year worldwide.

CONCEPT CHECK 43.1

1. Although pus is often seen simply as a sign of infection, it is also an indicator of immune defenses in action. Explain.
2. **MAKE CONNECTIONS** How do the molecules that activate the vertebrate TLR signal transduction pathway differ from the ligands in most other pathways, such as those shown in Concept 11.2 (pp. 210–214)?
3. **WHAT IF?** Suppose humans were the major host for a bacterial species. What temperature would you predict would be optimal for growth of this species? Explain.

For suggested answers, see Appendix A.

CONCEPT 43.2

In adaptive immunity, receptors provide pathogen-specific recognition

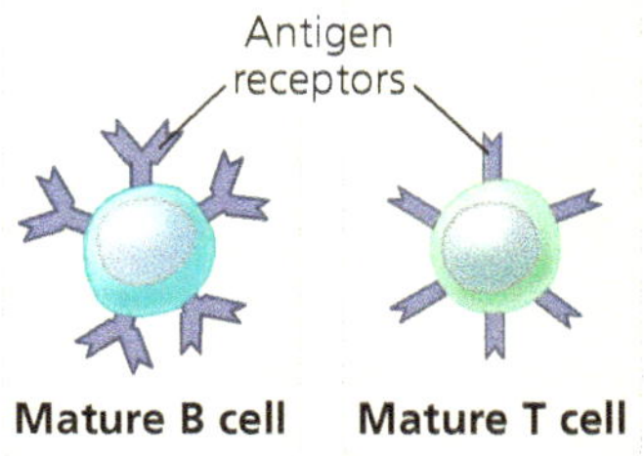

Vertebrates are unique in having adaptive immunity in addition to innate immunity. The adaptive response relies on T cells and B cells, which are types of white blood cells called **lymphocytes**. Like all blood cells, lymphocytes originate from stem cells in the bone marrow. Some lymphocytes migrate from the bone marrow to the **thymus**, an organ in the thoracic cavity above the heart (see Figure 43.7). These lymphocytes mature into **T cells**. Lymphocytes that remain and mature in the bone marrow develop as **B cells**. (Lymphocytes of a third type remain in the blood and become the natural killer cells active in innate immunity.)

Any substance that elicits a response from a B cell or T cell is called an **antigen**. In adaptive immunity, recognition occurs when a B cell or T cell binds to an antigen, such as a bacterial or viral protein, via a protein called an **antigen receptor**. An antigen receptor is specific enough to bind to just one part of one molecule from a particular pathogen, such as a species of bacteria or strain of virus. Although the cells of the immune system produce millions of different antigen receptors, all of the antigen receptors made by a single B or T cell are identical. Infection by a virus, bacterium, or other pathogen triggers activation of B and T cells with antigen receptors specific for parts of that pathogen. B and T cells are shown here with only a few antigen receptors, but there are actually about 100,000 antigen receptors on the surface of a single B or T cell.

Antigens are usually foreign and are typically large molecules, either proteins or polysaccharides. Many antigens protrude from the surface of foreign cells or viruses. Other antigens, such as toxins secreted by bacteria, are released into the extracellular fluid.

The small, accessible portion of an antigen that binds to an antigen receptor is called an **epitope**, or *antigenic determinant*. An example is a group of amino acids in a particular protein. A single antigen usually has several different epitopes, each binding a receptor with a different specificity. Because all antigen receptors produced by a single B cell or T cell are identical, they bind to the same epitope. Each B cell or T cell thus displays *specificity* for a particular epitope, enabling it to respond to any pathogen that produces molecules containing that same epitope.

The antigen receptors of B cells and T cells have similar components, but they encounter antigens in different ways. We'll consider the two processes in turn.

Antigen Recognition by B Cells and Antibodies

Each B cell antigen receptor is a Y-shaped molecule consisting of four polypeptide chains: two identical **heavy chains** and two identical **light chains**, with disulfide bridges linking the chains together (**Figure 43.9**). A transmembrane region near one end of each heavy chain anchors the receptor in the cell's plasma membrane. A short tail region at the end of the heavy chain extends into the cytoplasm.

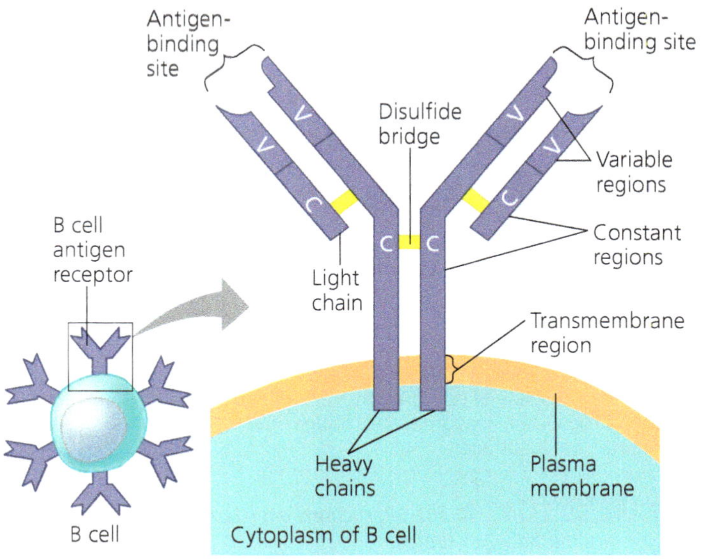

▲ **Figure 43.9 The structure of a B cell antigen receptor.**

The light and heavy chains each have a *constant (C) region*, where amino acid sequences vary little among the receptors on different B cells. The C region includes the cytoplasmic tail and transmembrane region of the heavy chain and all of the disulfide bridges. Within the two tips of the Y shape, the light and heavy chains each have a *variable (V) region*, so named because its amino acid sequence varies extensively from one B cell to another. Together, parts of a heavy-chain V region and a light-chain V region form an asymmetrical binding site for an antigen. As shown in Figure 43.9, each B cell antigen receptor has two identical antigen-binding sites.

The binding of a B cell antigen receptor to an antigen is an early step in B cell activation, leading eventually to formation of cells that secrete a soluble form of the receptor (**Figure 43.10a**). This secreted protein is called an **antibody**, or **immunoglobulin (Ig)**. Antibodies have the same Y-shaped organization as B cell antigen receptors, but they are secreted rather than membrane bound. It is the antibodies, rather than the B cells themselves, that actually help defend against pathogens. Antibodies have distinct functions, as we'll see later.

The antigen-binding site of a membrane-bound receptor or antibody has a unique shape that provides a lock-and-key fit for a particular epitope. Many noncovalent bonds between an epitope and the binding surface provide a stable and specific interaction. Differences in the amino acid sequences of variable regions provide the variation in binding surfaces that enables this highly specific binding.

B cell antigen receptors and antibodies bind to intact antigens in the blood and lymph. As illustrated in **Figure 43.10b** for antibodies, they can bind to antigens on the surface of pathogens or free in body fluids. The antigen receptors of T cells function quite differently, as we'll see next.

▼ **Figure 43.10 Antigen recognition by B cells and antibodies.**

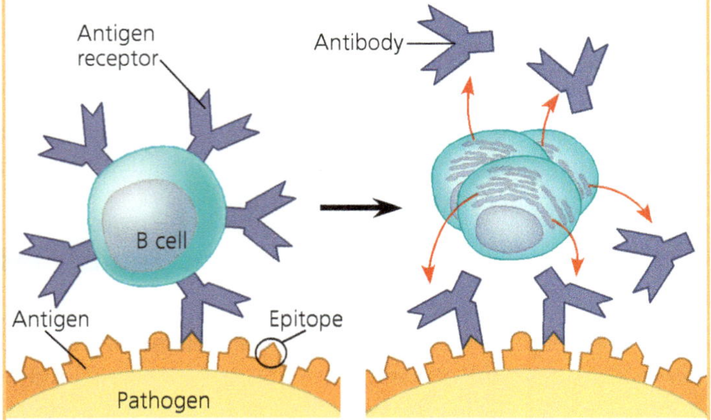

(a) B cell antigen receptors and antibodies. An antigen receptor of a B cell binds to an epitope, a particular part of an antigen. Following binding, the B cell gives rise to cells that secrete a soluble form of the antigen receptor. This soluble receptor, called an antibody, is specific for the same epitope as the original B cell.

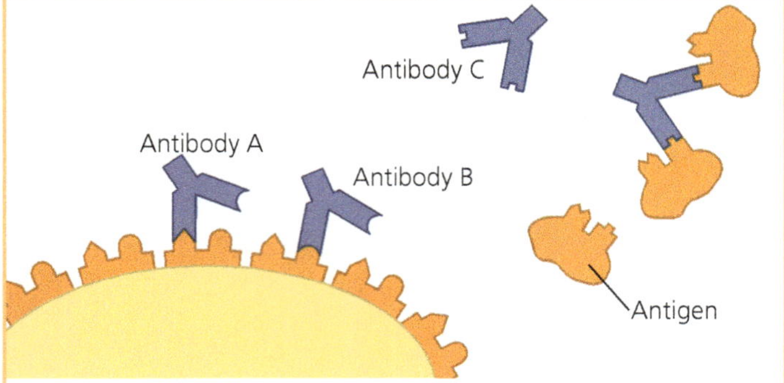

(b) Antigen receptor specificity. Different antibodies can recognize distinct epitopes on the same antigen. Furthermore, antibodies can recognize free antigens as well as antigens on a pathogen's surface.

MAKE CONNECTIONS *The interactions depicted here involve a highly specific binding between antigen and receptor, as shown in Figure 5.19 (p. 81). How is this similar to the enzyme-substrate interaction shown in Figure 8.14 (p. 154)?*

Antigen Recognition by T Cells

For a T cell, the antigen receptor consists of two different polypeptide chains, an *α chain* and a *β chain*, linked by a disulfide bridge (**Figure 43.11**). Near the base of the T cell antigen receptor (often called simply a T cell receptor) is a transmembrane region that anchors the molecule in the cell's plasma membrane. At the outer tip of the molecule, the variable (V) regions of α and β chains together form a single antigen-binding site. The remainder of the molecule is made up of the constant (C) regions.

Although T cell and B cell antigen receptors have many features in common, they function in fundamentally different ways. Whereas the antigen receptors of B cells bind to epitopes of intact antigens circulating in body fluids, those of

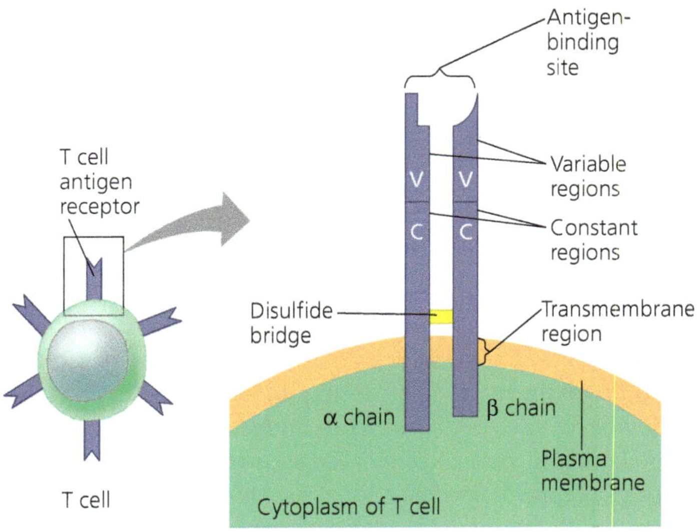

▲ **Figure 43.11 The structure of a T cell antigen receptor.**

T cells bind only to fragments of antigens that are displayed, or *presented*, on the surface of host cells. The host protein that displays the antigen fragment on the cell surface is called an **MHC (major histocompatibility complex) molecule**.

Recognition of protein antigens by T cells begins when a pathogen or part of a pathogen either infects or is taken in by a host cell (**Figure 43.12a**). Inside the host cell, enzymes in the cell cleave the antigen into smaller peptides. Each peptide, called an *antigen fragment*, then binds to an MHC molecule inside the cell. Movement of the MHC molecule and bound antigen fragment to the cell surface results in **antigen presentation**, the display of the antigen fragment in an exposed groove of the MHC protein. **Figure 43.12b** shows a close-up view of antigen presentation, which advertises the fact that a host cell contains a foreign substance. If the cell displaying an antigen fragment encounters a T cell with the right specificity, the antigen receptor on the T cell can bind to both the antigen fragment and the MHC molecule. This interaction of an MHC molecule, an antigen fragment, and an antigen receptor is necessary for a T cell to participate in an adaptive immune response, as we'll see later.

▼ **Figure 43.12 Antigen recognition by T cells.**

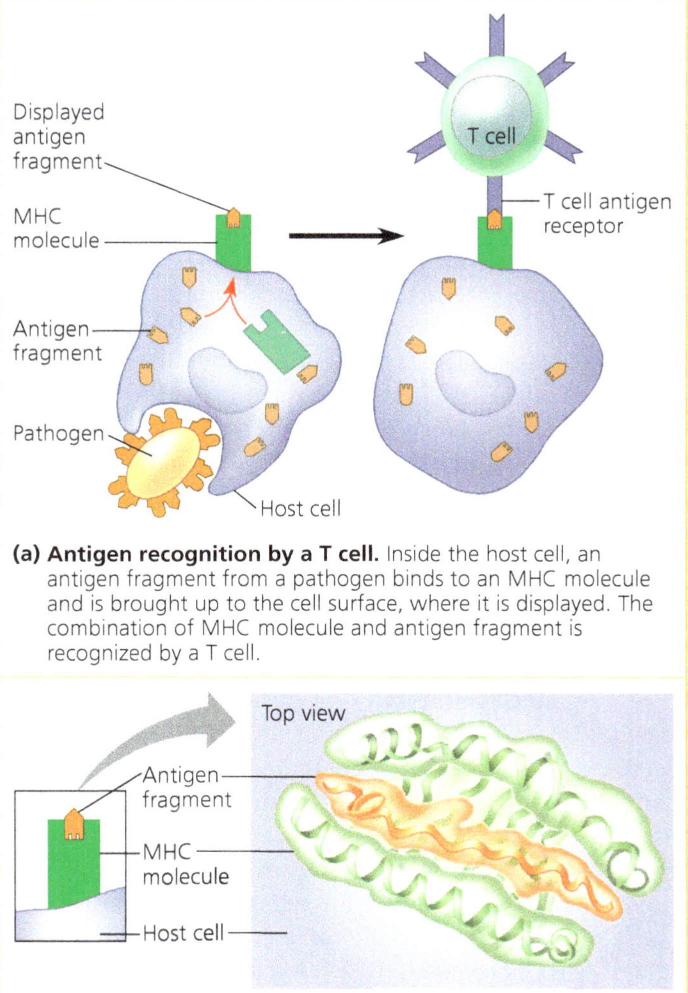

(a) Antigen recognition by a T cell. Inside the host cell, an antigen fragment from a pathogen binds to an MHC molecule and is brought up to the cell surface, where it is displayed. The combination of MHC molecule and antigen fragment is recognized by a T cell.

(b) A closer look at antigen presentation. As shown in this ribbon model, the top of the MHC molecule cradles an antigen fragment, like a bun holding a hot dog. An MHC molecule can display many different antigen fragments, but the antigen receptor of a T cell is specific for a single antigen fragment.

B Cell and T Cell Development

Now that you know how B cells and T cells recognize antigens, let's consider four major characteristics of adaptive immunity. First, there is an immense diversity of lymphocytes and receptors, enabling the immune system to detect pathogens never before encountered. Second, adaptive immunity normally has self-tolerance, the lack of reactivity against an animal's own molecules and cells. Third, cell proliferation triggered by activation greatly increases the number of B and T cells specific for an antigen. Fourth, there is a stronger and more rapid response to an antigen encountered previously, due to a feature known as *immunological memory*.

Receptor diversity and self-tolerance arise as a lymphocyte matures. Proliferation of cells and the formation of immunological memory occur later, after a mature lymphocyte encounters and binds to a specific antigen. We'll consider these four characteristics in the order in which they develop.

Generation of B and T Cell Diversity

Each person makes more than 1 million different B cell antigen receptors and 10 million different T cell antigen receptors. Yet there are only about 20,000 protein-coding genes in the human genome. How, then, do we generate such remarkable diversity in antigen receptors? The answer lies in combinations. Think of selecting a car with a choice of three interior colors and six exterior colors. There are 18 (3×6) color combinations to consider. Similarly, by combining variable elements, the immune system assembles many different receptors from a much smaller collection of parts.

To understand the origin of receptor diversity, let's consider an immunoglobulin (Ig) gene that encodes the light chain of both secreted antibodies (immunoglobulins) and membrane-bound B cell antigen receptors. Although we'll analyze only a single Ig light-chain gene, all B and T cell antigen receptor genes undergo very similar transformations.

The capacity to generate diversity is built into the structure of Ig genes. A receptor light chain is encoded by three gene segments: a variable (V) segment, a joining (J) segment, and a constant (C) segment. The V and J segments together encode the variable region of the receptor chain, while the C segment encodes the constant region. The light-chain gene contains a single C segment, 40 different V segments, and 5 different J segments. These alternative copies of the V and J segments are arranged within the gene in a

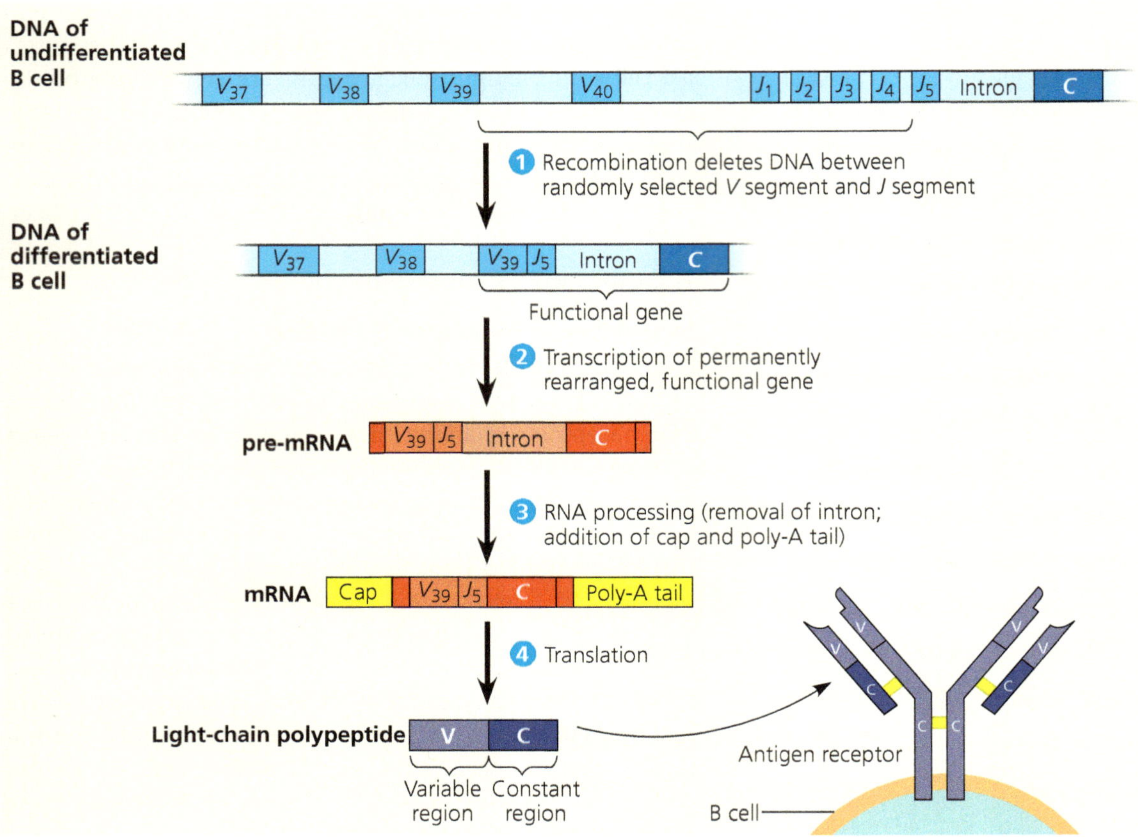

Figure 43.13 Immunoglobulin (antibody) gene rearrangement. The joining of randomly selected *V* and *J* gene segments (V_{39} and J_5 in this example) results in a functional gene that encodes the light-chain polypeptide of a B cell antigen receptor. Transcription, splicing, and translation result in a light chain that combines with a polypeptide produced from an independently rearranged heavy-chain gene to form a functional receptor. Mature B cells (and T cells) are exceptions to the generalization that all nucleated cells in the body have exactly the same DNA.

MAKE CONNECTIONS

Both alternative splicing (see Figure 18.13 on p. 363) and joining of V and J segments by recombination generate diverse gene products from a limited set of gene segments. How do these processes differ?

series **(Figure 43.13)**. Because a functional gene is built from one copy of each type of segment, the pieces can be combined in 200 different ways (40 $V \times 5\ J \times 1\ C$). The number of different heavy-chain combinations is even greater, resulting in even more diversity.

Assembling a functional Ig gene requires rearranging the DNA. Early in B cell development, an enzyme complex called *recombinase* links one light-chain *V* gene segment to one *J* gene segment. This recombination event eliminates the long stretch of DNA between the segments, forming a single exon that is part *V* and part *J*. Because there is only an intron between the *J* and *C* DNA segments, no further DNA rearrangement is required. Instead, the *J* and *C* segments of the RNA transcript will be joined when splicing removes the intervening RNA (see Figure 17.11 to review RNA splicing).

Recombinase acts randomly, linking any one of the 40 *V* gene segments to any one of the 5 *J* gene segments. Heavy-chain genes undergo a similar rearrangement. In any given cell, however, only one allele of a light-chain gene and one allele of a heavy-chain gene are rearranged. Furthermore, the rearrangements are permanent and are passed on to the daughter cells when the lymphocyte divides.

After both the light- and heavy-chain genes have rearranged, antigen receptors can be synthesized. The rearranged genes are transcribed, and the transcripts are processed for translation. Following translation, the light chain and heavy chain assemble together, forming an antigen receptor (see Figure 43.13). Each pair of randomly rearranged heavy and light chains results in a different antigen-binding site. For the total population of B cells in a human body, the number of such combinations has been calculated as 3.5×10^6. Furthermore, mutations introduced during *VJ* recombination add additional variation, making the number of possible antigen-binding specificities even greater.

Origin of Self-Tolerance

How does adaptive immunity distinguish self from nonself? Because antigen receptor genes are randomly rearranged, some immature lymphocytes produce receptors specific for epitopes on the organism's own molecules. If these self-reactive lymphocytes were not eliminated or inactivated, the immune system could not distinguish self from nonself and would attack body proteins, cells, and tissues. Instead, as lymphocytes mature in the bone marrow or thymus, their antigen receptors are tested for self-reactivity. Some B and T cells with receptors specific for the body's own molecules are destroyed by *apoptosis*, which is a programmed cell death (see Chapter 11). The remaining self-reactive lymphocytes are typically rendered nonfunctional, leaving only those that react to foreign molecules. Since the body normally lacks mature lymphocytes that can react against its own components, the immune system is said to exhibit *self-tolerance*.

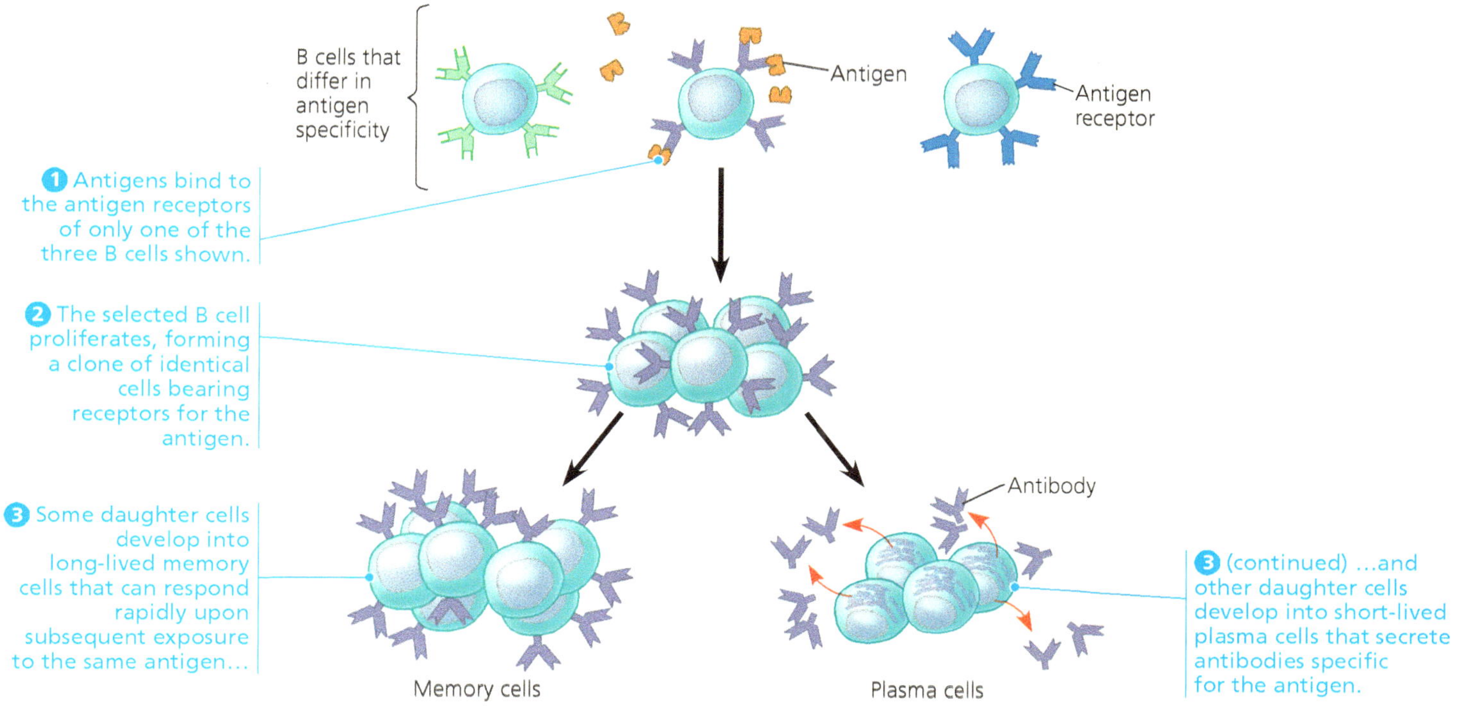

▲ **Figure 43.14 Clonal selection.** This figure illustrates clonal selection, using B cells as an example. In response to a specific antigen and to immune cell signals (not shown), one B cell divides and forms a clone of cells. The remaining B cells, which have antigen receptors specific for other antigens, do not respond. The clone of cells formed by the selected B cell gives rise to memory B cells and antibody-secreting plasma cells. T cells also undergo clonal selection, generating memory T cells and effector T cells (cytotoxic T cells and helper T cells).

Proliferation of B Cells and T Cells

Despite the enormous variety of antigen receptors, only a tiny fraction are specific for a given epitope. So how is adaptive immunity so effective? To begin with, an antigen is presented to a steady stream of lymphocytes in the lymph nodes (see Figure 43.7) until a match is made. A successful match then triggers changes in cell number and activity for the lymphocyte to which an antigen has bound.

The binding of an antigen receptor to an epitope initiates events that activate the lymphocyte. Once activated, a B cell or T cell undergoes multiple cell divisions. For each activated cell, the result of this proliferation is a clone, a population of cells that are identical to the original cell. Some cells from this clone become **effector cells**, short-lived cells that take effect immediately against the antigen and any pathogens producing that antigen. The effector forms of B cells are plasma cells, which secrete antibodies. The effector forms of T cells are helper T cells and cytotoxic T cells, whose roles we'll explore in Concept 43.3. The remaining cells in the clone become **memory cells**, long-lived cells that can give rise to effector cells if the same antigen is encountered later in the animal's life.

Figure 43.14 summarizes the proliferation of a lymphocyte into a clone of cells in response to binding to an antigen, using B cells as an example. This process is called **clonal selection** because an encounter with an antigen *selects* which lymphocyte will divide to produce a *clonal* population of thousands of cells specific for a particular epitope.

Immunological Memory

Immunological memory is responsible for the long-term protection that a prior infection or vaccination provides against many diseases, such as chickenpox. This type of protection was noted almost 2,400 years ago by the Greek historian Thucydides. He observed that individuals who had recovered from the plague could safely care for those who were sick or dying, "for the same man was never attacked twice—never at least fatally."

Prior exposure to an antigen alters the speed, strength, and duration of the immune response. The production of effector cells from a clone of lymphocytes during the first exposure to an antigen is the basis for the **primary immune response**. The primary response peaks about 10–17 days after the initial exposure. During this time, selected B cells and T cells give rise to their effector forms. If an individual is exposed again to the same antigen, the response is faster (typically peaking only 2–7 days after exposure), of greater magnitude, and more prolonged. This is the **secondary immune response**, a hallmark of adaptive, or acquired, immunity. Because selected B cells give

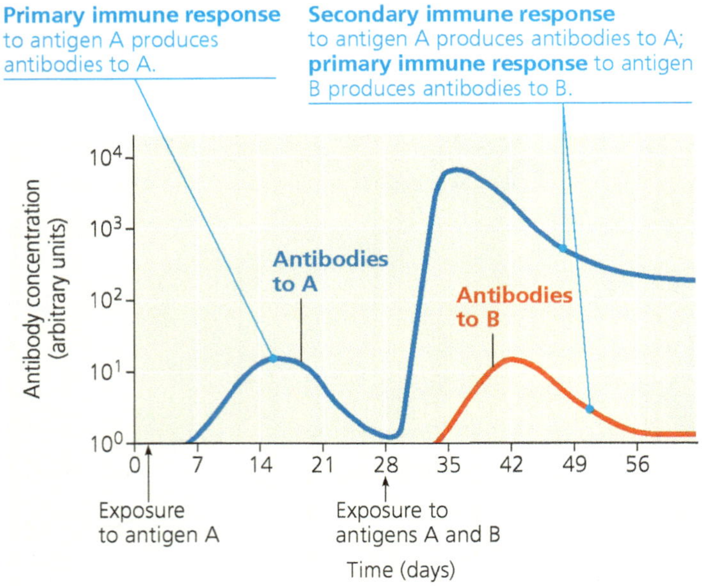

▲ **Figure 43.15 The specificity of immunological memory.** Long-lived memory cells generated in the primary response to antigen A give rise to a heightened secondary response to the same antigen, but do not affect the primary response to a different antigen (B).

rise to antibody-secreting effector cells, measuring the concentrations of specific antibodies in blood over time distinguishes the primary and secondary immune responses (**Figure 43.15**).

The secondary immune response relies on the reservoir of T and B memory cells generated following initial exposure to an antigen. Because these cells are long-lived, they provide the basis for immunological memory, which can span many decades. (Effector cells have much shorter life spans, which is why the immune response diminishes after an infection is overcome.) If an antigen is encountered again, memory cells specific for that antigen enable the rapid formation of clones of thousands of effector cells also specific for that antigen, thus generating a greatly enhanced immune defense.

Although the processes for antigen recognition, clonal selection, and immunological memory are similar for B cells and T cells, these two classes of lymphocytes fight infection in different ways and in different settings, as we'll explore next.

CONCEPT CHECK 43.2

1. **DRAW IT** Sketch a B cell antigen receptor. Label the V and C regions of the light and heavy chains. Label the antigen-binding sites, disulfide bridges, and transmembrane region. Where are these features located relative to the V and C regions?
2. Explain two advantages of having memory cells when a pathogen is encountered for a second time.
3. **WHAT IF?** If both copies of a light-chain gene and a heavy-chain gene recombined in each (diploid) B cell, how would this affect B cell development?

For suggested answers, see Appendix A.

CONCEPT 43.3

Adaptive immunity defends against infection of body fluids and body cells

Having considered how clones of lymphocytes arise, we now explore how these cells help fight infections and minimize damage by pathogens. The activities of B and T lymphocytes produce a humoral immune response and a cell-mediated immune response. The **humoral immune response** occurs in the blood and lymph, which were long ago called body humors (fluids). In the humoral response, antibodies help neutralize or eliminate toxins and pathogens in the blood and lymph. In the **cell-mediated immune response**, specialized T cells destroy infected host cells. Both responses include a primary immune response and a secondary immune response, with memory cells enabling the secondary response.

Helper T Cells: A Response to Nearly All Antigens

A type of T cell called a **helper T cell** triggers both the humoral and cell-mediated immune responses. Helper T cells themselves do not carry out those responses. Instead, signals from helper T cells initiate production of antibodies that neutralize pathogens and activate T cells that kill infected cells.

Two requirements must be met for a helper T cell to activate adaptive immune responses. First, a foreign molecule must be present that can bind specifically to the antigen receptor of the T cell. Second, this antigen must be displayed on the surface of an **antigen-presenting cell**. The antigen-presenting cell can be a dendritic cell, macrophage, or B cell.

When host cells are infected, they too display antigens on their surface. What then distinguishes an antigen-presenting cell? The answer lies in the existence of two classes of MHC molecules. Most body cells have only class I MHC molecules, but antigen-presenting cells have both class I and class II MHC molecules. The class II molecules provide a molecular signature by which an antigen-presenting cell is recognized.

A helper T cell and the antigen-presenting cell displaying its specific epitope have a complex interaction (**Figure 43.16**). The antigen receptors on the surface of the helper T cell bind to the antigen fragment and to the class II MHC molecule displaying that fragment on the antigen-presenting cell. At the same time, an accessory protein on the helper T cell surface binds to the class II MHC molecule, helping keep the cells joined. As the two cells interact, signals in the form of cytokines are exchanged in both directions. For example, the cytokines secreted from a dendritic cell act in combination with the antigen to stimulate the helper T cell, causing it to produce its own set of cytokines. Also, extensive contact between the cell surfaces enables further information exchange.

The different types of antigen-presenting cells interact with helper T cells in distinct contexts. Antigen presentation

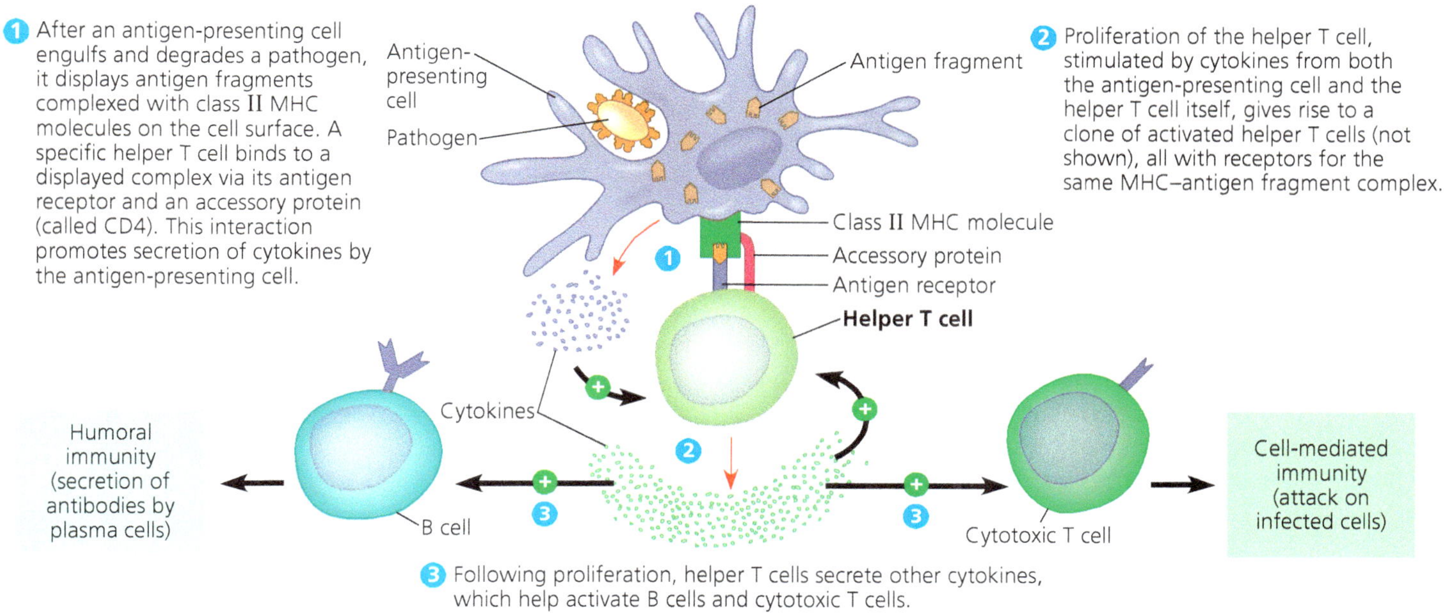

▲ **Figure 43.16 The central role of helper T cells in humoral and cell-mediated immune responses.** In this example, a helper T cell responds to a dendritic cell displaying a microbial antigen.

by a dendritic cell or macrophage activates a helper T cell. The helper T cell then proliferates, forming a clone of activated helper T cells. The B cells present antigens to *already* activated helper T cells, which in turn activate the B cells themselves. Activated helper T cells also help stimulate cytotoxic T cells, as we'll discuss next.

Cytotoxic T Cells: A Response to Infected Cells

In the cell-mediated immune response, **cytotoxic T cells** are the effector cells. The term *cytotoxic* refers to their use of toxic gene products to kill infected cells. To become active, they require signaling molecules from helper T cells as well as interaction with a cell that presents an antigen. Once activated, cytotoxic T cells can eliminate cells that are infected by viruses or other intracellular pathogens.

Fragments of foreign proteins produced in infected host cells associate with class I MHC molecules and are displayed on the cell surface, where they can be recognized by cytotoxic T cells (**Figure 43.17**). As with helper T cells, cytotoxic T cells have an accessory protein that binds to the MHC molecule, helping keep the two cells in contact while the T cell is activated.

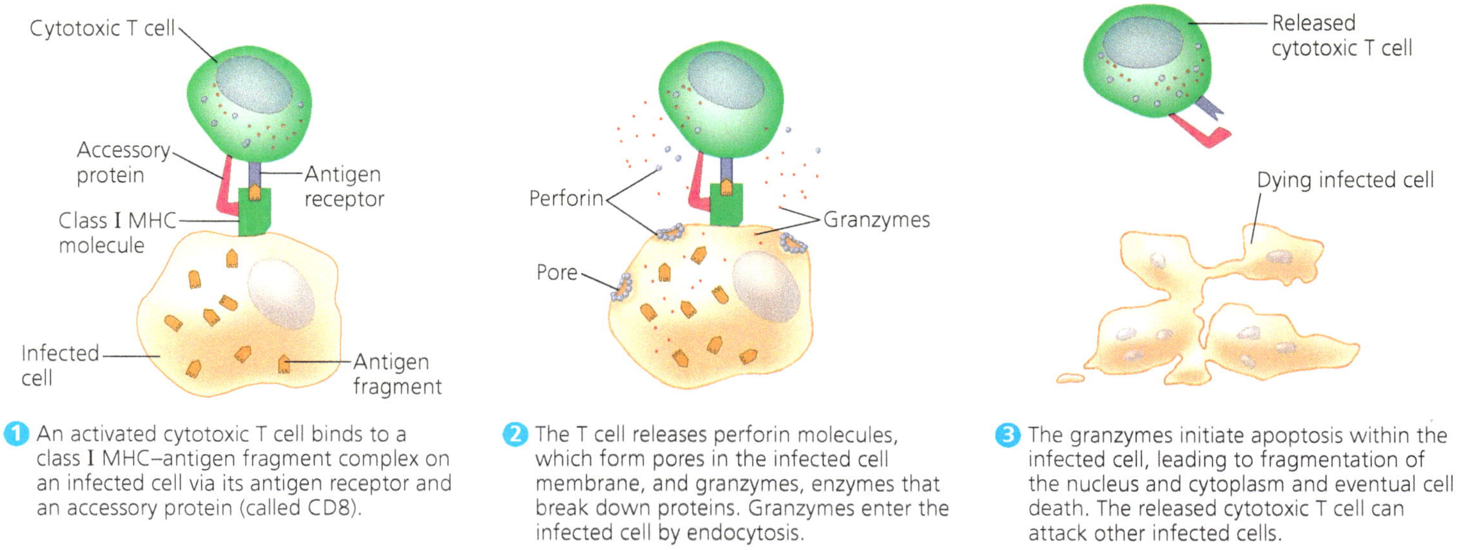

▲ **Figure 43.17 The killing action of cytotoxic T cells on an infected host cell.** An activated cytotoxic T cell releases molecules that make pores in an infected cell's membrane and enzymes that break down proteins, promoting the cell's death.

The targeted destruction of an infected host cell by a cytotoxic T cell involves the secretion of proteins that disrupt membrane integrity and trigger apoptosis (see Figure 43.17). The death of the infected cell not only deprives the pathogen of a place to reproduce, but also exposes cell contents to circulating antibodies, which mark them for disposal. After destroying an infected cell, the cytotoxic T cell can move on and kill other cells infected with the same pathogen.

B Cells and Antibodies: A Response to Extracellular Pathogens

The secretion of antibodies by clonally selected B cells is the hallmark of the humoral immune response. We'll explore how B cells become activated before investigating how antibodies function.

Activation of B Cells

Activation of the humoral immune response typically involves B cells and helper T cells as well as proteins on the surface of pathogens. As depicted in **Figure 43.18**, B cell activation by an antigen is aided by cytokines secreted from helper T cells that have encountered the same antigen. Stimulated by both an antigen and cytokines, the B cell proliferates and differentiates into memory B cells and antibody-secreting effector cells called **plasma cells**.

The pathway for antigen processing and display in B cells differs from that in other antigen-presenting cells. A macrophage or dendritic cell can present fragments from a wide variety of protein antigens, whereas a B cell presents only the antigen to which it specifically binds. When an antigen first binds to receptors on the surface of a B cell, the cell takes in a few foreign molecules by receptor-mediated endocytosis (see Figure 7.22). The class II MHC protein of the B cell then presents an antigen fragment to a helper T cell. This direct cell-to-cell contact is usually critical to B cell activation (see step 2 in Figure 43.18).

B cell activation leads to a robust humoral immune response: An activated B cell gives rise to thousands of identical plasma cells. These plasma cells stop expressing a membrane-bound antigen receptor and begin producing and secreting antibodies (see step 3 in Figure 43.18). Each plasma cell secretes approximately 2,000 antibodies every second of the cell's 4- to 5-day life span. Furthermore, most antigens recognized by B cells contain multiple epitopes. An exposure to a single antigen therefore normally activates a variety of B cells, with different plasma cells producing antibodies directed against different epitopes on the common antigen.

Antibody Function

Antibodies do not kill pathogens, but by binding to antigens, they mark pathogens in various ways for inactivation or

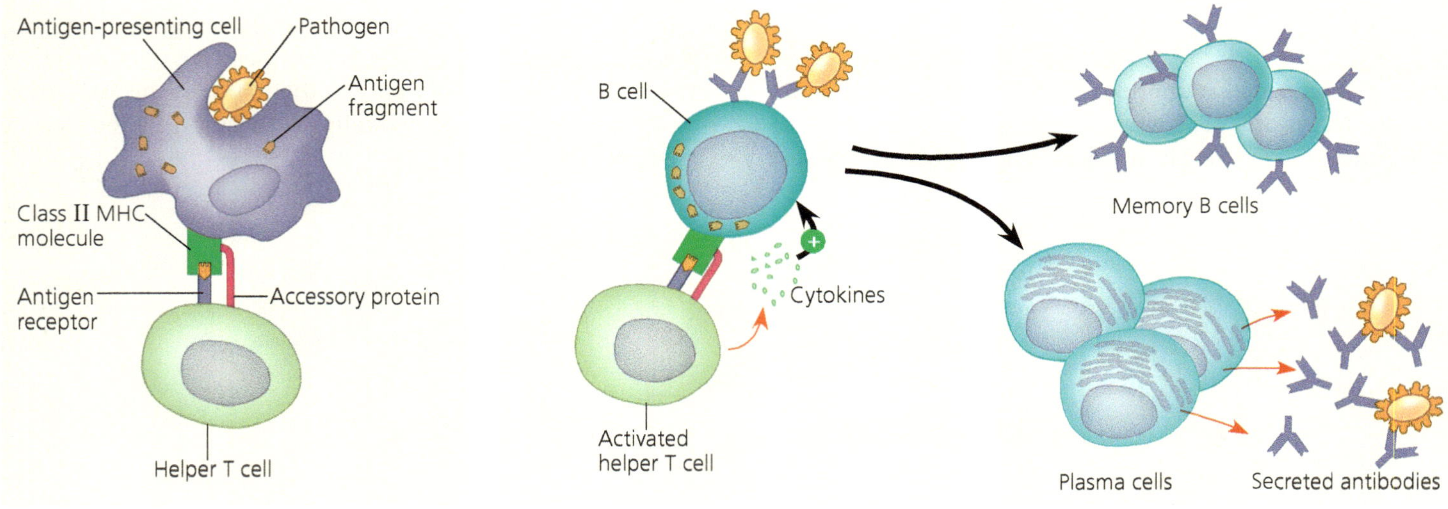

1 After an antigen-presenting cell engulfs and degrades a pathogen, it displays an antigen fragment complexed with a class II MHC molecule. A helper T cell that recognizes the complex is activated with the aid of cytokines secreted from the antigen-presenting cell.

2 When a B cell with receptors for the same epitope internalizes the antigen, it displays an antigen fragment on the cell surface in a complex with a class II MHC molecule. An activated helper T cell bearing receptors specific for the displayed fragment binds to the B cell. This interaction, with the aid of cytokines from the T cell, activates the B cell.

3 The activated B cell proliferates and differentiates into memory B cells and antibody-secreting plasma cells. The secreted antibodies are specific for the same antigen that initiated the response.

▲ **Figure 43.18 Activation of a B cell in the humoral immune response.** Most protein antigens require activated helper T cells to trigger a humoral response. A macrophage (shown here) or a dendritic cell can activate a helper T cell, which in turn can activate a B cell to give rise to antibody-secreting plasma cells.

? *What function do cell-surface antigen receptors play for memory B cells?*

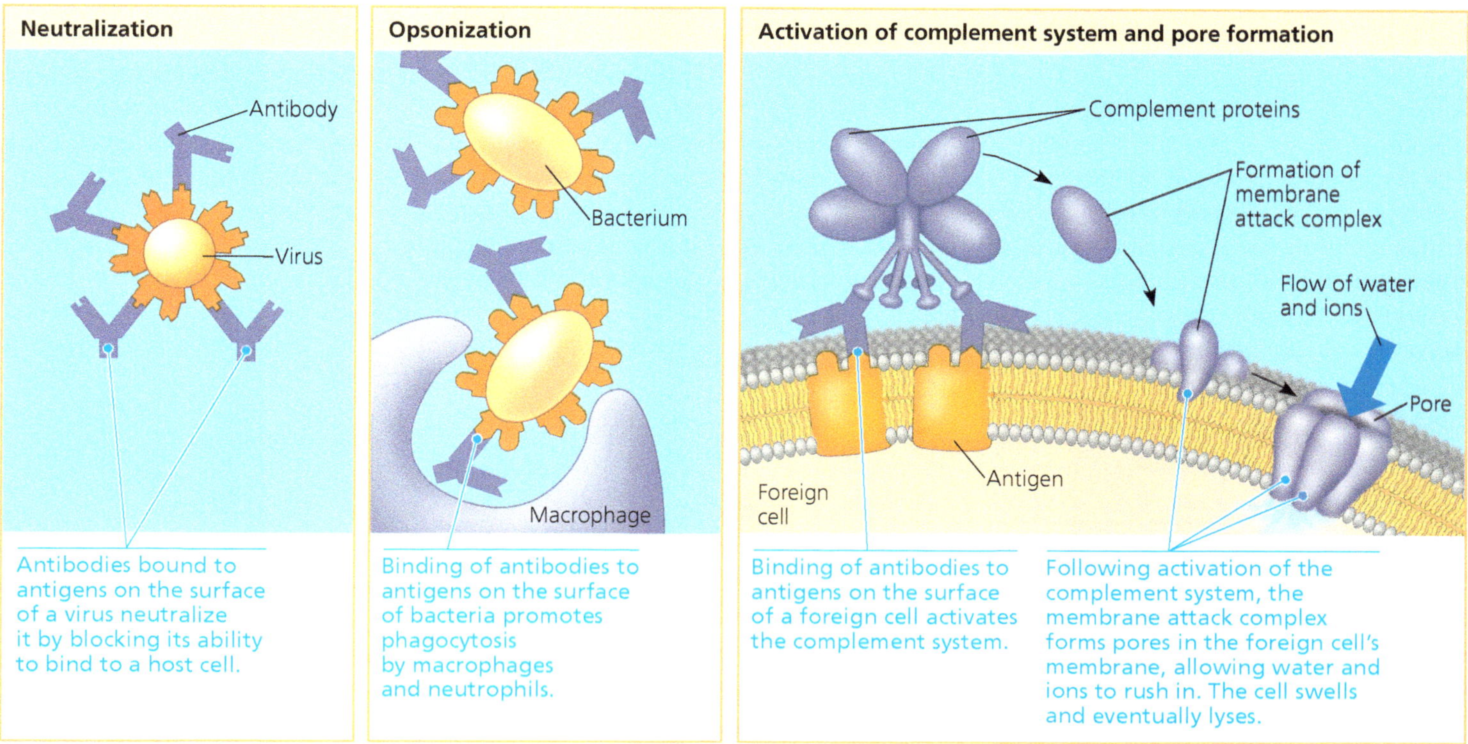

▼ **Figure 43.19 Antibody-mediated mechanisms of antigen disposal.**

destruction. In the simplest of these activities, *neutralization*, antibodies bind to viral surface proteins (**Figure 43.19**, left). The bound antibodies prevent infection of a host cell, thus neutralizing the virus. Similarly, antibodies sometimes bind to toxins released in body fluids, preventing the toxins from entering body cells. In another process, called *opsonization*, antibodies bound to antigens on bacteria present a readily recognized structure for macrophages or neutrophils and therefore increase phagocytosis (**Figure 43.19**, middle). Because each antibody has two antigen-binding sites, antibodies sometimes also facilitate phagocytosis by linking bacterial cells, virus particles, or other foreign substances into aggregates.

Antibodies sometimes work together with the proteins of the complement system to dispose of pathogens. (The name *complement* reflects the fact that these proteins increase the effectiveness of antibody-directed attacks on bacteria.) Binding of a complement protein to an antigen-antibody complex on a foreign cell (or an enveloped virus) triggers a cascade in which each protein of the complement system activates the next protein. Ultimately, activated complement proteins generate a *membrane attack complex* that forms a pore in the membrane of the foreign cell. Ions and water rush into the cell, causing it to swell and lyse (**Figure 43.19**, right).Whether activated as part of innate defenses or as part of adaptive defenses, this cascade of complement protein activity results in the lysis of foreign cells and produces factors that promote inflammation or stimulate phagocytosis.

When antibodies facilitate phagocytosis (see Figure 43.19, middle), they also help fine-tune the humoral immune response. Recall that phagocytosis enables macrophages and dendritic cells to present antigens to and stimulate helper T cells, which in turn stimulate the very B cells whose antibodies contribute to phagocytosis. This positive feedback between innate and adaptive immunity contributes to a coordinated, effective response to infection.

Although antibodies are the cornerstone of the response in body fluids, there is also a mechanism by which they can bring about the death of infected body cells. When a virus uses a cell's biosynthetic machinery to produce viral proteins, these viral products can appear on the cell surface. If antibodies specific for epitopes on these viral proteins bind to the exposed proteins, the presence of bound antibody at the cell surface can recruit a natural killer cell. The natural killer cell then releases proteins that cause the infected cell to undergo apoptosis.

B cells can express five different forms of immunoglobulin (Ig). For a given B cell, each form or *class* has an identical antigen-binding specificity, but a distinct heavy-chain C region. The B cell antigen receptor, known as IgD, is membrane bound. The other four classes consist of soluble antibodies. IgM is the first class of soluble antibody produced. IgG, which follows next, is the most abundant antibody in blood. We will learn more about the function of IgG, as well as the two remaining antibody classes (IgA and IgE), as we further explore the role of antibodies in immunity and disease.

Summary of the Humoral and Cell-Mediated Immune Responses

As noted earlier, both the humoral and cell-mediated responses can include primary and secondary immune responses. Memory cells of each type—helper T cell, B cell, and cytotoxic T cell—enable the secondary response. For example, when body fluids are reinfected by a pathogen encountered previously, memory B cells and memory helper T cells initiate a secondary humoral response. **Figure 43.20** reviews the events that initiate humoral and cell-mediated immune responses, highlights the central role of the helper T cell, and serves as a helpful summary of adaptive immunity.

Active and Passive Immunization

Our discussion of adaptive immunity has to this point focused on **active immunity**, the defenses that arise when a pathogen infects the body and prompts a primary or secondary immune response. In contrast, a different type of immunity results when the IgG antibodies in the blood of a pregnant female cross the placenta to her fetus. The transferred antibodies can immediately react with any pathogens for which they are specific. This protection is called **passive immunity** because the antibodies provided by the mother guard against pathogens that have never infected the newborn. Because passive immunity does not involve the recipient's

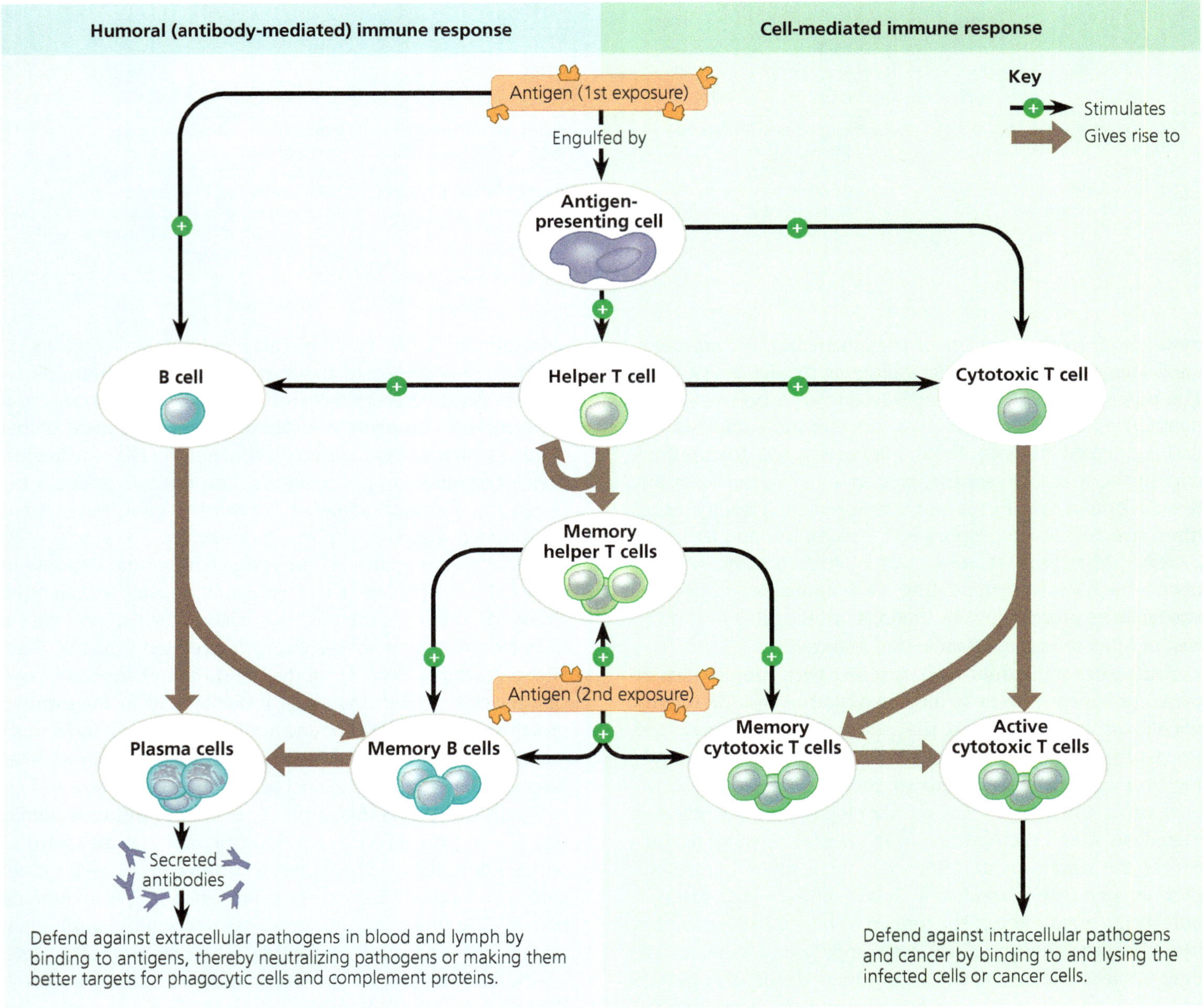

▲ **Figure 43.20 An overview of the adaptive immune response.**

? *Identify each black or brown arrow as representing part of the primary or secondary response.*

ANIMATION **BioFlix** Visit the Study Area at **www.masteringbiology.com** for the BioFlix® 3-D Animation on Immunology.

B and T cells, it persists only as long as the transferred antibodies last (a few weeks to a few months).

After giving birth, a nursing mother continues to transfer protection against disease to her infant. IgA antibodies present in breast milk provide additional passive immunity to the infant's digestive tract while the infant's immune system develops. Later in life, IgA functions in active immunity: IgA antibodies secreted in tears, saliva, and mucus protect the mucous membranes of both males and females.

Both active immunity and passive immunity can be induced artificially. Active immunity can develop from the introduction of antigens into the body through **immunization**. In 1796, Edward Jenner noted that milkmaids who had cowpox, a mild disease usually seen only in cows, did not contract smallpox, a far more dangerous disease. In the first documented immunization (or **vaccination**, from the Latin *vacca*, cow), Jenner used the cowpox virus to induce adaptive immunity against the closely related smallpox virus. Today, many sources of antigen are used to make vaccines, including inactivated bacterial toxins, killed pathogens, parts of pathogens, weakened pathogens that generally do not cause illness, and even genes encoding microbial proteins. Because all of these agents induce a primary immune response and immunological memory, an encounter with the pathogen from which the vaccine was derived triggers a rapid and strong secondary immune response (see Figure 43.15).

Vaccination programs have been successful against many infectious diseases that once killed or incapacitated large numbers of people. A worldwide vaccination campaign led to eradication of smallpox in the late 1970s. In industrialized nations, routine active immunization of infants and children has dramatically reduced the incidence of sometimes devastating diseases, such as polio, measles, and whooping cough. Unfortunately, not all pathogens are easily managed by vaccination. Furthermore, some vaccines are not readily available in impoverished areas of the globe.

Misinformation about vaccine safety and disease risk has led some parents to refuse to immunize their children with available, effective vaccines. The consequence has been a substantial and growing public health problem. Consider measles as just one example. Side effects of immunization are remarkably rare: Fewer than one in a million children suffer a significant allergic reaction to the measles vaccine. The disease, however, is quite dangerous: Roughly one out of every 1,000 patients develop *encephalitis*, an inflammation of the brain. Worldwide, measles kills more than 200,000 people each year. Sadly, declines in measles vaccination rates in parts of the United Kingdom, Russia, and the United States have recently resulted in a number of measles outbreaks and significant numbers of preventable deaths.

In artificial passive immunization, antibodies from an immune animal are injected into a nonimmune animal. For example, humans bitten by venomous snakes are sometimes treated with antivenin, serum from sheep or horses that have been immunized against the venom of one or more species of venomous snakes. When injected immediately after a snakebite, the antibodies in antivenin can neutralize toxins in the venom before the toxins do massive damage.

Antibodies as Tools

The power of antibody specificity and antigen-antibody binding has been harnessed in research, diagnosis, and therapy. Some antibody tools are *polyclonal*: They are the products of many different clones of plasma cells, each specific for a different epitope (**Figure 43.21**). Antibodies that an animal produces after exposure to a microbial antigen are polyclonal. In contrast, other antibody tools are *monoclonal*: They are prepared from a single clone of B cells grown in culture. The **monoclonal antibodies** produced by such a culture are identical and specific for the same epitope on an antigen.

Monoclonal antibodies have provided the basis for many recent advances in medical diagnosis and treatment. For example, home pregnancy kits use monoclonal antibodies to detect human chorionic gonadotropin (hCG). Because hCG is produced as soon as an embryo implants in the uterus (see Chapter 46), the presence of this hormone in a woman's urine is a reliable indicator for a very early stage of pregnancy. In the clinic, monoclonal antibodies are being used to treat many human diseases. For this type of therapy, researchers use mouse B cell clones to identify antibodies specific for an epitope on diseased cells. Next, the mouse antibody genes are altered to code for antibodies that appear less foreign to the human adaptive immune defenses. Scientists then use the "humanized" genes to produce large amounts of antibody for injecting into patients.

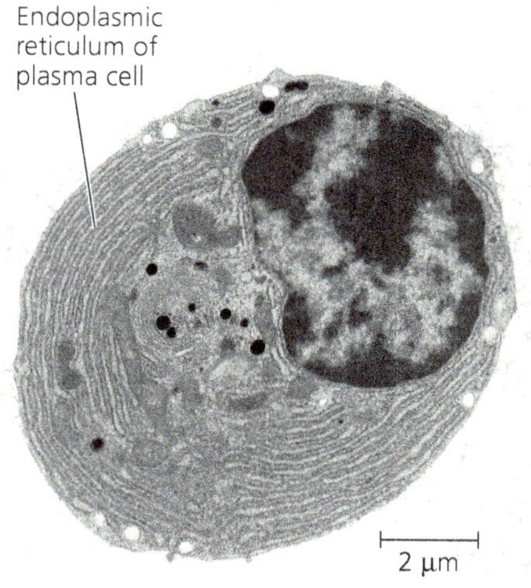

▲ **Figure 43.21 A plasma cell.** A plasma cell contains abundant endoplasmic reticulum, a common feature of cells dedicated to making proteins for secretion (TEM).

Immune Rejection

Like pathogens, cells from another person can be recognized as foreign and attacked by immune defenses. For example, skin transplanted from one person to a genetically nonidentical person will look healthy for a week or so but will then be destroyed (rejected) by the recipient's immune response. Keep in mind that the body's rejection of transplanted tissues or organs or of an incompatible blood transfusion is the expected reaction of a healthy immune system exposed to foreign antigens. (It remains a largely unanswered question why a pregnant woman does not reject her fetus as nonself tissue.)

Blood Groups

To avoid a blood transfusion being recognized as foreign by the recipient's immune system, the ABO blood groups of the donor and recipient must be taken into account. As discussed in Chapter 14, red blood cells are designated as type A if they have the type A carbohydrate on their surface. Similarly, the type B carbohydrate is found on type B red blood cells; both A and B carbohydrates are found on type AB red blood cells; and neither carbohydrate is found on type O red blood cells (see Figure 14.11).

To understand how ABO blood groups affect transfusions, let's consider the immune response of someone with type A blood. It turns out that certain bacteria normally present in the body have epitopes very similar to the A and B carbohydrates. By responding to the bacterial epitope similar to the B carbohydrate, a person with type A blood makes antibodies that will react with the type B carbohydrate. No antibodies are made against the bacterial epitope similar to the type A carbohydrate because lymphocytes reactive with the body's own molecules are inactivated or eliminated during development. If the person with type A blood receives a transfusion of type B blood, that person's anti-B antibodies cause an immediate and devastating transfusion reaction. The transfused red blood cells undergo lysis, which can lead to chills, fever, shock, and kidney malfunction. By the same token, anti-A antibodies in the donated type B blood will act against the recipient's type A red blood cells. Although such interactions prevent type O individuals from receiving transfusions of any other blood type, the recent discovery of enzymes that can cleave the A and B carbohydrates from red blood cells may eliminate this problem.

Tissue and Organ Transplants

In the case of tissue and organ transplants, or grafts, MHC molecules stimulate the immune response that leads to rejection. Each vertebrate species has many alleles for each MHC gene, enabling presentation of antigen fragments that vary in shape and net electrical charge. This diversity of MHC molecules almost guarantees that no two people, except identical twins, will have exactly the same set. Thus, in the vast majority of graft and transplant recipients, some MHC molecules on the donated tissue are foreign to the recipient. To minimize rejection, physicians use donor tissue bearing MHC molecules that match those of the recipient as closely as possible. In addition, the recipient takes medicines that suppress immune responses (but also leave the recipient more susceptible to infections).

Transplants of bone marrow from one person to another can also cause an immune reaction, but for a different reason. Bone marrow transplants are used to treat leukemia and other cancers as well as various hematological (blood cell) diseases. Prior to receiving transplanted bone marrow, the recipient is typically treated with radiation to eliminate his or her own bone marrow cells, thus destroying the source of abnormal cells. This treatment effectively obliterates the recipient's immune system, leaving little chance of graft rejection. However, lymphocytes in the donated marrow may react against the recipient. This *graft versus host reaction* is limited if the MHC molecules of the donor and recipient are well matched. Bone marrow donor programs continually seek volunteers because the great variability of MHC molecules makes a diverse pool of donors essential.

CONCEPT CHECK 43.3

1. If a child were born without a thymus, what cells and functions would be deficient? Explain.
2. Treatment of antibodies with a particular protease clips the heavy chains in half, releasing the two arms of the Y-shaped molecule. How might the antibodies continue to function?
3. **WHAT IF?** Suppose that a snake handler bitten by a particular venomous snake species was treated with antivenin. Why might the same treatment for a second such bite have different results?

For suggested answers, see Appendix A.

CONCEPT 43.4

Disruptions in immune system function can elicit or exacerbate disease

Although adaptive immunity offers significant protection against a wide range of pathogens, it is not fail-safe. In this last section of the chapter, we'll first examine the problems that arise when adaptive immunity is blocked or misregulated. We'll then turn to some of the evolutionary adaptations of pathogens that diminish the effectiveness of host immune responses.

Exaggerated, Self-Directed, and Diminished Immune Responses

The highly regulated interplay among lymphocytes, other body cells, and foreign substances generates an immune response that provides extraordinary protection against many pathogens. When allergic, autoimmune, or immunodeficiency disorders disrupt this delicate balance, the effects are frequently severe and sometimes life-threatening.

Allergies

Allergies are exaggerated (hypersensitive) responses to certain antigens called **allergens**. The most common allergies involve antibodies of the IgE class. Hay fever, for instance, occurs when plasma cells secrete IgE antibodies specific for antigens on the surface of pollen grains (**Figure 43.22**). Some IgE antibodies attach by their base to mast cells in connective tissues. Pollen grains that enter the body later attach to the antigen-binding sites of these IgE antibodies. This attachment links adjacent IgE molecules, inducing the mast cell to release histamine and other inflammatory chemicals from granules (vesicles). Acting on a variety of cell types, these signals bring about the typical allergy symptoms: sneezing, runny nose, teary eyes, and smooth muscle contractions that can result in breathing difficulty. Drugs called antihistamines diminish allergy symptoms (and inflammation) by blocking receptors for histamine.

An acute allergic response sometimes leads to *anaphylactic shock*, a whole-body, life-threatening reaction that can occur within seconds of exposure to an allergen. Anaphylactic shock develops when widespread release of mast cell contents triggers abrupt dilation of peripheral blood vessels, causing a precipitous drop in blood pressure, as well as constriction of bronchioles. Death may occur within minutes due to lack of blood flow and the inability to breathe. Allergic responses to bee venom or penicillin can lead to anaphylactic shock in people who are extremely allergic to these substances. Likewise, people very allergic to peanuts, fish, or shellfish can die from ingesting only tiny amounts of these allergens, which trigger reactions through interactions with mast cells on the surface of the digestive tract. People with severe hypersensitivities often carry syringes containing the hormone epinephrine, which counteracts this allergic response (see Figure 45.8).

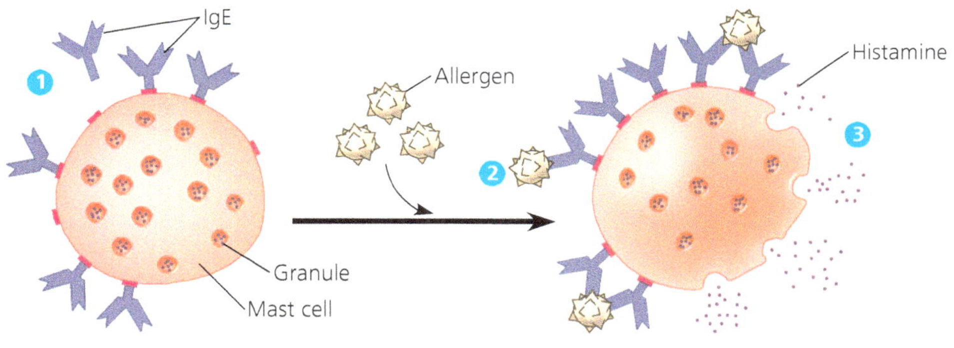

1 IgE antibodies produced in response to initial exposure to an allergen bind to receptors on mast cells.

2 On subsequent exposure to the same allergen, IgE molecules attached to a mast cell recognize and bind the allergen.

3 Cross-linking of adjacent IgE molecules triggers release of histamine and other chemicals, leading to allergy symptoms.

▲ **Figure 43.22 Mast cells, IgE, and the allergic response.** In this example, pollen grains act as the allergen.

Autoimmune Diseases

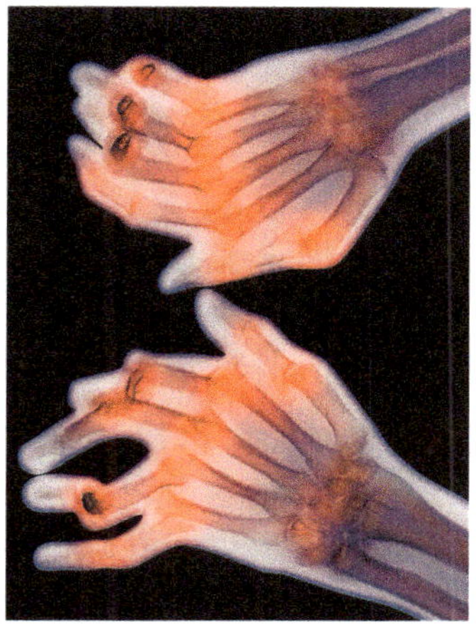

▲ **Figure 43.23 X-ray of hands deformed by rheumatoid arthritis.**

In some people, the immune system is active against particular molecules of the body, causing an **autoimmune disease**. Such a loss of self-tolerance has many forms. In *systemic lupus erythematosus*, commonly called *lupus*, the immune system generates antibodies against histones and DNA released by the normal breakdown of body cells. These self-reactive antibodies cause skin rashes, fever, arthritis, and kidney dysfunction. Another autoimmune disease, *rheumatoid arthritis*, leads to damage and painful inflammation of the cartilage and bone of joints (**Figure 43.23**). In *type 1 diabetes mellitus*, the insulin-producing beta cells of the pancreas are the targets of autoimmune cytotoxic T cells. The most common chronic neurological disorder in developed countries is the autoimmune disease *multiple sclerosis*. In this disease, T cells infiltrate the central nervous system. The result is destruction of the myelin sheath that surrounds parts of many neurons (see Figure 48.12), leading to muscle paralysis through a disruption in neuron function.

Gender, genetics, and environment all influence susceptibility to autoimmune disorders. For example, members of certain families show an increased susceptibility to particular autoimmune disorders. In addition, many autoimmune diseases afflict females more often than males. Women are two to three times more likely than men to suffer from multiple sclerosis and rheumatoid arthritis and nine times more likely to develop lupus. The cause of this sex bias, as well as the rise in autoimmune disease frequency in industrialized countries, is an area of active research and debate. Clearly, much remains to be learned about these often devastating disorders.

Exertion, Stress, and the Immune System

Many forms of exertion and stress influence immune system function. Consider, for example, susceptibility to the common cold and other infections of the upper respiratory tract. Moderate exercise improves immune system function and significantly reduces the risk of these infections. In contrast, exercise to the point of exhaustion leads to more frequent infections and to more severe symptoms. Studies of marathon runners support the conclusion that exercise intensity is the critical variable. On average, such runners get sick less often than their more sedentary peers during training, a time of moderate exertion, but have a marked increase in illness in the period immediately following the grueling race itself. Similarly, psychological stress has been shown to disrupt immune system regulation by altering the interplay of the hormonal, nervous, and immune systems (see Figure 45.21). Recent research also confirms that rest is important for immunity: Adults who averaged fewer than 7 hours of sleep a night got sick three times as often when exposed to a cold virus as individuals who averaged at least 8 hours of sleep.

Immunodeficiency Diseases

A disorder in which an immune system response to antigens is defective or absent is called an **immunodeficiency**. An *inborn immunodeficiency* results from a genetic or developmental defect in the immune system. An *acquired immunodeficiency* develops later in life following exposure to chemical or biological agents. Whatever its cause and nature, an immunodeficiency can lead to frequent and recurrent infections and increased susceptibility to certain cancers.

Inborn immunodeficiencies result from defects in the development of various immune system cells or defects in the production of specific proteins, such as antibodies or the proteins of the complement system. Depending on the specific genetic defect, either innate or adaptive defenses—or both—may be impaired. In severe combined immunodeficiency (SCID), functional lymphocytes are rare or absent. Lacking an adaptive immune response, SCID patients are susceptible to infections, such as pneumonia and meningitis, that can cause death in infancy. Treatments include bone marrow and stem cell transplantation.

Exposure to certain agents can cause immunodeficiencies that develop later in life. Drugs used to fight autoimmune diseases or prevent transplant rejection suppress the immune system, leading to an immunodeficient state. Certain cancers also suppress the immune system, especially Hodgkin's disease, which damages the lymphatic system. Acquired immunodeficiencies range from temporary states that may arise from physiological stress to the devastating **acquired immunodeficiency syndrome (AIDS)**, which is caused by the human immunodeficiency virus (HIV). We will discuss AIDS further in the next section, which focuses on how pathogens escape the adaptive immune response.

Evolutionary Adaptations of Pathogens That Underlie Immune System Avoidance

EVOLUTION Just as immune systems that ward off pathogens have evolved in animals, mechanisms that thwart immune responses have evolved in pathogens. Using human pathogens as examples, we'll examine some common mechanisms: antigenic variation, latency, and direct attack on the immune system.

Antigenic Variation

One mechanism for escaping the body's defenses is for a pathogen to alter how it appears to the immune system. Immunological memory is a record of the foreign epitopes an animal has encountered. If the pathogen that expressed those epitopes no longer does so, it can reinfect or remain in a host without triggering the rapid and robust response that memory cells provide. Such changes in epitope expression, which are called *antigenic variation*, are regular events for some viruses and parasites. The parasite that causes sleeping sickness (trypanosomiasis) provides one example. By periodically switching at random among 1,000 different versions of the protein found over its entire surface, this pathogen can persist in the body without facing an effective adaptive immune response **(Figure 43.24)**.

Antigenic variation is the major reason the influenza, or "flu," virus remains a major public health problem. As it replicates in one human host after another, the human influenza virus mutates. Because any change that lessens recognition by the immune system provides a selective advantage, the virus steadily accumulates such alterations. These changes in the surface proteins of the influenza virus are the reason that a new flu vaccine must be manufactured and distributed each year. Of much greater danger, however, is the fact that the human virus occasionally exchanges genes with influenza viruses that infect domesticated animals, such as pigs or chickens. When this occurs, influenza can take on such a radically different appearance that none of the memory cells in

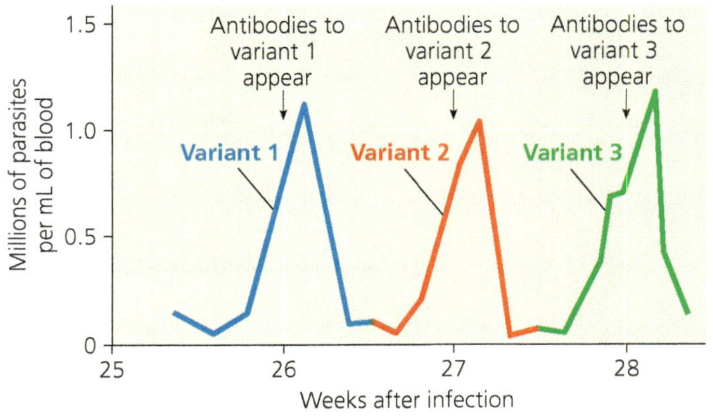

▲ **Figure 43.24 Antigenic variation in the parasite that causes sleeping sickness.** Blood samples taken from a patient during a chronic infection of sleeping sickness reveal cyclic variation in the surface coat protein of the parasite. The infection has become chronic because this weekly variation allows the parasite to evade the adaptive immune response.

the human population recognize the new strain. Such an event led to the influenza outbreak of 1918–1919, which killed more than half a million people in the United States (see Figure 19.9). Worldwide more than 20 million people died, a greater number than had died in World War I.

In 2009, an influenza virus called H1N1 appeared that contained a novel combination of genes from flu viruses that normally circulate in pigs, birds, and humans. The rapid spread of this flu across the human population caused a *pandemic*, an outbreak of worldwide proportions. Fortunately, a rapidly developed H1N1 vaccine soon provided public health officials with an excellent means of slowing the spread of this virus and reducing the impact of the outbreak.

Latency

After infecting a host, some viruses enter a largely inactive state called *latency*. Because such dormant viruses cease making most viral proteins and typically produce no free virus particles, they do not trigger an adaptive immune response. Nevertheless, the viral genome persists in the nuclei of infected cells, either as a separate small DNA molecule or as a copy integrated into the host genome. Latency typically persists until conditions arise that are favorable for viral transmission or unfavorable for host survival, such as when the host is infected by another pathogen. Such circumstances trigger the synthesis and release of virus particles that can infect new hosts.

Herpes simplex viruses, which establish themselves in human sensory neurons, provide a good example of latency. The type 1 virus causes most oral herpes infections, whereas the type 2 virus is responsible for most cases of genital herpes. Because sensory neurons express relatively few MHC I molecules, the infected cells are inefficient at presenting viral antigens to circulating lymphocytes. Stimuli such as fever, emotional stress, or menstruation reactivate the virus to reproduce and infect surrounding epithelial tissues. Activation of the type 1 virus can result in blisters around the mouth that are inaccurately called "cold" sores. The type 2 virus can cause genital sores, but people infected with either type 1 or type 2 virus often lack any apparent symptoms. Infections of the type 2 virus, which is sexually transmitted, pose a serious threat to the babies of infected mothers and can increase transmission of HIV, the virus that causes AIDS.

Attack on the Immune System: HIV

The human immunodeficiency virus (HIV), the pathogen that causes AIDS, both escapes and attacks the adaptive immune response. Once introduced into the body, HIV infects helper T cells with high efficiency. To infect these cells, the virus binds specifically to the CD4 accessory protein (see Figure 43.16). However, HIV also infects some cell types that have low levels of CD4, such as macrophages and brain cells. In the cell, the HIV RNA genome is reverse-transcribed, and the product DNA is integrated into the host cell's genome (see Figure 19.8). In this form, the viral genome can direct production of new virus particles.

Although the body responds to HIV with an immune response sufficient to eliminate most viral infections, some HIV invariably escapes. One reason HIV persists is antigenic variation. The virus mutates at a very high rate during replication. Altered proteins on the surface of some mutated viruses reduce interaction with antibodies and cytotoxic T cells. Such viruses survive, proliferate, and mutate further. The virus thus evolves within the body. The continued presence of HIV is also helped by latency. When the viral DNA integrates into the chromosome of a host cell but does not produce new virus proteins or particles, it is shielded from the immune system by the host cell. This inactive, or latent, viral DNA is also protected from antiviral agents currently used against HIV because they attack only actively replicating viruses.

Over time, an untreated HIV infection not only avoids the adaptive immune response but also abolishes it **(Figure 43.25)**. Viral reproduction and cell death triggered by the virus lead to loss of helper T cells, impairing both humoral and cell-mediated immune responses. The result is a progression to AIDS, characterized by a susceptibility to infections and cancers that a healthy immune system would usually defeat. For example, *Pneumocystis carinii*, a common fungus that does not cause disease in healthy individuals, can result in severe pneumonia in people with AIDS. Likewise, the Kaposi's sarcoma herpesvirus causes a cancer among AIDS patients that is extremely rare in individuals not infected with HIV. Such opportunistic diseases, as well as nerve damage and body wasting, are the primary causes of death in AIDS patients, not the HIV virus itself.

At present, HIV infection cannot be cured, although certain drugs can slow HIV reproduction and the progression to

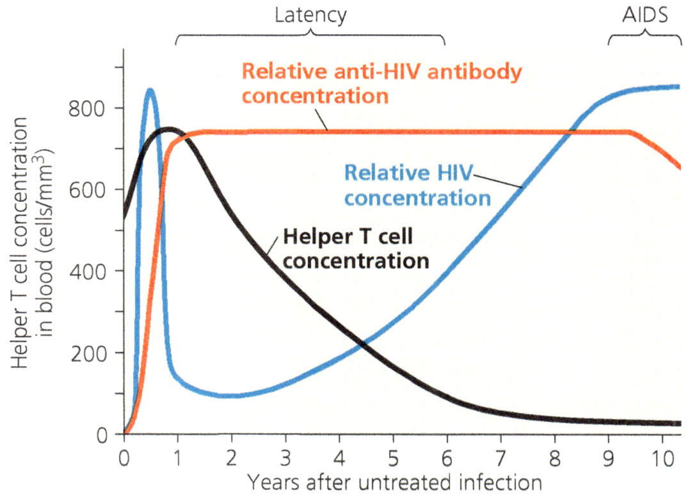

▲ **Figure 43.25 The progress of an untreated HIV infection.**

AIDS. Unfortunately, mutations that occur in each round of viral reproduction can generate strains of HIV that are drug resistant. The impact of such viral drug resistance can be reduced by the use of a combination of drugs; viruses newly resistant to one drug can be defeated by another. However, the appearance of strains resistant to multiple drugs reduces the effectiveness of such multidrug "cocktails" in some patients. Frequent mutations in genes for HIV surface antigens also have hampered efforts to develop an effective vaccine. Worldwide, the AIDS epidemic continues to grow. In 2008, approximately 2 million people died of AIDS, and the disease is now the leading cause of death in Africa.

Transmission of HIV requires the transfer of virus particles or infected cells from person to person via body fluids such as semen, blood, or breast milk. Unprotected sex (that is, without a condom) and transmission via HIV-contaminated needles (typically among intravenous drug users) account for the vast majority of HIV infections. The virus can enter the body through the mucosal linings of the vagina, vulva, penis, or rectum during intercourse or via the mouth during oral sex. The likelihood of transmission is increased by factors that may damage these linings, especially other sexually transmitted infections that cause ulcers or inflammation.

People infected with HIV can transmit the disease in the first few weeks of infection, *before* they express HIV-specific antibodies that can be detected in a blood test (see Figure 43.25). Currently, 10–50% of all new HIV infections appear to be caused by recently infected individuals.

Cancer and Immunity

When adaptive immunity is inactivated, the frequency of certain cancers increases dramatically. For example, the risk of developing Kaposi's sarcoma is 20,000 times greater for untreated AIDS patients than for healthy people. This observation was unanticipated. If the immune system recognizes only nonself, it should fail to recognize the uncontrolled growth of self cells that is the hallmark of cancer. It turns out, however, that viruses are involved in about 15–20% of all human cancers. Because the immune system can recognize viral proteins as foreign, it can act as a defense against viruses that can cause cancer and against cancer cells that harbor viruses.

Scientists have identified six viruses that can cause cancer in humans. The Kaposi's sarcoma herpesvirus is one such virus. Hepatitis B virus, which can trigger liver cancer, is another. A vaccine directed against hepatitis B virus that was introduced in 1986 was demonstrated to be the first vaccine to help prevent a specific human cancer. Rapid progress on virus-induced cancers continues. In 2006, the release of a vaccine against cervical cancer, specifically human papillomavirus (HPV), marked a major victory against a disease that afflicts more than half a million women worldwide every year (**Figure 43.26**).

▼ Figure 43.26

IMPACT

Vaccinating Against Cervical Cancer

In the 1970s, Harald zur Hausen, working in Heidelberg, Germany, proposed that human papillomavirus (HPV) causes cervical cancer. Many scientists were skeptical that cancer could result from infection by HPV, the most common sexually transmitted pathogen. However, after more than a decade of work, zur Hausen isolated two particular types of HPV from patients with cervical cancer. He quickly made samples available to other scientists, leading to development of highly effective vaccines against cervical cancer. In 2008, zur Hausen shared the Nobel Prize in Physiology or Medicine for his discovery. This computer graphic image of an HPV particle illustrates the abundant copies of the capsid protein (yellow) that is used as the antigen in vaccination.

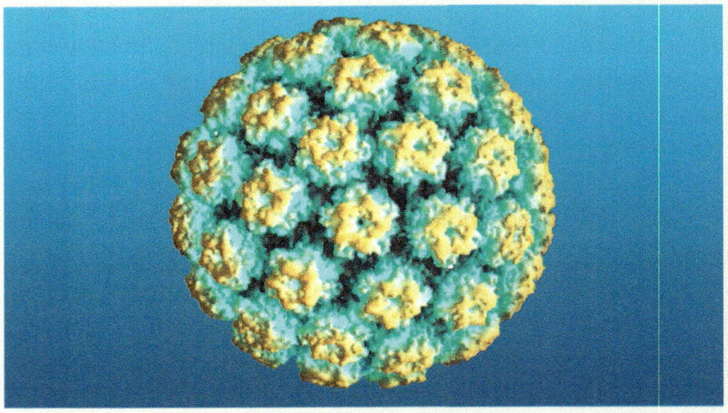

WHY IT MATTERS Cervical cancer kills more than 4,000 women annually in the United States and is the fifth-most common cause of cancer deaths among women worldwide. Administering an HPV vaccine, either Gardasil or Cervarix, to preteen girls and young women greatly reduces their chance of being infected with the HPV viruses that cause most cervical cancers.

FURTHER READING L. R. Badenet et al., Human papillomavirus vaccine: Opportunity and challenge, *New England Journal of Medicine* 356:1990–1991 (2007).

WHAT IF? Suppose you tracked the health of women infected with the types of HPV that cause cancer. Why might only a fraction of such women develop cervical cancer? (*Hint*: Refer to Figure 18.25 on p. 376 and the accompanying text.)

CONCEPT CHECK 43.4

1. In myasthenia gravis, antibodies bind to and block certain receptors on muscle cells, preventing muscle contraction. Is this disease best classified as an immunodeficiency disease, an autoimmune disease, or an allergic reaction? Explain.
2. People with herpes simplex type 1 viruses often get mouth sores when they have a cold or similar infection. How might this location benefit the virus?
3. **WHAT IF?** How would a macrophage deficiency likely affect a person's innate and adaptive defenses?

For suggested answers, see Appendix A.

43 CHAPTER REVIEW

SUMMARY OF KEY CONCEPTS

CONCEPT 43.1

In innate immunity, recognition and response rely on traits common to groups of pathogens (pp. 930–935)

- In both invertebrates and vertebrates, **innate immunity** is mediated by physical and chemical barriers as well as cell-based defenses. Activation of innate immune responses relies on recognition proteins specific for broad classes of **pathogens**. In insects, pathogens that penetrate barrier defenses are ingested by cells in the hemolymph that also release antimicrobial peptides.
- In vertebrates, intact skin and mucous membranes form barriers to pathogens. Mucus produced by membrane cells, the low pH of the skin and stomach, and degradation by **lysozyme** also deter pathogens. Microbes that penetrate barrier defenses are ingested by phagocytic cells, including **macrophages** and **dendritic cells**. Additional cellular defenses include **natural killer cells**, which can induce the death of virus-infected cells. **Complement system** proteins, **interferons**, and other antimicrobial peptides also act against microbes. In the **inflammatory response**, **histamine** and other chemicals released from cells at the injury site promote changes in blood vessels that allow fluid, more phagocytic cells, and antimicrobial peptides to enter tissues.
- Pathogens sometimes evade innate immune defenses. For example, some bacteria have an outer capsule that prevents recognition, while others are resistant to breakdown within lysosomes.

? *In what ways does innate immunity protect the mammalian digestive tract?*

CONCEPT 43.2

In adaptive immunity, receptors provide pathogen-specific recognition (pp. 935–940)

- **Adaptive immunity** relies on **lymphocytes** that arise from stem cells in the bone marrow and complete their maturation in the bone marrow (**B cells**) or in the **thymus** (**T cells**). Lymphocytes have cell-surface **antigen receptors** for foreign molecules. All receptor proteins on a single B or T cell are the same, but there are millions of B and T cells in the body that differ in the foreign molecules that their receptors recognize. Upon infection, B and T cells specific for the pathogen are activated. Some T cells help other lymphocytes; others kill infected host cells. B cells called **plasma cells** produce soluble receptor proteins called **antibodies**, which bind to foreign molecules and cells. The activated lymphocytes called **memory cells** defend against future infections by the same pathogen.
- Recognition of foreign molecules involves the binding of variable regions of receptors to an **epitope**, a small region of an antigen. B cells and antibodies recognize epitopes on the surface of antigens circulating in the blood or lymph. T cells recognize protein epitopes in small antigen fragments (peptides) that are presented on the surface of host cells, complexed with cell-surface proteins called **MHC (major histocompatibility complex) molecules**.
- The four major characteristics of B and T cell development are the generation of cell diversity, self-tolerance, proliferation, and immunological memory.

The following figure uses B cells to illustrate clonal selection:

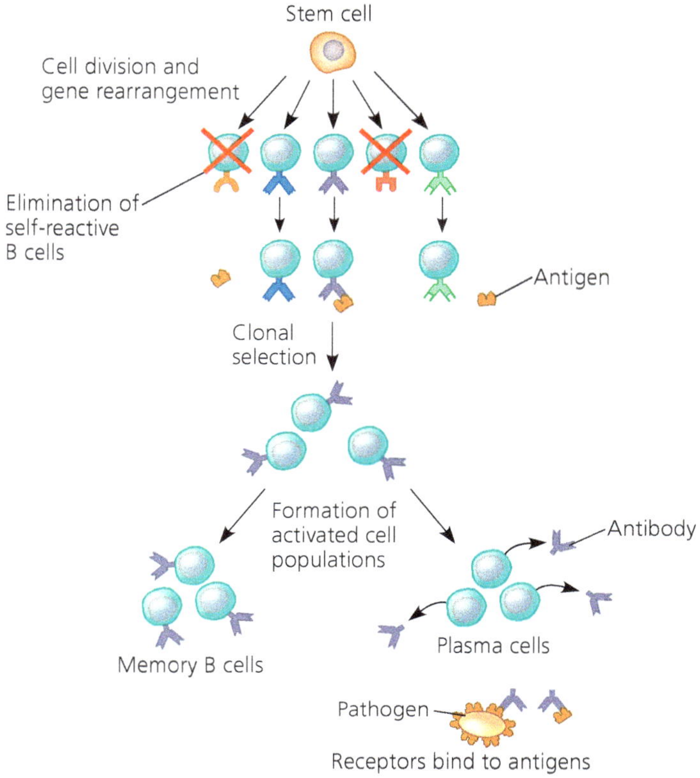

? *Why is the adaptive immune response to an initial infection slower than the innate response?*

CONCEPT 43.3

Adaptive immunity defends against infection of body fluids and body cells (pp. 940–946)

- **Helper T cells** interact with antigen fragments displayed by class II MHC molecules on the surface of dendritic cells, macrophages, and B cells (**antigen-presenting cells**). Activated helper T cells secrete **cytokines** that stimulate other lymphocytes as part of the response to nearly all antigens. **Cytotoxic T cells** bind to a complex of an antigen fragment and a class I MHC molecule on infected host cells. In the **cell-mediated immune response**, activated cytotoxic T cells secrete proteins that initiate destruction of infected cells. All T cells have an accessory protein that enhances binding to MHC–antigen fragment complexes.

  In the **humoral immune response**, B cell antigen receptors and antibodies bind to extracellular foreign substances in blood and lymph.The binding of antibodies helps eliminate antigens by phagocytosis and complement-mediated lysis. The five major antibody classes differ in distribution and function.
- **Active immunity** develops in response to infection or to immunization with a nonpathogenic form or part of a pathogen. Active immunity includes a response to and immunological memory for that pathogen. **Passive immunity**, which provides immediate, short-term protection, is conferred naturally when IgG crosses the placenta from mother to fetus or when IgA passes from mother to infant in breast milk. It also can be conferred artificially by injecting antibodies into a nonimmune person.

- Tissues or cells transferred from one person to another are subject to immune rejection. In tissue grafts and organ transplants, MHC molecules stimulate rejection. Lymphocytes in bone marrow transplants may cause a graft versus host reaction.

? *Is immunological memory after a natural infection fundamentally different from immunological memory after vaccination? Explain.*

CONCEPT 43.4

Disruptions in immune system function can elicit or exacerbate disease (pp. 946–950)

- Disruption of normal immune system regulation or function can result in an exaggerated, self-directed, or diminished response. In localized allergies, IgE attached to **mast cells** induces the cells to release histamine and other mediators that cause vascular changes and allergic symptoms. Loss of self-tolerance can lead to **autoimmune diseases**, such as multiple sclerosis. Inborn **immunodeficiencies** result from defects that interfere with innate, humoral, or cell-mediated defenses. **AIDS** is an acquired immunodeficiency caused by HIV.
- Antigenic variation, latency, and direct assault on the immune system allow some pathogens to thwart immune responses. HIV infection destroys helper T cells, leaving the patient prone to disease. Immune defense against cancer appears to primarily involve action against viruses that can cause cancer, as well as against cancer cells that harbor viruses.

? *Is being infected with HIV the same as having AIDS? Explain.*

TEST YOUR UNDERSTANDING

LEVEL 1: KNOWLEDGE/COMPREHENSION

1. Which of these is *not* part of insect immunity?
   a. enzyme activation of microbe-killing chemicals
   b. activation of natural killer cells
   c. phagocytosis by hemocytes
   d. production of antimicrobial peptides
   e. a protective exoskeleton

2. An epitope associates with which part of an antigen receptor or antibody?
   a. the disulfide bridge
   b. the heavy-chain constant regions only
   c. variable regions of a heavy chain and light chain combined
   d. the light-chain constant regions only
   e. the tail

3. Which statement best describes the difference in responses of effector B cells (plasma cells) and cytotoxic T cells?
   a. B cells confer active immunity; cytotoxic T cells confer passive immunity.
   b. B cells kill pathogens directly; cytotoxic T cells kill host cells.
   c. B cells secrete antibodies against a pathogen; cytotoxic T cells kill pathogen-infected host cells.
   d. B cells carry out the cell-mediated response; cytotoxic T cells carry out the humoral response.
   e. B cells respond the first time a pathogen is present; cytotoxic T cells respond subsequent times.

LEVEL 2: APPLICATION/ANALYSIS

4. Which of the following statements is *not* true?
   a. An antibody has more than one antigen-binding site.
   b. An antigen can have different epitopes.
   c. A pathogen makes more than one antigen.
   d. A lymphocyte has receptors for multiple different antigens.
   e. A liver cell makes one class of MHC molecule.

5. Which of the following should be the same in identical twins?
   a. the set of antibodies produced
   b. the set of MHC molecules produced
   c. the set of T cell antigen receptors produced
   d. the susceptibility to a particular virus
   e. the set of immune cells eliminated as self-reactive

LEVEL 3: SYNTHESIS/EVALUATION

6. Vaccination increases the number of
   a. different receptors that recognize a pathogen.
   b. lymphocytes with receptors that can bind to the pathogen.
   c. epitopes that the immune system can recognize.
   d. macrophages specific for a pathogen.
   e. MHC molecules that can present an antigen.

7. Which of the following would *not* help a virus avoid triggering an adaptive immune response?
   a. having frequent mutations in genes for surface proteins
   b. infecting cells that produce very few MHC molecules
   c. producing proteins very similar to those of other viruses
   d. infecting and killing helper T cells
   e. building the viral shell from host proteins

8. **DRAW IT** Consider a pencil-shaped protein with two epitopes, Y (the "eraser" end) and Z (the "point" end). They are recognized by antibodies A1 and A2, respectively. Draw and label a picture showing the antibodies linking proteins into a complex that could trigger endocytosis by a macrophage.

9. **MAKE CONNECTIONS** Contrast Lamarck's idea for the inheritance of acquired characteristics, discussed on pp. 454–455 of Concept 22.1, with the clonal selection of lymphocytes.

10. **EVOLUTION CONNECTION**
    Describe one invertebrate defense mechanism and discuss how it is an evolutionary adaptation retained in vertebrates.

11. **SCIENTIFIC INQUIRY**
    A diagnostic test for tuberculosis (TB) involves injecting antigen (from the bacterium that causes TB) under the skin and then waiting a few days for a reaction to appear. This test is *not* useful for diagnosing TB in AIDS patients. Why?

12. **WRITE ABOUT A THEME**
    The Genetic Basis of Life Among all nucleated body cells, only B and T cells lose DNA during their development and maturation. In a short essay (100–150 words), discuss the relationship between this loss and the theme of DNA as heritable biological information, focusing on similarities between cellular and organismal generations.

For selected answers, see Appendix A.

MasteringBIOLOGY www.masteringbiology.com

1. MasteringBiology® Assignments

Tutorial Acquired Immunity
Activities The Inflammatory Response • Immune Responses • The Adaptive Immune Response • Discovery Channel Video: Vaccines • HIV Reproductive Cycle • Discovery Channel Video: Fighting Cancer
Questions Student Misconceptions • Reading Quiz • Multiple Choice • End-of-Chapter

2. eText

Read your book online, search, take notes, highlight text, and more.

3. The Study Area

Practice Tests • Cumulative Test • ***BioFlix*** 3-D Animations • MP3 Tutor Sessions • Videos • Activities • Investigations • Lab Media • Audio Glossary • Word Study Tools • Art

APPENDIX A Answers

Chapter 2

Figure Questions

Figure 2.2 The most significant difference in the results would be that the two *Cedrela* saplings inside each garden would show similar amounts of dying leaf tissue because a poisonous chemical released from the *Duroia* trees would presumably reach the saplings via the air or soil and would not be blocked by the insect barrier. The *Cedrela* saplings planted outside the gardens would not show damage unless *Duroia* trees were nearby. Also, any ants present on the unprotected *Cedrela* saplings inside the gardens would probably not be observed making injections into the leaves. However, formic acid would likely still be found in the ants' glands, as it is for most species of ants. **Figure 2.9** Atomic number = 12; 12 protons, 12 electrons; 3 electron shells; 2 valence electrons **Figure 2.16** One possible answer:

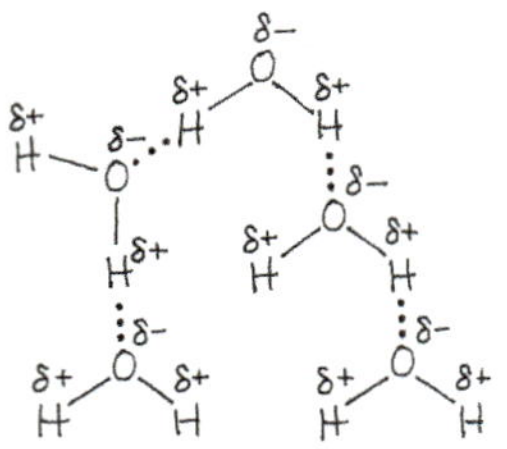

Figure 2.19 The plant is submerged in water (H_2O), in which the CO_2 is dissolved. The sun's energy is used to make sugar, which is found in the plant and can act as food for the plant itself, as well as for animals that eat the plant. The oxygen (O_2) is present in the bubbles.

Concept Check 2.1

1. Table salt (sodium chloride) is made up of sodium and chlorine. We are able to eat the compound, showing that it has different properties from those of a metal (sodium) and a poisonous gas (chlorine). **2.** Yes, because an organism requires trace elements, even though only in small amounts **3.** A person with an iron deficiency will probably show fatigue and other effects of a low oxygen level in the blood. (The condition is called anemia and can also result from too few red blood cells or abnormal hemoglobin.) **4.** Variant ancestral plants that could tolerate the toxic elements could grow and reproduce in serpentine soils. (Plants that were well adapted to nonserpentine soils would not be expected to survive in serpentine areas.) The offspring of the variants would also vary, with those most capable of thriving under serpentine conditions growing best and reproducing most. Over many generations, this probably led to the serpentine-adapted species we see today.

Concept Check 2.2

1. 7 **2.** $^{15}_{7}N$ **3.** 9 electrons; two electron shells; 1*s*, 2*s*, 2*p* (three orbitals); 1 electron is needed to fill the valence shell. **4.** The elements in a row all have the same number of electron shells. In a column, all the elements have the same number of electrons in their valence shells.

Concept Check 2.3

1. Each carbon atom has only three covalent bonds instead of the required four. **2.** The attraction between oppositely charged ions, forming ionic bonds **3.** If you could synthesize molecules that mimic these shapes, you might be able to treat diseases or conditions caused by the inability of affected individuals to synthesize such molecules.

Concept Check 2.4

1.

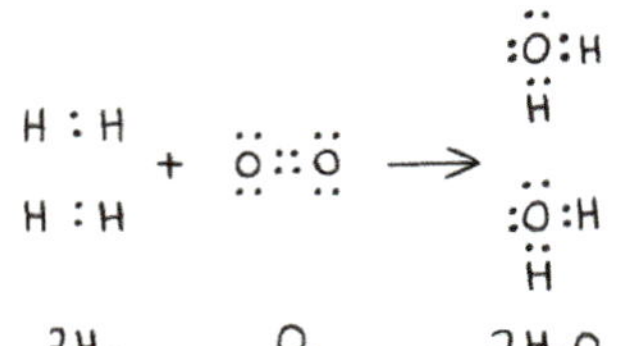

2. At equilibrium, the forward and reverse reactions occur at the same rate. **3.** $C_6H_{12}O_6 + 6\ O_2 \rightarrow 6\ CO_2 + 6\ H_2O$ + Energy. Glucose and oxygen react to form carbon dioxide and water, releasing energy. We breathe in oxygen because we need it for this reaction to occur, and we breathe out carbon dioxide because it is a by-product of this reaction. (This reaction is called cellular respiration, and you will learn more about it in Chapter 9.)

Summary of Key Concepts Questions

2.1 Iodine (part of a thyroid hormone) and iron (part of hemoglobin in blood) are both trace elements, required in minute quantities. Calcium and phosphorus (components of bones and teeth) are needed by the body in much greater quantities.
2.2

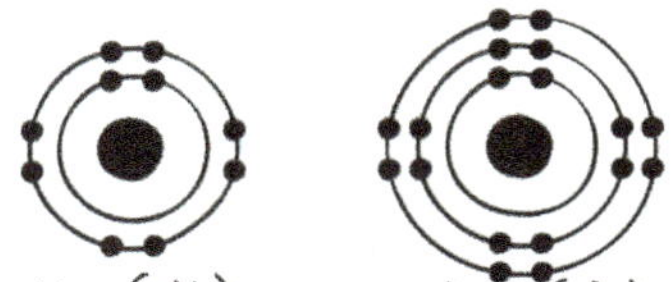

Both neon and argon have completed valence shells, containing 8 electrons. They do not have unpaired electrons that could participate in chemical bonds. **2.3** Electrons are shared equally between the two atoms in a nonpolar covalent bond. In a polar covalent bond, the electrons are drawn closer to the more electronegative atom. In the formation of ions, an electron is completely transferred from one atom to a much more electronegative atom. **2.4** The concentration of products would increase as the added reactants were converted to products. Eventually, an equilibrium would again be reached in which the forward and reverse reactions were proceeding at the same rate and the relative concentrations of reactants and products returned to where they were before the addition of more reactants.

Test Your Understanding

1. a **2.** e **3.** b **4.** a **5.** d **6.** b **7.** c **8.** e

9. a. 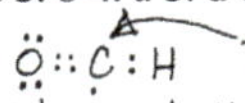 This structure doesn't make sense because the valence shell of carbon is incomplete; carbon can form 4 bonds.

b. 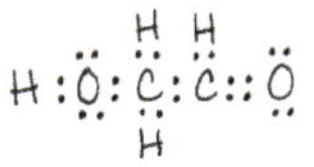 This structure makes sense because all valence shells are complete, and all bonds have the correct number of electrons.

c. 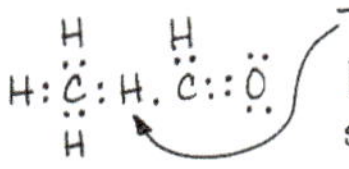 This structure doesn't make sense because H has only 1 electron to share, so it cannot form bonds with 2 atoms.

d. This structure doesn't make sense for several reasons:

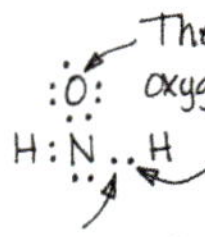
The valence shell of oxygen is incomplete; oxygen can form 2 bonds.
H has only 1 electron to share, so it cannot form a double bond.

Nitrogen usually makes only 3 bonds. It does not have enough electrons to make 2 single bonds, make a double bond, and complete its valence shell.

Chapter 3

Figure Questions

Figure 3.2 One possible answer:

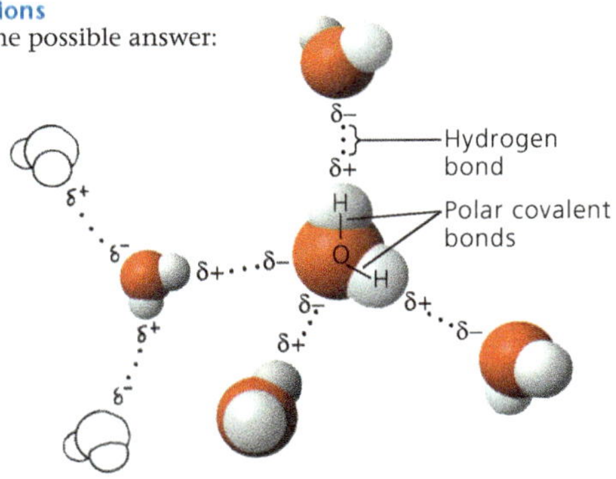

Figure 3.6 Without hydrogen bonds, water would behave like other small molecules, and the solid phase (ice) would be denser than liquid water. The ice would sink to the bottom and would no longer insulate the whole body of water, which would eventually freeze because the average annual temperature at the South Pole is –50°C. The krill could not survive. **Figure 3.7** Heating the solution would cause the water to evaporate faster than it is evaporating at room temperature. At a certain point, there wouldn't be enough water molecules to dissolve the salt ions. The salt would start coming out of solution and re-forming crystals. Eventually, all the water would evaporate, leaving behind a pile of salt like the original pile. **Figure 3.12** By causing the loss of coral reefs, a decrease in the ocean's carbonate concentration would have a ripple effect on noncalcifying organisms. Some of these organisms depend on the reef structure for protection, while others feed on species associated with reefs.

Concept Check 3.1

1. Electronegativity is the attraction of an atom for the electrons of a covalent bond. Because oxygen is more electronegative than hydrogen, the oxygen atom in H_2O pulls electrons toward itself, resulting in a partial negative charge on the oxygen atom and partial positive charges on the hydrogen atoms. Atoms in neighboring water molecules with opposite partial charges are attracted to each other, forming a hydrogen bond. **2.** The hydrogen atoms of one molecule, with their partial positive charges, would repel the hydrogen atoms of the adjacent molecule. **3.** The covalent bonds of water molecules would not be polar, and water molecules would not form hydrogen bonds with each other.

Concept Check 3.2

1. Hydrogen bonds hold neighboring water molecules together. This cohesion helps the chain of water molecules move upward against gravity in water-conducting cells as water evaporates from the leaves. Adhesion between water molecules and the walls of the water-conducting cells also helps counter gravity. **2.** High humidity hampers cooling by suppressing the evaporation of sweat. **3.** As water freezes, it expands because water molecules move farther apart in forming ice crystals. When there is water in a crevice of a boulder, expansion due to freezing may crack the boulder. **4.** A liter of blood would contain 7.8×10^{13} molecules of ghrelin (1.3×10^{-10} moles per liter $\times$ 6.02×10^{23} molecules per mole). **5.** The hydrophobic substance repels water, perhaps helping to keep the ends of the legs from becoming coated with water and breaking through the surface. If the legs were coated with a hydrophilic substance, water would be drawn up them, possibly making it more difficult for the water strider to walk on water.

Concept Check 3.3

1. 10^5, or 100,000 **2.** $[H^+] = 0.01\ M = 10^{-2}\ M$, so pH = 2. **3.** $CH_3COOH \rightarrow CH_3COO^- + H^+$. CH_3COOH is the acid (the H^+ donor), and CH_3COO^- is the base (the H^+ acceptor). **4.** The pH of the water should decrease from 7 to about 2; the pH of the acetic acid solution will decrease only a small amount, because the reaction shown for question 3 will shift to the left, with CH_3COO^- accepting the influx of H^+ and becoming CH_3COOH molecules.

Summary of Key Concepts Questions

3.1

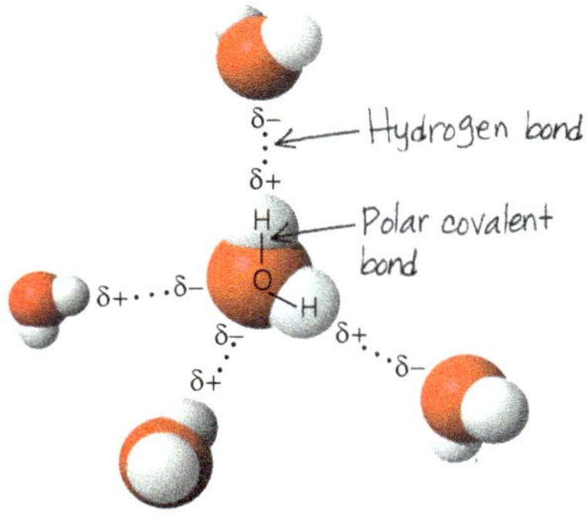

Each water molecule can make four hydrogen bonds with neighboring molecules. **3.2** Ions dissolve in water when polar water molecules form a hydration shell around them. Polar molecules dissolve as water molecules form hydrogen bonds with them and surround them. Solutions are homogeneous mixtures of solute and solvent. Colloids form when particles that are too large to dissolve remain suspended in a liquid. **3.3** CO_2 reacts with H_2O to form carbonic acid (H_2CO_3), which dissociates into H^+ and bicarbonate (HCO_3^-). Although the carbonic acid–bicarbonate reaction is a buffering system, adding CO_2 drives the reaction to the right, releasing more H^+ and lowering pH. The excess protons combine with CO_3^{2-} to form bicarbonate, lowering the concentration of carbonate available for the formation of calcium carbonate (calcification) by corals.

Test Your Understanding

1. d **2.** b **3.** c **4.** e **5.** c **6.** a **7.** e **8.** d

9.

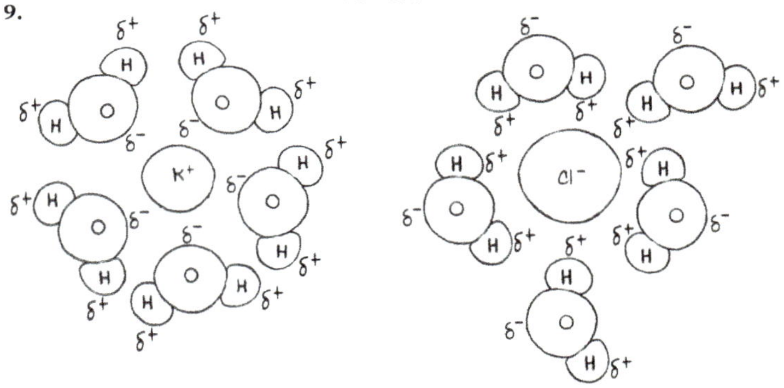

10. Both global warming and ocean acidification are caused by increasing levels of carbon dioxide in the atmosphere, the result of burning fossil fuels. **11.** Due to intermolecular hydrogen bonds, water has a high specific heat (the amount of heat required to increase the temperature of water by 1°C). When water is heated, much of the heat is absorbed in breaking hydrogen bonds before the water molecules increase their motion and the temperature increases. Conversely, when water is cooled, many H bonds are formed, which releases a significant amount of heat. This release of heat can provide some protection against freezing of the plants' leaves, thus protecting the cells from damage.

Chapter 4

Figure Questions

Figure 4.2 Because the concentration of the reactants influences the equilibrium (as discussed in Chapter 2), there might have been more HCN relative to CH_2O, since there would have been a higher concentration of the reactant gas containing nitrogen.

Figure 4.4

Figure 4.6 The tails of fats contain only carbon-hydrogen bonds, which are relatively nonpolar. Because the tails occupy the bulk of a fat molecule, they make the molecule as a whole nonpolar and therefore incapable of forming hydrogen bonds with water.

Figure 4.7

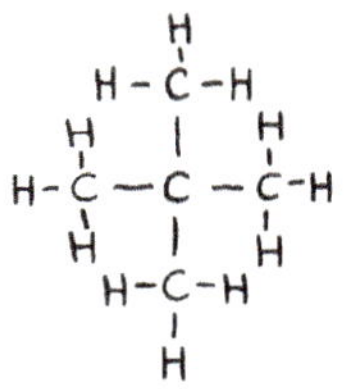

Figure 4.9 Molecule b, because there are not only the two electronegative oxygens of the carboxyl group, but also an oxygen on the next (carbonyl) carbon. All of these oxygens help make the bond between the O and H of the —OH group more polar, thus making the dissociation of H^+ more likely.

Concept Check 4.1

1. Prior to Wöhler's experiment, the prevailing view was that only living organisms could synthesize "organic" compounds. Wöhler made urea, an organic compound, without the involvement of living organisms. **2.** The spark provided energy needed for the inorganic molecules in the atmosphere to react with each other. (You'll learn more about energy and chemical reactions in Chapter 8.)

Concept Check 4.2

1.

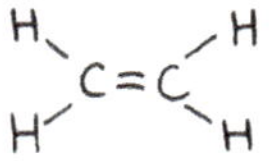

2. The forms of C_4H_{10} in (b) are structural isomers, as are the butenes in (c). **3.** Both consist largely of hydrocarbon chains. **4.** No. There is not enough diversity in the atoms. It can't form structural isomers because there is only one way for three carbons to attach to each other (in a line). There are no double bonds, so *cis-trans* isomers are not possible. Each carbon has at least two hydrogens attached to it, so the molecule is symmetrical and cannot have enantiomers.

Concept Check 4.3

1. It has both an amino group (—NH_2), which makes it an amine, and a carboxyl group (—COOH), which makes it a carboxylic acid. **2.** The ATP molecule loses a phosphate, becoming ADP. **3.** A chemical group that can act as a base has been replaced with a group that can act as an acid, increasing the acidic properties of the molecule. The shape of the molecule would also change, likely changing the molecules with which it can interact. The original cysteine molecule has an asymmetric carbon in the center. After replacement of the amino group with a carboxyl group, this carbon is no longer asymmetric.

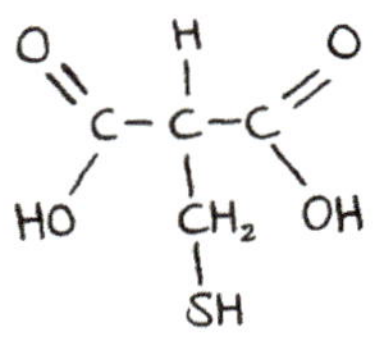

Summary of Key Concepts Questions

4.1 Miller showed that organic molecules could form under the physical and chemical conditions estimated to have been present on early Earth. This abiotic synthesis of organic molecules would have been a first step in the origin of life. **4.2** Acetone and propanal are structural isomers. Acetic acid and glycine have no asymmetric carbons, whereas glycerol phosphate has one. Therefore, glycerol phosphate can exist as forms that are enantiomers, but acetic acid and glycine cannot. **4.3** The methyl group is nonpolar and not reactive. The other six groups are called functional groups. They are each hydrophilic, increasing the solubility of organic compounds in water, and can participate in chemical reactions.

Test Your Understanding

1. b **2.** b **3.** d **4.** d **5.** a **6.** b **7.** a **8.** The molecule on the right; the middle carbon is asymmetric.
9. •Si• (with a dot above and below) Si has 4 valence electrons, the same number as carbon. Therefore, silicon would be able to form long chains, including branches, that could act as skeletons for large molecules. It would clearly do this much better than neon (with no valence electrons) or aluminum (with 3 valence electrons).

Chapter 5

Figure Questions

Figure 5.3 Glucose and fructose are structural isomers.
Figure 5.4

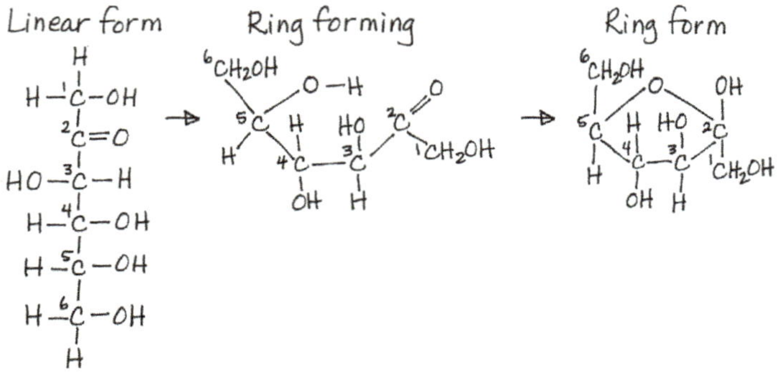

Note that the oxygen on carbon 5 lost its proton and that the oxygen on carbon 2, which used to be the carbonyl oxygen, gained a proton. Four carbons are in the fructose ring, and two are not. (The latter two carbons are attached to carbons 2 and 5, which are in the ring.) The fructose ring differs from the glucose ring, which has five carbons in the ring and one that is not. (Note that the orientation of this fructose molecule is flipped relative to that of the one in Figure 5.5b.)

Figure 5.5

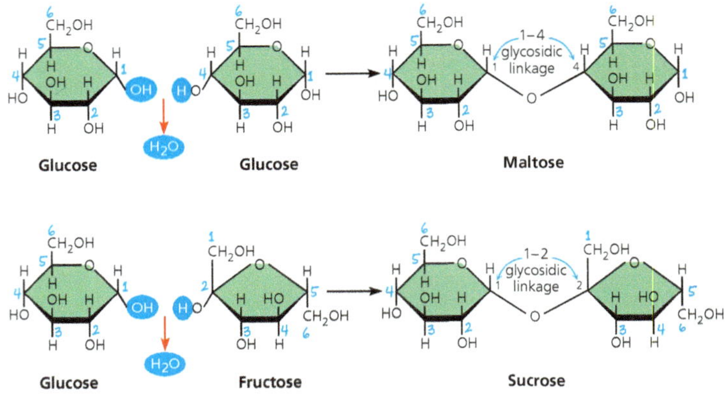

Figure 5.12

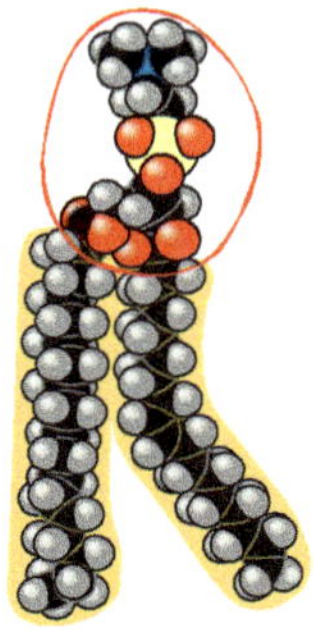

Figure 5.14

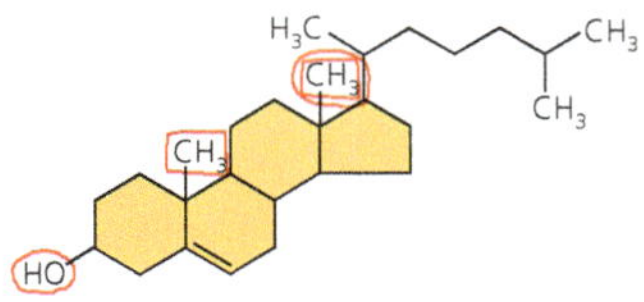

Figure 5.17

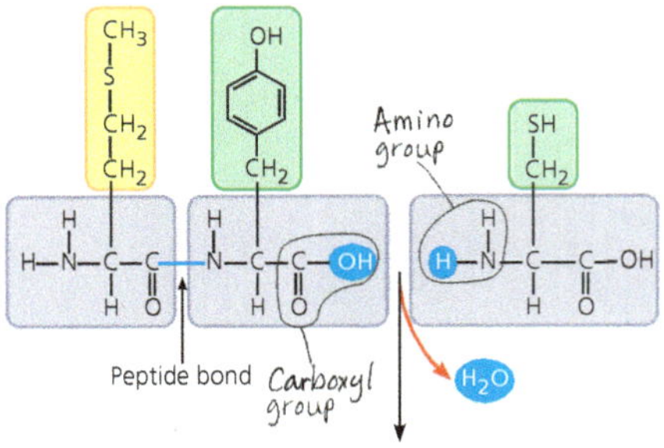

Figure 5.21 The R group on glutamic acid is acidic and hydrophilic, whereas that on valine is nonpolar and hydrophobic. Therefore, it is unlikely that valine can participate in the same intramolecular interactions that glutamic acid can. A change in these interactions causes a disruption of molecular structure.
Figure 5.24 The spirals are α helices.

Concept Check 5.1

1. The four main classes are proteins, carbohydrates, lipids, and nucleic acids. Lipids are not polymers. **2.** Nine, with one water molecule required to hydrolyze each connected pair of monomers **3.** The amino acids in the fish protein must be released in hydrolysis reactions and incorporated into other proteins in dehydration reactions.

Concept Check 5.2

1. $C_3H_6O_3$ **2.** $C_{12}H_{22}O_{11}$ **3.** The antibiotic treatment is likely to have killed the cellulose-digesting prokaryotes in the cow's stomach. The absence of these prokaryotes would hamper the cow's ability to obtain energy from food and could lead to weight loss and possibly death. Thus, prokaryotic species are reintroduced, in appropriate combinations, in the gut culture given to treated cows.

Concept Check 5.3

1. Both have a glycerol molecule attached to fatty acids. The glycerol of a fat has three fatty acids attached, whereas the glycerol of a phospholipid is attached to two fatty acids and one phosphate group. **2.** Human sex hormones are steroids, a type of hydrophobic compound. **3.** The oil droplet membrane could consist of a single layer of phospholipids rather than a bilayer, because an arrangement in which the hydrophobic tails of the membrane phospholipids were in contact with the hydrocarbon regions of the oil molecules would be more stable.

Concept Check 5.4

1. The function of a protein is a consequence of its specific shape, which is lost when a protein becomes denatured. **2.** Secondary structure involves hydrogen bonds between atoms of the polypeptide backbone. Tertiary structure involves interactions between atoms of the side chains of the amino acid subunits. **3.** These are all nonpolar amino acids, so you would expect this region to be located in the interior of the folded polypeptide, where it would not contact the aqueous environment inside the cell.

Concept Check 5.5

1.

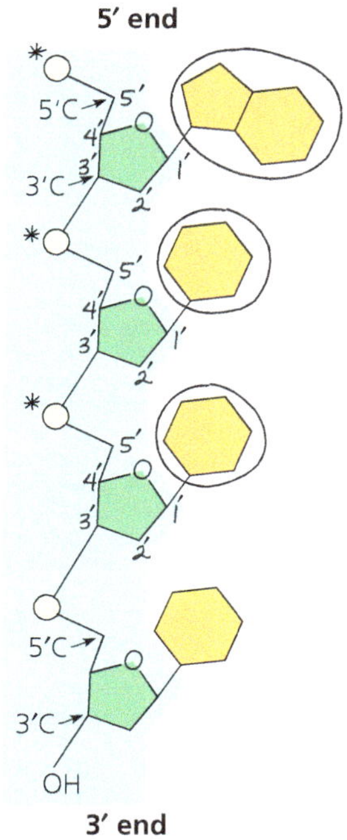

2. 5′–T A G G C C T–3′
3′–A T C C G G A–5′

3. a. Mismatch
5′–T A (A) G C C T–3′
3′–A T (C) C G G A–5′

b. 3′–A T T C G G A–5′

Summary of Key Concepts Questions

Concept 5.1 The polymers of carbohydrates, proteins, and nucleic acids are built from three different types of monomers: monosaccharides, amino acids, and nucleotides, respectively. **Concept 5.2** Both starch and cellulose are polymers of glucose, but the glucose monomers are in the α configuration in starch and the β configuration in cellulose. The glycosidic linkages thus have different geometries, giving the polymers different shapes and thus different properties. Starch is an energy-storage compound in plants; cellulose is a structural component of plant cell walls. Humans can hydrolyze starch to provide energy but cannot hydrolyze cellulose. Cellulose aids in the passage of food through the digestive tract. **Concept 5.3** Lipids are not polymers because they do not exist as a chain of linked monomers. They are not considered macromolecules because they do not reach the giant size of many polysaccharides, proteins, and nucleic acids. **Concept 5.4** A polypeptide, which may consist of hundreds of amino acids in a specific sequence (primary structure), has regions of coils and pleats (secondary structure), which are then folded into irregular contortions (tertiary structure) and may be noncovalently associated with other polypeptides (quaternary structure). The linear order of amino acids, with the varying properties of their side chains (R groups), determines what secondary and tertiary structures will form to produce a protein. The resulting unique three-dimensional shapes of proteins are key to their specific and diverse functions. **Concept 5.5** The complementary base pairing of the two strands of DNA makes possible the precise replication of DNA every time a cell divides, ensuring that genetic information is faithfully transmitted. In some types of RNA, complementary base pairing enables RNA molecules to assume specific three-dimensional shapes that facilitate diverse functions.

Test Your Understanding

1. d **2.** a **3.** b **4.** a **5.** b **6.** c **7.** d

8.

	Monomers or Components	Polymer or larger molecule	Type of linkage
Carbohydrates	Monosaccharides	Polysaccharides	Glycosidic linkages
Lipids	Fatty acids	Triacylglycerols	Ester linkages
Proteins	Amino acids	Polypeptides	Peptide bonds
Nucleic acids	Nucleotides	Polynucleotides	Phosphodiester linkages

9.

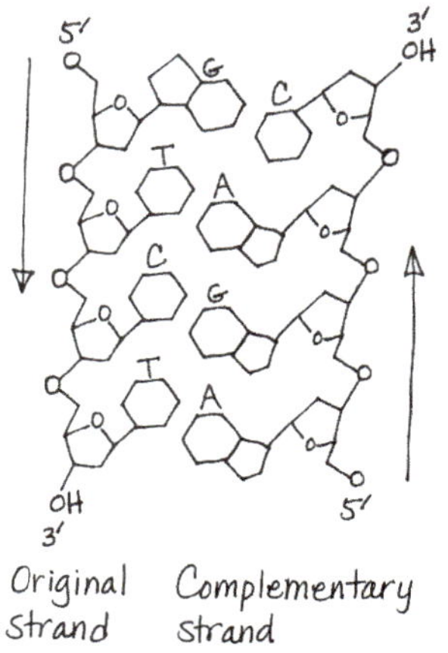

Chapter 6

Figure Questions

Figure 6.6 A phospholipid is a lipid, consisting of a glycerol molecule joined to two fatty acids and one phosphate group. Together, the glycerol and phosphate end of the phospholipid form the "head," which is hydrophilic, while the hydrocarbon chains on the fatty acids form hydrophobic "tails." The presence in a single molecule of both a hydrophilic and a hydrophobic region makes the molecule ideal as the main building block of a membrane. **Figure 6.9** The DNA in a chromosome dictates synthesis of a messenger RNA (mRNA) molecule, which then moves out to the cytoplasm. There, the information is used for the production, on ribosomes, of proteins that carry out cellular functions. **Figure 6.10** Any of the bound ribosomes (attached to the endoplasmic reticulum) could be circled, because any could be making a protein that will be secreted. **Figure 6.22** Each centriole has 9 sets of 3 microtubules, so the entire centrosome (two centrioles) has 54 microtubules. Each microtubule consists of a helical array of tubulin dimers (as shown in Table 6.1).

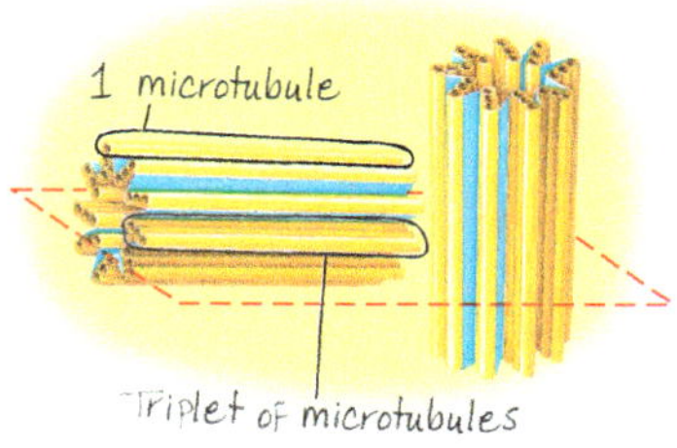

Figure 6.24

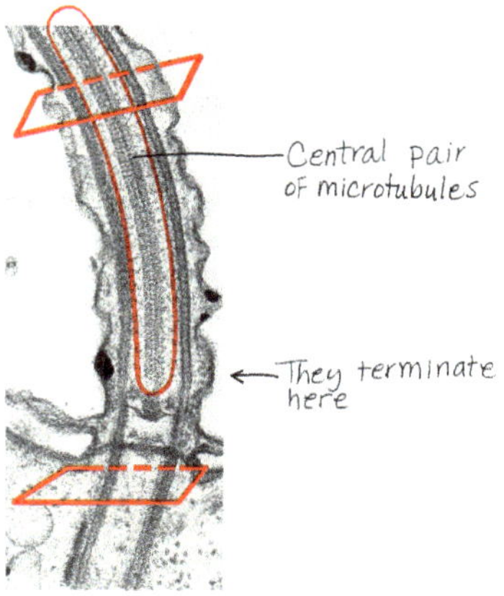

The two central microtubules terminate above the basal body, so they aren't present at the level of the cross section through the basal body, indicated by the lower red rectangle in (a). **Figure 6.29** The microtubules would reorient, and based on the earlier results, the cellulose synthase proteins would also change their path, orienting along the repositioned microtubules. (This is, in fact, what was observed.)

Concept Check 6.1

1. Stains used for light microscopy are colored molecules that bind to cell components, affecting the light passing through, while stains used for electron microscopy involve heavy metals that affect the beams of electrons passing through. **2.** (a) Light microscope, (b) scanning electron microscope

Concept Check 6.2

1. See Figure 6.8.
2.

This cell would have the same volume as the cells in columns 2 and 3 but proportionally more surface area than that in column 2 and less than that in column 3. Thus, the surface-to-volume ratio should be greater than 1.2 but less than 6. To obtain the surface area, you would add the area of the six sides (the top, bottom, sides, and ends): 125 + 125 + 125 + 125 + 1 + 1 = 502. The surface-to-volume ratio equals 502 divided by a volume of 125, or 4.0.

Concept Check 6.3

1. Ribosomes in the cytoplasm translate the genetic message, carried from the DNA in the nucleus by mRNA, into a polypeptide chain. **2.** Nucleoli consist of DNA and the ribosomal RNA (rRNA) made according to its instructions, as well as proteins imported from the cytoplasm. Together, the rRNA and proteins are assembled into large and small ribosomal subunits. (These are exported through nuclear pores to the cytoplasm, where they will participate in polypeptide synthesis.) **3.** No. Each chromosome is present whether its chromatin is relatively diffuse (when the cell is not dividing) or condensed (when the cell is dividing).

Concept Check 6.4

1. The primary distinction between rough and smooth ER is the presence of bound ribosomes on the rough ER. Both types of ER make phospholipids, but membrane proteins and secretory proteins are all produced on the ribosomes of the rough ER. The smooth ER also functions in detoxification, carbohydrate metabolism, and storage of calcium ions. **2.** Transport vesicles move membranes and substances they enclose between other components of the endomembrane system. **3.** The mRNA is synthesized in the nucleus and then passes out through a nuclear pore to be translated on a bound ribosome, attached to the rough ER. The protein is synthesized into the lumen of the ER and perhaps modified there. A transport vesicle carries the protein to the Golgi apparatus. After further modification in the Golgi, another transport vesicle carries it back to the ER, where it will perform its cellular function.

Concept Check 6.5

1. Both organelles are involved in energy transformation, mitochondria in cellular respiration and chloroplasts in photosynthesis. They both have multiple membranes that separate their interiors into compartments. In both organelles, the innermost membranes—cristae, or infoldings of the inner membrane, in mitochondria, and the thylakoid membranes in chloroplasts—have large surface areas with embedded enzymes that carry out their main functions. **2.** Yes. Plant cells are able to make their own sugar by photosynthesis, but mitochondria in these eukaryotic cells are the organelles that are able to generate energy from sugars, a function required in all cells. **3.** Mitochondria and chloroplasts are not derived from the ER, nor are they connected physically or via transport vesicles to organelles of the endomembrane system. Mitochondria and chloroplasts are structurally quite different from vesicles derived from the ER, which are bounded by a single membrane.

Concept Check 6.6

1. Both systems of movement involve long filaments that are moved in relation to each other by motor proteins that grip, release, and grip again adjacent polymers. **2.** Dynein arms, powered by ATP, move neighboring doublets of microtubules relative to each other. Because they are anchored within the organelle and with respect to one another, the doublets bend instead of sliding past each other. Synchronized bending of the nine microtubule doublets brings about bending of both structures. **3.** Such individuals have defects in the microtubule-based movement of cilia and flagella. Thus, the sperm can't move because of malfunctioning or nonexistent flagella, and the airways are compromised because cilia that line the trachea malfunction or don't exist, and so mucus cannot be cleared from the lungs.

Concept Check 6.7

1. The most obvious difference is the presence of direct cytoplasmic connections between cells of plants (plasmodesmata) and animals (gap junctions). These connections result in the cytoplasm being continuous between adjacent cells. **2.** The cell would not be able to function properly and would probably soon die, as the cell wall or ECM must be permeable to allow the exchange of matter between the cell and its external environment. Molecules involved in energy production and use must be allowed entry, as well as those that provide information about the cell's environment. Other molecules, such as products synthesized by the cell for export and the by-products of cellular respiration, must be allowed to exit. **3.** The parts of the protein that face aqueous regions would be expected to have polar or charged (hydrophilic) amino acids, while the parts that go through the membrane would be expected to have nonpolar (hydrophobic) amino acids. You would predict polar or charged amino acids at each end (tail), in the region of the cytoplasmic loop, and in the regions of the two extracellular loops. You would predict nonpolar amino acids in the four regions that go through the membrane between the tails and loops.

Summary of Key Concepts Questions

6.1 Both light and electron microscopy allow cells to be studied visually, thus helping us understand internal cellular structure and the arrangement of cell components. Cell fractionation techniques separate out different groups of cell components, which can then be analyzed biochemically to determine their function. Performing microscopy on the same cell fraction helps to correlate the biochemical function of the cell with the cell component responsible. **6.2** The separation of different functions in different organelles has several advantages. Reactants and enzymes can be concentrated in one area instead of spread throughout the cell. Reactions that require specific conditions, such as a lower pH, can be compartmentalized. And enzymes for specific reactions are often embedded in the membranes that enclose or partition an organelle. **6.3** The nucleus contains the genetic material of the cell in the form of DNA, which codes for messenger RNA, which in turn provides instructions for the synthesis of proteins (including the proteins that make up part of the ribosomes). DNA also codes for ribosomal RNA, which is combined with proteins in the nucleolus into the subunits of ribosomes. Within the cytoplasm, ribosomes join with mRNA to build polypeptides, using the genetic information in the mRNA. **6.4** Transport vesicles move proteins and membranes synthesized by the rough ER to the Golgi for further processing and then to the plasma membrane, lysosomes, or other locations in the cell, including back to the ER. **6.5** According to the endosymbiont theory, mitochondria originated from an oxygen-using prokaryotic cell that was engulfed by an ancestral eukaryotic cell. Over time, the host and endosymbiont evolved into a single organism. Chloroplasts originated when at least one of these eukaryotic cells containing mitochondria engulfed and then retained a photosynthetic prokaryote. **6.6** Inside the cell, motor proteins interact with components of the cytoskeleton to move cellular parts. Motor proteins may "walk" vesicles along microtubules. The movement of cytoplasm within a cell involves interactions of the motor protein myosin and microfilaments (actin filaments). Whole cells can be moved by the rapid bending of flagella or cilia, which is caused by the motor-protein-powered sliding of microtubules within these structures. Cell movement can also occur when pseudopodia form at one end of a cell (caused by actin polymerization into a filamentous network), followed by contraction of the cell toward that end; this is powered by interactions of microfilaments with myosin. Interactions of motor proteins and microfilaments in muscle cells can propel whole organisms. **6.7** A plant cell wall is primarily composed of microfibrils of cellulose embedded in other polysaccharides and proteins. The ECM of animal cells is primarily composed of collagen and other protein fibers, such as the glycoprotein fibronectins. These fibers are embedded in a network of carbohydrate-rich proteoglycans. A plant cell wall provides structural support for the cell and, collectively, for the plant body. In addition to giving support, the ECM of an animal cell allows for communication of environmental changes into the cell.

Test Your Understanding

1. b **2.** d **3.** b **4.** e **5.** a **6.** d **7.** c **8.** See Figure 6.8.

Chapter 7

Figure Questions

Figure 7.2

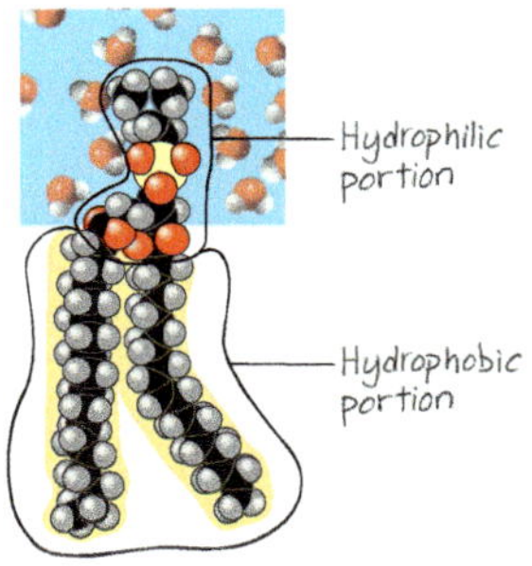

The hydrophilic portion is in contact with an aqueous environment (cytosol or extracellular fluid), and the hydrophobic portion is in contact with the hydrophobic portions of other phospholipids in the interior of the bilayer. **Figure 7.7** You couldn't rule out movement of proteins within membranes of the same species. You might propose that the membrane lipids and proteins from one species weren't able to mingle with those from the other species because of some incompatibility. **Figure 7.10** A transmembrane protein like the dimer in (f) might change its shape upon binding to a particular ECM molecule. The new shape might enable the interior portion of the protein to bind to a second, cytoplasmic protein that would relay the message to the inside of the cell, as shown in (c). **Figure 7.11** The shape of a protein on the HIV surface is likely to be complementary to the shape of the receptor (CD4) and also to that of the co-receptor (CCR5). A molecule that was a similar shape to the HIV surface protein could bind CCR5, blocking HIV binding. (Another alternative would be a molecule that bound to CCR5 and changed the shape of CCR5 so it could no longer bind HIV.)

Figure 7.12

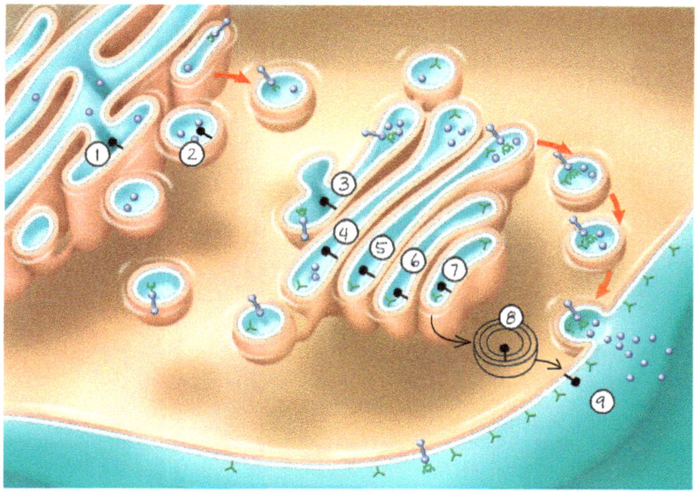

The protein would contact the extracellular fluid. **Figure 7.14** The orange dye would be evenly distributed throughout the solution on both sides of the membrane. The solution levels would not be affected because the orange dye can diffuse through the membrane and equalize its concentration. Thus, no additional osmosis would take place in either direction. **Figure 7.19** The diamond solutes are moving into the cell (down), and the round solutes are moving out of the cell (up); both are moving against their concentration gradient.

Concept Check 7.1

1. They are on the inner side of the transport vesicle membrane. **2.** The grasses living in the cooler region would be expected to have more unsaturated fatty acids in their membranes because those fatty acids remain fluid at lower temperatures. The grasses living immediately adjacent to the hot springs would be expected to have more saturated fatty acids, which would allow the fatty acids to "stack" more closely, making the membranes less fluid and therefore helping them to stay intact at higher temperatures. (Cholesterol could not be used to moderate the effects of temperature on membrane fluidity because it is not found within plant cell membranes.)

Concept Check 7.2

1. O_2 and CO_2 are both nonpolar molecules that can easily pass through the hydrophobic interior of a membrane. **2.** Water is a polar molecule, so it cannot pass very rapidly through the hydrophobic region in the middle of a phospholipid bilayer. **3.** The hydronium ion is charged, while glycerol is not. Charge is probably more significant than size as a basis for exclusion by the aquaporin channel.

Concept Check 7.3

1. CO_2 is a nonpolar molecule that can diffuse through the plasma membrane. As long as it diffuses away so that the concentration remains low outside the cell, it will continue to exit the cell in this way. (This is the opposite of the case for O_2, described in this section.) **2.** The water is hypotonic to the plant cells, so the plant cells take up water. Thus, the cells of the vegetable remain turgid rather than plasmolyzing, and the vegetable (for example, lettuce or spinach) remains crisp and not wilted. **3.** The activity of *Paramecium caudatum*'s contractile vacuole will decrease. The vacuole pumps out excess water that accumulates in the cell; this accumulation occurs only in a hypotonic environment.

Concept Check 7.4

1. The pump uses ATP. To establish a voltage, ions have to be pumped against their gradients, which requires energy. **2.** Each ion is being transported against its electrochemical gradient. If either ion were transported down its electrochemical gradient, this *would* be considered cotransport. **3.** The internal environment of a lysosome is acidic, so it has a higher concentration of H^+ than does the cytoplasm. Therefore, you might expect the membrane of the lysosome to have a proton pump such as that shown in Figure 7.20 to pump H^+ into the lysosome.

Concept Check 7.5

1. Exocytosis. When a transport vesicle fuses with the plasma membrane, the vesicle membrane becomes part of the plasma membrane.
2.

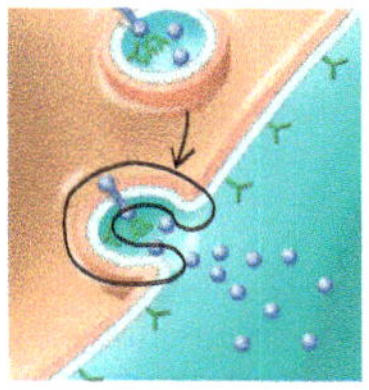

3. The glycoprotein would be synthesized in the ER lumen, move through the Golgi apparatus, and then travel in a vesicle to the plasma membrane, where it would undergo exocytosis and become part of the ECM.

Summary of Key Concepts Questions

7.1 Plasma membranes define the cell by separating the cellular components from the external environment. This allows conditions inside cells to be controlled by membrane proteins, which regulate entry and exit of molecules and even cell function (see Figure 7.10). The processes of life can be carried out inside the controlled environment of the cell, so membranes are crucial. In eukaryotes, membranes also function to subdivide the cytoplasm into different compartments where distinct processes can occur, even under differing conditions such as pH. **7.2** Aquaporins are channel proteins that greatly increase the permeability of a membrane to water molecules, which are polar and therefore do not readily diffuse through the hydrophobic interior of the membrane. **7.3** There will be a net diffusion of water out of a cell into a hypertonic solution. The free water concentration is higher inside the cell than in the solution (where water molecules are not free, but are clustered around the higher concentration of solute particles). **7.4** One of the solutes moved by the cotransporter is actively transported against its concentration gradient. The energy for this transport comes from the concentration gradient of the other solute, which was established by an electrogenic pump that used energy to transport the other solute across the membrane. **7.5** In receptor-mediated endocytosis, specific molecules act as ligands when they bind to receptors on the plasma membrane. The cell can acquire bulk quantities of those molecules when a coated pit forms a vesicle and carries the bound molecules into the cell.

Test Your Understanding

1. b **2.** c **3.** a **4.** d **5.** b
6. (a)

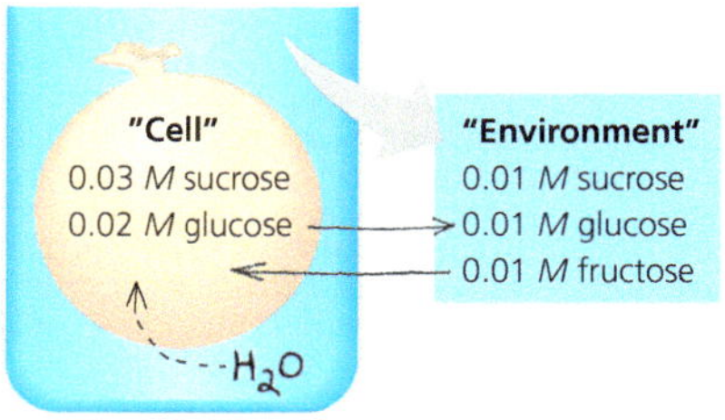

(b) The solution outside is hypotonic. It has less sucrose, which is a nonpenetrating solute. (c) See answer for (a). (d) The artificial cell will become more turgid. (e) Eventually, the two solutions will have the same solute concentrations. Even though sucrose can't move through the membrane, water flow (osmosis) will lead to isotonic conditions.

Chapter 8

Figure Questions

Figure 8.12

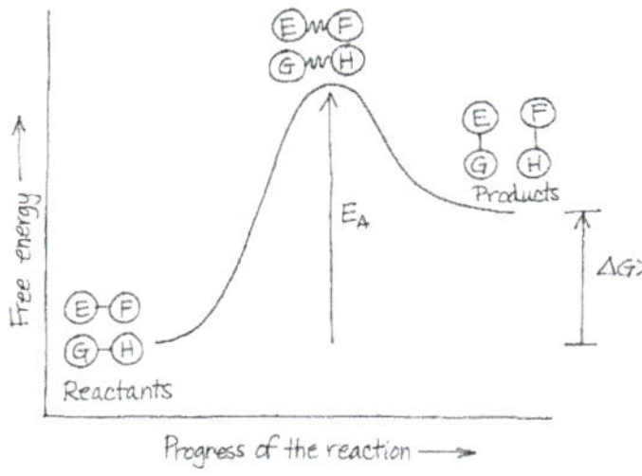

Figure 8.16

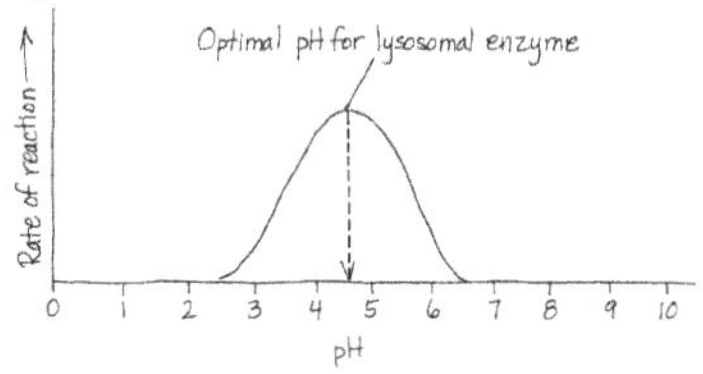

Figure 8.20 Because the affinity of the caspase for the inhibitor is very low (as is expected of an allosterically inhibited enzyme), the inhibitor is likely to diffuse away. Because no additional source of the inhibitory compound is present and the concentration of inhibitor is very low, the inhibitor is unlikely to bind again to the allosteric binding site once the covalent linkage is broken. Therefore, the activity of the enzyme would most likely be normal. (In fact, this is what the researchers observed when they broke the disulfide linkage.)

Concept Check 8.1

1. The second law is the trend toward randomization, or increasing entropy. When the concentrations of a substance on both sides of a membrane are equal, the distribution is more random than when they are unequal. Diffusion of a substance to a region where it is initially less concentrated increases entropy, making it an energetically favorable (spontaneous) process as described by the second law. This explains the process seen in Figure 7.13. **2.** The apple has potential energy in its position hanging on the tree, and the sugars and other nutrients it contains have chemical energy. The apple has kinetic energy as it falls from the tree to the ground. Finally, when the apple is digested and its molecules broken down, some of the chemical energy is used to do work, and the rest is lost as thermal energy. **3.** The sugar crystals become less ordered (entropy increases) as they dissolve and become randomly spread out in the water. Over time, the water evaporates, and the crystals form again because the water volume is insufficient to keep them in solution. While the reappearance of sugar crystals may represent a "spontaneous" increase in order (decrease in entropy), it is balanced by the decrease in order (increase in entropy) of the water molecules, which changed from a relatively compact arrangement as liquid water to a much more dispersed and disordered form as water vapor.

Concept Check 8.2

1. Cellular respiration is a spontaneous and exergonic process. The energy released from glucose is used to do work in the cell or is lost as heat. **2.** When the H^+ concentrations are the same, the system is at equilibrium and can do no work. Hydrogen ions can perform work only if their concentrations on each side of a membrane differ—in other words, when a gradient is present. This is consistent with Figure 7.20, which shows that an energy input (provided by ATP hydrolysis) is required to establish the concentration gradient (the H^+ gradient) that can in turn perform work. **3.** The reaction is exergonic because it releases energy—in this case, in the form of light. (This is a nonbiological version of the bioluminescence seen in Figure 8.1.)

Concept Check 8.3

1. ATP usually transfers energy to endergonic processes by phosphorylating (adding phosphate groups to) other molecules. (Exergonic processes phosphorylate ADP to regenerate ATP.) **2.** A set of coupled reactions can transform the first combination into the second. Since this is an exergonic process overall, ΔG is negative and the first combination must have more free energy (see Figure 8.9). **3.** Active transport: The solute is being transported against its concentration gradient, which requires energy, provided by ATP hydrolysis.

Concept Check 8.4

1. A spontaneous reaction is a reaction that is exergonic. However, if it has a high activation energy that is rarely attained, the rate of the reaction may be low. **2.** Only the specific substrate(s) will fit properly into the active site of an enzyme, the part of the enzyme that carries out catalysis. **3.** In the presence of malonate, increase the concentration of the normal substrate (succinate) and see whether the rate of reaction increases. If it does, malonate is a competitive inhibitor. **4.** If lactose wasn't present in the environment as a source of food and the fucose-containing disaccharide was available, bacteria that could digest the latter would be better able to grow and multiply than those that could not.

Concept Check 8.5

1. The activator binds in such a way that it stabilizes the active form of an enzyme, whereas the inhibitor stabilizes the inactive form. **2.** An inhibitor that binds to the active site of the enzyme you want to inhibit could also bind to and block the enzymes with similar structures, causing significant side effects. For this reason, you would be better off choosing to screen chemical compounds that bind allosterically to the enzyme in question, because allosteric regulatory sites are less likely to share similarity with other enzymes.

Summary of Key Concepts Questions

8.1 The process of "ordering" a cell's structure is accompanied by an increase in the entropy or disorder of the universe. For example, an animal cell takes in highly ordered organic molecules as the source of matter and energy used to build and maintain its structures. In the same process, however, the cell releases heat and the simple molecules of carbon dioxide and water to the surroundings. The increase in entropy of the latter process offsets the entropy decrease in the former. **8.2** A spontaneous reaction has a negative ΔG and is exergonic. For a chemical reaction to proceed with a net release of free energy ($-\Delta G$), the enthalpy or total energy of the system must decrease ($-\Delta H$), and/or the entropy or disorder must increase (yielding a more negative term, $-T\Delta S$). Spontaneous reactions supply the energy to perform cellular work. **8.3** The free energy released from the hydrolysis of ATP may drive endergonic reactions through the transfer of a phosphate group to a reactant molecule, forming a more reactive phosphorylated intermediate. ATP hydrolysis also powers the mechanical and transport work of a cell, often by powering shape changes in the relevant motor proteins. Cellular respiration, the catabolic breakdown of glucose, provides the energy for the endergonic regeneration of ATP from ADP and Ⓟ$_i$. **8.4** Activation energy barriers prevent the complex molecules of the cell, which are rich in free energy, from spontaneously breaking down to less ordered, more stable molecules. Enzymes permit a regulated metabolism by binding to specific substrates and forming enzyme-substrate complexes that selectively lower the E_A for the chemical reactions in a cell. **8.5** A cell tightly regulates its metabolic pathways in response to fluctuating needs for energy and materials. The binding of activators or inhibitors to regulatory sites on allosteric enzymes stabilizes either the active or inactive form of the subunits. For example, the binding of ATP to a catabolic enzyme in a cell with excess ATP would inhibit that pathway. Such types of feedback inhibition preserve chemical resources within a cell. If ATP supplies are depleted, binding of ADP to the regulatory site of catabolic enzymes would activate that pathway.

Test Your Understanding

1. b **2.** c **3.** b **4.** a **5.** c **6.** e **7.** c

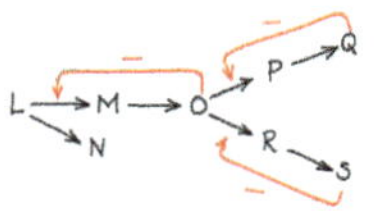

9.

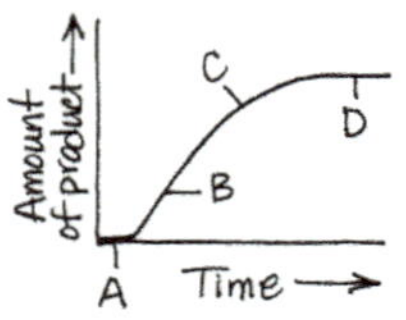

A. The substrate molecules are entering the cells, so no product is made yet.
B. There is sufficient substrate, so the reaction is proceeding at a maximum rate.
C. As the substrate is used up, the rate decreases (the slope is less steep).
D. The line is flat because no new substrate remains and thus no new product appears.

Chapter 9

Figure Questions

Figure 9.7 Because there is no external source of energy for the reaction, it must be exergonic, and the reactants must be at a higher energy level than the products. **Figure 9.9** The removal would probably stop glycolysis, or at least slow it down, since it would push the equilibrium for step 5 toward the left. If less (or no) glyceraldehyde 3-phosphate were available, step 6 would slow down (or be unable to occur). **Figure 9.15** At first, some ATP could be made, since electron transport could proceed as far as complex III, and a small H^+ gradient could be built up. Soon, however, no more electrons could be passed to complex III because it could not be reoxidized by passing its electrons to complex IV. **Figure 9.16** First, there are 2 NADH from the oxidation of pyruvate plus 6 NADH from the citric acid cycle (CAC); 8 NADH $\times$ 2.5 ATP/NADH = 20 ATP. Second, there are 2 $FADH_2$ from the CAC; 2 $FADH_2 \times$ 1.5 ATP/$FADH_2$ = 3 ATP. Third, the 2 NADH from glycolysis enter the mitochondrion through one of two types of shuttle. They pass their electrons either to 2 FAD, which become $FADH_2$ and result in 3 ATP, or to 2 NAD^+, which become NADH and result in 5 ATP. Thus, 20 + 3 + 3 = 26 ATP, or 20 + 3 + 5 = 28 ATP from all NADH and $FADH_2$.

Concept Check 9.1

1. Both processes include glycolysis, the citric acid cycle, and oxidative phosphorylation. In aerobic respiration, the final electron acceptor is molecular oxygen (O_2); in anaerobic respiration, the final electron acceptor is a different substance. **2.** $C_4H_6O_5$ would be oxidized and NAD^+ would be reduced.

Concept Check 9.2

1. NAD^+ acts as the oxidizing agent in step 6, accepting electrons from glyceraldehyde 3-phosphate, which thus acts as the reducing agent. **2.** Since the overall process of glycolysis results in net production of ATP, it would make sense for the process to slow down when ATP levels have increased substantially. Thus, we would expect ATP to allosterically inhibit phosphofructokinase.

Concept Check 9.3

1. NADH and $FADH_2$; they will donate electrons to the electron transport chain. **2.** CO_2 is released from the pyruvate that is the end product of glycolysis, and CO_2 is also released during the citric acid cycle. **3.** In both cases, the precursor molecule loses a CO_2 molecule and then donates electrons to an electron carrier in an oxidation step. Also, the product has been activated due to the attachment of a CoA group.

Concept Check 9.4

1. Oxidative phosphorylation would eventually stop entirely, resulting in no ATP production by this process. Without oxygen to "pull" electrons down the electron transport chain, H^+ would not be pumped into the mitochondrion's intermembrane space and chemiosmosis would not occur. **2.** Decreasing the pH means addition of H^+. This would establish a proton gradient even without the function of the electron transport chain, and we would expect ATP synthase to function and synthesize ATP. (In fact, it was experiments like this that provided support for chemiosmosis as an energy-coupling mechanism.) **3.** One of the components of the electron transport chain, ubiquinone (Q), must be able to diffuse within the membrane. It could not do so if the membrane were locked rigidly into place.

Concept Check 9.5

1. A derivative of pyruvate, such as acetaldehyde during alcohol fermentation, or pyruvate itself during lactic acid fermentation; oxygen **2.** The cell would need to consume glucose at a rate about 16 times the consumption rate in the aerobic environment (2 ATP are generated by fermentation versus up to 32 ATP by cellular respiration).

Concept Check 9.6

1. The fat is much more reduced; it has many —CH_2— units, and in all these bonds the electrons are equally shared. The electrons present in a carbohydrate molecule are already somewhat oxidized (shared unequally in bonds), as quite a few of them are bound to oxygen. **2.** When we consume more food than necessary for metabolic processes, our body synthesizes fat as a way of storing energy for later use. **3.** Glycogen is a storage polysaccharide in liver and muscle cells. When energy is needed, glucose units are hydrolyzed from glycogen. Glycolysis in the cytosol breaks down glucose to two pyruvate molecules, which are transported into the mitochondrion. Here they are further oxidized, ultimately producing the needed ATP. **4.** AMP will accumulate, stimulating phosphofructokinase, and thus increasing the rate of glycolysis. Since oxygen is not present, the cell will convert pyruvate to lactate in lactic acid fermentation, providing a supply of ATP. **5.** When oxygen is present, the fatty acid chains containing most of the energy of a fat are oxidized and fed into the citric acid cycle and the electron transport chain. During intense exercise, however, oxygen is scarce in muscle cells, so ATP must be generated by glycolysis alone. A very small part of the fat molecule, the glycerol backbone, can be oxidized via glycolysis, but the amount of energy released by this portion is insignificant compared to that released by the fatty acid chains. (This is why moderate exercise, staying below 70% maximum heart rate, is better for burning fat—because enough oxygen remains available to the muscles.)

Summary of Key Concepts Questions

9.1 Most of the ATP produced in cellular respiration comes from oxidative phosphorylation, in which the energy released from redox reactions in an electron transport chain is used to produce ATP. In substrate-level phosphorylation, an enzyme directly transfers a phosphate group to ADP from an intermediate substrate. All ATP production in glycolysis occurs by substrate-level phosphorylation; this form of ATP production also occurs at one step in the citric acid cycle. **9.2** The oxidation of the three-carbon sugar, glyceraldehyde 3-phosphate, yields energy. In this oxidation, electrons and H^+ are transferred to NAD^+, forming NADH, and a phosphate group is attached to the oxidized substrate. ATP is then formed by substrate-level phosphorylation when this phosphate group is transferred to ADP. **9.3** The release of six molecules of CO_2 represents the complete oxidation of glucose. During the processing of two pyruvates to acetyl CoA, the fully oxidized carboxyl group (—COO^-) is given off as CO_2. The remaining four carbons are released as CO_2 in the citric acid cycle as citrate is oxidized back to oxaloacetate. **9.4** The flow of H^+ through the ATP synthase complex causes the rotor and attached rod to rotate, exposing catalytic sites in the knob portion that produce ATP from ADP and Ⓟ$_i$. ATP synthases are found in the inner mitochondrial membrane, the plasma membrane of prokaryotes, and membranes within chloroplasts. **9.5** Anaerobic respiration yields more ATP. The 2 ATP produced by substrate-level phosphorylation in glycolysis represent the total energy yield of fermentation. NADH passes its "high-energy" electrons to pyruvate or a derivative of pyruvate, recycling NAD^+ and allowing glycolysis to continue. Anaerobic respiration uses an electron transport chain to capture the energy of the electrons in NADH via a series of redox reactions; ultimately, the electrons are transferred to an electronegative molecule other than oxygen. And additional molecules of NADH are produced in anaerobic respiration as pyruvate is oxidized. **9.6** The ATP produced by catabolic pathways is used to drive anabolic pathways. Also, many of the intermediates of glycolysis and the citric acid cycle are used in the biosynthesis of a cell's molecules.

Test Your Understanding

1. d **2.** c **3.** c **4.** a **5.** e **6.** a **7.** b
8.

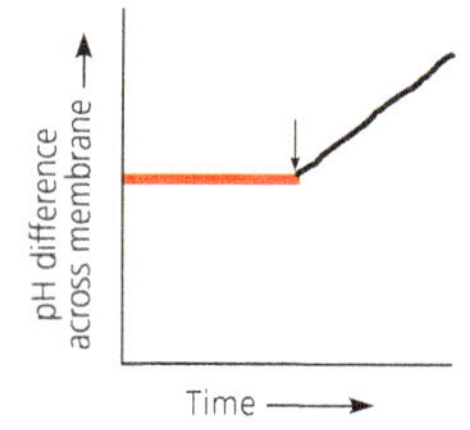

Chapter 10

Figure Questions

Figure 10.3 Situating containers of algae near sources of CO_2 emissions makes sense because algae need CO_2 to carry out photosynthesis. The higher their rate of photosynthesis, the more plant oil they will produce. At the same time, algae would be absorbing the CO_2 emitted from industrial plants or from car engines, reducing the amount of CO_2 entering the atmosphere. **Figure 10.10** Red, but not violet-blue, wavelengths would pass through the filter, so the bacteria would not congregate where the violet-blue light normally comes through. Therefore, the left "peak" of bacteria would not be present, but the right peak would be observed because the red wavelengths passing through the filter would be used for photosynthesis. **Figure 10.12** In the leaf, most of the chlorophyll electrons excited by photon absorption are used to power the reactions of photosynthesis.
Figure 10.16 The person at the top of the photosystem I tower would not turn and throw his electron into the bucket. Instead, he would throw it onto the top of the ramp right next to the photosystem II tower. The electron would then roll down the ramp, get energized by a photon, and return to him. This cycle would continue as long as light was available. (This is why it's called cyclic electron flow.)
Figure 10.19

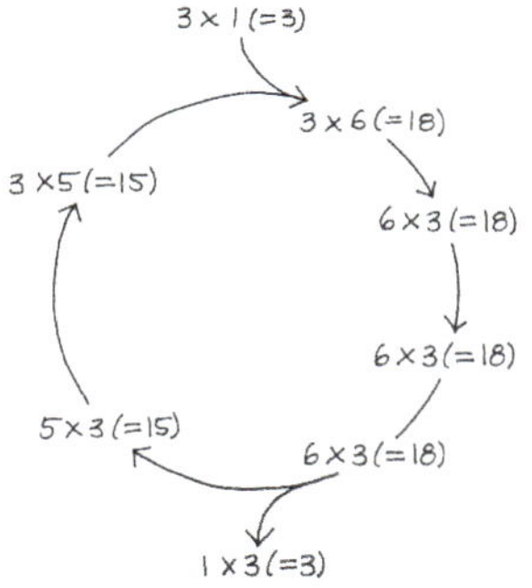

Three carbon atoms enter the cycle, one by one, as individual CO_2 molecules, and leave the cycle in one three-carbon molecule (G3P) per three turns of the cycle.
Figure 10.22

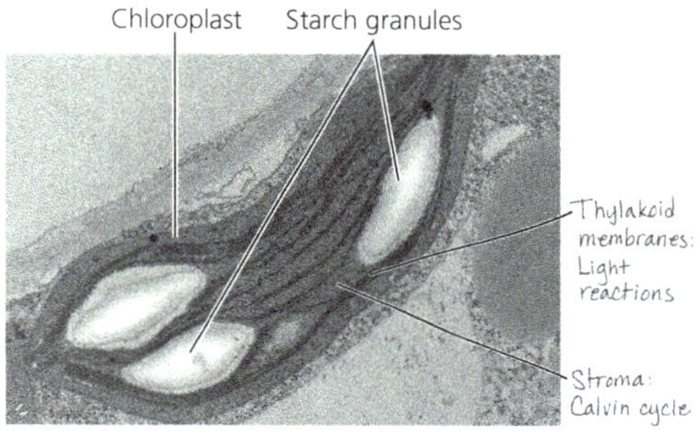

The photosystems that carry out the light reactions are embedded in the thylakoid membranes, and the ATP and NADPH that are formed are released into the stroma. There, they are used for the reactions of the Calvin cycle, which produces G3P. Excess sugar molecules that are not used by the plant can be converted to glucose, then stored in the form of starch.

Concept Check 10.1

1. CO_2 enters leaves via stomata, and water enters via roots and is carried to leaves through veins. **2.** Using ^{18}O, a heavy isotope of oxygen, as a label, researchers were able to confirm van Niel's hypothesis that the oxygen produced during photosynthesis originates in water, not in carbon dioxide. **3.** The light reactions could *not* keep producing NADPH and ATP without the $NADP^+$, ADP, and Ⓟ$_i$ that the Calvin cycle generates. The two cycles are interdependent.

Concept Check 10.2

1. Green, because green light is mostly transmitted and reflected—not absorbed—by photosynthetic pigments **2.** In chloroplasts, light-excited electrons are trapped by a primary electron acceptor, which prevents them from dropping back to the ground state. In isolated chlorophyll, there is no electron acceptor, so the photoexcited electrons immediately drop back down to the ground state, with the emission of light and heat. **3.** Water (H_2O) is the initial electron donor; $NADP^+$ accepts electrons at the end of the electron transport chain, becoming reduced to NADPH. **4.** In this experiment, the rate of ATP synthesis would slow and eventually stop. Because the added compound would not allow a proton gradient to build up across the membrane, ATP synthase could not catalyze ATP production.

Concept Check 10.3

1. 6, 18, 12 **2.** The more potential energy a molecule stores, the more energy and reducing power is required for the formation of that molecule. Glucose is a valuable energy source because it is highly reduced, storing lots of potential energy in its electrons. To reduce CO_2 to glucose, much energy and reducing power are required in the form of large numbers of ATP and NADPH molecules, respectively. **3.** The light reactions require ADP and $NADP^+$, which would not be formed in sufficient quantities from ATP and NADPH if the Calvin cycle stopped. **4.** In glycolysis, G3P acts as an intermediate. The 6-carbon sugar fructose 1,6-bisphosphate is cleaved into two 3-carbon sugars, one of which is G3P. The other is an isomer called dihydroxyacetone phosphate, which can be converted to G3P by an isomerase. Because G3P is the substrate for the next enzyme, it is constantly removed, and the reaction equilibrium is pulled in the direction of conversion of dihydroxyacetone phosphate to more G3P. In the Calvin cycle, G3P acts as both an intermediate and a product. For every three CO_2 molecules that enter the cycle, six G3P molecules are formed, five of which must remain in the cycle and become rearranged to regenerate three 5-carbon RuBP molecules. The one remaining G3P is a product, which can be thought of as the result of "reducing" the three CO_2 molecules that entered the cycle into a 3-carbon sugar that can later be used to generate energy.

Concept Check 10.4

1. Photorespiration decreases photosynthetic output by adding oxygen, instead of carbon dioxide, to the Calvin cycle. As a result, no sugar is generated (no carbon is fixed), and O_2 is used rather than generated. **2.** Without PS II, no O_2 is generated in bundle-sheath cells. This avoids the problem of O_2 competing with CO_2 for binding to rubisco in these cells. **3.** Both problems are caused by a drastic change in Earth's atmosphere due to burning of fossil fuels. The increase in CO_2 concentration affects ocean chemistry by decreasing pH, thus affecting calcification by marine organisms. On land, CO_2 concentration and air temperature are conditions that plants have become adapted to, and changes in these characteristics have a strong effect on photosynthesis by plants. Thus, alteration of these two fundamental factors could have critical effects on organisms all around the planet, in all different habitats. **4.** C_4 and CAM species would replace many of the C_3 species.

Summary of Key Concepts Questions

10.1 CO_2 and H_2O are the products of respiration; they are the reactants in photosynthesis. In respiration, glucose is oxidized to CO_2 as electrons are passed through an electron transfer chain from glucose to O_2, producing H_2O. In photosynthesis, H_2O is the source of electrons, which are energized by light, temporarily stored in NADPH, and used to reduce CO_2 to carbohydrate. **10.2** The action spectrum of photosynthesis shows that some wavelengths of light that are not absorbed by chlorophyll *a* are still effective at promoting photosynthesis. The light-harvesting complexes of photosystems contain accessory pigments such as chlorophyll *b* and carotenoids, which absorb different wavelengths and pass the energy to chlorophyll *a*, broadening the spectrum of light useful for photosynthesis.
10.3

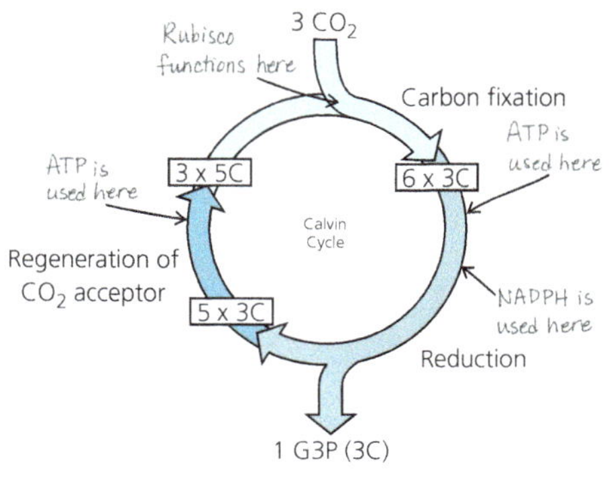

Appendix A Answers

In the reduction phase of the Calvin cycle, ATP phosphorylates a 3-carbon compound, and NADPH then reduces this compound to G3P. ATP is also used in the regeneration phase, when five molecules of G3P are converted to three molecules of the 5-carbon compound RuBP. Rubisco catalyzes the first step of carbon fixation—the addition of CO_2 to RuBP. **10.4** Both C_4 and CAM photosynthesis involve initial fixation of CO_2 to produce a 4-carbon compound (in mesophyll cells in C_4 plants and at night in CAM plants). These compounds are then broken down to release CO_2 (in the bundle-sheath cells in C_4 plants and during the day in CAM plants). ATP is required for recycling the molecule that is used initially to combine with CO_2. These pathways avoid the photorespiration that consumes ATP and reduces the photosynthetic output of C_3 plants when they close stomata on hot, dry, bright days. Thus, hot, arid climates would favor C_4 and CAM plants.

Test Your Understanding

1. d **2.** b **3.** c **4.** d **5.** c **6.** b **7.** d

9.

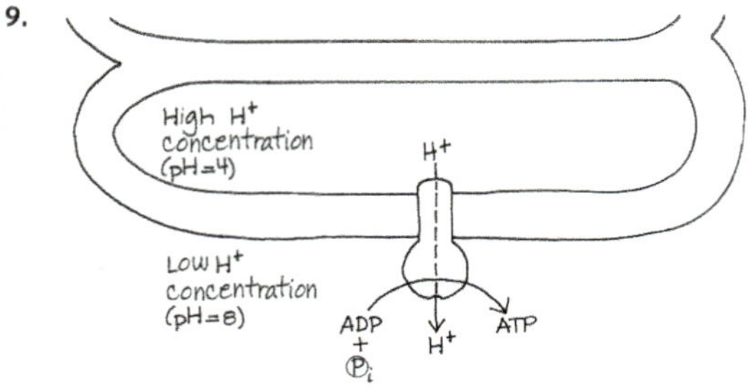

The ATP would end up outside the thylakoid. The thylakoids were able to make ATP in the dark because the researchers set up an artificial proton concentration gradient across the thylakoid membrane; thus, the light reactions were not necessary to establish the H^+ gradient required for ATP synthesis by ATP synthase.

Chapter 11

Figure Questions

Figure 11.6 Epinephrine is a signaling molecule; presumably, it binds to a cell-surface receptor protein. **Figure 11.7** Figure 7.1 shows a potassium channel, which, according to the description on p. 135, opens in response to an electrical stimulus, allowing potassium ions to rush out of the cell. Thus, it is a voltage-gated ion channel. **Figure 11.8** When the receptor is actively transmitting a signal to the inside of the cell, it is bound to a G protein. To determine a structure corresponding to that state, it might work to crystallize the receptor in the presence of many copies of the G protein. (In fact, the researchers planned to try this approach next. Another research group also used this approach successfully with a related G protein-coupled receptor the following year.) **Figure 11.9** The testosterone molecule is hydrophobic and can therefore pass directly through the lipid bilayer of the plasma membrane into the cell. (Hydrophilic molecules cannot do this.) **Figure 11.10** The active form of protein kinase 2 **Figure 11.11** The signaling molecule (cAMP) would remain in its active form and would continue to signal. **Figure 11.17** In the model, the directionality of growth is determined by the association of Fus3 with the membrane near the site of receptor activation. Thus, the development of shmoos would be severely compromised, and the affected cell would likely resemble the ΔFus3 and Δformin cells. **Figure 11.18** The signaling pathway shown in Figure 11.14 leads to the splitting of PIP_2 into the second messengers DAG and IP_3, which produce different responses. (The response elicited by DAG is mentioned but not shown.) The pathway shown for cell B is similar in that it branches and leads to two responses.

Concept Check 11.1

1. The two cells of opposite mating type (**a** and α) each secrete a certain signaling molecule, which can only be bound by receptors carried on cells of the opposite mating type. Thus, the **a** mating factor cannot bind to another **a** cell and cause it to grow toward the first **a** cell. Only an α cell can "receive" the signaling molecule and respond by directed growth (see Figure 11.17 for more information). **2.** The secretion of neurotransmitter molecules at a synapse is an example of local signaling. The electrical signal that travels along a very long nerve cell and is passed to the next nerve cell can be considered an example of long-distance signaling. (Note, however, that local signaling at the synapse between two cells is necessary for the signal to pass from one cell to the next.) **3.** Glucose 1-phosphate is not generated, because the activation of the enzyme requires an intact cell, with an intact receptor in the membrane and an intact signal transduction pathway. The enzyme cannot be activated directly by interaction with the signaling molecule in the test tube. **4.** Glycogen phosphorylase acts in the third stage, the response to epinephrine signaling.

Concept Check 11.2

1. NGF is water-soluble (hydrophilic), so it cannot pass through the lipid membrane to reach intracellular receptors, as steroid hormones can. Therefore, you'd expect the NGF receptor to be in the plasma membrane—which is, in fact, the case. **2.** The cell with the faulty receptor would not be able to respond appropriately to the signaling molecule when it was present. This would most likely have dire consequences for the cell, since regulation of the cell's activities by this receptor would not occur appropriately. **3.** Binding of a ligand to a receptor changes the shape of the receptor, altering the ability of the receptor to transmit a signal. Binding of an allosteric regulator to an enzyme changes the shape of the enzyme, either promoting or inhibiting enzyme activity.

Concept Check 11.3

1. A protein kinase is an enzyme that transfers a phosphate group from ATP to a protein, usually activating that protein (often a second type of protein kinase). Many signal transduction pathways include a series of such interactions, in which each phosphorylated protein kinase in turn phosphorylates the next protein kinase in the series. Such phosphorylation cascades carry a signal from outside the cell to the cellular protein(s) that will carry out the response. **2.** Protein phosphatases reverse the effects of the kinases. **3.** The signal that is being transduced is the information that a signaling molecule is bound to the cell-surface receptor. Information is transduced by way of sequential protein-protein interactions that change protein shapes, causing them to function in a way that passes the signal along. **4.** The IP_3-gated channel opens, allowing calcium ions to flow out of the ER, which raises the cytosolic Ca^{2+} concentration.

Concept Check 11.4

1. At each step in a cascade of sequential activations, one molecule or ion may activate numerous molecules functioning in the next step. **2.** Scaffolding proteins hold molecular components of signaling pathways in a complex with each other. Different scaffolding proteins would assemble different collections of proteins, leading to different cellular responses in the two cells. **3.** A malfunctioning protein phosphatase would not be able to dephosphorylate a particular receptor or relay protein. As a result, the signaling pathway, once activated, would not be able to be terminated. (In fact, one study found altered protein phosphatases in cells from 25% of colorectal tumors.)

Concept Check 11.5

1. In formation of the hand or paw in mammals, cells in the regions between the digits are programmed to undergo apoptosis. This serves to shape the digits of the hand or paw so that they are not webbed. **2.** If a receptor protein for a death-signaling molecule were defective such that it was activated even in the absence of the death signal, this would lead to apoptosis when it wouldn't normally occur. Similar defects in any of the proteins in the signaling pathway, which would activate these relay or response proteins in the absence of interaction with the previous protein or second messenger in the pathway, would have the same effect. Conversely, if any protein in the pathway were defective in its ability to respond to an interaction with an early protein or other molecule or ion, apoptosis would not occur when it normally should. For example, a receptor protein for a death-signaling ligand might not be able to be activated, even when ligand was bound. This would stop the signal from being transduced into the cell.

Summary of Key Concepts Questions

11.1 A cell is able to respond to a hormone only if it has a receptor protein on the cell surface or inside the cell that can bind to the hormone. The response to a hormone depends on the specific cellular activity that a signal transduction pathway triggers within the cell. The response can vary for different types of cells. **11.2** Both GPCRs and RTKs have an extracellular binding site for a signaling molecule (ligand) and an α helix region of the polypeptide that spans the membrane. GPCRs usually trigger a single transduction pathway, whereas the multiple activated tyrosines on an RTK dimer may trigger several different transduction pathways at the same time. **11.3** A protein kinase is an enzyme that adds a phosphate group to another protein. Protein kinases are often part of a phosphorylation cascade that transduces a signal. A second messenger is a small, nonprotein molecule or ion that rapidly diffuses and relays a signal throughout a cell. Both protein kinases and second messengers can operate in the same pathway. For example, the second messenger cAMP often activates protein kinase A, which then phosphorylates other proteins. **11.4** In G protein-coupled pathways, the GTPase portion of a G protein converts GTP to GDP and inactivates the G protein. Protein phosphatases remove phosphate groups from activated proteins, thus stopping a phosphorylation cascade of protein kinases. Phosphodiesterase converts cAMP to AMP, thus reducing the effect of cAMP in a signal transduction pathway. **11.5** The basic mechanism of controlled cell suicide evolved early in eukaryotic evolution, and the genetic basis for these pathways has been conserved during animal evolution. Such a mechanism is essential to the development and maintenance of all animals.

Test Your Understanding

1. c **2.** d **3.** a **4.** b **5.** a **6.** d **7.** c **8.** c **9.** This is one possible drawing of the pathway. (Similar drawings would also be correct.)

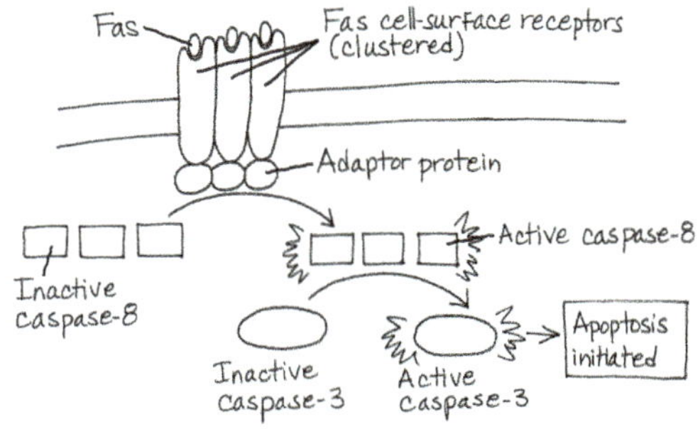

Chapter 12

Figure Questions

Figure 12.4

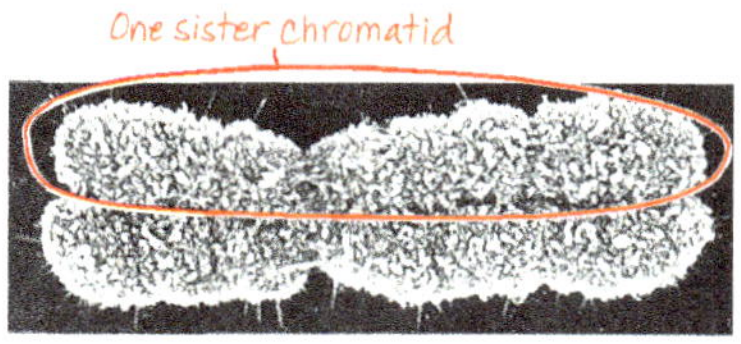

Circling the other chromatid instead would also be correct. **Figure 12.5** The chromosome has four arms. **Figure 12.7** 12; 2; 2; 1
Figure 12.8

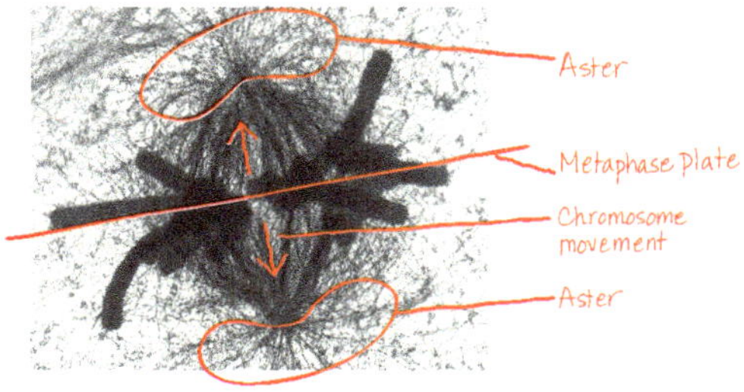

Figure 12.9 The mark would have moved toward the nearer pole. The lengths of fluorescent microtubules between that pole and the mark would have decreased, while the lengths between the chromosomes and the mark would have remained the same. **Figure 12.14** In both cases, the G_1 nucleus would have remained in G_1 until the time it normally would have entered the S phase. Chromosome condensation and spindle formation would not have occurred until the S and G_2 phases had been completed. **Figure 12.16** The cell would divide under conditions where it was inappropriate to do so. If the daughter cells and their descendants also ignored the checkpoint and divided, there would soon be an abnormal mass of cells. (This type of inappropriate cell division can contribute to the development of cancer.) **Figure 12.17** Passing the G_2 checkpoint in the diagram corresponds to the beginning of the "Time" axis of the graph, and entry into the mitotic phase (yellow background on the diagram) corresponds to the peaks of MPF activity and cyclin concentration on the graph (see the yellow M banner over the peaks). During G_1 and S phase in the diagram, Cdk is present without cyclin, so on the graph both cyclin concentration and MPF activity are low. The curved purple arrow in the diagram shows increasing cyclin concentration, seen on the graph during the end of S phase and throughout G_2 phase. Then the cell cycle begins again. **Figure 12.18** The cells in the vessel with PDGF would not be able to respond to the growth factor signal and thus would not divide. The culture would resemble that without the added PDGF. **Figure 12.21** The intracellular estrogen receptor, once activated, would be able to act as a transcription factor in the nucleus, turning on genes that may cause the cell to pass a checkpoint and divide. The HER2 receptor, when activated by a ligand, would form a dimer, and each subunit of the dimer would phosphorylate the other. This would lead to a series of signal transduction steps, ultimately turning on genes in the nucleus. As in the case of the estrogen receptor, the genes would code for proteins necessary to commit the cell to divide.

Concept Check 12.1

1. 2 **2.** 39; 39; 78

Concept Check 12.2

1. 6 chromosomes, duplicated; 12 chromatids **2.** Following mitosis, cytokinesis results in two genetically identical daughter cells in both plant cells and animal cells. However, the mechanism of dividing the cytoplasm is different in animals and plants. In an animal cell, cytokinesis occurs by cleavage, which divides the parent cell in two with a contractile ring of actin filaments. In a plant cell, a cell plate forms in the middle of the cell and grows until its membrane fuses with the plasma membrane of the parent cell. Material inside the cell plate thus becomes the new cell wall. **3.** They elongate the cell during anaphase. **4.** During eukaryotic cell division, tubulin is involved in spindle formation and chromosome movement, while actin functions during cytokinesis. In bacterial binary fission, it's the opposite: Tubulin-like molecules are thought to act in daughter cell separation, and actin-like molecules are thought to move the daughter bacterial chromosomes to opposite ends of the cell. **5.** Microtubules made up of tubulin in the cell provide "rails" along which vesicles and other organelles can travel, based on interactions of motor proteins with tubulin in the microtubules. In muscle cells, actin in microfilaments interacts with myosin filaments to cause muscle contraction.
6. From the end of S phase in interphase through the end of metaphase in mitosis

Concept Check 12.3

1. The nucleus on the right was originally in the G_1 phase; therefore, it had not yet duplicated its chromosome. The nucleus on the left was in the M phase, so it had already duplicated its chromosome. **2.** A sufficient amount of MPF has to exist for a cell to pass the G_2 checkpoint; this occurs through the accumulation of cyclin proteins, which combine with Cdk to form MPF. **3.** Most body cells are in a nondividing state called G_0. **4.** Both types of tumors consist of abnormal cells, but their characteristics are different. A benign tumor stays at the original site and can usually be surgically removed; the cells have some genetic and cellular changes from normal, non-tumor cells. Cancer cells from a malignant tumor have more significant genetic and cellular changes, can spread from the original site by metastasis, and may impair the functions of one or more organs. **5.** The cells might divide even in the absence of PDGF. In addition, they would not stop when the surface of the culture vessel was covered; they would continue to divide, piling on top of one another.

Summary of Key Concepts Questions

12.1 The DNA of a eukaryotic cell is packaged into structures called *chromosomes*. Each chromosome is a long molecule of DNA, which carries hundreds to thousands of genes, with associated proteins that maintain chromosome structure and help control gene activity. This DNA-protein complex is called *chromatin*. The chromatin of each chromosome is long and thin when the cell is not dividing. Prior to cell division, each chromosome is duplicated, and the resulting sister *chromatids* are attached to each other by proteins at the centromeres and, for many species, all along their lengths (sister chromatid cohesion). **12.2** Chromosomes exist as single DNA molecules in G_1 of interphase and in anaphase and telophase of mitosis. During S phase, DNA replication produces sister chromatids, which persist during G_2 of interphase and through prophase, prometaphase, and metaphase of mitosis. **12.3** Checkpoints allow cellular surveillance mechanisms to determine whether the cell is prepared to go to the next stage. Internal and external signals move a cell past these checkpoints. The G_1 checkpoint, called the "restriction point" in mammalian cells, determines whether a cell will complete the cell cycle and divide or switch into the G_0 phase. The signals to pass this checkpoint often are external—such as growth factors. Passing the G_2 checkpoint requires sufficient numbers of active MPF complexes, which in turn orchestrate several mitotic events. MPF also initiates degradation of its cyclin component, terminating the M phase. The M phase will not begin again until sufficient cyclin is produced during the next S and G_2 phases. The signal to pass the M phase checkpoint is not activated until all chromosomes are attached to kinetochore fibers and are aligned at the metaphase plate. Only then will sister chromatid separation occur.

Test Your Understanding

1. b **2.** a **3.** c **4.** c **5.** a **6.** b
7. See Figure 12.7 for a description of major events.

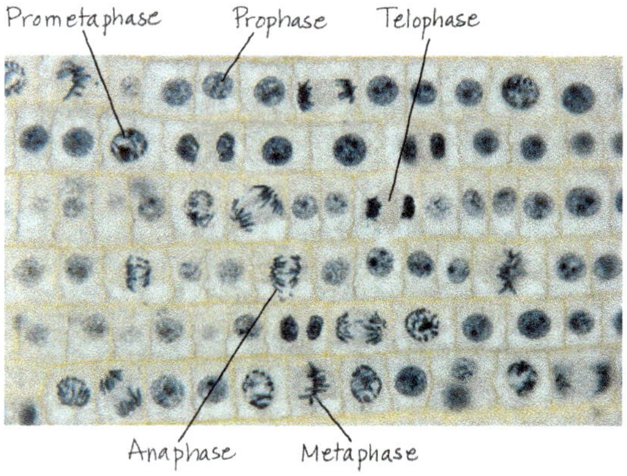

Only one cell is indicated for each stage, but other correct answers are also present in this micrograph.
8. a **9.** e
10.

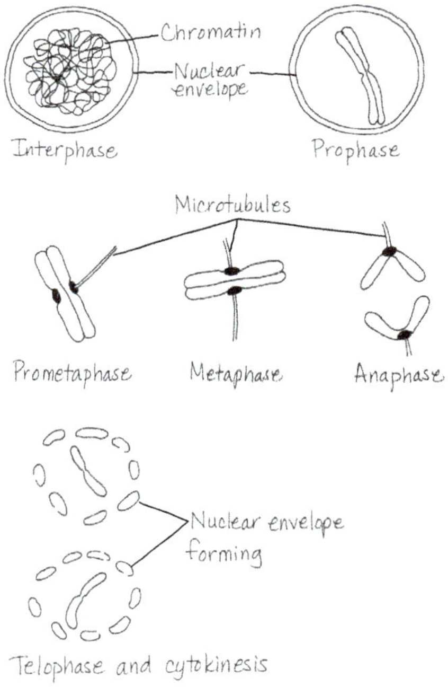

Chapter 27

Figure Questions

Figure 27.10 It is likely that the expression or sequence of genes that affect glucose metabolism may have changed; genes for metabolic processes no longer needed by the cell also may have changed. **Figure 27.11** Transduction results in horizontal gene transfer when the host and recipient cells are members of different species. **Figure 27.16** Thermophiles live in very hot environments, so it is likely that their enzymes can continue to function normally at much higher temperatures than do the enzymes of other organisms. At low temperatures, however, the enzymes of thermophiles may not function as well as the enzymes of other organisms. **Figure 27.18** From the graph, plant uptake can be estimated as 0.7, 0.6, and 0.95 (mg K) for strains 1, 2, and 3, respectively. These values average to 0.75 mg K. If bacteria had no effect, the average plant uptake of potassium for strains 1, 2, and 3 should be close to 0.5 mg K, the value observed for plants grown in bacteria-free soil.

Concept Check 27.1

1. Adaptations include the capsule (shields prokaryotes from host's immune system) and endospores (enable cells to survive harsh conditions and to revive when the environment becomes favorable). **2.** Prokaryotic cells generally lack the internal compartmentalization of eukaryotic cells. Prokaryotic genomes have much less DNA than eukaryotic genomes, and most of this DNA is contained in a single ring-shaped chromosome located in the nucleoid rather than within a true membrane-bounded nucleus. In addition, many prokaryotes also have plasmids, small ring-shaped DNA molecules containing a few genes. **3.** Plastids such as chloroplasts are thought to have evolved from an endosymbiotic photosynthetic prokaryote. More specifically, the phylogenetic tree shown in Figure 26.21 indicates that plastids are closely related to cyanobacteria. Hence, we can hypothesize that the thylakoid membranes of chloroplasts resemble those of cyanobacteria because chloroplasts evolved from a cyanobacterium endosymbiont.

Concept Check 27.2

1. Prokaryotes have extremely large population sizes, in part because they have short generation times. The large number of individuals in prokaryotic populations makes it likely that in each generation there will be thousands of individuals that have new mutations at any particular gene, thereby adding considerable genetic diversity to the population. **2.** In transformation, naked, foreign DNA from the environment is taken up by a bacterial cell. In transduction, phages carry bacterial genes from one bacterial cell to another. In conjugation, a bacterial cell directly transfers plasmid or chromosomal DNA to another cell via a mating bridge that temporarily connects the two cells. **3.** The population that includes individuals capable of conjugation would probably be more successful, since some of its members could form recombinant cells whose new gene combinations might be advantageous in a novel environment. **4.** Yes. Genes for antibiotic resistance could be transferred (by transformation, transduction, or conjugation) from the nonpathogenic bacterium to a pathogenic bacterium; this could make the pathogen an even greater threat to human health. In general, transformation, transduction, and conjugation tend to increase the spread of resistance genes.

Concept Check 27.3

1. A phototroph derives its energy from light, while a chemotroph gets its energy from chemical sources. An autotroph derives its carbon from a form of CO_2, while a heterotroph gets its carbon from organic nutrients such as glucose. Thus, there are four nutritional modes: photoautotrophic, photoheterotrophic (unique to prokaryotes), chemoautotrophic (unique to prokaryotes), and chemoheterotrophic. **2.** Chemoheterotrophy; the bacterium must rely on chemical sources of energy, since it is not exposed to light, and it must be a heterotroph if it requires a source of carbon other than CO_2 (or a related compound, such as bicarbonate). **3.** If humans could fix nitrogen, we could build proteins using atmospheric N_2 and hence would not need to eat high-protein foods such as meat, fish, or soy. Our diet would, however, need to include a source of carbon, along with minerals and water. Thus, a typical meal might consist of carbohydrates as a carbon source, along with fruits and vegetables to provide essential minerals (and additional carbon).

Concept Check 27.4

1. Before molecular systematics, taxonomists classified prokaryotes according to phenotypic characters that did not clarify evolutionary relationships. Molecular comparisons—of DNA in particular—indicate key divergences in prokaryotic lineages. **2.** By not requiring that organisms be cultured in the laboratory, genetic prospecting has revealed an immense diversity of previously unknown prokaryotic species. Over time, the ongoing discovery of new species by genetic prospecting is likely to alter our understanding of prokaryotic phylogeny greatly. **3.** At present, all known methanogens are archaea in the clade Euryarchaeota; this suggests that this unique metabolic pathway probably arose in ancestral species within Euryarchaeota. Since Bacteria and Archaea have been separate evolutionary lineages for billions of years, the discovery of a methanogen from the domain Bacteria would suggest that adaptations that enabled the use of CO_2 to oxidize H_2 may have evolved twice—once in Archaea (within Euryarchaeota) and once in Bacteria. (It is also possible that a newly discovered bacterial methanogen could have acquired the genes for this metabolic pathway by horizontal gene transfer from a methanogen in domain Archaea. However, horizontal gene transfer is not a likely explanation because of the large number of genes involved and because gene transfers between species in different domains are rare.)

Concept Check 27.5

1. Although prokaryotes are small, their large numbers and metabolic abilities enable them to play key roles in ecosystems by decomposing wastes, recycling chemicals, and affecting the concentrations of nutrients available to other organisms **2.** Cyanobacteria produce oxygen when water is split in the light reactions of photosynthesis. The Calvin cycle incorporates CO_2 from the air into organic molecules, which are then converted to sugars.

Concept Check 27.6

1. Sample answers: eating fermented foods such as yogurt, sourdough bread, or cheese; receiving clean water from sewage treatment; taking medicines produced by bacteria **2.** No. If the poison is secreted as an exotoxin, live bacteria could be transmitted to another person. But the same is true if the poison is an endotoxin—only in this case, the live bacteria that are transmitted may be descendants of the (now-dead) bacteria that produced the poison. **3.** Some of the many different species of prokaryotes that live in the human gut compete with one another for resources (in the food you eat). Because different prokaryotic species have different adaptations, a change in diet may alter which species can grow most rapidly, thus altering species abundance.

Summary of Key Concepts Questions

27.1. Prokaryotes are small, they have short generation times, and their populations can consist of trillions of individuals. As a result, populations of prokaryotes can evolve substantially in short periods of time, enabling them to adapt to a wide range of environments. Specific structural features that enable prokaryotes to thrive in diverse environments include their cell walls (which provide shape and protection), flagella (which function in directed movement), and ability to form endospores (which enable tolerance of harsh conditions). Prokaryotes also possess biochemical adaptations for growth in varied conditions, such as those that enable them to tolerate extremely hot or salty environments. **27.2** Prokaryotes reproduce extremely rapidly, and their populations can number in the trillions. As a result, even though mutations are rare, every day thousands of offspring are produced that have new mutations at particular gene loci. In addition, even though prokaryotes reproduce asexually and hence the vast majority of offspring are genetically identical to their parent, the genetic variation of their populations can be increased by transduction, transformation, and conjugation. Each of these (nonreproductive) processes can increase genetic variation by transferring DNA from one cell to another—even among cells that are of different species. **27.3** Prokaryotes have an exceptionally broad range of metabolic adaptations. As a group, prokaryotes perform all four modes of nutrition (photoautotrophy, chemoautotrophy, photoheterotrophy, and chemoheterotrophy), whereas eukaryotes perform only two of these (photoautotrophy and chemoheterotrophy). Prokaryotes are also able to metabolize nitrogen in a wide variety of forms (again unlike eukaryotes), and they frequently cooperate with other prokaryotic cells of the same or different species. **27.4** Phenotypic criteria such as shape, motility, and nutritional mode do not provide a clear picture of the evolutionary history of the prokaryotes. In contrast, molecular data have elucidated relationships among major groups of prokaryotes. Molecular data have also allowed researchers to sample genes directly from the environment; using such genes to construct phylogenies has led to the discovery of major new groups of prokaryotes. **27.5** Prokaryotes play key roles in the chemical cycles on which life depends. For example, prokaryotes are important decomposers, breaking down corpses and waste materials, thereby releasing nutrients to the environment where they can be used by other organisms. Prokaryotes also convert inorganic compounds to forms that other organisms can use. With respect to their ecological interactions, many prokaryotes form life-sustaining mutualisms with other species. In some cases, such as hydrothermal vent communities, the metabolic activities of prokaryotes provide an energy source on which hundreds of other species depend; in the absence of the prokaryotes, the community collapses. **27.6** Human well-being depends on our associations with mutualistic prokaryotes, such as the many species that live in our intestines and digest food that we cannot. Humans also can harness the remarkable metabolic capabilities of prokaryotes to produce a wide range of useful products. Negative effects of prokaryotes result primarily from bacterial pathogens that cause disease.

Test Your Understanding

1. e **2.** a **3.** d **4.** d **5.** b **6.** a

8. (a)

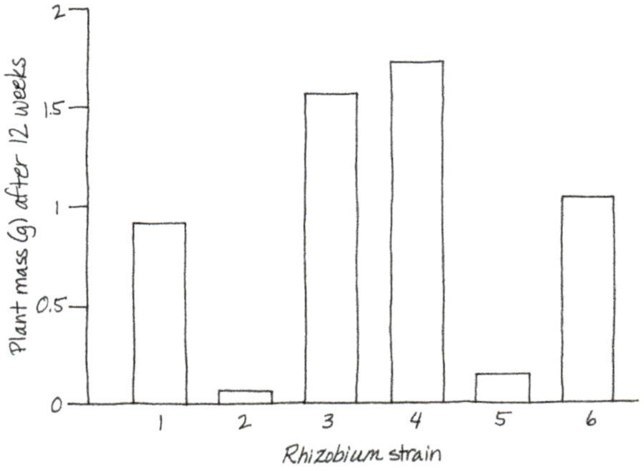

(b) Some *Rhizobium* strains are much more effective at promoting plant growth than other *Rhizobium* strains; the most ineffective strains have little positive effect (plant growth with these strains differs little from plant growth in the absence of *Rhizobium*). The ineffective strains may transfer relatively little nitrogen to their plant host, limiting plant growth.

Chapter 28

Figure Questions

Figure 28.10 Merozoites are produced by the asexual (mitotic) cell division of haploid sporozoites; similarly, gametocytes are produced by the asexual cell division of merozoites. Hence, it is likely that individuals in these three stages have the same complement of genes and that morphological differences between them result from changes in gene expression. **Figure 28.16** The sperm cells in the diagram are produced by the asexual (mitotic) division of cells in a single male gametophyte, which was itself produced by the asexual (mitotic) division of a single zoospore. Thus, the sperm cells are all derived from a single zoospore and so are genetically identical to one another. However, the male gametophyte that produced the sperm developed from one zoospore, while the female gametophyte that produced the egg developed from a different zoospore. Zoospores are produced by meiosis, so each zoospore differs genetically from the others. Hence, the sperm cells (which can be traced back to one zoospore) differ genetically from the egg cell (which can be traced back to another zoospore). **Figure 28.22** The following stage should be circled: step 6, where a mature cell undergoes mitosis and forms four or more daughter cells. In step 7, the zoospores eventually grow into mature haploid cells, but they do not produce new daughter cells. Likewise, in step 2, a mature cell develops into a gamete, but it does not produce new daughter cells. **Figure 28.23** If the assumption is correct, then their results indicate that the DHFR-TS gene fusion may be a derived trait shared by members of four supergroups of eukaryotes (Excavata, Chromalveolata, Rhizaria, and Archaeplastida). However, if the assumption is not correct, the presence or absence of the gene fusion may tell little about phylogenetic history. For example, if the genes fused multiple times, groups could share the trait because of convergent evolution rather than common descent. If the genes were secondarily split, a group with such a split could be placed (incorrectly) in Unikonta rather than its correct placement in one of the other four supergroups. **Figure 28.28** If diatom populations decrease in size as global temperatures warm, less carbon dioxide would be "pumped" from surface waters to the deep ocean floor. Atmospheric carbon dioxide levels might increase as a result, thereby potentially causing further warming. If this process continues, a positive-feedback loop would result in which warming causes diatom populations to drop, thereby causing additional warming, further drops in diatom populations, and so on.

Concept Check 28.1

1. Sample response: Protists include unicellular, colonial, and multicellular organisms; photoautotrophs, heterotrophs, and mixotrophs; species that reproduce asexually, sexually, or both ways; and organisms with diverse physical forms and adaptations. **2.** Strong evidence shows that eukaryotes acquired mitochondria after an early eukaryote first engulfed and then formed an endosymbiotic association with an alpha proteobacterium. Similarly, chloroplasts in red and green algae appear to have descended from a photosynthetic cyanobacterium that was engulfed by an ancient heterotrophic eukaryote. Secondary endosymbiosis also played an important role: Various protist lineages acquired plastids by engulfing unicellular red or green algae. **3.** The modified tree would look as follows:

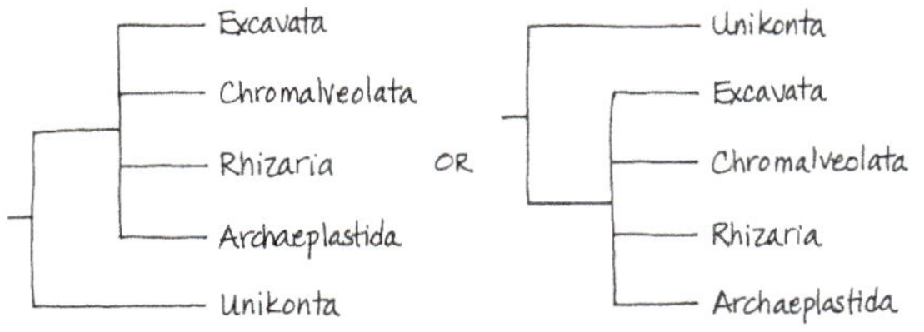

Concept Check 28.2

1. Their mitochondria do not have an electron transport chain and so cannot function in aerobic respiration. **2.** Since the unknown protist is more closely related to diplomonads than to euglenids, it must have originated after the diplomonads and parabasalids diverged from the euglenozoans. In addition, since the unknown species has fully functional mitochondria—yet both diplomonads and parabasalids do not—it is likely that the unknown species originated *before* the last common ancestor of the diplomonads and parabasalids.

Concept Check 28.3

1. Some DNA data indicate that Chromalveolata is a monophyletic group, but other DNA data fail to support this result. In support of monophyly, for many species in the group, the structure of their plastids and the sequence of their plastid DNA suggest that the group originated by a secondary endosymbiosis event (in which a red alga was engulfed). However, other species in the group lack plastids entirely, making the secondary endosymbiosis hypothesis difficult to test. **2.** The plastid DNA would likely be more similar to the chromosomal DNA of cyanobacteria based on the well-supported hypothesis that eukaryotic plastids (such as those found in the eukaryotic groups listed) originated by an endosymbiosis event in which a eukaryote engulfed a cyanobacterium. If the plastid is derived from the cyanobacterium, its DNA would be derived from the bacterial DNA. **3.** Figure 13.6b. Algae and plants with alternation of generations have a multicellular haploid stage *and* a multicellular diploid stage. In the other two life cycles, either the haploid stage or the diploid stage is unicellular.

Concept Check 28.4

1. Because foram tests are hardened with calcium carbonate, they form long-lasting fossils in marine sediments and sedimentary rocks. **2.** Convergent evolution. The different organisms have come to display similar morphological adaptations over time owing to their similar lifestyles. **3.** During photosynthesis, aerobic algae produce O_2 and use CO_2. O_2 is produced as a by-product of the light reactions, while CO_2 is used as an input to the Calvin cycle (the end products of which are sugars). Aerobic algae also perform cellular respiration, which uses O_2 as an input and produces CO_2 as a waste product.

Concept Check 28.5

1. Many red algae contain an accessory pigment called phycoerythrin, which gives them a reddish color and allows them to carry out photosynthesis in relatively deep coastal water. Also unlike brown algae, red algae have no flagellated stages in their life cycle and must depend on water currents to bring gametes together for fertilization. **2.** *Ulva*'s thallus contains many cells and is differentiated into leaflike blades and a rootlike holdfast. *Caulerpa*'s thallus is composed of multinucleate filaments without cross-walls, so it is essentially one large cell. **3.** Red algae have no flagellated stages in their life cycle and hence must depend on water currents to bring their gametes together. This feature of their biology might increase the difficulty of reproducing on land. In contrast, the gametes of green algae are flagellated, making it possible for them to swim in thin films of water. In addition, a variety of green algae contain compounds in their cytoplasm, cell wall, or zygote coat that protect against intense sunlight and other terrestrial conditions. Such compounds may have increased the chance that descendants of green algae could survive on land.

Concept Check 28.6

1. Amoebozoans have lobe-shaped pseudopodia, whereas forams have threadlike pseudopodia. **2.** Slime molds are fungus-like in that they produce fruiting bodies that aid in the dispersal of spores, and they are animal-like in that they are motile and ingest food. However, slime molds are more closely related to gymnamoebas and entamoebas than to fungi or animals. **3.** Support. Unikonts lack the unique cytoskeletal features shared by many excavates (see Concept 28.2). Thus, if the unikonts were the first group of eukaryotes to diverge from other eukaryotes (as shown in Figure 28.23), it would be unlikely that the eukaryote common ancestor had the cytoskeletal features found today in many excavates. Such a result would strengthen the case that many excavates share cytoskeletal features because they are members of a monophyletic group, the Excavata.

Concept Check 28.7

1. Because photosynthetic protists constitute the base of aquatic food webs, many aquatic organisms depend on them for food, either directly or indirectly. (In addition, a substantial percentage of the oxygen produced in photosynthesis on Earth is made by photosynthetic protists.) **2.** Protists form mutualistic and parasitic associations with other organisms. Examples include photosynthetic dinoflagellates that form a mutualistic symbiosis with coral polyps, parabasalids that form a mutualistic symbiosis with termites, and the oomycete *Phytophthora ramorum*, a parasite of oak trees. **3.** Corals depend on their dinoflagellate symbionts for nourishment, so coral bleaching would be expected to cause the corals to die. As the corals die, less food will be available for fishes and other species that eat coral. As a result, populations of these species may decline, and that, in turn, might cause populations of their predators to decline.

Summary of Key Concepts Questions

28.1 Sample response: Protists, plants, animals, and fungi are similar in that their cells have a nucleus and other membrane-bounded organelles, unlike the cells of prokaryotes. These membrane-bounded organelles make the cells of eukaryotes more complex than the cells of prokaryotes. With respect to differences between protists and other eukaryotes, most protists are unicellular, unlike animals, plants, and most fungi. Protists also have greater nutritional diversity than other eukaryotes. **28.2** Unique cytoskeletal features are shared by many excavates. In addition, some members of Excavata have an "excavated" feeding groove for which the group was named. DNA evidence does not strongly support or refute Excavata as a group. Overall, evidence for the group is relatively weak. **28.3** We can infer that the common ancestor of the group had a plastid, in this case of red algal origin. Thus, we would expect that members of Chromalveolata either would have plastids or would have lost their plastids over the course of evolution. **28.4** The main subgroups are the radiolarians, the forams, and the cercozoans. They are united as a clade by DNA similarities. **28.5** Red algae, green algae, and land plants are placed in the same supergroup because considerable evidence indicates that these organisms all descended from the same ancestor, an ancient heterotrophic protist that acquired a cyanobacterial endosymbiont. **28.6** The unikonts are a diverse group of eukaryotes that includes many protists, along with animals and fungi. Most of the protists in Unikonta are amoebozoans, a clade of amoebas that have lobe- or tube-shaped pseudopodia (as opposed to the threadlike pseudopodia of rhizarians). Other protists in Unikonta include several groups that are closely related to fungi and several other groups that are closely related to animals. **28.7** Sample response: Ecologically important protists include photosynthetic dinoflagellates that provide essential sources of energy to their symbiotic partners, the corals that build coral reefs. Other important protistan symbionts include those that enable termites to digest wood and *Plasmodium*, the pathogen that causes malaria. Photosynthetic protists such as diatoms are among the most important producers in aquatic communities; as such, many other species in aquatic environments depend on them for food.

Test Your Understanding

1. d **2.** b **3.** c **4.** e **5.** d **6.** d

7. 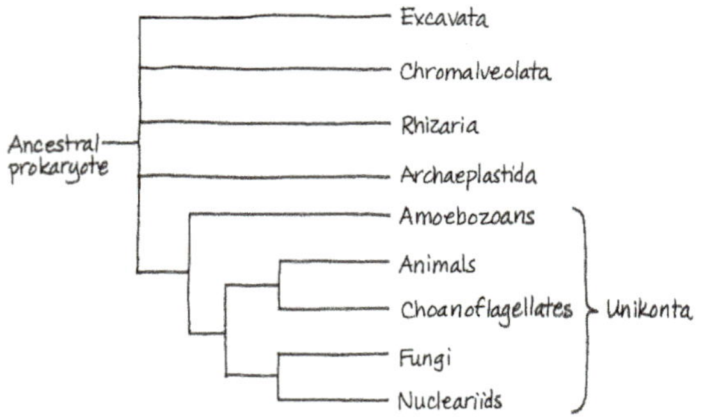

Pathogens that share a relatively recent common ancestor with humans should also share metabolic and structural characteristics with humans. Because drugs target the pathogen's metabolism or structure, developing drugs that harm the pathogen but not the patient should be most difficult for pathogens with whom we share the most recent evolutionary history. Working backward in time, we can use the phylogenetic tree to determine the order in which humans shared a common ancestor with pathogens in different taxa. This process leads to the prediction that it should be hardest to develop drugs to combat animal pathogens, followed by choanoflagellate pathogens, fungal and nucleariid pathogens, amoebozoans, other protists, and finally prokaryotes.

Chapter 43

Figure Questions

Figure 43.5 The seemingly inactive peptides might offer protection against pathogens other than those studied. Also, some antimicrobial peptides might work best in combination. **Figure 43.6** Cell-surface TLRs recognize pathogens identifiable by surface molecules, whereas TLRs in vesicles recognize pathogens identifiable by internal molecules after the pathogens are broken down. **Figure 43.10** Part of the enzyme or antigen receptor provides a structural "backbone" that maintains overall shape, while interaction occurs at a surface with a close fit to the substrate or antigen. The combined effect of multiple noncovalent interactions at the active site or binding site is a high-affinity interaction of tremendous specificity. **Figure 43.13** After gene rearrangement, a lymphocyte and its daughter cells make a single version of the antigen receptor. In contrast, alternative splicing is not heritable and can give rise to diverse gene products in a single cell. **Figure 43.18** These receptors enable memory cells to present antigen on their cell surface to a helper T cell. This presentation of antigen is required to activate memory cells in a secondary immune response. **Figure 43.20** Primary response: arrows extending from Antigen (1st exposure), Antigen-presenting cell, Helper T cell, B cell, Plasma cells, Cytotoxic T cell, and Active cytotoxic T cells; secondary response: arrows extending from Antigen (2nd exposure), Memory helper T cells, Memory B cells, and Memory cytotoxic T cells. **Figure 43.26** The loss of growth control that characterizes cancer involves many changes in gene regulation. HPV and other viruses can bring about some of these changes, but other mutations must occur in an infected cell to transform the cell to a cancerous state.

Concept Check 43.1

1. Because pus contains white blood cells, fluid, and cell debris, it indicates an active and at least partially successful inflammatory response against invading microbes. **2.** Whereas the ligand for the TLR receptor is a foreign molecule, the ligand for many signal transduction pathways is a molecule produced by the animal itself. **3.** Bacteria with a human host would be likely to grow optimally at normal body temperature or, if fever were often induced, at a temperature a few degrees higher.

Concept Check 43.2

1. See Figure 43.9. The transmembrane regions lie within the C regions, which also form the disulfide bridges. In contrast, the antigen-binding sites are in the V regions. **2.** Generating memory cells ensures both that a receptor specific for a particular epitope will be present and that there will be more lymphocytes with this specificity than in a host that had never encountered the antigen. **3.** If each B cell produced two different light and heavy chains for its antigen receptor, different combinations would make four different receptors. If any one was self-reactive, the lymphocyte would be eliminated in the generation of self-tolerance. For this reason, many more B cells would be eliminated, and those that could respond to a foreign antigen would be less effective at doing so due to the variety of receptors (and antibodies) they express.

Concept Check 43.3

1. A child lacking a thymus would have no functional T cells. Without helper T cells to help activate B cells, the child would be unable to produce antibodies against extracellular bacteria. Furthermore, without cytotoxic T cells or helper T cells, the child's immune system would be unable to kill virus-infected cells. **2.** Since the antigen-binding site is intact, the antibody fragments could neutralize viruses and opsonize bacteria. **3.** If the handler developed immunity to proteins in the antivenin, another injection could provoke a severe immune response. The handler's immune system might also now produce antibodies that could neutralize the venom.

Concept Check 43.4

1. Myasthenia gravis is considered an autoimmune disease because the immune system produces antibodies against self molecules (certain receptors on muscle cells). **2.** A person with a cold is likely to produce oral and nasal secretions that facilitate viral transfer. In addition, since sickness can cause incapacitation or death, a virus that is programmed to exit the host when there is a physiological stress has the opportunity to find a new host at a time when the current host may cease to function. **3.** A person with a macrophage deficiency would have frequent infections. The causes would be poor innate responses, due to diminished phagocytosis and inflammation, and poor adaptive responses, due to the lack of macrophages to present antigens to helper T cells.

Summary of Key Concepts Questions

43.1 Lysozyme in saliva destroys bacterial cell walls; the viscosity of mucus helps trap bacteria; acidic pH in the stomach kills many bacteria; and the tight packing of cells lining the gut provides a physical barrier to infection. **43.2** Sufficient numbers of cells to mediate an innate immune response are always present, whereas an adaptive response requires selection and proliferation of an initially very small cell population specific for the infecting pathogen. **43.3** No. Immunological memory after a natural infection and after vaccination are very similar. There may be minor differences in the particular antigens that can be recognized in a subsequent infection. **43.4** No. AIDS refers to a loss of immune function that can occur over time in an individual infected with HIV, not to the viral infection itself. For individuals infected with HIV, certain multidrug combinations ("cocktails") or rare genetic variations usually prevent progression to AIDS.

Test Your Understanding

1. b **2.** c **3.** c **4.** d **5.** b **6.** b **7.** c

8. One possible answer:

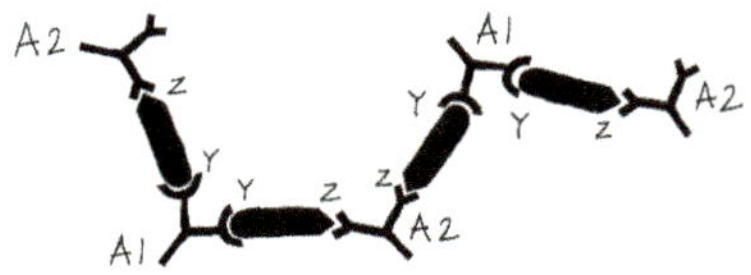

APPENDIX

B Periodic Table of the Elements

Atomic number (number of protons) → 6
Element symbol → C
Atomic mass (number of protons plus number of neutrons averaged over all isotopes) → 12.01

Metals | Metalloids | Nonmetals

Groups: Elements in a vertical column have the same number of electrons in their valence (outer) shell and thus have similar chemical properties.

Periods: Each horizontal row contains elements with the same total number of electron shells. Across each period, elements are ordered by increasing atomic number.

Representative elements: Groups 1–2 (1A: Alkali metals; 2A: Alkaline earth metals) and Groups 13–18 (7A: Halogens; 8A: Noble gases). Transition elements: Groups 3–12.

Period number	1 Group 1A	2 Group 2A	3 3B	4 4B	5 5B	6 6B	7 7B	8 8B	9 8B	10 8B	11 1B	12 2B	13 Group 3A	14 Group 4A	15 Group 5A	16 Group 6A	17 Group 7A	18 Group 8A
1	1 H 1.008																	2 He 4.003
2	3 Li 6.941	4 Be 9.012											5 B 10.81	6 C 12.01	7 N 14.01	8 O 16.00	9 F 19.00	10 Ne 20.18
3	11 Na 22.99	12 Mg 24.31											13 Al 26.98	14 Si 28.09	15 P 30.97	16 S 32.07	17 Cl 35.45	18 Ar 39.95
4	19 K 39.10	20 Ca 40.08	21 Sc 44.96	22 Ti 47.87	23 V 50.94	24 Cr 52.00	25 Mn 54.94	26 Fe 55.85	27 Co 58.93	28 Ni 58.69	29 Cu 63.55	30 Zn 65.41	31 Ga 69.72	32 Ge 72.64	33 As 74.92	34 Se 78.96	35 Br 79.90	36 Kr 83.80
5	37 Rb 85.47	38 Sr 87.62	39 Y 88.91	40 Zr 91.22	41 Nb 92.91	42 Mo 95.94	43 Tc (98)	44 Ru 101.1	45 Rh 102.9	46 Pd 106.4	47 Ag 107.9	48 Cd 112.4	49 In 114.8	50 Sn 118.7	51 Sb 121.8	52 Te 127.6	53 I 126.9	54 Xe 131.3
6	55 Cs 132.9	56 Ba 137.3	57* La 138.9	72 Hf 178.5	73 Ta 180.9	74 W 183.8	75 Re 186.2	76 Os 190.2	77 Ir 192.2	78 Pt 195.1	79 Au 197.0	80 Hg 200.6	81 Tl 204.4	82 Pb 207.2	83 Bi 209.0	84 Po (209)	85 At (210)	86 Rn (222)
7	87 Fr (223)	88 Ra (226)	89† Ac (227)	104 Rf (261)	105 Db (262)	106 Sg (266)	107 Bh (264)	108 Hs (269)	109 Mt (268)	110 Ds (271)	111 Rg (272)	112 Cn (285)	113 — (284)	114 — (289)	115 — (288)	116 — (293)	117 — (294?)	118 — (294)

*Lanthanides	58 Ce 140.1	59 Pr 140.9	60 Nd 144.2	61 Pm (145)	62 Sm 150.4	63 Eu 152.0	64 Gd 157.3	65 Tb 158.9	66 Dy 162.5	67 Ho 164.9	68 Er 167.3	69 Tm 168.9	70 Yb 173.0	71 Lu 175.0
†Actinides	90 Th 232.0	91 Pa 231.0	92 U 238.0	93 Np (237)	94 Pu (244)	95 Am (243)	96 Cm (247)	97 Bk (247)	98 Cf (251)	99 Es 252	100 Fm 257	101 Md 258	102 No 259	103 Lr 260

Name (Symbol)	Atomic Number	Name (Symbol)	Atomic Number	Name (Symbol)	Atomic Number	Name (Symbol)	Atomic Number	Name (Symbol)	Atomic Number
Actinium (Ac)	89	Copernicium (Cn)	112	Iridium (Ir)	77	Palladium (Pd)	46	Sodium (Na)	11
Aluminum (Al)	13	Copper (Cu)	29	Iron (Fe)	26	Phosphorus (P)	15	Strontium (Sr)	38
Americium (Am)	95	Curium (Cm)	96	Krypton (Kr)	36	Platinum (Pt)	78	Sulfur (S)	16
Antimony (Sb)	51	Darmstadtium (Ds)	110	Lanthanum (La)	57	Plutonium (Pu)	94	Tantalum (Ta)	73
Argon (Ar)	18	Dubnium (Db)	105	Lawrencium (Lr)	103	Polonium (Po)	84	Technetium (Tc)	43
Arsenic (As)	33	Dysprosium (Dy)	66	Lead (Pb)	82	Potassium (K)	19	Tellurium (Te)	52
Astatine (At)	85	Einsteinium (Es)	99	Lithium (Li)	3	Praseodymium (Pr)	59	Terbium (Tb)	65
Barium (Ba)	56	Erbium (Er)	68	Lutetium (Lu)	71	Promethium (Pm)	61	Thallium (Tl)	81
Berkelium (Bk)	97	Europium (Eu)	63	Magnesium (Mg)	12	Protactinium (Pa)	91	Thorium (Th)	90
Beryllium (Be)	4	Fermium (Fm)	100	Manganese (Mn)	25	Radium (Ra)	88	Thulium (Tm)	69
Bismuth (Bi)	83	Fluorine (F)	9	Meitnerium (Mt)	109	Radon (Rn)	86	Tin (Sn)	50
Bohrium (Bh)	107	Francium (Fr)	87	Mendelevium (Md)	101	Rhenium (Re)	75	Titanium (Ti)	22
Boron (B)	5	Gadolinium (Gd)	64	Mercury (Hg)	80	Rhodium (Rh)	45	Tungsten (W)	74
Bromine (Br)	35	Gallium (Ga)	31	Molybdenum (Mo)	42	Roentgenium (Rg)	111	Uranium (U)	92
Cadmium (Cd)	48	Germanium (Ge)	32	Neodymium (Nd)	60	Rubidium (Rb)	37	Vanadium (V)	23
Calcium (Ca)	20	Gold (Au)	79	Neon (Ne)	10	Ruthenium (Ru)	44	Xenon (Xe)	54
Californium (Cf)	98	Hafnium (Hf)	72	Neptunium (Np)	93	Rutherfordium (Rf)	104	Ytterbium (Yb)	70
Carbon (C)	6	Hassium (Hs)	108	Nickel (Ni)	28	Samarium (Sm)	62	Yttrium (Y)	39
Cerium (Ce)	58	Helium (He)	2	Niobium (Nb)	41	Scandium (Sc)	21	Zinc (Zn)	30
Cesium (Cs)	55	Holmium (Ho)	67	Nitrogen (N)	7	Seaborgium (Sg)	106	Zirconium (Zr)	40
Chlorine (Cl)	17	Hydrogen (H)	1	Nobelium (No)	102	Selenium (Se)	34		
Chromium (Cr)	24	Indium (In)	49	Osmium (Os)	76	Silicon (Si)	14		
Cobalt (Co)	27	Iodine (I)	53	Oxygen (O)	8	Silver (Ag)	47		

APPENDIX C The Metric System

Measurement	Unit and Abbreviation	Metric Equivalent	Metric-to-English Conversion Factor	English-to-Metric Conversion Factor
Length	1 kilometer (km)	= 1,000 (10^3) meters	1 km = 0.62 mile	1 mile = 1.61 km
	1 meter (m)	= 100 (10^2) centimeters	1 m = 1.09 yards	1 yard = 0.914 m
		= 1,000 millimeters	1 m = 3.28 feet	1 foot = 0.305 m
			1 m = 39.37 inches	
	1 centimeter (cm)	= 0.01 (10^{-2}) meter	1 cm = 0.394 inch	1 foot = 30.5 cm
				1 inch = 2.54 cm
	1 millimeter (mm)	= 0.001 (10^{-3}) meter	1 mm = 0.039 inch	
	1 micrometer (μm) (formerly micron, μ)	= 10^{-6} meter (10^{-3} mm)		
	1 nanometer (nm) (formerly millimicron, mμ)	= 10^{-9} meter (10^{-3} μm)		
	1 angstrom (Å)	= 10^{-10} meter (10^{-4} μm)		
Area	1 hectare (ha)	= 10,000 square meters	1 ha = 2.47 acres	1 acre = 0.405 ha
	1 square meter (m^2)	= 10,000 square centimeters	1 m^2 = 1.196 square yards	1 square yard = 0.8361 m^2
			1 m^2 = 10.764 square feet	1 square foot = 0.0929 m^2
	1 square centimeter (cm^2)	= 100 square millimeters	1 cm^2 = 0.155 square inch	1 square inch = 6.4516 cm^2
Mass	1 metric ton (t)	= 1,000 kilograms	1 t = 1.103 tons	1 ton = 0.907 t
	1 kilogram (kg)	= 1,000 grams	1 kg = 2.205 pounds	1 pound = 0.4536 kg
	1 gram (g)	= 1,000 milligrams	1 g = 0.0353 ounce	1 ounce = 28.35 g
			1 g = 15.432 grains	
	1 milligram (mg)	= 10^{-3} gram	1 mg = approx. 0.015 grain	
	1 microgram (μg)	= 10^{-6} gram		
Volume (solids)	1 cubic meter (m^3)	= 1,000,000 cubic centimeters	1 m^3 = 1.308 cubic yards	1 cubic yard = 0.7646 m^3
			1 m^3 = 35.315 cubic feet	1 cubic foot = 0.0283 m^3
	1 cubic centimeter (cm^3 or cc)	= 10^{-6} cubic meter	1 cm^3 = 0.061 cubic inch	1 cubic inch = 16.387 cm^3
	1 cubic millimeter (mm^3)	= 10^{-9} cubic meter		
		= 10^{-3} cubic centimeter		
Volume (liquids and gases)	1 kiloliter (kL or kl)	= 1,000 liters	1 kL = 264.17 gallons	
	1 liter (L or l)	= 1,000 milliliters	1 L = 0.264 gallons	1 gallon = 3.785 L
			1 L = 1.057 quarts	1 quart = 0.946 L
	1 milliliter (mL or ml)	= 10^{-3} liter	1 mL = 0.034 fluid ounce	1 quart = 946 mL
		= 1 cubic centimeter	1 mL = approx. 1/4 teaspoon	1 pint = 473 mL
			1 mL = approx. 15–16 drops (gtt.)	1 fluid ounce = 29.57 mL
				1 teaspoon = approx. 5 mL
	1 microliter (μL or μl)	= 10^{-6} liter (10^{-3} milliliters)		
Time	1 second (s or sec)	= 1/60 minute		
	1 millisecond (ms or msec)	= 10^{-3} second		
Temperature	Degrees Celsius (°C) (Absolute zero, when all molecular motion ceases, is −273.15°C. The Kelvin [K] scale, which has the same size degrees as Celsius, has its zero point at absolute zero. Thus, 0 K = −273.15°C.)		°F = 9/5°C + 32	°C = 5/9 (°F − 32)

D A Comparison of the Light Microscope and the Electron Microscope

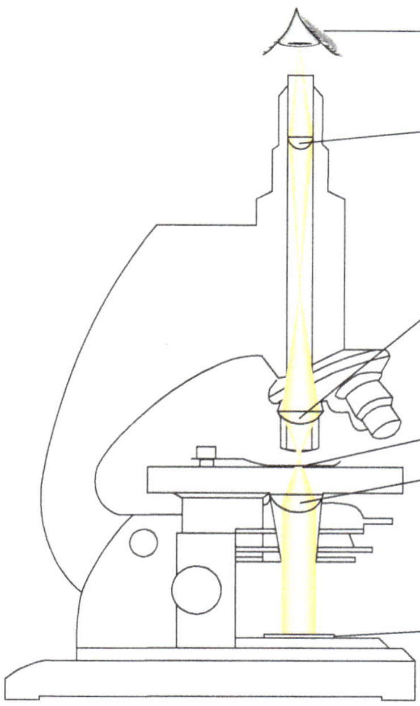

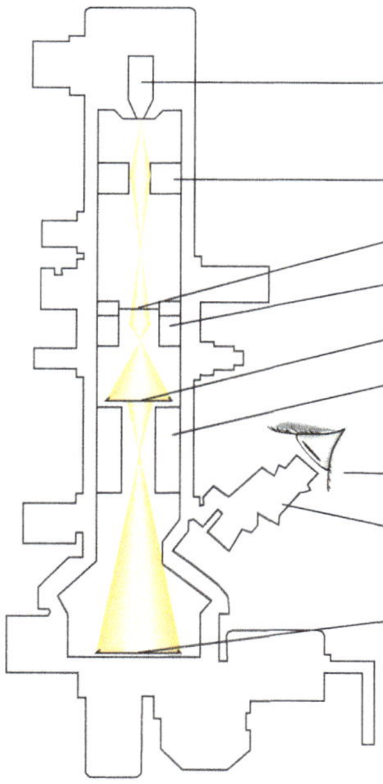

Light Microscope

In light microscopy, light is focused on a specimen by a glass condenser lens; the image is then magnified by an objective lens and an ocular lens, for projection on the eye, digital camera, digital video camera, or photographic film.

Electron Microscope

In electron microscopy, a beam of electrons (top of the microscope) is used instead of light, and electromagnets are used instead of glass lenses. The electron beam is focused on the specimen by a condenser lens; the image is magnified by an objective lens and a projector lens for projection on a digital detector, fluorescent screen, or photographic film.

APPENDIX **E** Classification of Life

This appendix presents a taxonomic classification for the major extant groups of organisms discussed in this text; not all phyla are included. The classification presented here is based on the three-domain system, which assigns the two major groups of prokaryotes, bacteria and archaea, to separate domains (with eukaryotes making up the third domain).

Various alternative classification schemes are discussed in Unit Five of the text. The taxonomic turmoil includes debates about the number and boundaries of kingdoms and about the alignment of the Linnaean classification hierarchy with the findings of modern cladistic analysis. In this review, asterisks (*) indicate currently recognized phyla thought by some systematists to be paraphyletic.

DOMAIN BACTERIA

- Proteobacteria
- Chlamydia
- Spirochetes
- Gram-positive Bacteria
- Cyanobacteria

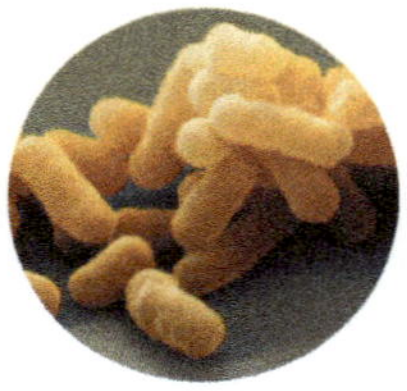

DOMAIN ARCHAEA

- **Korarchaeota**
- **Euryarchaeota**
- **Crenarchaeota**
- **Nanoarchaeota**

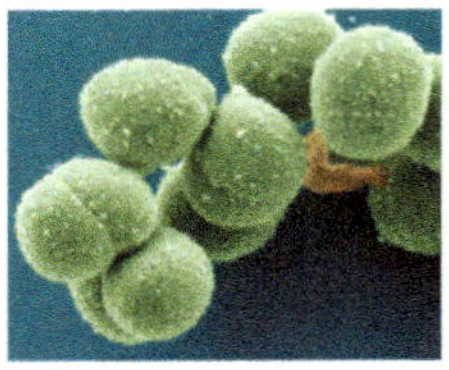

DOMAIN EUKARYA

In the phylogenetic hypothesis we present in Chapter 28, major clades of eukaryotes are grouped together in the five "supergroups" listed below in blue type. Formerly, all the eukaryotes generally called protists were assigned to a single kingdom, Protista. However, advances in systematics have made it clear that Protista is in fact polyphyletic: Some protists are more closely related to plants, fungi, or animals than they are to other protists. As a result, the kingdom Protista has been abandoned.

Excavata

- Diplomonadida (diplomonads)
- Parabasala (parabasalids)
- Euglenozoa (euglenozoans)
  - Kinetoplastida (kinetoplastids)
  - Euglenophyta (euglenids)

Chromalveolata

- Alveolata (alveolates)
  - Dinoflagellata (dinoflagellates)
  - Apicomplexa (apicomplexans)
  - Ciliophora (ciliates)
- Stramenopila (stramenopiles)
  - Bacillariophyta (diatoms)
  - Chrysophyta (golden algae)
  - Phaeophyta (brown algae)
  - Oomycota (water molds)

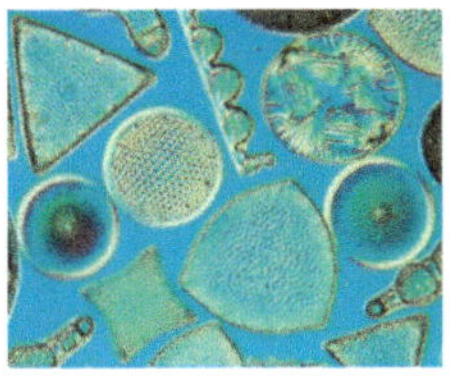

Archaeplastida

- Rhodophyta (red algae)
- Chlorophyta (green algae: chlorophytes)
- Charophyta (green algae: charophytes)
- Plantae

Phylum Hepatophyta (liverworts)	Nonvascular plants (bryophytes)	
Phylum Bryophyta (mosses)		
Phylum Anthocerophyta (hornworts)		
Phylum Lycophyta (lycophytes)	Seedless vascular plants	
Phylum Pterophyta (ferns, horsetails, whisk ferns)		
Phylum Ginkgophyta (ginkgo)	Gymnosperms	Seed plants
Phylum Cycadophyta (cycads)		
Phylum Gnetophyta (gnetophytes)		
Phylum Coniferophyta (conifers)		
Phylum Anthophyta (flowering plants)	Angiosperms	

Rhizaria

- Radiolaria (radiolarians)
- Foraminifera (forams)
- Cercozoa (cercozoans)

DOMAIN EUKARYA, continued

Unikonta

- Amoebozoa (amoebozoans)
  - Myxogastrida (plasmodial slime molds)
  - Dictyostelida (cellular slime molds)
  - Gymnamoeba (gymnamoebas)
  - Entamoeba (entamoebas)
- Nucleariida (nucleariids)
- Fungi
  - *Phylum Chytridiomycota (chytrids)
  - *Phylum Zygomycota (zygomycetes)
  - Phylum Glomeromycota (glomeromycetes)
  - Phylum Basidiomycota (club fungi)
  - Phylum Ascomycota (sac fungi)

- Choanoflagellata (choanoflagellates)
- Animalia
  - Phylum Porifera (sponges)
  - Phylum Ctenophora (comb jellies)
  - Phylum Cnidaria (cnidarians)
    - Hydrozoa (hydrozoans)
    - Scyphozoa (jellies)
    - Cubozoa (box jellies and sea wasps)
    - Anthozoa (sea anemones and most corals)
  - Phylum Acoela (acoel flatworms)
  - Phylum Placozoa (placozoans)

<u>Lophotrochozoa (lophotrochozoans)</u>

- Phylum Kinorhyncha (kinorhynchs)
- Phylum Platyhelminthes (flatworms)
  - Catenulida (chain worms)
  - Rhabditophora (planarians, flukes, tapeworms)
- Phylum Nemertea (proboscis worms)
- Phylum Ectoprocta (ectoprocts)
- Phylum Brachiopoda (brachiopods)
- Phylum Phoronida (phoronids)
- Phylum Rotifera (rotifers)
- Phylum Cycliophora (cycliophorans)
- Phylum Mollusca (molluscs)
  - Polyplacophora (chitons)
  - Gastropoda (gastropods)
  - Bivalvia (bivalves)
  - Cephalopoda (cephalopods)
- Phylum Annelida (segmented worms)
  - Polychaeta (polychaetes)
  - Oligochaeta (oligochaetes)
- Phylum Acanthocephala (spiny-headed worms)

<u>Ecdysozoa (ecdysozoans)</u>

- Phylum Loricifera (loriciferans)
- Phylum Priapula (priapulans)
- Phylum Nematoda (roundworms)
- Phylum Arthropoda (This survey groups arthropods into a single phylum, but some zoologists now split the arthropods into multiple phyla.)
  - Subphylum Chelicerata (horseshoe crabs, arachnids)
  - Subphylum Myriapoda (millipedes, centipedes)
  - Subphylum Hexapoda (insects, springtails)
  - Subphylum Crustacea (crustaceans)
- Phylum Tardigrada (tardigrades)
- Phylum Onychophora (velvet worms)

<u>Deuterostomia (deuterostomes)</u>

- Phylum Hemichordata (hemichordates)
- Phylum Echinodermata (echinoderms)
  - Asteroidea (sea stars, sea daisies)
  - Ophiuroidea (brittle stars)
  - Echinoidea (sea urchins and sand dollars)
  - Crinoidea (sea lilies)
  - Holothuroidea (sea cucumbers)
- Phylum Chordata (chordates)
  - Subphylum Cephalochordata (cephalochordates: lancelets)
  - Subphylum Urochrodata (urochordates: tunicates)
  - Subphylum Craniata (craniates)
    - Myxini (hagfishes)
    - Vertebrates:
      - Petromyzontida (lampreys)
      - Chondrichthyes (sharks, rays, chimaeras)
      - Actinopterygii (ray-finned fishes)
      - Actinistia (coelacanths)
      - Dipnoi (lungfishes)
      - Amphibia (amphibians)
      - Reptilia (tuataras, lizards, snakes, turtles, crocodilians, birds)
      - Mammalia (mammals)

Credits

Photo Credits

Cover Image "Succulent I" © 2005 Amy Lamb, www.amylamb.com

Unit Opening Interviews **UNIT 1** Justina Thorsen, Showcase Reflections; **UNIT 2** Brian Wilson, Princeton University; **UNIT 3** William K. Sacco, Yale ITS Photo + Design Services; **UNIT 4** Steve Bonnel, Bonnel Photography; **UNIT 5** Jacob Mailman; **UNIT 6** Michael Starghill, Michael Starghill Photography; **UNIT 7** River Healey, Rusty Healey Photography; **UNIT 8** Marsha Miller, Director of Photography, University of Texas at Austin.

About the Author Photos River Healey, Rusty Healey Photography.

Detailed Contents **UNIT I** Photo by T. Naeser, from Patrick Cramer Laboratory, Gene Center Munich, Ludwig-Maximilians-Universität München, Munich, Germany; **UNIT II** Biophoto Associates/Photo Researchers; **UNIT III** From: Multicolor Spectral Karyotyping of Human Chromosomes. E. Schröck, S. du Manoir, T. Veldman, B. Schoell, J. Wienberg, M. A. Ferguson-Smith, Y. Ning, D. H. Ledbetter, I. Bar-Am, D. Soenksen, Y. Garini, T. Ried. *Science*. 1996 Jul 26;273(5274):494-7.; **UNIT IV** © Ted Daeschler/Academy of Natural Sciences/VIREO; **UNIT V** Yufeng Zhou/iStockphoto; **UNIT VI** PD photo: John Walker (www.fourmilab.ch); **UNIT VII** David Wall/Alamy; David Doubilet/Getty Images; Kenneth Catania; **UNIT VIII** Daniel Mosquin.

Chapter 1 **1.1** Amy Lamb Studio; **1.2** Walter Teague; **1.3.2** R.Dirscherl/FLPA; **1.3.3** Kim Taylor and Jane Burton/Dorling Kindersley; **1.3.4** Malcolm Schuyl/FLPA; **1.3.5** Frans Lanting/Corbis; **1.3.6** Michael & Patricia Fogden/Corbis; **1.3.7** Joe McDonald/Corbis; **1.3.8** ImageState/International Stock Photography Ltd.; **1.4.2** WorldSat International/Photo Researchers, Inc.; **1.4.3** Bill Brooks/Alamy; **1.4.4** Linda Freshwaters Arndt/Alamy; **1.4.5** Michael Orton/Photographer's Choice/Getty Images; **1.4.6** Ross M. Horowitz/Getty Images; **1.4.7** Photodisc/Getty Images; **1.4.9** Jeremy Burgess/SPL/Photo Researchers; **1.4.11** John Durham/Photo Researchers; **1.4.13** Electron micrograph by Wm. P. Wergin, courtesy of E. H. Newcomb, University of Wisconsin; **1.5.2** James Balog/Aurora Creative/Getty Images; **1.6.2** Anup Shah/Nature Picture Library; **1.6.3** Anup Shah/Nature Picture Library; **1.7.2** Photodisc/Getty Images; **1.7.3** Janice Sheldon; **1.8.2** S. C. Holt/Biological Photo Service; **1.8.3** Steve Gschmeissner/Photo Researchers; **1.9.2** Conly L. Rieder; **1.9.3** Conly L. Rieder; **1.10.2** Camille Tokerud/Stone/Getty Images; **1.11.2** Photodisc/Getty Images; **1.12** Roy Kaltschmidt, Lawrence Berkeley National Laboratory; **1.15.2** Oliver Meckes/Nicole Ottawa/Photo Researchers; **1.15.3** Eye of Science/Photo Researchers; **1.15.4** Kunst & Scheidulin/AGE Fotostock; **1.15.5** Peter Lilja/Taxi/Getty Images; **1.15.6** Anup Shah/Nature Picture Library; **1.15.7** D. P. Wilson/Photo Researchers; **1.16.2** VVG/SPL/Photo Researchers; **1.16.3** W. L. Dentler/Biological Photo Service; **1.16.4** OMIKRON/Photo Researchers; **1.17** Photo by Dede Randrianarisata, from Kristi Curry Rogers, Macalester College, St. Paul, MN; **1.18** ARCHIV/Photo Researchers; **1.19.2** Michael P. Fogden/Bruce Coleman/Alamy; **1.19.3** Matt T. Lee; **1.19.4** Hal Horwitz/Corbis; **1.21** Frank Greenaway/Dorling Kindersley; **1.23.2** Karl Ammann/Corbis; **1.23.3** Tim Ridley/Dorling Kindersley, Courtesy of the Jane Goodall Institute, Clarendon Park, Hampshire; **1.25.2** Breck P. Kent; **1.25.3** Barry Mansell/Nature Picture Library; **1.26.2** David Pfennig; **1.26.3** David Pfennig; **1.28** Gary L. Firestone's Lab researchers meeting in the Dept. of Molecular and Cell Biology, University of California at Berkeley. Photo: Seelevel.com, Pearson Science; **1.29** Tim Sharp/AP Images; **p. 25, top right** James Balog/Aurora Creative/Getty Images; **p. 25, bottom left** Anup Shah/Nature Picture Library; **p. 25, bottom left** Anup Shah/Nature Picture Library; **p.26, top** PhotoDisc/Getty Images; **p. 26, 2nd from top** S. C. Holt/ Biological Photo Service; **p. 26, 3rd from top** Steve Gschmeissner/Photo Researchers.

Chapter 2 **2.1** Martin Dohrn/BBC Natural History Unit; **2.2.3** Martin Dohrn/BBC Natural History Unit; **2.3.2** Chip Clark; **2.4.2** C. Michael Hogan; **2.4.3** Rick York and California Native Plant Society (www.cnps.org); **2.4.4** Andrew Alden; **2.6.2** Clayton T. Hamilton, Stanford University; **2.7.2** National Library of Medicine; **p. 41, left** Jerry Young/Dorling Kindersley; **2.19** Nigel Cattlin/Photo Researchers; **p. 45** Rolf Nussbaumer/Nature Picture Library.

Chapter 3 **3.1** Alexander/Fotoloia; **3.3.2** N.C. Brown Center for Ultrastructure Studies, SUNY-Environmental Science & Forestry, Syracuse, NY; **3.4** iStockphoto; **3.6.2** Jan van Franeker, IMARES, Alfred Wegener Institute for Polar and Marine Research; **3.9** NASA/JPL-Caltech/University of Arizona/Texas A&M University; **3.10.2** Jakub Semeniuk/iStockphoto; **3.10.3** Feng Yu/iStockphoto; **3.10.4** Monika Wisniewska/iStockphoto; **3.10.5** Beth Van Trees/Shutterstock; **3.12.2** From "Coral Reefs Under Rapid Climate Change and Ocean Acidification." O. Hoegh-Guldberg, et al. *Science* 14 December 2007: 318 (5857):1737–1742. Photos by Ove Hoegh-Guldberg, Centre for Marine Studies, The University of Queensland; **3.12.3** From "Coral Reefs Under Rapid Climate Change and Ocean Acidification." O. Hoegh-Guldberg, et al. *Science* 14 December 2007: 318 (5857):1737–1742. Photos by Ove Hoegh-Guldberg, Centre for Marine Studies, The University of Queensland; **3.12.4** From "Coral Reefs Under Rapid Climate Change and Ocean Acidification." O. Hoegh-Guldberg, et al. *Science* 14 December 2007: 318 (5857):1737–1742. Photos by Ove Hoegh-Guldberg, Centre for Marine Studies, The University of Queensland.

Chapter 4 **4.1** Shutterstock; **4.6.2** David M. Phillips/Photo Researchers.

Chapter 5 **5.1** Photo by T. Naeser, from Patrick Cramer Laboratory, Gene Center Munich, Ludwig-Maximilians-Universität München, Munich, Germany; **5.6.2** John N. A. Lott/Biological Photo Service; **5.6.3** H. Shio and P. B. Lazarow; **5.8.2** Alexey Repka/iStockphoto; **5.8.3** John Durham/Photo Researchers; **5.8.4** Biophoto Associates/Photo Researchers; **5.9.2** F. Collet/Photo Researchers; **5.9.3** Corbis; **5.11.2** Dorling Kindersley; **5.11.3** Dorling Kindersley; **5.15.2** Andrey Stratilatov/Shutterstock; **5.15.3** Nina Zanetti; **5.15.4** Nina Zanetti; **5.19.2** Reproduced by permission from Tulip WR, Varghese JN, Laver WG, Webster RG, Colman PM. Refined crystal structure of the influenza virus N9 neuraminidase-NC41 Fab complex. *J Mol Biol.* September 5; 227(1):122–48. Copyright © 1992 by Elsevier Science Ltd. **5.20.3** Dieter Hopf/AGE Fotostock; **5.20.8** Monika Wisniewska/iStockphoto; **5.21.2** Eye of Science/Photo Researchers; **5.21.3** Eye of Science/Photo Researchers; **5.23.2** Reprinted by permission from *Nature*. P. B. Sigler from Z. Xu, A. L. Horwich, and P. B. Sigler. 388:741–750 Copyright © 1997 Macmillan Magazines Limited; **5.24.2** Dave Bushnell; **5.24.4** Dave Bushnell; **p. 90** Dorling Kindersley.

Chapter 6 **6.1** Eye of Science/Photo Researchers; **6.3.2** Elisabeth Pierson, FNWI-Radboud University Nijmegen, Pearson Science; **6.3.3** Elisabeth Pierson, FNWI-Radboud University Nijmegen, Pearson Science; **6.3.4** Elisabeth Pierson, FNWI-Radboud University Nijmegen, Pearson Science; **6.3.5** Elisabeth Pierson, FNWI-Radboud University Nijmegen, Pearson Science; **6.3.6** Michael W. Davidson/The Florida State University Research Foundation; **6.3.7** Karl Garsha, Beckman Institute for Advanced Science and Technology, University of Illinois; **6.3.9** Macrophage fluorescently stained for tubulin (yellow), actin (red) and the nucleus (DAPI, blue). Top part of the image: data recorded with a widefield microscope and visualized with the Simulated Fluorescence Process (SFP) volume rendering algorithm. Bottom part: the same dataset, deconvolved using Huygens Professional (Scientific Volume Imaging, Hilversum, The Netherlands) and again rendered with the SFP algorithm. Data courtesy Dr. James G. Evans, Whitehead Institute, MIT Boston MA, USA; **6.3.11** From "Motile Cilia of Human Airway Epithelia Are Chemosensory." Alok S. Shah, Yehuda Ben-Shahar, T.s O. Moninger, J. N. Kline, M. J. Welsh. Science Express on 23 July 2009. *Science* 28 August 2009:325 (5944): 1131–1134 (cover). Pseudocolored scanning electron micrograph by Tom Moninger (epithelia generated by Phil Karp); **6.3.12** William Dentler/Biological Photo Service; **6.3.13** From "STED microscopy reveals that synaptotagmin remains clustered after synaptic vesicle exocytosis." Katrin I. Willig, Silvio O. Rizzoli, Volker Westphal, Reinhard Jahn & . . . Stefan W. Hell. *Nature*, 440 (13) April 2006. doi:10.1038/nature04592, Letters; **6.5.2** S. C. Holt/Biological Photo Service; **6.6.2** Daniel Friend; **6.8.4** S. Cinti/Photo Researchers; **6.8.6** SPL/Photo Researchers; **6.8.8** A. Barry Dowsett/Photo Researchers; **6.8.10** Biophoto Associates/Photo Researchers; **6.8.12** SPL/Photo Researchers; **6.8.14** From "Flagellar Microtubule Dynamics in *Chlamydomonas:* Cytochalasin D Induces Periods of Microtubule Shortening and Elongation; and Colchicine Induces Disassembly of the Distal, but Not Proximal, Half of the Flagellum." W. L. Dentler and C. Adams. *The Journal of Cell Biology*, 117(6): 1289–1298, Copyright © 1992 by The Rockefeller University Press; **6.9.2** Reproduced by permission from L.Orci and A.Perelet, Freeze-Etch Histology. (Heidelberg: Springer-Verlag, 1975). Copyright ©1975 by Springer-Verlag GmbH & Co KG; **6.9.3** Reproduced by permission from A. C. Faberge, *Cell Tiss. Res.* 151 Copyright © 1974 by Springer-Verlag GmbH & Co KG; **6.9.4** Reprinted by permission from *Nature* 323. U. Aebi et al. Copyright © 1996: 560–564, figure 1a. Used with permission. Macmillan Magazines Limited; **6.10.2** D. W. Fawcett/Photo Researchers; **6.11.2** R. Bolender, D. Fawcett/Photo Researchers; **6.12.2** Don W. Fawcett/Photo Researchers; **6.13.2** Daniel S. Friend; **6.13.3** Daniel S. Friend; **6.14.2** E. H. Newcomb; **6.17.2** Daniel S. Friend; **6.17.3** From "The shape of mitochondria and the number of mitochondrial nucleoids during the cell cycle of *Euglena gracilis*." Y. Hayashi and K. Ueda. *Journal of Cell Science*, 93:565–570, Copyright © 1989 by Company of Biologists; **6.18.2** Courtesy of W.P. Wergin and E.H. Newcomb, University of Wisconsin/Biological Photo Service; **6.18.3** Franz Grolig, Philipps-University Marburg, Germany. Image acquired with the confocal microscope Leica TCS SP2; **6.19.2** From S. E. Fredrick and E. H. Newcomb, *The Journal of Cell Biology* 43 (1969):343. Provided by E. H. Newcomb; **6.20** Albert Tousson, High Resolution Imaging Facility, University of Alabama at Birmingham; **6.21.2** Dr. Bruce J. Schnapp; **p. 113, left** Mary Osborn; **p. 113, middle** Frank Solomon; **p. 113, right** Mark S. Ladinsky and J. Richard McIntosh, University of Colorado; **6.22.2** Kent L. McDonald; **6.23.2** Biophoto Associates/Photo Researchers; **6.23.3** Oliver Meckes & Nicole Ottawa/Eye of Science/Photo Researchers; **6.24.2** OMIKRON/Science Source/Photo Researchers; **6.24.3** W. L. Dentler/Biological Photo Service; **6.24.4** Linck RW, Stephens RE. Functional protofilament numbering of ciliary, flagellar, and centriolar microtubules. *Cell Motil Cytoskeleton.* 2007 Jul;64(7):489–95; cover. Micrograph by D. Woodrum Hensley; **6.26.2** From Hirokawa Nobutaka, *The Journal of Cell Biology* 94 (1982):425 by copyright permission of The Rockefeller University Press; **6.27.2** Clara Franzini-Armstrong, University of Pennsylvania; **6.27.3** M. I. Walker/Photo Researchers; **6.27.4** Michael Clayton, University of Wisconsin-Madison; **6.28.2** G. F. Leedale/Photo Researchers; **6.29.2** David Ehrhardt; **6.29.3** From "Visualization of cellulose synthase demonstrates functional association with microtubules." A. R. Paredez, C. R. Somerville, D. W. Ehrhardt. *Science.* 2006 Jun 9;312(5779):1491–5. Epub 2006 Apr 20; **6.31.2** Micrograph by W. P. Wergin, provided by E. H. Newcomb; **6.32.2** From Douglas J. Kelly, *The Journal of Cell Biology* 28 (1966): 51. Fig.17. Reproduced by copyright permission of The Rockefeller University Press; **6.32.3** From Douglas J. Kelly, *The Journal of Cell Biology* 28 (1966):51 by copyright permission of The Rockefeller University Press; **6.32.5** Reproduced by permission from L. Orci and A. Perrelet, Freeze-Etch Histology. (Heidelberg: Springer-Verlag). Copyright © 1975 by Springer-Verlag GmbH & Co KG; **6.32.7** From C. Peracchia and A. F. Dulhunty, *The Journal of Cell Biology* 70 (1976):419 by copyright permission of The Rockefeller University Press; **6.33.2** Lennart Nilsson/Scanpix; **p. 123, top** Courtesy E. H. Newcomb; **p. 123, bottom** From S. E. Fredrick and E. H. Newcomb, *The Journal of Cell Biology* 43 (1969): 343.

Chapter 7 **7.1** Roderick Mackinnon; **7.4.3** D. W. Fawcett/Photo Researchers; **7.4.4** D. W. Fawcett/Photo Researchers; **7.16.2.** Michael Abbey/Photo Researchers; **7.22.3** H. S. Pankratz, T.C. Beaman & P. Gerhardt/Biological Photo Service; **7.22.6** D. W. Fawcett/Photo Researchers; **7.22.9** M. M. Perry and A. B. Gilbert, *J. Cell Science* 39 (1979) 257. Copyright 1979 by The Company of Biologists Ltd. **7.22.10** M. M. Perry and A. B. Gilbert, *J. Cell Science* 39 (1979) 257. Copyright 1979 by The Company of Biologists Ltd.

Chapter 8 **8.1** Photoshot/NHPA; **8.2.2** Jupiter Images; **8.3.2** Robert N. Johnson/RnJ Photography; **8.3.3** Robert N. Johnson/RnJ Photography; **8.4.1** Brandon Blinkenberg/iStockphoto; **8.4.2** Bridget Lazenby/iStockphoto; **8.14.2** Thomas A. Steitz, Yale University; **8.14.3** Thomas A. Steitz, Yale University; **8.20.3** Scheer JM, Romanowski MJ, Wells JA. A common allosteric site and mechanism in caspases. *Proc Natl Acad Sci U S A.* 2006 May 16;103(20):7595–600; Fig. 4a; **8.22.2** Nicolae Simionescu.

Chapter 9 **9.1** Anup Shah/Nature Picture Library.

Chapter 10 **10.1** Bob Rowan, Progressive Image/Corbis; **10.2.2** Jean-Paul Nacivet/AGE Fotostock; **10.2.3** Lawrence Naylor/Photo Researchers; **10.2.4** M. I. Walker/Photo Researchers; **10.2.5** Susan M. Barns; **10.2.6** National Library of Medicine; **10.3** Robert Clark Photography, robertclark.com; **10.4.2** Image courtesy Andreas Holzenburg and Stanislav Vitha, Dept. of Biology and Microscopy & Imaging Center, Texas A&M University; **10.4.3** E.H Newcomb & WP Wergin/Biological Photo Service; **10.12.2** Christine L. Case, Skyline College; **10.13.2** From "Architecture of the photosynthetic oxygen-evolving center."; . . . K. N. Ferreira,T. M. Iverson, K. Maghlaoui, J. Barber, S. Iwata. *Science.* 2004 Mar 19;303(5665):1831–8. Epub 2004 Feb 5; **10.21.2** David Muench/Corbis; **10.21.3** Dave Bartruff/Corbis.

Chapter 11 **11.1** Winfried Wisniewski/Corbis; **11.3.2** A. Dale Kaiser, Stanford University; **11.8.2** From "High-resolution crystal structure of an engineered human beta2-adrenergic G protein-coupled receptor." V. Cherezov, D. M. Rosenbaum, M. A. Hanson, S. G. F. Rasmussen, F. S. Thian, T. S. Kobilka, H-J. Choi, P. Kuhn, W. I. Weis, B. K. Kobilka, R. C. Stevens. *Science.* 2007 Nov 23;318(5854):1258–65. Epub 2007 Oct 25; **11.17.3** Matheos D, Metodiev M, Muller E, Stone D, Rose MD. Pheromone-induced polarization is dependent on the Fus3p MAPK acting through the formin Bni1p. *J Cell Biol.* 2004 April; 165(1);99–109; Fig.9. Reproduced by copyright permission of The Rockefeller University Press; **11.20.2** Gopal Murti/Photo Researchers; **11.22.2** *Development* 127, 5245–5252 (2000). Mesenchymal cells engulf and clear apoptotic footplate cells in macrophageless PU.1 null mouse embryos. William Wood, Mark Turmaine, Roberta Weber, Victoria Camp, Richard A. Maki, Scott R. McKercher and Paul Martin; **11.22.3** *Development* 127, 5245–5252 (2000). Mesenchymal cells engulf and clear apoptotic footplate cells in macrophageless PU.1 null mouse embryos. William Wood, Mark Turmaine, Roberta Weber, Victoria Camp, Richard A. Maki, Scott R. McKercher and Paul Martin; **11.22.4** *Development* 127, 5245–5252 (2000). Mesenchymal cells engulf and clear apoptotic footplate cells in macrophageless PU.1 null mouse embryos. William Wood, Mark Turmaine, Roberta Weber, Victoria Camp, Richard A. Maki, Scott R. McKercher and Paul Martin.

Chapter 12 **12.1** Jan-Michael Peters/Silke Hauf; **12.2.2** Biophoto Associates/Photo Researchers; **12.2.3** C.R. Wyttenbach/Biological Photo Service; **12.2.4** Biophoto/Science Source/Photo Researchers; **12.3.2** John Murray; **12.4.2** Biophoto/Photo Researchers; **12.5.2** Biophoto/Photo Researchers; **12.7.2** Conly L. Rieder; **12.7.3** Conly L. Rieder; **12.7.4** Conly L. Rieder; **12.7.6** Conly L. Rieder; **12.7.7** Conly L. Rieder; **12.7.8** Conly L. Rieder; **12.8.2** J. Richard McIntosh, University of Colorado at Boulder; **12.8.3** Reproduced by permission from Matthew Schibler, from *Protoplasma* 137. Copyright © 1987: 29–44 by Springer-Verlag GmbH & Co KG; **12.10.2** Don W. Fawcett/Photo Researchers; **12.10.3** B. A. Palevitz, Courtesy of E. H. Newcomb, University of Wisconsin; **12.11.2** Elisabeth Pierson, FNWI-Radboud University Nijmegen, Pearson Science; **12.11.3** Elisabeth Pierson, FNWI-Radboud University Nijmegen, Pearson Science; **12.11.4** Elisabeth Pierson, FNWI-Radboud University Nijmegen, Pearson Science; **12.11.5** Elisabeth Pierson, FNWI-Radboud University Nijmegen, Pearson Science; **12.11.6** Elisabeth Pierson, FNWI-Radboud University Nijmegen, Pearson Science; **12.18.2** Guenter Albrecht-Buehler; **12.19.2** Lan Bo Chen; **12.19.3** Lan Bo Chen; **12.21** Anne Weston, LRI, CRUK, Wellcome Images; **p. 245** USDA/ARS/Agricultural Research Service.

Chapter 13 **13.1** Steve Granitz/WireImage/Getty Images; **13.2.2** Roland Birke/OKAPIA/Photo Researchers; **13.2.3** SuperStock; **13.3.2** Veronique Burger/Phanie Agency/Photo Researchers; **13.3.3** CNRI/Photo Researchers; **13.11.2** Mark Petronczki and Maria Siomos; **13.12.2** John Walsh, Micrographia.com.

Chapter 14 **14.1** Mendel Museum, Augustinian Abbey, Brno; **14.14.2** Altrendo nature/Getty Images; **14.14.3** PictureNet Corporation/Corbis; **14.15.2** Photodisc/Getty Images; **14.15.3** Photodisc/Getty Images; **14.15.4** Anthony Loveday; **14.15.5** Anthony Loveday; **14.16.2** Rick Guidotti and Diane McLean/Positive Exposure; **14.17.2** Michael Ciesielski Photography; **14.18** Douglas C. Pizac/AP Images; **14.19.2** CNRI/Photo Researchers; **p. 285** Norma Jubinville.

Chapter 15 **15.1** David C. Ward; **15.3.2** From: "Learning to Fly: Phenotypic Markers in *Drosophila*." A poster of common phenotypic markers used in *Drosophila* genetics. Jennifer Childress, Richard Behringer, and Georg Halder. 2005. *Genesis* 43(1). Cover illustration; **15.3.3** From: "Learning to Fly: Phenotypic Markers in *Drosophila*." A poster of common phenotypic markers used in *Drosophila* genetics. Jennifer Childress, Richard Behringer, and Georg Halder. 2005. *Genesis* 43(1). Cover illustration; **15.5** Andrew Syred/Photo Researchers; **15.8.2** Dave King/Dorling Kindersley; **15.15.2** Lauren Shear/SPL/Photo Researchers; **15.15.3** CNRI/SPL/Photo Researchers; **15.18** Geoff Kidd/Photo Researchers; **p. 304** James K. Adams, Biology, Dalton State College, Dalton, Georgia.

Chapter 16 **16.1** National Institutes of Health; **16.3.2** Oliver Meckes/Photo Researchers; **16.6.2** Courtesy of the Library of Congress; **16.6.3** From "The Double Helix" by James D. Watson, Atheneum Press, N.Y., 1968, p. 215. © 1968. Courtesy CSHL Archive; **16.12.2** Jerome Vinograd; **16.12.3** From D. J. Burks and P. J. Stambrook, *The Journal of Cell Biology 77* (1978). 762, fig. 6 by copyright permission of The Rockefeller University Press. Photo provided by P. J. Stambrook; **16.21** Peter Lansdorp; **16.22.2** S. C. Holt/Biological Photo Service; **16.22.3** Victoria E. Foe; **16.22.4** Barbara Hamkalo; **16.22.5** From J. R. Paulsen and U. K. Laemmli, *Cell* 12 (1977):817–828; **16.22.6** Biophoto/Photo Researchers; **16.23.2** From: Multicolor Spectral Karyotyping of Human Chromosomes. E. Schröck, S. du Manoir, T. Veldman, B. Schoell, J. Wienberg, M. A. Ferguson-Smith, Y. Ning, D. H. Ledbetter, I. Bar-Am, D. Soenksen, Y. Garini, T. Ried. *Science.* 1996 Jul 26;273(5274):494–7; **16.23.3** From: The new cytogenetics: blurring the boundaries with molecular biology; . . . M. R. Speicher, N. P. Carter. *Nat Rev Genet.* 2005 Oct;6(10):782–92; **p. 324** Thomas A. Steitz, Yale University.

Chapter 17 **17.1** Deutscher Fotodienst GmbH; **17.6.2** Keith V. Wood, University of California, San Diego; **17.6.3** AP Images; **17.16.2** Thomas Steitz; **17.17.2** Joachim Frank; **17.21.2** B. Hamkalo and O. Miller, Jr; **17.25.2** Reprinted with permission from O. L. Miller , Jr., B. A. Hamkalo, and C. A Thomas, Jr., *Science* 169 (1970):392. Copyright © 1970 American Association for the Advancement of Science.

Chapter 18 **18.1** Reproduced by permission from Cook O, Biehs B, Bier E. Brinker and optomotor-blind act coordinately to initiate development of the L5 wing vein primordium in *Drosophila*. *Development.* 2004 May: 131(9):2113–24; **18.12.2** From: The new cytogenetics: blurring the boundaries with molecular biology; M. R. Speicher, N. P. Carter. *Nat Rev Genet.* 2005 Oct;6(10):782–92; **18.16.2** Mike Wu; **18.16.3** Hans Pfletschinger/Peter Arnold/PhotoLibrary; **18.20.2** F. R. Turner, Indiana University; **18.20.3** F. R. Turner, Indiana University; **18.21.2** Wolfgang Driever, University of Freiburg, Freiburg, Germany; **18.21.3** Wolfgang Driever, University of Freiburg, Freiburg, Germany; **18.22.2** Ruth Lehmann, The Whitehead Institution; **18.22.3** Ruth Lehmann, The Whitehead Institution; **18.26** Roy Kaltschmidt, Lawrence Berkeley National Laboratory.

Chapter 19 **19.1** Science Photo Library/Photo Researchers; **19.2.2** Peter von Sengbusch/Botanik; **19.2.3** Peter von Sengbusch/Botanik; **19.2.4** Peter von Sengbusch/Botanik; **19.3.2** Robley C. Williams/Biological Photo Service; **19.3.3** R.C. Valentine and H.G. Pereira, "Antigens and Structure of the Adenovirus," *Journal of Molecular Biology* 13: 13–20 (1965); **19.3.4** Hazel Appleton, Health Protection Agency Centre for Infections/Photo Researchers; **19.3.5** Robley C. Williams/ Biological Photo Service; **19.8.2** C. Dauguet/Institute Pasteur/Photo Researchers; **19.8.3** C. Dauguet/Institute Pasteur/Photo Researchers; **19.8.4** C. Dauguet/Institute Pasteur/Photo Researchers; **19.8.5** C. Dauguet/Institute Pasteur/Photo Researchers; **19.8.6** C. Dauguet/Institute Pasteur/Photo Researchers; **19.9.2** NIBSC/Photo Researchers; **19.9.3** Seo Myung-gon/AP Images; **19.9.4** National Museum of Health and Medicine/Armed Forces Institute of Pathology; **19.10.2** Dennis E. Mayhew; **19.10.3** Thomas A. Zitter; **19.10.4** A. Vogler/Shutterstock.

Chapter 20 **20.1** Reproduced with permission from R.F. Service, *Science* (1998) 282:396–3999. Copyright 1998 American Association for the Advancement of Science. Incyte Pharmaceuticals, Inc., Palo Alto, CA; **20.5.2** L. Brent Selinger, Pearson Science; **20.9.2** Repligen Corporation; **20.14** Ethan Bier; **20.15.2** Reproduced with permission from R.F. Service, *Science* (1998) 282:396–3999. Copyright 1998 American Association for the Advancement of Science. Incyte Pharmaceuticals, Inc., Palo Alto, CA; **20.20** Pat Sullivan/AP Images; **20.24.2** Brad DeCecco Photography; **20.24.3** Brad DeCecco Photography; **20.25.2** Steve Helber/AP Images.

Chapter 21 **21.1** Karen Huntt/Corbis; **21.5.2** From "The genetic landscape of a cell." M. Costanzo, et al. *Science.* 2010 Jan 22;327(5964):425–31; **21.6** GeneChip Human Genome U133 Plus 2.0 Array, image courtesy of Affymetrix; **21.8.2** AP Images; **21.8.3** Courtesy of Virginia Walbot, Stanford University; **21.11.2** Courtesy of O. L. Miller Jr., Dept. of Biology, University of Virginia; **21.17.2** Shu W, Cho JY, Jiang Y, Zhang M, Weisz D, Elder GA, Schmeidler J, De Gasperi R, Sosa MA, Rabidou D, Santucci AC, Perl D, Morrisey E, Buxbaum JD. Altered ultrasonic vocalization in mice with a disruption in the *Foxp2* gene. : *Proc Natl Acad Sci U S A.* 2005 Jul 5; 102(27):9643–8; Fig. 3. Image supplied by Joseph Buxbaum; **21.17.3** Shu W, Cho JY, Jiang Y, Zhang M, Weisz D, Elder GA, Schmeidler J, De Gasperi R, Sosa MA, Rabidou D, Santucci AC, Perl D, Morrisey E, Buxbaum JD. Altered ultrasonic vocalization in mice with a disruption in the *Foxp2* gene. : *Proc Natl Acad Sci U S A.* 2005 Jul 5;102(27):9643–8; Fig. 3. Image supplied by Joseph Buxbaum; **21.17.4** Shu W, Cho JY, Jiang Y, Zhang M, Weisz D, Elder GA, Schmeidler J, De Gasperi R, Sosa MA, Rabidou D, Santucci AC, Perl D, Morrisey E, Buxbaum JD. Altered ultrasonic vocalization in mice with a disruption in the *Foxp2* gene. : *Proc Natl Acad Sci U S A.* 2005 Jul 5; 102(27):9643–8; Fig. 3. Image supplied by Joseph Buxbaum; **21.17.5** Joe McDonald/Corbis.

Chapter 22 **22.1** Olivier Grunewald; **22.2.2** PD image: Skeleton of the "Rhinoceros unicorne de Java" in the Paris Museum of Natural History. From G. Cuvier, Recherches sur les ossemens fossiles, *Atlas*, pl. 17. 1836; **22.2.3** The Natural History Museum/Alamy; **22.2.4** Wayne Lynch/AGE Fotostock; **22.2.5** Neg./Transparency no. 330300. Courtesy Dept. of Library Services, American Museum of Natural History; **22.4** Michael S. Yamashita/Corbis; **22.5.2** George Richmond/ARCHIV/Photo Researchers; **22.5.3** National Maritime Museum; **22.6.2** Michel Gunther/PhotoLibrary; **22.6.3** David Hosking/FLPA; **22.6.4** David Hosking/Alamy; **22.7** Darwin's Tree of life sketch, MS.DAR.121:p.36. Reproduced with the permission of the Cambridge University Library; **22.9.2** Gerard Schulz/Naturphoto; **22.9.3** Robert Sarno/iStockphoto; **22.9.4** Paul Rapson/Alamy; **22.9.5** Izaokas Sapiro/Shutterstock; **22.9.6** YinYang/iStockphoto; **22.9.7** floricica buzlea/iStockphoto; **22.10** Laura Jesse, Extension Entomologist, Iowa State University; **22.11.2** Richard Packwood/Oxford Scientific/Jupiter Images; **22.12.2** E. S. Ross, California Academy of Sciences; **22.12.3** Mark Taylor/Nature Picture Library; **22.13.1** Scott P. Carroll; **22.16.2** Dr. Keith Wheeler/Photo Researchers; **22.16.3** Lennart Nilsson/Scanpix; **22.18.2** Visible Earth (http://visibleearth.nasa.gov/), NASA; **22.19.2** Chris Linz, Thewissen lab, Northeastern Ohio Universities College of Medicine; **22.19.3** Chris Linz, Thewissen lab, Northeastern Ohio Universities College of Medicine; **22.19.4** Chris Linz, Thewissen lab, Northeastern Ohio Universities College of Medicine; **22.19.5** Chris Linz, Thewissen lab, Northeastern Ohio Universities College of Medicine.

Chapter 23 **23.1** Rosemary B. Grant; **23.3.2** Erick Greene, University of Montana; **23.3.3** Erick Greene, University of Montana; **23.4.2** Janice Britton-Davidian, ISEM, UMR 5554 CNRS, Universite Montpellier II. Reprinted by permission from *Nature,* Vol. 403, 13 January 2000, p. 158. © 2000 Macmillan Magazines Ltd.; **23.4.3** Janice Britton-Davidian, ISEM, UMR 5554 CNRS, Universite Montpellier II. Reprinted by permission from *Nature*, Vol. 403, 13 January 2000, p. 158. © 2000 Macmillan Magazines Ltd.; **23.4.4** Steve Gorton/Dorling Kindersley; **23.5.2** New York State Department of Environmental Conservation; **23.6.2** Gary Schultz/Photoshot; **23.6.3** James L. Davis/ProWildlife; **23.11.2** William Ervin/SPL/Photo Researchers; **23.12.2** Jan Visser; **23.14.2** John Visser/Photoshot; **23.15** Dave Blackey/PhotoLibrary; **23.16.2** Allison M. Welch; **23.19** Merlin D. Tuttle, Bat Conservation International, www.batcon.org.

Chapter 24 **24.1** Mark Jones/AGE Fotostock; **24.2.2** Malcolm Schuyl/Alamy; **24.2.3** Wave Rf/PhotoLibrary; **24.2.4** Robert Kneschke/iStockphoto; **24.2.5** Justin Horrocks/iStockphoto; **24.2.6** Photodisc/Getty Images; **24.2.7** Photodisc/Getty Images; **24.2.8** Photodisc/Getty Images; **24.2.9** Masterfile; **24.3.a2** Joe McDonald/Photoshot; **24.3.b2** Joe McDonald/Corbis; **24.3.c2** USDA/APHIS/Animal and Plant Health Inspection Service; **24.3.d2** Stephen Krasemann/Photo Researchers; **24.3.e1** Michael Dietrich/

imagebroker/Alamy; **24.3.f2** Takahito Asami; **24.3.g2** William E. Ferguson; **24.3.h2** Charles W. Brown; **24.3.i2** Photodisc/Getty Images; **24.3.j2** Corbis; **24.3.k2** DawnYL/Fotolia; **24.3.12** Kazutoshi Okuno; **24.4.2** CLFProductions/Shutterstock; **24.4.3** Boris Karpinski/Alamy; **24.4.4** Troy Maben/AP Images; **24.6.2** John Shaw/Bruce Coleman/Photoshot; **24.6.3** Michael Fogden/Bruce Coleman/Photoshot; **24.6.4** Corbis; **24.7.2** From: Morphology, performance, fitness: functional insight into a post-Pleistocene radiation of mosquitofish. R. B. Langerhans. *Biology Letters* 2009;5(4):488–491; **24.7.3** From: Morphology, performance, fitness: functional insight into a post-Pleistocene radiation of mosquitofish. R. B. Langerhans. *Biology Letters* 2009;5(4):488–491; **24.8.2** Visible Earth (http://visibleearth.nasa.gov/), NASA; **24.8.3** Arthur Anker; **24.12** Ole Seehausen; **24.13.2** Jeroen Speybroeck, Research Institute for Nature and Forest, Belgium; **24.13.3** Jeroen Speybroeck, Research Institute for Nature and Forest, Belgium; **24.16.2** Ole Seehausen; **24.16.3** Ole Seehausen; **24.16.4** Ole Seehausen; **24.18** Jason Rick; **24.20.2** Reprinted by permission from *Nature* Bradshaw HD, Schemske DW. Allele substitution at a flower colour locus produces a pollinator shift in monkeyflowers. *Nature*. 2003 November 12; 426(6963):176–8 Copyright © 2003. Macmillan Magazines Limited; **24.20.3** Reprinted by permission from *Nature* Bradshaw HD, Schemske DW. Allele substitution at a flower colour locus produces a pollinator shift in monkeyflowers. *Nature*. 2003 November 12; 426(6963):176–8 Copyright © 2003. Macmillan Magazines Limited; **24.20.4** Reprinted by permission from *Nature* Bradshaw HD, Schemske DW. Allele substitution at a flower colour locus produces a pollinator shift in monkeyflowers. *Nature*. 2003 November 12; 426(6963):176–8 Copyright © 2003. Macmillan Magazines Limited; **24.20.5** Reprinted by permission from *Nature* Bradshaw HD, Schemske DW. Allele substitution at a flower colour locus produces a pollinator shift in monkeyflowers. *Nature*. 2003 November 12; 426(6963):176–8 Copyright © 2003. Macmillan Magazines Limited.

Chapter 25 **25.1** Gerhard Boeggemann; **p. 507.UN1** Rebecca Hunt; **25.2.2** UPI Photo/Landov; **25.3.2** Courtesy of F. M. Menger and Kurt Gabrielson, Emory University; **25.3.3** M. Hanczyc; **25.4.10** Specimen No 12478, Markus Moser, Staatliches Museum fur Naturkunde Stuttgart; **25.4.2** S. M. Awramik/Biological Photo Service; **25.4.3** Sinclair Stammers/Photo Researchers; **25.4.4** Andrew H. Knoll, Harvard University; **25.4.5** Lisa-Ann Gershwin, University of California-Berkeley, Museum of Paleontology; **25.4.6** Chip Clark; **25.4.7** Ted Daeschler/Academy of Natural Sciences/VIREO; **25.4.8** Roger Jones; **25.4.9** Seelevel.com; **25.11.2** Shuhai Xiao, Tulane University; **25.11.3** Shuhai Xiao, Tulane University; **25.20.2** Gerald D. Carr; **25.20.3** Gerald D. Carr; **25.20.4** Gerald D. Carr; **25.20.5** Gerald D. Carr; **25.20.6** Gerald D. Carr; **25.20.7** Bruce G. Baldwin; **25.21.2** Jean Kern; **25.21.3** Jean Kern; **25.22** Juniors Bildarchiv/Alamy; **25.23** Reprinted from "The Origin of Form" by Sean B. Carroll, *Natural History*, November 2005. Burke, A.C. 2000, "*Hox*" Genes and the Global Patterning of the Somitic Mesoderm. In Somitogenesis. C. Ordahl (ed.) "Current Topics in Developmental Biology", Vo. 47. Academic Press; **25.25.1** Oxford Scientific/PhotoLibrary; **25.25.3** Shapiro MD, Marks ME, Peichel CL, Blackman BK, Nereng KS, Jonsson B, Schluter D, Kingsley DM. Genetic and developmental basis of evolutionary pelvic reduction in threespine sticklebacks. *Nature*. Erratum. 2006 Feb 23;439(7079):1014; Fig. 1; **25.25.4** Shapiro MD, Marks ME, Peichel CL, Blackman BK, Nereng KS, Jonsson B, Schluter D, Kingsley DM. Genetic and developmental basis of evolutionary pelvic reduction in threespine sticklebacks. *Nature*. Erratum. 2006 Feb 23;439(7079):1014; Fig. 1.

Chapter 26 **26.1** Ken Griffiths/NHPA/Photoshot; **26.2.2** Ryan McVay/Photodisc/Getty Images; **26.2.3** Neil Fletcher/Dorling Kindersley; **26.2.4** Dorling Kindersley; **26.17.2** Ed Heck; **26.17.3** Courtesy Dept. of Library Services, American Museum of Natural History.

Chapter 27 **27.1** Bonnie K. Baxter, Great Salt Lake Institute, Westminster College, Utah; **27.2.2** CDC; **27.2.3** Dr. Kari Lounatmaa/Photo Researchers; **27.2.4** Stem Jems/Photo Researchers; **27.3.2** L. Brent Selinger, Pearson Science; **27.4.2** Dr. Immo Rantala/Photo Researchers; **27.5.2** Kwangshin Kim/Photo Researchers; **27.6.2** Julius Adler; **27.7.2** S. W. Watson; **27.7.3** Norma J. Lang/Biological Photo Service; **27.8.2** Huntington Potter, Byrd Alzheimer's Institute and University of South Florida, David Dressler, Oxford University and Balliol College; **27.9.2** H.S. Pankratz, T.C. Beaman/Biological Photo Service; **27.12.2** Charles C. Brinton, Jr., University of Pittsburgh; **27.14.2** Susan M. Barns; **27.16** Jack Dykinga/Stone/Getty Images; **27.17.3** L. Evans Roth/Biological Photo Service; **27.17.5** Yuichi Suwa; **27.17.7** National Library of Medicine; **27.17.9** From "Scanning electron microscopy of fruiting body formation by myxobacteria." P. L. Grilione and J. Pangborn. *J. Bacteriol.* 1975 December; 124(3): 1558–1565; **27.17.13** Photo Researchers; **27.17.15** Moredon Animal Health/SPL/Photo Researchers; **27.17.17** CNRI/SPL/Photo Researchers; **27.17.19** Culture Collection CCALA, Institute of Botany, Academy of Sciences Dukelska, Czech Republic; **27.17.21** Paul Hoskisson, Strathclyde Institute of Pharmacy and Biomedical Sciences, Glasgow, Scotland; **27.17.22** David M. Phillips/Photo Researchers; **27.18.2** Pascale Frey-Klett; **27.19** Ken Lucas/Biological Photo Service; **27.20.2** Scott Camazine/Photo Researchers; **27.20.3** David M. Phillips/Photo Researchers; **27.20.4** James Marshall/The Image Works; **27.21** Metabolix; **27.21.3** Courtesy of Exxon Mobil Corporation; **27.21.4** Seelevel.com.

Chapter 28 **28.1** Brian S. Leander; **28.3.3** Joel Mancuso, University of California, Berkeley; **28.3.5** M. I. Walker/NHPA/Photoshot; **28.3.7** Howard Spero, University of California-Davis; **28.3.8** NOAA; **28.3.10** Kim Taylor/Nature Picture Library; **28.3.11** David J. Patterson/micro*scope; **28.3.13** Tom Stack/PhotoLibrary; **28.4.2** David M. Phillips/The Population Council/Photo Researchers; **28.5.2** David J. Patterson; **28.6.2** Meckes/Ottawa/Photo Researchers; **28.7.2** D. J. Patterson, L. Amaral-Zettler, M. Peglar and T. Nerad, http://micro*scope.mbl.edu; **28.8.2** Guy Brugerolle, Universitad Clearmont, Ferrand; **28.9.2** Virginia Institute of Marine Science; **28.10.2** Masamichi Aikawa, Tokai University School of Medicine, Japan; **28.11.2** M. I. Walker/Photo Researchers; **28.12.2** Centers for Disease Control & Prevention; **28.13.2** Steve Gschmeissner/Photo Researchers; **28.14.2** Stephen Durr; **28.15.2** Colin Bates; **28.16.2** J. Robert Waaland/Biological Photo Service; **28.17.2** Fred Rhoades; **28.18.2** Robert Brons/Biological Photo Service; **28.19.2** Eva Nowack; **28.20.2** D. P. Wilson, Eric & David Hosking/Photo Researchers; **28.20.3** Michael D. Guiry; **28.20.4** Biophoto Associates/Photo Researchers; **28.20.6** Michael Yamashita/IPN/Aurora Photos; **28.20.7** David Murray/Dorling Kindersley; **28.21.2** Laurie Campbell/NHPA; **28.21.3** Marine Sciences, University of Puerto Rico; **28.22.2** William L. Dentler; **28.24.2** George Barron; **28.24.3** Mikel Tapia (www.argazkik.com); **28.25.2** Robert Kay; **28.25.3** Robert Kay; **28.26.2** Kevin Carpenter and Patrick Keeling.

Chapter 29 **29.1** Martin Rugner/AGE Fotostock; **29.2.2** S. C. Mueller and R. M. Brown, Jr; **29.3.2** Natural Visions; **29.3.3** Linda Graham, University of Wisconsin-Madison; **29.5.11** Ed Reschke; **29.5.12** CDC; **29.5.3** Linda Graham, University of Wisconsin-Madison; **29.5.4** Karen S. Renzaglia; **29.5.6** Alan S. Heilman; **29.5.7** Michael Clayton, University of Wisconsin-Madison; **29.5.9** David John Jones (mybitoftheplanet.com); **29.6.2** Charles H. Wellman; **29.6.3** Charles H. Wellman; **29.8.2** Laurie Knight (www.laurieknight.net); **29.9.2** Linda Graham, University of Wisconsin-Madison; **29.9.3** Alvin E. Staffan/National Audubon Society/Photo Researchers; **29.9.4** Hidden Forest; **29.9.6** Hidden Forest; **29.9.8** Tony Wharton, Frank Lane Picture Agency/Corbis; **p. 609** From "Mosses and Other Bryophytes, an Illustrated Glossary" (2006) by Bill and Nancy Malcolm; **29.11.1** Brian Lightfoot/AGE Fotostock; **29.11.2** Chris Lisle/Corbis; **29.15.2** Jody Banks, Purdue University; **29.15.3** Murray Fagg, Australian National Botanic Gardens; **29.15.4** Helga & Kurt Rasbach; **29.15.6** Jon Meier/iStockphoto; **29.15.8** Milton Rand/Tom Stack & Associates; **29.15.9** Francisco Javier Yeste Garcia (www.flickr.com/photos/fryega/); **29.15.10** Francisco Javier Yeste Garcia (www.flickr.com/photos/fryega/); **29.16.2** The Open University; **p. 616, top** Ed Reschke; **p. 616, bottom** Michael Clayton.

Chapter 30 **30.1** National Museum of Natural History, Smithsonian Institution; **30.5.1** Johannes Greyling/iStockphoto; **30.5.3** CC-BY-SA photo: Kurt Stueber (www.biolib.de); **30.5.4** Jeroen Peys/iStockphoto; **30.5.5** Michael Clayton; **30.5.6** Thomas Schoepke; **30.5.7** Bob Gibbons/FLPA; **30.5.8** Raymond Gehman/Corbis; **30.5.9** Adam Jones/Getty Images; **30.5.10** David Muench/Corbis; **30.5.11** Gunter Marx Photography/Corbis; **30.5.12** Royal Botanic Gardens Sydney; **30.5.13** Jaime Plaza/ AP Images; **30.5.14** Mario Verin/Photolibrary; **30.8.2** Dave King/Dorling Kindersley; **30.8.3** Andy Crawford/Dorling Kindersley; **30.8.4** Dave King/Dorling Kindersley; **30.8.5** Maria Dryfhout/iStockphoto; **30.8.6** Peter Rees/Getty Images; **30.9.2** PIXTAL/AGE Fotostock; **30.9.3** Hans Dieter Brandl, Frank Lane Picture Agency/Corbis; **30.9.4** Scott Camazine/Photo Researchers; **30.9.5** Derek Hall/Dorling Kindersley; **30.11.2** David L. Dilcher; **30.13.1** Howard Rice/Dorling Kindersley; **30.13.2** Jack Scheper, Floridata.com; **30.13.3** Stephen McCabe; **30.13.4** Andrew Butler/Dorling Kindersley; **30.13.6** Eric Crichton/Dorling Kindersley; **30.13.7** John Dransfield; **30.13.8** Dorling Kindersley; **30.13.10** Terry W. Eggers/Corbis; **30.13.11** CC-BY-SA photo: Artslave; **30.13.12** Matthew Ward/Dorling Kindersley; **30.13.13** Tony Wharton, Frank Lane Picture Agency/Corbis; **30.13.14** Howard Rice/Dorling Kindersley; **30.13.15** Gerald D. Carr; **30.14** kkaplin/Shutterstock; **30.15** D. Wilder; **30.16.2** NASA's Earth Observatory; **30.16.3** NASA's Earth Observatory.

Chapter 31 **31.1** Georg Müller; **31.2.2** Fred Rhoades/Mycena Consulting; **31.2.3** Hans Reinhard/Taxi/Getty Images; **31.2.4** George Barron, University of Guelph, Canada; **31.4.2** © N. Allin & G.L. Barron, University of Guelph/Biological Photo Service; **31.6.2** Biophoto Associates/Photo Researchers; **31.6.3** Popovaphoto/dreamstime; **31.7.2** Stephen J. Kron; **31.9.2** Dirk Redecker, Robin Kodner, and Linda E. Graham. Glomalean Fungi from the Ordovician. *Science* 15 September 2000; 289: 1920–1921; **31.10.2** CDC; **31.11.2** John Taylor; **31.11.3** Ray Watson; **31.11.5** Reproduced by permission from Kiers ET, van der Heijden MG. Mutualistic stability in the arbuscular mycorrhizal symbiosis: exploring hypotheses of evolutionary cooperation. *Ecology*. 2006 July: 87(7):1627–36: Fig. 1a. Image by Marcel van der Heijden, Swiss Federal Research Station for Agroecology and Agriculture. Copyright © 2006, Ecological Society of America; **31.11.6** Frank Young/Papilio/Corbis; **31.11.7** Phil Dotson/Photo Researchers; **31.12.2** William E. Barstow; **31.13.2** Antonio D'Albore/iStockphoto; **31.13.3** Alena Kubátová (http://botany.natur.cuni.cz/cs/sbirka-kultur-hub-ccf); **31.13.4** Ed Reschke/Peter Arnold/PhotoLibrary; **31.13.5** George Barron; **31.14.2** G.L. Barron/Biological Photo Service; **31.15.2** M. F. Brown/Biological Photo Service; **31.16.2** Viard/Jacana/Photo Researchers; **31.16.3** Douglas Adams/iStockphoto; **31.17.2** Fred Spiegel; **31.18.2** Fletcher and Baylis/Photo Researchers; **31.18.3** Michael Fogden/Photolibrary; **31.18.4** Frank Paul/Alamy; **31.19.2** Biophoto Associates/Photo Researchers; **31.20** University of Tennessee Entomology and Plant Pathology; **31.22** Mark Bowler/Photo Researchers; **31.23.2** Ralph Lee Hopkins/Getty Images; **31.23.3** Geoff Simpson/naturepl.com; **31.23.4** Wild-Worlders of Europe/Benvie/naturepl.com; **31.24.2** Eye of Science/Photo Researchers; **31.25.2** Alamy; **31.25.3** Peter Chadwick/Dorling Kindersley; **31.25.4** Hecker-Sauer/AGE Fotostock; **31.26.2** Vance T. Vredenburg, San Francisco State University; **31.27.2** Christine Case.

Chapter 32 **32.1** Jeff Hunter/Image Bank/Getty Images; **p. 656** Biological Photo Service; **32.4.2** © The Museum Board of South Australia 2004 Photographer: Dr. J. Gehling; **32.4.3** © The Museum Board of South Australia 2004 Photographer: Dr. J. Gehling; **32.5** J. Sibbick/The Natural History Museum, London; **32.6.2** Wikramanayake AH, Hong M, Lee PN, Pang K, Byrum CA, Bince JM, Xu R, Martindale MQ. An ancient role for nuclear beta-catenin in the evolution of axial polarity and germ layer segregation. *Nature*. 2003 Nov 27;426(6965):446–50; Fig. 2, 3 and 4; **32.6.3** Wikramanayake AH, Hong M, Lee PN, Pang K, Byrum CA, Bince JM, Xu R, Martindale MQ. An ancient role for nuclear beta-catenin in the evolution of axial polarity and germ layer segregation. *Nature*. 2003 Nov 27;426(6965):446–50; Fig. 2, 3 and 4; **32.6.4** Wikramanayake AH, Hong M, Lee PN, Pang K, Byrum CA, Bince JM, Xu R, Martindale MQ. An ancient role for nuclear beta-catenin in the evolution of axial polarity and germ layer segregation. *Nature*. 2003 Nov 27;426(6965):446–50; Fig. 2, 3 and 4; **32.6.5** Wikramanayake AH, Hong M, Lee PN, Pang K, Byrum CA, Bince JM, Xu R, Martindale MQ. An ancient role for nuclear beta-catenin in the evolution of axial polarity and germ layer segregation. *Nature*. 2003 Nov 27; 426(6965):446–50; Fig. 2, 3 and 4; **32.12** Kent Wood/Photo Researchers; **32.13.2** Hecker/Sauer/AGE Fotostock.

Chapter 33 **33.1** C. Wolcott Henry III/National Geographic/Getty Images; **33.3.1** Andrew J. Martinez/Photo Researchers; **33.3.2** Robert Brons/Biological Photo Service; **33.3.3** Teresa (Zubi) Zuberbühler; **33.3.4** Stephen Dellaporta; **33.3.5** Gregory G. Dimijian/Photo Researchers; **33.3.6** Ed Robinson/Pacific Stock/Photolibrary; **33.3.7** Hecker/Sauer/AGE Fotostock; **33.3.8** W. I. Walker/Photo Researchers; **33.3.9** Kåre Telnes/Image Quest Marine; **33.3.10** PD image: Anilocra/Taken by Neil Campbell, University of Aberdeen, Scotland, UK; **33.3.11** Erling Svensen/UWPhoto ANS; **33.3.12** Peter Funch; **33.3.13** photonimo/iStockphoto; **33.3.14** Peter Batson/Image Quest Marine; **33.3.15** Reinhart Mobjerg Kristensen; **33.3.16** Erling Svensen/

UWPhoto ANS; **33.3.17** Andrew Syred/Photo Researchers; **33.3.18** Reproduced with permission from A. Eizinger and R. Sommer, Max Planck Institut fur entwicklungsbiologie, Tubingen. Copyright 2000 American Association for the Advancement of Science. Cover 278(5337) 17 Oct 97; **33.3.19** Thomas Stromberg; **33.3.20** Tim Flach/Stone/Getty Images; **33.3.21** Heather Angel/Natural Visions; **33.3.22** Robert Harding World Imagery/Alamy; **33.3.23** Robert Brons/Biological Photo Service; **33.4.2** Andrew J. Martinez/Photo Researchers; **33.7.2** Andrew J. Martinez/Photo Researchers; **33.7.3** Robert Brons/Biological Photo Service; **33.7.4** Commonwealth of Australia (GBRMPA); **33.7.5** Neil G. McDaniel/Photo Researchers; **33.8.2** Robert Brons/Biological Photo Services; **33.9.2** Ed Robinson/Pacific Stock/PhotoLibrary; **33.11.2** CDC; **33.12.2** Eye of Science/Photo Researchers; **33.13.2** W. I. Walker/Photo Researchers; **33.14.2** Hecker/Sauer/AGE Fotostock; **33.14.3** Kåre Telnes/Image Quest Marine; **33.16** Jeff Foott/Tom Stack and Associates; **33.17.2** Amruta Bhelke/Dreamstime.com; **33.17.3** Corbis; **33.19** H. W. Pratt/Biological Photo Service; **33.21.2** photonimo/iStockphoto; **33.21.3** Mark Conlin/Image Quest Marine; **33.21.4** Jonathan Blair/Corbis; **33.22.2** Photograph courtesy of the U.S. Bureau of Fisheries (1919), and Illinois State Museum; **33.22.3** © Zoological Society of London (ZSL); **33.23.2** Peter Batson/Image Quest Marine; **33.24.2** A.N.T./NHPA/Photoshot; **33.25** Astrid & Hanns-Frieder Michler/Photo Researchers; **33.26.2** Reproduced with permission from A. Eizinger and R. Sommer, Max Planck Institut fur entwicklungsbiologie, Tubingen. Copyright 2000 American Association for the Advancement of Science. Cover 278(5337) 17 Oct 97; **33.27.2** SPL/Photo Researchers; **33.28** Collection of Dan Cooper; **33.29.2** Grenier JK, Garber TL, Warren R, Whitington PM, Carroll S. Evolution of the entire arthropod *Hox* gene set predated the origin and radiation of the onychophoran/arthropod clade. *Curr Biol.* 1997 Aug 1;7(8):547–53; Fig. 3c; **33.31** Mark Newman/FLPA; **33.32.2** Tim Flach/Stone/Getty Images; **33.32.3** Andrew Syred/Photo Researchers; **33.32.4** Eric Lawton/iStockphoto; **33.34.2** Premaphotos/Nature Picture Library; **33.34.3** Tom McHugh/Photo Researchers; **33.36** Meul/ARCO/Nature Picture Library; **33.37.2** John Shaw/Tom Stack and Associates; **33.37.3** John Shaw/Tom Stack and Associates; **33.37.4** John Shaw/Tom Stack and Associates; **33.37.5** John Shaw/Tom Stack and Associates; **33.37.6** John Shaw/Tom Stack and Associates; **33.38.2** Dr. John Brackenbury/Photo Researchers; **33.38.3** Perry Babin; **33.38.4** PREMAPHOTOS/Nature Picture Library; **33.38.5** Dante Fenolio/Photo Researchers; **33.38.6** John Cancalosi/Nature Picture Library; **33.38.7** Hans Christoph Kappel/Nature Picture Library; **33.38.8** Michael & Patricia Fogden/CORBIS; **33.38.9** CC-BY-SA photo: Bruce Marlin (www.cirrusimage.com/fly_whale-tail.htm); **33.39.2** Maximilian Weinzierl/Alamy; **33.39.3** Peter Herring/Image Quest Marine; **33.39.4** Peter Parks/Image Quest Marine; **33.40.2** Andrey Nekrasov/Image Quest Marine; **33.41** Daniel Janies; **33.42** Jeff Rotman/Photo Researchers; **33.43** Robert Harding World Imagery/Alamy; **33.44** Jurgen Freund/Nature Picture Library; **33.45** Hal Beral/Corbis.

Chapter 34 **34.1** Xian-guang H, Aldridge RJ, Siveter DJ, Siveter DJ, Xiang-hong F. New evidence on the anatomy and phylogeny of the earliest vertebrates. *Proceedings of the Royal Society of London-B-Biological Sciences,* Sept. 22, 2002; 269 (1503) 1865–1869; Fig. 1c; **34.4.2** Oxford Scientific/PhotoLibrary; **34.5.2** Robert Brons/Biological Photo Service; **34.8.2** Nanjing Institute of Geology and Palaeontology; **34.9.2** Tom McHugh/Photo Researchers; **34.10.2** A Hartl/AGE Fotostock; **34.10.3** Marevision/AGE Fotostock; **34.14.2** The Field Museum, #GEO82014; **34.15.2** Carlos Villoch/Image Quest Marine; **34.15.3** Masa Ushioda/Image Quest Marine; **34.15.4** Andy Murch/V&W/Image Quest Marine; **34.17.2** James D. Watt/Image Quest Marine; **34.17.3** Jez Tryner/Image Quest Marine; **34.17.4** George Grall/Getty Images; **34.17.5** Fred McConnaughey/Photo Researchers; **34.18.2** From "The oldest - articulated osteichthyan reveals mosaic gnathostome characters." M. Zhu et al. *Nature.* 2009 Mar 26;458(7237):469–74; **34.19** Arnaz Mehta; **34.20.2** © Ted Daeschler/ Academy of Natural Sciences/VIREO; **34.20.3** © Ted Daeschler/Academy of Natural Sciences/VIREO; **34.20.4** © Kalliopi Monoyios; **34.22.2** Alberto Fernandez/AGE Fotostock America, Inc. **34.22.3** Michael Fogden/Bruce Coleman/Photoshot; **34.22.4** Michael Fogden/Bruce Coleman/Photoshot; **34.23.2** John Cancalosi/Peter Arnold/PhotoLibrary; **34.23.3** Stephen Dalton/Photo Researchers; **34.23.4** Hans Pfletschinger/Peter Arnold/PhotoLibrary; **34.24** Michael Fogden/OSF/PhotoLibrary; **34.27** Nobumichi Tamura; **34.28** Michael Fogden/OSF/PhotoLibrary; **34.29.2** Natural Visions/Alamy; **34.29.3** Matt T. Lee; **34.29.5** Medford Taylor/National Geographic Image Collection; **34.29.6** Carl & Ann Purcell/Corbis; **34.30.2** Visceralimage/dreamstime; **34.30.3** Janice Sheldon; **34.32** Russell Mountford/Alamy; **34.33** DLILLC/Corbis; **34.34** Yufeng Zhou/iStockphoto; **34.35.1** McPHOTO/AGE Fotostock; **34.35.2** paolo barbanera/AGE Fotostock; **34.36** Gianpiero Ferrari/FLPA; **34.36** Gianpiero Ferrari/FLPA; **34.38.2** D. Parer and E. Parer Cook/Auscape International Proprietary Ltd. **34.38.3** Mervyn Griffiths/Commonwealth Scientific and Industrial Research Organization; **34.39.2** John Cancalosi/Alamy; **34.39.3** Wells Bert & Babs/OSF/PhotoLibrary; **34.42** Frans Lanting/Corbis; **34.44.2** Kevin Schafer/AGE Fotostock America, Inc. **34.44.3** J & C Sohns/Photolibrary; **34.45.2** Morales/AGE Fotostock America, Inc. **34.45.3** Anup Shah/ImageState/Alamy Images; **34.45.4** T. J. Rick/Nature Picture Library; **34.45.5** E. A. Janes/AGE Fotostock America, Inc. **34.45.6** Frans Lanting/Corbis; **34.47** Fossilized bone, partial skeleton, of *Ardipithecus ramidus,* articulated, with bones laid in their approximate positions. Housed in National Museum of Ethiopia, Addis Ababa. Photo © T. White 2009, From "*Ardipithecus ramidus* and the Paleobiology of Early Hominids." T. White et al. *Science.* 2009 Oct 2;326(5949):75–86; **34.48.2** John Reader/SPL/Photo Researchers; **34.48.3** John Gurche Studios; **34.49.2** Alan Walker @ National Museums of Kenya. Printed with permission; **34.51** David L Brill/Brill Atlanta; **34.52** C. Henshilwood & F. d'Errico/Professor Chris Henshilwood.

Chapter 35 **35.1** PD photo: John Walker (www.fourmilab.ch); **35.3** Robert & Linda Mitchell; **35.4.2** CC-BY-SA photo: Forest & Kim Starr; **35.4.3** Rob Walls/ Alamy; **35.4.4** YinYang/iStockphoto; **35.4.5** Robert Holmes/Corbis; **35.4.6** Geoff Tompkinson/Science Photo Library/Photo Researchers; **35.5.2** Donald Gregory Clever; **35.5.3** Gusto Productions/SPL/Photo Researchers; **35.5.4** Dorling Kindersley; **35.5.5** Aflo Foto Agency/Alamy; **35.7.2** Neil Cooper/Alamy; **35.7.3** Martin Ruegner/ jupiterimages; **35.7.4** Mike Zens/Corbis; **35.7.5** Jerome Wexler/Photo Researchers; **35.7.6** Kathy Piper/iStockphoto; **35.9.2** Purdue Extension Entomology; **35.10.2** Brian Capon; **35.10.4** © Clouds Hill Imaging/www.lastrefuge.co.uk; **35.10.6** Graham Kent, Pearson Science; **35.10.7.** Graham Kent, Pearson Science; **35.10.9** N.C. Brown Center for Ultrastructure Studies, SUNY-Environmental Science & Forestry, Syracuse, NY; **35.10.11** Graham Kent; **35.10.12** Reproduced with permission from "Plant Cell Biology on CD," by B E S Gunning, www.plantcellbiologyonCD.com; **35.10.13** Professor Ray F. Evert; **35.13.2** From "*Arabidopsis* TCP20 links regulation of growth and cell division control pathways." C. Li et al. *Proc Natl Acad Sci U S A.* 2005 Sep 6;102(36):12978–83. Epub 2005 Aug 25. Photo: Peter Doerner; **35.14.2** Natalie B. Bronstein; **35.14.3** Ed Reschke; **35.14.4** Ed Reschke; **35.15.2** Michael Clayton; **35.16.2** Michael Clayton; **35.17.2** Ed Reschke; **35.17.3** Ed Reschke; **35.18.2** Ed Reschke; **35.18.3** Ed Reschke; **35.19.2** Michael Clayton; **35.19.3** Alison W. Roberts; **35.21.2** Dr. Edward R. Cook; **35.23** California Historical Society Collection (CHS-1177), University of Southern California on behalf of the USC Specialized Libraries and Archival Collections; **p. 756** Reproduced by permission from Janet Braam, *Cell* 60 (9 February 1990): Cover. Copyright ©1990 Cell Press. Image courtesy of Elsevier Sciences Ltd. **35.25.2** From "Microtubule plus-ends reveal essential links between intracellular polarization and localized modulation of endocytosis during division-plane establishment in plant cells." P. Dhonukshe et al. *BMC Biol.* 2005 Apr 14;3:11; **35.26.2** From "The tangled-1 mutation alters cell division orientations throughout maize leaf development without altering leaf shape." L. G. Smith et al. *Development.* 1996 Feb;122(2):481–9; **35.26.3** From "The tangled-1 mutation alters cell division orientations throughout maize leaf development without altering leaf shape." L. G. Smith et al. *Development.* 1996 Feb;122(2):481–9; **35.28** From figure 1a in U. Mayer et al., *Development* 117 (1): 149–162. © 1993 The Company of Biologists Ltd. **35.29.2** From "Microtubule plus-ends reveal essential links between intracellular polarization and localized modulation of endocytosis during division-plane establishment in plant cells." P. Dhonukshe et al. *BMC Biol.* 2005 Apr 14;3:11; **35.29.3** B. Wells and K. Roberts; **35.30.2** Reproduced by permission from figure 1 in D. Hareven et al, *Cell* 84 (5): 735–744. Copyright © 1996. by Elsevier Science Ltd. **35.30.3** Reproduced by permission from figure 1 in D. Hareven et al, *Cell* 84 (5): 735–744. Copyright © 1996. by Elsevier Science Ltd. **35.31.2** Reproduced by permission from Figure 2g in Hung et al, *Plant Physiology* 117:73–84. Copyright © 1998 by the American Society of Plant Biologists. Image courtesy of John Schiefelbein/Univesity of Michigan; **35.32.2** Dr. Gerald D. Carr, PhD; **35.33.2** Dr. E. M. Meyerowitz and John Bowman, *Development* 112 1991:1–231.2. Division of Biology, California Institute of Technolgy; **35.33.3** Dr. E. M. Meyerowitz and John Bowman, *Development* 112 1991:1–231.2. Division of Biology, California Institute of Technolgy.

Chapter 36 **36.1** Peggy Heard/FLPA/Alamy; **36.3.2** Rolf Rutishauser; **36.3.3** Rolf Rutishauser; **36.5.2** Dr. Jeremy Burgess/SPL/Photo Researchers; **p. 770, top** Nigel Cattlin/Holt Studios International/Photo Researchers; **p. 770, bottom** Nigel Cattlin/Holt Studios International/Photo Researchers; **36.11** Scott Camazine/Photo Researchers; **36.14.2** Graham Kent; **36.14.3** Graham Kent; **36.16.2** Mlane/Dreamstime.com; **36.16.3** Kate Shane, Southwest School of Botanical Medicine; **36.16.4** Frans Lanting/Corbis; **36.16.5** Natalie Bronstein; **36.16.6** Andrew de Lory/Dorling Kindersley; **36.16.7** Danita Delimont/Alamy; **36.19.2** M. H. Zimmermann, courtesy of Professor P. B. Tomlinson, Harvard University; **36.19.3** M. H. Zimmermann, courtesy of Professor P. B. Tomlinson, Harvard University; **36.19.4** M. H. Zimmermann, courtesy of Professor P. B. Tomlinson, Harvard University; **36.20.2** From "A coiled-coil interaction mediates cauliflower mosaic virus cell-to-cell movement." L. Stavolone et al. *Proc Natl Acad Sci U S A.* 2005 Apr 26;102(17):6219–24. Epub 2005 Apr 18.

Chapter 37 **37.1** Chris Mattison/Alamy; **37.2.2** USDA/ARS/Agricultural Research Service; **37.4** National Oceanic and Atmospheric Administration NOAA; **37.5** U.S. Geological Survey, Denver; **37.6** Kevin Horan/Stone/Getty Images; **37.9.2** White et al., *Plant Physiology,* June 2003; **37.11.2** Scimat/Photo Researchers; **37.11.3** E. H. Newcomb and S. R. Tandon/Biological Photo Service; **37.13.2** Hugues B. Massicotte, University of Northern British Columbia Ecosystem Science and Management Program, Prince George, B.C., Canada; **37.13.3** Mark Brundrett (http://mycorrhizas.info); **37.13.4** Mark Brundrett (http://mycorrhizas.info); **37.14.2** Elizabeth J. Czarapata/The Park People; **37.15.1** Wolfgang Kaehler/Corbis; **37.15.2** Ruud de Man/iStockphoto; **37.15.3** Kevin Schafer/Corbis; **37.15.4** Gary W. Carter/Corbis; **37.15.5** Kim Taylor and Jane Burton/Dorling Kindersley; **37.15.6** Biophoto Associates/Photo Researchers; **37.15.7** Philip Blenkinsop/Dorling Kindersley; **37.15.8** Paul A. Zahl/Photo Researchers; **37.15.9** Fritz Polking, Frank Lane Picture Agency/Corbis.

Chapter 38 **38.1** Pierre-Michel Blais; **38.3.2** Ed Reschke/Peter Arnold/PhotoLibrary; **38.3.3** Ed Reschke/Peter Arnold/PhotoLibrary; **38.3.4** David Scharf/Peter Arnold/PhotoLibrary; **38.4.2** Marianne Wiora; **38.4.3** Stephen Dalton/NHPA Limited/Photoshot; **38.4.4** Bjorn Rorslett Photographe; **38.4.5** Bjorn Rorslett Photographe; **38.4.7** Doug Backlund; **38.4.8** Martin Pieter Heigan; **38.4.9** Rolf Nussbaumer/Nature Picture Library; **38.4.10** Merlin D. Tuttle, Bat Conservation International, www.batcon.org; **38.5** Photo: © W. Barthlott/W. Rauh; **38.11.2** Kevin Schafer/Alamy; **38.11.3** Nature Production; **38.11.4** Brian Gordon Green/National Geographic Image Collection; **38.11.5** Steve Bloom Images/Alamy; **38.11.6** Aaron McCoy/Botanica/Photolibrary; **38.11.7** California Department of Food and Agriculture's Plant Health and Pest Prevention Services; **38.11.8** Kim A. Cabrera Photographer; **38.11.9** Steve Shattuck/CSIRO Entomology; **38.11.10** Alan Williams/ Alamy; **38.12.11** Dennis Frates/Alamy; **38.13.2** Marcel E. Dorken; **38.13.3** Marcel E. Dorken; **38.13.4** Nobumitsu Kawakubo, Gifu University, Japan; **38.14.2** Bruce Iverson, Photomicrography; **38.14.3** Bruce Iverson, Photomicrography; **38.14.4** Meriel G. Jones, University of Liverpool, UK; **38.15.2** Sinclair Stammers/Photo Researchers; **38.16.2** Andrew McRobb/Dorling Kindersley; **38.16.3** Andrew McRobb/Dorling Kindersley; **38.17** John Van Hasselt/Corbis.

Chapter 39 **39.1** *Plant Physiol.* 1999 Jul;120(3): Cover. By permission of the American Society of Plant Physiologists. Illustration by Niemeyer MI and Fernandez MC; **39.2** Natalie B. Bronstein; **39.7.2** "Regulation of Polar Auxin transport ATPIN1 in - *Arabidopsis* Vascular Tissue," by Leo Galweiler, et al. *Science* 18 DEC 1998, vol. 282; **39.7.3** "Regulation of Polar Auxin transport ATPIN1 in *Arabidopsis* Vascular Tissue," by Leo Galweiler, et al. *Science* 18 DEC 1998, vol. 282; **39.9.2** Malcolm B. Wilkins, University of Glasgow, Glasgow, Scotland, U.K. **39.9.3** Malcolm B. Wilkins, University of Glasgow, Glasgow, Scotland, U.K. **39.9.4** Malcolm B. Wilkins, University of Glasgow, Glasgow, Scotland, U.K. **39.10.2** Dr. Richard Amasino; **39.10.3** Fred Jensen; **39.12.2** Mia Molvray; **39.12.3** Karen E. Koch; **39.14.2** Kurt Stepnitz, DOE

Plant Research Laboratory, Michigan State University; **39.14.3** Joe Kieber, University of North Carolina; **39.15.2** Ed Reschke; **39.16.3** Malcolm B. Wilkins, University of Glasgow, Glasgow, Scotland, U.K. **39.16.4** Malcolm B. Wilkins, University of Glasgow, Glasgow, Scotland, U.K. **39.17.2** Malcolm B. Wilkins, University of Glasgow, Glasgow, Scotland, U.K. **39.17.3** Malcolm B. Wilkins, University of Glasgow, Glasgow, Scotland, U.K. **39.17.4** Malcolm B. Wilkins, University of Glasgow, Glasgow, Scotland, U.K. **39.17.5** Malcolm B. Wilkins, University of Glasgow, Glasgow, Scotland, U.K. **39.17.6** Malcolm B. Wilkins, University of Glasgow, Glasgow, Scotland, U.K. **39.20.2** Malcolm B. Wilkins, University of Glasgow, Glasgow, Scotland, U.K. **39.20.3** Malcolm B. Wilkins, University of Glasgow, Glasgow, Scotland, U.K. **39.24.2** Michael Evans, Ohio State University; **39.24.3** Michael Evans, Ohio State University; **39.24.4** Michael Evans, Ohio State University; **39.24.5** Michael Evans, Ohio State University; **39.25** Reproduced by permission from Janet Braam, *Cell* 60 (9 February 1990): Cover. Copyright © 1990 Cell Press. Image courtesy of Elsevier Sciences Ltd. **39.26.2** Martin Shields/Photo Researchers; **39.26.3** Martin Shields/Photo Researchers; **39.26.4** From K. Esau. "Anatomy of Seed Plants," 2nd ed. (New York: John Wiley and Sons, 1977), fig. 19.4, p.358; **39.26.5** From K. Esau. "Anatomy of Seed Plants," 2nd ed. (New York: John Wiley and sons, 1977), fig. 19.4, p.358; **39.27.2** J. L. Basq and M. C. Drew; **39.27.3** J. L. Basq and M. C. Drew; **39.29.2** New York State Agricultural Experiment Station (NYSAES)/Cornell.

Chapter 40 **40.1** JOEL SARTORE/National Geographic Stock; **40.2.2** Peter Aitken/Peter Arnold Images/ PhotoLibrary; **40.2.3** Duncan Usher/Alamy; **40.2.4** Frank Greenaway/Dorling Kindersley/Getty Images; **40.4.2** Susumu Nishinaga/Photo Researchers; **40.4.3** Eye of Science/Photo Researchers; **40.4.4** Susumu Nishinaga/Photo Researchers; **40.5.7** CNRI/SPL/Photo Researchers; **40.5.11** Dr. Gopal Murti/SPL/Photo Researchers; **40.5.13** Chuck Brown/Photo Researchers; **40.5.15** Nina Zanetti; **40.5.17** Nina Zanetti; **40.5.19** Nina Zanetti; **40.5.21** Alamy; **40.5.24** Nina Zanetti; **40.5.26** Ed Reschke/Peter Arnold Images/PhotoLibrary; **40.5.28** Manfred Kage/Peter Arnold/Photolibrary; **40.5.31** Ulrich Gartner; **40.5.33** Thomas Deerinck/National Center for Microscopy and Imaging Research, University of California, San Diego; **40.10.2** Patricio Robles Gil/naturepl.com; **40.10.3** Matt T. Lee; **40.13** Robert Ganz; **40.18** Jeff Rotman/Alamy.

Chapter 41 **41.1** Michael deYoung/Corbis; **41.2** Roland Seitre/Peter Arnold/PhotoLibrary; **41.3** Stefan Huwiler/Rolf Nussbaumer Photography/Alamy; **41.5.2** cameilia/Shutterstock; **41.6.2** Hervey Bay Whale Watch (www.herveybaywhalewatch.com.au); **41.6.4** Thomas Eisner; **41.6.5** Photo Lennart Nilsson/Scanpix; **41.6.6** Gunter Ziesler/Peter Arnold/PhotoLibrary; **41.11.2** Visuals Unlimited/Corbis; **41.17.2** Fritz Polking/Peter Arnold, Inc./Alamy; **41.17.3** EyeWire Collection/Photodisc/Getty Images; **41.22** Photo courtesy of The Jackson Laboratory, Bar Harbor, Maine; **41.23** Wolfgang Kaehler/Corbis.

Chapter 42 **42.1** Stephen Dalton/Photo Researchers; **42.2.2** Reinhard dirscheri/PhotoLibrary; **42.2.3** Eric Grave/Photo Researchers; **42.10.2** From "Human Histology Photo CD." Image courtesy Indigo Instruments (www.indigo.com); **42.10.3** Photo Lennart Nilsson/Scanpix; **42.16** Biophoto Associates/Photo Researchers; **42.18.2** Eye of Science/Photo Researchers; **42.22.2** Peter Batson/Image Quest Marine; **42.22.3** Olgysha/Shutterstock; **42.22.4** Jez Tryner/Image Quest Marine; **42.24.2** Prepared by Dr. Hong Y. Yan, University of Kentcky and Dr. Peng Chai, University of Texas; **42.25.2** Motta & Macchiarelli/Anatomy Dept., Univ. La Sapienza, Rome/Photo Researchers; **42.27.2** Hans-Rainer Duncker, University of Giessen, Germany.

Chapter 43 **43.1** Biology Media/Science Source/Photo Researchers; **43.4** Dominique Ferrandon; **43.21.2** Steve Gschmeissner/Photo Researchers; **43.23** CNRI/Photo Researchers; **43.26** The Laboratory of Structural Cell Biology, headed by Stephen C. Harrison, Harvard Medical School/HHMI.

Chapter 44 **44.1** David Wall/Alamy; **44.4** Mark Conlin/Image Quest Marine; **44.5.2** Dr. John Crowe, University of California, Davis; **44.5.3** Dr. John Crowe, University of California, Davis; **44.9** AFP/Getty Images; **44.14.5** Steve Gschmeissner/Photo Researchers; **44.17** John Cancalosi/Peter Arnold Images/PhotoLibrary; **44.18** Michael Fodgen/OSF/PhotoLibrary.

Chapter 45 **45.1** Ralph A. Clevenger/Corbis; **p. 974** Stuart Wilson/Photo Researchers; **45.3** Volker Witte; **45.18.2** Astier/BSIP/Photo Researchers; **45.19.2** Photoshot Holdings Ltd/Alamy; **45.19.3** Jurgen & Christine Sohns/FLPA.

Chapter 46 **46.1** David Doubilet/Getty Images; **46.2** David Wrobel; **46.4** Chris Wallace Photography, photographersdirect.com; **46.5.2** P. de Vries, courtesy of David Crews; **46.6** Andy Sands/naturepl.com; **46.7** John Cancalosi/Peter Arnold/PhotoLibrary; **46.17.2** Photo Lennart Nilsson/Scanpix; **46.17.3** Photo Lennart Nilsson/Scanpix; **46.17.4** Photo Lennart Nilsson/Scanpix.

Chapter 47 **47.1** Photo Lennart Nilsson/Scanpix; **47.4.2** Vacquier VD, Payne JE. Methods for quantitating sea urchin sperm-egg binding. *Exp Cell Res.* 1973 Nov;82(1):227–35; **47.4.3** Vacquier VD, Payne JE. Methods for quantitating sea urchin sperm-egg binding. *Exp Cell Res.* 1973 Nov;82(1):227–35; **47.4.4** Vacquier VD, Payne JE. Methods for quantitating sea urchin sperm-egg binding. *Exp Cell Res.* 1973 Nov;82(1):227–35; **47.4.5** Vacquier VD, Payne JE. Methods for quantitating sea urchin sperm-egg binding. *Exp Cell Res.* 1973 Nov;82(1):227–35; **47.4.7** Hafner, M., Petzelt, C., Nobiling, R., Pawley, J., Kramp, D. and G. Schatten. Wave of Free Calcium at Fertilization in the Sea Urchin Egg Visualized with Fura-2. *Cell Motil. Cytoskel.*, 9:271–277 (1988); **47.4.8** Hafner, M., Petzelt, C., Nobiling, R., Pawley, J., Kramp, D. and G. Schatten. Wave of Free Calcium at Fertilization in the Sea Urchin Egg Visualized with Fura-2. *Cell Motil. Cytoskel.*, 9:271–277 (1988); **47.4.9** Hafner, M., Petzelt, C., Nobiling, R., Pawley, J., Kramp, D. and G. Schatten. Wave of Free Calcium at Fertilization in the Sea Urchin Egg Visualized with Fura-2. *Cell Motil. Cytoskel.*, 9:271–277 (1988); **47.4.10** Hafner, M., Petzelt, C., Nobiling, R., Pawley, J., Kramp, D. and G. Schatten. Wave of Free Calcium at Fertilization in the Sea Urchin Egg Visualized with Fura-2. *Cell Motil. Cytoskel.*, 9:271–277 (1988); **47.6.2** George von Dassow; **47.6.3** George von Dassow; **47.6.4** George von Dassow; **47.6.5** George von Dassow; **47.7.2** Jürgen Berger/Max Planck Institute for Developmental Biology, Tübingen, Germany; **47.7.3** Andrew J. Ewald, Johns Hopkins Medical School; **47.9.2** Charles A. Ettensohn; **47.13.2** Huw Williams; **47.13.3** Thomas Poole, SUNY Health Science Center; **47.14.2** Dr. Keith Wheeler/Photo Researchers; **47.17.2** Hiroki Nishida, *Developmental Biology* 121 (1987): 526. Reprinted by permission of Academic Press; **47.17.3** Hiroki Nishida, *Developmental Biology* 121 (1987): 526. Reprinted by permission of Academic Press; **47.18.2** J. E. Sulston and H. R. Horvitz, *Dev. Biol.* 56 (1977):110–156; **47.19.2** Adapted from Strome (International Review of Cytology 114: 81–123, 1989); **47.20.2** Adapted from Strome (International Review of Cytology 114: 81–123, 1989); **47.20.3** Adapted from Strome (International Review of Cytology 114: 81–123, 1989); **47.20.4** Adapted from Strome (International Review of Cytology 114: 81–123, 1989); **47.20.5** Adapted from Strome (International Review of Cytology 114: 81–123, 1989); **47.24.2** Kathryn W. Tosney, University of Michigan; **47.25.2** Dennis Summerbell.

Chapter 48 **48.1** Marinethemes.com; **48.2.2** David Fleetham/Alamy; **48.6.2** Thomas Deerinck; **48.13.2** Bear, Connors, and Paradiso, "Neuroscience: Exploring the Brain" © 1996, p. 43; **48.16.2** Edwin R. Lewis, University of California at Berkeley.

Chapter 49 **49.1** Brainbow mouse cerebellum. Image by Tamily Weissman, Harvard University. The Brainbow mouse was produced by Livet J, Weissman TA, Kang H, Draft RW, Lu J, Bennis RA, Sanes JR, Lichtman JW. *Nature* (2007) 450:56–62; **49.6.2** N. Kedersha/Photo Researchers; **49.9** Larry Mulvehill/Corbis; **49.14.2** From "A functional MRI study of happy and sad affective states induced by classical music."; . . . M. T. Mitterschiffthaler et al. *Hum Brain Mapp.* 2007 Nov;28(11):1150–62; **49.16.2** Marcus E. Raichle, M.D., Washington University Medical Center; **p. 1075** From "Dr. Harlow's Case of Recovery from the passage of an Iron Bar through the Head." H. Bigelow. *Am. Journal of the Med. Sci.* July 1850;XXXIX. Images from the History of Medicine (NLM); **49.21** Image by Sebastian Jessberger. Fred H. Gage, Laboratory of Genetics LOG-G, The Salk Institute for Biological Studies; **49.24.2** Martin M. Rotker/Photo Researchers.

Chapter 50 **50.1** Kenneth Catania; **50.6.1** CSIRO; **50.6.2** R. A. Steinbrecht; **50.7.2** James Gerholdt/Photolibrary; **50.7.3** Splashdown Direct/ OSF/PhotoLibrary; **50.9.2** From Richard Elzinga, "Fundamentals of Entomology" 3rd. © 1987, p. 185. Reprinted by permission of Prentice-Hall, Upper Saddle River, NJ; **50.10.2** SPL/Photo Researchers; **50.16.2** USDA/APHIS Animal and Plant Health Inspection Service; **50.17.3** STEVE GSCHMEISSNER/SPL/Photo Researchers; **50.21** From "Gene therapy for red-green colour blindness in adult primates."; . . . K. Mancuso et al. *Nature.* 2009 Oct 8; 461(7265):784–7. Photo: Neitz Laboratory; **50.26.2** Clara Franzini-Armstrong, University of Pennsylvania; **50.27.2** Courtesy of Dr. H. E. Huxley; **50.27.3** Courtesy of Dr. H. E. Huxley; **50.27.4** Courtesy of Dr. H. E. Huxley; **50.33** George Cathcart Photography, photographersdirect.com; **50.38** Dave Watts/NHPA/Photo Researchers; **50.39** Vance A. Tucker.

Chapter 51 **51.1** Michael Nichols/National Geographic/Getty Images; **51.3** Susan Lee Powell; **51.5.2** Kenneth Lorenzen, UC Davis; **51.7.2** Thomas McAvoy/Life Magazine/Getty Images; **51.7.3** Operation Migration Inc.; **51.9.2** Lincoln Brower, Sweet Briar College; **51.9.3** Lincoln Brower, Sweet Briar College; **51.9.4** Lincoln Brower, Sweet Briar College; **51.11** Clive Bromhall/OSF/PhotoLibrary; **51.12.2** Alissa Crandall/Corbis; **51.12.3** Richard Wrangham; **51.15.2** Matt T. Lee; **51.15.3** David Osborn/Alamy; **51.15.4** Bill Schmoker; **51.16.2** James D Watt/Image Quest Marine; **51.17** Courtesy of Gerald S. Wilkinson; from G.S. Wilkinson and G. N. Dodson, in J. Choe and B. Crespi (eds)., "The Evolution of Mating Systems in Insects and Arachnids," Cambridge University Press, Cambridge (1997), pp. 310–328; **51.18** Cyril Laubscher/Dorling Kindersley; **51.21** Martin Harvey/Peter Arnold Images/PhotoLibrary; **51.22** Erik Svensson, Lund University, Sweden; **51.23.2** Robert Pickett/Corbis; **51.24** Lowell L. Getz and Lisa Davis; **51.25** Rory Doolin; **51.27** Jennifer Jarvis; **51.29.2.** Stephen J. Krasemann/Peter Arnold/PhotoLibrary.

Chapter 52 **52.1** Dr. Paul A. Zahl/Photo Researchers; **52.2.2** James D. Watt/Stephen Frink Collection/Alamy; **52.2.3** Gianni Tortoli/Photo Researchers; **52.2.4** Robyn Mackenzie/shutterstock; **52.2.5** B. Tharp/Photo Researchers; **52.2.6** Yann Arthus-Bertrand/ Corbis; **52.2.7** NASA/Goddard Space Flight Center; **52.11.1** JTB Photo Communications, Inc./Alamy; **52.11.2** imagebroker/Alamy; **52.12.2** Frans Lanting/Corbis; **52.12.4** Gordon Whitten/Corbis; **52.12.6** Wolfgang Kaehler/Corbis; **52.12.8** The California Chaparral Institute (http://californiachaparral.net). Photo supplied by Richard Halsey; **52.12.10** All Canada Photos/SuperStock; **52.12.12** Shutterstock; **52.12.14** Kennan Ward/Corbis; **52.12.16** Darrell Gulin/Corbis; **52.16.1** Allen Russell/PhotoLibrary; **52.16.2** AfriPics.com/Alamy; **52.16.3** David Tipling/Nature Picture Library; **52.16.4** Ron Watts/Corbis; **52.16.5** Photononstop/SuperStock; **52.16.6** James Randklev/Image Bank/Getty Images; **52.16.7** Stuart Westmorland/Corbis; **52.16.8** Stuart Westmorland/Corbis; **52.16.9** Digital Vision/Getty Images; **52.16.10** William Lange/Woods Hole Oceanographic Institution; **52.17.2** Geoff Dann/Dorling Kindersley; **52.19.2** Peter Llewellyn/Alamy; **52.21** Daniel Mosquin.

Chapter 53 **53.1** Arpat Ozgul; **53.2** Todd Pusser/Naturepl.com; **53.4.2** Bernard Castelein/Nature Picture Library/Alamy; **53.4.3** Frans Lanting/Corbis; **53.4.4** Niall Benvie/Corbis; **53.8.2** Hansjoerg Richter/iStockphoto; **53.11** Photodisc/White/PhotoLibrary; **53.12** Stone Nature Photography/Alamy; **53.13.2** H. Willcox/Wildlife Picture/Peter Arnold/Photolibrary; **53.14.2** Jean Louis Batt/Taxi/Getty Images; **53.14.3** Christine Osborne/Corbis; **53.14.4** Edward Parker/Alamy; **53.17.2** fotoVoyager/iStockphoto; **53.17.3** Adrian Bailey/Aurora Photos; **53.17.4** Joe Raedle/Getty Images; **53.17.5** Patrick Clayton, www.fisheyeguyphotography.com; **53.17.6** JOZSEF SZENTPETERI/NGS Image Collection; **53.17.7** Andrew Syred/Photo Researchers; **53.17.8** Nicholas Bergkessel, Jr./Photo Researchers; **53.19.2** Joe McDonald/Corbis; **53.20.2** Robert Kay; **53.21.2** Niclas Fritzén.

Chapter 54 **54.1** Hal Beral VWPics/SuperStock; **54.2.2** Joseph T. Collins/Photo Researchers; **54.2.3** National Museum of Natural History/Smithsonian Institution; **p. 1196** Frank W Lane/FLPA; **54.5.2** Barry Mansell/Nature Picture Library; **54.5.3** Fogden/Corbis; **54.5.4** Stephen J. Krasemann/Photo Researchers; **54.5.5** Robert Pickett/Papilio/Alamy; **54.5.6** Edward S. Ross, California Academy of Sciences; **54.5.7** © James K. Lindsey; **54.6** Douglas Faulkner/Photo Researchers; **54.7.2** Fogden/Corbis; **54.7.3** Dan Janzen, Department of Biology, University of Pennsylvania; **54.8** Peter Johnson/Corbis; **54.9.2** Sally D. Hacker; **54.12** Cedar Creek Ecosystem Science Reserve, University of Minnesota; **54.17.2** Genny Anderson; **54.19** SuperStock; **54.21.2** Ron Landis Photography, www.ronlandisphotography.co; **54.21.3** Scott T. Smith/Corbis; **54.22.2** Charles Mauzy/Corbis; **54.22.3** Keith Boggs; **54.22.4** Glacier Bay National Park Photo/Glacier Bay National Park and Preserve; **54.22.5** Terry Donnelly, Mary Liz Austin; **54.24.2** R. Grant Gilmore, Dynamac Corporation; **54.24.3** Lance Horn, National Undersea Research Center, University of North Carolina-Wilmington; **54.29** Nelish Pradhan, Bates College, Lewiston, ME; **54.30** Josh Spice.

Chapter 55 **55.1** Hassan Basagic; **55.2** Stone Nature Photography/Alamy; **55.3** Justus de Cuveland/AGE Fotostock; **55.16.2** Hubbard Brook Research Foundation/USDA Forest Service; **55.16.3** USDA Forest Service; **55.17.2** Mark Gallagher; **55.17.3** Mark Gallagher; **55.18.2** U.S. Department of Energy; **55.19.2** Stewart Rood, University of Lethbridge; **55.19.3** Photo provided by Kissimmee Division staff, South Florida Water Management District (WPB); **55.19.4** Tim Day, Xcluder Pest Proof Fencing Company; **55.19.5** Daniel H. Janzen, University of Pennsylvania; **55.19.6** Bert Boekhoven; **55.19.7** Jean Hall/Holt Studios/Photo Researchers; **55.19.8** Kenji Morita/Environment Division, Tokyo Kyuei Co., Ltd.

Chapter 56 **56.1** Stephen J Richards; **56.2** Wayne Lawler/Ecoscene/Corbis; **56.4.2** Neil Lucas/Nature Picture Library; **56.4.3** Mark Carwardine/Still Pictures/Peter Arnold/PhotoLibrary; **56.4.4** Nazir Foead; **56.5** Merlin D. Tuttle, Bat Conservation International, www.batcon.org; **56.6** Scott Camazine/Photo Researchers; **56.7** Michael Edwards/Getty Images; **56.8.2** Bruce Cowell, www.brucecowellphotographer.com; **56.8.3** Robert Ginn/PhotoEdit Inc. **56.9** Benezeth Mutayoba, photo provided by the University of Washington; **56.10** Richard Vogel/Liaison/Getty Images; **56.13.2** William Ervin/SPL/Photo Researchers; **56.14** Craighead Environmental Research Institute; **56.15.2** Tim Thompson/Corbis; **56.15.3** Chuck Bargeron, University of Georgia; **56.15.4** William D. Boyer, USDA Forest Service; **56.16.2** Yann Arthus-Bertrand/Corbis; **56.16.3** James P. Blair/National Geographic Image Collection; **56.17** R. O. Bierregaard, Jr., Biology Dept., University of North Carolina, Charlotte; **56.18** SPL/Photo Researchers; **56.21.2** Edwin Giesbers/naturepl.com; **56.22.2** Mark Chiappone and Steven Miller, Center for Marine Science, University of North Carolina-Wilmington, Key Largo, FL; **56.23** Nigel Cattlin/Photo Researchers; **56.24.2** NASA; **56.24.3** NASA; **56.26** Erich Hartmann/Magnum Photos; **56.28** Prof. William H. Schlesinger; **56.31.2** NASA; **56.31.3** NASA; **56.33.2** Serge de Sazo/Photo Researchers; **56.33.3** Hilde Jensen, University of Tubingen/Nature Magazine/AP Photo; **56.33.4** Gabriel Rojo/Nature Picture Library; **56.33.5** Titus Lacoste/Getty Images. Appendix A **p. A-5** OMIKRON/Science Source/Photo Researchers; **p. A-8** John N. A. Lott/Biological Photo Service; **p. A-10, top** Biophoto/Photo Researchers; **p. A-10, 2nd from top** Conly L. Rieder; **p. A-11** USDA/ARS/Agricultural Research Service; **p. A-35** Peter Kitin; Appendix E **p. E-1, top left** Dr. Kari Lounatmaa/Photo Researchers; **p. E-1, bottom left** Eye of Science/Photo Researchers; **p. E-1, middle** M. I. Walker/NHPA/Photoshot; **p. E-1 right** Kathy Piper/iStockphoto; **p. E-2, left** Douglas Adams/iStockphoto; **p. E-2, right** McPHOTO/AGE Fotostock.

Illustration and Text Credits

4.6b, 9.9, 17.17b and **c** are adapted from C. K. Matthews and K. E. van Holde, *Biochemistry*, 2nd ed. Copyright © 1996 Pearson Education, Inc., publishing as Pearson Benjamin Cummings. **4.7, 6.6b, 11.7, 11.12, 17.11, 18.25, 20.8, 21.9, and 21.10** are adapted from W. M. Becker, J. B. Reece, and M. F. Poenie, *The World of the Cell*, 3rd ed. Copyright © 1996 Pearson Education, Inc., publishing as Pearson Benjamin Cummings. **Table 6.1a** Adapted from W. M. Becker, L. J. Kleinsmith, and J. Hardin, *The World of the Cell*, 4th ed. p. 753. Copyright © 2000 Pearson Education, Inc., publishing as Pearson Benjamin Cummings. **6.8** and **6.23a** and cell organelle drawings in **6.11** and **6.12** are adapted from illustrations by Tomo Narashima in E. N. Marieb, *Human Anatomy and Physiology*, 5th ed. Copyright © 2001 Pearson Education, Inc., publishing as Pearson Benjamin Cummings. **6.9a, 50.12,** and **50.13** are also from *Human Anatomy and Physiology*, 5th ed. Copyright © 2001 Pearson Education, Inc., publishing as Pearson Benjamin Cummings. **30.4, 30.13i,** and **39.13** are adapted from M. W. Nabors, *Introduction to Botany*, Copyright © 2004 Pearson Education, Inc., publishing as Pearson Benjamin Cummings. **42.30a, 46.16, 49.8, 49.10, 50.26, and 50.30** are adapted from E. N. Marieb, *Human Anatomy and Physiology*, 4th ed. Copyright © 1998 Pearson Education, Inc., publishing as Pearson Benjamin Cummings. **42.30a** From Campbell et al., *Biology: Concepts and Connections*, 6th ed., fig. 22.10, p. 462. Copyright © 2009 Pearson Education, Inc., publishing as Pearson Benjamin Cummings. **43.8** Adapted from Gerard J. Tortora, Berdell R. Funke, and Christine L. Case. 1998. *Microbiology: An Introduction*, 6th ed. Copyright © 1998 Pearson Education, Inc., publishing as Pearson Benjamin Cummings. **44.8** and **51.8** are adapted from L. G. Mitchell, J. A. Mutchmor, and W. D. Dolphin. *Zoology*. Copyright © 1988 Pearson Education, Inc., publishing as Pearson Benjamin Cummings.

Chapter 1 **1.25** Map provided courtesy of David W. Pfennig, University of North Carolina at Chapel Hill; **1.27** Data in bar graph based on D. W. Pfennig et al. 2001. Frequency-dependent Batesian mimicry. *Nature 410*:323.

Chapter 2 **2.2 (bottom)** Reprinted by permission of Macmillan Publishers Ltd. *Nature*. M.E. Frederickson et al. 'Devil's gardens' bedeviled by ants, *437*:495, 9/22/05. Copyright © 2005.

Chapter 3 **3.8a** Adapted from *Scientific American*, Nov. 1998, p.102.

Chapter 5 **5.12** Adapted from *Biology: The Science of Life*, 3rd ed. by Robert Wallace et al. Copyright © 1991. Reprinted by permission of Pearson Education, Inc; **5.15 and 5.20F** PDB ID 1CGD: J. Bella, B. Brodsky, and H. M. Berman. 1995. Hydration structure of a collagen peptide. *Structure 3*: 893–906; **5.18** Adapted from D. W. Heinz et al. 1993. How amino-acid insertions are allowed in an alpha-helix of T4 lysozyme. *Nature 361*:561; **5.20D** PDB ID 3GS0: Palaninathan, S.K., Mohamedmohaideen, N.N., Orlandini, E., Ortore, G., Nencetti, S., Lapucci, A., Rossello, A., Freundlich, J.S., Sacchettini, J.C. 2009. Novel transthyretin amyloid fibril formation inhibitors: synthesis, biological evaluation, and X-ray structural analysis. *Public Library of Science One 4*: e6290–e6290; **5.20G, 21.10b, Un42.1** PDB ID 2HHB: G. Fermi, M. F. Perutz, B. Shaanan, R. Fourme. 1984. The crystal structure of human deoxyhaemoglobin at 1.74 A resolution. *J.Mol.Biol. 175*: 159–174.

Chapter 7 **7.9** PDB ID 3HAO: N. H. Joh, A. Oberai, D. Yang, J. P. Whitelegge, J. U. Bowie. 2009. Similar energetic contributions of packing in the core of membrane and water-soluble proteins. *J. Am.Chem.Soc. 131*: 10846–10847.

Chapter 8 **8.18** PDB ID 3e1f: D H. Juers, B. Rob, M. L. Dugdale, N. Rahimzadeh, C. Giang, M. Lee, B. W. Matthews, R.E. Huber. 2009. Direct and indirect roles of His-418 in metal binding and in the activity of beta-galactosidase (E. coli). *Protein Sci. 18:* 1281–1292; **8.19** PDB ID 1MDY: P. C. Ma, M.A. Rould, H. Weintraub, C.O. Pabo. 1994. Crystal structure of MyoD bHLH domain-DNA complex: perspectives on DNA recognition and implications for transcriptional activation. *Cell* (Cambridge, Mass.) *77*: 451–459; **8.20** Figures 4a and 4e from "A common allosteric site and mechanism in caspases" by J.M. Scheer et al., in *Proceedings of the National Academy of Sciences, 103*, no. 20: 7595–7600, May 16, 2006. Copyright © 2006 National Academy of Sciences, U.S.A. Used by permission.

Chapter 9 **9.5** From *Molecular Biology of the Cell*, 4th edition, by Bruce Alberts et al., fig. 2.69, p. 92. Copyright © 2002 by Bruce Alberts, Alexander Johnson, Julian Lewis, Martin Raff, Keith Roberts, and Peter Walter. Used by permission

Chapter 10 **10.13b** Figure 1a from "Architecture of the photosynthetic Oxygen-evolving center" by K.N. Ferreira et al., in *Science, 303*, No. 5665: 1831–1838, March 19, 2004. Copyright © 2004, The American Association for the Advancement of Science. Reprinted with permission from AAAS; **10.15** Adapted from Richard and David Walker. Energy, Plants, and Man, fig. 4.1, p. 69. Sheffield: University of Sheffield. Oxygraphics http://www.oxygraphics.co.uk. © Richard Walker. Used with permission.

Chapter 12 **12.13** From *Molecular Biology of the Cell*, 4th edition, by Bruce Alberts et al., fig. 18.41, p. 1059. Copyright © 2002 by Bruce Alberts, Alexander Johnson, Julian Lewis, Martin Raff, Keith Roberts, and Peter Walter. Used by permission.

Chapter 17 **17.13** Figure 10-45 from *Principles of Cell and Molecular Biology*, 2nd edition, by Valerie M. Kish and Lewis J. Kleinsmith. Copyright © 1995 HarperCollins College Publishers. Reprinted by permission of Pearson Education.

Chapter 18 **18.15a** Figure 1d from "An Abundant Class of Tiny RNAs with Probable Regulatory Roles in *Caenorhabditis elegans*" by N.C. Lau et al., in *Science, 294*, No. 5543: 858–862, Oct. 1, 2001. Copyright © 2001, The American Association for the Advancement of Science. Reprinted with permission from AAAS.

Chapter 20 **20.10** Figure 15.24, p. 481, from *Genetics*, 5th ed., by Peter Russell. Copyright © 1998 Pearson Education, Inc., publishing as Pearson Benjamin Cummings. Used by permission of the publisher.

Chapter 21 **21.2** Adapted from a figure by Chris A. Kaiser and Erica Beade; **21.5b** Figure 2B from "The Genetic Landscape of a Cell" by M. Costanzo et al., in *Science, 327*, No. 5964: 425–431, Jan. 22, 2010. Copyright © 2010, The American Association for the Advancement of Science. Reprinted with permission from AAAS; **21.17** Adapted from an illustration by William McGinnis in Peter Radetsky, "The homeobox: Something very precious that we share with flies, from egg to adult." Bethesda, MD: Howard Hughes Medical Institute, 1992, p. 92. Reprinted by permission of William McGinnis; **21.18** Adapted from M. Akam, "*Hox* genes and the evolution of diverse body plans," *Philosophical Transactions of the Royal Society B*, 1995, *349:* 313–319, fig. 3. © Royal Society of London. Reprinted by permission.

Chapter 22 **22.8** © Utako Kikutani 2007. Used with permission; **22.13** Figure 4a from "Host Race Radiation in the Soapberry Bug: Natural History with the History" by S.P. Carroll and C. Boyd, in Evolution, Vol. 46 No. 4, p. 1060, Aug. 1992. Reproduced with permission of Blackwell Publishing Ltd.; **22.14** Reprinted from *Trends in Microbiology, 16*, Issue 8: 361–369. B.A. Diep and M. Otto, "The role of virulence determinants in community-associated MRSA pathogenesis," Copyright © 2008, with permission from Elsevier.

Chapter 23 **23.5** Figure 3 from "Genetic mechanisms for adapting to a changing environment" by D.A. Powers et al., from *Annual Review of Genetics, 25*, Dec. 1991. Copyright © 1991 by Annual Reviews. Reprinted by permission; **23.11** Figure 20.6 (maps only) from *Discover Biology*, Second Edition by Michael L. Cain, Hans Damman, Robert A. Lue & Carol Kaesuk Loon, Editors. Copyright © 2002 by Sinauer Associates, Inc. Used by permission of W.W. Norton & Company, Inc.; **23.12** Reprinted by permission from Macmillan Publishers Ltd.: *Nature*. E. Postma and A. J. van Noordijk, Gene flow maintains a large genetic difference in clutch size at a small spatial scale, *433*, 1/6/05. Copyright © 2005; **23.14** Adapted from many sources including D. J. Futuyma. 2005. *Evolution*, fig. 11.3. Sunderland, MA: Sinauer Associates and from R. L. Carroll, 1988. *Vertebrate Paleontology and Evolution*. W.H. Freeman & Co.; **23.16** Adapted from A. M. Welch et al. 1998. Call duration as an indicator of genetic quality in male gray tree frogs. *Science* 280:1928–1930; **23.17** Adapted from A. C. Allison. 1961. Abnormal hemoglobin and erythrocyte enzyme-deficiency traits. In *Genetic Variation in Human Populations*, ed. G.A. Harrison. Oxford: Elsevier Science and from S. I. Hay et al., A world malaria map: *Plasmodium falciparum* endemicity in 2007. *PLoS Medicine 6*: fig. 3, p. 291; **23.18** Figure 2a from "Frequency-Dependent Natural Selection in the Handedness of Scale-Eating Cichlid Fish" by Michio Hori in *Science, 260*, No. 5105: 216–219, April 9, 1993. Copyright © 1993, The American Association for the Advancement of Science. Reprinted with permission from AAAS; **Un 23.2** Data from R. K. Koehn and T. J. Hilbish. 1987. The adaptive importance of genetic variation. *American Scientist* 75: 134–141.

Chapter 24 **24.7** Figure 3 from "Ecological Speciation in *Gambusia* Fishes" by R.B. Langerhans et al., from *Evolution, 61*, No. 9, July 2007, published by The Society for the Study of Evolution. Copyright © 2007 R.B. Langerhans, M.E. Gifford, E.O. Joseph. Reprinted by permission; **24.9** From figure 2 in "Correspondence between sexual isolation and allozyme differentiation" *Proceedings of the National Academy of Science, 87:* 2715–2719, 1990, p. 2718. Copyright © 1990 Stephen G. Tilley, Paul A. Verrell, Steven J. Arnold. Used with permission; **24.10a** Adapted from D. M. B. Dodd, 1989. Reproductive isolation as a consequence of adaptive divergence in *Drosophila pseudoobscura*. *Evolution* 43: 1308–1311; **24.13** *Hybrid Zone and the Evolutionary Process* edited by R.G. Harrison (1993): Map of Bombina hybrid zone (p. 263) and figure 10.1 (p. 278) from chapter "Analysis of hybrid zones with bombina" by J. M. Szymura. By permission of Oxford University Press; **24.15** Reprinted by permission of Macmillan Publishers Ltd.: *Nature*. G. P. Saetre et al. A sexually selected character displacement in flycatchers reinforces premating isolation, *387*:589–591, fig. 2, 6/5/97. Copyright © 1997; **24.19b** From fig. 2 in L. H. Rieseberg et al. 1996. Role of Gene Interactions in Hybrid Speciation: Evidence from Ancient and Experimental Hybrids. *Science* 272: 741–745. Copyright © 1996. Reprinted with permission from AAAS.

Chapter 25 **25.2** Graph based on data from A. P. Johnson et al., The Miller Volcanic spark discharge experiment, *Science* 322:404 (2008); **25.3a** Graph from "Experimental Models of Primitive Cellular Compartments Encapsulation Growth and Division" by M.M. Hanczyc et al., in *Science, 302*, No. 5645: 618–622. Oct. 24, 2003. Copyright © 2003, The American Association for the Advancement of Science. Reprinted with permission from AAAS; **25.5** From Don L. Eicher, *Geologic Time*, 1st edition, © 1968. Printed and Electronically reproduced by permission of Pearson Education, Inc., Upper Saddle River, New Jersey; **25.6a–d** Adapted from many sources including D. J. Futuyma. 2005. *Evolution*, fig. 4.10. Sunderland, MA: Sinauer Associates and from R. L. Carroll, 1988. *Vertebrate Paleontology and Evolution*. W.H. Freeman & Co.; **25.6e** Adapted from Luo et al. 2001. A new mammalia form from the Early Jurassic and evolution of mammalian characteristics. *Science* 292: 1535; **25.7** Adapted from D. J. Des Marais. September 8, 2000. When did photosynthesis emerge on Earth? *Science* 289: 1703–1705; **25.8** Reprinted by permission from Macmillan Publishers Ltd.: *Nature*. L. R. Kump. The rise of atmospheric oxygen, *451*:277–278, 1/17/08. Copyright © 2008; **25.13** Map adapted from http://gelology.er.usgs.gov/eastern/plates.html; **25.15** Graph created from D. M. Raup and J. J. Sepkoski, Jr. 1982. Mass extinctions in the marine fossil record. *Science*. 215: 1501–1503 and J. J. Sepkoski, Jr. 1984. A kinetic model of Phanerozoic taxonomic diversity. III. Post-Paleozoic families and mass extinctions. *Paleobiology*. Vol 10, No. 2, pp. 246–267 in D. J. Futuyma, fig. 7.3a, p. 143 and fig. 7.6, p. 145, Sunderland, MA: Sinauer Associates; **25.17** From Mayhew, P. J. et al. 2008. A long-term association between global temperature and biodiversity, origination and extinction in the fossil record. *Proceedings of the Royal Society B, 275:* 47–53, fig. 3b. Reprinted by permission; **25.18** Figure 3 from "Anatomical and ecological constraints on Phanerozoic animal diversity in the marine realm" by R.K. Bambach et al., in *Proceedings of the National Academy of Sciences, 99,* no. 10: 6854–6859, May 14, 2002. Copyright © 2002 National Academy of Sciences, U.S.A. Used by permission; **25.19** Adapted from Hickman, Roberts, and Larson.1997, *Zoology*, 10th ed, Wm. C. Brown, fig. 31.1; **25.24** Reprinted by permission from Macmillan Publishers Ltd: *Nature*. M. Ronshaugen et al. Hox protein mutation and macroevolution of the insect body plan, *415*:914–917, fig. 1a. Copyright © 2002; **25.26** Adapted from M. Strickberger, 1990. *Evolution*, Boston: Jones & Bartlett.

Chapter 26 **26.6** Figure 1 from "Which Whales Are Hunted? A Molecular Genetic Approach to Monitoring Whaling" by C.S. Baker and S.R. Palumbi, in *Science, 265,* No. 5178:1538–1539, Sep. 9, 1994. Copyright © 1994, The American Association for the Advancement of Science. Reprinted with permission from AAAS; **26.12** With kind permission from Springer Science+Business Media: *Development Genes and Evolution*, "The evolution of the hedgehog gene family in chordates: insights from amphioxus hedgehog," vol. 209, 1999, pp. 40–47, Jan. 1999, S.M. Shimeld, fig. 3; **26.19** Figure 4.3c, p. 124 from *Molecular Markers, Natural History, and Evolution*, 2nd edition, by John Avise. Copyright © 2004 Sinauer Associates. Used by permission; **26.20** Figure from "Timing the Ancestor of the HIV-1 Pandemic Strains" by B. Korber et al., in *Science, 288,* No. 5472: 1789–1796, Jun 9, 2000. Copyright © 2000, The American Association for the Advancement of Science. Reprinted with permission from AAAS; **26.21** Figure 4.1, p. 45, "The three domains of life" by S.L. Baldauf et al., from *Assembling the Tree of Life,* edited by Joel Cracraft and Michael Donoghue. By permission of Oxford University Press, Inc. **26.22** Adapted from S. Blair Hedges. The origin and evolution of model organisms. *Nature Reviews Genetics 3*:838–848, fig. 1, p. 840.

Chapter 27 **27.10** Reprinted by permission from Macmillan Publishers Ltd.: *Nature*. V.S. Cooper and R.E. Lenski. The population genetics of ecological specialization in evolving *E. coli* populations, *407*:736–739, fig. 1. Copyright © 2000; **27.18** Graph created from data in C. Calvaruso et al. 2006. Root-associated bacteria contribute to mineral weathering and to mineral nutrition in trees: A budgeting analysis. *Applied and Environmental Microbiology* 72: 1258–1266.

Chapter 28 **28.2** Reprinted from *Trends in Genetics,18,* No. 11, J.M. Archibald and P.J. Keeling, "Recycled plastids: a 'green movement' in eukaryotic evolution," Copyright © 2002, with permission from Elsevier; **28.11** Figure 12.7, p. 350, from *Microbiology*, by R. W. Bauman. Copyright © 2004 Pearson Education, Inc., publishing as Benjamin Cummings; **28.23** Data from A. Stechman and T. Cavalier-Smith. 2002. Rooting the eukaryote tree by using a derived gene fusion, *Science* 297: 89–91; **28.28** Reprinted by permission from Macmillan Publishers Ltd.: *Nature*. M.J. Behrenfeld et al. Climate-driven trends in contemporary ocean productivity, *44:* 752–755, fig. 3. Copyright © 2006.

Chapter 29 **29.10** Source: R. D. Bowden. 1991. Inputs, outputs and accumulation of nitrogen in an early successional moss *(Polytrichum)* ecosystem. *Ecological Monographs* 61: 207–223; **29.14** Adapted from Raven et al. *Biology of Plants,* 6th ed., fig. 19.7, W. H. Freeman & Co.

Chapter 30 **30.12a** From P. R. Crane. 1985. Phylogenetic analysis of seed plants and the origin of angiosperms. *Annals of the Missouri Botanical Garden, 72*:716–793, fig. 11a, p. 738. Used by permission of Missouri Botanical Garden Press; **30.12b** Figure 2.3, p. 28, from *Phylogeny and Evolution of Angiosperms* by Douglas E. Soltis et al. Copyright © 2005 Sinauer Associates. Used by permission. **Table 30.1** Adapted from Randy Moore et al., *Botany,* 2nd ed. Dubuque, IA: Brown, 1998, Table 2.2, p. 37.

Chapter 31 **31.21** Figures 4 and 5 from "Fungal endophytes limit pathogen damage in a tropical tree" by A.E. Arnold et al., in *Proceedings of the National Academy of Sciences, 100,* no. 26: 15652–15653, December 23, 2003. Copyright © 2003 National Academy of Sciences, U.S.A. Used by permission; **31.26** Figure 1 from "Reversing introduced species effects: Experimental removal of introduced fish leads to rapid recovery of a declining frog" by Vance T. Vredenburg from *Proceedings of the National Academy of Sciences 101:* 7646–7650. Copyright © 2004 National Academy of Sciences, U.S.A. Used by permission.

Chapter 33 **33.22** C. Lydeard et al., The Global Decline of Nonmarine Mollusks, *BioScience*, Vol. 54, No. 4: 321–330. © 2004, American Institute of Biological Sciences. Used by permission. All rights reserved. (Updated data are from International Union for Conservation of Nature, 2008.); **33.29a** Reprinted from *Current Biology, 7,* Issue 8, J.K. Grenier, S. Carroll et al., "Evolution of the entire arthropod *Hox* gene set predated the origin and radiation of the onychophoran/arthropod clade," p. 551, fig. 2a, Copyright © 1987, with permission from Elsevier.

Chapter 34 **34.8b** From "Fossil sister group of craniates: Predicted and found" by J. Mallatt and J. Chen, from *Journal of Morphology, 258,* Issue 1, May 15, 2003. Copyright © 2003 Wiley-Liss, Inc. Reprinted with permission; **34.12** From *Vertebrates: Comparative Anatomy, Function, Evolution,* 3/e by Kenneth Kardong. © 2002 McGraw-Hill Science/Engineering/Mathematics. Reprinted by permission of The McGraw-Hill Companies, Inc. **34.18 (bottom)** Reprinted by permission of Macmillan Publishers Ltd.: *Nature*. M. Zhu et al. The oldest articulated osteichthyan reveals mosaic gnathostome characters, *458*:469–474. Copyright © 2009; **34.21 (left)** Reproduced by permission of the Royal Society of Edinburgh from Transactions of the Roya Society of Edinburgh: Earth Sciences, volume 87 (1996), pp. 363–421; **34.21 (right)** Reprinted by permission of Macmillan Publishers Ltd.: *Nature*. N.H. Shubin et al. The pectoral fin of *Tiktaalik roseae* and the origin of the tetrapod limb, *440*:768, fig. 4. Copyright © 2006; **34.37a** Adapted from many sources including D. J. Futuyma. 2005. *Evolution*, 1/e, fig. 4.10. Sunderland, MA: Sinauer Associates and from R. L. Carroll, 1988. *Vertebrate Paleontology and Evolution*. W.H. Freeman & Co. **34.47** Drawn from many photos of fossils. Some sources are *O. tugenensis* photo in Michael Balter, Early hominid sows division, *ScienceNow*, Feb. 22, 2001, © 2001 American Association for the Advancement of Science. *A. garhi,* and *H. neanderthalensis* adapted from *The Human Evolution Coloring Book. K platyops* drawn from photo in Meave Leakey et al., New hominid genus from eastern Africa shows diverse middle Pliocene lineages, *Nature*, March 22, 2001, 410: 433. *P. boisei* drawn from a photo by David Bill. *H. ergaster* drawn from a photo at www.inhandmuseum.com. *S. tchadensis* drawn from a photo in Michel Brunet et al., A new hominid from the Upper Miocene of Chad, Central Africa, *Nature*, July 11, 2002, 418: 147, fig. 1b; **34.50 (a/b)** Reprinted by permission of Macmillan Publishers Ltd.: *Nature*. I.V. Ovchinnikov et al. Molecular analysis of Neanderthal DNA from the northern Caucasus, *404:* 492, fig.3a and b. Copyright © 2000.

Chapter 35 **35.21** Figure 2b from "Mongolian Tree Rings and 20th-Century Warming" by G.C. Jacoby et al., in *Science, 273,* No. 5276: 771–773, Aug. 9, 1996. Copyright © 1996, The American Association for the Advancement of Science. Reprinted with permission from AAAS.

Chapter 39 **39.16 (top)** Adapted from M. Wilkins. 1988. *Plant Watching,* Facts of File Publ. **39.28** Figure "No Free Lunch" from "Plant Biology: New fatty acid-based signals: A lesson from the plant world" by Edward Farmer in *Science, 276,* No. 5314: 912–913, May 9, 1997. Copyright © 1997, The American Association for the Advancement of Science. Reprinted with permission from AAAS.

Chapter 40 **40.14** Figure 2 from "Thermoregulation in a brooding female Indian python, Python molurus bivittatus" by V.H. Hutchison et al., in *Science, 151,* No. 3711: 694–695, Feb. 11, 1966. Copyright © 1966, The American Association for the Advancement of Science. Reprinted with permission from AAAS; **40.15** Figure 7 from "Thermoregulation in Endothermic Insects" by Bernd Heinrich, in *Science, 185,* No. 4153: 747–756, August 30, 1974. Copyright © 1974, The American Association for the Advancement of Science. Reprinted with permission from AAAS; **40.21** Adapted from figures 2b and 2c from "The circadian clock stops ticking during deep hibernation in the European hamster" by F.G. Revel et al., in *Proceedings of the National Academy of Sciences, 104,* no. 34: 13816–13820, Aug. 21, 2007. Copyright © 2007 National Academy of Sciences, U.S.A. Used by permission.

Chapter 41 **41.9a** Figure 23.1 from *Human Anatomy and Physiology,* 8e, by Elaine Marieb and Katja Hoehn. Copyright © 2010 Pearson Education, Inc., publishing as Pearson Benjamin Cummings. Used by permission of the publisher; **41.9b** Figure 22-1 from Rhoades, *Human Physiology,* 3e. © 1996 Brooks/Cole, a part of Cengage Learning, Inc. Reproduced by permission. www.cengage.com/permissions; **41.21** Figure "Appetite Controllers" from "Cellular Warriors at the Battle of the Bulge" by Jean Marx, from *Science, 299:* p. 86, Feb. 7, 2003. Illustration by Kathleen Sutliff. Copyright © 2003, The American Association for the Advancement of Science. Reprinted with permission from AAAS.

Chapter 42 **42.20** Reprinted by permission from Macmillan Publishers Ltd.: *Nature*. D.J. Rader and A. Daugherty. Translating molecular discoveries into new therapies for atherosclerosis, *451*:904–913, fig. 1, 2/21/08. Copyright © 2008; **42.21** From J.C. Cohen et al., "Sequence variations in PCSK9, low LDL, and protection against coronary heart disease," *New England Journal of Medicine*. 2006 Mar 23; *354*:1264–72, fig. 1A. Copyright © 2006 Massachusetts Medical Society. Used by permission. All rights reserved; **42.26** Adapted from "Surface properties in relation to atelectasis and hyaline membrane disease" by M.E. Avery and J. Mead, from A.M.A. *American Journal of Diseases of Children 97*:517–523 (June 1959). Copyright © 1959 American Medical Association. Used by permission. All rights reserved.

Chapter 43 **43.5** From figures 2a and 4a in Phoebe Tzou et al., "Constitutive expression of a single antimicrobial peptide can restore wild-type resistance to infection in immuno-deficient *Drosophila* mutants," *PNAS, 99:* 2152–2157. Copyright © 2002 National Academy of Sciences, U.S.A. Used with permission; **43.7** Figures 20.4 and 20.5 from *Human Anatomy and Physiology,* 8e, by Elaine Marieb and Katja Hoehn. Copyright © 2010 Pearson Education, Inc., publishing as Pearson Benjamin Cummings. Used by permission of the publisher.

Chapter 44 **44.6** Kangaroo rat data adapted from K. B. Schmidt-Nielson. 1990. *Animal Physiology: Adaptation and Environment,* 4th ed., p. 339. Cambridge: Cambridge University Press; **44.7a** Adapted from K. B. Schmidt-Nielsen et al. 1958. Extrarenal salt excretion in birds. *American Journal of Physiology* 193: 101–107; **44.14B and 44.15** Figure 25.3b from *Human Anatomy and Physiology,* 8e, by Elaine Marieb and Katja Hoehn. Copyright © 2010 Pearson Education, Inc., publishing as Pearson Benjamin Cummings. Used by permission of the publisher; **44.21** Table 1 from "Requirement of human renal water channel aquaporin-2 for vasopressin-dependent concentration in urine" by P.M. Deen et al., in *Science, 264,* No. 5155: 92–95, April 1, 1994. Copyright © 1994, The American Association for the Advancement of Science. Reprinted with permission from AAAS; **44EOC** From W.S. Beck et al., *Life: An Introduction to Biology,* 3rd edition, copyright © 1991. Reprinted and electronically reproduced by permission of Pearson Education, Inc., Upper Saddle River, New Jersey.

Chapter 46 **46.9** Reprinted by permission of Macmillan Publishers Ltd.: *Nature*. R.R. Snook and D.J. Hosken, "Sperm death and dumping in *Drosophila*," *428*:939–941, fig. 2. Copyright © 2004.

Chapter 47 **47.16** Figures 1.10 and 8.25 from *Principles of Development* by Lewis Wolpert (1998). By permission of Oxford University Press, Inc. **47.17a** Figure 21-70 from *Molecular Biology of the Cell,* 4th ed, by Bruce Alberts, used with permission from Garland Science - Books, permission conveyed through Copyright Clearance Center, Inc. Adapted for use from "Cell commitment and gene expression in the axolotl embryo" by T.J. Mohun from *Cell 22:* 9–15 (1980), by permission of the author;

47.17b Reprinted from *Developmental Biology, 121,* Issue 2, Hiroki Nishida, "Cell lineage analysis in ascidian embryos by intracellular injection of a tracer enzyme: III. Up to the tissue restricted stage," p. 526, Copyright © 1987, with permission from Elsevier; **47.18** From *Molecular Biology of the Cell,* 4th edition, by Bruce Alberts et al., fig. 21.17, p. 1172. Copyright © 2002 by Bruce Alberts, Alexander Johnson, Julian Lewis, Martin Raff, Keith Roberts, and Peter Walter. Used by permission; **47.23 (experiment, results left)** Figures 1.10 and 8.25 from *Principles of Development* by Lewis Wolpert (1998). By permission of Oxford University Press, Inc. **47.23 (results, right)** Figure 15.12, p. 604, from *Developmental Biology,* 5th edition, by Gilbert et al. Copyright © 1997 Sinauer Associates. Used by permission; **47.26** Figure 23.1 from *Human Anatomy and Physiology,* 8e, by Elaine Marieb and Katja Hoehn. Copyright © 2010 Pearson Education, Inc., publishing as Pearson Benjamin Cummings. Used by permission of the publisher.

Chapter 48 **48.11** Adapted from G. Matthews, 2003. *Cellular Physiology of Nerve and Muscle,* 4th edition, fig. 6-2d, p. 61. Cambridge, MA: Blackwell Scientific Publications. Reprinted by permission of Wiley Blackwell; **48.18** Table 1 from "Opiate Receptor: Demonstration in nervous tissue" by C.B. Pert and S.H. Snyder in *Science, 179,* No. 4077: 1011–1014, March 9, 1973. Copyright © 1973, The American Association for the Advancement of Science. Reprinted with permission from AAAS.

Chapter 49 **49.11** Adapted from L. M. Mukhametov. 1984. Sleep in marine mammals. In *Sleep Mechanisms,* by A. A. Borbély and J. L. Valatx (eds.). Munich: Springer-Verlag, pp 227–238; **49.12** Figure 2a from "Transplanted suprachiasmatic nucleus determines circadian period" by M.R. Ralph et al., in *Science, 247,* No. 4945: 975–978, Feb. 23, 1990. Copyright © 1990, The American Association for the Advancement of Science. Reprinted with permission from AAAS; **49.18** Adapted from E. D. Jarvis et al. 2005. Avian brains and a new understanding of vertebrate brain evolution. *Nature Reviews Neuroscience* 6: 151–159, fig. 1c; **49.22** From *Schizophrenia Genesis: The Origins of Madness* by I. I. Gottesman, and D. Wolfgram, fig. 10, p. 96. New York: Freeman. © 1991 by Irving I. Gottesman. Used with the permission of Worth Publishers.

Chapter 50 **50.17A** Figure 15.4(a) from *Human Anatomy and Physiology,* 8e, by Elaine Marieb and Katja Hoehn. Copyright © 2010 Pearson Education, Inc., publishing as Pearson Benjamin Cummings. Used by permission of the publisher; **50.17B** Figure 15.15 from *Human Anatomy and Physiology,* 8e, by Elaine Marieb and Katja Hoehn. Copyright © 2010 Pearson Education, Inc., publishing as Pearson Benjamin Cummings. Used by permission of the publisher; **50.23** Reprinted by permission from Macmillan Publishers Ltd.: *Nature.* K.L. Mueller et al. The receptors and coding logic for bitter taste, *434:*225–229, fig. 4b. Copyright © 2005; **50.24a** Figure 15.23(a) and (b) from *Human Anatomy and Physiology,* 8e, by Elaine Marieb and Katja Hoehn. Copyright © 2010 Pearson Education, Inc., publishing as Pearson Benjamin Cummings. Used by permission of the publisher; **50.34** Grasshopper adapted from Hickman et al. 1993. *Integrated Principles of Zoology,* 9th ed., Fig. 22.6, p. 518. New York: McGraw-Hill Higher Education. © 1993 The McGraw-Hill Companies; **50.40** Figure 4 from "Locomotion: Energy Cost of Swimming, Flying, and Running" by K. Schmidt-Nielsen, in *Science, 177,* No. 4045: 222–228, July 21, 1972. Copyright © 1972, The American Association for the Advancement of Science. Reprinted with permission from AAAS.

Chapter 51 **51.2b** Figure 20, p. 28, from *Study of Instinct* by N. Tinbergen (1989). By permission of Oxford University Press, Inc.; **51.4** Reprinted by permission from Macmillan Publishing Ltd.: *Nature Reviews: Genetics:* M.B. Sokolowski. Drosophila: Genetics meets behavior, *2:*881, fig. 1. Copyright © 2001; **51.10** Figure 3a from "Prospective and retrospective learning in honeybees" by M. Biurfa and J. Benard, in *International Journal of Comparative Psychology, 19,* Issue 3: 358–367, 2006; Reprinted with permission; **51.13** Figure 2a from "Evolution of foraging behavior in Drosophila by density-dependent selection" by M.B. Sokolowski et al., in *Proceedings of the National Academy of Sciences, 94,* no. 14: 7373–7377, July 8, 1997. Copyright © 2007 National Academy of Sciences, U.S.A. Used by permission; **51.19** K. Witte and N. Sawka. 2003. Sexual imprinting on a novel trait in the dimorphic zebra finch: sexes differ. *Animal Behaviour* 65: 195–203. Art adapted from http://www.uni-bielefeld.de/biologue/vhf/KW/Forschungsprojekte2.html; **51.23 (left, bottom)** With kind permission from Springer Science+Business Media: *Genetica,* "The inheritance of mating songs in two cryptic, sibling lacewing species," vol. 116, 2002, pp. 269–289, C. S. Henry, fig. 2; **51.26 (top)** Adapted from a photograph by Jonathan Blair in Alcock, 2002. *Animal Behavior,* 7th ed. Sinauer Associates, Inc., Publiishers; **51.26 (bottom)** Reprinted by permission from Macmillan Publishers Ltd.: *Nature.* P. Berthold et al. Rapid microevolution of migratory behaviour in a wild bird species, *360:*668, fig. 1, 12/17/92. Copyright © 1992.

Chapter 52 **52.7** Adapted from L. Roberts. 1989. How fast can trees migrate? *Science* 243: 736, fig. 2. © 1989 by the American Association for the Advancement of Science; **52.8** Reprinted by permission from Macmillan Publishers Ltd.: *Nature.* C. Parmesan et al. Poleward shift of butterfly species' ranges associated with regional warming, *399:*579–583, fig. 3. Copyright © 1999; **52.9** Adapted from Heinrich Walter and Siegmar-Walter Breckle. 2003. *Walter's Vegetation of the Earth,* fig. 16, p. 36. Springer-Verlag, © 2003; **52.17** Figure 1.7, p. 9 from *Kangaroos, Their Ecology and Management in the Sheep Rangelands of Australia,* edited by Graeme Caughley, Neil Shepherd, Jeff Short. © Cambridge University Press 1987. Reprinted with the permission of Cambridge University Press; **52.19** Map adapted from R. L. Smith. 1974. *Ecology and Field Biology,* fig. 11.19, p. 353. Harper and Row Publishers. Map updated from D. A. Sibley. 2000. National Audubon Society *The Sibley Guide to Birds,* Alfred A. Knopf: New York; **52.20** Data from W. J. Fletcher. 1987. Interactions among subtidal Australian sea urchins, gastropods and algae: effects of experimental removals. *Ecological Monographs* 57: 89–109; **Un. 52.2** Data from J. Clausen, D. D. Keck, and W. M. Hiesey. 1948. Experimental studies on the nature of species. III. Environmental responses of climatic races of *Achillea.* Carnegie Institution of Washington Publication 581.

Chapter 53 **53.5** Adapted from P. W. Sherman and M. L. Morton. 1984. "Demography of Belding's ground squirrels," *Ecology,* Vol. 65, No. 5, p. 1622, fig. 1a, 1984. Copyright © 1984 Ecological Society of America. Used by permission; **53.15** Adapted from J. T. Enright. 1976. Climate and population regulation: The biogeographer's dilemma. *Oecologia* 24: 295–310; **53.16** Figure 3.5b, page 59, from *Soay Sheep: Dynamics and Selection in an Island Population* edited by T.H. Clutton-Brock and J.M. Pemberton. Copyright © Cambridge University Press 2004. Reprinted with the permission of Cambridge University Press; **53.18** Data courtesy of Rolf O. Peterson, Michigan Technological University; **53.23** Data from U. S. Census Bureau International Data Base; **53.24** Data from U. S. Census Bureau International Data Base; **53.25** Data from U. S. Census Bureau International Data Base 2008; **53.26** Source: Used courtesy of UNEP/GRID-Arendel at http://maps.grida.no/go/graphic/energy_consumption_per_capita_2004; **Tables 53.1** and **53.2** Data from P. W. Sherman and M. L. Morton, "Demography of Belding's ground squirrels," *Ecology,* Vol. 65, No. 5, p. 1622, fig. 1a, 1984. Copyright © 1984 Ecological Society of America.

Chapter 54 **54.2** From "The anoles of La Palma: aspects of their ecological relationships" by A.S. Rand and E. E. Williams in *Breviora 327:* 1–19, 1969. Copyright © 1969 by the President and Fellows of Harvard College. Used with permission from the Museum of Comparative Zoology, Harvard University; **54.9** Data for graph from S. D. Hacker and M. D. Bertness. 1999. Experimental evidence for factors maintaining plant species diversity in a New England salt marsh. *Ecology* 80: 2064–2073; **54.11** Adapted from N. Fierer and R. B. Jackson. 2006. The diversity and biogeography of soil bacterial communities. *Proceedings of the National Academy of Sciences USA 103:* 626–631 fig. 1a; **54.14** Adapted from E. A. Knox. 1970. Antarctic marine ecosystems. In *Antarctic Ecology,* ed. M. W. Holdgate, 69–96. London: Academic Press; **54.15** Adapted from D. L. Breitburg et al. 1997. Varying effects of low dissolved oxygen on trophic interactions in an estuarine food web. *Ecological Monographs* 67: 490. Copyright © 1997 Ecological Society of America; **54.16** Adapted from B. Jenkins. 1992. Productivity, disturbance and food web structure at a local spatial scale in experimental container habitats. *Oikos* 65: 252. Copyright © 1992 Oikos, Sweden; **54.17** Adapted from R. T. Paine. 1966. Food web complexity and species diversity. *American Naturalist* 100: 65–75; **54.18** From J. A. Estes et al. 1998. Killer whale predation on sea otters linking oceanic and nearshore ecosystems. *Science 282:* 474, fig. 1. Copyright © 1998 by the American Association for the Advancement of Science; **54.20** Graph adapted from A. R. Townsend et al. 1997. The intermediate disturbance hypothesis, refugia, and diversity in streams. *Limnology and Oceanography* 42:938–949. **54.22** Adapted from R. L. Crocker and J. Major. 1955. Soil Development in relation to vegetation and surface age at Glacier Bay, Alaska. *Journal of Ecology* 43: 427–448; **54.23** Data from F. S. Chapin, III, et al. 1994. Mechanisms of primary succession following deglaciation at Glacier Bay, Alaska. *Ecological Monographs* 64: 149–175; **54.25** Adapted from D. J. Currie. 1991. Energy and large-scale patterns of animal- and plant-species richness. *American Naturalist* 137: 27–49; **54.26** Adapted from F. W. Preston. 1960. Time and space and the variation of species. *Ecology 41:* 611–627; **54.28** Adapted from F. W. Preston. 1962. The canonical distribution of commonness and rarity. *Ecology 43:*185–215, 410–432.

Chapter 55 **55.4** and **Un 55.1** Adapted from D. L. DeAngelis. 1992. *Dynamics of Nutrient Cycling and Food Webs.* New York: Chapman & Hall; **55.7** Adapted from National Oceanic and Atmospheric Administration's National Data Buoy Center Voluntary Observing Ship Project, www.vos.noaa.gov/MWL/dec_06/Images/OCP_fig4.jpg; **55.8** Figure 2 from "Nitrogen, phosphorus, and eutrophication in the coastal marine environment" by J.H. Ryther, in *Science, 171,* No. 3975: 1008–1013, Mar. 12, 1971. Copyright © 1971, The American Association for the Advancement of Science. Reprinted with permission from AAAS; **55.9** Fig. 4.1 p. 82 from *Communities and Ecosystems,* 1e by Robert H. Whittaker. Copyright © 1970 Robert H. Whittaker. Reprinted by permission of Pearson Education; **55.14** From *The Economy of Nature,* 5th edition by Robert E. Ricklefs. © 2001 by W. H. Freeman and Company. Used with the permission of Worth Publishers. **55.15a** Adapted from J.A. Trofymow et al., *The Canadian Intersite Decomposition Experiment: Project and Site Establishment Report,* Information Report BC-X-378, page 2, Natural Resources Canada, Canadian Forest Service (1998). Reproduced with permission from the Minister of Public Works and Government Services, Canada, 2010; **55.15b** Adapted from: T.R. Moore et al., Litter decomposition rates in Canadian forests, *Global Change Biology* 5: 75–82 (1999), copyright © 2001, 1998 Blackwell Science Ltd. Reproduced with permission from the Minister of Public Works and Government Services, Canada, 2010 and Wiley Blackwell; **55.18b** Data adapted from Wu, W-M et al. 2006. Pilot-scale in situ bioremediation of uranium in a highly contaminated aquifer. 2. Reduction of U(VI) and geochemical control of a U(VI) bioavailability. *Environ Sci. Technol. 40:* 3986–3995, fig. 1D; **Table 55.1** Data from Menzel and Ryther. 1961. *Deep Sea Ranch* 7: 276–281.

Chapter 56 **56.12** Figure 19.1 from *Ecology,* 5e by C.J. Krebs. Copyright © 2001 Pearson Education, Inc. Reprinted by permission; **56.13** Figure 2 from "Tracking the Long-Term Decline and Recovery of an Isolated Population" by R.L. Westemeier et al., in *Science, 282,* No. 5394: 1695–1698, Nov. 27, 1998. Copyright © 1998, The American Association for the Advancement of Science. Reprinted with permission from AAAS; **56.19** Reprinted by permission from Macmillan Publishers, Ltd.: *Nature.* N. Myers et al. Biodiversity hotspots for conservation priorities, *403:* 853–858, fig. 1, 2/24/00. Copyright © 2000; **56.20** Reprinted from W. D. Newmark, 1985. Legal and biotic boundaries of western North American national parks: a problem of congruence. *Biological Conservation* 33: 197–208, fig. 1, p. 199. © 1985, with permission from Elsevier; **56.21a** Map adapted from W. Purves and G. Orians, *Life, The Science of Biology,* 5th ed., fig. 55.23, p. 1239. © 1998 by Sinauer Associates, Inc. Reprinted with permission; **56.27** CO_2 data from www.esrl.noaa.gov/gmd/ccgg/trends. Temperature data from www.giss.nasa.gov/gistemps/graphs/Fig.A.lrg.gif; **56.29** Data from ozonewatch.gsfc.nasa.gov/facts/history/html; **56.32** Data from Instituto Nacional de Estadistica y Censos de Costa Rica and Centro Centroamericano de Poblacion, Universidad de Costa Rica.

Glossary

Pronunciation Key

Pronounce		
ā	as in	ace
a/ah		ash
ch		chose
ē		meet
e/eh		bet
g		game
ī		ice
i		hit
ks		box
kw		quick
ng		song
ō		robe
o		ox
oy		boy
s		say
sh		shell
th		thin
ū		boot
u/uh		up
z		zoo

′ = primary accent
′ = secondary accent

5′ cap A modified form of guanine nucleotide added onto the 5′ end of a pre-mRNA molecule.

A site One of a ribosome's three binding sites for tRNA during translation. The A site holds the tRNA carrying the next amino acid to be added to the polypeptide chain. (A stands for aminoacyl tRNA.)

ABC hypothesis A model of flower formation identifying three classes of organ identity genes that direct formation of the four types of floral organs.

abiotic (ā′-bī-ot′-ik) Nonliving; referring to the physical and chemical properties of an environment.

abortion The termination of a pregnancy in progress.

abscisic acid (ABA) (ab-sis′-ik) A plant hormone that slows growth, often antagonizing the actions of growth hormones. Two of its many effects are to promote seed dormancy and facilitate drought tolerance.

absorption The third stage of food processing in animals: the uptake of small nutrient molecules by an organism's body.

absorption spectrum The range of a pigment's ability to absorb various wavelengths of light; also a graph of such a range.

abyssal zone (uh-bis′-ul) The part of the ocean's benthic zone between 2,000 and 6,000 m deep.

acanthodian (ak′-an-thō′-dē-un) Any of a group of ancient jawed aquatic vertebrates from the Silurian and Devonian periods.

accessory fruit A fruit, or assemblage of fruits, in which the fleshy parts are derived largely or entirely from tissues other than the ovary.

acclimatization (uh-klī′-muh-tī-zā′-shun) Physiological adjustment to a change in an environmental factor.

acetyl CoA Acetyl coenzyme A; the entry compound for the citric acid cycle in cellular respiration, formed from a fragment of pyruvate attached to a coenzyme.

acetylcholine (as′-uh-til-kō′-lēn) One of the most common neurotransmitters; functions by binding to receptors and altering the permeability of the postsynaptic membrane to specific ions, either depolarizing or hyperpolarizing the membrane.

acid A substance that increases the hydrogen ion concentration of a solution.

acid precipitation Rain, snow, or fog that is more acidic than pH 5.2.

acoelomate (uh-sē′-lō-māt) A solid-bodied animal lacking a cavity between the gut and outer body wall.

acrosomal reaction (ak′-ruh-sōm′-ul) The discharge of hydrolytic enzymes from the acrosome, a vesicle in the tip of a sperm, when the sperm approaches or contacts an egg.

acrosome (ak′-ruh-sōm) A vesicle in the tip of a sperm containing hydrolytic enzymes and other proteins that help the sperm reach the egg.

actin (ak′-tin) A globular protein that links into chains, two of which twist helically about each other, forming microfilaments (actin filaments) in muscle and other kinds of cells.

action potential An electrical signal that propagates (travels) along the membrane of a neuron or other excitable cell as a nongraded (all-or-none) depolarization.

action spectrum A graph that profiles the relative effectiveness of different wavelengths of radiation in driving a particular process.

activation energy The amount of energy that reactants must absorb before a chemical reaction will start; also called free energy of activation.

activator A protein that binds to DNA and stimulates gene transcription. In prokaryotes, activators bind in or near the promoter; in eukaryotes, activators generally bind to control elements in enhancers.

active immunity Long-lasting immunity conferred by the action of B cells and T cells and the resulting B and T memory cells specific for a pathogen. Active immunity can develop as a result of natural infection or immunization.

active site The specific region of an enzyme that binds the substrate and that forms the pocket in which catalysis occurs.

active transport The movement of a substance across a cell membrane against its concentration or electrochemical gradient, mediated by specific transport proteins and requiring an expenditure of energy.

adaptation Inherited characteristic of an organism that enhances its survival and reproduction in a specific environment.

adaptive immunity A vertebrate-specific defense that is mediated by B lymphocytes (B cells) and T lymphocytes (T cells). It exhibits specificity, memory, and self-nonself recognition. Also called acquired immunity.

adaptive radiation Period of evolutionary change in which groups of organisms form many new species whose adaptations allow them to fill different ecological roles in their communities.

addition rule A rule of probability stating that the probability of any one of two or more mutually exclusive events occurring can be determined by adding their individual probabilities.

adenosine triphosphate *See* ATP (adenosine triphosphate).

adenylyl cyclase (uh-den′-uh-lil) An enzyme that converts ATP to cyclic AMP in response to an extracellular signal.

adhesion The clinging of one substance to another, such as water to plant cell walls by means of hydrogen bonds.

adipose tissue A connective tissue that insulates the body and serves as a fuel reserve; contains fat-storing cells called adipose cells.

adrenal gland (uh-drē′-nul) One of two endocrine glands located adjacent to the kidneys in mammals. Endocrine cells in the outer portion (cortex) respond to adrenocorticotropic hormone (ACTH) by secreting steroid hormones that help maintain homeostasis during long-term stress. Neurosecretory cells in the central portion (medulla) secrete epinephrine and norepinephrine in response to nerve signals triggered by short-term stress.

adrenocorticotropic hormone (ACTH) A tropic hormone that is produced and secreted by the anterior pituitary and that stimulates the production and secretion of steroid hormones by the adrenal cortex.

aerobic respiration A catabolic pathway for organic molecules, using oxygen (O_2) as the final electron acceptor in an electron transport chain and ultimately producing ATP. This is the most efficient catabolic pathway and is carried out in most eukaryotic cells and many prokaryotic organisms.

age structure The relative number of individuals of each age in a population.

aggregate fruit A fruit derived from a single flower that has more than one carpel.

AIDS (acquired immunodeficiency syndrome) The symptoms and signs present during the late stages of HIV infection, defined by a specified reduction in the number of T cells and the appearance of characteristic secondary infections.

alcohol fermentation Glycolysis followed by the reduction of pyruvate to ethyl alcohol, regenerating NAD^+ and releasing carbon dioxide.

aldosterone (al-dos′-tuh-rōn) A steroid hormone that acts on tubules of the kidney to regulate the transport of sodium ions (Na^+) and potassium ions (K^+).

algae A diverse grade of photosynthetic protists, including unicellular and multicellular forms. Algal species are included in three of the five eukaryote supergroups (Chromalveolata, Rhizaria, and Archaeplastida).

alimentary canal (al′-uh-men′-tuh-rē) A complete digestive tract, consisting of a tube running between a mouth and an anus.

allele (uh-lē′-ul) Any of the alternative versions of a gene that may produce distinguishable phenotypic effects.

allergen An antigen that triggers an exaggerated immune response.

allopatric speciation (al′-uh-pat′-rik) The formation of new species in populations that are geographically isolated from one another.

allopolyploid (al′-ō-pol′-ē-ployd) A fertile individual that has more than two chromosome sets as a result of two different species interbreeding and combining their chromosomes.

allosteric regulation The binding of a regulatory molecule to a protein at one site that affects the function of the protein at a different site.

alpha (α) helix (al′-fuh hē′-liks) A coiled region constituting one form of the secondary structure of proteins, arising from a specific pattern of hydrogen bonding between atoms of the polypeptide backbone (not the side chains).

alternation of generations A life cycle in which there is both a multicellular diploid form, the sporophyte, and a multicellular haploid form, the gametophyte; characteristic of plants and some algae.

alternative RNA splicing A type of eukaryotic gene regulation at the RNA-processing level in which different mRNA molecules are produced from the same primary transcript, depending on which RNA segments are treated as exons and which as introns.

altruism (al′-trū-iz-um) Selflessness; behavior that reduces an individual's fitness while increasing the fitness of another individual.

alveolate (al-vē′-uh-let) A protist with membrane-bounded sacs (alveoli) located just under the plasma membrane.

alveolus (al-vē′-uh-lus) (plural, **alveoli**) One of the dead-end air sacs where gas exchange occurs in a mammalian lung.

Alzheimer's disease (alts′-hī-merz) An age-related dementia (mental deterioration) characterized by confusion and memory loss.

amacrine cell (am′-uh-krin) A neuron of the retina that helps integrate information before it is sent to the brain.

amino acid (uh-mēn′-ō) An organic molecule possessing both a carboxyl and an amino group. Amino acids serve as the monomers of polypeptides.

amino group A chemical group consisting of a nitrogen atom bonded to two hydrogen atoms; can act as a base in solution, accepting a hydrogen ion and acquiring a charge of 1+.

aminoacyl-tRNA synthetase An enzyme that joins each amino acid to the appropriate tRNA.

ammonia A small, toxic molecule (NH_3) produced by nitrogen fixation or as a metabolic waste product of protein and nucleic acid metabolism.

ammonite A member of a group of shelled cephalopods that were important marine predators for hundreds of millions of years until their extinction at the end of the Cretaceous period (65.5 million years ago).

amniocentesis (am′-nē-ō-sen-tē′-sis) A technique associated with prenatal diagnosis in which amniotic fluid is obtained by aspiration from a needle inserted into the uterus. The fluid and the fetal cells it contains are analyzed to detect certain genetic and congenital defects in the fetus.

amniote (am′-nē-ōt) Member of a clade of tetrapods named for a key derived character, the amniotic egg, which contains specialized membranes, including the fluid-filled amnion, that protect the embryo. Amniotes include mammals as well as birds and other reptiles.

amniotic egg An egg that contains specialized membranes that function in protection, nourishment, and gas exchange. The amniotic egg was a major evolutionary innovation, allowing embryos to develop on land in a fluid-filled sac, thus reducing the dependence of tetrapods on water for reproduction.

amoeba (uh-mē′-buh) A protist grade characterized by the presence of pseudopodia.

amoebocyte (uh-mē′-buh-sīt′) An amoeba-like cell that moves by pseudopodia and is found in most animals. Depending on the species, it may digest and distribute food, dispose of wastes, form skeletal fibers, fight infections, or change into other cell types.

amoebozoan (uh-mē′-buh-zō′-an) A protist in a clade that includes many species with lobe- or tube-shaped pseudopodia.

amphibian Member of the tetrapod class Amphibia, including salamanders, frogs, and caecilians.

amphipathic (am′-fē-path′-ik) Having both a hydrophilic region and a hydrophobic region.

amplification The strengthening of stimulus energy during transduction.

amygdala (uh-mig′-duh-luh) A structure in the temporal lobe of the vertebrate brain that has a major role in the processing of emotions.

amylase (am′-uh-lās′) An enzyme that hydrolyzes starch (a glucose polymer from plants) and glycogen (a glucose polymer from animals) into smaller polysaccharides and the disaccharide maltose.

anabolic pathway (an′-uh-bol′-ik) A metabolic pathway that consumes energy to synthesize a complex molecule from simpler molecules.

anaerobic respiration (an-er-ō′-bik) A catabolic pathway in which inorganic molecules other than oxygen accept electrons at the "downhill" end of electron transport chains.

analogous Having characteristics that are similar because of convergent evolution, not homology.

analogy (an-al′-uh-jē) Similarity between two species that is due to convergent evolution rather than to descent from a common ancestor with the same trait.

anaphase The fourth stage of mitosis, in which the chromatids of each chromosome have separated and the daughter chromosomes are moving to the poles of the cell.

anatomy The structure of an organism.

anchorage dependence The requirement that a cell must be attached to a substratum in order to initiate cell division.

androgen (an′-drō-jen) Any steroid hormone, such as testosterone, that stimulates the development and maintenance of the male reproductive system and secondary sex characteristics.

aneuploidy (an′-yū-ploy′-dē) A chromosomal aberration in which one or more chromosomes are present in extra copies or are deficient in number.

angiosperm (an′-jē-ō-sperm) A flowering plant, which forms seeds inside a protective chamber called an ovary.

angiotensin II A peptide hormone that stimulates constriction of precapillary arterioles and increases reabsorption of NaCl and water by the proximal tubules of the kidney, increasing blood pressure and volume.

anhydrobiosis (an-hī′-drō-bī-ō′-sis) A dormant state involving loss of almost all body water.

animal pole The point at the end of an egg in the hemisphere where the least yolk is concentrated; opposite of vegetal pole.

anion (an′-ī-on) A negatively charged ion.

anterior Pertaining to the front, or head, of a bilaterally symmetrical animal.

anterior pituitary A portion of the pituitary that develops from nonneural tissue; consists of endocrine cells that synthesize and secrete several tropic and nontropic hormones.

anther In an angiosperm, the terminal pollen sac of a stamen, where pollen grains containing sperm-producing male gametophytes form.

antheridium (an-thuh-rid′-ē-um) (plural, **antheridia**) In plants, the male gametangium, a moist chamber in which gametes develop.

anthropoid (an′-thruh-poyd) Member of a primate group made up of the monkeys and the apes (gibbons, orangutans, gorillas, chimpanzees, bonobos, and humans).

antibody A protein secreted by plasma cells (differentiated B cells) that binds to a particular antigen; also called immunoglobulin. All antibodies have the same Y-shaped structure and in their monomer form consist of two identical heavy chains and two identical light chains.

anticodon (an′-tī-kō′-don) A nucleotide triplet at one end of a tRNA molecule that base-pairs with a particular complementary codon on an mRNA molecule.

antidiuretic hormone (ADH) (an′-tī-dī-yū-ret′-ik) A peptide hormone, also known as vasopressin, that promotes water retention by the kidneys. Produced in the hypothalamus and released from the posterior pituitary, ADH also functions in the brain.

antigen (an′-ti-jen) A substance that elicits an immune response by binding to receptors of B cells, antibodies, or of T cells.

antigen presentation The process by which an MHC molecule binds to a fragment of an intracellular protein antigen and carries it to the cell surface, where it is displayed and can be recognized by a T cell.

antigen receptor The general term for a surface protein, located on B cells and T cells, that binds to antigens, initiating adaptive immune responses. The antigen receptors on B cells are called B cell receptors, and the antigen receptors on T cells are called T cell receptors.

antigen-presenting cell A cell that upon ingesting pathogens or internalizing pathogen proteins generates peptide fragments that are bound by class II MHC molecules and subsequently displayed on the cell surface to T cells. Macrophages, dendritic cells, and B cells are the primary antigen-presenting cells.

antiparallel Referring to the arrangement of the sugar-phosphate backbones in a DNA double helix (they run in opposite $5' \rightarrow 3'$ directions).

aphotic zone (ā′-fō′-tik) The part of an ocean or lake beneath the photic zone, where light does not penetrate sufficiently for photosynthesis to occur.

apical bud (ā′-pik-ul) A bud at the tip of a plant stem; also called a terminal bud.

apical dominance (ā′-pik-ul) Tendency for growth to be concentrated at the tip of a plant shoot, because the apical bud partially inhibits axillary bud growth.

apical ectodermal ridge (AER) A thickened area of ectoderm at the tip of a limb bud that promotes outgrowth of the limb bud.

apical meristem (ā′-pik-ul mār′-uh-stem) Embryonic plant tissue in the tips of roots and buds of shoots. The dividing cells of an apical meristem enable the plant to grow in length.

apicomplexan (ap′-ē-kom-pleks′-un) A protist in a clade that includes many species that parasitize animals. Some apicomplexans cause human disease.

apomixis (ap′-uh-mik′-sis) The ability of some plant species to reproduce asexually through seeds without fertilization by a male gamete.

apoplast (ap′-ō-plast) Everything external to the plasma membrane of a plant cell, including cell walls, intercellular spaces, and the space within dead structures such as xylem vessels and tracheids.

apoptosis (ā-puh-tō′-sus) A type of programmed cell death, which is brought about by activation of enzymes that break down many chemical components in the cell.

aposematic coloration (ap′-ō-si-mat′-ik) The bright warning coloration of many animals with effective physical or chemical defenses.

appendix A small, finger-like extension of the vertebrate cecum; contains a mass of white blood cells that contribute to immunity.

aquaporin A channel protein in the plasma membrane of a plant, animal, or microorganism cell that specifically facilitates osmosis, the diffusion of free water across the membrane.

aqueous solution (ā′-kwē-us) A solution in which water is the solvent.

arachnid A member of a major arthropod group, the chelicerates. Arachnids include spiders, scorpions, ticks, and mites.

arbuscular mycorrhiza (ar-bus′-kyū-lur mī′-kō-rī′-zuh) Association of a fungus with a plant root system in which the fungus causes the invagination of the host (plant) cells' plasma membranes.

arbuscular mycorrhizal fungus A symbiotic fungus whose hyphae grow through the cell wall of plant roots and extend into the root cell (enclosed in tubes formed by invagination of the root cell plasma membrane).

Archaea (ar′-kē′-uh) One of two prokaryotic domains, the other being Bacteria.

Archaeplastida (ar′-kē-plas′-tid-uh) One of five supergroups of eukaryotes proposed in a current hypothesis of the evolutionary history of eukaryotes. This monophyletic group, which includes red algae, green algae, and land plants, descended from an ancient protist ancestor that engulfed a cyanobacterium. *See also* Excavata, Chromalveolata, Rhizaria, and Unikonta.

archegonium (ar-ki-gō′-nē-um) (plural, **archegonia**) In plants, the female gametangium, a moist chamber in which gametes develop.

archenteron (ar-ken′-tuh-ron) The endoderm-lined cavity, formed during gastrulation, that develops into the digestive tract of an animal.

archosaur (ar′-kō-sōr) Member of the reptilian group that includes crocodiles, alligators and dinosaurs, including birds.

arteriole (ar-ter′-ē-ōl) A vessel that conveys blood between an artery and a capillary bed.

artery A vessel that carries blood away from the heart to organs throughout the body.

arthropod A segmented ecdysozoan with a hard exoskeleton and jointed appendages. Familiar examples include insects, spiders, millipedes, and crabs.

artificial selection The selective breeding of domesticated plants and animals to encourage the occurrence of desirable traits.

ascocarp The fruiting body of a sac fungus (ascomycete).

ascomycete (as′-kuh-mī′-sēt) Member of the fungal phylum Ascomycota, commonly called sac fungus. The name comes from the saclike structure in which the spores develop.

ascus (plural, **asci**) A saclike spore capsule located at the tip of a dikaryotic hypha of a sac fungus.

asexual reproduction The generation of offspring from a single parent that occurs without the fusion of gametes (by budding, division of a single cell, or division of the entire organism into two or more parts). In most cases, the offspring are genetically identical to the parent.

assisted migration The translocation of a species to a favorable habitat beyond its native range for the purpose of protecting the species from human-caused threats.

assisted reproductive technology A fertilization procedure that generally involves surgically removing eggs (secondary oocytes) from a woman's ovaries after hormonal stimulation, fertilizing the eggs, and returning them to the woman's body.

associative learning The acquired ability to associate one environmental feature (such as a color) with another (such as danger).

aster A radial array of short microtubules that extends from each centrosome toward the plasma membrane in an animal cell undergoing mitosis.

astrocyte A glial cell with diverse functions, including providing structural support for neurons, regulating the interstitial environment, facilitating synaptic transmission, and assisting in regulating the blood supply to the brain.

atherosclerosis A cardiovascular disease in which fatty deposits called plaques develop in the inner walls of the arteries, obstructing the arteries and causing them to harden.

atom The smallest unit of matter that retains the properties of an element.

atomic mass The total mass of an atom, which is the mass in grams of 1 mole of the atom.

atomic nucleus An atom's dense central core, containing protons and neutrons.

atomic number The number of protons in the nucleus of an atom, unique for each element and designated by a subscript to the left of the elemental symbol.

ATP (adenosine triphosphate) (a-den′-ō-sēn trī-fos′-fāt) An adenine-containing nucleoside triphosphate that releases free energy when its phosphate bonds are hydrolyzed. This energy is used to drive endergonic reactions in cells.

ATP synthase A complex of several membrane proteins that functions in chemiosmosis with adjacent electron transport chains, using the energy of a hydrogen ion (proton) concentration gradient to make ATP. ATP synthases are found in the inner mitochondrial membranes of eukaryotic cells and in the plasma membranes of prokaryotes.

atrial natriuretic peptide (ANP) (ā′-trē-ul na′-trē-yū-ret′-ik) A peptide hormone secreted by cells of the atria of the heart in response to high blood pressure. ANP's effects on the kidney alter ion and water movement and reduce blood pressure.

atrioventricular (AV) node A region of specialized heart muscle tissue between the left and right atria where electrical impulses are delayed for about 0.1 second before spreading to both ventricles and causing them to contract.

atrioventricular (AV) valve A heart valve located between each atrium and ventricle that prevents a backflow of blood when the ventricle contracts.

atrium (ā′-trē-um) (plural, **atria**) A chamber of the vertebrate heart that receives blood from the veins and transfers blood to a ventricle.

autocrine Referring to a secreted molecule that acts on the cell that secreted it.

autoimmune disease An immunological disorder in which the immune system turns against self.

autonomic nervous system (ot′-ō-nom′-ik) An efferent branch of the vertebrate peripheral nervous system that regulates the internal environment; consists of the sympathetic, parasympathetic, and enteric divisions.

autopolyploid (ot′-ō-pol′-ē-ployd) An individual that has more than two chromosome sets that are all derived from a single species.

autosome (ot′-ō-sōm) A chromosome that is not directly involved in determining sex; not a sex chromosome.

autotroph (ot′-ō-trōf) An organism that obtains organic food molecules without eating other organisms or substances derived from other organisms. Autotrophs use energy from the sun or from oxidation of inorganic substances to make organic molecules from inorganic ones.

auxin (ôk′-sin) A term that primarily refers to indoleacetic acid (IAA), a natural plant hormone that has a variety of effects, including cell elongation, root formation, secondary growth, and fruit growth.

average heterozygosity (het′-er-ō-zī-gō′-si-tē) The percentage, on average, of a population's loci that are heterozygous in members of the population.

avirulent Describing a pathogen that can mildly harm, but not kill, the host.

axillary bud (ak′-sil-ār-ē) A structure that has the potential to form a lateral shoot, or branch. The bud appears in the angle formed between a leaf and a stem.

axon (ak′-son) A typically long extension, or process, of a neuron that carries nerve impulses away from the cell body toward target cells.

B cells The lymphocytes that complete their development in the bone marrow and become effector cells for the humoral immune response.

Bacteria One of two prokaryotic domains, the other being Archaea.

bacterial artificial chromosome (BAC) A large plasmid that acts as a bacterial chromosome and can carry inserts of 100,000 to 300,000 base pairs (100–300 kb).

bacteriophage (bak-tēr′-ē-ō-fāj) A virus that infects bacteria; also called a phage.

bacteroid A form of the bacterium *Rhizobium* contained within the vesicles formed by the root cells of a root nodule.

balancing selection Natural selection that maintains two or more phenotypic forms in a population.

bark All tissues external to the vascular cambium, consisting mainly of the secondary phloem and layers of periderm.

Barr body A dense object lying along the inside of the nuclear envelope in cells of female mammals, representing a highly condensed, inactivated X chromosome.

basal angiosperm A member of one of three clades of early-diverging lineages of flowering plants. Examples are *Amborella*, water lilies, and star anise and its relatives.

basal body (bā′-sul) A eukaryotic cell structure consisting of a "9 + 0" arrangement of microtubule triplets. The basal body may organize the microtubule assembly of a cilium or flagellum and is structurally very similar to a centriole.

basal metabolic rate (BMR) The metabolic rate of a resting, fasting, and nonstressed endotherm at a comfortable temperature.

basal taxon In a specified group of organisms, a taxon whose evolutionary lineage diverged early in the history of the group.

base A substance that reduces the hydrogen ion concentration of a solution.

basidiocarp Elaborate fruiting body of a dikaryotic mycelium of a club fungus.

basidiomycete (buh-sid′-ē-ō-mī′-sēt) Member of the fungal phylum Basidiomycota, commonly called club fungus. The name comes from the club-like shape of the basidium.

basidium (plural, **basidia**) (buh-sid′-ē-um, buh-sid′-ē-ah) A reproductive appendage that produces sexual spores on the gills of mushrooms (club fungi).

Batesian mimicry (bāt′-zē-un mim′-uh-krē) A type of mimicry in which a harmless species looks like a species that is poisonous or otherwise harmful to predators.

behavior Individually, an action carried out by muscles or glands under control of the nervous system in response to a stimulus; collectively, the sum of an animal's responses to external and internal stimuli.

behavioral ecology The study of the evolution of and ecological basis for animal behavior.

benign tumor A mass of abnormal cells with specific genetic and cellular changes such that the cells are not capable of surviving at a new site and generally remain at the site of the tumor's origin.

benthic zone The bottom surface of an aquatic environment.

benthos (ben′-thōz) The communities of organisms living in the benthic zone of an aquatic biome.

beta (β) pleated sheet One form of the secondary structure of proteins in which the polypeptide chain folds back and forth. Two regions of the chain lie parallel to each other and are held together by hydrogen bonds between atoms of the polypeptide backbone (not the side chains).

beta oxidation A metabolic sequence that breaks fatty acids down to two-carbon fragments that enter the citric acid cycle as acetyl CoA.

bicoid A maternal effect gene that codes for a protein responsible for specifying the anterior end in *Drosophila melanogaster*.

bilateral symmetry Body symmetry in which a central longitudinal plane divides the body into two equal but opposite halves.

bilaterian (bī′-luh-ter′-ē-uhn) Member of a clade of animals with bilateral symmetry and three germ layers.

bile A mixture of substances that is produced in the liver and stored in the gallbladder; enables formation of fat droplets in water as an aid in the digestion and absorption of fats.

binary fission A method of asexual reproduction by "division in half." In prokaryotes, binary fission does not involve mitosis, but in single-celled eukaryotes that undergo binary fission, mitosis is part of the process.

binomial The two-part, latinized format for naming a species, consisting of the genus and specific epithet; a binomen.

biodiversity hot spot A relatively small area with numerous endemic species and a large number of endangered and threatened species.

bioenergetics (1) The overall flow and transformation of energy in an organism. (2) The study of how energy flows through organisms.

biofilm A surface-coating colony of one or more species of prokaryotes that engage in metabolic cooperation.

biofuel A fuel produced from dry organic matter or combustible oils produced by plants.

biogenic amine A neurotransmitter derived from an amino acid.

biogeochemical cycle Any of the various chemical cycles, which involve both biotic and abiotic components of ecosystems.

biogeography The study of the past and present geographic distribution of species.

bioinformatics The use of computers, software, and mathematical models to process and integrate biological information from large data sets.

biological augmentation An approach to restoration ecology that uses organisms to add essential materials to a degraded ecosystem.

biological clock An internal timekeeper that controls an organism's biological rhythms. The biological clock marks time with or without environmental cues but often requires signals from the environment to remain tuned to an appropriate period. *See also* circadian rhythm.

biological magnification A process in which retained substances become more concentrated at each higher trophic level in a food chain.

biological species concept Definition of a species as a group of populations whose members have the potential to interbreed in nature and produce viable, fertile offspring, but do not produce viable, fertile offspring with members of other such groups.

biology The scientific study of life.

biomanipulation An approach that applies the top-down model of community organization to alter ecosystem characteristics. For example, ecologists can prevent algal blooms and eutrophication by altering the density of higher-level consumers in lakes instead of by using chemical treatments.

biomass The total mass of organic matter comprising a group of organisms in a particular habitat.

biome (bī′-ōm) Any of the world's major ecosystem types, often classified according to the predominant vegetation for terrestrial biomes and the physical environment for aquatic biomes and characterized by adaptations of organisms to that particular environment.

bioremediation The use of organisms to detoxify and restore polluted and degraded ecosystems.

biosphere The entire portion of Earth inhabited by life; the sum of all the planet's ecosystems.

biotechnology The manipulation of organisms or their components to produce useful products.

biotic (bī-ot′-ik) Pertaining to the living factors—the organisms—in an environment.

bipolar cell A neuron that relays information between photoreceptors and ganglion cells in the retina.

bipolar disorder A depressive mental illness characterized by swings of mood from high to low; also called manic-depressive disorder.

birth control pill A chemical contraceptive that inhibits ovulation, retards follicular development, or alters a woman's cervical mucus to prevent sperm from entering the uterus.

blade (1) A leaflike structure of a seaweed that provides most of the surface area for photosynthesis. (2) The flattened portion of a typical leaf.

blastocoel (blas′-tuh-sēl) The fluid-filled cavity that forms in the center of a blastula.

blastocyst (blas′-tuh-sist) The blastula stage of mammalian embryonic development, consisting of an inner cell mass, a cavity, and an outer layer, the trophoblast. In humans, the blastocyst forms 1 week after fertilization.

blastomere An early embryonic cell arising during the cleavage stage of an early embryo.

blastopore (blas′-tō-pōr) In a gastrula, the opening of the archenteron that typically develops into the anus in deuterostomes and the mouth in protostomes.

blastula (blas′-tyū-luh) A hollow ball of cells that marks the end of the cleavage stage during early embryonic development in animals.

blood A connective tissue with a fluid matrix called plasma in which red blood cells, white blood cells, and cell fragments called platelets are suspended.

blue-light photoreceptor A type of light receptor in plants that initiates a variety of responses, such as phototropism and slowing of hypocotyl elongation.

body cavity A fluid- or air-filled space between the digestive tract and the body wall.

body plan In multicellular eukaryotes, a set of morphological and developmental traits that are integrated into a functional whole—the living organism.

Bohr shift A lowering of the affinity of hemoglobin for oxygen, caused by a drop in pH. It facilitates the release of oxygen from hemoglobin in the vicinity of active tissues.

bolus A lubricated ball of chewed food.

bone A connective tissue consisting of living cells held in a rigid matrix of collagen fibers embedded in calcium salts.

book lung An organ of gas exchange in spiders, consisting of stacked plates contained in an internal chamber.

bottleneck effect Genetic drift that occurs when the size of a population is reduced, as by a natural disaster or human actions. Typically, the surviving population is no longer genetically representative of the original population.

bottom-up model A model of community organization in which mineral nutrients influence community organization by controlling plant or phytoplankton numbers, which in turn control herbivore numbers, which in turn control predator numbers.

Bowman's capsule (bō′-munz) A cup-shaped receptacle in the vertebrate kidney that is the initial, expanded segment of the nephron where filtrate enters from the blood.

brachiopod (bra′-kē- uh-pod′) A marine lophophorate with a shell divided into dorsal and ventral halves; also called a lamp shell.

brain Organ of the central nervous system where information is processed and integrated.

brainstem A collection of structures in the vertebrate brain, including the midbrain, the pons, and the medulla oblongata; functions in homeostasis, coordination of movement, and conduction of information to higher brain centers.

branch point The representation on a phylogenetic tree of the divergence of two or more taxa from a common ancestor. A branch point is usually shown as a dichotomy in which a branch representing the ancestral lineage splits (at the branch point) into two branches, one for each of the two descendant lineages.

brassinosteroid A steroid hormone in plants that has a variety of effects, including inducing cell elongation, retarding leaf abscission, and promoting xylem differentiation.

breathing Ventilation of the lungs through alternating inhalation and exhalation.

bronchiole (brong′-kē-ōl′) A fine branch of the bronchi that transports air to alveoli.

bronchus (brong′-kus) (plural, **bronchi**) One of a pair of breathing tubes that branch from the trachea into the lungs.

brown alga A multicellular, photosynthetic protist with a characteristic brown or olive color that results from carotenoids in its plastids. Most brown algae are marine, and some have a plantlike body (thallus).

bryophyte (brī′-uh-fīt) An informal name for a moss, liverwort, or hornwort; a nonvascular plant that lives on land but lacks some of the terrestrial adaptations of vascular plants.

budding Asexual reproduction in which outgrowths from the parent form and pinch off to live independently or else remain attached to eventually form extensive colonies.

buffer A solution that contains a weak acid and its corresponding base. A buffer minimizes changes in pH when acids or bases are added to the solution.

bulk feeder An animal that eats relatively large pieces of food.

bulk flow The movement of a fluid due to a difference in pressure between two locations.

bundle-sheath cell In C_4 plants, a type of photosynthetic cell arranged into tightly packed sheaths around the veins of a leaf.

C_3 plant A plant that uses the Calvin cycle for the initial steps that incorporate CO_2 into organic material, forming a three-carbon compound as the first stable intermediate.

C_4 plant A plant in which the Calvin cycle is preceded by reactions that incorporate CO_2 into a four-carbon compound, the end product of which supplies CO_2 for the Calvin cycle.

calcitonin (kal′-si-tō′-nin) A hormone secreted by the thyroid gland that lowers blood calcium levels by promoting calcium deposition in bone and calcium excretion from the kidneys; nonessential in adult humans.

callus A mass of dividing, undifferentiated cells growing in culture.

calorie (cal) The amount of heat energy required to raise the temperature of 1 g of water by 1°C; also the amount of heat energy that 1 g of water releases when it cools by 1°C. The Calorie (with a capital C), usually used to indicate the energy content of food, is a kilocalorie.

Calvin cycle The second of two major stages in photosynthesis (following the light reactions), involving fixation of atmospheric CO_2 and reduction of the fixed carbon into carbohydrate.

CAM plant A plant that uses crassulacean acid metabolism, an adaptation for photosynthesis in arid conditions. In this process, carbon dioxide entering open stomata during the night is converted to organic acids, which release CO_2 for the Calvin cycle during the day, when stomata are closed.

Cambrian explosion A relatively brief time in geologic history when many present-day phyla of animals first appeared in the fossil record. This burst of evolutionary change occurred about 535–525 million years ago and saw the emergence of the first large, hard-bodied animals.

cAMP *See* cyclic AMP (cAMP).

canopy The uppermost layer of vegetation in a terrestrial biome.

capillary (kap′-il-ār′-ē) A microscopic blood vessel that penetrates the tissues and consists of a single layer of endothelial cells that allows exchange between the blood and interstitial fluid.

capillary bed A network of capillaries in a tissue or organ.

capsid The protein shell that encloses a viral genome. It may be rod-shaped, polyhedral, or more complex in shape.

capsule (1) In many prokaryotes, a dense and well-defined layer of polysaccharide or protein that surrounds the cell wall and is sticky, protecting the cell and enabling it to adhere to substrates or other cells. (2) The sporangium of a bryophyte (moss, liverwort, or hornwort).

carbohydrate (kar′-bō-hī′-drāt) A sugar (monosaccharide) or one of its dimers (disaccharides) or polymers (polysaccharides).

carbon fixation The initial incorporation of carbon from CO_2 into an organic compound by an autotrophic organism (a plant, another photosynthetic organism, or a chemoautotrophic prokaryote).

carbonyl group (kar-buh-nēl′) A chemical group present in aldehydes and ketones and consisting of a carbon atom double-bonded to an oxygen atom.

carboxyl group (kar-bok′-sil) A chemical group present in organic acids and consisting of a single carbon atom double-bonded to an oxygen atom and also bonded to a hydroxyl group.

cardiac cycle (kar′-dē-ak) The alternating contractions and relaxations of the heart.

cardiac muscle A type of striated muscle that forms the contractile wall of the heart. Its cells are joined by intercalated disks that relay the electrical signals underlying each heartbeat.

cardiac output The volume of blood pumped per minute by each ventricle of the heart.

cardiovascular system A closed circulatory system with a heart and branching network of arteries, capillaries, and veins. The system is characteristic of vertebrates.

carnivore An animal that mainly eats other animals.

carotenoid (kuh-rot′-uh-noyd′ An accessory pigment, either yellow or orange, in the chloroplasts of plants and in some prokaryotes. By absorbing wavelengths of light that chlorophyll cannot, carotenoids broaden the spectrum of colors that can drive photosynthesis.

carpel (kar′-pul) The ovule-producing reproductive organ of a flower, consisting of the stigma, style, and ovary.

carrier In genetics, an individual who is heterozygous at a given genetic locus for a recessively inherited disorder. The heterozygote is generally phenotypically normal for the disorder but can pass on the recessive allele to offspring.

carrying capacity The maximum population size that can be supported by the available resources, symbolized as *K*.

cartilage (kar′-til-ij) A flexible connective tissue with an abundance of collagenous fibers embedded in chondroitin sulfate.

Casparian strip (ka-spār′-ē-un) A water-impermeable ring of wax in the endodermal cells of plants that blocks the passive flow of water and solutes into the stele by way of cell walls.

catabolic pathway (kat′-uh-bol′-ik) A metabolic pathway that releases energy by breaking down complex molecules to simpler molecules.

catalyst (kat′-uh-list) A chemical agent that selectively increases the rate of a reaction without being consumed by the reaction.

catastrophism (kuh-tas′-truh-fiz′-um) The principle that events in the past occurred suddenly and were caused by different mechanisms than those operating today. *See* uniformitarianism.

catecholamine (kat′-uh-kōl′-uh-mēn) Any of a class of neurotransmitters and hormones, including the hormones epinephrine and norepinephrine, that are synthesized from the amino acid tyrosine.

cation (cat′-ī-on) A positively charged ion.

cation exchange A process in which positively charged minerals are made available to a plant when hydrogen ions in the soil displace mineral ions from the clay particles.

cDNA library A gene library containing clones that carry complementary DNA (cDNA) inserts. The library includes only the genes that were transcribed in the cells whose mRNA was isolated to make the cDNA.

cecum (sē′-kum) (plural, **ceca**) The blind pouch forming one branch of the large intestine.

cell body The part of a neuron that houses the nucleus and most other organelles.

cell cycle An ordered sequence of events in the life of a cell, from its origin in the division of a parent cell until its own division into two. The eukaryotic cell cycle is composed of interphase (including G_1, S, and G_2 subphases) and M phase (including mitosis and cytokinesis).

cell cycle control system A cyclically operating set of molecules in the eukaryotic cell that both triggers and coordinates key events in the cell cycle.

cell division The reproduction of cells.

cell fractionation The disruption of a cell and separation of its parts by centrifugation at successively higher speeds.

cell plate A membrane-bounded, flattened sac located at the midline of a dividing plant cell, inside which the new cell wall forms during cytokinesis.

cell wall A protective layer external to the plasma membrane in the cells of plants, prokaryotes, fungi, and some protists. Polysaccharides such as cellulose (in plants and some protists), chitin (in fungi), and peptidoglycan (in bacteria) are important structural components of cell walls.

cell-mediated immune response The branch of adaptive immunity that involves the activation of cytotoxic T cells, which defend against infected cells.

cellular respiration The catabolic pathways of aerobic and anaerobic respiration, which break down organic molecules and use an electron transport chain for the production of ATP.

cellular slime mold A type of protist characterized by unicellular amoeboid cells and aggregated reproductive bodies in its life cycle.

cellulose (sel′-yū-lōs) A structural polysaccharide of plant cell walls, consisting of glucose monomers joined by β glycosidic linkages.

Celsius scale (sel′-sē-us) A temperature scale (°C) equal to $^5/_9$(°F – 32) that measures the freezing point of water at 0°C and the boiling point of water at 100°C.

central canal The narrow cavity in the center of the spinal cord that is continuous with the fluid-filled ventricles of the brain.

central nervous system (CNS) The portion of the nervous system where signal integration occurs; in vertebrate animals, the brain and spinal cord.

central vacuole In a mature plant cell, a large membranous sac with diverse roles in growth, storage, and sequestration of toxic substances.

centriole (sen′-trē-ōl) A structure in the centrosome of an animal cell composed of a cylinder of microtubule triplets arranged in a 9 + 0 pattern. A centrosome has a pair of centrioles.

centromere (sen′-trō-mēr) In a duplicated chromosome, the region on each sister chromatid where they are most closely attached to each other by proteins that bind to specific DNA sequences; this close attachment causes a constriction in the condensed chromosome. (An uncondensed, unduplicated chromosome has a single centromere, identified by its DNA sequence.)

centrosome (sen′-trō-sōm) A structure present in the cytoplasm of animal cells that functions as a microtubule-organizing center and is important during cell division. A centrosome has two centrioles.

cephalization (sef′-uh-luh-zā′-shun) An evolutionary trend toward the concentration of sensory equipment at the anterior end of the body.

cercozoan An amoeboid or flagellated protist that feeds with threadlike pseudopodia.

cerebellum (sār′-ruh-bel′-um) Part of the vertebrate hindbrain located dorsally; functions in unconscious coordination of movement and balance.

cerebral cortex (suh-rē′-brul) The surface of the cerebrum; the largest and most complex part of the mammalian brain, containing nerve cell bodies of the cerebrum; the part of the vertebrate brain most changed through evolution.

cerebral hemisphere The right or left side of the cerebrum.

cerebrospinal fluid (suh-rē′-brō-spī′-nul) Blood-derived fluid that surrounds, protects against infection, nourishes, and cushions the brain and spinal cord.

cerebrum (suh-rē′-brum) The dorsal portion of the vertebrate forebrain, composed of right and left hemispheres; the integrating center for memory, learning, emotions, and other highly complex functions of the central nervous system.

cervix (ser′-viks) The neck of the uterus, which opens into the vagina.

chaparral A scrubland biome of dense, spiny evergreen shrubs found at midlatitudes along coasts where cold ocean currents circulate offshore; characterized by mild, rainy winters and long, hot, dry summers.

chaperonin (shap′-er-ō′-nin) A protein complex that assists in the proper folding of other proteins.

character An observable heritable feature that may vary among individuals.

character displacement The tendency for characteristics to be more divergent in sympatric populations of two species than in allopatric populations of the same two species.

checkpoint A control point in the cell cycle where stop and go-ahead signals can regulate the cycle.

chelicera (kē-lih′-suh-ruh) (plural, **chelicerae**) One of a pair of clawlike feeding appendages characteristic of chelicerates.

chelicerate (kē-lih-suh′-rāte) An arthropod that has chelicerae and a body divided into a cephalothorax and an abdomen. Living chelicerates include sea spiders, horseshoe crabs, scorpions, ticks, and spiders.

chemical bond An attraction between two atoms, resulting from a sharing of outer-shell electrons or the presence of opposite charges on the atoms. The bonded atoms gain complete outer electron shells.

chemical energy Energy available in molecules for release in a chemical reaction; a form of potential energy.

chemical equilibrium In a chemical reaction, the state in which the rate of the forward reaction equals the rate of the reverse reaction, so that the relative concentrations of the reactants and products do not change with time.

chemical reaction The making and breaking of chemical bonds, leading to changes in the composition of matter.

chemiosmosis (kem′-ē-oz-mō′-sis) An energy-coupling mechanism that uses energy stored in the form of a hydrogen ion gradient across a membrane to drive cellular work, such as the synthesis of ATP. Under aerobic conditions, most ATP synthesis in cells occurs by chemiosmosis.

chemoautotroph (kē′-mō-ot′-ō-trōf) An organism that obtains energy by oxidizing inorganic substances and needs only carbon dioxide as a carbon source.

chemoheterotroph (kē′-mō-het′-er-ō-trōf) An organism that requires organic molecules for both energy and carbon.

chemoreceptor A sensory receptor that responds to a chemical stimulus, such as a solute or an odorant.

chiasma (plural, **chiasmata**) (kī-az′-muh, kī-az′-muh-tuh) The X-shaped, microscopically visible region where crossing over has occurred earlier in prophase I between homologous nonsister chromatids. Chiasmata become visible after synapsis ends, with the two homologs remaining associated due to sister chromatid cohesion.

chitin (kī′-tin) A structural polysaccharide, consisting of amino sugar monomers, found in many fungal cell walls and in the exoskeletons of all arthropods.

chlorophyll (klōr′-ō-fil) A green pigment located in membranes within the chloroplasts of plants and algae and in the membranes of certain prokaryotes. Chlorophyll *a* participates directly in the light reactions, which convert solar energy to chemical energy.

chlorophyll *a* A photosynthetic pigment that participates directly in the light reactions, which convert solar energy to chemical energy.

chlorophyll *b* An accessory photosynthetic pigment that transfers energy to chlorophyll *a*.

chloroplast (klōr′-ō-plast) An organelle found in plants and photosynthetic protists that absorbs sunlight and uses it to drive the synthesis of organic compounds from carbon dioxide and water.

choanocyte (kō-an′-uh-sīt) A flagellated feeding cell found in sponges. Also called a collar cell, it has a collar-like ring that traps food particles around the base of its flagellum.

cholesterol (kō-les′-tuh-rol) A steroid that forms an essential component of animal cell membranes and acts as a precursor molecule for the synthesis of other biologically important steroids, such as many hormones.

chondrichthyan (kon-drik′-thē-an) Member of the class Chondrichthyes, vertebrates with skeletons made mostly of cartilage, such as sharks and rays.

chordate Member of the phylum Chordata, animals that at some point during their development have a notochord; a dorsal, hollow nerve cord; pharyngeal slits or clefts; and a muscular, post-anal tail.

chorionic villus sampling (CVS) (kōr′-ē-on′-ik vil′-us) A technique associated with prenatal diagnosis in which a small sample of the fetal portion of the placenta is removed for analysis to detect certain genetic and congenital defects in the fetus.

Chromalveolata One of five supergroups of eukaryotes proposed in a current hypothesis of the evolutionary history of eukaryotes. Chromalveolates may have originated by secondary endosymbiosis and include two large protist clades, the alveolates and the stramenopiles. *See also* Excavata, Rhizaria, Archaeplastida, and Unikonta.

chromatin (krō′-muh-tin) The complex of DNA and proteins that makes up eukaryotic chromosomes. When the cell is not dividing, chromatin exists in its dispersed form, as a mass of very long, thin fibers that are not visible with a light microscope.

chromosome (krō′-muh-sōm) A cellular structure carrying genetic material, found in the nucleus of eukaryotic cells. Each chromosome consists of one very long DNA molecule and associated proteins. (A bacterial chromosome usually consists of a single circular DNA molecule and associated proteins. It is found in the nucleoid region, which is not membrane bounded.) *See also* chromatin.

chromosome theory of inheritance A basic principle in biology stating that genes are located at specific positions (loci) on chromosomes and that the behavior of chromosomes during meiosis accounts for inheritance patterns.

chylomicron (kī′-lō-mī′-kron) A lipid transport globule composed of fats mixed with cholesterol and coated with proteins.

chyme (kīm) The mixture of partially digested food and digestive juices formed in the stomach.

chytrid (kī′-trid) Member of the fungal phylum Chytridiomycota, mostly aquatic fungi with flagellated zoospores that represent an early-diverging fungal lineage.

ciliate (sil′-ē-it) A type of protist that moves by means of cilia.

cilium (sil′-ē-um) (plural, **cilia**) A short appendage containing microtubules in eukaryotic cells. A motile cilium is specialized for locomotion or moving fluid past the cell; it is formed from a core of nine outer doublet microtubules and two inner single microtubules (the "9 + 2" arrangement) ensheathed in an extension of the plasma membrane. A primary cilium is usually nonmotile and plays a sensory and signaling role; it lacks the two inner microtubules (the "9 + 0" arrangement).

circadian rhythm (ser-kā′-dē-un) A physiological cycle of about 24 hours that persists even in the absence of external cues.

***cis-trans* isomer** One of several compounds that have the same molecular formula and covalent bonds between atoms but differ in the spatial arrangements of their atoms owing to the inflexibility of double bonds; formerly called a geometric isomer.

citric acid cycle A chemical cycle involving eight steps that completes the metabolic breakdown of glucose molecules begun in glycolysis by oxidizing acetyl CoA (derived from pyruvate) to carbon dioxide; occurs within the mitochondrion in eukaryotic cells and in the cytosol of prokaryotes; together with pyruvate oxidation, the second major stage in cellular respiration.

clade (klayd) A group of species that includes an ancestral species and all of its descendants.

cladistics (kluh-dis′-tiks) An approach to systematics in which organisms are placed into groups called clades based primarily on common descent.

class In Linnaean classification, the taxonomic category above the level of order.

cleavage (1) The process of cytokinesis in animal cells, characterized by pinching of the plasma membrane. (2) The succession of rapid cell divisions without significant growth during early embryonic development that converts the zygote to a ball of cells.

cleavage furrow The first sign of cleavage in an animal cell; a shallow groove around the cell in the cell surface near the old metaphase plate.

climate The long-term prevailing weather conditions at a given place.

climograph A plot of the temperature and precipitation in a particular region.

cline A graded change in a character along a geographic axis.

clitoris (klit′-uh-ris) An organ at the upper intersection of the labia minora that engorges with blood and becomes erect during sexual arousal.

cloaca (klō-ā′-kuh) A common opening for the digestive, urinary, and reproductive tracts found in many nonmammalian vertebrates but in few mammals.

clonal selection The process by which an antigen selectively binds to and activates only those lymphocytes bearing receptors specific for the antigen. The selected lymphocytes proliferate and differentiate into a clone of effector cells and a clone of memory cells specific for the stimulating antigen.

clone (1) A lineage of genetically identical individuals or cells. (2) In popular usage, an individual that is genetically identical to another individual. (3) As a verb, to make one or more genetic replicas of an individual or cell. *See also* gene cloning.

cloning vector In genetic engineering, a DNA molecule that can carry foreign DNA into a host cell and replicate there. Cloning vectors include plasmids and bacterial artificial chromosomes (BACs), which move recombinant DNA from a test tube back into a cell, and viruses that transfer recombinant DNA by infection.

closed circulatory system A circulatory system in which blood is confined to vessels and is kept separate from the interstitial fluid.

cnidocyte (nī′-duh-sīt) A specialized cell unique to the phylum Cnidaria; contains a capsule-like organelle housing a coiled thread that, when discharged, explodes outward and functions in prey capture or defense.

cochlea (kok′-lē-uh) The complex, coiled organ of hearing that contains the organ of Corti.

codominance The situation in which the phenotypes of both alleles are exhibited in the heterozygote because both alleles affect the phenotype in separate, distinguishable ways.

codon (kō′-don) A three-nucleotide sequence of DNA or mRNA that specifies a particular amino acid or termination signal; the basic unit of the genetic code.

coefficient of relatedness The fraction of genes that, on average, are shared by two individuals.

coelom (sē′-lōm) A body cavity lined by tissue derived only from mesoderm.

coelomate (sē′-lō-māt) An animal that possesses a true coelom (a body cavity lined by tissue completely derived from mesoderm).

coenocytic fungus (sē′-no-si′-tic) A fungus that lacks septa and hence whose body is made up of a continuous cytoplasmic mass that may contain hundreds or thousands of nuclei.

coenzyme (kō-en′-zīm) An organic molecule serving as a cofactor. Most vitamins function as coenzymes in metabolic reactions.

coevolution The joint evolution of two interacting species, each in response to selection imposed by the other.

cofactor Any nonprotein molecule or ion that is required for the proper functioning of an enzyme. Cofactors can be permanently bound to the active site or may bind loosely and reversibly, along with the substrate, during catalysis.

cognition The process of knowing that may include awareness, reasoning, recollection, and judgment.

cognitive map A neural representation of the abstract spatial relationships between objects in an animal's surroundings.

cohesion The linking together of like molecules, often by hydrogen bonds.

cohesion-tension hypothesis The leading explanation of the ascent of xylem sap. It states that transpiration exerts pull on xylem sap, putting the sap under negative pressure or tension, and that the cohesion of water molecules transmits this pull along the entire length of the xylem from shoots to roots.

cohort A group of individuals of the same age in a population.

coitus (kō′-uh-tus) The insertion of a penis into a vagina; also called sexual intercourse.

coleoptile (kō′-lē-op′-tul) The covering of the young shoot of the embryo of a grass seed.

coleorhiza (kō′-lē-uh-rī′-zuh) The covering of the young root of the embryo of a grass seed.

collagen A glycoprotein in the extracellular matrix of animal cells that forms strong fibers, found extensively in connective tissue and bone; the most abundant protein in the animal kingdom.

collecting duct The location in the kidney where processed filtrate, called urine, is collected from the renal tubules.

collenchyma cell (kō-len′-kim-uh) A flexible plant cell type that occurs in strands or cylinders that support young parts of the plant without restraining growth.

colloid A mixture made up of a liquid and particles that (because of their large size) remain suspended rather than dissolved in that liquid.

colon (kō′-len) The largest section of the vertebrate large intestine; functions in water absorption and formation of feces.

commensalism (kuh-men′-suh-lizm) A symbiotic relationship in which one organism benefits but the other is neither helped nor harmed.

communication In animal behavior, a process involving transmission of, reception of, and response to signals. The term is also used in connection with other organisms, as well as individual cells of multicellular organisms.

community All the organisms that inhabit a particular area; an assemblage of populations of different species living close enough together for potential interaction.

community ecology The study of how interactions between species affect community structure and organization.

companion cell A type of plant cell that is connected to a sieve-tube element by many plasmodesmata and whose nucleus and ribosomes may serve one or more adjacent sieve-tube elements.

competitive exclusion The concept that when populations of two similar species compete for the same limited resources, one population will use the resources more efficiently and have a reproductive advantage that will eventually lead to the elimination of the other population.

competitive inhibitor A substance that reduces the activity of an enzyme by entering the active site in place of the substrate, whose structure it mimics.

complement system A group of about 30 blood proteins that may amplify the inflammatory response, enhance phagocytosis, or directly lyse extracellular pathogens.

complementary DNA (cDNA) A double-stranded DNA molecule made *in vitro* using mRNA as a template and the enzymes reverse transcriptase and DNA polymerase. A cDNA molecule corresponds to the exons of a gene.

complete digestive tract A digestive tube that runs between a mouth and an anus; also called an alimentary canal.

complete dominance The situation in which the phenotypes of the heterozygote and dominant homozygote are indistinguishable.

complete flower A flower that has all four basic floral organs: sepals, petals, stamens, and carpels.

complete metamorphosis The transformation of a larva into an adult that looks very different, and often functions very differently in its environment, than the larva.

compound A substance consisting of two or more different elements combined in a fixed ratio.

compound eye A type of multifaceted eye in insects and crustaceans consisting of up to several thousand light-detecting, focusing ommatidia.

concentration gradient A region along which the density of a chemical substance increases or decreases.

conception The fertilization of an egg by a sperm in humans.

condom A thin, latex rubber or natural membrane sheath that fits over the penis to collect semen.

conduction The direct transfer of thermal motion (heat) between molecules of objects in direct contact with each other.

cone A cone-shaped cell in the retina of the vertebrate eye, sensitive to color.

conformer An animal for which an internal condition conforms to (changes in accordance with) changes in an environmental variable.

conidium (plural, **conidia**) A haploid spore produced at the tip of a specialized hypha in ascomycetes during asexual reproduction.

conifer Member of the largest gymnosperm phylum. Most conifers are cone-bearing trees, such as pines and firs.

conjugation (kon′-jū-gā′-shun) (1) In prokaryotes, the direct transfer of DNA between two cells that are temporarily joined. When the two cells are members of different species, conjugation results in horizontal gene transfer. (2) In ciliates, a sexual process in which two cells exchange haploid micronuclei but do not reproduce.

connective tissue Animal tissue that functions mainly to bind and support other tissues, having a sparse population of cells scattered through an extracellular matrix.

conodont An early, soft-bodied vertebrate with prominent eyes and dental elements.

conservation biology The integrated study of ecology, evolutionary biology, physiology, molecular biology, and genetics to sustain biological diversity at all levels.

contraception The deliberate prevention of pregnancy.

contractile vacuole A membranous sac that helps move excess water out of certain freshwater protists.

control element A segment of noncoding DNA that helps regulate transcription of a gene by serving as a binding site for a transcription factor. Multiple control elements are present in a eukaryotic gene's enhancer.

controlled experiment An experiment in which an experimental group is compared with a control group that varies only in the factor being tested.

convection The mass movement of warmed air or liquid to or from the surface of a body or object.

convergent evolution The evolution of similar features in independent evolutionary lineages.

convergent extension A process in which the cells of a tissue layer rearrange themselves in such a way that the sheet of cells becomes narrower (converges) and longer (extends).

cooperativity A kind of allosteric regulation whereby a shape change in one subunit of a protein caused by substrate binding is transmitted to all the other subunits, facilitating binding of additional substrate molecules to those subunits.

copepod (cō′-puh-pod) Any of a group of small crustaceans that are important members of marine and freshwater plankton communities.

coral reef Typically a warm-water, tropical ecosystem dominated by the hard skeletal structures secreted primarily by corals. Some coral reefs also exist in cold, deep waters.

corepressor A small molecule that binds to a bacterial repressor protein and changes the protein's shape, allowing it to bind to the operator and switch an operon off.

cork cambium (kam′-bē-um) A cylinder of meristematic tissue in woody plants that replaces the epidermis with thicker, tougher cork cells.

corpus callosum (kor′-pus kuh-lō′-sum) The thick band of nerve fibers that connects the right and left cerebral hemispheres in mammals, enabling the hemispheres to process information together.

corpus luteum (kor′-pus lū′-tē-um) A secreting tissue in the ovary that forms from the collapsed follicle after ovulation and produces progesterone.

cortex (1) The outer region of cytoplasm in a eukaryotic cell, lying just under the plasma membrane, that has a more gel-like consistency than the inner regions due to the presence of multiple microfilaments. (2) In plants, ground tissue that is between the vascular tissue and dermal tissue in a root or eudicot stem.

cortical nephron In mammals and birds, a nephron with a loop of Henle located almost entirely in the renal cortex.

corticosteroid Any steroid hormone produced and secreted by the adrenal cortex.

cotransport The coupling of the "downhill" diffusion of one substance to the "uphill" transport of another against its own concentration gradient.

cotyledon (kot′-uh-lē′-dun) A seed leaf of an angiosperm embryo. Some species have one cotyledon, others two.

countercurrent exchange The exchange of a substance or heat between two fluids flowing in opposite directions. For example, blood in a fish gill flows in the opposite direction of water passing over the gill, maximizing diffusion of oxygen into and carbon dioxide out of the blood.

countercurrent multiplier system A countercurrent system in which energy is expended in active transport to facilitate exchange of materials and generate concentration gradients.

covalent bond (kō-vā′-lent) A type of strong chemical bond in which two atoms share one or more pairs of valence electrons.

craniate A chordate with a head.

crassulacean acid metabolism (CAM) An adaptation for photosynthesis in arid conditions, first discovered in the family Crassulaceae. In this process, a plant takes up CO_2 and incorporates it into a variety of organic acids at night; during the day, CO_2 is released from organic acids for use in the Calvin cycle.

crista (plural, **cristae**) (kris′-tuh, kris′-tē) An infolding of the inner membrane of a mitochondrion. The inner membrane houses electron transport chains and molecules of the enzyme catalyzing the synthesis of ATP (ATP synthase).

critical load The amount of added nutrient, usually nitrogen or phosphorus, that can be absorbed by plants without damaging ecosystem integrity.

crop rotation The practice of planting nonlegumes one year and legumes in alternating years to restore concentrations of fixed nitrogen in the soil.

cross-fostering study A behavioral study in which the young of one species are placed in the care of adults from another species.

crossing over The reciprocal exchange of genetic material between nonsister chromatids during prophase I of meiosis.

cross-pollination In angiosperms, the transfer of pollen from an anther of a flower on one plant to the stigma of a flower on another plant of the same species.

crustacean (kruh-stā′-shun) A member of a subphylum of mostly aquatic arthropods that includes lobsters, crayfishes, crabs, shrimps, and barnacles.

cryptic coloration Camouflage that makes a potential prey difficult to spot against its background.

culture A system of information transfer through social learning or teaching that influences the behavior of individuals in a population.

cuticle (kyū′-tuh-kul) (1) A waxy covering on the surface of stems and leaves that prevents desiccation in terrestrial plants. (2) The exoskeleton of an arthropod, consisting of layers of protein and chitin that are variously modified for different functions. (3) A tough coat that covers the body of a nematode.

cyclic AMP (cAMP) Cyclic adenosine monophosphate, a ring-shaped molecule made from ATP that is a common intracellular signaling molecule (second messenger) in eukaryotic cells. It is also a regulator of some bacterial operons.

cyclic electron flow A route of electron flow during the light reactions of photosynthesis that involves only photosystem I and that produces ATP but not NADPH or O_2.

cyclin (sī′-klin) A cellular protein that occurs in a cyclically fluctuating concentration and that plays an important role in regulating the cell cycle.

cyclin-dependent kinase (Cdk) A protein kinase that is active only when attached to a particular cyclin.

cystic fibrosis (sis′-tik fī-brō′-sis) A human genetic disorder caused by a recessive allele for a chloride channel protein; characterized by an excessive secretion of mucus and consequent vulnerability to infection; fatal if untreated.

cytochrome (sī′-tō-krōm) An iron-containing protein that is a component of electron transport chains in the mitochondria and chloroplasts of eukaryotic cells and the plasma membranes of prokaryotic cells.

cytogenetic map A map of a chromosome that locates genes with respect to chromosomal features distinguishable in a microscope.

cytokine (sī′-tō-kīn′) Any of a group of small proteins secreted by a number of cell types, including macrophages and helper T cells, that regulate the function of other cells.

cytokinesis (sī′-tō-kuh-nē′-sis) The division of the cytoplasm to form two separate daughter cells immediately after mitosis, meiosis I, or meiosis II.

cytokinin (sī′-tō-kī′-nin) Any of a class of related plant hormones that retard aging and act in concert with auxin to stimulate cell division, influence the pathway of differentiation, and control apical dominance.

cytoplasm (sī′-tō-plaz′-um) The contents of the cell bounded by the plasma membrane; in eukaryotes, the portion exclusive of the nucleus.

cytoplasmic determinant A maternal substance, such as a protein or RNA, that when placed into an egg influences the course of early development by regulating the expression of genes that affect the developmental fate of cells.

cytoplasmic streaming A circular flow of cytoplasm, involving interactions of myosin and actin filaments, that speeds the distribution of materials within cells.

cytoskeleton A network of microtubules, microfilaments, and intermediate filaments that extend throughout the cytoplasm and serve a variety of mechanical, transport, and signaling functions.

cytosol (sī′-tō-sol) The semifluid portion of the cytoplasm.

cytotoxic T cell A type of lymphocyte that, when activated, kills infected cells as well as certain cancer cells and transplanted cells.

dalton A measure of mass for atoms and subatomic particles; the same as the atomic mass unit, or amu.

data Recorded observations.

day-neutral plant A plant in which flower formation is not controlled by photoperiod or day length.

decapod A member of the group of crustaceans that includes lobsters, crayfishes, crabs, and shrimps.

decomposer An organism that absorbs nutrients from nonliving organic material such as corpses, fallen plant material, and the wastes of living organisms and converts them to inorganic forms; a detritivore.

deductive reasoning A type of logic in which specific results are predicted from a general premise.

deep-sea hydrothermal vent A dark, hot, oxygen-deficient environment associated with volcanic activity on or near the seafloor. The producers in a vent community are chemoautotrophic prokaryotes.

de-etiolation The changes a plant shoot undergoes in response to sunlight; also known informally as greening.

dehydration reaction A chemical reaction in which two molecules become covalently bonded to each other with the removal of a water molecule.

deletion (1) A deficiency in a chromosome resulting from the loss of a fragment through breakage. (2) A mutational loss of one or more nucleotide pairs from a gene.

demographic transition In a stable population, a shift from high birth and death rates to low birth and death rates.

demography The study of changes over time in the vital statistics of populations, especially birth rates and death rates.

denaturation (dē-nā′-chur-ā′-shun) In proteins, a process in which a protein loses its native shape due to the disruption of weak chemical bonds and interactions, thereby becoming biologically inactive; in DNA, the separation of the two strands of the double helix. Denaturation occurs under extreme (noncellular) conditions of pH, salt concentration, or temperature.

dendrite (den′-drīt) One of usually numerous, short, highly branched extensions of a neuron that receive signals from other neurons.

dendritic cell An antigen-presenting cell, located mainly in lymphatic tissues and skin, that is particularly efficient in presenting antigens to helper T cells, thereby initiating a primary immune response.

density The number of individuals per unit area or volume.

density dependent Referring to any characteristic that varies with population density.

density independent Referring to any characteristic that is not affected by population density.

density-dependent inhibition The phenomenon observed in normal animal cells that causes them to stop dividing when they come into contact with one another.

deoxyribonucleic acid (DNA) (dē-ok′-sē-rī′-bō-nū-klā′-ik) A nucleic acid molecule, usually a double-stranded helix, in which each polynucleotide strand consists of nucleotide monomers with a deoxyribose sugar and the nitrogenous bases adenine (A), cytosine (C), guanine (G), and thymine (T); capable of being replicated and determining the inherited structure of a cell's proteins.

deoxyribose (dē-ok′-si-rī′-bōs) The sugar component of DNA nucleotides, having one fewer hydroxyl group than ribose, the sugar component of RNA nucleotides.

depolarization A change in a cell's membrane potential such that the inside of the membrane is made less negative relative to the outside. For example, a neuron membrane is depolarized if a stimulus decreases its voltage from the resting potential of −70 mV in the direction of zero voltage.

dermal tissue system The outer protective covering of plants.

desert A terrestrial biome characterized by very low precipitation.

desmosome A type of intercellular junction in animal cells that functions as a rivet, fastening cells together.

determinate cleavage A type of embryonic development in protostomes that rigidly casts the developmental fate of each embryonic cell very early.

determinate growth A type of growth characteristic of most animals and some plant organs, in which growth stops after a certain size is reached.

determination The progressive restriction of developmental potential in which the possible fate of each cell becomes more limited as an embryo develops. At the end of determination, a cell is committed to its fate.

detritivore (deh-trī′-tuh-vōr) A consumer that derives its energy and nutrients from nonliving organic material such as corpses, fallen plant material, and the wastes of living organisms; a decomposer.

detritus (di-trī′-tus) Dead organic matter.

deuteromycete (dū′-tuh-rō-mī′-sēt) Traditional classification for a fungus with no known sexual stage.

deuterostome development (dū′-tuh-rō-stōm′) In animals, a developmental mode distinguished by the development of the anus from the blastopore; often also characterized by radial cleavage and by the body cavity forming as outpockets of mesodermal tissue.

development The events involved in an organism's changing gradually from a simple to a more complex or specialized form.

diabetes mellitus (dī′-uh-bē′-tis mel′-uh-tus) An endocrine disorder marked by an inability to maintain glucose homeostasis. The type 1 form results from autoimmune destruction of insulin-secreting cells; treatment usually requires daily insulin injections. The type 2 form most commonly results from reduced responsiveness of target cells to insulin; obesity and lack of exercise are risk factors.

diacylglycerol (DAG) (dī-a′-sil-glis′-er-ol) A second messenger produced by the cleavage of the phospholipid PIP_2 in the plasma membrane.

diaphragm (dī′-uh-fram′) (1) A sheet of muscle that forms the bottom wall of the thoracic cavity in mammals. Contraction of the diaphragm pulls air into the lungs. (2) A dome-shaped rubber cup fitted into the upper portion of the vagina before sexual intercourse. It serves as a physical barrier to the passage of sperm into the uterus.

diapsid (dī-ap′-sid) Member of an amniote clade distinguished by a pair of holes on each side of the skull. Diapsids include the lepidosaurs and archosaurs.

diastole (dī-as′-tō-lē) The stage of the cardiac cycle in which a heart chamber is relaxed and fills with blood.

diastolic pressure Blood pressure in the arteries when the ventricles are relaxed.

dicot A term traditionally used to refer to flowering plants that have two embryonic seed leaves, or cotyledons. Recent molecular evidence indicates that dicots do not form a clade; species once classified as dicots are now grouped into eudicots, magnoliids, and several lineages of basal angiosperms.

differential gene expression The expression of different sets of genes by cells with the same genome.

differentiation The process by which a cell or group of cells become specialized in structure and function.

diffusion The spontaneous movement of a substance down its concentration or electrochemical gradient, from a region where it is more concentrated to a region where it is less concentrated.

digestion The second stage of food processing in animals: the breaking down of food into molecules small enough for the body to absorb.

dihybrid (dī′-hī′-brid) An organism that is heterozygous with respect to two genes of interest. All the offspring from a cross between parents doubly homozygous for different alleles are dihybrids. For example, parents of genotypes *AABB* and *aabb* produce a dihybrid of genotype *AaBb*.

dihybrid cross A cross between two organisms that are each heterozygous for both of the characters being followed (or the self-pollination of a plant that is heterozygous for both characters).

dikaryotic (dī′-kār-ē-ot′-ik) Referring to a fungal mycelium with two haploid nuclei per cell, one from each parent.

dinoflagellate (dī′-nō-flaj′-uh-let) Member of a group of mostly unicellular photosynthetic algae with two flagella situated in perpendicular grooves in cellulose plates covering the cell.

dinosaur Member of an extremely diverse clade of reptiles varying in body shape, size, and habitat. Birds are the only extant dinosaurs.

dioecious (dī-ē′-shus) In plant biology, having the male and female reproductive parts on different individuals of the same species.

diploblastic Having two germ layers.

diploid cell (dip′-loyd) A cell containing two sets of chromosomes ($2n$), one set inherited from each parent.

diplomonad A protist that has modified mitochondria, two equal-sized nuclei, and multiple flagella.

directional selection Natural selection in which individuals at one end of the phenotypic range survive or reproduce more successfully than do other individuals.

disaccharide (dī-sak′-uh-rīd) A double sugar, consisting of two monosaccharides joined by a glycosidic linkage formed by a dehydration reaction.

dispersal The movement of individuals or gametes away from their parent location. This movement sometimes expands the geographic range of a population or species.

dispersion The pattern of spacing among individuals within the boundaries of a population.

disruptive selection Natural selection in which individuals on both extremes of a phenotypic range survive or reproduce more successfully than do individuals with intermediate phenotypes.

distal tubule In the vertebrate kidney, the portion of a nephron that helps refine filtrate and empties it into a collecting duct.

disturbance A natural or human-caused event that changes a biological community and usually removes organisms from it. Disturbances, such as fires and storms, play a pivotal role in structuring many communities.

disulfide bridge A strong covalent bond formed when the sulfur of one cysteine monomer bonds to the sulfur of another cysteine monomer.

DNA (deoxyribonucleic acid) (dē-ok′-sē-rī′-bō-nū-klā′-ik) A double-stranded, helical nucleic acid molecule, consisting of nucleotide monomers with a deoxyribose sugar and the nitrogenous bases adenine (A), cytosine (C), guanine (G), and thymine (T); capable of being replicated and determining the inherited structure of a cell's proteins.

DNA ligase (lī′-gās) A linking enzyme essential for DNA replication; catalyzes the covalent bonding of the 3′ end of one DNA fragment (such as an Okazaki fragment) to the 5′ end of another DNA fragment (such as a growing DNA chain).

DNA methylation The presence of methyl groups on the DNA bases (usually cytosine) of plants, animals, and fungi. (The term also refers to the process of adding methyl groups to DNA bases.)

DNA microarray assay A method to detect and measure the expression of thousands of genes at one time. Tiny amounts of a large number of single-stranded DNA fragments representing different genes are fixed to a glass slide and tested for hybridization with samples of labeled cDNA.

DNA polymerase (puh-lim′-er-ās) An enzyme that catalyzes the elongation of new DNA (for example, at a replication fork) by the addition of nucleotides to the 3′ end of an existing chain. There are several different DNA polymerases; DNA polymerase III and DNA polymerase I play major roles in DNA replication in *E. coli.*

DNA replication The process by which a DNA molecule is copied; also called DNA synthesis.

domain (1) A taxonomic category above the kingdom level. The three domains are Archaea, Bacteria, and Eukarya. (2) A discrete structural and functional region of a protein.

dominant allele An allele that is fully expressed in the phenotype of a heterozygote.

dominant species A species with substantially higher abundance or biomass than other species in a community. Dominant species exert a powerful control over the occurrence and distribution of other species.

dopamine A neurotransmitter that is a catecholamine, like epinephrine and norepinephrine.

dormancy A condition typified by extremely low metabolic rate and a suspension of growth and development.

dorsal Pertaining to the top of an animal with radial or bilateral symmetry.

dorsal lip The region above the blastopore on the dorsal side of the amphibian embryo.

double bond A double covalent bond; the sharing of two pairs of valence electrons by two atoms.

double circulation A circulatory system consisting of separate pulmonary and systemic circuits, in which blood passes through the heart after completing each circuit.

double fertilization A mechanism of fertilization in angiosperms in which two sperm cells unite with two cells in the female gametophyte (embryo sac) to form the zygote and endosperm.

double helix The form of native DNA, referring to its two adjacent antiparallel polynucleotide strands wound around an imaginary axis into a spiral shape.

Down syndrome A human genetic disease usually caused by the presence of an extra chromosome 21; characterized by developmental delays and heart and other defects that are generally treatable or non-life-threatening.

Duchenne muscular dystrophy (duh-shen′) A human genetic disease caused by a sex-linked recessive allele; characterized by progressive weakening and a loss of muscle tissue.

duodenum (dū′-uh-dēn′-um) The first section of the small intestine, where chyme from the stomach mixes with digestive juices from the pancreas, liver, and gallbladder as well as from gland cells of the intestinal wall.

duplication An aberration in chromosome structure due to fusion with a fragment from a homologous chromosome, such that a portion of a chromosome is duplicated.

dynamic stability hypothesis The concept that long food chains are less stable than short chains.

dynein (dī′-nē-un) In cilia and flagella, a large motor protein extending from one microtubule doublet to the adjacent doublet. ATP hydrolysis drives changes in dynein shape that lead to bending of cilia and flagella.

E site One of a ribosome's three binding sites for tRNA during translation. The E site is the place where discharged tRNAs leave the ribosome. (E stands for exit.)

ecdysozoan Member of a group of animal phyla identified as a clade by molecular evidence. Many ecdysozoans are molting animals.

ecdysteroid A steroid hormone, secreted by the prothoracic glands, that triggers molting in arthropods.

echinoderm (i-kī′-nō-derm) A slow-moving or sessile marine deuterostome with a water vascular system and, in larvae, bilateral symmetry. Echinoderms include sea stars, brittle stars, sea urchins, feather stars, and sea cucumbers.

ecological footprint The aggregate land and water area required by a person, city, or nation to produce all of the resources it consumes and to absorb all of the wastes it generates.

ecological niche (nich) The sum of a species' use of the biotic and abiotic resources in its environment.

ecological species concept A definition of species in terms of ecological niche, the sum of how members of the species interact with the nonliving and living parts of their environment.

ecological succession Transition in the species composition of a community following a disturbance; establishment of a community in an area virtually barren of life.

ecology The study of how organisms interact with each other and their environment.

ecosystem All the organisms in a given area as well as the abiotic factors with which they interact; one or more communities and the physical environment around them.

ecosystem ecology The study of energy flow and the cycling of chemicals among the various biotic and abiotic components in an ecosystem.

ecosystem engineer An organism that influences community structure by causing physical changes in the environment.

ecosystem service A function performed by an ecosystem that directly or indirectly benefits humans.

ecotone The transition from one type of habitat or ecosystem to another, such as the transition from a forest to a grassland.

ectoderm (ek′-tō-durm) The outermost of the three primary germ layers in animal embryos; gives rise to the outer covering and, in some phyla, the nervous system, inner ear, and lens of the eye.

ectomycorrhiza (ek′-tō-mī′-kō-rī′-zuh) Association of a fungus with a plant root system in which the fungus surrounds the roots but does not cause invagination of the host (plant) cells' plasma membranes.

ectomycorrhizal fungus A symbiotic fungus that forms sheaths of hyphae over the surface of plant roots and also grows into extracellular spaces of the root cortex.

ectoparasite A parasite that feeds on the external surface of a host.

ectopic Occurring in an abnormal location.

ectoproct A sessile, colonial lophophorate; also called a bryozoan.

ectothermic Referring to organisms for which external sources provide most of the heat for temperature regulation.

Ediacaran biota (ē′-dē-uh-keh′-run bī-ō′-tuh) An early group of soft-bodied, multicellular eukaryotes known from fossils that range in age from 565 million to 550 million years old.

effective population size An estimate of the size of a population based on the numbers of females and males that successfully breed; generally smaller than the total population.

effector cell (1) A muscle cell or gland cell that performs the body's response to stimuli as directed by signals from the brain or other processing center of the nervous system. (2) A lymphocyte that has undergone clonal selection and is capable of mediating an adaptive immune response.

egg The female gamete.

egg-polarity gene A gene that helps control the orientation (polarity) of the egg; also called a maternal effect gene.

ejaculation The propulsion of sperm from the epididymis through the muscular vas deferens, ejaculatory duct, and urethra.

ejaculatory duct In mammals, the short section of the ejaculatory route formed by the convergence of the vas deferens and a duct from the seminal vesicle. The ejaculatory duct transports sperm from the vas deferens to the urethra.

electrocardiogram (ECG or EKG) A record of the electrical impulses that travel through heart muscle during the cardiac cycle.

electrochemical gradient The diffusion gradient of an ion, which is affected by both the concentration difference of an ion across a membrane (a chemical force) and the ion's tendency to move relative to the membrane potential (an electrical force).

electrogenic pump An active transport protein that generates voltage across a membrane while pumping ions.

electromagnetic receptor A receptor of electromagnetic energy, such as visible light, electricity, or magnetism.

electromagnetic spectrum The entire spectrum of electromagnetic radiation, ranging in wavelength from less than a nanometer to more than a kilometer.

electron A subatomic particle with a single negative electrical charge and a mass about 1/2,000 that of a neutron or proton. One or more electrons move around the nucleus of an atom.

electron microscope (EM) A microscope that uses magnets to focus an electron beam on or through a specimen, resulting in a practical resolution of a hundredfold greater than that of a light microscope using standard techniques. A transmission electron microscope (TEM) is used to study the internal structure of thin sections of cells. A scanning electron microscope (SEM) is used to study the fine details of cell surfaces.

electron shell An energy level of electrons at a characteristic average distance from the nucleus of an atom.

electron transport chain A sequence of electron carrier molecules (membrane proteins) that shuttle electrons down a series of redox reactions that release energy used to make ATP.

electronegativity The attraction of a given atom for the electrons of a covalent bond.

electroporation A technique to introduce recombinant DNA into cells by applying a brief electrical pulse to a solution containing the cells. The pulse creates temporary holes in the cells' plasma membranes, through which DNA can enter.

element Any substance that cannot be broken down to any other substance by chemical reactions.

elimination The fourth and final stage of food processing in animals: the passing of undigested material out of the body.

embryo sac (em′-brē-ō) The female gametophyte of angiosperms, formed from the growth and division of the megaspore into a multicellular structure that typically has eight haploid nuclei.

embryonic lethal A mutation with a phenotype leading to death of an embryo or larva.

embryophyte Alternate name for land plants that refers to their shared derived trait of multicellular, dependent embryos.

emergent properties New properties that arise with each step upward in the hierarchy of life, owing to the arrangement and interactions of parts as complexity increases.

emigration The movement of individuals out of a population.

enantiomer (en-an′-tē-ō-mer) One of two compounds that are mirror images of each other and that differ in shape due to the presence of an asymmetric carbon.

endangered species A species that is in danger of extinction throughout all or a significant portion of its range.

endemic (en-dem′-ik) Referring to a species that is confined to a specific geographic area.

endergonic reaction (en′-der-gon′-ik) A nonspontaneous chemical reaction, in which free energy is absorbed from the surroundings.

endocrine gland (en′-dō-krin) A ductless gland that secretes hormones directly into the interstitial fluid, from which they diffuse into the bloodstream.

endocrine system The internal system of communication involving hormones, the ductless glands that secrete hormones, and the molecular receptors on or in target cells that respond to hormones; functions in concert with the nervous system to effect internal regulation and maintain homeostasis.

endocytosis (en′-dō-sī-tō′-sis) Cellular uptake of biological molecules and particulate matter via formation of vesicles from the plasma membrane.

endoderm (en′-dō-durm) The innermost of the three primary germ layers in animal embryos; lines the archenteron and gives rise to the liver, pancreas, lungs, and the lining of the digestive tract in species that have these structures.

endodermis In plant roots, the innermost layer of the cortex that surrounds the vascular cylinder.

endomembrane system The collection of membranes inside and surrounding a eukaryotic cell, related either through direct physical contact or by the transfer of membranous vesicles; includes the plasma membrane, the nuclear envelope, the smooth and rough endoplasmic reticulum, the Golgi apparatus, lysosomes, vesicles, and vacuoles.

endometriosis (en′-dō-mē-trē-ō′-sis) The condition resulting from the presence of endometrial tissue outside of the uterus.

endometrium (en′-dō-mē′-trē-um) The inner lining of the uterus, which is richly supplied with blood vessels.

endoparasite A parasite that lives within a host.

endophyte A fungus that lives inside a leaf or other plant part without causing harm to the plant.

endoplasmic reticulum (ER) (en′-dō-plaz′-mik ruh-tik′-yū-lum) An extensive membranous network in eukaryotic cells, continuous with the outer nuclear membrane and composed of ribosome-studded (rough) and ribosome-free (smooth) regions.

endorphin (en-dōr′-fin) Any of several hormones produced in the brain and anterior pituitary that inhibit pain perception.

endoskeleton A hard skeleton buried within the soft tissues of an animal.

endosperm In angiosperms, a nutrient-rich tissue formed by the union of a sperm with two polar nuclei during double fertilization. The endosperm provides nourishment to the developing embryo in angiosperm seeds.

endospore A thick-coated, resistant cell produced by some bacterial cells when they are exposed to harsh conditions.

endosymbiont theory The theory that mitochondria and plastids, including chloroplasts, originated as prokaryotic cells engulfed by an ancestral eukaryotic cell. The engulfed cell and its host cell then evolved into a single organism.

endosymbiosis A process in which a unicellular organism (the "host") engulfs another cell, which lives within the host cell and ultimately becomes an organelle in the host cell. *See also* endosymbiont theory.

endothelium (en′-dō-thē′-lē-um) The simple squamous layer of cells lining the lumen of blood vessels.

endothermic Referring to organisms that are warmed by heat generated by their own metabolism. This heat usually maintains a relatively stable body temperature higher than that of the external environment.

endotoxin A toxic component of the outer membrane of certain gram-negative bacteria that is released only when the bacteria die.

energetic hypothesis The concept that the length of a food chain is limited by the inefficiency of energy transfer along the chain.

energy The capacity to cause change, especially to do work (to move matter against an opposing force).

energy coupling In cellular metabolism, the use of energy released from an exergonic reaction to drive an endergonic reaction.

enhancer A segment of eukaryotic DNA containing multiple control elements, usually located far from the gene whose transcription it regulates.

enteric division One of three divisions of the autonomic nervous system; consists of networks of neurons in the digestive tract, pancreas, and gallbladder; normally regulated by the sympathetic and parasympathetic divisions of the autonomic nervous system.

entropy A measure of disorder, or randomness.

enzymatic hydrolysis The process in digestion that splits macromolecules from food by the enzymatic addition of water.

enzyme (en′-zīm) A macromolecule serving as a catalyst, a chemical agent that increases the rate of a reaction without being consumed by the reaction. Most enzymes are proteins.

enzyme-substrate complex A temporary complex formed when an enzyme binds to its substrate molecule(s).

epicotyl (ep′-uh-kot′-ul) In an angiosperm embryo, the embryonic axis above the point of attachment of the cotyledon(s) and below the first pair of miniature leaves.

epidemic A general outbreak of a disease.

epidermis (1) The dermal tissue system of nonwoody plants, usually consisting of a single layer of tightly packed cells. (2) The outermost layer of cells in an animal.

epididymis (ep′-uh-did′-uh-mus) A coiled tubule located adjacent to the mammalian testis where sperm are stored.

epigenetic inheritance Inheritance of traits transmitted by mechanisms not directly involving the nucleotide sequence of a genome.

epinephrine (ep′-i-nef′-rin) A catecholamine that, when secreted as a hormone by the adrenal medulla, mediates "fight-or-flight" responses to short-term stresses; also released by some neurons as a neurotransmitter; also known as adrenaline.

epiphyte (ep′-uh-fīt) A plant that nourishes itself but grows on the surface of another plant for support, usually on the branches or trunks of trees.

epistasis (ep′-i-stā′-sis) A type of gene interaction in which the phenotypic expression of one gene alters that of another independently inherited gene.

epithelial tissue (ep′-uh-thē′-lē-ul) Sheets of tightly packed cells that line organs and body cavities as well as external surfaces.

epithelium An epithelial tissue.

epitope A small, accessible region of an antigen to which an antigen receptor or antibody binds; also called an antigenic determinant.

EPSP *See* excitatory postsynaptic potential.

equilibrium potential (E_{ion}) The magnitude of a cell's membrane voltage at equilibrium; calculated using the Nernst equation.

erythrocyte (eh-rith'-ruh-sīt) A blood cell that contains hemoglobin, which transports oxygen; also called a red blood cell.

erythropoietin (EPO) (eh-rith'-rō-poy'-uh-tin) A hormone that stimulates the production of erythrocytes. It is secreted by the kidney when body tissues do not receive enough oxygen.

esophagus (eh-sof'-uh-gus) A muscular tube that conducts food, by peristalsis, from the pharynx to the stomach.

essential amino acid An amino acid that an animal cannot synthesize itself and must be obtained from food in prefabricated form.

essential element A chemical element required for an organism to survive, grow, and reproduce.

essential fatty acid An unsaturated fatty acid that an animal needs but cannot make.

essential nutrient A substance that an organism cannot synthesize from any other material and therefore must absorb in preassembled form.

estradiol (es'-truh-dī'-ol) A steroid hormone that stimulates the development and maintenance of the female reproductive system and secondary sex characteristics; the major estrogen in mammals.

estrogen (es'-trō-jen) Any steroid hormone, such as estradiol, that stimulates the development and maintenance of the female reproductive system and secondary sex characteristics.

estrous cycle (es'-trus) A reproductive cycle characteristic of female mammals except humans and certain other primates, in which the nonpregnant endometrium is reabsorbed rather than shed, and sexual response occurs only during mid-cycle at estrus.

estuary The area where a freshwater stream or river merges with the ocean.

ethylene (eth'-uh-lēn) A gaseous plant hormone involved in responses to mechanical stress, programmed cell death, leaf abscission, and fruit ripening.

etiolation Plant morphological adaptations for growing in darkness.

euchromatin (yū-krō'-muh-tin) The less condensed form of eukaryotic chromatin that is available for transcription.

eudicot (yū-dī'-kot) Member of a clade that contains the vast majority of flowering plants that have two embryonic seed leaves, or cotyledons.

euglenid (yū'-glen-id) A protist, such as *Euglena* or its relatives, characterized by an anterior pocket from which one or two flagella emerge.

euglenozoan Member of a diverse clade of flagellated protists that includes predatory heterotrophs, photosynthetic autotrophs, and pathogenic parasites.

Eukarya (yū-kar'-ē-uh) The domain that includes all eukaryotic organisms.

eukaryotic cell (yū'-ker-ē-ot'-ik) A type of cell with a membrane-enclosed nucleus and membrane-enclosed organelles. Organisms with eukaryotic cells (protists, plants, fungi, and animals) are called eukaryotes.

eumetazoan (yū'-met-uh-zō'-un) Member of a clade of animals with true tissues. All animals except sponges and a few other groups are eumetazoans.

eurypterid (yur-ip'-tuh-rid) An extinct carnivorous chelicerate; also called a water scorpion.

Eustachian tube (yū-stā'-shun) The tube that connects the middle ear to the pharynx.

eutherian (yū-thēr'-ē-un) Placental mammal; mammal whose young complete their embryonic development within the uterus, joined to the mother by the placenta.

eutrophic lake (yū-trōf'-ik) A lake that has a high rate of biological productivity supported by a high rate of nutrient cycling.

eutrophication A process by which nutrients, particularly phosphorus and nitrogen, become highly concentrated in a body of water, leading to increased growth of organisms such as algae or cyanobacteria.

evaporation The process by which a liquid changes to a gas.

evaporative cooling The process in which the surface of an object becomes cooler during evaporation, a result of the molecules with the greatest kinetic energy changing from the liquid to the gaseous state.

evapotranspiration The total evaporation of water from an ecosystem, including water transpired by plants and evaporated from a landscape, usually measured in millimeters and estimated for a year.

evo-devo Evolutionary developmental biology; a field of biology that compares developmental processes of different multicellular organisms to understand how these processes have evolved and how changes can modify existing organismal features or lead to new ones.

evolution Descent with modification; the idea that living species are descendants of ancestral species that were different from the present-day ones; also defined more narrowly as the change in the genetic composition of a population from generation to generation.

evolutionary tree A branching diagram that reflects a hypothesis about evolutionary relationships among groups of organisms.

Excavata One of five supergroups of eukaryotes proposed in a current hypothesis of the evolutionary history of eukaryotes. Excavates have unique cytoskeletal features, and some species have an "excavated" feeding groove on one side of the cell body. *See also* Chromalveolata, Rhizaria, Archaeplastida, and Unikonta.

excitatory postsynaptic potential (EPSP) An electrical change (depolarization) in the membrane of a postsynaptic cell caused by the binding of an excitatory neurotransmitter from a presynaptic cell to a postsynaptic receptor; makes it more likely for a postsynaptic cell to generate an action potential.

excretion The disposal of nitrogen-containing metabolites and other waste products.

exergonic reaction (ek'-ser-gon'-ik) A spontaneous chemical reaction, in which there is a net release of free energy.

exocytosis (ek'-sō-sī-tō'-sis) The cellular secretion of biological molecules by the fusion of vesicles containing them with the plasma membrane.

exon A sequence within a primary transcript that remains in the RNA after RNA processing; also refers to the region of DNA from which this sequence was transcribed.

exoskeleton A hard encasement on the surface of an animal, such as the shell of a mollusc or the cuticle of an arthropod, that provides protection and points of attachment for muscles.

exotoxin (ek'-sō-tok'-sin) A toxic protein that is secreted by a prokaryote or other pathogen and that produces specific symptoms, even if the pathogen is no longer present.

expansin Plant enzyme that breaks the cross-links (hydrogen bonds) between cellulose microfibrils and other cell wall constituents, loosening the wall's fabric.

exponential population growth Growth of a population in an ideal, unlimited environment, represented by a J-shaped curve when population size is plotted over time.

expression vector A cloning vector that contains a highly active bacterial promoter just upstream of a restriction site where a eukaryotic gene can be inserted, allowing the gene to be expressed in a bacterial cell. Expression vectors are also available that have been genetically engineered for use in specific types of eukaryotic cells.

external fertilization The fusion of gametes that parents have discharged into the environment.

extinction vortex A downward population spiral in which inbreeding and genetic drift combine to cause a small population to shrink and, unless the spiral is reversed, become extinct.

extracellular digestion The breakdown of food in compartments that are continuous with the outside of an animal's body.

extracellular matrix (ECM) The meshwork surrounding animal cells, consisting of glycoproteins, polysaccharides, and proteoglycans synthesized and secreted by the cells.

extraembryonic membrane One of four membranes (yolk sac, amnion, chorion, and allantois) located outside the embryo that support the developing embryo in reptiles and mammals.

extreme halophile An organism that lives in a highly saline environment, such as the Great Salt Lake or the Dead Sea.

extreme thermophile An organism that thrives in hot environments (often 60–80°C or hotter).

extremophile An organism that lives in environmental conditions so extreme that few other species can survive there. Extremophiles include extreme halophiles ("salt lovers") and extreme thermophiles ("heat lovers").

F factor In bacteria, the DNA segment that confers the ability to form pili for conjugation and associated functions required for the transfer of DNA from donor to recipient. The F factor may exist as a plasmid or be integrated into the bacterial chromosome.

F plasmid The plasmid form of the F factor.

F_1 generation The first filial, hybrid (heterozygous) offspring arising from a parental (P generation) cross.

F_2 generation The offspring resulting from interbreeding (or self-pollination) of the hybrid F_1 generation.

facilitated diffusion The passage of molecules or ions down their electrochemical gradient across a biological membrane with the assistance of specific transmembrane transport proteins, requiring no energy expenditure.

facilitation An interaction in which one species has a positive effect on the survival and reproduction of another species without the intimate association of a symbiosis.

facultative anaerobe (fak′-ul-tā′-tiv an′-uh-rōb) An organism that makes ATP by aerobic respiration if oxygen is present but that switches to anaerobic respiration or fermentation if oxygen is not present.

family In Linnaean classification, the taxonomic category above genus.

fast block to polyspermy The depolarization of the egg plasma membrane that begins within 1–3 seconds after a sperm binds to an egg membrane protein. The depolarization lasts about 1 minute and prevents additional sperm from fusing with the egg during that time.

fast-twitch fiber A muscle fiber used for rapid, powerful contractions.

fat A lipid consisting of three fatty acids linked to one glycerol molecule; also called a triacylglycerol or triglyceride.

fate map A territorial diagram of embryonic development that displays the future derivatives of individual cells and tissues.

fatty acid A carboxylic acid with a long carbon chain. Fatty acids vary in length and in the number and location of double bonds; three fatty acids linked to a glycerol molecule form a fat molecule, also known as a triacylglycerol or triglyceride.

feces (fē′-sēz) The wastes of the digestive tract.

feedback inhibition A method of metabolic control in which the end product of a metabolic pathway acts as an inhibitor of an enzyme within that pathway.

fermentation A catabolic process that makes a limited amount of ATP from glucose (or other organic molecules) without an electron transport chain and that produces a characteristic end product, such as ethyl alcohol or lactic acid.

fertilization (1) The union of haploid gametes to produce a diploid zygote. (2) The addition of mineral nutrients to the soil.

fetus (fē′-tus) A developing mammal that has all the major structures of an adult. In humans, the fetal stage lasts from the 9th week of gestation until birth.

fiber A lignified cell type that reinforces the xylem of angiosperms and functions in mechanical support; a slender, tapered sclerenchyma cell that usually occurs in bundles.

fibroblast (fī′-brō-blast) A type of cell in loose connective tissue that secretes the protein ingredients of the extracellular fibers.

fibronectin An extracellular glycoprotein secreted by animal cells that helps them attach to the extracellular matrix.

filament In an angiosperm, the stalk portion of the stamen, the pollen-producing reproductive organ of a flower.

filtrate Cell-free fluid extracted from the body fluid by the excretory system.

filtration In excretory systems, the extraction of water and small solutes, including metabolic wastes, from the body fluid.

fimbria (plural, **fimbriae**) A short, hairlike appendage of a prokaryotic cell that helps it adhere to the substrate or to other cells.

first law of thermodynamics The principle of conservation of energy: Energy can be transferred and transformed, but it cannot be created or destroyed.

fission The separation of an organism into two or more individuals of approximately equal size.

fixed action pattern In animal behavior, a sequence of unlearned acts that is essentially unchangeable and, once initiated, usually carried to completion.

flaccid (flas′-id) Limp. Lacking turgor (stiffness or firmness), as in a plant cell in surroundings where there is a tendency for water to leave the cell. (A walled cell becomes flaccid if it has a higher water potential than its surroundings, resulting in the loss of water.)

flagellum (fluh-jel′-um) (plural, **flagella**) A long cellular appendage specialized for locomotion. Like motile cilia, eukaryotic flagella have a core with nine outer doublet microtubules and two inner single microtubules (the "9 + 2" arrangement) ensheathed in an extension of the plasma membrane. Prokaryotic flagella have a different structure.

florigen A flowering signal, probably a protein, that is made in leaves under certain conditions and that travels to the shoot apical meristems, inducing them to switch from vegetative to reproductive growth.

flower In an angiosperm, a specialized shoot with up to four sets of modified leaves, bearing structures that function in sexual reproduction.

fluid feeder An animal that lives by sucking nutrient-rich fluids from another living organism.

fluid mosaic model The currently accepted model of cell membrane structure, which envisions the membrane as a mosaic of protein molecules drifting laterally in a fluid bilayer of phospholipids.

follicle (fol′-uh-kul) A microscopic structure in the ovary that contains the developing oocyte and secretes estrogens.

follicle-stimulating hormone (FSH) A tropic hormone that is produced and secreted by the anterior pituitary and that stimulates the production of eggs by the ovaries and sperm by the testes.

follicular phase That part of the ovarian cycle during which follicles are growing and oocytes maturing.

food chain The pathway along which food energy is transferred from trophic level to trophic level, beginning with producers.

food vacuole A membranous sac formed by phagocytosis of microorganisms or particles to be used as food by the cell.

food web The interconnected feeding relationships in an ecosystem.

foot (1) The portion of a bryophyte sporophyte that gathers sugars, amino acids, water, and minerals from the parent gametophyte via transfer cells. (2) One of the three main parts of a mollusc; a muscular structure usually used for movement. *See also* mantle, visceral mass.

foraging The seeking and obtaining of food.

foram (foraminiferan) An aquatic protist that secretes a hardened shell containing calcium carbonate and extends pseudopodia through pores in the shell.

forebrain One of three ancestral and embryonic regions of the vertebrate brain; develops into the thalamus, hypothalamus, and cerebrum.

fossil A preserved remnant or impression of an organism that lived in the past.

founder effect Genetic drift that occurs when a few individuals become isolated from a larger population and form a new population whose gene pool composition is not reflective of that of the original population.

fovea (fō′-vē-uh) The place on the retina at the eye's center of focus, where cones are highly concentrated.

fragmentation A means of asexual reproduction whereby a single parent breaks into parts that regenerate into whole new individuals.

frameshift mutation A mutation occurring when nucleotides are inserted in or deleted from a gene and the number inserted or deleted is not a multiple of three, resulting in the improper grouping of the subsequent nucleotides into codons.

free energy The portion of a biological system's energy that can perform work when temperature and pressure are uniform throughout the system. The change in free energy of a system (ΔG) is calculated by the equation $\Delta G = \Delta H - T\Delta S$, where ΔH is the change in enthalpy (in biological systems, equivalent to total energy), T is the absolute temperature, and ΔS is the change in entropy.

frequency-dependent selection Selection in which the fitness of a phenotype depends on how common the phenotype is in a population.

fruit A mature ovary of a flower. The fruit protects dormant seeds and often aids in their dispersal.

functional group A specific configuration of atoms commonly attached to the carbon skeletons of organic molecules and involved in chemical reactions.

G protein A GTP-binding protein that relays signals from a plasma membrane signal receptor, known as a G protein-coupled receptor, to other signal transduction proteins inside the cell.

G protein-coupled receptor (GPCR) A signal receptor protein in the plasma membrane that responds to the binding of a signaling molecule by activating a G protein. Also called a G protein-linked receptor.

G_0 phase A nondividing state occupied by cells that have left the cell cycle, sometimes reversibly.

G_1 phase The first gap, or growth phase, of the cell cycle, consisting of the portion of interphase before DNA synthesis begins.

G_2 phase The second gap, or growth phase, of the cell cycle, consisting of the portion of interphase after DNA synthesis occurs.

gallbladder An organ that stores bile and releases it as needed into the small intestine.

game theory An approach to evaluating alternative strategies in situations where the outcome of a particular strategy depends on the strategies used by other individuals.

gametangium (gam′-uh-tan′-jē-um) (plural, **gametangia**) Multicellular plant structure in which gametes are formed. Female gametangia are called archegonia, and male gametangia are called antheridia.

gamete (gam′-ēt) A haploid reproductive cell, such as an egg or sperm. Gametes unite during sexual reproduction to produce a diploid zygote.

gametogenesis The process by which gametes are produced.

gametophore (guh-mē′-tō-fōr) The mature gamete-producing structure of a moss gametophyte.

gametophyte (guh-mē′-tō-fīt) In organisms (plants and some algae) that have alternation of generations, the multicellular haploid form that produces haploid gametes by mitosis. The haploid gametes unite and develop into sporophytes.

gamma-aminobutyric acid (GABA) An amino acid that functions as a CNS neurotransmitter in the central nervous system of vertebrates.

ganglia (gang′-glē-uh) (singular, **ganglion**) Clusters (functional groups) of nerve cell bodies in a centralized nervous system.

ganglion cell A type of neuron in the retina that synapses with bipolar cells and transmits action potentials to the brain via axons in the optic nerve.

gap junction A type of intercellular junction in animal cells, consisting of proteins surrounding a pore that allows the passage of materials between cells.

gas exchange The uptake of molecular oxygen from the environment and the discharge of carbon dioxide to the environment.

gastric juice A digestive fluid secreted by the stomach.

gastrovascular cavity A central cavity with a single opening in the body of certain animals, including cnidarians and flatworms, that functions in both the digestion and distribution of nutrients.

gastrula (gas′-trū-luh) An embryonic stage in animal development encompassing the formation of three layers: ectoderm, mesoderm, and endoderm.

gastrulation (gas′-trū-lā′-shun) In animal development, a series of cell and tissue movements in which the blastula-stage embryo folds inward, producing a three-layered embryo, the gastrula.

gated channel A transmembrane protein channel that opens or closes in response to a particular stimulus.

gated ion channel A gated channel for a specific ion. The opening or closing of such channels may alter a cell's membrane potential.

gel electrophoresis (ē-lek′-trō-fōr-ē′-sis) A technique for separating nucleic acids or proteins on the basis of their size and electrical charge, both of which affect their rate of movement through an electric field in a gel made of agarose or another polymer.

gene A discrete unit of hereditary information consisting of a specific nucleotide sequence in DNA (or RNA, in some viruses).

gene annotation Analysis of genomic sequences to identify protein-coding genes and determine the function of their products.

gene cloning The production of multiple copies of a gene.

gene expression The process by which information encoded in DNA directs the synthesis of proteins or, in some cases, RNAs that are not translated into proteins and instead function as RNAs.

gene flow The transfer of alleles from one population to another, resulting from the movement of fertile individuals or their gametes.

gene pool The aggregate of all copies of every type of allele at all loci in every individual in a population. The term is also used in a more restricted sense as the aggregate of alleles for just one or a few loci in a population.

gene therapy The introduction of genes into an afflicted individual for therapeutic purposes.

gene-for-gene recognition A widespread form of plant disease resistance involving recognition of pathogen-derived molecules by the protein products of specific plant disease resistance genes.

genetic drift A process in which chance events cause unpredictable fluctuations in allele frequencies from one generation to the next. Effects of genetic drift are most pronounced in small populations.

genetic engineering The direct manipulation of genes for practical purposes.

genetic map An ordered list of genetic loci (genes or other genetic markers) along a chromosome.

genetic profile An individual's unique set of genetic markers, detected most often today by PCR or, previously, by electrophoresis and nucleic acid probes.

genetic recombination General term for the production of offspring with combinations of traits that differ from those found in either parent.

genetic variation Differences among individuals in the composition of their genes or other DNA segments.

genetically modified (GM) organism An organism that has acquired one or more genes by artificial means; also known as a transgenic organism.

genetics The scientific study of heredity and hereditary variation.

genome (jē′-nōm) The genetic material of an organism or virus; the complete complement of an organism's or virus's genes along with its noncoding nucleic acid sequences.

genome-wide association study A large-scale analysis of the genomes of many people having a certain phenotype or disease, with the aim of finding genetic markers that correlate with that phenotype or disease.

genomic imprinting A phenomenon in which expression of an allele in offspring depends on whether the allele is inherited from the male or female parent.

genomic library A set of cell clones containing all the DNA segments from a genome, each within a plasmid, BAC, or other cloning vector.

genomics (juh-nō′-miks) The study of whole sets of genes and their interactions within a species, as well as genome comparisons between species.

genotype (jē′-nō-tīp) The genetic makeup, or set of alleles, of an organism.

genus (jē′-nus) (plural, **genera**) A taxonomic category above the species level, designated by the first word of a species' two-part scientific name.

geographic variation Differences between the gene pools of geographically separate populations or population subgroups.

geologic record The division of Earth's history into time periods, grouped into three eons—Archaean, Proterozoic, and Phanerozoic—and further subdivided into eras, periods, and epochs.

germ layer One of the three main layers in a gastrula that will form the various tissues and organs of an animal body.

gestation (jes-tā′-shun) Pregnancy; the state of carrying developing young within the female reproductive tract.

gibberellin (jib′-uh-rel′-in) Any of a class of related plant hormones that stimulate growth in the stem and leaves, trigger the germination of seeds and breaking of bud dormancy, and (with auxin) stimulate fruit development.

glans The rounded structure at the tip of the clitoris or penis that is involved in sexual arousal.

glia (glial cells) Cells of the nervous system that support, regulate, and augment the functions of neurons.

global climate change Increase in temperature and change in weather patterns all around the planet, due mostly to increasing atmospheric CO_2 levels from the burning of fossil fuels. The increase in temperature, called global warming, is a major aspect of global climate change.

global ecology The study of the functioning and distribution of organisms across the biosphere and how the regional exchange of energy and materials affects them.

glomeromycete (glō′-mer-ō-mī′-sēt) Member of the fungal phylum Glomeromycota, characterized by a distinct branching form of mycorrhizae called arbuscular mycorrhizae.

glomerulus (glō-mār′-yū-lus) A ball of capillaries surrounded by Bowman's capsule in the nephron and serving as the site of filtration in the vertebrate kidney.

glucagon (glū′-kuh-gon) A hormone secreted by pancreatic alpha cells that raises blood glucose levels. It promotes glycogen breakdown and release of glucose by the liver.

glucocorticoid A steroid hormone that is secreted by the adrenal cortex and that influences glucose metabolism and immune function.

glutamate An amino acid that functions as a neurotransmitter in the central nervous system.

glyceraldehyde 3-phosphate (G3P) (glis′-er-al′-de-hīd) A three-carbon carbohydrate that is the direct product of the Calvin cycle; it is also an intermediate in glycolysis.

glycogen (glī′-kō-jen) An extensively branched glucose storage polysaccharide found in the liver and muscle of animals; the animal equivalent of starch.

glycolipid A lipid with one or more covalently attached carbohydrates.

glycolysis (glī-kol′-uh-sis) A series of reactions that ultimately splits glucose into pyruvate. Glycolysis occurs in almost all living cells, serving as the starting point for fermentation or cellular respiration.

glycoprotein A protein with one or more covalently attached carbohydrates.

glycosidic linkage A covalent bond formed between two monosaccharides by a dehydration reaction.

gnathostome (na′-thu-stōm) Member of the vertebrate subgroup possessing jaws.

golden alga A biflagellated, photosynthetic protist named for its color, which results from its yellow and brown carotenoids.

Golgi apparatus (gol′-jē) An organelle in eukaryotic cells consisting of stacks of flat membranous sacs that modify, store, and route products of the endoplasmic reticulum and synthesize some products, notably noncellulose carbohydrates.

gonads (gō′-nadz) The male and female sex organs; the gamete-producing organs in most animals.

grade A group of organisms that share the same level of organizational complexity or share a key adaptation.

graded potential In a neuron, a shift in the membrane potential that has an amplitude proportional to signal strength and that decays as it spreads.

Gram stain A staining method that distinguishes between two different kinds of bacterial cell walls; may be used to help determine medical response to an infection.

gram-negative Describing the group of bacteria that have a cell wall that is structurally more complex and contains less peptidoglycan than the cell wall of gram-positive bacteria. Gram-negative bacteria are often more toxic than gram-positive bacteria.

gram-positive Describing the group of bacteria that have a cell wall that is structurally less complex and contains more peptidoglycan than the cell wall of gram-negative bacteria. Gram-positive bacteria are usually less toxic than gram-negative bacteria.

granum (gran′-um) (plural, **grana**) A stack of membrane-bounded thylakoids in the chloroplast. Grana function in the light reactions of photosynthesis.

gravitropism (grav′-uh-trō′-pizm) A response of a plant or animal to gravity.

gray matter Regions of dendrites and clustered neuron cell bodies within the CNS.

green alga A photosynthetic protist, named for green chloroplasts that are similar in structure and pigment composition to those of land plants. Green algae are a paraphyletic group, some of whose members are more closely related to land plants than they are to other green algae.

greenhouse effect The warming of Earth due to the atmospheric accumulation of carbon dioxide and certain other gases, which absorb reflected infrared radiation and reradiate some of it back toward Earth.

gross primary production (GPP) The total primary production of an ecosystem.

ground tissue system Plant tissues that are neither vascular nor dermal, fulfilling a variety of functions, such as storage, photosynthesis, and support.

growth An irreversible increase in size or biomass.

growth factor (1) A protein that must be present in the extracellular environment (culture medium or animal body) for the growth and normal development of certain types of cells. (2) A local regulator that acts on nearby cells to stimulate cell proliferation and differentiation.

growth hormone (GH) A hormone that is produced and secreted by the anterior pituitary and that has both direct (nontropic) and tropic effects on a wide variety of tissues.

guard cells The two cells that flank the stomatal pore and regulate the opening and closing of the pore.

gustation The sense of taste.

guttation The exudation of water droplets from leaves, caused by root pressure in certain plants.

gymnosperm (jim′-nō-sperm) A vascular plant that bears naked seeds—seeds not enclosed in protective chambers.

hair cell A mechanosensory cell that alters output to the nervous system when hairlike projections on the cell surface are displaced.

half-life The amount of time it takes for 50% of a sample of a radioactive isotope to decay.

Hamilton's rule The principle that for natural selection to favor an altruistic act, the benefit to the recipient, devalued by the coefficient of relatedness, must exceed the cost to the altruist.

haploid cell (hap′-loyd) A cell containing only one set of chromosomes (n).

Hardy-Weinberg principle The principle that frequencies of alleles and genotypes in a population remain constant from generation to generation, provided that only Mendelian segregation and recombination of alleles are at work.

haustorium (plural, **haustoria**) (ho-stōr′-ē-um, ho-stōr′-ē-uh) In certain symbiotic fungi, a specialized hypha that can penetrate the tissues of host organisms.

heart A muscular pump that uses metabolic energy to elevate the hydrostatic pressure of the circulatory fluid (blood or hemolymph). The fluid then flows down a pressure gradient through the body and eventually returns to the heart.

heart attack The damage or death of cardiac muscle tissue resulting from prolonged blockage of one or more coronary arteries.

heart murmur A hissing sound that most often results from blood squirting backward through a leaky valve in the heart.

heart rate The frequency of heart contraction (in beats per minute).

heat The total amount of kinetic energy due to the random motion of atoms or molecules in a body of matter; also called thermal energy. Heat is energy in its most random form.

heat of vaporization The quantity of heat a liquid must absorb for 1 g of it to be converted from the liquid to the gaseous state.

heat-shock protein A protein that helps protect other proteins during heat stress. Heat-shock proteins are found in plants, animals, and microorganisms.

heavy chain One of the two types of polypeptide chains that make up an antibody molecule and B cell receptor; consists of a variable region, which contributes to the antigen-binding site, and a constant region.

helicase An enzyme that untwists the double helix of DNA at replication forks, separating the two strands and making them available as template strands.

helper T cell A type of T cell that, when activated, secretes cytokines that promote the response of B cells (humoral response) and cytotoxic T cells (cell-mediated response) to antigens.

hemoglobin (hē′-mō-glō′-bin) An iron-containing protein in red blood cells that reversibly binds oxygen.

hemolymph (hē′-mō-limf′) In invertebrates with an open circulatory system, the body fluid that bathes tissues.

hemophilia (hē′-muh-fil′-ē-uh) A human genetic disease caused by a sex-linked recessive allele resulting in the absence of one or more blood-clotting proteins; characterized by excessive bleeding following injury.

hepatic portal vein A large vessel that conveys nutrient-laden blood from the small intestine to the liver, which regulates the blood's nutrient content.

herbivore (hur′-bi-vōr′) An animal that mainly eats plants or algae.

herbivory An interaction in which an organism eats parts of a plant or alga.

heredity The transmission of traits from one generation to the next.

hermaphrodite (hur-maf′-ruh-dīt′) An individual that functions as both male and female in sexual reproduction by producing both sperm and eggs.

hermaphroditism (hur-maf′-rō-dī-tizm) A condition in which an individual has both female and male gonads and functions as both a male and female in sexual reproduction by producing both sperm and eggs.

heterochromatin (het′-er-ō-krō′-muh-tin) Eukaryotic chromatin that remains highly compacted during interphase and is generally not transcribed.

heterochrony (het′-uh-rok′-ruh-nē) Evolutionary change in the timing or rate of an organism's development.

heterocyst (het′-er-ō-sist) A specialized cell that engages in nitrogen fixation in some filamentous cyanobacteria; also called a *heterocyte*.

heterokaryon (het′-er-ō-kār′-ē-un) A fungal mycelium that contains two or more haploid nuclei per cell.

heteromorphic (het′-er-ō-mōr′-fik) Referring to a condition in the life cycle of plants and certain algae in which the sporophyte and gametophyte generations differ in morphology.

heterosporous (het-er-os′-pōr-us) Referring to a plant species that has two kinds of spores: microspores, which develop into male gametophytes, and megaspores, which develop into female gametophytes.

heterotroph (het′-er-ō-trōf) An organism that obtains organic food molecules by eating other organisms or substances derived from them.

heterozygote advantage Greater reproductive success of heterozygous individuals compared with homozygotes; tends to preserve variation in a gene pool.

heterozygous (het′-er-ō-zī′-gus) Having two different alleles for a given gene.

hexapod An insect or closely related wingless, six-legged arthropod.

hibernation A long-term physiological state in which metabolism decreases, the heart and respiratory system slow down, and body temperature is maintained at a lower level than normal.

high-density lipoprotein (HDL) A particle in the blood made up of thousands of cholesterol molecules and other lipids bound to a protein. HDL scavenges excess cholesterol.

hindbrain One of three ancestral and embryonic regions of the vertebrate brain; develops into the medulla oblongata, pons, and cerebellum.

histamine (his′-tuh-mēn) A substance released by mast cells that causes blood vessels to dilate and become more permeable in inflammatory and allergic responses.

histone (his′-tōn) A small protein with a high proportion of positively charged amino acids that binds to the negatively charged DNA and plays a key role in chromatin structure.

histone acetylation The attachment of acetyl groups to certain amino acids of histone proteins.

HIV (human immunodeficiency virus) The infectious agent that causes AIDS. HIV is a retrovirus.

holdfast A rootlike structure that anchors a seaweed.

holoblastic (hō′-lō-blas′-tik) Referring to a type of cleavage in which there is complete division of the egg; occurs in eggs that have little yolk (such as those of the sea urchin) or a moderate amount of yolk (such as those of the frog).

homeobox (hō′-mē-ō-boks′) A 180-nucleotide sequence within homeotic genes and some other developmental genes that is widely conserved in animals. Related sequences occur in plants and yeasts.

homeostasis (hō′-mē-ō-stā′-sis) The steady-state physiological condition of the body.

homeotic gene (hō-mē-o′-tik) Any of the master regulatory genes that control placement and spatial organization of body parts in animals, plants, and fungi by controlling the developmental fate of groups of cells.

hominin (hō′-mi-nin) A member of the human branch of the evolutionary tree. Hominins include *Homo sapiens* and our ancestors, a group of extinct species that are more closely related to us than to chimpanzees.

homologous chromosomes (hō-mol′-uh-gus) A pair of chromosomes of the same length, centromere position, and staining pattern that possess genes for the same characters at corresponding loci. One homologous chromosome is inherited from the organism's father, the other from the mother. Also called homologs, or a homologous pair.

homologous structures Structures in different species that are similar because of common ancestry.

homology (hō-mol′-ō-jē) Similarity in characteristics resulting from a shared ancestry.

homoplasy (hō′-muh-play′-zē) A similar (analogous) structure or molecular sequence that has evolved independently in two species.

homosporous (hō-mos′-puh-rus) Referring to a plant species that has a single kind of spore, which typically develops into a bisexual gametophyte.

homozygous (hō′-mō-zī′-gus) Having two identical alleles for a given gene.

horizontal cell A neuron of the retina that helps integrate the information that is sent to the brain.

horizontal gene transfer The transfer of genes from one genome to another through mechanisms such as transposable elements, plasmid exchange, viral activity, and perhaps fusions of different organisms.

hormone In multicellular organisms, one of many types of secreted chemicals that are formed in specialized cells, travel in body fluids, and act on specific target cells in other parts of the body, changing the target cells' functioning. Hormones are thus important in long-distance signaling.

hornwort A small, herbaceous, nonvascular plant that is a member of the phylum Anthocerophyta.

host The larger participant in a symbiotic relationship, often providing a home and food source for the smaller symbiont.

host range The limited number of species whose cells can be infected by a particular virus.

human chorionic gonadotropin (hCG) (kōr′-ē-on′-ik gō-na′-dō-trō′-pin) A hormone secreted by the chorion that maintains the corpus luteum of the ovary during the first three months of pregnancy.

Human Genome Project An international collaborative effort to map and sequence the DNA of the entire human genome.

humoral immune response (hyū′-mer-ul) The branch of adaptive immunity that involves the activation of B cells and that leads to the production of antibodies, which defend against bacteria and viruses in body fluids.

humus (hyū′-mus) Decomposing organic material that is a component of topsoil.

Huntington's disease A human genetic disease caused by a dominant allele; characterized by uncontrollable body movements and degeneration of the nervous system; usually fatal 10 to 20 years after the onset of symptoms.

hybrid Offspring that results from the mating of individuals from two different species or from two true-breeding varieties of the same species.

hybrid zone A geographic region in which members of different species meet and mate, producing at least some offspring of mixed ancestry.

hybridization In genetics, the mating, or crossing, of two true-breeding varieties.

hydration shell The sphere of water molecules around a dissolved ion.

hydrocarbon An organic molecule consisting only of carbon and hydrogen.

hydrogen bond A type of weak chemical bond that is formed when the slightly positive hydrogen atom of a polar covalent bond in one molecule is attracted to the slightly negative atom of a polar covalent bond in another molecule or in another region of the same molecule.

hydrogen ion A single proton with a charge of 1+. The dissociation of a water molecule (H_2O) leads to the generation of a hydroxide ion (OH^-) and a hydrogen ion (H^+); in water, H^+ is not found alone but associates with a water molecule to form a hydronium ion.

hydrolysis (hī-drol′-uh-sis) A chemical reaction that breaks bonds between two molecules by the addition of water; functions in disassembly of polymers to monomers.

hydronium ion A water molecule that has an extra proton bound to it; H_3O^+, commonly represented as H^+.

hydrophilic (hī′-drō-fil′-ik) Having an affinity for water.

hydrophobic (hī′-drō-fō′-bik) Having no affinity for water; tending to coalesce and form droplets in water.

hydrophobic interaction A type of weak chemical interaction caused when molecules that do not mix with water coalesce to exclude water.

hydroponic culture A method in which plants are grown in mineral solutions rather than in soil.

hydrostatic skeleton A skeletal system composed of fluid held under pressure in a closed body compartment; the main skeleton of most cnidarians, flatworms, nematodes, and annelids.

hydroxide ion A water molecule that has lost a proton; OH^-.

hydroxyl group (hī-drok′-sil) A chemical group consisting of an oxygen atom joined to a hydrogen atom. Molecules possessing this group are soluble in water and are called alcohols.

hymen A thin membrane that partly covers the vaginal opening in the human female. The hymen is ruptured by sexual intercourse or other vigorous activity.

hyperpolarization A change in a cell's membrane potential such that the inside of the membrane becomes more negative relative to the outside. Hyperpolarization reduces the chance that a neuron will transmit a nerve impulse.

hypersensitive response A plant's localized defense response to a pathogen, involving the death of cells around the site of infection.

hypertension A disorder in which blood pressure remains abnormally high.

hypertonic Referring to a solution that, when surrounding a cell, will cause the cell to lose water.

hypha (plural, **hyphae**) (hī′-fuh, hī′-fē) One of many connected filaments that collectively make up the mycelium of a fungus.

hypocotyl (hī′-puh-cot′-ul) In an angiosperm embryo, the embryonic axis below the point of attachment of the cotyledon(s) and above the radicle.

hypothalamus (hī′-pō-thal′-uh-mus) The ventral part of the vertebrate forebrain; functions in maintaining homeostasis, especially in coordinating the endocrine and nervous systems; secretes hormones of the posterior pituitary and releasing factors that regulate the anterior pituitary.

hypothesis (hī-poth′-uh-sis) A testable explanation for a set of observations based on the available data and guided by inductive reasoning. A hypothesis is narrower in scope than a theory.

hypotonic Referring to a solution that, when surrounding a cell, will cause the cell to take up water.

imbibition The physical adsorption of water onto the internal surfaces of structures.

immigration The influx of new individuals into a population from other areas.

immune system An animal body's system of defenses against agents that cause disease.

immunization The process of generating a state of immunity by artificial means. In active immunization, also called vaccination, an inactive or weakened form of a pathogen is administered, inducing B and T cell responses and immunological memory. In passive immunization, antibodies specific for a particular microbe are administered, conferring immediate but temporary protection.

immunodeficiency A disorder in which the ability of an immune system to protect against pathogens is defective or absent.

immunoglobulin (Ig) (im′-yū-nō-glob′-yū-lin) Any of the class of proteins that function as antibodies. Immunoglobulins are divided into five major classes that differ in their distribution in the body and antigen disposal activities.

imprinting In animal behavior, the formation at a specific stage in life of a long-lasting behavioral response to a specific individual or object. *See also* genomic imprinting.

***in situ* hybridization** A technique using nucleic acid hybridization with a labeled probe to detect the location of a specific mRNA in an intact organism.

***in vitro* fertilization (IVF)** (vē′-trō) Fertilization of oocytes in laboratory containers followed by artificial implantation of the early embryo in the mother's uterus.

***in vitro* mutagenesis** A technique used to discover the function of a gene by cloning it, introducing specific changes into the cloned gene's sequence, reinserting the mutated gene into a cell, and studying the phenotype of the mutant.

inclusive fitness The total effect an individual has on proliferating its genes by producing its own offspring and by providing aid that enables other close relatives to increase production of their offspring.

incomplete dominance The situation in which the phenotype of heterozygotes is intermediate between the phenotypes of individuals homozygous for either allele.

incomplete flower A flower in which one or more of the four basic floral organs (sepals, petals, stamens, or carpels) are either absent or nonfunctional.

incomplete metamorphosis A type of development in certain insects, such as grasshoppers, in which the young (called nymphs) resemble adults but are smaller and have different body proportions. The nymph goes through a series of molts, each time looking more like an adult, until it reaches full size.

indeterminate cleavage A type of embryonic development in deuterostomes in which each cell produced by early cleavage divisions retains the capacity to develop into a complete embryo.

indeterminate growth A type of growth characteristic of plants, in which the organism continues to grow as long as it lives.

induced fit Caused by entry of the substrate, the change in shape of the active site of an enzyme so that it binds more snugly to the substrate.

inducer A specific small molecule that binds to a bacterial repressor protein and changes the repressor's shape so that it cannot bind to an operator, thus switching an operon on.

induction The process in which one group of embryonic cells influences the development of another, usually by causing changes in gene expression.

inductive reasoning A type of logic in which generalizations are based on a large number of specific observations.

inflammatory response An innate immune defense triggered by physical injury or infection of tissue involving the release of substances that promote swelling, enhance the infiltration of white blood cells, and aid in tissue repair and destruction of invading pathogens.

inflorescence A group of flowers tightly clustered together.

ingestion The first stage of food processing in animals: the act of eating.

ingroup A species or group of species whose evolutionary relationships we seek to determine.

inhibin A hormone produced in the male and female gonads that functions in part by regulating the anterior pituitary by negative feedback.

inhibitory postsynaptic potential (IPSP) An electrical change (usually hyperpolarization) in the membrane of a postsynaptic neuron caused by the binding of an inhibitory neurotransmitter from a presynaptic cell to a postsynaptic receptor; makes it more difficult for a postsynaptic neuron to generate an action potential.

innate behavior Animal behavior that is developmentally fixed and under strong genetic control. Innate behavior is exhibited in virtually the same form by all individuals in a population despite internal and external environmental differences during development and throughout their lifetimes.

innate immunity A form of defense common to all animals that is active immediately upon exposure to pathogens and that is the same whether or not the pathogen has been encountered previously.

inner cell mass An inner cluster of cells at one end of a mammalian blastocyst that subsequently develops into the embryo proper and some of the extraembryonic membranes.

inner ear One of three main regions of the vertebrate ear; includes the cochlea (which in turn contains the organ of Corti) and the semicircular canals.

inositol trisphosphate (IP_3) (in-ō′-suh-tol) A second messenger that functions as an intermediate between certain signaling molecules and a subsequent second messenger, Ca^{2+}, by causing a rise in cytoplasmic Ca^{2+} concentration.

inquiry The search for information and explanation, often focusing on specific questions.

insertion A mutation involving the addition of one or more nucleotide pairs to a gene.

insulin (in′-suh-lin) A hormone secreted by pancreatic beta cells that lowers blood glucose levels. It promotes the uptake of glucose by most body cells and the synthesis and storage of glycogen in the liver and also stimulates protein and fat synthesis.

integral protein A transmembrane protein with hydrophobic regions that extend into and often completely span the hydrophobic interior of the membrane and with hydrophilic regions in contact with the aqueous solution on one or both sides of the membrane (or lining the channel in the case of a channel protein).

integrin In animal cells, a transmembrane receptor protein with two subunits that interconnects the extracellular matrix and the cytoskeleton.

integument (in-teg′-yū-ment) Layer of sporophyte tissue that contributes to the structure of an ovule of a seed plant.

integumentary system The outer covering of a mammal's body, including skin, hair, and nails, claws, or hooves.

intercalated disk (in-ter′-kuh-lā′-ted) A specialized junction between cardiac muscle cells that provides direct electrical coupling between the cells.

interferon (in′-ter-fēr′-on) A protein that has antiviral or immune regulatory functions. Interferon-α and interferon-β, secreted by virus-infected cells, help nearby cells resist viral infection; interferon-γ, secreted by T cells, helps activate macrophages.

intermediate disturbance hypothesis The concept that moderate levels of disturbance can foster greater species diversity than low or high levels of disturbance.

intermediate filament A component of the cytoskeleton that includes filaments intermediate in size between microtubules and microfilaments.

internal fertilization The fusion of eggs and sperm within the female reproductive tract. The sperm are typically deposited in or near the tract.

interneuron An association neuron; a nerve cell within the central nervous system that forms synapses with sensory and/or motor neurons and integrates sensory input and motor output.

internode A segment of a plant stem between the points where leaves are attached.

interphase The period in the cell cycle when the cell is not dividing. During interphase, cellular metabolic activity is high, chromosomes and organelles are duplicated, and cell size may increase. Interphase often accounts for about 90% of the cell cycle.

intersexual selection Selection whereby individuals of one sex (usually females) are choosy in selecting their mates from individuals of the other sex; also called mate choice.

interspecific competition Competition for resources between individuals of two or more species when resources are in short supply.

interspecific interaction A relationship between individuals of two or more species in a community.

interstitial fluid The fluid filling the spaces between cells in most animals.

intertidal zone The shallow zone of the ocean adjacent to land and between the high- and low-tide lines.

intracytoplasmic sperm injection (ICSI) The fertilization of an egg in the laboratory by the direct injection of a single sperm.

intrasexual selection Selection in which there is direct competition among individuals of one sex for mates of the opposite sex.

introduced species A species moved by humans, either intentionally or accidentally, from its native location to a new geographic region; also called non-native or exotic species.

intron (in′-tron) A noncoding, intervening sequence within a primary transcript that is removed from the transcript during RNA processing; also refers to the region of DNA from which this sequence was transcribed.

invasive species A species, often introduced by humans, that takes hold outside its native range.

inversion An aberration in chromosome structure resulting from reattachment of a chromosomal fragment in a reverse orientation to the chromosome from which it originated.

invertebrate An animal without a backbone. Invertebrates make up 95% of animal species.

ion (ī′-on) An atom or group of atoms that has gained or lost one or more electrons, thus acquiring a charge.

ion channel A transmembrane protein channel that allows a specific ion to diffuse across the membrane down its concentration or electrochemical gradient.

ionic bond (ī-on′-ik) A chemical bond resulting from the attraction between oppositely charged ions.

ionic compound A compound resulting from the formation of an ionic bond; also called a salt.

IPSP *See* inhibitory postsynaptic potential.

iris The colored part of the vertebrate eye, formed by the anterior portion of the choroid.

isomer (ī′-sō-mer) One of several compounds with the same molecular formula but different structures and therefore different properties. The three types of isomers are structural isomers, *cis-trans* isomers, and enantiomers.

isomorphic Referring to alternating generations in plants and certain algae in which the sporophytes and gametophytes look alike, although they differ in chromosome number.

isopod A member of one of the largest groups of crustaceans, which includes terrestrial, freshwater, and marine species. Among the terrestrial isopods are the pill bugs, or wood lice.

isotonic (ī′-sō-ton′-ik) Referring to a solution that, when surrounding a cell, causes no net movement of water into or out of the cell.

isotope (ī′-sō-tōp′) One of several atomic forms of an element, each with the same number of protons but a different number of neutrons, thus differing in atomic mass.

iteroparity Reproduction in which adults produce offspring over many years; also known as repeated reproduction.

joule (J) A unit of energy: 1 J = 0.239 cal; 1 cal = 4.184 J.

juxtaglomerular apparatus (JGA) (juks′-tuh-gluh-mār′-yū-ler) A specialized tissue in nephrons that releases the enzyme renin in response to a drop in blood pressure or volume.

juxtamedullary nephron In mammals and birds, a nephron with a loop of Henle that extends far into the renal medulla.

karyogamy (kār′-ē-og′-uh-mē) In fungi, the fusion of haploid nuclei contributed by the two parents; occurs as one stage of sexual reproduction, preceded by plasmogamy.

karyotype (kār′-ē-ō-tīp) A display of the chromosome pairs of a cell arranged by size and shape.

keystone species A species that is not necessarily abundant in a community yet exerts strong control on community structure by the nature of its ecological role or niche.

kidney In vertebrates, one of a pair of excretory organs where blood filtrate is formed and processed into urine.

kilocalorie (kcal) A thousand calories; the amount of heat energy required to raise the temperature of 1 kg of water by 1°C.

kin selection Natural selection that favors altruistic behavior by enhancing the reproductive success of relatives.

kinetic energy (kuh-net′-ik) The energy associated with the relative motion of objects. Moving matter can perform work by imparting motion to other matter.

kinetochore (kuh-net′-uh-kōr) A structure of proteins attached to the centromere that links each sister chromatid to the mitotic spindle.

kinetoplastid A protist, such as a trypanosome, that has a single large mitochondrion that houses an organized mass of DNA.

kingdom A taxonomic category, the second broadest after domain.

***K*-selection** Selection for life history traits that are sensitive to population density; also called density-dependent selection.

labia majora A pair of thick, fatty ridges that encloses and protects the rest of the vulva.

labia minora A pair of slender skin folds that surrounds the openings of the vagina and urethra.

labor A series of strong, rhythmic contractions of the uterus that expels a baby out of the uterus and vagina during childbirth.

lactation The continued production of milk from the mammary glands.

lacteal (lak′-tē-ul) A tiny lymph vessel extending into the core of an intestinal villus and serving as the destination for absorbed chylomicrons.

lactic acid fermentation Glycolysis followed by the reduction of pyruvate to lactate, regenerating NAD^+ with no release of carbon dioxide.

lagging strand A discontinuously synthesized DNA strand that elongates by means of Okazaki fragments, each synthesized in a $5' \rightarrow 3'$ direction away from the replication fork.

lancelet Member of the clade Cephalochordata, small blade-shaped marine chordates that lack a backbone.

landscape An area containing several different ecosystems linked by exchanges of energy, materials, and organisms.

landscape ecology The study of how the spatial arrangement of habitat types affects the distribution and abundance of organisms and ecosystem processes.

large intestine The portion of the vertebrate alimentary canal between the small intestine and the anus; functions mainly in water absorption and the formation of feces.

larva (lar′-vuh) (plural, **larvae**) A free-living, sexually immature form in some animal life cycles that may differ from the adult animal in morphology, nutrition, and habitat.

larynx (lār′-inks) The portion of the respiratory tract containing the vocal cords; also called the voice box.

lateral geniculate nucleus One of a pair of structures in the brain that are the destination for most of the ganglion cell axons that form the optic nerves.

lateral inhibition A process that sharpens the edges and enhances the contrast of a perceived image by inhibiting receptors lateral to those that have responded to light.

lateral line system A mechanoreceptor system consisting of a series of pores and receptor units along the sides of the body in fishes and aquatic amphibians; detects water movements made by the animal itself and by other moving objects.

lateral meristem (mār′-uh-stem) A meristem that thickens the roots and shoots of woody plants. The vascular cambium and cork cambium are lateral meristems.

lateral root A root that arises from the pericycle of an established root.

lateralization Segregation of functions in the cortex of the left and right cerebral hemispheres.

law of conservation of mass A physical law stating that matter can change form but cannot be created or destroyed. In a closed system, the mass of the system is constant.

law of independent assortment Mendel's second law, stating that each pair of alleles segregates, or assorts, independently of each other pair during gamete formation; applies when genes for two characters are located on different pairs of homologous chromosomes or when they are far enough apart on the same chromosome to behave as though they are on different chromosomes.

law of segregation Mendel's first law, stating that the two alleles in a pair segregate (separate from each other) into different gametes during gamete formation.

leading strand The new complementary DNA strand synthesized continuously along the template strand toward the replication fork in the mandatory $5' \rightarrow 3'$ direction.

leaf The main photosynthetic organ of vascular plants.

leaf primordium A finger-like projection along the flank of a shoot apical meristem, from which a leaf arises.

learning The modification of behavior based on specific experiences.

lens The structure in an eye that focuses light rays onto the photoreceptors.

lenticel (len′-ti-sel) A small raised area in the bark of stems and roots that enables gas exchange between living cells and the outside air.

lepidosaur (leh-pid′-uh-sōr) Member of the reptilian group that includes lizards, snakes, and two species of New Zealand animals called tuataras.

leptin A hormone produced by adipose (fat) cells that acts as a satiety factor in regulating appetite.

leukocyte (lū′-kō-sīt′) A blood cell that functions in fighting infections; also called a white blood cell.

Leydig cell (lī′-dig) A cell that produces testosterone and other androgens and is located between the seminiferous tubules of the testes.

lichen The mutualistic association between a fungus and a photosynthetic alga or cyanobacterium.

life cycle The generation-to-generation sequence of stages in the reproductive history of an organism.

life history The traits that affect an organism's schedule of reproduction and survival.

life table An age-specific summary of the survival pattern of a population.

ligament A fibrous connective tissue that joins bones together at joints.

ligand (lig′-und) A molecule that binds specifically to another molecule, usually a larger one.

ligand-gated ion channel A transmembrane protein containing a pore that opens or closes as it changes shape in response to a signaling molecule (ligand), allowing or blocking the flow of specific ions; also called an ionotropic receptor.

light chain One of the two types of polypeptide chains that make up an antibody molecule and B cell receptor; consists of a variable region, which contributes to the antigen-binding site, and a constant region.

light microscope (LM) An optical instrument with lenses that refract (bend) visible light to magnify images of specimens.

light reactions The first of two major stages in photosynthesis (preceding the Calvin cycle). These reactions, which occur on the thylakoid membranes of the chloroplast or on membranes of certain prokaryotes, convert solar energy to the chemical energy of ATP and NADPH, releasing oxygen in the process.

light-harvesting complex A complex of proteins associated with pigment molecules (including chlorophyll *a*, chlorophyll *b*, and carotenoids) that captures light energy and transfers it to reaction-center pigments in a photosystem.

lignin (lig′-nin) A hard material embedded in the cellulose matrix of vascular plant cell walls that provides structural support in terrestrial species.

limiting nutrient An element that must be added for production to increase in a particular area.

limnetic zone In a lake, the well-lit, open surface waters far from shore.

linear electron flow A route of electron flow during the light reactions of photosynthesis that involves both photosystems (I and II) and produces ATP, NADPH, and O_2. The net electron flow is from H_2O to $NADP^+$.

linkage map A genetic map based on the frequencies of recombination between markers during crossing over of homologous chromosomes.

linked genes Genes located close enough together on a chromosome that they tend to be inherited together.

lipid (lip′-id) Any of a group of large biological molecules, including fats, phospholipids, and steroids, that mix poorly, if at all, with water.

littoral zone In a lake, the shallow, well-lit waters close to shore.

liver A large internal organ in vertebrates that performs diverse functions, such as producing bile, maintaining blood glucose level, and detoxifying poisonous chemicals in the blood.

liverwort A small, herbaceous, nonvascular plant that is a member of the phylum Hepatophyta.

loam The most fertile soil type, made up of roughly equal amounts of sand, silt, and clay.

lobe-fin Member of the vertebrate clade Sarcopterygii, osteichthyans with rod-shaped muscular fins, including coelacanths, lungfishes, and tetrapods.

local regulator A secreted molecule that influences cells near where it is secreted.

locomotion Active motion from place to place.

locus (lō′-kus) (plural, **loci**) A specific place along the length of a chromosome where a given gene is located.

logistic population growth Population growth that levels off as population size approaches carrying capacity.

long-day plant A plant that flowers (usually in late spring or early summer) only when the light period is longer than a critical length.

long-term memory The ability to hold, associate, and recall information over one's lifetime.

long-term potentiation (LTP) An enhanced responsiveness to an action potential (nerve signal) by a receiving neuron.

loop of Henle The hairpin turn, with a descending and ascending limb, between the proximal and distal tubules of the vertebrate kidney; functions in water and salt reabsorption.

lophophore (lof′-uh-fōr) In some lophotrochozoan animals, including brachiopods, a crown of ciliated tentacles that surround the mouth and function in feeding.

lophotrochozoan Member of a group of animal phyla identified as a clade by molecular evidence. Lophotrochozoans include organisms that have lophophores or trochophore larvae.

low-density lipoprotein (LDL) A particle in the blood made up of thousands of cholesterol molecules and other lipids bound to a protein. LDL transports cholesterol from the liver for incorporation into cell membranes.

lung An infolded respiratory surface of a terrestrial vertebrate, land snail, or spider that connects to the atmosphere by narrow tubes.

luteal phase That portion of the ovarian cycle during which endocrine cells of the corpus luteum secrete female hormones.

luteinizing hormone (LH) (lū′-tē-uh-nī′-zing) A tropic hormone that is produced and secreted by the anterior pituitary and that stimulates ovulation in females and androgen production in males.

lycophyte (lī′-kuh-fīt) An informal name for a member of the phylum Lycophyta, which includes club mosses, spike mosses, and quillworts.

lymph The colorless fluid, derived from interstitial fluid, in the lymphatic system of vertebrates.

lymph node An organ located along a lymph vessel. Lymph nodes filter lymph and contain cells that attack viruses and bacteria.

lymphatic system A system of vessels and nodes, separate from the circulatory system, that returns fluid, proteins, and cells to the blood.

lymphocyte A type of white blood cell that mediates immune responses. The two main classes are B cells and T cells.

lysogenic cycle (lī′-sō-jen′-ik) A type of phage replicative cycle in which the viral genome becomes incorporated into the bacterial host chromosome as a prophage, is replicated along with the chromosome, and does not kill the host.

lysosome (lī′-suh-sōm) A membrane-enclosed sac of hydrolytic enzymes found in the cytoplasm of animal cells and some protists.

lysozyme (lī′-sō-zīm) An enzyme that destroys bacterial cell walls; in mammals, found in sweat, tears, and saliva.

lytic cycle (lit′-ik) A type of phage replicative cycle resulting in the release of new phages by lysis (and death) of the host cell.

macroclimate Large-scale patterns in climate; the climate of an entire region.

macroevolution Evolutionary change above the species level. Examples of macroevolutionary change include the origin of a new group of organisms through a series of speciation events and the impact of mass extinctions on the diversity of life and its subsequent recovery.

macromolecule A giant molecule formed by the joining of smaller molecules, usually by a dehydration reaction. Polysaccharides, proteins, and nucleic acids are macromolecules.

macronutrient An essential element that an organism must obtain in relatively large amounts. *See also* micronutrient.

macrophage (mak′-rō-fāj) A phagocytic cell present in many tissues that functions in innate immunity by destroying microbes and in acquired immunity as an antigen-presenting cell.

magnoliid Member of the angiosperm clade that is most closely related to the combined eudicot and monocot clades. Extant examples are magnolias, laurels, and black pepper plants.

major depressive disorder A mood disorder characterized by feelings of sadness, lack of self-worth, emptiness, or loss of interest in nearly all things.

major histocompatibility complex (MHC) molecule A host protein that functions in antigen presentation. Foreign MHC molecules on transplanted tissue can trigger T cell responses that may lead to rejection of the transplant.

malignant tumor A cancerous tumor containing cells that have significant genetic and cellular changes and are capable of invading and surviving in new sites. Malignant tumors can impair the functions of one or more organs.

Malpighian tubule (mal-pig′-ē-un) A unique excretory organ of insects that empties into the digestive tract, removes nitrogenous wastes from the hemolymph, and functions in osmoregulation.

mammal Member of the class Mammalia, amniotes that have hair and mammary glands (glands that produce milk).

mammary gland An exocrine gland that secretes milk to nourish the young. Mammary glands are characteristic of mammals.

mandible One of a pair of jaw-like feeding appendages found in myriapods, hexapods, and crustaceans.

mantle One of the three main parts of a mollusc; a fold of tissue that drapes over the mollusc's visceral mass and may secrete a shell. *See also* foot, visceral mass.

mantle cavity A water-filled chamber that houses the gills, anus, and excretory pores of a mollusc.

map unit A unit of measurement of the distance between genes. One map unit is equivalent to a 1% recombination frequency.

marine benthic zone The ocean floor.

mark-recapture method A sampling technique used to estimate the size of animal populations.

marsupial (mar-sū′-pē-ul) A mammal, such as a koala, kangaroo, or opossum, whose young complete their embryonic development inside a maternal pouch called the marsupium.

mass extinction The elimination of a large number of species throughout Earth, the result of global environmental changes.

mass number The sum of the number of protons and neutrons in an atom's nucleus.

mast cell A vertebrate body cell that produces histamine and other molecules that trigger inflammation in response to infection and in allergic reactions.

mate-choice copying Behavior in which individuals in a population copy the mate choice of others, apparently as a result of social learning.

maternal effect gene A gene that, when mutant in the mother, results in a mutant phenotype in the offspring, regardless of the offspring's genotype. Maternal effect genes, also called egg-polarity genes, were first identified in *Drosophila melanogaster*.

matter Anything that takes up space and has mass.

maximum likelihood As applied to molecular systematics, a principle that states that when considering multiple phylogenetic hypotheses, one should take into account the hypothesis that reflects the most likely sequence of evolutionary events, given certain rules about how DNA changes over time.

maximum parsimony A principle that states that when considering multiple explanations for an observation, one should first investigate the simplest explanation that is consistent with the facts.

mechanoreceptor A sensory receptor that detects physical deformation in the body's environment associated with pressure, touch, stretch, motion, or sound.

medulla oblongata (meh-dul′-uh ōb′-long-go′-tuh) The lowest part of the vertebrate brain, commonly called the medulla; a swelling of the hindbrain anterior to the spinal cord that controls autonomic, homeostatic functions, including breathing, heart and blood vessel activity, swallowing, digestion, and vomiting.

Medusa (plural, medusae) (muh-dū′-suh) The floating, flattened, mouth-down version of the cnidarian body plan. The alternate form is the polyp.

megapascal (MPa) (meg′-uh-pas-kal′) A unit of pressure equivalent to about 10 atmospheres of pressure.

megaphyll (meh′-guh-fil) A leaf with a highly branched vascular system, characteristic of the vast majority of vascular plants. *See* microphyll.

megaspore A spore from a heterosporous plant species that develops into a female gametophyte.

meiosis (mī-ō′-sis) A modified type of cell division in sexually reproducing organisms consisting of two rounds of cell division but only one round of DNA replication. It results in cells with half the number of chromosome sets as the original cell.

meiosis I The first division of a two-stage process of cell division in sexually reproducing organisms that results in cells with half the number of chromosome sets as the original cell.

meiosis II The second division of a two-stage process of cell division in sexually reproducing organisms that results in cells with half the number of chromosome sets as the original cell.

melanocyte-stimulating hormone (MSH) A hormone produced and secreted by the anterior pituitary with multiple activities, including regulating the behavior of pigment-containing cells in the skin of some vertebrates.

melatonin A hormone that is secreted by the pineal gland and that is involved in the regulation of biological rhythms and sleep.

membrane potential The difference in electrical charge (voltage) across a cell's plasma membrane due to the differential distribution of ions. Membrane potential affects the activity of excitable cells and the transmembrane movement of all charged substances.

memory cell One of a clone of long-lived lymphocytes, formed during the primary immune response, that remains in a lymphoid organ until activated by exposure to the same antigen that triggered its formation. Activated memory cells mount the secondary immune response.

menopause The cessation of ovulation and menstruation marking the end of a human female's reproductive years.

menstrual cycle (men′-strū-ul) In humans and certain other primates, a type of reproductive cycle in which the nonpregnant endometrium is shed through the cervix into the vagina; also called the uterine cycle.

menstrual flow phase That portion of the uterine (menstrual) cycle when menstrual bleeding occurs.

menstruation The shedding of portions of the endometrium during a uterine (menstrual) cycle.

meristem (mār′-uh-stem) Plant tissue that remains embryonic as long as the plant lives, allowing for indeterminate growth.

meristem identity gene A plant gene that promotes the switch from vegetative growth to flowering.

meroblastic (mār′-ō-blas′-tik) Referring to a type of cleavage in which there is incomplete division of a yolk-rich egg, characteristic of avian development.

mesoderm (mez′-ō-derm) The middle primary germ layer in a triploblastic animal embryo; develops into the notochord, the lining of the coelom, muscles, skeleton, gonads, kidneys, and most of the circulatory system in species that have these structures.

mesohyl (mez′-ō-hīl) A gelatinous region between the two layers of cells of a sponge.

mesophyll (mez′-ō-fil) Leaf cells specialized for photosynthesis. In C_3 and CAM plants, mesophyll cells are located between the upper and lower epidermis; in C_4 plants, they are located between the bundle-sheath cells and the epidermis.

messenger RNA (mRNA) A type of RNA, synthesized using a DNA template, that attaches to ribosomes in the cytoplasm and specifies the primary structure of a protein. (In eukaryotes, the primary RNA transcript must undergo RNA processing to become mRNA.)

metabolic pathway A series of chemical reactions that either builds a complex molecule (anabolic pathway) or breaks down a complex molecule to simpler molecules (catabolic pathway).

metabolic rate The total amount of energy an animal uses in a unit of time.

metabolism (muh-tab′-uh-lizm) The totality of an organism's chemical reactions, consisting of catabolic and anabolic pathways, which manage the material and energy resources of the organism.

metagenomics The collection and sequencing of DNA from a group of species, usually an environmental sample of microorganisms. Computer software sorts partial sequences and assembles them into genome sequences of individual species making up the sample.

metamorphosis (met′-uh-mōr′-fuh-sis) A developmental transformation that turns an animal larva into either an adult or an adult-like stage that is not yet sexually mature.

metanephridium (met′-uh-nuh-frid′-ē-um) (plural, **metanephridia**) An excretory organ found in many invertebrates that typically consists of tubules connecting ciliated internal openings to external openings.

metaphase The third stage of mitosis, in which the spindle is complete and the chromosomes, attached to microtubules at their kinetochores, are all aligned at the metaphase plate.

metaphase plate An imaginary structure located at a plane midway between the two poles of a cell in metaphase on which the centromeres of all the duplicated chromosomes are located.

metapopulation A group of spatially separated populations of one species that interact through immigration and emigration.

metastasis (muh-tas′-tuh-sis) The spread of cancer cells to locations distant from their original site.

methanogen (meth-an′-ō-jen) An organism that produces methane as a waste product of the way it obtains energy. All known methanogens are in domain Archaea.

methyl group A chemical group consisting of a carbon bonded to three hydrogen atoms. The methyl group may be attached to a carbon or to a different atom.

microclimate Climate patterns on a very fine scale, such as the specific climatic conditions underneath a log.

microevolution Evolutionary change below the species level; change in the allele frequencies in a population over generations.

microfilament A cable composed of actin proteins in the cytoplasm of almost every eukaryotic cell, making up part of the cytoskeleton and acting alone or with myosin to cause cell contraction; also known as an actin filament.

micronutrient An essential element that an organism needs in very small amounts. *See also* macronutrient.

microphyll (mī′-krō-fil) In lycophytes, a small leaf with a single unbranched vein. *See* megaphyll.

micropyle A pore in the integuments of an ovule.

microRNA (miRNA) A small, single-stranded RNA molecule, generated from a hairpin structure on a precursor RNA transcribed from a particular gene. The miRNA associates with one or more proteins in a complex that can degrade or prevent translation of an mRNA with a complementary sequence.

microspore A spore from a heterosporous plant species that develops into a male gametophyte.

microtubule A hollow rod composed of tubulin proteins that makes up part of the cytoskeleton in all eukaryotic cells and is found in cilia and flagella.

microvillus (plural, **microvilli**) One of many fine, finger-like projections of the epithelial cells in the lumen of the small intestine that increase its surface area.

midbrain One of three ancestral and embryonic regions of the vertebrate brain; develops into sensory integrating and relay centers that send sensory information to the cerebrum.

middle ear One of three main regions of the vertebrate ear; in mammals, a chamber containing three small bones (the malleus, incus, and stapes) that convey vibrations from the eardrum to the oval window.

middle lamella (luh-mel′-uh) In plants, a thin layer of adhesive extracellular material, primarily pectins, found between the primary walls of adjacent young cells.

migration A regular, long-distance change in location.

mineral In nutrition, a simple nutrient that is inorganic and therefore cannot be synthesized in the body.

mineralocorticoid A steroid hormone secreted by the adrenal cortex that regulates salt and water homeostasis.

minimum viable population (MVP) The smallest population size at which a species is able to sustain its numbers and survive.

mismatch repair The cellular process that uses specific enzymes to remove and replace incorrectly paired nucleotides.

missense mutation A nucleotide-pair substitution that results in a codon that codes for a different amino acid.

mitochondrial matrix The compartment of the mitochondrion enclosed by the inner membrane and containing enzymes and substrates for the citric acid cycle, as well as ribosomes and DNA.

mitochondrion (mī′-tō-kon′-drē-un) (plural, **mitochondria**) An organelle in eukaryotic cells that serves as the site of cellular respiration; uses oxygen to break down organic molecules and synthesize ATP.

mitosis (mī-tō′-sis) A process of nuclear division in eukaryotic cells conventionally divided into five stages: prophase, prometaphase, metaphase, anaphase, and telophase. Mitosis conserves chromosome number by allocating replicated chromosomes equally to each of the daughter nuclei.

mitotic (M) phase The phase of the cell cycle that includes mitosis and cytokinesis.

mitotic spindle An assemblage of microtubules and associated proteins that is involved in the movement of chromosomes during mitosis.

mixotroph An organism that is capable of both photosynthesis and heterotrophy.

model organism A particular species chosen for research into broad biological principles because it is representative of a larger group and usually easy to grow in a lab.

molarity A common measure of solute concentration, referring to the number of moles of solute per liter of solution.

mold Informal term for a fungus that grows as a filamentous fungus, producing haploid spores by mitosis and forming a visible mycelium.

mole (mol) The number of grams of a substance that equals its molecular weight in daltons and contains Avogadro's number of molecules.

molecular clock A method for estimating the time required for a given amount of evolutionary change, based on the observation that some regions of genomes evolve at constant rates.

molecular mass The sum of the masses of all the atoms in a molecule; sometimes called molecular weight.

molecular systematics A scientific discipline that uses nucleic acids or other molecules to infer evolutionary relationships between different species.

molecule Two or more atoms held together by covalent bonds.

molting A process in ecdysozoans in which the exoskeleton is shed at intervals, allowing growth by the production of a larger exoskeleton.

monoclonal antibody (mon′-ō-klōn′-ul) Any of a preparation of antibodies that have been produced by a single clone of cultured cells and thus are all specific for the same epitope.

monocot Member of a clade consisting of flowering plants that have one embryonic seed leaf, or cotyledon.

monogamous (muh-nog′-uh-mus) Referring to a type of relationship in which one male mates with just one female.

monohybrid An organism that is heterozygous with respect to a single gene of interest. All the offspring from a cross between parents homozygous for different alleles are monohybrids. For example, parents of genotypes *AA* and *aa* produce a monohybrid of genotype *Aa*.

monohybrid cross A cross between two organisms that are heterozygous for the character being followed (or the self-pollination of a heterozygous plant).

monomer (mon′-uh-mer) The subunit that serves as the building block of a polymer.

monophyletic (mon′-ō-fī-let′-ik) Pertaining to a group of taxa that consists of a common ancestor and all of its descendants. A monophyletic taxon is equivalent to a clade.

monosaccharide (mon′-ō-sak′-uh-rīd) The simplest carbohydrate, active alone or serving as a monomer for disaccharides and polysaccharides. Also known as simple sugars, monosaccharides have molecular formulas that are generally some multiple of CH_2O.
monosomic Referring to a diploid cell that has only one copy of a particular chromosome instead of the normal two.
monotreme An egg-laying mammal, such as a platypus or echidna. Like all mammals, monotremes have hair and produce milk, but they lack nipples.
morphogen A substance, such as Bicoid protein in *Drosophila*, that provides positional information in the form of a concentration gradient along an embryonic axis.
morphogenesis (mōr′-fō-jen′-uh-sis) The cellular and tissue-based processes by which an animal body takes shape.
morphological species concept A definition of species in terms of measurable anatomical criteria.
moss A small, herbaceous, nonvascular plant that is a member of the phylum Bryophyta.
motor neuron A nerve cell that transmits signals from the brain or spinal cord to muscles or glands.
motor protein A protein that interacts with cytoskeletal elements and other cell components, producing movement of the whole cell or parts of the cell.
motor system An efferent branch of the vertebrate peripheral nervous system composed of motor neurons that carry signals to skeletal muscles in response to external stimuli.
motor unit A single motor neuron and all the muscle fibers it controls.
movement corridor A series of small clumps or a narrow strip of quality habitat (usable by organisms) that connects otherwise isolated patches of quality habitat.
MPF Maturation-promoting factor (or M-phase-promoting factor); a protein complex required for a cell to progress from late interphase to mitosis. The active form consists of cyclin and a protein kinase.
mucus A viscous and slippery mixture of glycoproteins, cells, salts, and water that moistens and protects the membranes lining body cavities that open to the exterior.
Müllerian mimicry (myū-lār′-ē-un) Reciprocal mimicry by two unpalatable species.
multifactorial Referring to a phenotypic character that is influenced by multiple genes and environmental factors.
multigene family A collection of genes with similar or identical sequences, presumably of common origin.
multiple fruit A fruit derived from an entire inflorescence.
multiplication rule A rule of probability stating that the probability of two or more independent events occurring together can be determined by multiplying their individual probabilities.
muscle tissue Tissue consisting of long muscle cells that can contract, either on its own or when stimulated by nerve impulses.
mutagen (myū′-tuh-jen) A chemical or physical agent that interacts with DNA and can cause a mutation.
mutation (myū-tā′-shun) A change in the nucleotide sequence of an organism's DNA or in the DNA or RNA of a virus.
mutualism (myū′-chū-ul-izm) A symbiotic relationship in which both participants benefit.
mycelium (mī-sē′-lē-um) The densely branched network of hyphae in a fungus.
mycorrhiza (mī′-kō-rī′-zuh) (plural, **mycorrhizae**) A mutualistic association of plant roots and fungus.
mycosis (mī-kō′-sis) General term for a fungal infection.
myelin sheath (mī′-uh-lin) Wrapped around the axon of a neuron, an insulating coat of cell membranes from Schwann cells or oligodendrocytes. It is interrupted by nodes of Ranvier, where action potentials are generated.
myofibril (mī′-ō-fī′-bril) A longitudinal bundle in a muscle cell (fiber) that contains thin filaments of actin and regulatory proteins and thick filaments of myosin.
myoglobin (mī′-uh-glō′-bin) An oxygen-storing, pigmented protein in muscle cells.
myosin (mī′-uh-sin) A type of motor protein that associates into filaments that interact with actin filaments to cause cell contraction.
myotonia (mī′-uh-tō′-nī-uh) Increased muscle tension, characteristic of sexual arousal in certain human tissues.
myriapod (mir′-ē-uh-pod′) A terrestrial arthropod with many body segments and one or two pairs of legs per segment. Millipedes and centipedes are the two major groups of living myriapods.
NAD^+ Nicotinamide adenine dinucleotide, a coenzyme that cycles easily between oxidized (NAD^+) and reduced (NADH) states, thus acting as an electron carrier.
$NADP^+$ Nicotinamide adenine dinucleotide phosphate, an electron acceptor that, as NADPH, temporarily stores energized electrons produced during the light reactions.
natural family planning A form of contraception that relies on refraining from sexual intercourse when conception is most likely to occur; also called the rhythm method.
natural killer cell A type of white blood cell that can kill tumor cells and virus-infected cells as part of innate immunity.
natural selection A process in which individuals that have certain inherited traits tend to survive and reproduce at higher rates than other individuals *because of* those traits.
negative feedback A form of regulation in which accumulation of an end product of a process slows the process; in physiology, a primary mechanism of homeostasis, whereby a change in a variable triggers a response that counteracts the initial change.
negative pressure breathing A breathing system in which air is pulled into the lungs.
nematocyst (nem′-uh-tuh-sist′) In a cnidocyte of a cnidarian, a capsule-like organelle containing a coiled thread that when discharged can penetrate the body wall of the prey.
nephron (nef′-ron) The tubular excretory unit of the vertebrate kidney.
neritic zone The shallow region of the ocean overlying the continental shelf.
nerve A fiber composed primarily of the bundled axons of PNS neurons.
nerve net A weblike system of neurons, characteristic of radially symmetrical animals, such as hydras.
nervous system The fast-acting internal system of communication involving sensory receptors, networks of nerve cells, and connections to muscles and glands that respond to nerve signals; functions in concert with the endocrine system to effect internal regulation and maintain homeostasis.
nervous tissue Tissue made up of neurons and supportive cells.
net ecosystem production (NEP) The gross primary production of an ecosystem minus the energy used by all autotrophs and heterotrophs for respiration.
net primary production (NPP) The gross primary production of an ecosystem minus the energy used by the producers for respiration.
neural crest In vertebrates, a region located along the sides of the neural tube where it pinches off from the ectoderm. Neural crest cells migrate to various parts of the embryo and form pigment cells in the skin and parts of the skull, teeth, adrenal glands, and peripheral nervous system.
neural plasticity The capacity of a nervous system to change with experience.
neural tube A tube of infolded ectodermal cells that runs along the anterior-posterior axis of a vertebrate, just dorsal to the notochord. It will give rise to the central nervous system.
neurohormone A molecule that is secreted by a neuron, travels in body fluids, and acts on specific target cells, changing their functioning.
neuron (nyūr′-on) A nerve cell; the fundamental unit of the nervous system, having structure and properties that allow it to conduct signals by taking advantage of the electrical charge across its plasma membrane.
neuropeptide A relatively short chain of amino acids that serves as a neurotransmitter.
neurotransmitter A molecule that is released from the synaptic terminal of a neuron at a chemical synapse, diffuses across the synaptic cleft, and binds to the postsynaptic cell, triggering a response.
neutral theory The hypothesis that much evolutionary change in genes and proteins has no effect on fitness and therefore is not influenced by natural selection.
neutral variation Genetic variation that does not provide a selective advantage or disadvantage.
neutron A subatomic particle having no electrical charge (electrically neutral), with a mass of about 1.7×10^{-24} g, found in the nucleus of an atom.
neutrophil The most abundant type of white blood cell. Neutrophils are phagocytic and tend to self-destruct as they destroy foreign invaders, limiting their life span to a few days.

nitric oxide (NO) A gas produced by many types of cells that functions as a local regulator and as a neurotransmitter.

nitrogen cycle The natural process by which nitrogen, either from the atmosphere or from decomposed organic material, is converted by soil bacteria to compounds assimilated by plants. This incorporated nitrogen is then taken in by other organisms and subsequently released, acted on by bacteria, and made available again to the nonliving environment.

nitrogen fixation The conversion of atmospheric nitrogen (N_2) to ammonia (NH_3). Biological nitrogen fixation is carried out by certain prokaryotes, some of which have mutualistic relationships with plants.

nociceptor (nō′-si-sep′-tur) A sensory receptor that responds to noxious or painful stimuli; also called a pain receptor.

node A point along the stem of a plant at which leaves are attached.

node of Ranvier (ron′-vē-ā′) Gap in the myelin sheath of certain axons where an action potential may be generated. In saltatory conduction, an action potential is regenerated at each node, appearing to "jump" along the axon from node to node.

nodule A swelling on the root of a legume. Nodules are composed of plant cells that contain nitrogen-fixing bacteria of the genus *Rhizobium*.

noncompetitive inhibitor A substance that reduces the activity of an enzyme by binding to a location remote from the active site, changing the enzyme's shape so that the active site no longer effectively catalyzes the conversion of substrate to product.

nondisjunction An error in meiosis or mitosis in which members of a pair of homologous chromosomes or a pair of sister chromatids fail to separate properly from each other.

nonequilibrium model A model that maintains that communities change constantly after being buffeted by disturbances.

nonpolar covalent bond A type of covalent bond in which electrons are shared equally between two atoms of similar electronegativity.

nonsense mutation A mutation that changes an amino acid codon to one of the three stop codons, resulting in a shorter and usually nonfunctional protein.

norepinephrine A catecholamine that is chemically and functionally similar to epinephrine and acts as a hormone or neurotransmitter; also known as noradrenaline.

norm of reaction The range of phenotypes produced by a single genotype, due to environmental influences.

Northern blotting A technique that enables specific nucleotide sequences to be detected in samples of mRNA. It involves gel electrophoresis of RNA molecules and their transfer to a membrane (blotting), followed by nucleic acid hybridization with a labeled probe.

northern coniferous forest A terrestrial biome characterized by long, cold winters and dominated by cone-bearing trees.

no-till agriculture A plowing technique that minimally disturbs the soil, thereby reducing soil loss.

notochord (nō′-tuh-kord′) A longitudinal, flexible rod made of tightly packed mesodermal cells that runs along the anterior-posterior axis of a chordate in the dorsal part of the body.

nuclear envelope In a eukaryotic cell, the double membrane that surrounds the nucleus, perforated with pores that regulate traffic with the cytoplasm. The outer membrane is continuous with the endoplasmic reticulum.

nuclear lamina A netlike array of protein filaments that lines the inner surface of the nuclear envelope and helps maintain the shape of the nucleus.

nucleariid Member of a group of unicellular, amoeboid protists that are more closely related to fungi than they are to other protists.

nuclease An enzyme that cuts DNA or RNA, either removing one or a few bases or hydrolyzing the DNA or RNA completely into its component nucleotides.

nucleic acid (nū-klā′-ik) A polymer (polynucleotide) consisting of many nucleotide monomers; serves as a blueprint for proteins and, through the actions of proteins, for all cellular activities. The two types are DNA and RNA.

nucleic acid hybridization The process of base pairing between a gene and a complementary sequence on another nucleic acid molecule.

nucleic acid probe In DNA technology, a labeled single-stranded nucleic acid molecule used to locate a specific nucleotide sequence in a nucleic acid sample. Molecules of the probe hydrogen-bond to the complementary sequence wherever it occurs; radioactive, fluorescent, or other labeling of the probe allows its location to be detected.

nucleoid (nū′-klē-oyd) A non-membrane-bounded region in a prokaryotic cell where the DNA is concentrated.

nucleolus (nū-klē′-ō-lus) (plural, **nucleoli**) A specialized structure in the nucleus, consisting of chromosomal regions containing ribosomal RNA (rRNA) genes along with ribosomal proteins imported from the cytoplasm; site of rRNA synthesis and ribosomal subunit assembly. *See also* ribosome.

nucleosome (nū′-klē-ō-sōm′) The basic, bead-like unit of DNA packing in eukaryotes, consisting of a segment of DNA wound around a protein core composed of two copies of each of four types of histone.

nucleotide (nū′-klē-ō-tīd′) The building block of a nucleic acid, consisting of a five-carbon sugar covalently bonded to a nitrogenous base and one or more phosphate groups.

nucleotide excision repair A repair system that removes and then correctly replaces a damaged segment of DNA using the undamaged strand as a guide.

nucleotide-pair substitution A type of point mutation in which one nucleotide in a DNA strand and its partner in the complementary strand are replaced by another pair of nucleotides.

nucleus (1) An atom's central core, containing protons and neutrons. (2) The organelle of a eukaryotic cell that contains the genetic material in the form of chromosomes, made up of chromatin. (3) A cluster of neurons.

nutrition The process by which an organism takes in and makes use of food substances.

obligate aerobe (ob′-lig-et ār′-ōb) An organism that requires oxygen for cellular respiration and cannot live without it.

obligate anaerobe (ob′-lig-et an′-uh-rōb) An organism that only carries out fermentation or anaerobic respiration. Such organisms cannot use oxygen and in fact may be poisoned by it.

ocean acidification Decreasing pH of ocean waters due to absorption of excess atmospheric CO_2 from the burning of fossil fuels.

oceanic pelagic zone Most of the ocean's waters far from shore, constantly mixed by ocean currents.

odorant A molecule that can be detected by sensory receptors of the olfactory system.

Okazaki fragment (ō′-kah-zah′-kē) A short segment of DNA synthesized away from the replication fork on a template strand during DNA replication. Many such segments are joined together to make up the lagging strand of newly synthesized DNA.

olfaction The sense of smell.

oligodendrocyte A type of glial cell that forms insulating myelin sheaths around the axons of neurons in the central nervous system.

oligotrophic lake A nutrient-poor, clear lake with few phytoplankton.

ommatidium (ōm′-uh-tid′-ē-um) (plural, **ommatidia**) One of the facets of the compound eye of arthropods and some polychaete worms.

omnivore An animal that regularly eats animals as well as plants or algae.

oncogene (on′-kō-jēn) A gene found in viral or cellular genomes that is involved in triggering molecular events that can lead to cancer.

oocyte A cell in the female reproductive system that differentiates to form an egg.

oogenesis (ō′-uh-jen′-uh-sis) The process in the ovary that results in the production of female gametes.

oogonium (ō′-uh- gō′-nē-em) (plural, **oogonia**) A cell that divides mitotically to form oocytes.

oomycete (ō′-uh-mī′-sēt) A protist with flagellated cells, such as a water mold, white rust, or downy mildew, that acquires nutrition mainly as a decomposer or plant parasite.

open circulatory system A circulatory system in which fluid called hemolymph bathes the tissues and organs directly and there is no distinction between the circulating fluid and the interstitial fluid.

operator In bacterial and phage DNA, a sequence of nucleotides near the start of an operon to which an active repressor can attach. The binding of the repressor prevents RNA polymerase from attaching to the promoter and transcribing the genes of the operon.

operculum (ō-per′-kyuh-lum) In aquatic osteichthyans, a protective bony flap that covers and protects the gills.

operon (op′-er-on) A unit of genetic function found in bacteria and phages, consisting of a promoter, an operator, and a coordinately regulated cluster of genes whose products function in a common pathway.

opisthokont (uh-pis'-thuh-kont') Member of the diverse clade Opisthokonta, organisms that descended from an ancestor with a posterior flagellum, including fungi, animals, and certain protists.

opposable thumb A thumb that can touch the ventral surface of the fingertips of all four fingers.

opsin A membrane protein bound to a light-absorbing pigment molecule.

optic chiasm The place where the two optic nerves meet and axons representing distinct sides of the visual field are segregated from one another before reaching the brain.

optimal foraging model The basis for analyzing behavior as a compromise between feeding costs and feeding benefits.

oral cavity The mouth of an animal.

orbital The three-dimensional space where an electron is found 90% of the time.

order In Linnaean classification, the taxonomic category above the level of family.

organ A specialized center of body function composed of several different types of tissues.

organ identity gene A plant homeotic gene that uses positional information to determine which emerging leaves develop into which types of floral organs.

organ of Corti The actual hearing organ of the vertebrate ear, located in the floor of the cochlear duct in the inner ear; contains the receptor cells (hair cells) of the ear.

organ system A group of organs that work together in performing vital body functions.

organelle (ōr-guh-nel') Any of several membrane-enclosed structures with specialized functions, suspended in the cytosol of eukaryotic cells.

organic chemistry The study of carbon compounds (organic compounds).

organismal ecology The branch of ecology concerned with the morphological, physiological, and behavioral ways in which individual organisms meet the challenges posed by their biotic and abiotic environments.

organogenesis (ōr-gan'-ō-jen'-uh-sis) The process in which organ rudiments develop from the three germ layers after gastrulation.

orgasm Rhythmic, involuntary contractions of certain reproductive structures in both sexes during the human sexual response cycle.

origin of replication Site where the replication of a DNA molecule begins, consisting of a specific sequence of nucleotides.

orthologous genes Homologous genes that are found in different species because of speciation.

osculum (os'-kyuh-lum) A large opening in a sponge that connects the spongocoel to the environment.

osmoconformer An animal that is isoosmotic with its environment.

osmolarity (oz'-mō-lār'-uh-tē) Solute concentration expressed as molarity.

osmoregulation Regulation of solute concentrations and water balance by a cell or organism.

osmoregulator An animal that controls its internal osmolarity independent of the external environment.

osmosis (oz-mō'-sis) The diffusion of free water across a selectively permeable membrane.

osteichthyan (os'-tē-ik'-thē-an) Member of a vertebrate clade with jaws and mostly bony skeletons.

outer ear One of three main regions of the ear in reptiles (including birds) and mammals; made up of the auditory canal and, in many birds and mammals, the pinna.

outgroup A species or group of species from an evolutionary lineage that is known to have diverged before the lineage that contains the group of species being studied. An outgroup is selected so that its members are closely related to the group of species being studied, but not as closely related as any study-group members are to each other.

oval window In the vertebrate ear, a membrane-covered gap in the skull bone, through which sound waves pass from the middle ear to the inner ear.

ovarian cycle (ō-vār'-ē-un) The cyclic recurrence of the follicular phase, ovulation, and the luteal phase in the mammalian ovary, regulated by hormones.

ovary (ō'-vuh-rē) (1) In flowers, the portion of a carpel in which the egg-containing ovules develop. (2) In animals, the structure that produces female gametes and reproductive hormones.

oviduct (ō'-vuh-duct) A tube passing from the ovary to the vagina in invertebrates or to the uterus in vertebrates, where it is also known as a fallopian tube.

oviparous (ō-vip'-uh-rus) Referring to a type of development in which young hatch from eggs laid outside the mother's body.

ovoviviparous (ō'-vō-vī-vip'-uh-rus) Referring to a type of development in which young hatch from eggs that are retained in the mother's uterus.

ovulation The release of an egg from an ovary. In humans, an ovarian follicle releases an egg during each uterine (menstrual) cycle.

ovule (o'-vyūl) A structure that develops within the ovary of a seed plant and contains the female gametophyte.

oxidation The complete or partial loss of electrons from a substance involved in a redox reaction.

oxidative phosphorylation (fos'-fōr-uh-lā'-shun) The production of ATP using energy derived from the redox reactions of an electron transport chain; the third major stage of cellular respiration.

oxidizing agent The electron acceptor in a redox reaction.

oxytocin (ok'-si-tō'-sen) A hormone produced by the hypothalamus and released from the posterior pituitary. It induces contractions of the uterine muscles during labor and causes the mammary glands to eject milk during nursing.

P generation The true-breeding (homozygous) parent individuals from which F_1 hybrid offspring are derived in studies of inheritance; P stands for "parental."

P site One of a ribosome's three binding sites for tRNA during translation. The P site holds the tRNA carrying the growing polypeptide chain. (P stands for peptidyl tRNA.)

***p53* gene** A tumor-suppressor gene that codes for a specific transcription factor that promotes the synthesis of proteins that inhibit the cell cycle.

paedomorphosis (pē'-duh-mōr'-fuh-sis) The retention in an adult organism of the juvenile features of its evolutionary ancestors.

pain receptor A sensory receptor that responds to noxious or painful stimuli; also called a nociceptor.

paleoanthropology The study of human origins and evolution.

paleontology (pā'-lē-un-tol'-ō-jē) The scientific study of fossils.

pancreas (pan'-krē-us) A gland with exocrine and endocrine tissues. The exocrine portion functions in digestion, secreting enzymes and an alkaline solution into the small intestine via a duct; the ductless endocrine portion functions in homeostasis, secreting the hormones insulin and glucagon into the blood.

pandemic A global epidemic.

Pangaea (pan-jē'-uh) The supercontinent that formed near the end of the Paleozoic era, when plate movements brought all the landmasses of Earth together.

parabasalid A protist, such as a trichomonad, with modified mitochondria.

paracrine Referring to a secreted molecule that acts on a neighboring cell.

paralogous genes Homologous genes that are found in the same genome as a result of gene duplication.

paraphyletic (pār'-uh-fī-let'-ik) Pertaining to a group of taxa that consists of a common ancestor and some, but not all, of its descendants.

parareptile A basal group of reptiles, consisting mostly of large, stocky quadrupedal herbivores. Parareptiles died out in the late Triassic period.

parasite (pār'-uh-sīt) An organism that feeds on the cell contents, tissues, or body fluids of another species (the host) while in or on the host organism. Parasites harm but usually do not kill their host.

parasitism (pār'-uh-sit-izm) A symbiotic relationship in which one organism, the parasite, benefits at the expense of another, the host, by living either within or on the host.

parasympathetic division One of three divisions of the autonomic nervous system; generally enhances body activities that gain and conserve energy, such as digestion and reduced heart rate.

parathyroid gland One of four small endocrine glands, embedded in the surface of the thyroid gland, that secrete parathyroid hormone.

parathyroid hormone (PTH) A hormone secreted by the parathyroid glands that raises blood calcium level by promoting calcium release from bone and calcium retention by the kidneys.

parenchyma cell (puh-ren'-ki-muh) A relatively unspecialized plant cell type that carries out most of the metabolism, synthesizes and stores organic products, and develops into a more differentiated cell type.

parental type An offspring with a phenotype that matches one of the true-breeding parental (P generation) phenotypes; also refers to the phenotype itself.

Parkinson's disease A progressive brain disease characterized by difficulty in initiating movements, slowness of movement, and rigidity.

parthenogenesis (par'-thuh-nō'-jen'-uh-sis) A form of asexual reproduction in which females produce offspring from unfertilized eggs.

partial pressure The pressure exerted by a particular gas in a mixture of gases (for instance, the pressure exerted by oxygen in air).

passive immunity Short-term immunity conferred by the transfer of antibodies, as occurs in the transfer of maternal antibodies to a fetus or nursing infant.

passive transport The diffusion of a substance across a biological membrane with no expenditure of energy.

pathogen An organism, virus, viroid, or prion that causes disease.

pattern formation The development of a multicellular organism's spatial organization, the arrangement of organs and tissues in their characteristic places in three-dimensional space.

peat Extensive deposits of partially decayed organic material often formed primarily from the wetland moss *Sphagnum*.

pedigree A diagram of a family tree with conventional symbols, showing the occurrence of heritable characters in parents and offspring over multiple generations.

pelagic zone The open-water component of aquatic biomes.

penis The copulatory structure of male mammals.

PEP carboxylase An enzyme that adds CO_2 to phosphoenolpyruvate (PEP) to form oxaloacetate in mesophyll cells of C_4 plants. It acts prior to photosynthesis.

pepsin An enzyme present in gastric juice that begins the hydrolysis of proteins.

pepsinogen The inactive form of pepsin secreted by chief cells located in gastric pits of the stomach.

peptide bond The covalent bond between the carboxyl group on one amino acid and the amino group on another, formed by a dehydration reaction.

peptidoglycan (pep'-tid-ō-glī'-kan) A type of polymer in bacterial cell walls consisting of modified sugars cross-linked by short polypeptides.

perception The interpretation of sensory system input by the brain.

pericycle The outermost layer in the vascular cylinder, from which lateral roots arise.

periderm (pār'-uh-derm') The protective coat that replaces the epidermis in woody plants during secondary growth, formed of the cork and cork cambium.

peripheral nervous system (PNS) The sensory and motor neurons that connect to the central nervous system.

peripheral protein A protein loosely bound to the surface of a membrane or to part of an integral protein and not embedded in the lipid bilayer.

peristalsis (pār'-uh-stal'-sis) (1) Alternating waves of contraction and relaxation in the smooth muscles lining the alimentary canal that push food along the canal. (2) A type of movement on land produced by rhythmic waves of muscle contractions passing from front to back, as in many annelids.

peristome A ring of interlocking, tooth-like structures on the upper part of a moss capsule (sporangium), often specialized for gradual spore discharge.

peritubular capillary One of the tiny blood vessels that form a network surrounding the proximal and distal tubules in the kidney.

permafrost A permanently frozen soil layer.

peroxisome (puh-rok'-suh-sōm') An organelle containing enzymes that transfer hydrogen atoms from various substrates to oxygen (O_2), producing and then degrading hydrogen peroxide (H_2O_2).

petal A modified leaf of a flowering plant. Petals are the often colorful parts of a flower that advertise it to insects and other pollinators.

petiole (pet'-ē-ōl) The stalk of a leaf, which joins the leaf to a node of the stem.

pH A measure of hydrogen ion concentration equal to $-\log [H^+]$ and ranging in value from 0 to 14.

phage (fāj) A virus that infects bacteria; also called a bacteriophage.

phagocytosis (fag'-ō-sī-tō'-sis) A type of endocytosis in which large particulate substances or small organisms are taken up by a cell. It is carried out by some protists and by certain immune cells of animals (in mammals, mainly macrophages, neutrophils, and dendritic cells).

pharyngeal cleft (fuh-rin'-jē-ul) In chordate embryos, one of the grooves that separate a series of pouches along the sides of the pharynx and may develop into a pharyngeal slit.

pharyngeal slit (fuh-rin'-jē-ul) In chordate embryos, one of the slits that form from the pharyngeal clefts and communicate to the outside, later developing into gill slits in many vertebrates.

pharynx (fār'-inks) (1) An area in the vertebrate throat where air and food passages cross. (2) In flatworms, the muscular tube that protrudes from the ventral side of the worm and ends in the mouth.

phase change A shift from one developmental phase to another.

phenotype (fē'-nō-tīp) The observable physical and physiological traits of an organism, which are determined by its genetic makeup.

pheromone (fār'-uh-mōn) In animals and fungi, a small molecule released into the environment that functions in communication between members of the same species. In animals, it acts much like a hormone in influencing physiology and behavior.

phloem (flō'-em) Vascular plant tissue consisting of living cells arranged into elongated tubes that transport sugar and other organic nutrients throughout the plant.

phloem sap The sugar-rich solution carried through a plant's sieve tubes.

phosphate group A chemical group consisting of a phosphorus atom bonded to four oxygen atoms; important in energy transfer.

phospholipid (fos'-fō-lip'-id) A lipid made up of glycerol joined to two fatty acids and a phosphate group. The hydrocarbon chains of the fatty acids act as nonpolar, hydrophobic tails, while the rest of the molecule acts as a polar, hydrophilic head. Phospholipids form bilayers that function as biological membranes.

phosphorylated intermediate A molecule (often a reactant) with a phosphate group covalently bound to it, making it more reactive (less stable) than the unphosphorylated molecule.

photic zone (fō'-tic) The narrow top layer of an ocean or lake, where light penetrates sufficiently for photosynthesis to occur.

photoautotroph (fō'-tō-ot'-ō-trōf) An organism that harnesses light energy to drive the synthesis of organic compounds from carbon dioxide.

photoheterotroph (fō'-tō-het'-er-ō-trōf) An organism that uses light to generate ATP but must obtain carbon in organic form.

photomorphogenesis Effects of light on plant morphology.

photon (fō'-ton) A quantum, or discrete quantity, of light energy that behaves as if it were a particle.

photoperiodism (fō'-tō-pēr'-ē-ō-dizm) A physiological response to photoperiod, the relative lengths of night and day. An example of photoperiodism is flowering.

photophosphorylation (fō'-tō-fos'-fōr-uh-lā'-shun) The process of generating ATP from ADP and phosphate by means of chemiosmosis, using a proton-motive force generated across the thylakoid membrane of the chloroplast or the membrane of certain prokaryotes during the light reactions of photosynthesis.

photoreceptor An electromagnetic receptor that detects the radiation known as visible light.

photorespiration A metabolic pathway that consumes oxygen and ATP, releases carbon dioxide, and decreases photosynthetic output. Photorespiration generally occurs on hot, dry, bright days, when stomata close and the O_2/CO_2 ratio in the leaf increases, favoring the binding of O_2 rather than CO_2 by rubisco.

photosynthesis (fō'-tō-sin'-thi-sis) The conversion of light energy to chemical energy that is stored in sugars or other organic compounds; occurs in plants, algae, and certain prokaryotes.

photosystem A light-capturing unit located in the thylakoid membrane of the chloroplast or in the membrane of some prokaryotes, consisting of a reaction-center complex surrounded by numerous light-harvesting complexes. There are two types of photosystems, I and II; they absorb light best at different wavelengths.

photosystem I (PS I) A light-capturing unit in a chloroplast's thylakoid membrane or in the membrane of some prokaryotes; it has two molecules of P700 chlorophyll *a* at its reaction center.

photosystem II (PS II) One of two light-capturing units in a chloroplast's thylakoid membrane or in the membrane of some prokaryotes; it has two molecules of P680 chlorophyll *a* at its reaction center.

phototropism (fō′-tō-trō′-pizm) Growth of a plant shoot toward or away from light.

phragmoplast (frag′-mō-plast′) An alignment of cytoskeletal elements and Golgi-derived vesicles that forms across the midline of a dividing plant cell.

phyllotaxy (fil′-uh-tak′-sē) The pattern of leaf attachment to the stem of a plant.

PhyloCode Proposed system of classification of organisms based on evolutionary relationships: Only groups that include a common ancestor and all of its descendants are named.

phylogenetic species concept A definition of species as the smallest group of individuals that share a common ancestor, forming one branch on the tree of life.

phylogenetic tree A branching diagram that represents a hypothesis about the evolutionary history of a group of organisms.

phylogeny (fī-loj′-uh-nē) The evolutionary history of a species or group of related species.

phylum (fī′-lum) (plural, **phyla**) In Linnaean classification, the taxonomic category above class.

physical map A genetic map in which the actual physical distances between genes or other genetic markers are expressed, usually as the number of base pairs along the DNA.

physiology The processes and functions of an organism.

phytochrome (fī′-tuh-krōm) A type of light receptor in plants that mostly absorbs red light and regulates many plant responses, such as seed germination and shade avoidance.

phytoremediation An emerging technology that seeks to reclaim contaminated areas by taking advantage of some plant species' ability to extract heavy metals and other pollutants from the soil and to concentrate them in easily harvested portions of the plant.

pilus (plural, **pili**) (pī′-lus, pī′-lī) In bacteria, a structure that links one cell to another at the start of conjugation; also known as a sex pilus or conjugation pilus.

pineal gland (pī′-nē-ul) A small gland on the dorsal surface of the vertebrate forebrain that secretes the hormone melatonin.

pinocytosis (pī′-nō-sī-tō′-sis) A type of endocytosis in which the cell ingests extracellular fluid and its dissolved solutes.

pistil A single carpel or a group of fused carpels.

pith Ground tissue that is internal to the vascular tissue in a stem; in many monocot roots, parenchyma cells that form the central core of the vascular cylinder.

pituitary gland (puh-tū′-uh-tār′-ē) An endocrine gland at the base of the hypothalamus; consists of a posterior lobe, which stores and releases two hormones produced by the hypothalamus, and an anterior lobe, which produces and secretes many hormones that regulate diverse body functions.

placenta (pluh-sen′-tuh) A structure in the pregnant uterus for nourishing a viviparous fetus with the mother's blood supply; formed from the uterine lining and embryonic membranes.

placental transfer cell A plant cell that enhances the transfer of nutrients from parent to embryo.

placoderm A member of an extinct group of fishlike vertebrates that had jaws and were enclosed in a tough outer armor.

planarian A free-living flatworm found in ponds and streams.

plasma (plaz′-muh) The liquid matrix of blood in which the blood cells are suspended.

plasma cell The antibody-secreting effector cell of humoral immunity. Plasma cells arise from antigen-stimulated B cells.

plasma membrane The membrane at the boundary of every cell that acts as a selective barrier, regulating the cell's chemical composition.

plasmid (plaz′-mid) A small, circular, double-stranded DNA molecule that carries accessory genes separate from those of a bacterial chromosome; in DNA cloning, used as vectors carrying up to about 10,000 base pairs (10 kb) of DNA. Plasmids are also found in some eukaryotes, such as yeasts.

plasmodesma (plaz′-mō-dez′-muh) (plural, **plasmodesmata**) An open channel through the cell wall that connects the cytoplasm of adjacent plant cells, allowing water, small solutes, and some larger molecules to pass between the cells.

plasmodial slime mold (plaz-mō′-dē-ul) A type of protist that has amoeboid cells, flagellated cells, and a plasmodial feeding stage in its life cycle.

plasmodium A single mass of cytoplasm containing many diploid nuclei that forms during the life cycle of some slime molds.

plasmogamy (plaz-moh′-guh-mē) In fungi, the fusion of the cytoplasm of cells from two individuals; occurs as one stage of sexual reproduction, followed later by karyogamy.

plasmolysis (plaz-mol′-uh-sis) A phenomenon in walled cells in which the cytoplasm shrivels and the plasma membrane pulls away from the cell wall; occurs when the cell loses water to a hypertonic environment.

plastid One of a family of closely related organelles that includes chloroplasts, chromoplasts, and amyloplasts. Plastids are found in cells of photosynthetic eukaryotes.

plate tectonics The theory that the continents are part of great plates of Earth's crust that float on the hot, underlying portion of the mantle. Movements in the mantle cause the continents to move slowly over time.

platelet A pinched-off cytoplasmic fragment of a specialized bone marrow cell. Platelets circulate in the blood and are important in blood clotting.

pleiotropy (plī′-o-truh-pē) The ability of a single gene to have multiple effects.

pluripotent Describing a cell that can give rise to many, but not all, parts of an organism.

point mutation A change in a single nucleotide pair of a gene.

polar covalent bond A covalent bond between atoms that differ in electronegativity. The shared electrons are pulled closer to the more electronegative atom, making it slightly negative and the other atom slightly positive.

polar molecule A molecule (such as water) with an uneven distribution of charges in different regions of the molecule.

polarity A lack of symmetry; structural differences in opposite ends of an organism or structure, such as the root end and shoot end of a plant.

pollen grain In seed plants, a structure consisting of the male gametophyte enclosed within a pollen wall.

pollen tube A tube that forms after germination of the pollen grain and that functions in the delivery of sperm to the ovule.

pollination (pol′-uh-nā′-shun) The transfer of pollen to the part of a seed plant containing the ovules, a process required for fertilization.

poly-A tail A sequence of 50–250 adenine nucleotides added onto the 3′ end of a pre-mRNA molecule.

polygamous Referring to a type of relationship in which an individual of one sex mates with several of the other.

polygenic inheritance (pol′-ē-jen′-ik) An additive effect of two or more genes on a single phenotypic character.

polymer (pol′-uh-mer) A long molecule consisting of many similar or identical monomers linked together by covalent bonds.

polymerase chain reaction (PCR) (puh-lim′-uh-rās) A technique for amplifying DNA *in vitro* by incubating it with specific primers, a heat-resistant DNA polymerase, and nucleotides.

polynucleotide (pol′-ē-nū′-klē-ō-tīd) A polymer consisting of many nucleotide monomers in a chain. The nucleotides can be those of DNA or RNA.

polyp The sessile variant of the cnidarian body plan. The alternate form is the medusa.

polypeptide (pol′-ē-pep′-tīd) A polymer of many amino acids linked together by peptide bonds.

polyphyletic (pol′-ē-fī-let′-ik) Pertaining to a group of taxa derived from two or more different ancestors.

polyploidy (pol′-ē-ploy′-dē) A chromosomal alteration in which the organism possesses more than two complete chromosome sets. It is the result of an accident of cell division.

polyribosome (polysome) (pol′-ē-rī′-buh-sōm′) A group of several ribosomes attached to, and translating, the same messenger RNA molecule.

polysaccharide (pol′-ē-sak′-uh-rīd) A polymer of many monosaccharides, formed by dehydration reactions.

polytomy (puh-lit′-uh-mē) In a phylogenetic tree, a branch point from which more than two descendant taxa emerge. A polytomy indicates that the evolutionary relationships between the descendant taxa are not yet clear.

pons A portion of the brain that participates in certain automatic, homeostatic functions, such as regulating the breathing centers in the medulla.

population A group of individuals of the same species that live in the same area and interbreed, producing fertile offspring.

population dynamics The study of how complex interactions between biotic and abiotic factors influence variations in population size.

population ecology The study of populations in relation to their environment, including environmental influences on population density and distribution, age structure, and variations in population size.

positional information Molecular cues that control pattern formation in an animal or plant embryonic structure by indicating a cell's location relative to the organism's body axes. These cues elicit a response by genes that regulate development.

positive feedback A form of regulation in which an end product of a process speeds up that process; in physiology, a control mechanism in which a change in a variable triggers a response that reinforces or amplifies the change.

positive pressure breathing A breathing system in which air is forced into the lungs.

posterior Pertaining to the rear, or tail end, of a bilaterally symmetrical animal.

posterior pituitary An extension of the hypothalamus composed of nervous tissue that secretes oxytocin and antidiuretic hormone made in the hypothalamus; a temporary storage site for these hormones.

postzygotic barrier (pōst′-zī-got′-ik) A reproductive barrier that prevents hybrid zygotes produced by two different species from developing into viable, fertile adults.

potential energy The energy that matter possesses as a result of its location or spatial arrangement (structure).

predation An interaction between species in which one species, the predator, eats the other, the prey.

pregnancy The condition of carrying one or more embryos in the uterus.

prepuce (prē′-pyūs) A fold of skin covering the head of the clitoris or penis.

pressure potential (Ψ_P) A component of water potential that consists of the physical pressure on a solution, which can be positive, zero, or negative.

prezygotic barrier (prē′-zī-got′-ik) A reproductive barrier that impedes mating between species or hinders fertilization if interspecific mating is attempted.

primary cell wall In plants, a relatively thin and flexible layer that surrounds the plasma membrane of a young cell.

primary consumer An herbivore; an organism that eats plants or other autotrophs.

primary electron acceptor In the thylakoid membrane of a chloroplast or in the membrane of some prokaryotes, a specialized molecule that shares the reaction-center complex with a pair of chlorophyll *a* molecules and that accepts an electron from them.

primary growth Growth produced by apical meristems, lengthening stems and roots.

primary immune response The initial adaptive immune response to an antigen, which appears after a lag of about 10 to 17 days.

primary oocyte (ō′-uh-sīt) An oocyte prior to completion of meiosis I.

primary producer An autotroph, usually a photosynthetic organism. Collectively, autotrophs make up the trophic level of an ecosystem that ultimately supports all other levels.

primary production The amount of light energy converted to chemical energy (organic compounds) by the autotrophs in an ecosystem during a given time period.

primary structure The level of protein structure referring to the specific linear sequence of amino acids.

primary succession A type of ecological succession that occurs in an area where there were originally no organisms present and where soil has not yet formed.

primary transcript An initial RNA transcript from any gene; also called pre-mRNA when transcribed from a protein-coding gene.

primary visual cortex The destination in the occipital lobe of the cerebrum for most of the axons from the lateral geniculate nuclei.

primase An enzyme that joins RNA nucleotides to make a primer during DNA replication, using the parental DNA strand as a template.

primer A short stretch of RNA with a free 3′ end, bound by complementary base pairing to the template strand and elongated with DNA nucleotides during DNA replication.

primitive streak A thickening along the future anterior-posterior axis on the surface of an early avian or mammalian embryo, caused by a piling up of cells as they congregate at the midline before moving into the embryo.

prion An infectious agent that is a misfolded version of a normal cellular protein. Prions appear to increase in number by converting correctly folded versions of the protein to more prions.

problem solving The cognitive activity of devising a method to proceed from one state to another in the face of real or apparent obstacles.

producer An organism that produces organic compounds from CO_2 by harnessing light energy (in photosynthesis) or by oxidizing inorganic chemicals (in chemosynthetic reactions carried out by some prokaryotes).

product A material resulting from a chemical reaction.

production efficiency The percentage of energy stored in assimilated food that is not used for respiration or eliminated as waste.

progesterone A steroid hormone that prepares the uterus for pregnancy; the major progestin in mammals.

progestin Any steroid hormone with progesterone-like activity.

progymnosperm (prō′-jim′-nō-sperm) An extinct seedless vascular plant that may be ancestral to seed plants.

prokaryotic cell (prō′-kār′-ē-ot′-ik) A type of cell lacking a membrane-enclosed nucleus and membrane-enclosed organelles. Organisms with prokaryotic cells (bacteria and archaea) are called prokaryotes.

prolactin A hormone produced and secreted by the anterior pituitary with a great diversity of effects in different vertebrate species. In mammals, it stimulates growth of and milk production by the mammary glands.

proliferative phase That portion of the uterine (menstrual) cycle when the endometrium regenerates and thickens.

prometaphase The second stage of mitosis, in which the nuclear envelope fragments and the spindle microtubules attach to the kinetochores of the chromosomes.

promiscuous Referring to a type of relationship in which mating occurs with no strong pair-bonds or lasting relationships.

promoter A specific nucleotide sequence in the DNA of a gene that binds RNA polymerase, positioning it to start transcribing RNA at the appropriate place.

prophage (prō′-fāj) A phage genome that has been inserted into a specific site on a bacterial chromosome.

prophase The first stage of mitosis, in which the chromatin condenses into discrete chromosomes visible with a light microscope, the mitotic spindle begins to form, and the nucleolus disappears but the nucleus remains intact.

prostaglandin (pros′-tuh-glan′-din) One of a group of modified fatty acids secreted by virtually all tissues and performing a wide variety of functions as local regulators.

prostate gland (pros′-tāt) A gland in human males that secretes an acid-neutralizing component of semen.

protease An enzyme that digests proteins by hydrolysis.

proteasome A giant protein complex that recognizes and destroys proteins tagged for elimination by the small protein ubiquitin.

protein (prō′-tēn) A biologically functional molecule consisting of one or more polypeptides folded and coiled into a specific three-dimensional structure.

protein kinase An enzyme that transfers phosphate groups from ATP to a protein, thus phosphorylating the protein.

protein phosphatase An enzyme that removes phosphate groups from (dephosphorylates) proteins, often functioning to reverse the effect of a protein kinase.

proteoglycan (prō′-tē-ō-glī′-kan) A large molecule consisting of a small core protein with many carbohydrate chains attached, found in the extracellular matrix of animal cells. A proteoglycan may consist of up to 95% carbohydrate.

proteomics (prō′-tē-ō′-miks) The systematic study of the full protein sets (proteomes) encoded by genomes.

protist An informal term applied to any eukaryote that is not a plant, animal, or fungus. Most protists are unicellular, though some are colonial or multicellular.

protocell An abiotic precursor of a living cell that had a membrane-like structure and that maintained an internal chemistry different from that of its surroundings.

proton (prō′-ton) A subatomic particle with a single positive electrical charge, with a mass of about 1.7×10^{-24} g, found in the nucleus of an atom.

proton pump An active transport protein in a cell membrane that uses ATP to transport hydrogen ions out of a cell against their concentration gradient, generating a membrane potential in the process.

protonema (plural, **protonemata**) A mass of green, branched, one-cell-thick filaments produced by germinating moss spores.

protonephridium (prō'-tō-nuh-frid'-ē-uhm) (plural, **protonephridia**) An excretory system, such as the flame bulb system of flatworms, consisting of a network of tubules lacking internal openings.

proton-motive force The potential energy stored in the form of a proton electrochemical gradient, generated by the pumping of hydrogen ions (H^+) across a biological membrane during chemiosmosis.

proto-oncogene (prō'-tō-on'-kō-jēn) A normal cellular gene that has the potential to become an oncogene.

protoplast The living part of a plant cell, which also includes the plasma membrane.

protoplast fusion The fusing of two protoplasts from different plant species that would otherwise be reproductively incompatible.

protostome development In animals, a developmental mode distinguished by the development of the mouth from the blastopore; often also characterized by spiral cleavage and by the body cavity forming when solid masses of mesoderm split.

provirus A viral genome that is permanently inserted into a host genome.

proximal tubule In the vertebrate kidney, the portion of a nephron immediately downstream from Bowman's capsule that conveys and helps refine filtrate.

pseudocoelomate (sū'-dō-sē'-lō-māt) An animal whose body cavity is lined by tissue derived from mesoderm and endoderm.

pseudogene (sū'-dō-jēn) A DNA segment very similar to a real gene but which does not yield a functional product; a DNA segment that formerly functioned as a gene but has become inactivated in a particular species because of mutation.

pseudopodium (sū'-dō-pō'-dē-um) (plural, **pseudopodia**) A cellular extension of amoeboid cells used in moving and feeding.

pterophyte (ter'-uh-fīt) An informal name for a member of the phylum Pterophyta, which includes ferns, horsetails, and whisk ferns and their relatives.

pterosaur Winged reptile that lived during the Mesozoic era.

pulmocutaneous circuit A branch of the circulatory system in many amphibians that supplies the lungs and skin.

pulmonary circuit The branch of the circulatory system that supplies the lungs.

pulse The rhythmic bulging of the artery walls with each heartbeat.

punctuated equilibria In the fossil record, long periods of apparent stasis, in which a species undergoes little or no morphological change, interrupted by relatively brief periods of sudden change.

Punnett square A diagram used in the study of inheritance to show the predicted genotypic results of random fertilization in genetic crosses between individuals of known genotype.

pupil The opening in the iris, which admits light into the interior of the vertebrate eye. Muscles in the iris regulate its size.

purine (pyū'-rēn) One of two types of nitrogenous bases found in nucleotides, characterized by a six-membered ring fused to a five-membered ring. Adenine (A) and guanine (G) are purines.

pyrimidine (puh-rim'-uh-dēn) One of two types of nitrogenous bases found in nucleotides, characterized by a six-membered ring. Cytosine (C), thymine (T), and uracil (U) are pyrimidines.

quantitative character A heritable feature that varies continuously over a range rather than in an either-or fashion.

quaternary structure (kwot-er-nār-ē) The particular shape of a complex, aggregate protein, defined by the characteristic three-dimensional arrangement of its constituent subunits, each a polypeptide.

R plasmid A bacterial plasmid carrying genes that confer resistance to certain antibiotics.

radial cleavage A type of embryonic development in deuterostomes in which the planes of cell division that transform the zygote into a ball of cells are either parallel or perpendicular to the vertical axis of the embryo, thereby aligning tiers of cells one above the other.

radial symmetry Symmetry in which the body is shaped like a pie or barrel (lacking a left side and a right side) and can be divided into mirror-imaged halves by any plane through its central axis.

radiation The emission of electromagnetic waves by all objects warmer than absolute zero.

radicle An embryonic root of a plant.

radioactive isotope An isotope (an atomic form of a chemical element) that is unstable; the nucleus decays spontaneously, giving off detectable particles and energy.

radiolarian A protist, usually marine, with a shell generally made of silica and pseudopodia that radiate from the central body.

radiometric dating A method for determining the absolute age of rocks and fossils, based on the half-life of radioactive isotopes.

radula A straplike scraping organ used by many molluscs during feeding.

***ras* gene** A gene that codes for Ras, a G protein that relays a growth signal from a growth factor receptor on the plasma membrane to a cascade of protein kinases, ultimately resulting in stimulation of the cell cycle.

ratite (rat'-īt) Member of the group of flightless birds.

ray-finned fish Member of the class Actinopterygii, aquatic osteichthyans with fins supported by long, flexible rays, including tuna, bass, and herring.

reabsorption In excretory systems, the recovery of solutes and water from filtrate.

reactant A starting material in a chemical reaction.

reaction-center complex A complex of proteins associated with a special pair of chlorophyll *a* molecules and a primary electron acceptor. Located centrally in a photosystem, this complex triggers the light reactions of photosynthesis. Excited by light energy, the pair of chlorophylls donates an electron to the primary electron acceptor, which passes an electron to an electron transport chain.

reading frame On an mRNA, the triplet grouping of ribonucleotides used by the translation machinery during polypeptide synthesis.

receptacle The base of a flower; the part of the stem that is the site of attachment of the floral organs.

receptor potential An initial response of a receptor cell to a stimulus, consisting of a change in voltage across the receptor membrane proportional to the stimulus strength.

receptor tyrosine kinase (RTK) A receptor protein spanning the plasma membrane, the cytoplasmic (intracellular) part of which can catalyze the transfer of a phosphate group from ATP to a tyrosine on another protein. Receptor tyrosine kinases often respond to the binding of a signaling molecule by dimerizing and then phosphorylating a tyrosine on the cytoplasmic portion of the other receptor in the dimer. The phosphorylated tyrosines on the receptors then activate other signal transduction proteins within the cell.

receptor-mediated endocytosis (en'-dō-sī-tō'-sis) The movement of specific molecules into a cell by the inward budding of vesicles containing proteins with receptor sites specific to the molecules being taken in; enables a cell to acquire bulk quantities of specific substances.

recessive allele An allele whose phenotypic effect is not observed in a heterozygote.

reciprocal altruism Altruistic behavior between unrelated individuals, whereby the altruistic individual benefits in the future when the beneficiary reciprocates.

recombinant chromosome A chromosome created when crossing over combines DNA from two parents into a single chromosome.

recombinant DNA A DNA molecule made *in vitro* with segments from different sources.

recombinant type (recombinant) An offspring whose phenotype differs from that of the true-breeding P generation parents; also refers to the phenotype itself.

rectum The terminal portion of the large intestine, where the feces are stored prior to elimination.

red alga A photosynthetic protist, named for its color, which results from a red pigment that masks the green of chlorophyll. Most red algae are multicellular and marine.

redox reaction (rē'-doks) A chemical reaction involving the complete or partial transfer of one or more electrons from one reactant to another; short for **red**uction-**ox**idation reaction.

reducing agent The electron donor in a redox reaction.

reduction The complete or partial addition of electrons to a substance involved in a redox reaction.

reflex An automatic reaction to a stimulus, mediated by the spinal cord or lower brain.

refractory period (rē-frakt'-ōr-ē) The short time immediately after an action potential in which the neuron cannot respond to another stimulus, owing to the inactivation of voltage-gated sodium channels.

regulator An animal for which mechanisms of homeostasis moderate internal changes in a particular variable in the face of external fluctuation of that variable.

regulatory gene A gene that codes for a protein, such as a repressor, that controls the transcription of another gene or group of genes.

reinforcement In evolutionary biology, a process in which a process in which natural selection strengthens prezygotic barriers to reproduction, thus reducing the chances of hybrid formation. Such a process is likely to occur only if hybrid offspring are less fit than members of the parent species.

relative abundance The proportional abundance of different species in a community.

relative fitness The contribution an individual makes to the gene pool of the next generation, relative to the contributions of other individuals in the population.

renal cortex The outer portion of the vertebrate kidney.

renal medulla The inner portion of the vertebrate kidney, beneath the renal cortex.

renal pelvis The funnel-shaped chamber that receives processed filtrate from the vertebrate kidney's collecting ducts and is drained by the ureter.

renin-angiotensin-aldosterone system (RAAS) A hormone cascade pathway that helps regulate blood pressure and blood volume.

repetitive DNA Nucleotide sequences, usually noncoding, that are present in many copies in a eukaryotic genome. The repeated units may be short and arranged tandemly (in series) or long and dispersed in the genome.

replication fork A Y-shaped region on a replicating DNA molecule where the parental strands are being unwound and new strands are being synthesized.

repressor A protein that inhibits gene transcription. In prokaryotes, repressors bind to the DNA in or near the promoter. In eukaryotes, repressors may bind to control elements within enhancers, to activators, or to other proteins in a way that blocks activators from binding to DNA.

reproductive isolation The existence of biological factors (barriers) that impede members of two species from producing viable, fertile offspring.

reproductive table An age-specific summary of the reproductive rates in a population.

reptile Member of the clade of amniotes that includes tuataras, lizards, snakes, turtles, crocodilians, and birds.

residual volume The amount of air that remains in the lungs after forceful exhalation.

resource partitioning The division of environmental resources by coexisting species such that the niche of each species differs by one or more significant factors from the niches of all coexisting species.

respiratory pigment A protein that transports oxygen in blood or hemolymph.

response (1) In cellular communication, the change in a specific cellular activity brought about by a transduced signal from outside the cell. (2) In feedback regulation, a physiological activity triggered by a change in a variable.

resting potential The membrane potential characteristic of a nonconducting excitable cell, with the inside of the cell more negative than the outside.

restriction enzyme An endonuclease (type of enzyme) that recognizes and cuts DNA molecules foreign to a bacterium (such as phage genomes). The enzyme cuts at specific nucleotide sequences (restriction sites).

restriction fragment A DNA segment that results from the cutting of DNA by a restriction enzyme.

restriction fragment length polymorphism (RFLP) A single nucleotide polymorphism (SNP) that exists in the restriction site for a particular enzyme, thus making the site unrecognizable by that enzyme and changing the lengths of the restriction fragments formed by digestion with that enzyme. A RFLP can be in coding or noncoding DNA.

restriction site A specific sequence on a DNA strand that is recognized and cut by a restriction enzyme.

reticular formation (re-tik′-yū-ler) A diffuse network of neurons in the core of the brainstem that filters information traveling to the cerebral cortex.

retina (ret′-i-nuh) The innermost layer of the vertebrate eye, containing photoreceptor cells (rods and cones) and neurons; transmits images formed by the lens to the brain via the optic nerve.

retinal The light-absorbing pigment in rods and cones of the vertebrate eye.

retrotransposon (re′-trō-trans-pō′-zon) A transposable element that moves within a genome by means of an RNA intermediate, a transcript of the retrotransposon DNA.

retrovirus (re′-trō-vī′-rus) An RNA virus that replicates by transcribing its RNA into DNA and then inserting the DNA into a cellular chromosome; an important class of cancer-causing viruses.

reverse transcriptase (tran-skrip′-tās) An enzyme encoded by certain viruses (retroviruses) that uses RNA as a template for DNA synthesis.

reverse transcriptase–polymerase chain reaction (RT-PCR) A technique for determining expression of a particular gene. It uses reverse transcriptase and DNA polymerase to synthesize cDNA from all the mRNA in a sample and then subjects the cDNA to PCR amplification using primers specific for the gene of interest.

Rhizaria (rī-za′-rē-uh) One of five supergroups of eukaryotes proposed in a current hypothesis of the evolutionary history of eukaryotes; a morphologically diverse protist clade that is defined by DNA similarities. *See also* Excavata, Chromalveolata, Archaeplastida, and Unikonta.

rhizobacterium A soil bacterium whose population size is much enhanced in the rhizosphere, the soil region close to a plant's roots.

rhizoid (rī′-zoyd) A long, tubular single cell or filament of cells that anchors bryophytes to the ground. Unlike roots, rhizoids are not composed of tissues, lack specialized conducting cells, and do not play a primary role in water and mineral absorption.

rhizosphere The soil region close to plant roots and characterized by a high level of microbiological activity.

rhodopsin (rō-dop′-sin) A visual pigment consisting of retinal and opsin. Upon absorbing light, the retinal changes shape and dissociates from the opsin.

rhythm method A form of contraception that relies on refraining from sexual intercourse when conception is most likely to occur; also called natural family planning.

ribonucleic acid (RNA) (rī′-bō-nū-klā′-ik) A type of nucleic acid consisting of a polynucleotide made up of nucleotide monomers with a ribose sugar and the nitrogenous bases adenine (A), cytosine (C), guanine (G), and uracil (U); usually single-stranded; functions in protein synthesis, gene regulation, and as the genome of some viruses.

ribose The sugar component of RNA nucleotides.

ribosomal RNA (rRNA) (rī′-buh-sō′-mul) RNA molecules that, together with proteins, make up ribosomes; the most abundant type of RNA.

ribosome (rī′-buh-sōm′) A complex of rRNA and protein molecules that functions as a site of protein synthesis in the cytoplasm; consists of a large and a small subunit. In eukaryotic cells, each subunit is assembled in the nucleolus. *See also* nucleolus.

ribozyme (rī′-buh-zīm) An RNA molecule that functions as an enzyme, such as an intron that catalyzes its own removal during RNA splicing.

RNA interference (RNAi) A technique used to silence the expression of selected genes. RNAi uses synthetic double-stranded RNA molecules that match the sequence of a particular gene to trigger the breakdown of the gene's messenger RNA.

RNA polymerase An enzyme that links ribonucleotides into a growing RNA chain during transcription, based on complementary binding to nucleotides on a DNA template strand.

RNA processing Modification of RNA primary transcripts, including splicing out of introns, joining together of exons, and alteration of the 5′ and 3′ ends.

RNA splicing After synthesis of a eukaryotic primary RNA transcript, the removal of portions of the transcript (introns) that will not be included in the mRNA and the joining together of the remaining portions (exons).

rod A rodlike cell in the retina of the vertebrate eye, sensitive to low light intensity.

root An organ in vascular plants that anchors the plant and enables it to absorb water and minerals from the soil.

root cap A cone of cells at the tip of a plant root that protects the apical meristem.

root hair A tiny extension of a root epidermal cell, growing just behind the root tip and increasing surface area for absorption of water and minerals.

root pressure Pressure exerted in the roots of plants as the result of osmosis, causing exudation from cut stems and guttation of water from leaves.
root system All of a plant's roots, which anchor it in the soil, absorb and transport minerals and water, and store food.
rooted Describing a phylogenetic tree that contains a branch point (often, the one farthest to the left) representing the most recent common ancestor of all taxa in the tree.
rough ER That portion of the endoplasmic reticulum with ribosomes attached.
round window In the mammalian ear, the point of contact where vibrations of the stapes create a traveling series of pressure waves in the fluid of the cochlea.
***r*-selection** Selection for life history traits that maximize reproductive success in uncrowded environments; also called density-independent selection.
rubisco (rū-bis′-kō) Ribulose bisphosphate (RuBP) carboxylase, the enzyme that catalyzes the first step of the Calvin cycle (the addition of CO_2 to RuBP).
ruminant (rū′-muh-nent) An animal, such as a cow or a sheep, with multiple stomach compartments specialized for an herbivorous diet.
S phase The synthesis phase of the cell cycle; the portion of interphase during which DNA is replicated.
saccule In the vertebrate ear, a chamber in the vestibule behind the oval window that participates in the sense of balance.
salicylic acid (sal′-i-sil′-ik) A signaling molecule in plants that may be partially responsible for activating systemic acquired resistance to pathogens.
salivary gland A gland associated with the oral cavity that secretes substances that lubricate food and begin the process of chemical digestion.
salt A compound resulting from the formation of an ionic bond; also called an ionic compound.
saltatory conduction (sol′-tuh-tōr′-ē) Rapid transmission of a nerve impulse along an axon, resulting from the action potential jumping from one node of Ranvier to another, skipping the myelin-sheathed regions of membrane.
sarcomere (sar′-kō-mēr) The fundamental, repeating unit of striated muscle, delimited by the Z lines.
sarcoplasmic reticulum (SR) (sar′-kō-plaz′-mik ruh-tik′-yū-lum) A specialized endoplasmic reticulum that regulates the calcium concentration in the cytosol of muscle cells.
saturated fatty acid A fatty acid in which all carbons in the hydrocarbon tail are connected by single bonds, thus maximizing the number of hydrogen atoms that are attached to the carbon skeleton.
savanna A tropical grassland biome with scattered individual trees and large herbivores and maintained by occasional fires and drought.
scaffolding protein A type of large relay protein to which several other relay proteins are simultaneously attached, increasing the efficiency of signal transduction.
scanning electron microscope (SEM) A microscope that uses an electron beam to scan the surface of a sample, coated with metal atoms, to study details of its topography.
schizophrenia (skit′-suh-frē′-nē-uh) A severe mental disturbance characterized by psychotic episodes in which patients have a distorted perception of reality.
Schwann cell A type of glial cell that forms insulating myelin sheaths around the axons of neurons in the peripheral nervous system.
science An approach to understanding the natural world.
scion (sī′-un) The twig grafted onto the stock when making a graft.
sclereid (sklār′-ē-id) A short, irregular sclerenchyma cell in nutshells and seed coats. Sclereids are scattered throughout the parenchyma of some plants.
sclerenchyma cell (skluh-ren′-kim-uh) A rigid, supportive plant cell type usually lacking a protoplast and possessing thick secondary walls strengthened by lignin at maturity.
scrotum A pouch of skin outside the abdomen that houses the testes; functions in maintaining the testes at the lower temperature required for spermatogenesis.
second law of thermodynamics The principle stating that every energy transfer or transformation increases the entropy of the universe. Usable forms of energy are at least partly converted to heat.
second messenger A small, nonprotein, water-soluble molecule or ion, such as a calcium ion (Ca^{2+}) or cyclic AMP, that relays a signal to a cell's interior in response to a signaling molecule bound by a signal receptor protein.
secondary cell wall In plant cells, a strong and durable matrix that is often deposited in several laminated layers around the plasma membrane and provides protection and support.
secondary consumer A carnivore that eats herbivores.
secondary endosymbiosis A process in eukaryotic evolution in which a heterotrophic eukaryotic cell engulfed a photosynthetic eukaryotic cell, which survived in a symbiotic relationship inside the heterotrophic cell.
secondary growth Growth produced by lateral meristems, thickening the roots and shoots of woody plants.
secondary immune response The adaptive immune response elicited on second or subsequent exposures to a particular antigen. The secondary immune response is more rapid, of greater magnitude, and of longer duration than the primary immune response.
secondary oocyte (ō′-uh-sīt) An oocyte that has completed the first of the two meiotic divisions.
secondary production The amount of chemical energy in consumers' food that is converted to their own new biomass during a given time period.
secondary structure Regions of repetitive coiling or folding of the polypeptide backbone of a protein due to hydrogen bonding between constituents of the backbone (not the side chains).
secondary succession A type of succession that occurs where an existing community has been cleared by some disturbance that leaves the soil or substrate intact.
secretion (1) The discharge of molecules synthesized by a cell. (2) The discharge of wastes from the body fluid into the filtrate.
secretory phase That portion of the uterine (menstrual) cycle when the endometrium continues to thicken, becomes more vascularized, and develops glands that secrete a fluid rich in glycogen.
seed An adaptation of some terrestrial plants consisting of an embryo packaged along with a store of food within a protective coat.
seed coat A tough outer covering of a seed, formed from the outer coat of an ovule. In a flowering plant, the seed coat encloses and protects the embryo and endosperm.
seedless vascular plant An informal name for a plant that has vascular tissue but lacks seeds. Seedless vascular plants form a paraphyletic group that includes the phyla Lycophyta (club mosses and their relatives) and Pterophyta (ferns and their relatives).
selective permeability A property of biological membranes that allows them to regulate the passage of substances across them.
self-incompatibility The ability of a seed plant to reject its own pollen and sometimes the pollen of closely related individuals.
semelparity Reproduction in which an organism produces all of its offspring in a single event; also known as big-bang reproduction.
semen (sē′-mun) The fluid that is ejaculated by the male during orgasm; contains sperm and secretions from several glands of the male reproductive tract.
semicircular canals A three-part chamber of the inner ear that functions in maintaining equilibrium.
semiconservative model Type of DNA replication in which the replicated double helix consists of one old strand, derived from the parental molecule, and one newly made strand.
semilunar valve A valve located at each exit of the heart, where the aorta leaves the left ventricle and the pulmonary artery leaves the right ventricle.
seminal vesicle (sem′-i-nul ves′-i-kul) A gland in males that secretes a fluid component of semen that lubricates and nourishes sperm.
seminiferous tubule (sem′-i-nif′-er-us) A highly coiled tube in the testis in which sperm are produced.
senescence (se-nes′-ens) The growth phase in a plant or plant part (as a leaf) from full maturity to death.
sensitive period A limited phase in an animal's development when learning of particular behaviors can take place; also called a critical period.
sensor In homeostasis, a receptor that detects a stimulus.
sensory adaptation The tendency of sensory neurons to become less sensitive when they are stimulated repeatedly.

sensory neuron A nerve cell that receives information from the internal or external environment and transmits signals to the central nervous system.

sensory reception The detection of a stimulus by sensory cells.

sensory receptor An organ, cell, or structure within a cell that responds to specific stimuli from an organism's external or internal environment.

sensory transduction The conversion of stimulus energy to a change in the membrane potential of a sensory receptor cell.

sepal (sē′-pul) A modified leaf in angiosperms that helps enclose and protect a flower bud before it opens.

septum (plural, **septa**) One of the cross-walls that divide a fungal hypha into cells. Septa generally have pores large enough to allow ribosomes, mitochondria, and even nuclei to flow from cell to cell.

serial endosymbiosis A hypothesis for the origin of eukaryotes consisting of a sequence of endosymbiotic events in which mitochondria, chloroplasts, and perhaps other cellular structures were derived from small prokaryotes that had been engulfed by larger cells.

serotonin (ser′-uh-tō′-nin) A neurotransmitter, synthesized from the amino acid tryptophan, that functions in the central nervous system.

set point In homeostasis in animals, a value maintained for a particular variable, such as body temperature or solute concentration.

seta (sē′-tuh) (plural, **setae**) The elongated stalk of a bryophyte sporophyte.

sex chromosome A chromosome responsible for determining the sex of an individual.

sex-linked gene A gene located on either sex chromosome. Most sex-linked genes are on the X chromosome and show distinctive patterns of inheritance; there are very few genes on the Y chromosome.

sexual dimorphism (dī-mōr′-fizm) Differences between the secondary sex characteristics of males and females.

sexual reproduction A type of reproduction in which two parents give rise to offspring that have unique combinations of genes inherited from both parents via the gametes.

sexual selection A form of selection in which individuals with certain inherited characteristics are more likely than other individuals to obtain mates.

Shannon diversity An index of community diversity symbolized by H and represented by the equation $H = -(p_A \ln p_A + p_B \ln p_B + p_C \ln p_C + \ldots)$, where A, B, C . . . are species, p is the relative abundance of each species, and ln is the natural logarithm.

shared ancestral character A character, shared by members of a particular clade, that originated in an ancestor that is not a member of that clade.

shared derived character An evolutionary novelty that is unique to a particular clade.

shoot system The aerial portion of a plant body, consisting of stems, leaves, and (in angiosperms) flowers.

short tandem repeat (STR) Simple sequence DNA containing multiple tandemly repeated units of two to five nucleotides. Variations in STRs act as genetic markers in STR analysis, used to prepare genetic profiles.

short-day plant A plant that flowers (usually in late summer, fall, or winter) only when the light period is shorter than a critical length.

short-term memory The ability to hold information, anticipations, or goals for a time and then release them if they become irrelevant.

sickle-cell disease A recessively inherited human blood disorder in which a single nucleotide change in the β-globin gene causes hemoglobin to aggregate, changing red blood cell shape and causing multiple symptoms in afflicted individuals.

sieve plate An end wall in a sieve-tube element, which facilitates the flow of phloem sap in angiosperm sieve tubes.

sieve-tube element A living cell that conducts sugars and other organic nutrients in the phloem of angiosperms; also called a sieve-tube member. Connected end to end, they form sieve tubes.

sign stimulus An external sensory cue that triggers a fixed action pattern by an animal.

signal In animal behavior, transmission of a stimulus from one animal to another. The term is also used in the context of communication in other kinds of organisms and in cell-to-cell communication in all multicellular organisms.

signal peptide A sequence of about 20 amino acids at or near the leading (amino) end of a polypeptide that targets it to the endoplasmic reticulum or other organelles in a eukaryotic cell.

signal transduction The linkage of a mechanical, chemical, or electromagnetic stimulus to a specific cellular response.

signal transduction pathway A series of steps linking a mechanical, chemical, or electrical stimulus to a specific cellular response.

signal-recognition particle (SRP) A protein-RNA complex that recognizes a signal peptide as it emerges from a ribosome and helps direct the ribosome to the endoplasmic reticulum (ER) by binding to a receptor protein on the ER.

silent mutation A nucleotide-pair substitution that has no observable effect on the phenotype; for example, within a gene, a mutation that results in a codon that codes for the same amino acid.

simple fruit A fruit derived from a single carpel or several fused carpels.

simple sequence DNA A DNA sequence that contains many copies of tandemly repeated short sequences.

single bond A single covalent bond; the sharing of a pair of valence electrons by two atoms.

single circulation A circulatory system consisting of a single pump and circuit, in which blood passes from the sites of gas exchange to the rest of the body before returning to the heart.

single nucleotide polymorphism (SNP) A single base-pair site in a genome where nucleotide variation is found in at least 1% of the population.

single-lens eye The camera-like eye found in some jellies, polychaete worms, spiders, and many molluscs.

single-strand binding protein A protein that binds to the unpaired DNA strands during DNA replication, stabilizing them and holding them apart while they serve as templates for the synthesis of complementary strands of DNA.

sinoatrial (SA) node A region in the right atrium of the heart that sets the rate and timing at which all cardiac muscle cells contract; the pacemaker.

sister chromatids Two copies of a duplicated chromosome attached to each other by proteins at the centromere and, sometimes, along the arms. While joined, two sister chromatids make up one chromosome. Chromatids are eventually separated during mitosis or meiosis II.

sister taxa Groups of organisms that share an immediate common ancestor and hence are each other's closest relatives.

skeletal muscle A type of striated muscle that is generally responsible for the voluntary movements of the body.

sliding-filament model The idea that muscle contraction is based on the movement of thin (actin) filaments along thick (myosin) filaments, shortening the sarcomere, the basic unit of muscle organization.

slow block to polyspermy The formation of the fertilization envelope and other changes in an egg's surface that prevent fusion of the egg with more than one sperm. The slow block begins about 1 minute after fertilization.

slow-twitch fiber A muscle fiber that can sustain long contractions.

small interfering RNA (siRNA) One of multiple small, single-stranded RNA molecules generated by cellular machinery from a long, linear, double-stranded RNA molecule. The siRNA associates with one or more proteins in a complex that can degrade or prevent translation of an mRNA with a complementary sequence. In some cases, siRNA can also block transcription by promoting chromatin modification.

small intestine The longest section of the alimentary canal, so named because of its small diameter compared with that of the large intestine; the principal site of the enzymatic hydrolysis of food macromolecules and the absorption of nutrients.

smooth ER That portion of the endoplasmic reticulum that is free of ribosomes.

smooth muscle A type of muscle lacking the striations of skeletal and cardiac muscle because of the uniform distribution of myosin filaments in the cells; responsible for involuntary body activities.

social learning Modification of behavior through the observation of other individuals.

sociobiology The study of social behavior based on evolutionary theory.

sodium-potassium pump A transport protein in the plasma membrane of animal cells that actively transports sodium out of the cell and potassium into the cell.

soil horizon A soil layer with physical characteristics that differ from those of the layers above or beneath.
solute (sol′-yūt) A substance that is dissolved in a solution.
solute potential (Ψ_S) A component of water potential that is proportional to the molarity of a solution and that measures the effect of solutes on the direction of water movement; also called osmotic potential, it can be either zero or negative.
solution A liquid that is a homogeneous mixture of two or more substances.
solvent The dissolving agent of a solution. Water is the most versatile solvent known.
somatic cell (sō-mat′-ik) Any cell in a multicellular organism except a sperm or egg or their precursors.
somite One of a series of blocks of mesoderm that exist in pairs just lateral to the notochord in a vertebrate embryo.
soredium (plural, **soredia**) In lichens, a small cluster of fungal hyphae with embedded algae.
sorus (plural, **sori**) A cluster of sporangia on a fern sporophyll. Sori may be arranged in various patterns, such as parallel lines or dots, which are useful in fern identification.
Southern blotting A technique that enables specific nucleotide sequences to be detected in samples of DNA. It involves gel electrophoresis of DNA molecules and their transfer to a membrane (blotting), followed by nucleic acid hybridization with a labeled probe.
spatial learning The establishment of a memory that reflects the environment's spatial structure.
spatial summation A phenomenon of neural integration in which the membrane potential of the postsynaptic cell is determined by the combined effect of EPSPs or IPSPs produced nearly simultaneously by different synapses.
speciation (spē′-sē-ā′-shun) An evolutionary process in which one species splits into two or more species.
species (spē′-sēz) A population or group of populations whose members have the potential to interbreed in nature and produce viable, fertile offspring, but do not produce viable, fertile offspring with members of other such groups.
species diversity The number and relative abundance of species in a biological community.
species richness The number of species in a biological community.
species-area curve The biodiversity pattern that shows that the larger the geographic area of a community is, the more species it has.
specific heat The amount of heat that must be absorbed or lost for 1 g of a substance to change its temperature by 1°C.
spectrophotometer An instrument that measures the proportions of light of different wavelengths absorbed and transmitted by a pigment solution.
sperm The male gamete.
spermatheca (sper′-muh-thē′-kuh) In many insects, a sac in the female reproductive system where sperm are stored.
spermatogenesis The continuous and prolific production of mature sperm cells in the testis.
spermatogonium (plural, **spermatogonia**) A cell that divides mitotically to form spermatocytes.
sphincter (sfink′-ter) A ringlike band of muscle fibers that controls the size of an opening in the body, such as the passage between the esophagus and the stomach.
spiral cleavage A type of embryonic development in protostomes in which the planes of cell division that transform the zygote into a ball of cells are diagonal to the vertical axis of the embryo. As a result, the cells of each tier sit in the grooves between cells of adjacent tiers.
spliceosome (splī′-sō-sōm) A large complex made up of proteins and RNA molecules that splices RNA by interacting with the ends of an RNA intron, releasing the intron and joining the two adjacent exons.
spongocoel (spon′-jō-sēl) The central cavity of a sponge.
spontaneous process A process that occurs without an overall input of energy; a process that is energetically favorable.
sporangium (spōr-an′-jē-um) (plural, **sporangia**) A multicellular organ in fungi and plants in which meiosis occurs and haploid cells develop.
spore (1) In the life cycle of a plant or alga undergoing alternation of generations, a haploid cell produced in the sporophyte by meiosis. A spore can divide by mitosis to develop into a multicellular haploid individual, the gametophyte, without fusing with another cell. (2) In fungi, a haploid cell, produced either sexually or asexually, that produces a mycelium after germination.
sporocyte A diploid cell, also known as a spore mother cell, that undergoes meiosis and generates haploid spores.
sporophyll (spō′-ruh-fil) A modified leaf that bears sporangia and hence is specialized for reproduction.
sporophyte (spō-ruh-fīt′) In organisms (plants and some algae) that have alternation of generations, the multicellular diploid form that results from the union of gametes. The sporophyte produces haploid spores by meiosis that develop into gametophytes.
sporopollenin (spōr-uh-pol′-eh-nin) A durable polymer that covers exposed zygotes of charophyte algae and forms the walls of plant spores, preventing them from drying out.
stabilizing selection Natural selection in which intermediate phenotypes survive or reproduce more successfully than do extreme phenotypes.
stamen (stā′-men) The pollen-producing reproductive organ of a flower, consisting of an anther and a filament.
standard metabolic rate (SMR) Metabolic rate of a resting, fasting, and nonstressed ectotherm at a particular temperature.
starch A storage polysaccharide in plants, consisting entirely of glucose monomers joined by α glycosidic linkages.
start point In transcription, the nucleotide position on the promoter where RNA polymerase begins synthesis of RNA.
statocyst (stat′-uh-sist′) A type of mechanoreceptor that functions in equilibrium in invertebrates by use of statoliths, which stimulate hair cells in relation to gravity.
statolith (stat′-uh-lith′) (1) In plants, a specialized plastid that contains dense starch grains and may play a role in detecting gravity. (2) In invertebrates, a dense particle that settles in response to gravity and is found in sensory organs that function in equilibrium.
stele (stēl) The vascular tissue of a stem or root.
stem A vascular plant organ consisting of an alternating system of nodes and internodes that support the leaves and reproductive structures.
stem cell Any relatively unspecialized cell that can produce, during a single division, one identical daughter cell and one more specialized daughter cell that can undergo further differentiation.
steroid A type of lipid characterized by a carbon skeleton consisting of four fused rings with various chemical groups attached.
sticky end A single-stranded end of a double-stranded restriction fragment.
stigma (plural, **stigmata**) The sticky part of a flower's carpel, which receives pollen grains.
stimulus In feedback regulation, a fluctuation in a variable that triggers a response.
stipe A stemlike structure of a seaweed.
stock The plant that provides the root system when making a graft.
stoma (stō′-muh) (plural, **stomata**) A microscopic pore surrounded by guard cells in the epidermis of leaves and stems that allows gas exchange between the environment and the interior of the plant.
stomach An organ of the digestive system that stores food and performs preliminary steps of digestion.
stramenopile A protist in which a "hairy" flagellum (one covered with fine, hairlike projections) is paired with a shorter, smooth flagellum.
stratum (strah′-tum) (plural, **strata**) A rock layer formed when new layers of sediment cover older ones and compress them.
striated muscle Muscle in which the regular arrangement of filaments creates a pattern of light and dark bands.
strigolactones A class of plant hormone that inhibits shoot branching, triggers the germination of parasitic plant seeds, and stimulates the association of plant roots with mycorrhizal fungi.
strobilus (strō-bī′-lus) (plural, **strobili**) The technical term for a cluster of sporophylls known commonly as a cone, found in most gymnosperms and some seedless vascular plants.
stroke The death of nervous tissue in the brain, usually resulting from rupture or blockage of arteries in the head.
stroke volume The volume of blood pumped by a heart ventricle in a single contraction.
stroma (strō′-muh) The dense fluid within the chloroplast surrounding the thylakoid membrane and containing ribosomes and DNA; involved in the synthesis of organic molecules from carbon dioxide and water.

stromatolite Layered rock that results from the activities of prokaryotes that bind thin films of sediment together.

structural isomer One of several compounds that have the same molecular formula but differ in the covalent arrangements of their atoms.

style The stalk of a flower's carpel, with the ovary at the base and the stigma at the top.

substrate The reactant on which an enzyme works.

substrate feeder An animal that lives in or on its food source, eating its way through the food.

substrate-level phosphorylation The enzyme-catalyzed formation of ATP by direct transfer of a phosphate group to ADP from an intermediate substrate in catabolism.

sugar sink A plant organ that is a net consumer or storer of sugar. Growing roots, shoot tips, stems, and fruits are examples of sugar sinks supplied by phloem.

sugar source A plant organ in which sugar is being produced by either photosynthesis or the breakdown of starch. Mature leaves are the primary sugar sources of plants.

sulfhydryl group A chemical group consisting of a sulfur atom bonded to a hydrogen atom.

suprachiasmatic nucleus (SCN) A group of neurons in the hypothalamus of mammals that functions as a biological clock.

surface tension A measure of how difficult it is to stretch or break the surface of a liquid. Water has a high surface tension because of the hydrogen bonding of surface molecules.

surfactant A substance secreted by alveoli that decreases surface tension in the fluid that coats the alveoli.

survivorship curve A plot of the number of members of a cohort that are still alive at each age; one way to represent age-specific mortality.

suspension feeder An aquatic animal, such as a sponge, clam, or baleen whale, that feeds by sifting small organisms or food particles from the water.

sustainable agriculture Long-term productive farming methods that are environmentally safe.

sustainable development Development that meets the needs of people today without limiting the ability of future generations to meet their needs.

swim bladder In aquatic osteichthyans, an air sac that enables the animal to control its buoyancy in the water.

symbiont (sim′-bē-ont) The smaller participant in a symbiotic relationship, living in or on the host.

symbiosis An ecological relationship between organisms of two different species that live together in direct and intimate contact.

sympathetic division One of three divisions of the autonomic nervous system; generally increases energy expenditure and prepares the body for action.

sympatric speciation (sim-pat′-rik) The formation of new species in populations that live in the same geographic area.

symplast In plants, the continuum of cytoplasm connected by plasmodesmata between cells.

synapse (sin′-aps) The junction where a neuron communicates with another cell across a narrow gap via a neurotransmitter or an electrical coupling.

synapsid Member of an amniote clade distinguished by a single hole on each side of the skull. Synapsids include the mammals.

synapsis (si-nap′-sis) The pairing and physical connection of duplicated homologous chromosomes during prophase I of meiosis.

systematics A scientific discipline focused on classifying organisms and determining their evolutionary relationships.

systemic acquired resistance A defensive response in infected plants that helps protect healthy tissue from pathogenic invasion.

systemic circuit The branch of the circulatory system that supplies oxygenated blood to and carries deoxygenated blood away from organs and tissues throughout the body.

systems biology An approach to studying biology that aims to model the dynamic behavior of whole biological systems based on a study of the interactions among the system's parts.

systole (sis′-tō-lē) The stage of the cardiac cycle in which a heart chamber contracts and pumps blood.

systolic pressure Blood pressure in the arteries during contraction of the ventricles.

T cells The class of lymphocytes that mature in the thymus; they include both effector cells for the cell-mediated immune response and helper cells required for both branches of adaptive immunity.

taproot A main vertical root that develops from an embryonic root and gives rise to lateral (branch) roots.

tastant Any chemical that stimulates the sensory receptors in a taste bud.

taste bud A collection of modified epithelial cells on the tongue or in the mouth that are receptors for taste in mammals.

TATA box A DNA sequence in eukaryotic promoters crucial in forming the transcription initiation complex.

taxis (tak′-sis) An oriented movement toward or away from a stimulus.

taxon (plural, **taxa**) A named taxonomic unit at any given level of classification.

taxonomy (tak-son′-uh-mē) A scientific discipline concerned with naming and classifying the diverse forms of life.

Tay-Sachs disease A human genetic disease caused by a recessive allele for a dysfunctional enzyme, leading to accumulation of certain lipids in the brain. Seizures, blindness, and degeneration of motor and mental performance usually become manifest a few months after birth, followed by death within a few years.

technology The application of scientific knowledge for a specific purpose, often involving industry or commerce but also including uses in basic research.

telomerase An enzyme that catalyzes the lengthening of telomeres in eukaryotic germ cells.

telomere (tel′-uh-mēr) The tandemly repetitive DNA at the end of a eukaryotic chromosome's DNA molecule. Telomeres protect the organism's genes from being eroded during successive rounds of replication. *See also* repetitive DNA.

telophase The fifth and final stage of mitosis, in which daughter nuclei are forming and cytokinesis has typically begun.

temperate broadleaf forest A biome located throughout midlatitude regions where there is sufficient moisture to support the growth of large, broadleaf deciduous trees.

temperate grassland A terrestrial biome that exists at midlatitude regions and is dominated by grasses and forbs.

temperate phage A phage that is capable of replicating by either a lytic or lysogenic cycle.

temperature A measure of the intensity of heat in degrees, reflecting the average kinetic energy of the molecules.

template strand The DNA strand that provides the pattern, or template, for ordering, by complementary base pairing, the sequence of nucleotides in an RNA transcript.

temporal summation A phenomenon of neural integration in which the membrane potential of the postsynaptic cell in a chemical synapse is determined by the combined effect of EPSPs or IPSPs produced in rapid succession.

tendon A fibrous connective tissue that attaches muscle to bone.

terminator In bacteria, a sequence of nucleotides in DNA that marks the end of a gene and signals RNA polymerase to release the newly made RNA molecule and detach from the DNA.

territoriality A behavior in which an animal defends a bounded physical space against encroachment by other individuals, usually of its own species.

tertiary consumer (ter-shē-ār′-ē) A carnivore that eats other carnivores.

tertiary structure The overall shape of a protein molecule due to interactions of amino acid side chains, including hydrophobic interactions, ionic bonds, hydrogen bonds, and disulfide bridges.

testcross Breeding an organism of unknown genotype with a homozygous recessive individual to determine the unknown genotype. The ratio of phenotypes in the offspring reveals the unknown genotype.

testis (plural, **testes**) The male reproductive organ, or gonad, in which sperm and reproductive hormones are produced.

testosterone A steroid hormone required for development of the male reproductive system, spermatogenesis, and male secondary sex characteristics; the major androgen in mammals.

tetanus (tet′-uh-nus) The maximal, sustained contraction of a skeletal muscle, caused by a very high frequency of action potentials elicited by continual stimulation.

tetrapod A vertebrate clade whose members have limbs with digits. Tetrapods include mammals, amphibians, and birds and other reptiles.

thalamus (thal′-uh-mus) An integrating center of the vertebrate forebrain. Neurons with cell bodies in the thalamus relay neural input to specific areas in the cerebral cortex and regulate what information goes to the cerebral cortex.

thallus (plural, **thalli**) A seaweed body that is plantlike, consisting of a holdfast, stipe, and blades, yet lacks true roots, stems, and leaves.

theory An explanation that is broader in scope than a hypothesis, generates new hypotheses, and is supported by a large body of evidence.

thermal energy *See* heat.

thermocline A narrow stratum of abrupt temperature change in the ocean and in many temperate-zone lakes.

thermodynamics (ther′-mō-dī-nam′-iks) The study of energy transformations that occur in a collection of matter. *See* first law of thermodynamics; second law of thermodynamics.

thermoreceptor A receptor stimulated by either heat or cold.

thermoregulation The maintenance of internal body temperature within a tolerable range.

theropod Member of a group of dinosaurs that were bipedal carnivores.

thick filament A filament composed of staggered arrays of myosin molecules; a component of myofibrils in muscle fibers.

thigmomorphogenesis A response in plants to chronic mechanical stimulation, resulting from increased ethylene production. An example is thickening stems in response to strong winds.

thigmotropism (thig-mo′-truh-pizm) A directional growth of a plant in response to touch.

thin filament A filament consisting of two strands of actin and two strands of regulatory protein coiled around one another; a component of myofibrils in muscle fibers.

threatened species A species that is considered likely to become endangered in the foreseeable future.

threshold The potential that an excitable cell membrane must reach for an action potential to be initiated.

thrombus A fibrin-containing clot that forms in a blood vessel and blocks the flow of blood.

thylakoid (thī′-luh-koyd) A flattened, membranous sac inside a chloroplast. Thylakoids often exist in stacks called grana that are interconnected; their membranes contain molecular "machinery" used to convert light energy to chemical energy.

thymus (thī′-mus) A small organ in the thoracic cavity of vertebrates where maturation of T cells is completed.

thyroid gland An endocrine gland, located on the ventral surface of the trachea, that secretes two iodine-containing hormones, triiodothyronine (T_3) and thyroxine (T_4), as well as calcitonin.

thyroxine (T_4) One of two iodine-containing hormones that are secreted by the thyroid gland and that help regulate metabolism, development, and maturation in vertebrates.

Ti plasmid A plasmid of a tumor-inducing bacterium (the plant pathogen *Agrobacterium*) that integrates a segment of its DNA (T DNA) into a chromosome of a host plant. The Ti plasmid is frequently used as a vector for genetic engineering in plants.

tidal volume The volume of air a mammal inhales and exhales with each breath.

tight junction A type of intercellular junction between animal cells that prevents the leakage of material through the space between cells.

tissue An integrated group of cells with a common structure, function, or both.

tissue system One or more tissues organized into a functional unit connecting the organs of a plant.

Toll-like receptor (TLR) A membrane receptor on a phagocytic white blood cell that recognizes fragments of molecules common to a set of pathogens.

tonicity The ability of a solution surrounding a cell to cause that cell to gain or lose water.

top-down model A model of community organization in which predation influences community organization by controlling herbivore numbers, which in turn control plant or phytoplankton numbers, which in turn control nutrient levels; also called the trophic cascade model.

topoisomerase A protein that breaks, swivels, and rejoins DNA strands. During DNA replication, topoisomerase helps to relieve strain in the double helix ahead of the replication fork.

topsoil A mixture of particles derived from rock, living organisms, and decaying organic material (humus).

torpor A physiological state in which activity is low and metabolism decreases.

torsion In gastropods, a developmental process in which the visceral mass rotates up to 180°, causing the animal's anus and mantle cavity to be positioned above its head.

totipotent (tō′-tuh-pōt′-ent) Describing a cell that can give rise to all parts of the embryo and adult, as well as extraembryonic membranes in species that have them.

trace element An element indispensable for life but required in extremely minute amounts.

trachea (trā′-kē-uh) The portion of the respiratory tract that passes from the larynx to the bronchi; also called the windpipe.

tracheal system In insects, a system of branched, air-filled tubes that extends throughout the body and carries oxygen directly to cells.

tracheid (trā′-kē-id) A long, tapered water-conducting cell found in the xylem of nearly all vascular plants. Functioning tracheids are no longer living.

trait One of two or more detectable variants in a genetic character.

trans fat An unsaturated fat, formed artificially during hydrogenation of oils, containing one or more *trans* double bonds.

transcription The synthesis of RNA using a DNA template.

transcription factor A regulatory protein that binds to DNA and affects transcription of specific genes.

transcription initiation complex The completed assembly of transcription factors and RNA polymerase bound to a promoter.

transcription unit A region of DNA that is transcribed into an RNA molecule.

transduction (1) A process in which phages (viruses) carry bacterial DNA from one bacterial cell to another. When these two cells are members of different species, transduction results in horizontal gene transfer. (2) In cellular communication, the conversion of a signal from outside the cell to a form that can bring about a specific cellular response; also called *signal transduction*.

transfer RNA (tRNA) An RNA molecule that functions as a translator between nucleic acid and protein languages by carrying specific amino acids to the ribosome, where they recognize the appropriate codons in the mRNA.

transformation (1) The conversion of a normal animal cell to a cancerous cell. (2) A change in genotype and phenotype due to the assimilation of external DNA by a cell. When the external DNA is from a member of a different species, transformation results in horizontal gene transfer.

transgenic Pertaining to an organism whose genome contains a gene introduced from another organism of the same or a different species.

translation The synthesis of a polypeptide using the genetic information encoded in an mRNA molecule. There is a change of "language" from nucleotides to amino acids.

translocation (1) An aberration in chromosome structure resulting from attachment of a chromosomal fragment to a nonhomologous chromosome. (2) During protein synthesis, the third stage in the elongation cycle, when the RNA carrying the growing polypeptide moves from the A site to the P site on the ribosome. (3) The transport of organic nutrients in the phloem of vascular plants.

transmission The passage of a nerve impulse along axons.

transmission electron microscope (TEM) A microscope that passes an electron beam through very thin sections stained with metal atoms and is primarily used to study the internal ultrastructure of cells.

transpiration The evaporative loss of water from a plant.

transport epithelium One or more layers of specialized epithelial cells that carry out and regulate solute movement.

transport protein A transmembrane protein that helps a certain substance or class of closely related substances to cross the membrane.

transport vesicle A small membranous sac in a eukaryotic cell's cytoplasm carrying molecules produced by the cell.

transposable element A segment of DNA that can move within the genome of a cell by means of a DNA or RNA intermediate; also called a transposable genetic element.

transposon A transposable element that moves within a genome by means of a DNA intermediate.

transverse (T) tubule An infolding of the plasma membrane of skeletal muscle cells.

triacylglycerol (trī-as′-ul-glis′-uh-rol) A lipid consisting of three fatty acids linked to one glycerol molecule; also called a fat or triglyceride.

triiodothyronine (T_3) (trī′-ī-ō′-dō-thī′-rō-nēn) One of two iodine-containing hormones that are secreted by the thyroid gland and that help regulate metabolism, development, and maturation in vertebrates.

trimester In human development, one of three 3-month-long periods of pregnancy.

triple response A plant growth maneuver in response to mechanical stress, involving slowing of stem elongation, thickening of the stem, and a curvature that causes the stem to start growing horizontally.

triplet code A genetic information system in which a set of three-nucleotide-long words specify the amino acids for polypeptide chains.

triploblastic Possessing three germ layers: the endoderm, mesoderm, and ectoderm. Most eumetazoans are triploblastic.

trisomic Referring to a diploid cell that has three copies of a particular chromosome instead of the normal two.

trochophore larva (trō′-kuh-fōr) Distinctive larval stage observed in some lophotrochozoan animals, including some annelids and molluscs.

trophic efficiency The percentage of production transferred from one trophic level to the next.

trophic structure The different feeding relationships in an ecosystem, which determine the route of energy flow and the pattern of chemical cycling.

trophoblast The outer epithelium of a mammalian blastocyst. It forms the fetal part of the placenta, supporting embryonic development but not forming part of the embryo proper.

tropic hormone A hormone that has an endocrine gland or cells as a target.

tropical dry forest A terrestrial biome characterized by relatively high temperatures and precipitation overall but with a pronounced dry season.

tropical rain forest A terrestrial biome characterized by relatively high precipitation and temperatures year-round.

tropics Latitudes between 23.5° north and south.

tropism A growth response that results in the curvature of whole plant organs toward or away from stimuli due to differential rates of cell elongation.

tropomyosin The regulatory protein that blocks the myosin-binding sites on actin molecules.

troponin complex The regulatory proteins that control the position of tropomyosin on the thin filament.

true-breeding Referring to organisms that produce offspring of the same variety over many generations of self-pollination.

tubal ligation A means of sterilization in which a woman's two oviducts (fallopian tubes) are tied closed to prevent eggs from reaching the uterus. A segment of each oviduct is removed.

tube foot One of numerous extensions of an echinoderm's water vascular system. Tube feet function in locomotion and feeding.

tumor-suppressor gene A gene whose protein product inhibits cell division, thereby preventing the uncontrolled cell growth that contributes to cancer.

tundra A terrestrial biome at the extreme limits of plant growth. At the northernmost limits, it is called arctic tundra, and at high altitudes, where plant forms are limited to low shrubby or matlike vegetation, it is called alpine tundra.

tunicate Member of the clade Urochordata, sessile marine chordates that lack a backbone.

turgid (ter′-jid) Swollen or distended, as in plant cells. (A walled cell becomes turgid if it has a lower water potential than its surroundings, resulting in entry of water.)

turgor pressure The force directed against a plant cell wall after the influx of water and swelling of the cell due to osmosis.

turnover The mixing of waters as a result of changing water-temperature profiles in a lake.

turnover time The time required to replace the standing crop of a population or group of populations (for example, of phytoplankton), calculated as the ratio of standing crop to production.

twin study A behavioral study in which researchers compare the behavior of identical twins raised apart with that of identical twins raised in the same household.

tympanic membrane Another name for the eardrum, the membrane between the outer and middle ear.

uniformitarianism The principle that mechanisms of change are constant over time. *See* catastrophism.

Unikonta (yū′-ni-kon′-tuh) One of five supergroups of eukaryotes proposed in a current hypothesis of the evolutionary history of eukaryotes. This clade, which is supported by studies of myosin proteins and DNA, consists of amoebozoans and opisthokonts. *See also* Excavata, Chromalveolata, Rhizaria, and Archaeplastida.

unsaturated fatty acid A fatty acid that has one or more double bonds between carbons in the hydrocarbon tail. Such bonding reduces the number of hydrogen atoms attached to the carbon skeleton.

urea A soluble nitrogenous waste produced in the liver by a metabolic cycle that combines ammonia with carbon dioxide.

ureter (yū-rē′-ter) A duct leading from the kidney to the urinary bladder.

urethra (yū-rē′-thruh) A tube that releases urine from the mammalian body near the vagina in females and through the penis in males; also serves in males as the exit tube for the reproductive system.

uric acid A product of protein and purine metabolism and the major nitrogenous waste product of insects, land snails, and many reptiles. Uric acid is relatively nontoxic and largely insoluble.

urinary bladder The pouch where urine is stored prior to elimination.

uterine cycle The changes that occur in the uterus during the reproductive cycle of the human female; also called the menstrual cycle.

uterus A female organ where eggs are fertilized and/or development of the young occurs.

utricle In the vertebrate ear, a chamber in the vestibule behind the oval window that opens into the three semicircular canals.

vaccination *See* immunization.

vaccine A harmless variant or derivative of a pathogen that stimulates a host's immune system to mount defenses against the pathogen.

vacuole (vak′-yū-ōl′) A membrane-bounded vesicle whose specialized function varies in different kinds of cells.

vagina Part of the female reproductive system between the uterus and the outside opening; the birth canal in mammals. During copulation, the vagina accommodates the male's penis and receives sperm.

valence The bonding capacity of a given atom; usually equals the number of unpaired electrons required to complete the atom's outermost (valence) shell.

valence electron An electron in the outermost electron shell.

valence shell The outermost energy shell of an atom, containing the valence electrons involved in the chemical reactions of that atom.

van der Waals interactions Weak attractions between molecules or parts of molecules that result from transient local partial charges.

variation Differences between members of the same species.

vas deferens In mammals, the tube in the male reproductive system in which sperm travel from the epididymis to the urethra.

vasa recta The capillary system in the kidney that serves the loop of Henle.

vascular cambium A cylinder of meristematic tissue in woody plants that adds layers of secondary vascular tissue called secondary xylem (wood) and secondary phloem.

vascular plant A plant with vascular tissue. Vascular plants include all living plant species except liverworts, mosses, and hornworts.

vascular tissue Plant tissue consisting of cells joined into tubes that transport water and nutrients throughout the plant body.

vascular tissue system A transport system formed by xylem and phloem throughout a vascular plant. Xylem transports water and minerals; phloem transports sugars, the products of photosynthesis.

vasectomy The cutting and sealing of each vas deferens to prevent sperm from entering the urethra.

vasocongestion The filling of a tissue with blood, caused by increased blood flow through the arteries of that tissue.

vasoconstriction A decrease in the diameter of blood vessels caused by contraction of smooth muscles in the vessel walls.

vasodilation An increase in the diameter of blood vessels caused by relaxation of smooth muscles in the vessel walls.

vector An organism that transmits pathogens from one host to another.

vegetal pole The point at the end of an egg in the hemisphere where most yolk is concentrated; opposite of animal pole.

vegetative reproduction Cloning of plants by asexual means.

vein (1) In animals, a vessel that carries blood toward the heart. (2) In plants, a vascular bundle in a leaf.

ventilation The flow of air or water over a respiratory surface.

ventral Pertaining to the underside, or bottom, of an animal with radial or bilateral symmetry.

ventricle (ven′-tri-kul) (1) A heart chamber that pumps blood out of the heart. (2) A space in the vertebrate brain, filled with cerebrospinal fluid.

venule (ven′-yūl) A vessel that conveys blood between a capillary bed and a vein.

vernalization The use of cold treatment to induce a plant to flower.

vertebrate A chordate animal with a backbone, including sharks and rays, ray-finned fishes, coelacanths, lungfishes, amphibians, reptiles, and mammals.

vesicle (ves′-i-kul) A membranous sac in the cytoplasm of a eukaryotic cell.

vessel A continuous water-conducting micropipe found in most angiosperms and a few nonflowering vascular plants.

vessel element A short, wide water-conducting cell found in the xylem of most angiosperms and a few nonflowering vascular plants. Dead at maturity, vessel elements are aligned end to end to form micropipes called vessels.

vestigial structure A feature of an organism that is a historical remnant of a structure that served a function in the organism's ancestors.

villus (plural, **villi**) (1) A finger-like projection of the inner surface of the small intestine. (2) A finger-like projection of the chorion of the mammalian placenta. Large numbers of villi increase the surface areas of these organs.

viral envelope A membrane, derived from membranes of the host cell, that cloaks the capsid, which in turn encloses a viral genome.

viroid (vī′-royd) A plant pathogen consisting of a molecule of naked, circular RNA a few hundred nucleotides long.

virulent Describing a pathogen against which an organism has little specific defense.

virulent phage A phage that replicates only by a lytic cycle.

virus An infectious particle incapable of replicating outside of a cell, consisting of an RNA or DNA genome surrounded by a protein coat (capsid) and, for some viruses, a membranous envelope.

visceral mass One of the three main parts of a mollusc; the part containing most of the internal organs. *See also* foot, mantle.

visible light That portion of the electromagnetic spectrum that can be detected as various colors by the human eye, ranging in wavelength from about 380 nm to about 750 nm.

vital capacity The maximum volume of air that a mammal can inhale and exhale with each breath.

vitamin An organic molecule required in the diet in very small amounts. Many vitamins serve as coenzymes or parts of coenzymes.

viviparous (vī-vip′-uh-rus) Referring to a type of development in which the young are born alive after having been nourished in the uterus by blood from the placenta.

voltage-gated ion channel A specialized ion channel that opens or closes in response to changes in membrane potential.

vulva Collective term for the female external genitalia.

water potential (Ψ) The physical property predicting the direction in which water will flow, governed by solute concentration and applied pressure.

water vascular system A network of hydraulic canals unique to echinoderms that branches into extensions called tube feet, which function in locomotion and feeding.

wavelength The distance between crests of waves, such as those of the electromagnetic spectrum.

wetland A habitat that is inundated by water at least some of the time and that supports plants adapted to water-saturated soil.

white matter Tracts of axons within the CNS.

wild type The phenotype most commonly observed in natural populations; also refers to the individual with that phenotype.

wilting The drooping of leaves and stems as a result of plant cells becoming flaccid.

wobble Flexibility in the base-pairing rules in which the nucleotide at the 5′ end of a tRNA anticodon can form hydrogen bonds with more than one kind of base in the third position (3′ end) of a codon.

xerophyte A plant adapted to an arid climate.

X-linked gene A gene located on the X chromosome; such genes show a distinctive pattern of inheritance.

X-ray crystallography A technique used to study the three-dimensional structure of molecules. It depends on the diffraction of an X-ray beam by the individual atoms of a crystallized molecule.

xylem (zī′-lum) Vascular plant tissue consisting mainly of tubular dead cells that conduct most of the water and minerals upward from the roots to the rest of the plant.

xylem sap The dilute solution of water and dissolved minerals carried through vessels and tracheids.

yeast Single-celled fungus. Yeasts reproduce asexually by binary fission or by the pinching of small buds off a parent cell. Many fungal species can grow both as yeasts and as a network of filaments; relatively few species grow only as yeasts.

yolk Nutrients stored in an egg.

zero population growth (ZPG) A period of stability in population size, when additions to the population through births and immigration are balanced by subtractions through deaths and emigration.

zona pellucida The extracellular matrix surrounding a mammalian egg.

zone of polarizing activity (ZPA) A block of mesoderm located just under the ectoderm where the posterior side of a limb bud is attached to the body; required for proper pattern formation along the anterior-posterior axis of the limb.

zoned reserve An extensive region that includes areas relatively undisturbed by humans surrounded by areas that have been changed by human activity and are used for economic gain.

zoonotic pathogen A disease-causing agent that is transmitted to humans from other animals.

zoospore Flagellated spore found in chytrid fungi and some protists.

zygomycete (zī′-guh-mī′-sēt) Member of the fungal phylum Zygomycota, characterized by the formation of a sturdy structure called a zygosporangium during sexual reproduction.

zygosporangium (zī′-guh-spōr-an′-jē-um) In zygomycete fungi, a sturdy multinucleate structure in which karyogamy and meiosis occur.

zygote (zī′-gōt) The diploid cell produced by the union of haploid gametes during fertilization; a fertilized egg.

Index

NOTE: A page number in regular type indicates where a topic is discussed in text (topic may also be in a figure on that page); a **bold** page number indicates where a term is bold and defined; an *f* following a page number indicates a figure (topic may also be discussed in text on that page); a *t* following a page number indicates a table (topic may also be discussed in text on that page).

C

D

E

F

Index

Q

R